Wörterbuch Gebäudetechnik

Englisch – Deutsch

Dictionary Building Services

English – German

Herbert Bucksch

Wörterbuch Gebäudetechnik

Englisch – Deutsch

Dictionary Building Services

English – German

Bibliografische Information Der Deutschen Bibliothek

Die Deutsche Bibliothek verzeichnet diese Publikation in der Deutschen Nationalbibliografie; detaillierte bibliografische Daten sind im Internet über <http://dnb.ddb.de> abrufbar.

ISBN 3-8167-4739-6

Herstellung und Layout: Georgia Zechlin
Satz: Späth Mediendesign, Birenbach
Umschlaggestaltung: Markus Bauer
Druck: Konrad Tritsch GmbH, Ochsenfurt-Hohestadt

Für den Druck des Buches wurde chlor- und säurefreies Papier verwendet.

Fraunhofer-Informationszentrum Raum und Bau IRB
Postfach 80 04 69, 70504 Stuttgart
Telefon (07 11) 970-25 00
Telefax (07 11) 970-25 99
E-Mail: irb@irb.fraunhofer.de
http://www.irbbuch.de

PREFACE

The building industry of today is increasingly influenced by internationalization. Building projects are put out to tender, planned, controlled and carried out Europe-wide and even world-wide. Due to the complexity of modern buildings, technical building equipment and building service engineering become more and more important. At the same time many technical disciplines from civil engineering to mechanical and electrical engineering, up to automation- and controlling techniques are involved here, all with different specialized terminologies. This demands a great deal of communication between the participating persons.

This specialized dictionary offers technical terms for building services engineering from A to Z, from "absorption factor" to "zone air conditioning". Offering more than 35.000 entries this dictionary covers the entire terminology of the building services field. To each keyword the common synonyms are given, considering both British and American versions. Many terms offer additional short explanations in German or English for better understanding and correct use. The dictionary includes not only technical terms from the construction field, but also from heating-, ventilation and air condition technology, computer terminology as well as from control and regulation techniques and thus meets the requirements of a large circle of users in this interdisciplinary field.

Vorwort

Das Baugeschehen der Gegenwart ist geprägt durch eine zunehmende Internationalisierung. Bauprojekte werden europa- und sogar weltweit ausgeschrieben, geplant, gesteuert und ausgeführt. Bei der Komplexität moderner Bauwerke bekommt dabei die technische Gebäudeausrüstung ein immer größeres Gewicht. Gleichzeitig überschneiden sich in diesem Bereich viele Fachdisziplinen von der Bautechnik über den Maschinenbau und die Elektrotechnik bis hin zur Automatisierungs- und Steuerungstechnik mit ihren jeweils unterschiedlichen Fachterminologien. Dies stellt hohe Ansprüche an die Kommunikation zwischen den Planenden und Ausführenden.

Dieses Wörterbuch liefert Fachbegriffe aus dem weiten Feld der Gebäudetechnik von A bis Z, von »Absorptionsgrad« bis »Zonenklimatisierung«. Mit über 35.000 Stichwörtern erschließt dieses Fachwörterbuch die gesamte Begriffsterminologie der Gebäudetechnik. Zu jedem Stichwort werden die gebräuchlichen Synonyme aufgeführt, wobei britische und amerikanische Versionen berücksichtigt werden. Viele Begriffe haben zum besseren Verständnis und zur richtigen Verwendung zusätzlich kurze inhaltliche Erläuterungen auf Deutsch oder Englisch. Das Wörterbuch enthält nicht nur Fachbegriffe aus dem bautechnischen Bereich, sondern auch aus der Heizungs-, Lüftungs- und Klimatechnik, der EDV sowie der Steuerungs- und Regeltechnik und erfüllt somit die Ansprüche eines großen Benutzerkreises in diesem interdisziplinären Arbeitsfeld.

EXPLANATORY NOTES

All terms including proper names, abbreviations and word combinations are compiled in alphabetical order. In addition certain terms are accompanied by others despite the alphabetical order, that are used synonymously in this technical jargon.

fault, malfunction, breakdown, interference, error Fehler *m*, Störung *f*

Use of brackets:

(a) Square brackets:

1) To define or explain a term in German or English, e.g.

discharge air Abluft *f* [*Die aus einem Raum abströmende Luft*]

fuse board [*A board carrying fuses for a number of circuits in an electric installation*] Sicherungs(verteiler)tafel *f*

2) To inform about a term, its origin, specialised field etc. in German or English, e.g.

oil burner Ölbrenner *m* [*DIN 4787, Teil 1*]

operating theatre [*hospital*] Operationsraum *m*

(b) Round brackets:

1) For terms or part of terms which can be omitted, e.g.

exhaust (ventilation) system = exhaust system

Fuß(bade)wanne = Fußwanne

2) For omitting letters, e.g.

tile(d) stove = tile stove

Gasheiz(ungs)ofen = Gasheizofen

3) For abbreviations, e.g.

(Brit.) = Great Britain; **(US)** = USA

The comma separates synonyms, e.g.

inspection, examination Inspektion *f*

light barrier Lichtschranke, Lichtsperre *f*

The semicolon separates equivalent definitions, but alternatively used either in the USA or Great Britain.

elevator (US); lift (Brit.) Aufzug *m*

The gender of the German words is indicated by:

m = masculine

f = feminine

n = neuter, e.g.

Aufzug *m* = der Aufzug

m (pl) = The term is given in its plural form, with the singular being masculine, *f* (pl) and *n* (pl) apply accordingly.

Abbreviations:

B.S. = British Standard

A.S.T.M. = American Society for Testing Materials

DIN = Deutsches Institut für Normung

Hinweise zur Benutzung

Alle Begriffe, einschließlich Eigennamen, Abkürzungen und Wortkombinationen sind alphabetisch geordnet. Ergänzend sind bei manchen Begriffen weitere, außerhalb der alphabetischen Reihenfolge stehende Wörter aufgeführt, die in der Fachsprache synonym verwendet werden, z. B.:

fault, malfunction, breakdown, interference, error Fehler *m*, Störung *f*

Verwendung der Klammern:

(a) Eckige Klammern:

1) Erklärung oder Definition eines Wortes in Deutsch oder Englisch, z. B.:

discharge air Abluft *f* [*Die aus einem Raum abströmende Luft*]

fuse board [*A board carrying fuses for a number of circuits in an electric installation*] Sicherungs(verteiler)tafel *f*

2) Erläuternde Hinweise zu Herkunft, Sachgebiet, etc. in Deutsch oder Englisch, z. B.:

oil burner Ölbrenner *m* [*DIN 4787, Teil 1*]

operating theatre [*hospital*] Operationsraum *m*

(b) Runde Klammern:

1) Auslassungen von Wörtern, z. B.:

exhaust (ventilation) system = exhaust system

Fuß(bade)wanne = Fußwanne

2) Auslassungen von Buchstaben, z. B.:

tile(d) stove = tile stove

Gasheiz(ungs)ofen = Gasheizofen

3) Abkürzungen, z. B.:

(Brit.) = Großbritannien; **(US)** = USA

Das Komma trennt Synonyme, z. B.:

inspection, examination Inspektion *f*

light barrier Lichtschranke, Lichtsperre *f*

Das Semikolon trennt Wörter gleicher Definition, die in USA und Großbritannien unterschiedlich verwendet werden, z. B.:

elevator (US); lift (Brit.) Aufzug *m*

Das Geschlecht der deutschen Wörter wird wie folgt angegeben:

m = männlich

f = weiblich

n = sächlich, z. B.:

Aufzug *m* = der Aufzug

m (pl) = Das Wort ist im Plural angegeben, die Einzahl ist männlich; f(pl) und n(pl) sind sinngemäß zu verstehen.

Abkürzungen:

B.S. = British Standard

A.S.T.M. = American Society for Testing Materials

DIN = Deutsches Institut für Normung

A

A → address field

AAA → advance(d) application architecture

AAC → analogue(-to)-analogue converter

AA cooling, air-air cooling [*Self-cooling by natural circulation of air*] Luftselbstkühlung, Selbstluftkühlung *f*

AA/FA cooling, air-air/forced-air cooling Luftselbstkühlung mit Zwangsluftkühlung, Selbstluftkühlung *f* mit Zwangslüftung

A AND NOT B gate → AND NOT gate

AB → audio broadcast(ing)

AB, arbitration bus Zuteilungsbus *m*

abandoned call [*A call in which the caller cancels the call after a connection has been made, but before conversation takes place*] abgebrochene Ruf, aufgegebene Ruf *m*

abattoir waste Schlachthofabfall *m*

abat-vent [*French*] Deckmatte an Gewächshäusern, Gewächshausdeckmatte *f*

abat-vent [*French. Louver (US)/louvre (Brit.) which is placed in an exterior wall or masonry wall opening to permit light and air to enter but to break the wind*] Windladen *m*, Schutzbrett *n*, unbewegliche Jalousie *f*

abbreviated dialing (US); abbreviated dialling [*A method comprising special circuits which require less than the usual number of dial(l)ing operations to connect to one or more subscribers*] Kurzwahl, verkürzte Wahl, abgekürzte Wahl *f*

abbreviation, ABR Abkürzung *f*, Kurzzeichen *n*

ABC → automatic background control

a-b component transformation Clarktransformation *f*

ABCS → automatic building construction system

ABES → active bass extension system

ability → power

ablutionary appliance → wash appliance

ABM → asynchronous balanced mode

AB method → arbitration bus method

abnormal statement [*An element of a FORTRAN V (UNIVAC) program(me) which specifies that certain function subroutines must be called every time they are referred to*] ano(r)male Anweisung, regelwidrige Anweisung *f*

abode time Aufenthaltszeit *f* [*Gütesicherung*]

to abort abbrechen [*Programm*]

abort(ion) [*A predefined software-controlled or operator-initiated action resulting in the immediate cessation of an activity*] Abbruch *m*, Abbrechen *n*

abort(ion) of dialling (in), AOD Wahlabbruch *m*

above-atmospheric pressure → overpressure

above board Hauptspeichererweiterungsplatine *f* für expandierten Speicher [*von Intel*]

above finished floor(ing), AFF über (Fuß)Boden (-höhe)

above-freezing specific heat spezifische Wärme *f* über dem Gefrierpunkt

above-ground hydrant, surface hydrant, pillar hydrant, above-grade hydrant, post hydrant, out-of-ground hydrant [*B.S. 336*] Überflurhydrant, Oberflächenhydrant, oberirdische Hydrant, obererdige Hydrant, Säulenhydrant *m*

above-resonance hochabgestimmt

ABR, automatic baud rate detection, autobaud rate detection [*A process by which a receiving device determines the speed, code level, and stop bits of incoming data by examining the first character – usually a preselected sign-on character. ABR allows the receiving device to accept data from a variety of transmitting devices operating at different data rates without needing to configure the receiver for each specific data range in advance*] automatische Baudratenerkennung *f*, automatische Baudratennachweis *m*

ABR → abbreviation

abrasion resistance, abrasion strength Abriebfestigkeit *f*, Abriebwiderstand *m*

abrupt abrupt, unsanft, plötzlich, schroff, sprunghaft, sprungweise

abrupt-characteristic thermal detector, abrupt-characteristic thermal detecting device Temperaturmelder *m* mit Sprungverhalten, Temperaturmeldegerät *n* mit Sprungverhalten [*VDE 0660, T. 302*]

ABS → absent subscriber

absence code Abwesenheitskode, Abwesenheitscode *m* [*Zeiterfassung*]

absent subscriber, ABS abwesende Teilnehmer, nichtanwesende Teilnehmer *m*

absent subscriber's service, ABS service Fernsprechauftragsdienst, Fernsprechkundendienst *m*

A

absolute address, absolute ADRS, machine address, machine ADRS [*A reference to a storage location that has a fixed displacement from absolute memory location zero*] Absolutadresse, absolute Adresse *f*

absolute addressing [*Computer programming. The practice of using the actual number representing the memory location of the data to be processed as the address portion of the coded instruction*] Absolutadressierung, absolute Adressierung *f*

absolute alarm Grenzwertüberschreitungsalarm *m*, Gefahr(en)meldung *f* bei Grenzwertüberschreitung

absolute chiller → absolute refrigerating machine

absolute dielectric constant absolute Dielektrizitätskonstante, Verschiebungskonstante *f*

absolute electrometer [*A very precise type of attracted disk electrometer in which the attraction between the disks is balanced against the force of gravity*] Spannungswaage *f*

absolute field strength level, absolute field intensity level, absolute magnetic force level absolute Feldstärkepegel *m*

absolute filter, HEPA filter, high efficiency particulate air filter [*It has an efficiency of at least 99.90%, capable of dealing with particle sizes down to 0.01 micrometer*] Feinstfilter, Schwebstofffilter, HEPA-Filter, Hochleistungsschwebstofffilter *m, n* [*DIN 24184*]

absolute humidity; vapor concentration, vapor density (US) [*British spelling is "vapour". In a mix(ture) of water vapor and dry air, the mass of water vapor in a specific volume of the mix(ture)*] absolute (Luft)Feuchtigkeit, absolute (Luft)Feuchte *f*

absolute jump Absolutsprung, absolute Sprung *m*

absolute level absolute Pegel, Messpegel *m* [*Logarithmierte Verhältnis des Wertes einer physikalischen Größe zu einem definierten Bezugswert gleicher Dimension*]

absolute limit(ing) value absolute Grenzwert, Absolutgrenzwert *m* [*DIN 41 848*]

absolute loader → real loader

absolute power level absolute Leistungspegel *m* [*Logarithmierte Verhältnis des Leistungswertes zu einem definierten Leistungsbezugswert*]

absolute pressure [*Pressure above a perfect vacuum; the sum of ga(u)ge pressure and atmospheric pressure*] Absolutdruck, absolute Druck *m*, p_{abs}

absolute programming, one-level code, single-level code, specific code [*A method of source coding that uses no programming language but rather uses specific memory addresses and machine-coded (bit-pattern) operators*] absolute Programmierung *f*

absolute refrigerating machine, absolute chiller [*See remark under "chiller"*] Absolutkälteanlage, Absolutkältemaschine *f*

absolute viscosity, dynamic viscosity [*Force per unit area required to produce unit relative velocity between two parallel areas of fluid unit distance apart. It is denoted by the Greek letter μ (My)*] Absolutviskosität, dynamische Viskosität, absolute Viskosität, dynamische Zähflüssigkeit, absolute Zähflüssigkeit, Absolutzähflüssigkeit *f*

absolute voltage level absolute Spannungspegel *m* [*Logarithmierte Verhältnis des Spannungswertes zu einem definierten Spannungsbezugswert*]

absorbed real power → input power

absorbed voltage verbrauchte Spannung *f*

absorbent, absorptive, absorbing absorbierend, aufsaugend, aufzehrend, einsaugend, aufnehmend

absorbent, absorbing agent, absorbing material, absorbing medium, absorbing substance [*An example of an absorbent is table salt, which changes from a solid to a liquid as it absorbs moisture*] Absorptionsmittel *n*, Absorptionsstoff *m*, Absorbens *n*, absorbierende Stoff, aufnehmende Stoff *m*, Aufsaugmittel, Einsaugmittel *n*, einsaugende Stoff *m*

absorbent liquid cooler → absorption liquid cooler

absorber → (shock) absorber

absorber [*A material or device that takes up and dissipates radiated energy; may be used to shield an object from the energy, prevent reflection of the energy, determine the nature of the radiation, or selectively transmit one or more components of the radiation*] Absorber *m*

absorber-evaporator Absorberverdampfer *m*

absorption [*Process whereby a porous material extracts one or more substances from an atmosphere, a mix(ture) of gases, or a mix(ture) of liquids*] Absorption, Aufsaugung, Absorbierung, Aufzehrung, Einsaugung *f*

absorption, dielectric soak, dielectric absorption dielektrische Absorption *f*

absorption ammonia refrigerator → ammonia absorption refrigerator

absorption chiller, absorption refrigerating system [*Refrigeration is created in this system by evapora-tion of a refrigerant in a heat exchanger (evaporator), with the vapo(u)r then absorbed by an absorbent medium from which it is subsequently expelled by heating at a higher partial vapo(u)r* *pressure (in a generator) and condensed by cooling in another heat exchanger (condenser)*] Absorptionskälteanlage, Kälteanlage *f*

absorption coefficient, absorption factor, coefficient of absorption, factor of absorption [*In acoustics, the ratio of the sound energy absorbed by a surface of a medium or material to the sound energy incident on the surface*] Absorptionsgrad, Schallschluckgrad *m*

absorption coefficient, absorption ratio [*In optics, the fraction of incident energy that is absorbed per unit pathlength*] Absorptionskoeffizient *m*, Absorptionsverhältnis *n*

absorption coefficient [*In physics. If a flux through a material decreases with distance "x" in proportion to "e^{-ax}", then "a" is called the absorption coefficient*] Absorptionskoeffizient *m*

absorption constant, absorptivity [*Analytical chemistry. A constant used in Beer's law that is a relative measure of the absorption intensity of a compound and is defined as the absorbance per unit concentration per unit length of light path*] Absorptionskonstante *f*

absorption cooling Absorptionskühlung *f*

absorption counter Kühl(ungs)theke *f* mit Absorptionskühlmaschine

absorption cryogenic machine Absorptionskältemaschine, Absorptionskryomaschine *f*

absorption cycle [*In refrigeration, the process whereby a circulating refrigerant, for example ammonia, is evaporated by heat from an aqueous solution at elevated pressure and subsequently reabsorbed at low pressure, displacing the need for a compressor*] kompressorlose Kälteerzeugung *f*

absorption dehumidifier Absorptionsentfeuchter *m*

absorption diffusion heat pump → diffusion absorption heat pump

absorption discontinuity Absorptionssprung *m*

absorption display, absorption merchandiser, absorption showcase Vitrine *f* mit Absorptionskältemaschine

absorption edge Absorptionskante *f*

absorption fading [*A variation in the absorption of radio waves propagated through the ionosphere due to changes in the densities of ionization*] Absorptionsschwund *m*

absorption frequency meter [*A frequency meter working on the principle that maximum power is absorbed by a calibrated tuned circuit when this resonates with the unknown frequency. In waveguide applications the tuned circuit is a cavity resonator*] Absorptionsfrequenzmesser *m*

absorption heating machine Absorptionsheiz(ungs)maschine *f*

absorption heat pump Absorptionswärmepumpe *f*, AWP

absorption hygrometer, chemical hygrometer Absorptionsfeuchtemesser, Absorptionsfeuchtigkeitsmesser *m*

absorption law Absorptionsgesetz *n*

absorption liquid, absorbent liquid Absorptionsflüssigkeit *f*, absorbierende Flüssigkeit, flüssige Absorptionsmittel, flüssige Absorbens *n*

absorption liquid cooler, absorbent liquid cooler Absorptionsflüssigkeitskühler *m*

absorption loss [*That part of the transmission loss due to the dissipation or conversion of either sound energy or electromagnetic energy into other forms of energy, either within the medium or attendant upon a reflection*] Absorptionsverlust *m*

absorption machine Absorptionsmaschine *f*

absorption of sound → acoustic absorption

absorption peak [*Spectroscopy. The wavelength at which the degree of absorption by a particular molecular substance is at a maximum*] Absorptionsspitze *f*

absorption ratio → absorption coefficient

absorption refrigerating system → absorption chiller

absorption refrigeration [*1. Refrigeration process whereby a secondary fluid absorbs the refrigerant, and in so doing, gives up heat, then releases the refrigerant, during which it absorbs heat. 2. Refrigeration in which cooling is effected by the expansion of the liquid refrigerant into gas and absorption of the gas by water. The refrigerant is reused after the water evaporates.*] Absorptionskälteerzeugung *f*

absorption refrigerator, absorption refrigerating unit, absorption refrigeration unit [*See remark under "refrigerating unit"*] Absorptionskälteaggregat *n*, Absorptionskälteeinheit *f* [*Hat eine kleinere Leistung als eine Absorptionskälteanlage*]

absorption resistance Absorptionswiderstand, Schallschluckwiderstand *m*

absorption solution Absorptionslösung *f*

absorption (sound) attenuator, absorption (sound) damping device, absorption (sound) muffler, absorption sound absorber Absorptions(schall)dämpfer *m*

absorption spectrophotometer, difference spectrophotometer [*An instrument used to measure*

A

the relative intensity of absorption spectral lines and bands] Absorptionsspektralphotometer *n*, Absorptionsspektrophotometer [*„Fotometer" ist synonym*]

absorption system [*Equipment intended or installed for the purpose of heating or cooling air by an absorption unit by direct or indirect means, and discharging such air in any room or space*] Absorptionssystem *n*

absorption system [*Refrigerating system in which the gas evolved in the evaporator is taken up by an absorber*] Absorptionssystem *n*

absorption unit [*Factory-tested assembly of component parts producing refrigeration for comfort cooling by the application of heat. The definition applies to absorption units that also produce comfort heating*] Absorptionseinheit *f*

absorption unit → sabin

absorptive attenuator, resistive attenuator [*A length of waveguide containing a dissipative material*] Absorptionsabschwächer *m*, absorbierende Dämpfungsglied, absorbierende Abschwächer *n*

absorptivity → absorption constant

ABS service → absent subscriber's service

abstract automata theory [*The mathematical theory which characterizes automata by three sets: input signals, internal states, and output signals; and two functions: input function and output function*] abstrakte Automattheorie *f*

abstract code, pseudocode abstrakte Kode, abstrakte Code, Pseudokode, Pseudocode *m*

abstracting [*The collection of items and quantities, involved in a building contract in preparation for a bill of quantities. The descriptions and quantities are arranged in the recognized sequence in works sections and reduced to the appropriate units of measurement*] Ausziehen *n* von Mengen, Mengenauszug *m*

abstracting [*Technique(s) whereby, on the basis of statistical properties, a subject of the sentences in a document is selected as representative of the general content of that document*] Ausziehen *n*, *Zusammenfassung f*

abstract syntax [*The specification of application layer data or application-protocol-control information by using notation rules which are independent of the encoding technique used to represent them (ISO 8822)*] darstellungsunabhängige Syntax, abstrakte Syntax *f*

A-bus, primary bus A-Bus *m*, Erstbus, Primärbus

AC → automatic ((open) loop) control

AC → alternating current

A/C → air conditioning

AC adaptor, alternating current adaptor, a.c. adaptor Wechselstromanpasser *m*

ACAU → automatic calling and answering unit

a.c. balancer → static balancer

a.c. ballast, alternating current ballast, AC ballast Wechselstromvorschaltgerät *n*

ACC → advance(d) control circuit

ACC → area communication controller

ACC → adaptive ((closed) loop) control

ACC → (storage) battery

ACC, allocatable channel zuweisbare Kanal, anweisbare Kanal *m*

AC capacitor, alternating current capacitor Wechselspannungskondensator *m*

to accelerate → to quicken

accelerating function → ramp function

acceleration condensor, speeding-up condenser Beschleunigungskondensator, Hochlaufkondensator *m*

acceleration due to gravity → gravitation(al) acceleration

acceleration (gain) constant Beschleunigungsfaktor, integrale Übertragungsfaktor, Hochlauffaktor *m* [*Systemtechnik*]

acceleration generator Beschleunigungsgeber, Hochlaufgeber *m* [*Systemtechnik*]

acceleration jogging, accelerating jogging Hochlauftippen, Beschleunigungstippen *n*

acceleration level, accelerating level Beschleunigungspegel, Hochlaufpegel *m*

acceleration machine, accelerating machine, accelerator [*In electronics, a machine used to accelerate charged particles to very high energies*] Beschleunigungsmaschine *f*, Beschleuniger *m*, Hochlaufmaschine *f*

acceleration memory, accelerating memory Beschleunigungsspeicher, Hochlaufspeicher *m*

acceleration pickup, accelerating pickup Beschleunigungsaufnehmer, Hochlaufaufnehmer *m*

acceleration pressure loss, accelerating pressure loss Beschleunigungsdruckverlust, Hochlaufdruckverlust *m*

acceleration(-rate) controller Beschleunigungsregler, Hochlaufregler, Sanftanlaufregler *m*

acceleration torque Beschleunigungsmoment *n*

accelerator → acceleration machine

accelerator card Beschleunigerkarte *f*

accent light Objektlicht, Akzentlicht *n*

accelerogram [*A graphic representation of the data recorded by an accelograph*] Beschleunigungsoszillogramm , Beschleunigungsschwingungsbild *n*

accelerograph [*An accelerator having provisions for recording the acceleration of a point on the earth during an earthquake or for recording any other type of acceleration*] Beschleunigungsschreiber *m*

accelerometer [*An instrument which measures acceleration or gravitational force capable of improving acceleration*] Beschleunigungsmesser *m*

accent lighting, accent illumination [*Any directional lighting which emphasizes a particular object or draws attention to a particular area*] Objektbeleuchtung, Akzentbeleuchtung *f*

accent luminaire, accent light(ing) fitting Akzentleuchte *f*, Akzentbeleuchtungskörper *m*, Objektleuchte *f*, Objektbeleuchtungskörper *m*

accentuator (circuit) [*Electronics. An electrical network or device that is used for preemphasis; i.e. to amplify a particular band of frequencies, such a device includes filter, equalizer and tone controller*] Anheber, Tonanheber *m*

to accept annehmen, übernehmen, aufnehmen

to accept abnehmen [*Güteprüfung*]

acceptability [*The status of equipment, products, devices and procedures that meet or exceed minimum standards*] Annehmbarkeit, Abnehmbarkeit *f* [*Güteprüfung*]

acceptable output, acceptable performance, acceptable service [*Performance of a component or system that meets specified design parameters under actual load*] Gutleistung, abnehmbare Leistung *f* [*Maschine; Gerät; Anlage*]

acceptable (quality) level, A(Q)L, accepted (quality) level Abnahmegutwert, Gutwert, Gütewert, Abnahme(güte)wert *m* [*Anteil schlechter Bauelemente in einem bestimmten Lieferlos, der vom Kunden noch abgenommen wird*]

acceptable (quality) limit, accepted (quality) limit Abnahmegutwertgrenze, Gutwertgrenze, Abnahme(güte)wertgrenze, Gütewertgrenze *f*

acceptable (radio) interference [*A level of (radio) interference that would normally be classed as harmful but, by agreement between the authorities operat-ing the radio service in question, is deemed to be acceptable. Usually such a level is only acceptable for an interim period until the performance can be up-graded or until a temporary service is withdrawn*] hinnehmbare (Funk)Störung *f*

acceptance Aufnahme *f*

acceptance Annahme *f*

acceptance Abnahme *f* [*Güteprüfung*]

acceptance angle, incident angle, incidence angle, angle of acceptance, angle of incidence Einfall(s)winkel, Öffnungswinkel *m*

acceptance batch Abnahmemenge *f* [*VDE 0715*]

acceptance certificate → certificate of acceptance

acceptance check(ing) Abnahmekontrolle *f* [*Güteprüfung*]

acceptance number Gutzahl, Abnahmezahl *f* [*Güteprüfung*]

acceptance test Abnahmeprüfung *f*, Abnahmeversuch *m*, Abnahmeprobe *f*

acceptance tolerance Abnahmetoleranz *f* [*Güteprüfung. Zulässige Abweichung vom vorgegebenen* Maß]

accept data state, ACDS Datenübernahmezustand *m* der Senke [*Ein Schnittstellenzustand*]

accepted power → power acceptance

accepting SS user, acceptor, accepting session service user annehmende Benutzer *m* des Dienstes der Kommunikationssteuerung

accept(ing) switch Quittierschalter *m*

acceptor Akzeptor, Akzepter *m* [*Halbleiter*]

acceptor energy level Akzeptorenergieniveau, Akzepterenergieniveau *n*, Akzeptorenergiehöhe, Akzepterenergiehöhe *f* [*Halbleiter*]

acceptor idle state, AIDS (Daten)Senkeruhezustand, Ruhezustand *m* der (Daten)Senke [*programmierbare Messgeräte DIN IEC 625*]

acceptor not ready state, ANRS Nichtbereitzustand *m* der (Daten)Senke [*programmierbare Messgeräte DIN IEC 625*]

acceptor (of data), data acceptor [*Any device capable of accepting data in a controlled manner; it is used in British Standard Interface Specifications to refer to devices which take data from source*] (Daten)Senke *f* [*programmierbare Messgeräte*]

acceptor ready state, ACRS Bereitzustand *m* der (Daten)Senke [*programmierbare Messgeräte DIN IEC 625*]

acceptor wait(ing) for new cycle state, AWNS Wartezustand *m* der (Daten)Senke [*programmierba-*

A

re Messgeräte DIN IEC 625]

to accept pulses verarbeiten von Impulsen

to access zugreifen [*Speicher*]

access [*In a local area network the ability to work with files. Usually used in connection with files stored on network disks. Various access rights may be assigned to users*] Zugriff *m*

access Zutritt *m*; Zugang *m* [*frühere Benennung*]

access Anbindung *f*

access ... → cleaning ...

accessable (US) → available

access aperture → access opening

access arm [*In automatic data processing, the mechanical device which positions the read/write head on a magnetic storage unit*] Zugriffsarm *m*

access authority, access authorization, access privilege, access entitlement Zutrittsberechtigung, Zugangsberechtigung *f* [*frühere Benennung*]

access authorization, access authority, access privilege, access entitlement Zugriffsberechtigung *f*

access capability Anschlusskapazität, Anschlussfähigkeit *f*, Anschlussvermögen *n* [*ISDN*]

access card Zutrittskarte, Eintrittskarte; Zugangskarte *f* [*frühere Benennung*]

(access) chamber [*deprecated*] **→ manhole**

access channel Anschlusskanal *m* [*ISDN*]

access contention Zugriffskonflikt *m* [*Konflikt zwischen Endeinrichtungen am passiven Bus, wenn diese gleichzeitig im D-Kanal senden wollen*]

access contention resolution Auflösung *f* des Zugriffskonfliktes, Zugriffskonfliktauflösung [*In der Schicht 1*]

access control Zugriffsschutz *m*

access control unit, ACU Adressenberechtigungseinheit *f* [*Baugruppen in GSM-Netzen*]

access cover → access panel

access database Zugriffsdatenbank *f*

access delay Zutrittsverzug *m*, Zutrittsverzögerung *f*; Eintrittsverzug *m*, Eintrittsverzögerung *f*, Zugangsverzug *m*, Zugangsverzögerung *f* [*frühere Benennungen*]

access denial Zutrittsverweigerung, Eintrittsverweigerung, Zugangsverweigerung *f* [*frühere Benennung*]

access door → cleaning door

access door [*A temporary door fabricated during construction of a building to allow access to concealed areas, such as crane spaces*] Bautür *f*

access enable, ACEN Zugriffsfreigabe *f*

access entitlement → access authority

access eye → cleaning eye

access for fire department vehicles, fire department vehicles access Feuerwehrzufahrt *f*

access frequency Zugriffshäufigkeit *f*

access granted decision Zutrittsgenehmigungsentscheid, Eintrittsgenehmigungsentscheid; Zugangsgenehmigungsentscheid *m* [*frühere Benennung*]

access granting, access permission Zugriffserteilung, Zugriffsgenehmigung *f*

access granting, access permission Zutrittserteilung, Zutrittsgenehmigung, Eintrittserteilung, Eintrittsgenehmigung; Zugangserteilung, Zugangsgenehmigung *f* [*frühere Benennungen*]

access gully → cleaning gully

access hole Zugang(s)loch *n* [*gedruckte Schaltung*]

(access) hopper Einwurfstation *f* [*Müllschlucker*]

accessibility (US) → availability

accessibility [*The network elements, such as ports or trunks, that can be reached by a user group*] Zugreifbarkeit, Greifbarkeit *f*

accessibility [*See remark under "accessible"*] Zugänglichkeit, Erreichbarkeit *f* [*von Geräten*]

accessible [*As applied to equipment. Equipment is accessible when close approach is not prevented by (locked) doors, elevation, or other effective means*] zugänglich, erreichbar

accessible-hermetic motorcompressor, semihermetic motorcompressor halbhermetische Motorverdichter, Deckelmotorverdichter, halbhermetischer Motorkompressor, Deckelmotorkompressor *m*

accessible-hermetic (refrigerant) compressor, semihermetic (refrigerant) compressor, half-hermetic (refrigerant) compressor [*Its housing is sealed by one or more gasketed joints and is provided with means of access for servicing internal parts in the field*] halbhermetische (Kältemittel)Verdichter, halbhermetische (Kältemittel)Kompressor *m*

accessible to hand, manually accessible handzugänglich, handerreichbar, greifbar

accessing [*database*] Zugreifen *n*

access intent Zugriffsabsicht *f*

accession number Zugriffszahl *f* [*z.B. zur Literatursuche*]

access key Ordnungswort, Passwort *n*

access ladder Einsteig(e)leiter, Einstiegsleiter, Zugangsleiter *f*

access level Zugriffsebene *f*

A

access line, local line, local loop, access loop [*The connection between a subscriber's facility and a public network – either a PDN, public switched network, or public telephone network*] Verbindungsleitung, Anschlussleitung *f*

access management [*The array of physical, hardware, and software techniques used to limit access to a computing system to authorized personnel, program or network nodes*] Zugriffsbegrenzung *f*

access method Zugriffsverfahren *n*, Zugriffsmethode *f* [*Je nach Speicheraufbau ist Zugriff nur unter bestimmten Voraussetzungen möglich; zum Beispiel lässt ein RAM-Speicher einen wahlfreien Zugriff zu, ein Magnetbandspeicher nur einen sequentiellen*]

access mode [*A programming clause in COBOL which is required when using a random-access device so that a specified record may be read out of or written into a mass storage bin*] Zugriffsart *f*

access motion, access movement Zugriffsbewegung *f*

access opening, access port, access aperture, access orifice Zutrittsöffnung; Zugangsöffnung *f* [*frühere Benennung*]

accessories of a building, appurtenances of a building, building accessories, building appurtenances Beiwerk *n*, Zubehöre; Zubehörden *f* [*Schweiz*]

accessory, appurtenance [*A part or assembly that aids the effectiveness of machinery and equipment without changing its function*] Zubehör *n* [*selten: m*]

accessory box, appurtenance box Einbaudose *f* für Installationsgeräte

accessory item, appurtenance item Zubehörteil *m, n*

access panel [*IEC 825*] Abdeckung *f* [*Abnehmbare Teil des Schutzgehäuses eines Lasergerätes. VDE 0837*]

access panel Zugangstürchen *n* [*Badewanne*]

access panel, access cover Montagedeckel *m*

access path, access route [*A sequence of steps, as from a directory to subdirectory or from one network node to another, that will read a desired file, network node or user*] Zugang(s)weg *m*

access permission → access granting

access plate [*A remov(e)able plate bolted on the side of a tank for a hot water system, etc.*] Zugangsplatte *f*

access point Zutrittsstelle, Zugangsstelle *f* [*frühere Benennung*] [*Fernmeldetechnik*]

access prefix Vorwahlziffer *f* [*Fernmeldetechnik*]

access privilege → access authorization

access protection Zugriffsschutz *m*

access protocol Zugriffsprotokoll *n*

access readiness Zugriffsbereitschaft *f*

(access) refusal list, black list Zutrittssperrliste, Eintrittssperrliste, Sperrliste; Zugangssperrliste *f* [*frühere Benennung*]

access relay set Anschaltesatz *m* [*Fernmeldetechnik*]

access release Zutrittsfreigeber, Zugangsfreigeber *m*

access request, ACRQ Zugriffsanforderung *f*

access right Zugriffsrecht *n*

access right Zutrittsberechtigung, Eintrittsberechtigung, Zugangsberechtigung *f* [*frühere Benennung*] [*Intrusionsschutz*]

access route → access path

access security Zutrittssicherung; Zugangssicherung *f* [*frühere Benennung*]

access speed Zugriffsgeschwindigkeit *f*

access surveillance Zutrittsüberwachung, Eintrittsüberwachung, Zugangsüberwachung *f* [*frühere Benennung*] [*Intrusionsschutz*]

access system controller, access system regulator Zutrittssystemregler, Zugangssystemregler *m* [*frühere Benennung*]

access table Zugriffstabelle *f*

access time [*Computer engineering. The time interval between a request for data to be retrieved from memory and the time the data is available for processing*] Zugriffszeitspanne *f*

access time [*The time interval between generation of a mass storage address by a control unit or input/output controller and the initiation of data transfer be-tween that location and the main memory*] Zugriffszeitspanne *f*

access time, read time [*The time interval between generation of a read signal from the external interface of a storage device and the first bit or bit pattern to be read back out from the storage medium*] Zugriffszeitspanne, Lesezeitspanne *f*

access tree Zugriffsbaum *m*

access vector Zugriffsvektor *m*

access zone Zutrittszone, Zugangszone *f* [*frühere Benennung*]

accidental arc Störlichtbogen, Fehlerlichtbogen *m*

accidental contact protection cap Berührungsschutzabdeckung, Kontaktschutzabdeckung *f* [*Abdeckung gegen zufälliges Berühren*]

A

accidental earth (Brit.); accidental ground (US) Erdschluss *m*

accident hazard Unfallgefährdung *f*

accident prevention equipment Unfallverhütungsgerät(e) *n*, (*f*)

accidentproof unfallsicher

accident safety Unfallsicherheit *f*

accident security Unfallsicherung *f*

accompanied fuse, accompanied fuze Teilbereichssicherung *f*

according to modular coordination modulgerecht, maßordnungsgerecht

according to scale maßstäblich, maßstab(s)gerecht

account Zugangsberechtigung *f* [*z. B. zu einem Rechner*]

account acquisition Rechnungserfassung *f*

accrued compilation → total compilation

accrued outage → total outage

to accumulate (an)sammeln, anhäufen

accumulated frost, collected frost, deposited frost Reif(belag) *m*, Reifschicht *f*

accumulated heat, stored heat Speicherwärme *f*

accumulated pitch error, total pitch error Gesamtteilungsfehler, Sammelteilungsfehler *m* [*z. B. Elektromaschine; optische Gitter*]

accumulating elapsed time meter, elapsed time accumulating meter Summierzeitzähler, Zeitsummierzähler *m*

accumulation (An)Sammlung, Anhäufung *f*

accumulative latching mechanical system Rastmechanismus *m* mit Fremdauslösung [*Drucktaster*]

accumulation layer Anreicherungsschicht, Anreicherungslage *f* [*Halbleiter*]

accumulator [*A device that accumulates or stores something*] Akkumulator, Sammler *m*

accumulator → low-pressure (refrigerant) receiv-eaccumulator cell; battery [*misnomer*] Akkumulatorzelle, Sammlerzelle *f*

accumulator (register) [*A special(-purpose) register and associated equipment within the arithmetic/ logic unit of the processor, used for temporary storage of operands and intermediate results of arithmetic and logic operations*] Akkumulator, Spezialregister *m*, Sonderregister, Sammlerregister *n*

accumulator shift instruction [*A computer instruction that causes the word stored in a register to be displaced a specified number of bit positions to the left or right*] Stellenverschiebungsbefehl *m*

accumulator switch, battery switch Zellenschalter *m*

accuracy, trueness [*The extent to which a measurement approaches the true value of the measured quantity*] Genauigkeit, Maßhaltigkeit *f*

accuracy of the mean, mean accuracy, trueness of the mean, mean trueness Treffgenauigkeit *f*, Richtigkeit [*DIN 55 350 Teil 13. Ausmaß der Übereinstimmung zwischen dem Erwartungswert und dem wahren Wert*]

accuracy to size, size accuracy, trueness to size, size trueness Maßhaltigkeit *f*

accurate within one cycle periodengenau

ACD → automatic call distribution

a.c.-d.c. device, universal device, a.c.-d.c. appliance, universal appliance [*Also written: AC/DC and ac/dc*] Wechselstrom-Gleichstrom-Gerät *n*

ac/dc receiver, universal receiver, AC/DC receiver [*A radio receiver designed to operate from either an alternating or direct current power line*] Allstromempfänger *m*

AC/DC switch, universal switch, alternating current/direct current switch, a.c./d.c. switch Wechselspannung(s)-Gleichspannung(s)-Schalter, Allstromschalter, Universalschalter *m*

a.c. disconnector, A.C. disconnector, alternating current disconnector [*IEC 129*] Wechselstromtrennschalter *m* [*VDE 0670, Teil 2*]

ACDS → accept data state

a.c. dump, alternating current dump, A.C. dump Ausfall *m* der Wechselstromversorgung, Wechselstromausfall, Versagen *n* der Wechselstromversorgung, Wechselstromversagen *n*

ACE → automatic calling equipment

a.c. earthing switch, alternating current earthing switch (Brit.); a.c. grounding switch, alternating current grounding switch (US) [*IEC 129*] Wechselstromerdungsschalter, Erdungswechselstromschalter *m* [*VDE 0670, Teil 2*]

ACEN → access enable

a.c. generator → alternator

achromatic colour (Brit.); achromatic color (US) [*A colo(u)r that has no hue or saturation but only brightness, such as white, black and various shades of gray*] achromatische Farbe, unbunte Farbe *f*

achromat(ic lens), antispectroscopic lens Achromat *m*, achromatische Linse, unbunte Linse *f* [*DIN 19 040*]

achromatic locus, achromatic region [*Optics. The region in a chromaticity diagram that includes points representing chromaticities that occur commonly* *and may therefore be acceptable reference standards of illumination*] Unbuntbereich, Farblosbereich *m* [*selten: n*]

achromatism, achromaticity [*The quality or fact of being achromatic; a lack of colo(u)r*] Farblosigkeit, Unbuntheit, Achromasie *f*

acid-base indicator [*Analytical chemistry. Any substance that reveals the pH or degree of acidity or basicity of a solution through a colo(u)r change*] pH-Indikator, pH-Anzeiger *m*

acid dew point Säuretaupunkt *m*

acidic → acidulous

acidic fermentation, sour fermentation, acidulous fermentation saure Gärung *f*

acid level indicator Säurestandanzeiger *m*

acid number, acid value [*A measure of the free acid content of a substance, such as oil, resin, or wax, as determined by the number of milligrams of potassium hydroxid(e), KOH, required to neutralize one gram of the substance*] Säurezahl *f*, Säurewert *m*

acid slop sink Säureausguss *m*

acid strength, acid stability Säurefestigkeit, Säurebeständigkeit *f*

to acidulate, to make acidic, to make sour ansäuern

acidulated, made acidic, made sour angesäuert

acidulous, acidic, sour sauer

acid value → acid number

to acknowledge, to confirm rückmelden, quittieren, bestätigen

acknowledge [*Telecommunications. A statement transmitted by a sensing location to a receiving location requesting confirmation that a previous message has been received and understood*] Bestätigungsaufforderung, Quittungsaufforderung, Rückmeldungsaufforderung, Rückmeldeaufforderung *f*

acknowledged data transfer → data-transfer-acknowledged

acknowledged service → confined service

acknowledg(e)ment, confirmation, ACK [*A control character used (with NAK) in BSC communications protocol to indicate that the previous transmission block was correctly received and that the receiver is ready to accept the next block. Also used as a ready reply in other communications protocols, such as Hewlett-Packard's ENQ/ACK protocol and the ETX/ACK method of flow control*] Bestätigung, Quittung, Rückmeldung *f*

acknowledg(e)ment key, confirmation key, confirming key, ACK key Bestätigungstaste, Rückmeldungstaste, Quittungstaste, Rückmeldetaste *f*

acknowledg(e)ment light-emitting diode, confirmation light-emitting diode, confirming light-emitting diode, acknowledg(e)ment LED, confirmation LED, confirming LED, ACK light-emitting diode Bestätigungsleuchtdiode, Quittungsleuchtdiode, Rückmeldungsleuchtdiode *f*

acknowledg(e)ment proof, confirmation proof, ACK proof, confirming proof Bestätigungsnachweis, Quittungsnachweis, Rückmeldungsnachweis *m* [*Nachweis für eine Leit- und/oder Inselzentrale, der diesen Geräten eine firmenneutrale Datenkommunikation bestätigt*]

acknowledg(e)ment sequence, confirmation sequence, confirming sequence, ACK sequence Bestätigungsreihenfolge, Quittungsreihenfolge, Rückmeldungsreihenfolge *f*

acknowledg(e)ment signal, confirmation signal, confirming signal, ACK signal Bestätigungssignal, Quittungssignal, Rückmeldungssignal *n*

acknowledg(e)ment telegram, confirmation telegram, ACK telegram, confirming telegram Quittungstelegramm, Rückmeldungstelegramm, Bestätigungstelegramm *n* [*Positive Rückmeldung einer Leitzentrale auf eine Spontanmeldung einer Inselzentrale*]

acknowledg(e)ment tone, confirmation tone, ACK tone, confirming tone Bestätigungston, Quittierungston, Rückmeldungston, Quittierton, Rückmeldeton *m*

acknowledg(e)ment window, confirmation window, confirming window, ACK window Bestätigungsfenster, Quittungsfenster, Rückmeldungsfenster *n*

acknowledging, confirming Quittieren, Bestätigen, Rückmelden *n*

aclastic [*Optics. Relating to or describing something that does not refract light*] lichtrefraktionswidrig

a-contact → make contact (element)

a.c.-operated, AC-operated, alternating-current-operated wechselstrombetätigt, wechselstrombedient, wechselstrombetrieben

acoustic [*Of or relating to sound or the scientific study of sound. Many compound terms beginning with "acoustic" may also begin with "acoustical", "sonic" or "sound"*] akustisch

acoustic absorber, silencer, muffler, sound absorber, sonic absorber Schalldämpfer *m* [*DIN EN ISO 11 654*]

acoustic absorption, sonic absorption, absorption of sound, sound absorption (Schall)Absorp-

A

tion, (Schall)Schluckung *f* [*Abschwächung der Luftschallreflexion in geschlossenen Räumen durch schallschluckende Wand- oder Deckenverkleidungen, die in halligen Räumen die Nachhallzeit verkürzen und eine Schallsenkung bewirken*]

acoustic absorption coefficient, sound absorption coefficient, acoustic absorptivity, sound absortivity, sonic absorptivity [*The ratio of sound energy absorbed to that arising at a surface or medium. "Factor" may synonymously be used for "coefficient"*] Schallschluckgrad, Schallschluckkoeffizient, Schallabsorptionsgrad, Schallabsorptionskoeffizient *m*

acoustic acuity → **acuity**

acoustical analysis, sound analysis Akustikanalyse, Schallanalyse *f*

acoustic alarm, audible alarm, sound alarm, sonic alarm Höralarm, akustische Alarm, Schallalarm, Akustikalarm *m*

acoustic alarm detector, audible alarm detector, sound alarm detector, sonic alarm detector [*"Detecting device" and "detection device" may syn-onymously be used for "detector"*] Höralarmmelder, Schallalarmmelder, akustische Alarmmelder, Akustikalarmmelder *m* [*„Meldegerät" kann synonym für „Melder"verwendet werden*]

acoustic alarm device, audible alarm device, sound alarm device, sonic alarm device Höralarmgeber, Akustikalarmgeber, Schallalarmgeber, akustische Alarmgeber *m* [*Elektronische Sirenen, Motorsirenen, Hupen, Lautsprecher, Glocken, Schnellschlagwecker, Klingeln, Pfeifen, usw.*]

acoustic alarm unit, sound alarm unit, audible alarm unit, sonic alarm unit akustische Alarmanlage, Schallalarmanlage, Höralarmanlage, Akustikalarmanlage *f*

acoustical correction, sound correction Akustikkorrektur, Schallkorrektur, Akustikberichtigung, Schallberichtigung *f*

acoustical engineering [*The practical application of the science of acoustics, including such elements as noise and vibration control, sound reproduction and broadcasting, and the use of sound instruments to make measurements and to examine or process various materials*] Akustiktechnik *f*

acoustical environment Akustikumgebung, Akustikumwelt *f* , Akustikmilieu *n*

acoustically lined akustisch verkleidet, akustisch bekleidet

acoustical material Akustik(werk)stoff *m*, Akustikmaterial *n*

acoustical treatment Akustikbehandlung *f*

acoustic attenuation, sonic attenuation, sound attenuation, sonic damping, sound damping, sonic deadening, sound deadening, sonic abatement, sound abatement, sonic muffling, sound muffling, acoustic damping, acoustic deadening, acoustic abadement [*Process in which sound energy is absorbed or otherwise diminished in intensity*] Schallschwächung, Schalldämpfung, akustische Schwächung, akustische Dämpfung *f*

acoustic background, sonic background, sound background Schallhintergrund, akustischer Hintergrund *m*

acoustic barrier, sound barrier, sonic barrier, transonic barrier [*Aviation. A sharp, sudden increase in aerodynamic drag of an aircraft approaching the speed of sound*] Schallgrenze *f*

acoustic burglar(y) protection, sonic burglar(y) protection, sound burglar(y) protection akustische Einbruchschutz, akustische Einbrecherschutz, akustische Wertschutz, Schall-Einbruchschutz, Schall-Einbrecherschutz, Schall-Wertschutz *m*

acoustic calibrator Eichschallquelle *f*

acoustic card, soundcard, sonic card Klangerzeuger(karte), elektronische Klangerzeuger(karte) *m*, (*f*)

acoustic center (US); acoustic centre (Brit.) [*For a sound transmitter or receiver, the point from which sound appears to emanate or diverge. "Sound" and "sonic" may synonymously be used for "acoustic"*] Schallzentrum, akustische Zentrum *n*

acoustic communication, sonic communication, sound communication Schallkommunikation, akustische Kommunikation *f*

acoustic (computer) memory [*An early type of (computer) memory using an acoustic delay line composed of a material such as mercury or quartz*] akustische (Rechner)Speicher *m*

acoustic construction [*A structure which reduces sound levels at the source and is used to impede the passage of airborne noise and impact sound transmission by means of its surface, its discontinuous construction or its density*] Akustikbau *m*

acoustic control, sound control Schallbeherrschung *f*

acoustic coupler [*A kind of modem that converts electrical signals into audio signals, enabling data to be transmitted over the public telephone network via a conventional telephone handset; it also converts the audio signals back into electrical signals at the receiving end*] Akustikkopp(e)ler, Akustikkupp(e)ler, akustische Kopp(e)ler, akustische Kupp(e)ler *m*

acoustic coverage, sound coverage, sonic coverage Schallbedeckung, Akustikbedeckung *f*

acoustic decay, sound decay, sonic decay Schallzerfall, akustischer Zerfall *m*

acoustic delay, sonic delay, sound delay [*An electronically or physically induced increase in the time between the transmission of a sound and reception of the sound, so as to produce some desired effect*] Schallverzögerung, akustische Verzögerung *f*, Schallverzug, akustische Verzug *m*

acoustic delay line, sonic delay line, sound delay line, acoustic storage [*Electronics. A transmission line that delays acoustic signals. The delay is created by the sound pulses being reflected through some liquid or solid material; e.g. a quartz bar*] Schallverzögerungsleitung, akustische Verzögerungsleitung *f*

acoustic device, sonic device, sounder, sound(ing) device [*Colloquially called "alarm". "Appliance" may synonymously be used for "device"*] Schallgeber, Hörmelder *m*, Akustikgerät *n*, akustische Melder *m*, Schallgerät *n* [*z.B. eine Sirene*]

acoustic device circuit, sonic device circuit, sounder circuit, sound(ing) device circuit Hörmelderschaltung, akustische Melderschaltung, Schallgeberschaltung *f*

acoustic distortion, sonic distortion, sound distortion [*Distortion in a sound-reproducing system*] Schallverzerrung, akustische Verzerrung *f*

acoustic elevator arrival signal (US); acoustic lift arrival signal (Brit.) akustische (Aufzug)Ankunftsmeldung *f*

acoustic energy, sound energy, sonic energy [*Of a part of a medium. The total energy in this part of the medium less the energy that would exist with no sound waves present*] Schallenergie, akustische Energie *f* [*In einem Mediumteil. Kinetische und potentielle Energie des Schalles*]

acoustic energy density, sound energy density, sonic energy density [*At a point in a sound field. The sound energy in an infinitesimal part of the medium surrounding the point divided by the volume of that part of the medium. Symbol E. Note: The term 'sound energy density' may be qualified by the terms 'instantaneous', 'maximum', 'peak', 'average', etc. When the term 'average' is used this may refer to time average (at a point) or, by extension, to the space average (at an instant)*] Schallenergiedichte *f* [*Quotient aus Schallenergie und zugehörigem Volumen. Räumliche Dichte der Schallenergie*]

acoustic energy flux, sound energy flux, sonic energy flux [*In a specified direction. The average rate of flow of sound energy for one period through a specified area normal to that direction. Symbol P. Note 1. In an isotropic medium of density p for a plane or spherical free progressive wave having a phase velocity c, the sound energy flux through an area S is (p2 S/pc) cos q, where S is the r.m.s. sound pressure and q the angle between the direction of propagation and the normal to S. Note 2. For an individual wave it is customary to define the area S parallel to the wave front.*] Schallenergiefluss *m*

acoustic energy reflection coefficient, sonic energy reflection coefficient, sound energy reflection coefficient [*Of a surface or material at a given frequency and under specified conditions. The ratio which the sound energy reflected from the surface or material bears to that incident upon it under those conditions*] Schallreflexionsgrad *m*

acoustic environment, sonic environment, sound environment akustische Umwelt, Schallumwelt *f*

acoustic external alarm, external acoustic alarm Schallexternalarm, akustische Externalarm, örtliche Alarm *m*

acoustic fidelity, sound fidelity, sonic fidelity Schalltreue, akustische Treue, Akustiktreue *f*

acoustic flux, sound flux, sonic flux Schallfluss, akustische Fluss, Akustikfluss *m*

acoustic frequency, sound frequency, sonic frequency Schallfrequenz, akustische Frequenz, Akustikfrequenz *f*

acoustic generation, sound generation, sonic generation Schallerzeugung, Schallentstehung, Schallbildung *f*

acoustic heated ceiling → **heated acoustic ceiling**

acoustician, acoustics expert Schallfachmann, Akustikfachmann, Akustiker *m*

acoustic impedance, sound impedance, sonic impedance [*It is given by complex ratio of sound pressure on surface to sound flux through surface, and has imaginary (reactance) and real (resistance) components, respectively. Unit is the acoustic(al) ohm*] akustische Impedanz, Schallimpedanz *f*, akustische Scheinwiderstand, akustische Standwert, Schallscheinwiderstand *m*

acoustic incidence angle, sound incidence angle, sonic incidence angle Schalleinfall(s)winkel, akustische Einfall(s)winkel *m*

acoustic indication, sound indication, sonic indication akustische Anzeige, Schallanzeige *f*

A

acoustic inlet, sound inlet, sonic inlet Einsprechöffnung *f*

acoustic insulation, sound insulation, sonic insulation [*Diminishing sound energy that passes through a material or strikes its surface*] Schalldämmung, akustische Dämmung *f*

acoustic intensity, sound intensity, sonic intensity [*The average acoustic power transported across a unit area, usually expressed in watts per square meter*] Schallintensität, Schallstärke, akustische Intensität, akustische Stärke *f*

acoustic isolation, sound isolation, sonic isolation [*Between rooms. Note: The distinction between sound insulation and sound isolation is not carefully preserved in international practice*] Schalltrennung, akustische Trennung *f*

acoustic level, sound level, sonic level [*A weighted value of the sound pressure level as determined by a sound level meter*] Schallpegel *m* [*Bei einer Frequenz oder in einem Frequenzbereich: Logarithmierte Vehältnis einer Schallfeldgröße oder einer Schallenergiegröße zu einer gleichartigen Bezugsgröße. Der Wert der Bezugsgröße ist gegebenenfalls anzugeben. Anmerkung: Schallpegel werden im Allgemeinen in dB angegeben. Wenn nichts anderes angegeben, sind im Zähler unter der Schallfeldgröße der Effektivwert dieser Größe und unter der Schallenergiegröße der zeitlich gemittelte Wert dieser Größe zu verstehen. Der Frequenzbereich ist gegebenenfalls durch Wortzusätze (z.B. Oktavpegel, Terzpegel, Pegel in der Bandbreite 1 Hz) anzugeben*]

acoustic level meter, sound level meter, sonic level meter [*An instrument designed to measure a frequency-weighted value of the sound pressure level. It consists of a microphone, amplifier and indicating instrument having a declared performance in respect of directivity, frequency response, rectification characteristic and ballistic response*] Schallpegelmesser *m*, Schallpegelmessgerät *n*

acoustic level recorder, sound level recorder, sonic level recorder Schallpegelschreiber *m*

acoustic model [*A scale model of a room, building, or other object used to measure quantities of sound distribution and noise control*] Akustikmodell *n*

acoustic noise [*An unwanted sound that is subjected to some form of noise control*] Stör(ungs)schall *m*

acoustic noise [*Any sound that is audible to human hearing*] Hörschall *m*

acoustic panel ceiling, sound panel ceiling, sonic panel ceiling Akustiktafeldecke, Akustikpaneldecke *f*

acoustic particle detection [*A specimen is irradiated with an acoustic signal, and certain reflections of the signal indicate the presence of a particle or particles*] akustische Teilchenerkennung *f*, akustische Teilchennachweis *m*

acoustic penetration, sound penetration, sonic penetration Schalleindringung *f*

acoustic phonon, sound phonon, sonic phonon [*A quantum of excitation of an acoustic mode of vibration*] Schallquantum, akustische Phonon, Schallphonom, akustische Quantum *n*

acoustic pickup, sound pickup, sonic pickup akustische Aufnehmer *m*, akustische Dose *f*, Schallaufnehmer *m*, Schalldose *f*

acoustic power, sound power, sonic power [*The power due to sound energy that is emitted from a sound source, normally measured in watts and calculated with one of several equations, depending on the medium and its theoretical model*] Schallleistung *f* [*ISO 3741*]

acoustic power level, sound power level, sonic power level Schallintensitätspegel, Schallleistungspegel *m* [*ISO 3741*]

acoustic pressure, sonic pressure, sound pressure [*At a point in a sound field. The alternating component of the pressure at the point. The unit is the Newton's square-metre. Note. The term 'sound pressure' may be qualified by the terms 'instantaneous', 'peak', 'maximum', 'r.m.s.', etc. The unqualified term 'sound pressure', symbol p, is frequently taken to imply r.m.s. value*] Schalldruck, Akustikdruck *m* [*Durch die Schallschwingung hervorgerufener Wechseldruck*]

acoustic pressure level, sound pressure level, SPL, sonic pressure level [*The sound pressure level of a sound, in decibels, is equal to 20 times the logarithm to the base 10 of the ratio of the r.m.s. sound pressure. In case of doubt, the reference sound pressure should be stated. In the absence of any statement to the contrary, the reference sound pressure in air is taken to be* 2×10^{-5} N/m^2 *and in water* $0.1\ N/m^2$. *Note. The use of the micropascal* ($1 \times 10^{-6}\ N/m^2$) *has been suggested as the reference level for sound pressure in underwater work which would have the practical advantage of generally avoiding the use of negative values in expressing sound levels*] Schalldruckpegel, Akustikdruckpegel *m*

acoustic pressure reflection coefficient, sound (pressure) reflection coefficient, pressure reflection coefficient, sonic (pressure) reflection coefficient [*Of a surface or material at a*

A

given frequency and under specified conditions. The ratio which the reflected sound pressure from a surface or material bears to the incident sound pressure under those conditions; in general a complex number] Schall-Reflexionssfaktor *m*

acoustic privacy, sound privacy, sonic privacy Stille *f*

acoustic probe, sound probe, sonic probe akustische Sonde, Schallsonde, Akustiksonde *f*

acoustic propagation, (sound) propagation, sonic propagation [*The wave process whereby sound energy is transferred from one part of a medium to another*] Schallausbreitung, Schallfortpflanzung *f*

acoustic propagation constant, (sound) propagation constant, sonic propagation constant Schallfortpflanzungskonstante, Schallausbreitungskonstante *f*

acoustic pulse, sound pulse, sonic pulse Schallimpuls, akustische Impuls, Akustikimpuls *m*

acoustic radiating ability, sound radiating ability, sonic radiating ability Schall(ab)strahlungsvermögen *n*, Schall(ab)strahlungsfähigkeit *f*, Akustik(ab)strahlungsvermögen *n*, Akustik(ab)strahlungsfähigkeit *f*

acoustic radiation, sound radiation, sonic radiation [*Infrasonic, sonic, or ultrasonic waves propagating through a solid, liquid, or gaseous medium*] Akustik(ab)strahlung, Schall(ab)strahlung *f*

acoustic radiation pressure, radiation sound pressure [*A unidirectional, steady-state pressure exerted upon a surface exposed to a sound wave*] Akustik(ab)strahlungsdruck, Schall(ab)strahlungsdruck *m*

acoustic radiator [*A vibrating surface that produces sound waves, such as a loudspeaker cone or a headphone diaphragm*] Akustikstrahler, Schallabstrahler *m*

acoustic radio glass break detector akustische Funkglasbruchmelder *m*, akustische Funkglasbruchmeldegerät *n* [*Er reagiert auf den Schall zerbrechenden Glases und dient der Überwachung von Glasflächen*]

acoustic radiometer, sound radiometer, sonic radiometer Schall(ab)strahl(ungs)druckmesser, Akustik(ab)strahl(ungs)druckmesser *m*

acoustic reflection, sound reflection, sonic reflection [*Strong sound reflections are generally called 'echoes'*] Schallreflexion *f*

acoustic refraction, sound refraction, sonic refraction [*The change in the direction of sound as it travels through a medium due to differences in temperature, pressure, and other characteristics of the medium, or a change in the medium*] Schallbrechung, Schallrefraktion *f*

acoustic reinforcement system, sound-reinforcement system, sonic reinforcement system Schallverstärkungssystem *n* [*Mikrophon(e), Verstärker, Lautsprecher*]

acoustic resistance, sound resistance, sonic resistance [*The real component of the acoustic impedance due to characteristics of the medium; the real part of the ratio of sound pressure at a specific point in the medium to practicle velocity in a free plane. Symbol R_a*] Schallresistenz *f*, Schallwiderstand *m*

acoustics, sound science, sonic science, science of sound [*The science relating to sounds, their production, transmission, and their effects on the organs of hearing. For the builder, it usually means controlling the transmission of noises, prevention of sounds passing through barriers of various types, assurance that speech sounds will be heard intelligibly, and assurance that music will be hard satisfactorily*] Akustik, Schalllehre *f*

acoustics [*The totality of those physical characteristics of an auditorium or room which affect an individual's perception, and his judgement, of the quality of speech or music produced in the room*] Akustik *f*

acoustic scattering, sound scattering, sonic scattering Schallstreuung, Akustikstreuung *f*

acoustics consultant Schallberater, Akustikberater *m*

acoustic screen, sound screen, sonic screen Schallblende, Akustikblende *f*

acoustic screening, acoustic shielding, sound screening, sound shielding, sonic screening, sonic shielding Schallabschirmung, Akustikabschirmung *f*

acoustics engineer, sound engineer Schallingenieur, Akustikingenieur *m*

acoustic shadow, sound shadow, sonic shadow Schallschatten, Akustikschatten *m*

acoustic shielding → acoustic screening

acoustic signal, sound signal, audible signal, sonic signal Hörsignal, akustische Signal, Schallsignal, Akustiksignal *n*

acoustic signal converter, sound signal converter, sonic signal converter, audible signal converter akustische Signalwandler, akustische Signalumsetzer, Schallsignalwandler, Schallsignalumsetzer *m* [*Alarmgabe über einen (Druckkammer)Lautsprecher (elektronisch erzeugter Signalton) oder über eine Sirene. DIN VDE 0833 Teil 3*]

A

acoustic signal message, sound signal message, sonic signal message, audible signal message Hörmeldung, akustische Meldung, Schallmeldung, Akustikmeldung *f*

acoustic signal output, audible signal output, sonic signal output, sound signal output akustische Signal(aus)gabe *f*

acoustic signal processing, sound signal processing, sonic signal processing [*The extraction of useful information from acoustic signals, using electronic technique, such as a fast Fourier transformer*] Hörsignalverarbeitung, Schallsignalverarbeitung, akustische Signalverarbeitung *f*

acoustic signal transmitter, sound signal transmitter, sonic signal transmitter, audible signal transmitter akustische Signalgeber, Hörsignalgeber, Schallsignalgeber *m*

acoustic sound, audible sound Hörschall *m*

acoustic source, sound source, sonic source Schallquelle *f*

acoustic spectrum, sonic spectrum, sound spectrum [*The range of acoustic frequencies, extending from subsonic and ultrasonic frequencies, that is, approximately from zero to at least 1 megahertz*] Schallspektrum, akustische Spektrum *n*

acoustic storage → acoustic delay line

acoustic surface wave → surface acoustic wave

acoustic thermometer, sonic thermometer, sound thermometer [*One that evaluates temperatures from the measurement of the speed of sound in a gas; used for very low temperatures*] Akustikthermometer, Schallthermometer *n*

acoustic tone signals → audio tone signals

acoustic track, sound track, sonic track Tonspur *f*

acoustic track film, sound track film, sonic track film Ton(spur)film *m*

acoustic transmission, sound transmission, sonic transmission [*The movement of acoustic energy by wave motion through a medium*] Schallübertragung, Schalltransmission *f*

acoustic transmission coefficient, sound transmission coefficient, acoustic transmissivity, sonic transmission coefficient, sound transmissivity, sonic transmissivity [*The ratio of transmitted to incident sound energy at an interface in a sound medium, the value depends on the angle of incidence of the sound*] Schalltransmissionskoeffizient, Schallübertragungskoeffizient *m*

acoustic transmission loss, (sound) transmission loss, TL (US); sound reduction index (Brit.); sound reduction factor [*deprecated*] [*Of a surface, partition or device at a given frequency and under specified conditions. Ten times the logarithm to the base 10 of the reciprocal of the sound transmission coefficient under these conditions. Symbol R. The unit is the decibel (dB). Note 1. Unless otherwise specified, the sound fields on both sides of the surface, partition or device are assumed to be reverberant. Note 2. In American usage, this term is called 'sound transmission loss'. "Sonic" may synonymously be used for "acoustic"*] Schalldämmmaß *n*, Schalldämmzahl *f*, Schallschutzmaß *n*, Dämmzahl *f*, Dämmmaß, Schutzmaß *n*

acoustic treatment [*The use of sound-absorbing materials to give a room a desired degree of freedom from echo and reverberation*] Raumakustik *f*

acoustic warning, sound warning, sonic warning Warnton *m*, Tonwarnung *f*

acoustic wave, sound wave, sonic wave Akustikwelle *f*

a.c. output voltage, alternating current output voltage Ausgang(s)wechselspannung *f* [*Stromrichter*]

AC-parametric test dynamische Verhaltensprüfung *f*

AC(P)H, air (ex)changes per hour, hourly air (ex)changes, air renewals per hour, hourly air renewals, air turnover per hour, hourly air turnover Luftwechsel je Stunde, Luftwechsel per Stunde, Stundenluftwechsel *m*

a.c. power conversion, alternating current power conversion Wechselstromumrichten *n*

AC-pressurization method → alternating method

acquisition [*The act or fact of acquiring something*] Erfassung *f*, Erfassen *n*

acquisition time Erfassungszeit *f*

ACR, air conditioning refrigeration Klimawärmeentzug *m*, Klimakälteerzeugung *f*

a.c. regulator, alternating-current regulator, AC regulator Wechselstromsteller *m*

acronym Kurzbezeichnung *f*

across-the-line (motor) starter, across-the-line (motor) switch [*When engaged, it impresses full line voltage on the motor windings*] Direktanlasser, Direktstarter *m*

ACRQ → access request

ACRS → acceptor ready state

a.c. signalling, alternating current signalling (Brit.); a.c. signaling, alternating current signal-ing (US) [*The use of alternating current signals or tones to accomplish transmission of infor-*

mation and/or control signals] Wechselstromsignalisieren *n*, Wechselstromsignalgabe *f*, Wechselstrommelden *n*

a.c. switch, AC switch, alternating-current switch Wechselstromschalter *m*

actinograph [*An actinometer that is equipped to record its measurements*] Strahlenmessschreiber, Strahlungsmessschreiber *m*

actinometer [*An instrument used to measure the intensity of radiation, especially that coming from a source causing photochemical reactions, such as the sun*] Strahlungsmesser, Strahlenmesser *m*

actinometry [*The measurement of the intensity of radiation from a body, especially an astronomical body*] Strahlenmessung, Strahlungsmessung *f*

action diagram [*Computer programming. A software development tool that graphically illustrates the hierarchical structure and process flow of the pro-gram(me) being developed*] Wirk(ungs)plan *m*

action flow Wirk(ungs)ablauf *m*

action interface Wirk(ungs)schnittstelle *f*

action interstice Wirk(ungs)spalt *m* [*DIN 8580*]

action medium Wirk(ungs)medium *n* [*DIN 8580*]

action pair Wirk(ungs)paar *n* [*DIN 8580*]

action path Wirkungsweg *m*

action period [*The period of time during which data in a Williams tube storage device can be read or new data can be written into this storage*] Funktionszeit, Operationszeit, Betriebszeit, Arbeitszeit, Wirk(ungs)zeit *f*

action temperature → opening temperature

to activate, to make active [*To start some process or motion*] aktivieren, in Tätigkeit setzen

to activate, to make active scharfschalten, aktivieren

activated carbon, activated charcoal, active charcoal, active carbon [*A powered, granular, or pelleted form of amorphous charcoal characterized by very large surface area per unit volume because of an enormous number of fine pores. Widely used in filter-ing systems for absorbing gases and solids*] Aktiv(holz)kohle, aktivierte (Holz)Kohle *f*

activation, activating Aktivierung *f*, Aktivieren *n*

activation, activating Scharfschalten *n*

active [*Being a source of electrical energy*] wirkend, betriebsam, aktiv

active air lock aktive Luftschleuse *f* [*Luftschleuse mit Anschluss an eine RLT-Anlage*]

active area Beschriftungsfläche, Schreibfläche *f* [*Kode*]

active bass extension system, ABES aktive Basslautsprecher *m*

active component, active device, active element [*Electronics. A component that produces more power in the output signal than is present in the input signal*] Wirkelement *n*, Wirkkomponente *f*

active (computer) storage (device) → associative (computer) memory

active conductor induzierte Leiter *m*

active current, watt(ful) current, true current [*Electricity. The current component that is in phase with the voltage in an alternating current; it equals the average power divided by the effective voltage*] Wattstrom *m*

active data link channel state Arbeitszustand *m* des Übermittlungsabschnittes

active data master file → active master (data) file

active device → active component

active electric network [*A network with at least one active component*] Wirk(ungs)netz(werk) *n*

active electronic element, electronic active element aktive elektronische Bauelement, elektronische aktive Bauelement, steuerbare Bauelement *n*

active element → active component

active energy Wirk(ungs)energie *f*

active end-of-line unit aktive Endglied *n* [*Datenübertragung*]

active energy meter Wirk(ungs)energiezähler, Wattstundenzähler, Wirkleistungszähler *m*

active filter [*Electronics. A filter with active components, such as amplifiers, that aid in signal selection and rejection. This filter usually employs negative feedback and tuning elements in the feedback path*] Aktivfilter *m, n*

active fire defense; active fire defence (Brit.) [*It includes (1) the installation of alarm and detector systems to give warnings of an outbreak. (2) The installation of equipment for automatic fire extinction (sprinklers, CO_2 flooding systems). (3) The provision of first aid fire fighting appliances. (4) The provision of facilities to assist the fire services*] mechanisierte Brandschutz, mechanisierte Feuerschutz *m*

active line Beschriftungszeile, Schreibzeile *f* [*Kode*]

active logic [*Electronics. A logic system that employs active components for the purpose of pulse*

A

width modulation, signal amplification, and signal phase shift] aktive Logik *f*

active master (data) file, active data master file [*It contains elements that are accessed or updated frequently*] aktive Hauptdatei, aktive Generaldatei *f*

active material [*An energy-storing material, such as lead oxid(e), used in the plates of a storage battery*] aktive Masse *f*

active memory → principal memory

active page Beschriftungsseite *f*, Schreibseite [*Kode*]

active part → live part

active position, AP Schreibstelle, Beschriftungsstelle *f* [*Kode*]

active-power control band Wirkleistungs-Regelbereich *m [selten: n]* [*Generatorsatz*]

active-power meter Wattmeter, KWmeter Wirkleistungsmesser, Wattmesser, Leistungsmesser, Wattzähler *m*

active-power relay Wirkleistungsrelais *n*

active program (US); active programme (Brit.) [*A program(me) that is in current use or that is regularly used*] betriebene Programm, Aktivprogramm *n*

active qualified area, qualified active area spezifische Beschriftungsfläche, spezifische Schreibfläche *f* [*Kode*]

active return loss Echodämpfung, Rückflussdämpfung *f*

active selected area, selected active area ausgewählte Schreibfläche, ausgewählte Beschriftungsfläche *f* [*Kode*]

active sensor [*See remark under "sensor"*] aktive Sensor, aktive Fühler, aktive Detektor *m*, aktive Detektionselement *n*

active signalling point (Brit.); active signaling point (US) aktive Meldepunkt *m* [*Er ist freigegeben, um bei Zustandsänderungen gemeldet zu werden*]

active (signal) transducer [*Transducer whose output signal is dependent on power that is controlled by one or more actuating signals*] aktive (Signal)Umsetzer, aktive (Signal)Wandler *m*

active solar energy, active sun energy [*Energy from the sun's radiation that is directly put to use, such as by collection in a solar roof panel*] Nutzsonnenenergie, Sonnennutzenergie *f*

active two-terminal network aktives Zweipolnetzwerk *n*

actor → ac(tua)tor

actual, real tatsächlich, wirklich, aktuell, echt, real

actual value [*IEC 50(351)*]**, measure value, instantaneous value, real value** Istwert *m*, Echtwert, aktuelle Wert, wirkliche Wert, tatsächliche Wert, Realwert, Momentwert [*Der Wert, den eine veränderliche Größe im betrachteten Zeitpunkt hat. E DIN 19 226, Teil 5*]

to actuate → to release

to actuate, to excite erregen

actuating cycle [*actuator, IEC 337-1*] Betätigungsspiel *n* [*Bedien(ungs)teil, Hilfsstromschalter, HSS; VDE 0660, Teil 200*]

actuating device, actuating appliance [*These devices are essentially motor-controlled dampers and valves, and may be hydraulic or pneumatic cylinders, air diaphragms, and electric motors, frequently with reversible and adjustable speeds*] Betätigungsgerät *n*

actuating drive → servo(-controlled) drive

actuating element Betätigungselement *n*, Betätiger *m*

actuating inhibitor Betätigungshemmteil *m, n*, Betätigungshemmer *m*

actuating mechanism Betätigungsmechanismus *m*

actuating pulse Betätigungsimpuls *m*

actuating quantity [*IEC 137-2B*] Betätigungsmenge *f* [*VDE 0660, Teil 204*]

actuating series [*IEC 337-2A*] Betätigungsreihe *f* [*VDE 0660, Teil 202*]

actuating signal, actuation signal, servo-signal [*ANSI C81.5. Command signal to switch on*] Stellsignal, Servosignal, Auslösesignal, Betätigungssignal *n*

actuating system, driving system [*Any control system that supplies and transmits the elastic, hydraulic, pneumatic, or other energy to drive a mechanism or system*] (An)Triebssystem *n*

actuating travel, actuating path Betätigungsweg, Betätigungspfad *m*, Betätigungsbahn *f*

actuating voltage, operating voltage Betätigungsspannung *f*

act(uat)or [*Mechanical engineering. Any mechanical device that is moved a predetermined distance to operate or control another mechanical device*] Aktor *m*, handelnde Anlagenteil *m, n*

ACU → automatic ((open) loop) control unit

ACU → access control unit

acuity, audible acuity, acoustic acuity Hörschärfe, Hörempfindlichkeit *f*

acuity, visual acuity [*The ability of the eye to distinguish between details that are very close together. Visual acuity increases as lighting levels increase*] Sehschärfe, Sehempfindlichkeit *f*

acusto-optic, A-O [*It refers to the interaction of optical and acoustical waves*] akustisch-optisch

acute accent Akut *m* Betonungszeichen

acute angle spitze Winkel, Spitzwinkel *m* [*Er hat weniger als 90°*]

acyclic machine, homopolar generator, homopolar machine, unipolar machine [*A direct-current generator in which the poles presented to the armature are all of the same polarity, so that the voltage generated in active conductors has the same polarity at all times; a pure direct current is thus produced, without commutation*] azyklische Maschine, Unipolarmaschine *f*

A/D → analog(ue)/digital

AD → analog(ue) device

AD → automatic device

ADAC → analog(ue)-digital-analog(ue) converter

adaline → adaptive linear element

to adapt anpassen

adaptability Anpassungsfähigkeit *f*, Anpassungsvermögen *n*, Anpassbarkeit *f*

adap(ta)tion [*Changing in response to some condi-tion*] Anpassung *f*

adapter, adaptor Anpasser, Adapter *m*, Passstück *n*

adapter to modify operation(s) → operation(s)-modifying adapter

adaptive [*Computer programming. Of a system capable of some form of alteration or adjustment in response to existing conditions*] adaptiv, anpassungsfähig, lernfähig

adaptive colour shift (Brit.); adaptive color shift (US) Farbwand(e)lung *f* [*Licht. Änderung des Farbeindruck(e)s unter dem Einfluss der Farbumstimmung*]

adaptive differential pulse code modulation, ADPCM Komprimierungsverfahren *n* für Datenübertragung auf PCM-Basis

adaptive linear element, adaline, linear adaptive element lineare adaptive Element, adaptive lineare Element, anpassungsfähige lineare Element, lineare anpassungsfähige Element, lineare lernfähige Element, lernfähige lineare Element *n* [*Neutronenschaltung*]

adaptive transform acoustic coding (method) → ATRAC

adaptometer Anpassungsmesser *m*

ADB → address bus

ADC → analog(u)e(-to)-digital converter

ADCON → address constant

A/D converter → analog(ue)(-to)-digital converter

to add addieren, summieren, zusammenzählen

to add hinzufügen

added output power hinzugefügte Ausgangsleistung *f* [*Diode*]

adding circuit, adder (circuit) [*A circuit in which two or more signals are combined to give an output-signal amplitude that is proportional to the sum of the input-signal amplitudes*] Additionsschaltung, Addierschaltung, Summenschaltung *f*

add(ing) counter Addierzähler, Summierzähler, Additionszähler, Zusammenzähler *m*

add(ing) element Addierglied, Additionsglied, Summierglied *n*

add(ing) for (the) 60 Hz operation Einbau *m* für 60 Hz Betrieb

add(ing) instruction (programmierbare) Summierbefehl, (programmierbare) Addierbefehl *m*, (programmierbare) Additionsbefehl [*Siehe Anmerkung unter „Befehl"*]

add(ing) machine [*A device which performs the arithmetical operations of addition and subtraction*] Addiermaschine, Additionsmaschine, Summiermaschine *f*

add(ing) mechanism Addierwerk, Additionswerk, Summierwerk *n*

additional capacitor, add-on capacitor, supplementary capacitor Zusatzkondensator, Ergänzungskondensator *m*

additional change key, add-on change key Mehrschlüssel, Vielschlüssel *m*

additional energy, supplementary energy Ergänzungsenergie, Zusatzenergie *f*

additional gun electrode, add-on gun electrode, additional acceleration electrode, add-on acceleration electrode Nachbeschleunigungselektrode *f*

additional heating, add-on heating Zusatzheizung, Ergänzungsheizung *f*

additional kit, additional assembly, additional set, add-on kit, add-on assembly, add-on set Nachrüst(bau)satz *m*, Nachrüstkombination, Nachrüstgruppe *f*, Nachrüstaggregat *n*, Nachrüstgarnitur *f*

A

additional memory, add-on memory Peripheriespeicher *m*

additional operation, add-on operation Zuschaltbetrieb *m [DIN 41 745]*

additional ventilation, add-on ventilation Zusatzlüftung, Zusatzbe- und -entlüftung *f*

additive shading Aufaddieren *n* [*Farbanteile eines Pixels auf ein anderes*]

add-on [*A circuit or system which increases the memory capacity of a computer*] Anbauelement, Zusatzelement, Nachrüstelement, Ergänzungselement *n*

to address, to select anwählen

address, ADRS [*The number or name that uniquely identifies a register, memory location, or storage device in a computer*] Adresse *f* [*DIN 4430*]

addressability [*The capability of Direct Broadcast Satellite (DBS) or cable communications system to allow computer control of program(me) distribution from a central location*] Adressierbarkeit *f*

addressable addressierbar

address allocation, address allotment, address assignment Adresszuordnung, Adresszuteilung, Adressenweisung *f*

address bits Netzwerkadresse *f* der Empfangsstation [*ISO-OSI-Schichtenmodell*]

address bus, ADB Adress(en)bus *m*, ADB [*Mikrorechner. Dieser Bus umfasst bei 8-Bit-Rechnern meist 16 Leitungen so dass 2^{16} = 65536 unterschiedliche Adressen übertragen werden können*]

address code → address mark

address constant, ADCON, presumptive address, reference address [*Computer programming. A number used as the base or starting point from which subsequent addresses are to be modified*] Basisadresse, Grundadresse *f*

address control module Adress(en)steuer(ungs)-modul, Adress(en)steuer(ungs)baustein *m*

address counter [*Computer engineering. A binary incrementing counter in a direct-memory access controller that is used to control data block transfer by storing the destination*] Adress(en)zähler *m*

address decoding Adress(en)entschlüsselung, Adress(en)dekodierung *f*

address delimiter Adress(en)begrenzungszeichen *n*

address dot Vorlage *f* [*Adress(en)verwaltung*]

addressed memory (voll)adressierte Speicher *m*

address error Adress(en)fehler *m*

address field [*Computer programming. The specific portion of a machine instruction that contains the operand, the address of the operand, or the information needed to derive the address of the operand*] Adress(en)feld *n*

address format [*Computer programming. The arrangement of the fields and entries in the address por-tion of an instruction*] Adress(en)format *n*

address function Adress(en)funktion *f*

address generation [*Computer programming. The process of obtaining an effective memory address by adding the contents of the base register to the relative address specified in the computer instruction*] Adress(en)bildung *f*

address generator Adress(en)bildner *m*

address identifier Adress(en)kennzeichen *n*, Adresskennzeichner *m*

addressing Anlegen *n* einer Adresse, Adressieren

addressing domain Adressier(ungs)bereich *m [selten: n]*

addressing mode [*Computer programming. A rule for interpreting or modifying the address field of an instruction to identify the operand; types include immediate, direct, indirect and register*] Adressier(ungs)art *f*

addressing module, adressing block Adressier(ungs)modul *m*, Adressier(ungs)baustein *m*

address instruction Adress(en)befehl *m*

address label Adress(en)etikette *f* [*„Etikett n" in der Schweiz und in Österreich, sonst veraltet*]

address latch Adress(en)(signal)speicher *m*

address latch enable, ALE Adress(en)(signal)speicherfreigabe *f*

addressless instruction adress(en)lose Befehl *m*

address management Adress(en)verwaltung *f*

address mark, address code Adress(en)kode, Adress(en)code *m* [*Spezieller Kode, der am Anfang besonderer Felder auf einer Plattenspur benutzt wird*]

address modification Adress(en)änderung *f*

address of virtual information, virtual information address virtuelle Adresse *f* [*Eine intern existierende Adresse die sich wie verdrahtete analoge oder binäre Adressen verarbeiten lässt*]

address parameter [*A number or code that defines the position of information in the memory of a computer. It is also used to control the destination of transmitted data*] Adress(en)parameter *m* [*Zahl oder Memocode, die oder der die Stelle einer Information im Rechner bezeichnet. Ebenso auch verwendet wie die Senke einer übertragenen Information*]

address part Adress(en)teil *m, n* [*Operandenteil oder Bestandteil desselben, der für Operandenadressen oder Befehlswörter vorgesehen ist. DIN 44 300 Teil 4*]

address pointer Adress(en)verweis *m*

address register, AR, memory address register [*A register wherein the address part of an instruction is stored by a computer*] Adress(en)register *n*

address resolution Adress(en)auflösung *f*

address resolution protocol, ARP Adress(en)auflösungsprotokoll *n*

address resolution request Adress(en)auflösungsanfrage *f*

address selection Adress(en)anwahl, Adress(en)-ansteuerung *f*

address setup time Adress(en)vorbereitungszeit *f*

address signal [*A signal sent in the forward direction containing some or all of the address informations*] Adress(en)signal *n*

address substitution Adress(en)(aus)tausch *m* [*Die Adresse wird durch den Adressenteil des durch sie selbst bezeichneten Befehls ausgetauscht*]

address unit Adress(en)werk *n* [*Rechner*]

add time [*The time required by a computer to perform an addition, not including the time needed to obtain the addends from storage and put the sum back into storage*] Addierzeit, Additionszeit, Summier(ungs)zeit *f*

ADE → **automatic design engineering**

adhesive Kleber *m*, Kleb(e)mittel, Haft(ungs)mittel *n*

adhesive film Kleb(e)folie, Haft(ungs)folie *f*

adhesive flange Kleb(e)flansch, Haft(ungs)-flansch *m*

adhesive strength, adhesive stability Haft(ungs)-festigkeit, Haft(ungs)beständigkeit, Haft(ungs)-kraft, Kleb(e)festigkeit, Kleb(e)beständigkeit, Kleb(e)kraft, Lagenbindung, Schichtenbindung *f*

adhesive tag [*shoplifting prevention*] Kleb(e)etikett, Haft(ungs)etikett *n*

adhesive tape Kleb(e)band, Haft(ungs)band *n*

adiabatic [*Term applied to a process in which there is no heat flow between a substance or system and their surroundings*] adiabatisch, wärmedicht, ohne Wärmeaustausch *m*

adiabatic change [*Temperature change within a substance caused only by its own expansion or compression*] adiabatische Wechsel, wärmedichte Wechsel *m*

adiabatic combustion adiabatische Verbrennung, wärmedichte Verbrennung *f* [*Es erfolgt keine Wärmeabgabe nach außen*]

adiabatic compression [*Thermodynamics. A reduction in the volume of a system that does not involve heat transfer either into or out of the system*] adiabatische Verdichtung, adiabatische Kompression, wärmedichte Verdichtung, wärmedichte Kompression *f*

adiabatic cooling [*A process in which the temperature of a system, or a portion thereof, is reduced without any heat being exchanged between the system, or a portion thereof, and its surroundings*] adiabatische Kühlung, Transpirationskühlung, wärmedichte Kühlung *f*

adiabatic demagnetization, Giaque-Debye method, (para)magnetic cooling [*A method of cooling paramagnetic salts to temperatures of 10^{-3} K; the sample is cooled to boiling point of medium in a strong magnetic field, thermally isolated, and then removed from the field to demagnetize it*] Giaque-Debye-Methode *f*, Giaque-Debye-Verfahren *n*

adiabatic efficiency adiabatische Wirkungsgrad, wärmedichte Wirkungsgrad *m*

adiabatic ellipse [*A plot of the speed of sound as a function of the speed of flow for the adiabatic flow of a gas, which forms one quadrant of an ellipse*] adiabatische Ellipse, wärmedichte Ellipse *f*

adiabatic expansion [*Thermodynamics. An increase in the volume of a system that does not involve any transfer of heat either into or out of the system*] adiabatische Expansion, adiabatische Ausdehnung, adiabatische Entspannung, wärmedichte Expansion, wärmedichte Ausdehnung, wärmedichte Entspannung *f*

adiabatic flow [*Fluid mechanics. A flow in which it can be assumed that the rate of energy transfer both into and out of the control volume is effectively zero*] adiabatische Strömung, wärmedichte Strömung *f*

adiabatic humidification adiabatische Befeuchtung, Befeuchtung mit Umwälz(ungs)wasser, wärmedichte Befeuchtung, adiabatische Anfeuchtung, wärmedichte Anfeuchtung, Anfeuchtung mit Umwälz(ungs)wasser *f*

adiabatic humidifier adiabatische Befeuchter, adiabatische Anfeuchter, wärmedichte Anfeuchter, wärmedichte Befeuchter *m* [*Es wird umlaufendes Wasser benutzt*]

adiabatic saturation [*Air flows through a water spray*] adiabatische Sättigung, wärmedichte Sättigung *f*

A

adiabatic throttling adiabatische Dross(e)lung, wärmedichte Dross(e)lung *f*

adiabatic washer adiabatische Wäscher(r) *f*, (*m*) [*Die Wassertemperatur während des Betriebes ist unter 20°C um zu vermeiden dass sich Legionellen vermehren*]

adiathermic, adiathermanous [*Not capable of transmitting radiant heat*] strahlungswärmeundurchlässig

ADIF → **analog(ue) data interchange format**

adjacent channel, neighbouring channel (Brit.); neighbouring channel (US) Nachbarkanal, benachbarte Kanal *m*

adjacent channel interference Nachbarkanalstörung *f* [*Sie entsteht durch übersprechen benachbarter Kanäle*]

adjacent channel selectivity Trennschärfe *f* gegen den Nachbarkanal

to adjust umstecken, regeln

to adjust abgleichen [*Messgerät*]

to adjust, to vary, to control, to set verstellen, justieren, (ein)stellen

adjustability, controllability, variability Verstellbarkeit, Justierbarkeit, (Ein)Stellbarkeit *f*

adjustable, variable, controllable justierbar, verstellbar, (ein)stellbar

adjustable regelbar, abgleichbar [*Messgerät*]

adjustable (attached) (metal) fin [*See remark under "(attached) (metal) fin"*] Leit(blech)lamelle, Leitflosse *f* [*Sie ist senkrecht und waag(e)recht verstellbar; der Luftstrom wird in jede Richtung gelenkt*]

adjustable charging device, adjustable charging appliance, adjustable feeding device, adjustable feeding appliance Regulierschüttvorrichtung *f* [*Kesselbrennstoffzufuhr*]

adjustable clearance, variable clearance, controllable clearance [*The adjustable distance separating two objects or parts*] justierbare Spiel *n*, justierbare Freiheit *f*, verstellbare Spiel *n*, verstellbare Freiheit *f*, (ein)stellbare Spiel *n*, (ein)stellbare Freiheit *f*

adjustable key → **extension key**

adjustable roller lever umstellbare Schwenkhebel *m* [*Endschalter*]

adjustable spot light, search(ing) spot light Suchscheinwerfer *m*

adjuster, adjusting device, adjustment device Einsteller, Steller, Versteller, Justierer *m*, Einstellgerät, Stellgerät, Justiergerät, Verstellgerät *n* [*Eine Baueinheit, mit der ein bestimmter Wert ihres Ausgangssignals festgelegt und aufrecht erhalten wird. E DIN 19226, Teil 5*]

adjusting Verstellen, Justieren, Einstellen, Stellen *n*

adjusting assembly, adjusting kit, adjusting set Verstell(ungs)satz, Justier(ungs)satz, Einstellungssatz, Stell(ungs)satz, Einstellsatz *m* [*„Kombination", „Gruppe", „Aggregat", „Garnitur" und „Bausatz" können synonym für „Satz" verwendet werden*]

adjusting blade Verstell(ungs)zunge, Einstel-l(ungs)zunge, Justier(ungs)zunge *f*

adjusting ((closed) loop) control, adjusting feedback control, adjusting closed control, adjusting regulation Folgereg(e)lung *f*

adjusting collar, adjusting ring (Ein)Stell(ungs)-ring, Verstell(ungs)ring, Justier(ungs)ring *m*

adjusting gauge; adjusting gage (US) (Ein)Stell-(ungs)lehre *f*, Justier(ungs)regler *m*

adjusting nut (Ein)Stell(ungs)mutter, Verstell-(ungs)mutter, Justier(ungs)mutter *f*

adjusting roller plunger, adjusting plunger with roller Verstell(ungs)rollenhebel, Justier(ungs)rollenhebel, (Ein)Stell(ungs)rollenhebel *m* [*Positionsschalter*]

adjusting straight(-way) valve → **straight(-way) adjusting valve**

adjusting valve (Ein)Stell(ungs)ventil, Verstellungsventil, Justier(ungs)ventil *n*

adjusting voltage (Ein)Stell(ungs)spannung, Verstell(ungs)spannung, Justier(ungs)spannung *f*

adjustment, balancing, compensation, compensating Abgleich(ung) *m*, (*f*), Abgleichen *n*

adjustment → **correction**

adjustment Verstellung, Justierung, (Ein)Stellung *f*

adjustment characteristic (curve) Verstell(ungs)-kennlinie, Verstell(ungs)kurve, (Ein)Stell(ungs)-kennlinie, (Ein)Stell(ungs)kurve, Justier(ungs)-kennlinie, Justier(ungs)kurve *f*

adjustment clause Gleitklausel *f*

adjustment control, adjustment element Verstellungselement, Justier(ungs)element, (Ein)-Stell(ungs)element, Verstellelement *n*

adjustment curve Eichkurve *f*

adjustment lamp Justierlampe, (Ein)Stell(ungs)-lampe, Verstell(ungs)lampe *f*

adjustment lever (Ein)Stell(ungs)hebel, Verstell-(ungs)hebel, Justier(ungs)hebel *m*

ADM, asynchronous disconnected mode unabhängige Wartemodus *m*, unabhängige Wartebetriebsart *f*, Frei-Wartemodus *m*, Frei-Wartebetriebsart *f*

administration management domain, ADMD öffentliche Versorgungsbereich *m [selten: n] [Betreut vom jeweiligen Telegraphieträger, also beispielsweise der Telekom]*

admissible relative humidity, relative admissible humidity zulässige relative Feuchte, zulässige relative Feuchtigkeit, relative zulässige Feuchte, relative zulässige Feuchtigkeit *f*

admission Beaufschlagung *f*

admission Zutritt *m*

admission of heat, heat admission Wärmezutritt *m*

admission of light, light admission Lichtzutritt *m*

admission of steam, steam admission Dampfzutritt *m*

admitted (for use) by the local government(s) bauaufsichtlich zugelassen

ADP → automatic data processing

ADPCM → adaptive differential pulse code modulation

ADPE → automatic data processing equipment

ADPI, air diffusion performance index Bewertungszahl *f* für Luftausbreitung [*nach dem Luftauslass*]

A/D processor → analog(ue)(-to)-digital processor

ADPS, automatic data processing system [*A system employed to perform automatic data processing functions*] automatische Datenverarbeitungssystem *n*

ADR → automatic data retrieval

ADRS → address

ADSL → asymmetric digital subscriber line

adsorbent, adsorptive agent [*Material which has the ability to cause molecules of gases, liquids, or solids to adhere to its surface without changing the adsorbent physically or chemically. Certain commercially available solid materials, such as silica gel, activated carbon, and activated alumina, have this property*] Adsorptionsmittel, Anlagerungsmittel *n*, Adsorptionsstoff, Anlagerungsstoff *m*

adsorbent charged element Element *n* mit Adsorptionsmittelfüllung, Patrone *f* mit Adsorptionsmittelfüllung

adsorption, surface adherence, surface retention [*Of a material in extracting one or more substances present in an atmosphere or mix(ture) of gases and liquids, unaccompanied by physical or chemical change*] Adsorption, Anlagerung *f*

adsorption cold Adsorptionskälte, Anlagerungskälte *f*

adsorption dehumidification Adsorptionsentfeuchtung, Anlagerungsentfeuchtung *f*

adsorption dehumidifier Adsorptionsentfeuchter, Anlagerungsentfeuchter *m*

adsorption device, adsorber [*It dehumidifies air by bringing it into contact with a solid adsorbing substance*] Anlagerungsmittel, Adsorptionsmittel *n*

adsorption dryer, adsorption drier Adsorptionstrockner, Anlagerungstrockner *m*

adsorption drying Adsorptionstrocknung, Anlagerungstrocknung *f*

adsorption indicator [*Analytical chemistry. An indicator such as fluorescein, that shows a colo(u)r change at the endpoint in a titration*] Adsorptionsanzeiger, Anlagerungsanzeiger *m*

ADT → advance(d) data transfer

adulterated [*Condition of food. Term describing food that bears or contains poisonous or deleterious substances rendering it injurious to health*] verdorben, ungenießbar, nicht genießbar, schlecht

adumbration [*Optics. The fact of being darkened*] Dunkelmachen *n*

adumbration [*Optics. The fact of being overshadowed*] Überschattung *f*

to advance [*To move forward or along some route or path*] voreilen, vorrücken, vorlaufen

to advance [*readout*] erhöhen

advance Voreilen, Vorrücken, Vorlaufen *n*

advance angle Vorgabewinkel *m* [*Parallelschalten*]

advanced fortgeschritten, fortschrittlich

advance(d) application architecture, AAA fortschrittliche Anwendungsarchitektur *f* [*Betriebssystemerweiterung Windows*]

advance(d) code [*readout*] Erhöhungskode, Erhöhungscode *m*

advance(d) coding [*readout*] Erhöhungskodierung, Erhöhungscodierung *f*

advance(d) control circuit, ACC Steuerschalt(ungs)kreis *m* [*Futurebus*]

advance(d) data transfer, ADT erhöhte Datentransfer *m*

advance(d) low energy house Nullheiz(ungs)-energiehaus *n*

advance(d) opening, early opening [*valve*] vorzeitige Öffnen *n*, frühe Öffnen

advance(d) production system → APS

advance(d) solid logic technique → ASLT

A

advance(d) technology → **AT**

advance(d) warning, early warning Frühwarnung, Vorwarnung *f*

advance gas cooled reactor schnelle Brüter *m*

advance interval Voreildauer *f* [*Impuls*]

adverse element Schadelement *n*

adversely corrosive stark korrosiv

adverse state, AS Umkehr(ungs)zustand *m*

adverse state detector → **ASD**

adverse weather, bad weather Schlechtwetter *n*

advertisement curtain, advertising curtain Werbevorhang, Reklamevorhang *m*

advertising light(ing), advertisement light(ing) Reklamebeleuchtung, Werbebeleuchtung *f*

advertising luminaire, advertisement luminaire Reklameleuchte, Werbeleuchte *f*

advisory development plan Vorprojektplan *m*

ADX → **automatic data exchange**

AE → **application entity**

aeration, aerating [*1. Exposing a substance, or area, to the action of air. 2. Causing air to bubble through. 3. Introducing air into a liquid by spraying, stirring, or similar method. 4. Supplying or infusing with air, as in sand, soil or water*] Belüften *n*, Belüftung *f*

aeration tank, aeration basin, aerating tank, aerating basin Belüftungsbecken *n*

aerator, aeration device, aerating device [*Equipment used for aeration*] Belüfter *m*, Belüftungsvorrichtung *f*

aerial; antenna (US); Ae, Aerial [*obsolete*] Antenne *f*

aerial cable Luftkabel *n*

aerial characteristic; antenna characteristic (US) Antennenkennwert *m* [*Messgröße, mit der Antenneneigenschaften beschrieben werden, z. B. der Fußpunktwiderstand und die Arbeitsfrequenz*]

aerial (conductor) Luftleiter *m*

aerial gallery; antenna gallery (US) Antennengang *m*

aerial junction box; antenna junction box (US) Antennenanschlussdose, Antennenabzweigdose, Antennenverbindungsdose *f*

aerial mast; antenna mast (US) Antennenmast *m*

aerial plug-in point; antenna plug-in point (US) Antennenanschluss *m* [*In einer Wohnung an eine Gemeinschaftsantenne*]

aerial (radio)frequency power, aerial RF power [*"Aerial" is called antenna in the USA. Radio-frequency power delivered to an antenna*] Antennen-Hochfrequenzleistung *f*

aerial receptacle, aerial socket; antenna receptacle, antenna socket (US) Antennensteckdose *f*

aerial response characteristic; antenna response characteristic (US) Antennenempfangs-Charakteristik *f*

aerial response curve; antenna response curve (US) Antennenempfangskurve *f*

aerial tower; antenna tower (US) Antennenturm *m*

aerial transformer; antenna transformer (US) Antennenübertrager *m*

aerobic bacterium Aerobier *m*, luftbedürftige Bakterie *f*

aerobic decomposition, aerobic decay aerobe Zersetzung *f*, aerobe Zerfall *m*

aerometer → **airometer**

aerodynamic Aerodynamik *f*

aerostatic bearing aerostatische Lager, Luftlager *n*

(a)esthetically pleasing → **esthetically pleasing**

AF → **audio frequency**

AF band → **audio frequency band**

AFC → **automatic frequency ((open) loop) control**

AFF → **above finished floor(ing)**

affinity law Affinitätsgesetz *n*

AFG → **analog(ue) frequency generator**

AFI, authority and format identifier Zuteilungsstellen- und Formatkennungsgerät *n*

AFM → **atomic force microscope**

AF power → **audio frequency power**

AF signal level → **audio frequency signal level**

AFT → **audio frequency transformer**

afterburner Nachbrenner *m*

aftercondenser Nachkondensator *m*

aftercooler, secondary cooler [*A device for removing the heat of compression of the air or gas after compression is completed. An aftercooler is one of the most effective means of removing moisture from compressed air*] Nachkühler, Sekundarkühler *m*

aftercooling Nachkühlung *f*

A

after-flush [*The small amount of water remaining in a cistern after a WC pan is flushed. This water trickles down to remake the water seal in the pan*] Tröpfel(spül)wasser *n*

afterheat Abschaltwärme, Nachwärme *f*

afterheater → (air) afterheater

afterheating electric energy, electric afterheating energy [*"Current" may synonymously be used for "energy"*] Nachheiz(ungs)energie *f*, Nachheiz(ungs)strom *m*

afterignition Nachzündung *f*

afterinjection Nach(ein)spritzen, Nachinjizieren *n*

afterpressure gauge; afterpressure gage (US) Nachdruckmesser *m*, Nachmanometer *n*

aftertouch Druckdynamik *f* [*Nicht verwechseln mit „Anschlag(s)dynamik"*]

aftertreatment unit Nachbehandlungseinheit *f*

AF transformer → audio frequency transformer

AF transmitter → audio frequency transmitter

against payment entgeltlich, gegen Bezahlung

AGC → automatic gain ((closed) loop) control

age coating Innenschwärzung *f* [*Glühlampe*]

age distribution Alter(s)verteilung *f*

age hardening Aushärtung *f*, Aushärten *n*

ag(e)ing by cooling, cooling ag(e)ing Kühl(ungs)alterung *f*, Kühl(ungs)altern *n*

ag(e)ing room Alterungsraum *m*

age of air → air age

AGFC, automatic gain and ((closed) loop) frequency control automatische Verstärkungs- und Frequenzreg(e)lung *f*

aggregate crosstalk(ing) → babble

aggregation Aggregation, Zusammensetzung *f* [*Von aktivierten Regeln in der unscharfen Logik*]

aggregation operator, aggregator Aggregationsoperator *m* [*MAX- oder SUM-Operator in der unscharfen Logik*]

agitation, agitating, stirring Rühren *n*

agitation, agitating Einblasen *n* [*Kältemitteldampf in Flüssigkältemittel*]

agitator, stirrer, agitating device, stirring device, agitating appliance, stirring appliance [*It creates turbulent motion in a fluid*] Rührer *m*, Rührwerk *n*, Rührapparat *m*, Rührgerät *n*, Rührvorrichtung *f*

Ah → ampere-hour

a.h.m. → amperehourmeter

A-HOE → amplitude-holographic optical equipment

AI → artificial intelligence

AI → analog(ue) input

aid Hilfsmittel *n*

aid, help Hilfe *f*

AIDS → acceptor idle state

Aiken Code Aikenkode *m* [*Symmetrischer 2-4-2-1 Binärkode für Dezimalziffern. Auch Reflex-Kode genannt, weil sich die Ziffern 4 und 5, 3 und 6 usw. entgegengesetzt entsprechen*]

air ... → (compressed-)air ...

air accumulation Luftansammlung, Luftwalze *f*

air adjuster [*Please refer to "adjuster" for synonymous terms*] Lufteinsteller *m*

air-admittance valve → vacuum breaker

(air) afterheater (Luft)Nacherhitzer, (Luft)Nach(er)wärmer *m*

(air) afterheating (Luft)Nach(er)wärmen, (Luft)-Nacherhitzen *n*

air afterprocessing, air aftertreatment Luftnachveredelung, Luftnachbehandlung, Luftnachaufbereitung *f* [*Teilbehandlung der Luft in einer RLT-Anlage, die in Strömungsrichtung nach der Lufthauptbehandlung erfolgt*]

air age, age of air [*Time of passage of air from one point to another within an indoor airspace*] Luftalter *n*

air-agitating system [*Combination consisting of a blower, distributing piping, and flexibly connected fittings for delivering air to the water in ice cans to agitate the water and promote production of clear ice*] Lufteinblassystem *n* zur Eisherstellung

air agitation Vermischung *f* durch Lufteinblasen in eine Flüssigkeit

air agitation zone Vermischungszone *f* durch Luft-einblasen in eine Flüssigkeit

air-air cooling → AA cooling

air-air/forced-air cooling → AA/FA cooling

air/air heat exchanger, air-to-air heat exchange unit, air-to-air heat interchanger, air/air heat exchange unit [*It transfers heat from an exhaust airstream to a separated supply airstream. Note: Fixed plate rotary wheels, heat pipes, runaround coil loops, and shell-and-tube are the most common types. See remark under "heat exchanger"*] Luft/Luft-Wärme(aus)tauscher [*In der Industrie so genannt*], Luft/Luft-Wärmeübertrager *m*

A

air/air heat pump, air-to-air heat pump Luft-Luft-Wärmepumpe *f* [*Siehe Anmerkung unter „Wärmepumpe"*]

air/air (thermal) resistance, air-to-air (thermal) resistance [*The total thermal resistance of all thermal components occurring through a structure from the inside surface layer to the outside surface layer*] Wärmedurchgangswiderstand *m*

air and water cooled condenser, combination condenser, combined condenser luft- und wassergekühlte Kondensator, Luft-Wasser-Kondensator *m*

air and water cooling → cooling with air and water

air and water induction unit → (combined) air and water induction unit

air annulus space, ring air space, annulus air space, air ring space Luftringraum, Ringluftraum *m*

air aspirator valve → air input valve

air atomizer Luftzerstäuber *m*

air-atomizing burner, oil-atomizing burner [*Burner in which the oil is atomized by compressed air which is forced into and through one or more streams of oil, breaking the oil into a fine spray*] Ölzerstäuberbrenner *m*

air baffle(r) → air deflector

air baffle(r) plate, air deflector plate Luftleitblech, Luftführungsblech *n*, Luftschikane *f*

air-balancing, air-equalizing, air-compensating Luftausgleich *m*

(air) bellows (Blase)Balg *m*

air blast, air jet Luftstrahl *m*

air-blast circuit-breaker (switch), air-jet circuit-breaker (switch) [*An electric switch that, as it opens, uses a high-pressure gas blast to break the arc*] Druckluftleistungs(trenn)schalter, Pressluftleistungs(trenn)schalter, pneumatische Leistungs(trenn)schalter *m*

air-blast cooling, air-jet cooling Anblasekühlung, Luftfremdkühlung, Fremdkühlung *f* mit Luft

air-blast lift, air-jet lift [*sewer*] Drucklufttheber, Presslufttheber, pneumatische Heber *m*

air-blast outlet, air-jet outlet Luftstrahlauslass, Luftstrahlabgang, Luftstrahlablass, Luftstrahlaustritt *m*

air bleed(er) Entlüfter *m*

air blow → air throw

(air) blower Gebläse *n*, Bläser *m*

air(borne) dust Flugstaub, Luftstaub *m*

air(borne) impurity Luftverunreinigung *f*

air(borne) insulation margin M_a [*ISO/R 717*] (Luft)Schallschutzmaß *n*, LSM [*DIN 52 210, Blatt 4*]

air(borne) intense noise → intense air(borne) noise

air(borne) noise Luftgeräusch *n* [*Es breitet sich in der Luft aus*]

air(borne) particles, air(borne) particulates [*Solid or liquid impurities from whatever source*] Luftverunreinigungs-Teilchen *f*

air(borne) sound [*It reaches the point of detection by radiation through the air*] (Luft)Schall *m* [*DIN 1320*]

air(borne) sound insulation index I_a [*ISO/R 717*] bewertete (Luft)Schalldämmmaß R_w *n* [*DIN 52210, Blatt 4*]

air bottle [*A vessel connected to a high point of a hot water system to collect air from it*] Lufttopf *m*

air box → air (distribution) box

air (break) circuit-breaker Leistungsluftschalter, Luft(leistungs)schalter *m*

air (break) disconnector Lufttrennschalter, Trennluftschalter, Trennschalter *m* in Luft

air (break) switch [*A switch in which the breaking of electric circuit takes place in air*] Luftunterbrechungsschalter *m*

air (break) switch disconnector Luftlasttrenner, Luftlasttrennschalter, Lasttrenner *m* in Luft

to airbrush [*To alter a piece of artwork with an airbrush, as by painting out some elements of the artwork*] spritzretuschieren

airbrush [*A small pencil-shaped gun that uses compressed air to fine-spray paint for adjusting tonal quality of drawings and photographs*] Spritzretusche *f*

air buildup → air stagnation

air bunching [*Accumulation of air around a ventilator due (usually) to the inadequacy of the extractor*] Luftstau *m*

air bypass duct → bypass (air) duct

air calculation → air(flow) calculation

air capacity Luftleistung *f*

(air) ceiling grill(e), ceiling air grill(e) [*See remark under "(air) grill(e)"*] (Luft)Zwischendeckengitter, Zwischendeckenluftgitter *n*

air central distributor → air manifold

air central heating, central air heating Luftzentralheizung, Zentralluftheizung *f* [*In der Diele jeder Wohnung ist ein Wärmeaustauscher vorhanden*]

A

air central heating plant, air central heating installation, central air heating plant, central air heating installation Luftzentralheiz(ungs)anlage *f*

air centrifugal dehumidifier Zentrifugalluftentfeuchter *m [DIN 30 600. Reg.-Nr. 06101]*

air chamber Luftkammer *f*

air change, air renewal, air turnover, air exchange Luftwechsel *m*, Lufterneuerung *f*, Luftaustausch, Luftumsatz *m*

air change effectiveness, air turnover effectiveness, air renewal effectiveness, air exchange effectiveness Luftwechselwirksamkeit, Luftumsatzwirksamkeit, Lufterneuerungswirksamkeit, Luftaustauschwirksamkeit *f*

air change loss, air renewal loss, air turnover loss, air exchange loss [*A source of heat dissipation peculiar to buildings due to the cold outside air infiltrating into the building and requiring heat to warm it to the inside temperature*] Luftwechselverlust, Lufterneuerungsverlust, Luftaustauschverlust, Luftumsatzverlust *m*

air change number Luftwechselzahl *f*

air choke Luftdrossel *f*

air circuit Luftkreislauf *m*

air-circulated umlüftet

air circulation ratio [*The volume of air circulated within a closed space per unit time divided by the volume of the space*] Luftumwälz(ungs)verhältnis, Luftzirkulationsverhältnis, Luftumlaufverhältnis *n*

air cleaner, air purifier [*Device used to remove dirt, lint, and airborne impurities from the air*] Luftreiniger *m*

air cleaning, air purification Luftreinigung *f*, Luftsäuberung

air cleanliness, air purity Luftreinheit *f*

air/cloth ratio, feet per minute [*Value used in dust collector calculations to measure air velocity through the bag filter media. Note: Calculated by dividing active cubic feet per minute of airflow by the effective area (square feet) of a cloth*] Luft/-(Filter)Gewebe-Verhältnis *n*

air cock [*A small valve used to release air trapped in a hot water radiator*] Luft(ablass)hahn, Entlüftungshahn *m*

air coil [*Coil used in refrigeration as a condenser or evaporator to remove heat or pick it up*] Luft-(rohr)schlange, Luftkühl(ungs)(rohr)schlange *f*

air collection (pipe)line, air collecting (pipe)-line Luftsammel(rohr)leitung, Sammelluft(rohr)-leitung *f*

air collection tank, air collecting tank Luftsammelbehälter, Sammelluftbehälter *m*

air collector Luftkollektor *m [Eine Komponente der Solarenergienutzung]*

air column Luftsäule *f*

air-compensating → air-balancing

air component → air constituent

air composition Luftzusammensetzung *f*, Luftaufbau *m*

air compressor Luftverdichter, Luftkompressor *m*

air condenser Luftverflüssiger, Luftkondensator, luftgekühlte Verflüssiger, luftgekühlte Kondensator *m [In erster Linie zur Verflüssigung des dampfförmigen Kältemittels in Kältemaschinen, seltener zum Niederschlagen von Wasserdampf zum Ausnutzen der aufgenommenen Kondensierungswärme für Luftheizung]*

to air-condition [*1. To supply an enclosure with air conditioning. 2. To treat air by means of air conditioning*] klimatisieren, klimaregeln

(air-)conditioned klimageregelt, klimatisiert

(air-)conditioned air heating Klima-Luftheizung *f*

(air-)conditioned zone klimageregelte Zone, klimatisierte Zone, Klimazone *f*

air conditioner [*A factory-made encased unit which is designed to deliver conditioned air to an enclosed space without the use of ducts; usually mounted in a window or in an opening in a wall, or as a console*] Klimagerät, Raumklimagerät *n*

air conditioning, climate control, A/C, a/c, AC, A.C., ac, a.c. [*The control of the atmospheric state within an enclosure as to temperature, moisture content and purity. Not to be confused with "air treatment"*] Klimareg(e)lung, Klimatisierung, Klimaaufbereitung *f*

air conditioning automation Klimaautomation *f*

air conditioning block Klimablock *m [Zentraler Klimablock aus Baukastenelementen (= Sektionen): Misch- und Filtersektion, Wärmeübertragungssektion, Befeuchtersektion und Lüftersektion. Zusammenwirken mit Ventilatorkonvektoren oder Induktionsgeräten in den Räumen]*

air conditioning ceiling, air-conditioned ceiling Klimadecke *f*

air conditioning central heating, central air conditioning heating Klimazentralheizung *f*

air conditioning (central) station Klimazentrale, Klimastation, Zentralklimaanlage *f*

A

air conditioning chiller, air conditioning refrigerating system [*See remark under "chiller"*] Klima-(Kompressor)Kälteanlage, Klima-Verdichterkälteanlage *f*

air conditioning ((closed) loop) control Klimareg(e)lung *f*

air conditioning cold water, air conditioning chilled water Klimakaltwasser *n*

air conditioning component Klimakomponente *f*, Klimabestandteil *m* [*z.B. eine Kühl(ungs)decke*]

air conditioning condensate, air conditioning condensation water, air conditioning condens(at)ing water Klimareg(e)lungskondenswasser, Klimareg(e)lungskondensat *n*

air conditioning convector Klimakonvektor *m*

air conditioning cooler Klimakühler *m*

air conditioning duct Klimakanal *m*

air conditioning ductwork, air conditioning duct system Klimakanalsystem *n*

air conditioning duty, air conditioning load Klimalast, Klimaleistung *f*

air conditioning engineer Klimaingenieur *m*
air conditioning engineering Klimatechnik *f*

air conditioning evaporator Klimaverdampfer *m*

air conditioning fan Klimaventilator *m*

air conditioning field Klimabranche *f*, Klimasektor *m*

air conditioning induction equipment Klimainduktionsgerät(e) *n*, (*f*)

air conditioning installation, air conditioning plant [*An installation of full control over the internal air supply to give specified comfort or other conditions. It is a form of mechanical ventilation that has an extra ability to cool and heat the air, and to change humidity*] Klimaanlage, raumlufttechnische Anlage *f* [*DIN 1946*]

air conditioning luminaire, air conditioning light(ing) fitting, air conditioning light(ing) fixture Klimaleuchte *f*, Klimabeleuchtungskörper *m*

air conditioning ((open) loop) control Klimasteuerung *f*

air conditioning outfit Klimaausrüstung, Klimaausstattung, Klimabestückung *f*

air conditioning product Klimaerzeugnis, Klimaprodukt *n*

air conditioning refrigerating system → **air conditioning chiller**

air conditioning refrigeration → **ACR**

air conditioning return duct Klimarückluftkanal *m*

air conditioning supply duct Klimazuluftkanal *m*

air conditioning terminal Klimaendgerät *n*

air conditioning timer Klimazeitschalter *m*

air conditioning unit Klimaeinheit *f*

air constituent, air component Luftbestandteil *m*

air consumption Luftverbrauch *m*

air contactor → **air (break) contactor**

air contaminant Luftverunreinigungsmittel, Luftverschmutzungsmittel *n*

air conveyance efficiency Lufttransportwirkungsgrad, Luftförderungswirkungsgrad *m*

air-cooled luftgekühlt

air cooler [*When the evaporator cools air directly, the plant is said to operate on the direct expansion system as distinct from the case in which a secondary refrigerant is cooled and circulated through the air cooler*] Luftkühler *m*, Luftkühlgerät *n*, LK

air cooling [*The cooling of hot bodies by a stream of cold air*] Luftkühlung *f*

air cooling and venting installation Lüftungs- und Luftkühl(ungs)anlage *f*

air cooling apparatus Luftkühl(ungs)apparat *m* [*Ventilator, Motor, Wärmeaustauscher usw. sind in einem transportablen Gehäuse zusammengebaut. Je nach Größe werden unterschieden Unterfensterkühl(ungs)geräte und Schrankkühl(ungs)geräte. Die Geräte sind entweder mit einem Kälteteil versehen, der bei einigen Typen als Wärmepumpe ausgelegt ist, oder werden an Zentralkälteanlagen angeschlossen. Nicht verwechseln mit Luftkühl(ungs)anlage (air cooling system)*]

air cooling coil Luftkühl(ungs)schlange *f*

air cooling equipment Luftkühl(ungs)gerät(e) *n*, (*f*)

air cooling evaporator Verdampfer zur Kühlung von Luft, Luftkühlungsverdampfer *m*

air cooling installation, air cooling plant Luftkühl(ungs)anlage, Luftkühl(ungs)installation *f* [*Die umgewälzte Raumluft wird gekühlt; wird Außenluft angesaugt heißt die Anlage Lüftungs- und Luftkühl(ungs)anlage (air cooling and venting installation)*]

air cooling power Kühl(ungs)stärke *f* der Luft [*Gemessen mit dem Katathermometer*]

air cooling unit Luftkühl(ungs)einheit *f*

air-core [*It is not magnetic, as one of plastic or fibre*] Luftkern *m*

air-core coil [*Electronics. A coil that is wound on a nonmagnetic core*] Luftkernspule *f*

air-core transformer [*Electronics. A transformer that has two or more air-core coils*] Luftkerntransformator *m*

air current, air stream, air flow Luftstrom *m*, Luftströmung *f*

air curtain [*A high velocity current of warm or cool air across a door or gate (opening)*] Luftschleier, Luftvorhang *m*, Luftschürze *f*

air curtain cooling Luftschleierkühlung, Luftvorhangkühlung, Luftschürzenkühlung *f*

air curtain door Luftschleiertür, Luftvorhangtür, Luftschürzentür *f*

air curtain gate Luftschleiertor, Luftvorhangtor, Luftschürzentor *n*

air curtain merchandiser, air curtain showcase Luftschleiervitrine, Luftschleierschautruhe, Luftvorhangvitrine, Luftvorhangschautruhe, Luftschürzenvitrine, Luftschürzenschautruhe *f*

air curtain unit [*Mechanical air-moving device designed to produce an air curtain*] Luftschleiereinheit, Luftvorhangeinheit, Luftschürze(n)einheit *f*

air cycle Luftspiel *n*, Luftzyklus *m*

air deficiency, air shortage Luftmangel *m*, Luftknappheit *f*

air deficiency safety device, air shortage safety device Luftmangelsicherung, Luftknappheitsicherung *f*

air deflector, air baffle(r) Luftschikane *f*, Luftumlenker, Luft(ab)lenker, Luftstrombrecher *m*, Luft(ab)lenkvorrichtung *f*

air deflector plate → air baffle(r) plate

air defrosting, air thawing Luftentfrosten, Luftabtauen *n*

air degradation Luftverschlechterung *f*

air dehumidification Luftentfeuchtung, Luftfeuchteentziehung *f* [*Die Luftentfeuchtung wird entweder durch Außenluftbeimischung oder durch Unterkühlen der Luft erreicht*]

air dehumidifier Luftentfeuchter *m*, LE [*DIN 30 600. Reg.-Nr. 06099*]

air density [*The weight of air, pounds per cubic foot*] Luftdichte *f*

air dew point Lufttaupunkt, Taupunkt der Luft *m*

air diffusion entrainment ratio → induction ratio

air diffusion performance index → ADPI

air diffuser, air diffusor [*A device, often a louver, designed to introduce air from an air-conditioning duct in a ceiling and mix it with air already in the room*] Luftdiffusor *m*

air diffuser, air diffusor [*activated sludge process*] Luftdiffusor *m*

air dilution [*A hazardous gas or vapo(u)r released into a confined space where there is a minimum rate of dilution by ventilation air to maintain the environment within the contamination limits specified for protection against health hazard or fire hazard*] Luftverdünnung *f*

air direction Luftrichtung *f*

(air) director (Luft)Geschränk *n*

(air) director hood (Luft)Geschränkhaube *f*, (Luft)Geschränkkasten *m*

air dirt Luftschmutz *m*

air discharge coefficient [*Ratio of the net area at vena contracta of air flowing through an orifice to the total free area of the opening*] Luftablasskoeffizient, Luftaustrittskoeffizient *m*

air discharge opening Luftaustritt(s)öffnung, Abluftöffnung *f*

air discharge outlet → discharge (air) outlet

air discharge shaft → discharge (air) shaft

air discharge valve → discharge (air) valve

air distribution Luftverteilung *f*

air (distribution) box Luft(verteil(ungs))kasten *m*

air (distribution) duct Luft(verteil(ungs))kanal *m*

air distributor Luftverteiler *m*

air door Lufttür *f*

(air) draft, (air) draught [*A device to control the flow of air, as in a stove or fireplace*] (Luft)Zug *m*

airdraulic [*Combining pneumatic and hydraulic action for operation*] drucklufthydraulisch, presslufthydraulisch, pneumatisch hydraulisch

air-dried luftgetrocknet

air drop [*Vertical distance between the base of an air outlet and the bottom of the airstream at the end of the air throw. Note: Drop occurs from the natural expansion of the airstream*] Luftabfall *m*

air-dry lufttrocken

air drying Lufttrocknung *f*

air drying installation, air drying plant Lufttrocknungsanlage *f*

air duct [*See remark under "duct"*] Luftkanal *m*

air duct heater, air duct heating device, air duct heating appliance Luftkanal(er)wärmer, Luftkanalerhitzer *m*

A

air duct humidity controller, air duct hygrostat Luftkanalfeuchteregler, Luftkanalfeuchtigkeitsregler *m*

air duct manufacturer, air duct producer Luftkanalhersteller *m*

air duct noise Luftkanalgeräusch *n*

air duct system [*Ducts, elbows, connectors, plenums, and fans assembled to form a continuous passageway for the distribution of air*] Luftkanalsystem *n*

(air) ductwork, (air) trunking [*A system of ducts for distribution and extraction of air*] Luftkanalsystem *n*

air dumping [*Rapid falling of cold air that occurs when a variable air volume (VAV) box or other device reduces airflow, and the supply air leaves the diffuser at very low velocity*] Kaltluftsturz *m*

air eliminator [*A device in a steam distribution system that opens when air or other noncondensables reach it, and closes if water or steam is present in the vent body, for the purposes of removing the noncondensables from the system*] Luftabführung, Entlüftung *f* [*Dampfrohrleitung*]

air emulsification, air emulsifying Luftemulgierung, Luftemulgation *f*

air engineering Lufttechnik *f*

air enthalpy, air (total) heat content, air total heat Luftwärmeinhalt *m*, Luftgesamtwärme, Luftenthalpie *f*

air entrainment Lufteinschluss *m*

(air) entry → (air) inlet

air envelope Lufthülle *f*

air environment [*Telecommunications. The total airborne paraphernalia of any communications system*] Luftgerät(e), Luftausrüstung *f*

air-equalizing → air-balancing

air equipment Luftgerät(e) *n*, (*f*)

air excess, combustion air excess Luftüberschuss, Verbrennungsluftüberschuss *m* [*Je weiter ein Brennstoff von der gasförmigen Phase entfernt ist, um so schwieriger ist eine einwandfreie Vermischung Brennstoff-Verbrennungsluft. Man benötigt stets mehr Verbrennungsluft als theoretisch notwendig ist. Das Verhältnis von tatsächlich zugeführter Luftmenge zur theoretisch notwendigen Luftmenge, nennt man den Luftüberschuss. Wenn also zur Verbrennung von 1 m^3 Ferngas 4,2 m^3 Verbrennungsluft notwendig sind, werden bei Luftüberschuss 1,5 = 4,2 mal 1,5 = 6,3 m^3 Luft zugeführt. Trotz der damit verbundenen Erhöhung des Abgasverlustes wird aus Sicherheitsgründen mit einem gewissen Luftüberschuss gefahren. Der kleinste Luftüberschuss (1,05 bis 1,35) wird bei Brenngasen, der größte Luftüberschuss (> 1,4) bei festen Brennstoffen erreicht*]

air (ex)change → air change

air (ex)changes per hour → AC(P)H

air (ex)changes rate, air renewal rate, air turnover rate [*Airflow in volume units per hours divided by the building space volume in identical volume units; normally expressed in air (ex)changes per hour (AC(P)H)*] Luftwechselrate *f*

air exfiltration → air spillage

air exhaust, air extraction Sauglüftung *f* [*Die Luft wird mittels eines Ventilators aus einem Raum abgesaugt und ins Freie geblasen, während aus dem benachbarten Raum Luft durch Türen und andere Öffnungen aus dem Freien durch die Fenster nachströmt*]

air exhaust hood, air extraction hood Sauglüftungshaube *f*

air exhaust plant, air exhaust installation, air extraction plant, air extraction installation Entlüftungsanlage, Entlüftungsinstallation *f*

air exit grid → exit air grid

air exit valve → exit air valve

air extraction rate, air exhaust rate Entlüftungsleistung, Abluftleistung *f*

air filter → (dust) air filter

air filter fan Luftfilterventilator *m*

air filtering Luftfiltern *n*, Luftfilterung *f*

air filter monitoring Luftfilterüberwachung *f*

(air) fire → (air) heater

(air) floor grill(e), floor air grill(e) [*See remark under "(air) grill(e)"*] (Luft)Deckengitter, Deckenluftgitter *n*

air floor heating → air (under)floor heating

air floor outlet → floor air outlet

air flow, air stream, air current Luftstrom *m*, Luftströmung *f*

air flue [*A vertical duct to convey vitiated air from a room to the outside air*] Schlechtluftabzug(kanal) *m*

air flush(ing) Luftspülung *f*

air for combustion → combustion air

air fractionation [*Separation of the various components of air through distillation at very low temperature*] Luftzerlegung *f*

air free of contaminants → clean air

air freshener, fresh air generator Frischlufterzeuger *m*

(air/(fuel)oil) emulsion burner Luft-(Heiz)Öl-Emulsionsbrenner *m*

air furnace Windofen *m*

air gap [*The space existing between two elements or parts*] Luftlücke *f*

air gas → producer gas

air gas mix(ture), gas air mix(ture) Luft-Gas-Gemisch *n*, Luft-Gas-Mischung *f*, Gas-Luft-Gemisch *n*, Gas-Luft-Mischung *f*

air gas ratio, gas air ratio [*The ratio of the volume of primary air to the volume of gas in a mix(ture)*] Luft-Gas-Verhältnis, Gas-Luft-Verhältnis *n*

air gauge; air gage (US) [*A device used to compare a machined surface to a standard surface by assessing the rate of air passage between them*] Luftspaltlehre *f*

air gravity heating → gravity air heating

(air) grill(e) [*A covering for any area through which air passes. Not to be confused with "grid"*] (Luft)Gitter *n*

air gun → spray gun

air hazard Luftgefährdung *f*

air heat Luftwärme *f*

air heat content → air enthalpy

(air-)heated luftbeheizt, lufterwärmt

(air) heater, (air) heating appliance, (air) heating apparatus, (air) heating device, (air) fire [*Heating of the air may be carried out through steam air heaters, water air heaters, electric(al) air heaters and gas air heaters*] Luftheiz(ungs)gerät *n*, Luftheiz(ungs)apparat, Luftheizer, Lufterhitzer, Luft(er)wärmer *m*

(air) heater furnace, (air) heating furnace Warmluftofen *m*

(air) heater pump Lufterhitzerpumpe, Luftheiz(ungs)pumpe, Luft(er)wärmerpumpe *f*

(air) heating, (air) warming [*It may be carried out through the medium of steam, water, electricity or gas*] Luft(er)wärmung, Lufterhitzung, Luftheizung *f*

(air) heating assembly, (air) warming assembly Luftheiz(ungs)aggregat, Luft(er)wärmungsaggregat, Lufterhitzungsaggregat *n*

(air) heating/cooling installation, (air) heating/ cooling plant Luftheiz(ungs)- und -kühl(ungs)anlage, Teilklima(tisierungs)anlage *f* [*Die Raumluft wird je nach Anforderung oder Jahreszeit geheizt oder gekühlt. Die Anlage arbeitet mit Außenluftzusatz*]

(air) heating device, (air) heating appliance, (air) heating apparats, air heater, air fire Luftheiz(ungs)gerät *n*, Luftheizer, Lufterhitzer, Luftheiz(ungs)apparat, Luft(er)wärmer *m*

(air) heating installation, (air) heating plant Luftheiz(ungs)anlage *f* [*Die Anlage fördert, entweder mit natürlichem Auftrieb oder zwangsweise mit Ventilator, wie eine Belüftungsanlage Luft in den zu beheizenden Raum, jedoch mit dem Unterschied, dass die Luft über Raumtemperatur auf etwa 30° C bis 70° C erwärmt wird. Zusätzlich zugeführte Wärme dient zur Deckung des Wärmeverlustes des Raumes, so dass sich die Luft im Raum allmählich auf Raumtemperatur abkühlt*]

(air) heating solar collector Luftkollektor *m* [*solare Lufterwärmung*]

(air) heating system Luftheiz(ungs)system, Luft-Wärme-System *n*

(air) heating unit, (air) heater [*Unit which includes means for air heating, and which may also include means for other air-handling unit functions*] Lufterhitzer, Luft(er)wärmer, Luftheizer *m*

air (horse)power [*The theoretical minimum power required for air delivery by a fan, blower, compressor or vacuum pump with no losses*] Luft-PS *f*

air hose Luftschlauch, flexible Luftkanal *m*

air humidification Luftbefeuchtung, Luftanfeuchtung *f*

air humidification/dehumidification plant, air humidification/dehumidification installation Luftbe- und –entfeuchtungsanlage *f*, Teilklimaanlage [*Die Raumluft wird je nach Anforderung befeuchtet oder entfeuchtet. Die Anlage arbeitet mit Außenluftzusatz. Die Anlage kann die Bezeichnung FHK (Filtern, Heizen, Kühlen) haben*]

air humidification device, air humidifying device Luftbefeuchtungsvorrichtung, Luftanfeuchtungsvorrichtung *f*

air humidification efficiency, air humidifying efficiency Luftbefeuchtungswirkungsgrad, Luftanfeuchtungswirkungsgrad *m*

air humidification filter, air humidifying filter Luftbefeuchtungsfilter, Luftanfeuchtungsfilter, Smogfilter *m*, *n*

air humidification system, air humidifying system Luftbefeuchtungssystem, Luftanfeuchtungssystem *n* [*Die Raumluft wird durch geeignete Mittel befeuchtet. Luftbefeuchtungsanlagen haben Außenluftzusatz*]

air humidifier, air humidificator Luftbefeuchter, Luftanfeuchter *m*, LB, LA

A

air humidifying load, air humidification load Luftbefeuchtungslast, Luftanfeuchtungslast *f* [*Dampfmassestrom, der dem Raum zugeführt werden muss, um einen angestrebten Raumluftzustand zu halten*]

air hygrometer Luftfeuchtemesser, Luftfeuchtigkeitsmesser *m*

air indicator → **air(flow) indicator**

(air) inlet, (air) entry, (air) ingress [*Device or opening through which air is discharged into a conditioned space. Gri(ll)es, registers, diffusers, and slots may be used as air inlets*] Lufteinlass, Lufteintritt, Luftzutritt *m*

(air) inlet cock (Luft) Einlasshahn *m*

(air) inlet core area [*Total plane area of the portion of a grill(e), face, or register bounded by a line tangent to the outer edges of the outer openings through which air can pass*] (Luft)Einlasskernfläche *f*

(air) inlet duct (Luft)Einlasskanal *m*

(air) inlet free area [*1. Total area through which air can pass in a grill(e) face or register. 2. Actual open area between the fins of grill(e) or register*] (Luft-)Einlassfreifläche *f*

(air) inlet grill(e), (air) inlet grating, (air) inlet grate (Luft)Einlass(abdeck)rost *m*, (Luft)Einlass(abdeck)gitter *n*

(air) inlet opening (Luft)Einlassöffnung *f*

(air) inlet system (Luft)Einlasssystem *n*

(air) inlet valve [*A valve with mica flap to allow the fresh air free access to the drains*] (Luft)Einlassschieber *m*

air input, air intake Belüftung *f*

air input filter, air intake filter Belüftungsfilter, Zuluftfilter *m, n*

air input valve, air intake valve Belüftungsventil, Zuluftventil *n*

air installation, air plant Luftanlage *f*

air-insulated luftisoliert

air insulation [*Term applied when the insulation for part of an electric(al) circuit is provided by atmospheric air, e.g., a high voltage transmission line, which is suspended between transmission towers (pylons), is insulated for the section between the towers by atmospheric air*] Luftisolierung *f*

air intake → **air input**

air ion Luftion *n*

air jacket Luft(führungs)mantel *m*

air jet → **air blast**

air-jet ... → **air-blast ...**

air large space heating → **large space (warm) air heating**

air leak(age) Luftundichtigkeit, Luftundichtheit *f*

air leaking outward → **air spillage**

air line [*A hose, duct, or pipe that delivers air, compressed or otherwise, to a pneumatic tool or other such devices*] Luftleitung *f*

air line component Luftleitungsbauteil *m, n*

air lock Luftschleuse *f* [*Raum zur verkehrsmäßigen Verbindung bei luftmäßiger Trennung von Bereichen unterschiedlichen Luftstandards mit mindestens zwei gegenseitig verriegelten Türen*]

air lock(ing) [*The stoppage of the flow of water in a hot water system due to the presence of a bubble of air trapped in it*] Luftsperre *f*

air lock strip (US); weatherstrip [*A narrow strip used around the edges of doors and windows to prevent the passage of air and water through the joints*] Fugen(ab)dicht(ungs)streifen *m*

air lock stripping (US) → **weatherstripping**

air magnetic circuit-breaker Magnetblasschalter *m*

air main Lufthauptleitung, Hauptluftleitung *f*

air management Luftführung *f*

air manifold, air central distributor, central air distributor Luftverteiler(rohr) *m, (n)*, Luftverteil(ungs)rohr, Luftverteil(er)stück, Luftverteilungsstück *n*, Luftverteil(er)batterie, Luftverteilungsbatterie *f*

air mass [*Totality of gas molecules in a defined volume of air*] Luftmasse *f*

air mass ratio [*Ratio of the mass of atmosphere in the actual earth-sun path to the mass which would exist at the sea level if the sun were directly overhead*] Luftmassenverhältnis *n*

air/media ratio [*Same as air/cloth ratio, but used where noncloth bag media, such as paper cartridges, are the filter media*] Luft/Nicht(filter)gewebe-Verhältnis *n*

air meter [*It measures the rate of flow of air and other gases*] Luftströmungsmesser *m*

(air) mixing (Luft)Mischen *n*

(air) mixing box, (air) mixing plenum, mixed air box, mixed air plenum [*A compartment in which a recirculating air supply is mixed with incoming out-door air before being discharged to an air-conditioning system*] (Luft)Mischkasten, Mischluftkasten *m*, (Luft)Mischkammer, Mischluftkammer *f*

air mixing damper (Luft)Mischklappe, Mischluftklappe *f*

(air) mixing tube (Luft)Mischröhre, Mischluftröhre *f*

air mix(ture) Luftmischung *f*, Luftgemisch *n*

air monitoring [*Ongoing or continuous air sampling, especially in terms of pollution level and radio-activity*] Luftreinhaltungsüberwachung *f*

air motor [*1. Device which convects compressed air into mechanical force. 2. Air-operated device that is used primarily for opening or closing dampers. 3. Pneumatic operator, a final control device that assumes a position as directed by an input-pressure signal*] Luftmotor *m*

air need → air requirement

air nozzle → vent(ilation) nozzle

(air oil) emulsion burner (Luft-Öl-)Emulsionsbrenner *m*

airometer, aerometer [*It is used to measure the weight and density of air and other gases*] Aerometer *n*

air-only (air) conditioning plant, air-only (air) conditioning installation Nur-Luft-Klima(tisierungs)anlage *f*

air-only system Nur-Luft-System *n*, Nur-Luft-Anlage *f* [*Energiezufuhr an den Versorgungsstellen nur durch Zuluft. DIN 1946, Teil 1*]

air opening, air port Luftöffnung *f*

air outlet Luftauslass, Luftaustritt *m*

air outlet grating Luftauslassgitter, Luftaustritt(s)gitter *n*

air output Luftleistung *f* [*m^3/Stunde; m^3/Sekunde; m^3/Minute*]

air panel heating, air radiant heating, panel air heating, radiant air heating Luftstrahlungsheizung *f*

air pass(age), air passageway Luftdurchgang *m*

air path Luftweg *m*

air permeability, air perviousness Luftdurchlässigkeit *f*

air-permeable, air-pervious luftdurchlässig

airphone (öffentliche) Flugzeugfunktelefon *n*

air pipe [*A by-pass to prevent the unsealing of traps*] Luftrohr *n*

air pipeline, air piping Luftrohrleitung *f*

air plant → air installation

air plate duct Luftblechkanal *m*

air pocket [*The restriction or the stopping of the flow of water by air trapped in piping*] Luftsack *m*, Lufttasche *f*

air pollutant, air polluting agent [*Unwanted airborne constituent that may reduce acceptability of the air: dust, fog, fume, mist, smog and smoke*] Luftverunreiniger, Luftverschmutzer *m*

air pollution Luftverschmutzung, Luftverunreinigung *f*

air pollution monitoring Luftreinhaltungsüberwachung *f*

air pollution prevention, air pollution control Luftreinhaltung *f*

air port → air opening

air precleaner, air prefilter Luftvorfilter *m, n*

(air) preheater (Luft)Vor(er)wärmer, (Luft)Vorheizer, (Luft)Vorerhitzer, Wärmeaustauschkörper, Luvo *m* [*Er bringt die Außenluft von der Außentemperatur auf Taupunkttemperatur*]

(air) preheating (Luft)Vor(er)wärmen, (Luft)Vorerhitzen *n*

air premixing Luftvormischen *n*, Luftvormischung *f*

air-premixing burner, premixing air burner Injektorbrenner, Saugbrenner, Brenner mit Luftvormischung *m*

air pressure Luftdruck *m*

air-pressure atomization Luftdruckzerstäubung *f*

air-pressure atomizing burner Luftdruckzerstäuberbrenner *m*

air-pressure metering Luftdruckmessung *f*

air-pressure monitor Luftdruckwächter *m*

air-pressure-reducing valve Luftdruckminderventil, Luftdruckreduzierventil *n*

air-pressure sensor Luftdruckfühler, Luftdrucksensor *m*

air pretreatment Luftvorbehandlung *f* [*Luftteilbehandlung in Strömungsrichtung vor der Lufthauptbehandlung. DIN 1946, Teil 1*]

air processing Luftveredelung, Luftaufbereitung *f*

airproof, airtight luftdicht

air pump → (compressed-)air pump

air purging → air scavenging

air purification, air cleaning Luftreinigung, Luftsäuberung *f*

air purifier → air cleaner

air purity, air cleanliness Luftreinheit *f*

air quality Luftgüte *f*

air quality ((closed) loop) control Luftgütereg(e)lung *f*

air quality ((closed) loop) controller, air quality regulator Luftgüterregler *m*

A

air quality index Luftgüteindex *m*

air quality pickup, air quality measured value transmitter Luftgütemesswertgeber *n*

air quality sensor Luftgütefühler *m*

air quantity Luftmenge *f* [*Ausgedrückt als Masse in kg oder als Volumen in* m^3 *oder in kmol*]

air (quantity) adjuster blade Luft(mengen)verstell(ungs)zunge *f*

air radiant heating → **air panel heating**

(air-raid) house shelter, house air-raid shelter Haus(luft)schutzraum *m*

(air-raid) shelter Schutzraum, Luftschutzraum *m*

air rate, rate of airflow Luftrate *f*, Luftdurchsatz, bezogene Luftvolumenstrom *m*, Luftdurchflussmenge *f* [*Die je Person erforderliche Luftmenge*]

air ratio Luftverhältnis *n*

air recirculation, air return Luftrückführung, Luftrückkehr *f*

air recooling [*Cooling of air that has been previously heated*] Luftrückkühlung *f*

air-reducing valve Luftminder(ungs)ventil *n*

(air) register [*Slotted frame for control of the direction of air delivered to the space and its flowrate*] (Luft)Register *n*

air removal Entlüftung *f*

air removal disk valve, air removal disc valve Entlüftungstellerventil, Entlüftungsscheibenventil, Tellerentlüftungsventil, Scheibenentlüftungsventil *n*

air renewal → **air change**

air requirement, air need Luftbedarf *m*

air ring space → **air annulus space**

air room thermostat → **(compressed-)air room thermostat**

air rush, rush of air Luftstoß *m*

air sample Luftprobe *f*

air scavenging, air purging Luftspülen *n*

air scrubber → **air washer**

air scrubbing → **air washing**

air seal(ing) Luft(ab)dichtung *f*

air-seasoned freiluftgetrocknet [*Holz*]

air secondary (water) supply → **(compressed-) air secondary water supply**

air separation [*Removing air from another fluid stream*] Luft(ab)scheidung, Luftabsonderung *f*

air separation unit, air separator [*A device that uses an air current to separate a material from another of greater density or particles from others of greater size*] Luft(ab)scheider *m*

air shaft, air well, light shaft, light well [*An unroofed shaft in the middle of a building. It gives light and air to windows that open on to the shaft*] Lichtschacht, Luftschacht *m*

air shaft grate, air well grate, light shaft grate, light well grate [*"Grill(e)" and "grating" may synonymously be used for "grate"*] Lichtschacht(abdeck)rost *m*, Lichtschacht(abdeck)gitter *n*, Luftschacht(abdeck)rost *m*, Luftschacht(abdeck)gitter *n*

air shaft security, air well security, light shaft security, light well security Lichtschachtsicherung, Luftschachtsicherung *f*

air shaft ventilation Schachtlüftung *f*

air shock, air water hammer Luft-Wasserschlag, Luft-Wasserstoß, Luftdruckschlag, Luftdruckstoß *m* [*Wasserrohrleitung*]

air shortage, air deficiency Luftmangel *m*, Luftknappheit *f*

air shortage safety device → **air deficiency safety device**

air shower Luftdusche *f* [*Ein Gerät, das einen Luftstrom erzeugt, dessen Temperatur, relative Feuchte und Geschwindigkeit solche Werte aufweisen, die die Wärmebilanz der angeblasenen Person günstig beeinflusst*]

air side Luftseite *f*

air side [*heat pump*] → **cold side**

air siphon, air syphon [*An inclined tube built into a wall to prevent damps*] Luftsiphon *m*

air sound → **air(borne) sound**

air space, cavity, void Hohlraum *m*

air speed [*The velocity of air as measured while passing through a ventilation duct, register or outlet*] Durchströmungsgeschwindigkeit *f*

air spillage, air exfiltration, air leaking outward Luftleckage *f*, Luftlecken *n*

air spring [*A spring in which the energy storage element is air confined in a container that includes an elastomeric bellows or diaphragm*] Luftfeder *f*

air stagnation, air buildup Luft(auf)stau *m*

air stratification Luftschichtung *f*

air stream → **air flow**

air suction Absaugung, Luftabsaugung *f* [*Örtliche maschinelle Luftabführung zur Senkung des Raumbelastungsgrades. DIN 1946, Teil 1*]

air suction opening Luft(an)saugöffnung *f*

air supply [*Supplying air by natural or mechanical means to any space. Such air may or may not have been conditioned*] Belüften *n*, Belüftung *f*, Luftzufuhr, Lufteinführung *f*

air switch → **air (break) switch**

air switch disconnector → **air (break) switch disconnector**

air syringe, chip blower Spanbläser, Spänebläser *m*

air tank Luftbehälter *m*

air temperature level Lufttemperaturniveau *n*, Lufttemperaturstand, Lufttemperaturpegel *m*

air terminal Fangvorrichtung *f*, Fänger *m* [*Blitzschutz*]

air-termination network, air-terminations Fangleitung(s)maschennetz *n*, Dachleitungen, Fangvorrichtung *f* [*Blitzschutz*]

air-termination rod Fangstange, Blitzschutzstange *f*

air test [*A test for the soundness of drain(age) systems and soil, waste and ventilating pipe systems by applying air pressure internally*] Abdrücken *n*

air thawing → **air defrosting**

air throughput Luftdurchsatz *m*

air throw, air blow, blow throw [*In air distribution, the distance an air stream travels from an outlet to a position at which air motion along the axis reduces to a velocity of 50 ft/minute*] Luftstromweite, Blasweite *f*

airtight → **airproof**

air tightness [*Qualitative term describing the integrity of the building envelope relative to air permeation; the resistance of the building envelope to the flow of air and entrained moisture*] Luftdichtheit *f*

air-to-air ... → **air/air ...**

air-to-refrigerant-to-air system Luft-Kältemittel-Luft-System *n*

air-to-refrigerant-to-water-to-air system Luft-Kältemittel-Wasser-Luft-System *n*

air (total) heat content → **air enthalpy**

air-to-water heat exchanger, air-to-water heat exchange unit, air-to-water heat interchanger Luft-Wasser-Wärme(aus)tauscher [*In der Industrie so genannt*], Luft-Wasser-Wärmeübertrager *m* [*Siehe Anmerkung unter „Wärme(aus)tauscher"*]

air-to-water-to-refrigerant-to-water system Luft-Wasser-Kältemittel-Wasser-System *n*

air transfer [*Transfer of air from one place to another*] Luftübertragung *f*

air trap Luftfalle *f*

(air) trunking → **(air) ductwork**

air tube → **carrier tube**

air tube conveyor, air tube system, pneumatic conveyor, pneumatic system, carrier tube conveyor, carrier tube system Rohrpost *f* [*DIN 6651*]

air turbo luminaire Druckleuchte *f*

(air) turbulence device [*oil burner*] (Luft)Wirbelvorrichtung *f*, (Luft)Wirbler *m*

air turnover → **air change**

air turnover ... → **air change ...**

air type Luftart *f* [*Bezeichnung der Luft nach dem jeweiligen Ort innerhalb einer RLT-Anlage oder einer Lüftungseinrichtung (z.B. Außenluft, Zuluft, vorbehandelte Luft, Abluft, Umluft, Fortluft)*]

air (under)floor heating, air (under)floor warming Luft(fuß)bodenheizung, Luft-FB-Heizung, (Fuß)Bodenluftheizung *f*

air valve → **ventilation valve**

air-vane relay Windfahnenrelais *n*

air vapor-mix(ture) (US); air vapour mix(ture) (Brit.) Luft-Wasserdampf-Gemisch *n*, Luft-Wasserdampf-Mischung *f*

air velocity [*Rate of motion of air in a given direction, measured as distance per unit time*] Luftgeschwindigkeit, Strömungsgeschwindigkeit *f* der Luft

air velocity detector → **(duct) air velocity detector**

air velocity measured-value transmitter Luftgeschwindigkeits-Messwertgeber *m*

air-vent [*An opening designed to allow air to enter or escape from a confined space, as in a building or mechanical system*] Luftdurchlass *m*

air-vent measuring station Luftdurchlassmessstation *f*

air vessel Luftgefäß *n*

air vitiation Luftverschlechterung *f*

(air) volume act(uat)or (Luft)Volumenstromsteller *m* [*Bauelement zur Steuerung eines (Luft)Volumenstromes (mit Vordruckeinfluss). DIN 1946, Teil 1*]

(air) volume flow (Luft)Volumenstrom *m* [*Quotient aus gefördertem Luftvolumen und Zeit*]

(air) volume flow ((closed) loop) controller (Luft)Volumenstromregler *m* [*Bauelement zu Einhaltung eines konstanten (Luft)Volumenstromes (ohne Vordruckeinfluss). DIN 1946, Teil 1*]

(air) volume system (Luft)Volumenstromsystem *n*

(air) wall cooler → **wall air cooler**

A

(air) wall grill(e), wall air grill(e) [*See remark under "(air) grill(e)"*] (Luft)Wandgitter, Wandluftgitter *n*

(air) wall inlet → wall air inlet

(air) wall outlet → wall air outlet

air warming → air heating

air washer, air scrubber [*A chamber in an air-conditioning system in which air is cleaned by being passed through water sprays or over wet plates which remove dust and smells*] Sprühbefeuchter *m*, Luftwäsche(r) *f*, (*m*), Düsenkammer *f*

air washing, air scrubbing Sprühbefeuchtung *f*, Luftwaschen *n*, Sprühanfeuchtung *f*

air-waste gas chimney kombinierte Schornstein, KLA-Schornstein, Luft-Abgas-Schornstein, LA-Schornstein *m* [*DVGW-G 627. DVGW = Deutscher Verein des Gas- und Wasserfaches*]

air water air conditioning plant, air water air conditioning installation Luft-Wasser-Klima(tisierungs)anlage *f*

air water atomizer → (compressed-)air water atomizer

air water hammer → air shock

air water heat pump Luft/Wasser-Wärmepumpe, L-W-Wärmepumpe *f*, L-W-WP [*Siehe Anmerkung unter „Wärmepumpe"*]

air water jet [*A high-pressure jet of air and water for cleaning surfaces*] Luft-Wasser-Strahl *m*

air water system Luft-Wasser-System *n* [*Energiezufuhr an den Versorgungsstellen durch Zuluft und durch Wasser; auch in Verbindung mit Raumheiz(ungs-) und Kühl(ungs)flächen, soweit diese mit der raumlufttechnischen Anlage in einem funktionellen Zusammenhang stehen (z.B. gekoppelte Reg(e)lung)*]

air water tank → (compressed-)air water tank

airway [*A space between roof insulation and roof boards for movement of air*] Dachlüftungsraum *m*

air well → air shaft

air well grate → air shaft grate

air well security → air shaft security

air whirl(pool) Luftsprudler *m*

air withdrawal slot Luftabzugsschlitz *m*

AL → acceptable (quality) level

alarm alarmieren

alarm [*colloquial*] **→ sounder**

alarm [*1. An annunciation, either audible or visual or both, which alerts an operator to an off-normal condition that may require corrective action. 2. An abnormal condition detected by a device or controller that implements a rule or logic specifically designed to look for that condition*] Alarm *m*

alarm acceptance, alarm reception Alarmempfang *m*, Alarmannahme *f*

alarm actuation → alarming

alarm alert Alarmbereitschaft *f*

alarm annunciation board, alarm annunciation panel, mimic board, mimic panel, synoptic board, synoptic panel Synoptiktafel *f*

alarm annunciator, alarm panel Alarmtableau *n*, Alarmtafel *f*

alarm automatic mode → alarm processing

(alarm) bell Wecker *m*, Alarmklingel *f*

alarm block Alarmsperre *f*, Alarmblock *m*

alarm blocking Alarmblockierung, Alarmsperrung, Alarmsperre *f*

alarm bus Alarmbus *m*

alarm buzzer Alarmsummer *m*

alarm cable Alarmkabel *n*

(alarm) call point, alarm point [*A place where an alarm system, such as a fire alarm, can be manually operated*] Alarm(melde)stelle, Handalarm(melde)stelle *f*

alarm characteristic Alarmmerkmal, Alarmkriterium *n*

alarm circuit Alarmkreis *m*

alarm condition → alarm state

alarm connector Alarmsteckerausgang *m*

alarm contact Alarmkontakt *m*

alarm(contact)shunt Alarm(kontakt)überbrückung *f*

alarm control center (US); alarm control centre (Brit.) Alarmzentrale *f*

alarm-controlled alarmgesteuert

alarm counter Alarmzähler *m* [*Vorrichtung in einer Alarmzentrale, die bei jedem Alarm ein Zählwerk betätigt*]

alarm current, current draw on alarm Alarmstrom *m*

alarm data Alarmdaten *f* [*Alle festgehaltenen oder eingegebenen Daten über unzulässige Zustände, z.B. Zutrittskontrolle. Aufgrund der Alarmdaten erfolgt eine Meldung an die hilfeleistende(n) Stelle(n)*]

alarm data base, alarm data bank Alarmdatenbank *f*

alarm decision Alarmentscheidung *f*

alarm delay Alarmverzögerung *f*, Alarmverzug *m*

alarm delay time Alarmverzögerungszeit, Alarmverzugszeit *f*

alarm deviation Alarm *m* bei unzulässiger Regelabweichung

alarm device Alarmgeber *m*, Alarmvorrichtung *f*, Alarmgerät *n*

alarm display Alarm(sicht)anzeige *f*

alarm door lock Alarmtürschloss *n*

alarm duration Alarmdauer *f*

alarm electronics Alarmelektronik *f*

alarm equipment Alarmgerät(e) *n*, (*f*)

alarm (equipment) installer Alarm(geräte)errichter *m*

alarm (e)valuation, alarm interpretation Alarmauswertung, Alarmbewertung *f*

alarm event Alarmereignis *n*

alarm execution → alarm processing

alarm float Alarmschwimmer *m*

alarm foil Alarmfolie *f* [*Dünner Metallstreifen zur Durchbruch(s)überwachung. Nicht verwechseln mit „Alarmsicherheitsfolie"*]

alarm for initiation control, initiation control alarm Wirk(ungs)alarm *m*

alarm fuse Alarmsicherung *f*

alarm gauge; alarm gage (US) [*A device that gives off a warning signal if the steam pressure or water level in a boiler reaches a crucial level*] Alarmlehre, Warnlehre, Standlehre *f*

alarm generation, alarm release Alarmauslösung *f*

alarm glass → signal glass

alarm (glass) pane, signal (glass) pane Alarmglasscheibe, Signalglasscheibe, Verbundglasscheibe mit Alarmdrahteinlage *f*

alarm glazing, signal glazing Alarmverglasung, Signalverglasung *f*

alarm horn Alarmhupe *f*

alarm identification Alarmidentifizierung, Alarmkenn(zeichn)ung *f*

alarm identifier Alarmkenner *m* [*Leittechnik*]

alarm image, alarm picture Alarmbild *n*

alarm indicator Alarmanzeiger, Alarmgeber *m* [*Vorrichtung mit der Dritte auf eine Notlage aufmerksam gemacht und zur Hilfeleistung aufgefordert werden*]

alarm (indicator) lamp, alarm light Alarmanzeigelampe, Alarmgeberlampe *f*, optische Alarmsignal *n*

alarming, alarm release, alarm actuation, alarm generation, alarm activation, alarm signalling; alarm signaling (US) Alarmierung, Alarmieren *n*, Alarmgabe, Alarmauslösung, Alarmaktivierung *f*

alarm input Alarmeingabe *f*

alarm installation → alarm plant

alarm interpretation → alarm (e)valuation

alarm key Alarmtaste *f*

alarm line Alarmleitung *f*

alarm locating, alarm localizing Alarmlokalisierung *f*

alarm lock Alarmschloss, Sicherheitsschloss *n*

alarm loop Alarmschleife *f*

alarm luminaire Alarmleuchte *f*

alarm management Alarmverwaltung, Alarmbehandlung *f*

alarm management system, AMS Alarmverwaltungssystem, Alarmbehandlungssystem *n*

alarm memory Alarmspeicher *m*

alarm modem [*See remark under "modem"*] Alarmmodem, Alarm-Modulator/Demodulator *m*

alarm ((open) loop) control Alarmsteuerung *f*

alarm organization Alarmorganisation *f* [*Organisatorische Festlegung der bei einem Alarmeingang zu veranlassenden Maßnahmen und deren zeitlicher Ablauf*]

alarm output Alarmausgang *m* [*Vorrichtung zum Auslösen und Absetzen einer Alarmmeldung*]

alarm override (control), override alarm (control) Alarmübersteuerung *f*

alarm overview Alarmübersicht *f*

alarm panel → alarm annunciator

alarm path Alarmpfad, Alarmweg *m*

alarm picture → alarm image

alarm plan, alarm schedule Alarmplan *m* [*Teil eines Notfallplanes, in dem ereignisbezogene Sofortmaßnahmen niedergelegt sind*]

alarm plant, alarm installation Alarmanlage *f*

alarm point → alarm (call) point

alarm pressure switch Alarmdruckschalter *m*

alarm printer Alarm(wert)drucker *m*

alarm probability Alarmwahrscheinlichkeit *f*

alarm processing, alarm execution, alarm automatic mode Alarmabarbeiten *n*, Alarmabarbeitung *f*, Alarmverarbeiten *n*, Alarmverarbeitung, Alarmbearbeitung *f*, Alarmbearbeiten *n*

alarm pushbutton Alarmdruckknopf *m*

A

alarm readiness Alarmbereitschaft *f*

alarm receiver Alarmempfänger *m*

alarm receiving centre; alarm receiving center (US) Alarmempfangszentrum *n*

alarm reception → alarm acceptance

alarm reception system → alarm acceptance system

alarm recording Alarmaufzeichnen *n*, Alarmaufzeichnung *f*

alarm relay Alarmrelais *n*

alarm relaying Alarmweiterleitung *f*

alarm release → alarming

alarm remote transmission, remote alarm transmission Alarmfernübertragung, Fernalarmübertragung *f*

alarm repetition Alarmwiederholung *f*

alarm reset(ing) Alarmrückstellung, Alarmrücksetzung *f*

alarm routing board Alarmwegewahlplatine, Alarmwegewahlleiterplatte *f*, Alarmwegewahlprint *m*

alarm safety foil Alarmsicherheitsfolie *f* [*Mehrschichtiges Polyestersicherheitsfolienlaminat mit Meldeteilen. Nicht verwechseln mit „Alarmfolie"*]

alarm schedule → alarm plan

alarm sensor Alarmpunkt, physikalische Alarmauslöser, Alarmfühler *m*

alarm sequence Alarmfolge *f*

alarm service Alarmdienst *m*

alarm setpoint Alarmsollwert, Alarmvorgabewert *m*

alarm signalling → alarming

alarm signal message Alarmmeldung *f*

alarm signal message system Gefahrenmeldeanlage, Gefahrenmeldungsanlage, GMA *f* [*Eine Fernmeldeanlage zum zuverlässigen Melden von Gefahren für Personen und Sachen. Zu einer GMA gehören Vorrichtungen für Eingabe, leitungsgeführte oder nichtleitungsgeführte Übertragung, Verarbeitung und Ausgabe von Meldungen und Energieversorgung*]

alarm siren Alarmsirene *f*

alarm site Alarmort *m*

alarm source Alarmquelle *f*

alarms report Alarmbericht *m*

alarm state, alarm status, alarm condition Alarmzustand, Alarmstatus *m*

alarm statistics Alarmstatistik *f*

alarm switch Alarmschalter *m*

alarm switch(ing)-off unit Alarmabschalteinheit, Alarmabstelleinheit *f*

alarm system Alarmsystem *n*, Alarmgeber *m*

alarm termination Alarmbeendigung *f*

alarm tester, alarm testing device Alarmanzeige(r)prüfer *m*

alarm threshold Alarmschwelle *f*

alarm timing Alarmzeitabstimmung *f*

alarm tone Alarmton *m*

alarm transceiver Alarm-Sende-Empfangs-Gerät *n*

alarm transmission Alarmübertragung *f*

alarm transmitter Alarmsender *m*

alarm valuation → alarm (e)valuation

alarm verification Alarmbewahrheitung, Alarmverifikation *f*, Nachweis *m* der Alarmrichtigkeit

alarm whistle Alarmpfeife *f*

alarm wire Alarmdraht *m* [*Ein elektrisch leitender Silberdraht flächig über die Glasscheibe verteilt*]

alarm wire glass Alarmdrahtglas *n*

alarm wire paper hanging Alarmdrahtpapiertapete *f*

alarm wire insert Alarmdrahteinlage *f*

alarm wire loop Alarmdrahtschleife *f*

alarm wire textile hanging Alarmdrahttextiltapete *f*

albuminoid nitrogen Albuminoidstickstoff, Eiweißstickstoff *m*

ALC → automatic level ((closed) loop) control

ALE → address latch enable

to alert ankündigen [*Fernsprechruf*]

to alert → to alarm

alert scharf [*Stromkreis*]

alerting signal [*Specific signal that is applied to subscriber access lines to indicate an incoming call*] Rufankündigungssignal *n*

algebraic product algebraische Produkt *n*, *t*-Norm *f* [*unscharfe Logik*]

algebraic sum algebraische Summe, *t*-Konorm *f* [*unscharfe Logik*]

ALGOL, algorithmic language, Algol problemorientierte Programmiersprache *f* für mathematisch-naturwissenschaftliche Anwendungen

algorithm Algorithmus *m* [*1) Nach DIN 66 201 Teil 1 ist der Regelalgorithmus eine Vorschrift zur Berechnung der Werte einer oder mehrerer Stellgrößen aus den Werten einer oder mehrerer Regeldifferenzen. Die zwischen den Reglereingangs- und*

Reglerausgangsgrößen gültige Rechenvorschrift wird Regelalgorithmus genannt. DIN 19 225.
2) Eine vollständig festgelegte Folge von Anweisungen, nach der Eingabedaten verarbeitet und Ausgabedaten erzeugt werden, heißt Algorithmus. DIN 19 226 Teil 1]

algorithmic geometry algorithmische Geometrie *f*

algorithmic language → ALGOL

algorithmic pattern generator, A(L)PG Bitmustererzeuger *m*

ALI → application layer interface

alias [*Computer programming. An alternate name for a certain file, device, data element, or the like*] Alternativbenennung, Parallelbenennung *f*

alias frequency Aliasfrequenz, Faltungsfrequenz, Frequenz *f* mit anderem Namen

aliasing Umklappen, Falten, Überlappen, Überlagern *n* [*Störende Erscheinung, wenn die Abtastfrequenz kleiner als die doppelte Tonfrequenz ist; dadurch entsteht ein Überlappen. Diese Mehrdeutigkeit führt bei der Rückwandlung in den analogen Bereich zu Tönen, die im Original nicht vorhanden waren (Verzerrungen). Vermeiden wird das Überlappen durch ein Filter*]

aliasing filter [*misnormer*] **→ anti-aliasing filter**

alien frequency Fremdfrequenz *f*

alight [*lamp; luminaire*] leuchtend, brennend

aligned-grid tube (Brit.); alined-grid tube (US) Gitterabschalt(ungs)röhre *f*

alignment; alinement (US) Ausfluchtung *f*

alignment; alinement (US) [*Telecommunication. Adjustment of preset tuned circuits to give optimum performance*] Abgleich *m*

ALIS → automatic link setup

alive, live, energized, under tension, in circuit spannungsführend, spannungsdurchflossen, spannungsenthaltend, stromführend, stromdurchflossen, stromenthaltend

alkaline-metal thermoelectric converter Parabolspiegel-Wärmekraftmaschine *f* [*Sie wandelt Sonnenenergie in elektrischen Strom um*]

alkaline-metal vapour lamp (Brit.); alkaline-metal vapor lamp (US) Akali-Metalldampflampe *f*

alkaline (storage) battery, alkaline accumulator, alkaline reversible cell, alkaline secondary cell [*A storage battery in which the electrolyte consists of an alkaline solution, usually potassium hydroxid(e)*] alkalische Sammler *m*, alkalische Akku(mulator) *m*

all-air conditioning system, all-air conditioning plant, all-air conditioning installation Nur-Luft-Klimaanlage *f* [*Die Luft wird in einer Zentrale aufbereitet und durch Kanäle in die zu klimatisierenden Räume gefördert, in denen außer einer statischen Steuerung keine weitere Nachbehandlung stattfindet*]

Allan's double-ported (slide) valve, Allan's dual-ported (slide) valve, Allan's twin-ported (slide) valve Allansche (Doppel)Schieber *m*

allbox Tabellenoption *f* des Dokumentaufbereitungs-kommandos [*Jeder Tabelleneintrag wird eingerahmt*]

all-concrete floor heating Vollbetondeckenheizung *f*

all-electric ((closed) loop) control Ganzelektrikreg(e)lungsvorrichtung *f*

all-electric ((open) loop) control Ganzelektriksteuer(ungs)vorrichtung *f*

Allen (head) screw Inbusschraube *f*

Allen key, hexagonal key (Innen)Sechskantschlüssel *m*

all-fuel boiler → universal boiler

all-gas burner, universal gas burner Allgasbrenner *m* [*Er kann über die Düse der jeweiligen Gasart angepasst werden*]

all-gas heating appliance, universal gas heating appliance, universal gas heater, all-gas heater Allgasheiz(ungs)gerät *n* [*Gasheiz(ungs)gerät für alle Gasarten, z. B. Stadtgas, Erdgas, Propan/Butan*]

all-gas range, universal gas range Allgasherd *m*

all-gas room heater, universal gas room heater Allgasraumheizer *m*

all-gas small room heater, universal gas small room heater Allgaskleinraumheizer *m*

all-glass door Ganzglastür, Vollglastür, Allglastür, Nurglastür *f*

all-glass escape door Ganzglasfluchttür, Vollglasfluchttür, Allglasfluchttür, Nurglasfluchttür *f*

all-glass luminaire, all-glass light(ing) fitting Allglasleuchte, Ganzglasleuchte, Nurglasleuchte, Vollglasleuchte *f*, Allglasbeleuchtungskörper, Ganzglasbeleuchtungskörper, Nurglasbeleuchtungskörper, Vollglasbeleuchtungskörper *m*

all-hose system Nurschlauchsystem *n*

alligator clip Krokodilklemme *f*

alligator wrench Alligatorzange *f*, gezahnte Rohrschlüssel *m*

A

Allihn condenser [*A condenser whose condensing surface is a series of interconnected glass bulbs*] Allihnkondensator *m*

all-insulated sheathed cable berührungssichere Mantelkabel, vollisolierte Mantelkabel, schutzisolierte Mantelkabel *n*

all-insulated wiring berührungssichere Verdrahtung, vollisolierte Verdrahtung, schutzisolierte Verdrahtung *f*

all-in system Komplettsystem *n*

all-in tariff Einheitstarif *m*

all-metal piston pump Ganzmetallkolbenpumpe, Vollmetallkolbenpumpe *f*

allocatable channel → ACC

to allocate, to allot, to assign zuordnen, zuweisen, zuteilen, belegen

allocation, assignment, allotment Zuordnung, Zuweisung, Zuteilung, Belegung *f*

allocation counter Zuordnungszähler *m*

allocator Zuordner, Zuweiser, Zuteiler *m*

all-or-nothing relay Schaltrelais, Kipprelais *n*

to allot → to allocate

alloter [*A telephone term referring to a distributor, associated with the finder control group relay assembly, which allots an idle line-function in preparation for an additional call*] Anrufordner *m*

allowed transition erlaubte Übergang *m* [*zwischen Energieniveaus*]

alloy transistor Legierungstransistor *m* [*Ein Transistor, bei dem durch Anlegen von Elektroden an den als Basiszone dienenden Halbleiterkörper gleichzeitig die entgegengesetzte Dotierung der Emitter- und Kollektorzone erzielt wird*]

all-pole → (in) all poles

all-purpose ... → universal .. .

allround .. → complete ...

allrounder, universal device, all-purpose device Universalgerät, Allzweckgerät *n*

all-speed governor, universal governor, all-purpose governor Universaldrehzahlregler, Allzweckdrehzahlregler *m*

all-weather slatted blind Allwetterjalousie *f*

ALP → azimuth locking pin

A(L)PG → algorithmic pattern generator

alphabet Alphabet *n* [*In vereinbarter Reihenfolge geordneter Buchstabenvorrat. DIN 44 300*]

alphabetic interpreting Lochschriftübersetzung *f*

alphabetic shift [*Automatic data processing. The status of a card punch when the program(me) control is off (star wheels are raised)*] alphabetische Umschaltung, alphabetische Versetzung *f*

alphabetic string → character string

alphabetic translation, alphabetic translating Lochstreifenübersetzung *f*

alpha channel Alphakanal *m* [*Ein Graustufenkanal zum Maskieren. Ein geändertes Objekt wird mit diesem Kanal nahtlos in ein Bild eingesetzt*]

allotter relay [*A telephone term referring to a relay of the line-finder circuit whose function is to preallot an idle line-finder to the next incoming call from the line, and to guard relays*] Anrufordnerrelais *n*

alpha channel mask Alphakanalmaske, Rubinschicht *f*

alphageometric alphageometrisch

alphamosaic alphamosaisch

alpha(nu)meric, AN alpha(nu)merisch, AN [*Sich auf einen Zeichenvorrat beziehend, der mindestens aus Dezimalziffern und Buchstaben besteht. DIN 44 300, Teil 2*]

alpha(nu)meric character set [*The set of characters including A–Z, the numerals 0-9 and some special characters, e.g. (:, -,). Punctuation marks, control characters, mathematical symbols and Greek characters may also be considered to be alpha(nu)meric*] alpha(nu)merische Zeichensatz *m*

alpha(nu)meric display unit, alpha(nu)meric display device Zeichenbildschirmeinheit, alpha(nu)merische Anzeigeeinheit *f* [*Es werden Zahlen und Buchstaben angezeigt*]

alpha(nu)meric instruction [*It can be read equally well with alphabetic or numeric kinds of fields of data*] alpha(nu)merische Befehl *m*

alpha(nu)meric keyboard, ANKB alpha(nu)merische Tastatur *f*

alpha(nu)meric menu alpha(nu)merische Menü *n* [*Ein von einem Überwachungs- und Bedienergerät zur Verfügung gestelltes Menü, von welchem Funktionen und Eingabe funktionsbezogener Zahlen, Ziffern oder Buchstaben ausgewählt werden können*]

alpha(nu)meric reader [*A device capable of reading alphabetic, numeric, and special characters and punctuation marks*] alpha(nu)merische Leser *m*

alphaphotographic alphaphotographisch

alpha ray Alphastrahl *m*

alpha system [*A signal(l)ing system in which alphabetic characters delineate the signal(l)ing code to be used*] Alphaanlage *f*

to alter → to change

alternate mark inversion code, inversion alternate mark code, AMI code alternierende Flankenpulskode, alternierende Flankenpulscode, ALP *m*

alternate mark inversion method, AMI method alternierende Flankenpulsverfahren *n*, alternierende Flankenpulsmethode *f*, AFP

alternate recipient allocation, alternate recipient assignment Ersatzempfängerzuordnung, Ersatzempfängerzuweisung, Ersatzempfängerzuteilung *f*

alternate refrigerant, replacement refrigerant [*It can accomplish the same task as the one being replaced, but may require that refrigeration or air-conditioning equipment be redesigned or replaced*] Ersatzkältemittel, Alternativkältemittel *n*

alternate track [*In disk storage, a track that is automatically substituted for a defective primary track*] Ersatzspur *f*

alternating ((closed) loop) control Ablösereg(e)lung *f*

alternating current, AC, a.c. Wechselstrom *m*

alternating current... → a.c....

alternating current generator → alternator

alternating double filtration [*wastewater*] zweistufige Wechselbetriebs-Tropfkörperbehandlung *f*

alternating method, AC-pressurization method pulsierende Druckmethode *f*, pulsierende Druckverfahren *n* [*Luftwechselmessung durch Überdruck/Unterdruck in Gebäuden*]

alternative assisted heat pump installation, alternative assisted heat pump plant, assisted alternative heat pump installation, assisted alternative heat pump plant bivalente alternative Wärmepumpenanlage *f*, alternative bivalente Wärmepumpenanlage [*Die Wärmepumpe kann bei Außenlufttemperaturen um den Umschaltpunkt über Thermostat oder Raumsteuer(ungs)anlage abgeschaltet werden*]

alternative energy Alternativenergie, alternative Energie *f*

alternative energy source [*A nondepletable source alternative to energy derived from combustible waste or heat recovery processes*] alternative Energiequelle, Alternativenergiequelle *f*

alternat(iv)ely → electively

alternator, synchronous generator, a.c. generator, alternating current generator Wechselstromgenerator, Synchrongenerator, Drehstromgenerator *m*, Wechselstrommaschine, Drehstrommaschine *f*, Wechselstromerzeuger, Drehstromerzeuger *m*

altitude Höhe(nlage) *f* [*über Gelände*]

ALU → arithmetic (and) logic unit

aluminium (convector) radiator, aluminium convector heater (Brit.) [*US spelling is "aluminum". See remark under "radiator"*] Alu(minium)radiator, Alu(minium)gliederheiz(ungs)körper, Alu(minium)heiz(ungs)körper *m* [*Fehlbenennung*]

aluminium (lightning) conductor (Brit.); aluminum (lightning) conductor (US) Alu(minium)-(blitz)ableiter *m*

AM → amplitude modulation

amateur radio, ham radio [*A radio used for two-way radio communications by private individuals as leisure-time activity*] Amateurfunk *m*

ambient, surrounding [*Encompassing on all sides*] umgebend

ambient air, surrounding air [*Atmosphere exterior to refrigerated or conditioned space*] Umgebungsluft *f*

ambient condition, surrounding condition Umgebungsbedingung, Umgebungsgegebenheit *f*

ambient heat, surrounding heat Umgebungswärme *f*

ambient humidity, surrounding humidity Umgebungsfeuchte, Umgebungsfeuchtigkeit *f*

ambient illumination, ambient lighting, surround-ing illumination, surrounding lighting [*Background uncontrollable light level at a location*] Umgebungsbeleuchtung *f*

ambient noise, surrounding noise Umgebungsgeräusch *n*

ambient noise level, surrounding noise level Umgebungsgeräuschpegel *m*

ambient pressure, surrounding pressure Umgebungsdruck *m*

ambient standard temperature → standard ambient temperature

ambient temperature, surrounding temperature [*1. Temperature of the medium surrounding an object. In a domestic or commercial system having an air-cooled condenser, it is the temperature of the air entering the condenser.*
2. Temperature of the medium, such as air, water, or earth, into which the heat of equipment is dissipated.
3. For self-ventilated equipment, the ambient temperature is the average temperature of the air in the immediate neighbo(u)rhood of the equipment.
4. For air- or gas-cooled equipment with forced ventilation or secondary water cooling, the ambient temperature is taken as that of the ingoing air or cooling gas.

5. For self-ventilated enclosed (including oil-immersed) equipment considered as a complete unit, the ambient temperature is the average temperature of the air outside of the enclosure in the immediate neighbo(u)rhood of the equipment.] Umgebungstemperatur *f*

ambiophony Ambiophonie *f* [*Stereophonie mit zusätzlichen Raummikrophonen*]

AMC → access management controller

amendment record → change record

American Society of Heating and Ventilating Engineers → ASHVE

American Society of Heating, Refrigerating and Air-Conditioning Engineers, Inc. → ASHRAE

American Standard Code for Information Interchange → ASCII

AMF → assemble macro feature

AMI code → alternate mark inversion code

AMI method → alternate mark inversion method

ammeter, ampere meter Strommesser *m*, Amperemeter *n*, Amperemesser *m*

ammonia Ammoniak *n*, NH_3

ammonia absorption refrigerator, absorption am-monia refrigerator [*An absorption-cycle refrigerator which uses ammonia as the circulating refrigerant*] Ammoniakabsorptionskälteaggregat, Absorptionsammoniakkälteaggregat *n*, Ammoniakabsorptionskälteeinheit, Absorptionsammoniakkälteeinheit *f*

ammoniacal nitrogen Ammoniakstickstoff *m*

ammonia chlorin(e) treatment Chlorgas-Ammoniakbehandlung *f*

ammonia compressor [*A device that decreases the volume of a quantity of gaseous ammonia by the amplification of pressure; used in refrigeration systems*] Ammoniakverdichter, Ammoniakkompressor *m*

ammonia condenser [*A device in an ammonia refrigerating system that raises the pressure of the ammonia gas in the evaporating coil, conditions the ammonia, and delivers it to the condensing system*] Ammoniakkondensator, Ammoniakverflüssiger *m*

ammonia liquor Ammoniakwasser *n*

ammonia refrigerant → (anhydrous) ammonia refrigerant

ammonia valve [*A valve that is resistant to corrosion by ammonia*] ammoniakbeständige Ventil *n*

amorphous semiconductor, glass semiconductor amorphe Halbleiter, Glashalbleiter *m*

amount of light, light amount Lichtmenge *f*

amount of time, time amount Zeitaufwand *m*

ampacity; current(-carrying) capacity [*obsolete*] [*A wire's, or cable's, ability to carry current safely, without undue heating*] Strombelastbarkeit *f* [*VDE 0298, T. 2*]

amperage → current intensity

ampere-hour, Ah, amp-hr [*An alternative to the Coulomb as a measure of the quantity of electricity. The unit is used for rating the capacity of a rechargeable battery*] Amperestunde, Stromstärkestunde *f*

amperehourmeter, a.h.m., AHM, Ah meter Amperestundenzähler, Amperestundenmesser *m*

amplification [*The production of an output of greater magnitude than the input*] Verstärkung *f*

amplification factor [*electron-beam tube*] Verstärkungsfaktor *m*, Leerlaufverstärkung *f*

amplification principle Verstärkungsgrundsatz *m*

amplification ratio Verstärkungsverhältnis *n*

amplifier [*A device capable of increasing the magnitude or power level of a physical quantity, such as an electric current or a hydraulic mechanical force, that is varying with time, without distorting the wave shape of the quantity*] Verstärker *m*

amplifier block Verstärkerblock *m*

amplifier cabinet, amplifier cubicle Verstärkerschrank *m*

amplifier circuit Verstärkerschaltung *f*

amplifier frequency response Verstärkerfrequenzgang *m*

amplifier with negation indicator, negation indicator amplifier [*IEC 117-15*] negierende Verstärker *m* [*DIN 40 700, Teil 14*]

to amplify verstärken

amplitude Amplitude, Schwingungsweite *f*, Ausschlag *m* [*In der Bewegungslehre, insbesondere bei Schwingungen, die größte Abweichung aus der Mittellage*]

amplitude build up Amplitudenaufbau *m*, Amplitudenaufschauk(e)lung *f*

amplitude signalling Amplitudentastung *f* [*Digitale Modulation, bei der die Amplitude des Trägersignals durch ein digitales Nutzsignal beeinflusst wird*]

amplitude control → ((closed) loop) amplitude control

amplitude-discrete amplitudendiskret

amplitude discriminator [*A circuit that produces a specified output pulse when and only when it* *receives an input pulse whose amplitude exceeds an assigned value*] Amplitudendiskriminator *m*

amplitude distortion Klirrverzerrung, nichtlineare Verzerrung, Amplitudenverzerrung *f*

amplitude excursion Amplitudenauslenkung *f*

amplitude fading (out) Amplitudenschwund *m*, Amplitudenabklingen, Amplitudenschwinden *n*

amplitude (frequency) response Amplitudenfrequenzgang, Frequenzgang der Amplitude, Frequenzgang G(f) *m* [*VDE 0432, Teil 3*]

amplitude grid Amplitudenraster *m*

amplitude-holographic optical equipment, A-HOE holographisch-optische Bauelement, optisch-holographische Bauelement *n*

amplitude jitter Amplitudenzittern *n*

amplitude limiter [*A transducer, such as a shunt-polarized diode between resistors, in which the output above a critical value of the input does not vary; used particularly to limit the amplitude of a radio, telephone, and recording device*] Amplitudenbegrenzer *m*

amplitude-log frequency curve Amplitudengang *m* [*Reg(e)lung; DIN 19229*]

amplitude modulation, AM Amplitudenmodulation, AM *f*

amplitude range [*Ratio, usually expressed in decibels, of the upper and lower limits of program(me) amplitudes which contain all significant energy contributions*] Amplitudenbereich *m [selten: n]*

amplitude scale Amplitudenmaßstab *m*

amplitude selector → pulse-height selector

amplitude shift key(ing), ASK [*A method of transmitting binary coded messages in which a sinusoidal carrier is pulsed so that one of the binary states is represented by the presence of the carrier while the other is represented by its absence*] Amplitudenumtastung *f*

AMS → alarm management system

AN → alpha(nu)meric

analog(ue) [*Pertaining to representation by means of continuously variable physical quantities*] analog, stufenlos, stetig veränderlich [*Gegensatz: digital*]

analog(ue) [*A physical variable which remains similar to another variable in so far as the proportional relationships are the same over some specified range; for example, a temperature may be represented by a voltage which is its analog*] Analoggröße *f*

analog(ue) clock [*A clock of the traditional type, in which time is represented by the position of hands that rotate in a dial, rather than by a numerical display*] Zeigeruhr, Zifferblattuhr *f*

analog(ue) ((closed) loop) control, analog(ue) closed control, analog(ue) regulation, analog(ue) modulation control analoge Reg(e)lung, analoge Regulierung *f* [*Siehe Anmerkung unter „Reg(e)lung"*]

analog(ue) comparator [*1. A comparator that checks digital values to determine whether they are within predetermined upper and lower limits. 2. A comparator that produces high and low digital output signals when the sum of two analog voltages is positive and negative, respectively*] Analogvergleicher *m*

analog(ue) computer [*A computer in which quantities are represented by physical variables; problem parameters are translated into equivalent mechanical or electrical circuits as an analog(ue) for the physical phenomenon being investigated*] Analogrechner *m*

analog(ue) control → analog(ue) ((closed) loop) control

analog(ue) control → analog(ue) ((open) loop) control

analog(ue) controller Analogregler *m* [*DIN 19225. Die Eingangs- und Ausgangsgrößen sind analoge Signale*]

analog(ue) data Analogiedaten, analoge Daten *f* [*Sie bestehen nur aus kontinuierlichen Funktionen. DIN 44300, Teil 2*]

analgo(ue) data interchange format, ADIF Datenaustauschformat *n* für programmierbare Mess- und Prüfzwecke

analog(ue) device, AD, analog(ue) appliance [*A control device that operates with variables represented by continuously measured voltages or other quantities*] Analoggerät *n*

analgo(ue) digital, A/D analog-digital

analog(ue)-digital-analog(ue) converter, ADAC Analog-Digital-Analog-Umsetzer *m*

analog(ue)-digital conversion Analog-Digital-Umsetzung *f*

analog(ue)-digital converter → analog(ue)(-to)-digital converter

analog(ue) digital signal analysis → digital analog(ue) signal analysis

analog(ue) electric(al) ((closed) loop) control system Analog-Elektro-Reg(e)lungssystem, Elektro-Analog-Reg(e)lungssystem *n*

A

analog(ue)-electric(al) ((open) loop) control system Analog-Elektro-Steuer(ungs)system, Elektro-Analog-Steuer(ungs)system *n*

analog(ue) engineering Analogtechnik, analoge elektronische Schaltungstechnik, elektronische analoge Schaltungstechnik *f*

analog(ue) frequency generator, AFG analoge Frequenzerzeuger *m*, AFG

analog(ue) input, AI Analogeingabe *f*, Analogeingang *m*, AE [*Physikalischer Eingang eines Gerätes des Gebäudebetriebssystems zur Umwandlung eines stetigen Signals in eine elektronische Größe. Seine Auflösung ist größer als diejenige eines Digitaleingangs. Typische Eingangsauflösungen sind: = oder > 10 bit. DIN 19 226 Teil 5, DIN V 32 734*]

analog(ue) input configuration, analog(ue) input circuit analoge Eingangsschaltung *f* [*Sie dient dem Anschluss der analogen Prozess- und Bedienungssignale an eine Steuerung*]

analog(ue) input/output Analog-Ein-/Ausgang *m* [*DIN 19 226 T.5, DIN V 32734. Für Signale mit kontinuierlichem Werteverlauf und dabei Punkt für Punkt unterschiedlicher Informationszuordnung. Verwendung in der Gebäudeautomation für die physikalischen Grundfunktionen Messen und Stellen.*]

analog(ue) input router Analogeingaberangierer *m*

analog(ue) input signal Analogeingangssignal *n*

analog(ue) input unit Analogeingangseinheit *f*

analog(ue) line Analoglinie *f*

analog(ue) measuring data analoge Messdaten *f*

analog(ue) measuring instrument Analogmessgerät, Analogmesser *m*, Messgerät mit analoger Ausgabe *n*, Messer *m* mit analoger Ausgabe

analog(ue) modulating control → analog(ue) ((closed) loop) control

analog(ue) ((open) loop) control analoge Steuerung *f* [*DIN 19 226 Teil 5; DIN 19 237. Sie arbeitet innerhalb der Signalverarbeitung vorwiegend mit analogen Signalen*]

analog(ue) output, AO analoge Wertausgabe, Analogausgabe *f*, analoge Ausgang, Analogausgang, AA *m* [*Physikalischer Ausgang, dessen Auflösung größer ist als diejenige eines Digitalausgangs. Typische Ausgangsauflösungen sind = oder > 8 bit. DIN 19 226 Teil 5, DIN V 32 734*]

analog(ue) output unit, AO unit Analogausgabeeinheit *f* [*V DIN 19 237*]

analog(ue) recording [*Electronics. A method of recording machine in which the recording signal varies in a manner analogous to the original signal*] Analogaufzeichnung *f*

analog(ue) regulation → analog(ue) ((closed) loop) control

analog(ue) ring line module Analogringlinienmodul *m*

analog(ue) signal analoge Signal, Analogsignal *n* [*1. Ein Signal mit kontinuierlichem Wertebereich des Informationsparameters. DIN 19 226 Teil 5. 2. Ein Signal bei dem die Werteverläufe der Signalparameter eine Nachricht oder Daten darstellen, die nur aus kontinuierlichen Funktionen besteht bzw. bestehen. DIN 44 300 Teil 2*]

analog(ue) (signal) voltage [*Electronics. A voltage representing an analog(ue) signal*] Analogsignalspannung *f*

analog(ue) standardized signal Analogeinheitssignal, Einheitsanalogsignal *n* [*z.B. 0/4 – 20 mA*]

analog(ue)(-to)-analog(ue) converter, AAC Analog-Analog-Umsetzer, Analog-Analog-Wandler, A/A-Umsetzer, A/A-Wandler *m*

analog(ue)(-to)-digital computer, hybrid computer [*A computer designed to handle both analog and digital data*] Analog-Digital-Rechner, Hybridrechner, A/D-Rechner *m*

analog(ue)(-to)-digital converter, ADC, digitizer, A/D converter Analog-Digital-Umsetzer, A/D-Umsetzer, Analog-Digital-Wandler, A/D-Wandler, ADU, Hybridumsetzer, Hybridwandler *m* [*Anordnung, mit der eine analoge physikalische Größe, z.B. eine Amplitude, in eine digitale Darstellung umgesetzt wird. DIN 19 226, Teil 5*]

analog(ue)(-to)-digital processor, A/D processor Analog-Digital-Prozessor, A/D-Prozessor *m*

analog(ue)(-to)–frequency converter [*A converter in which an analog input in some form other than frequency is converted to a proportional change in frequency*] Analog-Frequenz-Umsetzer, Analogfrequenzwandler *m*

analog(ue) value [*Instantaneous value of an analog signal*] Analogwert *m* [*Momentanwert eines Analogsignals*]

analog(ue)-value indicator analoge Messwertanzeiger, Analogmesswertanzeiger *m*

analog(ue) watch [*A watch of the traditional type, in which time is represented by the position of hands that rotate on a dial, rather than by a numerical display*] Armbandzeigeruhr, Zeigerarmbanduhr *f*

to anchor verankern

anchorage fixing, anchorage fastening, anchoring [*A fixing which is achieved by means of a devi-*

ce pass-ing through the minor component of an assembly and anchoring into, without completely piercing, the base component] Verankern, Ankerbefestigen *n*, Verankerung *f*

ancillary heating surface Nachschaltheiz(ungs)fläche, Heiz(ungs)nachschaltfläche *f*

to AND verknüpfen nach UND

AND [*Mathematics. A logical operation used to connect elements in an AND function; if A is true and B is true, then (and only then) A AND B is true*] UND

ANDASP → automatic noise dependent audio signal processing

AND-before-OR logic UND-vor-ODER-Verknüpfung, UND-vor-ODER-Logik *f*

AND branch UND-Aufspaltung [*DIN 19237*], UND-Verzweigung *f*

AND circuit UND-Schaltung, Konjunktionsschaltung *f*

AND dependency UND-Abhängigkeit *f*

AND element, AND gate [*A logic element having one output and two or more inputs, and having the prop-erty that the output will stand at in defined logic 1-state if, and only if, all the inputs stand at their de-fined logic 1-states*] UND-Glied *n*, UND-Tor(schaltung), (*f*)

AND input converter UND-Eingangsstufe *f*

AND (instruction) [*Computer programming. An instruction that computes the bitwise logical AND function on two computer words*] UND(-Anweisung) *n*, (*f*)

AND NOT gate, A AND NOT B gate [*A coincidence circuit that performs the logic operation AND NOT, under which a result is true only if statement A is true and statement B not*] UND-NICHT-Gatter(schaltung), UND-NICHT-Tor(schaltung)*n*, (*f*), UND-NICHT-Glied *n*

AND operation, ANDing, AND relation, AND function UND-Operation, UND-Verknüpfung, UND-Funktion, Konjunktion *f*

AND-OR gate, AND-OR element [*It produces a pre-scribed output condition when several possible com-bined input signals are applied; exhibits are characteristics of the AND gate and the OR gate*] UND-ODER-Gatter(schaltung), UND-ODER-Tor(schaltung) *n*, (*f*), UND-ODER-Glied *n*

AND-OR progression matrix UND-ODER-Weiterschaltmatrix *f*

anechoic, nonreverberant, free (nach)hallfrei, (nach)hallwidrig, echofrei, echowidrig, reflexionsfrei, reflexionswidrig, reflektionsfrei, reflektionswidrig, schalltot

anechoic field → free field

anechoic room, nonreverberant room, free-field room, anechoic chamber, free (field) chamber, nonreverberant chamber (nach)hallfreie Raum, (nach)hallwidrige Raum, echofreie Raum, echowidrige Raum, reflexionsfreie Raum, reflexionswidrige Raum, reflektionsfreie Raum, reflektionswidrige Raum, schalltote Raum, Schallschluckraum *m* [*ISO 3475*]

anechoic termination, free termination, nonreverberant termination [*Device placed at the end of a test duct to prevent excessive reflection of the sound waves back into the test duct where they would interfere with the waves to be measured*] (nach)hallfreie Ende, (nach)hallwidrige Ende, reflexionfreie Ende, reflexionswidrige Ende, echofreie Ende, echowidrige Ende, reflektionsfreie Ende, reflektionswidrige Ende, schalltote Ende *n*

anemometer [*Instrument for measuring the velocity of air speed. Usually used at the face of air supply grilles, diffusors and ventilation ducts*] Luftgeschwindigkeitsmesser *m*

anemometer, wind speed meter Windgeschwindigkeitsmesser *m*, Anemometer *n*

aneroid [*Containing or using no liquid*] flüssigkeitsfrei

aneroid (barometer) [*A simple form of barometer which measures variations in atmospheric pressure by detecting changes on the sides of a sealed thin metal box. It is used for approximate measurements of altitude*] Aneroid(barometer), Dosenbarometer, Kapselbarometer *n*

aneroid capsule, bellows [*Thin, disc-shaped capsule, usually metallic, partially evacuated and sealed, held extended by a spring, which expands and contracts with changes in atmospheric or gas pressure*] Aneroidkapsel, Aneroiddose *f*

aneroid diaphragm [*A thin plate, usually metal, covering the end of an aneroid capsule and moving axially as the ambient gas pressure increases or decreases*] Aneroidmembran(e) *f*

aneroid liquid level meter [*A mechanism to meas-ure liquid depth by pressure of the fluid against a bellows which in turn acts on a manometer or signal transmitter*] Aneroidflüssigkeitstandmesser *m*

angel echo Engelecho *n* [*Echoimpulse, die ohne Geräteursache und ohne Doppelreflexionen auf dem Anzeigegerät erscheinen*]

angels Engel *f* [*Radiostörungen*]

A

angle backpressure valve, angle check(ing) valve [*A backpressure valve with its outlet opening at right angles to its inlet opening*] Winkelrückschlagventil, Rückschlagwinkelventil *n*

angle collar, bevel collar [*A cast-iron pipe fitting with a socket at each end for jointing the spigot ends of two pipes not in the same alignment*] Winkelmanschette *f*

angled, angling (ab)gewinkelt, wink(e)lig

angle(d) (bath)tub, angling (bath)tub (amerikanische) Winkel(bade)wanne, Spar(bade)wanne *f*

angle(d) bolt-nab, angle(d) strike (plate), angle(d) striking plate [*See remark under "bolt-nab"; "angling" may synonymously be used for "angle(d)"*] Winkelschließblech *n*, Winkelschließkappe *f*

angle(d) roller lever, angling roller lever Winkelrollenhebel, Rollenwinkelhebel *m*

angle(d) striking plate, angling striking plate Winkelschließblech *n* [*Tür*]

angle(d) valve, angling valve Eck((ab)sperr)ventil, Winkel((ab)sperr)ventil *n*

angle(-entry) plug Winkelstecker, Stecker mit seitlicher Einführung, Seiteneinführungsstecker *m*

angle equation, triangle equation Dreiecksgleichung *f*

angle flange Winkelflansch *m*

angle frame connection Winkelrahmenverbindung *f*

angle hinge Winkelband *n*, Winkelscharnier *n*

angle in decimal degree Winkel *m* in dezimaler Schreibweise

angle iron, iron angle Winkeleisen *n*, Eisenwinkel *m*

angle (lock) striking plate, angle (lock) strike plate, angle (lock) strike(r), angled striking plate, angling striking plate [*See remark under "(lock) striking plate"*] Winkelschließblech *n*

angle modulation [*Modulation in which the electrical angle of a carrier wave is the characteristic that is varied by the modulating signal. The term thus embraces both phase modulation and frequency modulation*] Winkelmodulation *f*

angle of acceptance → acceptance angle

angle of incline, incline angle, angle of inclination, inclination angle Steigungswinkel, Neigungswinkel, Schrägungswinkel *m*

angle pressure reducing valve, pressure reducing angle valve Druckmindereckventil, Eckdruckminderventil *n*

angle reciprocating compressor, angle piston compressor [*There are vertical and horizontal cylinders and a common crank*] Winkelkolbenverdichter, Winkelkolbenkompressor *m*

angle seat(ing) → oblique seat(ing)

angle section Winkelprofil, Eckprofil *n*

angle stop valve, stop angle valve Eck(ab)sperrventil, (Ab)Sperreckventil *n*

angle(-type) non-return valve, non-return angle(-type) valve Eckrückschlagventil, Rückschlageckventil *n*

angle(-type) return (flow) temperature limiter Eckrücklauftemperaturbegrenzer *m*

angle vertical → principal vertical

angling → angled

angular acceleration Winkelbeschleunigung, Drehbeschleunigung *f*

angular characteristic frequency, characteristic angular frequency Kennkreisfrequenz, charakteristische Kreisfrequenz *f*

angular corner frequency → corner angular frequency

angular degree Winkelgrad *m*

angular distance Winkelabstand *m*

angular distortion [*Distortion in a map projection because of nonconformity*] Winkelverzerrung *f*

angular fishplate Winkellasche *f*

angular frequency Kreisfrequenz *f*

angular handle, angular (hand)grip Winkel(hand)griff *m*

angular hinge → cleaning hinge

angularity Wink(e)ligkeit *f*

angular momentum, moment of momentum, mo-mentum moment [*In mechanics, the cross product of vector from a specified reference point to a particle, with the particle's linear momentum*] Winkeldrehmoment *n*, Drehimpuls, Drall *m*, Impulsmoment *n*, Scherung *f*

angular momentum operator, moment of momentum operator [*In quantum mechanics, any vector operator satisfying communication rules of the type* $[J_x, J_y] = iJ_z$] Drehimpulsoperator, Winkeldrehmomentoperator, Dralloperator, Impulsmomentoperator, Scherungsoperator *m*

angular perspective [*A form of plane linear perspec-tive in which some of the principal lines of the picture are either parallel or perpendicular to the picture plane and some are oblique*] Winkelperspektive *f*

angular rate, angular speed, angular velocity [*Change of direction per unit time, as of a target on a radar screen, without regard to the direction* *of the rotation axis; in other words, the magnitude of the angular velocity sector*] Winkelgeschwindigkeit *f*

anharmonic ratio Doppelverhältnis, unharmonische Verhältnis, anharmonische Verhältnis, nicht harmonische Verhältnis *n*

(anhydrous) ammonia refrigerant, R 717 Ammoniakkältemittel *n*, R 717, NH_3

anhysteretic curve anhysteretische Magnetisier(ungs)kurve *f*, ideale Magnetisier(ungs)kurve

anion exchanger Anionen(aus)tauscher *m*

ANKB → alpha(nu)meric keyboard

ANL → automatic noise limiter

announcement Ansage *f*

announcement [*An "alert" to an operator. Not to be confused with alarm, the operator is informed of a change in a system. This change could require an alarm but not necessarily*] Mitteilung, Ankündigung *f*

announcer Mitteilungszeichen, Ankündigungszeichen *n* [*Kode*]

annoyance → nuisance

annoyance alarm → nuisance alarm

annual average, yearly average Jahresmittel *n*, Jahresdurchschnitt *m*

annual cycle thermal energy storage, yearly cycle thermal energy storage Jahreswärmespeicherung, Wärmejahresspeicherung *f*

annual schedule, yearly schedule, annual plan, yearly plan Jahresplan *m*

annular-gap arcing chamber, ring-gap arcing chamber Ringspaltlöschkammer *f*

annular slide valve, ring slide valve Ringschieber *m*

annulus [*The area enclosed by two concentric circles*] Ringraum *m*

annulus air space → air annulus space

annulus earth electrode → ring earth electrode

annunciating signal, annunciation signal Meldesignal *n*

annunciation Meldung *f* [*Sie kennzeichnet einen Zustand oder eine Zustandsänderung der Steuerung oder der zu steuernden Vorrichtung. Das zugehörige Signal heißt „Meldesignal". E DIN 19 226, Teil 5 und V DIN 19 227*]

annunciation device, annunciation appliance Meldegerät, Meldungsgerät *n*, Melder *m*

annunciator, panel, indicator annunciator, indicator panel, display annunciator, display panel [*"Board" may synonymously be used for "panel"*] Anzeigetableau, Anzeigefeld *n*, Anzeigetafel *f*, Tableau, Signaltafel [*Alarmierungsvorrichtung zur optischen oder akustischen selektiven Signalisation von Meldungen. Eine Anzeigetafel wird meistens als parallele Anzeige der Meldergruppen zur Brandmelderzentrale eingesetzt*]

anode, plate Anode, positive Elektrode *f*, Pluspol *m*

anode current, plate current Anodenstrom *m*

anode efficiency, plate efficiency Anodenwirkungsgrad *m*

anode ignition voltage, plate ignition voltage Anodenzündspannung *f*

anode impedance, plate impedance [*Total impedance between anode and cathode exclusive of the electron stream*] Anodenimpedanz *f*, Anodenscheinwiderstand *m*

anode input power, plate input power zugeführte Anodenleistung *f*

anode modulation, plate modulation [*Modulation produced by introducing the modulating signal into the anode circuit of any tube in which the carrier is present*] Anodenmodulation *f*

anode mud, anode slime, plate mud, plate slime Anodenschlamm *m*

anode reactor, plate reactor Anodendrossel *f*

anode resistance, plate resistance Anodenwiderstand *m*

anodic cleaning, anodic pickling, reverse current cleaning, reverse current pickling [*The removal of a foreign substance from a metallic surface by electrolysis with the metal as the anode*] anodische Metallreinigung *f*

ANRS → acceptor not ready state

answer(ing) cord [*Cord nearest the face of the switchboard which is used for answering subscriber's calls and incoming trunks*] Abfrageschnur, Antwortschnur, Teilnehmerschnur *f*

answering jack [*Jack on which a station calls in and is answered by an operator*] Abfrageklinke, Antwortklinke, Teilnehmerklinke *f*

answering lamp [*Telephone switchboard lamp that lights when an answer cord is plugged into a line jack; the lamp goes out when the call is completed*] Abfrage- und Schlusslampe, Antwort- und Schlusslampe *f*

answering set Anrufbeantworter *m*

answering signal [*A signal sent in the backward direction when a call is answered. Such a signal may also indicate, or inhibit, call charging. If a specific answer signal capable of inhibiting char-*

ging does not exist in a system, charging can be prevented by omitting the backward signal when the call is answered] Antwortsignal *n*

antecedent part Regel-WENN-Teil, WENN-Teil *n, m* einer Regel [*unscharfe Logik*]

antenna (US) → aerial

antenna ... (US) → aerial ...

ante room (control) function [*When this function is activated it is impossible for two doors to be simultaneously unlocked and opened, even when the correct entry code is entered*] Schleusenfunktion *f*

antialiasing Herstellung *f* weicher Objektkanten ohne Treppenstufen, Kantenglättung *f* bei der Bildberechnung

antialiasing filter; aliasing filter [*misnomer*] Unterdrückungsfilter Verhinderungsfilter *m, n,* [*Filter, das den Frequenzbereich oberhalb der halben Abtastfrequenz extrem stark unterdrückt*]

antiblocking function Blockier(ungs)(schutz)-funktion *f*

antibugging device, debugging device, debugger Antilauschvorrichtung *f*, „Entwanzer", „Wanzenjäger" *m*

anticapillary kapillarbrechend, antikapillar

anticatalytic effect Hemmeffekt, antikatalytische Effekt *m* [*Vorgang beim Löschen, wenn dem Feuer die bei einem Brand durch Wärme entstehenden und mit dem Luftsauerstoff zu Oxiden reagierenden Radikalen durch die Löschpulverteilchen ABC entzogen werden*]

anticipating control [*One which, by artificial means, is actuated sooner than it would be without such means to produce a smaller differential of the controlled variable*] Frühsteuerung *f*

anti-clash key Entwirrungstaste *f*

anticlimbing coat(ing) Antikletterbeschichtung *f*, Antikletterüberzug *m*

anticlockwise direction counting Zählsinn links, linke Zählsinn, Linkszählsinn *m*

anticooling protection Auskühl(ungs)schutz *m* [*Gebäude*]

anticreep tongue → creep stop

antidiffusion voltage Antidiffusionsspannung *f* [*Diode*]

antidistortion Entzerrung *f*

antidrill feature Anbohrsicherung *f*

antidrill protection Anbohrschutz *m*

antidrumming pad Antidröhnmatte *f*

antifreeze agent, antifrost agent Gefrierschutzmittel, Frostschutzmittel *n* [*Zusätze zum Wasser sind: Alkohol, Chlorkalzium, Glyzerin, Glykol, Glysantin, Kochsalzlösung und andere*]

antifreeze function, antifrost function Gefrierschutzfunktion, Frostschutzfunktion *f* [*Eine Funktion, die in allen Betriebsarten (ausser Handbetrieb) aktiv ist, um ein Einfrieren der HLK-Anlage durch Bereitstellung besonderer Ausgangssignale an die entsprechenden Anlagenkomponenten zu verhindern.*]

antifreeze switch Gefrierschutzschalter, Frostschutzschalter *m*

antifreeze thermostat, antifrost thermostat Gefrierschutzthermostat, Frostschutzthermostat *m*

antifrost circuit (arrangement), antifrost circuity Frostschutzschaltung, Gefrierschutzschaltung *f* [*als Stromkreis*]

antifrost convector Frostschutzkonvektor, Gefrierschutzkonvektor *m*

antiglare system Blendschutzsystem *n*

antiinterference filter → noise filter

antiintrusion wire Einbruchschutzdraht *m*

antiliming, lime removal Entkalkung *f*, Kalkentzug *m*

antimist (heated) mirror beschlagfreie Spiegel *m*

antimisting [*mirror*] Beschlagfreiheit *f*

antimonial lead pipe, hard lead pipe Antimonbleirohr, Hartbleirohr, Bleisparrohr, Sparbleirohr *n*

antimonial lead pressure pipe, hard lead pressure pipe Antimonbleidruckrohr, Hartbleidruckrohr, Bleispardruckrohr, Sparbleidruckrohr *n*

antinode, loop of oscillation, oscillation loop [*A point, line, or surface in a standing-wave system at which some characteristic of the wave has maximum amplitude*] Wellenbauch, Schwingungsbauch *m* [*Physik*]

antipanic lock Antipanikschloss *n*

antipanic lock door release Antipanikschlosstüröffner *m*

antipassback Doppelzutrittssperre, Doppeleintrittssperre, Doppelzugangssperre *f* [*frühere Benennung*]

antipicking device Aufsperrsicherung *f* [*Schloss*]

antipollution device Reinhaltungsvorrichtung *f*

antiresonant circuit, parallel resonant circuit [*A circuit in which an alternating current voltage is applied across a capacitor and a coil in parallel*] Entkopp(e)lungskreis, Saugkreis, Parallelschwingkreis *m*

anti-rotation lug slot → slot for anti-rotation lug

antiscald device Verbrühungsschutz *m* [*Badebatterie*]

antiscald (type) mixer valve verbrüh(ungs)freie Mischventil *n*

antiscale (agent) Kesselsteinverhütungsmittel, Kesselsteinverhinderungsmittel *n*

antisiphon(age) → back-siphon(age)

antisiphon(age) pipe, puff pipe, A.S.P., backsyphon(age) pipe [*An air pipe in a plumbing system which admits air downstream of a water seal in a w.c., sink or washbasin, thus preventing the creation of a vacuum in the wake of water flushed down the drain and avoiding the breakage of a seal which such a vacuum could cause. The top of this pipe may be connected to a ventilation pipe*] Rohrbelüfter *m* [*DIN 3266*]

anti-siphon(age) pipe with leakage fitting, puff pipe with leakage fitting, A.S.P. with leakage fitting, backsiphon(age) pipe with leakage fitting Rohrbe- und -entlüfter *m* [*DIN 3266*]

antisiphon(age) trap, deep seal trap, backsiphon(age) trap [*A type of drainage trap below a sink or washbasin. It contains a large volume of water which makes it difficult for the water seal to be broken. It can be a replacement for an antisiphonage. "Preventer" may synonymously be used for "trap"*] (Wasser)(Geruch)Verschluss *m* mit großer (Wasser)Verschlusshöhe [*ungefähr 75 mm*]

antismudge ring [*A frame attached around the perimeter of a ceiling-mounted air diffuser, which minimizes the formation of rings of dirt on the ceiling*] schmutzabweisende Ring *m*

antispectroscopic lens → achromatic lens

antistatic antistatisch [*elektrische Aufladung aufhebend*]

antitampering cap Behördenkappe *f*

antitheft, theft-protected diebstahlsicher, diebstahlgeschützt

antitheft lock Behördenschloss *n* [*Einsteckschloss nach DIN 18251, Klasse 4*]

antitwisting plate Sicherungsblech *n*

antivacuum valve Rohrbelüfter *m*

antivibration mounting vibrationsfreie Montage, schwingungsfreie Montage *f*

antiwindup Vernachlässigung *f* des Integralanteils bei Erreichen der Stellgrößenbegrenzung

AO → analog(ue) output

A-O → acusto-optic

AOC → auxiliary output chip

AOD → abort of dialling (in)

AOQ → mean outgoing quality

AOQL → mean outgoing quality limit

AO unit → analog(ue) output unit

AP → active position

AP → application protocol

APC → automatic phase ((closed) loop) control

APC → automatic power control

APCVD → atmospheric pressure chemical vapor (US)/vapour (Brit.) deposition

APD → avalanche photodiode

APDU → application protocol data unit

aperiodic → deadbeat

aperiodic... → deadbeat...

aperture → port

aperture time Öffnungszeit *f*

apex Scheitel *m*

APG → algorithmic pattern generator

API → application programme interface (Brit.)

APL, A programming language Programmiersprache *f* [*Sie enthält sehr viele Sonderbefehle und Operationen, speziell für mathematische Aufgaben in der Produktionstechnik*]

apparatus [*A compound unit designed to carry out a specific function*] Apparat *m*

apparatus dewpoint (temperature) → (conditioning) apparatus dewpoint (temperature)

apparent final drying rate, final apparent drying rate scheinbare Endtrocknungsgeschwindigkeit *f*

apparent power, apparent current [*Product of the volts and amperes of a circuit. This product generally is divided by 1.000 and designed in kilovoltamperes (kVA). It comprises both real and reactive power*] Scheinstrom *m*, Scheinleistung, scheinbare Leistung *f*, scheinbare Strom *m* [*Kürzel: (s)*]

appearance (Erscheinungs)Bild *n*, Aussehen *n*

to append anhängen, hinzufügen [*Zeichen an eine Zeichenkette oder Benennung am Ende einer Liste*]

appliance → device

appliance... → device...

appliance description language → DDL

applicable, usable anwendbar, einsetzbar, verwendbar, benutzbar, brauchbar, gebrauchsfähig, verwendungsfähig, anwendungsfähig

application, use, duty Anwendung, Verwendung, Benutzung, Einsetzung *f*, Gebrauch, Einsatz *m*

application entity, AE [*The aspects of an application process pertinent to OSI (ISO 7498)*] Anwendungsinstanz *f* [*E DIN ISO 7498*]

A

application error Anwendungsfehler *m* [*Falsche Meldertyp; falsche Meldereinstellung; falsche Melderstandort usw.*]

application guideline Anwendungsrichtlinie *f*

application layer, layer 7 Anwendungsschicht, Anwendungsebene *f*, sieb(en)te Schicht, sieb(en)te Ebene, Schicht 7, Ebene 7, Anwenderebene, Anwenderschicht [*OSI. Auf dieser Schicht werden alle Details spezifiziert, die sich mit dem Datenaustausch befassen*]

application layer interface, ALI, layer 7 interface Anwendungsschichtschnittstelle *f*

application object Kommunikationsobjekt *n*

application-orien(ta)ted multi(ple)-function unit anwendungsorientierte Kombinationsglied *n*, betriebsspezifische Kombinationsglied [*z. B. Zähler, Register, Addierer, Vergleicher usw. V DIN 19 237*]

application process [*An element within a real open system that performs the information processing for a particular application*] Anwendungsprozess *m* [*E DIN ISO 7498*]

application program (US); application programme (Brit.); application routine Anwendungsprogramm *n* [*Für Aufgaben des Anwenders geschrieben*]

application protocol, AP Anwendungsprotokoll *n* [*E DIN ISO 7498*]

application programme interface (Brit.); application program interface (US); API, application programming interface, application routine interface Anwendungsprogramm(ier)schnittstelle, Softwareschnittstelle *f* für Anwendungsprogramme

application protocol data unit, APDU Anwendungsprotokolldateneinheit *f*

application server Spezialrechner *m* im Netzwerkbetrieb, Sonderrechner *m* im Netzwerkbetrieb

application service, AS Anwendungsdienst *m* [*E DIN ISO 7598*]

application sharing gemeinsame Zugriff *m*

application-specific controller, unitary controller, ASC (US); terminal-equipment controller anwendungsspezifische Steuer(ungs)- und Regel(ungs)einheit *f*, ASR [*Eine nicht frei programmierbare Steuer(ungs)- und Regel(ungs)einheit in der eine oder mehrere Anwendungen gespeichert sind*]

application-specific integrated circuit, ASIC anwendungsspezifische integrierte Schaltung, integrierte anwendungsspezifische Schaltung *f*

to apply, to impress aufschalten [*Strom; Signal*]

apportioned operating cost(s) Umlage *f* [*Umlagen sind auf die Mieter umgelegte und im allgemeinen mit der Miete erhobene Betriebskosten*]

apportionment Aufteilung *f*, Zuteilung

appraisement (of value) → **(e)valuation**

approach Ansatz *m*

approach section [*Flow safety device of a gas boiler*] Anlaufstrecke *f*

appropriate equipment(s) Sachgerät(e) *n*, (*f*)

approval Zulassung, Genehmigung *f*

approval authority Genehmigungsbehörde, Zulassungsbehörde *f*

approval label Zulassungsetikett, Genehmigungsetikett *n*

approval seal → **seal of approval**

approved [*Approved as applied to a material, device, equipment, or method of construction means approv-ed by the building official under the provisions of a code, or approved by other authority designated by law to give approval in the matter in question*] genehmigt, zugelassen, anerkannt, gutgeheißen

approved [*Term indicating conformance to appropriate standards and good public practice*] unbeeinträchtigt

approximate dimension Ungefährdimension, Ungefährabmessung *f*

approximate measure (An)Näherungsmaß, Ungefährmaß, unscharfe Maß *n*

approximate reasoning angenäherte Schließen, angenäherte Folgern, unscharfe Schließen, unscharfe Folgern, ungefähre Schließen, ungefähre Folgern *n*

approximation (method) (An)Näherung(sverfahren) *f*, (*n*) [*Mathematisches Verfahren zur (An)Näherung eines Kurvenverlaufs an Stützstellen*]

appurtenance → **accessory**

appurtenance box → **accessory box**

appurtenance item → **accessory item**

appurtenances of a building → **accessories of a building**

A programming language → **APL**

apron induction unit → **spandrel (wall) induction unit**

APS → **auxiliary power supply**

APS, advance(d) production system fortschrittliche Fertigungssystem, fortschrittliche Herstellungssystem, fortschrittliche Produktionssystem *n*

APT → **Automatic Programming Tool**

APT → **automatic picture transmission**

APU → **arithmetic processing unit**

A(Q)L → acceptable (quality) level

aquastat [*Thermostat designed for use in water*] Wasserthermostat, automatische Wassertemperaturregler *m*

aquatic chemistry, water chemistry Wasserchemie *f*

aquatic physics, water physics Wasserphysik *f*

aqueous chlorine solution wäss(e)rige Chlorlösung *f*

AR → address register

arbiter [*A computer unit that determines the priority sequence in which two or more processor inputs are connected to a single functional unit such as a memory or multiplier*] Zuteiler, Verwalter *m*

arbitration Arbitrierung, Entscheidung über Zugriffsrechte, Zugriffs(recht)entscheidung *f*

arbitration bus → AB

arbitration bus method, AB method Zuteilungsbusverfahren *n*, Zuteilungsbusmethode *f*

arc (Kreis)Bogen *m* [*Ein graphisches Objekt*]

arc [*A discharge of electric current across an air gap*] Lichtbogen *m*

ARC → automatic remote ((closed) loop) control

arch(ed) door Bogentür *f*

archeria schmale Dachfenster, Schlitzfenster *n*

Archimedes number Archimedeszahl, Ar-Zahl *f*

architectural acoustics Raumakustik *f*

architectural (builders') hardware → finish(ing) (builders') hardware

architectural daylighting [*Introducing daylight into the built environment*] bauliche Tageslichtbeleuchtung *f*

architectural diagramm, architectural plan Installationsplan *m* [*DIN 40 719, Blatt 1*]

architectural-electrical plan, electrical-architectural plan [*Architectural plan on which electrical work is shown*] Elektrobauplan, Bauelektroplan *m*

architectural fountain → water fountain

architectural hygiene, sanitary science as applied to buildings Bauhygiene *f*

architectural ironmongery → (builders') iron supplies

architecture Architektur, Baukunst *f*

archive Archiv *n*

archived (data) file archivierte Datei, archivierte Daten *f*

archiving of historic(al) data, historic(al) data archiving, long-term storage Datenhistorisierung, Archivierung historischer Daten, historische Datenlangzeitspeicherung, historische Langzeitdatenspeicherung *f*

archiving, filing, off-line storage, back-up restore Archivierung *f*, Archivieren *n*, Datenauslagerung, Datensicherung *f* [*Sie fasst mehrere Dateien komprimiert in einer einzigen Datei zusammen, einem so genannten Archiv. Die Daten lassen sich mit dieser Methode sehr effektiv packen, da Zeit eine eher untergeordnete Rolle spielt. Dafür sind die in einem Archiv liegenden Dateien erst nach manuellem Entpacken wieder zugänglich. Die Archivierung eignet sich deshalb vorrangig für das langfristige und platzsparende Speichern von Daten, die nicht täglich benötigt werden*]

arc(ing) chute, arc(ing) chamber Lichtbogenkammer *f*

arc(ing) contact Lichtbogenkontakt *m*, Lichtbogenberührung *f*

arc(ing) interference, accidental arc, arc(ing) fault, internal fault Störlichtbogen *m*

arc(ing) time Lichtbogenzeit *f*

arc lamp, electric arc lamp (Elektro)Bogenlampe *f*

arc minute Bogenminute *f*

arc-over Lichtbogenüberschlag, Überschlag *m*

arc-over ground(ing) fault (US); arc-over earth(ing) fault (Brit.) Lichtbogenerdschluss, aussetzende Erdschluss *m*

arc-over voltage Lichtbogenüberschlagspannung, Überschlagspannung *f*

arc quencher, arc quenching device, arc quenching appliance Lichtbogenlöscher *m*, Lichtbogenlöschvorrichtung *f*

arc quenching nozzle Lichtbogenlöschdüse *f*

arc quenching sleeve Lichtbogenlöschmantel *m*

arc second Bogensekunde *f*

arc spit Brennfleck, Lichtbogenbrennfleck *m*

arc-suppression coil, Petersen coil [*IEC 281*] [*A grounding reactor, used in alternating current power transmission systems, which is designed to limit the current flowing to ground at the location of a fault almost to zero by setting up a reactive current to ground flowing from the lines*] Petersenspule, Erdschlusslöschspule, Esp *f* [*VDE 0532, Teil 20*], Erdschlusslöscher *m*, Löschspule *m*

arc-through [*Of a gas tube, a loss of control resulting in the flow of a principal electron stream in the normal direction during a scheduled nonconducting period*] Durchzündung *f*

Arcton → halogenated hydrocarbon refrigerant

arc voltage Lichtbogenspannung, Brennspannung *f* [*Lampe*]

A

area, superficial content (Ober)Flächeninhalt *m*, Fläche *f*

area care, area preventive maintenance Flächenpflege, Flächenwartung *f*

area communication controller, ACC Gebietsvermittlungsrechner, Vermittlungsrechner *m* eines Gebietes [*z.B. zwischen Datenfunk und Datex-P (ein Telekomprodukt)*]

area content Flächeninhalt *m*

area drainage Flächenentwässerung *f*

area efficiency Flächenwirkungsgrad *m* [*Wärmeübertragung*]

area heating Flächenheizung *f* [*Eine in Decke, Wand oder (Fuß)Boden liegende Heiz(ungs)fläche*]

area management Flächenbewirtschaftung *f*

area monitoring Flächenüberwachung *f*

area of surface → **surface area**

area potential equalization Flächenpotentialausgleich *m*

area-related flächenbezogen

area restriction Flächenbeschränkung *f*

area rug Teppichboden *m*

area search [*A computer search that examines only those records which satisfy some broad criteria*] flächenhafte Suche *f*

area weight Flächenmasse *f*; Flächengewicht *n* [*frühere Benennung*]

arithmetically left shifting arithmetische Linksverschiebung *f* [*Eine Bit-Manipulation*]

arithmetic (and) logic unit, ALU [*The register or portion of hardware of a computer or processor in which binary arithmetic and logic(al) functions are performed*] (logische) Rechenwerk *n*, logische Recheneinheit, Rechen- und Steuereinheit, arithmetisch-logische Einheit, Arithmetik- und Logikeinheit *f* [*Funktionseinheit innerhalb eines digitalen Rechensystems, die Rechenoperationen ausführt. DIN 44 300*]

arithmetic expression arithmetische Ausdruck, arithmetische Term *m*

arithmetic logic unit, ALU arithmetische logische Einheit, logische arithmetische Einheit *f*

arithmetic mean, arithmetic average arithmetische Mittel(wert) *n*, (*m*), arithmetische Durchschnitt(swert) *m*

arithmetic mean temperature difference, arithmetic average temperature difference [*In a parallel flow or counterflow heat exchanger, arithmetic(al) mean of the temperature difference between the two fluids at both ends of the exchanger*] arithmetische durchschnittliche Temperaturunterschied, durchschnittliche arithmetische Temperaturunterschied *m*

arithmetic operation Rechenoperation *f*

arithmetic overflow arithmetische Überlauf *m* [*Ein Ergebnis größer als die aufnehmende Stelle bei einer arithmetischen Operation*]

arithmetic processing unit, APU arithmetische Verarbeitungseinheit *f*

arithmetic register, A register [*In automatic data processing a specific memory location reserved for intermediate results of arithmetic operations*] arithmetische Register *n*

arithmetic shift(ing) [*In automatic data processing, a shift(ing) of the digits of a number, expressed in a positional notation system, in the register without changing the sign of the number*] Stellen(wert)verschiebung *f*

arithmetic statement arithmetische Anweisung *f*

arithmetic unit, AU Rechenwerk *n*, Arithmetikeinheit, Arithmetik und Logik *f* [*Eine auf diesem Chip realisierte Funktion des Mikroprozessors*]

arity Wertigkeit *f* [*Die auf einen Operator wirkende Anzahl der Operanden*]

to arm scharf schalten [*Gefahrenmeldeanlage*]

ARM → **asynchronous response mode**

armature [*The revolving part of an electric motor or dynamo on which the coils are wound*] Anker *m*

armature reversal via contactors Ankerumschaltung *f* mit Schützen

armature rotational speed Ankerdrehzahl *f*

armature voltage isolating amplifier Ankerspannungstrennverstärker *m*

armature winding Ankerwick(e)lung *f*

arm barrier Armschranke, Armsperre *f*

armed scharf (geschaltet)

armed bewaffnet

arming Triggersperre *f* [*digitale Messgerät*]

armored cable (US); armoured cable (Brit.) Panzerkabel *n*

armoured door (Brit.); armored door (US) Panzertür *f*

armoured pipe (Brit.); armored pipe (US) Panzerrohr *n*

armoured plastic pipe (Brit.); armored plastic pipe (US) Kunststoffpanzerrohr, Panzerkunststoffrohr *n*

armoured wire (Brit.); armored wire (US) Panzerdraht, Rohrdraht *m*

armour(ing); armor(ing) (US) Panzerung *f* [*Hülle aus Metallbändern oder -drähten zum äußeren mechanischen Schutz, z.B. eines erdverlegten Kabels*]

arm rest Armauflage *f*, Armstütze *f*

Aron circuit → double wattmeter method

ARP → address resolution protocol

to arrange, to dispose, to lay out anordnen

arrangement, disposition, layout Anordnung *f*

arras, handwoven tapestry handgewobene Tapete *f*

array Feld *n* [*Reihung von gleichartigen Elementen, wobei auf die Komponenten mit Hilfe eines Index zugegriffen wird. Felder gibt es in den meisten Programmsprachen. Das Schlüsselwort lautet „array" oder manchmal „DIMENSION (DIM)"*]

to arrest, to divert ableiten [*Blitz*]

to arrest aufhalten, hemmen, sperren, blockieren, arretieren, feststellen

arrester, lightning arrester Ableiter, Blitzableiter *m*

arresting Hemmen, Sperren, Blockieren, Aufhalten, Festhalten, Arretieren *n*

arris, edge, salient corner [*Two plane surfaces meet at this corner*] Kante, vorspringende Ecke, Vorsprungecke *f*

arris offset → edge offset

arris-punched card → edge-punched card

arrival lamp → (elevator) arrival lamp

arrow key Pfeiltaste *f*

arson, fire-raising, incendiarism Brandstiftung *f*

arson attack Brandanschlag *m*

arsonist, incendiary, fire setter Brandstifter *m*; Brandleger [*Österreich*]

article security, product security Artikelsicherung, Produktsicherung, Erzeugnissicherung *f* [*Sicherung gegen Diebstahl und Beschädigung*]

articulation loss of consonants, consonant articulation loss [*(loud)speaker*] Konsonantenverlust, Mitlaut(e)verlust *m*

artificial cooling künstliche Kühlen *n*, künstliche Kühlung, Fremdkühlung *f*, Fremdkühlen *n*

artificial intelligence, AI [*Capability of a device to perform functions that normally are associated with human intelligence, such as reasoning, learning, adaptation and self-improvement*] künstliche Wissensverarbeitung; künstliche Intelligenz *f*, KI [*Fehlbenennungen*]

artificial light [*It is obtained from a lamp as opposed to natural light*] künstliche Licht, Kunstlicht *n*

artificial mouth [*Small loudspeaker mounted in a shaped baffle which is proportioned to simulate the acoustical constants of the human mouth; used for calibrating and testing close-talking microphones*] künstliche Mund *m*

artificial network [*EN 5006*] Ersatznetz(werk) *n*, Netz(werk)nachbildung *f*

artificial reverberation künstliche Nachhall *m*

artificial voice, synthetic voice [*It is produced by a multiple tone generator; used to produce a voice reply in some real-time computer applications*] künstliche Stimme *f*

artificial warmth künstliche Wärme *f*

Arton shunt → universal shunt

artwork Unterlagen, Druckvorlage, gedruckte Schaltung *f*, Schaltplan *m*

artwork design, artworking Schaltplanentflechtung *f*

AS → adverse state

AS → application service

AS → automation station

asbestos rope Asbeststrick *m*

as-built drawing, record drawing, as-constructed drawing [*Construction drawing revised to show significant changes made during the construction process, usually based on marked-up prints, drawings, and other data furnished by the contractor or anyone who prepared the original drawings*] Bestand(s)zeichnung *f*, Bestand(s)plan *m*

ASC → automatic sensitivity ((closed) loop) control

ASC → application-specific controller

ASCII, American Standard Code for Information Interchange ASCII-Format *n*, amerikanische Standardkode *m* für Datenübertragung, binäre Zeichencode [*Acht Bit (= 1 Byte) langer Code für alle druckbaren Zeichen; er wird z. B. von der Tastatur erzeugt und an den Rechner übergeben, der dadurch erkennt, welche Taste gedrückt wurde*]

ascending... → rising...

ASD, adverse state detector Umkehr(ungs)zustandsmelder, Umkehr(ungs)statusmelder *m*, Umkehr(ungs)zustandsmeldegerät, Umkehr(ungs)statusmeldegerät *n*

to ash einäschern

ash collector Asche(n)sammler *m*

ash container Asche(n)behälter *m*

ash drop door Asche(n)falltür *f*

ash (fusing) cone Asche(n)(schmelz)kegel *m*

A

ashpan [*A metal receptacle beneath a grating for collection and removal of ashes*] Asche(n)kübel, Asche(n)kasten *m*, Asche(n)kiste *f*

ashpit [*An area beneath a fireplace or burner designed to collect and hold ash before it is removed*] Asche(n)grube *f*, Asche(n)raum *m*

ashpit refuse unverbrannte Kohle *f* [*Sie ist durch den Rost gefallen*]

ash pocket Asche(n)sack *m*

ASHRAE, American Society of Heating and Refrigerating and Air-Conditioning Engineers Inc. [*1791. Tullie Circle, Atlanta, Georgia 30329*] Amerikanische Gesellschaft *f* der Heizungs-, Kälte- und Klimaingenieure, Gesellschaft der Heizungs-, Kälte- und Klimaingenieure der USA

ash removal → **deashing**

ash scraper Asche(n)kratzer *m*

ash softening Asche(n)erweichen *n*

ash stop Asche(n)schieber *m*

ASHVE, American Society of Heating and Ventilating Engineers Inc. Amerikanische Gesellschaft *f* der Heizungs- und Lüftungsingenieure, Gesellschaft der Heizungs- und Lüftungsingenieure der USA

Asiatic w.c. [*deprecated*] → **squatting w.c.**

ASIC → **application-specific integrated circuit**

ASK → **amplitude shift key(ing)**

ASLT, advance(d) solid logic technique ultraschnelle integrierte Schaltung, integrierte ultraschnelle Schaltung *f*

ASN → **mean sample number**

A.S.P. → **antisiphonage pipe**

aspergus negri [*Latin*] → **mould fungus**

aspiration psychrometer, aspirated psychrometer [*A psychrometer in which the ventilation is provided by a suction fan*] Aspirations-Psychrometer *n*, Aspirations-Luftfeuchtigkeitsmesser *m*

A.S.P. with leakage fitting → **antisiphonage pipe with leakage fitting**

as required nach Bedarf, wie erforderlich, erforderlichenfalls

to assemble, to mount zusammenbauen, montieren

to assemble [*See remark under „assembler"*] übersetzen, assemblieren

assembled on site, mounted on site, locally assembled, locally mounted zusammengebaut am Standort, montiert am Standort

assembled representation zusammenhängende Darstellung *f* [*DIN 40 719, Blatt 1*]

assemble macro feature, AMF zeitkritische Programmsequenz *f* der Programmsprache C

assembler → **fitter**

assembler, assembly routine; assembler program, assembly program (US); assembler programme, assembly programme (Brit.); symbolic assembly system [*A programming language designed to convert symbolic instruction into a form suitable for execution on a computer*] Assembl(ier)er, Übersetzer(programm) *m*, (*n*) (für Assemblersprachen), maschinenorientierte Programmiersprache, maschinennahe Programmiersprache, Maschinensprache *f*

assembly Bauform *f*

assembly, assembling Zusammenbau(en) *m*, (*n*), Montage *f*, Montieren *n*

assembly, assembling [*In automatic data processing, the automatic translation into machine language of a computer program(me) written in symbolic language*] Übersetzen *n*, Assemblieren

assembly, set, kit, package Satz *m*, Kombination, Gruppe *f*, Aggregat *n*, Bausatz *m*, Garnitur *f*

assembly bracket Anreihlasche, Reihenlasche *f*

assembly construction method, kit construction method, set construction method Satzbauweise, Kombinationsbauweise, Gruppenbauweise, Aggregatbauweise, Garniturbauweise *f* [*Bauelemente im gemeinsamen Gehäuse; zusammen transportierbar und versetzbar*]

assembly drawing, mounting drawing [*A drawing that displays all the parts of a construction and indicates the joints or connections between them*] Zusammenbauzeichnung, Montagezeichnung, Fertigungszeichnung *f*

assembly group → **subassembly**

assembly language [*A low-level computer language one step above the binary machine language*] Assemblersprache *f*

assembly line, assembling line Montageband *n*

assembly list [*In automatic data processing, a printed list which is the byproduct of an assembly procedure; it lists in logical instruction sequence all details of a routine, showing the coded and symbolic notation next to the actual notations established by the assembly procedure; this listing is highly useful in the debugging of a routine*] Assemblerliste *f*, Assemblerverzeichnis *n*

assembly model Bauwerkteilmodell *n*

assembly model structure Einheit *f* zum Lokalisieren der Bauwerkteilmodelle

assembly place → **place of assembly**

assembly routine → **assembler**

assembly unit [*In automatic data processing, a de-vice which performs the function of associating and joining several parts together in a program(me)*] Assemblereinheit *f*

assembly unit [*In automatic data processing, a portion of a program(me) which is capable of being assembled into a larger program(me)*] Assemblerteil *m, n*

assessment (of value) → (e)valuation

to assign → to allocate

to assign manually, to assign by hand vermitteln von Hand [*Telefon*]

assignment → allocation

assignment of point address [*Linking of point address to plant schematic diagram*] Adresszuordnung *f* [*Zuordnung einer Information zu einem Anlagenschema*]

assignment to I/O equipment [*Linking of automation station channels to a transducer/actuator*] Zu-ordnung *f* zu Ein/Ausgabegeräten [*Zuordnung der Automationsstation zu Fühlern und Stellgliedern*]

to assign parameters → to parameterize

assisted alternative heat pump installation → alternative assisted heat pump installation

assisted heat pump installation, assisted heat pump plant bivalente Wärmepumpenanlage *f* [*Sie ist mit einem unabhängigen Wärmeerzeuger gekoppelt*]

assisted parallel heat pump installation, assisted parallel heat pump plant, parallel assisted heat pump installation, parallel assisted heat pump plant bivalente parallele Wärmepumpenanlage *f*, parallele bivalente Wärmepumpenanlage [*Die Wärmepumpe wird neben dem Heizkessel betrieben sodass eine zweite Wärmeerzeugung lediglich die notwendige Temperaturspitze auf die Vorlauftemperatur setzt*]

Assmann psychrometer [*A special form of the aspiration psychrometer in which the thermometric elements are well shielded from radiation*] Aßmannsche Psychrometer *n*, Aßmannsche Luftfeuchtigkeitsmesser *m*

associated mode, associated signalling; associated signaling (US) assoziierte Zeichengabe *f* [*Signalisierungssystem Nr. 7*]

associative (computer) memory, active (computer) memory, associative (computer) storage (device), active (computer) storage (device) [*A content-addressable memory in which stored data is identified by properties of its own value, rather than by an address as in conventional (computer) memories*] inhalt(s)adressierbare (Rechner)Speicher, Assoziativ(rechner)speicher, Verknüpfungs(rechner)speicher *m*

associativity Assoziativität *f* [*Verbindung voneinander abhängiger Daten*]

assorting Sortieren, Auslesen, Trennen *n* [*Verschiedenartige Elemente einer Menge nach ihrer Verschiedenartigkeit nach vorgegebenen Sorten, Arten, Merkmalen oder Merkmalwerten trennen. DIN 1319, Teil 1*]

assorting needle Sortier(ungs)nadel, Auslesenadel *f* [*Zum Aussortieren gesuchter Karten aus einer Kartei von Randlochkarten. Diese Nadeln werden an bestimmten Lochstellen durch das Kartenpaket gestochen, wodurch alle nicht gesuchten Karten festgehalten und die gesuchten aus dem Paket herausgeschüttet werden*]

assumed decimal point, virtual decimal point [*In automatic data processing, for a decimal number stored in a computer or appearing on a printout, a position in the number at which place values change from positive to negative powers of 10, but to which no location is assigned or at which no printed character appears, as opposed to an actual decimal point*] virtuelle Dezimalpunkt *m*

astatic [*Without orientation or directional characteristics; having a tendency to change position*] astatisch

astatic measuring element, astatic measuring instrument, astatic measuring mechanism [*"Measurement" may synonymously be used for "measuring"*] astatische Messwerk, Messwerk *n* mit magnetischer oder elektrischer Abschirmung

asymmetric digital subscriber line, ADSL asymmetrische digitale Teilnehmerleitung, digitale asymmetrische Teilnehmerleitung *f*

asymmetric encoding asymmetrische Verschlüsselung *f* [*Das Konzept der asymmetrischen Kryptographie wurde 1976 von Whitfiled Diffie und Martin Hellman vorgeschlagen, um das Problem des Schlüsselmanagements (der Begriff umfasst das Erzeugen, die Übertragung und das Speichern des Schlüssels) zu lösen. Nach ihrem Konzept hat jeder Beteiligte zwei Schlüssel: Einen öffentlichen und einen privaten, der geheim bleibt. Es ist damit nicht mehr notwendig, dass Sender und Empfänger ein gemeinsames Geheimnis haben: Jede Kommunikation umfasst nur öffentliche Schlüssel, private werden nie übertragen oder geteilt. Der Absender chiffriert mit dem öffentlichen Schlüssel, das Chiffrat kann jedoch nur mit dem privaten Schlüssel dechiffriert werden, der sich ausschließlich in der Hand des Empfängers befindet. Darüber hinaus ist asymmetrische Kryptographie auch zur Authentifikation (digitale Unterschriften) verwendbar.*]

A

asynchronous balanced mode, ABM Betriebsart *f* des bitorientierten synchron arbeitenden Sicherungsprotokolls [*In Zweipunktverbindungen und Punkt-zu-Mehrpunktverbindungen*]

asynchronous counter, ripple counter asynchrone Zähler *m*, Asynchronzähler, seriengetakt(et)e Zähler [*Zähler, dessen Zählglieder durch serielle Übertragungssignale in zeitlicher Folge schalten. V DIN 19237*]

asynchronous (data) communication [*Data communication in which the timing of information transmission is not related to any system frequency*] asynchrone (Daten)Kommunikation *f*

asynchronous disconnected mode → ADM

asynchronous ((open) loop) control asynchrone Steuerung *f* [*Eine ohne Taktsignal arbeitende Steuerung. DIN 19226. Teil 5*]

asynchronous heating → (electric) asynchronous heating

asynchronous response mode, ARM Spontanbetrieb *m*

AT, advance(d) technology verbesserte Technologie, fortgeschrittene Technologie *f* [*Mikrorechnergeneration mit 16/16-Bit-Architektur (intern/extern)*]

AT-Bus → IDE

ATE → automatic test equipment

ATF, automatic track following automatische Spurnachführung *f* [*Bei digitalem Audioband und 8-mm-Video*]

ATM → automatic (bank) teller (machine)

atmosphere, atmospheric pressure [*The pressure exerted by the earth's atmosphere under certain standard conditions*] Atmosphäre(ndruck) *f, (m)*

atmosphere, earth's atmosphere [*The layer of gases that surrounds the earth*] Atmosphäre, Lufthülle *f*

atmospheric all-gas burner atmosphärische Allgasbrenner *m*

atmospheric (cooling) tower, natural draft cooling tower, natural draught cooling tower [*One in which air movement through the tower occurs by aspiration or natural convection, as opposed to mechanical means*] Naturzugkühlturm *m*

atmospheric electricity atmosphärische Elektrizität *f*

atmospheric evaporation condenser atmosphärische Verdunstungskondensator, atmosphärische Verdunstungsverflüssiger *m*

atmospheric freeze drying, atmospheric lyophilization [*Process in which the solid phase of the solvent is sublimed at atmospheric pressure*] atmosphärische Gefriertrocknung *f*

atmospheric (gas) burner, blowerless (gas) burner [*Air for combustion is supplied by natural draft and the inspirating force created by gas velocity through orifices*] atmosphärische (Gas)Brenner, (Gas)Brenner ohne Gebläse, gebläselose (Gas)-Brenner *m* [*DIN 4788 Teil 1*]

atmospheric gas(-fired) furnace atmosphärische Gasfeuerstätte, atmosphärische Gasfeuerung(sanlage) *f*

(atmospheric ozone impact) Ozonloch *n*

atmospheric pressure, barometric pressure [*The pressure at any point in the atmosphere due solely to the weight of the atmospheric gas above the point concerned*] Atmosphärendruck, Barometerdruck *m*, Pam_b

atmospheric pressure chemical vapour deposition (Brit.); atmospheric pressure chemical vapor deposition (US); APCVD Schichtabscheidung *f* bei atmosphärischem Druck [*aus der Gasphase*]

atmospheric valve Luftventil, atmosphärische Ventil *n*

atomic force microscope, AFM, nuclear force microscope Kernkraftmikroskop *n*, Atomkraftmikroskop

atomic waste, nuclear waste radioaktive Abfall (-stoff), radioaktive Müll *m*

atomization; atomisation (Brit.) Zerstäubung *f*, Zerstäuben *n*

atomization air humidification; atomisation air humidification (Brit.) Zerstäubungsluftbefeuchtung, Zerstäubungsluftanfeuchtung *f*

atomization air humidifier; atomisation air humidifier (Brit.) Zerstäubungsluftbefeuchter, Zerstäubungsluftanfeuchter *m*

atomization disk air humidifier, atomization disk air humidifier [*British spelling = atomisation*] Scheiben-Zerstäubungsluftbefeuchter, Scheiben-Zerstäubungsluftanfeuchter *m*

atomization nozzle; atomisation nozzle (Brit.) Zerstäubungsdüse *f*

to atomize; to atomise (Brit.) zerstäuben

atomized oil → oil mist

atomized oil separator → oil mist separator

atomizer; atomiser (Brit.) Zerstäuber *m*

atomizing air; atomising air (Brit.) Zerstäubungsluft, zerstäubte Luft *f*

atomizing burner; atomising burner (Brit.) Zerstäubungsbrenner, Zerstäuberbrenner *m* [*Nicht verwechseln mit „Schichtungsbrenner"*]

atomizing cone; atomising cone (Brit.) Zerstäuberkegel, Zerstäubungskegel *m*

atomizing drier, atomizing dryer; atomising drier, atomising dryer (Brit.) Zerstäubungstrockner, Zerstäubertrockner *m*

atomizing evaporator; atomising evaporator (Brit.) Zerstäubungsverdampfer, Zerstäuberverdampfer *m*

atomizing gun; atomising gun (Brit.) Zerstäubungspistole, Zerstäuberpistole *f*

atomizing head; atomising head (Brit.) Zerstäubungskopf, Zerstäuberkopf *m*

atomizing humidifier; atomising humidifier (Brit.) Zerstäubungsbefeuchter, Zerstäubungsanfeuchter, Zerstäuberbefeuchter, Zerstäuberanfeuchter *m*

atomizing medium; atomising medium (Brit.) [*Supplementary fluid, such as air or steam, that assists in breaking down a liquid into a finely divided state*] Zerstäubermittel, Zerstäubungsmittel *n*

atomizing nozzle; atomising nozzle (Brit.) Zerstäuberdüse, Zerstäubungsdüse *f*

atomizing oil burner, oil atomizing burner; atomising oil burner, oil atomising burner (Brit.) [*See remark under "atomizing burner"*] Ölzerstäubungsbrenner, Ölzerstäuberbrenner, Zerstäubungsölbrenner, Zerstäuberölbrenner *m* [*DIN 4787, Teil 1*]

atomizing tip; atomising tip (Brit.) Zerstäuberspitze, Zerstäubungsspitze *f*

at option → optional

ATPG → automatic test program generator (US)

ATRAC (method), adaptive transform acoustic coding (method) Datenreduktionsverfahren *n*, Datenreduktionsmethode *f* [*Minidisk von Sony*]

at-rest pressure, pressure at-rest Ruhedruck *m*, statische Überdruck [*Nicht strömendes oder nicht fließendes Medium*]

AT-slot Steckplatz *m*

attachable anbaubar, anmontierbar

attached anmontiert, angebaut, angebracht

(attached) (metal) fin [*Thin piece of metal soldered to a pipe, tube, or other surface in order to increase the heat transfer area*] Lamelle, Flosse, Blechlamelle, Blechflosse *f*

attaching Anmontieren, Anbauen, Anbringen *n*

attachment Anmontierung *f*, Anbau *m*, Anbringung, Anlage, Beilage *f*, Anhang *m*

attachment, attached file [*It is attached to an E-mail*] Dateianhang *m*, angehängte Datei *f*

attachment Anschließung *f*, Anschluss *m*

attack checking glazing angriffshemmende Verglasung *f* [*DIN 52 290, Teil 1, 11.88*]

attack-resistant, attack-resisting, attack-proof angriffsbeständig, angriffsfest

attack time (Ton)Einschwingzeit *f*

to attend anwesend sein

to attend, to operate, to control bedienen, betätigen

attendance check(ing) → reconnaissance squad check(ing)

attendance detector, attendance detection device, attendance detecting device Anwesenheitsmeldegerät, Präsenzmeldegerät *n*, Anwesenheitsmelder, Präsenzmelder *m* [*Zur Erfassung der Anwesenheit von Personen*]

attendance management → reconnaissance squad check(ing)

attendance monitoring Anwesenheitsüberwachung, Präsenzüberwachung *f*

attendance signal message Anwesenheitsmeldung, Präsenzmeldung *f*

attendance time recording Anwesenheitszeitaufzeichnung, Präsenzzeitaufzeichnung *f*

attendant Wärter *m*

attendant → (human) operator

attendant-operated, attendant-controlled, attended bedienergesteuert, führerbedient, führergesteuert [*Aufzug*]

attendant-operated elevator, attendant-controlled elevator, attended elevator (US) [*"Elevator" is called "lift" in British English*] Führeraufzug, Bedieneraufzug *m*

attended (elevator) operation (US); attended (lift) operation (Brit.) (Aufzug)Führerbetrieb, (Aufzug)Bedienerbetrieb *m*

attenuation [*Physics. A reduction in amplitude, density, or energy as the result of such effects as friction, absorption or scattering*] Abschwächung, Dämpfung *f*

attenuation [*Electricity. An exponential reduction of signal amplitude due to distance*] Abklingen *n*

attenuation → acoustic attenuation

attenuation coefficient [*Electromagnetism. The fraction of a beam of electromagnetic energy that is lost by scatter or absorption in passing through a given thickness of a material, used in relation to light*] Dämpfungskoeffizient, Abschwächungskoeffizient *m*

attenuation compensator, attenuation equalizer [*Electronics. A device that equalizes transmission*

A

loss for a line or circuit of interest] Dämpfungsausgleicher, Abschwächungsausgleicher *m*

attenuation constant, attenuation factor [*A wave with space-time dependence exp [i(kx – wt)], where k = Kp + ik$_i$ is attenuated as it propagates according to the factor exp (=k$_i$x) and k$_i$ is called the attenuation constant*] Abschwächungsfaktor, Dämpfungsfaktor *m*, Abschwächungskonstante, Dämpfungskonstante *f*

attenuation network [*Arrangement of circuit elements, usually impedance elements, inserted in circuit to introduce a known loss or to reduce the impedance level without reflections*] Dämpfungsnetz(werk), Abschwächungsnetz(werk) *n*

attenuation peak Dämpfungsspitze, Abschwächungsspitze *f*

attenuation response characteristic Dämpfungs-Charakteristik, Abschwächungs-Charakteristik *f*

attenuation response curve Dämpfungskurve, Abschwächungskurve *f*

attenuator [*An adjustable or fixed transducer for reducing the amplitude of a wave without introducing appreciable distortion*] Abschwächer *m*, Abschwächungsglied, Dämpfungsglied *n*, Teiler, Wellendämpfer, Dämpfer, Wellenabschwächer *m*

attic central furnace [*One designed specifically for installation in an attic, normally unoccupied*] Dachbodenwarmluftfeuerung, Warmluftdachbodenfeuerung *f*

attic fan Dachbodenventilator *m*

attractive force, attractive power Anziehungskraft *f*

(attractively-)styled formschön, stilvoll

attribute Attribut *n*, Beifügung, Ergänzung *f*, Zusatz *m* [*Nichtgraphische Beifügung an ein graphisches Element*]

attribute [*A data item containing informations about a variable*] Attribut *n*, Beifügung, Ergänzung *f*, Zusatz *m*

at zero voltage → grounded

AU → arithmetic unit

audibility, loudness Hörbarkeit, Hörsamkeit *f*

audibility range → range of audibility

audible ... → acoustic ...

audible acuity → acuity

audio... [*A combining form meaning "hearing" or "sound" as in "audio frequency power"*] Audio..., Ton..., Schall...

audio amplification, audio frequency amplification Niederfrequenzverstärkung *f*

audio broadcast(ing), AB Hörfunk *m*, Rundfunk [*Gegensatz: Fernsehen*]

audio channel Tonfrequenzkanal *m*

audio circuit override Audiostromkreisübersteuerung *f*

audio clarity Hördeutlichkeit, Hörklarheit, Hörverständlichkeit *f*

audio component [*Telecommunications. The part of any wave or signal whose frequencies are within the audible range*] Hörbestandteil *m*, Hörkomponente *f*

audio conference, telephone conference Fernsprechkonferenz *f*, Telefonkonferenz [*alte Schreibung: Telephon*]

audio data interleaving geschachtelte Audiodatenadressierung, verschachtelte Audiodatenadressierung, Audiodatenverwürfelung, verwürfelte Audiodatenadressierung *f*

audio feedback howl(ing) Rückkopp(e)lungspfeifen *n*

audio file format Audiodateiformat *n*

audio frequency, AF [*Any frequency that can be detected by a normal human ear. This audible range extends from about 20 to 20.000 Hz. In a communication system intelligible speech can be obtained if a frequency range from about 300 to 3400 Hz is reproduced; any frequency in this range is called voice frequency (VF)*] Tonfrequenz *f*, Hörfrequenz, TF, Audiofrequenz, AF, Niederfrequenz, NF, HF

audio frequency amplification → audio amplification

audio frequency band, AF band Tonfrequenzband *n*, Hörfrequenzband, Audiofrequenzband, AF-Band, Niederfrequenzband, TF-Band, NF-Band, HF-Band

audio frequency meter [*Any of various instruments used to measure the frequencies of sound waves that are audible to the human ear*] Hörfrequenzmesser *m*, AF-Messer, Audiofrequenzmesser, Niederfrequenzmesser, NF-Messer, Tonfrequenzmesser, TF-Messer, HF-Messer

audio frequency noise, AF noise [*Unwanted disturb-ance in the audiofrequency range*] Tonfrequenzgeräusch, TF-Geräusch, Hörfrequenzgeräusch, Audiofrequenzgeräusch, AF-Geräusch, Niederfrequenzgeräusch, HF-Geräusch, NF-Geräusch *n*

audio frequency oscillator [*An oscillator that produces AF signals C20–20.000 hertz*] Tonfrequenzoszillator, AF-Oszillator, Hörfrequenzoszillator, TF-Oszillator, Audiofrequenzoszillator, Niederfrequenzoszillator, HF-Oszillator, NF-Oszillator *m*

audio frequency power, AF power Tonfrequenzleistung, TF-Leistung, Hörfrequenzleistung, Audiofrequenzleistung, AF-Leistung, Niederfrequenzleistung, HF-Leistung, NF-Leistung *f*

audio frequency range [*Acoustics. The portion of the acoustic energy spectrum, from about 20 to 20,000 vibrations per second, that can be detected by a human with normal hearing*] Tonfrequenzbereich, Hörfrequenzbereich, Audiofrequenzbereich, AF-Bereich, TF-Bereich, Niederfrequenzbereich, NF-Bereich, HF-Bereich *m* [*selten: n*]

audio frequency signal level, AF signal level Tonfrequenzpegel, TF-Pegel, NF-Pegel Hörfrequenzpegel, Audiofrequenzpegel, AF-Pegel, Niederfrequenzpegel *m*

audio frequency spectrum, AF spectrum [*Continuous range of frequencies extending from the lowest to the highest audio frequency*] Tonfrequenzspektrum, AF-Spektrum, Hörfrequenzspektrum, TF-Spektrum, Audiofrequenzspektrum, Niederfrequenzspektrum, NF-Spektrum *n*

audio frequency transformer, AF transformer, AFT [*An iron-core transformer designed for use with audio frequency circuits*] Tonfrequenztransformator, TF-Transformator, Hörfrequenztransformator, Audiofrequenztransformator, Niederfrequenztransformator, AF-Transformator, NF-Transformator *m*

audio frequency transmitter, AF transmitter Tonfrequenzsender, TF-Sender, Hörfrequenzsender, Audiofrequenzsender, Niederfrequenzsender, AF-Sender, NF-Sender *m*

audiology [*The scientific study of measurement of hearing ability*] Gehörlehre, Gehörkunde *f*

audiometer [*An instrument used to measure the intensity of sound*] Schallstärkemesser *m*, Schallstärkemessgerät *n*

audiometer [*Acoustical engineering. An instrument that generates sounds of known frequency and intensity; used to measure an individual's hearing ability, especially the level at which a sound becomes audible at a given frequency*] Hörfähigkeitsmesser *m*, Hörbarkeitsmesser

audiometry [*The scientific measurement of hearing ability through the use of audiometers and similar devices*] Hörbarkeitsmessung, Hörfähigkeitsmessung *f*

audio-suppression Sprechsperrung, Sprechunterdrückung *f*

audio tone signals, acoustical tone signals akustische Schwingungen *f* [*Ein Modem verwandelt digitale Rechnerdaten in telefonisch weiterleitende akustische Schwingungen*]

audiovisual [*Telecommunications. Of or relating to the use of electrical, chemical, optical, and mechan-ical media in the reproduction of audible signals, and visual image*] audiovisuell, zugleich hör- und sichtbar

audit Auswerten *n* örtlich erfasster Aufzeichnungen

auditory sensation area, aural sensation area, auditory impression area, aural impression area [*The region enclosed by the curves defining the threshold of pain and the threshold of hearing as a function of frequency*] Hörempfindungsbereich, Wahrnehmungsbereich *m* [*selten: n*]

augmented heat exchange surface → extended heat exchange surface

authenticity certificate → certificate of authenticity

authority and format identifier, AFI Zuteilungsstellen- und Formatkennung *f*

authority level → level of authority

authority (of a valve), valve authority [*The ratio of the pressure drop across a control valve to the pressure drop across the entire system including the control valve*] Schieberautorität *f*

to authorize berechtigen, autorisieren, befugen

authorized dealer Vertragshändler *m*

authorized person Berechtigte, Befugte *m* [*authorized male person*], Berechtigte, Befugte *f* [*authorized female person*]

authorized security representative → security delegate

auto... → automatic...

auto-consumption, self-consumption Selbstverbrauch, Eigenverbrauch *m*

auto-correction, self-correction Selbstkorrektur, Eigenkorrektur *f*

auto-design Nachprojektieren *n* [*Leittechnik. Automatischer Ablauf gespeicherter Eingaben*]

autodial(l)ing [*Automatic telephone destination dial(l)ing, connection, and link confirmation by a modem equipment*] automatische Wählbetrieb *m* [*Automatische Fernwahl, Leitungsaufbau und Verbindungsbestätigung durch eine MODEM-Einheit*]

autofocus automatische Scharfeinstellung, automatische Schärfeeinstellung *f*

automata theory Automatentheorie *f*

to automate [*To operate or control any system by means of automation*] automatisieren

automatic, self-acting, automated, self-working [*Of a mechanism that performs a required act at a predetermined time to certain conditions*] automatisch, automatisiert, selbsttätig

automatic acknowledg(e)ment, automatic ACK, automatic confirmation [*See remark under "ac-*

A

knowledg(e)ment"] automatische Bestätigung, automatische Quittung *f*

automatic act(uat)or [*See remark under „act(uat)or"*] Selbststellglied *n*

automatic air escape valve; automatic air vent (US) automatische Luft(ablass)ventil *n*

automatic air(flow) control damper, automatic airstream control damper Luftstromklappenautomat *m*, automatische Luftstromklappe *f*

automatic air (room) heater, automatic air (room) heating unit automatische Luft(raum)heizer *m*, automatische Raumluftheizer

automatic air vent [*See remark under „air vent"*] Luftdurchlassautomat *m*

automatically fired boiler automatisch gefeuerte Kessel, Automatikkessel *m*

automatic answer, auto-answer automatische Antwort, automatische Beantwortung *f*

automatic appliance → automatic device

automatic backflow(ing) shutter automatische Rückstauverschluss, automatische Rückflussverschluss *m* [*Vorrichtung zum automatischen Abschluss einer anschließenden Leitung(sstrecke) bei Rückstau*]

automatic background ((closed) loop) control, automatic brightness ((closed) loop) control automatische Helligkeitsreg(e)lung *f*

automatic background ((open) loop) control, automatic brightness ((open) loop) control, ABC automatische Helligkeitssteuerung *f*

automatic ball valve Automatikkugelventil, Kugelselbstschlussventil *n*

automatic (bank) teller (machine), ATM Bankautomat, Geldautomat *m*

automatic batchmeter Dosierautomat, Automatikdosierer *m*

automatic baud, autobaud automatische Geschwindigkeitserkennung *f* [*Telekom-Modem MDG 96-HX-21*]

automatic baud rate detection → ABR

automatic beverage vending machine Getränkeautomat, automatische Getränkespender *m*

automatic brightness ((closed) loop) control → automatic background ((closed) loop) control

automatic building construction system, ABCS automatische Hochbausystem *n*

automatic building (services) control system automatische Gebäudeleitsystem *n*

automatic burner [*Burner equipped with an automatic appliance control system*] Automatikbrenner, automatische Brenner, Brennerautomat *m*

automatic call automatische Wahl *f* [*Funktion einer Datenübertragungsvorrichtung, um nach Aufforderung einer EDV-Anmeldung eine Wählerverbindung herzustellen*]

automatic callback automatische Rückruf *m*

automatic callback system Rückrufautomatik *f*, automatische Rückrufsystem *n*

automatic call distribution, ACD, auto-call distribution automatische (An)Rufverteilung *f* bei digitalen Nebenstellenanlagen

automatic calling and answering unit, computer-assisted calling and answering unit, ACAU rechnergestützte Fernsprechvermittlung, automatische Fernsprechvermittlung *f*

automatic calling equipment, ACE automatische Anrufgerät(e) *n*, (*f*)

automatic central ((open) loop) control, central automatic ((open) loop) control automatische Zentralsteuerung *f*

automatic character recognition automatische Zeichenerkennung *f*

automatic chargeover [*IEC 292-3*] automatische Umschalten *n* [*VDE 0660, Teil 301*]

automatic check(ing) automatische Kontrolle *f*

automatic chiller, automatic refrigerating system Kälteautomat *m*, automatische Kältesystem *n*

automatic circuit breaker, automatic c.b. [*IEC 157-1*]

automatic power circuit breaker (US); automatic power breaker automatische Leistungsausschalter, (Leistungs)Selbstausschalter, Sicherungsautomat *m* [*VDE 0660. Teil 101*]

automatic circuity Automatikschaltung *f*

automatic ((closed) loop) control, automatic regulation automatische Reg(e)lung, Automatikreg(e)lung, Selbstreg(e)lung, selbsttätige Reg(e)lung *f*

automatic closing device, self-closing device [*It ensures that a door will return to its closed position after being opened*] Türautomatikschließer *m*

automatic closing device, self-closing device [*A mechanism designed to ensure that an open fire door will close and latch in the event of fire*] Brandtürautomatikschließer, Feuertürautomatikschließer *m*

automatic closing valve → flow closing valve

automatic clothes washer Waschautomat *m*, automatische Waschmaschine *f*

automatic coal heating installation, automatic coal heating plant Automatik-Kohle(n)-heiz(ungs)anlage *f*

automatic coding, autocoding automatische Kodierung *f*, automatische Codierung, Selbstkodierung, Selbstcodierung

automatic coin changing device Geldwechselautomat *m*, Münz(en)wechselautomat

automatic computer [*It carries out a special set of operations without human intervention*] Automatikrechner, automatische Rechner, Rechnerautomat *m*

automatic confirmation → automatic acknowledg(e)ment

automatic connection [*Ability of electronic switching equipment to make a connection between users without human intervention*] Selbstanschluss, automatische Anschluss *m*

automatic contact protection → positive contact protection

automatic control → automatic ((closed) loop) control

automatic control → automatic ((open) loop) control

automatic controller [*A device in which a signal from the sensing element is compared with a signal representing the set value and which operates in such a way as to reduce the deviation. This it does by providing a control signal dependent on the deviation, for transmission to the correction unit which adjusts the value of the correcting condition. NOTE. The two functions of the automatic controller, namely to determine the deviation and to generate the control signal dependent on the deviation, are in many controllers carried out by two separate parts of the controller, the comparing element and the controlling element respectively.*] Automatikregler, Selbstregler *m*

automatic control multi(ple)-loop system → multi(ple)-loop automatic control system

automatic control switch automatische Hilfsstromschalter *m*

automatic control unit → automatic ((open) loop) control unit

automatic cooker Kochautomat, Automatikkocher, automatische Kocher *m*

automatic coupler, automatic synchronizer, check synchronizer, paralleling device Parallelschaltgerät *n*, Synchronisiergerät, Synchromat *m*

automatic cutout [*A device, usually operated by centrifugal force or by an electromagnet, that automatically shorts part of a circuit at a particular time*] Sicherungsautomat, Kleinautomat, Automat *m*

automatic data exchange, ADX automatische Datenaustausch *m*

automatic data processing, ADP, datamation automatische Datenverarbeitung, ADV *f*

automatic data processing equipment, ADPE, datamation equipment automatische Datenverarbeitungsgerät(e) *n*, (*f*)

automatic data processing system → ADPS

automatic data retrieval, ADR automatische Datenauffinden *n*

automatic defrosting [*Defrosting automatically at set intervals, based on time or temperature conditions, or both*] automatische Entfrostung, automatische Enteisung *f*, automatische Abtauen *n*

automatic degausser [*Electronics. A circuit that automatically demagnetizes a colo(u)r television picture tube*] automatische Entmagnetisierer, Entmagnetisier(ungs)automat *m*

automatic demand (matching) unit automatische Lastanforderungseinheit, Lastanforderungsautomatik *f*

automatic design engineering, ADE automatische Konstruktionstechnik *f*

automatic detector → vibratory detector

automatic device, automatic appliance, AD Automationsgerät *n*, AG [*Geräteeinheit mit festem Funktionsumfang, wobei die Parameter für die jeweilige Anwendung angepasst werden können, z. B. digitale Einzelraumregler, Kommunikationsbaugruppe, EA-Baugruppe usw.*]

automatic dialer (US); automatic dialler (Brit.) [*A device in which a telephone number up to a maximum of 14 digits can be stored in a memory and then activated, directly into the line by the caller's pressing a button*] automatische Wählspeichergerät, automatische Speicherwählgerät *n*

automatic dialling unit (Brit.); automatic dialing unit (US) [*Device capable of generating dial(l)ing digits automatically*] automatische Wählgerät *n*, Wählautomat *m*

automatic dictionary, autodictionary [*Any table which establishes one-to-one correspondence between two sets of characters*] automatische Wörterbuch *n*, Wörterbuchautomat *m*

automatic dish washer Geschirrspülautomat *m*, automatische Geschirrspülmaschine *f*

automatic diverter (assembly) automatische Umsteller, Selbstumsteller [*Bad(e)batterie*] *m*

automatic door bottom closer → automatic thresh-old closer

automatic door closer Automatiktürschließer, Türschließautomat, Selbsttürschließer *m*

A

automatic door control [*Türfeststellanlage*]; **automatic gate control** [*Torfeststellanlage*] Feststellanlage *f* [*Ein Gerät oder eine Gerätekombination geeignet die Funktion von Schließmitteln für bewegliche Raumabschlüsse kontrollierend unwirksam zu machen. Eine Feststellanlage besteht aus mindestens einem Brandmelder, einer Auslösevorrichtung und einer Energieversorgung. Beim Ansprechen der Auslösevorrichtung im Brandfall oder anderweitiger Auslösung werden offengehaltene Feuerschutzabschlüsse durch ihre Schließmittel selbsttätig geschlossen*]

automatic door release ((open) loop) control automatische Türöffnersteuerung *f*

automatic double bolts zweifach wirkende Fallenriegel, doppelt wirkende Fallenriegel *m*

automatic early fire recognition automatische Brandfrüherkennung, automatische Feuerfrüherkennung *f*

automatic electromagnetic valve → self-closing solenoid valve

automatic electronic machine, electronic automat-ic machine Elektronikautomat *m*, elektronische Automatikmaschine *f*

automatic energy control unit Energiesparautomat *m*

automatic error correction automatische Fehlerberichtigung, automatische Fehlerkorrektur *f*

automatic escalator cleaning machine Fahrtreppenreinigungsautomat, Rolltreppenreinigungsautomat *m*

automatic exchange, automatic switching system, machine switching system [*A telephone, teletypewriter, or data-transmission exchange in which communication between subscribers is effected, without the intervention of an operator, by devices set in operation by the originating subscriber's instrument*] Wähl(er)amt *n*

automatic expansion valve [*1. Expansion valve de-signed to maintain a constant pressure in the evaporator regardless of superheat. It is seldom used in air-conditioning where loads generally fluctuate and cannot be used on systems with multi(ple)-valve installations. 2. Device used to control the flow of a refrigeration system by pressure*] automatische Expansionsventil, automatische Ausdehnungsventil *n*

auto(matic) feed Automatikvorschub, automatische Vorschub *m*

automatic fire detector, automatic fire detecting device, automatic fire detecting appliance automatische Brandmelder, automatische Feuermelder, Brandmeldeautomat, Feuermeldeautomat *m*

automatic fire protection system automatische Brandschutzsystem, automatische Feuerschutzsystem *n*

automatic firing device, automatic firing appliance Feuerungsautomat *m*, automatische Feuerungsgerät *n*, automatische Feuerungsvorrichtung *f*

automatic fixed firefighting system → automatic stationary firefighting system

(automatic) flow control valve [*One that shuts automatically when the circular pump stops, thereby preventing gravity circulation*] (automatische) Absperrschieber *m*

automatic flush(ing) Zeitspülung, automatische Spülung *f*

automatic flush(ing) cistern [*A water tank system which provides automatically for the periodic flushing of urinals and other pumping fixtures, or of pipes having too small a slope to drain effectively*] Spül(ungs)kastenautomat, automatische Spül(ungs)kasten, Zeitspül(ungs)kasten *m*

automatic flushometer (US); automatic flush(ing) valve (Brit.) (Druck)Spül(ungs)automat, automatische (Druck)Spüler *m*

automatic forwarded indication automatische Weiterleitungsanzeige, Anzeige der automatischen Weiterleitung *f*

automatic frequency ((open) loop) control, AFC [*A device that automatically maintains the frequency of an alternating voltage within specified limits. Error-operated, the device normally consists of two parts, a frequency discriminator that compares actual with desired frequencies, and a reactor controlled by the discriminator in such a way as to correct any frequency error*] automatische Frequenzsteuerung *f*

automatic gain and ((closed) loop) frequency control → AGFC

automatic gain ((closed) loop) control, AGC, automatic volume control, AVC [*In a receiver, a system whereby the gain of the RF and IF stages is automatically adjusted with the object of minimizing changes in the level of the output signal with changes in the level of the received signal*] automatische Verstärkungsreg(e)lung *f*, AVR

automatic gas detector, automatic gas detecting device, automatic gas detection device automatische Gasmelder *m*, automatische Gasdetektionsmittel *n*, Gasmeldeautomat *m*, automatische Gasmeldegerät *n*

automatic gas heater Gasheiz(ungs)automat *m*

automatic gate control → **automatic door control**

automatic heater Heiz(ungs)automat, automatische Heizer *m*

automatic hot air generator, automatic warm air generator Warmluftautomat, automatische Warmlufterzeuger *m*

automatic humidifier automatische Befeuchter, Befeuchtungsautomat, automatische Anfeuchter, Anfeuchtungsautomat *m*

automatic incinerator Verbrennungsautomat *m*, automatische Verbrennungsanlage *f*

automatic indexing, autoindexing [*Selection of key words from a document by computer for use as index entries*] automatische Indizieren *n*, automatische Indizierung *f*

automatic indicating light, automatic indicating lamp, automatic indicator light, automatic indicator lamp, automatic signal light, automatic signal lamp, automatic pilot (light), automatic pilot lamp automatische Meldelampe *f*, Meldelampeautomat *m*

automatic ironer Bügelautomat *m*, automatische Bügeleisen *n*

automatic level ((closed) loop) control, ALC automatische Standreg(e)lung *f*

automatic link setup, ALIS Funkprozessor *m* zum Aufbau automatischer Verbindungen

automatic load limitation [*relay*] automatische Leistungsbegrenzung *f*

automatic magnet closing valve, magnet automatic closing valve Magnetselbstschlussventil *n*, Selbstschlussmagnetventil

automatic meter reading automatische Zählerablesung *f*, automatische Zählerableser *m*

automatic message accounting, automatic toll ticketing [*System whereby toll calls are automatically recorded and timed, and toll tickets printed, under control of the calling telephone's dial pulses and without the intervention of an operator*] automatische Gesprächsberechnung *f*

automatic message switching center (US); automatic message switching centre (Brit.) [*A center in which messages are automatically routed according to information in them*] automatische Mitteilungsumschalt(ungs)zentrale *f*

automatic noise dependent audio signal processing, ANDASP automatische rauschabhängige Tonsignalverarbeitung *f*

automatic noise limiter, ANL [*A circuit employed in radio transmitter/receivers, now extensively used in citizen band transceivers to reduce noise or interference such as that caused by motor-car ignition systems*] automatische Rauschbegrenzer *m*

automatic oil burner automatische Ölbrenner *m* [*DIN 4787, Teil 1. Ein Brenner, der mit selbsttätig wirkenden Zünd-, Flammenüberwachungs- und Steuereinrichtungen ausgerüstet ist. Das Zünden, die Flammenüberwachung sowie das Ein- und Ausschalten des Brenners erfolgen ohne Einwirkung durch das Bedienungspersonal. Die Wärmeleistung des Ölbrenners kann während des Betriebes selbsttätig geregelt oder von Hand gesteuert werden.*]

automatic opening system Öffnungsautomatik *f*, automatische Öffnungssystem *n*

automatic ((open) loop) control, AC [*The controls include three basic types of devices*] automatische Steuerung, Selbststeuerung, Automatiksteuerung, selbsttätige Steuerung *f*

automatic ((open) loop) control unit, ACU automatische Steuer(ungs)einheit, Selbststeuer(ungs-)einheit *f* [*Sie führt als zielgerichtet wirkendes System von Gliedern die automatische Steuerung eines Objektes aus*]

automatic ((open) loop) control valve Steuer(ungs)ventilautomat *m*, automatische Steuer(ungs)ventil, Selbststeuer(ungs)ventil *n*

automatic outdoor air damper, automatic outside air damper Außenluftklappenautomat *m*, automatische Außenluftklappe *f*

automatic (output) volume control [*Process or means by which a substantially constant output volume is automatically maintained in a system or transducer*] automatische Ausgangsvolumenreg(e)lung *f*

automatic overflow valve automatische Überströmventil, automatische Schlabberventil *n*

automatic phase ((closed) loop) control, APC automatische Phasenreg(e)lung, Selbstphasenreg(e)lung, automatische Phasenkorrektur, Selbstphasenkorrektur *f*

automatic picture transmission, APT automatische Bildübertragung *f*

automatic poll Sendeaufruf *m*

automatic power circuit breaker (US) → **automat-ic circuit breaker**

automatic power control, APC automatische Stromverbrauchsoptimierung *f*

automatic power protection device, automatic power protection appliance Leistungsschutzautomat *m*

A

automatic pressure controller [*See remark under "automatic controller"*] Druckwächter *m*

automatic primary starter → automatic stator-circuit starter

automatic priority system Vorrangautomatik *f*, Prioritätsautomatik

automatic programming [*The preparation of machine-language instructions by use of a computer*] automatische Programmieren, Selbstprogrammieren *n*

Automatic Programming Tool, APT [*A computer language used to program numerically controlled machine tools*] automatische Programmierer *m* für numerisch gesteuerte Werkzeugmaschinen

automatic push-button control [*A method of control by means of buttons at the landings and in the lift-car, the momentary pressure of which will cause the lift-car to start and automatically stop at the landing corresponding to the button pressed*] Druckknopf-Innensteuerung *f* für Selbstfahrer(aufzug)

automatic range → autorange

automatic ranging, autoranging [*Automatic switching of a multirange meter from its lowest to the next higher range, with the switching process repeated until a range is reached for which the full-scale value is not exceeded*] Bereichsautomatikschaltung, Automatikbereichsschaltung *f*

automatic ranging unit, autoranging unit [*data processing*] Bereichsautomatik *f*

automatic rate-and-price indicator automatische Börsenkursanzeiger, Börsenkursanzeigeautomat *m*

automatic reclose Kurzunterbrechungsvorrichtung, Wiedereinschalt(ungs)automatik *f*

automatic reclosure, automatic reclosing, rapid automatic reclosure, rapid automatic reclosing Kurzunterbrechung, KU, automatische Wiedereinschaltung *f*; automatische Wiederschließen *n* [*VDE 0670, Teil 101*]

automatic reclosure lockout, automatic reclosing lockout Wiedereinschalt(ungs)sperre *f* [*Kurzunterbrechung*]

automatic recognition automatische Erkennung, Selbsterkennung *f*

automatic refeeding automatische Nachspeisen *n*, automatische Nachspeisung *f*, Selbstnachspeisen *n*, Selbstnachspeisung *f*

automatic refrigerating machine → automatic chiller

automatic regulation → automatic ((closed) loop) control

automatic remote ((closed) loop) control, ARC, automatic ((closed) loop) telecontrol automatische Fernreg(e)lung, Selbstfernreg(e)lung, Fernautomatikreg(e)lung, Fernselbstreg(e)lung *f*

automatic repeat request automatische Wiederholungsaufforderung *f*

automatic restart Selbstanlauf *m*, Selbstanfahren *n*, automatische Anlauf *m*, automatische Anfahren *n* [*Fähigkeit eines Systems nach Netzausfall wieder Strom aufzunehmen*]

automatic return of water, automatic water return Dampfschleife *f*, Wasserkreislauf *m*

automatic return trap Rückspeiseautomat, automatische Rückspeiser *m* [*Dampfheizung*]

automatic rotor starter Läuferselbstanlasser, automatische Läuferanlasser, Selbstläuferanlasser *m*

automatic routine [*It is executed independently of manual operations, but only if certain conditions occur within a program(me) or record, or during some other process*] automatische Prozedur, automatische Behandlungsweise *f*

automatic safety function automatische Sicherheitsfunktion *f* [*Eine Hardware- oder Softwarefunktion, die darauf abzielt, infolge der Erfassung eines Ereignisses oder einer Folge von Ereignissen, das System automatisch in einen „sicheren Zustand" zu überführen*]

automatic (safety) isolation, automatic (safety) disconnection automatische Freischaltung *f*, automatische Freischalten *n* [*VDE 0100. T.200*]

(automatic) safety shut-off valve, (automatic) safety shutdown valve, (automatic) safety stop valve, (automatic) safety disconnection valve [*A valve that is actuated by the safety control system so as to admit and stop gas flow automatically*] Sicherheitsabsperrventil, SAV, Absperrsicherheitsventil *n*

automatic scanning receiver [*A receiver which can automatically and continuously sweep across a preselected frequency, either to stop when a signal is toned or to plot signal occupancy within the frequency spectrum being swept*] automatische Abtastempfänger, Abtastempfängerautomat *m*

automatic sensitivity ((closed) loop) control, automatic sensitiveness ((closed) loop) control, automatic susceptibility ((closed) loop) control, ASC automatische Empfindlichkeitsreg(e)lung, Selbstempfindlichkeitsreg(e)lung *f*

automatic shut-off (switch) [*A switch in a tape recorder which automatically stops the recorder when the tape ends or breaks*] Abstellautomat, Ausschaltautomat, automatische Abstellschalter, automatische Ausschalter *m*

automatic smoke alarm service [*It consists of equipment designed and installed, whereby abnormal products of combustion other than heat result in the transmission of a distinctive smoke alarm signal to a signal reception system*] automatische Rauchalarmanlage *f*

(automatic) smoke (and fire) vent (cover) [*It is installed on a roof, which opens automatically when the heat exceeds 160 F (71.4 Grad C), thereby releasing the cover and venting the fire*] Rauch- und Feuerentlüftungsdeckel *m*

automatic solenoid valve → self-closing solenoid valve

automatic (solid fuel) stoker, automatic (solid fuel) stoking device, automatic (solid fuel) stok-ing appliance automatische Festbrennstoffbeschicker, automatische Festbrennstoffaufgeber *m*

automatic sprinkler automatische Sprinkler, Sprinklerautomat, Sprengerautomat, automatische Sprenger *m*

automatic sprinkler head automatische Sprinklerkopf, automatische Sprengerkopf *m*

automatic star-delta starter Stern-Dreieck-Schalt-Automat, automatische Stern-Dreieck-Schalter *m*

automatic starter Anlasserautomat, Automatikanlasser, Selbstanlasser, automatische Anlasser *m*

automatic start(up) device, automatic start(up) appliance, automatic starting device, automatic starting appliance Anlassautomatik *f*

automatic state Automation *f* [*DIN 19233. Die Benennung des durch Automatisierung erreichten Zustandes, bei dem die Vorgänge automatisch abgelaufen sind*]

automatic stationary firefighting system, automatic fixed firefighting system, stationary automatic firefighting system, fixed automatic firefighting system stationäre automatische Feuerlöschsystem, automatische stationäre Feuerlöschsystem *n*

automatic stator-circuit starter, automatic reduced-voltage starter, automatic primary starter Ständerselbstanlasser, Selbstständeranlasser, automatische Ständeranlasser *m*

automatic steam boiler → dwarf steam boiler

automatic stop, automatic halting [*Of a computer processing operation as the result of an error detected by a built-in checking device*] automatische Abstellung, Selbstabstellung *f*

automatic storage water heater, automatic storage water heating appliance automatische Speicherwassererhitzer, automatische Speicherwasser(er)wärmer, automatische Speicherwasserheizer *m*

automatic switching system → automatic exchange

automatic switch to night service, operator not working [*A PABX feature whereby the switch on to night service occurs automatically at any time when there is no operator's instrument plugged into the switchboard*] Nachtbetriebsumschaltautomatik *f*, automatische Nachtbetriebsumschalter *m*

automatic synchronizer → automatic coupler

automatic syphon, automatic siphon Siphonautomat, automatische Siphon *m*

automatic system Automatik *f*, automatische System *n*

automatic system for boiler operation (Heiz)Kessel-Automatik *f*

automatic tape punch [*It punches holes in a paper top upon reception of electronic signals from a central processing unit*] automatische Papierstreifenlocher, Papierstreifenlocherautomat *m*

automatic telegraph transmission [*Form of telegraphy in which telegraph signals are transmitted mechanically from a perforated tape*] Lochstreifentelegraphie, Lochstreifentelegrafie *f*

automatic telephone call transmitter automatische Telefonrufgerät, automatische Fernsprecherrufgerät *n*

automatic ((tele)phone) switchboard [*The connections are made by using remotely controlled switches*] automatische (Fernsprech)Schalttafel *f*

automatic telephony Wählertechnik *f*

automatic teller (machine) → automatic (bank) teller (machine)

automatic test circuit Prüfautomatik *f*

automatic test equipment, ATE Gerät(e) *n*, *(f)* zur automatischen Prüfung, automatische Prüfgerät(e), selbsttätige Prüfgerät(e)

automatic test program generator (US); automatic test programme generator (Brit.); ATPG Prüfprogrammgenerator *m* [*Programmpaket zur automatischen Erzeugung von Prüfprogrammen*]

automatic threshold Schwelle *f* der automatischen Verstärkungsreg(e)lung, Schwelle der selbsttätigen Verstärkungsreg(e)lung

automatic threshold closer, automatic door bottom closer automatische Türschließer, Türschließautomat, selbsttätige Türschließer *m*

automatic time control Zeitautomatik *f*

automatic time switch [*Combination of a switch with an electric(al) or spring-wound clock, arranged*

A

to turn an apparatus on and off at predetermined times] automatische Zeitschalter, Zeitschaltautomat, Selbstzeitschalter, Zeitselbstschalter *m*

automatic toll ticketing → automatic message accounting

automatic track following → ATF

automatic transfer switch [*In an electric(al) circuit, a switch which automatically transfers a specific load from the normal source to an energy source if the former fails or if the voltage of the normal source drops below a preset minimum*] Übertragungsschaltautomat *m*, automatische Übertragungsschalter *m*

automatic transformer (equipment) automatische Lastausgleichsanlage *f*

automatic transmitter Maschinengeber *m*

automatic valve Automatikventil, automatische Ventil *n*

automatic ventilation automatische Lüftung, automatische Be- und Entlüftung *f*

(automatic) voltage regulator, IR drop compensator, (A)VR [*A device that maintains the terminal voltage of a generator or other voltage source within required limits despite variations in input voltage or load*] Spannungskonstantregler, Spannungskonstanthalter, automatische Spannungsregler *m*

automatic volume control → automatic gain ((closed) loop) control

automatic volume expander [*engineering acoustics*] Dynamikdehner *m*

automatic warm air generator → automatic hot air generator

automatic wet-pipe (fire) sprinkler system, automatic wet-pipe fire-protection sprinkler system [*A fire-protection sprinkler system, all of whose parts are filled with water at sufficient pressure to provide an immediate continuous discharge if the system is actuated*] automatische Nass-Sprinkleranlage, automatische Nass-Sprengeranlage *f*

automation Automatisieren, Automatisierung; Automation *f* [*Automatische Überwachung, Steuerung oder Reg(e)lung eines Betriebsablaufes*]

automation (control) level network Automationsebenennetz(werk) *n*, AEN

automation (control) scheme Automationsschema *n*, MSR-Automationsschema [*MSR = Messen, Steuern, Regeln*]

automation equipment Automationsgerät(e) *n*, (*f*)

automation level MSR-Ebene, Automations(system)ebene; DDC-Unterstation *f* [*Nach VDI 3814 Blatt 1*] [*Gedachte Ebene eines Gebäudeautomationssystems, in der die meisten Automatisierungsvorgänge ablaufen. In dieser Ebene sind eigenständige in einem Gebäude erforderliche Automationsstationen und -geräte installiert. Funktionen sind: Schalten, Stellen, Melden, Messen, Zählen, Regeln, Steuern, Rechnen. Mit aktuellen Automationsstationen [DIN V 32 734) wird die Funktionalität dieser Ebene durch Funktionen der früheren Gruppen- oder Gesamtleitebene erweitert: Optimieren, Bedienen, Beobachten, Tendenzdarstellung*]

automation plant, automation installation Automatisierungsanlage *f* [*Der Teil einer Gesamtanlage, der zur Informationsgewinnung, -übertragung, -verarbeitung, -ausgabe und -nutzung zur Automatisierung von technologischen Prozessen dient; nicht als eigenständige Anlage, immer nur im Zusammenhang mit einer technologischen Anlage verwertbar; umfasst automatisierungstechnische Ausrüstung und Anlagendokumentation*]

automation station, AS Automationsstation *f*, AS [*Vorrichtung zur autarken Bearbeitung der zugeordneten Automatisierungsaufgabe nach Laden entsprechender Programme und Parameter. Sie besteht aus Grundeinheit, Zentraleinheit, Ein-/Ausgabeeinheiten und Kommunikationseinheiten zum Anschluss von Feldbus, Bedienungsstation(en), Leitstation(en) und Automationsgeräten. Verwendung beim Bau komplexer Systeme. DIN V 32 734, GAEB LB 071, VDMA-HKG, CEN TC247*]

automaton [*A robot which functions without step-by-step guidance by a human operator*] endliche Automat *m*

automonitor [*In automatic data processing, a computer program(me) used in debugging which instructs a computer to make a record of its own operations*] automatische Überwacher, Selbstüberwacher *m*

automotive vehicle → motor vehicle

autonomous autark, unabhängig, selb(st)ständig

autopolarity Eigenpolarität *f*

auto-power-off circuit Ausschaltautomatik, automatische Ausschaltung, Abschaltautomatik, automatische Abschaltung *f*

auto-protected range eigenfeste Bereich *m* [*selten: n*]

autoradiography, radioautography [*A technique for detecting radioactivity in a specimen by producing an image on a photographic film or plate*] Autoradiographie, Radioautographie, Autoradiografie, Radioautografie *f*

autorange, automatic range Messbereich(s)automatik *f*

autoranging → **automatic ranging**

autoranging unit → **automatic ranging unit**

autostability [*The ability of a device (such as a servomechanism) to hold a steady position, either by virtue of its shape and proportions, or by control of a servomechanism*] Eigenstabilität, Selbststabilität *f*

autotransductor [*In electromagnetism, a saturable reactor in which the same windings carry the main and control currents*] Autotransduktor, Spar(schaltungs)transduktor, Transduktor *m* in Sparschaltung

autotransformer starter, autostarter, reduced-voltage starter [*Motor starter having an autotransformer to furnish a reduced voltage for starting; includes the necessary switching mechanism*] Anlassspartransformator, Starterspartransformator, Autospartransformatoranlasser, Autospartransformatorstarter, Anlasser mit Spartransformator, Starter mit Spartransformator *m*

autowalk, passenger conveyor Fahrsteig, Rollsteig *m*

autowalk with upper and lower transition curve, passenger conveyor with upper and lower transition curve Fahrsteig mit oberem und unterem Horizontalauslauf, Rollsteig mit oberem und unterem Horizontalauslauf *m*

autowalk with upper transition curve, passenger conveyor with upper transition curve Fahrsteig mit oberem Horizontalauslauf, Rollsteig mit oberem Horizontalauslauf *m*

auxiliaries, auxiliary equipment Hilfsgeräte(e), Reservegerät(e) *n (f)*

auxiliary boiler → **donkey boiler**

auxiliary burner Hilfsbrenner, Reservebrenner *m*

auxiliary bus(bar) Hilfssammelschiene, Reservesammelschiene, Hilfsstromschiene, Reservestromschiene, Hilfsverteilerschiene, Reserveverteilerschiene *f*

auxiliary carry, auxiliary borrow Zusatzübertrag *m* [*Mikroprozessor*]

auxiliary channel [*In data transmission, a secondary channel whose direction of transmission is independ-ent of the primary channel and is controlled by a appropriate set of secondary control interchange circuits*] Hilfskanal *m*

auxiliary ((closed) loop) control Hilfsreg(e)lung *f*

auxiliary contact [*Contact, in a switching device, in addition to the main circuit contact, which functions with the movement of the latter*] Hilfskontakt *m*, Hilfsschaltglied, Hilfsschaltstück *n*

auxiliary controller Hilfsregler *m*

auxiliary (current) circuit Hilfsstromkreis *m*

auxiliary disconnector, auxiliary isolator [*obsolete*] Hilfstrenn(schalt)er *m*

auxiliary electrode, secondary electrode [*In an electrochemical cell, such as a battery, the electrode that transfers current to the test electrode*] Hilfselektrode, Sekundärelektrode, Zweitelektrode *f*

auxiliary energizing quantity Hilfsgröße *f* [*Relay. E VDE 0435, Teil 110*]

auxiliary energy → **auxiliary power**

auxiliary equipment, off-line equipment [*Computer engineering. Terms for any peripheral device not under direct control of a computer*] Hilfsgerät(e) *n*, *(f)*

auxiliary fan Hilfsventilator, Hilfslüfter *m*

auxiliary gate valve (US) → **auxiliary sluice valve**

auxiliary generator → **backup generator**

auxiliary heat → **backup heat**

auxiliary isolator [*obsolete*] → **auxiliary disconnector**

auxiliary lens Vorsatzlinse *f*

auxiliary memory → **external memory**

auxiliary output chip, AOC, secondary output chip integrierte Schaltung *f* mit zusätzlichen Ausgangsschaltungen

auxiliary output voltage, secondary output voltage Hilfsausgangsspannung *f*

auxiliary power, auxiliary energy [*See remark under "power"*] Hilfsenergie, Hilfskraft *f*

auxiliary power (basic) unit, auxiliary energy (basic) unit Baueinheit mit Hilfsenergie, Baueinheit mit Hilfskraft, Hilfsenergiebaueinheit, Hilfskraftbaueinheit *f* [*Die zur Bildung des Ausgangssignals benötigte Energie wird ganz oder teilweise von einer Hilfsenergiequelle geleistet. E DIN 19 226, Teil 5*]

auxiliary power drive, auxiliary energy drive Stellantrieb *m* [*Bauglied zur Betätigung des Stellgliedes durch Hilfsenergie*]

auxiliary power supply, APS, auxiliary energy supply Hilfsstromversorgung *f*

auxiliary relay [*1. A relay that supports another relay or device in the performance of a function. 2. A relay that is actuated by another relay*] Hilfsrelais *n*

auxiliary resistance heating, supplementary resistance heating [*Electric resistance heaters that supplement the heat from the heat pump*] Hilfswiderstandsheizung, Zusatzwiderstandsheizung *f*

A

aux(iliary) return, effect return Behelfsrückführung *f* [*Effektsignal*]

auxiliary rim lock [*A secondary or extra lock that is surface-mounted on a door to provide additional security*] Zusatzkastenschloss mit Umschweif, Umschweif-Zusatzkastenschloss *n*

auxiliary routine; auxiliary programme (Brit.); auxiliary program (US) Hilfs(dienst)programm, Diensthilfsprogramm *n* [*Dienstprogramm eines Rechners für bestimmte Arbeitsfunktionen, z. B. Fehlerbeseitigungs- oder Editierprogramme*]

auxiliary sluice valve, auxiliary parallel slide valve; auxiliary gate valve (US) Hilfsparallelschieber, Parallelhilfsschieber *m* [*Kesselarmatur*]

auxiliary source opening voltage Auslösespannung *f* der Hilfsspannungsquelle [*E VDE 0664, Teil 100*]

auxiliary sprinkler station Sprengernebenstation, Sprinklernebenstation *f* [*Nicht verwechseln mit „Sprinklerunterstation"*]

auxiliary storage (device) → auxiliary memory

auxiliary supply Hilfsversorgung *f*

auxiliary switch [*1. A switch in series or parallel with another switch. 2. A switch that is actuated by another switch*] Hilfsschalter *m*

auxiliary water supply [*It utilizes a low pressure source for sprinklers and standpipes*] Hilfswassereinspeisung *f* [*Sprinkleranlage*]

auxiliary winding Hilfswick(e)lung *f*

availability; accessibility (US) [*Of a switching network, a statement of the number of outlets (trunks) of a desired group to which an inlet has access via the network; i.e. the number of outlets that can be tested for the busy or idle condition*] Verfügbarkeit *f*, Be-triebsbereitschaft *f*

available; accessable (US) [*See remark under „availability"*] verfügbar, betriebsbereit

available for supply lieferbar

available pictorial line [*Portion of the length of the scanning line that can be used specifically for picture signals in a facsimile system*] verfügbare Bildzeile *f*

available power [*Maximum power obtainable from a given source by suitable adjustment of the load*] verfügbare Leistung *f*

available power gain [*Electronics. The ratio of the available power from the output part of a transducer to the available power from the source connected to the input part of the transducer*] verfügbare Leistungsverstärkung *f*

avalanche breakdown [*Electronics. In a semiconductor junction, a nondestructive breakdown caused by multiplicative ion liberation that creates large reverse current flow*] Lawinendurchbruch, Lawinendurchschlag *m*

avalanche breakdown voltage, avalanche BV Lawinendurchbruchspannung, Lawinendurchschlag(s)spannung *f*

avalanche diode [*Electronics. A silica diode that is capable of nondestructive breakdown, in which the junction breakdown voltage is nearly constant for different current levels*] Lawinendiode *f*

avalanche (diode) noise [*Electronics. Noise created in an avalanche diode at the beginning of avalanche breakdown*] Lawinendiodengeräusch *n*

avalanche (effect) [*Electronics. Terms for a condition in which a highly reverse-biased semiconductor junction generates large currents due to high energy electrons that ionize atoms and create new charge carriers with a snowball effect*] Lawinendurchbruch *m*

avalanche photodiode, APD Lawinenphotodiode, Lawinenfotodiode *f*

avalanche rectifier diode Lawinengleichrichterdiode *f*

avalanche transistor [*Electronics. A transistor that uses avalanche breakdown at high reverse voltage to produce chain generation of charge-carriers*] Lawinentransistor *m*

avalanche transit time diode ATT-Diode *f* [*Sie hat einen Lawinen- und Laufzeiteffekt, und kann daher als negativer Widerstand und im Zentimeterbereich als Schwingungserzeuger benutzt werden*]

avalanche voltage [*Electronics. The voltage across a P-N junction diode at which breakdown occurs*] Lawinendurchbruch(s)spannung *f*

AVC → automatic gain ((closed) loop) control

average [*Mathematics. The sum of a group of items, divided by the total number of items*] Mittel *n*

average Durchschnitt *m*

average ... → mean ...

averaging Mittelwertbildung *f* [*Mathematik*]

averaging Durchschnitt(s)bildung *f*

averaging operator Mittelwertoperator, mittelnde Operator *m* [*unscharfe Logik*]

Avogadro('s) law, Avogadro('s) hypothesis [*Under the same conditions of pressure and temperature, equal volumes of all gases contain equal numbers of molecules*] Gesetz *n* von Avogadro, Hypothese *f* von Avogadro [*1776-1856*]

Avogadro('s) number [*Physics. 6.02 x 10 ^{23}, the number of atoms or molecules in a mole*] Avogadrozahl *f*

avoidance Vermeidung *f*

AVR → **(automatic) voltage regulator**

awning Markise *f*

AWNS → **acceptor waiting for new cycle state**

axial condenser Axialkondensator *m*, Axialverflüssiger

axial(-flow) fan, axisymmetric(-flow) fan [*This fan is the modern development of a propeller fan and consists of a rotor in the air stream, having blades of efficient aerofoil action*] Axialventilator *m*, Axiallüfter

axial(-flow) fan condenser, axisymmetric(-flow) fan condenser Axialventilatorverflüssiger *m*, Axialventilatorkondensator, Axiallüfterverflüssiger, Axiallüfterkondensator

axial source level Sonarschallpegel *m*

axial thermostat valve body Axial-Thermostat-Ventilunterteil *n, m*

axial vacuum pump, vacuum axial pump Axialvaku-umpumpe *f*, Vakuumaxialpumpe

axial valve Axialventil *n*

axis Achse *f*, Mittellinie *f*

azeotropic mix(ture), azeotrope Azeotrop *n*, azeotrope Mischung *f*, azeotrope Gemisch *n* [*Mischung zweier Stoffe, die sich nach dem Mischen wie ein einziger Stoff verhalten; also gemeinsam verdampfen und auch verflüssigen. R 502 ist z. B. ein azeotropes Gemisch*]

azimuth angle Seitenwinkel *m*, Azimut *m, n*

azimuth locking pin, ALP Winkelkontrollstift, Azimutkontrollstift *m* [*Zur Bandführung in der digitalen Kompaktkassette*]

B

BA → **building automation**

babble [*1. Aggregate crosstalk(ing) from a large number of channels. 2. Unwanted disturbing sounds in a carrier or other multi(ple)-channel system which result from the aggregate crosstalk(ing) or mutual interference from other channels*] Gemurmel *n*, Gebabbel *n*, Geplapper *n*

baby spot(light) [*A small spot(light) usually equipped with a hood, used (as in the theatre) to concentrate light on an area or an object a small distance from the spot(light)*] Stufenlinsenkleinscheinwerfer *m*, Kleinstufenlinsenscheinwerfer

to back rückstauen

back → **mattock**

backboard [*A wood(en) or metal panel used for mounting miscellaneous apparatus. May be equipped with premounted connecting blocks, mounting brackets, and/or predrilled holes*] Montagetafel *f*

back burner plate → **burner back plate**

back case, back casing, back housing, back enclosure Hintergehäuse *n*

back contact [*Normally closed stationary contact on a relay that is opened when the relay is energized*] Ruhekontakt *m*

backdraft, backdraught Rückzug(luft) *m, (f)*, Rückluft(zug) *f, (m)*

backdrop manhole, tumbling bay manhole, manhole with tumbling bay Absturzschacht *m*, Absturzmannloch *n*

back elevation, rear elevation Rückaufriss *m*, Hinteraufriss

back enclosure → **back case**

back-end computer Nachrechner *m*

backflow(ing) Rückfließen *n*, Rückströmen, Rückstau *m*, Rückfluss *m* [*Fließen oder Strömen gegen das Leitungsgefälle*]

backflow(ing) barrier Rückstausperre *f*, Rückflusssperre, Rückström(ungs)sperre

backflow(ing) cooler Rückstaukühler, Rückflusskühler *m*, Rückström(ungs)kühler

backflow(ing) level Rückflussebene *f*, Rückstauebene, Rückström(ungs)ebene [*Ebene bis zu der Rückstau auftreten kann; in der Regel behördlich vorgeschrieben*]

backflow(ing) preventer, backflow(ing) prevention device, BFP [*Device designed to prevent reverse flow in a water system. Note: Term normally used where backpressure(-type) backflow(ing) is implied*] Rückstauverhinderer, Rückflussverhinderer *m*, Rückström(ungs)verhinderer [*Rückflussverhinderer sind Rückschlagklappen, Rückschlagventile, Rückflusssperren und Membran(e)flussverhinderer*]

backflow(ing) shutter Rückstauverschluss *m*, Rückflussverschluss, Rückström(ungs)verschluss [*Vorrich-*

B

tung zum manuellen oder automatischen Abschluss einer anschließenden Leitung(sstrecke) bei Rückstau]

backflow(ing) valve → **check(ing) valve**

backflush(ing) → **backwash(ing)**

backflush(ing) valve, backwash(ing) valve Rückspül(ungs)ventil *n*

background brightness Hintergrundhelligkeit *f*

background charge, background load Hintergrundlast *f*

background heating [*It is designed to be supplemented by another heat source*] Grundheizung *f*

background illumination → **back illumination**

background image, background picture Hintergrundbild *n*

background irradiance, background radiant flux density [*The irradiance at the entrance aperture of the infrared sensing system which is not radiated directly from the object being investigated*] indirekte Bestrahlungsstärke *f*, mittelbare Bestrahlungsstärke

background lighting → **back illumination**

background limited performance, BLIP untere Grenzwert *m* des detektierbaren Signals

background load → **background charge**

background luminance Hintergrundleuchtdichte, Leuchtdichte *f* des Leuchtschirmhintergrundes [*Oszilloskop*]

background memory → **background storage (device)**

background music, mood music Hintergrundmusik, funktionelle Musik *f*

background noise Hintergrundgeräusch *n*, Ruhegeräusch

background noise level Hintergrundgeräuschpegel *m*, Ruhegeräuschpegel

background program(me) Hintergrundprogramm *n* [*Hauptprogramm welches in einem Rechnersystem Unterbrechungen verarbeitet*]

background radiant flux density → **background irradiance**

background radiation Hintergrundstrahlung *f*

background storage (device), background memory Hintergrundspeicher *m*

background warmth → **warmth to touch**

back housing → **back case**

back illumination, rear illumination, backlighting, rear lighting, background lighting, background illumination Hintergrundbeleuchtung, Durchlichtbeleuchtung *f*

backing, jam Stau, Aufstau *m*

backing plate Aufspannplatte *f*

backing pump Vorpumpe *f*

backing storage (device), backing memory [*A computer storage (device) whose capacity is larger, but whose access time is slower, than that of the computer's main storage or immediate access storage; usually slower than main storage*] Ergänzungsspeicher, Hilfsspeicher, Zusatzspeicher *m*

backlash, lost motion, lost movement Spiel *n*, Luft *f*, tote Gang, Spielraum *m*

backlash, reversed oblique lash umgekehrte Schrägstrich, Gegenschrägstrich *m* [*Symbol:* \]

backlash flank clearance Flankenspiel *n*

backlight, rear light [*Graphic arts. A light that illuminates the subject of a photograph from behind, separating the subject from the background*] Hinterlicht *n*

back lighting → **back illumination**

back lobe [*The three-dimensional portion of the radiation pattern of a directional antenna that is directed away from the intended direction*] Hinterlappen, Hinterzipfel *m*, Hinterkeule *f*

backlog of orders Auftrag(s)bestand *m*

back mounting switch, rear mounting switch Zwischenbauschalter *m*

back panel, rear panel Rücktafel *f*

backplane Rückwand *f* [*z.B. bei einem Steckverbinder*]

backplate Grundblech *n*, Sockel *m* [*Lampenfassung*]

backplate Grundplatte *f* [*Leuchte*]

backplate (lamp)holder Grundblechfassung, Sockelfassung *f*

backpressure Gegendruck, Rückdruck, Staudruck, Rückschlag *m*

backpressure(-type) backflow(ing) Druckrückstau, Druckrückfluss *m*

backpressure valve → **check(ing) valve**

backscattering Rückstreuung *f* [*Lichttechnik; Lichtwellenleiter*]

backset [*The horizontal distance between the face on the centreline of a lock or latch to the centre of the keyhole, lock cylinder, or lock mechanism through which a bolt passes*] Dornmaß *n*

back-siphonage, anti-siphonage [*A form of liquid backflow into a supply system caused by unequal pressures or 'siphonage' developing in the pipes. Water authority regulations are designed to prevent the back-siphonage of polluted household water into main water supply*] Rücksaugen *n*

back-siphon(age) pipe → **anti-siphon(age) pipe**

back-siphon(age) pipe with leakage fitting → **anti-siphon(age) pipe with leakage fitting**

back-siphon(age) trap → **anti-siphon(age) trap**

back-spring lock → **bastard lock**

backstop Rücklaufsperre *f*

back-to-wall wandstehend [*z. B. (fuß)bodenmontierte Spülabort an der Wand stehend*]

back-to-wall pedestal Wandsäule *f*, Wandpiedestal *n*

backtwist Rückdrehung *f*

backup Bereitschaftreg(e)lung *f* [*Fällt der Regler aus, so übernimmt der parallel geschaltete bereite Regler in der Warte die Reg(e)lung*]

backup Unterstützung, Stützung, Reserve *f*

backup Kopieren *n* [*Von der Festplatte auf Disketten*]

backup → **(data) backup**

backup battery Stützbatterie *f*, Pufferbatterie, Unterstützungsbatterie, Reservebatterie

backup capacitor, backup (electrical) condenser Unterstützungskondensator *m*, Stützkondensator

backup controller Unterstützungsregler *m*, Stützregler, Reserveregler

backup copy Sicherheitskopie, Sicherungskopie, Ersatzkopie, Notkopie, Hilfskopie, Reservekopie *f* [*Programm- und Datenkopie für den Fall einer Zerstörung oder Verlust des Originals*]

backup file Sicherheitsdatei, Reservedatei, Sicherungsdatei *f*

backup fuse, backup fusible cutout, backup fuze Teilbereich(schmelz)sicherung, Vor(schmelz)sicherung, Kurzschluss(schmelz)sicherung, Reserve-(schmelz)sicherung, vorgeschaltete (Schmelz)Sicherung *f*

backup generator, auxiliary generator Hilfsgenerator *m*, Ersatzgenerator, Reservegenerator, Notgenerator, Sicherheitsgenerator, Sicherungsgenerator

backup heat, auxiliary heat Hilfswärme, Ersatzwärme, Notwärme, Reservewärme, Sicherheitswärme *f*

backup power operation, standby power operation Ersatzstrombetrieb, Notstrombetrieb, Reserve-strombetrieb *m* [*Funktionen in einem Gebäudebetriebssystem, die auch bei Ausfall der Netzspeisung zur Verfügung stehen müssen (z. B. bei Brand)*]

backup protection Leitungsdistanzschutz, vorgeordnete Schutz, ergänzende Kurzschlussschutz, Reserveschutz, Schutz, überlagernde Schutz, Vorsicherungsschutz, Schutz durch Vorsicherung(en) *m*

backup relay, protective relay, protection relay, protecting relay [*A relay designed to protect a power system in case a primary relay fails to operate as de-sired*] überlagernde Relais, Schutzrelais *n*

backup-restore → **archiving**

backup supply Reserveversorgung *f*, Ersatzversorgung

backup supply for retentive markers Haftmerkerpufferung *f*

backup system Reservesystem, Ersatzsystem *n* [*Systemeinheiten, die bei Ausfall einer Haupteinheit deren Funktion(en) übernehmen, damit das Gesamtsystem betriebsfähig bleibt*]

backup time, data support time Sicher(stell)ungszeit *f* [*Datenträger*]

backup time Puffer(ungs)zeit *f* [*Batteriespeicher; Kondensatorspeicher*]

back valve Rückschlagventil *n*

back view, rear view Rückansicht, Hinteransicht *f*

backward rückwärts

backward bladed centrifugal fan → **backward curved fan**

backward chaining Rück(wärts)verketten *n* [*Vom Ziel ausgehende(s) Verketten*]

backward channel Hilfskanal *m*, Rückwärtskanal [*Datenübertragung. E DIN 44 302*]

backward channel signal quality detector Hilfskanal-Empfangssignalpegel *m*

backward-curved centrifugal fan, BC centrifugal fan, backward-bladed centrifugal fan, BB centrifugal fan Ventilator *m* mit rückwärts gekrümmten Schaufeln

backward diode, unitunnel diode [*A semiconductor diode similar to a tunnel diode except that it has no forward tunnel current; used as a low-voltage rectifier*] Rückwärtsdiode *f*

backward-inclined fan, BI fan Ventilator *m* mit rückwärts geneigten Schaufeln, Lüfter *m* mit rückwärts geneigten Schaufeln

backward inclined (impeller) blade, backward sloped (impeller) blade [*fan*] rückwärts geneigte Schaufel *f*

backward raytracing → **raytracing**

backward search Suchlauf rückwärts, Rückwärtssuchlauf, Suchrückwärtslauf *m*

backward set-up message, BSM Verbindungsaufbaurückmeldung *f*

backward shift Bürstenrückschub *m*, Bürstenverschiebung *f* entgegen der Drehrichtung

backward tone Bestätigungs-Ton *m* [*Datenübertragung*]

backward wave amplifier, BWA Rückwärtsverstärker *m*

backward wave oscillator, BWO, carcinotron [*An electric device which amplifies microwave signals*

B

simultaneously over a wide band of frequencies and in which the travel(l)ing wave produced is reflected backward so as to sustain the wave oscillations] Karzinotron *n*, Rückwärtswellenoszillator *m*

backward wave tube, BWT Rückwärtswellenröhre *f*

backwash blower Rückspülgebläse *n*, Rückspülbläser *m*

backwash(ing), backflush(ing) [*Process of reversing water flow to clean a filter*] Rückspülung *f*, Rückspülen *n*

backwash(ing) valve → backflush(ing) valve

backwash protection filter, backwash protecting filter, backwash protective filter Rückspülschutzfilter *m, n*

back wave, spacing wave [*Signal emitted from a radio telegraph transmitter during spacing portions of the code characters*] negative Tastwelle *f*, Pausenwelle

BAC-net, building automation and control network Kommunikationsprotokoll *n* für Gebäudeautomation

bactericidal lamp bakterientötende Lampe, Entkeimungslampe *f*

bactericidal radiation bakterientötende Strahlung, Entkeimungsstrahlung *f*

bad Inkonsistenz *f* eines Dateisystems, Widersprüchlichkeit *f* eines Dateisystems

bad contact test [*CEE 14*] Glühkontaktprüfung, Glühkontaktprobe *f*, Glühkontaktversuch *m* [*VDE 0632*]

badge Abzeichen *n*

badge (card), identity card, identification card, ID card Ausweis(karte) *m*, (*f*), Kennkarte *f*, Ansteckausweis(karte) *m*, (*f*), Ansteckkennkarte *f*

badge (card) accessories, ID card accessories, identity card accessories, identification card accessories Ausweis(karten)zubehör, Kennkartenzubehör *n* [*selten: m*]

badge (card) check(ing), identity card check(ing), ID card check(ing), identification card check(ing) Ausweis(karten)kontrolle, Kennkartenkontrolle *f*

badge (card) holder, identity card holder, ID card holder, identification card holder Ausweis(karten)halter, Kennkartenhalter, Ausweis(karten)inhaber, Kennkarteninhaber *m*

badge (card) management system, identity card management system, ID card management system, identification card management system Ausweis(karten)system, Kennkartensystem, Ausweis(karten)führungssystem, Kennkartenführungssystem *n*

badge (card) number, identity card number, ID card number, identification card number Ausweis(karten)nummer, Kennkartennummer *f*

badge (card) reader, identity card reader, ID card reader, identification card reader Ausweis(karten)leser, Kennkartenleser *m*

badge (card) reader terminal, identity card reader terminal, ID card reader terminal, identification card reader terminal Ausweis(karten)leserterminal, Kennkartenleserterminal *n*

badge (card) scanner, ID card scanner, identity card scanner, identification card scanner Ausweis(karten)abtaster, Kennkartenabtaster *m* [*Zum Erfassen der Ausweiskartendaten*]

badge (card) system, identity card system, ID card system, identification card system Ausweis(karten)system, Kennkartensystem *n*

badge (card) with chip, ID card with chip, identity card with chip, identification card with chip Ausweis(karte) *m*, (*f*) mit integrierter Schaltung, Kennkarte *f* mit integrierter Schaltung

badge (card) with hologram (interferometer), identity card with hologram (interferometer), ID card with hologram (interferometer), identification card with hologram (interferometer) Ausweis(karte) *m*, (*f*) mit Hologramm, Kennkarte *f* mit Hologramm [*Es wird eine dreidimensionale Abbildung erzeugt*]

badge (card) with infrared (en)coding, identity card with infrared (en)coding, ID card with infrared (en)coding, identification card with infrared (en)coding Ausweis(karte) *m*, (*f*) mit Infrarotkodierung, Ausweis(karte) *m*, (*f*) mit Infrarotcodierung, Kennkarte *f* mit Infrarotkodierung, Kennkarte *f* mit Infrarotcodierung

baffle [*A section fitted between cladding panels to prevent a direct entry of rain*] (Ab)Dicht(ungs)profil *n*

baffle [*An artificial obstruction for checking or deflecting light flow*] Lichtsperre *f*

baffle [*Electricity. It is used for shielding a circuit breaker from gas or oil*] Abschirmung *f*

baffle [*A small acoustic screen attached to a microphone in order to reduce air flow that might cause local variations in the acoustic field around it*] Schallschirm *m*

baffle [*A partition in a loudspeaker cabinet that is used to reduce the interaction between the sound produced by the speaker and the walls of the cabinet*] Schallwand *f*

baffle screen [*A barrier or obstruction of any kind*] Blende *f*, Schutzschirm *m*

baffle [*Any partition or barrier that blocks any incoming sound*] Schallsperre *f*

baffle Schwappschutz *m* [*Batterie*]

baffle deflector, deflecting device, vane Schikane *f*, Umlenker *m*, Ablenker *m*, Prallteil *n*, Strombrecher *m*, Führ(ungs)wand *f*

baffle plate, deflector plate, deflecting plate Prallblech, Ablenkblech, Umlenkblech, Strombrecherblech, Schikaneblech, Führ(ungs)wandblech *n*

baffling, deflector system, deflecting system Schikaneanlage, Umlenk(er)anlage, Ablenk(er)anlage, Pralltteilanlage, Strombrecheranlage, Führungswandanlage *f*

bag check(ing) Taschenkontrolle *f*

bag filter Taschenfilter, Sackfilter *m, n*

baggage handling Gepäckabfertigung *f*

baglock Rucksackschloss, Felleisenschloss *n* [*veraltet*]

bag tag Kofferanhänger *m*

bain-marie [*French. It is used in commercial kitchens. It consists of a vessel of hot water within which another vessel containing food is placed to keep hot or to cook slowly*] Marienbad, Standbad, Wasserbad *n*

to balance auswuchten

to balance ausgleichen, abgleichen, symmetrieren

balance [*A condition or means of achieving equality*] Ausgleichen *n*, Abgleich *m*, Symmetrierung *f*

balance coil, interphase reactor [*An iron-core solenoid with adjustable taps near the center; used to convert a two-wire circuit to a three-wire circuit, the taps furnishing a neutral terminal for the latter*] Ausgleichsdrossel(spule), Abgleichsdrossel(spule), Symmetriedrossel(spule) *f*

balanced [*In communication practice the term usually signifies electrically alike and symmetrical with respect to a common reference point, usually ground, or arranged to provide conjugation between certain sets of terminals*] ausgeglichen, abgeglichen, symmetriert

balanced beam relay Waagebalkenrelais, Kipprelais *n*

balanced busbar protection Sammelschienen-Differenzialschutz, Differenzial-Sammelschienenschutz *m*

balanced circuit → balancing circuit

balanced converter → balun

balanced draft, balanced draught [*Two fans connected to a combustion unit, one to supply the combustion air and the other to induce draft*] ausgeglichene Zug, abgeglichene Zug, symmetrierte Zug *m*

balanced feeder [*radio*] Paralleldrahtspeiseleitung *f*

balanced line erdsymmetrische Leitung *f*

balanced protection relay Differenzialschutzrelais *n*

balanced single seat pressure reducing valve Druckminderer *m* mit Vordruckausgleich, Druckmindererventil *n* mit Vordruckausgleich

balanced (slide) valve → double-seated (slide) valve

balanced station, combined station Hybridstation *f*

balanced to earth, balanced to ground erdsymmetrisch, symmetrisch gegen Erde

balanced-to-earth current, balanced-to-ground current erdsymmetrische Strom *m*

balanced-to-earth voltage, balanced-to-ground voltage erdsymmetrische Spannung *f*

balanced(-to-unbalanced) converter → balun

balanced tree ausgeglichene Baum, ausgewogene Baum, höhenbalancierte Baum, AVL-Baum, abgeglichene Baum *m* [*Ein binärer Baum heißt so, wenn sich für jeden Knoten k die Höhen h der beiden Teilbäume von k um höchstens 1 unterscheiden. AVL = Adelson-Velskuj und Landis*]

balanced WC seat klappbare WC-Sitz, Klapp-WC-Sitz, klappbare Toilettensitz, klappbare Abortsitz, Klapp-Abortsitz *m*

balance loop → balancing loop

balance line symmetrische (Übertragungs)Leitung *f* [*Sie besteht aus zwei Adern, deren anliegende Spannung immer gleich groß und entgegengesetzt ist. Der Vorteil solcher Leitungen ist, dass sich Störspannungen gegenseitig auslöschen*]

balance measuring instrument Kompensationsmessgerät *n*

balance point impedance Kippimpedanz *f*, Kippscheinwiderstand *m*

balance rate of the line Streckenausgleich(s)wert *m*

balance relay, balancing relay Abgleichsrelais, Ausgleichsrelais *n*

balance tank, balance reservoir Ausgleichsbehälter, Abgleichsbehälter *n*

balance time Ausgleichszeit, Abgleichszeit *f*

balance-to-unbalance transformer Symmetrieübertrager *m*

balancing → adjustment

balancing bus, slack node Bilanzknoten *m* [*Netz*]

balancing cable, balancing rope [*lift*] Ausgleichsseil, Abgleichsseil *n*

balancing capacitor, balance capacitor Ausgleichs-kondensator, Abgleichskondensator *m*

balancing circuit, balanced circuit symmetrische Schaltung *f*, symmetrische Stromkreis, symmetrische Schaltkreis *m*, abgeglichene Schaltung *f*, abgeglichene Stromkreis, abgeglichene Schaltkreis *m*

balancing loop, balance loop Symmetrierungsschleife *f*

B

balancing pressure, balance pressure [*Pressure in a system or container equal to that outside*] Ausgleichsdruck *m*

balancing relay ((closed) loop) controller Abgleichsrelaisregler, Ausgleichsrelaisregler

balancing reservoir, service reservoir [*water supply*] Ausgleichsbehälter *m*, Abgleichsbehälter *m*

balancing resistor Symmetrierwiderstand, Abgleichswiderstand, Ausgleichswiderstand *m* [*Brücke*]

balancing valve Mengenabgleichsventil, Mengenausgleichsventil *n*

balcony drainage Balkonentwässerung *f*

balcony inlet, balcony drain [*A fitting intended to be interposed in a vertical rainwater pipe passing through a balcony, and providing an inlet for the drainage of rainwater from the balcony*] Balkoneinlauf, Balkonablauf *m*

ball angle adapter, ball angle adaptor Winkelschraubkopf *m*

ballast, control gear, ballast element [*Electrical device used with an electric-discharge lamp to obtain the necessary circuit conditions (voltage, current, wave-form) for starting and operating*] Vorschaltgerät, Vorschaltglied *n*

ballasted kompensiert, mit Vorschaltgerät, mit Vorschaltglied [*Leuchte*]

ballast frame, ballast support Vorschaltgerät(e)-träger, Vorschaltgliedträger *m* [*Leuchte*]

ballast hum Brummen *n* [*Vorschaltgerät*]

ballast resistor Vorschaltwiderstand *m* [*Leuchte*]

ballast tube Stromregel(ungs)röhre *f* [*Leuchte*]

ball bearing Kugellager *n*

ball-bearing (goods) gravity conveyor Kugellager-(Güter)Schwerkraftförderer *m*

ball-bearing hinge Kugellagergelenk *n*

ball-bearing motor Kugellagermotor *m*

ball-bearing support Kugellagerlagerung *f*

ball-blasted kugel(ab)gestrahlt

ball catch(er) [*A spring-loaded door catch which comprises a small ball, projecting from a mortise in the leading edge of the door, which engages in a hole in a striking plate set within the rebate of the door frame*] Kugelschnäpper, Kugelschnepper *m*

ball chain Kugelkette *f*

ball check(ing) valve → non-return ball valve

ball closing → ball shut(ting)-off

ball cock Kugelhahn *m* [*Siehe Anmerkung unter „Hahn"*]

ball float Kugelschwimmer *m*

ball float steam trap, ball float steam separator Kugelschwimmer-(Kondens)Wasserabscheider, Kugelschwimmerkondenstopf, Kugelschwimmerkondensatableiter, Kugelschwimmerwassersammler, Kugelschwimmerentwässerungstopf *m*

ball head Kugelkopf *m*

ball hinge, ball joint Kugelgelenk *n*

ball knob, globe knob Kugelknopf *m*

ball lamp, globe lamp Kugellampe *f*

ball lightning, globe lightning Kugelblitz *m*

ball mixer, globe mixer Kugelmischer *m*

ball point bonding, nailhead bonding Kugelbonden *n* [*Siehe Anmerkung unter „Bonden"*]

ball retainer, ball cage Kugelkäfig *m*

ball shut(ting)-off, ball closing, ball closure, ball closeoff Kugel(ab)sperrorgan *n*, Kugel(ab)sperrung *f*, Kugelschließung, Kugelschließe *f*

ball valve [*A valve on the supply pipe of an open-tapped water tank or a w.c. cistern, which is on the sur-face of the water. A float is fixed on the end of a rod which is hinged to the body of the valve*] Kugelventil *n*

ball valve → globe valve

balun, bazooka, line-balance converter, balanced
(-to-unbalanced) converter, BALUN [*A device used for matching an unbalanced coaxial transmission line or system to a balanced two-wire line or system*] Symmetrieglied *n*, Symmetrieübertrager *m*, EMI-Schleife *f*

balustrade [*escalator*] Balustrade *f*

balustrade illumination, balustrade lighting Balustradenbeleuchtung *f*

banana plug, split plug Bananenstecker *m*

band [*In automatic data processing, a set of circular or cyclic recording tracks on a storage device such as a magnetic drum, disk, or tape loop*] Band *n*

band, range of frequencies, frequency range, frequency domain [*It lies between two defined limits*] Band, Frequenzband *n*, (Frequenz)Bereich *m [selten: n]*

band-and-gudgeon hinge; band-and-hook hinge [*A gate hinge of heavy wrought iron, consisting of a long strap bent into a circle at one end and rotating on a vertical pin (or gudgeon). Strictly the gudgeon should be fixed to masonry. With any other fixing the hinge is more accurately described as a band-and-hook hinge*] Band-Haken-Scharnier, Haken-Band-Scharnier *n*

band armoured cable (Brit.); band armored cable (US) Bandpanzerkabel *n*

banded ore Banderz n

bandgap Bandlücke *f*

band of steps → step band

bandpass filter, BPF Bandpassfilter, Durchlassfilter *m, n* [*Siebschaltung, die ein begrenztes Frequenzband durchlässt, aber alle Frequenzen unterhalb und oberhalb sperrt. Eine Bandsperre hat die entgegengesetzte Funktion*]

bandpass response, flat top response [*Response characteristics in which a definite band of frequencies is transmitted uniformly*] Bandpassresonanz *f*

band rejection filter, band stop filter, band elimination filter [*An electric filter which transmits more or less uniformly at all frequencies of interest except for a band within which frequency components are largely attenuated*] Bandsperre *f*, Bandfilter *m, n*, Sperrfilter *m*, Filtersperre *f*, Filterband, Sperrband *n*

band selector, band switch [*A switch that selects any of the bands in which a receiver, signal generator, or transmitter is designed to operate and usually has two or more sections to make the required changes in all tuning circuits simultaneously*] Bereichsschalter *m* für Frequenz, Frequenzbereichsschalter, Wellenumschalter *m*

band sound-pressure level Schalldruckbandpegel, Bandschalldruckpegel *m*

bandwidth, BW Bandbreite *f*, Durchlassbereich *m* [*selten: n*] [*Der Frequenzbereich für eine gegebene Leitung. Allgemein gilt, je größer die Bandbreite, um so mehr Nachrichten können durch die Leitung innerhalb einer gegebenen Zeitspanne gesendet werden.*]

bandwidth concatenation factor, BW concatenation factor Bandbreite(n)verkettungsfaktor *m* [*Lichtwellenleiter*]

bang bang ((closed) loop) control, in-out ((clos-ed) loop) control Alles-oder-Nichts-Reg(e)lung, Ein-Aus-Reg(e)lung *f*

banjo coupling piece Schwenkverschraubung *f*

bank [*A number of similar electrical devices, such as resistors, connected together for use as a single device*] Bank *f*, Satz *m*, Gruppe, Batterie, Garnitur *f*

bank Schaltebene *f* [*Ventil*]

to bank (a burner boiler) abschalten unter Druck [*Brennerkessel*]

bank-and-wiper selector, two-motion selector, two-movement selector [*Switch in which electromagnetic ratches or other mechanisms are used, first, to move the wipers to a desired group of terminals, and second, to move the wipers over the terminals of the group to the desired bank contacts*] Heb(e)drehwähler, Hubdrehwähler *m*

bank (of contact), contact bank [*An assemblage of fused contacts over which one wiper or brush or more wipers or brushes move in order to establish electrical connections in automatic switching*] Kontaktbank, Kontaktgruppe *f*, Kontaktsatz *m*, Kontaktbatterie, Kontaktgarnitur *f*

bank ((open) loop) control → collective ((open) loop) control

to bank the fire, to cover the fire (zu)decken, ab-decken [*Kesselfeuer*]

bantam tube [*Vacuum tube having a standard octal base, but a considerably smaller glass tube than a standard glass tube*] (Sub)Miniaturröhre, Bantamröhre *f*

to bar verriegeln

bar Steg *m*, Segment *n*, Lamelle *f*

bar Balken *m* [*Graphik*]

bar Stab *m*

bar, lock bar Riegel, Schlossriegel *m*

barb bolt Widerhakenbolzen *m*

barbed wire tape, tape barbed wire Bandstacheldraht, Stachelbanddraht *m*

bar chart → bar graph

bar code Strichkode, Balkenkode, Strichcode, Balkencode *m* [*Informationskode aus verschieden starken Strichen, die mit Hilfe eines Lesegerätes gelesen werden*]

bar coded strichkodiert, balkenkodiert, strichcodiert, balkencodiert

bar-code label Strichkodeetikett, Balkenkodeetikett, Strichcodeetikett, Balkencodeetikett *n*

bar-code scanner, bar-code reader [*An optical scanning device that reads texts which have been convert-ed into a special bar code*] Strichkodeabtaster, Strichkodeleser, Balkenkodeabtaster, Balkenkodeleser *m*, Strichkodelesegerät, Balkenkodelesegerät *n* [*„...kode..." kann auch mit „c" geschrieben werden*]

bar-coding Strichkodierung, Balkenkodierung, Strichcodierung, Balkencodierung *f*

bar diagramm → bar graph

to bare abisolieren, blank machen [*Draht*]

to bare, to uncover freilegen, abdecken, entblößen, bloßlegen, blanklegen

bare unisoliert, nichtisoliert, blank

bare, uncovered, naked blank, freigelegt, entblößt, abgedeckt, nackt, bloß(gelegt), aufgedeckt

bare board test Prüfung *f* unbestückter Leiterplatten

bare resistance (-type) instantaneous (water) heater Blankwiderstand(s)durchlauf(wasser)heizer *m*

bar graph, bar chart, bar diagram [*Spaced rectangular bars of the same width erected from the same base line, their lengths representing their*

B

values. The bars may be either vertical or horizontal. Called "Gantt chart" in production planning and control, when the bars are horizontally arranged] Balkendiagramm *n*

bar handle, lock bar handle Riegelschwanz, Schlossriegelschwanz *m*

bar head, lock bar head Riegelkopf, Schlossriegelkopf *m* [*drehbare Riegel*]

baring, uncovering Aufdecken, Freilegen, Entblößen, Bloßlegen *n*, Entblößung, Aufdeckung, Freilegung *f*

Barkhausen(-Kurz) oscillator, positive-grid oscillator [*An oscillator of the retarding-field type in which the frequency of oscillation depends solely on the transit time of electrons oscillating about a highly positive grid before reaching the less positive anode*] Barkhausen(-Kurz)oszillator *m*

bar-mounting fuse base Reitersicherungssockel *m*, Reitersicherungsunterteil *m, n*

bar of closure, closure bar Vorlegestange *f* [*Tür*]

bar of damper winding, damper winding bar Dämpferstab *m*

barograph, recording barometer Luftdruckschreiber, Barograph, Barograf *m*

barometer [*An instrument that measures barometric pressure*] Barometer *n* [*Österreich und Schweiz: m*], Luftdruckmesser *m*

barometric draught regulator, barometric draft regulator, barometric damper [*It regulates the ve-locity exhaust of gases in a chimney or the like and thus controls the draft*] Zug(klappen)regler *m*

barometric effect [*Variation(s) in barometric pres-sure caused by altitude or weather changes*] Barometereffekt *m*, Barometerwirkung *f*

barometric pressure → atmospheric pressure

barrage jammer Breitbandstörsender *m*

barrage jamming [*The simultaneous jamming of a number of radio or radar bands of frequencies*] Breitband(sende)störung *f*

barrage reception [*A type of radio reception in which interference from certain directions is reduced by selecting from a system of directional antennas those which yield the greatest signal-to-interference ratio*] Breitbandempfang *m*

barrel lock Zylinderschloss *n*

barrel lug Rohrkabelschuh *m*

barrel pump Fasspumpe *f*

barrel steam boiler, steam barrel boiler, steam drum boiler, drum steam boiler zylinderförmige Dampfkessel *m*

barrel thread Mantelgewinde *n*

barrier, hold-off, shut-off Schranke, Barriere, Sperre *f*

barrier air Sperrluft *f*

barrier gate, grate(d) gate Gatter(tor) *n*

barrier (junction) → depletion layer

barring Sperren *n*

barring Sperrung *f* [*Ergebnis des Sperrens*]

barring rotation Sperr(ungs)drehung *f*

bar scale Maßstableiste *f*, Linearmaßstab *m*

bar soap, soap bar Stangenseife, Seifenstange *f*

bar winding [*An armature winding made up of a series of metallic bars connected at their ends*] Stabwick(e)lung *f*

BAS → building automation system

BAS controller → building automation system controller

bascule bolt → basquill bolt

bascule lock → basquill bolt

base Unterlage *f*

base Unterteil *m, n*

base, standing area Standfläche *f*

base [*Of a plane figure or of a solid*] Basis *f* [*einer ebenen Figur and eines Körpers*]

base, radix [*Of a power and of a logarithmic system*] Basis *f*

base, foot Basis *f*, Fuß *m*

base Sockel *m*

base [*Of a number system*] Basis, Grundzahl *f* [*Die Zahlen werden als Summe der Vielfachen von Potenzen dieser Grundzahl dargestellt*]

base (area) Grundfläche *f*

baseband [*1. Direct transmission method used for short distances (less than 10 miles); uses a bandwidth whose lowest frequency is zero (DC level) – that is, transmission of raw (carrierless) binary data. The transmission medium carries only one signal at a time. 2. Information signal that starts near PC and goes to a finite value. Video baseband is 30 Hz to 4.2 MHz, audio baseband is 30 Hz to 15 KHz*] Basisband, Grundband *n*

baseband frequency response [*Frequency response characteristics of the frequency band occupied by all of the signals used to modulate a transmitted carrier*] Basisbandfrequenzgang, Grundbandfrequenzgang *m*

baseband local area network, baseband LAN [*A local area network in which information is encoded, multiplexed, and transmitted without modulation of a carrier. The IBM Token Ring Network is an example*] lokale Basisbanddatennetz, lokale Grundbanddatennetz *n*

baseband modem, limited distance modem, short-haul modem [*A modem which does not apply a complex modulation scheme to the data before transmission, but which applies the digital input, or a simple transformation of it, to the transmission channel. This technique is available and only operates over short distances where signal amplification is not necessary*] Basisbandmodem, Grundbandmodem *n, m*

baseband signalling (Brit.); baseband signaling (US) [*Transmission of a signal at its original frequencies without modulation*] Basisbandsignalisierung, Basisbandsignalgabe, Grundbandsignalisierung, Grundbandsignalgabe *f*

baseband system [*It transmits at only one frequency and speed*] Basisbandsystem, Grundbandsystem *n*

baseband transmission Basisbandübertragung, Grundbandübertragung *f*

baseband transmission equipment Basisbandübertragungsgerät, Grundbandübertragungsgerät *n*

base bias [*The direct voltage applied to the majority-carrier contact (= base) of a bipolar transistor to ensure that signal excursions operate over the desired region of the transistor characteristic*] Basis(vor)-spannung, Grund(vor)spannung *f*

base(board), scrub board, mopboard, washboard, sanitary cove (US); skirting (board) (Brit.); base plate [*Scotland*] Sockel *m*, Fußleiste, Scheuerleiste *f*

base(board) convector, scrub board convector, mopboard convector, washboard convector, sanitary cove convector (US); skirting board convector (Brit.); base plate convector [*Scotland*] Sockelkonvektor, Fußleistenkonvektor, Scheuerleistenkonvektor, Heiz(ungs)leistenkonvektor *m*

base(board) heater, scrub board heater, mopboard heater, washboard heater, sanitary cove heater (US); skirting (board) heater (Brit.); base plate heater [*Scotland*] [*"Heating unit" may syno-nymously be used for "heater"*] Sockelheiz(ungs)körper, Fußleistenheiz(ungs)körper, Scheuerleistenheiz(ungs)körper *m*, Heiz(ungs)leiste *f*

base(board) heating Sockelheizung, Fußleistenheizung, Scheuerleistenheizung, Leistenheizung *f*

base(board) outlet Sockelauslass, Fußleistenauslass, Scheuerleistenauslass, Leistenauslass *m*

(base) cabinet → undercounter

base channel Unterlage-U-Profil *n*

base continuous current → continuous base current

base corner cabinet → corner base cabinet

base cup position [*lamp*] hängende Brennstellung *f*

base current Basisstrom, Grundstrom *m* [*Transistor*]

base-down position [*lamp*] stehende Brennstellung *f*

base electrode, base element [*An ohmic or majority carrier contact to the base region of a transistor*] Basiselektrode, Grundelektrode *f*

base frame Grundrahmen, Fundamentrahmen, Basisrahmen *m*

base group [*Twelve communications VF paths. A unit of frequency division multiplexing systems bandwidth allocation*] Basisgruppe, Grundgruppe *f*

base illumination, base lighting Grundbeleuchtung, Basisbeleuchtung *f*

base insulator [*Heavy-duty insulator used to support the weight of an antenna mast and to insulate the mast from the ground or some other surface*] (Antennen)Isoliersockel, Stützisolator, Fuß(punkt)isolator *m*

base lighting → base illumination

base-limit register Grenzdatenregister *n*

base line, BL [*The line traced on amplitude-modulat-ed indicators which corresponds to the power level of the weakest echo detected by the radar; it is retraced with every pulse transmitted by the radar but appears as a nearly continuous display on the screen*] Basis-linie, Grundlinie *f*

basis line, BL [*A horizontal or vertical line created by the movement of a scanning dot across a cartridge-ray tube*] Basislinie, Grundlinie *f*

baseload [*Minimum load of a power generator over a given period of time*] Tiefstlast, Mindestlast *f*

baseload Grundlast *f*

basement [*The lowest stor(e)y in a building that is partly or wholly underground*] Kellergeschoss *n*

basement-to-ground floor non-walk-in lift (Brit.); basement-to-ground floor non-walk-in elevator (US) vereinfachte (Klein)Güteraufzug, Unterflur(klein)güteraufzug *m*

base mode → basic mode

base-mounted switch Aufbauschalter *m*

base pin Sockelstift *m* [*Lampe*]

base plate Grundplatte *f*

base plate [*Scotland*] **→ base (board)**

base plate convector [*Scotland*] **→ base (board) convector**

base plate heater [*Scotland*] **→ base (board) heater**

base procedure → basic mode

B

base register → index register

base speed, base rpm., base rotations per minute Grunddrehzahl, Grundtourenzahl *f*

base-spreading resistance [*Resistance which is found in the base of any transistor and acts in series with it, generally a few ohms in value*] Basis(bahn)widerstand, Grund(bahn)widerstand *m*

base station, BS Basisstation, Grundstation *f*

base station controller, BSC Basisstationsregler, Grundstationsregler *m* [*Festnetzverwaltung*]

base station subsystem, BSS Basisstationsuntersystem, Grundstationsuntersystem *n* [*Festnetzverwaltung*]

base steel plate, bearing steel plate [*A steel plate used to distribute the load beneath a steel stanchion or heavy equipment item*] Basisstahlplatte, Stahlbasisplatte, Grundstahlplatte, Stahlgrundplatte *f*

base system, basic system Grundsystem, Basissystem *n* [*Gegensatz: Spitzensystem*]

base transceiver station, BTS Grundstation, Basisstation *f* [*Festnetzverwaltung*]

base unit Unterbau, Unterschrank *m* [*Spültisch*]

BASIC, Beginners All-purpose Symbolic Instruction Code universelle problemorientierte Programmiersprache *f* für Anfänger [*Dialogorientierte Kontakt zum Rechner, speziell für Anfänger geschaffen*]

basic address → address constant

basic assembly, basic set, basic kit Grund(bau)satz *m*, Grundkombination, Grundbaugruppe *f*, Grundbauaggregat *n*, Grundblock *m*, Grundgarnitur *f* [*„Basis“ kann synonym für „Grund“ verwendet werden*]

(basic) block → basic module

basic breaker Schaltereinsatz, Leistungsschaltereinsatz *m*

basic channel Basiskanal *m*, Grundkanal

basic circuit → elementary circuit

basic configuration Basisausbau, Grundausbau *m* [*elektronische Steuerung*]

basic controller Kernregler *m* [*Im nicht unterbrechbaren Regelkreis*]

basic current Nennstrom *m* [*Elektrizitätszähler*]

basic current Basisstrom, Grundstrom *m* [*Relais*]

basic data Grunddaten, Basisdaten *f*

basic data access method, BDAM [*Providing a non-queueing facility, BDAM is a method employed by some computers of gaining access to data held in a backing store. Although not a primary store the backing store has in fact a larger capacity and a slower access time*] Grunddatenzugriffsverfahren *n*, Grunddatenzugriffsmethode *f*, Basisdatenzugriffsverfahren *n*, Basisdatenzugriffsmethode *f*

basic design, basic type, basic version Basisaufbau *m*, Basis(bau)art *f*, Grundaufbau *m*, Grund(bau)art, Grundausführung, Grundbauweise, Basisausführung, Basisbauweise *f*

basic dimensions → basic size

basic disc operating system, BDOS [*An area in the memory map of a microcomputer using a control program(me) for microcomputers disk operating system (DOS), having routines used by the system to access its system disk drives. It thus provides for file management*] Platten-Grundbetriebssystem, Platten-Basisbetriebssystem *n*

basic equation → fundamental equation

basic equipment Grundgerät(e), Basisgerät(e) *n*, (*f*)

basic equipment → basic outfit

basic final cleaning → final basic cleaning

basic firebed [*furnace*] Basisglut, Grundglut *f*

basic flow Basisströmung, Grundströmung *f*

basic form Basisform, Grundform *f*

basic frequency → elementary frequency

basic function Basisfunktion, Grundfunktion *f* [*Eine der innerhalb eines Gerätes erforderlichen Funktionen die nicht speziell sind*]

basic grid system Basisrastersystem, Grundrastersystem *n* [*gedruckte Schaltungen*]

basic hardware elektronische Grundbauteile *f* und Grundgeräte *f* einer Datenverarbeitungsanlage

basic heating Basisheizung, Grundheizung *f*

basic indissoluble information unit Einheit *f* für den Informationsgehalt einer Nachricht

basic input/output system, BIOS, basic I/O system [*A low-level interface set of instructions between application software and hardware. BIOS is normally contained in read only memory (ROM) and consists of a set of functions that are performed when a software calls (invokes) a specific BIOS routine*] Daten-Ein- und -ausgabesystem *n*

basic instruction, presumptive instruction, unmodified instruction [*An instruction in a computer program(me) which is systematically changed by the program(me) to obtain the instructions which are actually carried out*] Basisbefehl, Grundbefehl *m*

basic insulation Basisisolierung [*VDE 0100, Teil 200*], Grundisolierung *f* [*DIN IEC 536*]

basic I/O system → basic input/output system

basic kit → basic assembly

basic load rating Tragzahl *f*

basic lock Basisschloss, Grundschloss *n*

basic logic element Basisfunktionsglied, Grundfunktionsglied *n* [*Es verwirklicht eine Grundfunktion. V DIN 19 237*]

basic logic function Basisfunktion, logische Basisfunktion, Grundfunktion *f*, logische Grundfunktion [*Verarbeitung von Signalen im Sinne von Verknüpfungs-, Zeit- und/oder Speicherfunktionen. V DIN 19 237*]

basic luminaire, basic light(ing) fitting Leuchtengrundmodell, Beleuchtungskörpergrundmodell, Leuchtenbasismodell, Beleuchtungskörperbasismodell *n*

basic mode, basic procedure, base mode, base procedure [*In ACF/VTAM, Release 1 and in VTAM, a mode of data transfer in which the application program(me) can communicate with non-SNA terminals without using SNA protocols*] Basismodus *m*, Basisbetriebsart *f*, Grundmodus *m*, Grundbetriebsart *f*, Basisverfahren *n*, Basismethode *f*, Grundverfahren *n*, Grundmethode *f*

basic modular unit → modular basic unit

basic module, basic block [*A sequence of program(me) statements such that if any one of them is executed, all of them are; a sequence of statements with a label (if any) only at the beginning and a branch (if any) only at the end*] Grundbaustein *m*, Grundmodul *n*, Basisbaustein *m*, Basismodul *n*

basic operating system, Bos Basisbetriebssystem, Betriebsbasissystem *n*

basic outfit, basic equipment Basisausstattung, Basisausrüstung, Grundausstattung, Grundausrüstung, Grundbestückung, Basisbestückung *f*

basic power pack, basic power supply unit Grundstromversorgungsgerät, Basisstromversorgungsgerät *n*

basic procedure → basic mode

basic research Grundlagenforschung *f*

basic set → basic assembly

basic setting Basiseinstellung, Grundeinstellung *f*

basic signal Leitsignal *n*

basic size, basic dimensions Basisgröße, Basisabmessungen, Basismaße, Grundgröße, Grundabmessungen, Grundmaße *f*

basic sound system Beschallungsbasissystem, Beschallungsgrundsystem *n*

basic spline, b-spline mathematische Form *f* von Freiformkurven und Freiformflächen

basic switch Basisschalter, Grundschalter, Schaltereinsatz *m* [*Hilfsstromschalter*]

basic type → basic design

(basic) unit, base unit Grundeinheit, Basiseinheit, Baueinheit *f*, Baustein *m* [*Ein nach Aufbau oder Zu-sammensetzung abgrenzbares Gebilde. E DIN 19 226, Teil 5 und E DIN 44 300, Teil 1*]

(basic) unit priority system Rangfolge *f* der Baueinheiten, Rangfolge der Bausteine [*Für Baueinheiten sollen bei Wortzusammensetzungen in aufsteigender Rangfolge folgende Nachworte verwendet werden: Bauelement (als Nachwort) ...bauelement, ...element (...bauteil), Baugruppe (als Nachwort) ...baugruppe, Bauteil (als Nachwort) ...gerät, ...anlage mit der Festlegung, dass das Bauelement die im betrachteten Zusammenhang kleinste angesprochene Baueinheit ist. Die in aufsteigender Folge genannten Baueinheiten sind jeweils Zusammenfassungen der rangnied-rigeren Baueinheiten. Kleine Baugruppen als Bestandteil einer größeren lassen sich nunmehr über die Funktion ansprechen. E DIN 19 226 T 5*]

(basic) unit without auxiliary power Baueinheit *f* ohne Hilfsenergie [*Die zur Bildung des Ausgangssignals benötigte Energie wird vom Eingangssignal aufgebracht. E DIN 19 226, Teil 5*]

basic value Basiswert, Grundwert *m*

basic values Basiswert(reih)e , Grundwert(reih)e*f*

basic version → basic design

basin → toilet pan

basin → (wash)basin

basin → bowl

basin branch Beckenabzweig *m*

basin mixer Beckenmischer *m*

basin urinal Beckenurinal *n*

basis, (set of) fundamental points, simplex of reference [*Of a linear space and of a number field*] Basis *f* [*Eines linearen Raumes und Zahlenkörpers*]

basis band [*The band of frequencies occupied by all transmitted signals used to modulate the radio wave that is produced by the transmitter in the absence of the signals*] Basisband *n*

basket winding, basket coil [*A crisscross coil winding in which successive turns are far apart except at points of crossing, giving low distributed capacitance*] Kettenwick(e)lung, Korbbodenwick(e)lung *f*

basquill (bolt), basquill lock, bascule bolt, bascule lock Schwengelverschluss, Riegel *m* mit einem Zuge, Zugstange(nverschluss) *f*, *(m)*, Basquillriegel, Riegelverschluss, Basküleriegel *m*, Baskülestange *f*, Ziehriegel, Basküleverschluss, Basquillenverschluss, Zugriegel *m*, Ziehstange(nver-

B

schluss) *f, (m)* [*Riegelverschluss, bei welchem der obere und der untere Riegel gleichzeitig mittels Hebel geschlossen werden*]

bass compensation, bass boost(ing) [*A circuit that emphasizes the low-frequency response of an audio amplifier at low volume levels to offset the lower sensitivity of the human ear to weak low frequencies*] Bassanhebung, Basshervorhebung, Bassentzerrung *f*

bass-cut Sprachfilter *m, n* [*Mikrophon*]

bass response [*A measure of the output of an electronic device or system of a function of an input of low audio frequencies*] Basswiedergabe *f*

bastard lock, back-spring lock Bastardschloss *n* [*mundartlich: Schnippschnapp(schloss) m, (n)*]

batch Charge *f*

batch [*Any specific group of electronic accounting machine (EAM) cards*] Stapel *m*

batch compiler [*It translates the complete program(me) at one time , sometimes making two or more passes through it*] Stapelübersetzer, Stapelkompilierer *m*

batch computing Stapelrechnen *n*

batch (data) file Stapeldatei, Stapeldaten *f*

batch dimension → batch size

batching → batch processing

batching, grouping [*Records for the purpose of pro-cessing them in a computer*] Gruppierung *f*

batch mode, batch procedure Stapelmodus *m*, Stapelbetriebsart *f*, Stapelverfahren *n*, Stapelmethode *f*

batch process [*A process that is not in continuous or mass production; operations are carried out with discrete quantities of material or a limited number of items*] diskontinuierliche Prozess, Chargenprozess, unstetige Prozess, unterbrochene Prozess *m*

batch processing, batching Stapelverarbeitung *f*, Stapelbetrieb *m* [*Im Gegensatz zur Echtzeitverarbeitung die Verarbeitung von über einen bestimmten Zeitraum gesammelten Daten in einem Schub durch eine Datenverarbeitungsanlage, z. B. Buchungen eines Tages*]

batch programming Stapelprogrammierung *f*

batch size, batch dimensions Chargengröße, Chargenabmessungen *f*

batch terminal Stapelendgerät *n*, Stapelstation *f*

bath [*general term*] Bad *n* [*allgemeine Benennung*]

bath → (bath)tub

bath → bath(room)

bat-handle switch [*A toggle switch having an actuating lever shaped like a baseball bat*] baseballschlägerförmige Kippschalter *m*

bath- and shower-mixer Bad- und Brausemischer *m*

bath convector Badkonvektor, Badheizkörper *m*

bather Badende *m, f*

bath facility, bath setup Bad(e)einrichtung *f*

bath filtration and chlorination plant, bath filtration and chlorination system Bad-Filter- und Chlorungsanlage *f*, Bad-Filter- und Chlorungssystem *n*

bathing establishment Badeanstalt *f*

bathometer Brunnenpfeife *f*

bath(room) Bad(eraum) *n, m*, Badezimmer *n*

bath(room) cabinet, bath(room) cubicle Bad(eraum)schrank, Badezimmerschrank *m*

bath(room) door lock Badtürschloss *n* [*Eintüriges Schloss, dessen Riegel durch Drehen eines Vierkantstiftes betätigt wird*]

bath(room) furniture Bad(eraum)möbel, Badezimmermöbel *f*

bath(room) outfit, bath(room) equipment Bad(eraum)ausstattung, Bad(eraum)ausrüstung, Bad(eraum)bestückung, Badezimmerausrüstung, Badezimmerausstattung, Badezimmerbestückung *f*

bath(room) (sectional) radiator [*See remark under "radiator"*] Badradiator *m*, Badgliederheiz(ungs)körper *m*; Badheiz(ungs)körper [*Fehlbenennung*]

bath setup → bath facility

bath/shower mixer (valve) Bad/Brause-Mischer *m*, Bad/Brause-Batterie *f*

bath stove Badeofen *m*

bath towel holder Badetuchhalter *m*

(bath)tub, bath(ing tub) (Bade)Wanne *f* [*DIN 18022. Fehlbenennung: Bad n*]

(bath)tub apron, bath(ing tub) apron (Bade)Wannenschürze *f*

(bath)tub (drain) trap, (bath(ing tub)) (drain) trap [*See remark under "trap"*] (Bade)Wannenverschluss *m*

(bath)tub faucet, bath(ing tub) faucet (US); (bath)tub (water) tap, bath(ing tub) (water) tap [*See remark under "water trap"*] (Bade)Wannenhahn *m*

(bath)tub (hand) grip, bath(ing tub) (hand) grip (Bade)Wannengriff, Griff *m*

(bath)tub (hand) rail, bath(ing tub) (hand) rail (Bade)Wannenhandlauf *m*

(bath)tub inlet, bath(ing tub) inlet (Bade)Wanneneinlauf *m*

(bath)tub lead trap, bath(ing tub) lead trap [*See remark under "trap"*] Blei(-Bade)wannen-Geruchsverschluss *m*

(bath)tub ledge, bath(ing tub) ledge (Bade)Wannenrand *m*

(bath)tub ledge fitting, bath(ing tub) ledge fitting (Bade)Wannenrandarmatur *f* [*Siehe Anmerkung unter „Armatur"*]

(bath)tub mixer valve, (bath)tub blender, (bath)tub mix(ing) valve, (bath)tub blending valve, (bath)tub filler, bath mixer [*B.S. 1415. "Bath(ing tub)" may synonymously be used for "(bath)tub"*] Bad(e)mischer *m*, Bad(e)(misch)batterie, (Bade)Wannen(misch)batterie *f*, (Bade)Wannenmischer *m*

(bath)tub mounting, bath(ing tub) mounting (Bade)Wannenmontage *f*

(bath)tub outlet, (bath)tub drain, bath(ing tub) outlet, bath(ing tub) drain (Bade)Wannenablauf *m*

(bath)tub rim, (bath)tub roll, bath(ing tub) rim, bath(ing tub) roll (Bade)Wannenwulst *m, f*

(bath)tub trap, bath(ing tub) trap [*See remark under "trap"*] (Bade)Wannen(wasser)(geruchs)verschluss *m*

(bath)tub water, bath(ing tub) water Badewasser *n*

(bath)tub with overflow, bath(ing tub) with overflow, overflow (bath)tub, overflow bath(ing tub) Überlauf(bade)wanne *f*

bath water heater Badwasserheizer, Badwarmwasserbereiter *m*

bat's wing burner, batwing burner [*A gas burner with a slit to give a fan-shaped flame*] Fledermaus(gas)brenner *m*

batten door, boarded door, ledged door Lattentür, belattete Tür *f*

battery → **(storage) battery**

battery [*misnomer*] → **accumulator cell**

battery backed [*That means that the computer contains a battery to provide power in the event of a mains failure, thus preventing the loss of data*] batteriegepuffert, batterieunterstützt

battery backup Batterie(unter)stützung *f*

battery charger Batterielader *m*, Batterieladegerät *n*

(battery) charge regulator (Speicher)Batterieladeregler, Laderegler *m*

battery compartment Batteriefach *n*

battery(-fed) clock Batterieuhr *f*

battery feeding Batterieeinspeisung *f*

battery housing, battery casing, battery case Batteriegehäuse *n*

battery lavatories (US); multi(ple)-(wash)-basin installation Waschbeckenreihe, Reihenwaschbecken *f*

battery of boilers → **boiler battery**

battery power, battery current, battery energy Batteriestrom *m*, Batterieenergie *f*

battery-powered, battery-fed batteriebetrieben, batteriegespeist

battery switch → **accumulator switch**

battery (type) storage tank Batterie(lager)behälter *m*

baud, unit pulse [*A unit of telegraph signal(l)ing speed equal to the number of code elements (pulses and spaces) per second or twice the number of pulses per second*] Baud *n*

Baudelot cooler, falling film (surface) cooler [*Liquid cooler in which the liquid to be cooled passes by gravity over a refrigerated surface*] Baudelotkühler, Querrohrsystemkühler *m*

Baud generator Baudrahmenteiler *m*

BAUD(ot) code [*A teleprinter code that uses a combination of five or six marking and spacing intervals of equal duration for each character*] Baudot-Kode *m*, Baudot-Alphabet *n*

BAUD(ot) rate [*In the data transmission of binary characters the baud is the unit of measurement for the step or transmission speed (Ys).*

$$Vs = \frac{characters}{seconds} \quad (baud)$$

For teletype TTY: 110 baud = 110 binary characters per second. One complete instruction frame can be transmitted using:
1 start bit, 8 data bits, 2 stop bits. Therefore the key character transmission speed is 10 key characters per second.] Baudzahl, Schrittgeschwindigkeit, Baudrate *f*

BAUD(ot) rate switch block Baudratenschalterblock, Baudzahlschalterblock *m*

bay lock Rohr(schlüssel)schloss, Schloss mit Rohrschlüssel *n*

bayonet cap, BC [*A type of connector for some domestic electric lamps. Two small pins are used to engage slots in the lampholder*] Bajonettsockel *m*

bayonet-locking connector Bajonettverschluss *m*

bazooka → **balun**

B.B. → **BH**

BBD → **bucket brigade device**

B box → **index register**

BBSY → **bus busy**

BC → **business computer**

BC → **blind copy**

BC → **bayonet cap**

BCC → **block check(ing) character**

BC centrifugal fan → **backward-curved centrifugal fan**

BCD → **binary-coded decimal**

BCD character, binary-coded decimal character binär kodierte Dezimalzeichen, kodierte Dezimalbinärzeichen *n*

BCD code → **binary-coded decimal code**

B

BC fan → backward-curved fan

BCF fire extinguisher, halon fire extinguisher Halonfeuerlöscher *m*

BCLR → bus clear

b-contact element → break contact element

BCR → byte count register

BDAM → basic data access method

BDOS → basic disc operating system

BDPS, beam deflection phase shifter [*An entrancement of electron beam automatic test equipment (ATE), in which the electron beam passes through an aperture and is deflected around a rectangular path. It can be used to monitor internal waveforms of devices for fault diagnosis*] Strahlablenkphasenschieber, Strahlablenkphasendreher *m*

bead thermistor [*A thermistor made by applying the semiconducting material to two wire leads as a viscous droplet, which cements the leads upon firing*] Perlthermistor *m*

beak Gasbrenner *m* mit runder Öffnung

beam [*A concentrated, unidirectional stream of particles or radiation, as of light or sound*] Bündelstrahl, gebündelte Strahl *m*

beam antenna (US); beam aerial [*An antenna, es-pecially in VHF or short-wave antenna, that is designed to transmit in a certain direction*] Richtantenne *f*

beam clamp Balkenklemme *f*

beam deflection phase shifter → BDPS

beam depth Balkenhöhe *f*

beamer kompakte Röhrenprojektor *m*

beam index Schnellantrittspunkt *m* [*DIN 54 119*]

beam irradiance, direct normal irradiance, beam radiant flux density [*Irradiance received from the sun without significant change of direction from the apparent position of the sun*] direkte Bestrahlungsstärke *f*

beam lever [*steam engine*] Kolbenhebel *m*

beam of electrons → electron beam

beam path [*beam smoke detector*] Strahl(ungs)weg *m*

beam power tube, beam tetrode [*An electron-beam tube which uses directed electron beams to provide most of its power-handling capability and in which the control grid and screen grid are essentially aligned*] Bündeltetrode, Strahltetrode, Tetrode mit Elektronenbündelung *f*

beam radiant flux density → beam irradiance

beam smoke detector, beam smoke detecting device, beam smoke detection device linienförmige Rauchmelder *m*, linienförmige Rauchmeldegerät *n* [*Er besteht aus einem getrennten Sender und Empfänger die an den gegenüberliegenden Seiten eines Raumes in höchstens 100 m Abstand angebracht werden. Der Sender strahlt Infrarotlicht aus, welches der Empfänger aufnimmt und auf Intensitätsgrad prüft. Wird der Sendestrahl von Rauch durchströmt, reduziert sich der Intensitätsgrad des Strahles und es wird bei einer bestimmten Mindestintensität ein Alarm ausgelöst*]

beam web Balkensteg *m*

to bear away by degrees (gradmäßig) abfallen

beard, key beard Bart, Schlüsselbart *m*

bearing ball [*A hardened steel ball used in combination with others in bearings, to lessen the friction*] Lagerkugel *f*

bearing length Auflage(n)länge *f*

bearing steel plate → base steel plate

beat [*Physics. The periodic variation in amplitude of a wave that is the superposition of two harmonic waves of slightly different frequencies*] Schwebung, Überlagerung, Schwebe *f*

beat frequency, BF [*Either of the two frequencies obtained when signals of two frequencies are combined, in a non-linear circuit equal to the sum or difference, respectively, of the original frequencies*] Schwebungsfrequenz, Überlagerungsfrequenz, Schwebefrequenz *f*

beat-frequency oscillator, BFO, heterodyne oscillator [*An oscillator in which a desired signal with frequency, such as the beat frequency produced by combining two different signal frequencies such as two different radio frequencies*] Überlagerungsoszillator, Schwebungsfrequenzoszillator, Schwebefrequenzoszillator *m*

beat tone [*Acoustical engineering. A musical tone produced when two tones within about fifteen cycles of each other are played together*] Schwebeton, Schwebungston, Überlagerungston *m*

to be called angerufen werden

to become extinguished, to extinguish ausgehen, ausgelöscht werden [*Flamme*]

to become warm → to warm

bed lift (Brit.); bed elevator (US) Bettenaufzug *m*

bedspring array → billboard array

beep kurze Summerzeichen *n*, Piepton, Warnton *m*

beeper → pager

Beginners All-purpose Symbolic Instruction Code → BASIC

beginning sequence → turn-on sequence

to begrease einfetten

Belfast sink, butler('s) sink, stoneware sink [*A deep-sided sink made of stoneware*] Steinzeugausguss(becken) *m*, *(n)*

bell Klingel, Glocke *f*

bell, hub; socket (Brit.) [*The portion of a pipe which is enlarged to receive the end of another pipe of the same diameter for the purpose of making a joint*] Glockenmuffe *f*

bell butt soldering Kelchlötung *f*

bell cord, bell string Klingelschnur *f*

bell current Klingelstrom *m*

bell (drain) trap, bell stench trap, bell U-bend, bell fixture trap, bell siphon trap, bell water trap, bell water stop, bell water barrier [*A trap, used in sanitary appliances, consisting of a metal bell suspended in the water forming the trap*] Glocken(wasser)-(geruch)verschluss *m*, Glockenwasservorlage *f* [*Die Benennungen „Glockensyphon", „Glockensiphon" und „Glockentraps" sind veraltet. DIN 1986*]

bell flush(ing) cistern Glockenspül(ungs)kasten *m*

bell key Klingeltaste *f*

bellows → (air) bellows

bellows → aneroid capsule

bellows (gas) meter → dry (gas) meter

bellows seal [*A boiler seal in the form of a bellows which prevents leakage of air or gas*] Balg(en)(ab)-dichtung *f*

bell pipe, hub pipe; socket pipe (Brit.) [*See remark under "bell"*] Glockenmuffenrohr *n*

bell push Klingelknopf *m*

bell shape Glockenform *f*

bell signal Klingelsignal *n*

bell string → bell cord

bell (toilet) cistern, bell flush(ing) cistern [*A flushing cistern with a siphonic action activated by the movement of a bell in the cistern*] Glocken-(Abort)Spülkasten, Glocken-Toilettenspülkasten, Glocken-Klosettspülkasten *m*

bell trap → bell (drain) trap

bell wire Klingeldraht *m*

below-grade hydrant → sunk hydrant

below-zero degree → degree below zero

belt drive Riemen(an)trieb *m*

belt(ed) cable Gürtelkabel *n*

belt guard, belt protection Riemenschutz *m*

belt pallet(-type) moving walk [*A moving walk with a series of connected and power-driven pallets to which a continuous belt treadway is fastened*] Palettenfahrsteig mit Band, Band-Palettenfahrsteig *m*

belt(-type) goods conveyor [*It consists of a con-tin-uous belt made from reinforced rubber, closely woven cotton material or wire mesh running over a number of rollers carried on ball bearings and driven by means of a drum pulley, at one end, which is connected to a motor through reduction drive. At the other end a similar pulley has to be provided designed to take on the stretch of the belt and to ensure the necessary tension*] Bandgüterförderer, Güterbandförderer *m*

belt (type) moving walk, belt treadway moving walk Förderbandfahrsteig, Beförderungsbandfahrsteig *m*

belt(-type) passenger conveyor, belt(-type) autowalk [*See remark under "autowalk". Instead of metal treads, a rubber tread is used with steel rod bracing*] Gummifahrsteig, Gummirollsteig *m*

B(E)M → facility management

benchmark Vergleichsprüfung *f* [*Für Datenverarbeitungsanlagen unter Praxisbedingungen*]

benchmarking Auswertbarkeit *f* der Daten, Datenauswertbarkeit

benchmarking (Gebäude)Nutzungskostenverfolgung *f* [*Ermittlung und Vergleich von Verhältniswerten bei den Gebäudenutzungskosten*]

bench potatoe peeler Tischkartoffelschälmaschine *f*

bend, (pipe) bend fitting, pipe bend [*For a pipe or channel, or a curve made in a pipe for changing the direction of the run*] (Rohr-)Bogen(stück) *m*, (*n*), Bogenrohr *n*, (Rohr)Krümmer(stück) *m*, (*n*), Krümmerrohr *m*

bend offset, offset (bend), swan-neck, swan'(s) neck, gooseneck, s-tube [*A pipe fitting used to connect two pipes whose axes are parallel but not in line*] Schwanenhals(verschluss) *m*, S-Stück *n*, Etagenbogen, Stufenbogen *m*, Sprungrohr, Springrohr *n*, Rohr-U-Bogen, Knieverschluss, Siphon(verschluss) *m*, Heber(rohr) *m*, (*n*)

beneficial use of rainwater → rainwater use

benefits Funktionalität *f*

bent bolt, bolt with a handle, handle bolt, toe bolt gekröpfte Riegel, Schwanzriegel, Riegel mit Angriff, Angriffsriegel, Riegel mit Schwanz *m*

BER → bit error rate

Beranek scale [*A scale which measures the subjective loudness of a noise; noises are arranged into six arbitrary categories: very quiet, quiet, moderately quiet, noisy, very noisy, and intolerably noisy*] Beranekskala *f*

Berkeley Internet domain server → domain name server

betty Brecheisen *n* [*Schlosserei*]

between-season heating period Übergangsheizperiode *f*

BeV [*A billion (10^9) electron volts, a unit used in the USA; international unit is GeV (giga electron volts), which has the same value*] Gigaelektronenvolt *n*

B

bevel Schmiege, Schräg(kant)e, Abkantung *f*

bevel chart, bevel diagram Schmiegediagramm *n*, Schmiegeplan *m*, Schrägendiagramm *n*, Schrägenplan *m*, Abkantungsdiagramm *n*, Abkantungsplan *m*, Schrägkantendiagramm *n*, Schrägkantenplan *m*

bevel collar → angle collar

bevelled (Brit.); beveled (US); cut (ab)geschrägt, (ab)gekantet, geschmiegt, schräg, schief

beverage dispenser Getränkespender *m*

beyond-the-horizon communication, (forward) scatter propagation, over-the-horizon communication Überhorizontübertragung *f*

BF → beat frequency

BFO → beat-frequency oscillator

BFP → backflow(ing) preventer

BFS → breadth-first-search

BG → bus grant

BGIN → bus grant in line

BH, boiler house, boiler building, B.B. (Heiz)-Kesselhaus, (Heiz)Kesselgebäude *n*

BI → binary input

bias Vormagnetisierung *f*

bias current → closed-circuit current

bias current door release → closed-circuit current door release

biaural sound zweiohrige Schall, doppelohrige Schall *m*

bibcock, bibb cock [*A water tap fed by a horizontal pipe from the wall instead of from below like a pillar tap. Bibcocks are commonly installed above Belfast sinks*] Zapfhahn, Schnauzenhahn *m*

BI block → binary input block

biconditional gate → equivalence gate

bicore (computer) memory, bicore (computer) storage (device) Doppel-Magnetschichten-Speicher *m* [*Die untere Schicht dient zum Speichern, die obere zum Lesen*]

bidet [*A vessel which can be bestridden for bathing purposes*] Bidet, Sitzwaschbecken *n*

bidet trap [*See remark under "trap"*] Bidetgeruchverschluss, Sitzwaschbeckengeruchverschluss *m*

bidirectional [*Having the ability to function in two directions, usually two opposite directions*] zweiseitig, doppelseitig

bidirectional counter, forward/backward counter, reversible counter, up-down counter, down-up counter [*A counter that has both an add and a substract input so as to count in either an increasing or a decreasing direction*] Vorwärts-Rückwärts-Zähler, Zweirichtungszähler, Doppelrichtungszähler, Um-kehrzähler, V-R-Zähler, R-V-Zähler, Rückwärts-Vorwärts-Zähler *m*

bidirectional diode thyristor → diac

bidirectional microphone [*A microphone that responds equally well to sounds reaching it from the front and rear, corresponding to sound incidences of 0° and 180°*] Mikrophon *n* mit Achtercharakteristik, zweiseitig gerichtete Mikrophon, Mikrofon mit Achtercharakteristik, zweiseitig gerichtete Mikrofon

bidirectional transistor Doppelrichtungstransistor, Zweirichtungstransistor *m*

BI fan → backward-inclined fan

bifilar electrometer [*An electrostatic voltmeter in which two conducting quartz fibers, stretched by a small weight or spring, are separated by their attraction in opposite directions toward two plate electrodes carrying the voltage to be measured*] Wulf-Elektrometer *n*

bifringent filter, Lyot filter, monochromatic filter [*optics*] doppeltbrechende Filter *m, n*

bifurcated fan, forked fan gegabelte Ventilator *m*

bijection, one-to-one function Bi(in)jektion *f* [*Diese Funktion stellt sowohl eine Injektion als auch eine Subjektion dar*]

bilateral transducer, bidirectional transducer [*A transducer capable of measuring in both positive and negative directions from a reference position*] bilaterale Wandler *m* [*DIN 19 226 Teil 5*]

BILBO → built-in logic block observer

billboard (US); boarding (Brit.) [*outdoor advertising*] Reklametafel, Werbetafel *f*

billboard array, bedspring array, mattress array [*A broadside antenna array consisting of stacked dipoles spaced $^1/_4$ to $^3/_4$ wavelength apart in front of a large sheet-metal reflector*] Querstrahler *m* mit Blechreflektor

billing system Abrechnungssystem *n*

bill of material(s) → BOM

bimetal(lic) ((closed) loop) controlled bimetallgeregelt

bimetal(lic) contact Bimetallkontakt *m*

bimetal(lic) contact thermometer, contact bimetal(lic) thermometer Bimetallanliegethermometer, Anliegebimetallthermometer *n*

bimetal(lic) element, bimetal(lic) strip [*Two dif-ferent metals fused together so that when heated, the strip will bend and open and close points (as on a thermostat)*] Bimetallelement *n*, Bimetallstreifen *m*

bimetal(lic) ignition fuse, bimetal(lic) ignition fuze Bimetallzündsicherung *f*

bimetal(lic) ((open) loop) controlled bimetallgesteuert

bimetal(lic) relay Bimetallrelais, thermische Überlastrelais *n*

bimetal(lic) thermometer, differential thermometer Bimetallthermometer *n*

binary [*In automatic data processing, possessing a property for which there exists two choices or conditions, one choice or condition excluding the other*] binär, zweiwertig, dyadisch [*E DIN 44300, Teil 2*]

binary adder (circuit) Binäraddierer, Binärsummierer *m*

binary (arithmetic) operation [*Any arithmetic op-eration in which all operands and results are in binary notation*] Binäroperation *f*

binary card → binary (punched) card

binary channel Binärkanal *m*

binary character Binärzeichen *n*

binary code [*A code in which each allowable position has one of two possible states, commonly 0 and 1; the binary number system is one of many binary codes*] Binärkode, Binärcode *m*

binary-coded binärkodiert, binärcodiert

binary-coded decimal, BCD binärkodierte Dezimalziffer, binärcodierte Dezimalziffer, dual verschlüsselte Dezimalziffer *f*

binary-coded decimal character → BCD character

binary-coded decimal code, BCD code binär codierte Dezimalcode, BDC, binär kodierte Dezimalkode *m*

binary-coded decimals, BCD Binärkode *m* für Dezimalziffern

binary-coded scale binär codierte Maßstab, binär kodierte Maßstab *m*

binary code value binäre Kodewert, binäre Codewert *m* [*Beliebige Informationen werden durch elektrische Signale übertragen. Dabei werden alle zu übertragenden Zeichen durch eine eindeutige Folge binärer Werte dargestellt, aus denen am Empfangsort die Nutzinformationen rekonstruiert werden*]

binary complementary code → complementary binary code

binary counter → binary scaler

binary digit → bit

binary digit system → double digit system

binary divider Binärteiler, Dualdividierer *m*

binary error correcting code binäre Fehlerkorrekturcode, binäre Fehlerberichtigungscode, binäre Fehlerkorrekturkode, binäre Fehlerberichtigungskode *m*

binary error detecting code, binary error recognition code binäre Fehlererkennungskode, binäre Fehlererkennungscode *m*

binary fixed point number binäre Festkommazahl *f*

binary image Binärbild, binäre Bild *n*

binary input, BI binäre Eingang, Binäreingang *m* [*Physikalischer Eingang der einen von zwei Zuständen „0" oder „1" erfasst. DIN 19 226 Teil 5, DIN V 32 734*]

binary input block, BI block Binäreingabe-Baugruppe *f*

binary input unit, BI unit binäre Eingabeeinheit, Binäreingabeeinheit *f* [*V DIN 19 237. Eingabeeinheit für binäre Eingabesignale*]

binary logic binäre Verknüpfung *f*

binary machine language, binary machine code, binary computer language, binary computer code binäre Maschinensprache, binäre Rechnerprogrammsprache, binäre Programmiersprache eines Rechners *f*

binary number Binärzahl *f* [*Zahl eines Zahlensystems, das aus nur zwei Ziffern besteht, z. B. 0 und 1*]

binary number system → binary system

binary ((open) loop) control binäre Steuerung *f* [*Eine binäre Steuerung ist eine innerhalb der Signalverarbeitung vorwiegend mit binären Signalen arbeitende Steuerung. Die binären Signale sind nicht Be-standteile zahlenmäßig dargestellter Informationen. E DIN 19 226 Teil 5, DIN 19 237. Anmerkung: Die binäre Steuerung verarbeitet binäre Eingangssignale vorwiegend mit Verknüpfungs-, Zeit- und Speichergliedern zu binären Ausgangssignalen. E DIN 19 226 Teil 5*]

binary operation Operation *f* zwischen scharfen Mengen

binary output, BO binäre Ausgang *m*, Binärausgang [*Physikalischer Ausgang den einen von zwei Zuständen „0" oder „1" zu signalisieren. DIN 19 226 Teil 5, DIN V 32 734*]

binary output unit, BO unit, binary signal output unit binäre Ausgabesignaleinheit, Binärausgabesignaleinheit *f* [*V DIN 19 237. Ausgabeeinheit für binäre Ausgabesignale*]

binary point [*See remark under "point"*] Binärkomma *n*

binary primary detector [*ANSI C 37,100*], **binary initial element, binary sensing element, binary transducing sensor, binary sensor switch** Binärsignalgeber *m*

binary (punched) card Binär(loch)karte *f*

binary pure code → pure binary code

binary relation Relation *f* zwischen scharfen Mengen [*unscharfe Logik*]

binary ripple counter Binär-Asynchronzähler *m*

B

binary run file Objektprogrammdatei *f*

binary scaler, binary counter, scale-of-two circuit [*A scaler that produces one output pulse for every two input pulses*] Binäruntersetzer, Binärzähler *m*, Frequenzhalbier(ungs)schaltung *f*

binary signal binäre Signal, Binärsignal *n* [*Ein digitales Signal mit nur zwei Wertbereichen des Informationsparameters. E DIN 19 226, Teil 5*]

binary signal Binärsignal, binäre Signal *n* [*Ein digitales Signal, bei dem die Signalparameterwerte eine Nachricht oder Daten darstellen, die nur aus Binärzeichen besteht bzw. bestehen. E DIN 44 300, Teil 2*]

binary signal output unit → binary output unit

binary synchronous communication(s), BSC, synchronous binary communication(s), BISYNC binär-synchrone (Daten)Übertragung(ssteuerung) *f* [*BSC/ BISYNC. Ein zeichenorientiertes synchrones Datenkommunikationsprotokoll in Halbduplex; von IBM 1964 eingeführt*]

binary system, binary number system binäre (Zahlen)System *n*, Dualsystem, Zweiersystem, duale Zahlensystem [*Ein Stellenwertsystem zur Basis 2, das über einen Vorrat von zwei Ziffern (0 und 1) verfügt*]

binary word [*A group of bits which occupies one storage address and is treated by the computer as a unit*] Binärwort, binäre Wort *n*

binary zero binäre Null, Binärnull *f*

BIND → domain name server

binding coal → matting coal

binding post [*A post attached to an electric cable, wire, or apparatus, for making a connection to it coveniently*] Klemm(en)bolzen, Anschlussbolzen *m*

binding screw Klemm(en)schraube, Anschlussschraube *f*

binistor [*Electronics. A semicondutor device having two stable states used in switching and storage circuits, depending largely on an external voltage supply for its bistable negative-resistance characteristic*] Binistor *m* [*Transistorähnliche PNPN-Flächen-Tetrode*]

binoculars Fernglas *n*

binominal array antenna (US); binominal aerial array [*Directional antenna array for reducing minor lobes and providing maximum response in two opposite directions*] Richtantenne *f* mit Nebenzipfelreduzierung, Binominalantenne *f*

binominal distribution, Bernoulli distribution Binominalverteilung, Bernoulliverteilung *f*

biofilter, trickling filter Tropfkörper *m*

biogas energy Biogasenergie *f*

biological computer Biorechner, biologische Rechner *m* [*Er hat organische Schaltelemente*]

(biological) fouling (Pflanzen)Anwuchs *m*

bionics [*The study of systems, particulary electronic systems, which function after the manner of living systems*] Bionik *f*

BIOS → basic input/output system

biparting door → centre-opening door (Brit.)

bipolar three-step controller, bipolar three-step regulator, bipolar three-position controller, bipolar three-position regulator, bipolar triple-step controller, bipolar triple-step regulator, bipolar triple-position controller, bipolar triple-position regulator bipolare Dreipunktregler *m* [*Er gibt innerhalb eines kleinen Bereiches um die Regelabweichung 0 kein Ausgangssignal ab. Bei Überschreiten des Be-reiches – je nach Vorzeichen der Regelabweichung – wird ein Ausgangssignal auf zwei Signalausgänge ge-geben. DIN 19 225*]

bipolar transistor Bipolartransistor *m* [*Entweder ein NPN- oder PNP-Typ. Diese Transistoren benutzen positive und negative Ladungsträger*]

biradiate zweistrahlig, doppelstrahlig

bird-mouth(ed) end Schnabelende *n*

bird screen Vogelschutzgitter *n*, Vogelschutzblende *f*

B-ISDN, broad-band integrated services digital network, wide-band integrated services digital network breitbandige dienstintegrierte digitale Netz *n*

bismuth soldering → soldering with bismuth

bistable multivibrator, trigger flip-flop, Eccles-Jordan multivibrator, trigger circuit, flip-flop (circuit), bistable element, FF, Eccles-Jordan circuit, trigger pair circuit [*A multivibrator in which either of the two active devices may remain conducting, with the other nonconducting, until the application of an external pulse. IEC 117-15*] Flip-Flop(-Schaltung) *m*, (*f*), Triggerschaltung, bistabile Kippschaltung, (bi-stabile) Kippstufe *f*, bistabile Trigger, Bivibrator, bistabile Kipper *m*, bistabile Kippglied *n* [*DIN 40 700*]

biswitch Fünfschichtdiode *f* [*Sie hat gleiche Eigenschaften in Sperr- und Durchlassrichtung*]

BISYNC → binary synchronous communication(s)

bit → bit(t)

bit, binary digit Binärzeichen, Dualzeichen *n*, Binärstelle *f*, Bit *n*, Dualstelle, zweiwertige Stelle *f* [*Stelle in einer Binärzahl, die nur den Wert 0 (Null) oder 1 (Eins) annimmt; in der niedrigstwertigen Stelle ist ein Bit die kleinste Informationseinheit*]

bit buffer unit [*A unit that terminates bit-serial communications lines coming from and going to technical control*] Bitpuffereinheit *f*

bit burst error [*A group of bits in which two consecutive erroneous bits are always separated by less than a given number of correct bits. In other words the erroneous bits are more closely spaced than is statically expected*] Bitbündelfehler *m*, Bitfehlerhäufung *f*

bit by bit bitweise

bit combination Bitverknüpfung, Bitkombination *f*

bit condition code Bitanzeige *f* [*programmierbare Steuer(ungs)gerät*]

bit condition code field, bit indicator field Bitanzeigefeld, Bitindikatorfeld *n*

bit density, record density [*Number of bits which can be placed, per unit length, area, or volume, on a storage medium; for example, bits per inch of magnetic tape*] Bitdichte, Aufzeichnungsdichte *f*

bit error Bitfehler *m*

bit error probability Bitfehlerwahrscheinlichkeit *f*

bit error rate, BER Bitfehlerquote *f* [*Das Verhältnis der falsch empfangenen Bits zu allen gesendeten Bits innerhalb eines Messzeitraumes*]

bi-threshold input Eingang *m* mit zwei Schwellenwerten, Doppelschwelleneingang *m*

bit indicator field → **bit condition code field**

bit key → **bit(ted) key**

bit map Bildelementgraphik, Pixelgraphik *f*

bit map memory, frame buffer Bildspeicher *m* [*Beim Rasterbildschirm der Speicher, in dem die Bildelemente gespeichert sind*]

bit map terminal Datensichtstation *f* für Ausgabe graphischer Darstellungen

bit-orient(at)ed protocol, BOP [*The most widely known BOP is IBM's synchronous data link control (SDLC). In this control, information is sent in frames, within each frame are fields which have specific functions*] bitorientierte Protokoll, bitweise Protokoll *n*

bit-orient(at)ed signalling (Brit.); bit-orient(-at)ed signaling (US); BOS [*A common signal(l)ing protocol used in AT & T's Digital Multiplexed Interface (DMI). In bit-orient(at)ed signal(l)ing, values of bits in the signal(l)ing channel represent physical signals on the E- and M-wires of an analog(ue) telephone. BOS is used only as an interim provision or systems immigrating between D4 framing (which uses AB signal(l)ing) and DMI (which uses message-orient(at)ed signal(l)ing according to CCITT common channel signal(l)ing system Nr. 2)*] bitorientierte Signalgabe, bitorientierte Signalisierung, bitweise Signalgabe, bitweise Signalisierung *f*

bit packet Bitpaket *n*

bit padding Bitstopfen *n*

bit position [*The position of a binary digit (= bit) in a word, generally numbered from the least significant bit*] Bitstelle *f*

bit processing Bitaufbereitung, Bitverarbeitung *f*

bit processor [*Integrated circuit design that enables the processing of 0 and 1*] Bitprozessor *m*

bit rate Bitrate, Bitfolgefrequenz *f*

bit/second Bit/Sekunde *n* [*Einheit der Übertragungsgeschwindigkeit; die Übertragungsgeschwindigkeit ist das Produkt aus Schrittgeschwindigkeit und Anzahl der Bits, die je Schritt übertragen werden. DIN 44 330*]

bit sequence → **bit string**

bit-serial process data highway interface system bitserielle Prozessbus-Schnittstellensystem *n*

bit significance → **bit valency**

bit skew Bitversatz *m*

bit slice Bitscheibe *n*

bits per inch → **bpi**

bits per second → **bps**

bit stream, bit flow Bitstrom, Bitfluss *m*

bit stream converter, bit flow converter Bitstromwandler, Bitfließwandler *m*

bit string, bit sequence [*A set of consecutive binary digits (= bits) representing data in coded form, in which the significance of each bit is determined by its position in the sequence and its relation to the other bits*] Bitreihe, Bitfolge, Bitkette *f*

bit system → **double digit system**

bit(t), key-bit(t), (key) step Schlüsselbart, Schlüsselkamm, Bart, Kamm *m*

bit(ted) key, step(ped) key Bartschlüssel, Kammschlüssel *m*

bit transfer area Bitübergabebereich *m*, Bitweitergabebereich, Bitübermitt(e)lungsbereich [*selten: n*]

bit transmission order → **order of bit transmission**

bit valency, bit significance Bitwertigkeit *f*

BI unit → **binary input unit**

BL → **base line**

black-and-white television → **monochrome television**

black-and-white video camera, monochrome video camera Schwarz-Weiß-Videokamera *f*

black-and-white (view(ing)) screen → **monochrome (view(ing)) screen**

blackbody, hohlraum, ideal radiator schwarze Körper, schwarze Strahler, Schwarzkörper, Schwarzstrahler, Hohlraumstrahler, Plankscher Körper, Plancksche Strahler, Hohlraumkörper *m* [*Auf der Suche nach einer vollkommen schwarzen Oberfläche ist Kirchhoff zum Hohlraum gelangt, der ein vollkommenes Absorptionsvermögen für alle Wellenlängen der Temperaturstrahlung besitzt. Er hat nach dem Kirchhoffschen Gesetz auch das größte Emissionsvermögen*]

blackbody radiation, hohlraum radiation, ideal radiator radiation Schwarz(körper)strahlung *f*

B

B

black box schwarze Kasten *m* [*Der Systemaufbau ist unwichtig, sondern nur Ein-/Ausgang(s)-Signale sind wichtig*]

black-bulb thermometer Schwarzkugelthermometer *n*

blackburst Schwarzbild *n* [*Insertschnitt*]

black disc, black disk Schallplatte *f* mit analoger Schallspeicherung

black level Schwarzpegel *m*

black list → **(access) refusal list**

blackout Verdunk(e)lung *f*

blackout → **total outage**

blackout curtain, blackout drape Verdunk(e)lungsvorhang *m*

black peak [*A peak excursion of the television picture signal in the black direction*] Schwarzspitze *f*, Schwarzmaximum *n*

black radiator, complete radiator Schwarzstrahler *m*

blade arrangement, blade disposition, blade layout [*fan*] Schaufelanordnung, Laufradanordnung *f*

blade damper Klappenzugregler *m*

blade terminal [*Flat, push-on (type) terminal connection (fast-on)*] Flachstecker *m*

blade wheel, vane wheel Schaufelrad, Laufrad *n* [*Ventilator*]

blading [*fan*] Beschaufelung *f*

blank (character), space (character) [*A character, either printed or appearing as a blank, used to denote a blank space among printed characters*] Leerzeichen *n*

blank commutation funkenfreie Kommutierung *f*

blank(et) Rohling *m*

blanketing Inertisierung *f* [*Verfahren, bei dem die Luft oder auch eine Verfahrenseinheit durch ein inertes Gas, z. B. Stickstoff oder Argon, ersetzt wird, um die Gefahr von Oxidation, Feuer oder Explosion zu verringern*]

blanking Austasten *n*

blanking interval, blanking gap Austast(ungs)-lücke *f*

blanking (off) plate [*Plate for blanking (off) unused or reserve spaces*] Verschlussplatte *f*

blanking signal [*The signal rendering the return trace invisible on the picture tube of a television receiver*] Austast(ungs)signal *n*

blank instruction → **no-operation instruction**

blank terminal Blindklemme *f*

blank tile Leermosaikbaustein *m* [*Siehe Anmerkung unter „Mosaikbaustein"*]

blank wall [*A wall without an opening*] Blindwand, Vollwand *f*

blast (air) [*Air under pressure, blown into a furnace*] Wind *m*, Gebläseluft, Bläserluft *f*

blast atomizer Injektionszerstäuber *m*

blast-effect inhibiting sprengwirkungshemmend

blast efficiency, blower efficiency Gebläsewirkungsgrad, Bläserwirkungsgrad *m*

blast-freeze tunnel → **tunnel freezer**

blast freezing → **jet freezing**

blasting [*Acoustics. The act of producing or especially reproducing sounds with excessive volumes*] Lärm-erzeugung *f*

blasting [*Distortion of sound due to overloading of any part of a radio transmitter, receiver or radio amplifier*] Dröhnen *n*

blast(ing) Sprengen *n*, Sprengung *f*

blast(ing) attack Sprengstoffanschlag *m*

blast(ing) effect Sprengwirkung *f*

blast steam pipe, steam blast pipe Dampfzugrohr, Zugdampfrohr *n*

blaze Feuersbrunst *f*, Großfeuer *n*, Großbrand *m*

bleaching croft → **croft**

bleed controller Ableit(ungs)regler *m*

bleeder [*deprecated*] → **blow-off pipe**

bleeder condenser, bleed-off condenser, bleeding condenser Beries(e)lungskondensator, Beries(e)lungsverflüssiger *m*

bleeder valve, bleed-off valve, bleeding valve Entnahmeventil *n*

bleed-off, deliberate wastage of water [*It is normally required in all open recirculating systems, including washers, to limit the condensation of solids. A bleed-off rate of about 1 % of the water circulation rate is generally satisfactory*] Wasserablassen *n*

bleed resistor Ableit(ungs)widerstand *m* [*Gerät*]

blended code optimization Optimierungsausgleich *m* zwischen verschiedenen Prozessorarchitekturen [*Ziel ist es, einen Binärkode zu erhalten, der auf allen 32-Bit-Intel-Prozessoren sehr gut arbeitet im Hinblick auf Programmausführungszeit bzw. erzielbare Rechenleistung*]

blended lamp → **mixed-light lamp**

blending Beimischen *n*, Beimischung *f*, Zumischen *n*, Zumischung *f*

blending pump, blender Beimischpumpe, Zumischpumpe *f*

blind copy, BC Blindkopie *f* [*Die Empfänger sehen nur ihre elektronische Postadresse*]

blind current → **reactive current**

blind flange [*Flange used to seal the end of a pipe*] Blindflansch *m*

blind lock, false lock, mock lock Blindschloss *n*

blind nut Blindmutter *f*

blind-power controller → reactive-power controller

(blind) slat (Jalousie)Lamelle *f*, (Jalousie)Stäbchen *n*

B line → index register

blinker lamp → Morse lamp

blinking cursor Blinkschreibmarke *f*

blinking frequency → flash(ing) frequency

BLIP → background limited performance

BLM → pressure zone microphone

Bloch wall, domain wall [*Solid-state physics. A transition layer with a finite thickness of a few hundred lattice constants, between adjacent ferromagnetic domains*] Blochwand, Domänenwand, Bereichswand *f*

to block [*To combine two or more information units into a single unit*] blockkoppeln

to block blockieren, sperren

block → block (of information)

block [*A (physical) unit of information for a logical record, which would normally be computed in bytes*] Block *m*

block → module

block, physical record Block *m* [*Eine technische Einheit von Speicherstellen*]

block check(ing) character, BCC Blockprüfzeichen, Blockkontrollzeichen *n*

block diagram → signal flow diagram

block diagram Übersichtsschalt(ungs)plan *m* [*DIN 40 719, Blatt 1*]

block diagram Blockdiagramm, Grobdiagramm, Ablaufdiagramm, Flussdiagramm *n*, Flussplan, grobe Ablaufplan, Blockplan, Grobplan *m*

blocked suction unloading Saugabsperrung *f*

block end → end of block

block end character → end-of-block character

block error probability Blockfehlerwahrscheinlichkeit *f*

block error rate, block error frequency Blockfehlerhäufigkeit *f*

(block) ice store basement (Block)Eiskeller, Eisblockkeller *m* [*Zur Lagerung von (Block)Eis*]

blocking [*Automatic data processing. Combining two or more computer records into one block*] Blockbildung, Kopp(e)lung *f*

blocking Blockieren, Sperren *n*

blocking Blockierung, Sperrung *f*

blocking cam Sperrnocken, Blockier(ungs)nocken *m*

blocking current Sperrstrom, Blockier(ungs)strom *m*

blocking diode Schutzdiode *f* [*Photovoltaik*]

blocking element Sperrelement, Blockier(ungs)element *n*

blocking out of light Abschatten *n*, Abschattung *f*

blocking period Sperrzeit, Blockier(ungs)zeit *f*

blocking protection Sperrschutz, Blockier(ungs)schutz *m*

blocking state region Sperrzone, Blockier(ungs)zone *f* [*Halbleiter*]

blocking voltage Sperrspannung, Blockier(ungs)spannung *f*

block input [*Automatic data processing. A block of computer words considered as a unit and intended or destined to be transferred from an internal storage medium to an external destination*] Blockeingabe *f*

block mark [*Automatic data processing. A special character that indicates the end of a data block*] Blockmarkierer, Datenblockmarkierer *m*

block (night) storage heater → night block storage heater

block of computer words → (computer) words block

block of grouped shops → shopping centre (Brit.)

block (of information) [*A group of information units (such as records, words, characters, or digits) that are transported or considered as a single unit by virtue of their being stored in successive storage locations; for example, a group of logical records constituting a physical record*] Block *m*

block security Blocksicherung *f* [*Fernwirktechnik*]

block-serial input, blockwise input blockweise Eingabe *f*

blocks interconnection → interconnection of blocks

block sorting [*Automatic data processing. A method of sorting a file, usually with punched card sorters, in which the file is first sorted according to the value of the digit in the highest digit position of the key; the resulting collections of records can next be sorted independently in smaller operations, and the separate sections then joined*] Blocksortieren *n*, Blocksortierung *f*

block switching diagram Übersichtsschaltplan *m* [*Vereinfachte Darstellung einer Schaltung mit ihren wesentlichen Teilen, die die Arbeitsweise und Gliederung einer elektrischen Vorrichtung zeigt*]

block transfer [*Automatic data processing. The movement of data in blocks instead of individual records*] Blockübertragung *f*

bloom(ing) Überstrahlung *f* (von weißen Punkten oder Flächen in benachbarte Bereiche) [*Unerwünschte Wirkung bei der Videoüberwachung*]

blooming Anlaufen *n* [*Oberfläche*]

B

blowback flap [*oil firing system*] Verpuffungsklappe *f*, Explosionsklappe

blowdown closure Ausblas(e)verschluss *m*

blowdown system Ausblas(e)system *n*

blowdown valve, purging valve, purge valve [*Device to allow liquid to flow out of a system*] Entwässerungsventil *n*

blower → (air) blower

blower air Gebläseluft, Bläserluft, Blasluft *f*

blower-assisted gebläseunterstützt, bläserunterstützt, blasunterstützt

blower boiler Gebläsekessel, Bläserkessel *m*

blower burner Gebläsebrenner, Bläserbrenner *m*

blower circuit Gebläseschaltung, Bläserschaltung *f*

blower control (switch) [*Temperature-actuated switch controlling only the on-off operation of an air-circulating blower*] Gebläseschalter, Bläserschalter *m*

blower convection Gebläsekonvektion, Bläserkonvektion *f* [*Konvektor mit Zwangsumwälzung*]

blower convector Gebläsekonvektor *m*, Bläserkonvektor [*Konvektor mit Zwangsumwälzung*]

blower door method Gebläsetürverfahren *n*, Gebläsetürmethode *f*, Bläsertürverfahren *n*, Bläsertürmethode *f* [*Ein in eine Außentür luftdicht eingebautes Gebläse erzeugt einen Druckunterschied zwischen dem Gebäudeinneren und der Umgebung um Lecks in der Gebäudehülle zu ermitteln*]

blower drier, blower dryer Gebläsetrockner, Bläsertrockner *m*

blower efficiency → blast efficiency

blower furnace [*See remark under "furnace"*] Gebläsefeuer(ungs)stätte, Gebläsefeuerung(sanlage), Bläserfeuer(ungs)stätte, Bläserfeuerung(sanlage) *f*

blower (gas) burner, gas blower burner Gasbrenner *m* mit Gebläse, Gebläsegasbrenner, Gasbläserbrenner, Bläsergasbrenner, Gasgebläsebrenner *m* [*DIN 4788, Teil 2*]

blower gas(-fired) furnace → gas(-fired) blower furnace

blowerless (gas) burner → atmospheric (gas) burner

blower motor Gebläsemotor, Bläsermotor *m*

blower shutoff, blower stop [*Approved manually reset, heat-actuated, thermostatic device installed in the duct system and arranged to shut off the supply of power to the blower whenever the temperature of the air at the point of installation reaches a point not in excess of 200° F. The device is non-adjustable above 200° F.*] Gebläseabsperrung, Bläserabsperrung *f*

blower unit heater [*Assembly of heating coils, finned tubes, or electric elements in a housing, but with quiet centrifugal-type fans to move air through the heating element*] Gebläseheiz(ungs)aggregat *n*, Gebläseheiz(ungs)einheit *f*, Bläserheiz(ungs)aggregat *n*, Bläserheiz(ungs)einheit *f*

blower wheel Gebläserad, Bläserrad *n*

blowing Blasen *n*

blowing through, purging, purge [*Removal of air from a water heating system*] Durchblasen *n*

blowlamp, soldering lamp, blowpipe lamp (Brit.); blowtorch (US) Lötlampe *f* [*DIN 8502*]

blown air geblasene Luft, Blasluft *f*

(blown) warm air hand dryer, (blown) warm air hand drier (Blas)Warmlufthändetrockner *m*

to blow off, to blow out abblasen, ausblasen

to blow off, to draw off ablassen, abblasen [*Schlamm bei der Kesselreinigung*]

blow-off → lift-off

blow-off coil, blowout coil [*A coil that produces a magnetic field in an electrical switching device for the purpose of lengthening and extinguishing an electric arc formed as the contact of the switching device part to interrupt the current*] Löschspule *f*

blow-off grill(e), blowout grill(e) (Aus)Blas(e)gitter, Abblas(e)gitter *n*

blow-off line, blowout line (Aus)Blas(e)leitung *f*

blow-off magnet, blowout magnet [*An electromagnet or permanent magnet used to deflect and extinguish the arc formed when a high-current circuit breaker or switch is opened*] (Aus)Blas(e)magnet, Abblas(e)magnet *m*

blow-off pipe, blowout pipe, relief pipe, purge pipe; bleeder [*deprecated*] [*A pipe attached to a gas main by which gas normally conveyed by the main can be discharged to the atmosphere*] (Aus)Blas(e)rohr, Abblas(e)rohr *n*

blow-off (pipe)line, blowout (pipe)line, relief (pipe)line, purge (pipe)line Abblas(e)(rohr)leitung, (Aus)Blas(e)(rohr)leitung *f*

blow-off pressure, blowout pressure Abblas(e)-druck, (Aus)Blas(e)druck *m* [*Dampf*]

blow-off recovery system, purge recovery system, purger, blowout recovery system [*System used to recover refrigerant from purged mix(ture) of gases and water*] Kältemittelrückgewinnungssystem *n*

blow-off safety valve → safety blow-off valve

blow-off tank → (steam) blow-off tank

blow-off valve → tap(ping) valve

blow-off valve, blowout valve [*Safety valve that maintains a predetermined pressure in a vessel by discharging excess gas or steam to the atmosphere*] Abblas(e)ventil, (Aus)Blas(e)ventil *n*

to blow off (water) → **to drain off (water)**

blow throw → **air throw**

blow-up → **bursting**

blow-up preventer, burst(ing) preventer Berstverhinderer *m*

blue flame Blauflamme *f*

blue jet technique dosierte Heiz(ungs)gaszugabe *f* an die Flamme

blue sky → **clear sky**

blue squares Blauraster *m*

blur → **lack of focus**

blurred image → **out-of-focus image**

BM(C) → **facility management**

BM(C)S information list → **building management system information list**

BMS → **building management system**

BO → **binary output**

board Platine *f* [*Auf ihr befinden sich alle Bauteile einer Datenverarbeitungsanlage in Form einer bestückten Leiterplatte*]

board Tafel *f*

boarded door → **batten door**

board information terminal integrierte Informations- und Kommunikationssystem *n*

board lamp-wire connector Platinenlüsterklemme *f*

board(-mounted) connector Steckverbinder *m* für Platinenmontage, Platinensteckverbinder *m*

board mounting Platinenmontage *f*

board of agrément, material(s) testing institute Materialprüfungsamt *n*, MPA, Materialprüfungsanstalt *f*, Materialprüfungsinstitut *n*

board test Platinenprüfung, Platinenprobe *f*, Platinenversuch *m*

bobbin Zugschnur, Ziehschnur *f* [*Kabelkanal*]

Bode diagram, frequency response diagram [*A diagram in which the phase shift or the gain of an amplifier, a servomechanism, or other device is plotted against frequency to show frequency response; logarithmic scales are customarily used for gain and frequency*] Bodediagramm, Frequenzkennliniendiagramm *n*

bod(il)y comfort → **comfort**

bod(il)y heat Körperwärme *f*

bod(il)y heat loss Körperwärmeverlust *m*

body current Körperstrom *m*

body gasket, body joint ring [*See remark under "gasket"*] Körper(ab)dicht(ungs)ring *m*, Körperflach(ab)dichtung *f*

body material Körper(werk)stoff *m*, Körpermaterial *n*

body model → **solid model**

body part Teil *m*, *n* [*Mitteilungsübermitt(e)lung*]

body part encryption indication Verschlüsselungsanzeige *f* über Mitteilungs-Teil

body scanner Körper(ab)taster *m*

body temperature Körpertemperatur *f*

body type Grundschrift *f* [*digitale Photosatz*]

to boil sieden

boiler → **water heater**

boiler Kessel, Heiz(ungs)kessel *m*

boiler accessories Kesselzubehör *n* [*selten: m*]

boiler base, boiler standing area Kesselstandfläche *f*

boiler base, boiler bottom Kesselsohle *f*, Kesselboden *m*

boiler battery, battery of boilers Kesselbatterie, Kesselreihe, Mehrkesselanlage, Vielkesselanlage *f*

boiler bearer, boiler cradle; boiler saddle (US) Kesselblock, Trag(e)block *m*

boiler blending pump Kesselbeimischpumpe *f*

boiler blower Kesselgebläse *n*, Kesselbläser *m*

boiler blow-off, boiler blowout [*Outlet on a boiler to permit emptying or discharge of sediment*] Kesselschlammablass, Kesselschlammauslass *m*

boiler body Kesselkörper *m*

boiler bracket; boiler lug (US) Kesselpratze, Trageplatte *f*

boiler building → **BH**

boiler burner Kesselbrenner *m*

boiler-burner set, boiler-burner kit, boiler-burner assembly Kessel-Brenner-(Bau)Satz *m*, Kessel-Brenner-Kombination, Kessel-Brenner-Gruppe *f*, Kessel-Brenner-Aggregat *n*, Kessel-Brenner-Garnitur *f*

boiler-burner unit Kesselbrennereinheit *f*

boiler capacity, boiler performance, boiler output Kesselleistung *f*

boiler circuit pump Kesselkreispumpe *f*

(boiler) clinker (Kessel)Schlacke *f*

boiler ((closed) loop) control → **((closed) loop) boiler control**

boiler compensation [*An operation which changes the operating temperature of a boiler usually according to the outside temperature*] Kesselausgleich *m*

boiler connection Kesselanschluss *m*, Kesselverbindung *f*

boiler control → **((closed) loop) boiler control**

B

boiler cradle → **boiler bearer**

boiler drain Kesseldrän *m*

boiler drainage cock Kesselentleer(ungs)hahn *m*

boiler efficiency Kesselwirkungsgrad *m* [*Quotient aus der an den Wärmeträger nutzbar abgegebenen Leistung und der zugeführten Leistung eines Kessels bezogen auf den unteren Heizwert H_u des Brennstoffes*]

boiler end plate Kesselrückwand *f*, Kesselrückblech *n*

boiler equipment, boiler outfit Kesselausrüstung(en) *f*

boiler family → **boiler series**

boiler feed(ing) Kesselspeisen *n*, Kesselspeisung *f*

boiler feed pump Kesselspeisepumpe *f*

boiler feedwater [*Water supplied to a boiler by pumping*] Kesselspeisewasser *n*

boiler fill(ing) cock Kesselfüll(ungs)hahn *m*

boiler fitting Kesselarmatur *f*, Kesselfitting *m*

boiler forward flow Kesselvorlauf *m*

boiler forward flow temperature Kesselvorlauftemperatur *f*

boiler foundation Kesselfundament *n*

boiler frame, boiler framing Kesselrahmen *m*, Kesselgerüst *n*

boiler furnace, steam boiler furnace, steam-generating furnace [*An enclosed space provided for the combustion of fuel to generate steam in a boiler*] Kesselfeuerung, Kesselfeuerstätte, Dampfkesselfeuerung, Dampfkesselfeuerstätte *f*

boiler heat Kesselhitze *f* [*Siehe Anmerkung unter „Hitze"*]

boiler heating area, boiler heating surface [*1. Portion of the surface of the heat transfer apparatus in contact with the fluid being heated and the gas or refractory being cooled, in which the fluid being heated forms part of the circulationg system. 2. The surface measured on the side receiving heat, including the boiler, water walls, water screens, and water floor*] Kesselheiz(ungs)fläche *f*

boiler horsepower [*It is equivalent to the evaporation of 34.5 lb of water per hour from 212° F or equal to a heat output of 970.3, 34.5 or 33.475 Btu/hour*] Kessel-PS *f*

boiler house → **BH**

boiler house pump sump Kesselhauspumpensumpf *m*

boiler jacket Kesselmantel *m*

boiler load(ing) Kesselbelastung, Kessellast *f*

boiler lug (US) → **boiler bracket**

boiler management system Kesselsteuerungssystem *n*

boiler material Kesselbaustoff, Kessel(werk)stoff *m*, Kesselmaterial *n*

boiler (mounting) room Kessel(aufstell)raum *m* [*VDI 2050*]

boiler ((open) loop) control, ((open) loop) boiler control Kesselsteuerung *f*

boiler operater → **stoker**

boiler pit Kesselgrube *f*, Kesselschacht *m*

boiler plant, boiler installation Kesselanlage, Kesselinstallation *f*

boiler plate Kesselblech *n*

boiler plug Kesselstopfen, Kesselstöpsel *m*

boiler power Kesselleistung *f*

boiler putty Kesselkitt *m*

boiler regulation → **((closed) loop) boiler control**

boiler return (flow) Kesselrücklauf *m*

boiler room → **boiler (mounting) room**

boiler saddle (US) → **boiler bearer**

boiler safety valve Kesselsicherheitsventil *n*

(boiler) scale (Kessel)Stein, Wasserstein *m*

(boiler) scale crust (Kessel)Steinschale, (Kessel-)Steinkruste, Wassersteinschale, Wassersteinkruste *f*

(boiler) scale formation, (boiler) scaling (Kessel-)Steinbildung, Wassersteinbildung *f*

(boiler) scale prevention (Kessel)Steinverhinderung, (Kessel)Steinverhütung, Wassersteinverhinderung, Wassersteinverhütung *f*

(boiler) scale prevention compound (Kessel)-Steinverhütungsmasse, (Kessel)Steinverhinderungsmasse *f*

(boiler) scale removal → **descaling**

boiler screw Kesselschraube *f*

boiler section Kesselglied *n*

boiler sensor Kesselfühler, Kesselsensor *m*

boiler series, boiler family Kesselbaureihe *f*

(boiler) shell plate (Kessel)Mantelblech *n*

boiler slag Kesselschlacke *f*

boiler standing area → **boiler base**

boiler steam Kesseldampf *m*

boiler supporting grid beam Kesseltragrostbalken *m*

boiler support structure, boiler substructure Kesselunterbau *m*

boiler switching system Kesselspeicherschaltfeld *n*

boiler temperature regulator, boiler thermostat Kesseltemperaturregler, Kesselthermostat *m*

boiler tube Sinderöhre *f* [*Sie liegt außerhalb des Kessels*]

B

boiler warmth Kesselwärme *f*

boiler water Kesselwasser *n*

boiler water flow temperature Kesselwasservorlauftemperatur *f*

boiler water level Kesselwasserstand *m*, Kesselwasserniveau *n*, Kesselwasserpegel *m*

boiler water takeoff Kesselwasserentnahme *f*

boiling Sieden *n*

boiling curve Siedekurve *f*

boiling delay Siedeverzug *m*, Siedeverzögerung *f*

boiling point, saturation temperature, bubble point [*Temperature at which the vapo(u)r pressure of a liquid equals the absolute external pressure at the liquid-vapo(u)r interface*] Siedepunkt *m*, Sättigungstemperatur *f*, Kochendpunkt *m*, Kochendtemperatur *f*

boiling storage water heater, storage boiling water heater Kochendwasserspeicher(gerät), Siedewasserspeicher(gerät) *m*, (*n*)

boiling water Kochendwasser, KoW, Siedewasser *n*

bold fett [*Zeichen*]

bold key Fett-Taste *f*

bollard luminaire Pollerleuchte *f*

bolometer, thermal detector, heat detector, thermal detecting device, heat detecting device, heat detection device, thermal detection device [*An instrument that measures the energy of electromagnetic radiation in certain wavelength regions by uti-lizing the change in resistance of a thin conductor caused by the heating effect of the radiation*] Wärmemelder *m*, Wärmemeldegerät *n*, thermische Melder *m*, thermische Meldegerät *n*, Thermomelder *m*, Thermomeldegerät *n*

to bolt verriegeln

to bolt verbolzen

bolt Falle *f*, Fallriegel *m* [*Fällt in das Schließblech, in die Schließkappe oder den Schließhaken ein und da-durch wird die Tür zugehalten*]

bolt Bolzen *m*

bolt Riegel *m*

bolt chisel → cross-cutting chisel

bolt clamp Riegelklampe *f* [*Innen auf dem Schlossblech aufgenietet*]

bolt clasp Riegelhaspel *f* [*selten: m*], Riegelhaken *m*

bolt contact Riegelkontakt *m*

bolt design Riegelkonstruktion *f*

to bolt down anbolzen

bolt eye Bolzenauge *n*, Schließenritze, Öse *f*

bolt handle Schubriegelgriff *m*

bolt hardware Riegelbeschlag *m*

bolt key, forelock Schließe *f*, Vorstecker, Splint, Bolzenkeil, Ansatz *m*

bolt lock Riegelschloss *n*

bolt lock Schließkappe *f*, Schließkloben *m*

bolt nab Schließblech *n* [*Blech mit Öffnungen, in welche die Schlossriegel eingreifen*]

bolt notch, notch of the bolt Einschnitt *m* am Riegel, Riegeleinschnitt *m* [*In die Einschnitte am Riegel greift die Zuhaltung ein*]

bolt of (a) door hinge, door hinge bolt Dorn *m* einer Türangel, Türangeldorn *m*

bolt-on connector Schraubverbinder *m*

bolt projection Riegelausschub *m*

bolt screw Bolzenschraube *f*

bolt shaft Riegelschaft *m*

bolt spring Fallenfeder, Riegelfeder *f* [*Feder in einem Schloss, welche die Falle dirigiert*]

bolt spring key → spring key of a bolt

bolt staple Schließhaken *m*

bolt telescopic lock → telescopic bolt lock

bolt with a handle → bent bolt

bolt without spring deutsche Falle, Fallklinke ohne Feder, federlose Fallklinke *f*

BOM, bill of material(s), material(s) bill Stückliste *f*

bomb attack Bombenanschlag, Bombenangriff *m* [*militärisch*]

bomb floor Bomben(schutz)decke *f*

bomb threat Bombendrohung *f*

Bommer (type) helical hinge Bommer-Band *n* [*Eigenname*]

to bonderize phosphatieren

bond(ing) Bindung *f*

bonding Bonden *n* [*Chipmontage in Gehäuse (Chipbonden) oder auf Chipträger und die Verbindung der Kontaktgebiete der Chips mit den nach außen führenden Kontakten des Gehäuses bzw. des Chipträgers (Drahtbonden)*]

bonding → keying-in

bond(ing) conductor [*IEE WR. A short conductor that earths the metal frame of electrical equipment*] Potentialausgleichsleiter *m*, PL

bond(ing) force, bond(ing) power Bindungskraft, Bindekraft *f*

bond(ing) strip → ground(ing) circuit (US)

bond(ing) wire Bonddraht *m*, Anschlussdraht [*Verbindung mit einem Teil eines Halbleiterelementes oder eines integrierten Schaltkreises, die durch Schweißen, Hartlöten oder Aufpressen unter Wärmeeinwirkung hergestellt wird*]

bondpad Chipkontaktgebiet *n*

B

bonnet → **hood**

bonnet air, hood air Haubenluft *f*

booked call (an)gemeldete (An)Ruf *m*

bookmark Lesezeichen *n*

book stack, bookshelf Buchregal, Bücherregal *n*

book theft protection Bücherdiebstahlschutz, Buchdiebstahlschutz *m*

boom [*Mechanical support for a microphone, used in a television studio to suspend the microphone within range of the persons and voices, but out of camera range*] Mikrophonanhänger, Mikrofonanhänger *m*

boom Blüte *f* [*Wirtschaft*]

to boost [*e.g. a signal*] verstärken

boost Verstärkung *f*

boost ((closed) loop) control Zusatzreg(e)lung *f*

boost converter Aufwärtsregler *m*

booster [*A machine which draws in air or gas which has already been compressed or is at pressure above that of the atmosphere and delivers it at a higher pressure*] Druckverstärker *m*

booster amplifier [*It is designed to boost the level of a signal from a preamplifier to provide power for driv-ing loudspeakers*] Zusatzverstärker *m*

booster compressor [*Machine for compressing air or gas from an initial pressure that is considerably above atmospheric pressure to a still higher pressure*] Verstärkerverdichter, Druckverdichter, Verstärkerkompressor, Druckkompressor, Zusatzverdichter, Zusatzkompressor, Verstärkungsverdichter, Verstärkungskompressor *m*

booster fan [*A fan used to stop up the static pressure in an air distribution system in order to serve a remote area, used only intermittently*] Verstärkerventilator, Zusatzventilator, Verstärkungsventilator *m*

booster heater, heating booster [*An auxiliary heating appliance fitted in a hot water system to raise the temperature of the hot water in a part of the system beyond its normal upper limit*] Verstärkerheizer, Verstärker(er)wärmer, Zusatzheizer, Zusatz(er)wärmer *m*

booster heating Zusatzheizung, Verstärkerheizung, Verstärkungsheizung *f*

booster relay Verstärkerrelais, Zusatzrelais, Verstärkungsrelais *n*

booster station [*See remark under "booster"*] Druckerhöhungsanlage, Druckverstärkeranlage, Druckverstärkungsanlage, Drucksteigerungsanlage *n*

booster transformer Zusatztransformator, Saugtransformator, Verstärkertransformator, Verstärkungstransformator *m*

boosting Verstärken *n*, Verstärkung *f*

boost(ing) period, preheat(ing) period [*The period immediately prior to the occupancy period during which plant is operated at its full rate capacity*] Vorwärmzeit *f*

boost ((open) loop) control Zusatzsteuerung *f*

bootleg unauthorisierte Tonträger *m*

to bootstrap laden durch Ureingabe, urladen [*Programm*]

bootstrap card Urladekarte *f*

bootstrap integrator, Miller generator [*A bootstrap sawtooth generator in which an integrating amplifier is used in the circuit*] Miller-Integrator *m*

bootstrapping, bootstrap loading Urladen *n*, Laden durch Ureingabe [*1. Eine Technik, beim Start eines Digitalrechners die ersten wenigen Befehle in den Speicher zu laden. Diese dienen dann dazu, ein um-fassendes Ladeprogramm von einem externen Speicher zu laden, das seinerseits den Betriebssystemkern einzulesen und zu starten gestattet. Früher war es nötig, die ersten Befehle von Hand an dem Bedienungspult einzugeben. Bei modernen Großrechnern steht dieses Urladeprogramm in einem Festprogrammspeicher zur Verfügung und ist durch Tastendruck zu aktivieren.*
2. Ein Technik, den Kompilierer für eine Programmiersprache von einer Rechenanlage auf eine andere zu übertragen, ohne dass er ganz neu geschrieben werden muss. Diese Technik ist anwendbar, wenn der Kompilierer in der eignen Sprache vorliegt. Nach Än-derung desjenigen Teiles, der den Maschinencode er-zeugt, wird die so geänderte Fassung mit der alten übersetzt, wobei ein Kompilierer für die Rechenanlage entsteht.]

bootstrap (progam) (US); bootstrap (programme) (Brit.); bootstrap loader Urladeprogramm, Urlader *m*, Vorlaufprogramm *n*, Rechnerstartprogramm [*Programm zum Starten eines Rechners. Es löscht meist den Hauptspeicher und ladet das Betriebssystem von einer Diskette oder Kassette.*]

bootstrap range Urladebereich *m* [*selten: n*]

boot-up Betriebssystemladen, Laden des Betriebssystems *n*

boot-up Rechnerstart, Rechnerbeginn *m*

BOP → **bit-orient(at)ed protocol**

borderline Grenzlinie *f*

bore [*Inside diameter of a hollow cylinder or of a hole*] Bohrung, Bohrweite *f*

borehole [*misnomer*] → **deep well**

boring pipe box Anbohrstelle *f*

borrowed air [*Room air drawn by fan or gravity ventilation from another room or space*] Fremdluft *f*

borrowed light [*Natural light that is brought from one space to another*] Fremdlicht *n*

borrowed water Fremdwasser *n*

BOS → bit-orient(at)ed signalling (Brit.)

Bos → basic operating system

boss branch Doppel-T-Stück, T-Doppel-Stück *n*

both sides locking lock Zweiseitenschloss, Doppelseitenschloss *n* [*Von zwei Seiten betätigte Schloss*]

bottle(d) gas, cylinder gas Flaschengas *n*

bottle (drain) trap, bottle stench trap, bottle U-bend, bottle fixture trap, bottle siphon trap, bottle water trap, bottle water stop, bottle water barrier [*See remark under "trap"*] Flaschen(wasser)(geruch)verschluss *m*, Flaschenwasservorlage *f* [*Die Benennungen „Flaschensiphon" und „Flaschentraps" sind veraltet. DIN 1986*]

bottle-trap(ped) outlet [*urinal*] Flaschenverschlussablauf *m*

bottom bolt Unterriegel *m* [*In die Schwelle greifender Schubriegel am stehenden (gewöhnlich geschlossenen) Flügel einer zweiflüg(e)ligen Tür*]

bottom (car) clearance [*Vertical distance between any obstruction in the pit, enclosure of the compen-sating device, buffer, and buffer supports, and the lowest point of the car understructure, exclusive of the safeties, car frame channels, guide shoes, and other accessory equipment attached to the underside of the platform, when the car floor is level with the bottom terminal landing*] Bodenfreihei, Fahrkorb-Bodenfreiheitt *f* [*Aufzug*]

bottom door closer Bodentürschließer *m*

bottom drainage Sohlenentwässerung *f*

bottoming cycles → organic Rankine cycles

bottom landing [*lift*] unterste Halt(estelle) *m, (f)*

bottom limit(ing) value → low limit(ing) value

bottom overtravel, bottom overshoot, bottom overrun [*Distance the car floor can travel below the level of the lower terminal landing until the weight of the car rests on the buffers; the resulting buffer compression is included*] untere Durchfahrt, untere Überfahrt *f*, untere Durchlauf(en) *m*, (*n*), untere Durchfahren, untere Überfahren *n* [*Aufzug*]

bottom plate, lower plate, cover plate Deckblech, Boden(blech) *m*, (*n*), Schlossblech *n*, Schlossdeckel *m*, Bodenplatte, Schlossplatte, Deckelplatte *f* [*Kastenschloss*]

bottom pressure Bodendruck *m* [*Er hängt von der Höhe des Flüssigkeitspegels über dem Gefäßboden und der Flüssigkeitsdichte ab*]

bottom sludge Bodenschlamm *m*

bottom swage, lower die Untergesenk *n* [*Schlosserei*]

bottom valve Bodenventil *n*

bottom view Unter(an)sicht *f* [*Unter der Vorder(an)sicht*]

bounce duration → duration of bounce

bounce-free, bouncing-free prellfrei

bouncing, bounce Prellen *n* [*Mehrfaches ungewolltes Betätigen eines Befehlsgebers bzw. falsche Eingaben über Tastaturen mit entsprechenden Fehlsignalen*]

bouncing suppression, bounce suppression Prellunterdrückung *f*

bound Begrenzung *f* [*Leiterplatte*]

bound Grenze *f* [*COBOL*]

boundary angle Randwinkel *m*

boundary layer [*Fluid mechanics. 1. A layer of fluid in the immediate vicinity of a bounding surface. 2. The region adjacent to the boundary dominated by shear stresses in a flow that is over a flat plate or through a pipe*] Grenzschicht, wandhohe Schicht *f* [*Rohrdurchfluss*]

boundary layer microphone → pressure zone microphone

boundary representation (method), B-Rep akkumulative Volumenmodellierung, Berandungsflächenmethode *f*, Berandungsflächenverfahren *n*, Berandungsdarstellung *f* [*Art der Volumenmodellierung, bei der die Volumina durch ihre Berandungsflächen definiert sind*]

boundary surfaces of the space Raumumschließungsflächen *f*

bounded difference begrenzte Differenz *f*, begrenzte Unterschied *m*, *t*-Norm *f* [*unscharfe Logik*]

bounded sum begrenzte Summe, *t*-Konorm *f* [*unscharfe Logik*]

boundless → infinite

bound variable gebundene Variable *f*

BO unit → binary output unit

Bourdon spring Röhrenfeder, Bourdonfeder *f* [*frühere Benennung*]

Bourdon tube, Bourdon (pressure) gauge; Bourdon pressure gage (US) [*Curved oval tube that tends to straighten when pressure is increased, and recurves with pressure decrease*] Röhrenfedermanometer *n*; Bourdonröhre *f* [*frühere Benennung*]

bow collector Bügelstromabnehmer *m*

bow key Rautenschlüssel, Ringschlüssel, Reifenschlüssel *m*

bowl, pan, basin Becken *n*

bowl → toilet bowl

bowl-fire Heizsonne *f*

bowl sink Beckenspüle *f*

bowl urinal → urinal bowl

bow of a key, key bow Schlüsselring, Schlüsselreif *m*

box Quader *m* [*Objektentwurf*]

box → (lock) box

box Tabellenoption *f* des Dokumentenaufbereitungskommandos

box air cooler Kastenluftkühl(ungs)gerät *n*

box-conveying belt Kastenförderband *n*

box-conveying continuous lift, box-conveying paternoster (lift) (Brit.); box-conveying continuous elevator, box-conveying paternoster (elevator) (US) Kastenpaternoster(aufzug), Kastenumlaufaufzug *m*

box-conveying plant, box-conveying installation Kastenförderanlage, Kastenförderinstallation *f*

box(ed) bolt nab → case(d) bolt-nab

box(ed) (bolt-)staple → case(d) bolt-staple

box(ed) nab → cased bolt-nab

box(ed) (roof) gutter, trough (roof) gutter, parallel (roof) gutter, parapet (roof) gutter [*A roof gutter of rectangular cross section used in valleys or parapets. The gutter may be made of wood and lined with flexible materials or bituminous materials*] Kasten(dach)rinne, Dachkastenrinne *f*

box equipment Kastengerät(e) *n, (f)*

box frame Gehäuse *n*, Kasten *m* [*Hinter den Fensterpfosten für die Rolladengewichte*]

box gutter Kasten(dach)rinne *f*

box heat pump Kastenwärmepumpe *f*

box lid Kastendeckel *m*

box of a hinge → hinge box

box staple → bolt lock

Boyle('s) law, Mariotte('s) law [*Product of the volume of a gas times its pressure is a constant at fixed temperature*] Boyle-Mariotte-Gesetz, Gesetz *n* von Boyle-Mariotte [*Robert Boyle, englischer Physiker und Chemiker 1627–1691; Edmé Mariotte, französischer Physiker 1620–1684*]

bp → boiling point

BPF → band pass filter

bpi, bits per inch, BPI Bits *f* je Zoll [*Maß für die Aufzeichnungsdichte, z. B. auf magnetischen Datenträgern; 1 Zoll = 25,4 mm*]

BR → brine return

bracket → (wall) bracket

bracket eckige Klammer *f*

bracket(-mount)ed, wall bracket(-mount)ed konsolenmontiert

brackish water [*Impure water with a lesser content of salt than seawater, but higher than that of potable water*] Brackwasser *n* [*Salz-/Süßwasser-Gemisch*]

braid Geflecht *n*

braided flexible cord Gummiaderschnur *f* [*DIN VDE 0282 Teil 801*]

braiding, hose braiding (Schlauch)Umspannung *f*

brainware Planungsaufwand *m* [*Realisierung einer Datenverarbeitung*]

brake horsepower, shaft horsepower Brems-PS *f*

brake releasing Bremslüftung *f*

brake (releasing) magnet Bremslüft(ungs)magnet *m*

braking controller → decelerator

braking radiation Bremsstrahlung *f*

Bramah('s) (detector) lock, (detector) lock of Bramah Bramahschloss *n*

branch [*Section of pipe from a main to a register or radiator*] Stutzen *m*

branch [*In ducts, pipings, or conduits; another section at an angle with the main*] (Ab)Zweig *m*

branch [*Portion of a network consisting of one, two or more terminal elements in a series*] (Ab)Zweig *m*

branch [*Computer. A set of instructions that are executed between two successive decision instructions*] Strang *m*

branch [*A transfer of control from one line of code to one other than the next logical instruction. Branches may be conditional, based on some defined circumstance, or unconditional*] Sprung *m*

branch (air) duct (Ab)Zweig(luft)kanal *m*

branch and junction box Richtungsänderungskasten *m*

branch bend [*A channel bend for building into the invert of manhole to receive a branch pipe*] (Ab-)Zweigbogen *m*

branch circuit Strom(ab)zweig *m*

branch-circuit panel (Ab)Zweigfeld *n*

branch destination → branch target

branch distributor [*See remark under "branch"*] Sprungverteiler *m*

branch duct (Ab)Zweigkanal *m*

branch fitting [*A fitting used to connect one or more branch pipes or channels to a main pipe or channel*] (Ab)Zweigpassstück *n*, (Ab)Zweigfitting *m*

branching Abzweigen *n*, Abzweigung *f*, Verzweigen *n*, Verzweigung *f*

branching instruction, jump instruction [*See remark under "instruction"*] Sprungbefehl *m*, Sprunginstruktion *f*

branch(ing) point Verzweigungsstelle, (Ab)Zweigstelle *f*, Verzweigungselement, (Ab)Zweigelement *n*

branch line, branch-off (Ab)Zweigleitung *f*

branch-off → branch line

branch overflow; branch nozzle (US) Überlaufstutzen *m*

branch prediction unit Verzweigungsvoraussagepuffer, (Ab)Zweigvoraussagepuffer *m*

(branch) service pipe [*The whole of the pipe (communication pipe + supply pipe = Endstrang + Grundstückstrang) between the main and customers premises*] (Ab)Zweigrohr, Stichrohr, Verzweigungsrohr *n*

branch sewer [*A sewer that receives sewage from a relatively small area*] (Ab)Zweigabwasserleitung *f*

branch switch (Ab)Zweigschalter *m*

branch target, branch destination [*See remark under "branch"*] Sprungziel *n*

branch terminal (Ab)Zweigklemme *f*

branch transmittance, branch gain [*The amplification of current or voltage in a branch of an electric(al) network; used in the representation of such a network by a signal-flow graph*] (Ab)Zweiggewinn *m*

branch water (pipe)line (Ab)Zweigwasserleitung *f*

brand mark Firmenkennzeichen *n*, Firmenmarke *f*

brashy [*wood*] spröde

brassware, brasswork Messingware(n) *f*

Braun tube → **cathode-ray tube**

to braze, to hard solder hartlöten, messinglöten

brazed, hard soldered hartgelötet, messinggelötet

brazed joint [*Gas-tight joint obtained by the joining of metal parts with alloys that melt at temperatures higher than 1000° F, but less than the melting temperatures of the joined parts*] Hartlötnaht, Messing-lötnaht *f*

brazing, hard soldering [*Joining by fusion using a spelter of brass*] Hartlöten, Messinglöten *n*, Hartlötung, Messinglötung *f*

breadth [*deprecated*] → **width**

breadth-first-search, BFS Breitendurchlauf *m* [*Durchlaufen eines Graphen bei dem jeder Knoten einmal erfasst wird*]

break → **interruption**

breakaway starting current → **locked-rotor current**

break-before-make contact [*One of a pair of contacts that interrupt one circuit before establishing another*] Umschaltkontakt *m* mit Unterbrechung, Unterbrechung(s)umschaltkontakt *m*

break circuit → **closed circuit**

break circuit contact → **closed-circuit contact**

break contact element [*IEC 337-1*], **b-contact element, normally closed contact, NC contact, ncc** Öffnungskontakt, Trennkontakt, Öffner *m*, Ausschaltglied *n* [*VDE 660, Teil 200*], Ruhekontakt [*frühere Benennung*] [*Schaltstück eines elektrischen Kontaktes, das beim Betätigen öffnet und im unbetätigten Zustand geschlossen ist. Gegensatz: Schließer*]

break contact opening time Öffneransprechzeit *f* [*monostabile Relais. DIN IEC 255, Teil 1-00*]

break cost(s) → **interruption cost(s)**

to break down → **to fail**

to break down aufgliedern, aufschlüsseln

breakdown Aufgliederung, Aufschlüss(e)lung *f*

breakdown Panne *f*, Ausfall *m*

breakdown → **fault**

breakdown diode, limiter diode [*Electronics. A Zener diode, avalanche diode, or any other semiconductor diode having a reverse-voltage breakdown mechanism based on either Zener or the avalanche effect*] Begrenzerdiode *f*

breakdown factor relative Kippmoment *n* [*elektrische Maschine*]

breakdown fuse, breakdown fuze Überspannungssicherung *f*

breakdown slip Kippschlupf *m*

breakdown voltage, Zener voltage, BV [*Electronics. The voltage produced at a particular current in the electrical breakdown region of a Zener diode*] Zenerspannung *f*

breaker arc-back Schalterrückzündung *f*

breaker interripping Schaltermitnahme *f*

breaker mechanism → **latching mechanism**

break frequency → **corner frequency**

break glass call point Einschlagbrandmelder, Einschlagfeuermelder *m*, Einschlagbrandmeldegerät, Einschlagfeuermeldegerät *n*, Glasbruchmelder *m*, Glasbruchmeldegerät *n*

break glass emergency cutoff switch Einschlagscheibe-Notausschalter *m*

to break in aufbrechen, gewaltsam eindringen, einbrechen

to break in einspielen [*Musik in eine elektronische Anlage*]

break(ing) → **cutout**

break(ing) arc Ausschalt(licht)bogen *m*

break(ing) capacity → **((closed) loop) break(ing) capacity**

break(ing) current Ausschaltstrom *m*

break(ing) current → **((closed) loop) break(ing) current**

break(ing) element → **break(ing) unit**

break(ing)-in → **burglary**

break(ing)-in... → **burglary...**

break(ing)-in relay Tastpausen-Empfangsrelais *n*

break(ing) key Abbruchtaste *f*

break(ing) through, cutout Durchbruch *m*

B

break(ing) unit, break(ing) element Ausschaltelement *n*

breaking up of structures → decomposition

break-in operation [*A method of radio communication in which it is possible for the receiving operator to interrupt or break into the transmission*] BK-Verkehr *m*

break of occupation, interruption of occupation Belegungspause, Belegungsunterbrechung *f* [*Gebäude*]

break of use, interruption of use, break of utilization, interruption of utilization Nutzungspause, Nutzungsunterbrechung *f*

to break out ausbrechen

breakpoint, interruption point, breakthrough (An)Haltepunkt, Unterbrechungspunkt, (Ablauf)Unterbrechungspunkt, Programm(ablauf)unterbrechungspunkt *m* [*Eine Stelle im Programm, an der vor Ausführung des dort gespeicherten Befehls der Programmablauf unterbrochen wird*]

breakpoint instruction, interruption instruction (An)Haltebefehl, (Ablauf)Unterbrechungsbefehl *m* [*Rechner*]

break setpoint, interruption setpoint (An)Haltesollwert, (Ablauf)Unterbrechungssollwert *m*

break setting, interruption setting Unterbrechungseinstellung, (An)Halteeinstellung *f*

break signal, interruption signal (An)Haltesignal, Unterbrechungssignal *n*

breakthrough → breakpoint

breakthrough, opening Durchbruch *m*, Durchführung *f* [*Durch eine Decke, Wand, Mauer oder (Fuß-) Boden*]

break value Speichergrenzwert, Grenzspeicherwert *m*

breast induction unit → spandrel (wall) induction unit

breathable, respirable atembar

to breathe, to respire atmen

breathing, respiration Atmen *n*, Atmung *f*

breathing air stream, respiration air stream Atemluftstrom, Atmungsluftstrom *m*

breathing valve, breather, respiration valve Atmungsventil *n*

breeches pipe, Y piece, trouser piece, trouser pipe, Y pipe, breeches piece Gabelrohr, Gabelstück, Hosenrohr, Hosenstück *n*, Rohrgabel *f*

breeching (US) → principal flue

breeder reactor Brutreaktor *m*

B register → index register

B-Rep → boundary representation (method)

brevity code [*It has as its sole purpose the shortening of messages rather than the concealment of their contents*] (Ab)Kürzungskode, (Ab)Kürzungscode *m*

brick chimney → brick fire(place)

brick chimney Ziegelschornstein, Ziegelkamin *m*

brick croft, brick yard Ziegelhof, Ziegelplatz *m*

brick fire(place), brick fireside, brick (grate) chimney [*See remark under "chimney"*] Ziegel(zimmer)kamin, Zimmerziegelkamin *m*

brick (grate) chimney → brick fire(place)

brick quarter → closer

brick yard → brick croft

bridge [*A device which connects two or more segments at the physical and data link layers. This device may also perform message filtering based upon MAC layer addresses*] Brücke *f*

bridge, interconnect(ing) server Brücke *f* [*Ein Gerät, mit dem der Zusammenschluss mehrerer LANs hergestellt wird, wobei die Kommunikation zwischen Datenendgeräten mit gleichen Protokollen an verschiedenen Netzwerken zustande kommt.*]

bridge amplifier Brückenverstärker *m*

bridge circuit Brückenschaltung *f*

bridge comparator [*See remark under "comparator"*] Brückenvergleicher *m*

bridge converter connection Stromrichterbrückenschaltung *f*

bridged überbrückt

bridge feedback Brückenrückkopp(e)lung *f*, Rückkopp(e)lung über Brückenschaltung

bridge hybrid → hybrid junction

bridge module Brückenbaustein *m*, Brückenmodul *n* [*Siehe Anmerkung unter „Baustein"*]

bridge rectifier Brückengleichrichter *m*

bridge resistor, bridge electrical resistor Brückenwiderstand(sgerät) *m, (n)*

bridge ward [*lock*] Eingerichte *n*

bridging Überbrückung *f*, Überbrücken *n*

Brigg's logarithm, common logarithm [*The exponent in the representation of a number as a power of 10*] Zehnerlogarithmus, dekadische Logarithmus, Logarithmus 10 *m*

bright chrome coated glanzverchromt

brightness [*obsolete*] **→ luminance**

brilliance [*The degree of brightness and clarity of the display of a cathode-ray tube*] Wiedergabebrillanz *f*

brilliance [*The degree to which the higher audio frequencies of an input sound are reproduced by a radio receiver, by a public address amplifier, or by a sound-recording play-back system*] helle Klangfarbe *f*

brine [*deprecated*] → **secondary coolant**

brine [*An aqueous solution of sodium chlorid(e) (= common salt), or calcium chlorid(e)*] Sole *f*, kochsalzhaltige Wasser *n*

brine/air heat pump Sole/Luft-Wärmepumpe *f*

brine coil Soleschlange *f*

brine cooler Solekühler *m*

brine cycle, brine circuit [*See remark under "cycle"*] Solekreislauf *m*

brine flow, BF Solevorlauf *m*

brine return, BR Solerücklauf *m*

brine tank Solebehälter, Soletank *m*

brine/water heat pump Sole/Wasser-Wärmepumpe *f*, S-W-WP [*Siehe Anmerkung unter „Wärmepumpe"*]

brise-soleil, pare-soleil [*French. A system of sunscreens, such as horizontal or vertical slats, that are fixed to the outside of a building*] (Sonnen)-Blendsystem *n*

brittle spröde

brittle fracture Sprödbruch *m*

brittle mica Sprödglimmer *m*

brittleness Sprödigkeit *f*, Sprödheit

broach, lock broach (Schloss)Dorn *m*

broad band, wide band [*As applied to data transmission, the terms denote transmission facilities capable of handling frequencies greater than those required for high-grade voice communication*] Breitband *n*

broad-band amplifier, wide-band amplifier [*It has an essentially flat response over a wide frequency range*] Breitbandverstärker *m*

broad-band feedback amplifier, wide-band feedback amplifier Breitbandrückkopp(e)lungsverstärker *m*

broad-band integrated services digital network → **B-ISDN**

broadbeam → **widebeam**

broadcast [*A television or radio transmission intended for public reception*] Sendung *f*

broadcast Nachricht *f* an alle Benutzer, Nachricht an alle Anwender, Mitteilung *f* an alle Benutzer, Mitteilung an alle Anwender, Information *f* an alle Benutzer, Information an alle Anwender, allgemeine Mitteilung, allgemeine Nachricht, allgemeine Information, Sammelaufruf *m*, Sammelnachricht [*Das Senden eines Signals von einer LAN-Station an alle anderen angeschlossenen Stationen, wobei alle Teilnehmer das gleiche Signal empfangen*]

broadcast band, standard broadcast band [*The band of frequencies extending form 535 to 1605 kilohertz, corresponding to assigned carrier frequencies that increase in multiples of 10 kHz between 540 and 1600 kHz for the United States of America*] Normalwellenband *n*, Normalwellenbereich, Rundfunkband, Rundfunkbereich *m* [*selten: n*]

(broadcast) station Sendestation Videotext *m*, Fernsehtext

(broadcast) videotext, television text Videotext *m*, Fernsehtext

broken chain device [*Escalator. It interrupts the power supply and applies the main drive brake in the event of one or both of the step chains break(s)*] Kettenbruchsicherung *f*

broken coke Brechkoks *m*

broken hinge(-band), double(d) hinge(-band) gebrochene Scharnierband *n*, gebrochene (Lappen)Band, gebrochene Gelenkband, Wiedergelenkband, Band mit Wiedergelenk

bronze plated bronziert

brown coal coke range Grude(koks)herd *m*, Braunkohlekoksherd

brush [*A conductive metal or carbon block used to make sliding electrical contact with a moving part*] Bürste *f*

brush(ed) fireproof treatment Anstrich-Feuerschutzbehandlung *f*

brush filter [*Air filter in which the filter medium consists of a screen of brushes*] Bürstenfilter *m, n*

brush interlock(ing) Bürstenverrieg(e)lung *f*

BS → **base station**

BSC → **base station controller**

BSC → **binary synchronous communication(s)**

BSI → **building supervisor integrated**

BSI technique → **building supervisor integrated technique**

BSM → **backward set-up message**

b-spline → **basic spline**

BSS → **base station subsystem**

B store → **index register**

BTS → **base transceiver station**

BU → **bus interface (unit)**

bubble, gas bubble (Gas)Blase *f*

bubble, liquid bubble (Flüssigkeits)Blase *f*

bubble (computer) memory → **(magnetic) bubble (computer) memory**

bubble flow, emulsion flow [*Form of two-phase flow where the liquid carries finely dispersed bubbles of gas*] Blasenströmung *f*, Blasenfließen *n*

bubble jet → **(ink-)jet**

bubble memory → **(magnetic) bubble (computer) memory**

bubble point → **boiling point**

bubble sprayer → **(ink-)spray plotter**

B

to buck absetzen [*Spannung*]

buck ((closed) loop) control Absetzreg(e)lung *f* [*Spannung*]

bucket Behälter *m* [*Zahlenfolge(n)sortierung*]

bucket bridge device, BBD, charge-coupled device, CCD Eimerkettenschaltung *f*, ladungsgekoppelte Bauelement *n* [*Elektronische Schaltung zur Speicherung und Weiterleitung von analogen oder digitalen Signalen nach dem Ladungstransportprinzip (CCD) aus der Gruppe der Ladungsverschiebeschaltungen*]

bucket sink, housemaid('s) sink, chambermaid('s) sink [*A sink fixed at or near floor level to facilitate the filling and emptying of buckets. It usually has a protective strip on the top of the front edge to prevent damage and a grid on which buckets can be rested. It may have a high splashback*] Spülausguss *m*

bucket sort Zahlenfolge(n)sortierung *f*

buck-tub Spülbank *f*, Waschstein (in einer Küche), Küchenspülbank, Küchenwaschstein *m*

to buffer [*data*] puffern [*Daten*]

bubble → **buffer solution**

buffer [*elevator car*] Aufsetzpuffer *m*

buffer Puffer(speicher) *m*, Zwischenspeicher [*Ein Speicher, der Daten vorübergehend aufnimmt, die von einer Funktionseinheit zu einer anderen übertragen werden. E DIN 44 300 T 6*]

buffer address [*The address of a location in a buffer*] Pufferadresse *f*

buffer amplifier [*It is used primarily to prevent changes of impedance in the circuit connected to its output from affecting the circuit connected to its input*] Pufferverstärker *m*

buffer(ed) computer [*A computer having a temporary storage device to compensate for differences in transmission speeds*] Pufferrechner *m*

buffer(ed) device [*A peripheral device, such as a printer or teleprinter, that is capable of serving as a buffer*] Puffergerät *n*

buffer(ed) I/O channel, buffer(ed) input/output channel [*A control channel for most peripheral de-vices that has associated temporary storage areas for input and output data; the channel controls the in-put/output operation(s), freeing the CPU for other processing*] Puffereingabe/Pufferausgabe-Kanal *m*

buffer(ed) terminal [*Any of a variety of user terminals that can accommodate differing data transmission rates by temporarily storing incoming data*] Pufferbenutzerstation, Benutzerpufferstation *f*

buffering [*telegraphy*] Geschwindigkeitsanpassung *f*, Puffern, Zwischenspeichern *n*

buffer management Pufferverwaltung *f* [*Speicher*]

buffer overflow Pufferüberlauf *m*

buffer (solution) Pufferlösung *f*

buffer zone [*An area of main memory set aside for temporary storage*] Pufferzone *f*

bug → **error**

bugging → **listening**

bug(ging device) → **listening device**

BU(I) → **bus interface (unit)**

to build [*To increase in received signal strength*] (sich) einschwingen, (sich) aufschaukeln

builder Hochbauer *m*

(builder) mason [*Scotland and Cornwall*] → **fixer (mason)**

builders' finish(ing) hardware → **finish(ing) (builders') hardware**

builders' iron mounting → **iron (builders') mounting**

(builders') iron supplies, (builders') ironmongery, (builders') iron furniture, architectural ironmongery [*Also loosely called "(builders') hardware"*] Kleineisenwaren *f*, Kleineisen(zeug) *n* [*Das zum „Eisen für Bauvorhaben" gehörende Eisen, das in Verbindung mit Baueisen oder auch einzeln in Bauvorhaben aufgeht: Beschlagteile, Nägel, Schrauben, Bolzen, Anker, Haken, Draht usw.*]

builders' mounting → **mounting**

building, facility Gebäude *n*, Objekt *n*

building access manager Gebäudeanbindungsbetreuer *m*

building accessories → **accessories of a building**

building acoustics Gebäudeakustik *f*, Gebäudeschalllehre *f*, Gebäudeschallkunde *f*

building air Gebäudeluft *f*

building air distribution Luftverteilung *f* im Gebäude [*Art der Luftleitungsführung einer raumlufttechnischen Anlage im Gebäude*]

building airtube conveyor, building airtube system, building pneumatic conveyor, building pneumatic system, building carrier tube conveyor, building carrier tube system Gebäuderohrpost *f* [*DIN 6651*]

building alarm system Gebäudealarmsystem *n*

building (and energy) management (and control) → **facility management**

building appurtenances → **accessories of a build-ing**

building area [*The maximum horizontal projected area of a building at or above grade, including all enclosed extensions*] Gebäudefläche *f*

building automatic system controller, BAS controller Gebäudeautomationssystemregler *m*

building automation, BA, building services automation Gebäudeautomation *f*, GA [*DIN 276 (1993), DIN V 32 734, GAEB LB071 Oberbegriff für die Summe aller Einrichtungen und der Software zur automa-tischen Überwachung, Steuerung, Reg(e)-lung und Betriebsoptimierung von Anlagen in einem oder mehreren Gebäuden; incl. Schaltschränken, Verkabelung der MSR-Technik sowie der Informationsübertragung. Die digitale Gebäudeautomation umfasst zusätzlich die zugeordneten Leit- und Managementebenen, mit weiteren Einrichtungen die aus der Sicht verteilter Unternehmen lokal, regional oder zentral angeordnet sind. Nicht zur GA gehören die „informationstech-nischen Gewerke" wie Gefahrenmeldeanlage (GMA = Brand, Einbruch), Zutrittskontrolle, Kamera-/Videoüberwachung (CCTV) sowie Personensucheinrichtungen (PSE). Diese Systeme können mit der GA kombiniert werden.*]

building automation and control network → BACnet

building automation controller Gebäudeautomationsregler *m*

building automation station Gebäudeautomationsstation, (Gebäude)Leitzentrale *f* [*frühere Benennung*]

building automation system, BAS, facility automation system, FAS Gebäudeautomationssystem *n*, GAS, Gebäudeleittechnik *f*, GLT [*frühere Benennungen*] [*Zusammenfassung der Automationseinrichtungen von gebäude- und sicherheitstechnischen Anlagen und Prozessen mittels entsprechender Kommunikationseinrichtungen zu einem System für die Betriebsführung in einem von mehreren Gebäuden*]

building biology, construction biology Baubiologie *f*

building bus engineering → bus engineering

building cabling Gebäudeverkabelung *f*

building carrier tube conveyor → building air-tube conveyor

building cavity Gebäudehohlraum *m*

building chimney Kamin, Gebäudeschornstein *m*

building climate Gebäudeklima *n*

building component → component

building connection Gebäudeanschluss *m*, Gebäudeverbindung *f*

building connection line Gebäudeanschlussleitung, Gebäudeverbindungsleitung *f*

building contract law, construction contract law Bauvertragsrecht *n*

building control → building (services) control

building control and monitoring engineering → building (monitoring and) control engineering

building controller Gebäuderegler *m*

building control system → building (services) control system

(building) conversion (Gebäude)Umbau *m*

building core → service core

building craft → building trade

building current supply → building power supply

building damage, construction damage Bauschaden *m*

building data Gebäudedaten *f*

building discharge pipe, building draining pipe, building drain(age) pipe [*A pipe which conveys the discharge from a sanitary appliance or appliances. It may also convey rainwater*] Gebäudeablaufrohr, Gebäudeabflussrohr, Gebäudeentwässerungsrohr *n*

building distribution board [*power engineering*] Gebäudeverteiler *m*, Gebäudeverteilungstafel *f*

building distribution (pipe) main → distribution (pipe) main

building door lock Gebäudetürschloss *n*

building drain Gebäudedrän *m*

building drainage Gebäudeentwässerung *f*

building dust, construction dust Bau(stellen-)staub *m*

building electric power system [*This system includes service equipment, switchboards, bus and heavy feeders, distribution panels, and motors and their control*] Gebäudeelektrosystem *n*

building electronics, construction electronics Bauelektronik, Elektronik im Bauwesen *f*

building electronics Gebäudeelektronik *f*

building electronic system Gebäudeelektroniksystem *n*

building energy consumption Gebäudeenergieverbrauch *m*

building energy monitoring Gebäudeenergieüberwachung *f*

building energy performance Gebäudeenergieleistung *f*

building energy saving(s) Gebäueenergieeinsparung(en) *f*

building engineer, construction engineer Bauingenieur *m*

building engineer → building (services) engineer

building engineering → building (services) engineering

building engineering, construction engineering, civil engineering Ingenieurbau *m*

building envelope Gebäudehülle *f*, Außenhaut *f*

building envelope monitoring Gebäudehülleüberwachung, Außenhautüberwachung *f* [*Überwachung von Gebäudeöffnungen auf Öffnen, Durchbruch und Eindringung*]

B

building envelope security Gebäudehüllesicherung, Außenhautsicherung *f* [*Mechanische Sicherungsmaßnahmen an Gebäuden mit einem Widerstands(zeit)wert, z. B. Gitter, Fenstersicherungen, Türsicherungen*]

building evacuation Gebäuderäumung, Gebäudeevakuierung *f*

(building) final circuit (Gebäude)Endstromkreis, Verbraucherstromkreis *m* [*VDE 0100, Teil 200*]

building fire Gebäudebrand *m*, Gebäudefeuer *n*

building grid Gebäuderaster *m*

building heat ... → building thermal heat ...

building heater, construction heater [*Portable oil-fired, coal-fired, coke-fired or electric appliance intended for temporary space heating during construction*] Bautrocknungsofen *m*

building heating forward flow Gebäudeheiz(ungs)-vorlauf *m*

building heating return (flow) Gebäudeheiz-(ungs)rücklauf *m*

building heating water Gebäudeheiz(ungs)wasser *n*

building illness syndrome → building related illness

building infrastructure Gebäudeinfrastruktur *f*

building installation Gebäudeinstallation *f*

building installation engineering Gebäudeinstallationstechnik *f* [*Mit dieser Benennung wird manchmal auch die Elektroinstallationstechnik bezeichnet*]

building integration Gebäudeintegration *f*

building interior Gebäudeinnere *n*

building inventory → building stock

building load profile Gebäudelastprofil *n*

building lock Gebäudeschloss *n*

building maintenance Gebäudeunterhaltung *f*

building management → facility management

building management product, facility management product Gebäudebetriebserzeugnis, Gebäudebetriebsprodukt *n*

building management system, BMS, facility management system, FMS Gebäudebetriebssystem *n* [*Umfassendes System für technische, kaufmännische und infrastrukturelle Gebäudebetriebsführung, „Objekt" kann synonym für „Gebäude" verwendet werden*]

building management system information list, facility management system information list, BM(C)S information list Gebäudebetriebssystem-Informationsliste *f*

building manager, (facility) manager Gebäudeverwalter, (Gebäude)Bewirtschafter, Gebäudebetreuer *m*

building metering → metering

building method, construction method Bauweise *f*

building micro-climate Gebäudemikroklima *n*

building monitoring Gebäudeüberwachung, Gebäudeaufsicht *f* [*Sicherheit*]

building monitoring, construction monitoring Bauaufsicht, Bauüberwachung *f*

building (monitoring and) control engineering, building control and monitoring engineering, building master control (and monitoring) engin-eering Gebäudeleittechnik *f*, GLT [*VDI 3814 Blatt 1 (1990), mit Blatt 2 von 3/93 geändert in Gebäude-automation. Bei der GLT mit der „potentialfreien" Klemmleiste bleibt die Selb(st)ständigkeit der einzelnen Anlagen BTA hinsichtlich ihrer Funktion erhalten, wogegen bei der digitalen Gebäudeautomation nach DIN 276 und DIN V 32 734 diese funktionale Selb(st)-ständigkeit nicht mehr gegeben ist, es sei denn, die BTA hat eine eigenständige MSR-Technik mit einer standardisierten Schnittstelle (z. B. gem. VDI 3814 Blatt 2) zur Gebäudeautomation.*]

building (monitoring and) control) system, build-ing control and monitoring system, building master control (and monitoring) system, facility integrator, facility integrating system Gebäudeleitsystem *n*

building occupancy Gebäudebelegung *f*

building operation Gebäudebetrieb *m*

(building) operator (Gebäude)Betreiber *m* [*Verantwortlicher für das Überwachen und Bedienen der technischen Anlagen mittels Gebäudeautomationssystem sowie zur Auswertung der Informationen*]

building-out network [*Network designed to be connected to a basic net(work) so that the combination will simulate the sending-end impedance, neglecting dissipation, of a line having a termination other than that for which the basic net(work) was designed*] Ergänzungsnetz(werk) *n*

building owner Bauherr(schaft) *m, (f)*

building owner Gebäudeeigentümer, (Gebäude-) Betreiber *m* [*Fehlbenennung*]

building physicist, construction physicist Bauphysiker *m*

building (pipe)line Gebäude(rohr)leitung *f*

building planner → facility planner

building planning, facility planning Gebäudeplanung *f*

building pneumatic conveyor → building airtube conveyor

building population Belegung(sstärke) *f* [*Gebäude*]

building possessor Gebäudebesitzer *m*

building power supply, building current supply Gebäudestromversorgung, Gebäudeelektrizitätsversorgung *f*

building preservation Gebäudeerhalt(ung) *m, (f)*

building related illness, sick building syndrome, SBS, building illness syndrome, tight building syndrome Gesundheits- und Behaglichkeitsbeschwerden *f*

building research, construction research Bauforschung *f*

building research centre, construction research centre (Brit.) [US spelling is "center", "station" may synonymously be used for "center"] Bauforschungsanstalt, Bauforschungszentrum *f*

(building) rubbish Gebäudemüll *m*

building safety Gebäudesicherheit *f*

building sanitation (engineering) Gebäude-Sanitärtechnik, Gebäude-Gesundheitstechnik *f*

building security Gebäudesicherung *f* [*Siehe Anmerkung unter „Sicherung"*]

(building) service chute [*A vertical or inclined tube or channel that conveys and controls the fall of ob-jects such as mail, laundry, and garbage to a lower level*] Betriebsrutsche *f*

building service connection Gebäudeanschluss *m*, Gebäudeverbindung *f*

building service (pipe)line Gebäudeanschluss(rohr)leitung *f*

building services, installation(s) Versorgungsausrüstung, (technische) Gebäudeausrüstung, (technische) Hausausrüstung, (T)GA, (T)HA, Haus(technik)anlagen, Installation, haustechnische Anlagen, gebäudetechnische Anlagen, (technische) Gebäudeausstattung, Technikbauten, betriebstechnische Anlagen, BTA *f* [*Heizungs-, Lüftungs-, Klima-, Kälte-, Gas-, Wasser-, Abwasser-, Stromversorgungsanlagen, nachrichtentechnische Anlagen, Aufzug- und Förderanlagen, Ver- und Entsorgungsanlagen in Wohn- und Gesellschaftsbauten, in Produktionsbetrieben und in der Landwirtschaft*]

building (services) automation → building automation

building services company, mechanical company Gebäudetechnikfirma, Haustechnikfirma *f*

building (services) control Gebäudeleitung, Gebäudebeherrschung, Hausleitung, Hausbeherrschung *f*

building (services) control engineering Gebäudeleittechnik, GLT, Hausleittechnik *f*

building (services) control system Gebäudeleitsystem, Hausleitsystem *n*

building (services) engineer, mechanical engineer Gebäudetechnikingenieur, Haustechnikingenieur *m*

building (services) engineering, mechanical engineering [*In building terms, the heating, cooling and ventilating of a building including the design of all related plant and equipment. Some specialist installations, such as lift installations, also fall into this category, but are usually designed and carried out by specialist engineers and subcontractors. It also covers some aspects of plumbing (the domestic hot water installation is usually undertaken by the mechanical engineer). Mechanical engineering sometimes merges with electrical engineering, for example, whilst a mechanical engineer would be responsible for the installation of electrically controlled equipment, the electrical engineer would be responsible for running a supply of the correct capacity to that equipment*] Installationstechnik, Gebäudetechnik, Haustechnik *f*

building (services) engineering research, mechan-ical engineering research, building services re-search Gebäudetechnikforschung, Haustechnikforschung, Installationstechnikforschung *f*

building sewage system; building wastewater system (US) Gebäudeabwassersystem, Hausabwassersystem *n*

building sewer Gebäudeabwasserleitung, Hausabwasserleitung *f*

building shaft → shaft

building skin Gebäudehülle, Gebäudehaut, Außenhaut, Haushaut, Haushülle, Außenhülle *f*

building skin monitoring Außenhautüberwachung, Gebäudehautüberwachung, Gebäudehüllenüberwachung *f* [*Einbruchmeldetechnik, bei der alle Türen, Fenster, Luken und sonstige Gebäudeöffnungen auf Durchbruch überwacht werden. Die Außenhautüberwachung meldet bereits den Versuch mit Hilfsmitteln wie z. B. Alarmfolien, Glasbruchmeldern oder Magnetkon-takten einer Intrusion, während sich der Eindringling noch im Außenbereich befindet*]

building skin security Außenhautsicherung, Gebäudehautsicherung, Gebäudehüllesicherung *f* [*Mechanische, an Gebäuden angebrachte und einen Widerstands(zeit)wert aufweisende Sicherung, z. B. Gitter, Fenstersicherung und Türsicherung*]

building sound control Gebäudeschallbeherrschung *f*

building specific gebäudetechnisch

(building) stack (Gebäude)Fallstrang *m* [*z. B. für Schmutzwasser*]

B

building stock, building inventory, on-hand build-ing quantity Gebäudebestand *m*

building structure → structure

building supervisor integrated technique, BSI technique GLT-integrierte Sicherheitstechnik *f*

building supply Gebäudeversorgung *f*

building supply line Gebäudeversorgungsleitung, Gebäudezuleitung *f*

building (supply) station Gebäudeversorgungsstation, Gebäudezuleitungsstation *f*

building temperature limitation [*It prevents room temperature during an unoccupied period from falling below or rising above acceptable limits. When the room temperature reaches those limits the HVAC plant cycles on and off, effectively maintaining the building temperature limitation set points*] Gebäudetemperaturbegrenzung *f*

building thermal load, building heat load Gebäudewärmelast *f*

building thermal loss, building heat loss Gebäudewärmeverlust *m*

building thermal requirement, building heat requirement, building thermal need, building heat need Gebäudewärmebedarf *m*

building thermal system, building heat system, building thermal plant, building heat plant, building thermal set, building heat set Gebäudewärmesystem *n*

building topology Gebäudetopologie *f*

building trade, building craft Baugewerbe, Bauhandwerk *n*

building use Gebäudenutzung *f*

building user Gebäudenutzer *m*

building ventilation, ventilation of building [*See remark under "ventilation"*] Gebäudelüftung, Gebäudebe- und -entlüftung, Gebäudeventilation *f*

building volume Gebäudevolumen, umbaute Volumen, Raum *m*

building wastewater system (US) → building sewage system

building water supply → water supply

building water supply line → water supply line

building wiring Gebäudeverdrahtung *f*

buildup → composition

buildup Stau *m* [*z. B. Wärmestau*]

built-in, in-built eingebaut, Einbau ...

built-in appliance, in-built appliance Einbaugerät *n* [*z. B. in einer Küche*]

built-in (bath(ing)) tub, in-built (bath(ing)) tub, encased (bath(ing)) tub Einbau(bade)wanne *f*

built-in box, in-built box Einbaukasten *m*

built-in burner, in-built burner Einbaubrenner *m*

built-in casing, in-built casing, built-in case, in-built case Einbaugehäuse *n* [*z. B. in eine Wand eingebaut*]

built-in ceiling luminaire, in-built ceiling luminaire Einbaudeckenleuchte, Deckeneinbauleuchte *f*

built-in cupboard → closet

built-in dishwasher, in-built dishwasher Einbau-Geschirrspüler *m*, Einbau-Geschirrspülmaschine *f*

built-in kitchen, in-built kitchen Einbauküche *f*

built-in lavatory (basin), in-built lavatory (basin), built-in (wash)basin, in-built (wash-) basin Einbauwaschbecken *n*

built-in lock → encased lock

built-in logic block observer, BILBO, in-built logic block observer Funktionsblock *m* [*Er kann auf Steuer(ungs)signal hin sowohl Prüfvektorgenerator als auch Prüfvektoranalysator sein*]

built-in luminaire, in-built luminaire Einbauleuchte *f*, Einbaubeleuchtungskörper *m*

built-in oven, in-built oven Einbaubackofen, Einbaubratofen *m*

built-in range, in-built range Einbau(küchen)-herd, Kücheneinbauherd *m*

built-in refrigerator, in-built refrigerator Einbaukühlschrank *m*, Kühleinbauschrank

built-in self test, in-built self test Schaltungsselbst(über)prüfung *f*

built in the wall wandeingebaut

built-in valve, in-built valve Einbauventil *n*

built-in vanity, in-built vanity Einbauwaschtisch *m*

bulb barometer Gefäßbarometer *n* [*Österreich und Schweiz: m*]

bulb socket Lampenfassung *f*

bulb sprinkler → (glass) bulb sprinkler

bulk Substrat *n* [*Feldeffekttransistor*]

bulk current Volumenstrom *m* [*Elektrizität*]

(bulk) dump(ing) Datenblockscheiden, Datenblockabwerfen *n*, Datenblockabwurf *m*, Datenblockscheidung *f*

bulk freezing, freezing of unpacked products Massengefrieren *n*

bulkhead coupling piece Schottverschraubung *f*

bulk meter, volume meter Mengenzähler, Volumenzähler *m*

bulk modulus of elasticity → modulus of volume elasticity

bulk movement → **mass transport(ation)**

bulk movement capacity → **mass transport(ation) capacity**

bulk power transmission Hochleistungsübertragung *f*

bullet-proof glass, bullet-resistant glass, bullet-resisting glass Panzerglas *n*

bullet-resistant, bullet-resisting, bullet-proof kugelsicher, kugelfest, schussfest, schusssicher, durchschussfest, durchschusssicher, beschussfest, beschusssicher

bump mapping Darstellung *f* tiefenstrukturierter Oberflächen durch ein Grundstufenbild [*Helle Bildteile springen vor, dunkle erwecken die Illusion von Tiefe*]

bunched conductor verwürgte Leiter, gebündelte Leiter *m*

buncher (resonator), first resonator, input (cavity) resonator [*The first resonator in a velocity-modulated tube, next to a cathode; here the faster electrons catch up with the slower ones to produce bunches of electrons*] Eingang(s)resonator *m*

bunch key Bundschlüssel *m*

bunch of keys → **key bunch**

bundle tube Bündelröhre *f*

bung Stopper, Stopfen *m*

Bunsen burner entleuchtete Brenner, Bunsenbrenner *m*

Bunsen (burner) flame Bunsen(brenner)flamme, entleuchtete Brennerflamme *f*

buoyancy-driven convection Auftriebskonvektion *f*

buoyant force velocity Auftriebsgeschwindigkeit *f*

burglar Einbrecher *m*

burglar bars Einbruchschutzstangen(gitter) *f, (n)*

burglar motion, burglar movement Einbrecherbewegung *f*

burglary, break(ing)-in Einbruch *m*

burglar(y) alarm, break(ing)-in alarm, intrusion alarm Einbruchalarm, Einbrecheralarm *m*

burglary attempt, break(ing)-in attempt Einbruchversuch *m*

burglar(y) detection, burglar(y) detecting, break(ing)-in detection, break(ing)-in detecting Einbruchmelden *n*, Einbruchdetektion *f*, Einbrucherkennen, Einbruchnachweisen *n*

burglar(y) detection plant, burglar(y) detection installation Einbruchmelde(r)anlage *f*, EMA

burglar(y) detection system, break(ing)-in detection system Einbruchmelde(r)system *n*

burglar(y) detector, burglar(y) detection device, burglar(y) detecting device [*"Break(ing)-in" may synonymously be used for "burglar(y)"*] Einbruchmelder *m*, Einbruchmeldegerät *n*, Einbruchdetektor, Einbruchmeldungsgeber *m*

burglar(y)-proof, burglar(y)-resistant, burglar(y)-resisting, break(ing)-in-proof, break(ing)-in-resistant, break(ing)-in-resisting einbruchsicher, einbrechersicher

burglar(y) protection, break(ing)-in protection Einbruchschutz, Wertschutz, Einbrecherschutz *m*

burglar(y) security Einbruchsicherung, Einbrechersicherung, Wertsicherung *f*

burglar(y) signal message plant, break(ing)-in signal message plant Einbruchmeldeanlage, Einbruchmeldungsanlage *f*, EMA [*Eine Gefahr(en)meldeanlage (GMA) zum automatischen Überwachen von Gegenständen auf unbefugte Wegnahme sowie Flächen und Räumen auf unbefugtes Eindringen*]

burial cable, buried cable, in-ground cable Erdkabel, Bodenkabel, erdverlegte Kabel, bodenverlegte Kabel *n* [*Nicht in einem in der Erde befindlichen Kanal oder Rohr verlegt*]

burial installation, buried installation, in-ground installation Erdeinbau *m*, Erdverlegung, Bodenverlegung, Bodeneinbau *f* [*Einbau in der Erde und nicht in einem in der Erde befindlichem Kanal oder Rohr*]

buried erdverlegt, erdinstalliert, bodenverlegt, bodeninstalliert [*Leitung im Untergrund ohne Kanal oder Rohr*]

buried cable Erdkabel *n* [*Im Erdreich verlegtes Kabel*]

buried layer Kollektorleitschicht, Kollektorleitlage, eingebettete Schicht, eingebettete Lage *f* [*Verfahren zur Verminderung des Kollektorbahnwiderstandes von Planartransistoren. In die Kollektorzone wird eine leitfähige Schicht parallel zur Kollektorsperrschicht eingebracht, die den Widerstand zwischen den Sperrschichten und dem Kollektoranschluss herabsetzt*]

burned-in address Token-Ring-Adresse *f* [*Jede Token-Ring-Karte ist hardwareseitig auf eine weltweit einmalige Adresse, bestehend aus einem zwölfstelligen Hexadezimalwert, eingestellt.*]

burned-out ausgebrannt, verbrannt

burner Brenner *m* [*Vorrichtung zur Aufbereitung, Luftbeimischung, Zündung und kontrollierter Verbrennung gasförmiger, flüssiger und fester Brennstoffe*]

burner aperture → **burner port**

burner back plate, back burner plate Brennerrückwand *f*

burner bank Brennerrampe *f*

burner blower Brennergebläse *n*, Brennerbläser *m*

burner boiler Brennerkessel *m*

burner chamber Brennerraum *m*, Brennerkammer *f*

burner ((closed) loop) control → **burner regulation**

B

burner controller [*A mechanism by means of which burners are automatically lit and extinguished at predetermined times*] Brennerregler *m*

burner crosspiece Brennergeschränk *n*

burner design, burner type, burner version Brenneraufbau *m*, Brennerausführung, Brenner(bau)art, Brennerbauweise *f*

burner fitting Brennerarmatur *f*

burner flame Brennerflamme *f*

burner head → **burner tip**

burner heat Brennerwärme *f*

burner heat load Brennerwärmelast *f*

burner hood Brennerhaube *f*

burner ignition [*1. Initiation of combustion. 2. Process or means of igniting a fuel-air mix(ture)*] Brennerzündung *f*

burner insert Brennereinsatz *m*

burner inspirator → **inspirator of burner**

burner (mixing) tube [*A tube connecting the gas injector and the burner ports, in which mixing of gas and air is completed. It frequently commences as a converging section leading in succession to a throat and a diverging section; it is then often erroneously called a "venturi tube"*] Brenner(misch)röhre, Mischbrennerröhre *f*

burner nozzle → **burner tip**

burner operation, burner run Brennerbetrieb *m*

burner operation Brennerbedienung *f*

burner port, burner aperture, burner orifice, flame port, flame aperture, flame orifice Brenneröffnung, Flammenöffnung *f*

burner pot Brennerschale *f*

burner power Brennerleistung *f*

burner pressure Brennerdruck *m*

burner pump Brennerpumpe *f*

burner rail → **gas manifold**

burner regulation, burner ((closed) loop) control Auf-Zu-Reg(e)lung, Brennerreg(e)lung *f* [*Durch den Raumthermostat geschaltet*]

burner run → **burner operation**

burner setting Brennereinstellung *f*

burner start(ing) (up) Brenneranfahren, Brenneranlauf(en) *(n)*, Brennerstart *m*

burner switch Brennerschalter *m*

burner system Brennersystem *n*

burner tip, burner nozzle, burner shroud, burner head Brennerkopf *m*

burner tube → **burner (mixing) tube**

burner tunnel [*See remark under "tunnel burner"*] Brennertunnel *m*

burner type → **burner design**

burner unit Brenner *m* als Baueinheit [*DIN 4787, Teil 1. Brenner als Baueinheit sind für sich als Einzelbrenner funktionsfähig und umfassen alle für den Betrieb erforderlichen Einrichtungen, wie Ölzerstäubungs-, Luftmisch- und Regelteil, gegebenenfalls interne Ölvorwärmeeinrichtungen einschließlich Öl-druckpumpe bei Öldruckzerstäubern, Verbrennungsluftgebläse sowie Feuerungsautomat (Flammenüberwachung), Zündeinrichtung und die erforderlichen Armaturen für Reg(e)lung und Sicherheitsabsperrung des Brenners*]

burner version → **burner design**

burn hole Brandloch, Brennloch *n* [*Entsteht zum Beispiel durch eine brennende Zigarette*]

burn-in Einbrennen *n*, Einbrand *m* [*Zur Verkürzung der Frühausfallphase*]

burning [*CD*] Brennen *n*

burning substance → **fuel**

burning time, combustion time, burning period, combustion period [*boiler*] Abbrandzeit, Abbrandperiode *f*

to burnish, to polish, to make shiny polieren

to burn out ausbrennen [*Kessel bei Teeransatz*]

burnout → **departure form nucleate boiling**

to burr-off abreifen [*Schlosserei. Mit dem Reifkloben den Reif entfernen*]

to burst, to disrupt (zer)springen, bersten, platzen

burst schnelle transiente Stör(ungs)größe, transiente schnelle Stör(ungs)größe *f*

burst abklingende Schwingung *f* [*Körperschallmessung*]

burst → **colour burst**

burst, data burst Paket, Bündel, Datenpaket, Datenbündel *n* [*Besteht in der Regel aus vier Doppelwörtern, also viermal 32 Bit*]

burst Prüfzyklenfolge *f*

burst impulsartige Informationsblockübertragung *f*

burst, discharge shock Entladungsstoß *m*

burst Rauschimpuls *m*

burst Farbsynchronsignal *n* [*Die dritte Synchroninformation die der Farbfernsehempfänger für den Gleichlauf der Bildwiedergabe mit der Abtastung im Studio benötigt*]

burst (Daten)Bündel(zer)störung *f*

burst Schauer *m*

burst error Bündelfehler, Fehlerhäufung *f*, Paketfehler *m* [*Gebündelte auftretender Fehler im Gegensatz zu zufälligen Einzelfehlern*]

burst firing Vielperiodensteuerung, Mehrperiodensteuerung, Schwingungspaketsteuerung [*Leistungselektronik*], Impulspaketsteuerung *f*

bursting, disruption, blow-up Bersten, (Zer-)Springen, (Zer)Platzen *n*

burst(ing) diaphragm, disruption diaphragm Berstmembran(e) *f*

burst(ing) disk, burst(ing) disc, disruption disk, disruption disc Berstscheibe *f* [*Dünne Membran(e), die beim Überschreiten des für sie ausgelegten Druckes bricht*]

burst(ing) mode Ping-Pong-Art , Zeitgetrenntart *f*

burst(ing) preventer → blow up preventer

bus → bus(bar)

bus Bus *m*; Hauptkabel *n* [*frühere Benennung*] [*Ein Bus (Funktionseinheit) ist ein System zur Übertragung von Daten zwischen mehreren Vorrichtungen (Teilnehmern, Stationen) über ein gemeinsames Leitungssystem. Dabei gelten für die übertragenen Da-ten, für den Anschluss der Einrichtungen und für den Austausch der Daten zwischen den Einrichtungen vereinbarte Bedingungen. Anmerkung: Die zu vereinbarenden Bedingungen – meist Protokoll genannt – werden in der Regel in mehreren Ebenen festgelegt. Auf der untersten Ebene werden dabei die technischen Eigenschaften der Signale und der Sammelleitung so-wie die Schnittstellen der Geräte vereinbart. Auf der darüberliegenden Ebene werden die Datenformate und die Bedingungen für deren Austausch festgelegt. Die höheren Vereinbarungsebenen dienen der Festlegung logischer Bedingungen beim Datenaustausch. E DIN 19 226 T 5*]

busable busfähig

bus access Buszugriff *m*

bus arbiter Buszuteiler *m*, Busverwalter

bus architecture [*A computer system configuration that uses a shared or multiplexed communication path between the processor and the peripheral de-vices, permitting multiple devices to transmit data at the same time*] Busarchitektur *f*

bus backplane Busrückwand *f*

bus(bar) [*A usually bare electrical conductor often in the form of a thick copper or alumin(i)um bar to which arc leads are clamped*] Sammelschiene, Stromschiene, Verteilerschiene *f*

bus(bar) bracket Sammelschienenhalter, Stromschienenhalter, Verteilerschienenhalter *m*

bus(bar) carrier Stromschienenträger, Sammelschienenträger, Verteilerschienenträger *m*

bus(bar) chamber Stromschienenkammer, Sammelschienenkammer, Verteilerschienenkammer *f*

bus(bar) conductor Sammelschienenleiter, Stromschienenleiter, Verteilerschienenleiter *m*

bus(bar) connection to circuit breaker Stromschienenanschluss *m* am Leistungsschalter, Sammelschienenanschluss am Leistungsschalter, Verteilerschienenanschluss am Leistungsschalter

bus(bar) crossunder Sammelschienenunterführung, Stromschienenunterführung, Verteilerschienenunterführung *f*

bus(bar) current Sammelschienenstrom, Verteilerschienenstrom, Stromschienenstrom *m*

bus(bar) current transformer Sammelschienenstromwandler, Verteilerschienenstromwandler, Stromschienenstromwandler *m*

bus(bar) grounding switch, bus(bar) earthing switch Sammelschienenerdungsschalter, Stromschienenerdungsschalter, Verteilerschienenerdungsschalter, Sammelschienenerder, Stromschienenerder, Verteilerschienenerder *m*

bus(bar) metering panel Sammelschienenmessfeld, Stromschienenmessfeld, Verteilerschienenmessfeld *n*

bus(bar) ring system Stromschienenringsystem, Sammelschienenringsystem, Verteilerschienenringsystem *n*

bus(bar) screw terminal Flachklemme *f* für Sammelschienenanschluss, Flachklemme für Stromschienenanschluss, Flachklemme *f* für Verteilerschienenanschluss

bus(bar) support, bus(bar) holder, bus(bar) grip Stromschienenhalter, Sammelschienenhalter, Verteilerschienenhalter *m*

bus(bar) system Stromschienensystem, Sammelschienensystem, Verteilerschienensystem *n*

bus(bar) trunking, bus(bar) duct Sammelschienenkanal, Stromschienenkanal, Verteilerschienenkanal *n*

bus(bar) unit Sammelschieneneinheit, Stromschieneneinheit, Verteilerschieneneinheit *f*

bus board, bus p.c.b. Busplatine *f*

bus busy, BBSY Bus belegt

bus card rack → bus subrack

bus clear, BCLR Bus frei

bus ((closed) loop) control Busreg(e)lung *f*

bus command Buskommando *n*

bus-compatible busverträglich

bus control → bus ((closed) loop) control

bus control → bus ((open) loop) control

bus converter Busumsetzer *m*

bus coupler, bus gateway Buskoppler, BK, Sammelschienenkoppelschalter, Stromschienenkoppelschalter, Verteilerschienenkoppelschalter *m*

bus current → bus(bar) current

bus driver Bus-Treiber *m* [*Schaltung, die durch geeignete Ansteuerung der Zentraleinheit diese mit dem Datenbus verbindet*]

B

bus duct [*An assembly of heavy bars of copper or alumin(i)um that acts as a conductor of large capacity*] Sammelschienenkanal, Stromschienenkanal, Verteilerschienenkanal *m*

bus enable, BUSEN Busfreigabe(signal) *f*, (*n*)

bus engineering, building bus engineering Gebäudesystemtechnik *f* [*Sie ist Teil der Gebäudetechnik. Der Begriff Gebäudesystemtechnik wurde von der EIBA (European Installation Bus Association) geprägt und kommt somit aus dem Bereich der Gebäudeinstallationstechnik. Die EIBA ist eine Vereinigung von Installationskomponenten-Anbietern, die den EIB unterstützen. In der Gebäudesystemtechnik sind alle Geräte über Bus miteinander verbunden.*]

bus extender Buserweiterer *m*

bus grant, BG Busbewilligung, Buszuteilung *f*

bus grant in line, BGIN Busbewilligungseingangsleitung, Buszuteilungseingangsleitung *f*

bush, grommet Tülle *f* [*Leitungseinführung*]

bushing [*cable accessory*] Durchführung *f*

bushing Buchse *f*

bushing terminal, lead-out terminal Durchführungsklemme *f*

business computer, office computer, BC Bürorechner, kommerzielle Rechner *m* [*PC der mittleren Da-tentechnik für Verwendung im Büro*]

business data processing kommerzielle Datenverarbeitung *f*

bus interface (unit), BU(I) Busschnittstelle *f* [*Vom Teilnehmer zum Datennetzwerk*]

bus interfacing, bus interface connection Busanschaltung *f*, Busschnittstellenanschluss *m*

bus line Sammelleitung, Busleitung, durchgehende Leitung *f*

(bus) master aktive (Bus)Teilnehmer, übergeordnet agierende (Bus)Teilnehmer *m* [*Feldbussystem*]

bus network, tree network [*A network in which LAN nodes are all connected to a single length of cable*] Busnetz, busförmige Netz *n*, Fernbus *m*

bus ((open) loop) control Bussteuerung *f*

bus p.c.b. → **bus board**

bus reactor [*A current-limiting reactor for connection between two different busses or two sections of the same bus for limiting and localizing disturbance due to a fault in either bus*] (Strom)Begrenzungsdrosselspule *f*

bus screening Busabschirmung *f*

bus section Busabschnitt *m*

bus segment [*A bus section that is electrically con-tinuous, with no intervening components, such as repeaters*] Bussegment *n*

(bus) slave passive (Bus)Teilnehmer, untergeordnete (Bus)Teilnehmer *m* [*Feldbussystem*]

(bus) station, (bus) user (Bus)Teilnehmer *m* [*An eine Busleitung angeschlossenes und mindestens einen Buskopp(e)ler enthaltendes Gerät*]

bus subrack, bus card rack Busbaugruppenträger *m*

bus technique Bustechnik *f* [*Übertragungsverfahren bei einer Gefahrenmeldeanlage*]

bus time monitoring Buszeitüberwachung *f*

bus topology Bus-Topologie *f*, Kettennetz *n*

(bus) transceiver Buskopp(e)ler, BK, Bustreiber *m* [*Ein Buskopp(e)ler verbindet den Bus mit der datenverarbeitenden Einrichtung. Er erfüllt die Funktionen des Sendens und des Empfanges der Daten und der Selbstüberwachung. E DIN 19 226 T 5*]

bus unit → **bus(bar) unit**

(bus) user → **(bus) station**

busway hanger → **track suspension device** [*IEC 570*]

busway system (US), bus(bar) trunking system [IEC 439-2] [*A standardized assembly of busbars rigidly supported by solid insulation and enclosed in a sheet-metal housing*] Schienenverteiler *m* [*VDE 0660, T. 500*]

busy, in use [*Of a telephone line which is not able to accept another incoming call*] besetzt, belegt, nicht frei

busy indicator Belegtanzeiger, Besetztanzeiger *m*

busy lamp [*It indicates a busy condition for a certain line or phone*] Besetztlampe, Belegtlampe *f*

busy signal [*An audible or visual signal which indicates that the called number of transmission path is unavailable*] Belegtsignal, Besetztsignal *n*

busy test [*A telephony test, made to find out whether certain facilities which may be desired, such as a subscriber line or trunk, are available for use*] Besetztprüfung, Belegtprüfung *f*

busy tone [*Interrupted low tone returned to the subscriber as an indication that the party's line is busy*] Besetztzeichen, Belegtzeichen *n*, Besetztton, Belegtton *m*

busy verification [*A feature in the switched telephone network that permits an attendant to verify the busy or idle state of station lines and/or to break into the conversation*] Besetztnachprüfung, Belegtnachprüfung *f*

butler('s) sink → **Belfast sink**

to butt stumpf (aneinander) stoßen

butt contact [*A hemispherically shaped contact designed to mate against a similarly shaped contact*] Druckkontakt *m*

butt-end Pressrest *m*

butterfly valve, butterfly dumper Drosselventil *n*

butt (hinge), (hinge and) loop, door loop, (door) hinge (Tür)Angel(band) *f*, (*n*), Einstemmband, Einsetzband, Band, Türband *n* [*DIN 401 und DIN 402*]

butt joint, square joint, straight joint [*A joint be-tween two materials or components that meet without overlapping*] Stumpffuge *f*, (An)Stoß(fuge) *m*, (*f*)

butt-jointed, square-jointed, straight-jointed stumpf gestoßen

button Knopf *m*

button → (push)button

button push Knopfdruck *m*

butt-welded stumpfgeschweißt

buying audience Kunde *f*

buying guide Einkauf(s)führer *m*

buying visitor Fachbesucher *m*

to buzz → to dither

buzz [*A low-pitched sound with high-frequency components resulting from nonsinusoidal voltage interfer-ence*] Brummton, Summton *m*, Brummen, Summen *n*

buzz, dither [*Control systems. A force with a controlled amplitude and frequency, continuously applied to a servomotor-driven device to prevent it from sticking in the null position*] Zittern, Bibbern *n*

buzzer Summer *m*, Brummer

buzzing sound Summen, Brummen *m*

BV → breakdown voltage

BW → bandwidth

BWA → backward wave amplifier

BWN → wide band network

BWO → backward wave oscillator

BWT → backward wave tube

BX (cable) [*A flexible, multi(ple)-conductor armo(u)red cable having an outer protective covering consisting of a helically wound steel strip; used for connections to electric equipment and in wiring houses*] bewehrte flexible Kabel, flexible bewehrte Kabel *n*

b, x directed control b, x-geführte Strategie *f* [*Gebäudeautomation*]

by-key Beischlüssel, Nachschlüssel *m*

bypass Nebenschluss *m* [*Parallele Stromführung zu einem Bauelement*]

bypass Umgehung, Umführung, Umleitung *f*

bypass Kurzschlussstrecke *f* [*Einrohrheizung*]

bypass air Umgehungsluft, Umführungsluft *f*

bypass (air) duct, air bypass duct Umführungs(luft)kanal, Umleitungs(luft)kanal, Umgehungs(luft)kanal, Luftumführungskanal, Luftumgehungskanal *m*

bypass channel [*A channel that is routed through a node without being demultiplexed*] Umgehungskanal, Umführungskanal, Umleitungskanal *m*

bypass ((closed) loop) control Umführungsreg(e)lung, Umgehungsreg(e)lung, Umleitungsreg(e)lung *f*

bypass flue (tube), bypass fire tube, bypass furnace tube Umführungsfeuerzug *m*, Umführungsfeuerrohr, Umführungsflammrohr *n*, Umgehungsfeuerzug *m*, Umgehungsfeuerrohr, Umgehungsflammrohr *n*

bypass humidity ((closed) loop) control Umführung(s)feuchtereg(e)lung, Umführung(s)feuchtigkeit(s)reg(e)lung *f*

bypass jumper Überbrückungsleiter *m*, Überbrückungsleitung *f*

bypass line Umgehungsleitung, Umführungsleitung *f*

bypass ((open) loop) control Umführungssteuerung, Umgehungssteuerung, Umleitungssteuerung *f*

bypass pipe Umführungsrohr, Umgehungsrohr, Umleitungsrohr *n*

bypass (pipe)line Umgehungs(rohr)leitung *f*

byte, eight-binary digit word n-Bit-Zeichen *n*, Bitgruppe *f*, Wort, Acht-Bitwort *n*, Biteinheit, Gruppe von 8 Bits *f*, Oktett *n* [*frühere Benennung*]

byte count register, BCR Bitgruppenzählerregister *n*

byte multiplexer channel [*Computer engineering. An approach to connecting peripheral devices to a processor and main memory using a shared channel that transmits one byte of data at a time and interleaves bytes of data from several devices; used with slower I/O devices*] Byte-Mehrfachkopp(e)ler, Byte-Vielfachkopp(e)ler *m*

byte serial transmission bitgruppen-serielle Übertragung *f*

byte timing Bitgruppentakt *m*

C

C Programmiersprache *f* [*Sie basiert auf der Sprache PASCAL, das heißt die meisten Anweisungen und die Struktur der Sprache sind genau so wie bei PASCAL. Sie erlaubt jedoch, falls erforderlich, wie eine Assemb-*

C

lersprache die speziellen Eigenheiten des Prozessors anzusprechen und jedes Bit gezielt zu bearbeiten. Dies ist häufig an den Übergangsstellen zwischen Rechner und Prozessor erforderlich, also beim Ansprechen der Analog-Digital- und Digital-Analog-Umsetzer, bei Zeitkoordinierungen usw.]

CA, cash adapter, cash adaptor, CM, cash master, TAV, teller assist unit BBA *m*, beschäftigtenbediente Banknotenautomat *m*, AKT *m*, automatische Kassentresor *m*

CA → computer access

cabin → enclosure

cabinet → (floor(-mounted)) (bath(room)) cabinet

cabinet → Schubladenschrank *m*

cabinet, display case, display casing Schautruhe, Vitrine *f*

cabinet → undercounter

cabinet distribution board, cubicle distribution board Schrankverteiler, Verteiler in Schrankbauweise *m*

cabinet door trash can, cubicle door trash can Schranktürmülleimer *m*

cabinet illumination, cabinet lighting, cubicle illumination, cubicle lighting Schrankbeleuchtung *f*

cabinet kitchen unit, cubicle kitchen unit Schrankküche *f*

cabinet lock, cubicle lock Schrankschloss *n*

cabinet (loud)speaker, cubicle loud(speaker) Gehäuselautsprecher *m*

cabinet suspension, cubicle suspension Schrankaufhängung *f*

cabinet switchboard, cubicle switchboard Schrankschalttafel, Schaltschranktafel *f*

cabinet(-type) lavatory (basin), cabinet(-type) (wash)basin [*See remark under "wash basin". "Cubicle" may synonymously be used for "cabinet"*] Schrank(wasch)becken *n*

cabinet unit, cubicle unit Schrankeinheit *f*

cabinet with inserts, cubicle with inserts bestückte Schrank *m*

cable [*Assembly of one or more conductors, usually within a protective sheath, so arranged that the conductors can be used separately or in groups*] Kabel, Elektrokabel, ELT-Kabel, elektrische (Leitungs)-Kabel *n*

cable access point, CAP Kabelanschlussstelle, Kabelverbindungsstelle *f*

cable article surveillance technique, cable article security technique, cable product surveillance technique, cable product security technique Kabelsicherungstechnik *f* [*Die Ware ist durch ein Kabel und eine Metallklammer mit einem Alarmgerät verbunden. Beim Lösen der Klammer (= Unterbrechung des Stromkreises) erfolgt Alarmgabe*]

cable assembly, cable set, cable harness Kabelsatz *m*

cable basement, cable cellar Kabelkeller *m*

cable box → cable (connecting) box

cable branch (line), (outgoing) cable feeder Kabel(ab)zweig *m*

cable branch terminal → cable tapping block

cable breakdown → cable fault

cable bundle Kabelbündel *n*

cable bushing Kabelbüchse *f*

cable ceiling heating (Elektro)Kabeldeckenheizung *f*

cable clutter Kabelwirrwarr *m*, Kabeldurcheinander *n*, Kabelsalat *m*

cable compound Kabelvergussmasse *f*

cable conductor → conductor

cable conduit → conduit

cable (connecting) box, cable connection box Kabel(anschluss)kasten, Kabelverbindungskasten *m*

cable (connecting) duct, cable connection duct Kabelübergang(s)formstück *n*

cable (connecting) sleeve, cable connection sleeve Kabel(verbindungs)muffe, Kabelanschlussmuffe *f*

cable connection Kabelanschluss *m*, Kabelverbindung *f*

cable connector Kabelverbindungsstück *n*, Kabelverbinde, Kabelanschlussstück *n*, Kabelübergang *m*

cable core Kabelseele *f*

cable cover(ing) hood Kabel(abdeck)haube *f*

cable design, cable type, cable version Kabelausführung, Kabelbauart, Kabelbauweise *f*

cable distribution box Kabelverteilerkasten, Kabelverteilungskasten *m*, KVK

cable distribution cabinet Kabelverteilerschrank, KVS, Kabelverteilungsschrank *m*

cable dowel Kabeldübel *m*

cable duct → cable trunking

cable electrode Kabelelektrode *f*

cable elevator (US) → cable lift

cable engineering Kabeltechnik *f*

cable entry, cable entrance, cable inlet, cable lead-in Kabeleingang *m*, Kabeleinführung, Kabelzuführung *f*

cable extension Kabelverlängerung *f*

cable eye Kabelschuh *m*, Kabelanschlusshülse *f*, Kabelanschlusslasche *f*, Kabelanschlussöse *f*

cable failure → cable fault

cable fanning out box Kabelauspleißkasten *m*

cable fault, cable malfunction, cable trouble, cable breakdown, cable failure, cable interference Kabelfehler *m*, Kabelstörung *f*

cable feeder → cable branch (line)

cable fire Kabelbrand *m*, Kabelfeuer *n*

cable floor heating, floor cable heating (Elektro-) Kabel(fuß)bodenheizung *f*

cable grommet Kabeltülle *f*

cable grounding (US); cable earthing (Brit.) Kabelerdung *f*

cable harness, cable set, cable assembly Kabelsatz *m*

cable holding spring Kabelhaltefeder *f*

cable inlet → cable entry

cable installation → cabling

cable interference → cable fault

cable jacket, cable sheath(ing) Kabelmantel *m*, Kabelummantelung *f*

cable junction box Kabelmuffe *f*

cable lead-in → cable entry

cable length compensator Kabelentzerrer *m*, Kabellängenausgleicher

cable lift, rope lift (Brit.); cable elevator, rope elevator (US) Seilaufzug *m*, Kabelaufzug

cable link Kabelüberführung *f*

cable loss Verlust *m* durch Kabel, Kabelverlust [*Der Anteil des RF-Signals (Radiofrequenz), der bei der Übertragung über ein Koax(ial)kabel verloren geht. Der Übertragungsverlust ist abhängig von den Medien und der Kabellänge. Beim Einsatz eines Koax(ial)kabels entstehen bei hohen Frequenzen größere Verluste als bei niedrigen Frequenzen, nach einer Logarithmus-Funktion.*]

cable main service → underground cable service

cable malfunction → cable fault

cable marker Kabelmerker *m*

cable monitoring Kabelüberwachung *f*

cable outgoing unit → outgoing cable unit

cable outlet Kabelabgang, Kabelausgang *m*

cable passage, cable penetration Kabeldurchführung *f*, Kabeldurchgang *m*

cable pipe Kabel(schutz)rohr *n*

cable product surveillance technique → cable article surveillance technique

(cable) pull box [*A metal box inserted into a conduit run for the purpose of providing a cable pulling point. Cable may be spliced in such a box*] (Kabel)Einzieh-kasten *m*, (Kabel)Einzugskasten

cable rack Kabelpritsche *f*, Kabelbahn *f*, Kabelgerät *n*

cable remote operation, cable teleoperation Kabelfernbedienung *f*

cable resistance Kabelwiderstand *m*

cable riser Kabelsteig(e)leitung *f*, Steig(e)kabelleitung

cable route, cable run, cabling path, cabling run, cabling route, cable path Kabeltrasse *f*, Kabelstrecke *f*, Kabelweg *m*

cable routing Kabel(trassen)führung, Kabeltraversierung *f*

cable schedule Kabelliste *f* [*Darstellung der Anschlusspunkte der elektrischen Einrichtungen und der daran anzuschließenden äußeren Kabel*]

cable service box Kabelanschlusskasten, Kabelverbindungskasten *m*

cable set → cable assembly

cable sheath(ing) → cable jacket

cable shielding, cable screening Kabelabschirmung *f*

cable sleeve → cable (connecting) sleeve

cable socket Kabelmuffe *m*

cable splice [*Connection between two or more separate lengths of cable. Conductors in one length are individually connected to conductors in the other length, and the protecting sheaths are so connected that protection is extended over the joint*] Kabelspleißung, Kabelverbindung *f*

cable stripping knife Kabelmesser *n*

cable system → cabling

cable tapping block, cable branch terminal Kabel(ab)zweigklemme *f*

cable teleoperation → cable remote operation

cable television, CATV, community antenna television Kabelfernsehen *n*

cable temperature sensor, cable temperature sensing element, cable temperature feeler [*See remark under "sensor"*] Kabeltemperaturfühler *m*, Kabeltemperatursensor *m*

cable termination Abschlusswiderstand *m* [*busförmige Verkabelung (Ethernet)*]

cable termination Kabelendverschluss *m*

cable tester Kabelprüfer *m*

cable tray [*An assembly of metalwork which is used to support insulated electric conductors; similar in function to a metal framework on the bottom and sides, with the top open*] Kabelpritsche *f*, Kabeltraggerüst *n*

cable trench Kabelgraben *m* [*Rechteckquerschnitt*]

cable trouble → cable fault

C

C

cable trunking, cable duct [*A fabricated system of linear enclosures of rectangular cross section, assembled together by connectors and having one removable side. Used for the protection of electrical cables*] Kabelkanal *m*

cable type → cable design

cable version → cable design

cable warning strip Kabelwarnstreifen *m*

cabling, cable system Verkabelung *f*, Kabelsystem *n*

cabling run → cable route

cache (memory), cache storage (Schnell)Pufferspeicher, (Schnell)Zwischenspeicher *m* [*Zwischen dem Arbeitsspeicher und der Zentraleinheit; dient zur Verkürzung der Zugriffszeit auf den Arbeitsspeicher*]

cache-on-a-stick → coast(-cache)

CAD/CAM Verkettung *f* von CAD, CAP, CAM und CAQ(A)

cadmium plated, cadmium coated kadmiert, kadmiumplattiert, kadmiumüberzogen, kadmiumbeschichtet

CAE → computer aided engineering

CAFM → computer-aided facility management

cage Dunstrohr-Windschutzhaube *f*

cage Käfig *m*

cage → (lift) car

cage... → (elevator) car...

CAI, computer-aided industry integrierte EDV-Einsatz *m* (in einem Unternehmen)

CAIFM → computer-aided and integrated facility management

caking coal → matting coal

CAL → computer-assisted learning

calculated value berechnete Wert *m* [*Ergebnis einer Berechnung, das bei der Bearbeitung in einem Gebäudeautomationssystem wie ein analoger Wert behandelt wird*]

calculated variable [*Variable which cannot actually be measured directly, but one which can be calculated by measuring other variables. Example: Measuring wet-bulb temperature and measuring dry-bulb temper-ature to determine enthalpy*] berechnete Veränderliche *f*

calculating punch [*A calculator having a card reader and a card punch*] elektronisch lochende Rechengerät *n*, Rechenlocher, Rechenstanzer *m*

calculation Berechnung *f*

calculation procedure Berechnungsvorgang *m*

calculation sheet Berechnungsblatt *n*, Berechnungsbogen *m*

calculator [*A portable or desktop device, primarily electronic, that is used to perform arithmetic and other numerical processing operations at the direction of an operator or a stored program(me)*] Rechenwerk, Rechengerät *n*, Rechenmaschine *f*, Rechner *m*

caldron Kleinkessel *m* [*Gliederkessel von 0,5 bis 5,0 m² Heiz(ungs)fläche*]

calender year clock (controller) Jahresuhr *f*

to calibrate [*1. To determine or mark a correct value using a meter or other measuring instrument. 2. To determine a measured value by adjusting for instrument error or environmental conditions*] kalibrieren, eichen

calibrating curve [*A line that plots the value for each reading calibrated from a meter or control dial*] Kalibrierkurve, Eich(ungs)kurve *f*

calibration Kalibrierung, Eichung *f*

caliduct [*1. A duct or pipe conveying hot air, hot water or steam for heating. 2. In the ancient Roman systems of furnace heating, a hot-air duct, usually of terracotta or built-up with brick partitions and tile facing*] Heiz(ungs)kanal *m*

to call aufrufen [*Aufrufen einer Koroutine oder Funktion an einer beliebigen Stelle im Programm*]

to call (an)rufen [*Fernsprecher*]

call Aufruf *m* [*Verwendung einer Koroutine oder Funktion an einer beliebigen Stelle im Programm*]

call Gespräch *n* [*Fernsprecher*]

call (An)Ruf *m* [*Fernsprecher*]

call accepted (An)Rufannahme *f* [*DIN 44 302, Teil 13*]

call address [*data processing*] Rufadresse *f*

call announcer [*Device for receiving pulses from an automatic telephone office and audibly reproducing the corresponding number in words, so that it may be heard by a manual operator*] akustische Rufnummernanzeiger *m*

callback Rückruf *m*, Rückfrage *f*

call bell (An)Rufglocke *f* [*Telefon*]

call billing Gebührenerfassung *f* [*Telefon*]

call booked by pre-arrangement Dauerrufanmeldung *f*

call booked the previous day Vortag(s)rufanmeldung *f*

call box → pay station

call button → call (push)button

call-by-call Anbieterauswahl *f* durch Netzvorwahl

call by name Namen(s)parameter *m*

call by reference, reference call Referenzparameter *m*, Verweisübergabe *f* auf eine Variable [*Eine Eigenschaft einer Programmiersprache*]

call by value, value call Wertparameter *m*, (Variablen)Wertübergabe *f* auf eine Funktion [*Eine Eigenschaft einer Programmiersprache*]

call by value-result Wert-Referenzparameter-Kombination *f*

call charge Gesprächsgebühr, Verbindungsgebühr, (An)Rufgebühr *f*

call-circuit key Dienstabfragetaste, DaT *f*

call-circuit key Dienstleitungstaste, D, DT *f*

call-circuit selector Dienstleitungswähler, DLW *m*

call collision (An)Rufzusammenstoß *m* [*Schnittstelle; Zeichengabe*]

call confirmation protocol, CCP Bestätigungsprotokoll *n* [*dezentrale Zeichengabe*]

call connected → CCO

call connection (An)Rufverbindung *f*

call deflection (An)Rufweiterschaltung *f* bei (An-) Rufzustellung

call distribution (An)Rufverteilung *f*

call diversion (An)Rufumleitung *f*

called line identity → CDI

Callendar('s) equation [*1. An equation of state for steam whose temperature is well above the boiling point at the existing pressure, but less than the critical temperature: (V – b) = (RT/p) – (s/Tn), where V is the volume, R is the gas constant, T is the tempera-ture, p is the pressure, n equals 10/3, and a and b are constants. 2. A very accurate equation relating tem-perature and resistance of platinum, according to which the temperature is the sum of a linear function of the resistance of platinum and a small correction term, which is a quadratic function of temperature.*] Callendarsche Zustandsgleichung *f*

Callendar('s) thermometer → platinum resistance thermometer

call header length (Auf)Ruflänge *f* [*programmierbare Steuer(ungs)gerät*]

call(ing) current, ringing current (An)Rufstrom *m*

call(ing) for heat → heat call(ing)

call(ing) for help, help call(ing) Hilferuf *m*

call(ing) lamp (An)Ruflampe, Fernsprechlampe, Telefonlampe *f*

call(ing) line, ringing line [*telephone*] (An)Rufleitung *f*

call(ing) line identification, ringing line identification Anschlusskennung *f*

call(ing) range [*communication cable*] (an)Rufbereich *m* [*selten: n*]

call(ing) sequence [*Computer programming. A set of program(me) instructions and data used to call a subroutine; it consists of an instruction that transfers control parameters to be passed to the subroutine, and one or more locations for the subroutine to transfer results back when its execution is complete*] (Auf-) Ruffolge *f*

call(ing) signal, ringing signal [*A call control signal transmitted over a circuit to indicate that a connection is desired*] (An)Rufsignal *n*

call(ing) station, ringing station (an)rufende Station *f*

call(ing) system, ringing system [*telephony*] (An-) Rufsystem *n*, (An)Rufanlage *f*

call in progress [*telephone*] laufende Gespräch *n*

call interception → interception of calls

cal(l)ipers, cal(l)ibers, cal(l)iper compasses Greifzirkel, Tastzirkel, Drehzirkel *m*

call key (An)Ruftaste *f*

call keyboard (An)Ruftastatur *f*

call not accepted (An)Rufabweisung *f* (An)Rufzurückweisung [*DIN 44 302, Teil 13*]

call number [*In computer operations, a set of characters identifying a subroutine, and containing information concerning parameters to be inserted in the subroutine, or information to be used in generating the subroutine, or information related to the operands*] Programmkennzahl, Rufzahl, Aufrufzahl *f*

call on hand bereitliegende Gesprächsanmeldung *f* [*Sie liegt am Fernplatz bereit*]

call ((open) loop) control method Verbindungssteuer(ungs)verfahren *n*, Verbindungssteuer(ungs)methode *f* [*Datenübertragung*]

callout device, callout apparatus anrufende Gerät *n* (An)Rufgerät

(call) pickup (An)Rufübernahme, Übernahme *f*

call point → (alarm) call point

call progress signal, CP (geschriebene) Netzmeldung *f*

call (push) button (An)Rufdruckknopf *m*

call receiver (An)Rufempfänger *m*

call redirection, call transfer (An)Rufumleitung, (An)Rufumlegung, Verbindungsumleitung, Verbindungsumlegung *f*

call request abgehende (An)Ruf *m* [*DIN 44 302, Teil 13*]

call statement (An)Rufanweisung *f*

call station Sprechstelle *f* [*Beschallungssystem*]

call station microphone Sprechstellenmikrophon, Sprechstellenmikrofon *n* [*Beschallungssystem*]

call storing lamp, call storage lamp [*telephony*] Wartelampe, WL *f*

call storing panel, call storage panel [*telephony*] Wartefeld, Wf *n*, Wartetafel, Wt *f*

C

call system (An)Rufsystem *n*

call tracing Fangen *n* [*Leistungsmerkmal*]

call transfer → call redirection

call waiting Anklopfen *n* [*Leistungsmerkmal*]

caloric kalorisch

caloric Wärmestoff *m*

caloric conductor → heat conductor

calorifere, air-stove Luftwärmofen *m*

calorific, heat-producing, heat-generating wärmeerzeugend

calorific boiler Brennwertkessel *m*

calorifier [*A type of heat exchanger that transfers heat from coils of hot water or steam pipes to cooler water surrounding the pipes*] Schlangenwärme(aus)tauscher *m*

calorimeter [*1. Device for measuring heat quantities, such as machine capacity, combustion heat, specific heat, vital heat, heat leakage, etc. 2. Device for meas-uring quality (or moisture content) of steam or other vapo(u)r 3. Equipment for measuring emitted or ab-sorbed heat quantities*] Kalorimeter, Wärmemessgerät *n*, Wärmemesser *m*

calorimetry, calorimetric measurement(s), meas-urement(s) of heat (capacities) Kalorimetrie, Wärmemessung(en) *f (pl)*

CAM, computer-aided manufacture rechnergestützte Herstellung, rechnergestützte Fabrikation, rechnergestützte Fertigung, rechnerunterstützte Herstellung, rechnerunterstützte Fabrikation, rechnerunterstützte Fertigung *f*

CAM → content-addressable (computer) memory camera adapter, camera adaptor [*An attachment to a camera that permits its use in a manner for which it was not designed*] Kameravorsatz, Photoapparatvorsatz, Fotoapparatvorsatz *m*

camera monitor(ing) set Überwachungskamera *f*

camera motion, camera movement Kamerafahrt *f*

camera-related kamerabezogen

cam(-operated) switch, camshaft switch Nockenschalter *m*

camping car Wohnwagen *m*

campus cabling Geländeverkabelung *f* [*Gegensatz: Gebäudeverkabelung*]

camset Nockensatz *m*

cam terminal Nockenklemme *f*

Canadian latch, Suffolk latch, Garden City latch (Brit.); lift latch (US); thumb latch, Norfolk latch (Daumen)Drückerfalle *f*

can be fitted by user nachrüstbar [*z. B. ein System*]

candela, cd [*The SI unit for luminous intensity; closely approximates to formerly accepted unit known as the "International Candle"*] Candela *f*, neue Kerze, NK *f*

candidate Kandidat *m* [*Eine Funktionsausführung wünschende Station*]

candidate protocol Kandidatenprotokoll *n*

candle filter Kerzenfilter *m, n*

candle jack Wachsstockleuchter *m*

candlepower, cp, CP [*The luminous intensity of a light source, expressed in candelas*] Kerzenstärke, Lichtstärke in Candela(s) *f*

canned rotor motorcompressor halboffene Motorverdichter *m*

canned rotor rotary motorcompressor halboffene Drehkolbenmotorverdichter *m*

canonical ((closed) loop) form Regelungsnormalform, kanonische Regelungsform *f*

canonical form kanonische Form, Normalform *f*

canonical ((open) loop) form Steuer(ungs)normalform, kanonische Steuer(ungs)form *f*

canopy hood, cooking canopy Dunstabzugshaube als Wandhaube, Wand-Dunstabzugshaube *f*

canteen service, canteen catering Speisenverteilung *f*, Kantinendienst *m*

cantilever valve → flapper valve

canvas Segeltuch *n*

CAP → common application programming interface

CAP → cable access point

CAP, computer-aided planning rechnergestützte Planung, rechnerunterstützte Planung *f*

capability (Leistungs)Fähigkeit *f*, (Leistungs)Vermögen *n*

capable of communication, with communication capability kommunikationsfähig

capable of real time operation echtzeittätig

capacitance; capacity [*obsolete*] [*The ratio of the charge on one of the conductors of a capacitor (there being an equal and opposite charge on the other conductor) to the potential difference between the conductors. Symbolized C*] Kapazität *f*

capacitance current Ladestrom *m*

capacitive kapazitiv [*Auf der Speicherfähigkeit eines Kondensators beruhend*]

capacitive coupling, capacitance coupling kapazitive Kopp(e)lung *f*

capacitive detector, capacitance detector, capacitive detection device, capacitive detecting device, capacitance detection device, capacitance detecting device Kapazitivmeldegerät *n*, Kapazitivmelder *m* [*Ein elektrisches Feld wird durch Eindringling verändert*]

capacitive feedback, capacitance feedback kapazitive Rückkopp(e)lung *f*

capacitive identification, capacitance identification kapazitive Kennung *f*

capacitor → condenser

capacitor bank, (electric) condenser bank [*A number of capacitors connected in series or in parallel*] Kondensationsblock, Kondensatorblock *m*

capacitor coil → capacitor roll

capacitor (loud)speaker, electrostatic (loud-) speaker, (electric) condenser (loud)speaker [*A loudspeaker in which the mechanical forces are produced by the action of electrostatic fields; in one type the fields are produced between a thin metal diaphragm and a rigid metal plate*] elektrostatische Lautsprecher *m*

capacitor microphone, electrostatic microphone, (electric) condenser microphone [*A microphone consisting essentially of a flexible metal diaphragm and a rigid metal plate that together form a two-plate air capacitor; sound waves set the diaphragm in vibration, producing capacitance variations that are converted into audio-frequency signals by a suitable amplifier circuit*] elektrostatische Mikrophon, Kondensatormikrophon *n* [*auch „Mikrofon" geschrieben*]

capacitor nominal current, nominal capacitor current, condenser nominal current, nominal condens-er current Kondensatornennstrom, Nennkondensatorstrom *m*

capacitor oil, condenser oil Kondensatoröl *n*

capacitor output, condenser output Kondensatorleistung *f*

capacitor plug, condenser plug Kondensatorstecker *m*

capacitor roll, capacitor coil, condenser roll, condenser coil Kondensatorwickel *m*

capacitor start(ing) and run(ning) motor [*Motor similar to the capacity start motor except that the capacitor and starting winding are designed to remain in the circuit at all times, thus eliminating the switch used to disconnect the starting winding*] Einphasenmotor *m* mit Kondensator für Anlauf und Betrieb

capacitor start(ing) motor [*1. Induction motor having a separate starting winding similar to the split-phase motor except that the capacitor start motor has an electrical condenser connected into the starting winding for added starting torque. 2. Motor that has a capacitor in the starting circuit*] Einphasenmotor mit Anlaufkondensator, Einphasenmotor *m* mit abschaltbarem Kondensator in der Hilfsphase

capacitor store Kondensatorspeicher *m* [*Für jedes zu speichernde Bit einen Kondensator enthaltender elektronischer Speicher*]

capacitor switching, condenser switching Kondensatorschalten, kapazitive Schalten *n*

capacitor transformer, condenser transformer kapazitive Wandler *m*

capacitor voltage divider, condenser voltage divider kapazitive Spannungsteiler, Spannungsteilerkondensator *m*

capacitor voltage transformer, condenser voltage transformer kapazitive Spannungswandler, Spannungswandlerkondensator *m*

capacity [*obsolete*] **→ capacitance**

capacity → power

capacity Fassungsvermögen *n*, Fassungsfähigkeit *f*

capacity calculation, service calculation, performance calculation Leistungsberechnung, Leistungsermittlung, Kapazitätsberechnung, Kapazitätsermittlung *f*

capacity characteristic, service characteristic, performance characteristic Leistungsmerkmal, Kapazitätsmerkmal *n*

capacity data, performance data, service data Leistungsdaten, Kapazitätsdaten, Leistungsangaben, Kapazitätsangaben *f* [*Maschine; Gerät; Anlage*]

capacity level, performance level, service level Leistungsstufe, Kapazitätsstufe *f* [*Maschine; Gerät; Anlage*]

capacity limitation, service limitation, performance limitation Leistungsbegrenzung, Kapazitätsbegrenzung *f*

capacity monitoring, service monitoring, perform-ance monitoring Leistungsüberwachung, Kapazitätsüberwachung *f*

capacity range, performance range, service range Leistungsbereich, Kapazitätsbereich *m* [*selten: n*] [*Maschine; Gerät; Anlage*]

capacity rating, performance rating, service rating Leistungseinstufung, Kapazitätseinstufung *f*

capacity reducer, performance reducer, service reducer [*In a compressor, a device, such as a clearance pocket, mov(e)able cylinder head, or suction bypass, by which compressor capacity can be adjusted without otherwise changing the operating conditions*] Leistungsabschwächer, Kapazitätsabschwächer *m*

capacity regulation, performance regulation, service regulation Leistungsreg(e)lung, Kapazitätsreg(e)lung *f*

capillary action, capillary attraction kapillare An-ziehung, Kapillaranziehung *f*

capillary column Kapillarsäule *f*

C

capillary condensation [*Condensation of an adsorb-ed vapo(u)r within the pores of the adsorbate*] Kapillarkondensation *f*

capillary hose Kapillarschlauch *m*

capillary liquid Kapillarflüssigkeit *f*

capillary restrictor Kapillardrossel(vorrichtung) *f*

capillary soldering Kapillarlöten *n*

capillary soldering fitting Kapillarlötfitting *m* [*Siehe Anmerkung unter „Lötfitting"*]

capillary tube Kapillarröhrchen, Haarröhrchen, Kapillarrohr, Haarrohr *n* [*Fehlbenennungen*]

capital character, capital letter Großbuchstabe *m*

capitalized value, income value Ertragswert *m*

cap nut Überwurfmutter *f*

cap rail Hutschiene *f*

CAPSC, computer-aided production scheduling and control Fertigungssteuerung *f*

cap seal(ing) valve [*Manual valve having the stem protected by a tightly fitting cap*] Kappen((ab)dichtungs)ventil *n*

cap shutoff valve Kappen(ab)sperrventil *n*

capstan Antriebsrolle *f* [*Magnetband*]

capstan tracking logic, CTL Bild-Zählimpuls *m*

capsule spring manometer Kapselfedermanometer, Balgfedermanometer, Kapselfederdruckmesser, Balgfederdruckmesser *n*

captain system → character and pattern telephone access information network system

caption Titel *m*, Überschrift, Unterschrift *f* [*nicht als Bestätigung*]

caption generator Titelerzeuger, Überschrifterzeuger, Unterschrifterzeuger *m* [*Nicht verwechseln mit „Unterschrift von Hand"*]

captive unverlierbar, nicht verlierbar

captive sponge rubber gasket unverlierbar eingeschäumte (Ab)Dichtung, eingeschäumte unverlierbare (Ab)Dichtung *f*

capture → (data) collection

capture effect [*The effect wherein a strong frequency-modulation signal in an FM receiver completely suppresses a weaker signal on the same or nearly the same frequency*] Übertönen *n*

CAQ(A) → computer-aided quality assurance

car [*general term*] Wagen *m*

CAR, computer-aided repair rechnergestützte Reparatur, rechnerunterstützte Reparatur *f*

CAR, computer aided robotics rechnergestützte Robotereinsatz, rechnerunterstützte Robotereinsatz *m*

CAR → computer-aided ((closed) loop) control

car... → (elevator) car...

caravan boiler Kofferkessel *m*

carbonate brine Karbonatsole *f*

carbonate (water) hardness vorübergehende (Wasser)Härte, temporäre (Wasser)Härte, zeitweilige (Wasser)Härte, Karbonathärte, KH *f*

carbon build up, carbon crust Kohlenstoffansatz *m*, Kohlenstoffkruste *f*

carbon copy, CC Durchschlag *m*, Zweitausfertigung *f*

carbon dioxid(e), CO_2 [*A colourless, odourless, tasteless gas about 1.5 times as dense as air*] Kohlendioxid *n*, Kohlendioxyd, Kohlensäure *f* [*Fehlbenennung*]

carbon dioxid(e) fire extinguisher Kohlendioxidfeuerlöscher *m*

carbon dioxid(e) fire-fighting plant, carbon diox-id(e) fire-fighting installation, CO_2 dioxid(e) fire-fighting plant, CO_2 dioxid(e) fire-fighting installation [*"Extinction" may synonymously be used for "fire-fighting"*] Kohlendioxidlöschanlage, Kohlendioxydlöschanlage, Kohlensäurelöschanlage *f* [*Fehlbenennung*]

carbon dioxid(e) ice → dry ice

carbon dioxid(e) lamp Kohlendioxidlampe, Kohlendioxydlampe, Kohlensäurelampe *f* [*Fehlbenennung*]

carbon film resistor Kohleschichtwiderstand *m*

carbonic acid freie Kohlendioxid *n*, Kohlensäure, freie Kohlendioxyd *f*

carbon steel Kohlenstoffstahl, gekohlte Stahl *m*

carbon steel pipe [*It owes its properties chiefly to the carbon content of the steel*] Kohlenstoffstahlrohr *n*

carbon tetrachlorid(e) → refrigerant-10

carbon tetrafluorid(e) → refrigerant-14

carcass → shell

carcinotron → backward wave oscillator

card [*An information-carrying medium, common to practically all computers, for the introduction of data and instructions into the computers either directly or indirectly*] Karte *f*

card blank(et) Kartenrohling *m*

card call box → card pay station

card code Kartenkode, Kartenschlüssel, Kartencode *m*

card column Kartenspalte *f*

card content Karteninhalt *m*

card deck Kartensatz *m*

card face, punch(ed) card face [*The printed side of a punch(ed) card or, if printing is on both sides, the side of chief importance*] (Loch)Kartenhauptseite *f*

card feeder [*In automatic data processing, a device that inserts cards into a machine one at a time*] Kartenspeiser *m*, Kartenvorschubgerät *n*

card field [*A specified group of card columns used for a particular category of data*] Kartenfeld *n*

card file Kartei *f*

card film Kartenfolie *f* [*aus Kunststoff*]

card holder Trag(e)tasche, Kartentrag(e)tasche *f*, Kartenetui *n*

card holder Kartenhalter *m* [*An der Wand montiert zum waag(e)rechten oder senkrechten Einstecken von Karten*]

(card) hopper [*A device that holds cards and makes them available to a card-feed mechanism*] Karten-(eingabe)fach, Karten(eingabe)magazin *n*

cardinal stimulus Definitionsvalenz *f*

card input Karteneingabe *f*

card interface Kartenschnittstelle *f*

cardioid microphone Richtmikrophon, Kardioidmikrophon *n* [*auch "...mikrofon"*]

card key Kartenschlüssel *m*

card loader *A programming routine which permits a deck of cards to be read into a memory*] Kartenlader *m*

card lock → (electronic) card lock

card lockout Kartensperrzeit *f*

card machine [*Any of several small computers which perform particular operations called for by instruction cards, concurrently with the reading of data cards*] Kartenmaschine *f*

card pay station, card call box, card public (tele)phone station öffentliche Kartentelephon *n*, öffentliche Kartenfernsprecher, öffentliche Kartentelefon *m*

card printer, card writer Kartendrucker *m*

card protective cover Kartenetui, Schutzetui *n*

card public (tele)phone station → card pay station

cards punch(ing machine), card puncher [*A com-puter output device that punches holes in a punch card upon reception of signals from the central processing unit*] Kartenstanzer, Kartenlocher *m*

card punching recorder Lochkartenstanzer *m*

card rack, subrack, mount(ing) rack, module frame Baugruppenträger *m*, BGT [*DIN 43350*]

card rack tier, subrack tier, mount(ing) rack tier, module frame tier Baugruppenträgerzeile, BGT-Zeile *f* [*DIN 43 350*]

card reader, punched-card reader (Loch-)Kartenleser *m*

card recognition Kartenerkennung *f*

card recorder → (punched) card recorder

card safety Kartensicherheit *f*

card security Kartensicherung *f*

card sorter, punched-card sorter [*A machine used to arrange punched cards into an appropriate sequence for further processing*] (Loch)Kartensortierer *m*

card stacker Kartenablage(fach) *f*, *(n)*

card storage box Kartenmagazin *n*

card-to-card conversion Karte/Karte-Umsetzung, Umsetzung *f* Karte/Karte

card-to-disk conversion, card-to-disc conversion [*A straight forward operating which consists in load-ing the data in a deck of cards onto a disk by means of utility program(me)*] Umsetzen von Karte auf Platte, Karte-Platte-Umsetzen *n*

card-to-tape conversion [*A straight forward operation which consists in loading the data in a deck of cards onto a magnetic tape by means of an utility program(me)*] Karte-Band-Umsetzen, Umsetzen *n* von Karte auf Band

card tracking Kartenbeobachtung, Kartenverfolgung *f*

care, preventive maintenance Pflege, Wartung, Bewahrung des Sollzustandes *f*

care → safe-keeping

care agent Pflegemittel *n*

care service Pflegedienst *m*

caretaker Hauswart *m*

Carnot engine [*It receives energy at a high level of temperature, converts a portion of the energy into work, and charges the remainder to a heat sink at low level of temperature*] Carnotmaschine *f*

Carnot (refrigeration) cycle [*It performs the reverse effect of the heat engine, because it transfers energy from a low level of temperature to a high level of temperature. It requires the addition of external work for its operation. The reversible processes in succession are: Isothermal expansion and heat addition, isentropic expansion, isothermal compression and heat rejection and isentropic compression*] Carnot-Prozess, Carnotsche Kreisprozess *m* [*rechts laufend*]

car park barrier Parkplatzschranke *f*

car park lift (Brit.); car park elevator (US) Garagenaufzug, Parkaufzug *m*

carpet Teppich *m*

carpet cleaner, carpet cleaning machine Teppichreiniger *m*, Teppichreinigungsmaschine *f*

carpet flooring Teppichboden *m*

carpet tile Teppichfliese *f*

carpet underlayment [*A padding material, laid directly on the floor, over which carpet is installed; usually manufactured of hair felt, foam rubber, hair felt and jute, sponge rubber, or some other combination of these materials*] Teppichunterlage *f*

car-port PKW-Unterstand *m*

C

carriage return, CR Zeilenschaltung *f* [*Funktionstaste an Terminals zur Ansteuerung einer Zeile*]

carriage return character, CR character Zeilenschaltungszeichen *n*

carriage return line feed, CRLF Zeilenschaltungsvorschub *m*

carriage tape, control tape [*Loop of paper tape to control the carriage operation of character printers*] Steuer(ungs)lochstreifen *m*

carrier Träger *m*

carrier Büchse, Kartusche *f* [*Rohrpost(anlage)*]

carrier basket → **collection basket**

carrier cable Trägerfrequenzkabel *n*

carrier current Trägerstrom *m*

carrier current telephony Trägerstromfernsprechen *n*

carrier damper Büchsenbremse, Kartuschenbremse *f* [*Rohrpost(anlage)*]

carrier detect, CD Trägererkennung *f* [*Die Identifizierung des Trägertons beim Verbindungsaufbau*]

carrier drop Trägerabfall *m* [*Modem kann die Verbindung nicht zwei Minuten halten*]

carrier frequency, resting frequency; center frequency (US); centre frequency (Brit.) [*The frequency generated by an unmodulated radio, radar, carrier communication, or other transmitter, or the average frequency of the emitted wave when modul-ated by a symmetrical signal*] Trägerfrequenz *f*

carrier frequency band [*A band of continuous frequencies that can be modulated with a signal*] Trägerfrequenzband *n*

carrier frequency cable Trägerfrequenzkabel *n*

carrier leak [*Carrier remaining after carrier suppression in a suppressed-carrier transmission system*] Trägerrest *m*

carrier level [*The level of an unmodulated point in a radio system, expressed in decibels in relation to some reference levels*] Trägerpegel, Trägerstand *m*

carrier noise, residual modulation [*Noise produced by undesired variation of a radio-frequency signal in the absence of any intended modulation*] Trägerrauschen *n*

carrier power output rating Trägerausgangsleistung *f*

carrier repeater [*telephony*] Trägerstromverstärker *m*

carrier sense multi(ple)-access → **CSMA**

carrier sense multi(ple)-access collision control → **CSMA/CC**

carrier sense multi(ple)-access/collision detection → **CSMA/CD**

carrier shift(ing) [*Transmission of radio-teletype writer messages by shifting the carrier frequency in one direction for a marking signal and in the opposite direction for a spacing signal*] Trägerfrequenzumtasten *n*

carrier signal Trägersignal *n*

carrier signalling (Brit.); carrier signaling (US) [*Method by which busy signals, ringing, or dial signal(l)ing relays are operated by the transmission of a carrier-frequency tone*] Trägerfrequenzruf *m*

carrier suppression Trägerunterdrückung *f*

carrier system [*A system permitting a number of simultaneous independent communications over the same circuit*] Träger(frequenz)system *n*

carrier telegraphy Trägerfrequenztelegraphie, Trägerfrequenztelegrafie *f*

carrier telephony Trägerfrequenztelephonie, Trägerfrequenztelefonie *f*

carrier-to-noise-ratio, CNR, C/N, $\frac{C}{N}$ [*The ratio of the magnitude of the carrier to that of a noise after selection and before any nonlinear process such as amplitude limiting and detection*] Träger/-Rausch-Verhältnis *n*

carrier transmission [*Transmission in which the transmitted electric wave is a wave resulting from the modulation of a single-frequency wave by a modu-lating wave*] Trägerfrequenzübertragung *f*

carrier tube, air tube, pneumatic tube Fahrrohr *n* [*Rohrpost*]

carrier tube conveyor → **air tube conveyor**

carrier wave Trägerwelle *f*

carry Übertrag *m* [*Signal oder Ausdruck, wenn die Summe zweier Ziffern mit demselben Stellenwert gleich oder größer als die Basis des verwendeten Zahlensystems ist*]

to carry away, to carry off abführen, wegführen

carrying bag Trag(e)tasche *f*

carrying capacity Tragfähigkeit *f*, Tragvermögen *n* [*Aufzug*]

carrying capacity [*escalator*] Personenförderleistung, Förderleistung *f* in Personen [*Rolltreppe*]

carrying case, carrying casing Trag(e)gehäuse *n*

carrying handle, carrying grip Trag(e)griff *m*

carrying rail Trag(e)schiene *f*

carrying rope, carrying cable Trag(e)seil *n*

carry lock Parallelübertragung *f* [*Durch eine zusätzliche logische Verknüpfung wird eine schnelle Übertragungsverarbeitung erreicht*]

carry lock ahead parallele Übertragungsbildung *f*

cartesian coordinates Cartesische Koordinaten *f* [*Rechtwink(e)liges Koordinatensysten zur räumlichen Bestimmung eines Punktes; bestehend aus X-, Y- und Z-Achse*]

to carton kartonieren [*In Kartons verpacken*]

cartoned kartonverpackt

cartridge Kassette, Bandkassette *f* [*zumeist „Magnetbandkassette"*]

cartridge Patrone *f*

cartridge disk → exchangeable disk

cartridge disk drive → exchangeable disk drive

cartridge filter → cellular filter

cartridge fuse; enclosed fuse, encapsulated fuse (US) [*May also be spelled "fuze". A fuse in an incombustible tube which protects the installation from fire; fuse melts or blows*] Patronensicherung, Sicherungspatrone *f*

CAS → column address strobe

CAS → column address select

cascade Kaskade *f* [*Beliebige Erweiterung der zu verarbeitenden Wortlänge durch Serienschaltungen mehrerer Prozessorbausteine (slices)*]

cascade amplifier, multistage amplifier Kaskadenverstärker *m*

cascade carry [*A carry process in which the addition of two numerals results in a sum numeral and a carry numeral that are in turn added together. This process being repeated until no new carriers are generated*] Kaskadenübertrag *m*

cascade circuit → connection-in-series

cascade compensation → series compensation

cascade connection, cascade arrangement, concatenation, tandem connection, tandem arrangement Kaskadenschaltung *f* [*Verschiebung des Stellwertes der Vorlauftemperatur in Abhängigkeit von der Raumtemperatur durch den Raumfühler*]

cascade controller Kaskadenregler *m*

cascaded kaskadiert

cascade(d) bridges Kaskadenbrückenschaltung *f*

cascade(d) ((closed) loop) control Kaskadenregelung *f*

cascade decay Kaskadenzerfall *m*

cascade(d) network [*It is divided into a number of linearly connected, contiguous segments*] Kaskadennetz(werk) *n*

cascade mixing preheater Kaskadenmischvorheizer *m* [*Das oben zugeführte Heizungsrücklaufwasser wird über Tassen geleitet, von denen es abtropft. Dabei mischt es sich intensiv mit dem im Gleichstrom ge-führten Dampf, der unter Abgabe seiner Verdampfungswärme kondensiert. Der Wasserspiegel wird durch einen Regler konstant gehalten*]

cascade refrigerating plant, cascade refrigerating installation Kaskadenkälteanlage, Kaskadenkälte-installation *f*

cascade refrigerating system [*One having two or more refrigerant circuits, each with a pressure imposing element, condenser, an evaporator of one circuit cools the condenser of another (lower temperature) circuit*] Kaskadenkältesystem *n*

cascade wind tunnel Kaskadenwindkanal *m*

case → (lock) box

case → jacket

case ... → casing ...

case(d) bolt-nab, box(ed) bolt-nab Schließkappe *f*

case(d) (bolt)-staple, box(ed) (bolt-)staple, case(d) catch, box(ed) catch [*The metal receiver for the bolt of a dead lock or rim lock*] Schließkappe *f*, Schließkloben *m*

casein glue Kaseinleim *m*

casein paint Kaseinfarbe *f*

casement fastener, casement turn, window casement fastener, window casement turn, cock spur fastener handle (Fenster)Vorreiber, (Fenster)-Einreiber, (einfache) Schiebefensterverschluss *m*

casement handle (Fenster)Flügelgriff *m*

casement hinge Angel, Fensterangel *f*

casement hook Stützhaken, Stützkegel *m* [*Fensterband*]

casement staple Fensterkrampe *f*

(casement) stay Kettelhaken, Sturmhaken, Fensterhaken *m*

casement turn → casement fastener

casement window Flügelfenster *n*

case of need, case of demand, case of requirement Bedarfsfall *m*

case pipe → jacket pipe

case(work) → housing

cash adapter → CA

cash box(ed) lock Kassenschloss *n*

cash handling Geldumgang *m* [*Physische Bearbeitung von Notengeld, Münzgeld, Sorten und Schecks*]

cash master → CA

cash pool Geldumgangszentrum *n*

cash register Registrierkasse *f*

casing, enclosure, housing, case Gehäuse *n* [*Funktionell erforderliche Hülle für Bauelemente. DIN 1946 Teil 1*]

casing → (lock) box

casing → jacket

casing → chest

casing ambient temperature, case ambient temperature, housing ambient temperature, enclosure ambient temperature Gehäuseumgebungstemperatur *f*

casing assembly, enclosure assembly, housing assembly, case assembly Gehäusezusammenbau *m*

C

casing base, enclosure base, case base, housing base Gehäuseunterkasten *m*

casing freezer → chest freezer

casing loss, case loss, housing loss, enclosure loss [*fan*] Gehäuseverlust *m*

casing outline, case outline, enclosure outline, housing outline Gehäusebauform *f*

casing panel, housing panel, enclosure panel, case panel Gehäusetafel *f*

casing pipe, case pipe, jacket pipe Mantelrohr *n*

casing station, housing station, enclosure station, case station Gehäusestation *f* [*Rohrpost(anlage)*]

casing strut, enclosure strut, case strut, housing strut Gehäusesteg *m*

casing tube, jacket tube, case tube Mantelröhre *f*

casing vessel → jacket vessel

cassette recorder Tonbandkassette *f*

cassette-ventilation system → (ceiling) cassette-ventilation system

casting alloy Gusslegierung *f*

casting lead → molten lead

casting resin Gießharz, Gussharz *n*

cast(-iron) bath(tub) Guss(eisen)(bade)wanne *f*

cast(-iron)boiler Guss(eisen)kessel *m*

cast(-iron)(convector) radiator [*See remark under "radiator"*] Guss(eisen)radiator, Guss(eisen)gliederheiz(ungs)körper, Guss(eisen)heiz(ungs)körper *m* [*Fehlbenennung*] [*DIN 4720*]

cast(-iron) hollow section → hollow cast(-iron) section

cast(-iron) pipe Guss(eisen)rohr *n*

cast(-iron) pressure pipe Guss(eisen)druckrohr *n* [*DIN 28 610 und 28 614*]

cast(-iron) section Gusseisenprofil *n*

cast(-iron) (sectional) boiler, sectional cast(-iron) boiler Guss(eisen)gliederheiz(ungs)kessel, Gliederguss(eisen)heiz(ungs)kessel, Guss(eisen)heiz(ungs)kessel in Gliederbauweise *m*

cast(-iron) (sectional) radiator, sectional cast (-iron) radiator [*See remark under "radiator"*] Gusseisenradiator, Gusseisengliederheiz(ungs)körper *m*

cast(-iron) sink Gusseisenspüle *f*

cast(-iron) slop sink, cast(-iron) service sink Gusseisenausguss(becken) *m, (n)*

cast(-iron) soil stack [*See remark under "soil stack"*] Guss(eisen)-Sanitärabfallrohr, Guss(eisen)-Abfallsanitärrohr *n* [*„Abfall" bedeutet hier „senkrecht"*]

cast(-iron) spout Guss(eisen)ablauf *m*

cast(-iron) valve Guss(eisen)ventil *n*

castor oil, ricinus oil Rizinusöl *n*

cast resin Gussharz *n*

casual access unbefugte Zutritt *m*

CAT, computer-aided testing rechnergestützte Güteprüfen, rechnerunterstützte Güteprüfen *n*

cataloged data set [*It is represented in an index, or hierarchy of indexes, in the system catalogue; the indexes provide the means for locating the data set*] katalogisierte Datensatz *m*

catalytic incineration katalytische Verbrennung *f*

cat(alyzer) Katalysator *m*

catastrophic degradation Spiegelzerstörung *f* durch hohe optische Leistungen [*cw, P >1MW/cm²*] [*Halbleiterlaser*]

catastrophic error [*A computer programming error that causes the computer to abandon program(me) compilation or execution*] Katastrophenfehler *m*

catastrophic failure, total failure Totalausfall, Gesamtausfall *m*

cat burglar Fassadenkletterer, Fassadeneinbrecher *m*

catch Trag(e)griff *m*

catch bolt → spring bolt

catched nab [*See remark under "nab"*] Schließkappe *f*

catch(er) Schnäpper, Schnepper *m*

catcher Sperrung *f* [*Schloss*]

catch(er) ball Schnäpperkugel, Schnepperkugel *f*

to catch in, to gear together eingreifen [*Einstriche des Schlüsselbartes*]

catching-in, gearing together Eingriff *m* [*Einstriche des Schlüsselbartes*]

catchlock Schnappschloss *n*

catchpot (fixture) trap, catchpot stench trap, catchpot drain trap, catchpot siphon trap [*A bottle trap with a removable lower part large enough to retain waste for examination or recovery*] (Auf)Fang-Flaschenwasserverschluss, (Auf)Fang-Flaschengeruchsverschluss *m*

catenary ((closed) loop) control Durchgangsreg(e)lung *f*

catenary wire Tragdraht *m* [*für Fahrdraht*]

catering (service) Kantinendienst *m*

catharometer → heat conductivity cell

cathode follower, grounded anode amplifier, grounded-plate amplifier Kat(h)odenfolger *m*

cathode glow Kat(h)odenglimmlicht *n*

cathode interface impedance, layer impedance [*The impedance between the cathode base and coat-ing in an electron tube, due to a high-resistivity layer or a poor mechanical bond*] Kat(h)oden-Zwischenschicht-Widerstand *m*

cathode ray Kat(h)odenstrahl *m*

(cathode-ray) oscillograph [*A photographic or other permanent record is produced by the cathode ray of the cathode-ray tube*] Kat(h)odenstrahloszillograph *m*

(cathode-ray) oscilloscope, CRO, scope [*A test instrument that uses a cathode ray tube to make visible on a fluorescent screen the instantaneous values and waveforms of electrical quantities that are rapidly varying as a function of time or another quantity*] Kat(h)odenstrahloszilloskop *n*

cathode-ray storage tube Kat(h)odenstrahlspeicherröhre, Braunsche Speicherröhre, Bildspeicherröhre *f*

cathode-ray tube, Braun tube, CRT, electron-ray tube Kat(h)odenstrahlröhre, Braunsche Röhre, Bildröhre *f*

cathode-ray tube controller, CRTC Kat(h)odenstrahlröhrenregler, Braunsche Röhrenregler, Bildröhrenregler *m*

cathode sheath Kat(h)odenglimmschicht *f*

cathode spot Kat(h)odenfleck *m*

(cathode) sputtering Kat(h)odenzerstäubung *f*

cation exchange Kationenaustausch *m*

CATV → cable television

CATV, community antenna television, community aerial television Gemeinschaftsantennefernsehen *n* [*Die logische Schleife eines Koaxialkabelsystems zum Übertragen von Fernseh- und anderen Signalen an angeschlossene Teilnehmer von einem logischen Punkt im Kabelsystem – dem Kabelkopf (headend)*]

catwalk Laufsteg *m*

CAU, controlled access unit steuerbare Token-Ring-leitungsverteiler *m*

CAV, constant angular velocity konstante Winkelgeschwindigkeit *f*

cavil → kevel

cavity → cavity resonator

cavity → air space

cavity ceiling heating → void ceiling heating

cavity oscillator [*An ultra-high frequency oscillator whose frequency is controlled by a cavity resonator*] Hohlraumoszillator *m*

cavity radiator [*A heated enclosure with a small opening which allows some radiation to escape or enter; the escaping radiation approximates that of a blackbody*] Hohlraumstrahler *m*

cavity (resonator), microwave cavity, resonant cavity, resonant chamber, resonant element, tuned cavity, waveguide resonator [*Electromagnetism. A metal-lined enclosure whose interior geometry is designed to induce and support resonance at certain frequencies, commonly in the microwave range*] Hohlraumresonator *m*

CAV system, constant air volume system [*The zone temperature is controlled by modulating the temperature of the air supplied to the zone*] Konstantvolumenstromanlage *f*, System *n* mit konstantem Volumenstrom [*Die Zonentemperatur wird durch Veränderung der Zulufttemperatur der Zone geregelt. Heizen und Kühlen der Zone erfolgen im allgemeinen durch eine Ventilator-Konvektor-Einheit*]

C

c.b. → circuit-breaker

C-battery → grid battery

CBDS, connectionless broadband data service verbindungslose Breitbanddatendienst *m*

c.b. for equipment → circuit-breaker for equipment

C/B ratio, saturation coefficient [*The ratio of the weight of water absorbed by a masonry unit during immersion in cold water to weight absorbed during immersion in boiling water; an indication of the prob-able resistance of brick to freezing and thawing*] Sättigungskoeffizient *m*

C.B system → central battery system

CBX → computerized branch exchange system

CC → crystal-controlled

CC → communication controller

CC → conversion command

CC → carbon copy

CCB → completion of calls for busy subscriber

CC-Bus → consumer's electronic bus

CCD → bucket-brigade device

CCDX → connecting circuit duplex

c-cell (US) → R 14 cell

C C_1 F_3 → R-13

CCI → charge-coupled imager

C CL_2 F_2 → R-12

C CL_3 F → R-11

CCO, call connected Anrufverbindung *f* geschaltet

CCP → console command processor

CCP, call confirmation protocol Bestätigungsprotokoll *n* [*dezentrale Zeichengabe*]

CCR → continuity check request

CCS → constant-current source

CCTV, closed-circuit television Kabelfernsehen *n*

CCW → counterclockwise

cd → candela

CD → compact disc

CD → carrier detect

C

CD → communication device

CDD → cooling degree-days

CD-DA → compact disc-digital audio

CDI, called line identity gerufene Anschlusskennung *f* [*zentrale Zeichengabe*]

CDR, CD recordable bespielbare digitale Schallplatte, digitale bespielbare Schallplatte *f*

CD-ROM, compact disk ROM, compact disc ROM Kompaktplatten-Nur-Lese-Speicher *m* [*Ein Audio- bzw. Videoplatte als Nur-Lese-Speicher mit Kapazitäten im Gigabytebereich*]

CDS → code and data separation

CE → chip enablement

ceiling [*The overhead surface of a room, usually a covering or decorative treatment used to conceal the floor above or the roof*] Decke, Unterdecke, nichttragende Decke *f*

ceiling air Deckenluft *f*

ceiling air conditioner, ceiling air conditioning unit Deckenklima(tisierungs)gerät *n* [*An der Decke aufgehängtes Klimagerät Teilen waag(e)recht nebeneinander angeordneten Teilen.*]

ceiling air cooler Deckenluftkühler *m*

ceiling air distributor Deckenluftverteiler *m*

ceiling air grill(e) → (air) ceiling grill(e)

ceiling air heater, ceiling air heat emitter, ceiling air heating appliance, ceiling air heating apparatus, ceiling air heating device Deckenluftheizer, Deckenlufterhitzer, Deckenluft(er)wärmer, Deckenluftheiz(ungs)apparat *m*, Deckenluftheiz(ungs)gerät *n*

ceiling air manifold, ceiling air distributor Deckenluftverteiler(rohr) *m*, *(n)*

ceiling (air) outlet, ceiling distribution outlet Decken(luft)auslass *m*

ceiling air supply Deckenluftzufuhr, Deckenluftversorgung, Deckenluftzuführung *f*

(ceiling) cassette-ventilation system Lüftungsdeckenkassettensystem, Deckenlüftungskassettensystem, Kassettendeckenlüftungssystem *n*

ceiling cavity → ceiling void

ceiling cooler stille Kühler, konvektive Deckenkühler, Deckenkühl(ungs)konvektor *m*

ceiling cover strip Deckendeckleiste *f*

ceiling detector, ceiling detecting device, ceiling detection device Deckenmelder *m*, Deckenmeldegerät *n*

ceiling diffuser, ceiling diffusor [*Supply-air diffuser designed for ceiling mounting*] Deckendiffusor *m*

ceiling evaporator Deckenverdampfer *m*

ceiling fan (unit) [*It is mounted in ceiling space between joists and discharges vertically, through a round duct through the roof connection*] Deckenventilator *m*

ceiling fan coil unit, ceiling forced convector Deckenkonvektor, Gebläsekonvektor *m* für Deckeneinbau

ceiling filter Deckenfilter *m*, *n*

ceiling finish Deckenoberflächengestaltung *f*

ceiling (fire) sprinkler [*See remark under "sprinkler"*] Deckensprinkler, Deckensprenger *m*

ceiling forced convector → ceiling fan coil unit

ceiling grill(e) → (air) ceiling grill(e)

ceiling guide rail Deckenführungsschiene *f*

ceiling hook [*Hook for use in ceilings, usually having two prongs*] Deckenhaken *m*

ceiling hung blower, ceiling suspended blower Deckenhängegebläse *n*, Deckenhängebläser *m*

ceiling illumination, ceiling lighting Deckenbeleuchtung *f*

ceiling installation → ceiling mounting

ceiling(-mounted) air diffusor, overhead air diffuser Deckenluftdiffusor *m*

ceiling(-mounted) equipment, overhead equipment Deckengerät(e) *n*, *(f)*

ceiling(-mounted) fan coil (unit), overhead fan coil (unit) Deckenklimatruhe *f*

ceiling(-mounted) lighting, overhead lighting, ceiling(-mounted) illumination, overhead illumination Deckenbeleuchtung *f*

ceiling(-mounted) luminaire, ceiling(-mounted) light(ing) fitting, overhead luminaire, overhead light(ing) fitting Deckenleuchte *f*, Deckenbeleuchtungskörper *m*

ceiling(-mounted) space heater, ceiling(-mount-ed) unit heater, overhead space heater, overhead unit heater Deckenraumheizer *m*

ceiling(-mounted) spray nozzle, overhead spray nozzle Deckensprühdüse *f* [*Brandbekämpfung*]

ceiling mounting, ceiling installation Deckenmontage *f*

ceiling outlet Deckenauslass *m*

ceiling panel Deckentafel *f*

ceiling panel heating, panel ceiling heating Deckentafelheizung, Tafeldeckenheizung *f*

ceiling panelling (Brit.); ceiling paneling (US) Decken(ver)täf(e)lung *f*

ceiling panel radiant heating, panel ceiling radiant heating Deckenstrahl(ungs)plattenheizung, Strahl(ungs)deckenplattenheizung *f*

ceiling pendant Hängeleuchte *f*

ceiling penetration Deckendurchführung *f* [*nichttragende Decke*]

ceiling plaster Deckenputz *m*

ceiling plaster panel Deckenputztafel *f*

ceiling plaster tubular heating → **tubular ceiling plaster heating**

ceiling radiant cable [*Electric cable embedded in the ceiling for heating*] (Decken)Strahl(ungs)heiz(ungs)-kabel *n*

ceiling radiant heating, radiant ceiling heating (Decken)Strahl(ungs)heizung, Strahl(ungs)deckenheizung *f*

ceiling radiation panel, radiant ceiling panel (Decken)Strah(lungs)heiz(ungs)platte *f*

ceiling radiator Deckenstrahler *m*

ceiling recessed floodlight → **recessed ceiling floodlight**

ceiling recessed reflector lamp → **recessed ceiling reflector lamp**

ceiling reflection [*Light reflection resulting in visibility reduction, or no contrast*] Schleierreflexion, Reflexkontrastminderung *f*

ceiling rose [*An electric(al) junction on the ceiling to which luminaires are attached or suspended*] Deckenrosette *f*

ceiling space Deckenraum *m*

ceiling supply air Deckenzuluft *f*

ceiling supply air distributor Deckenzuluftverteiler *m*

(ceiling) troffer (luminaire) (Decken)Langfeldleuchte, Wannenleuchte *f*, eingelassene Lichtband *n*

ceiling void, ceiling cavity, ceiling space Decken(hohl)raum *m*

cellar drain, cellar inlet Kellerablauf, Kellereinlauf *m* [*DIN 591*]

cellar draining, cellar drainage Kellerentwässerung *f*

cellar ejector (pump), cellar draining pump, cellar drainage pump Kellerentwässerungspumpe *f*

cell broadcasting Funkzelleninformationssystem *n*

cell computer Zellenrechner *m*

cell net(work) Zellennetz(werk) *n*

cell(-type) inclined-current air filter Zellenschrägstromluftfilter, Schrägstromzellenluftfilter *m, n*

cellular filter, cartridge filter [*Air filter of juxtaposed square or rectangular elements which can be easily dismantled for cleaning or replacement*] Patronenfilter, Zellenfilter *m, n*

cellular plastic, foam plastic Schaumkunststoff *m*

cellular plastic fire → **foam plastic fire**

cellular polyethylene, cellular PE verzellte Polyethylen, Zell-PE, verschäumte Polyethylen *n*, verzellte Polyäthylen, verschäumte Polyäthylen

cellular radio network → **CRN**

cellular storage heater Zellenspeicher(wasser)erhitzer *m*

cellulose nitrate, nitro-cellulose, nitro-cotton Nitrozellulose *f*

cellulose pad Zellulosevlies *n*; Zellulosevließ [*Österreich*] [*Filtermedium*]

15° (Celsius) calorie 15°-Kalorie *f* [*Die zur Erwärmung von 1 kg Wasser von 14,5 auf 15,5°C bei 760 Torr erforderliche Wärmemenge*]

cement-lime mortar → **gauged mortar**

cement mortar lining Zementmörtelauskleidung, Zementmörtelverkleidung, Zementmörtelbekleidung *f*

to center (US) → **to centre (Brit.)**

center (US) → **central station**

center (US); centre (Brit.) Nuss *f* [*Nussband*]

center frequency (US) → **carrier frequency**

center of area, center of figure (US); centre of area, centre of figure (Brit.); centroid [*For a plane figure, the centre of mass of a thin uniform plate having the same boundaries as a plane figure*] Flächenschwerpunkt *m*

center of mass, center of inertia (US); centre of mass, centre of inertia (Brit.); centroid Massenzentrum *n*

center of volume, center of figure (US); centre of volume, centre of figure (Brit.); centroid [*For a three-dimensional figure, the centre of mass of a homogeneous solid having the same boundaries as the figures*] Volumenschwerpunkt *m*

center positive (US); centre positive (Brit.) mittig positiv

centigrade temperature, Celsius temperature; centigrade scale [*obsolete*] [*Temperature scale used with the SI system in which the freezing point of water is 0°C, the triple point is 0.01°C, and the boiling point is 100°C*] Celsiustemperatur *f*

centimetric wave [*Radiation in the super-high-frequency band, having a wavelength between 1 and 10 centimeters, corresponding with frequencies between 3 and 30 gigahertz*] Zentimeterwelle *f*

central aerial, master aerial; central antenna, master antenna (US) Gemeinschaftsantenne *f*

central air conditioner, central air conditioning unit Zentralklima(tisierungs)gerät *n*

central air conditioning and ventilation plant → **central ventilation and air conditioning plant**

central air conditioning heating → **air conditioning central heating**

C

central air conditioning plant, central air conditioning installation Klimazentrale *f* [*Sie wird auf der Baustelle zusammengebaut*]

central air distributor → air manifold

central air heater Zentrallufterhitzer, Zentralluft(er)-wärmer, Zentralluftheizer *m*, Zentralluftheizgerät *n*

central (air) humidifier [*Device which humidifies air circulated through ducts in an air-conditioning system*] Zentral(luft)befeuchter, Zentral(luft)anfeuchter *m*

central annunciator board → central indicator board

central automatic ((open) loop) control → automatic central ((open) loop) control

centralbattery system, C.B. system [*A telephone or telegraph system which obtains all the energy for signal(l)ing (and for speaking in the case of the telephone) from a single battery of secondary cells locat-ed at the main exchange*] Zentralbatteriesystem *n*

central boiler installation, central boiler plant Kesselzentrale, Zentralkesselanlage *f*

central board → central panel

central (building) control and monitoring station → central (building) monitoring and control station

central burglar(y) detector station, central burglar(y) detecting device station, central burglar(y) detection device station [*"Breaking-in" may synonymously be used for "burglar(y)"*] Einbruchmeldezentrale, Einbruchdetektorzentrale, Einbruchmeldegerätzentrale, Einbruchmeldungsgeberzentrale, Einbruchnachweiszentrale *f*

central bus station Buszentrale *f* [*Zentrales Betriebsmittel einer Gefahrenmeldeanlage in Bus-Technik*]

central chamber Kammerzentrale *f* [*Kombination von Bauelementen in Kammerbauweise*]

central circulation air system, central circulating air system Zentral-Umluftsystem *n* [*Zentrale Umluftbehandlung für den Versorgungsbereich*]

central cold steam heating → cold steam central heating

central control and monitoring station → central monitoring and control station

central control facility zentrale Leiteinrichtung, Zentralleiteinrichtung *f*

central controller Zentralregler *m*

central control unit → primary control unit

central cryogenic station → cryogenic central station

central danger detection station → danger detection central station

central data processing [*The processing of all the data concerned with a given activity at one place, usually with fixed equipment within one building*] Zentraldatenverarbeitung *f*

central data station → data (processing) center

central dispatching system Einsatzleitsystem *n*

central display board → central indicator board

central distributor → header

central distributor valve → manifold valve

central domestic hot water, central domestic warm water [*C.P. (Code of Practice) 342*] Zentralbrauchwasser *n*

central electrical heating → electrical central heating

central electrical storage heating → electrical central storage heating

central emergency call station → emergency call central station

central emergency unlocking Zentralnotentrieg(e)-lung, Notzentralentrieg(e)lung *f*

central equipment Zentralgerät(e) *n, (f)*

central failure alarm system, central alarm signal system Störmeldezentrale *f*

central fan system [*Mechanical indirect system of heating, ventilating, or air conditioning where the air is treated or handled by equipment located outside the rooms served, usually at a central location, and conveyed to and from the rooms by means of a fan and a system of distributing ducts*] Zentralventilatorsystem *n*

central filter Zentralfilter *m, n*

central fire detection system Brandmelde(r)zentrale, Feuermelde(r)zentrale *f*

central furnace Zentralfeuerung, Zentralfeuerstätte *f*

central gas heating Gaszentralheizung, Zentralgasheizung *f*

central gas heating boiler → gas central heating boiler

central heating Zentralheizung, Sammelheizung *f* [*Fehlbenennung*]

central heating boiler Zentralheiz(ungs)kessel *m*

central heating boiler with warm water heater Zentralheiz(ungs)kessel *m* mit Warmwasserbereiter

central heating furnace Zentralheiz(ungs)feuerstätte, Zentralheiz(ungs)feuerung *f*

central heating installation, central heating plant Heiz(ungs)zentrale *f*

central heating room Zentralheiz(ungs)raum *m*

central heating station → heating central station

central heating system [*It consists of a centralized heat source, a distributing system and local heat emitting units. In general, the type of heating system is designated by the distributing medium and the main types are: low-pressure hot water, high-pressure hot water, steam, warm air, centralized heat*] Zentralheiz(ungs)system *n*

central heat pump station Wärmepumpen-Heiz(ungs)zentrale *f*

central heat supply station, heat supply central station Wärmeversorgungszentrale *f*

central high-pressure heating system → high-pressure (central heating) system

central hold-up detection installation → hold-up detection central installation

central hold-up detector station, central hold-up detecting device station, central hold-up detection device station Überfallmelderzentrale, Überfalldetektorzentrale, Überfallmelde(gerät)zentrale, Überfallmeldungsgeberzentrale *f*

central horizontal furnace → horizontal central furnace

central hot air heating → hot air central heating

central hot air heating system → hot air central heating system

central hot water heating, hot water central heat-ing Zentralwarmwasserheizung, Warmwasserzentralheizung *f*

central hot water (preparation) installation, hot water central (preparation) installation [*"Plant" and "station" may synonymously be used for "installation"*] zentrale Warmwasserbereitungsanlage *f*, WWB-Zentrale, Warmwasserzentrale *f* [*"...installation" kann synonym für "...anlage" verwendet werden*]

central humidifier → central (air) humidifier

central indicator board, central indicator panel, central annunciator board, central annunciator panel, central display board, central display panel Zentral(anzeige)tableau *n*, Zentral(anzeige)tafel *f* [*Siehe Anmerkung unter "Anzeigetableau"*]

central installation → central plant

central intelligence [*A system where algorithm proc-essing is only possible at the central station. The outstations are dormant when not in contact with the central station*] zentrale Intelligenz *f*

centralized air conditioning plant, centralized air conditioning installation zentral(isiert)e Klima(tisierungs)anlage, Klima(tisierungs)zentrale *f*

centralized alarm Sammelalarm, Zentralalarm *m*

centralized ((closed) loop) control zentrale Reg(e)lung *f*

centralized data processing Zentrale Datenaufbereitung *f*

centralized fault, centralized malfunction Sammelstörung *f*

centralized fault signal message detector, central-ized malfunction signal message detector, centralized fault signal message detecting device, centralized malfunction signal message detecting device Sammelstör(ungs)melder *m*, Sammelstör(ungs)meldegerät *n*

centralized ((open) loop) control zentrale Steuerung *f*, Zentralsteuerung

centralized system [*A system with a heating or cooling source and a ducted distribution network*] Zentralsystem *n*

central key system Zentralschlossanlage *f*, Zentralschlosssystem *n*

central location, central site Zentralstelle *f*

centrally ((closed) loop) controlled zentralgeregelt

centrally cooled zentralgekühlt

centrally ((open) loop) controlled zentralgesteuert

centrally supplied zentralversorgt

central master key, central main key, central principal key Zentralhauptschlüssel *m*

central memory → principal memory

central minibore heating → minibore central heating

central mix(ture) concrete Transportbeton *m*

central monitoring and control station, central control and monitoring station Leitzentrale *f*

central monitoring board, central monitoring panel Zentralüberwachungstableau *n*, Zentralüberwachungstafel *f*

central net(work) controller Zentralnetz(werk)-regler *m*

central oil-heated heating → oil-heated central heating

central ((open) loop) control panel, central ((open) loop) control board Zentralsteuer(ungs)-tafel *f*, Zentralsteuer(ungs)tableau *n*

central ((open) loop) control processor Zentralsteuer(ungs)prozessor *m*

central ((open) loop) control unit [*A computer that provides the primary control for an entire control system*] Zentralsteuer(ungs)rechner *n*

central panel, central board Zentraltafel *f*, Zentraltableau *n*

central plant, central installation Zentral(anlag)e *f*

central principal key system, central main key system, central master key system Zentralhauptschlüsselsystem *n*

C

central processing unit, CPU, frame, main frame, central processor (unit) Zentraleinheit *f*

central process (I/O) image zentrale Prozessabbild, zentrale PA *n* [*Es wird auf einer zentralen Vorrichtung geführt*]

central radio aerial, master radio aerial; central radio antenna, master radio antenna (US) Gemeinschaftsrundfunkantenne, Rundfunkgemeinschaftsantenne, Gemeinschaftsantenne, Gemeinschaftsradioantenne, Radiogemeinschaftsantenne *f*

central radio station, radio central station Funkzentrale, Zentralfunkstelle *f*

central receiver Zentralempfänger *m*

central recirculated air system Umluftzentralsystem, Zentralumluftsystem *n* [*System mit Umluftbehandlung zentral für den gesamten Versorgungsbereich*]

central refrigeration station, refrigeration central station Kältezentrale, Zentralkälteanlage *f*

central regulated air, regulated central air Zentralumluft *f* [*Abluft aus einem oder mehreren Versorgungsbereichen, die zentral der oder den RLT-Anlage(n) wieder zugeführt wird. DIN 1946, Teil 1*]

central release, central releasing zentrale Freischaltung, zentrale Abschaltung *f*

central roof air-conditioning installation, central roof air-conditioning plant [*See remark under "air conditioning installation"*] Dachklima(tisierungs)zentrale *f*

central roof station → roof (central) station

central safety station, safety central station Sicherheitszentrale *f*

central sanitary station, sanitary central station Sanitärzentrale *f*

central signal message Sammelmeldung *f*

central signal station Signalzentrale *f*

central stage → principal memory

central station Zentrale *f*

central (station) refrigeration [*See remark under "refrigeration"*] zentrale Kälteerzeugung *f*

central steam distributor → steam manifold

central steam heating → steam central heating

central steam plant, steam central plant, central steam installation, steam central installation Zentral-Dampfanlage, Dampf-Zentralanlage *f*

central storage → principal memory

central storage heating Zentralspeicherheizung *f*

central substation Unterzentrale *f* [*VDI 3814 Blatt 1. Die Unterzentrale (UZ) ist hierarchisch in der Gruppenleitebene angeordnet. Sie übernimmt einen Teil der erforderlichen Grund- und Verarbeitungsfunktionen durch programmierte Funktionsabläufe, die durch Informationen von den Unterstationen bzw. von der Leitzentrale (LZ) oder innerhalb der Unterzentrale selbst ausgelöst werden und selbständig ablaufen. Die Möglichkeit zum Anschluss von Ein- und Ausgabegeräten ist gegeben.*]

central supply zentrale Versorgung *f*

central switch(ing) cabinet, central switch(ing) cubicle Zentralschaltschrank *m*

central switch(ing) over zentrale Umschaltung *f*

central telephone system Telefonzentrale, Fernsprech(er)zentrale *f*

central terminal [*In a distributed processing environment, a device that controls communication between the user terminals and central processing unit*] Zentralterminal *n*

central T.V. aerial, master T.V. aerial; central T.V. antenna, master T.V. antenna (US) Gemeinschaftsfernsehantenne, Fernsehgemeinschaftsantenne *f*

central unit Zentraleinheit *f*

central ventilation, central vent(ilat)ing [*See remark under "ventilation"*] Zentrallüftung *f*

central ventilation and air conditioning plant, central ventilation and air conditioning installation, central air conditioning and ventilation plant, central air conditioning and ventilation installation Lüftungs- und Klima(tisierungs)zentrale, Klima(tisierungs)- und Lüftungszentrale *f*

central ventilation station, ventilation control centre (Brit.) [*US spelling is "center"*] Lüftungszentrale *f*

central ventilation system Zentrallüftungsanlage *f*

central warm air heating → hot air central heat-ing

central warm air heating system → hot air central heating system

central water distributor → water manifold

central water supply zentrale Wasserversorgung *f* [*DIN 4046*]

to centre (Brit.); to center (US) zentrieren

centre (Brit.); center (US) Nuss, Bandnuss *f* [*Nussband*]

centre bowl sink (Brit.); center bowl sink (US) Mittelbeckenspüle, Spüle *f* mit Mittelbecken

centre-break disconnector (Brit.); center-break disconnector (US) Drehtrennschalter, Trenndrehschalter *m*

centre feeder box (Brit.); center feeder box (US) Mitteneinspeisungskasten *m*

centre gutter (Brit.) → valley gutter

centre of area defuzzification; center of area defuccification (US); COA defuzzification Schwerpunktmethode *f*, Schwerpunktverfahren, Defussifizierungsverfahren *n*, Defussifizierungsmethode *f*

centre of largest area defuzzification; center of largest area defuzzification (US) Schwerpunkt *m* der größten Defuzzifizierungsfläche

centre of sums defuzzification; center of sums defuzzification (US); COS defuzzification Schwerpunktsummenmethode *f*, Schwerpunktsummenverfahren *n*

centre-opening door (Brit.); center-opening door (US); biparting door zentralöffnende und zentralschließende Schiebetür *f*

centre-to-centre distance (Brit.); center-to-center distance (US) Mittenabstand, Achs(en)abstand *m*

centrifugal air separator, centrifugal air separation unit [*See remark under "air separation unit"*] Zentrifugalluft(ab)scheider, Schleuderluft(ab)-scheider, Fliehkraftluft(ab)scheider *m*

centrifugal atomizer Zentrifugalzerstäuber, Schleuderzerstäuber, Fliehkraftzerstäuber *m*

centrifugal blower Zentrifugalgebläse, Schleudergebläse, Fliehkraftgebläse *n*, Schleuderbläser, Fliehkraftbläser, Zentrifugalbläser *m*

centrifugal brake operator Bremslüfter *m* [*elektromechanisch*]

centrifugal ((closed) loop) control Zentrifugalregler, Schleuderregler, Fliehkraftregler *m*

centrifugal compressor Kreiselkompressor, Kreiselverdichter *m*

centrifugal fan [*The air flows vertically through the impeller. Centrifugal fans are classified according to wheel blading: forward-curved or forward-bladed, backward-curved or backward-bladed and straight-bladed (radial)*] Zentrifugalventilator, Schleuderventilator, Fliehkraftventilator *m*

centrifugal (fan) blade Zentrifugalventilatorschaufel, Schleuderventilatorschaufel, Fliehkraftventilatorschaufel *f*

centrifugal fan unit heater, industrial blower heater [*See remark under "unit heater"*] Schleuderventilator-Gebläsekonvektor, Zentrifugalventilator-Gebläsekonvektor, Fliehkraftventilator-Gebläsekonvektor *m*

centrifugal freeze drying, centrifugal lyophilization Schleudergefriertrocknung, Zentrifugalgefrier-trocknung, Fliehkraftgefriertrocknung *f*

centrifugal governor [*See remark under "governor"*] Zentrifugalregler, Schleuderregler, Fliehkraftregler *m*

centrifugal humidifier Schleuderbefeuchter, Schleuderanfeuchter, Zentrifugalbefeuchter, Zentrifugalanfeuchter, Fliehkraftanfeuchter, Fliehkraftbefeuchter *m* [*DIN 30 600*]

centrifugal interceptor → **centrifugal separator**

centrifugal pipe fan Schleuderrohrventilator, Zentrifugalrohrventilator, Fliehkraftrohrventilator *m*

centrifugal psychrometer Schleuderpsychrometer, Zentrifugalpsychrometer, Fliehkraftpsychrometer *n*

centrifugal (refrigerant) compressor [*See remark under "(refrigerant) compressor"*] Zentrifugal(-Kältemittel)verdichter, Zentrifugal(-Kältemittel)kompressor, Schleuder(-Kältemittel)verdichter, Schleuder(-Kältemittel)kompressor, Fliehkraft(-Kältemittel)verdichter, Fliehkraft(-Kältemittel)kompressor *m*

centrifugal separator, centrifugal trap, centrifugal interceptor [*See remark under "interceptor"*] Zentrifugal(ab)scheider *m*

centrifugal switch Fliehkraftschalter, Zentrifugalschalter, Schleuderschalter *m*

(centrifugal) thruster Bremslüfter *m* [*hydraulisch*]

centrifugal trap → **centrifugal separator**

centroid → **center of area**

centroid → **center of volume**

centroid → **center of mass**

CER → **character error rate**

ceramic capacitor, ceramic (electric) condenser Keramikkondensator *m*

ceramic pedestal Keramiksäule *f* [*Waschbecken*]

ceramics Keramik *f*

ceramic slow combustion stove keramische Dauerbrandofen, Keramikdauerbrandofen *m*

ceramic vanity Keramikwaschtisch *m*

ceramic washbasin Keramikwaschbecken *n*

certificate (Prüf)Zeugnis *n*

certificate of acceptance, acceptance certificate Abnahme(prüf)zeugnis *n*

certificate of authenticity, authenticity certificate Echtheitszeugnis *n*

certification Zertifizierung, Zulassung *f*

certification mark Zulassungszeichen, Zertifizierungszeichen *n*

certification test Zulassungsprüfung, Zertifizierungsprüfung *f*

certified zertifiziert, zugelassen

cesspipe Abortabfallroh, Abtrittabfallrohr, Ab-fallabortrohr, Abfallabtrittrohrr *n*

CF → **control field**

CFM, cubic feet/minute, ft^3 per minute, cfm [*Used to designate the volume rate of air flow*] Kubikfuß/ Minute *m*

CFPP → **cold filter plugging point**

CG → **clock generator**

CGS → **continuous grain silicon**

cgs system, centimeter-gram-second system Zentimeter-Gramm-Sekunden-System *n*

C

chad, chip [*The piece of material removed when forming a hole or notch in a punched tape or punched card*] Span, Stanzrückstand *m*, Schuppe *f*, Stanzabfall *m*

chad tape, chip tape durchgestanzte Lochstreifen *m*

chain(-driven) elevator (US); chain(-driven) lift (Brit.) [*Elevator having its machine connected to a reversible motor, engine, or turbine by a chain*] Kettenaufzug *m*

chained list → linked list

chain grate stoker Wanderrost *m*

chain holder Kettenhalter *m*

chaining, linking (together) [*Of subprogram(me)s to form one master program(me)*] (Ver)Ketten, Aneinanderreihen *n*

chain insulator, suspension insulator Kettenisolator, Hängeisolator *m*

chainless kettenlos

chain printer, train printer [*An impact printer in which the printing characters are linked together in a chain or in a train of print slugs*] Kettendrucker *m*

chain pull Kettenzug *m*

chain pull(ing) door release Kettenzugtüröffner *m*

chain sampling plan, ChSP Kettenstichprobenplan *m*

chain shackle Kettenbügel *m*

chain tensioning device [*escalator*] Kettenspannvorrichtung *f*, Kettenspanner *m*

chair seating Bestuhlung *f*

chamber [*deprecated*] **→ manhole**

chamber (air) lock Kammer(luft)schleuse *f*

chamber-door lock, room-door lock Zimmertürschloss, Raumtürschloss *n*

chambermaid('s) sink → bucket sink

chamber superheater Kammerüberhitzer *m*

chamber type, chamber design Kammerbauweise, Kammerkonstruktion, Kammerbauart, Kammerausführung *f* [*Kombinationsart von Bauelementen in am Aufstell(ungs)ort aufgebauten Kammern, deren gemeinsames Transportieren und Versetzen aufgrund ihres konstruktiven Aufbaues nicht möglich ist*]

to change → to replace

to change, to switch schalten

to change, to alter (ab)ändern, umändern, verändern

change → replacement

change (Ab)Änderung *f*, Umänderung, Wechsel *m*

changeable, replaceable (aus)tauschbar, (aus-)wechselbar

changeable database driver → replaceable database driver

changeable part → replaceable part

changed verändert, abgeändert, umgeändert

change data, transaction data Bewegungsdaten, Änderungsdaten *f* [*Sie ändern sich ständig und werden in der Regel nur für einen bestimmten Prozess benötigt, z. B. Datum, Uhrzeit, Mengen usw.*]

change dump → differential dump

change file, transaction file Änderungsdatei, Bewegungsdatei *f*

change key, master key, pass key [*Key of change-keyed lock that differs from all others of the same series and will operate only its own lock (sometimes called room key). The term "change key" is used rather than "master key"*] Hauptschlüssel, Generalschlüssel *m*

change-key(ed) lock, master-key(ed) lock, pass-key(ed) lock Hauptschlüsselschloss, Generalschlüsselschloss *n*

change (lock) striking plate → replacement (lock) striking plate

change of state, COS, state change Zustandsänderung *f* [*EN V 1805-1 BACnet*]

change-of-state announcement Meldeanreiz *m* [*Fernwirktechnik*]

change of use, use change, utilization change, change of utilization Benutzungs(ver)änderung, Gebrauchs(ver)änderung *f*, Benutzungswechsel, Gebrauchswechsel, Nutzungswechsel *m*, Nutzungs(ver)änderung, Umnutzung *f*

change of value, value change, COV Wert(e)-(ver)änderung *f*, Wert(e)wechsel *m* [*CEN TC 247 WG 3 Teil 2 Entwurf*]

to change over umschalten, wechseln

changeover Umschaltung *f*, Wechsel *m*

changeover air-conditioning system Zweirohr-Klima(tisierungs)anlage *f* mit Umschaltung

changeover contact Umschalt(ungs)kontakt, Wechselkontakt *m*

changeover delay Umschalt(ungs)verzögerung, Wechselverzögerung, Umschaltverzögerung *f*, Wechselverzug *m*

changeover device, changeover appliance [*See remark under "device"*] Umschalt(ungs)gerät, Wechselgerät *n*, Umschalt(ungs)vorrichtung, Wechselschalt(ungs)vorrichtung *f*

changeover flap Umschalt(ungs)klappe, Wechselschalt(ungs)klappe *f*

changeover mechanism Umschalt(ungs)mechanismus, Wechselmechanismus *m*

changeover relay Umschalt(ungs)relais, Wechselrelais *n*

changeover rocker(-dolly) switch Wipp(en)umschalter, Wipp(en)wechselschalter *m*

changeover spray shower head, changeover shower rose [*See remark under "spray shower head"*] Umschalt(ungs)brausekopf, Wechselbrausekopf *m*

changeover surge Umschalt(ungs)stromstoß, Wechselstromstoß *m*

changeover valve Umschalt(ungs)ventil, Wechselventil *n*

changeover switch, double throw switch, dual throw switch, twin throw switch, two-throw switch, throw-over switch, transfer switch [*A means of moving an electric(al) circuit from one set of connections to another*] Umschalter, Wechsler, Wechselschalter *m*

changeover temperature [*That outside temperature at which the air-conditioning system is switched from summer to winter use, i.e., from chilled water to hot water*] Umschalt(ungs)temperatur, Wechseltemperatur *f*

change record, amendment record, transaction record [*A record that is used to alter information in a corresponding master record*] Änderungsaufzeichnung *f*

change tape, transaction tape [*A paper tape or magnetic tape carrying information that is to be used to update filed information; the latter is often on a master tape*] Änderungsband *n*

changing over Umschalten, Wechselschalten *n*

channel [*Portion of a spectrum assigned for the op-eration of a specific carrier and the minimum number of side bands necessary to convey intelligence; single path for transmitting electric signals*] Kanal *m*

channel [*The portion of storage media, such as a track on a magnetic tape or disk, that is available to a single character-reading state*] Kanal *m*

channel Kanal, Übertragungskanal *n* [*Eine spezielle Bandbreitenzuordnung in einem Radiofrequenzbereich (zum Beispiel in einem Breitband LAN) zur Übertragung eines Nachrichtensignals*]

channel [*A device associated with the input/output controller that transfers data between the processor and one or more peripheral devices; three types of channels are in use: selector channel, byte-multiplexer channel, and block-multiplexer channel*] Kanal *m*

channel [*A path along which digital or other information may flow in a computer*] Kanal *m*

channel, level [*One of the longitudinal rows of intelligence holes, punched along the length of paper tape*] Kanal *m*

channel, connection, line, link, path [*The specific path used for an electronic transmission*] Kanal *m*, Verbindung, Leitung *f*, Anschluss, Durchgang *m*

channel, U(-shap(ed)) section, section with U (-shap(ed)) cross section U-Profil *n*

channel [*In a field effect transistor, the conducting path between the source and the drain*] Strompfad, Stromweg *m*

channel adapter, input/output adapter [*Any of a variety of devices used to allow data transfer between devices that operate at differing transfer rates; rarely "adaptor"*] Kanalanpasser *m*

channel black, channel soot Kanalruß *m*

channel block generation [*EIB range*] Kanalblockbildung *f*

channel bundling Kanalbündelung *f*

channel capacity [*The rate at which data arrive on the channel, measured in bit/s*] Kanalkapazität *f*

channel (control) command (word) [*Computer programming. An instruction stored in a main memory or a special register that is decoded and executed by the I/O channels; usually contains an operation code (such as read or write), flags and options, the start-ing address and the word count of the data on which the operation is to be performed*] Kanalkommando *n*

channel designator → channel sequence number

channel effect [*Electronics. The leakage that occurs over the surface path between collector and emitter in some types of transistors*] Halbleiterelektron *n*

channel-end condition signal [*A signal that occurs when the data transmission is completed; the channel is considered busy until the CPU accepts this condition*] Datenübertragungsendesignal *n*

channel identification Kanalkennung *f*

channel miles [*A measurement, in miles, multiplied by the number of communication channels between two given points*] Kanalmeilen *f*

channel monitoring Kanalüberwachung *f* [*Fernwirktechnik*]

channel(-mounting) luminaire, track(-mounted) luminaire, track-suspended luminaire, trunking luminaire, channel-mounted luminaire Schienenleuchte *f*

channel program (US); channel programme (Brit.) [*Computer engineering. A list of instructions or channel commands stored in main memory that the channel retrieves and executes sequently to perform input and output operations*] Kanalprogramm *n*

channel pulse [*Telecommunications. An electrical pulse conveying channel information by modulation*

C

of that pulse, e.g. pulse-duration modulation] Kanalimpuls *m*

channel register Eingangsregister, Summandenzählwerk *n*

channel schedule [*Computer programming. A program(me) in a computer operating system that processes the list of requests for input and output operations in the appropriate order*] Kanalablaufsteuerung *f*

channel selector, CS [*Electricity. Any control, like a switch, that tunes in a desired channel*] Kanalwähler *m*, Kanalwählvorrichtung *f*

channel sequence number, channel designator [*Telecommunications. A unique identification assigned to a channel for reference purposes*] Kanalfolgenummer, Kanalablaufnummer *f*

channel skip (character) [*Computer engineering. A carriage-control character that instructs the printer to skip to the top of the next page or to carry out a vertical tab*] Kanalübersprungzeichen *n*

channel soot → channel black

channel status block Kanal-Status-Block *m*

channel (strip) [*Metal enclosure containing the ballast, starter, lampholder, and wiring for fluorescent tubes*] Kanal *m*

channel translator [*A device installed in a broadband LAN in order to raise the transmission frequency and to allocate the channels, the incoming signals to the headend being converted into outgoing signals*] Kanalumsetzer *m*

channel type Kanalart *f* [*Fernwirktechnik*]

channel urinal → urinal trough

channel wave-guide Streifenwellenleiter *m*

chaplet-hinge Paternosterband *n* [*Schlosserei*]

chapter Kapitel *n* [*Zusammenfassung einer Text- und Bilddatei in einem festgelegten Format*]

character Zeichen, alphanumerische Zeichen *n* [*Kleinste Einheit aus der sich Daten zusammensetzen (binäre Zeichen, Buchstaben, Ziffern und Sonderzeichen)*]

character and pattern telephone access information network system, captain Zeichen- und Bild-Zugriffssystem *n* über Fernsprechleitungen

character base line Zeichengrundlinie *f*

character code [*One of several standard sets of binary representations for the alphabet, numbers, common symbols, and control functions, common character codes include ASCII, Baudot, BCD and EBCDIC*] Zeichenkode, Zeichencode *m*

character density, record density, horizontal spacing [*The number of characters recorded per unit of length or area*] Zeichendichte *f*

character error rate, CER Zeichenfehlerrate, Zeichenfehlerhäufigkeit *f*

character fill(ing) Speicher(auf)füllung *f* [*Alle Speicherworte, d. h. alle Datenfelder eines Programms, werden mit dem gleichen Zeichen aufgefüllt*]

character font → character set

character-imaging zeichendarstellend

characteristic [*Inherent and measurable property of an object*] Kenn(zeichn)ung *f*, Merkmal *n*

characteristic [*A set of related values*] Kennwertreihe, Kennwerte *f*

characteristic → feature

characteristic angular frequency, angular characteristic frequency Kennkreisfrequenz, charakteristische Kreisfrequenz *f*

characteristic consumption value, consumption characteristic value Verbrauchskennwert, charakteristische Verbrauchswert *m*

characteristic curve Kennlinie, Kennkurve, charakteristische Linie, charakteristische Kurve *f*

characteristic curve basic form Kennliniengrundform, Kennkurvengrundform, charakteristische Liniengrundform, charakteristische Kurvengrundform *f*

characteristic curve tracer Kennlinienschreiber, Kennkurvenschreiber, charakteristische Linienschreiber, charakteristische Kurvenschreiber *m*

characteristic data, characteristics Kenndaten, charakteristische Daten *f*

characteristic equation Kenngleichung, charakteristische Gleichung *f*

characteristic function, indicator function charakteristische Mengenfunktion, Zugehörigkeitsfunktion, Kennmengenfunktion *f* [*unscharfe Logik*]

characteristic impedance, surge impedance Wellenwiderstand *m*, Kennimpedanz *f*, Schallwellenwiderstand *m*

characteristic line Kennlinie *f*

characteristic measure Kennmaß *n*

characteristic number → characteristic value

characteristic phenomenon Kennerscheinung, charakteristische Erscheinung *f*

characteristic power number, power characteristic number Leistungskennzahl, charakteristische Leistungszahl *f*

characteristic requirement number, characteristic need number, requirement characteristic number, need characteristic number Bedarf(s)-kennzahl, charakteristische Bedarf(s)zahl *f*

characteristic root → characteristic value

characteristics family → family characteristics

characteristic similitude value, similitude characteristic value Ähnlichkeitskennwert, charakteristische Ähnlichkeitswert *m*

characteristic value, characteristic root, characteristic number, eigenvalue latent root, proper value Kennwert, Bestimmungswert *m*

character mode Zeichenbetriebsart *f*

character-orient(at)ed computer [*A computer in which the location of individual characters, rather than words, can be addressed*] zeichenorientierte Rechner, zeichenweise Rechner *m*

character parity [*The addition of a redundant overhead bit to a character to provide error detection capability*] Zeichenparität, Zeichengleichstellung *f*

character path Zeichenweg, Zeichenpfad *m*

character printer → serial printer

character reader [*Any data-preparation device that can read, recognize, and convert written or printed characters into machine-processable form, such as an optical or magnetic reader*] Zeichenregistriergerät *n*

character reading Zeichenlesen *n*

character recognition [*Using a machine to sense and encode into a machine language the characters which are originally written or printed by human beings*] Zeichenerkennung *f*

character recorder [*A device capable of location identifying and translating into machine code the handwritten or printed data appearing on a source document*] Zeichenleser *m*

character sensing card Zeichenlochkarte *f*

character sequence Zeichenfolge *f*

character set, character font, set of characters, character collection, collection of characters [*A set of unique representations called characters, for example, the 26 letters of English alphabet, the Boolean 0 and 1, the set of signals in Morse code, and the 128 characters of the ASCII*] Zeichensatz, Zeichenvorrat *m*, endliche Zeichenmenge *f*, Zeichenrepertoire *n*

character skew Zeichenschiefe, Zeichenschrägstellung, Zeichenschräglage *f*

characters per inch, CPI Zeichen *f* je Zoll, Zeichen pro Zoll

characters per second, cps Zeichen *f* pro Sekunde, Zeichen je Sekunde

character string, string of characters [*A sequence of characters, which are regarded as a logical entity, in a computer memory or other storage device*] Zeichenfolge, Zeichenkette, Zeichenreihe, Zeichenreihung, Zeichenserie *f*

character subset, character subfont Teil-Zeichensatz, Teil-Zeichenvorrat, Zeichen-Teilvorrat, Zeichen-Teilsatz *m*, Zeichen-Teilmenge, Teil-Zeichenmenge *f*

character transfer rate Zeichenübertragungsgeschwindigkeit *f*

to charge, to feed (ein)speisen

to charge, to load laden, beschicken

charge, load Last *f*

charge [*Amount of fluid forced or drawn into a closed system, such as the refrigerant in a refrigerating system, the fluid in the bulb and power element of a thermostatic expansion valve, or oil in the crankcase of a compressor*] Ladung *f*

charge → quantity of electricity

charge → fee

charge acquisition → fee acquisition

charge amplifier Ladungsverstärker *m*

charge carrier Ladungsträger *m*

charge-coupled device → bucket brigade device

charge-coupled imager, CCI ladungsgekoppelte Bildwandler *m*

charge-coupled memory [*A volatile direct-acting memory (unit) made up of charge-coupled devices; used especially when memory contents are accessed serially*] ladungsgekoppelte Speicher *m*

charge coupling [*A means to store and transfer data based on the ability to store and transfer electric charges*] Ladungskopp(e)lung *f*

charge efficiency → volumetric efficiency

charge equalizer Ladungsausgleicher *m*

charge meter Ladungsmesser *m* [*Mit ihm lässt sich eine statische elektrische Aufladung erfassen*]

charge-on-ones recording, non-return-to-zero recording, NRZ recording Wechselschrift *f*

charge pump, charging pump Ladepumpe *f* [*z. B. beim Eisspeicher*]

charge regulator → (battery) charge regulator

charge simulation method, CSM Ersatzquellenverfahren *n*, Ersatzquellenmethode *f*

charge threshold value (Ver)Änderungsschwellenwert, Schwellen(ver)änderungswert *m*

charge transfer device, CTD [*Any device that relays stored charges positioned at predetermined locations, such as charge-coupled or charge-injection devices*] Ladungsverschiebungsschaltung *f*, Ladungsverschiebungselement, Ladungstransportschaltung, Ladungstransportelement *n*

charging cable Ladekabel *n*

charging capacitor Ladekondensator *m*

charging chute, fill(ing) chute [*underburning stove*] Füll(ungs)schacht *m*

charging ((closed) loop) control Ladereg(e)lung *f*

charging contactor, charging electric contactor Ladeschütz *n*

C

charging controller Laderegler *m*

charging current Ladestrom *m*

charging (fire) extinguisher Lade(feuer)löscher, Ladebrandlöscher *m*

charging ((open) loop) control Ladesteuerung *f*

charging period [*storage heater*] Ladedauer *f*

charging pump → **charge pump**

charging readiness Ladebereitschaft *f*

charging valve Ladeventil *n*

charging voltage Ladespannung *f*

Charles(') gun Charles-Strahlenerzeuger *m*

Charles(') law, Gay-Lussac('s) law [*The law that at constant pressure the volume of a fixed mass or quantity of gas varies directly with the absolute temperature; a close approximation*] Gay-Lussacsche Gesetz *n*, Gesetz von Gay-Lussac [*1802*]

charred verkohlt

chart, diagram, graph Diagramm *n*, Plan *m*, Schema, Bild *n* [*DIN 40 719. Die graphische Darstellung errechneter oder beobachteter Werte*]

chart speed accuracy Gleichlauf *m* [*des Registrierpapiers*]

chasing Stemmen *n* von Schlitzen

chasing stake Treibstöckchen *n* [*Schlosserei*]

chassis Einbaurahmen *m* [*Diskettenlaufwerk*]

chassis Blitzschutzmasse, Masse *f*

chassis ground (connection), chassis grounding (US); chassis earth (connection), chassis earthing (Brit.) Masseerde(r) *f*, (*m*), Masseerdanschluss *m*, Masseerd(verbind)ung *f*

to chat unterhalten

chatter-free prellfrei [*Relais*]

chatter(ing) Prellen *n* [*Relais*]

chatter(ing) suppression Prellunterdrückung *f* [*Relais*]

C H_3 Cl → **R-40**

C H Cl F_2 → **R-22**

to check kontrollieren, überprüfen, prüfen

check → **check (test)**

checkback signal Rückmeldesignal *n*

checkback signal message; checkback signaling message (US); checkback signalling message (Brit.) Rückmeldung *f* [*DIN 19 237*]

check(ing) Kontrollieren, Prüfen, Überprüfen *n*, Kontrolle *f*

check(ing) ball valve → **non-return ball valve**

check(ing) bit Kontrollbit, Prüfbit, Überprüfungsbit *m* [*Für die Paritätsprüfung verwendete binäre Prüfzahl*]

check(ing) bus Kontrollbus, Prüfbus, Überprüfungsbus *m*

check(ing) character [*A redundant character used to perform a check*] Kontrollzeichen, Prüfzeichen, Überprüfungszeichen *n*

check(ing) digit, check(ing) numeral Kontrollziffer, Prüfziffer, Überprüfungsziffer *f*

check(ing) display, check(ing) readout Kontrollanzeige, Prüfanzeige, Überprüfungsanzeige *f*

check(ing) field Kontrollfeld, Prüffeld, Überprüfungsfeld *n*

check(ing) indicator [*In automatic data processing, a console device, usually a light, informing the operator that an error has occurred*] Fehleranzeiger *m*

check(ing) indicator instruction [*A computer instruction which directs that a signal device is turned on to call the operator's attention to the fact that there is some discrepancy in the instruction now in use*] Fehleranzeigebefehl *m*

check(ing) information [*e.g. parity bit or check(ing) sum*] Kontrollinformation, Prüfinformation, Überprüfungsinformation *f*

check(ing) lamp Kontrolllampe, Prüflampe, Überprüfungslampe *f*

check(ing) light [*See remark under "check(ing) indicator"*] Fehleranzeigelampe *f*

check(ing) lighting, check(ing) illumination Kontrollbeleuchtung, Prüfbeleuchtung, Überprüfungsbeleuchtung *f*

check(ing) number [*A number denoting a specific type of hardware malfunction*] Kontrollnummer, Prüfnummer, Überprüfungsnummer *f*

check(ing) point [*That place in a routine at which the entire state of the computer (memory, registers, and so on), is written out on auxiliary storage (tape, disk, card) from which it may be read back into the computer if the program(me) is to be restarted later*] Kontrollpunkt, Prüfpunkt, Überprüfungspunkt *m*

check(ing) point reading Stechen *n*

check(ing) point recovery Kontrollpunktverfahren, Prüfpunktverfahren, Überprüfungspunktverfahren *n*, Kontrollpunktmethode, Prüfpunktmethode, Überprüfungspunktmethode *f*

check(ing) programme (Brit.); check(ing) program (US); check(ing) routine [*A computer program(me) which detects and determines the nature of errors in other program(me)s, particularly those that involve incorrect coding or punching of wrong* *characters*] Fehlerkontrollprogramm, Fehlerprüfprogramm, Fehlerüberprüfungsprogramm *n*

check(ing) readout → check(ing) display

check(ing) row [*A row (or one of two or more rows) on a paper tape which contains the cumulated sum of existing rows, column by column, resulting in either 1 or 0 by column, thus verifying that all rows have been properly read*] Kontrollreihe, Prüfreihe, Überprüfungsreihe *f*

check(ing) sum [*A sum of digits or numbers used in a summation check(ing)*] Kontrollsumme, Prüfsumme, Überprüfungssumme *f*

check(ing) symbol Kontrollsymbol, Prüfsymbol, Überprüfungssymbol *n*

check(ing) valve, backpressure valve, reflux valve, backflow(ing) valve, non-return valve [*An auto-matic valve which permits liquid to flow in only one direction. It is designed to close against backflow(-ing)*] Rückschlagventil, automatische Rückschlagventil, selbsttätige Rückschlagventil *n*

check(ing) word Kontrollwort, Überprüfungswort, Prüfwort *n*

checklist Abhakliste, Überprüfungsliste, Prüfliste, Kontrollliste *f*

check (test), checkout Kontrolle, Überprüfung, Prüfung *f*

chemical closet, chemical toilet [*A portable or fixed appliance containing a fluid, generally with disinfect-ant and deodorant properties, and used for the re-ception of human excrement*] Chemikalienabort, Chemikalienabtritt, Trockenabort, Trockenabtritt *m*, Chemikalienklosett, Trockenklosett *n*, Chemikalientoilette, Trockentoilette *f*

chemical dehumidification, chemical moisture removal chemische Entfeuchtung *f*

chemical dehumidifier chemische Entfeuchter *m*

chemical foam (fire-)extinguisher chemische Schaumfeuerlöscher, chemische Schaumbrandlöscher *m*

chemical hygrometer → absorption hygrometer

chemical light chemische (Not)Licht *n*, Leuchtstab *m*

chemical-mechanical chemomechanisch, chemisch-mechanisch

chemical substance [*See remark under "odo(u)rization"*] Geruchsstoff, Warnstoff *m*

chemical vapour deposition (Brit.); chemical vapor deposition (US); CVP Schichtabscheidung *f* [*aus der Gasphase*]

cherry Kugelsenker, Kugelknopf *m*, Kugelfräse *f* [*Schlosserei*]

chest, casing, case Truhe *f*

chest freezer, freezing chest, casing freezer, freez-ing casing, case freezer, freezing case [*Freezer in which access is gained by removing a lid*] Gefriertruhe *f*

CHI → computer-human interface

child(ren) (bath)tub Kinder(bade)wanne *f*

child(ren) security Kindersicherung *f*

to chill abschrecken

chill, cold(ness) [*As felt by the human body*] Kälte *f* [*Wärme bei Temperaturen unterhalb der Umgebungs-temperatur. Formelzeichen:* Q, q]

chilled cabinets, refrigerated cabinets Kühlmöbel *f* [*Kühlschränke und Kühltruhen*]

chilled surface, refrigerated surface Kälteoberfläche *f*

chilled water, refrigerated water [*Water used as a cooling medium (particularly in air-conditioning systems or in processes) at below ambient temperature*] Kaltwasser *n*, KW

chilled water cooling, refrigerated water cooling Kaltwasserkühlung, KW-Kühlung *f*

chilled water cooling coil, refrigerated water cool-ing coil Kaltwasserschlange, KW-Schlange *f*

chilled water (flow) switch, refrigerated water (flow) switch Kaltwasserschalter, KW-Schalter *m*

chilled water thermostat, refrigerated water thermostat Kaltwasserthermostat, KW-Thermostat *m*

chiller, refrigerating system, liquid chilling system, liquid chiller [*Complete, indirect refrigerating system of compressor, condenser, and evaporator with all operating and safety controls*] Kompressorkältesystem, Kältesystem, Verdichterkältesystem *n* [*Verdichter (= Kompressor), Verflüssiger (= Kondensator oder äußere Wärme(aus)tauscher) und Verdampfer (= innere Wärme(aus)tauscher) und mit einem Drosselorgan (Reduzierventil oder Kapillarröhre) zu einem Kältekreislauf zusammengeschaltet. Im allgemeinen Sprachgebrauch, in Normen, Merkblättern, Nomenklaturen, Klassifikationen usw. wird von „Kältemaschine" gesprochen bzw. geschrieben, obwohl „Kälteanlage" gemeint ist. Die Benennung „Kältemaschine" ist zwar sehr alt, ändert aber nichts daran, dass sie technisch falsch ist. „Kältemaschine" muss daher durch „Kälteanlage" ersetzt werden denn man bezeichnet auch eine Anlage für Raumklimatisierung als „Klimaanlage" und nicht als „Klimamaschine"*]

chill factor, wind chill factor [*Temperature of "calm" (4 mph) air that would produce the same chill effect on bare flesh as the combination of wind and temperature*] Frösteln *n*

C

chime Klangobjekt *n*

chime(s) Glockenspiel *n*

chimney Schornstein, Kamin *m*

chimney → fire(place)

chimney back [*The wall behind a fire(place). In the case of back-to-back fire(places) this can be one wall which forms a back to both*] Kaminrückwand, Kaminrückplatte, Kaminhinterwand, Kaminhinterplatte *f*

chimney bar Zungenstab *m*, Manteleisen, Rauchfangeisen *n*

chimney barometric draught regulator, chimney barometric draft regulator, chimney (barometric) damper Schornsteinzugregler, Kaminzugregler *m*

chimney board Kamintür *f*, Kaminvorsetzer *m*

chimney built back to back angebaute Kamin *m* [*An die Wand angebaut*]

chimney cleaning door → chimney soot door

chimney connection Schornsteinanschluss *m*, Schornsteinverbindung *f*

chimney coping, chimney cope, chimney capping Schornsteinkopf, Kaminkopf *m*

chimney corner Kaminhölle, Kaminnische *f*

chimney cover → chimney louvre

chimney draft, chimney draught Schornsteinzug, Kaminzug *m* [*Als Erscheinung, nicht als Bauteil*]

chimney draft boiler, chimney draught boiler Schornsteinzugkessel, Kaminzugkessel *m*

chimney dressing → chimney jambs

chimney effect, stack effect; flue effect [*deprecated*] [*Rising of air or gas in any vertical passage, induced when its density is lower than that of surrounding air or gas*] Schornsteinwirkung, Kaminwirkung, Steigwirkung *f*

chimney fan Schornsteinventilator, Kaminventilator *m*

chimney flue [*An incombustible and heat-resistant enclosed passage in a chimney to control and carry away products of combustion from a fireplace, furnace, or boiler to the outside air*] Schornsteinzug, Kaminzug *m*

chimney gas Schornsteingas, Kamingas *n*

chimney hood, chimney funnel Hals, Rauchfang, Schurz *m*, Kutte *f*

chimney jambs, chimney dressing [*The two vertical sides of a fire(place) opening*] Kamingewände *n*, Ka-mineinfassung *f*

chimney louvre (Brit.); chimney louver (US); chimney lantern, chimney luffer, chimney cover Schornsteinaufsatz, Kaminaufsatz *m*, Schornsteinhaube, Kaminhaube *f*

chimney outlet Schornsteinmündung, Kaminmündung *f*

chimney shaft [*A term usually applied to a large chimney from a boiler installation, which encloses the flue(s) from that installation; often constructed of precast concrete sections*] Schornsteinschaft, Kaminschaft *m*

chimney soot Schornsteinruß, Kaminruß *m*

chimney soot door, chimney cleaning door Schornsteinputztür, Schornsteinreinigungstür, Kaminputztür, Kaminreinigungstür *f*

chimney tube, flue opening Schornsteinröhre, Schornsteinhöhlung, Kaminröhre, Kaminhöhlung *f*

chimney with rectangular jambs französische Kamin *m*

Chinese (column) binary (representation) → column binary (representation)

chip Halbleiterbauelement *n*

chip → chad

chip blower → air syringe

chip bonding Chipbonden *n* [*Siehe Anmerkung unter „Bonden"*]

chip card Chipkarte *f*

chip enablement, CE [*Enablement of a chip by means of a chip-enable command (signal)*] Chipfreigabe *f*

chip tape → chad tape

chirping spektrale Verbreiterung *f* bei Amplitudenmodulation

chlorid(e) brine Chloridsole *f*

chlorination, chlorinating Chlorierung *f*, Chloren *n*

chlorinator [*A unit used to add chlorine to organic compounds or to sterrilize water with chlorine gas*] Chlorgasdosier(ungs)anlage *f* [*DIN 19 606*]

chlorine gas Chlorgas, gasförmige Chlor *n*

chlorine gas bulk flow Chlorgasmassenstrom *m*

chlorine gas cloud Chlorgaswolke *f*

chlorine gas inlet valve Chlorgaseinlassventil *n*

chlorine shutoff valve Chlor(ab)sperrventil *n*

chlorine solution Chlorlösung *f*

chloring Chloren *n*

chlorodifluoromethane → refrigerant-22

chloromethane → refrigerant-40

chlorotrifluoromethane → refrigerant-13

to choke verstopfen

choke (coil) Drossel(spule), Vorschalt(ungs)drossel(spule) *f*

choking calorimeter Drosselkalorimeter *n*, Drosselwärmemesser *m*

to chop zerhacken

to chop [*Electronics*] unterbrechen, trennen, ausschalten

chopper [*Electronics. A toothed disk that is used to interrupt a beam of light at regular intervals, or a de-vice of interrupting electrical signals*] Unterbrecher, Trenner, Ausschalter *m*

chopper [*Any device, such as a knife, that cuts an object into pieces*] Zerhacker *m*, Zerhackungsgerät *n*

chopper amplifier [*A circuit used to amplify low signals by chopping the slowly varying input into a succession of higher-frequency rectangular pulses, amplifying the pulses and filtering the output to restore the slowly varying signal*] Zerhackerverstärker, Zerhackungsverstärker *m*

chopper-bar recorder Punktschreiber *m* [*Messtechnik*]

chopper-bar recording Punktschreibung *f* [*Messtechnik*]

chopper-bar relay Feldbügelrelais *n*

chopper(-type) power supply unit → switched-mode power supply unit

chopping rate, chopping frequency, clock pulse rate, clock pulse frequency Taktfrequenz, Taktrate *f*

chorus Mehrstimmenwirkung *f*

CHP → combined heat and power

CHP plant → combined heat and power plant

chroma-keying Farbausstanzen *n*

chroma oscillator → color oscillator

chromaticity Farbart *f*

chromaticity signal amplifier Farbartsignalverstärker *m*

chrominance [*The difference between a given colo(u)r and a reference colo(u)r with the same degree of brightness*] Farbigkeit *f*

chrominance subcarrier oscillator → color oscillator

chronological, in correct time order zeitfolgerichtig, zeitfolgegemäß, zeitfolgerecht

chronological order, chronological sequence zeitliche Reihenfolge *f*

ChSP → chain sampling plan

Chubb('s) (detector) lock, (detector) lock of Chubb, Chubb ('s) lever-tumbler lock Chubbschloss, Mehrzuhaltungsschloss, Vielzuhaltungsschloss *n* [*Nach Jeremiah Chubb benanntes Sicherheitsschloss mit mehreren lamellenartigen, flachen Sperrzuhaltungen*]

church radiant heater Kirchenwärmestrahler *m*

churnrate, rate of change and turn Umnutzungsrate, Nutz(ungs)veränderungsrate *f* [*Gebäudenutzungsfläche*]

chute Rutsche *f*

chymol → grimmer

CI, coolant inlet, cooling agent inlet, cooling medium inlet Kühl(ungs)mitteleinlass, Kühl(ungs)mediumeinlass *m*

CI → comfort index

CI → counter input

CIB, computer integrated building rechnerintegrierte Gebäude *n*

CIC, circuit identification code Leitungskennzeichnungskode, Leitungskennzeichnungscode *m*, Nachrichtenkanaladresse *f* [*Signalisierungssystem Nr. 7*]

CID, connection identification Verbindungserkennung *f*

CIM → computer input microfilm

CIM, computer-integrated manufacturing [*It embraces the interaction of CAD, CAP, CAM and CAQA, thus permitting the integration of the technical or-ganizationed functions for product manufacture. It re-quires joint and overlapping use of a comprehensive data base*] rechnerintegrierte Herstellung (von Erzeugnissen), rechnerintegrierte Fertigung *f* (von Erzeugnissen)

CIMFFM → computer integrated manufacturing for facility management

cinder sifter Asche(n)klosett *n*, Asche(n)abort *m*

circle diagram Kreisortskurve *f*

circuit [*telephony*] Verbindung *f*

circuit → communication path [*DIN IEC 625. Programmable measuring apparatus*]

circuit → cycle

circuit, electric circuit [*An interconnection of electrical elements forming one or more complete paths for the flow of current*] Schaltung *f*, Schaltkreis, Stromkreis *m*

circuit algebra → logic(al) algebra

circuit angle Phasenwinkel *m* der Stromrichterschaltung, Schaltungs(phasen)winkel *m*

circuit arrangement, circuit disposition, circuit layout Schaltungsanordnung *f*

circuit-breaker, c.b. [*IEC 157-1*] Leistungsschalter, (Leistungs)Schutzschalter, (Leistungs)Selbstschalter *m* [*VDE 0660, Teil 101*]

C

circuit-breaker for equipment, c.b. for equipment, equipment circuit-breaker, equipment c.b. Geräte(leistungs)schutzschalter, Geräte(leistungs)-selbstschalter, GSS, Geräteleistungsschalter m

circuit-breaker for mesh-connected systems Maschennetzschalter *m*

circuit-breaker frame (Leistungs)Schutzschalterträger, (Leistungs)Selbstschalterträger m

circuit-breaker pole (Leistungs)Schutzschalterpol, (Leistungs)Selbstschalterpol m

circuit-breaker (switch) Leistungs(trenn)schalter *m*

circuit-breaker unit (Leistungs)Schutzschaltereinheit, (Leistungs)Selbstschaltereinheit *f*, (Leistungs-)Schutzschalter(abgangs)feld *n*, (Leistungs)-Schutzschalterabgang, (Leistungs)Selbstschalterabgang *m*

circuit-closing insulator Kurzschlussisolator *m*

circuit condition → circuit state

circuit design, circuit type, circuit version [*Electricity. The process of determining the components and connections of the electrical circuit or network*] Schalt(ungs)art, Schalt(ungs)ausführung, Schalt-(ungs)bauweise, Schalt(ungs)bauart *f*

circuit diagram, schematic diagram, elementary diagram, wiring diagram Stromlaufplan *m*, Stromlaufschema, Stromlaufbild, Schalt(ungs)bild, Schalt(ungs)schema, Verdrahtungsschema, Verdrahtungsbild *n*, Schalt(ungs)plan, Verdrahtungsplan *m* [*DIN 40 719 Blatt 1*]

circuit diagram ((open) loop) control → diagram ((open) loop) control [*IEC 113-2*]

circuit disposition → circuit arrangement

circuit element Stromkreiselement, Stromkreisglied, Schalt(ungs)element, Schalt(ungs)glied *n*

circuit engineering Schalt(ungs)technik *f*

circuit hysteresis Schalt(ungs)hysterese *f*

circuit identification code → CIC

circuit line Stromkreisleitung *f*

circuit logic Schalt(ungs)(kreis)logik *f*

circuit off-state interval Vorwärtssperrzeit *f*

circuit override Stromkreisübersteuerung *f*

circuit path Leiterbahn *f*

circuits, circuiting Beschaltung, Stromkreise *f*

circuit separation Stromkreistrennung *f*

circuit shift → cyclic(al) shift

circuit state, circuit condition Schaltbeschaffenheit *f*, Schaltzustand *m*

circuit strip Stromkreisleiste *f*

circuit supply (line) Stromkreis(zu)leitung *f*

circuit switched data net(work) → CSDN

circuit switched public data network, CSPDN leitungsvermittelnde öffentliche Datennetz, öffent-liche leitungsvermittelnde Datennetz *n*

circuit switching, CS leitungsvermittelte Kommunikation, Durchschaltevermittlung, Leitungsvermittlung, Wählvermittlung *f* [*Eine Verbindungsart von Kommunikationsteilnehmern*]

circuit symbol Schalt(ungs)zeichen *n*

circuit technique Schalt(ungs)technik *f* [*als betriebliche Anwendung*]

circuit terminating equipment Leitungsabschlussgerät(e) *n*, (*f*)

circuit tester Stromkreisprüfer *m*

circuit theory Schalt(ungs)lehre *f*

circuit with lock-out preventing closing [*IEC 157-1*] Leistungsschalter *m* mit (Einschalt)Sperre [*VDE 0660, Teil 101*]

circular (air) diffusor, round (air) diffusor, round (air) diffuser, circular (air) diffuser (Luft)Runddiffusor, (Luft)Kreisdiffusor *m*

circular (air) duct, round (air) duct Kreis(luft)-kanal, Luftkreiskanal *m*

circular air grid → draft regulating register

circular conductor Kreisleiter *m*

circular dial → dial

circular escalator, round escalator, circular moving staircase, round moving staircase [*It carries passengers through a graceful, sweeping arc*] Bogenrolltreppe, Bogenrfahrtreppe *f*

circular flow Kreisströmung *f*

circular interpolation, interpolation circular motion Kreisinterpolation, Zirkularinterpolation, zirkuläre Interpolation *f* [*Errechnung von Zwischenpunkten, um eine definierte Kreisbewegung näherungsweise auszuführen*]

circular interpolation motion, circular interpolation movement Kreisinterpolationsbewegung, Interpolationskreisbewegung *f*

circular left rotation zirkuläre Linksrotation, zirkuläre Linksdrehung *f* [*Eine Bitmanipulation*]

circular letter Rundschreiben *n*

circular measure, radian measure Bogenmaß *n*

circular movement, circular motion Kreisbewegung *f*

circular pendulum Kreispendel *n*

circular right rotation zirkuläre Rechtsrotation, zirkuläre Rechtsdrehung *f* [*Eine Bitmanipulation*]

circular scale → dial

circular shift → **cyclic(al) shift**

circular siphon, circular syphon Trompetenrohr *n*

to circulate umlaufen, umwälzen, zirkulieren

circulating Umwälzen, Zirkulieren, Umlaufen *n*

circulating air ag(e)ing, circulation air ag(e)ing, furnace air ag(e)ing Alterung *f* im Umluftofen, Umluft(ofen)alterung *f*

circulating air conditioning plant, circulating air conditioning installation [*See remark under "air conditioning installation"*] Umluft-Klimaanlage, Klima-Umluftanlage *f*

circulating air heater, circulation air heater [*Any device in a heating system that heats circulation air*] Umlufterhitzer *m*, Umluft(er)wärmer, Umluftheizer

circulating air induction system, circulation air induction system Induktions-Umluftsystem *n*, Umluft-Induktionssystem *n* [*Förderung der Umluft durch Induktion*]

circulating air plant, circulating air installation, circulation air plant, circulation air installation Umluftanlage *f*

circulating air system, circulation air system Umluftsystem *n* [*Es wird nur Umluft maschinell gefördert*]

circulating air treatment, circulation air treatment Umluftbehandlung *f*

circulating current, circulation current Kreisstrom *m*

circulating current cross-connection, circulation current cross-connection kreisstromführende Kreuzschaltung *f*

circulating current-free, circulation current-free kreisstromfrei

circulating fan, circulation fan [*Free-flow propeller fan designed to circulate the air in a room without any air duct*] Umwälzventilator, Zirkulationsventilator, Umlaufventilator, Umwälzlüfter, Zirkulationslüfter, Umlauflüfter *m*

circulating gas water heater, circulation gas water heater, circulating gas water heating appliance, circulation gas water heating appliance, gas water circulating heater, gas water circulation heater, gas water circulating heating appliance, gas water circulation heating appliance Umlauf-Gas-Wassererhitzer, Umlauf-Gas-Wasserheizer, Umlauf-Gas-Wasser(er)wärmer, Umlauf-Gas-(Warm)Wasserbereiter, Gas-Umlaufwassererhitzer, Gas-Umlaufwasser(er)wärmer, Gas-Umlauf(warm)wasserbereiter, Gas-Um-laufwasserheizer *m*

circulating gas water heating, circulation gas water heating, gas water circulation heating Umlauf-Gas-Wasserheizung, Gas-Umlaufwasserheizung *f*

circulating heating pump, circulation heating pump, heating circulating pump, heating circulation pump Heiz(ungs)umwälzpumpe, Heiz(ungs)-zirkulationspumpe, Heiz(ungs)umlaufpumpe *f*

circulating line, circulation line Umlaufleitung, Zirkulationsleitung, Umwälzleitung *f*

circulating means, circulation means [*A means of travel through a building, such as doors, corridors, stairs, elevators and escalators*] Personenverkehrsmittel *n*

circulating medium, circulation medium [*e.g. water*] Umlaufträger, Umwälzträger, Zirkulationsträger *m*

circulating pattern, circulation pattern [*The traffic pattern through an area or building*] Verkehrsmuster, Verkehrsbild *n*

circulating pipe, circulation pipe [*It forms part of the primary or secondary circuit of a hot water system*] Umlaufrohr, Zirkulationsrohr, Umwälzrohr *n*

circulating (pipe)line, circulating piping, circulation (pipe)line, circulation piping Umwälz(rohr)leitung, Umlauf(rohr)leitung, Zirkulations(rohr)leitung *f*

circulating pressure, circulation pressure Umlaufdruck, Umwälzdruck, Zirkulationsdruck *m*

circulating pump, circulation pump, circulator Umwälzpumpe, Zirkulationspumpe, Umlaufpumpe *f*

circulating scheme, circulation scheme [*In a building, a scheme providing for a smooth, economical, and functional flow of passenger traffic*] Personenverkehrsplan *m*

circulating spray (air) dehumidifier, circulation spray (air) dehumidifier Umlaufsprüh(luft)entfeuchter, Zirkulationssprüh(luft)entfeuchter, Umwälz(ungs)sprüh(luft)entfeuchter *m*

circulating system, circulation system Umwälzungssystem, Umlaufsystem, Zirkulationssystem, Umwälzsystem *n*

circulating (type) evaporator, circulation (type) evaporator [*Flooded evaporator comprising a low-pressure receiver, in which the unvaporized refrigerant returns to the evaporator inlet by gravity or by means of a pump or an injector*] Umlaufverdampfer, Zirkulationsverdampfer, Umwälz(ungs)-verdampfer *m*

circulating water, circulation water Umwälz(ungs)-wasser, Umlaufwasser, Zirkulationswasser *n*

circulating water heater, circulation water heater Umlaufwasserheizer, Umlaufwassererhitzer, Wasserumlaufheizer, Wasserumlauferhitzer *m*

circulating water quantity, circulation water quantity umlaufende Wassermenge *f*

circulating water system, circulation water system [*A system in which the same water circulates*

C

around a closed loop] Wasserumwälz(ungs)system, Wasserzirkulationssystem, Wasserumlaufsystem *n*

circulation [*The continuous flow of a fluid within a closed circuit*] Umwälzung, Zirkulation *f*, Umlauf *m*

circulator → circulating pump

circulatory (type) gas water heater, circulation (type) gas water heater Umlauf-Gaswasserheizer, Umlauf-Gaswasser(er)wärmer, Gas-Wasserumlaufheizer, Gas-Wasserumlauf(er)wärmer *m*

CISC computer → complex instruction set computer

cistern, rain water cistern Zisterne *f*, Regenwasserbehälter *m*

cistern, storage cistern [*A large watertight container used to store water supplies for a building. Cisterns are often installed in the roof and used to feed taps and the heating system*] Zisterne *f*

cistern → (WC) cistern

cistern gravity system Zisternen-Schwerkraftsystem *n* [*Der Wasserspeicher liegt über den Entnahmestellen*]

cistern pump(ing) system Zisternen-Pumpensystem *n*

city drainage, town drainage Stadtentwässerung, Netzentwässerung *f*

city water, town water Stadtwasser, Netzwasser *n*

civil engineering → building engineering

civil protection, civil defence; civil defense (Brit.) Zivilschutz *m*

CJ → cooling jacket

cladding skin Außenwandschale *f*

claim management Nachforderungswesen *n*

clamp Klammer, Schelle *f*

clamp Begrenzer, Klemme(r) *f*, (*m*) [*Er wird durch Hand- oder Software realisiert und verhindert, dass bestimmte Grenzen, z. B. Spannungen oder Ströme an einem Prüfling oder an einem Prüfautomaten überschritten werden*]

clamping Klemmen *n*

clamp(ing) sleeve Klemmhülse *f*

clamp(ing) terminal, terminal clamp Bügelklemme *f*, Klemmbügel *m*

clapper → (door) knocker

clarity [*received sound*] Deutlichkeit *f*

clasp and clamp → hasp and staple

clasp nail, clout nail, tack Flachkopfnagel *m*

classification letter → letter symbol

class of service, COS Dienstgüte *f*

class of traffic character, COT character Verkehrsklassenkennzeichen *n* [*dezentrale Zeichengabe*]

class tag Klassenetikett *n*

claw (type) clamp Tatzenklemme *f*

claxon, klaxon Elektromotorhupe *f*

to clean löschen [*In der Mac-Sprache „zappen" genannt*]

clean → noiseless

cleanable reinigungsbar

clean air, air free of contaminants, pure air Reinluft, Gutluft *f*

clean bench Reinheits-Arbeitsbereich *m* [*selten: n*], Reinraumbank *f* [*Arbeitsbereich mit hoher Reinheitsanforderung*]

clean channel → noiseless channel

cleaner filter [*Device to remove foreign matter from fuel*] Brennstofffilter *m, n*

cleaning Reinigen, Putzen *n*

cleaning Aufräumen *n*

cleaning cap Reinigungskappe, Putzkappe, Revisionskappe *f*

cleaning chamber, cleanout chamber, rodding chamber, inspection chamber, access chamber [*A chamber constructed over a subsidiary sewer or drain of not more than 9 in. diameter to permit inspection and access for rodding. "Pit" and "manhole" may be used synonymously for "chamber"*] Revisionskammer, Reinigungskammer, Putzkammer *f* [*„Schacht" kann synonym für „Kammer" verwendet werden*]

cleaning cloth, wiping cloth Reinigungstuch, Wischtuch *n*

cleaning cover, access cover, access plate, cleanout cover, cleanout plate, cleaning plate, rodding cover, rodding plate, inspection cover, inspection plate [*"Cap" may synonymously be used for "cover" and "plate"*] Schaulochdeckel, Reinigungslochdeckel, Putzlochdeckel, Revisionslochdeckel *m*

cleaning cradle → (window) cradle machine

cleaning door, cleanout door, rodding door, access door, inspection door Reinigungstür, Putztür, Revisionstür, Kontrolltür *f*

cleaning eye, rodding eye, access eye, inspection eye; cleanout (eye) (US) [*An opening in a drain closed by a plate held on to the opening by a bolt or wedge. It is often provided at pipe bends to enable the pipe to be rodded. "Hole" may synonymously be used for "eye"*] Revisionsöffnung, Schauöffnung, Kontrollöffnung, Reinigungsöffnung, Putzöffnung *f*, Schauloch, Revisionsloch, Putzloch *n*

C

cleaning eye, access eye, inspection eye; cleanout (eye) (US) [*A pipe fitting with a removable plug which provides access for inspection or cleaning of the pipe run*] Putzfitting, Reinigungsfitting, Revisionsfitting *m*

cleaning flap, cleanout flap Reinigungsklappe, Schauklappe, Revisionsklappe, Kontrollklappe *f*

cleaning gully, cleaning inlet, access gully, inspection gully, cleanout gully, access inlet, rodding inlet, inspection inlet, cleanout inlet [*A trapped gully with a rodding eye on the outlet permitting rodding from the surface of the ground*] Reinigungssinkkasten, Putzsinkkasten, Revisionssinkkasten *m*

cleaning hinge, offset hinge, angular hinge [*A long-legged window hinge which holds the open casement clear of the stile, thus allowing the outside of the glass to be cleaned from inside*] Winkelgelenk *n*

cleaning iron, cleanout iron Austreibeisen *n*

cleaning manhole, cleanout manhole Reinigungsschacht, Putzschacht, Revisionsschacht, Kontrollschacht, Inspektionsschacht *m*

cleaning opening, cleanout opening, cleaning hole, cleanout hole Putzöffnung, Reinigungsöffnung, Inspektionsöffnung, Revisionsöffnung *f*

cleaning plug, cleaning screw, cleanout plug, cleanout screw, access plug, access screw, rodding plug, rodding screw, inspection plug, inspection screw Putzstopfen, Reinigungsstopfen, Ausräumstopfen, Inspektionsstopfen *m*, Putzschraube, Reinigungsschraube, Ausräumschraube, Inspektionsschraube *f*

cleaning service Reinigungsdienst *m*

cleaning supplies (gelieferte) Reinigungsmittel *f*

cleaning tube, inspection tube, cleanout tube, rodding tube, access tube Inspektionsrohr, Reinigungsrohr, Putzrohr, Revisionsrohr *n*, Inspektionsstutzen, Revisionsstutzen, Reinigungsstutzen *m*, Putzstutzen [*Formstück in Abflussleitungen zur Reinigung derselben. DIN 1986*]

cleaning utensil Reinigungsgegenstand *m*

clean pipe, cleanout pipe, access pipe, rodding pipe, inspection pipe [*It is provided with an opening with a remov(e)able cover for injection and main-tenance*] Rohr *n* mit Putzöffnung, Putzöffnungsrohr, Reinigungsöffnungsrohr, Rohr mit Reinigungsöffnung

clean relay [*A relay with neither pole nor contact connected to a voltage source*] (fremd)spannungsfreie Relais *n*

clean room; white room [*obsolete*] [*An assembly room for precision products whose quality would be affected by dust, lint, or airborne pathogens; usually has smooth room surfaces to prevent dust collection; air precipitators or filters keep dust, lint, etc., to a specified minimum level*] Reinraum, (vollständig) staubfreie Raum *m*

clean room lock; white room lock [*obsolete*] Reinraumschleuse, reinraumgerechte Schleuse *f*

clean room wall; white room wall [*obsolete*] Reinraumwand, reinraumgerechte Wand *f*

clean water service reservoir, clean water service tank Reinwasserbehälter, Reinwassertank *m*

to clean with soda, to boil soda in the boiler auslaugen [*Kessel*]

to clear, to reset to zero löschen, räumen [*Rechner*]

to clear [*device*] zurücksetzen

clear fehlerfrei, ohne Fehler, fehlerlos, einwandfrei

clear, ready (betriebs)bereit, (betriebs)fertig

clearance → play

clearance area, clearance plane Sicherheitsbereich *m* [*selten: n*] [*Bereich, in den die Werkzeugspitze beim Erkennen einer bestimmten Anweisung zurückgezogen wird, z. B. beim Werkzeugwechsel, DIN 66 215*]

clearance between open contacts Schalt(ungs)strecke *f*

clearance to earth (Brit.); clearance to ground (US) Luftstrecke *f* zur Erde

clearance volume Totraum *m*

clear distance Sicherheitsabstand *m*, lichte Abstand [*Zur Vermeidung von Kollisionen in einem gegebenen Abstand von der Fläche*]

clear height Lichthöhe *f*, lichte Höhe

clear height above steps Kopffreiheit *f* [*Rolltreppe*]

clearing Auslösung *f*, Auslösen *n*

clearing Abschaltung *f* [*Kurzschluss*]

clear sky, cloudless sky, blue sky klare Himmel, wolkenlose Himmel, heitere Himmel, blaue Himmel *m*

clear sky radiation, cloudless sky radiation [*The loss of heat from a body by radiation that occurs when a night sky is dark and cloudless. This effect causes the temperature of an object, such as a roof, to fall below the surrounding air temperature*] Nacht-himmel-Strahlungswärmeverlust *m*

clear text, clear language Klartext, Reintext *m*

clear to send, CTS sendebereit, sendefertig

clear (work(ing) space, work(ing) clear space Frei(arbeits)raum, freie (Arbeits)Raum, Arbeitsfreiraum *m*

to cleat festklemmen

cleat Knagge(n) *f*, *(m)*

cleat, (electric) conduit cleat Rohrschelle, Schelle *f* [*Installationsrohr*]

C

cleat wiring [*Electric wiring on cleats or insulated supports which are mounted on a surface – e. g. a wall – leaving the wiring exposed; conduits or raceways are not used*] offene Leitungsverlegung *f* mit Schellenbefestigung

clenching, rivet bearing Stauchung *f*

to clench round, to clinch [*nail*] rundstauchen, umschlagen

clicket → (door) knocker

client Rechner *m* mit Auswahlfenster [*Auf dem Auswahlfenster werden alle für dieses Fenster zugelassenen Drucker angezeigt*]

client, data base client, data bank client Nutzer, Datenbanknutzer *m*

client Kunde *m*

client Datenempfänger *m*

client Netzwerkrechner *m* [*Er benutzt für bestimmte Aufgaben den Dienstanbieter (= Server)*]

client [*A client requests service from a server*] Dienstanforderer *m*, Dienstanfordernder, Applikator *m* [*DIN 19 245, Teil 2*]

client consultation, client consulting Kundenberatung *f*

client-server-relationship Kunden-Dienstleister-Verhältnis *n*

client-server system Rechnerverbund *m* [*Er hat einen Dienstanbieter und mehrere Rechner mit Auswahlfenstern*]

client signal Kundensignal *n*

client system Kundensystem *n*

client training Kundenschulung, Kundenausbildung *f*

climate control → air conditioning

climate (system) research Klimaforschung *f*

climatic envelope Klimahülle *f*

climatic (test) chamber Klima(prüf(ungs))raum *m*

climb milling Gleichlauffräsen *n* [*Werkzeugvorschub und Fräsrichtung haben den gleichen Drehsinn*]

climb-over protection, climb-over security Übersteig(e)schutz *m*, Übersteig(e)sicherung *f*

to clinch → to clench round

clinch Klinke, Klinkung *f* [*Nicht im Sinne von „Türklinke"*]

clinch-bolt Klink(ungs)bolzen *m*

clinch(ing) [*The turning over of a nail point projecting through the material*] Umbiegen, Verklinken *n*

clinker → (boiler) clinker

clip Schelle *f*

clipboard Zwischenablage *f* [*Zum Datenaustausch zwischen Programmen*]

clipboard Ablagefläche *f* [*im Rechnerspeicher*]

clip-on ammeter Zangenstrommesser *m*, Zangenstrommessgerät *n*

clip-on meter Zangenmesser *m*, Zangenmessgerät *n*

clipper, limiter, amplitude limiter, amplitude-limiting circuit, clipping circuit, limiter circuit, peak limiter [*Device whose output is zero or a fixed value for instantaneous input amplitudes up to a certain value, but a function of the input for amplitudes exceeding the critical value*] Schwellwertbegrenzer, Klipper *m*

clipper amplifier [*Amplifier designed to limit the instantaneous value of its output to a predetermined maximum*] Klipperverstärker, Schwellwertverstärker *m*

clipping Aussparen *n* [*Der nicht sichtbaren Objekte vor der Berechnung, so wird Rechenzeit gespart*]

clipping Schwellwertbestimmung *f* [*Videosignal*]

cloak hook Kleiderhaken *m*

to clock, to time, to cycle, to operate in the switching mode takten

clock Takt, Schrittpuls *m* [*Pulskodemodulation*]

clock → clock generator

clock Takt, feste Zeitabstand *m* [*Digitalsysteme verarbeiten Daten in ihren Baugruppen in festen Zeitabständen*]

clock accuracy Taktgenauigkeit *f*

clock channel Uhr(en)kanal *m*

clock(-controlled) switch → timing element

clock(-controlled] switching → time switching

clock dial (Uhren)Zifferblatt *n*

clocked, timed, cycled, operated in the switching mode getaktet, taktsynchron

clocked counter, timed counter, cycled counter, counter operated in the switching mode taktsynchrone Zähler, getaktete Zähler *m* [*Ein synchroner Zähler, dessen Weiterschalten zusätzlich von einem Taktsignal abhängt. V DIN 19 237*]

clock frequency Taktfrequenz *f* [*elektronische Steuerung. DIN 19 237*]

clock generator, CG, clock pulse generator, CPG, clock Taktgeber, Taktgenerator *m*

clock input Takteingabe *f*

clock jitter Taktzittern *n* [*Pulskodemodulation*]

clock line Taktleitung *f*

clock pulse, CP Taktimpuls *m*, Taktrate *f*

clock pulse amplifier Taktverstärker *m*

clock pulse generator → **clock generator**

clock pulse rate → **chopping rate**

clock pulse space Taktpause *f*

clock qualifier Taktkennzeichen *n*

clock relay Schaltuhr *f*

clock scan [*programmable controller*] Zeittaktabfrage *f*

clock speed Taktfrequenz *f* [*Rechner*]

clock switch Uhr(en)schalter *m*

clock system Uhr(en)system *n*

clock thermostat, clock temperature controller Uhr(en)thermostat *m*, Uhr(en)temperaturregler *m*

clockwise rechtsdrehend

clockwise direction counting Zählsinn *m* rechts, rechte Zählsinn, Rechtszählung *f*

clone Kopie *f*, Nachahmung *f*

close [*condition*] eng [*Zustand*]

close burning coal → **matting coal**

close circuit [*A complete path for current*] geschlossene Stromkreis *m*

close-controlled system [*IEC 50(351)*] → **plant**

close-coupled (flushing) cistern aufgesetzte Spülkasten *m*, Aufsetzspülkasten

close-coupled toilet → **single-piece toilet**

closed geschlossen, zu

closed geschlossen, druckfest [*z. B. Wasser(er)wärmer; Badeofen; Behälter*]

closed action flow geschlossene Wirkungsablauf *m*

closed action path geschlossene Wirkungsweg *m*

closed cascade(d) control → **cascade(d) closed control**

closed channel → **conduit**

closed circuit, break circuit Ruhestromschaltung *f*, geschlossene Stromschaltung

closed-circuit contact, break circuit contact Öffnungskontakt *m*, Öffner *m* [*Kontakt, der bei Betätigung öffnet und dadurch den Ruhestromkreis unterbricht. Öffnungskontakte sind Bestandteile der Gefahrenmeldeanlagen*]

closed-circuit cooling Kühlung *f* im geschlossenen Kreislauf, Kreislaufkühlung, Umlaufkühlung

closed-circuit current, bias current Ruhestrom *m*, Vor(magnetisierungs)strom, Stabilisierungsstrom

closed-circuit current door release, bias current door release Ruhestromtüröffner *m*

closed-circuit lubrication Umlaufschmierung *f*, Kreislaufschmierung

closed-circuit television → **CCTV**

closed-circuit television monitoring system, clos-ed-circuit TV monitoring system Fernsehüberwachungsanlage *f*

closed-circuit ventilation Umlauflüftung *f*, Kreislauflüftung

closed-circuit water-water cooling, CWW Wasserumlauf-Wasserkühlung *f*, WUW-Kühlung

closed-circuit working, closed-current working Ruhestrombetrieb *m* [*Fernwirktechnik*]

closed coil winding geschlossene Wick(e)lung *f*, unaufgeschnittene Wick(e)lung

closed condensate return → **non vented condensate return**

closed control → **((closed) loop) control**

closed control... → **((closed) loop) control...**

((closed) control) loop geschlossene Regel(ungs)schleife *f*, geschlossene Wirkungsweg *m*

closed-core transformer Transformator *m* mit geschlossenem Kern, Wandler *m* mit geschlossenem Kern, eingeschlossene Transformator, eingeschlossene Wandler

closed current Ruhestrom *m* [*Strom eines Stromkreises, der bei Gefahrenmeldeanlagen ständig in einer Meldelinie fließt und bei Unterbrechung oder Kurzschluss zur Alarmauslösung führt*]

closed cycle, closed system[*The fluid is used over and over again without the introduction of additional fluid*] geschlossene System *n*, geschlossene Kreislauf *m*

closed design → **closed type**

closed expansion vessel geschlossene Ausdehnungsgefäß *n* [*Wärmeträger steht unter dem Druck eines geschlossenen Gas- oder Dampfpolsters*]

closed flash tester Flammpunktprüfer *m* mit geschlossenem Tiegel

closed furnace Feuerstätte *f* mit geschlossenem Verbrennungsraum, Feuerstätte mit geschlossener Verbrennungskammer

closed gain → **((closed) loop) gain**

closed gas furnace Gasfeuerstätte *f* mit geschlossenem Verbrennungsraum, Gasfeuerstätte mit geschlossener Verbrennungskammer

closed hot water heater → **closed (warm) water heater**

closed humidity control → **((closed) loop) humid-ity control**

closed loop → **closed (control) loop**

C

((closed) loop) amplitude control, amplitude regulation Amplitudenreg(e)lung *f*

((closed) loop) boiler control, boiler ((closed) loop) control, boiler regulation Kesselreg(e)lung *f*

((closed) loop) break(ing) capacity [*IEC 265*] Ringausschalt(ungs)vermögen *n* [*VDE 0670, Teil 3*]

((closed) loop) break(ing) current [*IEC 265*] Ringausschalt(ungs)strom *m* [*VDE 0670, Teil 3*]

((closed) loop) cascade(d) control, cascade(d) ((closed) loop) control, cascade(d) regulation, cascade(d) closed control Kaskadenreg(e)lung *f*

((closed) loop) control, regulation, modulating control Reg(e)lung [*mit Rückführung*], Regulierung *f* [*mit Rückführung*] [*Ein Vorgang, bei dem eine Größe, die zu regelnde Größe (Regelgröße), fortlaufend erfasst, mit einer anderen Größe, der Führungsgröße, verglichen und abhängig vom Ergebnis dieses Vergleiches im Sinne einer Angleichung an die Führungsgröße beeinflusst wird. Der Wirkungsablauf findet in einem geschlossenen Kreis (Regel(ungs)kreis) statt*]

((closed) loop) control amplifier, regulation amplifier, modulating control amplifier Regel(ungs)verstärker, Reglungsverstärker *m*

((closed) loop) control block, regulation block Regel(ungs)block, Reglungsblock *m*

((closed) loop) control board → ((closed) loop) control panel

((closed) loop) control characteristic, regulation characteristic Regel(ungs)charakteristik, Reglungscharakteristik *f* [*Bei stabilisiertem Netzteil*]

((closed) loop) control circuit, feedback control circuit, servo loop, control loop, regulation circuit Regel(ungs)kreis, Reglungskreis *m* [*Er wird durch die Gesamtheit aller Glieder, die an dem Ablauf der Reglung teilnehmen, gebildet. Er besteht aus Regelstrecke und Regler*]

((closed) loop) control damper, regulation damper Regel(ungs)(stell)klappe, Reglungs(stell)-klappe *f*

((closed) loop) control (device), ((closed) loop) control appliance, regulation device, regulation appliance Regel(ungs)vorrichtung *f*, Regel(ungs)gerät *n*, Regel(ungs)anlage *f* [*Gesamtheit der Bauglieder, welche die aufgabengemäße Beeinflussung der Regel(ungs)strecke bewirkt*]

((closed) loop) control element, regulation element Regel(ungs)glied, Reglungsglied *n*

((closed) loop) control engineering, regulation engineering Regel(ungs)technik, Reglungstechnik *f*

((closed) loop) control equipment Regel(ungs)-gerät(e), Reglungsgerät(e) *n*, (*f*)

((closed) loop) control factor, regulation factor Regel(ungs)faktor *m*, Reglungsfaktor [*DIN 19226, Teil 5*]

((closed) loop) control function, regulation function Regel(ungs)funktion, Reglungsfunktion *f* [*Funktion der Regelvorrichtung als a) stetige oder quasistetige Regelfunktion, b) unstetige oder schaltende Re-gelfunktion. Je nach Zeitverhalten unterscheidet man: P-Regler = Proportionalregler, PI-Regler = Proportional-Integralregler, PID-Regler = Proportional-Integral-Differential-Regler, Regler mit anderen Zeitfunktionen. Die mathematische Grundform für das Zeitverhalten des Reglers wird Regelalgorithmus genannt.*]

((closed) loop) control gain, regulation gain Regel(ungs)verstärkung, Reglungsverstärkung *f*

((closed) loop) control gate valve, regulation gate valve Regel(ungs)schieber, Reglungsschieber *m*

((closed) loop) controllability Regelfähigkeit, Regelbarkeit *f*, Regelvermögen *n*

((closed) loop) controlled, regulated geregelt

((closed) loop) controlled variable, directly controlled variable, regulated variable Regelgröße, Reg(e)lungsgröße *f* [*Die Regelgröße x ist diejenige Größe der Strecke, die zum Zwecke des Regelns erfasst und über die Messvorrichtung der Regelvorrichtung zu-geführt wird. Sie ist die Ausgangsgröße der Strecke und Eingangsgröße der Messvorrichtung. E DIN 19 226, T 5*]

((closed) loop) controller station Reglerstation *m*

((closed) loop) controlling, regulating Regeln *n* [*1) Das Regeln ist ein Vorgang, bei dem fortlaufend eine Größe, die zu regelnde Größe (Regelgröße), erfasst, mit einer anderen Größe, der Führungsgröße, verglichen und abhängig vom Ergebnis dieses Vergleichs im Sinne einer Angleichung an die Führungsgröße beeinflusst wird (aus: DIN 19 226 Teil 1, Entwurf 05.84). Der sich dabei ergebende Wirkungsablauf findet in einem geschlossenen Kreis, dem Regelkreis, statt. DIN 19 222.*
2) Das Regeln ist ein Vorgang, bei dem eine Größe, die zu regelnde Größe (Regelgröße), fortlaufend erfasst, mit einer anderen Größe, der Führungsgröße, verglichen und im Sinne einer Angleichung an die Führungsgröße beeinflusst wird. Kennzeichen für das Regeln ist der geschlossene Wirkungsablauf, bei dem die Regelgröße im Wirkungsweg des Regelkreises fortlaufend sich selbst beeinflusst. E DIN 19 226, Teil 1]

((closed) loop) control matrix, regulation matrix Regel(ungs)matrix, Reglungsmatrix *f*

((closed) loop) control mode, regulation mode Regel(ungs)betriebsart, Regel(ungs)arbeitsart, Reglungsbetriebsart, Reglungsarbeitsart *f*

((closed) loop) control network Regel(ungs)-netz(-werk), Reglungsnetz(werk) *n*

((closed) loop) control of multivariable systems, regulation of multivariable systems Reg(e)lung *f* von Mehrgrößensystemen, Reg(e)-lung von Vielgrößensystemen

((closed) loop) control panel, ((closed) loop) control board, regulation board, regulation panel, Zentrale, Regelungstafel *f*

((closed) loop) control parameter, regulation parameter Regel(ungs)kennwert, Regel(ungs)parameter, Reglungskennwert, Reglungsparameter *m*

((closed) loop) control process, regulation process Regel(ungs)vorgang, Regulierungsvorgang, Reglungsvorgang *m*

((closed) loop) control property, regulation property Regel(ungs)eigenschaft *f*, Reglungseigenschaft

((closed) loop) control range, regulating range Regel(ungs)bereich, Reglungsbereich *m* [*selten: n*] [*Auslegungsbereich der technischen Anlage, innerhalb dessen die Regelgröße geregelt wird. DIN 41 745*]

((closed) loop) control signal, regulation signal Regel(ungs)signal, Reglungssignal *n*

((closed) loop) control stability, regulation stability Regel(ungs)stabilität, Reglungsstabilität *f*

((closed) loop) control station, regulation station Regel(ungs)station, Reglungsstation *f*

((closed) loop) control straight(-way) valve → straight(-way) ((closed) loop) control valve

((closed) loop) control system [*IEC 50 (351)*], **directly controlled member** [*IEC 27-2A*], **regulation system** [*"Architecture" may synonymously be used for "system"*] Regel(ungs)system *n* mit geschlossenem Wirkungsablauf, Regel(ungs)kette *f*, Regel(ungs)system mit Rückmeldung [*auch „Reg(e)lung"*]

((closed) loop) control system analog(ue) simulator, regulation system analog(ue) simulator Modellregel(ungs)kreis, Regel(ungs)modellkreis, Reglungsmodellkreis *m*

((closed) loop) control system malfunction [*IEC 50 (351)*], **directly controlled member malfunction** [*IEC 27-2A*], **regulation malfunction** (Funktions-)Störung *f* des Regel(ungs)systems, Regel(ungs)system(funktions)störung, Reglungssystem(funktions)störung *n*

((closed) loop) control valve Regel(ungs)ventil, Stellventil, Reglungsventil *n* [*Zur Reg(e)lung als Eingang(s)glied der Regelstrecke*]

((closed) loop) control variable, regulation variable Regel(ungs)größe *f* [*DIN 19 226, Teil 5*]

((closed) loop) dead time Regel(ungs)totzeit, Totregel(ungs)zeit, Reglungstotzeit, Totreglungszeit *f*

((closed) loop) digital control, digital ((closed) loop) control, digital regulation digitale Reg(e)lung, Digitalreg(e)lung *f*

((closed) loop) dynamic system dynamische Regel(ungs)system, dynamische Reglungssystem *n*

((closed) loop) error signal Regeldifferenz *f*, Regelunterschied *m* [*DIN 19 226, Teil 5*]

((closed) loop) feedforward control, closed feedforward control, feedforward regulation, feedforward ((closed) loop) control, feedforward closed control Reg(e)lung *f* mit Störgrößenaufschaltung, Vorwärtsreg(e)lung *f*

((closed) loop) flow temperature control → flow temperature ((closed) loop) control

((closed) loop) frequency response Frequenzgang *m* des geschlossenen Regelkreises

((closed) loop) function block Regel(ungs)baustein, Reglungsbaustein *m* [*programmierbare Regelgerät*]

((closed) loop) gain, operation(al) gain, regulation gain (Regel(ungs))Kreisverstärkung, Reglungskreisverstärkung *f*

((closed) loop) gas pressure control → gas pressure ((closed) loop) control

((closed) loop) humidity control, closed humidity control, humidity regulation Feuchtereg(e)lung, Feuchtigkeitsreg(e)lung, Feuchteregulierung, Feuchtigkeitsregulierung *f*

((closed) loop) infrared control → infrared ((closed) loop) control

((closed) loop) level control Niveaureg(e)lung, Standreg(e)lung *f* [*Flüssigkeit*]

((closed) loop) operation geschlossene Betrieb *m* [*Der Rechner gibt direkte Steuerwerte an den Prozess ab*]

((closed) loop) path geschlossene Schleifenweg *m*

((closed) loop) position control, ((closed) loop) positioning Lageregel(ungs)kreis *m*

((closed) loop) positioning valve → ((closed) loop) control valve

((closed) loop) power control, power ((closed) loop) control Leistungsregelung *f*

((closed) loop) process → control process

((closed) loop) process(ing) design Projektierung *f* des Regel(ungs)vorganges, Auslegung des Regel(ungs)vorganges

((closed) loop) project control, project regulation Projektreg(e)lung *f*

((closed) loop) pump control, pump regulation Pumpenreg(e)lung *f*

((closed) loop) radio space temperature control → radio space temperature ((closed) loop) control

C

((closed) loop) radio space temperature controller → radio space temperature ((closed) loop) controller

((closed) loop) room temperature control → room temperature ((closed) loop) control

((closed) loop) servo valve → ((closed) loop) control valve

(closed) loop structure [*In a binary control system*] Regel(ungs)kreisstruktur *f*

(closed) loop (system) Abschaltkreis *m*

((closed) loop) temperature control, closed temperature control, temperature (closed) loop control, temperature closed control, temperature regulation Temperaturreg(e)lung *f*

((closed) loop) time Regel(ungs)zeit, Reglungszeit *f*

((closed) loop) transfer function Übergabefunktion *f* des geschlossenen Regelkreises, Weitergabefunktion des geschlossenen Regelkreises, Übermitt(e)lungsfunktion des geschlossenen Regelkreises, Übertragungsfunktion des geschlossenen Regelkreises

((closed) loop) valve control Ventilreg(e)lung *f*

((closed) loop) valve-controlled ventilgeregelt

((closed) loop) variable [*See remark under "variable"*] Regel(ungs)größe *f* [*Ausgangsgröße der Regel(-ungs)strecke, wobei diese Größe aufgabengemäß geregelt wird*]

closed parallel current surface condenser geschlossene Oberflächenparallelstromkondensator *m*

closed (piping) system [*The circulating water or brine is completely enclosed and under pressure above atmospheric*] geschlossene Umwälzanlage *f*, geschlossene Umwälzsystem *n*

closed position Schließstellung, geschlossene Stellung *f*

closed raceway [*See remark under "raceway"*] geschlossene Leitungskanal *m*

closed room heater [*It burns only smokeless fuels since house coal would soon soot up the internal flueways and reduce efficiency*] geschlossene Raumheizer *m*

closed shell geschlossene Hülle *f*

closed shell-and-tube condenser, multi(ple)-shell condenser [*It consists of a number of closed shell-and-tube units*] geschlossene Rohrbündelverflüssiger, geschlossene Rohrbündelkondensator, geschlossene Bündelrohrverflüssiger, geschlossene Bündelrohrkondensator *m*

closed shop [*A data-processing center so organized that only professional programmers and operators have access to the center to meet the needs of users*] abgetrennte Rechenzentrum, geschlossene Rechenzentrum *n*

closed signal path control geschlossene Steuerung *f* [*Eine Steuerung, bei der die zu steuernden Größen eines Systems durch geeignete Steuer(ungs)-größen über einen geschlossenen Signal(fluss)weg zielgerichtet beeinflusst werden. Eine Form der geschlossenen Steuerung ist die Reg(e)lung*]

closed solar heating plant, closed solar heating installation, closed sun heating plant, closed sun heating installation geschlossene Sonnenheizungsanlage, geschlossene Solarheiz(ungs)anlage, geschlossene Sonnenheizanlage *f* [*Sie ist gegen Drucküberschreitung gesichert und hat keine ständige offene Verbindung zur Atmosphäre*]

closed subroutine; closed subprogram (US); closed subprogramme (Brit.) [*In automatic data processing, a subroutine that can be stored outside the main routine and can be connected to it by link-ages at one or more locations*] geschlossene Unterprogramm *n*

closed surface condenser geschlossene Oberflächenkondensator, Oberflächengefäßkondensator *m*

closed system [*In thermodynamics, a system which is isolated so that it cannot exchange matter or energy with its surrounding and can therefore attain a state of thermodynamic equilibrium*] isolierte System, geschlossene System *n*

closed tank (type) dishwashing machine Tank(geschirr)spülmaschine *f*

closed traverse geschlossene Polygonzug *m*

closed type, closed design, closed version geschlossene Bauart, geschlossene Bauweise *f*, geschlossene Aufbau *m*, geschlossene Ausführung, geschlossene Konstruktion *f*

closed (type) heating plant, closed (type) heating installation geschlossene Heiz(ungs)anlage, ge-schlossene Heiz(ungs)installation *f*

closed user group → CUG

closed version → closed type

closed vessel geschlossene Gefäß *n* [*z. B. Membran(druck)ausdehnungsgefäß*]

closed (warm) water heater, closed hot water heater, closed (warm) water heating appliance, closed hot water heating appliance geschlossene Wassererhitzer, geschlossene Wasser(er)wärmer, geschlossene Wasserheizer, geschlossene Warmwasserbereiter *m* [*Auslauf geschlossen*]

closed water heating geschlossene Wasserheizung *f*

close fit Feinpassung, Edelpassung *f*

close joint soldering Spaltlöten *n*

close mesh grating feinmaschige Rost *m*

close mesh wire feinmaschige Draht *m*

close miking Mikrophonierung *f* in nächster Nähe der Schallquelle

to close off → **to shut off**

closeoff → **shut(ting)-off**

close-pitch fluting [*radiator panel*] Feinprofilierung *f*

close protection Personenschutz *m*

closer → **shut(ting)-off device**

closer, quarter of a brick, brick quarter Viertelziegel *m*

close(-range) force Nahkraft *f*

close(-range) modem Nahmodem *n*

close sliding fit enge Schiebepassung *f*

closet, built-in cupboard, in-built cupboard [*A small enclosed storage area*] Wandschrank, Einbauschrank *m*

closet (US); small (private) room Kleinzimmer *n*

closet → **toilet**

closet lock Schrankschloss *n*

closet soil stack → **soil stack**

closet waste water pipe → **soil pipe**

closing → **shut(ting-)off**

closing authorization, closure authorization Schließberechtigung *f*

closing circuit, closure circuit Einschalt(strom)-kreis *m*

closing cock → **stopcock**

closing contact, closure contact Arbeitskontakt, Schließungskontakt, offene Kontakt *m* [*Er schließt bei Betätigung des Schalters*]

closing damper, closure damper Verschlussklappe *f*

closing device, closure device [*IEC 129*] Einschaltvorrichtung *f* [*VDE 0670, Teil 2*]

closing diaphragm pump → **diaphragm closing pump**

closing direction → **direction of closing**

closing force, closing power, closure force, closure power Schließ(ungs)kraft *f*

closing lockout, closure lockout Einschaltverrieg(e)lung, Einschaltsperre *f* [*Schaltgerät*]

closing mechanism, closure mechanism Einschaltmechanismus *m*

closing post, closure post Schließpfosten *m* [*Schiebetor*]

closing readiness, closure readiness Schließbereitschaft *f*

closing relay, closure relay Einschaltrelais *n*

closing sequence controller, closure sequence controller Schließfolgeregler *m*

closing signal, closure signal Schließsignal *n*

closing type, closing design, closing version, closure type, closure design, closure version Schließart *f* [*Schloss*]

closure → **shut(ting)-off**

closure bar → **bar of closure**

closure member Drosselkörper *m* [*Ventil*]

clothed human body bekleidete menschliche Körper *m*

clothes chute Kleiderrutsche *f*

clothes dryer, clothes drier Kleidertrockner *m*

clothes hamper Kleiderkorb *m*

clothes hook → **robe hook**

clothes washing machine → **household washing machine**

cloudless sky → **clear sky**

cloudless sky radiation → **clear sky radiation**

cloud point [*Temperature at which a clear liquid becomes hazy or cloudy due to the formation of crystals or particles, when tested under standardized conditions*] Erstarrungspunkt, Stockpunkt *m*

clout nail → **clasp nail**

cluster Verbund *m* von Prozessen, Prozessverbund

cluster Bausteinbündel *n*, funktionell zusammengehörige Bausteine *f*

cluster [*window 95*] Zuordnungseinheit *f*

clutch anchor Ankereisen *n*

clutch point [*In automatic data processing, the moment in time at which the clutch is engaged in a peripheral device such as a card punch*] Kupp(e)-lungspunkt *m*

CLV, constant linear velocity konstante lineare Geschwindigkeit, konstante Abtastgeschwindigkeit *f* [*Bei der digitalen Schallplatte wird die Spurabtastgeschwindigkeit für eine lange Spielzeit (optimale Speicherdichte) über den Abspielbereich konstant gehalten, die Drehzahl wird daher je nach Abspielhalbmesser verändert*]

CM → **CA**

CM → **common mode**

C

CMDR → **command reject**

CML → **current-mode logic**

CMMR → **common-mode-rejection ratio**

CMOS circuit → **complementary metal-oxid(e) semiconductor circuit**

CMR → **common-mode rejection**

CMRR → **common-mode rejection ratio**

CMTL → **current-mode transistor logic**

CMV → **common-mode voltage**

CMV ((closed) loop) control → **common-mode volt-age ((closed) loop) control**

CMV range → **common-mode voltage range**

$\frac{C}{N}$ → **carrier-to-noise-ratio**

CNC → **communication network coupler**

CNC → **computerized numerical ((open) loop) control**

C net(work) C-Netz *n* [*1986 in Deutschland eingeführtes zellulares Funknetz. Es arbeitet mit einer analogen Übertragung im Frequenzbereich um 450 Megahertz*]

CNR → **carrier-to-noise ratio**

CO → **control field**

CO → **coolant outlet**

coaching Schulung *f* des kundeneigenen Sicherheitsbeauftragten

coach screw Schlüsselschraube *f*

COA defuzzification → **centre of area defuzzification**

coal bath stove Kohlenbadeofen *m*

coal(-burning) boiler, coal-fired boiler Kohlenkessel *m*

coal(-burning) range, coal-fired range Kohle(n)-herd *m*

coal(-burning) stove, coal-fired stove Kohle(n)-ofen *m*

coal chute Kohlenrutsche *f*

coal domestic hot water heater, coal domestic warm water heater Kohlen-Brauchwasser(er)wärmer, Kohlen-Brauchwassererhitzer, Kohlen-Brauchwasserheizer *m*

coal dust furnace [*See remark under "furnace"*] Kohlenstaubfeuerung *f*

coal-fired kohlengefeuert

coal heating installation, coal heating plant Kohlenheiz(ungs)anlage *f*

coal stove Kohlenofen *m*

coal water heater [*See remark under "water heater"*] Kohlen-Wasserheizer, Kohlen-Wasser(er)wärmer, Kohlen-Wassererhitzer *m*

Coanda effect [*In fluid mechanics, the tendency of a gas or liquid coming out of a jet to travel close to the wall contour even if the wall's direction of curvature is away from the jet's axis; a factor in the operation of a fluidic element*] Coandaeffekt *m*, Coandawirkung *f*

coarse atomization; coarse atomisation (Brit.) Grobzerstäubung *f*

coarse (dust) filter Grob(staub)filter *m, n* [*DIN 24185*]

coarse (dust) filter cartridge Grob(staub)filterpatrone *f*

coarse (dust) filter fleece Grob(staub)filtervlies, Grob(staub)filtervließ *n* [*Österreich*]

coarse file, rough file, rubber (file) Grobfeile, Raufeile *f*

coarse-pitch fluting → **wide-pitch fluting**

coarse sludge Grobschlamm *m*

coarse suction filter (An)Sauggrobfilter, Grob(an)saugfilter *m, n*

coast(-cache), cache-on-a-stick Stangenschnellpufferspeicher, Schnellpufferstangenspeicher *m*

coated paper beschichtete Papier *n*

coated with frost → **frosted**

coat(ing) Überzug *m*, Beschichtung *f*

coaxial capacitor → **cylindrical capacitor**

coaxial connector, coaxial-entry plug Koaxialsteck(verbind)er *m*

coaxial heat exchanger, coaxial heat exchange unit, coaxial heat interchanger Koaxialwärme(aus)-tauscher, Koaxialwärmeübertrager *m* [*Siehe Anmerkung unter „Wärme(aus)tauscher (im weiteren Sinne)*]

coaxial (transmission) cable, coax cable, coaxial (transmission) line, concentric (transmission) cable, concentric (transmission) line [*One conductor is centered inside and insulated from an outer metal tube that serves as the second conductor*] Koaxialkabel, konzentrische (Zweileiter)Kabel *n*

coaxial tube Koaxialröhre *f*

COBOL, common business orien(ta)ted language [*A business data-processing language that can be given to a computer as a series of English statements describing a complete business operation*] COBOL, kaufmännische Programmiersprache, Programmiersprache für kommerzielle Datenverarbeitung, problemorientierte Symbolsprache *f*

cock [*Not to be confused with "faucet"*] Hahn *m* [*Absperr- und Drosselorgan für Rohrleitungen mit Be-wegung des Verschluss- oder Drosselstückes (Küken) quer zum Sitz (Querschalter). Es wird um die Achse des Hahnkükens auf ebenen Zylinder- oder Kegelsitzen gedreht*]

cock key → key of cock

cockle stove, Dutch tile stove Kachelofen *m* [*DIN 1294*]

cock plug Hahnküken *n*

cock(-type) water level indicator Hahnwasserstandanzeiger *m*

cock with screwed shank Schraubenhahn *m*

cock wrench Hahnschlüssel *m*

cocurrent design Mitstromausführung *f* [*Gegensatz: Gegenstromausführung*]

code Bauteilsatz *m* aus Kodierer und Entkodierer, Bauteilsatz aus Codierer und Entcodierer

code Kode, Code, Schlüssel *m* [*Vorschrift für die eindeutige Zuordnung der Zeichen eines Zeichenvorrates zu denjenigen eines anderen Zeichenvorrates. DIN 44 300*]

code and data separation, CDS Kode-Daten-Trennung, Code-Daten-Trennung *f* [*Daten werden für jede Aufgabe getrennt abgelegt, der Programmkode aber nur einmal*]

code card Kodekarte, Kodierkarte, Codekarte, Codierkarte, Schlüsselkarte *f*

code character Kodezeichen, Codezeichen, Codierzeichen, Kodierzeichen, Schlüsselzeichen *n*

code character set Kodezeichenvorrat, kodierte Zeichenvorrat, codierte Zeichenvorrat, verschlüsselte Zeichenvorrat, Schlüsselzeichenvorrat *m*

code conversion Kodewandlung, Codewandlung, Kodeumsetzung, Codeumsetzung, Schlüsselwandlung, Schlüsselumsetzung, Umkodierung, Umcodierung *f*

code converter Kodewandler, Codewandler, Kodierwandler, Kodierumsetzer, Kodeumsetzer *m*, Codeumsetzer, Schlüsselumsetzer, Schlüsselumwandler
[*1. Er setzt ein digitales Eingangssignal in ein anderes kodiertes digitales Ausgangssignal um. DIN 19 226, Teil 5. 2. Umsetzer, in dem die Zeichen eines Kode A Zeichen eines Kode B zugeordnet werden. DIN 44 300*]

coded kodiert, codiert, verschlüsselt

coded decimal code kodierte Dezimalkode *m*, kodierte Dezimalschlüssel, codierte Dezimalschlüssel, verschlüsselte Dezimalkode, verschlüsselte Dezimalcode *m* [*Dezimales Zahlensystem, in dem jede Dezimalzahl durch einen Kode ausgedrückt wird*]

code decimal digit, decimal-code digit [*One of ten arbitrarily selected patterns of 1 and 0 used to represent the decimal digits*] kodierte Dezimalziffer, codierte Dezimalziffer, verschlüsselte Dezimalziffer *f*

code decimal notation kodierte Dezimalschreibweise, codierte Dezimalschreibweise, verschlüsselte Dezimalschreibweise *f*

code diagnostic field → diagnostic code field

code disk, code disc [*Disk with small projections for operating contacts to give a certain predetermined code to a transmission*] Kodescheibe, Codescheibe, Kodierscheibe, Codierscheibe, Schlüsselscheibe *f*

code element Kodierglied, Kodierelement, Codierglied, Codierelement, Kodeglied, Kodeelement, Codeglied, Codeelement, Schlüsselglied, Schlüsselelement *n*

code error [*A surplus or lack of a bit or bits in a machine instruction*] Kodefehler, Codefehler, Kodierfehler, Codierfehler, Schlüsselfehler *m*

code extension Codeerweiterung, Kodeerweiterung, Schlüsselerweiterung *f*

code format Kodeformat, Codeformat, Schlüsselformat *n* [*Zuordnung der Lochkombinationen einer Spalte eines Lochstreifens zu einem bestimmten Zeichen*]

code generator Kodeerzeuger, Codeerzeuger, Kodegenerator, Codegenerator, Kodiererzeuger, Codiererzeuger, Kodiergenerator, Codiergenerator, Schlüsselerzeuger, Schlüsselgenerator *m*

code group [*A combination of letters or numerals or both, assigned to represent one or more words of plain text in a coded message*] Kodegruppe, Codegruppe, Gruppe von Kodeelementen, Gruppe von Codeelementen, Kodiergruppe, Codiergruppe, Schlüsselgruppe, Gruppe *f* von Schlüsselelementen

code hole [*One of the informational holes in perforated tape, as opposed to feed or other holes*] Kodeloch, Codeloch, Kodierloch, Codierloch, Schlüsselloch *n*

code inband signalling kodierte Im-Band-Kennzeichengabe, codierte Im-Band-Kennzeichengabe, verschlüsselte Im-Band-Kennzeichengabe *f*

code independent codeunabhängig, kodeunabhängig, schlüsselunabhängig

code input Kodeeingabe, Codeeingabe, Kodiereingabe, Codiereingabe, Schlüsseleingabe *f*

code inspection, code examination Kodeinspektion, Codeinspektion, Kodeuntersuchung, Codeuntersuchung, Schlüsselinspektion, Schlüsseluntersuchung *f* [*Kontrollfunktion der numerischen Steuerung, mit der falsche oder unzulässige Zeichen im Lochstreifen ausgeschlossen werden*]

code keyboard Kodetastatur, Codetastatur *f*

C

code letter → **letter symbol**

code listing Kodeaufstellung, Kodieraufstellung, Codeaufstellung, Codieraufstellung, Schlüsselaufstellung *f*

code lock Kodierschloss, Codierschloss, Kodeschloss, Codeschloss, Schlüsselschloss *n* [*Eingabe über Tastatur*]

code number Schlüsselzahl, Codezahl, Kodezahl, Codierzahl, Kodierzahl *f*

code of practice Merkblatt *n*

code position [*A location in a data-recording medium at which data may be entered, such as the intersection of a column and a row on a punch(ed) card, at which a hole may be punched*] Kodeposition *f*

code program (US); code programme (Brit.) [*A program(me) expressed in the required code for a computer*] Kodeprogamm, Codeprogramm, Kodierprogramm, Codierprogramm, Schlüsselprogramm *n*

coder [*A person who translates a sequence of computer instructions into codes acceptable to the machine*] Kodierer, Codierer, (Ver)Schlüsseler *m*

code selector Kodewähler, Codewähler, Kodierwähler, Codierwähler, Schlüsselwähler *m*

code set Kodevorrat, Codevorrat, Kodiervorrat, Codiervorrat, Schlüsselvorrat *m*

code signal [*A sequence of discrete conditions or events corresponding to a coded message*] Kodesignal, Codesignal, Kodiersignal, Codiersignal, Schlüsselsignal *n*

code storage Kodespeicher, Codespeicher, Kodierspeicher, Codierspeicher, Schlüsselspeicher *m*

code switch Kodeschalter, Codeschalter, Kodierschalter, Codierschalter, Schlüsselschalter *m*

code track Kodespur, Codespur, Schlüsselspur, Kodierspur, Codierspur *f*

code translation [*Conversion of a directory code or number into a predetermined code for controlling the selection of an outgoing trunk or line*] Kodeumsetzung, Codeumsetzung, Umschlüsselung, Kodierumsetzung, Codierumsetzung, Schlüsselumsetzung *f*

code word Kodewort, Codewort, Schlüsselwort, Kodierwort, Codierwort *n*

coding Kodieren, Codieren, Verschlüsseln *n*

coding sheet Kodierblatt, Codierblatt, Schlüsselblatt n

coefficient of absorption → **absorption coefficient**

coefficient of (heating) performance, CO(H)P, (heating) performance coefficient Leistungszahl *f* im Heiz(ungs)betrieb [*Verhältnis der Heiz(ungs)-leistung zur effektiven Leistungsaufnahme der Wärmepumpe*]

coefficient of performance, performance coefficient [*In the reversed Carnot cycle, the ratio of heat removed to energy expended in heat units*] Leistungszahl *f* der Maschine, Maschinenleistungszahl

coefficient of utilization, utilization coefficient [*The ratio between usable "lumens" and lamp lumens for particular combination of fixture and space*] (Aus)Nutzungskoeffizient *m*

COFF, common object file format gemeinsame Objektdatei *f*

CO_2 fire-fighting plant → **carbon dioxid(e) fire-fighting plant**

cog joint → **(double) cog joint**

coherent significant instant entsprechende Kennzeitpunkt *m*

CO(H)P → **coefficient of (heating) performance**

coil Blechrolle, Rolle *f* [*Von diesen Rollen wird das Blech maschinell abgespult und direkt in die Fertigungsstraße geführt*]

coil Spule *f*

coil, pipe coil, coiled pipe [*Cooling or heating element made of pipe or tube that may or may not be finned, formed into helical or serpentine shape*] (Rohr)Schlange *f*

coil cooler, serpentine cooler Schlangen(rohr)kühler, Rohrschlangenkühler *m*, Kühl(ungs)(rohr)schlange *f*

coil current Spulenstrom *m*

coil drain (Rohr)Schlangendrän *m*

coil-driven (loud)speaker → **dynamic (loud)-speaker**

coiled-coil filament Doppelwendelleuchtdraht *m*

coiled cord Spiralschnur, Federzugschnur *f*

coiled heat exchanger Rohrschlangenwärme(aus)-tauscher, Schlangen(rohr)wärme(aus)tauscher *m*

coiled tube Trompetenröhre *f*

coil freezing Rohrschlangeneinfrieren, Schlangenrohreinfrieren *n*

coil heating [*Heating by coils of pipes embedded in ceilings or walls, in the form of panels*] Rohrschlangenheizung, Schlangenrohrheizung *f*

coil linking flux Spulenfluss *m*

coil-loaded cable Pupinkabel, bespulte Kabel, pupinierte Kabel *n*

coil-loaded line Pupinleitung *f*, pupinierte Leitung, bespulte Leitung *f*

coil loading Bespulung, Pupinierung *f*

coil pressure Rohrschlangen(innen)druck, Schlangenrohr(innen)druck *m*

coil pressure drop Rohrschlangen(innen)druckabfall, Schlangenrohr(innen)druckabfall *m*

coil section Schlangenteil *m, n,* Rohrschlangenteil

coil temperature rise [*In a transformator is the difference between the temperature of the windings and insulation and the existing ambient temperature*] Spulentemperaturanstieg *m*

coil terminal Spulenanschluss *m*

coil voltage Spulenspannung *f*

coin automatic washer, coin automatic washing machine Münzwaschautomat *m,* automatische Münzwaschmaschine *f*

coin-box (tele)phone → public (switched) (tele-) phone coin box

coincidence Zusammentreffen *n* [*zweier oder mehrerer binärer Signale*]

coincidence Übereinstimmung *f* [*von Soll- und Istwert*]

coin contactor Münzkontaktgeber *m*

coin dryer, coin drier Münzwäschetrockner *m*

coin gas meter [*See remark under "gas meter"*] Münzgaszähler, Münzgasmesser *m*

coin meter Münzzähler, Münzmesser *m*

coin telephone booth Münzfernsprechzelle *f,* Münztelefonhäuschen, Münztelefonzelle, Münzfernsprechhäuschen *n*

coke(-burning) boiler, coke-fired boiler Koks-kes-sel *m*

coke furnace [*See remark under "furnace"*] Koksfeuerung(sanlage), Koksfeuerstätte *f*

coke furnace front Koksfeuer(ungs)geschränk *n,* Koksgarnitur *f*

coke grating [*fireplace*] Koksrost *m,* Koksgitter *n*

cold kalt [*im Sinne von niedriger Temperatur*]

cold [*Pertaining to electrical circuits that are disconnected from voltage supplies and at ground potential; opposed to hot, pertaining to carrying an electrical charge*] tot

cold Kälte *f*

cold air → cold (supply) air

cold air ((closed) loop) control damper Kaltluftregel(ungs)(stell)klappe *f*

cold (air) curtain Kalt(luft)schleier, Kalt(luft)-vorhang *m*

cold air distribution Kaltluftverteilung *f*

cold air duct Kaltluftkanal *m*

cold air flow, cold air current, cold air stream Kaltluftströmung *f,* Kaltluftstrom *m*

cold air machine Kaltluftmaschine *f*

cold air operation Kaltluftbetrieb *m*

cold air screening Kaltluftabschirmung *f*

cold beam lamp Kaltlichtspiegellampe *f*

cold bridge [*A ready path of heat transmittance through a structure. A part of relatively high U value when compared with the rest of the structure*] Kaltbrücke *f*

cold-cathode lamp → fluorescent lamp

cold chain Kaltkette *f*

cold chamber, refrigerator, cold room [*Insulated structure served by a refrigerating system for preservation of food. The term "refrigerator" is also applied to apparatus and plant for making ice*] Kälteraum *m,* Kältekammer *f*

cold curve Auslösekennlinie *f* ohne Vorlast [*Überlastrelais, DIN IEC 255, T.8*]

cold distribution Kälteverteilung *f*

cold draught, cold draft Kältezug *m*

cold-drawn steel wire kaltgezogene Stahldraht *m* [*DIN 177*]

cold drinking water → cold potable water

cold filter plugging point, CFPP Saugfiltrierbarkeitspunkt *m* [*Heizöl EL*]

cold finished steel blankgezogene Stahl *m*

cold gas heating pump Kaltgaswärmepumpe *f*

cold generation, cold production Kälteerzeugung *f*

cold generator Kälteerzeuger *m*

cold high-pressure water → high-pressure cold water

cold-hot resistant → hot-cold resistant

cold lap, cryolap Kaltschweißstelle, Kryoschweißstelle *f*

cold loop, cryoloop Kälteschleife, Kryoschleife *f*

cold(ness) → chill

cold (pipe)line, cryo(pipe)line Kälte(rohr)leitung, Kryo(rohr)leitung *f*

cold plunge pool, cold water basin Kaltwasserbecken, KW-Becken *n*

cold potable water, cold drinking water, cold drink-able water Brauchkaltwasser, Trinkkaltwasser, kalte sanitäre Brauchwasser, sanitäre kalte Brauchwasser, Kaltbrauchwasser, Kalttrinkwasser *n*

cold probe, cold explorer, cryoprobe, cryoexplorer Kryosonde, Kältesond *n*

cold pump water, pump cold water Pumpenkaltwasser *n,* PKW

C

cold recovery Kälterückgewinnung, Kältewiedergewinnung *f*

cold redundancy kalte Redundanz *f* [*Ein System, dessen Reserveeinheiten erst nach Ausfall der Grundeinheit arbeiten*]

cold restart(ing) Start *m* nach Spannungsausfall

cold roof → **ventilated flat roof**

cold room → **cold chamber**

cold room capacity, cold chamber capacity Kälteraumleistung, Kältekammerleistung *f*

cold solenoid, cryosolenoid Kryomagnetspule, Kältemagnetspule *f*

cold separation Kältetrennung *f*

cold side, air side [*heat pump*] kalte Seite, Luftseite *f*

cold source → **heat sink**

cold start(ing) device, cold start(ing) appliance Kaltstartgerät, Kaltanlassgerät *n*, Kaltstarter, Kaltanlasser *m*

cold start lamp Kaltstartlampe *f*

cold steam Kaltdampf *m*

cold steam central heating, central cold steam heating Kaltdampfzentralheizung, Zentralkaltdampfheizung *f*

cold steam cylinder Kaltdampfzylinder *m*

cold steam heating Kaltdampfheizung *f*

cold steam heating net(work) Kaltdampfheiz(ungs)netz, Kaltdampfheiznetz *n*

cold steam heating plant, cold steam heating installation Kaltdampfheiz(ungs)anlage *f* [*Als Wärmeträger dienen fluorierte Kohlenwasserstoffverbindungen, besonders das als Kältemittel R 114 gehandelte Tetrafluordichlorethan,* $C_2F_4Cl_2$]

cold (steel) finishing (Stahl)Blankziehen *n*

cold storage, cryogenic storage Kältespeicherung *f* **cold storage structure, cryogenic storage structure** Kühlhaus, Kühlgebäude *n*

cold storage vessel, cryogenic storage vessel Kältespeicher(behälter) *m*

cold store, refrigerator, cold storage building [*A building having a very low temperature, for the preservation of food. The temperature is lowered by various methods and agencies such as brine, ammonia, etc., and there are a number of patented meth-ods of insulating the store*] Kühlhaus, Kühlgebäude, Tiefkühlhaus, Tiefkühlgebäud *n*

cold store, refrigerator, cold storage room [*A room having a very low temperature, for the preservation of food. The temperature is lowered by various methods and agencies such as brine, ammonia. etc., and there are a number of patented methods of insulating the store*] Kühlraum, Tiefkühlraum *m*

cold supply, cryogenic supply, cryosupply Kälteversorgung, Kryoversorgung *f*

cold (supply) air [*General term used for supply air at 1.5 to 4.5°C*] Kalt(zu)luft *f*

cold supply plant, cold supply installation, cryogenic supply plant, cryogenic supply installation Kälteversorgungsanlage, Kälteversorgungsinstallation *f*

cold vapor (US); cold vapour (Brit.) Kaltdampf *m* [*Z.B. Ammoniak, Frigen und andere in den Verdampfern der Kältemaschinen*]

cold water Kaltwasser, KW, nicht erwärmte Brauchwasser *n* [*Temperatur je nach Jahreszeit 5°– 15°C. DIN 1946, Teil1*]

cold water assembly, cold water kit, cold water set Kaltwasser(bau)satz *m*, Kaltwassergarnitur, Kaltwasserkombination, Kaltwassergruppe *f*, Kaltwasseraggregat *n* [*„KW" kann synonym für „Kaltwasser" verwendet werden*]

cold water central distributor → **cold water manifold**

cold water cistern Kaltwasserzisterne, KW-Zisterne *f*

cold water connection Kaltwasseranschluss, KW-Anschluss *m*, Kaltwasserverbindung, KW-Verbindung *f*

cold water cooler Kaltwasserkühler, KW-Kühler *m*

cold water feed(ing) Kaltwasser(ein)speisung, KW-(Ein)Speisung *f*

cold water feed(ing) line Kaltwasser(ein)speiseleitung, Kaltwasserzuleitung, KW-(Ein)Speiseleitung, KW-Zuleitung *f*

cold water feed(ing) pipe Kaltwasser(ein)speiserohr, KW-(Ein)Speiserohr *n*

cold water feed(ing) tank Kaltwasser(ein)speisebehälter, KW-(Ein)Speisebehälter *m*

cold water flow temperature Kaltwasservorlauftemperatur, KW-Vorlauftemperatur *f*

cold water generator Kaltwassererzeuger, KW-Erzeuger *m*

cold water long-distance supply, long-distance cold water supply Kaltwasserfernversorgung, Kaltwasserfernbelieferung, KW-Fernversorgung, KW-Fernbelieferung *f*

cold water makeup line Kaltwassernachholleitung, KW-Nachholleitung *f*

cold water manifold, cold water central distributor Kaltwasserverteiler(rohr) *m*, *(n)*, Kaltwasserverteilerstück, KW-Verteilerstück *n*, KW-Verteiler(rohr)

cold water meter Kaltwasserzähler, KWZ *m*

cold water pipe Kaltwasserrohr, KW-Rohr *n*

cold water (pipe)line Kaltwasser(rohr)leitung, KW-(Rohr)Leitung *f*

cold water plunge pool Kaltwassertauchbecken, KW-Tauchbecken *n*

cold water point Kaltwasserstelle, KW-Stelle *f*

cold water production Kaltwassererzeugung, KW-Erzeugung *f*

cold water pump Kaltwasserpumpe, KW-Pumpe *f*

cold water service [*The piped cold water supply to a building to feed draw-off points and equipment in the building, either directly from the rising main or indirectly through a cold water storage tank*] Gebäudekaltwasserversorgung, Gebäude-KW-Versorgung *f*

cold water set → cold water assembly

cold water (storage) tank Kaltwasserlagerbehälter, KW-Lagerbehälter *m*

cold water supply [*Tap water supplied from the main, as distinct from the water stored for a domestic hot-water system, etc.*] Kaltwasserversorgung, Kaltwasserbelieferung, KW-Versorgung, KW-Belieferung *f*

cold water tap Kaltwasser(zapf)hahn, KW-(Zapf)-Hahn *f*

cold water throughput Kaltwasserdurchsatz, KW-Durchsatz *f*

cold water urinal connection Kaltwasserurinalanschluss, KW-Urinalanschluss, Urinal-KW-Anschluss, Urinal-Kaltwasseranschluss *f*

cold water valve Kaltwasserschieber, KW-Schieber *m*

cold water vapour (Brit.); cold water vapor (US) Kaltwasserdampf, KW-Dampf *m*

cold working, cryoworking Kalt(um)formung *f*

collapsed architecture eingeschränkte Schichtnutzung *f* [*Nur Nutzung der Schichten 1, 2, 3 und 7 des OSI-Referenzmodells*]

collar Ring *m*

collar Manschette *f*, Kragen, Bund *m*

to collate [*In automatic data processing, to combine two or more similarly ordered sets of values into one set that may not have the same order as the original sets*] einordnen

collating sequence [*In automatic data processing, the ordering of a set of items such that sets in that assigned order can be collected*] Einordnungsfolge *f*

collected frost → accumulated frost

collecting basket → collection basket

collecting flue, collection flue Sammelfuchs *m*

collecting line, collection line Sammelleitung *f*

collecting point, collection point Sammelstelle *f*

collecting unit, collector, collection unit [*In automatic data processing, an electro-mechanical device capable of performing singly or simultaneously the merging, sequence-checking, selection, and matching of punch(ed) cards*] Einordner *m*

collecting vessel, collection vessel Sammelgefäß *n*, Sammelbehälter *m*

collection → (data) collection

collection basket, carrier basket, collecting basket Auffangkorb, Büchsenkorb, Kartuschenkorb, Versandhülsenkorb *m* [*Rohrpost(anlage)*]

collection manhole, collecting manhole Sammelschacht *m*, Sammelmannloch *n*

collective addressing Sammeladressierung *f*

collective alarm Sammelalarm *m*

collective call → general call

collective detector, collective detecting device, collective detection device Sammelmelder *m*, Sammelmeldegerät *n*

collective heating system Sammelheizung *f*

collective message [*See remark under "message"*] Sammelnachricht, Sammelmitteilung *f*

collective ((open) loop) control, bank ((open) loop) control [*An improved form of automatic control in which calls from the lift car and lift-landings are registered, and are answered by the lift-car stopping in floor sequence at each lift landing for which a call has been registered, until all calls have had attention*] (rückführungslose) Sammelsteuerung *f*

collective operation(al) display, operation(al) collective display Betriebssammelanzeige, Sammelbetriebsanzeige *f*

collective security Sammelsicherung *f*

collector → (data) collector

collector → collecting unit

collector, interceptor Sammler *m* [*Leitungssammel-strang eines Entwässerungssektors*]

collector-base cut-off current Kollektor-Basis-Reststrom *m*

collector depletion layer, collector junction layer Kollektorsperrschicht, Kollektorgrenzschicht *f*

collector-emitter cut off current Kollektor-Emitter-Reststrom *m* I_{CEo}

collector-emitter cut off voltage Kollektor-Emitter-Restspannung *f* U_{CErest}

collector junction [*A semiconductor junction located between the base and collector electrodes of a transistor*] Kollektor(-Basis-Zonen)übergang *m*, Kollektorsperrschicht *f*

collector junction layer → collector depletion layer

C

collector ring, slip ring Schleifring, Stromsammelring *m*

collector voltage Kollektorspannung *f*

collision Zusammenstoß *m*, Kollision *f* [*Zusammentreffende Übertragungen, die entstehen, wenn zwei oder mehr Stationen gleichzeitig eine Übertragung starten*]

collision calculation Kollisionsberechnung *f* (EXAPT) [*Berechnung zur Vermeidung einer Kollision des Werkzeugs mit dem Werkstück oder mit anderen Geräteteilen*]

collision matrix → **scattering matrix**

colocated base station controller, colocated BSC direkt angebundene Basisstationssteuer(ungs)einheit, ortsgleiche Installation *f* von Basisstationssteuer(ungs)einheit und Hauptstation [*Festnetzbetrieb. Sie ist direkt an die Hauptstation angebunden*]

color difference signal (US); colour difference signal (Brit.) [*Electronics. A signal produced when the amplitude of a colo(u)r signal is reduced by an amount to the amplitude of the luminance signal*] Farbdifferenzsignal, Farbunterschiedssignal *n*

color(-subcarrier) oscillator (US); colour(-subcarrier) oscillator (Brit.); chroma oscillator, chrominance subcarrier oscillator [*Electronics. It produces a signal in colo(u)r television at the same frequency and phase as the incoming signal*] Farbträgeroszillator *m*

colour-corrected mercury lamp (Brit.); color-corrected mercury lamp (US) Quecksilberdampf-Niederdruck-Leuchtstofflampe, Niederdruck-Quecksilberdampf-Leuchtstofflampe *f*

colour burst (Brit.); color burst (US); (reference) burst [*The portion of the composite colo(u)r television signal consisting of a few cycles of a sine wave of chrominance subcarrier frequency*] Farbsynchronsignal *n*

colour cycling (Brit.); color cycling (US) Farb(registrier)animation *f*

coloured smoke (Brit.); colored smoke (US) Farbrauch *m*

colour flatbed scanner (Brit.); color flatbed scanner (US) Farbflachbettabtaster, Flachbettfarbabtaster *m*

colour graphics system (Brit.); color graphics system (US) Prozessleitsystem *n*

colour hygrometer (Brit.); color hygrometer (US) Farbhygrometer *n* [*Gewisse Salze, z. B. Kobaltbromid, auf saugfähiges Papier aufgetragen, verändern ihre Farbe mit der Aufnahme von Wasser aus der Luft. Sehr ungenau*]

colour (ink) style writer, colour (ink) style printer (Brit.); color (ink) style writer, color (ink) style printer (US) Farbtintenstrahl(druck)er *m*

colour-keyed (Brit.); color-keyed (US) farbmarkiert

colour killer (Brit.); color killer (US) Farbabschalter *m*

colour laser printer, colour laser writer (Brit.); color laser printer, color laser writer (US) Farblaserdrucker, Laserfarbdrucker *m*

colour perception (Brit.); color perception (US) Farbwahrnehmung *f*, Farbeindruck *m*

colour picker (Brit.); color picker (US) Farbrad *n*

colour print out (Brit.); color print out (US) Farbausdruck *m*

colour radiation pyrometer → **double colour pyrometer**

colour rendering (Brit.); color rendering (US) Farbwiedergabe *f*

colour rendering index (Brit.); color rendering index (US) Farbwiedergabeindex *m*

colour saturation (Brit.); color saturation (US) Farbsättigung *f*

colour scanner (Brit.); color scanner (US) Farbabtaster *m*

colour screen (Brit.); color screen (US) Farbbildschirm *m*

colour video camera (Brit.); color video camera (US) Farbvideokamera, Videofarbkamera *f*

colour video copy processor, color video printer (Brit.); color video copy processor, color video printer (US) Farbvideodrucker, Videofarbdrucker *m*

Colpitts (tank) circuit kapazitäre Dreipunktschaltung *f*

column Trennsäule *f* [*Chromatograph*]

column [*In automatic data processing, a vertical arrangement of characteristics or other expressions, usually referring to a specific print position an a printer or a vertical area on a card*] Spalte *f* [*DIN 40 719*]

column Umgebungsvariable *f* [*Sie enthält die festgelegte Anzahl Zeichen pro Ausgabezeile*]

column access strobe, CAS Speicherspaltensteuerungssignal *n*

column address Spaltenadresse *f*

column address select, CAS Spaltenadressauswahl *f*

column address strobe, CAS Spaltenadresseübernahmesignal *n*

column binary (punch(ed)) card [*A punch(ed) card containing data in column binary (representation)*] Binärspalten(loch)karte *f*

column binary (representation), Chinese (column) binary (representation) [*The binary representation of data on punch(ed) cards in which adjacent positions in a column correspond to adjacent bits of data*] Binärspaltendarstellung *f*

column heat exchanger Säulenwärme(aus)tauscher *m*

column inquiry Spaltenabfrage *f*

column luminaire, column light(ing) fitting Mastleuchte *f*, Mastbeleuchtungskörper *m*

column packing, column fill(ing) Säulenfüllung *f* [*Chromatograph*]

column printer Spaltendrucker *m*

column radiator Schmalsäuler *m* [*Radiator mit geringerer Bautiefe als DIN*]

column skip Spaltenübersprung *m*

column split [*automatic data processing*] Spaltenteilung *f*

column switching Trennsäulenumschaltung *f* [*Chromatograph*]

column vector Spaltenvektor *m*

COM → computer output on microfilm

coma, spot distortion Koma *n*, Lichtfleckverzerrung *f*

combatting → control

comb filter Kammfilter *m, n*

combinational circuit Schaltnetz *n* [*E DIN 44300/ 10.85*], kombinatorische Schaltung, Verknüpfungsschaltung *f*

combination condenser → air and water cooled condenser

combination control (device) [*A control (device) in which one or more control variables are monitored, such as a combination high- and low-pressure control for a refrigerant system*] Kombi(nations)-steuerung *f*

combination cylinder, combination tank [*A hot water cylinder combined with a cold water feed tank in one package unit*] Warmwasserspeicher *m* mit Kaltwasserspeicherbehälter

combination door release, door combination release Türöffnerkombination *f*

combination enclosure, combination casing, combination housing, combination case Anreihgehäuse *n*

(combination) filter/mixing box Filterteil *m, n* und Mischkasten *m* [*Lüftungsgerät*]

(combination) freezer-refrigerator unit → (combination) refrigerator-freezer unit

combination gas-oil burner → (gas) conversion burner

combination heating and cooling unit Heizungs-Kühlungs-(Klima)Gerät *n*

combination lock, permutation lock, puzzle lock Kombi(nations)schloss, Vexierschloss, Buchstabenschloss, Ziffernschloss *n*

combination range Kombi(nations)herd *m*

(combination) refrigerator-freezer unit, (combination) freezer-refrigerator unit Kühl-(schrank)-Gefrier(schrank)-Kombination, Kühlschranksäule *f*

combination sink Spültisch, Abwaschtisch *m*

combination sink cabinet and cooker unit, com-bined range and sink Herd-Spültisch-Kombination, Herd-Abwaschtisch-Kombination, Spültisch-Herd-Kombination, Abwaschtisch-Herd-Kombination *f*

combination tank → combination cylinder

combination water heater Kombi(nations)wasserheizer, Kombi(nations)(wasser)(er)wärmer *m*

to combine zusammenfassen, kombinieren

(combined) air and water induction unit Luft-Wasser-Induktions-(Klima)Gerät *n*

combined air conditioning plant, combined air conditioning installation, primary air central air conditioning plant, primary air central air condit-ioning installation kombinierte Klimaanlage, Primärluft-Zentralklimaanlage *f* [*Die angesaugte Luft wird vorbehandelt und in einer folgenden Anlage aufbereitet*]

combined air/waste gas chimney kombinierte Luft-Abgas-Schornstein, KLA-Schornstein, KLA-Kamin *m*

combined bar joist and sprinkler system Fachwerkdeckenträger *m* mit Sprinkleranlage

combined burner → (gas) conversion burner

combined check and stop valve Speisekopf *m* [*Kesselarmatur*]

combined crockery and slop sink → combined washup and slop sink

combined domestic water heating [*When the indirect domestic water heating is used for hot water supply, the same boiler may also be used for heating*] kombinierte Hauswasserheizung *f*

combined drain [*A pipe or channel that carries both sewage and storm water*] Misch(wasser)drän *m*

combined drainage → combined system (of sewerage)

combined drainage installation, combined drain-age plant, combined storm water and sewage in-stallation, combined storm water and sewage plant Mischentwässerungsanlage *f*

C

combined dry-pipe and pre-action sprinkler system kombinierte vorgesteuerte Trockensprinkleranlage *f*

combined fittings → (combined) hot and cold (water) tap

combined fuel burner Kombi(nations)brenner *m*

combined furnace kombinierte Feuerung, kombinierte Feuerstätte *f* [*DIN 4787, Teil 1. In einem Feuer(ungs)raum werden flüssige und andere Brennstoffe mit verschiedenen Brennern oder Mehrstoffbrennern verfeuert*]

combined ground resistance Gesamterdungswiderstand *m* [*VDE 0100, Teil 200*]

combined head → read/write head

combined head access aperture → read/write head access aperture

combined heat and power, CHP, c. h. p. Heizkraftkopp(e)lung, Kraft-Wärme-Kopp(e)lung, KWK *f*

combined heat and power plant, combined heat and power installation, CHP plant, CHP installation [*It makes use of waste heat from electricity generation for other purposes, such as district heating*] Heizkraftanlage *f*, Heizkraftwerk *n*

combined heating (system) kombinierte Heizung *f*

(combined) hot and cold (water) tap, (combined) hot and cold draw-off tap, combined fittings Warmwasser-Kaltwasser(zapf)hahn *m*

combined hot water and air heating Domothermheizung *f*

combined interface module Kombi(nations)-Anschaltung *f* [*MC-System; zwei Flachbaugruppen zum Anschluss von Speicherlaufwerken*]

(combined) radiant/convector electric heater Elektro-Strahlungs-Konvektor-Heizer, Strahlungs-Konvektor-Elektro-Heizer, Konvektor-Elektro-Strahlungs-Heizer *m*

combined range and sink → combination sink cabinet and cooker unit

combined right-angle disconnector and earth(ing) switch Winkeltrennerder, Trennwinkelerder *m*

combined services ceiling Decke *f* mit unterschiedlichen Leitungen

combined sewer [*It receives both stormwater and sewage*] Misch(wasser)kanal *m*

combined sewerage (system) → combined system (of sewerage)

combined sponge and soap holder, combined sponge and soap tray Schwamm- und Seifenschale, Seifen- und Schwammschale *f*

combined station, balanced station Hybridstation *f*

(combined) stay and casement fastener Kettelhaken *m* und Vorreiber, Sturmhaken und Vorreiber *m*

combined system intercepting sewer, combined system collecting sewer Misch(wasser)sammler *m*

combined system (of sewerage), combined system of sewers, combined sewerage (system), combin-ed drainage [*A system of sewers, carrying sewage and storm water*] Mischentwässerung, Mischkanalisation *f*, Mischsystem *n*

combined washup and slop sink, combined crockery and slop sink Geschirrspüle-Ausguss-Kombination *f*

combined water gebundene Wasser *n*

comb plate [*escalator*] Kammplatte *f*

comb strainer Spaltfilter *m, n*

combustibility (Ver)Brennbarkeit *f*

combustible (ver)brennbar

combustible gas Brenngas, (ver)brennbare Gas *n*

combustible refuse Brennmüll, (ver)brennbare Müll *m*

combustion Verbrennung *f*, Abbrand *m* [*des in einen Brennraum eingebrachten Brennstoffes*]

combustion air, air for combustion Verbrennungsluft *f*

combustion air blower Verbrennungsluftgebläse *n*

combustion air damper Verbrennungsluftklappe *f*

(combustion) air excess → air excess

combustion air fan Verbrennungsluftventilator *m*

combustion air input Verbrennungsluftzuführung *f*

combustion air intake, combustion air inlet Verbrennungslufteinlass *m*

combustion air limiter, combustion air limiting device Verbrennungsluftbegrenzer *m*

combustion air line Verbrennungsluftleitung *f*

combustion air opening, combustion air port Verbrennungsluftöffnung *f*

combustion air quantity Verbrennungsluftmenge *f*

combustion air room Verbrennungsluftraum *m*

combustion air supply Verbrennungsluftzufuhr, Verbrennungsluftversorgung, Verbrennungsluftzuführung *f*

combustion air throttle valve Verbrennungsluftdrosselventil *n*

combustion air volume flow Verbrennungsluftvolumenstrom *m* [*Quotient aus gefordertem Verbrennungsluftvolumen und Zeit*]

combustion appliance, combustion device Verbrennungsgerät *n*, Verbrennungsvorrichtung *f*

combustion calculation Verbrennungsrechnung *f*

combustion chamber [*1. Space forward of the mixing chamber where combustion is completed. 2. Space in a heating appliance provided for the combustion of fuel*] Brennraum, Verbrennungsraum *m*, Brennkammer, Verbrennungskammer *f*

combustion controller Verbrennungsregler *m*

combustion efficiency Verbrennungswirkungsgrad *m*

combustion engine heat pump Verbrennungsmotorwärmepumpe *f*, VWP

combustion fan Verbrennungsventilator *m*

combustion formula Verbrennungsformel *f*

combustion gas Brenngas *n*

combustion performance test verbrennungstechnische Prüfung *f*

combustion process Verbrennungsvorgang, Verbrennungsablauf *m*

combustion quality Verbrennungsgüte *f*

combustion residue Verbrennungsrückstand, Verbrennungsrest *m*

combustion substance → fuel

combustion temperature Verbrennungstemperatur *f*

combustion turbine Verbrennungsturbine *f*

to come on [*light*] aufleuchten [*Licht*]

comfort, human comfort, bod(il)y comfort Behaglichkeit, Wohligkeit *f*, Wohlbefinden *n*

comfortable behaglich, wohlig

comfortable feeling, feeling of comfort Behaglichkeitsgefühl, Wohl(igkeits)gefühl *n*

comfort air conditioning Behaglichkeitsklimareg(e)lung, Behaglichkeitsklimatisierung *f*

comfort air conditioning plant Behaglichkeit(s)klimaanlage *f* [*Gegensatz: Industrieklimaanlage*]

comfort chart, comfort diagram [*It shows curves of relative humidity and effective temperature superimposed upon rectangular coordinates of wet-bulb temperature and dry-bulb temperature*] Behaglichkeitsdiagramm *n*

comfort climate Behaglichkeitsklima *n*

comfort control [*Control of temperature, humidity, flow, and composition of air by using heating and air conditioning systems, fans, or other systems to increase the comfort of people in an enclosure*] Behaglichkeitssteuerung [*offener Wirkungsablauf*]; Behaglichkeitsreg(e)lung *f* [*geschlossener Wirkungsablauf*]

comfort cooling [*Cooling for comfort as opposed to cooling for storage or manufacture*] Behaglichkeitskühlung *f*

comfort cooling system [*All the equipment installed for the purpose of cooling air by mechanical means and discharging such air into any room or space. This definition does not include any evaporative cooler*] Behaglichkeitskühl(ungs)system *n*

comfort cooling unit [*Self-contained cooling system, factory assembled and tested, installed with or without conditioned air ducts and without connecting any refrigerant-containing parts. This definition does not include a portable comfort cooling unit or/and an absorption unit*] Behaglichkeitskühl(ungs)einheit *f*

comfort curve, curve of comfort [*A line drawn on a graph of air temperature versus some function of humidity (usually wet-bulb temperature or relative humidity) to show the varying conditions under which the average sedentary person feels the same degree of comfort*] Behaglichkeitskurve *f*

comfort engineering Behaglichkeitstechnik *f*

comfort equation Behaglichkeitsgleichung *f*

comfort heating equipment [*All warm-air furnaces, warm-air heaters, combustion product vents, comfort heating air distribution ducts and fans, steam and hot-water piping, together with all control devices and accessories installed as part of or in connection with any comfort heating system or appliance regulated by local codes*] Behaglichkeitsheiz(ung(s))gerät(e) *n*, (*f*)

comfort heating system [*Warm air heating system consisting of a heat exchanger enclosed in a casing, from which the heated air is distributed through ducts to rooms and areas. This system includes the circulat-ing air supply, conditioned air supply, and all accessory apparatus and equipment installed in connection therewith*] Behaglichkeitsheiz(ung(s))-system *n*

comfort index, temperature humidity index, CI, THI [*An index which gives a numerical value, in the general range of 70-80 reflecting outdoor atmospheric conditions of temperature and humidity as a measure of comfort (or discomfort) during the warm season of the year; equal to 15 plus 0.4 times the sum of the dry-bulb and wet-bulb temperatures in degrees Fahrenheit*] Behaglichkeitsindex, Unbehaglichkeitsindex *m*

comfort level Behaglichkeitsstufe *f*

comfort line [*Line on the comfort chart showing the relation between the effective temperature and the percentages of persons feeling comfortable*] Behaglichkeitslinie *f*

comfort office Wohlfühlbüro *n*, Behaglichkeitsbüro

comfort operation Behaglichkeitsbetrieb *m*

comfort requirement Behaglichkeitsanforderung *f*

comfort research Behaglichkeitsforschung *f*

C

C

comfort setting Behaglichkeitseinstellung *f*

comfort temperature Behaglichkeitstemperatur *f*

comfort ventilation, ventilation of human occupancy Behaglichkeitslüftung *f*

comfort zone [*The area on the psychrometric chart covered by combinations of temperatures and humidities at which, in subjective tests, more than 50 % of the subjects were comfortable*] Behaglichkeitszone *f*

coming signal message kommende Meldung *f* [*Eine Meldung, abgeleitet vom Wechsel des Eingabesignals in den zu meldenden Betriebszustand*]

comma [*A mark of punctuation (,) used to indicate a slight separation of sentence elements; commas are used to set off nonrestrictive or parenthetical elements, quotations, items in a series, etc.*] Komma *n*, Beistrich *m*

command Kommando *n*, interne Nachricht *f* [*Die Befehlsteile die nach ihrer Entschlüsselung vom Steuerwerk der Zentraleinheit an periphere Steuerwerke (Geräte- oder Kanalsteuereinheit) zur weiteren Entschlüsselung übermittelt werden, heißen Kommandos*]

command execution Kommandoausführung *f*

command history Kommandospeicher *m*

command language, job-control language [*A means by which a user can interact with a computer to describe to the system the requirements of a job, such as the amount of storage needed or input/output devices required, and actions to be taken, such as programs to be executed*] Kommandosprache *f*

command pass through Kommandodurchgabe *f*

command reject, CMDR Kommandorückweisung *f* [*Protokoll*]

command sequence, command string Kommandofolge *f*

command signal Kommandosignal, Kommandozeichen *n*

command variable → reference input variable

command verification Kommandobewahrheitung *f*, Nachweis *m* der Kommandorichtigkeit, Feststellung *f* der Kommandowahrheit

comment [*In automatic data processing, an expression identifying or explaining one or more steps in a routine, which has no effect on execution of the routine*] Bemerkung, Anmerkung *f*, Kommentar *m*

comment code [*In automatic data processing, one or more characters identifying a comment*] Bemerkungskode, Bemerkungscode, Anmerkungskode, Anmerkungscode, Kommentarcode, Kommentarkode *m*

commercial building Geschäftsgebäude, Geschäftshaus *n*

commercial centre autowalk, commercial centre passenger conveyor [*US spelling is "center". As opposed to a traffic installation autowalk*] Geschäftshausrollsteig, Geschäftshausfahrsteig, Geschäftsgebäuderollsteig, Geschäftsgebäudefahrsteig *m*

commercial centre escalator (Brit.); commercial center escalator (US) [*As opposed to a traffic installation escalator*] Geschäftshausrolltreppe, Geschäftshausfahrtreppe, Geschäftsgebäuderolltreppe, Geschäftsgebäudefahrtreppe *f*

commercial conveyance equipment Gewerbefördergerät(e) *n, (f)*

commercial cooling Gewerbekühlung *f*

commercial cooling furniture Gewerbekühl(ungs)möbel *f*

commercial dishwasher gewerbliche (Geschirr)-Spülmaschine *f*

commercial electrical furnace, electrical commercial furnace Gewerbeelektroofen *m*

commercial electrical kitchen, electrical commercial kitchen gewerbliche Elektroküche, Gewerbeelektroküche *f*

commercial freezer [*In cold storage, an insulated room kept below 30° F (-1° C)*] Gefrierraum, Frostraum *m*

commercial(-grade) handelsüblich

commercial kitchen Gewerbeküche *f*

commercial kitchen sink Gewerbeküchenspüle *f*

commercial lab(oratory) gewerbliche Labor(atorium) *n*

commercially produced maßstäblich hergestellt, hergestellt im Fertigungsmaßstab

commercial refrigerating system, commercial re-frigeration system [*It is assembled or installed in a building used for business or commercial purposes. See remark under "refrigerating system"*] Gewerbe-Kältemittelumwälzsystem *n*

commercial refrigerator [*Cabinet for keeping food or other items cool*] gewerbliche Kühlschrank *m*

commercial room gewerbliche Raum *m*

commercial timber Handelsholz *n*

to commission in Betrieb nehmen, in Betrieb setzen [*Anlage*]

commissioning Inbetriebnahme, Inbetriebsetzung, IBS *f*, Inbetriebnehmen, Inbetriebsetzen *n* [*Anlage*]

commissioning card Inbetriebnahmekarte *f*

commissioning engineer Inbetriebnahmeingenieur *m*

commode(-type) toilet, commode(-type) closet [*The term "lavatory" is also used as a euphemism for "toilet" and "closet"*] Sitzabort, Sitzabtritt *m*, Sitzklosett *n*, Sitztoilette *f*

common → grounded

common application programming interface, CAPI CAPI-Treiber *m*, Treiberprogramm *n* [*Schnittstelle zwischen ISDN-Software und ISDN-Karte*]

common area, common field [*In automatic data processing, an area of storage which two or more routines share*] gemeinsame Bereich *m [selten: n]*

Common Business Oriented Language → COBOL

common carrier Fernmeldegesellschaft *f*

common channel interference Gleichkanalstörung *f*

common control unit [*In automatic data processing, a control unit that is shared by more than onemachine*] gemeinsame Steuer(ungs)einheit *f*

common-drain amplifier, source-follower amplifier [*Electronics. A field-effect transistor amplifier in which the drain terminal is common to the input and the output*] Quelle-Folger-Verstärker *m*

common emitter Emitterschaltung *f* [*Transistorbetriebsschaltung in der der Emitter dem Eingangs- und Ausgangskreis angehört*]

common field → common area

common ground Betriebserde *f* [*Im Gegensatz zur Schutzerde ist die Betriebserde das gemeinsame Bezugspotential für alle Daten- und Steuer(ungs)leitungen innerhalb eines Rechnersystems*]

common language [*In automatic data processing, a machine-readable language that is common to a group of computers and associated equipment*] gemeinsame Sprache *f*

common lock gewöhnliche Schloss *n* [*Gegensatz: Sicherheitsschloss*]

common logarithm → Brigg(´s) logarithm

common mode, CM [*Having signals that are identical in amplitude and phase at both inputs, as in a differential operational amplifier*] Gleichtakt *m*

common-mode crosstalk(ing) Gleichtaktübersprechen *n*

common-mode gain Gleichtaktverstärkung *f*

common-mode input voltage Gleichtakteingang(s)spannung *f*

common-mode rejection, CMR, in-phase rejection Gleichtaktunterdrückung *f*

common-mode rejection ratio, CMRR Gleichtaktunterdrückung G *f*

common-mode signal, in-phase signal, I signal Gleichtaktsignal *n*

common-mode voltage, CMV, in-phase voltage [*A signal of the same polarity and voltage on both wires of a differential input, often caused by a common path*] Gleichtaktspannung *f*

common-mode voltage ((closed) loop) control, in-phase voltage ((closed) loop) control, CMV ((closed) loop) control Längsreg(e)lung *f* [*Spannungsreg(e)lung mittels einer zusätzlichen, variablen und phasengleichen Spannungskomponente*]

common-mode voltage range, CMV range, in-phase voltage range Gleichtakt-Spannungsbereich *m [selten: n]* [*Operationsverstärker*]

common neutral conductor [*A neutral conductor that is common to, or serves more than one circuit*] gemeinsame Neutralleiter, gemeinsame Mittelleiter, gemeinsame Nulleiter *m*

common object file format → COFF

(common) return Rückleiter *m*

common-sense [*language*] verständlich, allgemeingültig

common-status display, common-state display Zustandsanzeige *f* [*(Prozess)Leittechnik*]

common turn-off Summenlösung *f* [*Leistungselektronik*]

common variable gemeinsame Variable, gemeinsame Veränderliche *f*

common vent(ilation) stack, common vent(ilat)ing stack gemeinsame Lüftungsleitung *f* [*Zwei nebeneinander liegende Fallleitungen*]

communication [*Transmission of information from one point, person, or piece of equipment to another*] Kommunikation, Informationsübermittlung, (Nachrichten)Übermittlung *f*, Informationsaustausch, (Nachrichten)Austausch *m*

communication → data communication

communication cable Kommunikationskabel, Nachrichtenkabel, Fernmeldekabel *n*

communication centre (Brit.); communication center (US) Kommunikationszentrum, Nachrichtenzentrum *n*

communication channel Kommunikationskanal, Nachrichtenkanal, Fernmeldekanal *m*

communication controller, CC Datenfernverarbeitungssteuereinheit, Leitzentrale *f*

communication device, communication appliance Kommunikationsgerät, Nachrichtengerät *n*, Kommunikationsvorrichtung, Kommunikationseinheit, KE, Nachrichtenvorrichtung, Nachrichteneinheit *f*

communication driver Koppeltreiber *m*

communication electronics Nachrichtenelektronik, Kommunikationselektronik, Fernmeldeelektronik *f*

communication engineer Kommunikationsingenieur, Nachrichteningenieur, Fernmeldeingenieur *m*

C

communication engineering Kommunikationstechnik, Nachrichtentechnik, Fernmeldetechnik *f*

communication interface Kommunikationsschnittstelle, Nachrichtenschnittstelle, Fernmeldeschnittstelle *f*

communication interface board Schnittstelle *f* zwischen Zentraleinheit und Terminal(s)

communication interface service Kommunikationsschnittstellendienst, Nachrichtenschnittstellendienst, Fernmeldeschnittstellendienst *m*

communication line Kommunikationsleitung, Nachrichtenleitung, Fernmeldeleitung *f*

communication matrix Kommunikationsmatrix, Nachrichtenmatrix, Fernmeldematrix *f* [*Sie beschreibt die Kommunikationsbeziehungen der aktiven Netzteilnehmer*]

communication mechanism Kommunikationsmechanismus, Nachrichtenmechanismus, Fernmeldemechanismus *m*

communication memory Kommunikationsspeicher, Nachrichtenspeicher, Fernmeldespeicher *m*

communication net(work), communication lines [*Organization of stations capable of intercommunications but not necessarily on the same channel*] Kommunikationsnetz, Nachrichtennetz, Fernmeldenetz *n*

communication net(work) coupler, CNC, communication lines coupler Kommunikationsnetz(werk)-kopp(e)ler, Nachrichtennetz(werk)kopp(e)ler, Fernmeldenetz(werk)kopp(e)ler *m*

communication path [*DIN IEC 625. Programmable measuring apparatus*]; message route, circuit, line Übertragungsweg, Nachrichtenweg, Fernmeldeweg, Kommunikationsweg *m*, Leitung *f* [*programmierbare Messgerät*]

communication pipe [*A pipe between the Water Board's supply main and the consumer's stop valve, or the boundary of the consumer's site*] Grundstück-anschlussrohr, Grundstückverbindungsrohr *n*

communication program (US); communication programme (Brit.) Kommunikationsprogramm *n*

communication protocol → **protocol**

communication security Kommunikationssicherung, Nachrichtensicherung *f*

communication security system Kommunikationssicherungssystem, Nachrichtensicherungssystem *n*, Kommunikationssicherungsanlage, Nachrichtensicherungsanlage *f*

communication server Steuer(ungs)rechner *m* in Kommunikationsnetzen

communication service Kommunikationsdienst, Nachrichtendienst *m*, Kommunikationswesen, Nachrichtenwesen *n*

communication signal Nutzsignal, Kommunikationssignal *n* [*Gegensatz: Störsignal*]

communication system [*Electrical system whereby intelligence or signals may be transmitted to or through a central station, including telephone, telegraph, district messenger, fire and burglar alarm, watchman or sprinkler supervisory systems, and other central station systems of a similar nature, which commonly receive the power supply necessary for their operation from central office or local power sources, but not including radio communication equipment*] Kommunikationssystem, Nachrichtensystem, Übermittlungssystem, Fernmeldesystem *n*

communication throughput capacity [*Rate at which a communication link transmits data*] Kommunikationsgeschwindigkeit, Nachrichtengeschwindigkeit, Fernmeldegeschwindigkeit, Übermittlungsgeschwindigkeit *f*

communication topology Kommunikations-Topologie, Nachrichtentopologie, Fernmeldetopologie, Übermittlungs-Topologie *f*

communication unit Kommunikationseinheit, Nachrichteneinheit, Fernmeldeeinheit, Übermittlungseinheit *f*

community antenna television → **CATV**

(community) fire alarm headquarters Brandwache, Feuerwache *f*

commutating pole, compole, interpole Wendepol, Umkehrpol, Kommutierungspol *m*

commutating (pole) field, interpole field, compole field Wendepolfeld, Umkehrpolfeld, Kommutierungspolfeld *n*

commutator Stromumkehrer, Stromwender, Kommutator, Kollektor *m*

commutator bar copper Lamellenkupfer *n*

commutator pulse, position pulse, P pulse [*One of a series of pulses indicating the beginning or end of a signal representing a single binary digit in a computer word*] Kommutatorimpuls, Kollektorimpuls, Stromwenderimpuls, Stromumkehrimpuls *m*

commutator riser Stromwenderfahne, Kommutatorfahne, Kollektorfahne, Stromumkehrerfahne *f*

commutator switch → **sampling switch**

compact gedrungen, kompakt, gedrängt

compact controller, miniaturized controller Kompaktregler, gedrungene Regler, gedrängte Regler *m* [*Er enthält alle wesentlichen Teile in einem Gehäuse, ausgenommen Messgrößenaufnehmer und Stellglied. DIN 19 225*]

compact device Kompaktgerät, gedrungene Gerät, gedrängte Gerät *n*

compact disc, compact disc, CD, digital audio-disk, digital audio-disc, DAD digitale Schallplatte *f* [*Mit optischer Abtastung von Philips initiiert, in Zusammenarbeit mit Sony und anderen Firmen verbessert. DIN EN 60 908*]

compact disk-digital audio, compact disc-digital audio, CD-DA Festplatte *f* mit digitaler Schallspeicherung

compact disc-erasable, compact disk-erasable löschbare Festplatte *f*

compact disc read only memory, compact disk read only memory Festplatte *f* mit unveränderbaren Daten, Festplattenlesespeicher *m*

compact disk ROM → CD-ROM

compact heat exchanger, compact heat exchange unit, compact heat interchanger Hochleistungswärmeübertrager *m* [*Siehe Anmerkung unter „Wärme(aus)tauscher"*]

compact ice-storage system [*See remark under "ice-storage system"*] Kompakteisspeicher(anlage) *m, (f)*, Kompakteisspeichersystem *n*

compaction [*In automatic data processing, a technique for reducing the space required for data storage without losing any information control*] Zusammenfassung *f* in Gruppen, Gruppenbildung

compact machine Elektromaschine *f* mit Erregerelektronik, Kompaktmaschine

compact system, compact installation, compact plant Kompaktanlage, Kleinanlage *f*, Kompaktsystem, Kleinsystem *n*

companded kompandiert [*Amplitudenmodulation*]

compander, compressor expander [*Combination of a compressor at one point in a communication path for reducing the amplitude range of signals, followed by an expander at another point for a complementary increase in amplitude range*] Dynamikregler, Presser, Kompandor, Kompander *m* [*Kurzwellensprechverkehr*]

companding [*A process in which compression is followed by expansion; often used for noise reduction in equipment, in which case compression is applied before noise exposure and expansion after exposure*] Kompandieren *n*, Kompandierung *f*

companding [*electroacoustics*] Dynamikreg(e)lung *f*, Pressung/Dehnung der Dynamik

companding advantage Kompandorgewinn, Kompandorvorteil *m*

companding law Kompandorkennlinie *f*

company member, member company Firmenmitglied *n*, Mitgliedsfirma *f*

company's fuse, principal fuse, main fuse Hauptsicherung *f*

company's stopcock Wasserwerkabsperrhahn *m*

comparative tracking index, CTI Vergleichszahl *f* der Kriechwegbildung [*VDE 0303 Teil 1*], Kriechstromzahl, KZ *f*

comparator, comparing element, comparator device Differenzverstärker, Vergleichsglied *n*, Koinzidenzeinheit, Vergleichsvorrichtung *f*, Vergleicher, Komparator, Signalvergleicher *m* [*Vorrichtung, die z. B. Soll- und Istwert miteinander vergleicht und evtl. auftretende Abweichungen durch ein digitales oder analoges Signal verringert oder aufhebt*]

comparator check (test), comparator checkout Kontrolle *f* durch Vergleich, Überprüfung *f* durch Vergleich, Vergleichskontrolle, Vergleichsüberprüfung *f*, Kontrollvergleich, Überprüfungsvergleich *m*

comparator circuit Vergleichsschaltung *f* [*Messtechnik*]

comparing Vergleichen *n*

comparing control change → control change

comparison, computer comparison [*A computer operation in which two numbers are compared as to identity, relative magnitude, or sign*] Vergleich, Rechnervergleich *m*

comparison brush, comparing brush [*One of a set of metallic brushes which verifies that all the cards in a gangpunching operation have been properly punched*] Vergleichsbürste *f*

comparison cycle, comparing cycle Vergleichsprozess *m*

comparison element, comparing element Vergleichselement, Vergleichsglied *n*

comparison gate → (time) comparison gate

comparison indicator, comparing indicator [*One of the two registers, of which one is activated during the comparison of two quantities to indicate whether the first quantity is lower than, equal to, or greater than the second quantity*] Vergleichsanzeiger *m*

comparison scale, comparing scale Vergleichsskala *f*

compartment Abteil, Fach *n*

compartment → (measuring) compartment

compartmentation → fire-resistant compartmentation

compartment box [*A box for joining pipes in differ-ent directions. It may be 3-way, 4-way, 5-way, etc.*] (Rohr)Verzweigungskasten *m*

compartment for zone identification sheet(s) Dechiffrierfach, Dechiffrierabteil *n*

C

C

compartment(-type) switchgear abgeschottete Schaltanlage *f*

compartment valve Abschnittventil *n* [*Sprinkleranlage*]

compass [*An instrument that indicates direction, usually having a magnetic needle that swings freely on a pivot and points to the magnetic north*] Kompass *m*

compass azimuth Kompassazimut *m, n*

compass bearing Kompasspeilung *f*

compass direction Kompassrichtung *f*

compatibility Verträglichkeit, Übereinstimmung, Vereinbarkeit *f*

compatible current sinking logic stromziehende verträgliche Logik, verträgliche stromziehende Logik *f*

compatible data base interface verträgliche Datenbankschnittstelle, übereinstimmende Datenbankschnittstelle, vereinbare Datenbankschnittstelle *f*

compensating earth (Brit.); compensating ground (US) Ausgleichserde *f*

compensating loop → expansion loop

compensating piece → expansion piece

compensating rope [*elevator*] Ausgleichsseil *n*

compensation → adjustment

compensation Kompensation *f*, Ausgleich *m* [*Wechselseitige Aufhebung entgegengesetzter Kräfte*]

compensation → correction

compensation circuit Kompensationsschaltung, Ausgleichsschaltung *f*

compensation coil Kompensationsspule, Ausgleichsspule *f*

compensator Ausgleicher, (Aus)Dehn(ungs)ausgleicher, (Rohr)Kompensator *m* [*In langen Rohrleitungen zwischen Festpunkten eingebaut, um die Rohr-(aus)dehnungen bei Temperaturänderungen aufzunehmen*]

compensator → filter

compensator [*It is the most popular form of control for commercial buildings which are heated by radiators or similar wet systems. It varies the flow temperature as the weather changes*] Kompensator, Ausgleicher *m*

compensator Entzerrer *m*

compensatory AND kompensatorische UND(-Operator) *n, (m)* [*unscharfe Logik*]

compensatory operator kompensatorische Operator *m* [*unscharfe Logik*]

compensatory OR kompensatorische ODER(-Operator) *n, (m)* [*unscharfe Logik*]

competent fachmännisch

competitive edge, competitive advantage Wettbewerbsvorteil *m*

to compile kompilieren, übersetzen, zusammensetzen [*Programm*]

compiler, compiling routine Kompil(ier)er, Übersetzungsprogramm, kompilierende Programm *n* [*Ein Übersetzer, der Quellanweisungen, die in einer nicht-maschinenorientierten Programmiersprache abgefasst sind, in Zielanweisungen einer maschinenorientierten Programmsprache umwandelt (= kompiliert). E DIN 44 300, Teil 4*]

compiler-compiler Übersetzergenerator *m*

compiling Übersetzen, Wandeln, Programmübersetzen, Programmwandeln, Kompilieren *n*

complementary binary code, binary complement-ary code komplementäre Binärkode, Ergänzungsbinärkode, Binärergänzungskode *m*

complementary gate Komplementärgatter, Ergänzungsgatter *n*

complementary metal-oxid(e) semiconductor circuit, CMOS circuit komplementäre Metalloxid-Halbleiter-Schaltung *f* [*Schaltungstechnik mit hoher Packungsdichte. Versorgungsspannung in weitem Be-reich möglich (3-15 V). Großer Störabstand, geringe Leistungsaufnahme*]

complementary rectifier Ergänzungsgleichrichter *m*

complementary representation Ergänzungsdarstellung *f*

to complete durchschalten, schließen [*Stromkreis*]

complete circuit, full circuit Hin- und Rückleitung *f*, geschlossene Stromkreis, durchgeschaltete Stromkreis *m*

complete demineralization, full demineralization Vollentsalzung, Ganzentsalzung *f*

complete equipment Komplettgerät(e) *n, (f)*

completely encapsulated electronics, fully encapsulated electronics vergossene Elektronik, vollgekapselte Elektronik *f*

completely heated → fully heated

complete monitoring → total monitoring

complete operation, allround operation [*An operation which includes obtaining all operands from stor-age, performing the operation, returning resulting operands to storage, and obtaining the next instruction*] vollständige Rechenoperation *f*

complete outfit Vollausstattung, Komplettausstattung, Ganzausstattung *f*

complete protection, allround protection Vollschutz, Ganzschutz, Komplettschutz *m*

complete radiator → black radiator

complete routine, allround routine [*A routine, gen-erally supplied by a computer manufacturer, which does not have to be modified by the user before being applied*] vollständige Teilprogramm *n*

complete space heating, full space heating Vollraum(be)heizung *f*

complete treatment, full treatment Vollbehandlung *f*

completion date Fertigstellungstermin *m*

completion installation Fertigeinbau *m*

completion of calls for busy subscriber, CCB automatische Rückruf *m* wenn besetzt

complex instruction set computer, CISC computer Rechner *m* mit komplexem Befehlssatz

complex sound Tongemisch *n*, Tonmischung *f*, zusammengesetzte Ton *m*, gemischte Ton

compliance Übereinstimmung *f* [*z. B. eines Messgerätes mit den Anforderungen*]

compliance Federung *f*

compliance-controlled federgehemmt, federsteif [*Elektroakustik*]

compole → commutating pole

compole field → commutating (pole) field

component Komponente *f*, Bestandteil *m*

component, building component Bauteil *m, n*

component to be measured Messling *m*

compose-edit processor Textaufbereitungsprozessor *m*

composite insulation, compound insulation Verbunddämmung *f*

composite (material), compound material Verbund(werk)stoff *m*, Verbundmaterial *n*

composite picture signal, composite video signal, compound picture signal, compound video signal [*Television signal produced by combining a blank picture signal with the sync signal*] Bildaustastsynchronsignal, BAS-Signal, zusammengesetzte Videosignal *n*

composite plant, composite installation, compound plant, compound installation Verbundanlage, Verbundinstallation *f*

Composite System FRAS-System *n* [*elektronische Bildübertragung*]

composite system, compound system Verbundsystem *n*

composition, buildup Aufbau *m*, Zusammensetzung *f*

composition Komposition, Verkettung *f* [*von unscharfen Relationen*]

compositional inference rule Inferenzkompositionsregel, Kompositionsregel der Inferenz, Verkettungsregel der Inferenz *f* [*unscharfe Logik*]

compositional rule Kompositionsregel, Verkettungsregel *f*

compound compressor Verbundverdichter, Verbundkompressor *m*

compound material → composite (material)

compound palm leaf → palm tree frond

compound-wound motor [*Direct-current motor with relatively constant speed characteristics*] Kompoundmotor, Verbundmotor, Doppelschlusswicklung(s)motor *m*

to compress verdichten, komprimieren, zusammendrücken [*z. B. gasförmiges Kältemittel*]

(compressed-)air act(uat)or, pneumatic act(uat)or Druckluftstellglied, Pressluftstellglied, pneumatische Stellglied *n* [*Siehe Anmerkung unter „Stellglied"*]

(compressed-)air atomizer, pneumatic atomizer Druckluftzerstäuber, Pressluftzerstäuber, pneumatische Zerstäuber *m*

(compressed-)air burner, pneumatic burner Druckluftbrenner, Pressluftbrenner, pneumatische Brenner *m*

(compressed-)air ((closed) loop) controller, pneumatic ((closed) loop) controller Druckluftregler, Pressluftregler, pneumatische Regler *m*

(compressed-)air electric control (device), pneumatic electric control (device) [*1. Control (device) which interfaces pneumatic to electrical systems.*
2. Control (device) with a pneumatic input and electric output] pneumatisch-elektrische Steuerung(sgerät) *f, (n)*

(compressed-)air electric relay, pneumatic electric relay pneumatisch-elektrische Relais *n* [*Es öffnet oder schließt durch Druckluft einen elektrischen Kontakt wodurch elektrische Geräte, z. B. ein Ventilatormotor mit einem Druckluftregelsystem verbunden sind*]

(compressed-)air gun → spray gun

(compressed-)air lift, pneumatic lift [*A device of raising water from a well by the injection of compressed air*] Drucklufheber, Pressluftheber, pneumatische Heber *m*

(compressed-)air ((open) loop) controller, pneumatic ((open) loop) controller Steuer(ungs)-regler *m*

C

(compressed-)air pump, pneumatic pump Druckluftpumpe, Pressluftpumpe, pneumatische Pumpe *f*

(compressed-)air room thermostat, pneumatic room thermostat Druckluftraumthermostat, Pressluftraumthermostat, pneumatische Raumthermostat *f*

(compressed-)air secondary water supply, pneumatic secondary water supply [*It may or may not be automatic*] Druckluftwassernachspeisung, Pressluftwassernachspeisung, pneumatische Wassernachspeisung *f*

(compressed-)air shutt(ing)-off valve, pneumatic shutt(ing)-off valve Druckluft(ab)sperrventil, Pressluft(ab)sperrventil *n*

(compressed-)air sliding door drive, pneumatic sliding door drive Druckluft-Schiebetürantrieb, Pressluft-Schiebetürantrieb, pneumatische Schiebetürantrieb *m*

(compressed-)airtube conveyor, (compressed-) airtube system Druckluftrohrpost, Pressluftrohrpost, pneumatische Rohrpost *f* [*DIN 6651*]

(compressed-)air water atomizer, pneumatic water atomizer Druckluftwasserzerstäuber, Pressluftwasserzerstäuber, pneumatische Wasserzerstäuber *m*

(compressed-)air water atomizing nozzle, pneumatic water atomizing nozzle Druckluftwasserzerstäubungsdüse, Pressluftwasserzerstäubungsdüse, pneumatische Wasserzerstäubungsdüse *f*

(compressed-)air water tank, pneumatic water tank Druckluftwasserbehälter, Pressluftwasserbehälter, Druckluftwasserkessel, Pressluftwasserkessel, pneumatische Wasserkessel *m* [*Sprinkleranlage*]

compressed gas Druckgas, Pressgas *n*

compressed gas (fire) extinguisher → water (fire) extinguisher

compressibility Zusammendrückbarkeit, Verdichtbarkeit, Kompressibilität *f*

compressing → time compressing

compression Verdichtung, Kompression, Zusammendrückung *f*

compression cable lug Presskabelschuh *m*

compression cock, compression faucet [*It is equipped with a handle to screw a compressible disk down over the opening*] Verdichtungshahn, Kompressionshahn *m*

compression concrete cover(ing) → concrete compression cover(ing)

compression final temperature, final compression temperature Verdichtungsendtemperatur, Temperatur am Verdichterausgang, Kompressionsendtemperatur *f*

compression gland Stopfbuchse *f*

compression heat → heat of compression

compression ignition Verdichtungszündung *f*, Kompressionszündung

compression modulus → modulus of volume elasticity

compression ratio → pressure ratio

compression refrigerator [*A refrigerator, such as the common household type, that uses a compressor to change the refrigerant from vapo(u)r to liquid*] Kompressionskühlschrank *m*, Verdichtungskühlschrank

compression stroke Verdichtungshub, Kompressionshub *m*

compression work, compressor work Verdichtungsarbeit, Kompressionsarbeit *f* [*Energieaufwand innerhalb einer Zeitspanne zur Verdichtung der zu fördernden Gase*]

compressive load, pressure load Drucklast *f*

compressor [*A machine which draws in gas or air and delivers it at a higher pressure*] Verdichter, Kompressor *m*, Kompressionsmaschine, Verdichtungsmaschine *f*

compressor [*Transducer that to a given input amplitude range produces a smaller output range*] Umset-zerverdichter, Wandlerverdichter, Umsetzerkompressor, Wandlerkompressor *m*

compressor → (refigerant) compressor

compressor capacity modulation system Verdichterleistungsmodulationssystem, Kompressorleistungsmodulationssystem *n*

compressor chamber Verdichterkammer, Kompressorkammer *f*

compressor/condenser unit Kompressor-Kondensator-Einheit, Verdichter-Kondensator-Einheit, Verdichter-Verflüssiger-Einheit, *f*

compressor-controlled [*close(d) loop*] kompressorgeregelt

compressor-controlled [*open loop*] kompressorgesteuert

compressor cryogenic plant, compressor cryogenic installation Kompressorkälteanlage, Verdichterkälteanlage *f*

compressor cycle Kompressorkreislauf, Verdichterkreislauf *m*

compressor displacement [*Volume swept by the pistons or impellers of a reciprocating or rotary machine, usually stated in cubic feet per shaft revolution or in cubic feet per minute of some definite rpm*] Kompressorverdrängung, Verdichterverdrängung *f*

compressor efficiency Verdichterwirkungsgrad, Kompressorwirkungsgrad *m*

C

compressor expander → compander

compressor flywheel Kompressorschwungrad, Verdichterschwungrad *n*

compressor lubricating oil Kompressorschmieröl, Verdichterschmieröl *n*

compressor method Kompressorverfahren, Verdichterverfahren *n*, Kompressormethode, Verdichtermethode *f* [*Ein motorgetriebener Kompressor verdichtet das Kühlmittel*]

(compressor) piston displacement [*Volume swept by a compressor piston during one stroke, or one revolution of the crankshaft*] Kompressorkolbenverdrängung, Verdichterkolbenverdrängung *f*

compressor refrigerating effect [*Rate of heat removal by the refrigerant assigned to the compressor in a refrigerating system. This is equal to the product of the mass rate of refrigerant flow produced by the compressor and the difference in specific enthalpies of the refrigerant vapo(u)r at its thermodynamic state entering the compressor and refrigerant liquid at saturation temperature corresponding to the pressure of vapo(u)r leaving the compressor*] Kompressionskälterate, Verdichterkälterate *f*

compressor refrigerating plant, compressor refrigerating installation, compressor refrigeration plant, compressor refrigeration installation Kompressorkälteanlage, Verdichterkälteanlage *f*

compressor set Verdichtersatz, Kompressorsatz *m*

compressor simulation Verdichtersimulation, Kompressorsimulation *f*

compressor start(ing) Kompressorstart(en), Verdichterstart(en) *m, (n)*

compressor thermal protection Kompressorwärmeschutz, Verdichterwärmeschutz *m*

compressor unit [*A refrigerating component designed to compress a specific refrigerant vapo(u)r, consisting of compressor, prime mover, and regularly furnished accessories*] Verdichtereinheit, Kompressoreinheit *f*, Verdichtersatz, Kompressorsatz *m*

compressor valve Verdichterventil, Kompressorventil *n*

compressor work → compression work

computational linguistics, computer linguistics linguistische Datenverarbeitung, Rechnerlinguistik *f*

computer, (data) process(ing) machine [*A device that receives, processes, and presents data; the two types are analog and digital; Sometimes loosely called "device"*] Rechner *m*, Rechenmaschine, Rechenanlage, Datenverarbeitungsanlage *f*

computer access, CA Rechnerzugang, Rechnerzutritt *m*

computer-aided and integrated facility management, CAIFM Kombination *f* von Gebäudebewirtschaftungssystemen und Elementen aus Standard- oder Planungssoftware [*Das Merkmal ist die zusätzliche Möglichkeit zur Integration sämtlicher relevanter Daten, die im Zuge der Planung und Ausführung eines Bauvorhabens entstehen. Zu diesem Zweck werden Systeme mit Elementen aus Standard- oder Planungssoftware kombiniert. Für die Bewirtschaftung steht somit eine maximale Informationsgrundlage zur Verfügung.*]

computer-aided ((closed) loop) control, computer-aided feedback control, computer-aided closed control, computer-aided regulation, CAR rechnergestützte Reg(e)lung, rechnerunterstützte Reg(e)-lung, rechnergestützte Regulierung, rechnerunterstützte Regulierung, rechnergestützte geschlossene Wirkungsarbeit, rechnerunterstützte geschlossene Wirkungsarbeit *f*

computer-aided engineering, CAE [*It covers CAD and CAP*] rechnergestützte Ingenieurbüroarbeiten *f* oder Arbeiten im technischen Unternehmensbereich [*Die gesamte Tätigkeit des Ingenieurs wie sie in der HOAI definiert ist, von der ersten Abschätzung der Leistung einer technischen Anlage bis zum Abzeichnen der Stückliste. Rechnergestützte Maschineneinsatzplanung in der Fertigung und die Zeitvorgaben für die Montage gehören dazu*]

computer-aided facility management, CAFM, computer-aided building management rechnergestützte Gebäudebewirtschaftung, rechnerunterstützte Gebäudebewirtschaftung *f*

computer-aided industry → CAI

computer-aided learning rechnergestützte Lernen, rechnerunterstützte Lernen *f*

computer-aided manufacture → CAM

computer-aided monitoring rechnergestützte Überwachung, rechnerunterstützte Überwachung *f*

computer-aided planning → CAP

computer-aided production scheduling and control → CAPSC

computer-aided quality assurance, CAQ(A) rechnergestützte Qualitätssicherung, rechnergestützte Gütesicherung, rechnerunterstützte Qualitätssicherung, rechnerunterstützte Gütesicherung *f*

computer-aided repair, CAR rechnergestützte Reparatur, rechnerunterstützte Reparatur *f*

computer-aided robotics → CAR

computer-aided software engineering rechnerunterstützte Softwaretechnik, rechnergestützte Softwaretechnik *f*

computer-aided testing → CAT

C

computer-aided training, computer-based training rechnergestützte Ausbildung, rechnerunterstützte Ausbildung *f*

computer analyst, computer expert [*A person who defines a problem, determines exactly what is required in the solution, and defines the outlines of the machine solution; generally an expert in automatic data processing*] Programmierer *m*

computer application, computer use Rechnereinsatz, Rechnergebrauch *m*, Rechnerverwendung, Rechneranwendung, Rechnernutzung *f*

computer architecture Rechnerarchitektur *f*

computer-assisted calling and answering unit → automatic calling and answering unit

computer-assisted learning, CAL rechnergestützte Lernen, rechnerunterstützte Lernen *n*

computer-assisted program (US); computer-assisted programme (Brit.) rechnergestützte Programm, rechnerunterstützte Programm *n* [*Es dient zur Lösung bestimmter manuell aufwendiger Aufgaben. Dabei sind die zur Definition der Aufgabe erforderlichen Daten im allgemeinen sehr kurz und einfach. Alle weiteren Schritte bis zur Lösung führt dann dieses Programm mit Rechnerunterstützung aus*]

computer based message system rechnergestützte elektronische Mitteilungssystem, rechnerunterstützte elektronische Mitteilungssystem *n*

computer cabinet, computer cubicle Rechnerschrank *m*

computer card Rechnerkarte *f*

computer center (US) → electronic data-process-ing center

computer channel Rechnerkanal *m*

computer check(ing) routine Rechnerkontrollprogramm *n*

computer ((closed) loop) control Rechnerreglung, Rechnerregelung *f*

computer code → machine code

computer component [*Sometimes loosely called "device"*] Rechnerteil *m, n*, Rechnerkomponente *f*

computer console [*A part of a computer used for communication between a human operator and the computer*] Rechnerpult *n*, Rechnerkonsole *f*

computer control → computer ((closed) loop) control

computer control → computer ((open) loop) control

computer-controlled, process-controlled [*closed loop control*] rechnergeregelt, prozessgeregelt

computer-controlled, process-controlled [*open loop control*] rechnergesteuert, prozessgesteuert

computer design, computer type, computer version Rechneraufbau *m*, Rechnerkonstruktion, Rechnerausführung, Rechnerbauweise, Rechnerbauart *f* [*Nicht verwechseln mit „Rechnerstruktur"*]

computer display Rechnerbildschirm *m*

computer element Rechnerelement *n*

(computer) freak (Rechner)Begeisterte *m, f*

computer graphics [*The use of computers to generate graphics*] graphische Datenverarbeitung, Rechnergraphik *f*

computer-human interface, CHI Mensch-Maschine-Schnittstelle, Maschine-Mensch-Schnittstelle, Benutzerschnittstelle, Benutzungsschnittstelle *f*

computer input, data to be processed Eingabedaten, Rechnereingabedaten, Verarbeitungsdaten *f*

computer input microfilm, CIM Informationseingabemikrofilm *m*

computer installation Rechneranlage, Rechenanlage *f*

computer instruction Rechnerbefehl, Maschinenbefehl *m*

computer-integrated building, CIB rechnerintegrierte Gebäude *n*

computer-integrated manufacturing → CIM

computer integrated manufacturing for facility management, CIMFFM einheitliche digitale Schlüsselsystem, digitale einheitliche Schlüsselsystem *n* [*Für alle bei einer Anlage beteiligten Planer und Anlagenbauer. Das System eignet sich auch für Dienste und Instandhaltung, die dadurch auf der gleichen Datenbasis arbeiten*]

computer-integrated telephony rechnerintegrierte Telephonie, rechnerintegrierte Telefonie *f*

computer interface Rechnerschnittstelle *f*

computerized branch exchange → CBX

computerized branch exchange system, CBX system rechnergestützte Nebenstellenanlage *f* [*Kommunikationsanlage, die in ihrem Aufbau dem vom Telefon her bekannten Nebenstellenprinzip entspricht und die schon vorhandenen sternförmigen Leitungsnetze zur Sprach- und Datenübermittlung benutzt*]

computerized numerical ((open) loop) control, CNC rechnergestützte numerische Steuerung, numerische rechnergestützte Steuerung *f*

computerized private branch exchange → CPBX

computer language → machine code

computer linguistics → computational linguistics

computer literary Alphabetismus *m*

(computer) mapped input/output, (computer) mapped I/O, storage mapped input/output, storage mapped I/O (rechner)speicherorientierte Ein/Ausgabe, (rechner)speicherorientierte E/A *f*

computer master programme (Brit.); computer master program (US) Rechnerleitprogramm, Leitrechnerprogramm *n*

(computer) memory, (computer) storage (device) [*General terms for the computer equipment that holds information in any usually binary language in electrical or magnetic form. This equipment also receives information for storage and gives out the stored information for storage and later use. The word "memory" usually means storage inside the computer, while "storage" refers to magnetic drums, discs, cores, tapes, punched cards, or solid-state devices outside of the computer.*] (Rechner)-Speicher, rechnerische Speicher, Hauptspeicher, Informationsspeicher, Rechenspeicher *f*

(computer) memory access (Rechner)Speicherzugriff *m*

(computer) memory allocation, (computer) memory allotment, (computer) memory assignment (Rechner)Speicherzuordnung, (Rechner-) Speicherzuweisung, (Rechner)Speicherzuteilung *f*

(computer) memory array (Rechner)Speicherblock *m*

(computer) memory card (Rechner)Speicherkarte *f*

(computer) memory compaction (Rechner)Speicherkomprimierung *f*

(computer) memory construction (Rechner)Speicherbau *m*

(computer) memory cycle (Rechner)Speicherzyklus *m*

(computer) memory data register, MDR (Rechner-)Speicherdatenregister *n*

(computer) memory dump (Rechner)Speicherauszug *m*

(computer) memory element, (computer) memory cell (Rechner)Speicherelement *n*, (Rechner)Speicherzelle, kleinste (Rechner)Speichereinheit *f* [*E DIN 19226, Teil 5*]

(computer) memory extension, (computer) memory expansion (Rechner)Speichererweiterung *f*

(computer) memory filling (Rechner)Speicherfüllung *f*

(computer) memory guard (Rechner)Speicherverrieg(e)lung *f*

(computer) memory handling (Rechner)Speicherumgang *m*

(computer) memory hierarchy (Rechner)Speicherhierarchie *f*

(computer) memory interleave, (computer) memory interleaving (Rechner)Speicherverschränkung *f*

(computer) memory library (Rechner)Speicherbibliothek, (Rechner)Speicherbücherei *f*

(computer) memory location (Rechner)Speicherstelle *f*

(computer) memory map Rechnerhauptspeicherabbild, Haupt(rechner)speicherabbild *n*

(computer) memory module (Rechner)Speicherbaustein *m*, (Rechner)Speichermodul *n*

(computer) memory monitoring (Rechner)Speicherüberwachung *f*

(computer) memory pointer (Rechner)Basis(adress)register *n*

(computer) memory-programmable (rechner)-speicherprogrammierbar

(computer) memory protection (Rechner)Speicherschutz *m*

(computer) memory read signal, read signal (Rechner) Speicherlesesignal *n*

(computer) memory reference instruction (Rechner)Speicherreferenzbefehl *m*

(computer) memory settlement (Rechner)Speicherberechnung *f*

(computer) memory stack (Rechner)Speicherstapel *m*

(computer) memory-to(-computer)-memory instruction (Rechner)Speicher-(Rechner)Speicher-Befehl *m*

(computer) memory unit, MU (Rechner)Speichereinheit *f* [*Baueinheit von Digitalrechnern bestehend aus einer bestimmten Anzahl von Speicherzellen; mit einer Speichereinheit wird die Speicherkapazität dem Einsatzgebiet des Rechners angepasst*]

computer misuse Rechnermissbrauch *m*

computer mode Rechnerbetriebsart *f*, Rechnerbetriebsmodus *m*

computer module Rechnermodul *n*, Rechnerbaustein *m*

(computer) net(work) [*A collection of interconnected individual computer systems*] (Rechner)Netz(werk) *n*

computer ((open) loop) control Rechnersteuerung *f*

computer PABX interface, CPI Schnittstelle *f* zwischen privater Nebenstellenanlage und Rechner

computer output on microfilm, COM Drucken *n* auf Mikrofilm, Zeichnen auf Mikrofilm

C

C

computer power pack Rechnernetzgerät, Netzrechnergerät *n*

computer program (US); computer programme (Brit.) [*A schedule or plan that specifies actions which may or may not be taken, expressed in the form of a set of instructions suitable for execution by a computer*] Rechnerprogramm *n*

computer programming → programming

computer ratio → operating ratio

computer-readable rechnerlesbar, rechnerleserlich

computer room equipment Rechnerraumgerät(e) *n, (f)*

(computer) routine [*Computer programming. A sequence of instructions directing a computer to carry out a set of operations that form part of a program*] (Rechner)Prozedur, Routine *f*, Unterprogramm *n*

computer run Rechner(durch)lauf *m*

computer safety Rechnersicherheit *f*

computer science Informatik *f* [*Lehre vom Aufbau und Programm logischer Rechneranlagen*]

computer security Rechnersicherung *f*

computer security system Rechnersicherungssystem *n*

computer station [*Primary "point" of access to a B(E)MS. The computer system from which all operations are conducted*] Rechnerzentrale, Rechnerstation *f*

(computer) storage (device) → (computer) memory

computer structure Rechnerstruktur *f* [*Nicht verwechseln mit „Rechneraufbau"*]

computer supported telecommunications application, CSTA Standard *m* für Rechner-Telefon-Integration

computer (supported) telephony integration, computer (supported) telephoning, CTI rechnerintegrierte Telefon *n*, Integration *f* von Telefon und Rechner, Rechner-Telefon-Integration *f*

computer system [*A set of related but unconnected components (hardware) of a computer or data-processing system*] Rechnersystem *n*

computer telephony integration, CTI PC-gestützte Telefonie *f*

computer throughput Rechnerdurchsatz *m*

computer type → computer design

computer unit Rechnereinheit *f*

computer use → computer application

computer version → computer design

computer virus Rechnervirus *n* [*nicht fachsprachlich: m*] [*Ein Programm, das in der Lage ist, andere Programme zu infizieren und sich über einen Reproduktionsteil auf Speichermedien auszubreiten*]

computer word Rechnerwort, Maschinenwort *n*

(computer) words block, block of computer words [*See remark under "block"*] (Rechner)Wörterblock, Maschinenwörterblock *m*

computing Rechnen *n*

computing power Rechenleistung *f*

concealed, hidden verdeckt, verborgen

concealed, hidden unter Putz, putzverdeckt, putzverborgen, Unterputz...

concealed ... → hidden ...

concentrated heat, point heat Punktwärme, konzentrierte Wärme *f*

concentrating (solar) collector konzentrierte Sonnen(energie)kollektor, konzentrierte Solar(energie)kollektor *m*

concentration switch [*telephony*] Sammelschalter *m*

concentric (transmission) cable → coaxial (transmission) cable

conclusion DANN-Teil *m, n*, Konklusion *f*, Aussage *f*, Schlussfolgerung, Folgerung *f* [*Fuzzy-Logik*]

concrete konkret

concrete Beton *m*

concrete (air) duct Beton(luft)kanal, Betonlüftungskanal *m*

concrete benching Betonsohlen(ge)rinne, Sohlen(ge)rinne aus Beton *f, (n)*

concrete ceiling Beton(unter)decke, Unterbetondecke *f*

concrete compression cover(ing), compression concrete cover(ing) Betondruckschicht, Druckbetonschicht, Betondrucklage *f*, Druckbetonlage

concrete floor Beton(fuß)boden *m*

concrete floor heating → (electric) concrete floor heating

concrete floor tubular heating → tubular concrete floor heating

(concrete) foundation ground (connection) (US); (concrete) foundation earth (connection) (Brit.) (Beton)Fundamenterde(r) *f, (m)*, (Beton)-Fundamenterdung *f* [*In die Sohle eines Betonfundamentes eingebettete Leiter. VED 0100, Teil 200*]

concrete heating floor, heating concrete floor Betonheiz(ungs)decke *f*

concrete heating slab, heating concrete slab Betonheiz(ungs)platte *f*

concrete (in)fill(ing) Beton(aus)füllung *f*

concrete pipe [*B.S. 556*] Betonrohr, Zementrohr *n* [*DIN 4032*]

concrete repair Betonreparatur *f*

concrete sanitary building block module Beton-Sanitäinstallationsblock, Beton-Sanitär(raum)block *m*, Beton-Sanitärinstallationszelle, Beton-Sanitär(raum)zelle *f*

concrete slab with heating tubes → heating tubes concrete slab

concrete-surrounded pipes ceiling heating, concrete-surrounded tubes ceiling heating Beton-Rohrdeckenheizung, Rohrdeckenheizung mit Heizrohren im Beton *f* [*Nahtlose Rohre von 3/8", 1/2" oder 3/4" werden im Deckenbeton verlegt*]

concrete syntax [*Those aspects of the rules used in the formal specification of data which embody a specific representation of that data (ISO 7498)*] darstellungsabhängige Syntax, konkrete Syntax *f*

concrete under severe condition(s), highly stress-ed concrete hochbeanspruchte Beton *m*

concurrent, simultaneous zeitgleich, gleichzeitig

concurrent access, simultaneous access zeitgleiche Zugriff, gleichzeitige Zugriff *m* [*Zugriff zur gleichen Zeit von mindestens zwei Prozessen auf einen Datenpunkt mit gegensätzlichem Ziel*]

condensable vapour (Brit.); condensable vapor (US) [*A vapo(u)r that becomes liquid when temperature and/or pressure are changed appropriately*] verflüssigbare Wasserdampf, kondensierbare Wasserdampf *m*

condensate, condensation water, condens(at)-ing water, water of condensation [*Water supplied to cool a condenser or condensing unit*] Kondenswasser *n*

condensate collection vessel, condensate (collecting) vessel, condensate collector, condensation water collection vessel, condensation water (collecting) vessel, condensation water collector Kondensat(sammel)gefäß, Kondenswasser(sammel)gefäß *n*, Kondensatsammler, Kondenswassersammler, Kondensat(sammel)behälter, Kondenswassersammelbehälter *m* [*Zur Aufnahme des aus der Dampfheiz(ungs)anlage zurückfließenden Kondensats vor Rückführung in den oder die Kessel*]

condensate discharge Kondensatablauf, Kondenswasserablauf *m*

condensate discharging device, condensation water discharging device Kondensatableiter, Kondenswasserableiter *m*

condensate drain, condensation water drain Kondensatdrän, Kondenswasserdrän *m*

condensate drainage, condensation water drainage Kondensatentwässerung, Kondenswasserentwässerung *f*

condensate (drain) pan, condensation water (drain) pan Kondensatschale, Kondenswasserschale *f*

condensate dryer, condensate drier, condensation water drier, condensation water dryer Kondensat-trockner, Kondenswassertrockner *m*

condensate elevated tank → elevated condensate tank

condensate immersion pump, condensate submersible pump, condensation water immersion pump, condensation water submersible pump Kondensattauchpumpe, Kondenswassertauchpumpe *f*

condensate loop, condensation water loop Kondensatschleife, Kondenswasserschleife *f* [*Zwischen Dampfleitung und trockener Kondensatleitung darf keine direkte Verbindung bestehen, denn der Dampf würde nahezu ungehindert über die Be- und Entlüftung austreten. Die Dampfleitungen werden deshalb über Kondensatschleifen entwässert.*]

condensate meter, condensation water meter Kondensat(mengen)messer, Kondensat(mengen)-zähler, Kondenswasser(mengen)messer, Kondenswasser(mengen)zähler *m*

condensate pan, condensation water pan Kondensatschale, Kondensatwanne, Kondenswasserwanne, Kondenswasserschale *f*

condensate (pipe)line, condensate piping, condensation water (pipe)line, condensation water piping Kondensat(rohr)leitung, Kondenswasser-(rohr)leitung *f*

condensate pump, condensation water pump Kondensatpumpe, Kondenswasserpumpe *f*

condensate return feed, condensation water return feed Kondensatrückspeisung, Kondenswasserrückspeisung *f*

condensate return feeder, condensation water return feeder Kondensatrückspeiser, Kondenswasserrückspeiser *m*

condensate return feeder station, condensation water return feeder station Kondensatrückspeisestation, Kondenswasserrückspeisestation *f*

condensate return (flow), condensation water return (flow) Kondensatrücklauf, Kondenswasserrücklauf *m*, Kondensatrückführung, Kondenswasserrückführung *f*

C

condensate retrun pipe, condensation water return pipe Kondensatrücklaufrohr, Kondenswasserrücklaufrohr *n*

condensate return feeder, condensation water return feeder Kondensatrückspeiser, Kondenswasserrückspeiser *m*, Kondensatrückspeiseanlage, Kondenswasserrückspeiseanlage *f*

condensate return heating system, condensation water return heating system [*Steam heating system in which a return pipe carries condensate. The pipe usually is located below the level of the water line in the boiler, although not necessarily so*] Dampfheizungssystem *n* mit Kondensatrücklaufleitung

condensate return manifold, condensation water return manifold Kondensatrücklaufverteiler(rohr)-Kondenswasserrücklaufverteiler(rohr) *m, (n),*

condensate return (pipe)line, condensation water return (pipe)line, condensate return piping, condensation water return piping Kondensatrücklaufleitung, Kondenswasserrücklaufleitung *f*

condensate return pump, condensation water return pump Kondensatrücklaufpumpe, Kondenswasserrücklaufpumpe *f*

condensate separator, trap, condensation water separator (Kondens)Wasserabscheider *m*, Kondensattopf, Entwässerungstopf, Kondensatabscheider, (Kondens)Wassertopf *m* [*DIN 3680*]

condensate strainer, condensation water strainer [*In a boiler system, strainer that removes solid particles from the steam condensate before it is pumped back to the boiler*] Kondensatseiher, Kondenswasserseiher *m*

condensate trap, condensation water trap Kondensatfänger, Kondenswasserfänger *m*

condensate vessel, condensation water vessel Kondensatgefäß, Kondenswassergefäß *n*

condens(at)ing kondensierend, verflüssigend [*Kälteprozess*]

condensation [*Change of state of a vapo(u)r into a liquid by extracting heat from the liquid*] Kondensation, Verflüssigung, Kondensierung *f*, Verflüssigen, Kondensieren *n*

condensation → (surface) condensation

condensation boiler, condens(at)ing boiler Brennwertkessel, Kondensationskessel *m*

condensation heat → heat of condensation

condensation latent heat → latent heat of condensation

condensation output, condens(at)ing output, condensation performance, condens(at)ing performance Kondensationsleistung *f* [*Die Kondensationsleistung Q setzt sich aus der Kälteleistung* Q_o *und der Kompressionsleistung P zusammen:* $Q = Q_o \times P$ in kW. *Bei Nenntemperaturen gilt etwa:* $Q = 1.23 * Q_o$]

condensation power plant, condensation power station, condens(at)ing power plant, condens(at)ing power station Kondensationskraftwerk *n*

condensation pressure, condensating pressure Kondensationsdruck, Verflüssigungsdruck *m* [*Sättigungsdampfdruck im Verflüssiger*]

condensation set, condens(at)ing set Kondensatorsatz, Verflüssigersatz, Verflüssigungssatz *m* [*Kälteprozess. Kompressor mit Ventilator*]

condensation temperature, condens(at)ing temperature Verflüssigungstemperatur, Kondensationstemperatur *f* [*Temperatur, bei der das gasförmige Kältemittel bei Druck im Verflüssiger kondensiert*]

condensation waste heat, condens(at)ing waste heat Kondensationsabwärme, Verflüssigungsabwärme *f* [*Kälteprozess*]

condensed-mercury temperature kondensierte Quecksilbertemperatur *f*

condenser [*Optics. A device in which a lens or series of lenses collect light from one object and projects it evenly into another object, as in a microscope or a slide projector*] Kondensor, Lichtsammler, Lichtverstärker *m*

condenser [*Heat exchanger in which vapo(u)r is liquefied by the rejection of heat to a heat sink*] Verflüssiger, Kondensator *m*

condenser, capacitor, electrical condenser [*A device which consists essentially of two conductors (such as parallel metal plates) insulated from each other by a dielectric and which introduces capacitance into a circuit, stores electrical energy, blocks the flow of direct current, and permits the flow of alternating current to a degree dependent on the capacitor's capacitance and the current frequency. Symbolized C*] Kondensator *m* [*Symbol C*]

condenser air Kondensatorluft *f*

condenser blower Kondensatorgebläse *n*

condenser blow-off side, condenser blow-out side Verflüssigerausblas(e)seite, Kondensatorausblas(e)seite *f*

condenser brine [*deprecated*] **→ condenser second-ary coolant**

condenser calorimeter [*Apparatus for determining refrigerant flow rate, and subsequently the capacity of a condensing unit by measuring the heat input requir-ed to balance the refrigerating effect produced in the evaporator by the condensing unit*] Kondensatorkalorimeter *n*

condenser coil [*It is not enclosed in a pressure vessel*] Kondensatorschlange *f*

condenser cooling air Verflüssigerkühl(ungs)luft, Kondensatorkühl(ungs)luft *f*

condenser cooling water Kondensatorkühl(ungs)-wasser *n*

condenser fan Kondensatorventilator *m*

condenser heat Kondensatorwärme, Kondensationswärme *f*

condenser heat rejection effect, condenser heat removal effect, condenser duty [*The amount of heat energy rejected to its surrounding by the condenser of a system*] Kondensatorwärmeentzugswirkung, Kondensatorwärmeabführungswirkung *f*

condenser high-pressure cutout, high-pressure condenser cutout Kondensatorhochdruckausschaltung *f*

condenser oil Kondensatoröl *n*

condenser pipe Verflüssigerrohr, Kondensatorrohr *n*

condenser pressure Kondensatordruck *m*

condenser pressure ((closed) loop) control Kondensatordruckreg(e)lung *f*

condenser pressure ((closed) loop) controller Kondensatordruckregler *m*

condenser refrigerant Kondensatorkältemittel *n*

condenser secondary coolant; condenser brine [*deprecated*] [*See remark under "secondary coolant"*] Kondensatorsole *f*

condenser suction standpipe, standpipe for condenser suction Kondensatorsaugstutzen *m*

condenser tube Kondensatorröhre *f*

condenser water Kondensatorwasser *n*

condenser with open compressur offene Kondensator *m*

condensing, liquefying; liquifying [*obsolete*] [*high pressure refrigerant*] Kondensieren, Verflüssigen *n*

condensing furnace [*One that circulates the products of combustion and extracts available heat to a point that causes condensation to occur. Some of this latent heat of condensation is recovered as usable energy, resulting in higher operating efficiency*] Kondensationsfeuerung, Kondensationsfeuerstätte *f*

condensing heat rejection, condensing heat removal [*That portion of the total refrigerant heat-rejecting effect of a condenser which is used for condensing the entering refrigerant vapo(u)r to a saturated liquid at the entering refrigerant pressure*] Kondensationswärmeentzugsanteil, Kondensationswärmeabführungsanteil *m*

condensing mirror Kollimatorspiegel *m*

condensing nominal capacity → **nominal condensing capacity**

condensing steam kondensierende Dampf, Kondensationsdampf *m*

condensing unit [*A machine specially designed to condense refrigerant vapo(u)r to a liquid by compressing the vapo(u)r in a positive-displacement compressor and rejecting heat to a cooling medium. The unit consists usually of one or more positive-displacement compressors, motors, condensers, liquid receivers (when required), and necessary accessories, mounted an a common base*] Kondensatorsatz *m*

condensing unit [*split air-conditioning system*] Außeneinheit *f*

condensing unit calorimeter [*Apparatus for determining refrigerant flow rate, and subsequently the capacity of a condensing unit by measuring the heat input required to balance the refrigerating effect produced in the evaporator by the condensing unit*] Kondensatorsatzkalorimeter, Kondensatoraggregatkalorimeter *n* [*„Wärmemesser" ist synonym mit „Kalorimeter"*]

condensing unit with accessible hermetic motorcompressor halbhermetische Kondensatorsatz *m*

condensing unit with canned rotor motorcompressor Kondensatorsatz *m* mit halboffenem Motorverdichter

to condition, to treat [*air*] klimatisieren, aufbereiten, behandeln, klimaregeln

condition Bedingung, Gegebenheit, Voraussetzung *f*

condition, state Beschaffenheit *f*, Zustand *m*

conditional, C [*In automatic data processing, subject to the result of a comparison made during computation in a computer, or subject to human intervention*] bedingt, gegeben

conditional expression [*A COBOL language expression which is either true or false, depending upon the status of the variables, literals and logical operators within the expression*] bedingte Ausdruck *m*, gegebene Ausdruck

conditional instruction bedingte Befehl, gegebene Befehl *m*

conditional jump, conditional branch, conditional transfer, decision instruction, discrimination instruction [*A computer instruction that will cause the proper one of two or more addresses to be used in obtaining the next instruction, depending on some property of a numerical expression that may be the result of some previous instruction*] bedingte Sprung, gegebene Sprung *m*

conditionally stable circuit, conditionally stable system [*Electronics. A circuit or system that is stable when operating with specific values of input*

C

C

signal gain, but unstable for other values] bedingt stabile Schaltung *f*, bedingt stabile Schaltkreis, bedingt stabile Stromkreis *m*

condition change Zustandsänderung, Beschaffenheitsänderung *f*, Zustandswechsel, Beschaffenheitswechsel *m*

condition code Bedingungsschlüssel, Bedingungskode, Bedingungscode *m*

conditioned → **(air-)conditioned**

conditioned air [*Air treated to control its tempe-ra-ture, relative humidity, purity, pressure, and movement. Not to be confused with "treated air"*] klimageregelte Luft, klimatisierte Luft, Klimaluft *f*

conditioned air heating → **(air-)conditioned air heating**

conditioning, specific preparation [*The process of performing design and/or installation changes to make equipment compatible for use with other equipment or to bring it in line with specific standards*] Konditionierung, spezifische Vorbereitung *f*

conditioning, (specific) preparation [*Electronics. The process of modifying or adjusting communication transmission equipment to bring attenuation impedance and delay characteristics to within set limits*] Konditionierung, (spezifische) Vorbereitung *f*

(conditioning) apparatus dew-point (temperature) [*Of air leaving the conditioning apparatus, used interchangeably with maximum average coil surface temperature required to produce dew*] Klimaapparattaupunkt(temperatur) *m, (f)*

condition number Bedingungszahl *f*

condition observation → **state observation**

condition output Kriterienausgang *m* [*programmierbare Steuer(ungs)gerät*]

condition signal message Bedingungsmeldung *f* [*Eine Meldung aufgerufener und nicht erfüllter Freigaben bzw. Weiterschaltbedingungen (siehe DIN 19 237). Anmerkung: Nach Ablauf der Überwachzeit (siehe DIN 19 237) kann die Bedingungsmeldung zu einer Störmeldung führen*]

conducive to flow strömungsgünstig, strömungsfreundlich

conductance Wirkleitwert, Leitwert *m*, Konduktanz *f*

conductance → **(electrical) conductance**

conductance → **(heat) conductance**

conducting leitend

conducting layer leitende Schicht, Leit(ungs)-schicht *f*

conducting path, current path Strombahn *f*, Strompfad *m*

conducting-state current Durchlassstrom *m* [*Diode. DIN 41 781*]

conducting-state current loss Durchlassstromverlust *m* [*Diode. DIN 41 781*]

conducting-state voltage Durchlassspannung *f* [*Diode. DIN 41 781*]

conducting-state voltage current characteristic Durchlassspannungskennlinie *f* [*Diode. DIN 41 781*]

conducting substance, conduction substance Leiter *m*, Leit(ungs)substanz *f*

conducting track, conduction track Leiterbahn *f*

conduction Leitung *f*

conduction Ableitung *f* [*Blitzschutz*]

conduction capacity, conducting power, conducting ability, conducting capability Leit(ungs)-fähigkeit *f*, Leit(ungs)vermögen *n*

conduction heat gain Leit(ungs)wärmeabgabe *f*

conductive leitfähig, leitend

conductive exposed part → **exposed conductive part**

conductive heat transfer → **heat conduction**

conductively connected leitend verbunden, leitend angeschlossen

conductive part leitfähige Teil, leitende Teil *m, n* [*IEV 411-11-09. Teil, welches Strom leitet, auch wenn es nicht zum Übertragen des Betriebsstromes benutzt wird*]

conductive pattern Leitbild *n* [*gedruckte Schaltung*]

conductivity [*Ability of materials to transmit or conduct heat or electricity*] Leit(ungs)fähigkeit *f*, Leit(ungs)vermögen *n*

conductivity decay, conductivity decomposition Leit(ungs)fähigkeitsabfall, Leit(ungs)vermögensabfall *m*

conductivity electrode Leit(ungs)fähigkeitselektrode, Leit(ungs)vermögenselektrode *f*

conductivity pickup, conductivity measured-value transmitter Leit(ungs)fähigkeitsmesswertgeber, Leit(ungs)vermögensmesswertgeber *m*

conductor (US) → **downspout**

conductor, current conductor [*A substance which offers a low resistance to the passage of an electric current*] Leiter, Stromleiter *m*, stromführende Verbindung *f*

conductor, section of conductor, conductor section [*It melts when the current through it exceeds a rated value for a definite period of time*] Löschglied, Löschelement *n* [*Sicherung*]

conductor, wire conductor, cable conductor, conductor of the cable [*Wire or combination of wires, not insulated from one another, suitable for carrying an electric(al) current*] (Kabel)Ader *f*

conductor bar → terminal bar

conductor breakage, conductor rupture Leiterbruch *m*

conductor bundle, conductor assembly Leiterbündel *n*

conductor bushing Leiterbuchse *f*

conductor connected to ground (US) → ground(-ed) conductor

conductor cooled leitergekühlt

conductor element Leiterelement *n*

conductor gradient Leiterfeldstück *n*

conductor loop, current conductor loop Stromleiterschleife, Leiterstromschleife *f*

conductor of heat → heat conductor

conductor pair → pair of conductors

conductor rail, current conductor rail Stromschiene *f*

conductor rupture → conductor breakage

conductor section → conductor

conduit, closed channel [*For conducting the flow of a fluid*] geschlossene Leitung *f*

conduit, electrical conduit [*Circular raceway into which conductors are drawn. The terms include rigid conduit (metallic and nonmetallic) and flexible conduit*] Schutzrohr, Leitungs(schutz)rohr, Installationsrohr, Isolierrohr, Elektro(installations)rohr *n*

conduit, cable conduit Kabelkanal *m* [*Fernmeldetechnik*]

conduit coupling Rohrmuffe *f*, Rohrverbindungsstück *n*, Muffe *f* [*Installationsrohr*]

conduit nipple Installationsrohrnippel, Schutzrohrnippel, Isolierrohrnippel *m*

conduit union, electrical conduit union Installationsrohrverschraubung, Schutzrohrverschraubung *f*

conduit with heavy protection, conduit with high protection, electrical conduit with heavy protection, electrical conduit with high protection [*CEE 23*] Panzerrohr, Schutzrohr für schwere Beanspruchung *n*

conduit with light protection, electrical conduit with light protection [*CEE 23*] Installationsrohr *n* für leichte Beanspruchung, Schutzrohr für leichte Beanspruchung

conduit with medium protection, electrical conduit with medium protection [*CEE 23*] Installationsrohr *n* für mittlere Beanspruchung, Schutzrohr für mittlere Beanspruchung

cone diffusor, cone diffuser Kegeldiffusor *m*

cone holder Kegelhalter(ung) *m, (f)*

cone (loud)speaker [*A (loud)speaker employing a magnetic driving unit that is mechanically coupled to a paper or fibre cone*] Kegellautsprecher *m*

cone valve Kegelventil, V-Ventil *n*

conference call Sammelgespräch *n* [*Fernmeldetechnik*]

conference circuit, conference connection, conference communication Konferenzschaltung, *f*

conference dialling (Brit.); conference dialing (US) [*automatic telephony*] Konferenzwählen *n*

conference traffic Konferenzverkehr *m* [*Fernwirktechnik*]

conference unit Konferenzvorrichtung *f* [*Vorprogrammierte oder frei wählbare Teilnehmer werden zu einem Konferenzgespräch zusammengeschaltet und können miteinander in einer Wechsel- oder Gegensprechkonferenz sprechen*]

confidence coefficient [*The probability associated with a confidence interval; that is, the probability that the interval contains a given parameter or characteristic*] Vertrauenskoeffizient *m*

confidence interval [*An interval which has a specified probability of containing a given parameter or characteristic*] Vertrauensbereich *m [selten: n]*, Vertrauensspanne *f* [*DIN 55 350, Teil 24*]

confidence level [*The probability in acceptance sampling that the quality of accepted lots manufactured will be better than rejectable quality level (RQL); 90 % level indicates that accepted lots will be better than the RQL times in hundred*] Vertrauensniveau *n*, Vertrauenswert *m* [*DIN 55 350, Teil 24; Statistische Sicherheit, Aussagewahrscheinlichkeit für Vertrauensbereich*]

confidence limit Vertrauensgrenze *f* [*DIN 55 350, Teil 24*]

configurating Konfigurieren *n* [*Das Erstellen von Reg(e)lungs- und Steuer(ungs)funktionen aus Programmbausteinen bzw. das Zusammensetzen aus vorgegebenen Funktions- oder Baueinheiten*]

configuration, shape [*Form or figure as determined by the arrangement of parts*] Gebilde *n*, Konfiguration, Gestalt, Struktur, Form *f*, Aufbau *m*

C

configuration (Hardware)Ausstattung, Konfiguration, (Hardware)Zusammensetzung *f* [*Die Ausstattung einer Zentraleinheit mit Eingabe-Ausgabe-Einheiten und Direktzugriffsspeichern*]

configuration factor Himmelslichtquotient, gegenseitige Austauschfaktor *m* [*Beleuchtungstechnik*]

confined space Engraum, beschränkte Raum, beengte Raum, knappe Raum *m*

to confirm → to acknowledge

confirmation → acknowledg(e)ment

confirmed data transfer → data-transfer-acknowledged

confirmed service, acknowledged service Dienst mit Bestätigung, Bestätigungsdienst, Dienst mit Quittierung, Quittier(ungs)dienst, Dienst mit Rückmeldung, Rückmeldungsdienst, Rückmeldedienst *m*

confirming → acknowledging

confluent link Zuleitung *f* [*Fernmeldetechnik*]

conformance Übereinstimmung, Entsprechung *f*

confusion of sounds, sound confusion Schallgewirr(e) *n*, Schallverwirrung *f* [*volkstümlich: „Schallsalat“ m*]

congé, cove base; sanitary shoe (US) [*It is a mo(u)lding flush with the top of the wall above, but with a fillet between the cove and the floor. Hence any mo(u)lding of similar profile*] (Fuß)Bodenkehle *f*

congealed grease Fettschlamm *m*, Fangfett *n*

congestion, overloading Über(be)lastung *f* [*Netz*]

congestion tone Anhäufungszeichen, Andrangzeichen *n* [*Fernmeldetechnik*]

conical, tapered kegel(förm)ig, konisch, verjüngt

conical pendulum, tapered pendulum [*A weight attached to a cord or light rod, which is made to execute circular motion; the cord or light rod motions thus form the surface of a circular cone*] Kegelpendel *n*

conical (pipe) fitting thread, male (pipe) fitting thread, tapered (pipe) fitting thread Fittingaußengewinde, Rohrverbindungsstückaußengewinde, Formstückaußengewinde *n*

conic section parameter → parameter of a conic section

conjugate branches [*Of a network, any two branches such that an electromagnetic force inserted in one branch produces no current in the other branch*] (einander) untergeordnete (Ab)Zweige *f*

conjugate impedances [*Impedances having resistance components that are equal to the reactance components that are equal in magnitude but opposite in sign*] konjugierte Scheinwiderstände, konjugierte Impedanzen, konjugierte Widerstand(s)operatoren, konjugierte-komplexe Impedanzen, Blindwiderstände, Wechselstromwiderstände *f*

conjunction [*The connection of two statements by the word "and"*] Konjunktion, UND-Verknüpfung *f* [*logische Schaltung*]

conjunctive rule UND-verknüpfte Regel *f* [*unscharfe Logik*]

to connect einschalten

to connect anklemmen

to connect verbinden, anschließen

to connect zuschalten

connectable zuschaltbar

connectable anklemmbar

connectable anschließbar, anschlussbar, verbindbar

connected emergency load, emergency connected load Notstromanschlusslast *f*, Notstromanschlusswert *m*

connected in parallel parallel geschaltet

connected load [*Sum of all loads on an electric(al) circuit at a selected time*], Anschlusswert *m*, Flächenbelastung, Anschlusslast, Flächenlast, Flächenwert, Anschlussbelastung Anschlussleistung, Flächenleistung *f* [*Beleuchtungstechnik*]

connected load, installed load [*The volume of gas which would be used in unit time; all the appliances connected to a supply main were in full operation simultaneously*] Anschlusswert *m*, Anschlussleistung, Anschlusslast *f*

connected position [*IEC 439-1*] Betriebsstellung *f* [*Schalteinheit, VDE 0660, T.500*]

connected to ground (US) → grounded

connecting accessory, connection accessory Anschlusszubehör, Verbindungszubehör *n* [*selten: m*]

connecting arrangement, connection arrangement Anschlussanordnung, Verbindungsanordnung *f*

connecting box, connector box, connection box Klemmdose, Anschlussdose, Verbindungsdose *f* [*In ihr befinden sich übersichtlich angeordnet Klemmen für den Anschluss an die Lichtleitung (branch circuit)*]

connecting branch, connection branch Anschlussstutzen, Verbindungsstutzen, Stutzen *m*

connecting cable, connection cable, supply cable, service cable [*The supply authority's cable to the supplier's fuse box*] Anschlusskabel, Steckkabel, Verbindungskabel *n*

connecting cable duct, connection cable duct, serv-ice cable duct, supply cable duct Anschlusskabelkanal, Verbindungskabelkanal, Steckkabelkanal *m*

connecting cable (entrance) box, connection cable (entrance) box, service cable (entrance) box, supply cable (entrance) box Anschlusskabelkasten, Verbindungskabelkasten *m*

connecting circuit duplex, CCDX Verkehrswegplatine *f*

connecting component, connection component Anschlussteil, Verbindungsteil *m, n*

connecting contact, connection contact Anschlusskontakt, Verbindungskontakt *m*

connecting cord, connection cord Anschlussschnur, Verbindungsschnur *f*

connecting corridor, connection corridor Verbindungsflur, Anschlussflur *m*

connecting direction, connection direction Anschlussrichtung, Verbindungsrichtung *f*

connecting hose, connection hose Anschlussschlauch, Verbindungsschlauch *m*

connecting housing, connection housing Anschlussgehäuse, Verbindungsgehäuse *n*

connecting line, connection line Anschlussleitung, Verbindungsleitung *f*

connecting lug, connection lug Anschlussfahne, Anschlusslasche, Verbindungsfahne, Verbindungslasche *f*

connecting net(work), connection net(work) Anschlussnetzwerk, Durchschalt(e)netzwerk, Verbindungsnetzwerk *n*

connecting piece → connector

connecting pipe, connection pipe Anschlussrohr, Verbindungsrohr *n*

connecting (pipe) line, connection (pipe) line Anschluss(rohr)leitung, Verbindungs(rohr)leitung *f* [*Leitung vom Geruchverschluss eines Entwässerungsgegenstandes bis zur Fallleitung*]

connecting pipework, connection pipework Anschlussverrohrung, Verbindungsverrohrung *f*

connecting plate, connection plate Anschlussplatte, Verbindungsplatte *f*

connecting point, connection (point) Anschlussstelle, Verbindungsstelle *f*

connecting portion, connection portion Randkontakt *m* [*Leiterplatte*]

connecting post, connection post Leiterbrücke *f* [*In gedruckten Schaltungen*]

connecting pressure, connection pressure Anschluss-druck, Verbindungsdruck *m*

connecting relay, connection relay Durchschalt(e)relais, Anschlussrelais, Verbindungsrelais *n*

connecting strip, connection strip Anschlussleiste *f*, Verbindungsleiste

connecting (sub)assembly, connection (sub-) assembly Anschlussbaugruppe, Steckbaugruppe *f*, Verbindungsbaugruppe

connecting terminal unit, connection terminal unit [*IEC 23F.3*] Anschlussklemme, Verbindungsklemme *f* [*VDE 613*]

connecting thread, connection thread Anschlussgewinde *n*, Verbindungsgewinde

connecting type, connection type Anschlussart, Verbindungsart *f*

connecting wire, connection wire Schalt(ungs)-draht *m*

connecting wire, connection wire Anschlussdraht *m*, Verbindungsdraht

connecting wiring, connection wiring Anschlussverdrahtung *f*, Verbindungsverdrahtung

connecting work, connection work Anschlussarbeit(en) *f*, Verbindungsarbeit(en)

to connect in parallel parallel schalten

connection, junction [*Attachment of two or more component parts so that conduction can take place between them*] Anschluss *m*, Verbindung *f*

connection → channel

connection cabinet, terminal cabinet, connection cubicle, terminal cubicle Anschlussschrank *m*, Verbindungsschrank

connection carrying circulating circuit kreisstromführende Schaltung *f* [*Leistungselektrik*]

connection drop Spannungsabfall *m* [*Zwischen Litze und Bürste*]

connection endpoint Anschlussendpunkt *m*, Verbindungsendpunkt [*Kommunikationsnetz*]

connection facilities Anschlussteile *f*, Verbindungsteile [*EN 50014*]

connection for protective earth wire to VDE 0100 Anschlussstelle *f* für Schutzleitung nach VDE 0100, Verbindungsstelle für Schutzleitung nach VDE 0100

connection identification → CID

connection-in-parallel, parallel connection, shunt (connection) [*System in which flow is divided among two or more channels from a common starting point or header*] Parallelschaltung *f*, Nebenschlussschaltung

connection-in-series, series connection, series arrangement, series circuit, cascade circuit [*System in which flow through two or more chan-*

nels is in a single path entering after leaving the first or previous channel] Reihenschaltung, Serienschaltung, Kaskadenschaltung, Stromrichterschaltung *f*

C

connectionless broadband data service → CBDS

connection of units, units connection Einheitenzuschaltung *f*

connection orien(ta)ted data transfer anschluss-orientierte Datenaustausch, verbindungsorientierte Datenaustausch *m*

connection orien(ta)ted network service anschlussorientierte Netzdienst, verbindungsorientierte Netzdienst *m*

connection orien(ta)ted service, COTS anschluss-orientierte Dienst, verbindungsorientierte Dienst *m*

connection orien(ta)ted transport(ation) service → COTS

connection piece → connector

connection to ground (US) → grounding

connection (to the system) Zuschaltung *f*

connectivity Austauschfähigkeit *f*, Austauschvermögen *n* [*Die Fähigkeit einer relationalen Datenbank mit anderen Rechnern und Systemen Daten auszutauschen*]

connector Anschlussstecker, Verbindungsstecker, Steck(verbind)er *m*

connector, connection piece, connecting piece Anschlussstück, Verbindungsstück *n*

connector box → connecting box

connector (final switch), connector switch [*A switch, or relay group system, which finds the telephone line being called as a result of digits being dialed; it also has the function of causing interrupted ringing voltage to be placed on the called line or of returning a busy tone to the calling party if the line is busy*] Leitungswähler, Wahlschalter *m*

to (connect to) earth (Brit.); to (connect to) ground (US) erden [*An Erde legen, mit Erde verbinden*]

to connect (to the supply), to connect to the system, to connect to the load, to bring on to the supply, to bring on to the system, to bring on to the load zuschalten, aufschalten [*An das Netz schalten, an Spannung legen*]

consequent DANN-Teil einer Regel, Regel-DANN-Teil *m, n* [*unscharfe Logik*]

consequential damage Folgeschaden *m*

consequential damage cost(s) Folgeschadenkosten *f* [*Sie entstehen durch die zusätzlichen Instandsetzungskosten von Komponenten, die durch technische Kettenreaktionen von einem auslösenden Schaden betroffen sind*]

conservation of energy, energy conservation (Energie)Erhaltung *f*

conservation of energy theorem, conservation of energy law, energy conservation theorem, energy conservation law (Energie)Erhaltungssatz *m*, (Energie)Erhaltungsgesetz *n*

conservation of mass, mass conservation Masse(n)erhalt *m*

console [*A large cabinet for a radio or television receiver, standing on the floor rather than on a table*] Truhe *f*

console, control desk, control console [*The section of a computer that is used to control the machine manually, correct errors, manually revise the contents of storage, and provide communication in other ways between the operator or service engineer and the central processing unit*] Schaltungspult, (Schalt)Pult, Steuer(ungs)pult, Bedienungspult *n*, Konsole, (Stell)Warte *f*, (Schalt)Tisch, Steuer(ungs)tisch *m*

console air conditioner [*Packaged air conditioner mounted in a console, the condenser being water-cooled or air-cooled; the air intake passes through the wall*] Klimatruhe *f*

console command processor, CCP Bedienungspultbefehlsprozessor *m*

console typewriter, panel typewriter Bedienungspultschreibmaschine *f*

consonant articulation loss → articulation loss of consonants

conspicuous deutlich sichtbar, auffällig, auffallend [*Objekt; Lichtquelle*]

constant → switching element

constant → constant value

constant air volume system → CAV system

constant angular velocity → CAV

constant-current source, CCS, stabilized power supply Urstromquelle, Konstantstromquelle, Präzisionsstromquelle *f*, Stromspiegel *m*, Strombank *f* [*Für bestimmte Schaltungsanordnungen*]

constant directivity gleichbleibende Abstrahl(ungs)winkel *m* [*Beschallung*]

constant directivity (loud)speaker Hornlautsprecher, Lautsprecher mit Horn *m*

constant-inkjet method [*Du Pont*] → **continuous (ink) flow method**

constant linear velocity → **CLV**

constant pressure Gleichdruck, konstante Druck *m*

constant pressure valve Gleichdruckventil *n* [*Es ist ein in die Saugleitung eingebautes und den Verdampferdruck konstant haltendes druckgesteuertes Regulier(ungs)ventil*]

constants adressable memory Assoziativspeicher *m*

constant speed drive, CSD Gleichdrehzahlantrieb *m*

constant surface speed [*ISO 1056*] Schnittgeschwindigkeit *f* [*NC-Wegbedingung. DIN 66 025, Teil 2*]

constant (value), element [*Mathematical value that does not change during a particular process*] Konstante *f*, Konstantwert, Festwert *m*, Glied, Element *n*

constant volume controller, CVR Gleichvolumen(strom)regler *m*

to constrict → **to contract**

constriction, contraction, narrowing Einschnürung, Verengung, Einengung, Beengung *f*

construction ... → **building ...**

constructive solid geometry, CSG generative Volumenmodellierung *f*

consultancy talk, consultation talk Beratungsgespräch *n*

to consume fuel, to burn verbrennen

consumer Bedarfsträger, Verbraucher, Abnehmer *m*

consumer circuit Verbraucherschaltung, Bedarfsträgerschaltung, Abnehmerschaltung *f*

consumer-owned verbrauchereigen, abnehmereigen, bedarfsträgereigen

consumer('s) electronics Unterhaltungselektronik *f*

consumer('s) line Verbrauchsleitung *f* [*Die Verbrauchsleitungen in einem Gebäude unterteilt man in Verteilungsleitungen (wie bei Gas, meist im Keller), Steig(e)leitungen und Stockwerks- oder (Ab)-Zweigleitungen*]

consumer('s) main fuse, consumer('s) main fuze Hauptsicherung *f*

consumer('s) mains, consumer('s) net(work) Verbrauchernetz, Abnehmernetz, Bedarfsträgernetz *n*

consumer('s) receiver Aufbauempfänger *m* [*Rundsteuer(ungs)anlage, VDE 0420*]

consumer('s) terminal [*The demarcation between the electricity board's cable and the consumer's wiring; usually adjacent to the main switch*] Abnehmeranschluss, Verbraucheranschluss, Bedarfsträgeranschluss *m*

consumption characteristic value → **characteristic consumption value**

consumption excess Verbrauchsüberschuss, Mehrverbrauch, Überschussverbrauch *m*

consumption indicator Verbrauchsanzeiger *m*

consumption meter Verbrauchszähler, Verbrauchsmesser *m*

consumption per head Pro-Kopf-Verbrauch *m*

consumption (pipe)line Verbrauchs(rohr)leitung *f*

consumption point Verbrauchsstelle *f*

consumption statistics (Energie)Verbrauchsstatistik *f* [*(Energie)Verbrauchsdaten, die innerhalb des vom Gebäudemanagementsystem gesteuerten Gebäudes erfasst werden*]

consumption value Verbrauchswert *m*

contact Berührung *f*, Kontakt *m*

contact, contact member [*A part which is an electric conductor and which provides a low-resistance path for current flow upon mating with another conducting part which it is designed to operate*] Schaltglied, Schaltstück, Kontaktstück, Kontaktglied *n*, Kontakt *m*

contact arrangement → **contact layout**

contact assembly, contact group, group of contacts, assembly of contacts Kontaktstrang *m*, Kontaktgruppe *f* [*Zusammenfassung aller Kontakte eines elektromechanischen Relais oder Schalters. E VDE 0435, Teil 110*]

contact band → **band (of contact)**

contact bounce Schaltgliedprellen, Schaltstückprellen, Schaltkontaktprellen, Kontakt(stück)prellen *n*

contact break(ing) distance Schalt(ungs)strecke *f*

contact bridge, contact link Kontaktbrücke *f*, Kontaktbügel *m* [*Relais*]

contact carriage Kontaktschlitten *m*

contact clearance, contact distance Kontaktabstand *m*, Kontaktweite *f*

contact closing time → **contact holding time**

contact cooling [*Cooling by contact with a cold surface*] Kontaktkühlung *f*

contact current-closing rating Nenneinschaltstrom, Einschaltnennstrom *m*

contact drum dryer, contact drum drier Kontakttrommeltrockner, Trommelkontakttrockner *m*

contact drying Kontakttrocknung *f*

contact equipment (Schalt)Schütz(en)steuer(ungs)gerät(e) *n*, *(f)*

contact-fleeting on de-energization ausschaltwischend

contact-fleeting on energization einschaltwischend

contact-fleeting time → **fleeting contact time**

contact guard, contact protection, guard against contact, protection against contact Kontaktschutz *m*

contact holding time, contact closing time Kontaktschließzeit *f*

C

contact humidifier Kontaktbefeuchter, Kontaktanfeuchter, Rieselbefeuchter, Rieselanfeuchter *m*

(contact) icing [*Process of chilling in which finely crushed ice is placed in contact with the product in its unpacked or packed state*] Eiskühlung *f*, Eiskühlen *n*

contacting Berühren, Kontaktieren *n*

contact insert Kontakteinsatz *m* [*Steckvorrichtung*]

contact interrupted on de-energization ausschaltunterbrechend

contact interrupted on energization einschaltunterbrechend

contact interrupting rating, interrupting contact rating [*ASA C 37.1*] Nennausschaltleistung, Nennabschaltleistung *f* [*Relais*]

contact jaw Kontaktbacke *f*

contact layout, contact arrangement Kontaktanordnung *f*

contact lead-in Kontakteinführung *f*

contactless detector → **proximity detector**

contactless effect → **proximity effect**

contactless sensor → **proximity sensor**

contactless switch → **proximity switch**

contact load Kontaktlast *f*

contact loadability Kontaktbelastbarkeit *f*

contact lock Kontaktschloss *n*

contact maker, contactor, contact making device, electric contactor Kontaktgeber, Kontaktmacher, „Einschalter" *m*, kontaktgebende Glied, Kontaktgabewerk *n*, Kontaktvorrichtung *f*

contact making Kontaktgabe *f*, Kontaktgeben *n*

contact mat → **pressure mat**

contact material Schaltstückwerkstoff, Schaltstückmaterial, Kontaktwerkstoff *m*, Kontaktmaterial *n* [*Es wird an der Stromübergangsstelle eines Kontaktes verwendet*]

contact mechanism Schaltwerk *n*, Schaltmechanismus *m* [*Geräteschalter; VDE 0630*]

contact member → **contact**

contact monitoring Kontaktüberwachung *f* [*Überwachung mit Kontakten*]

contact multiplication Kontaktvervielfachung *f*

contact noise Kontaktrauschen, Schaltrauschen *n*

contact opening time Kontaktausschalt(ungs)zeit *f*

contactor (Schalt)Schütz(e) *m*, (*f*), (Schalt)Schützen *m*

contactor → **contact maker**

contactor assembly, contactor group Schützbaugruppe, Schützkombination *f*

contactor coil Schützspule *f*

contactor control → **contactor ((open) loop) control**

contactor control circuit Schützen-Steuer(ungs)-stromkreis *m*

contactor controller Schützsteuer(ungs)schalter, Schützsteuer(ungs)fahrschalter *m*, Schützsteuer-(ungs)walze *f*

contactor control system Schützsteuer(ungs)system *n*, Schützsteuer(ungs)anlage *f*

contactor disconnector Trennschütz(e) *m*, (*f*), Trennschützen *m*

contactor group → **contactor assembly**

contactor-interlocked → **interlocked with contactor**

contactor ((open) loop) control (Schalt)Schütz-(en)steuerung *f*

contactor panel Schütz(en)feld *n*

contactor relay [*IEC 337-1*] Hilfsschütz *n* [*VDE 0660, Teil 200*]

contactor servomechanism [*system engineering*] Regler *m* mit unstetigem Glied, Relaisregler

contactor servosystem [*system engineering*] Relaisfolgesystem *n*

contactor starter Schützanlasser *m*

contactor(-type) reversing controller, contactor (-type) reverser Schützumkehrschalter, Schützwendeschalter, Kraftschalter *m*, Kraftschaltgerät *n*

contactor(-type) starting control Anlassschütz-(en)steuerung *f*

contact parting, contact separation Kontakttrennung, Schaltstücktrennung, Kontaktunterbrechung, Schaltstückunterbrechung *f*

contact pin Kontaktstift, Schaltstückstift *m*

contact pitting Schaltstückfraß, Kontaktfraß, Kontaktabbrand, Schaltstückabbrand *m*

contact plate Bodenkontakt *m* [*Lampenfassung*]

contact plate of plug, contact plug plate Kontaktstöpselbrücke *f*

contact-protected, contact-proof, touch safe, safe from touch, safe to touch, contact-safe, proof to touch, proof to contact, safe to contact kontaktgeschützt, berühr(ungs)geschützt, berühr(ungs)sicher, kontaktsicher

contact protection → **contact guard**

contact rail Strom(abnehmer)schiene *f*

contact release Kontaktfreigabe *f*

contact separation → contact parting

contact series Kontaktreihe *f*, Stromführungsreihe

contact sheave, contact roller Kontaktrolle *f*, Stromführungsrolle

contact sound → impact sound

contact stroke Kontakthub *m*

contact tag Kontakt *m*

contact temperature Kontakttemperatur *f* [*Wenn sich zwei unterschiedlich temperierte Körper berühren, so stellt sich an der Berührungsfläche die Kontakttemperatur ein. Sie ist abhängig von den Wärmeeindring-koeffizienten beider Stoffe*]

contact track Kontaktstreifen *m* [*gedruckte Schaltung*]

contact transmitter Kontaktsender *m*

contact travel Schalt(stück)weg, Kontaktweg, Kontakthub, Schalt(stück)hub *m* [*Elektroapparat*]

contact voltage drop Kontaktspannungsabfall *m*

contact wafer Schaltplatine *f* [*Vielfachschalter*]

container Behälter *m*

container Innenbehälter *m* [*Heißwasserspeicher*]

container continuous lift, container paternoster (lift) (Brit.); container continuous elevator, container paternoster (elevator) (US) Behälterpaternoster(aufzug), Behälterumlaufaufzug *m*

container elevator (US); container lift (Brit.) Behälteraufzug *m*

containing iron → ferritic

containing nitrogen → nitrogenous

contaminant → pollutant

contaminated household water → polluted household water

contaminating signal Schmutzsignal *n*

contamination, pollution Verseuchung *f*

contamination, pollution Verschmutzung, Verunreinigung *f*

contamination control, pollution control Reinhaltung *f*

contamination layer Fremdschicht, Fremdlage *f* [*Isolierung*]

contaminator → pollutant

contaminator concentration → pollutant concentration

content-addressable inhalt(s)adressierbar

content-addressable (computer) memory, CAM inhalt(s)adressierbare (Rechner)Speicher *m*, Assoziativ(rechner)speicher, (Rechner)Speicher mit adressierbarem Inhalt

contention mode Überlappung, Auseinandersetzung *f*, Streit *m*, Konkurrenz(betrieb) *f*, (*m*) [*Zwischen zwei oder mehreren Endgeräten über den gleichzeitigen Einsatz eines gemeinsamen Kanals*]

contents-addressable memory, CAM inhalt(s)-adressierbare Speicher, Assoziativspeicher *m*

content (type) indication Inhalt(s)artanzeige *f* [*Mitteilungsübermitt(e)lung*]

context [*A set of data or information that completely describes a particular communication environment at a particular point in time*] Kontext, umgebende Text *m*

context management Verbindungsauf- und -abbau *m* [*Kommunikationsdienst*]

contingent fee Erfolgshonorar *n*

continuation Fortgang, Weitergang *m*

continuity check(ing) Durchgangskontrolle *f*

continuity check request, CCR Durchgangskontroll-anfrage *f* [*Ein Dienstelement der Anwendungsschicht im ISO/OSI-Schicht(en)modell*]

continuity equation → equation of continuity

continuity signal message Durchgangsmeldung *f* [*Siehe Anmerkung unter „Meldung"*]

continuous permanent, stetig, dauernd, fortlaufend, stufenlos, kontinuierlich, durchgängig

continuous(-action) controller, continuous (-action) regulator stetige Regler, Stetigregler, kontinuierliche Regler *m*

continuous alarm Daueralarm, Stetigalarm *m*

continuous arrangement (of a system) Durchgängigkeit *f* (eines Systems)

continuous band of steps Fahrstufenband, Rollstufenband *n*

continuous base current, base continuous current Basisgleichstrom *m*

continuous belt treadway, passenger-carrying surface [*moving walk*] Beförderungsband, Personenband *n*

continuous burning stove Dauerbrandofen *m*

continuous controller [*See remark under "controller"*] stetige Regler, kontinuierliche Regler, Stetigregler, Dauerregler *m*

continuous flow Durchfluss *m*

continuous flow appliance, continuous flow device Durchflussgerät *n*

C

continuous freeze drying, continuous lyophilization [*Process in which the product is continuously moved through the drying chamber*] Stetiggefrier-trocknung, Dauergefriertrocknung *f*

continuous gas heater → gas continuous heater

continuous gas water heater → no-storage gas water heater

continuous grain silicon, CGS Siliziumkristallschicht *f* [*Auf einer Glasplatte aufgebracht*]

continuous hot water heater → hot water contin-uous heater

continuous hunting [*automatic telephony*] Dauerwahl *f*, Durchdreher *m*

continuous (ink) flow method; constant-inkjet method [*Du Pont*] Dauer-Tintenstrahlverfahren *n*, Dauer-Tintenstrahlmethode *f* [*Die Tinte ist dauernd im Umlauf, wobei die Tintenmenge die nicht auf das Papier soll, nach dem Austritt aus dem Druckkopf über elektrostatische Felder abgelenkt und dem Tintenbehälter zugeführt wird*]

continuous (lavatory) range, continuous washbasin range [*The basins have overlaps to form a contin-uous range, as opposed to basins 50 mm or more apart*] Durchlaufwasch(becken)reihe, Durchlaufwasch(becken)batterie, Durchlaufreihenwaschanlage *f*, Durchlaufreihenwaschbecken *n*

continuous lift, paternoster (lift) (Brit.); contin-uous elevator, paternoster (elevator) (US) [*B. S. 2655*] Paternoster(aufzug), Umlaufaufzug, Paternosterfahrstuhl, Umlauffahrstuhl *m* [*DIN 15 307 und 15 308*]

continuous limit(ing) current → limit(ing) continuous current

continuous load Stetiglast, Dauerlast *f*

continuous(ly acting) ((closed) loop) control, continuous regulation stetige Reg(e)lung, dauernde Reg(e)lung, fortlaufende Reg(e)lung, kontinuierliche Reg(e)lung, stetig wirkende Reg(e)lung, dauernd wirkende Reg(e)lung, kontinuierlich wirkende Reg(e)lung, fortlaufend wirkende Reg(e)lung, Stetigreg(e)lung, Dauerreg(e)lung *f*

continuous(ly acting) ((open) loop) control stetige Steuerung, kontinuierliche Steuerung, dauernde Steuerung, fortlaufende Steuerung, stetig wirkende Steuerung, kontinuierlich wirkende Steuerung, fortlaufend wirkende Steuerung, dauernd wirkende Steuerung, Dauersteuerung *f*

continuously concurrent counter reading Stetig-Zählerablesung *f*

continuously variable slope delta modulation Deltamodulation *f* mit stetig veränderbarer Rampensteilheit, Deltamodulation mit stetig verstellbarer Rampensteilheit, Deltamodulation mit stetig einstellbarer Rampensteilheit, Deltamodulation mit stufenlos regelbarer Rampensteilheit *f*

continuous multi(ple)-cycle duty, continuous periodic duty Durchlaufschaltbetrieb, Stetigschaltbetrieb, Dauerschaltbetrieb *m*

continuous operating voltage Netzbetriebsspannung *f*

continuous operation, permanent operation, uninterrupted operation Dauerbetrieb, Stetigbetrieb *m*

continuous paper printer, endless paper printer Endlospapierdrucker *m*

continuous path control (system), contour(ing) control (system) Bahnsteuerung *f* [*Numerische Steuerungssystem, bei dem eine Bewegung in mehr als einer Achse gleichzeitig und kontinuierlich gesteuert wird. Zwischen den Bewegungen in den verschiedenen Achsen besteht ein Funktionszusammenhang*]

continuous periodic duty → continuous multi(ple)-cycle duty

continuous phase frequency shift keying → CPFSK

continuous pilot flame [*A small gas flame which on some gas water heaters is left always burning so as to ignite the main gas burners immediately the tap is opened*] Sparflamme *f*

continuous plotter Endloszeichenmaschine *f*

continuous range → continuous (lavatory) range

continuous record fortlaufende Aufzeichnung, stetige Aufzeichnung, dauernde Aufzeichnung, kontinuierliche Aufzeichnung, Registrierung, Daueraufzeichnung *f*

continuous recording fortlaufende Aufzeichnen, Registrieren, stetige Aufzeichnen, dauernde Aufzeichnen, kontinuierliche Aufzeichnen *n* [*DIN 19 222*]

continuous regulation → continuous(ly acting) ((closed) loop) control

continuous roll (type) filter, roller curtain (type) filter, roll (type) disposable media filter Roll-(band)filter *m*, *n*

continuous setting fortlaufende (Ein)Stellen, kontinuierliche (Ein)Stellen, stetige (Ein)Stellen, dauernde (Ein)Stelle *n*

continuous signal, maintained signal Stetigsignal, fortlaufende Signal, kontinuierliche Signal, stetige Signal, dauernde Signal, Dauersignal *n*

continuous spray (air) dehumidifier Durchlauf-Sprüh(luft)entfeuchter *m* [*DIN 30 600, Reg.-Nr. 06105*]

continuous spray (air) humidifier Durchlauf-Sprüh(luft)befeuchter *m* [*DIN 30 600, Reg.-Nr. 06105*]

continuous stationary Endlospapier *n* [*Schreiber; Drucker*]

continuous steam output Dampfdauerleistung, Dauerdampfleistung *f*

continuous stress, permanent stress Dauerbeanspruchung, Stetigbeanspruchung *f*

continuous surface geschlossene Oberfläche *f*

continuous switching Fortschalten, Weiterschalten *n*

continuous switching delay Fortschaltverzögerung, Weiterschaltverzögerung *f*, Fortschaltverzug, Weiterschaltverzug *m*

continuous-time zeitkontinuierlich

continuous tone Dauerton *m*

continuous (type) preheater Durchlaufvor(er)wärmer, Durchlaufvorerhitzer *m*

continuous washbasin range → continuous (lavatory) range

continuous water heater → no-storage water heater

continuous wave laser, CWL [*IEC 825*] Dauerstrichlaser, Stetigstrichlaser *m* [*VDE 0837*]

contour(ing) control (system) → continuous path control (system)

to contract verdingen, vergeben

to contract, to narrow, to constrict einschnüren, verengen, einengen, beengen

contracting Verdingung, Vergabe *f* [*Die Gesamtheit der zum Vertragsabschluss führenden Maßnahmen des Auftraggebers*]

contracting company Unternehmen *n*

contracting contract Verdingungsvertrag, Vergabevertrag *m*

contraction Einschnürung, Verengung, Verschwächung *f*

contraction degree Einschnürungsgrad, Verengungsgrad, Verschwächungsgrad *m*

contract management Vertragsverwaltung *f*

contraflow → counterflow

contra-running gegenläufig

contrast Kontrast *m*

to control, to adjust, to vary verstellen, justieren, (ein)stellen, umstellen

to control → to attend

control Regulierung, Reg(e)lung *f*

control Beherrschung *f*

control, combatting Bekämpfung, Abwehr *f*

control → ((closed) loop) control

control → ((open) loop) control

control agent, control medium [*It is manipulated by the controlled device. It may be air or gas flowing through a damper or gas, steam or water flowing through a valve; or an electric current*] Reg(e)lungsmedium, Regulierungsmedium *n*

control amplifier → ((closed) loop) control amplifier

control and indicating equipment [*EN 54*] Nebenmelde(r)zentrale *f* [*Brandbekämpfung*]

control and instrumentation technique, instrumentation and control technique, I & C Leittechnik, Mess- und Regeltechnik *f*

control and monitoring → monitoring and control

control and monitoring engineering → monitoring and control engineering

control and monitoring station → monitoring and control station

control and monitoring system → monitoring and control system

control and read-only memory, control and read-only store, CROM [*A read-only memory that also provides storage, sequencing, execution, and translation logic for a number of microinstructions*] Steuer(ungs)befehls- und Nur-Lese-Speicher, Registrier- und Nur-Lese-Speicher *m*

control and surveillance → monitoring and control

control and surveillance engineering → monitor-ing and control engineering

control and surveillance system → monitoring and control system

control assembly → control package

control block → ((closed) loop) control block

control block Steuer(ungs)satz *m* [*Er beinhaltet die kodierten numerischen Werte, die von der Maschine über die Steuerung in Schalt- und Bewegungsvorgänge umgewandelt werden*]

control board → control panel

control break [*In automatic data processing, a key change which takes place in a control data field, especially in the execution of a report program(me)*] Unterbrechung *f* des Programmverlaufs

C

control center attendant (US) → **control station attendant**

control chain → **((open) loop) control chain**

control change, comparing control change [*A change of function that occurs when successive records (such as those entered on punched cards) differ in the data entered in the control field; for example, a punch(ed)-card tabulator may change from adding to printing at the end of a series of items*] Gruppenwechsel *m*

control circuit → **((closed) loop) control circuit**

control clause [*COBOL*] Gruppensteuer(ungs)klausel *f*

control computer [*A computer which uses inputs from sensing devices and outputs connected to control mechanisms to control physical processes*] Leitrechner *m*

control console → **console**

control cycle [*Sequence of operations of functions under automatic control intended to maintain the desired conditions at all times*] automatische Reg(e)lungszyklus *m*

control data [*In automatic data processing, data used for identifying, selecting, executing, or modifying another set of data, a routine, a record, or the like*] Steuer(ungs)daten *f*

control decoder and sequencer Befehlskodierer *m* und Ablaufsteuerung *f* [*Mikroprozessor*]

control desk → **console**

control (device), (closed) loop control (device) Reg(e)lungsgerät *n*

control (device), (open) loop control (device) Steuer(ungs)gerät *n*

control equipment → **((closed) loop) control equipment**

control equipment → **((open) loop) control equipment**

control factor → **((closed) loop) control factor**

control field, CO, CF [*In automatic data processing, a constant location where information for control purposes is planned, such as a specified column on punched cards*] Steuer(ungs)feld *n*

control frequency response Führungsfrequenzgang *m* [*Systemtechnik*]

control gas valve → **((open) loop) control gas valve**

control gear Vorschaltgerät *n*

control instruction Steuer(ungs)befehl *m*

controllability Regelbarkeit *f*

controllability Steuerbarkeit *f*

controllability matrix Regelbarkeitsmatrix *f*

controllability matrix Steuerbarkeitsmatrix *f*

controllable regelbar

controllable steuerbar

controllable → **adjustable**

controllable clearance → **adjustable clearance**

controllable ((closed loop) canonical form Regel-(ungs)normalform *f*

controllable ((closed) loop) system regelbare System *n*

controllable key → **extension key**

controllable ((open) loop) canonical form Steuer(ungs)normalform *f*

controllable ((open) loop) system steuerbare System *n*

Control Language for Microcomputer → **CP/M**

controlled → **((closed) loop) controlled**

controlled access unit → **CAU**

controlled-avalanche-element Lawinenzelle *f* [*für sehr hohe Sperrspannungen*]

controlled-carrier modulation transmitter fremdgesteuerte Sender *m*

controlled member Regel(ungs)strecke, Reglungsstrecke *f* [*Systemtechnik*]

controlled not ready signal Antwortkennzeichen *n* für „gesteuert nicht bereit", Gesteuert/-nichtbereit-Kennzeichen, Nichtbetriebsbereitkennzeichen

controlled operation → **controller operation**

controlled switch Kipptriode, Vierschicht(en)triode *f*

controlled system [*IEC 50(351)*], **drectly controlled member** [*IEC 27-2A*], **plant, process, system** Strecke *f* [*1. Steuer(ungs)strecke, 2. Reg(e)lungsstrecke. Der Teil des Wirkungsweges, dem die beeinflussenden Glieder oder Teile eines Objekts oder einer Anlage, vereinbart für den jeweiligen Anwendungsfall, angehören*]

controlled system without self-regulation Strecke ohne Ausgleich, ausgleichlose Strecke *f*

controlled system with self-regulation → **self-regulation controlled system** [*IEC 50(351)*]

controlled transmitter fremdgesteuerte Sender *m*

controlled variable, controlled value [*closed loop control*] Regel(ungs)größe *f*

controller, CTRL Plattenkontrollprozessor *m*, Schaltung *f* zwischen Rechner und Platte, Schnittstelle [*Diese Benennung ist für diese Schaltung zu mehrdeutig*]

controller, CTRL [*Control systems. A device that directs the transmission of information over the data links of a network; it is controlled by a program(me) either within the device or a processor to which the controller is connected*] Regler *m*

controller, CTRL [*Computer engineering. In industrial control applications, the comparator and the control algorithms implemented within a hardware unit*] Regler *m*

controller, CTRL Regler *m*, Regel(ungs)einheit *f*, Regel(ungs)gerät, Regel(ungs)organ *n* [*Eine Vorrichtung zum Erhalten der eingestellten Regelgröße (z. B. Druck, Temperatur) nach dem vorgeschriebenen Sollwert*]

controller, CTRL Steuer(ungs)einheit *f*, Steuer(ungs)organ, Steuer(ungs)gerät *n*

controller automation station Reglerautomatisierungsstation *f*

controller command [*See remark under "command"*] Reglerkommando *n*

controller CPU, controller central processing unit, controller (main) frame [*See remark under "central processing unit"*] Reglerzentraleinheit *f*, Reglersteuerwerk *n*, Reglerverarbeitungsteil *m, n*

controller error signal [*The difference between the control point (actual value of the controlled variable) and the set point. This quantity may have a plus or minus sign*] Reglerfehlersignal *n*

controller family, controller series Reglerbaureihe, Reglerfamilie *f*

controller heating characteristic (curve) Reglerheiz(ungs)kennlinie *f*

controller input variable Reglereingang(s)größe *f* [*Am Reglereingang anliegende Größe. DIN 19225*]

controller interface Reglerschnittstelle *f*

controller (main)frame → **controller CPU**

controller memory, controller storage Reglerspeicher *m*

controller module Reglerbaustein *m*, Reglermodul *n* [*Siehe Anmerkung unter „Baustein"*]

controller network Reglernetz(werk) *n*

controller operation, controlled operation Reglerbetrieb *m*

controller output, CTRL output Reglerausgang *m*

controller output variable, controller output quantity, CTRL output variable, CTRL output quantity Reglerausgang(s)größe *f* [*Vom Reglerausgang abgegebene Größe. DIN 19 225*]

controller parameter Reglerparameter *m*

controller pressure probe Reglerdrucksonde *f*

controller section Reglerteil *m, n*

controller series → **controller family**

controller service not requested state, CSNS bedienungsaufrufende Zustand *m* der Steuer(ungs)-funktion [*programmierbare Messgerät; DIN IEC 625*]

controller service requested state, CSRS Bedienungsruf-Empfangszustand *m* der Steuer(ungs)-funktion [*programmierbare Messgerät; DIN IEC 625*]

controller setpoint Reglersollwert, Reglerführungswert *m*, Reglersollgröße, Reglerführungsgröße *f*

controller setting Einregulierung, Reglereinstellung *f*

controller set(ting) range Regler(ein)stell(ungs)-bereich, Einregulierungsbereich *m* [*selten: n*]

controller station → **((closed) loop) controller station**

controller storage → **controller memory**

controller submodule Reglersubmodul *n*, Reglerplatine *f*

controller switch Regelschalter *m*

controller synchronous wait state, CSWS synchronische Wartezustand *m* der Steuer(ungs)funktion [*programmierbare Messgerät; DIN IEC 625*]

controller transfer state → **CTRS**

controller unit Regel(ungs)vorrichtung *f*, Regel(ungs)gerät *n* [*Besteht aus Stellgerät (=Stellglied und Stellantrieb) und Regler*]

controller valve Reglerventil *n*

controller with disturbance variable feedforward [*control engineering*] Regler *m* mit Störgrößenaufschaltung

control level Leitebene, Führungsebene *f*

control limit Regel(ungs)grenze *f*

controlling Kostenverfolgung *f*

controlling Steuerung *f*

controlling Reg(e)lung *f*

controlling equipment Reg(e)lungsgerät(e) *n, (f)*

controlling equipment Steuer(ungs)gerät(e) *n, (f)*

controlling mechanism Regel(ungs)mechanismus, Reglungsmechanismus *m*

controlling mechanism Steuer(ungs)mechanismus *m*

control logic Steuer(ungs)logik *f*

control logic timing generator Steuer(ungs)signalzeitgeber, Steuer(ungs)signaltaktgeber *m*, Steuer(ungs)signalsynchronisiereinheit *f* [*Er fasst die beiden wichtigsten DRAM-Steuer(ungs)signale RAS (= row access strobe) und CAS (= column access strobe), die die Speicherzeilen und -spalten adressieren, zu einem Steuer(ungs)signal zusammen*]

control loop → **((closed) loop) control circuit**

C

control measuring unit Betätigungsmesseinheit *f*

control medium → **control agent**

control memory Speicher *m* einer Steuerung, Steuer(ungs)speicher [*Eine Funktionseinheit innerhalb einer Steuerung, die Programm (Programmspeicher) oder Daten (Datenspeicher) in digitaler Darstellung aufbewahrt. E DIN 19 226 Teil 5 und V DIN 19 237*]

control message Steuer(ungs)nachricht *f*, Steuer(ungs)mitteilung *f*

control message display [*A device, such as a console typewriter, on which control information, such as information on the progress of a running computer program(me), is displayed in ordinary language*] Steuer(ungs)nachrichtenanzeige, Steuer(ungs)mitteilungsanzeige *f*

control mode [*a.c. power controller*] Stellerbetrieb *m* [*Wechselstromsteller*]

control module [*In automatic data processing, the set of registers and circuity required to carry out a specific function*] Steuer(ungs)modul *n*, Steuerungsbaustein, Steuerbaustein *m*

control network → **((closed) loop) control network**

control network → **((open) loop) control network**

control of PAD service signals Steuerung *f* von PAD-Netzmeldungen

control package, control assembly, control set, control kit Steuer(ungs)garnitur, Steuer(ungs)kombination, Steuer(ungs)gruppe *f*, Steuer(ungs)satz *m*, Steuerungsaggregat *n*

control panel, control board Zentrale *f*, Bedien(er)feld, Bedienungsfeld *n*, Bedien(er)tafel, Bedienungstafel *f*

control panel → **((open) loop) control panel**

control passing Steuer(ungs)übergabe *f* [*programmierbares Messgerät*]

control path, signal control path Signalwirkungsweg, Signalflussweg *m*

control pedestal Steuer(ungs)säule *f*

control point [*radio*] Funkbetriebszentrale *f*

control point [*In automatic data processing, the numerical value of the controlled variable (speed, temperature, and so on) which, under any fixed set of operating conditions, an automatic controller operates to maintain*] Sollwert *m*

control point Einsatzleitstelle *f*

control power, regulating power Regel(ungs)leistung, Reg(e)lungsleistung *f*

control power Steuer(ungs)leistung *f*

control power element [*Actuator in an automatic control*] Steuer(ungs)leistungselement *n*

control power supply Steuer(ungs)leistungsversorgung *f*

control power transformer Steuer(ungs)transformator, Steuer(ungs)trafo *m*

control process, (closed) loop process Regelvorgang, Reg(e)lungsvorgang *m*

control process, (open) loop control process Steuer(ungs)vorgang *m*

control program (US) → **((open) loop) control program**

control pulse Ansteuer(ungs)signal *n* [*Leistungselektronik*]

control punch → **designation punch**

control punching, designation punching [*data processing*] Steuer(ungs)lochung, Leitlochung *f*

control quality Steuer(ungs)güte *f*

control quality Regelgüte, Reg(e)lungsgüte *f*

control range [*IEC 50(603)*] Wirkleistungsregelbereich *m* [*selten: n*] [*Generatorsatz*]

control reading Steuer(ungs)ablesung *f*

control register [*Any one of the registers in a com-puter used to control the execution of a computer program(me)*] Steuer(ungs)register *n*

control ROM → **control and read-only memory**

control room → **control station**

control room attendant → **control station attend-ant**

control sequence introducer → **CSI**

control set → **control package**

control setting Steuer(ungs)einstellung *f*

control signal Steuer(ungs)signal *n*

control signalling code, CSC Steuer(ungs)kode *m* zur Zeichengabe [*dezentrale Zeichengabe*]

control (slide) valve Steuer(ungs)schieber *m*

control stage Steuer(ungs)teil *m, n*

control state [*In automatic data processing, the oper-ating mode of a system which permits it to override its normal sequence of operation*] Steuer(ungs)zustand *m*

control station, control room; control center (US); control centre (Brit.) Steuer(ungs)warte, Steuer(ungs)station, Leitzentrale, Leitwarte, Steuer(ungs)zentrale, Leitstation *f*

control station attendant, control room attendant; control center attendant (US); control centre attendant (Brit.) Schaltwartenbediener, Steuer(ungs)wartenbediener, Steuer(ungs)stationsbediener, Leitzentralenbediener, Leitwartenbediener, Leitstationsbediener, Steuer(ungs)zentralenbediener *m*

control strategy Regel(ungs)strategie *f*, unterlegte Reg(e)lungsfunktion *f*

control strategy Steuer(ungs)strategie *f*, unterlegte Steuer(ungs)funktion *f*

control string Steuer(ungs)zeichenfolge *f* [*Kode*]

control supervisor [*The computer software which controls the processing of the system*] Ablaufüberwacher *m*

control switch [*IEC 337-1*] Steuerungsschalter *m* [*VDE 0660 T.200*]

control switchboard [*ANSI C 37.100*] Steuer(ungs)-tafel, (Schalt)Wartentafel *f*

control synchro → **control transformer**

control system earth, control system ground Regel(ungs)systemerde, Reg(e)lungssystemerde *f*

control system earth, control system ground Steuer(ungs)systemerde *f*

control system flowchart → **sequential function chart**

control system function chart → **sequential function chart**

control tap [*Part of a shower spray nozzle to vary the shape of volume of the spray*] Regulier(ungs)-hahn *m*

control tape → **carriage tape**

control terminal → **((open) loop) control terminal**

control time constant Steuer(ungs)zeitkonstante *f* [*DIN 41 745*]

control-to-load isolation → **electrical isolation**

control transfer Steuer(ungs)übergabe *f*

control unit, CU [*An electronic device containing data buffers and logical circuity; situated between the computer channel and the input/output device, and controlling data transfers and such operations as tape rewind*] Leitwerk [*DIN 44 300*], Steuer(ungs)werk *n*, Datenverarbeitungsanlage, Leiteinheit, Steuer(ungs)einheit *f* [*VDE 0660, Teil 200*]

control value Steuer(ungs)wert *m*

control voltage Steuer(ungs)spannung *f*

control voltage socket Steuer(ungs)spannungs-steckdose *f*

control wiring Steuer(ungs)verdrahtung *f*

control word [*A computer word specifying a certain action to be taken*] Leitwort, Steuer(ungs)-wort *n*

control zero Steuer(ungs)nullpunkt *m*

convected air Konvektionsluft *f*

convection, convectible heat transfer, convection flow, convective flow, convection current, convective current [*Transfer of heat from one place to another by virtue of warm air rising*] Konvektion, Wärmemitführung *f*, Konvektionsstrom *m*

convection circulation route Konvektionsumlaufweg *m*

convection cooling ceiling, convective cooling ceiling Konvektionskühl(ungs)decke *f*

convection current, convective current Konvektionsstrom *m*

convection dryer, convection drier Konvektions-trockner *m*

convection fan Konvektionsventilator *m*

convection furnace Konvektionsofen *m*

convection heat, heat of convection Konvektionswärme *f*

convection heater → **convector**

convection heating, convective heating, convector heating Konvektionsheizung, Konvektor-heizung *f*

convection port, convection orifice Konvektions-öffnung, Konvektoröffnung *f*

convection protection Konvektionsschutz *m*

convection shaft Konvektionsschacht, Konvektor-schacht *m*

convection space cooling, convective space cooling Konvektionsraumkühlung *f*

convective konvektiv

convector, convection heater [*Assembly of ferrous or nonferrous tubes with metal fins attached, mounted in a metal enclosure or a wall recess and covered in front, designed to transfer heat to the air, principally by convection*] Konvektor(heizkörper) *m*

convector cabinet, convector cubicle Konvektor-schrank *m*

convector casing, convector case, convector chest Konvektortruhe *f*

convector circuit breaker Konvektorausschalter, Konvektorunterbrecher *m*

convector core pipe Konvektorkernrohr *n*

convector fire [*deprecated*] → **radiant/convector gas fire**

convector heating panel, convector heater panel, convection heating panel, convective heating panel Konvektorplatte, Konvektionsplatte *f*

convector heating surface Konvektionsheiz(ungs)-fläche *f*

convector recess Konvektornische *f*

convector (room) heater, convector (room) heat-ing unit Konvektions(raum)heiz(ungs)körper, Konvektionszimmerheiz(ungs)körper, Konvektor(raum)heiz(ungs)körper, Konvektorzimmer-heiz(ungs)körper *m*

C

convector sanitary cove, convector mopboard, convector washboard, convector scrub board, convector base (board) (US); convector skirting (board) (Brit.); convector base plate (Scotland) Konvektor(fuß)leiste, Konvektorscheuerleiste, Konvektorsockelleiste *f*, Niedrigkonvektor *m*

convector type, convector design, convector version Konvektorbauweise, Konvektorausführung, Konvektor(bau)art *f*, Konvektoraufbau *m* [*Bei Raumgeräten, die Ausführung eines Wärme(aus)tauschers zur überwiegend konvektiven Wärmeübertragung an den Raum*]

convector wall Konvektorwand *f*

convector with primary air supply Konvektor *m* mit Primärluftzufuhr, Klimakonvektor, Induktionsgerät *n*

conventional milling Gegenlauffräsen *n*

conventional operating current vereinbarte Ansprechstrom, vereinbarte Auslösestrom *m* [*Schutzvorrichtung. VDE 0100, Teil 200*]

conversational compiler → interpreter

conversational mode → interactive mode

conversational service Dialogdienst *m* [*Ein Multimedia-Kommunikationsdienst*]

conversation meter Gespräch(s)zähler *m* [*Fernmeldetechnik*]

conversion Umrechnung *f*

conversion Umrüstung *f*

conversion → (building) conversion

conversion → (data) conversion

conversion, transformation [*Process of changing from one form of representation to another*] Konvertierung, Umsetzung, formelle Änderung, Umwandlung *f*, Umformen *n*

conversion assembly → conversion kit

conversion burner → (gas) conversion burner

conversion channel → (data) conversion channel

conversion command, CC Umsetz(ungs)kommando *n*

conversion cycle Umsetz(ungs)zyklus *m*

conversion end → end of conversion

conversion equipment → (data) conversion equipment

conversion error Umrechnungsfehler *m*

conversion factor [*IEC 146*] Stromrichtergrad, Leistungsumformergrad *m* [*Leistungselektronik*]

conversion factor Umrechnungsfaktor *m*

conversion indication Umsetz(ungs)anzeige *f*

conversion kit, conversion assembly, conversion set Umrüst(ungs)satz *m*, Umrüst(ungs)kombination, Umrüst(ungs)garnitur, Umrüst(ungs)gruppe *f*, Umrüst(ungs)aggregat *n*

conversion program (US); conversion programme (Brit.) [*A set of instructions which allows a program written for one system to be run on a different system*] Umsetz(ungs)befehlssatz, Umsetz(ungs)anweisungssatz *m*

conversion prohibition Umsetz(ungs)verbot *n* [*Mitteilungsübermittlung*]

conversion rate Umsetz(ungs)geschwindigkeit *f* [*Sie gibt die Anzahl der maximalen erreichbaren Wandlungen pro Sekunde an*]

conversion routine [*A flexible, self-contained, and generalized program(me) used for data conversion, which only requires specifications about a very few facts in order to be used by a programmer*] Umsetzungsprogramm *n*

conversion set → conversion kit

conversion table Umrechnungstabelle, Umrechnungstafel *f* [*Mathematik*]

conversion time [*The time required to read in data from one code into another code*] Umsetz(ungs)-(einlese)zeit *f*

to convert umsetzen, wandeln

to convert umstellen [*Von einer Betriebsweise auf eine andere*]

converter → (data) converter

converter Umsetzer, Wandler *m* [*Bauglied oder Gerät zum Umsetzen eines stetigen Signals (Eingangsgröße) in eine Impulsfolge (Ausgangsgröße) oder eine entsprechende Bitkombination und umgekehrt. Er wird erforderlich wenn die Steuer(ungs)- oder Regel(ungs)funktion von einem Digitalrechner wahrgenommen wird. E DIN 44300/10.85*]

converter, power converter, static converter Stromrichter, Umformer, Konverter, Umrichter, Umsetzer *m* [*Eine umlaufende Maschine oder umlaufender Maschinensatz zur Umwand(e)lung elektrischer Energie in solche anderer Stromart, Spannung oder Frequenz*]

converter amplifier → chopper amplifier

converter connection Stromrichterschaltung *f*

converter unit [*IEC 633*] Stromrichtergruppe *f*

converter unit sequence ((open) loop) control Stromrichtergruppenablaufsteuerung *f*

converter unit tap changer ((open) loop) control Stufenschaltersteuerung *f* der Stromrichtergruppe

to convert (to) umrechnen (in) [*Mathematik*]

convexity of fuzzy set Fuzzy-Mengen-Konvexität *f*

convexo-concave [*optics*] erhaben und hohl [*Auf einer Seite erhaben, auf der anderen hohl*]

convexo-convex [*optics*] erhaben auf beiden Seiten, beid(er)seits erhaben, beid(er)seitig erhaben

conveyance, transport(ation) Förderung *f*, Transport *m*

conveying system, conveying plant, conveying in-stallation Förderanlage, Transportanlage *f*, Fördersystem, Transportsystem *n*

conveying tube, conveyer tube Fahrrohr, Förderrohr, Transportrohr *n* [*Rohrpost*]

cooker Kocher *m*

cooker burner Kocherbrenner *m*

cooking area Kochbereich *m* [*selten: n*], Kochfläche *f*

cooking burner → **surface burner**

cooking canopy → **canopy hood**

cooking exhaust hood → **(kitchen) range exhaust hood**

cooking point Kochstelle *f*

cooking unit, cooker Kocher *m*, Kochgerät *n*

cool → **cool(ness)**

cool-air supply duct Kühlzuluftkanal, Zuluftkühlkanal *m*

coolant → **cooling medium**

coolant inlet, CI, cooling agent inlet, cooling medium inlet Kühl(ungs)mitteleinlass, Kühl(ungs)mediumeinlass *m*

coolant outlet, CO, cooling agent outlet, cooling medium outlet Kühl(ungs)mittelauslass, Kühl(ungs)mediumauslass *m*

coolant pump, cooling agent pump, cooling medium pump Kühl(ungs)mittelpumpe *f*

to cool back rückkühlen

to cool (down), to cool off abkühlen, auskühlen, kühlen [*Abführen fühlbarer Wärme*]

cooldown Abkühlung *f*

cooled freshwater gekühlte Süßwasser *n* [*z. B. für Klimaanlagen*]

cooler Kühler *m*

cooler battery Kühlerbatterie *f*

cooler case, cooler housing, cooler casing Kühlergehäuse *n*

cooler ((closed) loop) control, cooler regulation Kühlerreg(e)lung *f*

cooler fan Kühlerventilator *m*

cooler pressure drop Kühlerdruckabfall *m*

cooler surface Kühler(ober)fläche *f*

cooler surface temperature Kühler(ober)flächentemperatur *f*

cooler tube Kühlerröhre *f*

cooler valve Kühlerventil *n*

cooler vessel, cooling vessel Kühl(ungs)gefäß *n*

cooler work, cooling work Kühl(er)arbeit *f*

cooling [*Process of lowering the temperature of a substance to a specific temperature above freezing*] Kühlung *f*

cooling ag(e)ing → **ag(e)ing by cooling**

cooling agent → **cooling medium**

cooling agent pump → **coolant pump**

cooling air [*1. Cooling air used to lower the temperature of a space or products stored in a space. 2. Atmospheric air used to remove heat from a device, space, or system*] Kühl(ungs)luft *f*

cooling air-conditioning unit [*Specific air-treating combination consisting of means for ventilation, air circulation, air cleaning, and heat transfer, with control means for cooling*] Kühl(ungs)klimaeinheit *f*

cooling air curtain Kühl(ungs)luftschleier, Kühl(ungs)luftvorhang *m*

cooling air duct Kühl(ungs)(luft)kanal *m*

cooling air passage Kühl(ungs)(luft)weg *m*

cooling and heating heat pump [*System designed to utilise alternately or simultaneously the heat extracted at a low temperature and the heat rejected at a higher temperature for cooling and heating functions, respectively*] Kühl(ungs)-Heizungs-Wärmepumpe *f*

cooling appliance → **cooling device**

cooling area Kühl(ungs)fläche *f*

cooling bath [*Cold liquid in a tank for immersion cooling*] Kühl(ungs)bad *n*

cooling brine Kühl(ungs)sole *f*

cooling by makeup air, cooling by external air, cooling by exterior air, cooling by outside air freie Kühlung *f* [*Direkte oder indirekte Kühlung mittels Außenluft, deren Temperatur unterhalb der Raumlufttemperatur liegt*]

cooling capacity → **cooling output**

cooling ceiling Kühl(ungs)decke *f* [*Thermisch aktive Decke zur Raumkühlung*]

cooling ceiling cavity Kühl(ungs)deckenhohlraum *m*

cooling ceiling valve Kühl(ungs)deckenventil *n*

cooling chain Kühl(ungs)kette *f*

cooling circulation Kühl(ungs)kreislauf *m*

C

C

cooling climate Kühl(ungs)klima *n*

cooling coil → cooling (pipe) coil

cooling controller, cooling CTRL Kühl(ungs)regler *m*

cooling convector Kühl(ungs)konvektor *m*

cooling curve Kühl(ungs)kurve *f*

cooling cycle Kühl(ungs)kreislauf *m*

cooling degree-days, CDD [*Sum of the degree-days for cooling, using a common base of 75°F (24°C in Canada), is used with other factors to evaluate the energy requirements of a cooling season*] Kühl(ungs)gradtage *f*

cooling degree hour Kühl(ungs)gradstunde *f*

cooling device, cooling appliance Kühl(ungs)gerät *n*, Kühl(ungs)vorrichtung *f*

cooling drawer Kühl(ungs)schublade *f*

cooling drum → drum cooler

cooling duct Kühl(ungs)kanal *m*

cooling duty → cooling output

cooling efficiency Kühl(ungs)wirkungsgrad *m*

cooling element Kühl(ungs)element *n*

cooling façade Kühl(ungs)fassade *f*

cooling fin, cooling rib Kühl(ungs)rippe *f*

cooling fluid Kühl(ungs)strömungsmedium *n*

cooling heat → heat of cooling

cooling heat pump Kühl(ungs)wärmepumpe *f*

cooling installation → cooling plant

cooling jacket, CJ Kühl(ungs)mantel *m*

cooling limit(ing) temperature Kühl(ungs)grenztemperatur *f*

cooling liquid → liquid coolant

cooling load [*The total amount of heat energy that must be removed from a system by a cooling mechan-ism in a unit time, equal to the rate at which heat is generated by people, machinery, and processes, plus the net flow of heat into the system not associated with the cooling machinery*] Kühl(ungs)last *f*

cooling load calculation Kühl(ungs)lastberechnung *f*

cooling load peak Kühl(ungs)lastspitze *f*

cooling loss Kühl(ungs)verlust *m*

cooling medium, coolant, cooling agent [*Substance, used with or without a change of state, to lower the temperature of other bodies or substances*] Kühl(ungs)mittel, Kühl(ungs)medium *n*, Kühl(ungs)träger *m*

cooling mode Kühl(ungs)betrieb *m*

cooling-off, cooling-down [*Temperature reduction from some given higher value to a required setpoint*] Abkühlen *n*, Abkühlung *f*

cooling-off curve, cooling-down curve Abkühl(ungs)kurve *f*

cooling-off drying, cooling-down drying Abkühl(ungs)trocknung *f*

cooling-off heat, cooling-down heat Abkühl(ungs)wärme *f*

cooling-off load, cooling-down load Abkühlungslast, Abkühllast *f*

cooling-off loss, cooling-down loss Abkühl(ungs)verlust *m*

cooling-off period, cooling-down period [*The time taken for a measured temperature to reduce from some given higher value to a required setpoint*] Abkühl(ungs)zeitraum *m*

cooling-only unit Nurkühl(ungs)einheit *f*

cooling operation Kühl(ungs)betrieb *m*

cooling output, cooling capacity, cooling duty, cooling performance [*in Bthu*] Kühl(ungs)leistung *f* (in kcal/h) [*Von einem Kälteerzeuger oder Wärme(aus)tauscher abgeführte Wärme*]

cooling panel Kühl(ungs)tafel *f*

cooling performance → cooling output

cooling pipe Kühl(ungs)rohr *n*

cooling (pipe) coil Kühl(ungs)(rohr)schlange *f*

cooling plant, cooling installation Kühl(ungs)anlage *f*

cooling pond Kühl(ungs)teich *m*

cooling process Kühl(ungs)vorgang *m*

cooling rate [*The decrease in temperature in unit time*] Kühl(ungs)geschwindigkeit *f*

cooling recovery Kühl(ungs)rückgewinnung *f*

cooling rib → cooling fin

cooling rib(bed) pipe → rib(bed) cooling pipe

cooling section Kühl(ungs)strecke *f*

cooling surface [*Surface that is lower in temperature than the fluid being cooled by contact with it*] Kühl(ungs)(ober)fläche *f*

cooling surface temperature Kühl(ungs)(ober)flächentemperatur *f*

cooling tonnage Kühl(ungs)tonnage *f*

cooling tower [*Heat-transfer device, often tower-like, in which atmospheric air cools warm water, generally by direct contact (evaporation)*] Kühl(ungs)turm *m*

cooling tower (air) fan Kühl(ungs)turmventilator *m*

cooling tower basin Kühl(ungs)turmbecken *n*

cooling tower feed(ing) Kühl(ungs)turm(ein)speisung *f*

cooling tower packing, cooling tower fill(ing) [*The part of a crossflow, counterflow, or natural-draft cool-ing tower consisting of splash bars, vertical sheets of various configurations, or honeycomb assemblies, tile, or other materials which are placed within the cooling tower to effect heat and mass transfer be-tween the circulation water and the air flowing through the tower*] Kühl(ungs)turmpackung, Kühlungsturmfüllung, Kühlturmfüllung *f*

cooling tower pump Kühl(ungs)urmpumpe *f*

cooling tower starter Kühl(ungs)turmanlasser *m*

cooling tower water Kühl(ungs)turmwasser *n*

cooling unit Kühl(ungs)einheit *f*

cooling valve Kühl(ungs)ventil *n*

cooling vessel → cooler vessel

cooling ware Kühl(ungs)gut *n*

cooling washer Nassluftkühler *m*, Kühlwäscher(r) *f*, *(m)* [*Es wird gekühltes Wasser benutzt*]

cooling water Kühl(ungs)wasser *n*

cooling with air Kühlung *f* mit Luft

cooling with air and water, air and water cooling Kühlung *f* mit Luft und Wasser, Luft/Wasser-Kühlung, Wasser/Luft-Kühlung, Kühlung mit Wasser und Luft

cooling with water Kühlung *f* mit Wasser, Wasserkühlung

cool(ness) Kühle *f*

to cool off → to cool (down)

cool space air, cool room air Kühl(ungs)raumluft *f*

cool storage [*Techniques or systems used to store cooling capacity for later use*] Kühl(ungs)speicherung *f*

Coons [*proper name*] mathematische Vorschrift *f* zum Festlegen von Freiformkurven und -flächen

coordinated design formvollendete Gestaltung *f*

coordinated universal time → UTC

coordinates presetting → presetting of coordinates

coordinate storage, matrix storage [*A computer storage in which coordinates are used to address the locations or circuit elements*] Koordinatenspeicherung *f*

coordinating control Leitsteuerung *f* [*DIN 19237*]

coordinating level Leitebene *f*

coordinating spark gap abgestimmte Schutzfunkenstrecke, Pegelfunkenstrecke *f*

coordination flag, coordination marker Koordinierungsmerker *m*

coordination meeting Abstimmungsgespräch, Koordinationsgespräch *n*

co-phasal gleichphasig, phasengleich, phasenrichtig, in gleicher Phasenfolge, in richtiger Phasenfolge

copper-bearing steel kupferhaltige Stahl *m*

copper installation pipe Kupferinstallationsrohr *n* [*DIN 1786*]

copper jacket → copper sheath(ing)

copper (lightning) conductor Kupfer(blitz)ableiter *m*

copper manifold, copper central distributor Kupferverteiler(rohr), Kupferverteilungsrohr *m*, *(n)*, Kupferverteilerstück, Kupferverteilungsstück *n*, Kupferverteilerbatterie, Kupferverteilungsbatterie *f*

copper pipe ceiling heating, copper tube ceiling heating Kupferrohrdeckenheizung *f*

to copper (plate) kupferplattieren, verkupfern

copper plated, coppered kupferplattiert, verkupfert

copper plating, coppering Kupferplattierung, Verkupferung *f*

copper sheath(ing), copper jacket Kupfermantel *m*

copper sheet Kupferblech *n*

copper solder fitting Kupferlötfitting *m* [*Siehe Anmerkung unter „Fitting"*]

copper solvency, cuprosolvency Kupferlöslichkeit *f*

copper tube Kupferröhre *f*

copper tube ceiling heating Kupferröhrendeckenheizung, Deckenkupferröhrenheizung *f*

copper tube radiation heating Kupferröhrenstrahlungsheizung *f*

copper water tube Kupferwasserröhre, Kupfersiederöhre, Kupferverdampfungsröhre *f*

to copy kopieren [*An einem Platz lesen und an einem anderen Platz schreiben, wobei die Vorsprungsdaten an ihrem Platz erhalten bleiben und die Zieldaten mit ihnen übereinstimmen*]

copying paper Kopierpapier, Lichtpauspapier *n*

copy processor, printer, writer Drucker *m*

copy protection Kopierschutz, Lichtpausschutz *m*

cordless schnurlos, ohne Schnur

cordless digital telephony → digital cordless telephony

cord(-operated) switch → suspended switch

core → principal memory

core balance transformer Summenstromwandler, Stromsummenwandler *m*

core image library [*A collection of computer program(me)s residing on mass-storage device in ready-to-run form*] Speicherbildbibliothek *f*

C

core magnet Kernmagnet *m*

core memory → (magnetic) core memory

core network Kernnetz *n*

core of membership function, membership function core Kern *m*, unscharfe Mengentoleranz *f*, Toleranz einer unscharfen Menge

core pin Kernstift *m*

core screen Außenleitschicht *f* [*Kabel*]

core storage → (magnetic) core memory

corkboard thermal insulation, corkboard heat in-sulation, corkboard insulation of heat [*Cork granules, cleaned, compressed, and bonded by heat and used for thermal insulation*] Korksteinwärmedämmung *f*

cork slab insulation Korkplattendämmung *f*

cork tile flooring, cork tile floor (finish), cork tile floor cover(ing) Korkfliesen(fuß)boden *m*

corner angular frequency, angular corner frequency Eckkreisfrequenz, Kreiseckfrequenz *f*

corner base cabinet, base corner cabinet Eckunterschrank, Untereckschrank *m*

corner (bath)tub, corner bathing tub Eck(bade)-wanne *f*

corner burner Eckbrenner *m*

corner cabinet Eckschrank *m*

corner ((closed) loop) control valve Eckregulier(ungs)ventil *n*

corner fixing, corner fastening Eckbefestigung *f*

corner float valve Eckschwimm(er)ventil *n*

corner floor(-mounted) (bath(room)) cabinet, floor(-mounted) corner (bath(room)) cabinet Eck-(Bad(ezimmer))Hochschrank, Bad(ezimmer)-Eckhochschrank *m*

corner frequency, break frequency, knee frequency Eckfrequenz *f* [*Bodediagramm*]

cornering Abrundungsradius, Abrundungshalbmesser *m* [*Werkzeugbewegung bei unstetiger Änderung der Tangentenrichtung*]

corner lavatory, corner washbasin Eckwaschbecken *n*

corner mounting, corner installation Eckeinbau *m*, Eckinstallation, Eckmontage *f*

corner odd section, odd corner section Eckpassstück *n*

corner protection rail Eckschutzschiene *f*

corner reflector antenna (US); corner reflector aerial [*It consists of a primary radiating element and a dihedral corner reflector formed by the elements of the reflector*] Winkelreflektorantenne, Winkelspiegelantenne *f*

corner return valve → return corner valve

corner safety valve, safety corner valve Ecksicherheitsventil, Sicherheitseckventil *n*

corner soap holder Eckenseifenhalter *m*

corner stud → principal vertical

corner valve Eckventil *n*

corner vanity Eckwaschtisch *m*

corner vanity slab, vanity corner slab Eckwaschtischplatte, Waschtischeckplatte *f*

corner vertical → principal vertical

corner washbasin → corner lavatory

cornice (roof) gutter, cornice rainwater gutter (Ge)Sims(dach)rinne, (Ge)Simsregenrinne *f*

corporate network Netzverbund *m*

corrected value Korrekturwert, berichtigte Wert *m*

correction, adjustment, compensation Nachführung *f* [*Spannung; Frequenz*]

correctional facility Erziehungsanstalt *f*

correction of jump address, jump address correction [*See remark under "jump address"*] Sprungzielkorrektur, Sprungadressekorrektur *f*

correction time → settling time

corrective maintenance Instandsetzung *f*

corrective variable → secondary controlled variable

corrective variable feedforwarding → secondary controlled variable feedforwarding

correlation function Korrelationsfunktion, Zuordnungsfunktion *f*

correlator Zuordner *m*

corridor (air) duct Korridor(luft)kanal, Flur(luft)-kanal *m*

corridor kitchen [*It is satisfactory where doors are necessary at the end of the kitchen. This arrangement frequently has the advantage of the parallel walls being closer together than in the typical "U", but the disadvantage of a greater distance along the corridor*] zweiteilige Küche *f*

corridor sprinkler Korridorsprinkler, Flursprinkler, Korridorsprenger, Flursprenger *m*

corrosion check(ing) Korrosionskontrolle *f*

corrosion damage Korrosionsschaden *m*

corrosion hazard Korrosionsgefährdung *f*

corrosion sensitivity, corrosion sensitiveness, corrosion susceptibility Korrosionsanfälligkeit, Korrosionsempfindlichkeit *f*

corrosive vapour proof (Brit.); corrosive vapor proof (US) schwadenfest, schwadensicher

corrugated gewellt, wellig

corrugated pipe Wellrohr *n*

corrugated steel sheet Wellblech *n*

corrugated tube Wellröhre *f*

corrugation Welle *f*

corrugations Wellung *f*, Wellen *f*

corrugation top Wellenberg *m*

corrugation valley Wellental *n*

to corrupt verstümmeln [*Information*]

to corrupt (ver)fälschen [*Daten*]

COS → **class of service**

COS → **change of state**

COS defuzzification → **centre of sum defuzzification**

cost effective kostenwirksam

cost effectiveness Kostenwirksamkeit *f*

cost(s) control Kostensteuerung *f*

cost(s) safety Kostensicherheit *f*

COT character → **class of traffic character**

COTS, connection orien(ta)ted transport(ation) service verbindungsorientierte Datentransport *m* [*Gegensatz: paketorientierter Datentransport*]

Coulomb [*Quality of electricity that passes any point in an electric circuit in 1 second when the current is maintained constant at 1 A*] Coulomb C *n*

Coulomb attraction → **electrostatic attraction**

count Zählimpulseingang *m* [*Eingang an einem elektronischen Baustein, an dem Zählimpulse angelegt werden*]

to count down, to count inversely rückwärtszählen, abwärtszählen

countdown, counting down, inverse counting Abwärtszählfolge, Rückwärtszählfolge *f*

counted measurand Zählwert *m* [*Datenübertragung*]

counter [*A device used in counting; specific uses include: Engineering. A device that registers and totals the number of operations performed by a machine. Electrical engineering.*
1. A circuit in which an oscillator of known frequency increments a numerical output at regular intervals, so that the time be-tween two events can be determined by indicating the number of counts that have occurred.
2. A digital circuit, composed of a series of bistable devices such as flip-flops, that keeps track of the number of times a specific event occurs by switching from one state to the next each time a pulse is received.
3. A device that can change from one to the next of a sequence of distinguishable states after receiving a discrete input signal. Computer engineering.
1. A register whose contents go through a specified sequence of states, frequently representing consecutive binary or decimal integers. 2. An incrementing or decrementing register that is used to control repeated sets of instructions, such as a loop, or to represent a number of occurrences] Zähler *m*

counter Zählglied *n* [*Ein Schaltwerk, in dem eine Zahl gespeichert ist, zu der abhängig von einer Schaltgröße eine konstante Zahl, die Zähleinheit, addiert wird. E DIN 19226, Teil 5*]

counter → **scaler**

counter advance Zählerfortschaltung *f*

counterbalance → **counterweight**

counter-balanced chain versenkbare Kette *f* [*Waschtisch*]

counter cell Gegenzelle *f* [*Batterie*]

counter check(ing) Gegenkontrolle *f*, Nachkontrolle

counter circuit → **counting circuit**

counterclockwise, CCW linksdrehend, linksgängig, entgegen dem Uhrzeigersinn, im Gegenuhrzeigersinn

counter content Zählerstand *m*

counter coupling [*In automatic data processing, the technique of combining two or more counters into one counter of larger capacity in electromechanical devices by means of control panel wiring*] Zählerkopp(e)lung *f*

counter cross-flow Kreuzgegenstrom, Gegenkreuzstrom, Kreuzrückstrom, Rückkreuzstrom *m*

counter current → **counterflow**

counter (down) lighting, counter (down) illumination Verkaufstischbeleuchtung, Ladentischbeleuchtung *f*

counter excitation Gegenerregung *f*

counter flap hinge Nussband *n*

counterflow, contraflow, countercurrent [*In heat exchange between two fluids, the opposite direction of flow; i.e., the coldest portion of one fluid meeting the coldest portion of the other*] Gegenstrom, Rückstrom *m*

counterflow condenser, contraflow condenser, counter current condenser [*See remark under "condenser". Water-cooled condenser in which water and refrigerant flow in parallel paths but in opposite directions*] Gegenstromkondensator, Gegenstromverflüssiger, Rückstromkondensator, Rückstromverflüssiger *m*

counterflow (cooling) tower, contraflow (cooling) tower, counter current (cooling) tower [*One in which air, drawn in through air inlets at the tower perimeter (induced draft), or forced in (forced draft) at the base by the fan, flows upward through the fill material in a direction opposite to the falling hot water*] Gegenstromkühl(ungs)turm, Rückstromkühl(ungs)turm *m*

C

counterflow (heat) exchanger, counterflow interchanger, counter current (heat) exchanger, counter current interchanger, contraflow (heat) exchanger, contraflow interchanger Gegenstromwärme(aus)tauscher [*In der Industrie so genannt*], Gegenstromwärmeübertrager *m* [*Siehe Anmerkung unter „Wärme(aus)tauscher"*]

counter input, CI Zähl(er)eingabe *f*

counter light Ladentischlicht, Verkauf(s)tischlicht *n*

counter lighting, counter illumination Ladentischbeleuchtung, Verkaufstischbeleuchtung *f*

counter overflow Zählerüberlauf *m*

counter piece → mating piece

counterplate door release → door release counterplate

counterpoise → counterweight

counter reading, counter read operation Zählerablesung *f*, Zählerablesen *n*

counter-rotating gegendrehend

counter-rotating swirl Gegendreh(ungs)wirbel *m*

counter rotation Gegendrehung *f*

counterweight, counterbalance, counterpoise [*At weight that is designed to counterbalance the original load carried by an elevator, hoist, or similar mechanism, so that the engine must only drive against the additional, unbalanced load*] Gegengewicht *n*

counterweight guide rail Gegengewichtsführungsschiene *f*

counterweight hoistway; counterweight elevator shaft (US); counterweight lift shaft (Brit.) [*"Counterbalance" and "counterpoise" may synonymously be used for "counterweight"*] Gegengewichtfahrschacht, Gegengewichtaufzugschacht *m*

counterweight spring buffer, counterbalance spring buffer, counterpoise spring buffer [*Buffer that stores in a spring the kinetic energy of a descending counterweight*] Gegengewichtsfederpuffer *m*

counting Zählen *n*, Zählung *f* [*Das Ermitteln der An-zahl gleichartiger Elemente oder Ereignisse. DIN 19222. Nicht integrierend*]

counting accuracy Zählgenauigkeit *f*

counting changeover device → register changeover device

counting circuit, counter circuit [*A circuit that counts pulses by frequency-dividing techniques, by changing a capacitator in such a way as to produce a voltage proportional to the pulse count, or by other means*] Impulszählkreis *m*, Impulszählschaltung *f*

counting down → countdown

counting element Zählglied, Zählelement *n* [*Ein Schaltwerk, in dem eine Zahl gespeichert ist, zu der abhängig von einer Schaltgröße eine konstante Zahl, die Zähleinheit, addiert wird. E DIN 19226, Teil 5*]

counting input Zähleingang *m*

counting mechanism → register

counting mechanism constant → register constant

counting module → register module

counting range Zählbereich *m* [*selten: n*]

counting rate Zählrate *f* [*Die Anzahl der Impulse je Zeiteinheit, die im Dauerbetrieb vom Prozessrechensystem gezählt wird. DIN 66201, Teil 1*]

counting rate meter [*Nucleonics. An instrument that indicates the time rate of occurrence of input pulses to a radiation counter*] Impulsmesser *m*

counting up Vorwärtszählen, Aufwärtszählen *n*

to count inversely → to count down

to count up vorwärtszählen, aufwärtszählen

count value Zählwert *m*

coupling Kopp(e)lung *f* [*Verbindung von zwei Stromkreisen oder deren Teilen zur Energie- oder Impulsübertragung von einem Stromkreis zum anderen. Beispiel: Induktive oder kapazitive Kopp(e)lung bei kontaktlosen Ausweiskarten mit integrierendem Mikroprozessor*]

coupling module, coupling block Kopp(e)lungsmodul *n*, Kopp(e)lungsbaustein *m* [*Hardwareerzeugnis, das den Ausgang einer Unterstation oder eines Reglers mit dem (den) Peripheriegerät(en) verbindet. Es dient als elektromagnetischer Puffer/Filter und/oder ändert die Signalart, z. B. DC 0-10 V auf 4-20 mA*]

coupling piece Verschraubung *f*

coupling relay, crosspoint relay, interposing relay [*Coupling module for binary output*] Koppelrelais *n*

coupling torque Kupp(e)lungsdrehmoment *n*

course (US); shell ring, shell belt Kesselschuss, Mantelschuss *m*

courtesy mit freundlicher Genehmigung

COV → change of value

cove base → congé

cover Deckel *m*

cover → cover(ing)

coverage Flächen(ab)deckung *f*

covered (ab)gedeckt, bedeckt

covered (sectional) radiator [*See remark under "radiator"*] verkleidete Radiator, verkleidete Gliederheiz(ungs)körper *m*

covered (slide) valve Rahmenschieber, Dachschieber *m*

cover hinge Deckelscharnier *n*

cover(ing) Abdeckung *f* [*als Schutzmaßnahme*]

cover(ing) frame (Ab)Deckrahmen *m*

cover(ing) sheet (Ab)Deckblatt *n* [*Karte*]

cover(ing) sheet (Ab)Deckblech *n*

cover(ing) strip (Ab)Deckstreifen *m*

cover plate → bottom plate

cover plate Rosette *f*

to cover the fire → to bank the fire

cowled roof extractor, cowled roof exhaust(ing) unit, cowled roof extract(ion) unit, roof extract(ion) hood Dach(entlüftungs)haube *f*

cp → candlepower

CP → clock pulse

CPBX, computerized private branch exchange rechnergesteuerte private Nebenstellenanlage *f*

CPFSK, continous phase frequency shift keying kontinuierliche Phasenfrequenzumtastung *f*

CPG → clock generator

CPI → computer PABX interface

CPI → characters per inch

CP/M, Control Language for Microcomputer Betriebssystem *n* [*Es hat durch allgemeingültige Hardware-Definition erstmals die Software-Austauschbarkeit zwischen Geräten unterschiedlicher Hersteller geschafft*]

cps → characters per second

CPU → central processing unit

CR → pressure ratio

crackle, crackling Knistern, Geknister *n*, Knistergeräusch *n*

crack (method of) ventilation [*Leakage through the cracks around doors, windows and other openings*] Spalt(en)lüftung, Risslüftung *f*

cradle machine → (window) cradle machine

craft, trade Handwerk *n*

craftsman, tradesman Handwerker *m*

crane [*Scottish term*] **→ (water) tap**

crankcase heating Kurbelgehäuseheizung *f*

cranking Kurbelanlassen, Anwerfen *n*

crash Aufprall, Zusammenstoß *m*

crash [*Incapacity of a computer system brought about by a hardware or software fault*] Zusammenbruch, Absturz *m*, Arbeitsunfähigkeit *f*

CRC, cyclic redundancy check(ing) zyklische Blockprüfung *f* [*Ein Fehler-Prüf-Algorithmus in einem Da-tenpaket vor Übertragung. Der Empfänger prüft den CRC-Wert eines jeden empfangenen Paketes und entfernt die Prüffunktion vor der Freigabe des Paketes an die Empfangsstation. Bei fehlerhaftem CRC stehen zwei Möglichkeiten zur Verfügung: Entweder das Paket zurückweisen oder das beschädigte Paket mit Status des CRC-Fehlers zustellen.*]

CRCC → cycling redundancy check(ing) code

CR character → carriage return character

cread Empfängereinschalten *n* [*Eine Terminaleigenschaft bei der Datenfernübertragung*]

—cread, minus cread Empfängerabschalten *n* [*Eine Terminaleigenschaft bei der Datenfernübertragung*]

credit card [*Not to be confused with "badge"*] Kreditkarte *f*

credit card verifier Kreditkartenprüfer *m*

creep(ing), creepage [*A slow plastic deformation or movement of a material under stress*] Kriechen *n*

creep(ing) action, creepage action [*Slow make/break action of a switch mechanism in a controller, as differentiated from snap action or toggle action*] Kriechen *n*

creep(ing) distance, creepage distance Kriech(weg)länge *f*

creep(ing) resistance, creepage resistance Kriechstromfestigkeit *f*

creep (speed) Kriechgang *m* [*Niedrige Vorschubgeschwindigkeit, die in einem bestimmten Abstand vor Erreichen des Ortssollwertes eingestellt wird, um eine Position kleiner Einfahrtoleranz zu erreichen und da-bei möglichst einen Überlauf zu vermeiden*]

creep stop. anticreep tongue Haltezunge *f* [*Elektrizitätszähler*]

cremorne bolt → espagnolette (bolt)

crest factor Spitzenwertfaktor, Scheitelfaktor *m*

crime scene Tatort *m*

crimp(led) splice Sickenspleiße *m*

crisp [*fuzzy logic*] scharf

crispening Differenzierentzerrung, Versteilerung *f*, Versteilern *n* [*Impulsflanken*]

crisp input [*fuzzy logic*] scharfe Eingang(s)größe *f*

crisp logic scharfe Logik, Schaltalgebra *f*

C

C

crisp output [*fuzzy logic*] scharfe Ausgang(sgröße) *m, (f)*

crisp relation [*fuzzy logic*] scharfe Beziehung *f*

crisp set [*fuzzy logic*] scharfe Menge *f*

crisp value [*fuzzy logic*] scharfe Wert *m*

criss-cross arrangement Scherenanordnung *f* [*Rolltreppe; geneigte Fahrsteig*]

criterium Unterscheidungsmerkmal *n*

critical damped system, damped critical system Pt $_2$-System n mit $P = 1$ [*aperiodische Grenzfall*]

critical pressure [*For a specific fluid, the vapo(u)r pressure at which the liquid and vapo(u)r have identical properties*] kritische Druck *m*

critical refrigerant charge [*Compromise refrigerant quantity required by a system to maximize performance where a capillary or fixed restriction expansion device is used*] kritische Kältemittelfüllung *f*

critical velocity, critical speed kritische Geschwindigkeit *f*

CRLF → carriage return line feed

CRN, cellular radio network Mobilfunknetz *n*

CRO → (cathode-ray) oscilloscope

croft, bleaching croft Bleichplatz, Rasenbleichplatz *m*, Rasenbleiche *f*

CROM → control and read-only memory

crook → (door) crook

crooked shaft doorknob → doorknob with crooked shaft

cropping [*graphics*] Beschneiden *n*

cross nicht resident [*Auf einem anderen Rechner laufend*]

cross → cross (piece)

cross assembler Assemblierprogramm *n* [*Dieses Programm läuft auf Rechner 1 und übersetzt ein in der Assemblersprache des Rechners 2 geschriebenes Programm in die Maschinensprache des Rechners 2*]

cross bar Kreuzschiene *f*

cross bar system Kreuzschienensystem *n* [*Telekommunikation*]

crossbolt lock → lock with a cross bolt

cross bracing Diagonalversteifung *f*

cross-colour (Brit.); cross-color (US) Beeinflussung *f* zwischen Luminanz- und Chromainformationen [*Fernsehen; Video*]

cross connection [*In plumbing, an erroneous connection whereby a water supply pipe is inadvertently connected to a waste line causing contaminated water to mix with the supply water*] Falsch-Wasseranschluss, Wasser-Falschanschluss *m*

cross connection Kreuzschaltung *f*

cross counter-current heat exchanger, cross contra-current heat exchanger Kreuzgegenstromwärmeübertrager *m* [*Siehe Anmerkung unter „Wärme(aus)tauscher"*]

crosscurrent fan Querstromventilator *m*

crosscurrent heat exchanger, crosscurrent heat exchange unit, crosscurrent heat interchanger [*See German remark under "heat exchanger"*] Querstromwärme(aus)tauscher *m* (im weiteren Sinne)

cross-cutting chisel, bolt chisel Kreuzmeißel, Schlossmeißel *m*

cross detection Zweimelderabhängigkeit, Zweidetektorabhängigkeit, Zweimeldegerätabhängigkeit *f*

cross dimmer Überblendsteller *m*

crossed coil Kreuzspule *f*

cross feed(ing) Quervorschub *m*, Quervorschieben *n*

cross handle Kreuzgriff *m*, Handkreuz *n*, Griffkreuz

crosshatch (pattern) Schraffur *f*

cross-interleaved Reed-Solomon-Code Kreuzverschachtelung *f* von Digitalsignalen bei der digitalen Tonaufzeichnung

cross lighting, cross illumination [*Lighting an object from two opposite sides*] Kreuzbeleuchtung *f*

cross-linking [*organic chemistry*] Vernetzung *f*

cross-linking agent Vernetzungsmittel *n*

cross modulation gegenseitige Frequenzmodulation *f*

crossover Durchtritt *m*

crossover → frequency dividing network

crossover frequency Übergangsfrequenz *f* [*zwischen zwei Frequenzbereichen*]

crossover network, (loudspeaker) dividing network [*Electrical filter that separates the output signal from an amplifier into two or more separate frequency bands for a multispeaker system*] Kreuzungsnetz, Überkreuzungsnetz *n*

crosspiece Geschränk *n*

cross(piece) [*Pipe fitting with four branches in the same plane with right angles between them*] Kreuz(stück) *n*

crosspoint relay → coupling relay

cross reference Querverweis *m*

cross reference list Querverweisliste *f* [*Zusammenstellung, geordnet nach Operanden, mit Angaben, an welchen Programmstellen (Speicheradressen, Bausteinen) ein Operand verwendet wird, ggf. ergänzt durch Kommentar. VDI 2880 Blatt 3*]

cross-section(al) narrowing Querschnitt(s)verengung *f*

cross-section(al) variation, cross-section(al) change Querschnitt(s)änderung *f*, Querschnitt(s)wechsel *m*

cross tailed hinge [*Scotland*] → **double garnet**

to crosstalk übersprechen

crosstalk coupling, crosstalk loss Übersprechverlust *m*

crosstalk(ing) Übersprechen *n*

cross-vent(ilat)ed, transverse-vent(ilat)ed quergelüftet

cross ventilation, transverse ventilation, cross vent(ilat)ing, transverse vent(ilat)ing [*Type of ventilation with air supply and exhaust points at opposite sides of a ventilated space*] Querlüftung *f*

cross wall unit Querwandelement *n*

cross window ventilation, transverse window ventilation, transverse window vent(ilat)ing, cross window vent(ilat)ing Fensterquerlüftung, Quer-(fenster)lüftung *f* [*Fenster auf gegenüberliegenden Seiten*]

cross-wise air distribution [*Air is introduced at one side of a room and removed at the other*] Querluftverteilung *f*

cross zone alarm Zweigruppenalarm *m*

cross zoning Zweigruppen(alarm)abhängigkeit *f* [*Maßnahme zur Überprüfung der Richtigkeit von Brandalarmzuständen. Der Brandalarm wird erst nach Ansprechen je eines Melders aus zwei einander zugeordneten Meldergruppen eingeleitet. Mit Ansprechen des ersten Melders kann sowohl ein interner Alarm als auch eine Steuerfunktion eingeleitet werden. DIN VDE 0833-2*]

crowding Materialverdrängung, Werkstoffverdrängung *f* [*Stanzen*]

crow-foot earth electrode Strahlenerder *m*

crown → **soffit**

crown handle Kronengriff *m*

CRT → **cathode-ray tube**

CRTC → **cathode-ray tube controller**

CRT unit, CRT display (US); usual display unit, UDU, video terminal Bildschirmgerät *n*

crude sewage, raw sewage, untreated sewage; crude wastewater, raw wastewater, untreated wastewater (US) Rohabwasser *n* [*Siehe Anmerkung unter „Abwasser"*]

crust Schale *f*, Kruste *f*

crust formation → **incrustation**

cryo(elec)tronics [*It studies the design and functioning of electronic systems, circuits and devices at temperatures approaching absolute zero (0 K or –270°C), particularly as applied to superconductivity*] Kryoelektronik, Kälteelektronik *f*

cryogenic [*1. Of or relating to cryogenics. 2. Relating to or occurring at extremely low temperature*] kryogen

cryogen(ic agent) [*Any substance that is used to induce freezing*] Kryomittel, Kältemittel *n* [*DIN 8960, DIN 8962*]

cryogen(ic agent) collector Kryomittelsammler, Kältemittelsammler *m*

cryogenic bath Kryobad, Kältebad *n*

cryogenic cascade Kältekaskade, Kryokaskade *f*

cryogenic central station, central cryogenic station Kryozentrale, Kältezentrale *f*

cryogenic coil, cryocoil [*A coil cooled to extremely low temperatures so as to reduce its electrical resis-tivity*] Kryospule *f*, Kältespule

cryogenic conductor, cryoconductor, superconductor [*Any material that can exhibit superconductivity*] Supraleiter, supraleitende Stoff *m*

cryogenic cycle, cryocycle Kryokreislauf, Kältekreislauf *m*

cryogenic device, cryogenic appliance [*Any device operating at extremely low temperature*] Kryogerät, Kältegerät *n*

cryogenic district plant, cryogenic district installation, district cryogenic plant, district cryogenic installation Fernkryoanlage, Fernkälteanlage *f*

cryogenic energy Kryoenergie, Kälteenergie *f*

cryogenic engineering, cryoengineering [*It is involved with the design and development of low-temperature systems and components. In such activities the designer must be familiar with the properties of the fluids used to achieve these low temperatures as well as the physical properties of the components used to produce, maintain, and apply such temperatures*] Kryotechnik, Kältetechnik, Tieftemperaturtechnik *f*

cryogenic equipment Kryogerät(e), Kältegerät(e), kältetechnische Gerät(e) *n*, *(f)*

cryogenic film [*Computer. A coating of supercon-duct-ing material that could maintain or store a current indefinitely when kept at a temperature near absolute zero*] Kryofolie, Kältefolie *f*

cryogenic fitter, cryofitter Kryoanlagenmonteur, Kälteanlagenmonteur *m*

C

cryogen(ic fluid), cryofluid [*Any fluid that changes from a liquid to a gas at less than 110 k at standard pressure. These fluids are used in industry and in-clude: hydrogen R 707; helium R 704; nitrogen R 728; oxygen R 732*] Kryofluid, Kältefluid *n*

cryogenic gyro(scope), superconducting gyro(scope) [*It operates under very low temperatures, having a central rotating disk of superconduct-ing niobium which spins while in levitation*] Kryokreisel, Kältekreisel *m*

cryogenic liquid → liquid cryogen

cryogenic machine, cryomachine Kältemaschine, Kryomaschine *f*

cryogenic machine with evaporator and condenser, evaporator-condenser cryogenic machine Wasserkühl(ungs)aggregat *n*, Kältemaschine *f* mit Verdampfer und Kondensator, Kryomaschine mit Verdampfer und Kondensator

cryogenic medium Kälteträger, Kryoträger *m*

cryogenic meter, cryometer Kältezähler, Kryozähler *m*

cryogenic mix(ture) Kältemischung, Kryomischung *f*, Kältegemisch, Kryogemisch *n*

cryogenic period Kälteperiode, Kryoperiode *f*

cryogenic physics, cryophysics [*A branch of physics that is concerned with processes and phenomena at temperatures approaching absolute zero*] Kryophysik, Kältephysik *f*

cryogenic plant, cryogenic system, cryosystem, cryoplant Kryoanlage, Kälteanlage, kältetechnische Anlage, kryotechnische Anlage *f*

cryogenic power, cryopower Kryoleistung, Kälteleistung *f*

cryogenic product, cryoproduct kältetechnische Erzeugnis, kältetechnische Produkt, kryotechnische Produkt, kryotechnische Erzeugnis *n*

cryogenic process engineering, cryoprocess engineering Kryoverfahrenstechnik, Kälteverfahrenstechnik *f*

cryogenic pump, cryopump [*A pump capable of producing a very low vacuum (with pressures of about 10^{-8} mm Hg) through the cooling of surfaces to about 20 K*] Kryopumpe, Kältepumpe *f*

cryogenic production, production of low temperatures Kälteerzeugung *f*

cryogenic recovery, cryorecovery Kälterückgewinnung *f*

cryogenics, cryogenic science, cryoscience [*These terms are normally associated with low termperatures. However, the location on the temperature scale at which refrigeration generally ends and cryogenics begins has never been well defined. Most scientists and engineers working in this field restrict cryogenics to a temperature below 125 K because the normal boiling points of most of the permanent gases (e.g., helium, hydrogen, neon, nitrogen, argon, oxygen, and air) occur below this temperature. In contrast, most of the common refrigerants have boiling points above this temperature*] Kryowissenschaft, Tieftemperaturwissenschaft, Kältewissenschaft *f*

cryogenic storage → cold storage

cryogenic storage structure → cold storage structure

cryogenic storage vessel → cold storage vessel

cryogenic supply → cold supply

cryogenic supply plant → cold supply plant

cryogenic temperature, cryotemperature [*A temperature typical of cryogenic fluids, usually 77 K or lower*] Kryotemperatur, Kältetemperatur *f*

cryogenic transformer [*Electronics, A supercon-duct-ing transformer, especially designed to operate at temperatures approaching absolute zero (0 K or –270°C)*] Kryotransformator *m*

cryogrinding, freeze-grinding [*Grinding at low temperature of a substance which otherwise could not be ground or would be spoiled by the temperature rise resulting from grinding*] Kryomahlen, Káltemahlen *n*, Kryomahlung, Kältemahlung *f*

cryolap → cold lap

cryoloop → cold loop

cryomagnetic [*Relating to the production of extremely low temperatures, through the adiabatic demagetization of paramagnetic salts*] kryomagnetisch, kältemagnetisch

cryometer Kältethermometer *n*

cryopower → cryogenic power

cryoprobe → cold probe

cryorecovery → cryogenic recovery

cryoresistive (electric(al)) power transmission line [*Its conducting cables are cooled to the temperature of liquid nitrogen, 77.4 K or below (-196°C) (sometimes defined as -195°C), thereby causing a reduction of the conductor resistance by a factor of about 10, allowing increased transmission capability*] kryobeständige Übertragungsleitung, kältebeständige Übertragungsleitung *f*

cryosolenoid → cold solenoid

cryostat [*Batch operating apparatus in which a cryogenic liquid or solid evaporates to maintain a cryotemperature, which need not be constant but may vary in a predetermined fashion*] Kryostat *m*

cryotemperature → **cryogenic temperature**

cryosupply → **cold supply**

cryotrap [*Surface cooled below 120 K (-243°F) in order to condense vapo(u)rs*] Kryofalle, Kältefalle *f*

cryotron Kryotron, Tieftemperaturschaltelement *n*

cryotronics, cryoelectronics [*A field of engineering that studies the design and functioning of electronic systems, circuits and devices at temperatures approaching absolute zero (0 K or -270°C); particularly as applied to the phenomenon of superconductivity*] Kryoelektronik, Kälteelektronik *f*

cryowinding Kältewick(e)lung *f*

cryoworking → **cold working**

cryptanalysis[*Steps and operations performed in converting encrypted messages into plain text without previous knowledge of the key employed*] Kryptoanalyse *f*

cryptographic service provider, CSP Programmdatei *f* die ein kryptographisches Verfahren umsetzt

cryptography [*The science of preparing messages in a form which cannot be read by those not privy to the secrets of the form*] Kryptographie, Kryptografie *f*

cryptology [*The science of preparing messages in forms which are intended to be unintelligible to those not privy to the secrets of the form, and of decipher-ing such messages*] Kryptologie *f*

crypton(-filled) lamp Kryptonlampe *f*

crystal-controlled, CC quarzgesteuert

crystallized tin plate Weißblech *n* mit Metallmohr, Metallmohr-Weißblech

crystal mircophone → **piezoelectric microphone**

CS → **channel selector**

CS → **circuit switching**

CS Adressbussignal *n*

CS → **((open) loop) control signal**

CSC → **control signalling code**

CSD → **constant speed drive**

CSDN, circuit switched data net(work) Leitungsvermittlungsdatennetz *n*

CSF → **sequential function chart**

CSG → **constructive solid geometry**

CSI, control sequence introducer Einleitungszeichen *n* für Steuer(ungs)folge [*Kode*]

CSL → **current-sourcing logic**

CSL → **current-sinking logic**

CSM → **charge simulation method**

CSMA, carrier sense multi(ple)-access Vielfachzugriff *m* in lokalen Netzen, Mehrfachzugriff in lokalen Netzen [*Eine technische Leitungsvermittlung, die mehreren Netzstationen den gleichzeitigen Zugriff auf einen Kanal erlaubt*]

CSMA/CC, carrier sense multi(ple)-access collision control Zugriffsverfahren *n* mit Zusammenstoßverhinderung [*Eine technische Leitungsvermittlung, die mehreren Netzstationen die erfolgreiche Teilnahme auf einem Kanal erlaubt und über einen Sensor den Zusammenstoß verhindert. Über eine Erkennungsfunktion werden Kollisionen erkannt; das Paket wird erneut übertragen*]

CSMA/CD, carrier sense multi(ple)-access/collision detection [*It is a method used to control access to shared transmission medium, such as a coaxial cable bus to which a number of stations are connect-ed. A station wishing to transmit a message first senses (listens to) the medium and transmits the mes-sage only if the medium is quiet – no carrier present. Then, as the message is being transmitted, the station monitors the actual signal on the transmission medium and, if this is different from the signal being transmitted, a collision is said to have occurred and been detected. The station then ceases transmission and retries again later*] Ethernet *n* [*Steuer(ungs)verfahren für Vielfachzugriff mit Aktivitätsüberwachung und Kollisionserkennung in lokalen Netzwerken*]

CSNS → **controller service not requested state**

CSP → **cryptographic service provider**

CSPDN → **circuit switched public data network**

CSRS → **controller service requested state**

CSTA → **computer supported telecommunications application**

CSWS → **controller synchronous wait state**

CTC → **cartridge tape controller**

CTD → **charge transfer device**

CTI → **comparative tracking index**

CTI → **computer (supported) telephony integration**

CTL → **capstan tracking logic**

CTRL → **controller**

CTRL circuit generation → **controller circuit gener-ation**

CTRL output → **controller output**

CTRL output variable → **controller output variable**

CTRS, controller transfer state Übergabezustand *m* der Steuer(ungs)funktion [*programmierbare Messgerät, DIN IEC 625*]

CTS → **clear to send**

cu [*abbreviation*] → **crosstalk unit**

C

CU → **control unit**

cube ice Würfeleis *n*

cubic feet/minute → **CFM**

cubic measure Kubikmaß *n*

cudbear rote Indigo, Persio *n* [*Färberei*]

cue Überblendsignal *n*

CUG, closed user group geschlossene Teilnehmerbetriebsklasse, geschlossene Benutzergruppe, geschlossene Anwendergruppe *f* [*Leistungsmerkmal*]

culinary waste water → **dishwashing waste water**

cull Ausschuss *m*

culver house Taubenschlag *m*, Taubenhaus *n*

cumulative failure frequency distribution Ausfall-summenverteilung *f* [*DIN 40042*]

cup anemometer, cup wind speed meter [*It is composed of three or four vanes with uplike structures on their ends; wind speed is determined by the rate at which the vanes revolve around a central shaft*] Schalenkreuzanemometer *n*, Schalenkreuzwindgeschwindigkeitsmesser *m*

cupboard lock Schrankschloss *n*

cup dispenser Becherspender *m*

cupreous, containing copper kupferhaltig

cupreous, resembling copper kupferähnlich

cuprosolvency → **copper solvency**

cup wind speed meter → **cup anemometer**

current aktuell

current Strom, elektrische Strom *m*

current amplification, current gain [*Increase in signal current magnitude in transmission from one point to another, or the process thereof*] Stromverstärkung *f*

current annunciator [*A signal(l)ing apparatus which operates electromagnetically and serves to indicate visually, or visually and audibly, whether a current is flowing, has flowed, or has changed direction of flow in one or more circuits*] Strommelder *m*

current attenuation [*Decrease in signal current magnitude in transmission from one point to another, or the process thereof*] Stromdämpfung *f*, Stromabschwächung

current-balancing stromausgleichend

current-carrying → **alive**

current(-carrying) capacity [*obsolete*] → **ampacity**

current chopping Stromabriss *m*, Strom(ab)kippen *n*

current coil Stromspule *f*

current collapse → **current failure**

current conductor → **conductor**

current conductor loop → **conductor loop**

current connection Stromanschluss *m*, Stromverbindung *f*

current consumption → **power consumption**

current cut → **current failure**

current directory aktuelle Verzeichnis, zeitgemäße Verzeichnis *n*

current divider [*A device used to deliver a desired fraction of total current to a circuit*] Stromteiler *m*

current draw, current drain [*Taking current from a voltage source by a load*] Stromabnahme, Stromentnahme, Stromaufnahme *f*

current draw on alarm → **alarm current**

current dump → **dump**

current duration → **ON period**

current entry Stromeingang *m*

current failure, current collapse, current cut, current outage Stromausfall *m*

current-fed stromgespeist

current feedback [*Feedback introduced in series with the input circuit of an amplifier*] Stromrückkopp(e)lung *f*

current flow Strombahn *f*, Stromverlauf, Stromfluss *m*

current-flow-on-failure Ruhestromprinzip *n*

current from external source → **stray current**

current gain → **current amplification**

current grading Stromstaffelung *f*

(current) grid (Strom)Netz *n*

current harmonic Stromoberschwingung *f*

current input Stromaufnahme, Leistungsaufnahme, zugeführte Leistung *f*

current inrush (Einschalt)Stromstoß *m*

current intensity, current strength, amperage, amp. [*The amount of electric current in amperes*] Stromstärke *f*

current lead Stromführung *f*, Stromleitung

current level Strompegel *m*

current limitation, current limiting Strombegrenzung *f*

current limiter, current limiting device Strombegrenzer *m*

current limit(ing) acceleration Beschleunigungsausgleich, Hochlauf *m* an der Strombegrenzung

current limit(ing) circuit breaker strombegrenzende Leistungsschalter *m*

current limit(ing) fuse strombegrenzende Sicherung *f*

current limit(ing) resistor Strombegrenzungswiderstand *m*

current limit(ing) threshold Einsatzwert *m* der Strombegrenzung [*DIN 41 745*]

current line → (electric(al)) current line

current location counter laufende Adressenregister *n*

current loop [*peripheral(s)*] Linienstrom *m* [*Peripheriegerät(e)*]

current loop Stromschleife *f*

current loop interface [*peripheral(s)*] Linienstromschnittstelle *f* [*Peripheriegerät(e)*]

current loop receive Empfangsstromschleife *f*

current loop transmit Sendestromschleife *f*

current magnet Stromeisen *n* [*Elektrizitätszähler*]

current meter Stromzähler *m*

current-mode logic, CML Stromflusslogik(schaltung), Logik(schaltung) *f* mit Stromsteuerung

current-mode transistor logic, CMTL stromschaltende Transistorlogik *f*

current-operated protective device → residual current device

current outage → current failure

current output Stromausgang(s)leistung, (abgegebene) Stromleistung, Stromleistungsabgabe, Stromabgabe(leistung) *f*

current overload Stromüberlastung *f*

current passage, passage of current Stromdurchfluss, Stromdurchgang *m*

current path → conducting path

current period → ON period

current pulse Stromimpuls *m*

current pulse transmission Stromimpulsübertragung *f*

(current-)saving lamp (Energie)Sparlampe, Stromsparlampe *f*

current sensor Stromgeber *m*

current-sensor module Stromgeberbaugruppe *f*

current sink Stromsenke, stromziehende Schaltung *f*

current sinking logic, CSL stromziehende Schaltungstechnik, stromziehende Logik *f*, schaltende Logik nach M, Logik nach M schaltend [*Schaltungstechnik, bei der die Eingänge der Funktionsglieder bei positiver Versorgungsspannung und positiver Logik (nach DIN 40 700 Teil 14) auf den Signalwert 0 reagieren. Die Ausgänge ziehen bei diesem Signalwert Strom aus den Eingängen nachfolgender Funktionsglieder und leiten diesen Bezugspotential ab. Stromfluss ist entgegengesetzt zum Signalfluss. Signalwert 1 wird durch Spannung und keinen Stromfluss dargestellt (unbeschalteter Eingang wirkt wie mit Signalwert 1 beaufschlagt). V DIN 19 237*]

current source, current supply device Stromquelle, Stromversorgungsvorrichtung *f*

current sourcing logic, CSL stromliefernde Schaltungstechnik, stromliefernde Logik *f*, schaltende Logik nach P, Logik nach P schaltend [*Schaltungstechnik, bei der die Eingänge der Funktionsglieder bei positiver Versorgungsspannung und positiver Logik (nach DIN 40 700 Teil 14) auf den Signalwert 1 reagieren. Die Ausgänge liefern bei diesem Signalwert Strom in die Eingänge nachfolgender Funktionsglieder. Signalfluss und Stromfluss haben die gleiche Richtung. Signalwert 1 wird durch Spannung und Stromfluss dargestellt (unbeschalteter Eingang wirkt wie mit Signalwert 0 beaufschlagt). V DIN 19 237*]

current supply Stromversorgung, Stromlieferung *f*

current supply laufende Versorgung, laufende Lieferung *f*

current tariff Stromtarif *m*

current time → ON time

current transformation Stromwand(e)lung *f*

current transformer [*It has its primary winding connected in series with the circuit to be measured or controlled; the current is measured across the second-ary winding*] Stromwandler *m*

current type Stromart *f*

current-using apparatus elektrische Verbrauchsvorrichtung *f*, elektrische Verbrauchsmittel *n* [*VDE 0100, Teil 200*]

current weakening, weakening of current Stromschwächung *f*

current weakening technique Stromschwächungstechnik *f*, Stromschwächungsprinzip *n* [*Nach diesem Prinzip arbeiten die meisten Linien für Druckknopffeuermelder, Überfallmelder und Einbruchmelder*]

cursor, mov(e)able visible mark Einfügemarke, Vorlaufzeiger, Korrekturzeiger, Kursor *m*, Bildmarke, Schreibmarke *f*, Lichtzeichen *n*, Positionieranzeige(r) *f*, *(m)*, Schreibzeiger, Feldmarke, Reiter, Stellring, Cursor *m* [*Er wird über Tasten, Rollbügel oder ähnliches an die Stelle auf dem Bildschirm bewegt an der eine Eintragung oder Änderung vorgenommen wird. Meistens ein blinkender Unterstrich oder ein leuchtendes Rechteck*]

curtain rail, curtain rod Vorhangstange, Vorhangschiene, Schleierstange, Schleierschiene *f*

curvature Rundung *f*, Krümmung

curve Kurve *f*

curve [*A transparent plastic or pear-wood shape used by draughtsmen for drawing curves*] Kurvenlineal *n*

curve follower, graph follower Kurvenleser *m* [*Eingabegerät zum Umsetzen von Kurven in Signale. DIN 44 300*]

C

C

curve generator Kurvengenerator, Interpolator, Kurvenerzeuger *m* [*Vorrichtung, die den zu folgenden Weg und die Ganggeschwindigkeit für ein Schneidwerkzeug, einen Maschinenschlitten oder ein Maschinenelement berechnet, wenn dieser Vorrichtung die kodierte mathematische Form der Angaben eingegeben worden ist. Sie stellt über geglättete Kurven die Verbindung zwischen den einzelnen programmierten Punkten her*]

curve memory Kurvenspeicher *m*

curve of comfort → comfort curve

curve set point (value) → sliding set point (value)

curvilinear krummlinig

cushioned start(ing) → soft start(ing)

customer's complaint → notice of defects

customized, tailored, custom-engineered, custom-made, custom-fabricated, tailor-made kundenspezifisch, „maßgeschneidert", nach Kundenwunsch

customized form letter Serienbrief *m*

customizing Wunschanpassung *f* [*Anpassung der (CAD-)Benutzungsoberfläche an individuelle Wünsche*]

to cut schneiden

cut → bevelled

cut, failure, outage Ausfall *m*, Versagen *n*

to cut and paste montieren [*Text*]

cutaway view aufgeschnittene Ansicht *f*

to cut off, to cut away abschneiden, wegschneiden

to cut off → to switch off

cut off → switched off

cutoff → cutout

cutoff bias, cutout bias [*The direct-current bias voltage that must be applied to the grid of an electron tube to stop the flow of anode current*] Gittersperrung *f*

cutoff frequency, cutout frequency [*Frequency that is identified with the transition between a pass band and an adjacent attenuation band of a system of transducers. IEC 50(151)*] Grenzfrequenz, Eckfrequenz *f*

cutoff louvre, cutout louvre Schrägraster *m* [*Leuchte*]

cutoff pressure → cutout pressure

to cut out ausschneiden

cutout, cutting out, switching off, cutoff, switching out, cutting off, disconnection, interruption, deenergizing, break(ing) Abschaltung, Ausschaltung, Unterbrechung, Trennung *f*, Abschalten, Ausschalten, Unterbrechen, Trennen *n*

cutout → break(ing) through

cutout, hole Loch *n*

cutout Sicherungsautomat *m*

cutout Ausschnitt *m*

cutout (device), cut off (device), switch(ing) out device, cutoff switch, switch(ing) off device, cutout switch Abschalter, Ausschalter *m*, Abschaltvorrichtung, Ausschaltvorrichtung *f*

cutout key, cutoff key Unterbrechertaste(r) *f*, *(m)*

cutout point, switch(ing) out point, cutoff point, switch(ing) off point [*Predetermined value (pressure, temperature, etc.) at which operation ceases*] Ausschaltwert, Abschaltwert *m*

cutout pressure, cutoff pressure, switching out pressure, switching off pressure Abschaltdruck, Ausschaltdruck *m*

cutout timer, cutoff timer, switching off timer, switching out timer Abschaltzeitgeber, Ausschaltzeitgeber *m*

cutout torque, cutoff torque, switching out torque, switching off torque Abschaltdrehmoment, Ausschaltdrehmoment *n*

cut surface Schnittfläche *f* [*DIN 6580*]

cutting Schneiden *n*

cut(ting) Zuschnitt *m*

cut(ting) line Schnittverlauf *m* [*zeichnerische Darstellung*]

cutting out → cutout

cut(ting) plan Schnittaufteilung *f* [*EXAPT*]

cut(ting) surface Schnittfläche *f*, Schneid(e)fläche

cut(ting) torch Schneidbrenner *m*

cut(ting) variable Schnittgröße *f* [*DIN 6580*]

C_V-factor (Brit.) → flow factor

CVP → chemical vapour deposition

CVR → constant volume controller

CWL → continuous wave laser

CWW cooling → closed-circuit water-water cooling

to cycle periodisch ablaufen, periodisch durchlaufen, zyklisch ablaufen, zyklisch durchlaufen

to cycle → to clock

cycle [*Sequence of operations or functions*] Zyklus *m*

cycle, circuit [*A system of phases through which a working substance passes in an engine, (refrigerant) compressor, pump, turbine, power plant, or refrigeration system*] Kreislauf *m*

cycle periodische Vorgang *m*, Periode *f*

cycle [*Series of events that have a tendency to repeat*] Kreislauf *m*

cycled → clocked

cycled periodisch abgelaufen, periodisch durchgelaufen, zyklisch abgelaufen, zyklisch durchgelaufen

cycle guard valve Kreislaufüberwachungsventil *n* [*Absorptionskälteanlage*]

cycle index [*1. The number of times a cycle has been carried out by a computer. 2. The difference, or its negative, between the number of executions of a cycle which are desired and the number which have actually been carried out*] Schleifenindex, Programmschleifenindex *m*

cycle index counter [*A device that counts the number of times a given cycle of instructions in a computer program(me) has been carried out*] Schleifenindexzähler, Programmschleifenindexzähler *m*

cycle time, memory cycle Spielzeit, Zykluszeit, Gangzeit, Taktzeit *f* [*Diese Zeit ist bei einer Funktionseinheit durch die Zeitspanne zwischen dem Beginn zweier aufeinanderfolgender gleichartiger zyklisch wiederkehrender Vorgänge bestimmt. DIN 19226 Teil 5*]

to cycle (up) betriebsbereit schalten

cyclic(al) access zyklische Zugriff *m* [*Bei Magnetplatten-, Magnettrommel- und Laufzeitspeichern*]

cyclic(al) binary code zyklische Binärkode, zyklische Binärcode, Graykode, Graycode, reflektierte Binärkode, reflektierte Binärcode *m* [*Zyklischer Binärkode, bei dem bei einem Wechsel von einem Wert zum nächsten sich nur ein binärer Wert ändert*]

cyclic(al) code [*A code, such as a binary code, that changes only in one digit when going from one number to the number immediately following, and in that digit by only one unit*] Zykluskode, Zykluscode, zyklische Kode, zyklische Code *m*

cyclic(al) distortion zyklische Verzerrung, Zyklusverzerrung *f*

cyclic(al) memory [*A device, such as a magnetic drum, whose storage medium is arranged in such a way that information can be read into or extracted from individual locations at only certain fixed times in a basic cycle*] Zyklusspeicher, Umlaufspeicher *m*

cyclic(al) shift, circuit shift, circular shift, end-around shift, nonarithmetic shift, ring shift [*A computer shift in which the digits dropped off at one end of a word return at the other end of the word*] zyklische (Stellenwert)Verschiebung *f*

cyclic redundancy check(ing) → CRC

cyclic redundancy check(ing) code → CRCC

cyclic switching zyklische Schalten *n*

cycling periodische Ablauf(en), zyklische Ablauf(en), periodische Durchlauf(en), zyklische Durchlauf(en), Taktablauf(en) *m, (n)*

cycling → hunting

cycling redundancy check(ing) code, CRCC zyklische Redundanzprüf(ungs)kode *m*

cycloid [*The path traced by a point on the circumference of a circle, as the circle rolls along a straight line*] Zykloide *f*

cyclone (dust) collector Staubzyklon *m*

cyclone filter [*Funnel-shaped filter for removing particles from fluids by centrifugal means*] Filterzyklon *m*, Zyklonfilter *m, n*

cylinder → seek area

cylinder bore [*Finished internal diameter of a cylinder*] Zylinderbohrung *f*

cylinder core Zylinderkern *m*

cylinder dead bolt lock Zylinderriegelschloss, Riegelzylinderschloss *n*

cylinder detaching pin → detaching pin

cylinder discharge valve [*Valve in a compressor through which the gas leaves the cylinder*] Zylinderablassventil, Ablasszylinderventil *n*

cylinder gas → bottle(d) gas

cylinder head [*Upper part of a cylinder*] Zylinderkopf *m*

cylinder jacket Zylindermantel *m*

cylinder key Zylinderschlüssel *m*

cylinder knob Zylinderolive *f* [*Schloss*]

cylinder lock, key-in-knob lock Zylinderschloss *n*

cylinder-lock act(uat)or, cylinder-lock operator Schlüsselantrieb *m* mit Zylinderschloss, Zylinderschlossschlüsselantrieb

cylinder overflow Zylinderüberlaufbereich *m* [*selten: n*] [*Kommandosprache*]

cylinder padlock Zylinderhangschloss, Zylindervorhängeschloss *n*

cylinder rim (dead)latch lock Zylinderkastenfallenschloss, Kastenfallenzylinderschloss *n*

cylinder rim deadlock Zylinderkastenriegelschloss, Kastenriegelzylinderschloss *n*

cylinder rim deadlock with latch bolt and dead bolt and pull knob Zylinderkastenschloss *n* mit Falle und Riegel und Ziehknopf, Kastenzylinderschloss mit Falle und Riegel und Ziehknopf

cylinder rim lock Zylinderkastenschloss, Kastenzylinderschloss *n*

cylinder rim lock with automatic double bolts Zylinderkastenschloss *n* mit zweifach wirkendem Fallenriegel, Kastenzylinderschloss mit zweifach wirkendem Fallenriegel

cylinder suction valve [*Valve in a compressor through which the gas enters the cylinder*] Zylinder(an)saugventil *n*

cylinder surface lock, surface cylinder lock Aufliegerzylinderschloss, Zylinderaufliegerschloss, aufliegende Zylinderschloss *n*

cylindrical capacitor, coaxial capacitor [*A capacitor made of two concentric metal cylinders of the same length, with dielectric filling the space between the cylinders*] Zylinderkondensator *m*

cylindrical pad-lock Bolzenschloss *n* [*Ein Vorlegeschloss, von der Befestigungsweise des Bügels so genannt*]

cylindrical (pipe) fitting thread, female (pipe) fitting thread Fittinginnengewinde *n*, Formstückinnengewinde

cylindrical piston water meter Kapselwassermesser *m*

cylindrical vessel zylindrische Gefäß *n*

cylindrical wave [*Electromagnetism. A wave whose phase fronts form concentric surfaces of circular cylinders*] Zylinderwelle *f*

D

DA, demand assignment, need assignment bedarfsweise Zuteilung, Zuteilung nach Bedarf *f*

DA, design automation Konstruktionsautomatisierung *f*

DA → **destination address**

DAB → **digital audio broadcast(ing)**

DAC → **digital(-to)-analogue converter**

D/A conversion → **digital(-to)-analogue conversion**

D/A converter → **digital(-to)-analogue converter**

D-action → **derivative action**

D(-action) controller, D(-action) regulator, differential(-action) controller, differential(-action) regulator, derivative(-action) controller, derivative(-action) regulator D-Regler, Differentialregler, Vorhaltregler *m* [*Er benötigt eine größere Verstärkung als der P-Regler, damit bei plötzlichen Eingangsänderungen eine ausreichende Stellgröße am Ausgang vorhanden ist. Für den D-Regler gilt die Gleichung:*

$$u - u_0 = K_D \frac{d(w - r)}{dt} = K_D \frac{de}{dt}$$

KD ist der Kennwert des D-Reglers. E DIN 19 226 Teil 5]

Dahlander (pole changing) circuit → **delta-parallel-star circuit**

daily dry weather flow [*The total amount of foul water, together with infiltration, if any, discharged by a sewer or drain in dry weather in 24 hours*] tägliche Trockenwetterabfluss, Tagestrockenwetterabfluss *m*

daily log, day log Tagesprotokoll *n*

daily output, daily performance Tagesleistung *f*

daily pumpage Tagespumpenfördermenge *f*

daily schedule Tagesplan *m*

daily temperature range Tagestemperaturbereich *m* [*selten: n*]

daily work, day-to-day work tägliche Arbeit, Tagesarbeit *f*

daisy chain [*In automatic processing, a means of connecting devices (readers, printers, and so on) to a central processor by party-line input/output buses which join these devices by male and female connectors, the last female connector being shorted by a suitable line termination*] Melderkette, Fortschaltkette *f*

daisy chain device priority Rangreihenfolge *f* für Unterbrechungen

daisychaining Melderverkettung *f*

daisywheel Typenrad, Speichenrad *n*

daisy wheel printer [*A serial printer in which the printing element is a plastic hub that has a large number of flexible radial spokes, each spoke having one or more different raised printing characters; the wheel is rotated as it is moved horizontally step by step under computer control, and stops when a desired character is in a desired print position so a hammer can drive that character against an inked ribbon*] Speichenraddrucker, Typenraddrucker *m*

Dalton('s) law [*Each constituent of a mix(ture) of gases behaves thermodynamically as if it alone occupies the space. The sum of the individual pressures of the constituents equals the total pressures of the mix(ture)*] Gesetz *n* von Dalton, Daltongesetz

DAM → **direct access memory**

damage Schaden *m*

damage caused by natural hazards → **primary damage**

damaged schadhaft, beschädigt

damage data Schadensdaten *f*

damaged by fire, fire damaged brandbeschädigt, brandschadhaft, feuerbeschädigt, feuerschadhaft

damage-limiting, damage-restricting, damage-diminishing schadenbegrenzend, schadenmindernd

damage prevention Schadensverhütung *f*

damage probability Schadenswahrscheinlichkeit *f*

damage (protection) control [*A safety control function for switching plant or part thereof into a pre-defined and safe or protected state, in order to prevent harm to people and/or damage to the building, plant, etc. An acknowledg(e)ment facility shall be provided. Reset shall be manually instigated by an operator*] Sicherheitsschaltung *f*

damaging fire Schadensbrand *m*, Schadensfeuer *n*

Damköhler number V → **Reynolds number**

damp → **moist**

damped critical system → **critical damped system**

damped natural angular frequency, natural damped angular frequency Eigenkreisfrequenz *f*

damper Klappe, Stellklappe, Regelklappe *f*

damper → **draft regulator**

damper act(uat)or, damper final controlling element Klappenstellglied *n* [*Elektromechanisches Gerät, das ein elektrisches Signal in die geforderte Stellung einer Luftklappe umwandelt*]

damper (air) register [*A grill(e) having a damper for regulating the quantity of air passing through it*] Klappen(luft)register, Luftklappenregister *n*

damper ((closed) loop) control Klappenreg(e)lung, Stellklappenreg(e)lung *f*

damper drive Klappenantrieb, Stellklappenantrieb *m*

damper ((open) loop) control Klappensteuerung, Stellklappensteuerung *f*

damper winding bar → **bar of damper winding**

damping Dämpfen *n*, Dämpfung *f*

damping factor [*Physics. The ratio of the amplitude of a vibration to that of its succeeding vibration in an underdamped vibrational system*] Dämpfungsfaktor *m*

damping level Dämpfungspegel *m* [*Logarithmiertes Verhältnis des Spannungswertes oder Leistungswertes am Eingang eines Zweitors (= Vierpol) zu dem am Ausgang*]

damping magnet [*A permanent magnet used in conjunction with a disk or other moving conductor to produce a force that opposes motion of the conduction and thereby provides damping*] Bremsmagnet, Dämpfungsmagnet *m*

damping ratio Dämpfungsgrad *m*, relative Dämpfung *f*

damping reactor Dämpfungsdrossel *f*

damping resistor Dämpfungswiderstand *m*

damp proofing *see under* **waterproofing**

danger alarm Gefahrenalarm *m*

danger alarm system, danger alarm plant Gefahrenalarmsystem *n*, Gefahrenalarmanlage *f*

danger arrow Warnpfeil, Hochspannungs(warn)pfeil *m*

danger control Gefahrenabwehr, Gefahrenbekämpfung *f*

danger detection, danger recognition Gefahrenerkennung *f*, Gefahrennachweis *m*

danger detection central station, danger detecting central station, central danger detection station, central danger detecting station Gefahrenmeldezentrale, Gefahrenmeldungszentrale *f*

danger detection installation, danger detecting installation Gefahrenmeldungsanlage, Gefahrenmeldeanlage *f*

danger detector, danger detecting device, danger detection device Gefahrenmelder, Gefahrenmeldungsgeber, Gefahrendetektor *m*, Gefahrenmeldegerät *n*

danger detector computer, danger detecting device computer, danger detection device computer Gefahrenmelderechner, Gefahrendetektorrechner, Gefahrenmeldegerätrechner, Gefahrenmeldungsgeberrechner *m*

danger level Gefahrenstufe *f*, Gefahrenstand *m*

danger light Gefahrenfeuer *n*

dangerous gefährlich

danger recognition, danger detection Gefahrenerkennung *f*, Gefahrennachweis *m*

danger signal, red signal Gefahrensignal *n*

danger signal message, red signal message Gefahrensignalmeldung *f* [*Eine Stör(ungs)meldung bei unmittelbarer Gefahr*]

danger type Gefahrenart *f*

danger warning Gefahrenwarnung *f*

dark colored (US); dark coloured (Brit.) dunkelfarbig

dark current → **(electrode) dark current**

D

dark-lamp synchronizing Synchronisier(ungs)-Dunkelschaltung, Dunkel-Synchronisier(ungs)schaltung *f*

dark line defects dunkle parallel laufende Linien, parallel laufende dunkle Linien *f* [*Halbleiterlaser-Lebensdauer*]

dark location Dunkelstandort *m*

dark radiator Dunkelstrahler *m*

dark radiator heating system Dunkelstrahlerwärmesystem, Dunkelstrahlerheiz(ungs)system *n*

darting flame Stichflamme *f*

DAS → double attach station

to dash → to splash

dash-dotted strichpunktiert, punktgestrichelt

DAT → digital audio tape

DAT, digital audio taperecording digitale Tonbandaufzeichnung *f*

data Daten *f* [*Gebilde aus Zeichen oder kontinuierliche Funktionen, die aufgrund von bekannten oder unterstellten Abmachungen und vorrangig zum Zwecke der Verarbeitung Information darstellen. E DIN 44 300 T2*]

data acceptance Datenannahme *f*

data acceptor → acceptor (of data)

data access service Dateizugang(s)dienst *m*

data ACK → data transfer acknowledg(e)ment

data acknowledg(e)ment, DTACK Datenquittierung *f*

data acquisition, data gathering, data capture (Daten)Direkteingabe, (Daten)Erfassung *f*, (Daten-)Erfassen *n* [*Vorgang, bei dem durch Messen oder Zählen und gegebenenfalls Signalumformung analoge oder digitale Daten gewonnen werden. Prüfung (Plausibilitätskontrolle) und Aufbereitung (Linearisierung, Filterung) der eingehenden Signale können Bestandteile der Datenerfassung sein. DIN 1922*]

data acquisition panel → data gathering panel

data acquisition terminal, data gathering terminal, data capture terminal (Daten)Erfassungsstation *f*

data archives Datenarchiv *n*

data area, data zone Datenbereich *m* [*selten: n*], Datenzone *f*

data arrangement → data organization

data assurance [*Memory function that will retain data for a defined period in the event of power failure*] Datenpufferung *f*

data automation [*The use of electronic, electromechanical or mechanical equipment and associated techniques to automatically record, communicate and process data and to present the resultant information*] Datenautomation *f*

(data) backup, (data) re-store (Daten)Archivierung *f*

data base, data bank Datenbank *f* [*Es besteht aus Datenobjekten zusammen mit Beziehungen zwischen ihnen. Diese Datenobjekte genügen bestimmten Daten-Bauarten und bestimmten zusätzlichen Bedingungen. Die Beziehungen zwischen den konstituierenden Datenobjekten unterliegen ihrerseits bestimmten Bedingungen. Häufig werden die definierenden Daten-Bauarten und Bedingungen, die über die Zeit hinweg bestehen bleiben, Datenbank genannt. E DIN 44 300 T3*]

data base client → client

data base computer, data bank computer Datenbankrechner *m*

data base design, data bank design Datenbankentwurf *m*

data base driver, DD, data bank driver Datenbanktreiber *m*

data base interface, data bank interface Datenbankschnittstelle *f*

data base language, data bank language Datenbanksprache *f*

data base management system, DBMS, data bank management system, data base handler, data bank handler Datenbankprogramme *f* [*Die Programme, die es ermöglichen, in einem Rechensystem Datenbanken einzurichten, zu verwalten und zu nutzen. Dies schließt z. B. ein*
– mit Hilfe geeigneter Sprachen Datenbanken zu beschreiben,
– Speicher so zu organisieren, dass die Daten, die beschreibungsgemäß strukturiert sind, in vorgegebener Weise gespeichert werden können,
– Zugriffsberechtigungen zu beachten und Zugriffskonflikte zu lösen,
– die Konsistenz der Daten, auch im Störungsfall, zu gewährleisten,
– Daten strukturgerecht einzubringen und anwendungsgerecht bereitzustellen. E DIN 44 300 T4]

data base system, data bank system Datenbanksystem *n*

data block, DB Datenbaustein, Datenblock *m* [*Digitale Daten, von deren Zusammensetzung aus Zeichen im betrachteten Zusammenhang abstrahiert werden. E DIN 44 300, Teil 3*]

data block and text end, ETX Datenblock- und Textende *n*, Ende des Datenblocks und des Textes

data block end Datenblockende *n*, Ende des Datenblocks

data buffer [*A buffer register in the CPU used to hold a word before it is written to memory or after it is read from memory*] Datenpuffer *m*

data bug → data error

data burst → burst

data bus (bar), DB (bar) Datenbus *m*, Daten(sammel)schiene *f*, DB [*Er umfasst bei 8-Bit-Rechnern acht Leitungen, bei 15-Bit-Rechnern 16 Leitungen. Auf ihm können Daten von und zum Mikroprozessor transferiert werden*]

data bus (bar) driver, DB (bar) driver Datenbus-Treiber, Daten(sammel)schienentreiber, DB-Treiber *m*

data-bus-in, DB-in Datenbuseingabe, DB-Eingabe *f* [*Eine Gehäuseanschlussbezeichnung beim Mikroprozessor 8080. Sie zeigt an, dass Daten aus dem Bus in den Mikroprozessor übernommen werden*]

data cabinet, data cubicle Datenschrank *m*

data cable Datenkabel *n*

data cabling Verkabelung für Daten, Datenverkabelung *f*

data call Datenaufruf *m*

data call unit Datenaufrufeinheit *f*

data capture → data acquisition

data capture panel → data gathering panel

data capture terminal → data acquisition terminal

data card Datenkarte *f*

data carrier, data medium, storage device, storage medium, output Datenträger *m*, Datenaufbewahrungsmittel *n* [*E DIN 44 300 Teil 2*]

data carrier detect → DCD

data cartridge [*A tape cartridge used for nonvolatile and removable data storage in small digital systems*] Datenpatrone *f*

data catalog(ue) → data library

data catena → data chain

data cell drive (Magnet)Streifenspeicher *m*

data central station → data (processing) center

data centre; data center (US) Datenzentrale *f*, Datenzentrum *n*

data chain, data string, data catena [*Any combination of two or more data elements, data items, data codes, and data abbreviations in a prescribed sequence to yield meaningful information; for example, "data" consists of data elements year, month, and day*] Datenkette *f*

data chaining [*A technique used in scatter reading or scatter writing in which new storage areas are defined for use as soon as the current data transfer is completed*] Datenverkettung *f*

data channel [*A bidirectional data path between input/output devices and the main memory of a digital computer permitting one or more input/output operations to proceed concurrently with computation*] Datenkanal *m*

data channel received line signal Empfangssignal *n*

data channel received line signal detector, DcD Empfangssignaldetektor *m*

data circuit terminating equipment, DCE Datenübertragungsvorrichtung *f* [*Datenübertragungsvorrichtung, die am Standort des Anwenders installiert ist und alle für die Herstellung, Durchführung und Beendigung einer Verbindung erforderlichen Bauteile bietet. Die Einrichtung sichert die Signalumwandlung und -kodierung zwischen den Datengeräten und der gewöhnlichen Trägerleitung (Modem). DIN 44 302*]

data code [*A number, letter, character, symbol, or any combination thereof, used to represent a data item*] Datenkode, Datencode *m*

(data) collector (Daten)Sammler *m*

(data)communication, DC Datenübermitt(e)lung, Datenkommunikation *f*, Datenverkehr *m*

(data) communication equipment, DCE Datenkommunikationsgerät(e) *n*, *(f)*

data compaction → data reduction

data compression Datenkomprimierung, Datenkompression *f* [*Nicht verwechseln mit Datenverdichtung*]

data concentrator Datenkonzentrator *m*

data conferencing Datenkonferieren *n*

data confidentiality [*The property that information is not made available or disclosed to unauthorised individuals, entities, or processes*] Datenvertraulichkeit *f*

data connection Datenanschluss *m*, Datenverbindung *f*

data connector Datensteck(verbind)er *m*

data consistency Datenkonsistenz *f*

data control language, DCL Datensteuer(ungs)sprache *f* [*Sie wird in Protokollen benutzt*]

data control unit, DCU Datensteuer(ungs)einheit *f*

(data) conversion [*The changing of the representation of data from one form to another, as from binary to decimal, or from one physical recording medium to another, as from card to disk*] Datenumsetzung, Daten(um)wand(e)lung *f*

(data) conversion channel [*The channel, electronic or manual, through which data elements are transferred between data banks*] Datenumsetz(ungs)kanal, Daten(um)wand(e)lungskanal *m*

D

(data) conversion equipment [*Equipment used for conversion of data from one recording medium to another, as from card to tape*] Datenumsetz(ungs)-gerät(e), Daten(um)wand(e)lungsgerät(e) *n, (f)*

(data) converter [*A computer unit that changes numerical information from one form to another, as from decimal to binary or vice versa, from fixed-point to floating-point representation, or from punch(ed) cards to magnetic tape*] (Daten)Umsetzer *m* [*E DIN 44 300/10.85*]

data coupling Datenkopp(e)lung *f*

data courier Datenkurier *m*

data cubicle → data cabinet

data density Datendichte *f*

data descriptive language Datenbeschreibungssprache *f*

data design, data type, data version Datenbauart, Datenbauweise, Datenkonstruktion *f*, Datenaufbau *m* [*Ein Modell für digitale Daten, die ein und derselben Aufbau- oder Anordnungsvorschrift genügen. Diese digitalen Daten heißen Ausprägungen der Bauart. E DIN 44 300, Teil 3*]

data direction register → DDR

data directory, data dictionary Daten(inhalt)verzeichnis, Dateiverzeichnis *n*, Katalog *m* [*Protokollierung gemeinsam verwalteter Datenbestände*]

data display Datenanzeige, Datendarstellung *f*

data display channel Datenanzeigekanal, Datendarstellungskanal *m*[*Er regelt das Zusammenspiel von Graphikkarte und Monitor*]

data display terminal Datenanzeigegerät, Datendarstellungsgerät *n*, Datenanzeigestation, Datendarstellungsstation *f*

data division [*The section of a program(me) (written in the COBOL language) which describes each data item used for input, output, and storage*] Datenteil *m, n*

data element Datenelement *n* [*Eine als elementare Einheit betrachtete, umkehrbar eindeutige Beziehung zwischen den Elementen einer Menge von Objekten oder von Objekttypen oder von Objekteigenschaften einerseits und Elementen einer geeigneten Menge von Wörtern andererseits. E DIN 44 300 T 2*]

Data Encryption Standard → DES

data entering, entering of data Dateneinbringen, Dateneintragen *n*

data entry point → node

data error, data bug, data fault [*A deviation from correctness in data, which occurred prior to processing the data*] Datenfehler *m*

(data) (e)valuation → (data) interpretation

(data) (e)valuation system, (data) interpretation system [*Automatically operated equipment used to interpret data gathered by instrument installations*] (Daten)Auswertungssystem, (Daten)Bewertungssystem *n*

data exchange Datenaustausch *m*

data exchange DATEX-P 10 [*Öffentliches Datennetz der Deutschen Bundespost mit standardisierten Schnittstellen*]

data fault → data error

data field Datenfeld *n* [*Eine unmittelbar auf Zeichen gestützte Daten-Bauart für Bezeichner oder andere Datenobjekte, die, ebenso wie Bezeichner, Information darstellen, die in gegebenem Zusammenhang als elementar angesehen wird. E DIN 44 300 T 3*]

data file → data set

data file text, data set text Dateitext *m*

dataflow Datenfluss *m* [*Durchlauf von Daten durch Instanzen und Kanäle im Zuge auftragsgemäßer Verarbeitung. E DIN 44 300 Teil 1*]

dataflow chart → dataflow diagram

dataflow ((closed) loop) control Datenflussregelung *f*

dataflow computer Datenflussrechner *m*

dataflow diagram, dataflow chart, dataflow control diagram, dataflow sheet, dataflow plan [*See remark under "flow diagram"*] Datenfließbild, Datenflussbild, Datenflussdiagramm, Datenfließdiagramm, Datenablaufbild, Datenablaufdiagramm *n*, Datenablaufplan, Datenflussplan, Datenfließplan, *m*

dataflow language [*A model(l)ing language used to describe the information paths in program(me)s in which parallel or asynchronous processes occur*] Datenflusssprache *f*

dataflow ((open) loop) control Datenflusssteuerung *f*

dataflow plan → dataflow diagram

dataflow system, non-von Neumann architecture [*A computer system in which basic operations are initiated by the availability of the operands instead of by the sequential flow of control*] Datenflusssystem *n*

dataflow technique [*An approach to computer systems organization and design that specifies the movement of data through the system and the transformations that occur, as opposed to a traditional system design using process and control flow*] Datenflusstechnik *f*

data format Datenformat *n*

D

data formatting [*Structuring the presentation of data as numerical or alphabetic and specifying the size and type of each datum*] Datenformatierung *f*

data gathering → data acquisition

data gathering panel, DGP, data acquisition panel, data capture panel [*A device with no control functions, but which accepts signals, conditions them, and passes them to other devices via a cable or cables. This term is used by some manufacturers synonymously with dumb outstation, but this use is not preferred*] Datensammler *m*, Datensammelgerät *n*

data gathering terminal → data acquisition terminal

data headend (equipment) Datenendgerät(e) *n, (f)*

data hiding Geheimhaltung *f* von Datenstrukturen

data hierarchy [*A data structure that defines succeedingly detailed and lower-ranking subjects*] Datenhierarchie *f*

(data) identifier → identifier

data independence [*The abstractions of data from the processing so that changes in the logical or physical structure of the data have no effect on the application program(me)'s view of the data*] Datenunabhängigkeit *f*

data input bus, DIB Dateneingabebus *m*

data integrity Datenintegrität, Datenunversehrtheit *f* [*Keine Veränderung des Informationsgehalts einer Nachricht während eines Übertragungsvorgangs*]

data interception Lauschangriff *m*

data interchange [*The accessibility of data from one program(me) to a separate program(me), using software standards for data representations*] Datenaustausch *m*

data interface [*1. A common aspect of two or mor data systems involving the capability of intersystem communications. 2. A common boundary between automatic data-processing systems or parts of a single system. VDI 2880 Blatt 1*] Datenschnittstelle *f*

data interface unit, DIU Datenschnittstelleneinheit *f*, DSE [*Sie schaltet das Gebäudemanagementsystem mit einem Drittsystem zusammen. Solche Drittsysteme können HLK-Anlagen anderer Hersteller sein, oder Einrichtungen wie Aufzüge, Brandschutzanlagen und Sicherheitseinrichtungen*]

data interlocking, data interconnection Datenverbund *m*

(data) interpretation, (data) (e)valuation [*Interpreting data gathered by instrument installations*] (Daten)Auswertung, (Daten)Bewertung *f*

(data) interpretation system → (data) (e)valuation system

data inventory, data stock Datenbestand *m*

data item, datum [*A single unit of data; the smallest unit of named data*] Datenelementteil *m, n*, Datengliedteil, Datum *n*

data item information → information

data level [*The rank of a data element in a source language with respect to other elements in the same record*] Datenrang *m*

data library, data catalog(ue) [*It contains all the disks, tapes, documentations, user's manuals, and procedures that are relevant to using data resources*] Datenkatalog *m*

data link → data (transfer) link

data link connection gesicherte Systemanschluss *m*, gesicherte Systemverbindung *f* [*DIN ISO 7498*]

data link escape, DLE Datenübertragungsmodus *m* [*Einleitung eines transparenten Übertragungsmodus*]

data link layer, layer 2 Verbindungsschicht, Verbindungsebene *f*, zweite Schicht, zweite Ebene, (Daten)Sicherheitsschicht, (Daten)Sicherheit(s)-ebene, Schicht 2, Ebene 2 [*Auf dieser Schicht wird das Sprachsignal bei analoger Übertragung richtig verstärkt, bei digitaler Übertragung (Richtfunk) werden entsprechende Sprechkanäle bereitgestellt, das Analogsignal wird in PDM (Plus Code Modulation) gewandelt und eine bestimmte Datensicherheit gewährleistet*]

data link service Sicherungsdienst *m* [*E DIN ISO 7498*]

data loop Datenschleife, Datenringleitung *f*

data (magnetic) tape [*Computer engineering. A strip of flexible material that is coated with a magnetic film and used for data input, storage, and output*] (Daten)(Magnet)Band *n*

data management → input-output central system

data master file → master (data) file

datamation → automatic data processing

datamation equipment → automatic data processing equipment

data medium → data carrier

data memory [*See remark under "(computer) memory"*] Datenspeicher, Arbeitsspeicher *m* [*Teil der Zentraleinheit zur kurzfristigen Datenspeicherung. Zum Datenspeicher gehören die Merker und die Abbildungsregister*]

data misuse Datenmissbrauch *m*

data model [*A graphic database design tool used to represent the properties of stored data and their relationship to each other, independent of software, hardware, or performance considerations*] Datenmodell *n*

D

data movement statement [*COBOL*] zeichenfolgenverarbeitende Anweisung *f*

data name [*A symbolic name used to represent an item of data in a source program(me), in place of the address of the data item*] Datenname *m*

data network Daten(übermitt(e)lungs)netz *n*

data network identification code, DNIC Datennetzkennkode, Datennetzkenncode *m*

data networking Datenvernetzung *f*

data object Datenobjekt *n* [*Zusammenfassung von digitalen Daten zu einer Einheit. Diese Zusammenfassung folgt einer bekannten oder als bekannt unterstellten Daten-Bauart. E DIN 44 300 T 3*]

data origination [*The process of putting data in a form that can be read by a machine*] Datenlesbarmachen *n*

(data) output, data-out [*The output of data from a memory or into a further communication unit*] (Daten-) Ausgabe, (Daten)Abgabe *f*, (Daten)Ausgeben, (Daten)Auslesen, (Daten)Abgeben *n* [*Ein Vorgang, bei dem analoge oder digitale Daten aufgezeichnet, an-gezeigt oder an den Prozess gegeben werden. DIN 19 222*]

data overflow, data overrun Datenüberlauf *m*

data packet [*A relatively small unit of data transmitted over a pocket switching network as part of a message that is sent from one computer to another*] Datenpaket *n*

data path, data travel, data way Datenweg, Datenpfad *m*

data plotter [*An output device that provides hardcopy graphics or pictorial representations of computer-processed data, usually as line drawings*] Datenplotter, Datenstrichzeichner *m*

data plug Datenstecker *m*

data point, point Datenpunkt *m* [*Durch eine Benutzeradresse eindeutig identifizierte und Informationen enthaltende Einheit*]

data point input, point input Datenpunkteingabe, Punkteingabe *f*

data position Datenstelle *f* [*DIN 6763, Blatt 1*]

(data) preprocessing (Daten)Vorverarbeitung *f*

data privacy protection → protection of data privacy

(data) processing, DP [*Any operation or combination of operations on data, including everything that happens to data from the time they are observed or collected to the time they are destroyed*] (Daten)Verarbeitung, DV, (Daten)Bearbeitung *f* [*DIN 19 222*]

data (processing) center (US); data (processing) centre (Brit.); data central station, central data station [*An area containing automatic data processing equipment and personnel who centralize and control data processing services for others in the organization; usually managed as a separate organizational entity*] Daten(verarbeitungs)zentrale, Daten(verarbeitungs)warte *f*

(data) processing circuit (Daten)Verarbeitungsstromkreis *m*

(data) processing director (Daten)Verarbeitungskarte *f*

(data) processing equipment (Daten)Verarbeitungsgerät(e) *n, (f)*

(data) processing function (Daten)Verarbeitungsfunktion *f*

(data) processing inventory [*An identification of all major data-processing areas in any agency for the purpose of selecting and focus(s)ing upon those in which the use of automatic data-processing techniques appears to be potentially advantageous, establishing relative priorities and schedules for embarking on automatic data-processing studies, and identifying significant relationships among areas to pinpoint possibilities for the integration of systems*] (Daten)Verarbeitungskennzeichnung *f*

(data) processing machine → computer

(data) processing machine [*A component of a data-processing system, such as a card reader or tape unit*] (Daten)Verarbeitungsmaschine *f*

(data) processing system (Daten)Verarbeitungssystem, Rechensystem[*E DIN 44 300, Teil 5*] *n*

data process language; CL 1 [*deprecated*] netzwerkführige Sprache *f* [*Sie bietet eine einheitliche Schnittstelle zu verschiedenen Datenbanken an*]

(data) processor [*Any device capable of performing operations on data, for instance, a data calculator, an analog computer, or a digital computer*] (Daten)Verarbeitungsgerät *n*

(data) processor [*Person engaged in processing data*] (Daten)Verarbeiter *m*

data production Datenerzeugung, Datenherstellung *f*

data protection Datenschutz *m*

data purification [*The process of removing inaccurate or incorrect terms from a mass of data before automatic data processing is begun*] Datenreinigung, Datensäuberung *f*

data quantity Datenmenge *f*, Datenumfang *m*

data radio Datenfunk *m*

data ready Daten abfragebereit [*Signal*]

data rearrangement, data redisposition Datenum(an)ordnung *f*

data receptacle Datensteckdose *f*

data record [*An organized and identifiable set of data elements that are related in some way*] Datenaufnahme, Datenaufzeichnung, Datenregistrierung *f*

data recorder, data recording device [*It makes a continuous record of data, e.g. on a chart*] Datenschreiber *m*, Datenregistriergerät, Datenaufzeichnungsgerät, Datenaufnahmegerät *n*

data recording Datenaufzeichnen, Datenregistrieren, Datenaufnehmen *n*

(data) record structure (Daten)Satzaufbau, (Daten)Aufzeichnungsaufbau *m*, (Daten)Satzstruktur, (Daten)Aufzeichnungsstruktur *f*

data redisposition → data rearrangement

data reduction, data compaction [*The process of obtaining only the useful data from a larger set of data through computation, summarization, abstracting aggregation, or other abstracting methods*] Datenverdichtung *f*, Datenreduktion *f* [*Nicht verwechseln mit „Datenkomprimierung"*]

data redundancy, data redundance [*The replication of specific data item values in more than one file within a database*] Datenüberfluss, Datenüberschuss *m*

(data) remote processing → (data) teleprocessing

data remote transmission, data teletransmission, remote data transmission, teletransmission of data Datenfernübertragung, Ferndatenübertragung, DFÜ, FDÜ *f*

data representation [*The machine-readable code or format that is interpreted and produced by computer instructions*] Datendarstellung *f*

data resolution Datenauflösung *f* [*Auflösungsvermögen eines Steuerungs- bzw. Messsystems, kleinste Lageänderung, die von der Positionsmessvorrichtung noch erfasst werden kann*]

data response Datenreaktion *f*

(data) re-store → (data) backup

data retrieval, data recovery [*The process of searching for and selecting data in files or in a database stored in a computer*] Datenrettung, Datenauffindung, Datenwiedergewinnung *f*

data rule [*It must be met when processing data*] Daten(verarbeitungs)regel *f*

data safety Datensicherheit *f* [*Sachlage, bei der Daten unmittelbar oder mittelbar so weit wie möglich vor Beeinträchtigung bewahrt sind, und zwar unter Berücksichtigung verarbeitungsfremder Risiken wie auch im Verlauf auftrags- und ordnungsgemäßer Erbringung einer Datenverarbeitungsleistung.
Daten dürfen also
– weder bei datenverarbeitenden Prozessen oder auftragsbedingten Vor- und Nacharbeiten
– noch in Funktionseinheiten zur Abwicklung auftragsbedingter Arbeiten
– noch durch Handeln von an auftragsbedingten Arbeiten beteiligten Personen beeinträchtigt werden. E DIN 44 300 Teil 1*]

data search(ing) (run) Datensuche *f*, Datensuchen *n*, Datensuchlauf *m*

data security, data protection [*The application of procedures and safeguards to prevent deliberate or accidental data loss or damage, or the access of unauthorized persons to sensitive classified, or private data*] Datenschutz *m*, Datensicherung *f*

data serial datenseriell

data service Datendienst *m*

data service adapter, DSA Umsetzer *m* [*Er wandelt Faxtonsignale in digitales Format um*]

data set, data file Datei *f*, Datensatz *m* [*Ein Datenobjekt, das einer bestimmten Vorschrift genügt. Nach dieser Vorschrift werden Ausprägungen einer oder mehrerer Satzarten oder Satzgruppen oder Kombinationen davon zu einer Einheit zusammengefasst. Die Daten-Bauart der Datei wird u. a. geprägt von Gesichtspunkten des Sachbezuges oder der Verarbeitungsweise. Häufig wird diese Daten-Bauart zusammen mit den prägenden Gesichtspunkten, die über die Zeit hinweg bestehen bleiben, selbst Datei genannt. E DIN 44 300 T 3*]

data set ready Betrieb(s)bereitschaft *f* [*Schnittstellenleitung*]

data set text → data file text

data sheet [*A form for collecting or recording source data that is designed for ease and convenience of transcription to machine-readable format*] Datenblatt *n*, Datenbogen *m*

data shift technique Datenumschalt(ungs)technik *f*

data signal Datensignal *n*

data signal(l)ing rate → equivalent bit rate

data signal(l)ing rate selector, equivalent bit rate selector Übertragungsgeschwindigkeitswähler *m*

data signal quality detector Datensignalgütemelder *m* [*Schnittstellenleitung*]

data silo externe Datenspeicher *m*

data sink [*The equipment at the receiving location that accepts and stores data signals after transmission via a data communications channel*] Datensenke *f*

D

data source [*The equipment at the sending location that supplies the data signals to be transmitted over a data communications channel*] Datenquelle *f*

data stack → stack

data statement [*A statement in a source program(-me) that identifies an item of data in the program(-me) and specifies its format*] Datenanweisung *f*

data station [*A remove input/output device which handles a variety of transmissions to and from certain centralised computers*] Datenstation *f*

data station ((closed) loop) control Datenstation(s)reg(e)lung *f*

data station ((open) loop) control [*The control of a data station by means of a program(me) resident in the central computer*] Datenstation(s)steuerung *f*

data stock, data inventory Datenbestand *m*

data storage, data storing [*See remark under "(computer) memory"*] Datenspeicherung außerhalb des Rechners, Außendatenspeicherung *f*

data string → data chain

data string end → end-of-data string

data support time → backup time

data switching Datenvermittlung *f*

data switching exchange, DSE Datenvermittlung(sstelle) *f*

data syntax Datensyntax *f*

data system [*The means, either manual or automatic, of converting data into action or decision information, including the forms, procedures, and processes which together provide an organized and interrelated means of recording, communication, processing, and presenting information relative to a definable function or activity*] Datensystem *n*

data table [*Collection of data with each item uniquely identified either by same label or by its relative position*] Datentafel *f*

data tape → data (magnetic) tape

data telecommunication service Dateidienst *m*

data teleprocessing, data remote processing, remote data processing Datenfernverarbeitung, Ferndatenverarbeitung *f*

data teletransmission → data remote transmission

data terminal → terminal

data terminal equipment, DTE Datenendgerät(e) *n, (f)*

data theft Datendiebstahl *m*

data to be processed → computer input

data track Datenspur *f*

data traffic Datenverkehr *m*

data transcription, data translation [*Conveying data from its original state to a data-processing media*] Datenkonvertierung, Datenumsetzung *f*

data transfer, data transmission [*The technique used by the hardware manufacturer to transfer data from computer to storage device or vice versa; usually under specialised program(me) control*] Datentransfer *m*, Datenübermitt(e)lung, Datenübertragung, Datenübergabe, Datenweitergabe *f* [*Von Rechner zu Speicher und umgekehrt*]

data-transfer-acknowledged, data-transfer-confirmed, acknowledged data transfer, confirmed data transfer, DTC, DTA bestätigte Datentransfer, quittierte Datentransfer *m*, bestätigte Datenübermitt(e)lung, bestätigte Datenübergabe, quittierte Datenübermitt(e)lung, quittierte Datenübergabe, quittierte Datenweitergabe, bestätigte Datenweitergabe, quittierte Datenübertragung, bestätigte Datenübertragung *f* [*DIN 19 222*]

data transfer acknowledg(e)ment, data transfer confirmation, data ACK Datentransferbestätigung, Datentransferquittierung, Datenübergabebestätigung , Datenübergabequittierung, Datenweitergabebestätigung, Datenweitergabequittierung, Datenübermitt(e)lung(s)bestätigung, Datenübermittlung(s)quittierung, Datenübermittelung(s)quittierung *f*

data transfer bus, DTB Datenübermitt(e)lung(s)bus, Datentransferbus, DTB, Datenübergabebus, Datenweitergabebus *m*

data transfer device, data transfer appliance Datenübertragungsgerät *n*, Datenübertragungsvorrichtung *f*

data (transfer) link, DL Datenübermitt(e)lungsabschnitt, Koppelabschnitt, Datentransferabschnitt, Datenweitergabeabschnitt, Datenübergabeabschnitt *m*, Übermitt(e)lungsleitung, Datenübermitt(e)lungsstrecke, Koppelstrecke, Koppelleitung, Datentransferstrecke, Datentransferleitung, Datenweitergabestrecke, Datenweitergabeleitung, Datenübergabestrecke, Datenübergabeleitung *f*

data (transfer) link ((closed) loop) control Datenleitung(s)reg(e)lung *f*

data (transfer) link escape, DLE Datenübermitt(e)lungsumschaltung, Datenübergabeumschaltung, Datenweitergabeumschaltung, Datentransferumschaltung, Datenübertragungsumschaltung *f*

data-transfer-unacknowledged, data-transfer-unconfirmed, unacknowledged data transfer, unconfirmed data transfer, DTU unbestätigte Datentransfer, nichtquittierte Datentransfer *m*, unbestätigte Datenübergabe, unbestätigte Datenweitergabe, unbestätigte Datenübermitt(e)lung, nichtquittierte Datenübergabe, nichtquittierte Datenweitergabe, nichtquittierte Datenübermitt(e)lung *f*

data transmission unit Datenübertragungseinheit *f* [*z. B. Modem, ISDN-Karte, Telephon*]

data travel → data path

data type → object type

data type → data design

data typewriter Datenschreibmaschine *f*

data unit Dateneinheit *f* [*OSI*]

data updating Datenaktualisierung, Datennachführung *f*, Datenaktualisieren, Datennachführen *n*

data us(ag)e Datengebrauch *m*, Datenbenutzung *f*

data valid, DAV Daten *f* gültig [*Eine Funktion einer Schnittstellenleitung*]

data validation, data validity check(ing) [*The check(ing) of data for correctness, or the determination of compliance with applicable standards, rules and conventions*] Datengültigkeitskontrolle *f*

data version → data design

data waste, waste data Datenabfall, Datenmüll *m*, Abfalldaten, Mülldaten *f*

data way → data path

data wealth Datenfülle *f*, Datenreichtum *m*

data window Datenfenster *n*

data word, information word, DW [*A computer word that is part of the data which the computer is manipulating, in contrast with an instruction word*] Datenwort, DW *n*

data zone → data area

date Termin *m*

date Datum *n*

date code Datumkode, Datumcode, Datumschlüssel *m*

date management Terminbetreuung *f*

date time generator Datum-Uhrzeit-Geber *m*

date time group [*The data and time, expressed in digits and zone suffix, at which the message was prepared for transmission (expressed at six digits followed by the zone suffix; first pair of digits denoting the date, second pair the hours, third pair the minutes)*] Datum(-Zeit)gruppe *f*

datex circuit switching network Datex-C-Netz *n*

datex packet switching network Datex-P-Netz *n*

datum [*Any numerical or geometric quantity or value that serves as a base reference for other quantities or values (such as a point, line, or surface) in relation to which others are determined*] Bezugsgröße, gegebene Größe *f*

datum → data item

datum dead zone Totbereich *m* [*selten: n*], Totzone, tote Zone, neutrale Zone, Neutralzone *f* [*Bereich von Eingabewerten bei dem keine Systemreaktion auftritt*]

datum information → information

datum level Bezugsebene, Nullebene *f*

daughter bond Zusatzleiterplatte *f*

DaV → Reynold's number

DAV → data valid

dawn control → twilight control

dawn sensor → twilight sensor

day air temperature Tageslufttemperatur *f*

day charge Tagladung *f*

day economization [*A control function which permits heating or cooling plant to be turned off completely if the anticipated fall or rise in internal temperature does not exceed preselected limits within a period, usually 1 hour*] Tag(es)einsparung *f*

daylight Tageslicht *n*

daylight bath Tageslichtbad *n*

daylight engineering Tageslichttechnik *f*

daylighting, day(-time) illumination, daylight lighting, daylight illumination, day(-time) lighting Tageslichtbeleuchtung *f*

daylight-orien(ta)ted tageslichtorientiert

daylight quotient Tageslichtquotient *m* [*Prozentuales Verhältnis der Beleuchtungsstärke im Innenraum zur Beleuchtungsstärke im Außenraum*]

daylight saving time, DST Sommerzeit *f*

daylight sensor, daylight sensing element Tageslichtfühler, Tageslichtsensor *m*

daylight use Tageslichtnutzung *f*

day log → daily log

day mode, occupied mode [*heating*] Tagbetriebsmodus *m*

day/night room thermostat → double room thermostat

day/night safe system Tag-Nacht-Tresoranlage *f*, Tag-Nacht-Tresorsystem *n*

day-of-week dial, week day dial Wochenscheibe *f*

day optimization [*A control function which permits a heating or cooling plant to be turned off completely if the anticipated fall or rise in internal temperature does not exceed pre-selected limits within a period*] Tagesoptimierung *f*

dayroom Tagesraum *m* [*Klimatechnik*]

day setting Tageinstellung *f*

D

day time → **time of day**

day(-time) current Tagesstrom *m*

day(-time) illumination → **daylighting**

day timer Tagesschaltuhr *f*

day-to-day work → **daily work**

daywork Stundenlohnarbeit(en), Tagelohnarbeit(en) *f*

to dazzle blenden

dB → **decibel**

DB → **data block**

DB (bar) → **data bus (bar)**

DB (bar) driver → **data bus (bar) driver**

DB-in → **data-bus-in**

D bistable element, D(-type) flip-flop D-Kippglied, Binärkippglied, Kippbinärglied *n*

D bit → **delivery confirmation bit**

DBMS → **data base management system**

DBT → **distributor**

DC → **data communication**

DC → **digital coded**

DC → **(direct) digital control**

DC → **(electrode) dark current**

d.c., D.C., direct current Gleichstrom *m*

d.c. arc → **direct-current arc**

DC capacitor → **direct-current capacitor**

DC character → **device control character**

DcD → **data channel received line signal detector**

DCD, data carrier detect Signalisierung *f* des Datenempfangs

d.c.-d.c. converter → **direct-current-direct-current converter**

DCE → **data circuit terminating equipment**

DCE → **(data) communication equipment**

DCE clear indication Auslösemeldung *f* [*Datenübertragung*]

DCE common return Datenübertragungsrückleiter *m*, Datenübertragungsvorrichtung *f*

d.c. electrical quantity, direct-current electrical quantity Gleichstromgröße *f*

d.c. electricity meter, direct-current electricity meter Gleichstromzähler *m*

DC flipflop, pulse-triggered flipflop taktzustandgesteuerte Kippstufe, taktzustandgesteuerte bistabile Kippschaltung *f*, taktzustandgesteuerte bistabile Multivibrator *m*

DCL → **data control language**

DCL → **device clear**

D ((closed) loop) control parameter D-Reglerkennwert *m*

D-core Blechkern *m* mit 45° Schnitt, geblechte Kern mit 45° Schnitt

DC-parametric test Prüfung *f* typischer statischer Parameter

DC source → **direct-current source**

dc-to-ac inverter → **inverter**

DC to DC converter, direct current to direct current converter Gleichstromwandler *m*

DCU → **data control unit**

DD → **degree-day**

DD → **data base driver**

DD, direct drive Direktantrieb, vorlauffreie Antrieb *m*, direkte Einwirkung *f* [*Hilfsstromschalter*]

DD → **digital readout**

(D)DA → **(digital) differential analyser**

DDC → **direct digital controller**

DDC station → **direct digital control station**

d-delay [*contact element, IEC 337-2B*] Verzögerung "d" *f* [*Schaltglied, VDE 0660, T.203*]

DD(I) → **direct dialling (in)**

DDL, device description language, appliance description language, ADL Parametriersprache *f*

DDP, distributed data processing dezentrale Datenverarbeitung, verteilte Datenverarbeitung *f*

DDR, data direction register Datenrichtungsregister *n*

DDR → **digital data receiver**

DE → **design engineering**

to deactivate → **to switch off**

to deactivate inaktivieren, wirkungslos machen

deactivated → **unlocked**

deactivation → **(interlock) deactivation**

deactivation by (means of) key → **(interlock) deactivation by (means of) key**

dead blockiert, tot, gesperrt

dead tot, stromlos, nicht stromführend, spannungslos, nicht spannungsführend

dead band Unempfindlichkeitsbereich, Nichtempfindlichkeitsbereich, Totbereich *m*, *n*

deadbeat, aperiodic überschwingungsfrei, aperiodisch (gedämpft) [*Messgerät*]

deadbeat control, aperiodic control Reg(e)lung auf endliche Einstell(ungs)zeit, Reg(e)lung auf endliche Ausregel(ungs)zeit, aperiodische Reg(e)lung *f*

deadbeat oscillation, aperiodic oscillation endliche Schwingung, aperiodische Schwingung *f*

deadbeat response, aperiodic response [*Coming to rest without vibration or oscillation, as when the pointer of a meter moves to a new position without overshooting*] endliche Ausregel(ungs)zeit, endliche Einstell(ungs)zeit, aperiodische Ausregel(ungs)-zeit, aperiodische Einstell(ungs)zeit *f*

deadbeat response controller, aperiodic response controller Abtastregler mit endlicher Ausregel-(ungs)zeit, Abtastregler mit endlicher Einstell(ungs)-zeit *m* [*Abtastregler, die durch einen besonderen Regler-algorithmus und streckenabhängige Parametereinstellung mit endlicher Ausregel(ungs)zeit arbeiten. DIN 19 225*]

dead bolt Riegel, Schlossriegel *m*

dead bolt lock Riegelschloss *n*

dead earth, dead ground [*A connection between a line conductor and earth by means of a path of low resistance*] vollkommene Erdschluss, satte Erdschluss, komplette Erdschluss *m*

dead element → dead (time) element

to deaden entdröhnen

dead end [*A portion of a tapped coil through which no current is flowing at a particular band = switch position*] stromlose Ende *n*

dead end Nichtantriebsseite *f*

dead-end corridor Sackflur, Sackkorridor *m*

dead-end feeder Netzausläufer *m*, Ausläuferleitung, Stichleitung *f*

dead-end tension joint Abspannverbinder *m*

dead(ened) room schalltote Raum, Schallmessraum *m*

dead(ened) wall schalltote Wand, Schallmesswand *f*

dead(ened) window schalltote Fenster, Schallmessfenster *n*

dead-front assembly [*IEC 439-1*] Tafelbauform *f* [*VDE 0660, Teil 500*]

dead ground → dead earth

dead halt → (drop-)dead halt

dead interval stromlose Pause, spannungsfreie Pause, momentenfreie Pause, Pausezeit, Totzeit *f*

dead leg [*In a hot water system, a length of pipe that forms a spur from the circulating system. The water in the dead leg tends to remain stationary and therefore wastes heat*] Totstrang, Blindstrang, tote Strang, blinde Strang *m*, tote Leitung, Totleitung, Blindleitung, blinde Leitung *f*

dead letter unzustellbare Brief, nicht zustellbare Brief *m* [*Er wird im Absenderverzeichnis abgelegt*]

dead level → flush (with)

deadman ((closed) loop) control Totmannreg(e)lung *f*

deadman ((open) loop) control Totmannsteuerung *f*

deadman switch Totmannschalter *m*

dead replacement time → replacement dead time

dead short circuit satte Kurzschluss, vollkommene Kurzschluss, metallische Kurzschluss, widerstand(s)lose Kurzschluss, komplette Kurzschluss *m*

dead space circuit Totzonenschaltung *f*

dead space slide-in unit Totzoneneinschubeinheit *f*

dead-tank circuit-breaker Kessel(leistungs)schalter, Leistungsschalter mit Kessel am Erdpotential *m*

dead-tank oil circuit-breaker Kessel(leistungs)-ölschalter *m*

dead time, lag time [*Time interval between a change in the input variable and the initiation of the corresponding change in the output variable*] Totzeit, Pausenzeit, stromlose Pause, spannungslose Pause, momentenfreie Pause *f*

dead (time) block, lag (time) block Totzeitbaustein, T-Baustein *m*

dead (time) element, lag (time) element Totzeitglied, T-Glied *n* [*DIN 19 226*]

dead true höchstgenau, haargenau

dead water [*It must be run to waste before hot water is obtained if long draw-off pipes are involved*] unnütze Wasser, nutzlose Wasser *n*

dead water [*Water in a boiler, settling tank, or similar equipment which fails to circulate to the extent required for proper functioning of the equipment*] zirkulationsarme Wasser, umlaufarme Wasser *n*

deadweight load Ruhelast *f*

deadweight tester Kolbenmanometer *n*, Druckwaage *f*

dead zone Totzone, Unempfindlichkeitszone *f*

dead zone after echo, dead band after echo, neutral zone after echo, neutral band after echo tote Zone nach Echo, Echototzone, Echoimpuls-Einflusszone *f* [*DIN 54 119*]

deaeration [*Removing air from some other substance or object*] Entlüftung *f*

deaeration bore Entlüftungsbohrung *f*

deaeration line Entlüftungsleitung *f*

deaeration opening Entlüftungsöffnung *f*

deaeration pipe Entlüftungsrohr *n*

deaeration time Entlüftungszeit *f*

deaeration valve Entlüftungsventil *n*

D

deaerator [*Apparatus for removing air from some other substance or object*] Entlüfter *m*

deaerator (boiler) feedwater heater [*For removing oxygen*] Entlüfter-Kesselspeisewasser(er)wärmer, Entlüfter-Kesselspeisewassererhitzer *m*

to deash, to remove ash(es) abschüren, austragen von Asche, entaschen

deashing, ash removal Abschüren *n*, Aschenaustrag *m*, Entaschung, Aschenbeseitigung *f*

deathnium (centre) (Brit.); deathnium (center) (US) Haftstelle *f*, Rekombinationszentrum *n* [*Halbleiter*]

debatable time [*In the keeping of computer usage statistics, time that cannot be attributed with certainty to any one of various categories of computer use*] umstrittene Zeit *f*

deblocking [*ISO 7498*] Entblocken *n* [*OSI-System*]

to debounce entprellen [*Kontakt*]

debouncing Entprellung *f*, Entprellen *n* [*Kontakt*]

to debug [*To test for, locate and remove mistakes from a program(me) or malfunction from a computer*] entstören, ausprüfen, beseitigen von Fehlern

to debug [*To detect and remove secretly installed listening devices popularly known as bugs*] „entwanzen"

debugger → anti-bugging device

debugger Fehlersuchgerät *n* [*Softwarehilfsmittel als Bestandteil des Prüf(ungs)automatenbetriebssystems zum Prüfen von Prüf(ungs)programmen*]

debugging „Entwanzen" *n*

debugging Entstörung, Fehlerbeseitigung, Störbeseitigung, Ausprüfung *f* [*Programm*]

debugging statement [*Temporary instructions inserted into a program(me) being tested so as to pinpoint problem areas*] Entstörungsanweisung, Ausprüfungsanweisung, Fehlerbeseitigungsanweisung, Störbeseitigungsanweisung *f*

debunching [*Electronics. The tendency of a beam of electrons in a velocity-modulated tube to spread because of mutual repulsion*] Phasendefokussierung *f*

to deburr abgraten, entgraten

deburring tool Abgrater, Entgrater, Gratabnehmer *m*

decade [*Interval between any two quantities having the ratio of 10:1*] Dekade *f*

decade addressing → (direct) decade addressing

decade counter Dekadenzähler *m*

decade scaler [*Electronics. A scale-of-ten electronic counter that delivers one output pulse for each group of ten input pulses*] Dekadenteiler *m*

decade switch Dekadenschalter *m* [*Schalter mit zehn definierten Zuständen je Dekadenstelle. Mit einem Dekadenschalter können z. B. bestimmte Sollwerte manuell vorgegeben oder Werkzeugkorrekturen berücksichtigt werden*]

decade switching, decade circuit Dekadenschaltung *f*

decade switch ((open) loop) control Dekadenschaltersteuerung *f*

decalcification, lime removal Entkalkung, Kalkbeseitigung, Kalkentfernung *f*, Kalkentzug *m*

to decalcify, to delime, to unlime entkalken

decapsulation Entkapselung *f*

decay → decomposition

decay method, step-down method, dilution method, washout method (Spurengas)Abklingmethode *f*, (Spurengas)Abklingverfahren *n*

decelerating electrode [*Electronics. An electrode to which a potential is applied to reduce the velocity of the electrons in the beam of an electron-beam vacuum tube*] Bremselektrode *f*

deceleration time, stop time [*For a storage medium, such as magnetic tape, that must be physically moved in order for reading or writing to take place, the minimum time that must elapse between the completion of a reading or writing operation and the moment at that motion ceases*] Auslaufzeit, Rücklaufzeit, Bremszeit *f*

decelerator, braking controller Bremsregler *m*

decentralized, distributed dezentral, verteilt

decentralized data processing, distributed data processing [*The collection and maintenance of data at individual subdivisions of an organisation or at various geographical locations, each of which may be connected to any or all of the other units*] dezentrale Datenverarbeitung, verteilte Datenverarbeitung *f*

decentralized intelligence → distributed intelligence

deceptive alarm Täuschungsalarm *m* [*Signal, das durch äußere Einflüsse und nicht von einem Schadenfeuer kommenden Einwirkungen auf die Brandmelder verursacht wird, z.B. Zigaretten- oder Zigarren- oder Pfeifenrauch, Dampf, Staub, Insekten, Arbeiten mit Rauch- und/oder Wärmeentwicklung, elektrische Störeinflüsse, usw.*]

decibel, dB [*A dimensionless quantity, used to describe a comparison of power levels, with the number of decibels equal to 10 times the logarithm of the ratio of the power*] Dezibel, 1/10 Bel *n*

decibel meter Dezibelmesser *m*, Dezibelmessinstrument, Dezibelmessgerät *n*

decidability Entscheidbarkeit *f*

decimal adjustment Dezimalberichtigung, Dezimalanordnung, Dezimalkorrektur *f* [*Der Inhalt eines Registers wird entnommen und zur Schaffung des dezimalen Äquivalents der Binärzahl neu geordnet*]

decimal code Dezimalkode, Dezimalcode *m* [*Kode, bei dem jede Stelle einen von zehn möglichen Zuständen einnimmt*]

decimal-code digit → **code decimal digit**

decimal counter Dezimalzähler *m*

decimal digit Dezimalziffer *f*

decimal notation Dezimalschreibweise *f*

decimal number Dezimalzahl *f*

decimal (number) system Dezimal(zahlen)system, Zehnersystem *n* [*Stellenwertsystem zur Basis 10, d. h. mit 10 unterschiedlichen Ziffern (0 bis 9)*]

decimal point [*See remark under "point"*] Dezimalkomma *n*

decimal scaler Dezimalteiler *m*

to decipher entschlüsseln

decision box Entscheidungssymbol *n* [*Symbol in Flussdiagrammen zur Kennzeichnung von Programmverzweigungen*]

decision content Entscheidungsgehalt *m*

decision element, decision gate [*A circuit that performs a logical operation such as "and", "or", "not", or "except" on one or more binary digits of input information representing "yes" or "no" and that expresses the result in its output*] Entscheidungselement *n*

decision feedback Empfangsbestätigung *f* [*Fernwirktechnik*]

decision instruction → **conditional jump**

decision pattern Entscheidungsmuster *n*

decision table Entscheidungstabelle *f*

decision tree Entscheidungsbaum *m*

deck, set of punched cards Lochkartensatz *m*

declination of sun angle, sun angle declination [*Angle of the sun above or below the equatorial plane. It is plus if north of the plane and minus if south*] Sonnenwinkelneigung *f*

de-clinkering [*boiler*] Schlackenaustrag(en) *m, (n)*

to decode [*To translate coded characters into a more understandable form*] entkodieren, entcodieren, entschlüsseln, dekodieren, decodieren

decoder Dekodierer, Decodierer, Entschlüsseler, Entkodierer, Entcodierer *m*

decoder matrix Entschlüsselungsmatrix, Entkodier(ungs)matrix, Dekodier(ungs)matrix *f*

decoding, decryption Entkodieren, Entcodieren, Entschlüsseln, Dekodieren, Decodieren *n*

decomposition, breaking up of structures, spoilage, decay [*Process of chemical change*] Zerfall *m*

to decontaminate entseuchen

decontamination Dekontamination *f* [*Beseitigung der Radioaktivität aus dem Wasser*]

decontamination agent, decontaminating agent Entseuchungsmittel *n*

decontamination installation, decontaminating installation Entseuchungsanlage *f*

decontamination shower, decontaminating shower Entseuchungsdusche *f*

decorative chains cord, ornamental chains cord Lichtkettenleitung *f*

decorative furniture, ornamental furniture, ornamental garnishment, fancy mounting, decorative garnishment Dekor(ations)beschläge, Zierbeschläge, Ornamentbeschläge, Schmuckbeschläge *f*

decorative lock, ornamental lock Dekor(ations)-schloss, Zierschloss, Ornamentschloss, Schmuckschloss *n*

decoratively styled, ornamentally styled dekorativ gestaltet

decorative sprinkler, ornamental sprinkler Dekor(ations)sprinkler, Schmucksprinkler, Ziersprinkler, Ornamentsprinkler *m* [*„Sprenger" kann synonym für „Sprinkler" verwendet werden*]

to decouple entkoppeln

decoupled end entkoppelte Ausgang *m*, entkoppelte Ende *n* [*Ausgang eines Funktionsgliedes, vorbereitet für das rückwirkungsfreie Zusammenschalten mit gleichartigen Ausgängen anderer Funktionsglieder. DIN 19 237*]

decoupling Entkopp(e)lung *f*

to decrement [*To decrease the value of a variable*] dekrementieren, vermindern, verringern, herabsetzen

decrement [*The amount by which the value of a variable is decreased*] Dekrement *n*, Verminderung(sbetrag), Verringerung(sbetrag) *f, (m)*

decryption → **decoding**

dedicated anwendungsspezifisch, zweckbestimmt, fest zugeordnet [*Bezeichnung von eine Aufgabe erfüllenden Systemen, Komponenten und Programmen*]

dedicated 1+1 protection 1+1-Ersatzschaltung *f* [*Ringnetz. Die zu schützende Transportgröße wird gedoppelt und über zwei unabhängige (disjunktive)*

D

Wege geführt. Die für den Fall der Ersatzschaltung bereitgehaltene Transportgröße wird nicht für andere Zwecke genutzt]

(dedicated) special system, (D)SS [*A third party system that is interfaced to the building management system. The interface is often made via a data interface unit (DIU)*] anwendungsspezifische automatische Regel-/Steuergerät, zweckbestimmte automatische Regel-/Steuergerät, fest zugeordnete automatische Regel-/Steuergerät *n*

dedusting → dust removal

deemphasis [*Process that has an amplitude-frequency characteristic complementary to that used for preemphasis*] Höhenabsenkung *f* [*Videoaufzeichnung*]

to deenergize → to switch off

deenergizing → cutout

deenergizing circuit, switching off circuit Abschalt(ungs)kreis *m*

deep-freezing → rapid-freezing

deep ground(ing) rod, (US); deep earth(ing) rod (Brit.) Tiefstaberder, Tieferdungsstab *m* [*Blitzschutz. Ein Staberder in größerer Tiefe als 6m eingebracht, damit er eine gut leitende Bodenschicht erreicht*]

deep-level tiefliegend

deep manhole [*A manhole of such depth that an access shaft is required for reaching it*] Tiefmannloch, Schachtmannloch *n*

deep seal trap → anti-siphon(age) trap

deep suction pump Tiefsaug(e)pumpe, Saug(e)tiefpumpe *f*

deep well; borehole [*misnomer*] [*A well in which the water level is more than 25 ft. from ground level*] Tiefbrunnen *m*

defacement Unkenntlichmachen *n*

default [*Computer engineering. A value, option, course of action or attribute that is assumed by the program(me) or system in the absence of other specification or direction by the user*] Standardannahme *f*

to defeat außer Betrieb setzen, außer Betrieb nehmen [*System; Anlage*]

defeat Außerbetriebsetzung, Außerbetriebnahme *f*

defeater key → interlock deactivating key

defeating Entriegeln *n*, Entrieg(e)lung *f* [*Schaltgerät*]

defect → failure

defect acquisition Mängelerfassung *f*

defect alarm → failure alarm

defect indicator → failure indicator

defective sample fraction → sample fraction defective

defect notification, defect notice, failure notification, failure notice Mangelanzeige, Mängelanzeige *f*

defect type Mängelart, Mangelart *f*, Mängeltyp, Mangeltyp *m*

deference [*local area network*] Zurückstellung *f*

deferrable load zeitlich verlagerbare Last *f*

deferred addressing (iteriert-)indirekte Adressierung *f*

deferred processing [*The making of computer runs which are postponed until nonpeak periods*] ausgesetzte Aufbereitung *f*

deferred state Aktualisierungsstand *m* [*graphische Datenverarbeitung. DIN 66 752*]

deficiency of water → water shortage

defined rest bestimmte Ruhelage, bestimmte Ruhestellung *f* [*Stellglied*]

deflagration Verpuffung *f*

deflashing tool Gratabnehmer, Entgrater, Gratbeseitiger *m*

deflecting Ablenken *n*, Ablenkung *f*

deflecting defocus(s)ing Unschärfe *f* des Leuchtpunktes durch Ablenkung [*Kat(h)odenstrahlröhre*]

deflecting device → baffle

deflecting sensitivity, deflecting sensitiveness, deflecting susceptibility Ablenk(ungs)empfindlichkeit *f* [*z. B. einer Elektronenstrahlröhre*]

deflection Ablenkung *f*

deflection factor, deflection coefficient [*Oscilloscope, IEC 351-1*] Ablenk(ungs)koeffizient *m*

deflection sheave → sheave

deflector → baffle

deflector lock Sicherheitsschloss, Angeberschloss, Schloss mit Angeber *n*

deflector plate → baffle plate

deflector system → baffling

defluxing Flussmittelentfernung, Flussmittelbeseitigung, Entmagnetisierung, Neutralisierung *f*

defocus(s)ing Unscharfwerden *n* [*Optik*]

to defrost abtauen, entfrosten, enteisen

defrosting [*Planned process to remove or prevent accumulation of frost or ice on coils of refrigerating units*] Entfrostung, Abtauung, Enteisung *f*, Abtauen, Entfrosten, Enteisen *n*

defrosting clock Abtau(ungs)uhr, Entfrostungsuhr, Enteisungsuhr *f*

defrosting ((closed) loop) control Entfrostungsreg(e)lung, Abtau(ungs)reg(e)lung, Enteisungsreg(e)lung *f*

defrosting heat Entfrostungswärme, Enteisungswärme, Abtau(ungs)wärme *f*

defrosting heating Entfrostungsheizung, Enteisungsheizung, Abtau(ungs)heizung *f*

defrosting ((open) loop) control [*Act of controlling the refrigerating cycle or heating cycle to periodically melt the unacceptable accumulation of ice on evaporator tubes, windows, etc.*] Entfrostungssteuerung, Abtau(ungs)steuerung, Enteisungssteuerung *f*

defrosting suppression Entfrostungsunterdrückung, Enteisungsunterdrückung, Abtau(ungs)unterdrückung *f*

defuzzification Defussifizierung, Erzeugung eines scharfen Wertes *f*

to degas, to remove gas entgasen

degassifier Entgaser *m*

to degauss, to deperm, to demagnetize [*Computer engineering. To erase informations from a magnetic tape, disk, or drum by demagnetizing the oxid(e) surface*] löschen, entmagnetisieren

degaussing key, demagnetizing key, deperming key Entmagnetisierungstaste, Löschtaste *f* [*Bildschirmgerät*]

degenerative amplifier gegengekoppelte Verstärker, Gegenkopp(e)lungsverstärker *m*

degenerative delayed feedback → delayed negative feedback

degenerative feedback, negative feedback Gegenkopp(e)lung *f*

degenerative voltage Gegenkopp(e)lungsspannung *f*

degradation Verschlechterung *f*

degradation failure, degradation outage [*A failure in a device that arises from change in a parameter and causes the device to fall below minimum standards*] Verschlechterungsausfall, Verschlechterungsfehler *m*, Verschlechterungsversagen *n*

degraded performance, degraded output, degraded service Leistungsverschlechterung *f*

degreaser, degreasing agent [*A chemical, usually an organic liquid, that removes natural fats, oils, and waxes*] Entfetter *m*, Entfettungsmittel, Fettbeseitigungsmittel *n*

degreaser [*A machine used to remove grease or oil from an object*] Entfetter *m*, Entfettungsmaschine, Fettbeseitigungsmaschine *f*

degreasing, grease removal Entfetten *n*, Fettbeseitigung, Fettentfernung, Entfettung *f*

degree below zero, below-zero degree Kältegrad, Minusgrad *m*

degree-day, DD [*Unit based on temperature difference and time, used in estimating fuel consumption and specifying nominal heating load of a building in winter. For any one day, when the mean temperature is less than 65 °F, there exist as many degree days as there are degrees Fahrenheit difference in temperature between the mean temperature for the day and 65 °F*] Gradtag *m*, Heizgradtag [*Die englische Erklärung ist nicht auf deutsche Verhältnisse anwendbar. DIN 4108 (Wärmeschutz im Hochbau)*]

degree-days number → number of degree-days

degree hour Gradstunde *f*

degree of membership, membership degree Zugehörigkeitsgrad *m* [*Fuzzy-Logik*]

degree of priority, priority degree Dringlichkeitsstufe *f*, Dringlichkeitsgrad *m*

degree sign Gradzeichen *n* [*Kode*]

dehumidification coil, dehumidifying coil Entfeuchtungs(rohr)schlange *f*

dehumidification load, dehumidifying load Entfeuchtungslast *f* [*Dampfmassenstrom, der aus dem Raum abgeführt werden muss, um einen bestimmten Luftzustand aufrecht zu halten*]

to dehumidify entfeuchten

dehumidifying, dehumidification Entfeuchten *n*, Entfeuchtung *f* [*Verringern der absoluten Luftfeuchte. DIN 1946, Teil 1*]

dehydration, moisture removal [*Chemistry. The removal of water from a compound*] Dehydration, Dehydrierung *f*, Wasserentzug *m*, Entfeuchten *n*

dehydrator, dehumidifier [*1. Device in a refrigerating system that contains a desiccant. 2. Refrigeration device placed in the main air line of a pneumatic control system to reduce moisture*] Entfeuchter *m*

deicer, de-icer Entfroster, Enteiser *m*

deicer solution, de-icer solution Entfrosterlösung, Enteiserlösung, Enteisungslösung *f*

deicing, de-icing [*Intentional melting of an ice layer*] Enteisung, Entfrostung *f*

deinterleaving Entwürfelung *f* [*Siehe Anmerkung unter „geschachtelte Adressierung"*]

deion circuit breaker [*It is designed so that the current-interrupting arc is magnetically blown into a stack of insulated copper plates; the resulting short arcs in series become almost instantaneously deionized when the current falls to zero in the AC cycle, so the arc cannot reform*] Magnetblasschalter *m*

D

de-latching coil, latch release coil, unlatching coil Entklinkungsspule *f*

delay Laufzeit *f*

delay, lag Verzögerung *f*, Verzug *m*

delay angle → lag angle

delay bistable multivibrator → delay flipflop (circuit)

delay correction net(work) [*telegraphy*] Laufzeitentzerrer *m*

delay counter [*A register in a CPU that can temporarily delay a program(me) long enough to complete the execution of an operation*] Verzögerungsregister, Verzugsregister *n*

delay distortion → (envelope) delay distortion

delayed action ball valve, delayed action floating valve [*It is recommended for use in any application where it is desirable to have float controlled, full-flow delivery and/or rapid shut off*] Verzögerungsschwimmerventil, Verzögerungskugelventil, Verzugsschwimmerventil, Verzugskugelventil *n*

delayed and elastic negative feedback, delayed and elastic degenerative feedback verzögerte und nachgebende Rückführung, verzögerte und nachgebende Gegenkopp(e)lung *f* [*Rückführung, die anfangs nur langsam eingreift, wodurch infolge der hohen Verstärkung eine große Änderung der Stellgröße erfolgt. Die Rückführung wächst dann bis zu einem Höchstwert an, um anschließend wieder auf 0 abzusinken*]

delayed and stored command verzögerte und gespeicherte Kommando *n*

delayed automatic gain control, delayed agc, delayed automatic volume control verzögerte automatische Verstärkungsreg(e)lung, automatische verzögerte Verstärkungsreg(e)lung *f*

delayed bipolar three-step controller, delayed bipolar three-step regulator, delayed bipolar three-position controller, delayed bipolar three-position regulator, delayed bipolar triple-step controller, delayed bipolar triple-step regulator, delayed bipolar triple-position controller, delayed bipolar triple-position regulator Dreipunktschrittregler *m* [*Bipolarer Dreipunktregler, bei dem von beiden Reglerausgängen je eine verzögert wirkende Rückführung hinter den Vergleichern einwirkend angelegt ist. DIN 19 225*]

delayed combustion [*It occurs after gas exceeds the furnace volume of a boiler*] Verzugsverbrennung, Verzögerungsverbrennung *f*

delayed ignition, ignition delay Zündverzögerung, Verzögerungszündung, Verzugszündung *f*, Zündverzug *m*

delayed instruction Verzögerungsbefehl, Verzugsbefehl *m* [*nicht gespeichert*]

delayed negative feedback, delayed degenerative feedback, negative delayed feedback, degenerative delayed feedback verzögerte Rückführung, verzögerte Gegenkopp(e)lung *f* [*Rückführung, bei welcher ein mit der Zeit bis zu einem festen Wert wachsender Anteil der Ausgangsgröße auf den Eingang zurückgeführt wird, wodurch die wirksame Verstärkung im Laufe der Zeit auf einen bestimmten Wert absinkt*]

delay element of second order → second order delay element

delay flipflop (circuit), delay bistable multivibrator [*See remark under "bistable multivibrator"*] Verzögerungs-Triggerschaltung, D-Triggerschaltung, bistabile Verzögerungskippstufe *f*, bistabile Verzögerungsmultivibrator, bistabile D-Multivibrator *m*

delay fuse → (time-)delay fuse

delay interval relay → interval (time-)delay relay

delay line, DL Verzögerungsleitung, Verzugsleitung, Laufzeitkette, Ausgleichsleitung *f*

delay path Vorlaufweg, Schallweg zur Prüf(ungs)strecke *m*

delay relay → (time-)delay relay

delay time, equivalent dead time Verzugszeit, Verzögerungszeit *f*

delay timing, equivalent dead timing Verzögerungszeitabstimmung, Verzugszeitabstimmung *f*

delay unit [*A timed element within a control circuit*] Verzögerer *m*, Verzögerungseinheit *f*, Verzugseinheit

delay valve Verzögerungsventil, Verzugsventil *n*

D-element → derivative element

to delete löschen, tilgen, streichen

deleted gestrichen, getilgt, gelöscht

deleted representation [*In paper tape codes, the superposition of a pattern of holes upon another pattern of holes representing a character, to effectively remove or obliterate the latter*] endgültige Darstellung *f*

deleterious schädlich

deletion Löschung, Tilgung *f*

deletion record [*It removes and replaces an existing record when it is added to a file*] Lösch(ungs)aufzeichnung, Tilgungsaufzeichnung *f*

deletion time Lösch(ungs)zeit, Tilgungszeit *f*

deliberate wastage of water → bleed-off

to delime → to decalcify

delimiter Trenner *m* [*Unterteilung von Textdateien*]

delimiter [*A character that separates items of data*] Begrenzungszeichen, Abgrenzungszeichen *n*

delivery Übergabe *f*

delivery Auslieferung *f*

delivery Zustellung *f*

delivery agent Übergabesystemteil *m, n*

delivery confirmation Übergabebestätigung *f*

delivery pump Förderpumpe *f*

delivery scope → **scope of delivery**

delivery side, discharge side, pressure side [*compressor*] Druckseite *f*

delivery stroke [*piston*] Abgabehub *m*

delta/delta transformer Delta-Delta-Transformator, Delta-Delta-Trafo *m*

delta-parallel-star circuit, Dahlander (pole changing) circuit Dahlanderschaltung *f*

deluge shower (rose) → **drench shower (rose)**

deluge (sprinkler) system → **drench (sprinkler) system**

deluxe (bath)tub → **luxury (bath)tub**

deluxe fluorescent tube, luxury fluorescent tube Luxusleuchtstoffröhre *f*

deluxe shower, luxury shower Luxusdusche *f*

to demagnetize → **to degauss**

demagnetizer Entmagnetisierer *m*

demand Forderung *f*

demand, requirement, need Bedarf *m*, Nachfrage *f*

demand → **prompt note**

demand → **demand (load)**

demand assessment period → **reading-off period**

demand assignment → **DA**

demand (closed-loop) controlled, requirement (closed-loop) controlled bedarfsgeregelt

demand (data) processing, immediate (data) processing sofortige (Daten)Verarbeitung *f*

demand-dependent bedarfsabhängig

demanded position Sollposition, Sollstellung *f*

demanded value → **desired value**

demand factor Gleichzeitigkeitsfaktor, Bedarfsfaktor, Nachfragefaktor *m* [*Verhältnis des Leistungsbedarfs zur installierten Leistung*]

demand-guided → **requirement-guided**

demand interval clock signal Messperiodentakt *m*

demand (load) [*Electric power delivered by equipment; expressed in kilowatts, kilovolt-amperes, or other suitable unit at a given instant or averaged over any designated period of time*] Belastung *f*

demand note → **prompt note**

demand-orien(ta)ted bedarfsorientiert, bedarfsbezogen

demand period → **requirement period**

demand rate, demand price Leistungspreis *m*

demand-side-management → **(energy) demand-side-management**

demineralization, desalting, desalination, water demineralization, water desalting, water desalination (Wasser)Entsalzung *f*

demineralizer, water demineralizer (Wasser)Entsalzungsgerät *n*

demodulation Demodulation *f* [*Abtrennung der durch einen modulierten hochfrequenten Träger übertragenen niederfrequenten Schwingung in einem Empfänger*]

demonstration equipment Vorführ(ungs)gerät(e) *n*, (*f*)

to demount → **to disassemble**

demounted → **disassembled**

demounting → **disassembling**

demultiplexer Demultiplexer *m* [*Funktionseinheit, die ein an einem Eingang anliegendes Digital- oder Analogsignal durch Vorgabe einer Adresse (Adressensteuerung) auf unterschiedliche Ausgangskanäle (4, 8, 16) schaltet*]

denominator polynominal Nennerpolynom *n*

dense smoke Qualm *m*

density Dichte *f* [*Masse geteilt durch Volumen. So hat z. B. 1dm³ Wasser bei 40 °C eine Masse von 1kg. Die Wasserdichte ist somit 1kg/dm³*]

density bottle, specificgravity bottle, density flask, specificgravity flask [*A bottle-shaped instrument containing a thermometer, used to measure and compare the densities of liquids and solids*] Dichteflasche *f*

density ratio Dichteverhältnis *n*

deodorization Geruchsbeseitigung *f*

department(al) store passenger lift (Brit.); department(al) store passenger elevator (US) Kaufhaus-Fahrstuhl, Warenhaus-Fahrstuhl, Kaufhaus-Personenaufzug, Warenhaus-Personenaufzug *m*

departure from nucleate boiling, DNB, burnout Filmsieden *n* [*Eine Siedeform in Abhängigkeit des Dampfgehaltes. Bei Blasenströmung sammeln sich die Blasen überwiegend in der Nähe der beheizten Wand an, während im Kern der Strömung die Wasserphase vorherrscht. Damit ist der maximale Dampfgehalt in der Nähe der Heizfläche. Löst sich die Blasenschicht von der Wand, steigt bei gegebener Wärme-stromdichte die Wandtemperatur stark an und es entsteht Filmsieden*]

dependency Abhängigkeit *f*

D

dependent exchange [*automatic telephony*] Unterzentrale, Nebenzentrale *f*

depending on the effective direction wirkrichtungsabhängig

to deperm → **to degauss**

dephased phasenverschoben

dephlegmator, stripping column, partial condenser Teilkondensator *m*

depletion-FET Verarmungs-Feldeffekt-Transistor *m*

depletion layer, barrier (junction), barrier layer Verarmungsschicht, Sperrschicht, Sperrlage, Verarmungslage *f* [*Raumladungszone an einer Metall-Halbleiterschicht*]

depletion type Verarmungstyp *m* [*Feldeffekt-Transistor*]

deposited frost → **accumulated frost**

deposition → **arrangement**

depressurized, pressureless, unpressurized drucklos, ohne Druck

depth [*The vertical dimension below a horizontal reference datum*] Tiefe *f*

depth Bautiefe *f* [*z. B. bei einem Radiator*]

depth-adjustable, depth-variable, depth-controllable tiefenverstellbar, tiefenjustierbar, tiefen(ein)stellbar

depth channel Höhe „n" *f* [*U-Stahl*]

depth cueing Verschwinden *n* von Bildern in der Dunkelheit, Bildtiefensimulation *f*

depth of discharge, DOD Entladetiefe *f* [*Unterschied zwischen der Nennkapazität und dem Ladezustand einer Batterie*]

depth of (water) seal, depth of trap seal (Wasser-) Verschlusshöhe *f* [*Geruchsverschluss*]

DER, out of order gestört, außer Betrieb [*PAD-Netzmeldung*]

to derate, to underload lastmindern, unterbelasten [*Elektrotechnik*]

derating, underload(ing) Unterlast, Unterbelastung, Lastminderung, Lastabnahme *f*

derating factor Unterlast(ungs)faktor *m* [*Quotient aus Betriebsgröße und ihrem Nennwert, z. B. Betriebsspannung/Nennspannung oder Betriebsstrom/Nennstrom*]

derivation Ableitung, Herleitung *f*

derivative action → **rate action**

derivative(-action) controller → **D(-action) controller**

derivative(-action) element → **differentiator**

derivative(-action) time → **rate time**

derivative term Differentialterm, Differenzialterm, Differenzialterm *m*, Differentialglied *n*

derivative time, Td [*The time it takes the proportional term to match the derivative term contribution when the error changes linearly with time*] Differentialzeit, Differenzialzeit *f*

derivative unit → **differentiator**

DES, Data Encryption Standard Datenverschlüsselungsnorm *f* der USA

desalination, desalting Entsalzung *f*, Entsalzen *n*

desalination by freezing, freezing desalination Gefrierentsalzung *f*, Gefrierentsalzen *n*

to desalt, to desalinate entsalzen

to descale, to remove boiler scale entfernen von Kesselstein, beseitigen von Kesselstein, entkrusten

descaling, (boiler) scale removal, removal of (boiler) scale (Kessel)Steinbeseitigung, (Kessel-) Steinentfernung, Entkrustung, Krustenbeseitigung *f*

descaling Entzundern *n*, Entzunderung *f*

descending absteigend, abwärts

descending direction Abwärtsrichtung, absteigende Richtung *f*

descending sequence absteigende (Reihen)Folge, Abwärts(reihen)folge *f*

descrambler Entschlüsseler, Entwürf(e)ler *m* [*Er entschlüsselt einen Verwürf(e)ler*]

description Bezeichnung *f* [*Beschriftung im Zeichnungsschriftfeld*]

description Beschreibung *f*, Text *m*

description card Beschreibungskarte *f*

descriptive language, description language Beschreibungssprache *f*

descriptor [*A phrase or word used to identify a document in a computer-based information storage and retrieval system*] Beschreibungskennzeichen *n*, Beschreiber *m*

descriptor Hauptstichwort, Suchwort, Schlagwort, Kennwort, Schlüsselwort *n* [*Begriffsbeschreibendes Wort, um Daten in einer Datenbank zu finden*]

desiccant [*Absorbent or adsorbent, liquid or solid, that removes water or water vapo(u)r from a material*] wasserentziehende Mittel, wasserentziehende Material, Wasserentzugsmittel, Wasserentzugsmaterial *n*, wasserentziehende Stoff, Wasserentzugsstoff *m*

desiccant cooling Entfeuchtungskühlung mit adiabat(isch)er Verdunstung, Kühlung durch Entfeuchtung und adiabat(isch)er Verdunstung *f*

to design, to style ausgestalten, gestalten

to design konstruieren, entwerfen, auslegen

design, type, version Bauweise, Bauart, Konstruktion, Auslegung *f*, Aufbau, Entwurf *m*

design, styling Stilgebung, Bauform, Ausgestaltung, Gestaltung *f*

designation [*An item of data forming part of a computer record that indicates the type of record and thus determines how it is to be processed*] Bezeichnung *f*

designation punch, designation hole, control punch, control hole, function(al) punch, function(al) hole [*A hole in a punched card indicating what is the nature of the data on the card or which functions are to be performed by the computer*] Funktionsloch, Bezeichnungsloch *n*

design automation → **DA**

design calculation Entwurfsberechnung, Auslegungsberechnung, Konstruktionsberechnung *f*

design current (of a circuit), nominal current (of a circuit) Nennstrom *m* (eines Stromkreises) [*VDE 0110, Teil 200*]

design department Konstruktionsabteilung *f*

design dimensions → **design size**

design engineering, DE Konstruktionstechnik *f*

designer → **interior decorator**

designer Konstrukteur *m*

design external temperature, design outdoor temperature, design outside temperature, external design temperature, outdoor design temperature, outside design temperature [*For heating load calculations, the outdoor temperature that is equalled or exceeded during 97 $^1/_2$ % of the hours in December through march; for cooling load calculations, the outside temperature that is equalled or exceeded by 2 $^1/_2$% of the total hours of June to through September*] Außenberechnungstemperatur *f*

design flow temperature Auslegungsvorlauftemperatur, Entwurfsvorlauftemperatur, Konstruktionsvorlauftemperatur *f*

design handling, designing Konstruieren, Auslegen, Entwerfen *n*

design handling stage, designing stage Konstruktionsphase, Auslegungsphase, Entwurfsphase *f*

design heat loss, design thermal loss Entwurfswärmeverlust, Konstruktionswärmeverlust, Auslegungswärmeverlust *m*

designing → **design handling**

design(ing) engineer Konstrukteur *m*

design load Entwurfslast, Konstruktionslast, Auslegungslast *f*

design load peak, design peak load [*Instantaneous load that a system is expected to meet*] Entwurfsspitzenlast, Entwurfslastspitze *f*

design pressure Konstruktionsdruck, Entwurfsdruck, Auslegungsdruck *m*

design rating Entwurfsleistung, Konstruktionsleistung, Auslegungsleistung *f*

design return temperature Auslegungsrücklauftemperatur, Entwurfsrücklauftemperatur, Konstruktionsrücklauftemperatur *f*

design review Auslegungsprüfung, Konstruktionsprüfung, Entwurfsprüfung *f* [*DIN 55 350, Teil 16*]

design size, design dimensions Konstruktionsgröße, Konstruktionsmaße, Konstruktionsabmessungen *f* [*„Entwurfs..." und „Auslegungs..." können synonym für „Konstruktions..." verwendet werden*]

design standards and criteria manual Standardleistungshandbuch *n*, StLHB

design stress Auslegungsspannung, Konstruktionsspannung, Entwurfsspannung *f* [*mechanisch*]

design study Vorentwurf, Voraufbau *m*, Vorkonstruktion, Vorausführung, Vorbauweise, Vorbauart *f*

design temperature map Klimakarte *f* [*Zum Entwurf von Kälte-, Heizungs- und Klimaanlagen*]

design value Entwurfswert, Konstruktionswert *m*

design velocity Konstruktionsgeschwindigkeit, Entwurfsgeschwindigkeit, Auslegungsgeschwindigkeit *f*

design voltage [*Specific voltage for which a line or piece of equipment is designed. A reference level of voltage for identification, and not necessarily the precise level at which it operates. IEC 458*] Bauspannung *f* [*VDE 0712, 102*]

design working pressure [*1. Maximum allowable working pressure to which a vessel (Gefäß) is designed. 2. Maximum allowable working pressure for which a part of a system is designed*] Konstruktionsarbeitsdruck, Auslegungsarbeitsdruck, Entwurfsarbeitsdruck *m*

desired state, desired condition Sollzustand *m*, Sollbeschaffenheit *f*

desired value, demanded value Sollwert *m* [*Gewünschter Wert einer Größe zu einem bestimmten Zeitpunkt und bestimmten Bedingungen*]

desired value deviation Sollwertabweichung *f*

desk check(ing) Schreibtischkontrolle *f*

desk research Schreibtischforschung *f* [*Auswertung sekundär-statistischen Materials zur Markt- und Meinungsforschung*]

desk(top) (Schreib)Tisch *m*

D

desk(top) accessory (Schreib)Tischzubehör *n* [*selten: m*]

desk(top) calculator (Schreib)Tischrechenmaschine *f*

desk(top) computer → workstation (computer)

desk(top) data (Schreib)Tischdaten *f*

desk(top) data file (Schreib)Tischdatei *f*

desk(top) device, desk(top) appliance (Schreib-)Tischgerät *n*

desk(top) facsimile unit (Schreib)Tischfernkopierer *m*

desk(top) publishing → DTP

desk(top) writer, desk(top) printer (Schreib)Tischdrucker *m*

to desmoke → to remove smoke

destination, target Ziel *n* [*1. Datenempfänger; 2. Datensenke*]

destination address, target address, DA Zieladresse, Bestimmungsadresse *f*

destination network, target network Zielnetz, Bestimmungsnetz *n*

destination point Ziel-Zeichengabestelle, Zielkennung *f* [*Signalisierungssystem Nr. 7*], Zielzeichengabepunkt *m*

destination point code → DPC

destratification fan Entschichtungsventilator *m*

destructive fire Schadenfeuer *n*, Schadenbrand *m*

destructive reading [*Reading that partially or completely erases the stored information as it is being read*] (aus)löschende Lesen, zerstörende Lesen *n*

destructor [*A furnace for the combustion of household refuse collected by local authorities*] Haus(halts)müllverbrennungsofen *m*

desuperheater Enthitzer *m*

desuperheating Enthitzung *f*

detachable, remov(e)able abnehmbar [*Geräteteil*]

detachable act(uat)or, remov(e)able act(uat)or (Betätigungs)Vorsatz *m* [*Hilfsstromschalter*]

detachable plugboard, remov(e)able plugboard [*A control panel that can be removed from the computer or other system and exchanged for another without altering the positions of the plugs and cords*] abnehmbare Schalttafel *f*

detachable specular reflector, remov(e)able specular reflector Einsatzspiegel *m* [*Deckenleuchte*]

detached representation aufgelöste Darstellung *f* [*DIN 40 719, Blatt 1*]

detaching pin, cylinder detaching pin Auslösestift, Zylinderauslösestift *m* [*Schlosszubehör*]

detectability Erkennbarkeit Nachweisbarkeit, Erfassbarkeit *f*,

detectable erkennbar, nachweisbar, erfassbar

detecting instrument Indikatormessgerät *n* [*Zur annähernden Messung einer Größe und/oder des Größenvorzeichens*]

detection, recognition Erkennung, Erfassung *f*, Nachweis *m* [*Eines Ereignisses und dessen Meldung*]

detection Meldung *f* [*eines Ereignisses*]

detection system, detecting system Meldesystem *n*

detection threshold → recognition differential

detectivity Detektivität *f* [*Eine als Kehrwert der äquivalenten Rauschleistung (NEP) festgelegte Größe (D = 1/NEP). Sie dient als Gütemaß für die Nachweisgrenze von Photodetektoren*]

detectophone Lauschmikrophon, Abhörmikrophon *n*

detector, detection device, detecting device Melder, Meldungsgeber *m*, Meldegerät *n*

detector, detection device, detection appliance Suchgerät, Nachweisgerät *n*, Suchvorrichtung, Nachweisvorrichtung *f*, Sucher *m*

detector indicator system, detecting device indicator system, detection device indicator system [*sound level meter*] Gleichrichtungs- und Anzeigesystem *n* [*Schallpegelmesser*]

detector lock Angeberschloss, Entdeckerschloss, Schloss mit Angeber, Schloss mit Entdecker, Sicherheitsschloss *n*

(detector) lock of Bramah → Bramah('s) (detector) lock

(detector) lock of Chubb → Chubb('s) (detector) lock

detector power efficiency, detecting device efficiency, detection device efficiency [*IEC 147-1*] Leistungs-Richtwirkungsgrad *m* [*Diode*]

detent pin [*lock*] Anhalt(e)stift, Sperrstift *m*

determination Festlegung, Bestimmung *f*

detonation guard Detonationssicherung *f*

detrimental to health → insanitary

detritor, sand catch(er), sand collector Sandfang, Sandfänger *m*

detritus Fanggut, Fangmaterial *n*, Fangstoff *m*

detunable verstimmbar

developed length abgewickelte Länge, gestreckte Länge *f*

development Bebauung *f*

development system Entwicklungssystem *n* [*System zur rationellen Prüfung von Software und Hardware und deren Zusammenspiel beim Entwickeln eines Mikrorechnersystems*]

deviation [*ANSI C85.1*] → **system deviation** [*IEC 50(351)*]

deviation Abweichung *f*

deviation from linearity Linearitätsabweichung *f*

deviation level, level of deviation Abweichungsstand *m*

deviation limit(ing) → **limit(ing) deviation**

deviative ((closed) loop) control abweichende Reg(e)lung *f*

device [*An electronic element that cannot be divided without destroying its stated function; commonly applied to active elements such as transistors and transducers*] Element, Bauelement *n*

device [*In automatic data processing, a general-purpose term used, often indiscriminately, to refer to a computer component or the computer itself*] Einrichtung *f* [*E DIN 19 226, Teil 5*]

device → **computer**

device → **computer component**

device, appliance, piece of equipment, item of equipment, equipment piece, equipment (item) [*A mechanism, tool, or other piece of equipment designed for specific uses*] Gerät, Organ *n*, Vorrichtung *f*, Gegenstand *m*

device address, appliance address Geräteadresse, Vorrichtungsadresse *f*

device assignment, appliance assignment [*The use of a logical device number used in conjunction with an input/output instruction, and made to refer to a specific device*] Gerätezuordnung, Vorrichtungszuordnung *f*

device circuit, appliance circuit [*A brand circuit that supplies outlets specifically intended for appliances. Lighting is not supplied from such circuits and the number of outlets is generally limited*] Geräteschutz, Vorrichtungsschutz *m*

device clear, DCL, appliance clear Gerät *n* im Grundzustand, Vorrichtung *f* im Grundzustand

device control character, DC character, appliance control character Gerätesteuer(ungs)zeichen, Vorrichtungssteuer(ungs)zeichen *n* [*E DIN 44 300 Teil 2. Ein Übertragungssteuer(ungs)zeichen, das den Betrieb von am Übertragungsvorgang beteiligten Bauteilen beeinflusst*]

device description language → **DDL**

device error, appliance error Gerätefehler, Vorrichtungsfehler *m*

device load, appliance load Gerätelast, Vorrichtungslast *f*

device outlet, appliance outlet [*An outlet connected to an appliance circuit. It may be a single or duplex receptacle or an outlet box intended for direct connection to an appliance*] Gerätesteckdose, Vorrichtungssteckdose *f* [*z. B. an einem Elektroherd zum Anschluss von Küchengeräten*]

device under test, DUT, appliance under test Prüfling *m*

Dewar flask, Dewar vessel, vacuum bottle [*A silvered glass flask with double walls, the space between them being evacuated. It is used for the storage of liquid air*] Dewargefäß, Thermosgefäß *n*, Dewarflasche, Thermosflasche *f*

dewetting Entnetzen *n* [*Gegensatz: Benetzen*]

dew point → **100% relative humidity**

dew point air conditioning plant, dew point air conditioning installation Taupunkt(klima(tisierungs))anlage *f*

dew point barrier Taupunktsperre *f*

dew point ((closed) loop) control, dew point regulation Taupunktreg(e)lung *f*

dew point damper Taupunktklappe *f*

dew point determination Taupunktbestimmung, Taupunktfestlegung *f*

dew point gradient Taupunktgefälle *n*

dew point relay Taupunktrelais *n*

dew point sensor, dew point feeler Taupunktsensor, Taupunktfühler *m*

dew point system probe [*e. g. Dewcel (Foxboro-Yoxall). This probe detects the point at which equilibrium occurs between the vapo(u)r pressure exerted by the moisture diluted solution film of lithium chlorid(e) on the one hand and the partial pressure of the water in the air on the other*] Taupunktsystemsonde *f*

dexterity Geschicklichkeit, Behändigkeit *f*

dezincification Entzinkung *f*

D flip-flop → **D bistable element**

DFT → **design for testability**

DGP → **data gathering panel**

DI → **digital input**

DI → **comfort index**

diac, diode alternating current switch, bidirectional diode thyristor, diode ac switch, three-layer diode, triple-layer diode, trigger diode [*Electronics. A two-terminal, three-layer avalanche semiconductor diode characterized by its ability to conduct symmetrically with voltages of either polarity; used mainly in AC*

D

power-control applications] zweipolige Wechselstromschalter, Diac *m*, Zweirichtungs-Thyristordiode, Zweiwegschaltdiode *f* [*Der Diac entspricht im Aufbau der Antiparallelschaltung von 2 Vierschichtendioden*]

diagnostic code field, code diagnostic field Diagnosefeld *n*

D

diagnostic program (US); diagnostic programme (Brit.); diagnostic routine Fehlersuchprogramm, Diagnoseprogramm *n*

diagonal (bath)tub Diagonal(bade)wanne *f*

diagonal fan Diagonalventilator *m*

diagonal slating, drop-point slating Diagonal Schiefer(dach)deckung *f*

diagram → chart

diagram ((open) loop) control [*IEC 113-2*], **circuit diagram ((open) loop) control** Schaltplansteuerung *f*

to dial wählen [*Telefonnummer*]

to dial anwählen [*Datenübertragung*]

dial, dial plate, circular scale, circular dial Zifferblatt *n*

dial [*A separate scale or other device for initiating the value to which a control is set*] Skalenscheibe, Skalenplatte *f*

dial [*In automatic telephone switching, a type of calling device which, when wound up and released, generates pulses required for establishing connections*] Wähl(er)scheibe, Nummernscheibe, Zifferscheibe *f*

dial central office [*Telephone or teletypewriter office where necessary automatic equipment is located for connecting tow or more users together for communicating purposes*] Wählvermittlungsstelle *f*

dial circuit Uhrenlinie *f*

dial converter Wählumsetzer *m*

dial exchange [*A telephone exchange area in which all subscribers originate calls by dial(l)ing*] Wähl(er)-scheibenvermittlung, Nummern(scheiben)vermittlung, Zifferscheibenvermittlung *f*

dial gauge; dial gage (US) Messuhr *f*

dial impulse ratio Wählimpulsverhältnis *n*

dial-in (Netz)Einwählen *n*, (Netz)Einwahl *f*

dial(led) communication link → switched communication link

dial(led)-in telephone access, dial(led)(-in telephone) input Wählscheibeneingabe, Zifferscheibeneingabe *f*

dialling (Brit.); dialing (US) [*automatic telephony*] Wählen *n*, Wahl *f*

dialling directory (Brit.); dialing directory (US) Wählverzeichnis *n*

dialling keyboard, dialling keys (Brit.); dialing keyboard, dialing keys (US) Wähltastatur *f*

dialling net(work) (Brit.); dialing net(work) (US) Wählnetz *n*

dial lock Ringschloss, Buchstabenschloss *n*

dial mercury thermometer, pointer mercury thermometer Quecksilberzeigerthermometer, Zigerquecksilberthermometer *n*

dialog(ue) Dialog *m* [*Ein Ablauf, bei dem zur Abwick(e)-lung einer Arbeitsaufgabe der Benutzer – in einem Dialogschritt oder mehreren Dialogschritten – Daten eingibt und jeweils Rückmeldung über die Verarbeitung dieser Daten erhält. DIN 66 234, Teil 3*]

dialog(ue) [*The interaction or conversation between computer and user*] Dialog *m*, Wechselrede *f*

dial-out (Netz)Auswählen *n*, (Netz)Auswahl *f*

dial plate → dial

dial potentiometer Drehpotentiometer, Drehpotenziometer *n*

dial pulse, DP Pulswahl *f* [*Modem*]

dial segment → segment

dial selection rotary Nummernschalterwahl *f*

dial-up telephone network Fernsprech(wähl)netz, Telefon(wähl)netz *n*

dial thermometer, pointer thermometer [*One that indicates temperature by a pointer moving over a circular scale*] Zifferblatt-Thermometer, Zeigerthermometer *n*

dial time Wählzeit *f* [*Die Zeitspanne, während der das Programm nach dem Absenden des Wählbefehls auf eine positive oder negative Antwort des Modems auf den Verbindungswunsch, wie CARRIER oder BUSY, wartet. Bei Zeitüberschreitung wird die Anwahl abgebrochen*]

dial tone Amtston *m* [*Fernsprecher*]

dial up line, dial up connection Wählverbindung *f*

diamond(-shaped) handle, lozenge (-shaped) handle Brilliantgriff *m*

diaphragm, membrane Membran, Membrane *f* [*selten*]

diaphragm backflow preventer, membrane backflow preventer, diaphragm backflow preventing device, membrane backflow preventing device Membran(e)-rückflussverhinderer *m*

diaphragm ball valve, membrane ball valve, diaphragm globe valve, membrane globe valve [*A float-operated valve in which the flow of water is controlled by a diaphragm*] Membran(e)schwimmerventil *n*

diaphragm chamber, membrane chamber; measuring chamber [*deprecated*] Messraum *m* [*trockene Gaszähler*]

diaphragm ((closed) loop) controller Membran(e)regler *m*

diaphragm compressor, membrane compressor [*Type of oil-free compressor in which flexion of a diaphragm creates compression*] Membran(e)verdichter, Membran(e)kompressor *m*

diaphragm control, membrane control Blendensteuerung *f* [*Kamera*]

diaphragm dosing pump, membrane dosing pump, dosing diaphragm pump, dosing membrane pump Membran(e)dosierpumpe, Dosiermembran(e)pumpe *f*

diaphragm element, membrane element Plattenfedermesswerk, Membran(e)messwerk *n*

diaphragm (expansion) tank, diaphragm (expansion) vessel, membrane (expansion) tank, membrane (expansion) vessel [*In a sealed hot water system, a closed expansion tank divided by a rubber wall that expands with the water when the water is heated. The other side of the tank wall contains nitrogen gas which compresses easily and helps prevent the water pressure rising too much*] Membran(e)(druck)ausdehnungsgefäß *n* [*DIN 30 600, Reg.-Nr. 06173*]

diaphragm filter, membrane filter Membran(e)filter *m, n*

diaphragm globe valve → diaphragm ball valve

diaphragm keyboard, membrane keyboard, touch keyboard, diaphragm keypad, membrane keypad, touch keypad Folientastatur *f*, Folientastenfeld, Folientastenwerk *n*

diaphragm pressure tank, membrane pressure tank Membran(e)druckbehälter *m*

diaphragm pump, membrane pump Membran(e)pumpe *f*

diaphragm safety valve, membrane safety valve Membran(e)sicherheitsventil, Sicherheitsmembran(e)ventil *n*

diaphragm valve, membrane valve [*A valve whose action is controlled by fluid pressure on a diaphragm*] Membran(e)ventil *n*

diathermancy [*The property of a substance that allows radiant heat to pass through the substance*] Diathermie *f*

DIB → data input bus

dibit Bitpärchen *n*

di-box → direct injection box

dichlorodifluoromethan(e) → refrigerant-12

dichloromethan(e) → refrigerant-30

dichloro(mono)fluoromethan(e) → refrigerant-21

dichlorotetrafluoroethan(e) → refrigerant-114

dichotomizing search [*In automatic data processing, a procedure for searching an item in a set, in which, at each step, the set is divided into two parts, one part being then discarded if it can be logically shown that the item could not be in that part*] binäre Suche, eliminierende Suche, Intervallschachtelungssuche *f*

dichotomy [*In automatic data processing, a division into two subordinate classes; for example, all white and all nonwhite, or all zero and all nonzero*] Dichotomie, Zweiteilung *f*, Binarismus *m* (*in Begriffspaare*)

dictionary window Wörterbuchfenster *n*

die Siliziumplättchen *n*

to die away, to fade (out), to die down [*sound*] abklingen, verklingen, schwächer werden

dielectric constant (value), relative dielectric constant (value), relative permittivity, specific inductive capacity, SIC [*1. For an isotropic medium, the ratio of the capacitance of a capacitor filled with a given dielectric to that of the same capacitor having only a vacuum as dielectric. 2. More generally, 1 + yx, where y is 4ð in Gaussian and cgs electrostatic units or 1 in rationalized mks units, and x is the electric susceptibility tensor.*] dielektrische Konstantwert, Dielektrizitätskonstantwert *m*, dielektrische Konstante, Dielektrizitätskonstante *f*

dielectric soak → absorption

dielectric sensitiveness, dielectric susceptibility, dielectric sensitivity [*A dimensionless parameter measuring the ease of polarisation of a dielectric, equal (in meter-kilogram-second units) to the ratio of the polarisation to the product of the electric field strength and the vacuum permittivity*] dielektrische Aufnahmefähigkeit *f*, dielektrische Aufnahmevermögen *n*

diesel (engine) heat pump Diesel(motor)wärmepumpe, Diese(motor)-WP *f*

die sinking Gesenkfräsen *n* [*Nur mit numerischer Steuerung möglich*]

difference equation, differential equation Differenzialgleichung, Differentialgleichung *f*

difference in pressure → differential pressure

difference spectrophotometer → absorption spectrophotometer

differential (air) grill(e) Differentialluftgitter, Luftgitter mit veränderlichem Widerstand, Differenz(ial)luftgitter *n*

differential air pressure Differenz(ial)luftdruck, Differentialluftdruck *m*

differential air pressure detector, differential air pressure sensor Differentialluftdruckfühler, Differentialluftdruckdetektor, Differentialluftdrucksensor *m*, Differentialluftdruckdetektionsgerät *n*

D

D

differential analyser → (digital) differential analyzer

differential bolometer, differential thermal detector, differential thermal detection device, differential thermal detecting device [*See remark under "bolometer"*] (Thermo)Differentialmelder, (Thermo)Differentialdetektor *m*, (Thermo)Differentialmeldegerät *n* [*Er spricht auf einen vorbestimmten Temperaturanstieg pro Zeiteinheit (Steilheit der Temperaturkurve) an. Übliche Ansprechwerte sind 5 °C/min oder 10 °C/min*]

differential controller, differential CTRL, derivative(-action) controller, differential-action controller, rate-action controller [*1. Device used to maintain a given difference in pressure or temperature between two pipelines or spaces. 2. Controller relative to the difference between values of two variables*] Differenz(ial)regler, Differentialregler *m*, Differentialregelorgan, Differenz(ial)regelorgan *n*, Differentialregel(ungs)einheit, Differenz(ial)regel(ungs)-einheit *f*

differential detector, differential detecting device, differential detection device Differentialmelder, Differenz(ial)melder *m*, Differentialmeldegerät, Differenz(ial)meldegerät *n* [*Brandmelder*]

differential draught gauge; differential draft gage (US) Differentialzugmesser, Differenz(ial)zugmesser *m*

differential dump, change dump differentielle Speicherauszug *m* der Änderungen

differential element Differentialglied, D-Glied, Differen(ial)glied *n*

differential forward resistance Differentialdurchlasswiderstand, Differenz(ial)durchlasswiderstand, differentielle Durchgangswiderstand *m* [*DIN 41 760*]

differential gap Schaltdifferenz *f* [*Differenz zwischen oberem und unterem Umschaltwert*]

differential heat detector, differential heat detecting device, differential heat detection device Differentialwärmemelder, Wärmedifferentialmelder *m*, Differentialwärmemeldegerät, Wärmedifferentialmeldegerät *n* [*Dieser Melder spricht an wenn der Wärmeanstieg pro Zeiteinheit einen Wert für genügend lange Zeit überschreitet*]

differential linearity differentielle Linearität, differenzielle Linearität *f* [*Sie beschreibt den Höchstunterschied zwischen zwei be-nachbarten Digitalisierungsstufen abzüglich dem Idealwert von 1 Bit mit niedrigstem Stellenwert. DIN 44 472*]

differential linearity error differentielle Nichtlinearität, differenziale Nichtlinearität, differentiale Nichtlinearität *f*, differentielle Linearitätsfehler *m* [*DIN 44 472*]

differential line technique Differentiallinientechnik, Differenziallinientechnik *f*

differential magnetoresistor Differentialfeldplatte, Differenz(ial)-feldplatte *f*, magnetische Differentialwiderstand *m*

differential oil pressure switch → oil pressure cutout

differential oscilloscope probe Differentialtastkopf, Differenz(ial)tastkopf *m* [*Er hat zwei gegenüber einer Masse „schwebende" Messeingänge zum erdfreien Messen mit dem Oszilloskop*]

differential phase shift(ing) differentielle Phasenverschiebung, differenzielle Phasenverschiebung, differentiale Phasenverschiebung *f*

differential phase shift(ing) keying, DPSK differentielle Phasenumtastung, differenzielle Phasenumtastung, Phasenunterschied-Modulation *f*

differential piston compressor → stepped piston compound compressor

differential pressure, difference in pressure, pressure difference Druckdifferenz *f*, Differentialdruck, Differenz(ial)druck, Wirkdruck, Druckunterschied, Differenzialdruck *m* [*Die Differenz zweier Drücke p_1 und p_2 wird Druckdifferenz Äp = $p_1 - p_2$ oder auch, wenn sie selbst Messgröße ist, Differenzdruck $p_{1,2}$ genannt.*]

differential pressure controller Differentialdruckregler, Differenz(ial)druckregler, Druckdifferenz(ial)-regler, Wirkdruckregler, Druckdifferentialregler *m*

differential pressure gauge → manometer

differential pressure monitoring Differenz(ial)-drucküberwachung *f*

differential pressure overflow valve Differenz(ial)-drucküberströmventil *n*

differential pressure pickup, differential pressure measured value transmitter Differenz(ial)druckmesswertgeber, Differentialdruckmesswertgeber *m*

differential pressure sensor Differenz(ial)druckfühler, Differentialdruckfühler *m* [*z. B. Messblende*]

differential pressure switch Differentialdruckschalter, Differenzdruckschalter, Druckdifferenz(ial)-schalter, Wirkdruckschalter, Druckdifferentialschalter *m*

differential pressure vacuum steam heating, differential pressure steam vacuum heating Vakuum-Differenzdruck-Dampfheizung, Differenz(ial)druck-Vakuumdampfheizung, Differentialdruck-Vakuumdampfheizung *f* [*Sie arbeitet mit konstantem oder in Abhängigkeit vom Wärmebedarf abgestuftem Differenzdruck zwischen Dampf- und Kondensatnetz*]

differential temperature ((closed) loop) controller Temperaturdifferenz(ial)regler, Differenz(ial)temperaturregler *m*

differential thermal detector → **differential bolometer**

differential thermometer → **bimetal(lic) thermometer**

differential transducer [*A dual-input, single-output sensor that simultaneously senses two separate sources and provides an output proportional to the difference between them*] Differential(mess)umformer *m*

differential transformer, linear variable-differential transformer, LVDT [*A transformer that joins two or more signal sources to a common transmission line*] Differentialtransformator, Aufnehmer *m*

differential travel "H", on-off differential hysteresis Schalthysterese "H" *f* [*(An)Näherungsschalter*]

to differentiate → **to discriminate**

differentiating circuit Differenzierschaltung *f*

differentiator, derivative(-action) element, derivative(-action) unit Differenzierer *m* [*Glied, dessen Ausgangsgröße proportional zur Änderungsgeschwindigkeit der Eingangsgröße ist*]

diffractional pulse-height discriminator → **pulse-height selector**

diffraction grating, diffraction grate Beugungsgitter *n*

diffuse daylight, zenithal light diffuse Tageslicht, Zenitlicht, gestreute Tageslicht, zerstreute Tageslicht, diffundierte Tageslicht, unscharfe Tageslicht *n*

diffused lighting, diffused illumination [*Light that is not predominantly incident from any particular direction*] unscharfe Beleuchtung, diffuse Beleuchtung, gestreute Beleuchtung, zerstreute Beleuchtung, diffundierte Beleuchtung *f*

diffuse-field room → **reverberation room**

diffuse-field sensitivity Übertragungsfaktor *m* im diffusen Feld [*DIN IEC 651*]

diffuseness [*A measure of the directiveness of the light. Diffuse light comes from many directions*] Unschärfe, Zerstreuung *f*

diffuser, diffusor licht(zer)streuende Körper, lichtdiffundierende Körper, (Licht)Streukörper *m*

diffuser, diffusor Filterkerze *f* [*Beiderseits offene oder einseitig geschlossene keramische Hohlzylinder zur Filterung von Wasser oder Luft und zur feinblasigen Verteilung von Luft in Wasser*]

diffuser, diffusor [*An outlet which discharges the air in a spreading jet*] Diffusor *m*, Austritt(s)teil *m, n*

diffuser, diffusor [*Material placed in front of the lamp for the purpose of controlling the light flux produced*] licht(zer)streuende Material, lichtdiffundierende Material *n*

diffuse radiation [*That radiation which does not result from the direct incidence of solar rays on an object but the type of radiation experienced on overcast days*] diffuse Strahlung, gestreute Strahlung, zerstreute Strahlung, diffundierte Strahlung *f*

diffuser bowl, diffusor bowl, diffusing glass cover Abschlusswanne, streuende Leuchtenwanne, Leuchten(streu)wanne, Streuleuchtenwanne *f*, streuende Leuchtenglas, Leuchten(streu)glas, Streuleuchtenglas *n*

diffuse reflection [*A type of reflection in which the reflected light is spread out in all directions. The reflection of a matte finish surface*] diffuse Reflexion, gestreute Reflexion, zerstreute Reflexion, diffundierte Reflexion *f* [*„Rückstrahlung" kann synonym für „Reflexion" verwendet werden*]

diffuser luminaire, diffusor luminaire, diffusing luminaire, diffuser light(ing) fixture, diffusor light(ing) fixture, diffusing light(ing) fixture licht(zer)streuende Leuchte, lichtdiffundierende Leuchte *f*, licht(zer)streuende Beleuchtungskörper, lichtdiffundierende Beleuchtungskörper *m*

diffusing glass cover → **diffuser bowl**

diffusing lens Streulinse *f*

diffusing screen, diffusion screen Streuschirm, Weichzeichner *m*, Streuscheibe *f*

diffusing unit → **illumination diffuser**

diffusing wall → **illuminated wall**

diffusion absorption heat pump, absorption diffusion heat pump Diffusions-Absorptions-Wärmepumpe, DAWP, Absorptions-Diffusions-Wärmepumpe *f*

diffusion area Diffusionsfläche *f*

diffusion barrier Diffusionssperre *f*

diffusion coefficient → **mass diffusivity (ratio)**

diffusion flame Diffusionsflamme *f*

diffusion ignition burner Diffusionszündbrenner *m*

diffusion temperature difference [*Temperature difference between the air temperature at supply openings and design outdoor temperature*] Diffusionstemperaturunterschied *m*

diffusion-tight, diffusion-proof diffusionsdicht

diffusion transistor Diffusions-Transistor *m* [*Ein Flächentransistor, dessen Zonenfolge durch Eindiffundieren von Fremdstoffen in einen Halbleiter entstanden ist*]

diffusor → **diffuser**

diffusor efficiency, diffuser efficiency Diffusorwirkungsgrad *m*

digested sludge Faulschlamm *m*

digestion → **(sludge) digestion**

D

digestion plant → **(sludge) digestion plant**

digestion tank → **(sludge) digestion tank**

digit, numeral Ziffer *f* [*Kleinste Einheit in einem digitalen System. Kann im Dezimalsystem die Werte der zehn Dezimalziffern annehmen, in einem Binärsystem die Ziffern Null und Eins. Im Englischen bedeutet digit auch eine Stelle in einer Zahl*]

digit absorbing selector [*automatic telephony*] Nummer(n)schlucker, Zahl(en)schlucker, Ziffer(n)-schlucker *m*

digital [*A representation of a value by means of discretely variable physical quantities*] digital, ziffernmäßig

digital-analog(ue) conversion → **digital(-to)-analog(ue) conversion**

digital-analog(ue) converter → **digital(-to)-analog(ue) converter**

digital-analog(ue) input module → **digital(-to)-analog(ue) input module**

digital-analog(ue) signal analysis, analog(ue) digital signal analysis digitale Analogsignalverarbeitung, digitale Verarbeitung analoger Signale *f*

digital anemometer [*Generally a revolving vane anemometer the shaft of which provides impulses (electric, magnetic, etc.) which are picked up by an electronic circuit and transmitted to the indicating apparatus*] Digitalanemometer *n*, Digitalwindmesser *m*

digital audio broadcast(ing), DAB Digitalhörfunk, Digitalrundfunk *m*

digital audio tape, DAT digitale Audiosignalband *n*

digital audio taperecording → **DAT**

digital BA → **digital building (services) automation**

digital building (service) automation digitale Gebäudeautomation, digitale GA, DGA *f* [*Siehe Anmerkung unter „Gebäudeautomation"*]

digital cash digitale Zahlungsverkehr *m*

digital clock timer → **digital time switch**

digital ((closed) loop) control, digital feedback control, digital closed control, digital regulation Digitalreg(e)lung, digitale Reg(e)lung *f*

digital ((closed) loop) control circuit, digital feedback control circuit, digital regulation circuit, digital servo loop Digitalregel(ungs)schaltung, Digitalreglungsschaltung *f*

digital ((closed) loop) controller Digitalregler *m*

digital coded, DC digitalkodiert, digitalcodiert

digital compact controller, digital compact regulator, digital miniaturized controller, digital miniaturized regulator digitale Prozessregler, Kompaktprozessregler, Digitalkompaktregler, Kompaktdigitalregler, Prozesskompaktregler *m*

digital computer digitale Rechenanlage, digitale Datenverarbeitungsanlage *f*, digitale Rechner, Ziffernrechner, Digitalrechner *m* [*Die Gesamtheit der Baueinheiten, aus denen ein digitales Rechnersystem aufgebaut ist. E DIN 44 300, Teil 5*]

digital controller Digitalregler *m* [*Die Eingangsgröße oder die Ausgangsgröße oder beide sind digitale Signale. DIN 19 225*]

digital cordless telephony, cordless digital telephony schnurlose digitale Fernsprechverkehr, digitale schnurlose Fernsprechverkehr *m*

digital counter Digitalzähler, digitale Zähler *m*

digital data, discrete data [*Data represented by a sequence of code characters*] Digitaldaten, digitale Daten *f* [*E DIN 44 300, Teil 2*]

digital data processing system, discrete data processing system digitale Datenverarbeitungssystem, digitale Rechnersystem *n* [*E DIN 44 300, Teil 5*]

digital data receiver, DDR, discrete data receiver Digitaldatenempfänger *m*

digital data recorder, discrete data recorder [*An electronic device converting continuous electrical analogous signals into number (digital) values and records these values onto a data log via a high-speed typewriter*] digitale Datenschreiber, Digitaldatenschreiber *m*

(digital) differential analyzer, (D)DA, (digitized) differential analyser [*It uses numbers to represent analogous quantities*] (Digital)Differentialanalysator *m*

digital direct ((closed) loop) control → **direct digital ((closed) loop) control**

digital direct controller → **direct digital controller**

digital direct ((open) loop) control → **direct digital ((open) loop) control**

digital display → **digital readout**

digital domestic bus D2D-Bus *m* [*Zur Kopp(e)lung von Haushalt(s)geräten*]

digital dynamic ((closed) loop) control → **DDC**

digital engineering digitale Technik, Digitaltechnik *f* [*als Lehre*]

digital equipment Digitalgerät(e) *n*, (*f*)

digital freeze frame memories digitale Bildarchivierung *f*

digital frequency display digitale Frequenzanzeige *f*

digital heating controller Digitalheiz(ungs)regler *m*

digital image processing digitale Bildverarbeitung *f*

digital image stabilizer, DIS digitale Bildschirmstabilisator *m*

digital image storage system, digital picture storage system Digitalbildspeichersystem *n*

digital incremental controller, digital incremental regulator digital-inkrementale Regler *m* [*Regler, der an seinem Eingang mit wertdiskreten, zeitkontinuierlichen Binärsignalen arbeitet. DIN 19 225*]

digital indicator [*Elevator car position indicator with numerals on a rotating drum or readout slide viewed through a small lens. The numeral indicates the floor that the elevator car is passing*] Digitalanzeiger *m*

digital indicator (unit) [*A device that indicates the magnitude of a measured quantity in digital numbers, usually in decimal or binary form*] Digitalanzeige(r-)(einheit) *f*, (*m*), (*f*)

digital information processing digitale Informationsverarbeitung *f*

digital information(s) digitale Rechnerdaten *f*

digital ink digitale Tinte *f*

digital input, DI Digitaleingabe, DE *f*

digital input, DI Digitaleingang, DE *m*

digital input timer module, digital input timer block Digitaleingabe-Zeitbaustein *m*, Digitaleingabe-Zeitmodul *n*

digital integrated circuit, integrated digital circuit integrierte Digitalschaltung *f*

digital integrator [*In automatic data processing, a device for compute definite integrals in which increments in the input variables and output variables are represented by digital signals*] Digitalintegrator *m*

digitalization error, digitization error Umsetzungsfehler *m* [*DIN 44 472*]

digital keyboard, digital keypad Zifferntastatur *f*, Zifferntastenfeld, Zifferntastenwerk *n*

digital level digitale Pegel *m*

digital line unit, DLU digitale Leitungseinheit *f*

digital locking system digitale Schließ- und Öffnungssystem *n* [*Mit diesem System werden Türen berührungslos geöffnet und geschlossen*]

digital measuring step digitale Messschritt *m*

digital miniaturized controller → digital compact controller

digital modulation [*A method of placing digital traffic on a microwave system without use of modems, by transmitting the information in the form of discrete phase or frequency states determined by the digital signal*] Digitalmodulation *f*

digital monitoring Digitalüberwachung *f*

digital motion detector, digital motion detecting device, digital motion detection device Digitalbewegungsmeldegerät *n*, Digitalbewegungsmelder *m*

digital multi(ple)-contact switch → digital step(ping) switch

digital multiplier [*A multiplier that accepts two numbers in digital form and gives their product in the same digital form, usually by making repeated additions; the multiplying process is simpler if the numbers are in binary form wherein digits are represented by a 0 or 1*] Multipliziervorrichtung *f*

digital net(work) control → DNC

digital ((open) loop) control digitale Steuerung, Digitalsteuerung *f* [*Eine Steuerung wird digitale Steuerung genannt, wenn sie innerhalb der Signalverarbeitung mit digitalen Signalen arbeitet und vorwiegend zahlenmäßig dargestellte Informationen verarbeitet. E DIN 19 226 T 5, V DIN 19 237*]

digital paging system digitale Personenrufsystem *n*

digital parallel-coded controller, digital parallel-coded regulator digital-parallelkodierte Regler *m* [*Regler, der an seinem Eingang mit parallel anliegenden, wertdiskreten, zeitkontinuierlichen Signalen arbeitet. DIN 19 225*]

digital phase shifter [*In electronics, a device which provides a signal phase shift by the application of a control pulse; a reversal or phase shift requires a control pulse of opposite polarity*] Digitalphasenschieber *m*

digital picture storage system → digital image storage system

digital printer [*A printer that provides a permanent readable record of binary-coded decimal or other codes data in a digital form that may include some or all alphanumeric characters and special symbols along with numerals*] Digitaldrucker *m*

digital pulse duration modulation, DPDM digitale Impulslängenmodulation, digitale Schrittdauermodulation *f*

digital quantity Digitalmenge *f* [*Variable Menge, die nur bestimmte, diskrete, quantisierte Werte annehmen kann und durch eine Anzahl von Ziffern darstellbar ist*]

digital radio relay signal Digitalrichtfunksignal *n*

digital readout, digital display, DD, DR Digitalanzeige *f* [*Sie stellt den anzuzeigenden Wert nicht analog dar, also etwa durch einen Zeiger auf einer Skala, sondern in Ziffern als Flüssigkristallanzeige, Gasentladungsanzeige oder durch Leuchtdioden*]

digital recorder Digitalaufzeichner *m*

digital recording [*Magnetic recording in which the information is first coded in a digital form, generally with a binary code that uses two discrete values of residual flux*] Digitalaufzeichnung *f*

D

digital regulation → **digital ((closed) loop) control**

digital regulation circuit → **digital ((closed) loop) control circuit**

digital serial coded controller, digital serial coded regulator digital seriell kodierte Regler *m* [*Er arbeitet an einem Eingang mit einem seriell anliegendem, wertdiskreten und zeitdiskreten Binärsignal, bzw. mit bipolaren Signalen. DIN 19 225*]

digital serial computer → **serial digital computer**

digital servo loop → **digital ((closed) loop) control circuit**

digital signal digitale Signal, Digitalsignal *n* [*1) Ein Signal, dessen Informationsparameter eine endliche Anzahl von Wertebereichen annehmen kann, wobei jedem Wertebereich als Ganzem eine bestimmte Information zugeordnet ist; E DIN 19 226, T 5. 2) Ein Signal, bei dem die Werte der Signalparameter eine Nachricht oder Daten darstellen, die nur aus Zeichen besteht bzw. bestehen; E DIN 44 300, T 2.*]

digital signal processing, DSP digitale Signalverarbeitung, Digitalsignalverarbeitung *f*

digital signal processor, DSP digitale Signalprozessor, DSP, Digitalsignalprozessor *m* [*Er moduliert oder demoduliert ein Signal mit großer Genauigkeit, indem er das Ausgangssignal direkt aus dem Eingangssignal und den mathematischen Modulationsfunktionen errechnet. Seine große Rechenleistung ergibt nur unwesentliche Signalverzögerungen*]

digital simulation [*The representation of a system in a form acceptable to a digital computer as opposed to an analogue computer*] Digitalsimulation, Digitalvortäuschung, Digitalnachahmung *f*

digital statement Digitalanweisung *f*

digital step(ping) switch, digital step switching mechanism, digital multi(ple)-contact switch Digitalstufenschalter, Stufendigitalschalter *m*

digital subscriber line digitale Teilnehmeranschluss *m*

digital subscriber signal(l)ing system (No.1), DSS1 DSS1-Protokoll, normierte ISDN-Protokoll für Europa *n*

digital system monitoring digitale Systemüberwachung *f*

digital technique digitale Technik, Digitaltechnik *f* [*als betriebliche Anwendung*]

digital telecommunication digitale Nachrichtenübertragung *f*

digital telemetering [*In automatic data processing, a conversion of a continuous electrical analogue signal into a digital (number system) code prior to transmitting the signal to a receiver (such as digital readout, card punch, or tape)*] Digitalfernmessung, Ferndigitalmessung *f*

digital thumbwheel switch Zahlenrollensteller, Ziffernrollensteller *m*

digital timer-counter module, digital timer-counter block Digital-Zeit-Zählerbaustein *m*, Digital-Zeit-Zählermodul *n*

digital time switch, digital clock timer, digital clock relay Digitalschaltuhr, Schaltdigitaluhr *f*

digital(-to)-analogue conversion, D/A conversion Digital-Analog-Umsetzung, Digital-Analog-Wandlung, D-A-Umsetzung, D-A-Wandlung *f*

digital(-to)-analogue converter, DAC, D/A converter Digital-Analog-Wandler, D-A-Wandler, Digital-Analog-Umsetzer, D-A-Umsetzer, DAU *m* [*Er setzt ein digitales Eingangssignal in ein analoges Ausgangssignal um. DIN 19 226, Teil 5*]

digital(-to)-analogue input modul, digital(-to)-analogue input block Digital-Analogeingabebaustein *m*, Digital-Analogeingabemodul *n*

digital(-to)-digital interface Digital/Digital-Schnittstelle *f*

digital(-to)-time converter digitale Zeitumsetzer, Digitalzeitumsetzer *m*

digital transducer [*A transducer that measures physical quantities and transmits the information as coded digital signals rather than as continuously varying currents or voltages*] Digitalwandler *m*

digital (transmission) device digitale Übertragungsgerät, Digitalübertragungsgerät *n*

digital (transmission) link digitale Übertragungsabschnitt *m*

digital (transmission) system digitale Übertragungssystem *n*

digital universal computer → **universal digital computer**

digital value Digitalwert, Stellenwert, digitale Messwert *m*

digital value indicator digitale Messwertanzeiger *m* [*Siehe Anmerkung unter „Messwertanzeiger"*]

digital variable digitale Veränderliche *f*

digital versatile disc, digital versatile disk optische Videofilmspeicher *m* [*Er gibt Videofilme in aufgezeichneter Güte und mit sechskanaligem Konoton wieder*]

digital video effects → **DVE**

digital video motion detection digitale Videobewegungsmeldung, digitale Videobewegungserkennung *f*, digitale Videobewegungsnachweis *m*

digital video motion detector, digital video motion detecting device, digital video motion detection device Digitalvideobewegungsmeldegerät *n*, Digitalvideobewegungsmelder *m*

digital voltmeter, DVM Digitalvoltmeter, DVM *n*; Digitalspannungsmesser *m* [*Ein Instrument zum Messen elektrischer Spannungen*]

digit coding → numeric coding

digit delay element [*A logic element that introduces a delay of one digit period in the series of signals or pulses*] ziffernweise Verzögerungsvorrichtung *f*

digit emitter [*A character emitter limited to the twelfth-row pulses in a punch(ed) card*] Verteiler, Impulsgeber, (Loch)Kartenimpulsgeber *m*

digit filter [*A device used with punch(ed)-card equipment that detects the presence of a designation punch in a specified card column*] Ziffernfilter *m, n*

digitization error → digitalization error

to digitize [*To convert an analogue measurement of a quantity into a numerical value*] digital darstellen, ziffernmäßig darstellen, digitalisieren

digitizer → analog(ue)(-to)-digital converter

DIL, dual-in-line „Käfer" *m*; doppelt in Reihe [*Schaltungsphase mit in zwei Reihen angeordneten Anschlüssen*]

dilatation Ausdehnung *f* [*Physik*]

DIL package → dual-in-line package

dilute, weak [*solution*] schwach, dünn [*Lösung*]

dilution method → decay method

to dimension dimensionieren, abmessen, bemessen

dimension, distance [*A distance (e. g. between two points, lines, or planes). The definition relates to the geometric concept of a dimension. In ordinary usage, the word "dimension" is sometimes used to denote a specified size; this reference is made in a mechanical drawing to a labeled measure in a straight line of the width, height, or thickness of a part, the angular position of a line, or the location of a detail such as a hole or boss*] Abstand *m*, Entfernung, (Ab)weite, Distanz *f*

dimension Dimension, Abmessung *f*, Maß *n*

dimension analysis [*A technique that involves the study of dimensions of physical quantities, used primarily as a tool for obtaining information about physical systems too complicated for full mathematical solutions to be feasible*] Dimensionsanalyse, Abmessungsanalyse, Maßanalyse, Dimensionskontrolle, Abmessungskontrolle *f*

dimensional check(ing) Maßkontrolle *f*

dimensional constant [*A physical quantity whose numerical value depends on the units chosen for fundamental quantities but not on the system being considered*] Dimensionskonstante, Abmessungskonstante, Maßkonstante *f*

dimensional data Maßangaben, Maßdaten *f*

dimensional drawing Maßzeichnung *f*

dimensional stability, dimensional quality Maßhaltigkeit, Maßtreue *f*

dimension(ed) sketch Maßskizze *f*

dimensioning, sizing [*Assigning of dimensions to a mechanical drawing*] Dimensionieren, Bemessen, Bemaßen, Auslegen, Abmessen *n*

dimensioning factor → sizing factor

dimensioning table → sizing table

dimensionless group, nondimensioned group [*A combination of dimensional and dimensionless quantities possessing zero overall dimensions; an example is the Reynolds number*] dimensionslose Gruppe *f*

dimensionless number, nondimensioned parameter [*A ratio of various physical properties (such as density or heat capacity) and conditions (such as flow rate or weight) of such nature that the resulting number has no defining units of weight, rate, and so on*] dimensionslose Zahl *f*

dimension line Maßlinie *f*

dimension theory [*The study of abstract notions of the dimensions, which are topological invariants of a space*] Maßtheorie *f*

dimethylbenzene toluol mix(ture) → xylene-toluene mix(ture)

diminished radix complement Stellenkomplement *n* [*Numerale; E DIN 44 300/10.85*]

diminishing pipe → taper(ed) pipe

dimmed light gedämpfte Licht, Dämmer(ungs)-licht, Sparlicht *n*

dimmer [*Varying number of rheostats mounted in a bank and controlled by interlocking levers, used for dimming large rooms or spaces and stage lighting to various desired levels of illumination*] Dämmerlichteinsteller, Spar(beleuchtungs)einsteller, Dämmer(ungs)einsteller *m*

dimming switch, dimmer switch Dämmerlichtschalter, Dämmer(ungs)schalter, Spar(beleuchtungs)-schalter; Abblendschalter *m* [*Fehlbenennung, weil mit diesem Schalter auch aufgeblendet werden kann*]

diner Speisegast *m*

dining establishment Speiselokal, Speiserestaurant *n*, Speisegaststätte *f*

dinner elevator (US) → dumbwaiter

DIN rail → top-hat rail

diode [*Electron tube having two electrodes, a cathode and an anode*] Diode *f*

diode ac switch → trigger diode

diode alternating current switch → diac

D

D

diode decoupling unit Diodenentkopp(e)lungseinheit *f*

diode laser Diodenlaser *m*

diode logic, DL Diodenlogik *f*, DL

diode safety barrier Diodensicherheitssperre *f*

diode transistor logic, DTL Dioden-Transistor-Logik *f*, DTL

diode transistor continuity tester Dioden-Transistor-Durchgang(s)prüfer *m*

diode valve Diodenventil *n*

DIP → dual in-line package

dipole antenna (US); dipole (aerial) [*Straight radiator usually fed in the centre. Maximum radiation is produced in the plane normal to its axis*] Dipol(antenne) *m*, (*f*)

dip pipe, inverted siphon Düker *m*

dipstick, dip rod Peilstab, Messstab, Tauchstab, Eintauchstab *m*

dip switch → dimming switch

dip tank, immersion tank [*Liquid-holding process tank in which components to be treated can be immersed*] Tauchtank, Eintauchtank *m*

direct access, random access [*The facility to obtain data from a storage device, or to enter data into a storage device in such a way that the process depends only on the location of that data and not on a reference to data previously accessed*] Direktzugriff, direkte Zugriff *m*

direct access memory, DAM Direktzugriffsspeicher *m* [*Speicher mit der kürzestem Zugriffszeit, also ein bistabiler Multivibrator (flipflop) oder Ringkernspeicher. Magnettrommelspeicher und Magnetplattenspeicher werden auch als Direktzugriffsspeicher bezeichnet*]

direct-acting controller direkt wirkende Regler, unmittelbar wirkende Regler *m* [*Er arbeitet ohne Hilfskraft*]

direct-acting diaphragm valve, direct-acting membrane valve [*It closes with the admission of fluid pressure to a diaphragm and opens when pressure is reduced*] direkt wirkende Membran(e)ventil, unmittelbar wirkende Membran(e)ventil *n*

direct-acting thermostat [*1. Thermostat in which the output increases proportionally to an increase in the value of the sensed variable. 2. Pneumatic instrument which permits passage of control air when temperature of the controlled medium is higher than the temperature for which the instrument is set and cuts off the air when temperatures go below the temperature for which the instrument is set*] direkt wirkende Thermostat, direkt wirkende Temperaturregler, unmittelbar wirkende Thermostat, unmittelbar wirkende Temperaturregler *m*

direct address direkte Adresse, unmittelbare Adresse *f*

direct addressing direkte Adressierung, unmittelbare Adressierung *f* [*Die Adresse des Operanden ist Befehlsbestandteil*]

direct air cooler Direktluftkühler *m*

direct air cooling system Direktluftkühl(ungs)system *n* [*Kaltlufterzeugung durch Kältemaschinen an den Versorgungsstellen*]

direct air heating system Direktluftheiz(ungs)system *n* [*Warmlufterzeugung mittels Brennstoff, Strom oder Wärmepumpe an der Versorgungsstelle*]

direct allocation, direct assignment, direct allotment [*A system in which the storage locations and peripheral units to be assigned to use by a computer program(me) are specified when the program(me) is written, in contrast to dynamic allocation*] direkte Zuordnung, direkte Zuweisung, direkte Zuteilung, unmittelbare Zuordnung, unmittelbare Zuweisung, unmittelbare Zuteilung *f*

direct-axis subtransient reactance Anfangsreaktanz, subtransiente Reaktanz *f*, Anfangsbandwiderstand, subtransiente Bandwiderstand *m* [*Synchrongenerator*]

direct building (transfer) station, direct transfer station direkte Übergabestation, direkte Anschlussstation, direkte Gebäudeübergabestation, direkte Hausübergabestation, direkte Hausanschlussstation *f* [*„Unmittelbare“ kann synonym für „direkte“ verwendet werden*]

direct call Direktruf *m*

direct code [*A code in which instructions are written in the basic machine language*] Direktkode, Direktcode *m*

direct connection, galvanic connection galvanische Verbindung, direkte Verbindung *f*

direct-contact heat exchanger [*Apparatus in which the fluids exchanging are brought into contact with each other*] Direktkontaktwärme(aus)tauscher *m* [*Siehe Anmerkung unter „Wärme(aus)tauscher (im weiteren Sinne“)*]

direct control → direct ((open) loop) control

direct cooling [*The evaporator is in direct contact with the material or space cooled, or is located in air circulating passages communicating with such spaces*] direkte Kühlung, unmittelbare Kühlung *f*, direkte Kühlen, unmittelbare Kühlen *n*

direct current → d. c.

direct-current arc, d. c. arc Gleichstromlichtbogen *m*

direct-current cryogenic generator, direct-current cryoalternator, d. c. cryogenic generator, d. c. cryoalternator Gleichstrom-Kryogenerator, Gleichstromgenerator mit supraleitender Wick(e)lung *m*

direct-current-direct-current converter, d. c.-d. c. converter Gleichspannungswandler *m* [*Für die Bereitstellung einer beliebigen Gleichspannung aus einer anderen vorgegebenen Gleichspannung mit oder ohne Polaritätsumkehr*]

direct-current electrical quantity → d. c. electrical quantity

direct-current electricity meter → d. c. electricity meter

direct-current source, DC source Gleichstromquelle *f*

(direct) decade addressing Dekadenkonfiguration *f* [*Brandmelder*]

direct digital controller, digital direct controller, DDC digitale (Rechner)Regler *m* [*DIN 19 225*]

direct digital control station, DDC station direkte digitale Regel(ungs)station *f* [*Können alle Automatisierungsfunktionen einer BTA in der Einzelleitebene wahrgenommen werden, ohne dass eine übergeordnete Leitebene, z. B. Personal Computer (PS) mit ZLT-Grund- und Verarbeitungsfunktionen gemäß VDI-Richtlinie 3814 Bl 1, erforderlich ist, so wird der in DDC-Technik ausgeführte Verarbeitungsteil als „DDC-Station" bezeichnet. Ein Gerät zum Programmieren und Parametrieren der MSR-Funktionen von DDC-Stationen bildet keine übergeordnete Leitebene und macht die DDC-Station folglich nicht zur Unterstation einer ZLT-Anlage. Unterstationen einer ZLT-Anlage und DDC-Stationen sind nach den Merkmalen der ZLT aufgebaut und technisch identisch. Beide können die Verarbeitungsfunktionen für mehrere BTA übernehmen. [ZLT/DDC-86]*]

direct digital ((open) loop) control, digital direct ((open) loop) control, DDC [*Terms used to indicate that a plant is under the direct control of software in an outstation or central station and not through the intermediary of some analogue controller. Analogue inputs are therefore digitized before being used by the B(E)MS*] direkte Digitalsteuerung, Direktdigitalsteuerung, digitale direkte Steuerung, unmittelbare Digitalsteuerung, digitale unmittelbare Steuerung *f*

direct digital ((open) loop) operation → on-line ((open) loop) operation

direct distance dial(l)ing [*A telephone exchange service that allows a telephone user to dial subscribers outside the local area using a standard routing pattern from the local or end office*] Direktfernwählen *n*, Direktfernwahl *f*

direct (domestic) hot water system [*A system in which water passes through and is directly heated by a boiler to provide domestic hot water*] direkte Haushalt(s)warmwasseranlage *f*

direct domestic water heating [*The water drawn off is the same as that in the boiler, being heated "direct"*] direkte Hauswasserheizung *f*

direct drive → DD

direct earthing unmittelbare Erdung, direkte Erdung, widerstandslose Erdung, starre Erdung *f*

directed research gezielte Forschung, zweckgebundene Forschung *f* [*Gegensatz: Allgemeine Forschung*]

directed space air gerichtete Raumluft *f*

direct energy system Direkt-Energie-System *n* [*Wärme- bzw. Kälteerzeugung direkt an den Versorgungsstellen*]

direct evaporation [*See remark under "evaporation"*] Direktverdampfung *f*

direct evaporation fan Direktverdampfungsventilator *m*

direct evaporator Direktverdampfer *m*

(direct) exchange line → post-office line

direct exchange system direkte Austauschsystem, unmittelbare Austauschsystem *n*

direct expansion air cooler Direktexpansionsluftkühler *m*, Direktexpansionsluftkühlgerät *n*

direct expansion cooling direkte Ausdehn(ungs)kühlung, direkte Expansionskühlung *f*

direct feedback system [*A system in which electric(al) feedback is used directly, as in a tachometer*] direkte Rückkopp(e)lungssystem, unmittelbare Rückkopp(e)lungssystem *n*

direct file direkte Datei, unmittelbare Datei *f*

direct-fired direktbefeuert, unmittelbar befeuert

direct-fired furnace, open-flame furnace, inflame furnace; oven furnace [*deprecated*] [*A furnace in which stock is heated in a chamber through which flames and combustion products flow*] Direktfeuerung, Direktfeuerstätte *f*

direct floor heating Direkt(fuß)bodenheizung *f*

direct flow valve Durchflussventil *n*

direct glare [*1. Glare resulting from high luminaires, insufficiently shielded light sources in the field of view, or reflecting areas of high luminance. It is usually associated with bright areas, such as luminaires, ceilings and windows, that are outside the visual task or region being viewed. 2. Glare resulting from high brightness, insufficiently shielded light sources in the field of view, or reflecting areas of high brightness and large area*] Direktblendung, Infeldblendung *f*

direct heat drying Kontakttrocknung *f*

direct heating direkte Heizung, unmittelbare Heizung *f*, direkte Heizen, unmittelbare Heizen *n* [*Heizen ohne Zwischenschalten eines Mediums zwischen Wärmequelle und zu heizendem Medium. DIN 1946 Teil1*]

D

direct hot water (storage) cylinder, direct hot water storage tank [*The stored water is directly heated in the boiler rather than by contact with a heat exchanger as in the indirect cylinder*] Warmwasserspeicher *m* mit direkter Wasser(er)wärmung

direct humidifier Direktbefeuchter, Direktanfeuchter *m*

direct-indirect lighting system → general diffuse lighting system

direct induced draft, direct induced draught direkte Saugzug, unmittelbare Saugzug *m* [*Es wird die gesamte Rauchgasmenge in den Schornstein gefördert*]

direct injection box, di-box Transformatorkasten *m* für Audiosignale

direct insert subroutine, direct insert subprogram (US); direct insert subprogramme (Brit.) [*A body of coding or a group of instructions inserted directly into the logic of a program(me), often in multiple copies, whenever required*] Direkt-Einfügung(s)-Unterprogramm *n*

direct instruction [*An instruction containing the address of the operand on which the operation specified in the instruction is to be performed*] Direktbefehl *m*

direct interface Direktschnittstelle, Direktanpassungsschaltung *f* [*Siehe Anmerkung unter „Schnittstelle"*]

direct inward dialing (US); direct inward dialling (Brit.) [*The capability for dial(l)ing individual telephone extensions in a large organization directly from outside, without going through a central switchboard*] Durchwahl *f*, Durchwählen *n*

directional coupler Richtungskopp(e)ler *m* [*Ein passives Gerät, das in einem Verkabelungssystem RF-Quellen trennt oder zusammenführt*]

directional element gerichtete Glied, gerichtete Element *n*

directionality Richtungstreue *f*

directional lighting, directional illumination [*1. Illumination on the work plane or an object that is predominantly from a single direction.*
2. Lighting designed to illuminate a work plane or an object, predominantly from a preferred direction] Richtungsbeleuchtung *f*

directional ((open) loop) control Richtungssteuerung *f*

directional radio Richtfunk *m*

directional radio aerial; directional radio antenna (US) Richtfunkantenne *f*

directional sensitivity pattern Richtungs-Charakteristik *f* [*Akustik*]

directional vector Richtungsvektor *m*

direction of closing, closing direction Schließrichtung, Zu-Richtung *f*

direction of continuous rotation Dauerdrehrichtung, Drehdauerrichtung *f*

direction of opening → opening direction

direction of sound, sound direction Schallrichtung *f*

directive [*An instruction in a source program(me) that guides the compiler in making the translation to machine language, and is usually not translated into instructions in the object program(me)*] Übersetzungsbefehl *m*

directive Weisung *f*

directivity Richtfähigkeit *f*, Richtvermögen *n*

directivity Abstrahlwinkel *m* [*Beschallung*]

direct loop control → direct (open) loop control

directly close-controlled member [*IEC 27-2A*] **→ plant**

directly connected network [*A network that is accessible from a router without messages being released through an intervening router. A PTP is to a directly connected network if the PTP connection is currently active and no intervening router is used*] direkt angeschlossene Netz *n*

directly controlled member malfunction [*IEC 27-2A*] **→ ((closed) loop) control system malfunction** [*IEC 50(351)*]

directly controlled variable → ((closed) loop) controlled variable

direct measurement method, linear measurement method direkte Messverfahren, unmittelbare Messverfahren *n*, direkte Messmethode, unmittelbare Messmethode *f* [*Bei den direkten Messverfahren wird der Messwert einer Messgröße als Größenwert dieser Messgröße ausgegeben, meist unmittelbar durch die Anzeige eines Messgerätes. DIN 1319 Teil 1*]

direct memory access, dma, DMA; direct program access (US); direct programme access (Brit.) [*The use of special hardware for direct transfer of data to or from memory to minimize the interruptions caused by program(me)-controlled data transfers*] Direktspeicherzugriff, direkte (Rechner)Speicherzugriff, unmittelbare (Rechner)Speicherzugriff *m*

direct memory access controller, DMAC Direktspeicherzugriffsregler *m*

direct normal irradiance → beam irradiance

direct numerical ((open) loop) control, DNC [*The use of a computer to program(me), service, and log a process such as a machine-tool cutting operation*] direkte numerische Steuerung, numerische Direktsteuerung, unmittelbare numerische Steuerung *f* [*Nicht verkettete Maschinen*]

direct-on-line starter (Motor)Starter *m* zum direkten Einschalten

direct (open) loop control [*The control of one machine in a data-processing system by another, without human intervention*] direkte Steuerung, unmittelbare Steuerung *f*

direct operation unit, service operation unit, SOU direkte Bedien(ungs)element, DBE; Notbedien(ungs)element *n* [*frühere Benennung*]

director → (air) director

director Leitstation *f* [*Zur Steuerung eines Datennetzes*]

director [*Telephone switch which translates the digits dialed into the directing digits actually used to switch the call*] Register *n*

director hood → (air) director hood

director meter [*automatic telephony*] Umrechnerzählwerk, Umrechnungszählwerk *n*

directory Verzeichnis *n*

directory change Verzeichniswechsel *m*

directory name Verzeichnisname *m*

directory system agent, DSA Verzeichnissystemteil *m, n*

directory tree Verzeichnisbaum *m*

direct radiation [*It is incident upon the receiving object or absorber*] direkte Strahlung, Direktstrahlung, unmittelbare Strahlung *f*

direct radiator (loud)speaker [*A (loud)speaker in which the radiating element acts directly on the air, without a horn*] trichterlose Lautsprecher, Lautsprecher ohne Trichter *m*

direct read after write → DRAW

direct reading → local reading

direct recording [*Recording in which a record is produced immediately, without subsequent processing, in response to received signals*] Direktaufzeichnen *n*

direct return system [*Piping arrangement for a heating, air-conditioning, or refrigerating system, in which the heating or cooling fluid, after it has passed through a heat-exchange unit, is returned to the boiler or evaporator by the shortest direct path, resulting in considerable differences in the lengths of the several circuits composing the system*] Direktrücklaufsystem *n*

direct scan(ning) [*A scanning method in which the subject is illuminated at all times and only one elemental area of the subject is viewed at a time by the television camera*] Direktabtastung *f*

direct sound Originalton, O-Ton *m*

direct space-heating furnace [*Furnace intended for heating the space in which the furnace is located*] direkte Raumheiz(ungs)feuerung *f*

direct storage heating, storage direct heating Direktspeicherheizung, Speicherdirektheizung *f*

direct subventilation direkte Nebenlüftung, unmittelbare Nebenlüftung *f*

direct symbol recognition [*In automatic data processing, recognition by sensing the unique geometrical properties of symbols*] direkte Symbolerkennung, unmittelbare Symbolerkennung *f*

direct to network direkt zum Netzwerk

direct transfer station → direct building (transfer) station

direct view storage tube → DVST

dirt entrance, entrance of dirt Schmutzeintritt, Schmutzzutritt *m*

dirt pollution Verschmutzung, Schmutzbelastung *f*

dirt trap, strainer, dirt strainer, trapping sieve, dirt cleaner [*A device for separating injurious solid matter from liquid to prevent such matter from entering a pump, valve, tap, meter or pipework*] Schmutzfänger, Schmutzfang *m*, Fangsieb, Schmutzsieb *n*

dirty money [*Additional payment to operatives, e. g. labo(u)rers, for working in difficult or unpleasant or dirty conditions*] Schmutzzulage *f*, Schmutzgeld *n*

dirty (relay) contact [*It is pre-wired to a voltage or current source*] vorverdrahtete Kontakt *m*

DIS → digital image stabilizer

disability glare [*1. Glare resulting in reduced visual performance and visibility, often accompanied by discomfort. 2. Reflective light that impairs a viewer's ability to accurately discern an object, perceive colo(u)r, or to read*] Unfähigkeit(s)blendung *f*

to disable, to suppress entaktivieren, unterdrücken

disabled, suppressed entaktiviert, unterdrückt

disabled persons' access, disabled persons' entry Behindertenzutritt, Behindertenzugang, Behinderteneingang *m*

disablement → suppression

to disarm unscharf schalten [*Gefahrenmeldeanlage*]

to disassemble, to demount demontieren, abbauen, zerlegen, auseinandernehmen

disassembled, demounted demontiert, abgebaut, zerlegt, auseinandergenommen

disassembler Rückübersetzer *m* [*Er übersetzt ein Maschinenprogramm in ein Assemblerprogramm, wobei die beiden Programme die gleiche Bedeutung haben*

D

sollen. Dies ist jedoch nicht immer möglich, weil in Maschinenprogrammen kein Unterschied im Aufbau zwischen Daten und Befehlen besteht]

D

disassembling, demounting Demontieren, Abbauen, Zerlegen, Auseinandernehmen *n*

disaster alarm Katastrophenalarm *m*

disc → **disk**

to discharge abführen, ableiten, abziehen

to discharge entladen

to discharge ablaufen, abfließen, auslaufen

to discharge, to empty entleeren

discharge Entladung *f*

discharge Abfluss, Ablauf, Auslauf *m* [*Der einen Abflussquerschnitt durchfließende Flüssigkeitsstrom*]

discharge Abführung, Ableitung *f*, Abzug *m*

discharge, emptying Entleerung *f*, Entleeren *n*

discharge air Abluft *f* [*Die aus einem Raum abströmende Luft*]

discharge (air) collar [*fan*] Ausblas(e)(luft)ring *m*

discharge air outlet, air discharge outlet Abluftauslass *m*

discharge air shaft, air discharge shaft Abluftschacht *m*

discharge air valve, air discharge valve Abluftventil *n*

discharge ((closed) loop) controller Abflussregler, Ablaufregler, Auslaufregler *m*

discharge closing valve → **discharge shut(ting)-off valve**

discharge coil → **ground(ing) reactor**

discharge cross-section(al area), discharge area Abflussfläche, Ablauffläche *f*, Abflussquerschnitt(fläche), Ablaufquerschnitt(fläche) *m*, (*f*) [*Von der abfließenden Flüssigkeit erfüllte Schnitt senkrecht zu den Stromlinien*]

discharge current Entladestrom, Entladungsstrom *m*

discharge device, discharge appliance Entladevorrichtung *f*, Entladegerät *n*

discharge duct Abflusskanal, Ablaufkanal *m*

discharge end Druckseite *f* [*Pumpe*]

discharge end Auslaufseite, Austrittsseite *f*, Auslaufende, Austrittsende *n* [*Auslaufventil*]

discharge floating valve, floating discharge valve Ablaufschwimmerventil, Abflussschwimmerventil, Auslaufschwimmventil *n*

discharge flow Ablaufstrom, Auslaufstrom, Abflussstrom *m*

discharge installation Abführ(ungs)anlage, Ableitungsanlage, Abzuganlage, Ableitanlage *f*

discharge lamp [*A lamp in which the light is produced, either directly or by means of phosphors, by an electric discharge through a gas, a metal vapo(u)r or mix(ture) of several gases or vapo(u)rs*] Entladungslampe *f*

discharge mouth Ausmündung *f*

discharge noise, emptying noise Entleerungsgeräusch *n*

to discharge outdoors abführen ins Freie, ableiten ins Freie, abgeben ins Freie

discharge/overflow valve Ab- und Überlaufventil, Ab- und Überfließventil *n*

discharge path → **gap**

discharge plug Ablaufstopfen, Abflussstopfen, Auslaufstopfen *m*

discharge rate Entladerate *f* [*Der Betrag um welchen elektrischer Strom aus einer Batterie entnommen werden kann. Üblicherweise in Strom (Ampere) oder Zeit (Stunden) ausgedrückt*]

discharge (refrigerant) gas, refrigerant discharge gas Ablasskältemittelgas, Kältemittelablassgas *n*

discharge safety device [*water heater*] Ablaufsicherheitsvorrichtung *f*

discharge shaft air Schachtabluft *f*

discharge shock → **burst**

discharge shut(ting)-off valve, discharge closing valve, discharge closure valve [*Valve in the discharge passage of a compressor. It is frequently mounted directly on the compressor body or discharge manifold*] Ablass(ab)sperr(ungs)ventil, Ablassschließventil *n*

discharge side → **delivery side**

discharge valve Ablaufventil, Abflussventil, Auslaufventil, Ausflussventil *n*

discharge valve → **(gas) discharge valve**

discharge velocity Ablaufgeschwindigkeit, Abflussgeschwindigkeit, Auslaufgeschwindigkeit, Ausflussgeschwindigkeit *f*

discharge volume flow Ablaufvolumenstrom, Auslaufvolumenstrom, Ausflussvolumenstrom, Abflussvolumenstrom *m*

discharging pump → **discharge pump**

discharging resistor, discharge resistor Entladewiderstand, Ableitwiderstand *m*

discomfort Unbehaglichkeit *f*, Unbehagen *n* [*Rauminsasse*]

to disconnect → **to isolate**

to disconnect → **to switch off**

to disconnect trennen, abtrennen, lösen, unterbrechen

disconnected mode, DM Wartebetriebsart, abgebrochene Betriebsart *f [Datenübertragung]*

disconnectible abtrennbar, trennbar, lösbar

disconnecting Abtrennen, Trennen, Lösen, Unterbrechen, Abbrechen

disconnect(ing) switch → (switch) disconnector

disconnect(ing) trap → intercepting trap

disconnection Abtrennung, Trennung, Lösung, Unterbrechung *f*, Abbruch *m*

disconnection → cutout

disconnection on all poles allpolige Abschaltung, allpolige Trennung vom Netz *f*

disconnection on faults → shutdown on faults

disconnection on three poles dreipolige Abschalten *n*, dreipolige Trennung *f* vom Netz

disconnection switch → isolation switch

disconnection terminal Trennklemme, schaltbare Klemme *f*

disconnection under load Abschalten *n* unter Last, Abschaltung unter Last, Trennung (vom Netz) unter Last *f*

disconnection voltage peak Abschalt(ungs)überspannung, Trenn(ungs)überspannung *f*

disconnector → (switch) disconnector

disconnector fuse, disconnector fuze *[IEC 408]* Trenn(schalt)er mit Sicherung, Sicherungstrennschalter *m*

discontinuous ((closed) loop) control, intermittent ((closed) loop) control unstetige Reg(e)lung, diskontinuierliche Reg(e)lung, schubweise Reglung, intermittierende Reg(e)lung, unterbrochene Reg(e-)lung, schubweise Regelung *f*

discontinuous controller *[See remark under "controller"]* unstetige Regler, diskontinuierliche Regler, unterbrochene Regler *m*

discontinuous ((open) loop) control, intermittent ((open) loop) control diskontinuierliche Steuerung, unterbrochene Steuerung, unstetige Steuerung *f*

discontinuous operation, interrupted operation, intermittent operation Aussetzbetrieb, Sperrzeitenbetrieb *m [Heizung]*

discontinuous setting, intermittent setting (Ein-)Stellen *n* durch Schaltvorgänge

discord → dissonance

discrete camera versteckte Kamera *f*, versteckte Photoapparat, versteckte Fotoapparat *m*

discrete ((closed) loop) control diskrete Reg(e)lung (mit Rückführung), diskrete Regulierung (mit Rückführung) *f [Siehe Anmerkung unter „Reg(e)lung"]*

discrete data → digital data

discrete variable diskrete Zufallsgröße *f [DIN 55 350, Teil 21]*

to discriminate, to differentiate unterscheiden

discriminating anspruchsvoll

discriminating (circuit-)breaker, discriminative (circuit-)breaker Selektivschutzschalter *m*

discriminating earth-fault relay selektive Erdschlussrelais *n*

discriminating fuse-link, discriminating fuze-link träg-flinke Sicherungseinsatz *m*

discrimination, variance Selektion, Selektivität *f*

discrimination instruction → conditional jump

discrimination ratio, variance ratio Selektivitätsverhältnis, Selektionsverhältnis *n*

dish dryer, dish drier Geschirrtrockner *m*

dished end plate *[heat exchanger]* Kümpelboden *m*, schalenförmige Endplatte *f*

(dishes) rinsing water Geschirrspülwasser *n*

(dishes) washing water Geschirrwaschwasser *n*

dishwasher, dishwashing machine, washup machine Geschirrspülmaschine *f*, Geschirrspüler *m*

dishwashing, washing-up Geschirrspülen *n*

dishwashing basket, washup basket Geschirrspülkorb *m*

dishwashing room, washup room Geschirrspülraum *m*

dishwashing waste water, washup waste water, culinary waste water, kitchen waste water Küchenspülwasser, (Geschirr)Spülwasser; Spülicht *n [frühere Benennung]*

disincrustant, scale solvent Kesselsteinlösungsmittel *n*

disinfection Desinfektion, Desinfizierung, Entkeimung *f*

disinfection agent Desinfektionsmittel, Entkeimungsmittel *n*

disinfection agent dispenser Desinfektionsmittelspender, Entkeimungsmittelspender *m*

disjunction Disjunktion, ODER-Verknüpfung *f [Logische Operation, die zwei logische Größen durch den Operator ODER verknüpft]*

disjunctive rule Regel mit ODER verknüpft, ODER-verknüpfte Regel *f*

disk arrays, disc arrays zusammengeschaltete Festplatten *f [Sie sind im Audiobereich zur schnellen Datenübertragung konfiguriert]*

disk atomizer, disc atomizer Scheibenzerstäuber *m*

D

D

disk bursting, disk disruption, disc bursting, disc disruption Scheibenbersten *n* [*Siehe Anmerkung unter „Berstscheibe"*]

disk drive, disc drive Plattenspeicherlaufwerk *n*

diskette → floppy disk

diskette drive → floppy disk drive

disk file, disc file [*An organized collection of records held on a magnetic disk*] Festplattendatei, Plattendatei *f*

disk flake ice maker, disc flake ice maker Scheiben-Flockeneismaschine *f*

disk humidifier, disc humidifier Scheibenbefeuchter, Scheibenanfeuchter *m*

disk pack, disc pack, disk stack, disc stack [*In automatic data processing, a set of magnetic disks that can be removed from a disk drive as a unit*] Plattenstapel, Plattenturm *m* [*Leicht auswechselbare Plattensatz im hermetisch abgeschlossenen Gehäuse*]

disk space, disc space Plattenkapazität *f*

disk storage → (magnetic) disk storage

disk unit, disc unit Plattenlaufwerk *n*

disk water meter, disc water meter Scheibenwasserzähler *m*

dispatcher Verteiler, Abfertiger *m* [*Steuer(ungs)programm in Großrechenanlagen, das beim Stapelbetrieb die Anwenderprogramme aufruft oder bei der Vielfachprogrammierung dem parallel abzuarbeitenden Anwenderprogramm eine Zentraleinheitszeit zuteilt*]

dispatch tube Rohrpostrohr *n*

dispensing cock Zapfhahn *m*

dispensing point Zapfstelle *f*

dispensing valve Zapfventil *n*

dispersal, dispersion [*A data-processing operation in which grouped input items are distributed among a larger number of groups in the output*] Zerstreuung, Streuung *f*

dispersion Dispersion, Feinverteilung *f* [*Unter „Dispersionen" versteht man Systeme, bei denen eine Flüssigkeit, ein fester Stoff oder auch ein Gas außerordentlich fein in einem „Dispersionsmittel" verteil ist. Das Dispersionsmittel kann wiederum flüssig, fest oder gasförmig sein. Anstrichtechnisch von Bedeutung sind nur Dispersionen, in denen das Dispersionsmittel eine Flüssigkeit, z. B. Wasser, und die disperse Phase flüssig oder fest ist. Erstere nennt man „Emulsionen", letztere „Suspensionen". eine scharfe Grenze zwischen beiden Systemen besteht nicht. Bei dispergierten Teilchen sehr hoher Viskosität kann es zweifelhaft sein, ob eine Emulsion oder eine Suspension vorliegt. Vielfach wird „Dispersion" auf diese Systeme angewendet, z. B. bei den „Kunststoffdispersionen". In einigen Fällen nennt man sie auch in Anlehnung an den Kautschuklatex „Latex". Eine typische Suspension ist „Glassival"*]

dispersion element [*for fluids*] Dispersionselement *n*

dispersoid Dispersionskolloid, Feinverteilungskolloid *n*

displacement Verdrängung *f*

displacement [*In a compressor, the volume displaced per unit of time, usually expressed in cubic feet per minute; in a reciprocating compressor, the net area of the compressor piston multiplied by the length of the stroke and by the number of compression strokes per minute. The displacement rating of a multistage compressor is the displacement of the low-pressure cylinder only*] Verdrängung(svolumen) *f*, (*n*)

displacement, offset Distanzadresse *f* [*Sie kann eine positive oder negative ganze Zahl sein*]

displacement compressor Verdrängungsverdichter, Verdrängungskompressor *m*

displacement cooling Verdrängungskühlung *f*

displacement device, displacement appliance Verdränger *m*, Verdrängungsgerät *n*, Verdrängungsvorrichtung *f*

displacement flow(ing) Verdrängungsströmung *f* [*Verdrängung der Schadstoffe aus dem Aufenthaltsbereich zum Abluftsystem und dortige Absaugung ohne dass Zuluft und abgedrängte Luft sich nennenswert vermischen*]

displacement hot water storage tank, displacement warm water storage tank, displacement hot water storage vessel, displacement warm water storage vessel, displacement calorifier Verdrängungswarmwasserspeicher *m*

displacement law → (Wien's) displacement law

displacement meter, positive meter Verdrängungszähler *m* [*Die mit einem Zählwerk verbundene Messkammer wird durch das Strömungsmedium angetrieben*]

displacement outlet Verdrängungsauslass *m*

displacement pump Verdrängerpumpe, Verdrängungspumpe *f*

displacement (refrigerant) compressor, refrigerant displacement compressor Verdrängungs(kältemittel)verdichter, Verdrängungs(kältemittel)kompressor *m*

displacement piston Verdrängungskolben *m*

displacement space air flow Verdrängungsraumluftströmung *f*

displacement total flow meter → positive total flow meter

displacement ventilation, displacement vent(ilat)ing Verdrängungslüften *n*, Verdrängungslüftung *f*

to display, to indicate anzeigen [*Optisch wahrnehmbare Daten darstellen. DIN 19 226, Teil 5*]

display Darstellung, Anzeige, Bildanzeige *f* [*Auf einem reservierten Bereich des Monitors*]

display Anzeigeeinheit *f* [*Durch elektrische Signale steuerbare optische Symbol- oder Bilddarstellungseinheit. Man unterscheidet aktive (selbstleuchtende) und passive Anzeigen (Projektionsanzeigen). Bei letzteren ist eine fremde Lichtquelle nötig*]

display Einblendung *f*

display Anzeigefeld, Darstellungsfeld *n*

display → readout

displayable character abbildbare Zeichen *n* [*Bildschirm*]

display annunciator → annunciator

display case → cabinet

display change Bildwechsel *m* [*Bildschirmgerät*]

display console [*A cathode-ray tube or other display unit on which data being processed or stored in a computer can be presented in graphical or character form; sometimes equipped with a light pen with which the user can alter the information displayed*] graphische Bildschirmarbeitsplatz *m*, Bildschirmkonsole *f*

display file Bildaufbaudatei *f* [*Sie enthält die für den Bildaufbau am Bildschirm benötigten Daten*]

display generation Anzeigenerstellung, Sichtanzeigenerstellung, Bilderstellung *f*

display hierarchy Bildhierarchie *f* [*Bildschirmgerät*]

display image → screen image

displaying Anzeigen *n* [*Darstellen von Größen und Zuständen. DIN 19 222*]

display limits → display range

display loss, visibility factor [*The ratio of the minimum signal input detectable by ideal instruments connected to the output of a receiver, to the minimum signal power detectable by a human operator through a display connected to the same receiver*] Anzeigeverlust *m*

display mode [*oscilloscope*] Betriebsart *f* „Anzeigen"

display ((open) loop) control [*A unit in a computer system consisting of channels and associated control circuity that connect a number of (visual) display units with a central processor*] Anzeigensteuerung *f*

display panel → annunciator

display picture → screen image

display primary, receiver primary [*One of the primary colo(u)rs produced in a television receiver that, when mixed in proper proportions, serves to produce the other desired colo(u)rs*] Empfängerprimärwertigkeit, Primärempfängerwertigkeit *f*

display processor [*A section of a computer, or a microcomputer, which handles the routines required to display an output on a cathode-ray tube*] Anzeigeprozessor *m*

display range, display limits Anzeigebereich *m* [*selten: n*] [*(Bild)Schirm*]

(display-)screen (Bild)Schirm *m*

(display-)screen-compatible (bild)schirmgerecht, (bild)schirmverträglich

(display-)screen device, (display-)screen appliance (Bild)Schirmgerät *n*, (Bild)Schirmvorrichtung *f*

(display-)screen workstation (Bild)Schirmarbeitsplatz *m*

display storage screen → storage display screen

display typewriter Bildschirmschreibmaschine *f*

display unit Bildschirmeinheit *f*, Datensichtgerät *n*, optische Anzeiger *m*

disposable, throwaway wegwerfbar, Einweg..., Einmal...

disposable filter, throwaway filter Wegwerffilter, Einmalfilter, Einwegfilter *m, n*

disposable glass wool filter → glass wool disposable filter

disposable seal, throwaway seal Wegwerfverschluss, Einmalverschluss, Einwegverschluss *m*

disposal Entsorgung *f*

disposal chute Entsorgungsrutsche *f*

disposal engineering Entsorgungstechnik *f*

disposal facility, disposal setup Entsorgungseinrichtung *f*

disposal management Entsorgungsbewirtschaftung *f*

disposal plant, disposal installation Entsorgungsanlage *f*

disposal roof water → roof disposal water

disposal service Entsorgungsdienst *m*

disposal shaft Entsorgungsschacht *m*

disposal system Entsorgungssystem *n*

disposal unit, waste disposal unit, waste food grinder [*An electric-motor-driven device for grinding waste food and disposing of it through the plumbing drainage pipes; may be installed without a grease trap in a residence*] Küchenabfallzerkleinerer, Küchenabfallwolf *m*

to dispose → to arrange

to dispose entsorgen

disposition → arrangement

D

D

to disrupt → **to burst**

disruption → **bursting**

disruption diaphragm → **burst(ing) diaphragm**

disruption disk → **burst(ing) disk**

dissipation Dissipation *f* [*Übergang einer Energieform in Wärmeenergie*]

dissipation Dissipation, unerwünschte Rauschinformation *f* [*Die bedingte Entropie vom Eingang her gesehen; also der Teil der empfangenen Information, der nicht vom Sender ausgegangen ist. Die Dissipation erreicht ihr Maximum, wenn die Transinformation Null wird*]

dissipator → **heat sink**

dissonance, discord [*An unpleasant combination of harmonics heard when certain musical tones are played simultaneously*] Missklang *m*

distance → **dimension**

distance pin Distanzstift, Abstand(s)stift *m*

distance range Abstandsbereich, Distanzbereich *m* [*selten: n*]

distance sleeve Abstandshülse, Distanzhülse *f*

distance/velocity lag, transport(ation) lag [*In a control system, the delay caused by the amount of time required to transport material or propagate a signal or condition from one point to another*] Laufzeit *f*

distant heat, remote heat Fernwärme *f*

distant supply, remote supply Fernversorgung *f*

distant warning, remote warning Fernwarnung *f*

distillate fuel oil, light oil leichte Heiz(ungs)öl *n*

distillation Destillation, offene Verdampfung *f*

distinctive [*An alarm must, e. g., be distinctive, that is to say it must not be capable of being confused with any other alarm*] erkennbar, unterscheidbar, klar

distinctive letter → **letter symbol**

distortion [*Of a signal, an undesired change in waveform*] Verzerrung *f*

distortion → **(harmonic) distortion**

distributed → **decentralized**

distributed data processing → **decentralized data processing**

distributed data processor → **DDP**

distributed periphery, DP dezentrale Peripherie, DP *f*

distributed processing system [*A hierarchical network of computers connected by physical lines or satellites, each computer usually performing only a part of the calculations required by the whole system*] Rechnernetz *n*, Rechnerverbund(system) *m, (n)*

distributed queue dual bus → **DQDB**

distributed system verteilte System *n*

distributing three-way valve, distributing triple-way valve Verteilerdreiwegeventil, Dreiwegeverteilerventil *n*

distributing water pipe → **water distribution pipe**

distribution Verteilung *f*

distribution Vertrieb *m*

distribution board, distribution panel, distribution switchboard Verteiler(schalt)tafel *f*, Stromkreisverteiler *m*, Verteilungs(schalt)tafel

distribution box Verteilung(skasten), Verteilerkasten *f*, *(m)*

distribution cabinet, distribution cubicle Verteilungsschrank, Verteilerschrank *m*

distribution center (US) → **principal distribution board**

distribution connection Verteilungsverbindung, Verteilerverbindung, *f*, Verteilungsanschluss, Verteileranschluss *m*

distribution duct Verteilerkanal, Verteilungskanal *m*

distribution equipment Energieverteilungsgerät(e), Energieverteilergerät(e) *n*, *(f)*

distribution fuse board, distribution fuse frame Sicherungsverteiler *m*

distribution header → **header**

distribution line Verteilungsleitung, Verteilerleitung *f*

distribution logic Verteilungslogik, Verteilerlogik *f*

distribution net(work) Verteilungsnetz(werk), Verteilernetz(werk) *n*

distribution of heat, heat distribution, thermal distribution Wärmeverteilung *f*

distribution pipe Verteilungsrohr, Verteilerrohr *n*

distribution (pipe) main, building distribution (pipe) main Verteilungs(rohr)leitung, Gebäudeverteilungs(rohr)leitung *f*

distribution section Verteilerfeld, Verteilungsfeld *n*

distribution service Verteilungsdienst, Verteilerdienst *m* [*Ein Multimedia-Kommunikationsdienst*]

distribution storey (Brit.); distribution story (US); distribution floor Verteileretage, Verteilungsetage *f*, Verteilergeschoss, Verteilerstockwerk, Verteilungsgeschoss, Verteilungsstockwerk *n*

distribution system Verteilungssystem, Verteilersystem *n*

distribution trunk line Hauptverteilungsleitung, Verteilungshauptleitung, Hauptverteilerleitung, Verteilerhauptleitung *f*

distributor, DBT Verteiler *m*

district air conditioning [*Corresponding to "district heating" is the concept of centralized distribution of chilled water for air conditioning*] Fernklimatisierung, Fernklimareg(e)lung, Stadtklimareg(e)lung, Stadtklimatisierung *f*

district airtube conveyor, district airtube system, district pneumatic conveyor, district pneumatic system Fernrohrpost, Stadtrohrpost *f*

district cold water heating → **cold water district heating**

district cooling Fernkühlung, Stadtkühlung *f*

district cryogenic plant → **cryogenic district plant**

district heat Fernwärme, Stadtwärme *f*

district heat controller Fernwärmeregler, Stadtwärmeregler *m*

district heating [*The provision of heat to a large number of buildings covering a district or perhaps part of a town or a town centre from a central source by mains laid in ducts underground*] Stadtheizung, Fernheizung *f* [*Die Fernheizung ist eine über den Bereich eines Häuserblocks reichende Zentralheizung. Die Wärme der Stadtheizung wird in einem Kraftwerk erzeugt*]

district heating boiler Fernheiz(ungs)kessel, Stadtheiz(ungs)kessel *m*

district heating controller Fernheiz(ungs)regler, Stadtheiz(ungs)regler *m*

district heating duct Fernheiz(ungs)kanal, Stadtheiz(ungs)kanal *m*

district (heating) heat Fern(heiz(ungs))wärme, Stadt(heiz(ungs))wärme *f*

district heating manifold, district heating central distributor, town heating manifold, town heating central distributor Fernheiz(ungs)verteiler(rohr), Stadtheiz(ungs)verteiler(rohr) *m*, (*n*)

district heating network Fernheiz(ungs)netz(werk), Stadtheiz(ungs)netz(werk), Fernwärmenetz(werk), Stadtwärmenetz(werk) *n*

district heating (pipe)line Fernheiz(ungs)(rohr)leitung, Stadtheiz(ungs)(rohr)leitung *f*

district heating plant, district heating installation Fernwärmewerk, Fernheiz(ungs)werk, Heiz(ungs)werk, Stadtheiz(ungs)werk, Stadtwärmewerk *n* [*Es enthält hauptsächlich folgende Räume: Kesselraum mit zwei oder mehr Kesseln, Pumpenraum mit Wasseraufbereitung und Hauptverteilung, Schaltwarte mit Mess- und Regelgeräten, Werkstatt, Personalräume und Brennstofflager*]

district heating steam Fernheiz(ungs)dampf, Stadtheiz(ungs)dampf *m*

district heating substation Fernheiz(ungs)übergabestation, Stadtheiz(ungs)übergabestation *f*

district heating water Fernheiz(ungs)wasser, Stadtheiz(ungs)wasser *n*

district heat supply Fernwärmeversorgung, Stadtwärmeversorgung *f*

district (hot) water heating (system), hot water district heating (system) Warmwasserfernheizung, Warmwasserstadtheizung, Fernwarmwasserheizung, Stadtwarmwasserheizung *f*

district oil heating, oil district heating [*Oil is pumped into mains from central storage tanks, with branches serving the houses, blocks of flats, public buildings, etc., through meters*] Ölfernheizung, Ölstadtheizung *f*

district steam Ferndampf, Stadtdampf *m*

district steam heating, steam district heating [*See remark under "district heating"*] Dampffernheizung, Ferndampfheizung, Dampfstadtheizung, Stadtdampfheizung *f*

district steam heating network, steam district heating network Dampffernheiz(ungs)netz(werk), Ferndampfheiz(ungs)netz(werk), Dampfstadtheizungsnetz(werk), Stadtdampfheiz(ungs)netz(werk), Dampfstadtheiznetz(werk) *n*

district steam heating (pipe)line, steam district heating (pipe)line Dampffernheiz(ungs)(rohr)leitung, Ferndampfheiz(ungs)(rohr)leitung, Dampfstadtheiz(ungs)(rohr)leitung, Stadtdampfheiz(ungs)(rohr)leitung *f*

disturbance, interference [*The introduction of an unwanted command signal into a control system*] Störung *f*

disturbance current, interference current Stör(ungs)strom, Fremdstrom *m*

disturbance estimation, interference estimation Stör(ungs)schätzung *f*

disturbance field, interference field Stör(ungs)feld, Fremdfeld *n*

disturbance field strength, interference field strength (Funk)Stör(ungs)feldstärke, Fremdfeldstärke *f*

disturbance fit, interference fit [*Design engineering. A fit between the mating parts of a male and female assembly in which one of the parts is forced into the other to allow maximum metal overlap*] Übermaßpassung *f*, Festsitz *m*

disturbance frequency, interference frequency Stör(ungs)frequenz *f*

disturbance input matrix, interference input matrix Stör(ungs)matrix *f*

D

D

disturbance level, disturbance degree, interference level, interference degree Stör(ungs)pegel, Störungsgrad, Störgrad *m*

disturbance observation, interference observation Stör(ungs)beobachtung *f*

disturbance point, interference point Stör(ungs)stelle *f*, Stör(ungs)ort *m*

disturbance power, interference power Stör(ungs)leistung *f*

disturbance protection, interference protection Stör(ungs)schutz *m* [*DIN VDE 0100-420*]

disturbance radiation, interference radiation Stör(ungs)strahlung *f*

disturbance range, interference range Stör(ungs)bereich *m* [*selten: n*]

disturbance reducing conductor, interference reducing conductor stör(ungs)reduzierende Leiter *m*

disturbance rejection, disturbance suppression, interference rejection, interference suppression Stör(ungs)unterdrückung, Entstörung, Stör(ungs)beseitigung *f*

disturbance sensitivity, disturbance sensitiveness, disturbance susceptibility, interference sensitivity, interference sensitiveness, interference susceptibility Stör(ungs)empfindlichkeit *f*

disturbance signal, interference signal Stör(ungs)signal *n*

disturbance (suppression) filter, disturbance (suppression) eliminator, interference (suppression) filter, interference (suppression) eliminator [*Electronics. A circuit that is placed between a source of interference and a radio receiver in order to reduce or eliminate the interferences*] Stör(ungs)schutzfilter, Entstör(ungs)filter *m, n*

disturbance suppression symbol, disturbance reduction symbol, Funkschutzzeichen *n*

disturbance variable, interference variable [*An undesired command signal in a control system*] Störungsgröße, Störgröße *f* [*DIN 19 226*]

disturbance variable range, interference variable range Stör(ungs)bereich *m* [*selten: n*] [*DIN 19 226*]

disturbance vector, interference vector Stör(ungs)vektor *m*

to dither, to buzz bibbern [*umgangssprachlich*], zittern [*Signal*]

dither → buzz

DIU → data interface unit

diversified, unequal ungleich(mäßig)

diversified load, unequal load [*The diversified load of a group of apparatus is the sum of the individual demands of the group times the demand factor*] ungleich(mäßig)e Last *f*

to divert → to arrest

diverter Einspritzschaltung *f* [*Kühl(ungs)kreis*]

diverter [*electric machine*] Nebenschlussdämpfungswiderstand *m*

diverter (assembly) Umsteller *m* [*Bad(e)batterie*]

diverter (compressed-)air tube installation, diverter pneumatic tube installation Weichenrohrpost(anlage), Rohrpostweichenanlage *f*

diverter jet/shower, diverter spray/shower Umsteller *m* Strahl/Brause

diverter pulley, diverter sheave, groove(d) pulley, groove(d) sheave [*lift*] Ablenkrolle, Umlenkrolle *f*

diverter relay Nebenschlussrelais *n*

diverter spout (Bade)Wannenauslauf *m* mit Umsteller

diverter valve Einspritzschaltventil *n* [*Kühl(ungs)kreis*]

diverting valve [*Used for non storage heat exchangers, heating and cooling coils*] Ablenkschieber, Umlenkschieber *m*

divide-and-conquer method Teile-und-Herrsche-Verfahren *n*, Teile-und-Herrsche-Methode *f*

divided support disconnector geteilte Trennschalter *m*

divider Teiler *m*

divider Trennzeichen *n*

dividing amplifier Dividierverstärker *m* [*Ein Messverstärker (invertierender Operationsverstärker), dessen Ausgangsspannung proportional dem Quotienten zweier Eingangsspannungen n_1 und n_2 ist*]

dividing network → crossover network

division Unternehmensbereich *m* [*selten: n*]

division Teilung, Aufteilung *f*

DIY depot, do-it-yourself depot Heimwerkermarkt *m*

DL → data transfer link

DL → delay line

DL → diode logic

DLC → data transfer link ((closed) loop) control

DLC → data transfer link ((open) loop) control

DLE → data (transfer) link escape

DLL → dynamic link library

DLU → digital line unit

DM → disconnected mode

DMA → direct memory access

DMAC → direct memory access controller

D-MOS, double diffused metal-oxid(e) semiconductor doppelt diffudierende Transistor *m* [*Niedriges Rauschen und hohe Verstärkung bei hohen Frequenzen*]

DNB → departure from nucleate boiling

DNC, digital net(work) control Digitalnetzsteuerung *f* [*Mehrere numerische Steuerungen, denen ein Leitrechner die Programme zuteilt*]

DNC → direct numerical ((open) loop) control numerische Direktführung *f* [*Nicht verkettete Maschinen*]

DNIC → data network identification code

DNL, dynamic noise limiting dynamische Rauschbegrenzung *f*

DNS → domain name server

doctor's lock Außenschlüsselschloss *n* [*Es wird nur von außen mit Schlüssel geöffnet*]

document [*In automatic data processing, any record printed or otherwise, that can be read by man or machine*] Beleg *m*, Dokument *n*

documentation, document management [*In automatic data processing, the collection, organized and stored, of documents that describe the purpose, use, structure, details, data operation and requirements of a program(me), for the purpose of making thus information easily accessible to the user*] Belegverwaltung, Dokument(en)verwaltung, Dokumentation, Dokumentierung, Belegverwaltung *f*

document feeder Dokument(en)zuführung, Belegzuführung *f*

document flow [*In automatic data processing, the path taken by documents as they are processed through a record handling system*] Belegfluss, Dokument(en)-fluss *m*

document handler Belegbearbeiter, Dokument(en)-bearbeiter *m*

document handling [*Automatic data processing. In character recognition, the process of loading, feeding, transporting, and unloading a cut-form document that has been submitted for character recognition*] Belegbearbeitung, Dokument(en)bearbeitung *f*

document leading edge [*Automatic data processing. In character recognition that edge which is the foremost one encountered during the reading process and whose relative position defines the document's direction of travel*] Belegvorderkante, Dokument(en)-vorderkante *f*

document lift (Brit.); document elevator (US) Dokument(en)aufzug, Belegaufzug *m*

document management → documentation

document misregistration [*Automatic data processing. In character recognition, the improper state of appearance of a document, on site in a character reader, with respect to read or imaginary horizontal baselines*] Belegfehlregistrierung, Dokument(en)-fehlregistrierung *f*

document number [*In automatic data processing, the number given to a document by its originators to be used as a means for retrieval; this number will follow any one of various systems, such as chronological, subject area, or accession*] Belegnummer, Dokumentnummer *f*

document reader Belegleser, Dokument(en)leser *m*

document reference edge [*Automatic data processing. In character recognition that edge of a source document which provides the basis of all subsequent reading processes, insofar as it indicates the relative position of registration marks, and the impending text*] Belegbezugskarte, Dokument(en)bezugskarte *f*

documents cabinet, document cubicle Belegschrank, Dokument(en)schrank *m*

document shredder Reißwolf, Papierschnitzler *m*

documents management Belegverwaltung, Dokument(en)verwaltung *f*

document stacker Belegstapler, Dokument(en)-stapler *m*

docuterm [*In automatic data processing, a word or phrase descriptive of the subject matter or concept of an item of information and considered important for later retrieval of information*] Belegausdruck, Dokument(en)ausdruck *m*

DOD → depth of discharge

do-it-yourself depot → DIY depot

dolly → rocker (dolly)

dolly switch → rocker (dolly) switch

domain Domäne *f*, Bereich *m* [*selten: n*] [*Hierarchisch aufgebaute Adresse eines Rechners im Internet. Jeder Adresse ist außerdem eine eindeutige Zahlenkombination zugeordnet, die IP-Nummer. Diese wird bei der Datenübertragung anstatt des Rechnernamens verwendet*]

domain management Laden *n* und Lesen von Speicherbereichen

domain name server, domain name system, DNS, BIND, Berkeley Internet domain server System *n* zur Abbildung von Hostnamen auf IP-Adressen

domain wall → Bloch wall

domestic [*Refers to use in residential and small commercial buildings*] Haus...

domestic air conditioning, house air conditioning Hausklimareg(e)lung *f*, Hausklimatisierung

D

D

domestic air cooler, house air cooler Hausluftkühler *m*

domestic appliance, house appliance, domestic apparatus, house apparatus Hausgerät *n*, Hausapparat *m*

domestic application → domestic use

domestic automation, house automation Hausautomation *f* [*Siehe Anmerkung unter „Gebäudeautomation"*]

domestic bath, house bath [*Any of a wide range of domestic sanitary appliances used for ablution purposes*] Hausbad *n*

domestic boiler Hauskessel *m*

domestic consumption häusliche Verbrauch *m*

domestic current network, house current network Hausstromanlage *f*

domestic dishwasher Haushalt(s)-Geschirrspülmaschine *f*

domestic drainage häusliche Entwässerung *f*

domestic duty → domestic use

domestic electric hot water storage tank, electric domestic hot water storage tank elektrische Haushalt(s)warmwasserspeicher *m*

domestic electric refrigerator, electric domestic refrigerator Elektrohaushalt(s)kühlschrank *m*

domestic electronics, house electronics Hauselektronik *f*

domestic engineering → domestic (services) engineering

domestic equipment Hausgerät(e) *n*, (*f*)

domestic floor heating, house floor heating Haus(fuß)bodenheizung *f*

domestic garbage, food waste(s), kitchen garbage, kitchen waste(s), household waste(s) [*The portion of garbage resulting from the preparation, cooking, and dispensing of food in homes, hotels, clubs and restaurants*] Küchenabfall, Speiseabfall, Haushalt(s)-abfall, organische Abfall, Nahrungsabfall *m* [*„Müll" kann synonym für „Abfall" verwendet werden*]

domestic gas-fired(-type) incinerator Haushalt(s)-Gasverbrennungsanlage *f*

domestic gas boiler, gas domestic boiler, house gas boiler, gas house boiler Hausgaskessel *m*

domestic gas meter Haushalt(s)gaszähler, Wohnungsgaszähler *m*

domestic heating, house heating Hausheizung, häusliche Heizung *f*

domestic heating installation, house heating installation, domestic heating system, house heating system Hausheiz(ungs)anlage, Wohnungsheizungsanlage, Wohnungsheizanlage *f*

domestic heat meter Haushaltswärmezähler, Wohnungswärmezähler *m*

domestic heat pump, house heat pump Hauswärmepumpe *f* [*Bis zu etwa 50 kW Wärmeleistung*]

domestic hot water, potable hot water [*To be distinguished from hot water for heating*] erwärmte Trinkwasser; Brauchwasser *n* [*frühere Benennung*]

domestic hot water boiler, domestic warm water boiler Hauswarmwasserkessel *m*

domestic humidifier, house humidifier Haus(luft)befeuchter, Haus(luft)anfeuchter *m*

domestic incinerator → domestic (-type) incinerator

domestic insect, house insect Hausinsekt *n*

domestic outside(-type) incinerator [*Free-standing incinerator which does not form an integral part of a building or structure*] Einzelverbrenner *m*, Einzelverbrennungsanlage *f*

domestic range → (kitchen) range

domestic sanitaryware, house sanitaryware Haussanitärware, Hausgesundheitsware *f*

domestic sanitation, house sanitation Haussanitärtechnik, Hausgesundheitstechnik *f*

domestic (services) engineering, house (services) engineering Haustechnik *f*

domestic sewage, house sewage häusliche Abwasser, Hausabwasser, Haushalt(s)abwasser *n*

domestic sink Haushalt(s)küchenspüle *f*

domestic slow combustion range, house slow combustion range (Haus)Dauerbrandherd *m*

domestic solid fuel boiler → solid fuel domestic boiler

domestic storage tank Hauslagerbehälter, Hauslagertank *m*

domestic(-type) incinerator [*1. Incinerator with a combustion chamber of not over 5 ft^3 or a capacity exceeding a 25-lb/hour burning rate. 2. Direct-fed incinerator with a fire-box or charging compartment 5 ft^3 designed to incinerate materials that burn with no higher intensity than wood or paper*] Haushalt(s)-Verbrennungsanlage *f*

domestic use, domestic duty, domestic application, house use, house duty, house application Hausgebrauch *m*, Hausanwendung, Hausverwendung *f*

domestic warm water boiler → domestic hot water boiler

domestic waste(water), domestic wastes [*Waste (water) derived principally from dwellings, business buildings, institutions, and the like. It may or may not contain groundwater, surface water, and/or storm water*] Hausschmutzwasser *n*

domestic water, house water Hauswasser *n*

domestic water manifold, house water manifold Hauswasserverteiler *m*

domestic water meter, house water meter Hauswasserzähler *m* [*DIN 3260*]

domestic water meter manhole, house water meter manhole Hauswasserzählerschacht *m*

domestic water supply, house water supply Hauswasserversorgung *f*

domestic wiring, house wiring Hausverdrahtung *f*

domestic work(ing) room, house work(ing) room Hausarbeitsraum *m*

donkey boiler, auxiliary boiler Hilfskessel *m*

donkey pump, feedwater pump Speisewasserpumpe *f*

donor of the heat, heat donor Wärmebringer *m*

Do not disturb Bitte nicht stören [*Zurückweisen eines ankommenden Rufes zu der anrufenden Sprechstelle mit Aufmerksamkeitston an den anrufenden Teilnehmer*]

don't care unbedenklich

door access ((open) loop) control Türsteuerung *f*

door-activating device, door mechanism, door gear, door activator, door drive Tür(an)trieb, Türmechanismus *m*, Türbetätigungsvorrichtung *f*

door air curtain Türluftschleier, Türluftvorhang *m*

door air gap Türluftspalt *m*

door alarm device Türalarmvorrichtung *f*

door alarm system [*It signals the guard room when certain restricted areas have been entered or vacated by individuals*] Türalarmsystem *n*

door attendant → door operator

door bell Türklingel, Türschelle *f*

door breakthrough Türdurchbruch *m*

door buffer Türpuffer *m*

door chain Türkette *f*

door check [*A mechanical appliance for checking the closing of doors. It is usually fixed at the head*] Türschließwächter *m*

door chime(s) Türglockenspiel *n*

(door) clapper → (door) knocker

(door) clicket → (door) knocker

door closer, door closing device Türschließer *m*, Türschließvorrichtung *f*

door closing spring Türschließfeder *f*

door code unit Türkodegerät, Türcodegerät *n*

door coding Türkodierung, Türcodierung *f*

door combination release → combination door release

door contact ((open) loop) control Türkontaktsteuerung *f*

door contact switch Türkontaktschalter *m*

door controller Türregler *m*

door counterpressure Türgegendruck *m*

(door) crook, (door) hook, (door) hasp (Tür)Angelhaken *m*, (Tür)Angelhaspe, (Tür)Haspe *f*

door draught excluder, door draught seal, door draft seal, door draft excluder Tür-Zugluftschützer *m*

door drive → door-activating device

door fastener → door securer

door fastening bar, door fixing bar Türstange *f*

door furniture → door mountings

door furniture item, door furniture article, door mounting Türbeschlag *m*

door gasket → gasket

door gear → door-activating device

door grill(e) Türgitter *n* [*Ein Baubeschlag, der vorzugsweise an Haus- und Korridortüren angebracht wird, wenn diese Glasscheiben enthalten*]

(door) handle, (door) grip (Tür)Drücker *m*

door hardware [*See remark under "(builders´) iron supplies"*] Tür-Kleineisenwaren *f*, Tür-Kleineisen(-zeug) *n*

(door) hasp → (door) crook

door heater Türheizer, Tür(er)wärmer *m*

(door) hinge → butt (hinge)

door hinge bolt → bolt of (a) door hinge

door holder [*1. Device for fastening a door in open position. 2. Device that holds a door open at a selected position*] Türfeststeller, Türhalter *m*

(door) hook → (door) crook

door intercom(munication) system Türsprechanlage *f*

door interlock(ing) Türverrieg(e)lung *f*

door interlock(ing) switch Türverrieg(e)lungsschalter *m*

door joinery Türentischlerei, Türenschreinerei *f*

(door) keep (Tür)Öffnerfalle *f*

door key Türschlüssel *m*

door key plate [*Plate, either plain or ornamental, having one or more keyholes (but no knob socket) and adapted for attachment to the surface of a door*] Türschlüsselblech *n*

doorknob Türknopf *m*

D

doorknob capacitor, doorknob (electrical) condenser [*A high-voltage plastic-encased capacitor resembling a doorknob in size and shape*] Türknopfkondensator *m*

doorknob with crooked shaft, crooked shaft doorknob Türkrücke *f*

(door) knocker, (door) rapper, (door) clicket, (door) clapper (Tür)Klopfring, (Tür)Klopper, (Tür)Klopfer *m*

door latch; heck [*local term*] Türklinke *f*

door lead-in Türdurchführung *f*

door left open warning Offen-Tür-Warnung *f*

door letter box Türbriefkasten *m*

door lever handle → lever (door) handle

door lock Türschloss *n* [*DIN 18 251*]

door locking Türverrieg(e)lung *f*

(door) locking Abschließen, Türabschließen *n*

(door) lockings → lockings

(door) loop → butt (hinge)

door (loud)speaker Türlautsprecher *m*

door magnet(ic clamp) Tür(haft)magnet *m*

door mechanism → door-activating device

door monitoring Türüberwachung *f*

door mounting → door furniture item

door mountings, door furniture Türbeschläge *f*

door mounting staple → staple of door mounting

door mute, door silencer Türgeräuschdämpfer *m*

door opener Türöffner, Aufwerfer *m*

door opening angle Türöffnungswinkel, Aufwerfwinkel *m*

door ((open) loop) control Türsteuerung *f*

door operation Türbetätigung *f*

door operator, door attendant Türbediener *m* [*Personenaufzug*]

(door) pan, (door) socket(-ring), (door) sole (Tür)Angelpfanne *f*, (Tür)Angelring, (Tür)Angelreifen *m*

door plate [*A metal plate on the door of a house or apartment carrying the name of the occupant*] Türschild, Namen(s)schild *n*

door position Türstellung, Türlage *f*

door processing unit, DPU, door operation unit Türbetätigungseinheit *f*

door profile, door section, door shape, door unit Türprofil *n*

(door) pull handle (Tür)Ziehgriff, (Tür)Zuggriff *m*

(door) pull ring (Tür)Zugring, (Tür)Ziehring *m*

(door) rapper → (door) knocker

door reader Türleser *m*

door release Türauslöser, Türentsperrer, Türöffner *m*

door release counterplate, counterplate door release Türöffnergegenstück *n*

door release key Türöffnertaste(r) *f*, (*m*)

door release with unlocking Türöffner *m* mit Entrieg(e)lung

door scraper Abtretrost, Türrost *m*

door screen → door (wire) screen

door securer, door fastener Türzuhalter, Türzuwerfer *m*, Sperrstange *f*

door security Türsicherung *f*

door silencer → door mute

door slide rail Türlaufschiene, Türschiebeschiene *f*

(door) socket(-ring) → (door) pan

(door) sole → (door) pan

door speaker → door (loud)speaker

door spring Türfeder *f*

doorstop [*A projection on the floor to prevent the door from opening too far*] Tür(fuß)bodenanschlag *m*

door switch, DSW Türschalter *m*

door tidy [*A draught-excluding, spring-loaded metal plate installed inside a letter box to cover its slot*] Briefkastendeckel *m*

door track Türlaufschiene *f*

door type Türart *f*

(door) unlocking Aufschließen, Türaufschließen; Türöffnen *n* [*Fehlbenennung*]

door ventilation Türlüftung *f* [*Luftwechsel durch Öffnung von Türen*]

door ventilator Türlüfter *m*

door wing Türflügel *m*

door (wire) screen [*It is hinged in a door frame on the outside of the main door to exclude insects when the main door is open*] Türinsektenschutzgitter *n*

Doppler effect [*The change in the observed frequency of an acoustic or electromagnetic wave due to relative motion of source and observer*] Dopplerwirkung *f*, Dopplereffekt *m*

Doppler shift, Doppler frequency, frequency shift [*The amount of change in the observed frequency of a wave due to the Doppler effect*] Frequenzverschiebung *f*

dormant bolt → French bolt

dormitory Schlafsaal *m*

dosing diaphragm pump → diaphragm dosing pump

dosing electronics Dosier(ungs)elektronik *f*

dosing stroke Dosier(ungs)hub *m*

dosing system Dosier(ungs)system *n*

dosing tank Dosier(ungs)behälter *m*

dosing water Dosier(ungs)wasser *n*

dot Punkt *m* [*diakritische Zeichen*]

dot matrix Punktraster *m*, Punktmatrix *f*

dot-matrix printer → matrix printer

dot-matrix print head → matrix print head

dot pitch kürzeste Abstand *m* [*Von zwei gleichfarbigen Bildpunkten auf einem Bildschirm*]

dot printer, dot writer Punktdrucker *m*

dot resolution Punktauflösung *f*

dots per inch, DPI Bildpunktanzahl *f* pro Zoll, Punkte *f* pro Zoll [*Beschreibung der Auflösung, die ein Abtaster oder Drucker pro Zoll darstellt bzw. druckt*]

double-acting, dual-acting, twin-acting doppelt wirkend, zweifach wirkend [*z. B. Kompressor; Pumpe usw.*]

double action floor spring, dual action floor spring, twin action floor spring, two-action floor spring Pendelflügel-(Fuß)Boden(tür)schließer *m*

double address system, dual address system, twin address system, two-address system [*See remark under "address"*] Doppeladressensystem, Zwei-Adressensystem *n*

double (air) duct system, dual (air) duct system, twin (air) duct system Doppel-(Luft)Kanalsystem, Zwei-(Luft)Kanalsystem *n*

double angle bar, double angle (section) Doppelwinkel *m* [*Ein kaltgewalztes Profil in Form zweier zusammengesetzter stumpfer Winkel*]

double attach station, dual attach station, DAS, twin attach station Klasse-A-Station *f* [*Rückgratnetzwerk*]

double ball cock, dual ball cock, twin ball cock, two-ball cock Doppelkugelhahn *m*

double band pyrometer → double colour pyrometer

double bank (sectional) radiator, dual bank (sectional) radiator, twin bank (sectional) radiator, two-bank (sectional) radiator zweireihige Gliederheizkörper, doppelreihige Gliederheizkörper *m*

double beam oscilloscope, dual beam oscilloscope, twin beam oscilloscope, two-beam oscilloscope Zweistrahloszilloskop, Doppelstrahloszilloskop *n*

double bend, dual bend, twin bend, return bend Doppelbogenrohr, U-Rohr *n*

double bit error, dual bit error, twin bit error, two-bit error Doppelbitfehler, Zweibitfehler *m*

double-bit(ted) key, dual-bit(ted) key, twin-bit(ted) key, double ward key, dual ward key, twin ward key [*It has bittings on both sides, whereby either or both wings or sides of the key may actuate the tumblers*] Doppelbartschlüssel, Doppelkammschlüssel *m*

double-bit(ted) lock, dual-bit(ted) lock, twin-bit(ted) lock Doppelbartschloss, Doppelkammschloss *n*

double blade terminal, dual blade terminal, twin blade terminal, two-blade terminal Doppelflachanschlussklemme *f*

double bolt lock, dual bolt lock, twin bolt lock, two-bolt lock, gear motion mortice lock, gear motion mortise lock [*This combination of simple latch and deadlock is the most versatile of locks. It has a spring bolt operated by handles giving the convenience of a single latch (i. e. frequent opening and closing of a door), and in addition to a dead bolt for regular or occasional operation by a key, the security. The term "two-bolt" need not be used in a specification as this type will automatically be supplied if ordered simply as a "mortice lock" or "rim lock"*] Doppelriegelschloss, Zweiriegelschloss *n*

double bolt mortice lock, dual bolt mortice lock, twin bolt mortice lock, two-bolt mortice lock (Brit.) [*US spelling: mortise*] Doppelriegeleinsteckschloss, Einsteckdoppelriegelschloss, Zweiriegeleinsteckschloss, Einsteckzweiriegelschloss *n*

double bowl sink, dual bowl sink, twin bowl sink, two-bowl sink Doppelbeckenspüle, Spüle mit Doppelbecken *f*

double branch pipe, dual branch pipe, twin branch pipe Doppelabzweig(er) *m* [*Rohrformstück*]

double break, double interruption, dual break, dual interruption, twin break, twin interruption Doppelunterbrechung *f*

double break contacts, dual break contacts, twin break contact, two-break contacts [*See remark under "break contact"*] Doppelruhekontakt, Doppeltrennkontakt, Doppelöffnungskontakt, Doppelöffner, Doppeltrenner, Doppelunterbrecher, Doppelunterbrechungskontakt *m*

double break contact system, dual break contact system, twin break contact system, two-break contact system Kontaktfolge *f* mit Doppelunterbrechung

double break disconnector, dualbreak disconnector, twin-break disconnector Doppeltrennschalter, Zweifachtrennschalter, Trenndoppelschalter *m*

D

double bundle condenser, dual bundle condenser, twin bundle condenser, two-bundle condenser [*It contains two separate tube bundles allowing the option of either rejecting the heat to the cooling tower or to a building system requiring heat input*] Doppelbündelkondensator, Doppelbündelverflüssiger *m*

double burner range, dual burner range, twin burner range, two-burner range Doppelplattenherd, Zweiplattenherd *m*

double bus(bar) system, dual bus(bar) system, twin bus(bar) system, two-bus(bar) system Doppelsammelschienensystem, Doppelbussystem *n*

double button collective control, double button bank control, dual button collective control, dual button bank control, twin button collective control, twin button bank control, two-button collective control, two-button bank control Doppelknopfsammelsteuerung, Zweiknopfsammelsteuerung *f* [*Aufzug*]

double cable, duplex cable, dual cable, twin cable doppelad(e)rige Kabel, zweiad(e)rige Kabel, Duplexkabel, paarverseilte Kabel *n*

double cable sensor, dual cable sensor, twin cable sensor, two-cable sensor, duplex cable sensor [*It has separate transmit and receive cables*] Doppelkabelfühler, Zweikabelfühler *m*

double card, dual card, twin card Verbund(loch)-karte *f*

double chamber septic tank, dual chamber septic tank, twin chamber septic tank, two-chamber septic tank Zweikammerfaulgrube, Zweikammerkleinkläranlage *f* [*In Deutschland nur für Schmutzwasser zugelassen. DIN 4261*]

double channel adapter card, double channel adaptor card, dual channel adapter card, dual channel adaptor card, twin channel adapter card, twin channel adaptor card, two-channel adapter card, two-channel adaptor card Zweikanaladapterkarte, Doppelkanaladapterkarte *f*

double channel section, twin channel section, dual channel section, two-channel section Doppel-U-Profil *n*

double channel switch, dual channel switch, twin channel switch, two-channel switch Zweikanalschalter, Doppelkanalschalter *m*

double channel time switch, dual channel time switch, twin channel time switch, two-channel time switch [*See remark under "time switch"*] Zwei-Kanal-Schaltuhr, Doppel-Kanal-Schaltuhr *f*

double chime(s) unit, dual chime(s) unit, twin chime(s) unit, two-chime(s) unit Doppel(ton)-glockenspiel, Zweitonglockenspiel *n*

double circuit brake system, dual circuit brake system, twin circuit brake system, two-circuit brake system Doppelkreisbremse, Zweikreisbremse *f* [*Rolltreppe*]

double circuit (single-interruption) switch, dual circuit (single-interruption) switch, twin circuit (single-interruption) switch, two-circuit (single-interruption) switch Serienschalter *m*

(double) cog joint, dual cog joint, twin cog joint, two-cog joint Kreuzüberblattung, Überblatt(ungs)-kreuzung *f*

double colour pyrometer, dual colour pyrometer, twin colour pyrometer, two-colour pyrometer, colour radiation pyrometer, dual band pyrometer, twin band pyrometer, two-band pyrometer, ratio pyrometer [*US spelling is "color"*] Farbpyrometer, Verhältnispyrometer *n*

double compartment digestion pit, dual compartment digestion pit, twin compartment digestion pit, two-compartment digestion pit Zweikammer-(faul)grube, Zweikammerausfaulgrube, Doppelkammer(faul)grube, Doppelkammerausfaulgrube *f* [*Kleinkläranlage*]

double compartment septic tank, dual compartment septic tank, twin compartment septic tank, two-compartment septic tank Zweikammer-Grundstückskläranlage, Doppelkammer-Grundstückskläranlage, Zweikammer-Hauskläranlage, Doppelkammer-Hauskläranlage, Zweikammer-Kleinkläranlage, Doppelkammer-Kleinkläranlage *f* [*DIN 4261*]

double compartment sink (bowl), dual compartment sink (bowl), twin compartment sink (bowl), two-compartment sink (bowl) [*"Basin" may synonymously be used for "bowl"*] Doppel(becken)spüle, Zweibeckenspüle *f*

double compartment sink top, dual compartment sink top, twin compartment sink top, two-compartment sink top Doppel(becken)spültisch, Zweibeckenspültisch *m*

double component ((closed) loop) control, dual component ((closed) loop) control, twin component ((closed) loop) control, two-component ((closed) loop) control Zweikomponentenreg(e)lung, Aufschaltung auf Reglereingang *f*

double compressor, dual compressor, twin compressor, tandem compressor [*1. Two compressors driven by the stand engine or motor mounted on the same base or frame. 2. Assembly of two semihermetic compressor units having a common suction chamber*] Doppelverdichter, Doppelkompressor, Tandemverdichter, Tandemkompressor *m*

double compressor heat pump, dual compressor heat pump, twin compressor heat pump, two-com-

pressor heat pump Doppelkompressor-Wärmepumpe *f*, Doppelverdichter-Wärmepumpe, Tandemverdichter-Wärmepumpe, Tandemkompressor-Wärmepumpe

double current key, dual current key, twin current key, two-current key Doppelstromtaste *f*

double deaeration valve, dual deaeration valve, twin deaeration valve Doppelentlüftungsventil, Entlüftungsdoppelventil *n*

double deck elevator (US) → tandem elevator

double density disk, double density disc, dual density disk, dual density disc, twin density disk, twin density disc, two-tensity disk, two-density disc DD-Diskette *f*

double density recording, dual density recording, twin density recording, two-density recording Aufzeichnen *n* mit doppelter Dichte

double detector, dual detector, twin detector, double detecting device, dual detecting device, twin detecting device Dualmelder Doppelmelder, Dualdetektor, Doppeldetektor *m*, [*Er hat zwei physikalisch unterschiedlich wirkende Sensorsysteme. Er reagiert auf die Veränderungen infraroter Strahlungen und die der Wellenlängen im Ultraschallbereich*]

double detector dependence, dual detector dependence, twin detector dependence, two-detector dependence [*Two detectors in the same zone have responded*] Doppelmelderabhängigkeit *f*

double(d) hinge(-band) → broken hinge(-band)

double diffused metal-oxid(e) semiconductor → D-MOS

double digit system, dual digit system, twin digit system, binary digit system, bit system binäre Zahlensystem, duale Zahlensystem, Zweizahlensystem *n* [*Alle Zahlen werden aus den Ziffern 0 und 1 gebildet*]

double door bolt, dual door bolt, twin door bolt [*Two sliding bars, moving in opposite directions, to secure a door simultaneously at the top and bottom*] Doppeltürriegel, Doppelschubriegel, Doppelschiebe-riegel *m*

double door interlock, dual door interlock, twin door interlock, two-door interlock, interlocked doors Schleuse/Kabinenschleuse *f*

double drainer, double draining board, dual drainer, dual draining board, twin drainer, twin draining board [*See remark under "drainer"*] Doppelabtropffläche *f* [*Teil eines Spültisches*]

double D (slide) valve, dual D (slide) valve, twin D (slide) valve E-Schieber, Doppelmuschelschieber, Zweimuschelschieber *m*

double duct air-conditioning system, dual duct air-conditioning system, twin duct air-conditioning system, two-duct air-conditioning system [*System of a central plant that produces conditioned air at two temperature and humidity levels, to supply air through two independent duct systems to the points of usage where mixing may be carried out. "Plant" and "installation" may synonymously be used for "system"*] Doppelkanalklimaanlage, Zweikanalklimaanlage *f*

double duct high-pressure air conditioning system, dual duct high-pressure air conditioning system, twin duct high-pressure air conditioning system, two-duct high-pressure air conditioning system Zweikanal-Hochdruck-Klima(tisierungs)anlage, Hochdruck-Zweikanal-Klima(tisierungs)anlage *f* [*„System" kann synonym für „...anlage" verwendet werden*]

double duct mixing box, double duct mixing unit, double duct blending box [*See remark under "mixing box"; "twin", "dual" and "two-" may synonymously be used for "double"*] Doppelkanalmischkasten, Zweikanalmischkasten *m*

double duct system, dual duct system, twin duct system, two-duct system Doppelkanalanlage, Zweikanalanlage *f*, Doppelkanalsystem, Zweikanalsystem *n*[*System mit Zuluftleitungen unterschiedlicher Luftart oder unterschiedlichen Luftzustandes (z. B. Lufttemperatur) an den Versorgungsstellen*]

double effect steam (type) absorption machine, dual effect steam (type) absorption machine, twin effect steam (type) absorption machine, two-effect steam (type) absorption machine Dampf-Doppeleffektabsorptionsmaschine *f*

double fast-on connector, dual fast-on connector, twin fast-on connector Doppelflachsteckanschluss *m*

double feed tape carriage, dual feed tape carriage, twin feed tape carriage, two-feed tape carriage doppelte Lochbandvorschub, zweifache Lochbandvorschub *m*

double fence, dual fence, twin fence Doppelzaun *m*

double flush(ing) cistern, duo cistern, dual flush(ing) cistern, twin flush(ing) cistern Zweimengen(abort)spülkasten *m* [*Er kann mit zwei verschiedenen Wassermengen gespült werden und zwar für Fäkalausspülung und Urinausspülung*]

double frequency coil, dual frequency coil, twin frequency coil Doppelfrequenzspule, Zweifrequenzspule *f*

double frequency recording, dual frequency recording, twin frequency recording, two-frequency recording Zweifrequenzschrift, Wechseltaktschrift *f*

double frequency ripple, dual frequency ripple, twin frequency ripple, two-frequency ripple Brummfrequenz *f* bei Zweiweggleichrichtung

D

D

double fuel boiler, dual fuel boiler, twin fuel boiler, two-fuel boiler Doppelstoffkessel, Zweistoffkessel, Wechselstoffkessel *m*

double fuel burner, dual fuel burner, twin fuel burner, two-fuel burner [*Burner designed to burn either gas or oil but not both simultaneously*] Doppelstoffbrenner, Zweistoffbrenner, Wechselstoffbrenner *m*

double furniture vanity, dual furniture vanity, twin furniture vanity, two-furniture vanity Möbeldoppelwaschtisch, Doppelmöbelwaschtisch *m*

double garnet, dual garnet, twin garnet; cross tail(ed) hinge [*Scotland*] Kreuzband *n* [*Für eine schwere Tür*]

double gas heating appliance, double gas heater, dual gas heating appliance, dual gas heater, twin gas heating appliance, twin gas heater Doppelgasgerät *n* [*Gasheiz(ungs)gerät für zwei Gasarten*]

double globe valve, dual globe valve, twin globe valve Zweikugelventil, Doppelkugelventil *n*

double grip, dual grip, twin grip Doppel(hand)griff *m*

double grip (bath)tub, dual grip (bath)tub, twin grip (bath)tub, double grip bathing tub, dual grip bathing tub, twin grip bathing tub, two-grip (bath)tub, two-grip bathing tub Doppelgriff(bade)wanne, Zweigriff(bade)wanne *f*

double grip (bath)tub/shower installation, double grip bath(ing) tub/shower installation [*"Twin", "dual" and "two-..." may synonymously be used for double*] Zweigriff(mischer)-(Bade)Wannenfüll- und Brausebatterie, Doppelgriff(mischer)-(Bade)Wannenfüll- und Brausebatterie (oder Brausegarnitur) *f*

double grip (water) mixer, dual grip (water) mixer, twin grip (water) mixer, two-grip (water) mixer Doppelgriff(wasser)mischer, Zweigriff(wasser)-mischer *m*

double grip (water) mixing valve, double grip (water) mixer (valve) [*"twin", "dual" and "two-..." may synonymously be used for "double"*] Zweigriffmischventil, Doppelgriffmischventil, Zweigriffmischer(ventil), Zweigriffmischventil, Doppelgriffmischer(ventil) *m*, (*n*)

double group dependent circuit, dual group dependent circuit, twin group dependent circuit, two-group dependent circuit Zweigruppenabhängigkeitsschaltung *f*

double (hand)grip (bath)tub, dual (hand)grip (bath)tub, twin (hand)grip (bath)tub, two-(hand)grip (bath)tub Doppelgriff(bade)wanne, Zweigriff(bade)wanne *f*

double (hand)grip mixer, dual (hand)grip mixer, twin (hand)grip mixer, two-(hand)grip mixer Zweigriffmischer, Doppelgriffmischer *m* [*Bad*]

double handle mixing battery, double handgrip mixing battery, dual handle mixing battery, dual handgrip mixing battery, twin handle mixing battery, twin handgrip mixing battery, two-handle mixing battery, two-handgrip mixing battery Zweigriff(misch)batterie, Doppelgriff(misch)batterie *f*

double head circulating pump, dual head circulating pump, twin head circulating pump, two-head circulating pump [*Two single-head pumps in a common housing*] Doppelumwälzpumpe, Umwälzdoppelpumpe *f*

double hole bath fitting, dual hole bath fitting, twin hole bath fitting, two-hole bath fitting Zwei-Loch-Badarmatur, Doppelloch-Badarmatur *f*

double hole mixer (valve), double hole blender, double hole mix(ing) valve, double hole blending valve [*"dual", "twin" and "two" may synonymously be used for "double"*] Zwei-Loch-Mischbatterie, Misch-Zweiloch-Batterie *f*

double inlet fan, dual inlet fan, twin inlet fan Doppel(einlass)ventilator *m*

double interruption → double break

double jacket, dual jacket, twin jacket Doppelmantel *m*

double key, dual key, twin key Doppeltaste *f*

double key, dual key, twin key Nachschlüssel, Doppelschlüssel *m*

double lamp luminaire, dual lamp luminaire, twin lamp luminaire, two-lamp luminaire [*"light(ing) fitting" may synonymously be used for "luminaire"*] zweilampige Leuchte, doppellampige Leuchte, Zweilampenleuchte, Doppellampenleuchte *f*

double-legged manometer → U(-shape(d)) tube manometer

double length arithmetic, dual length arithmetic, twin length arithmetic, two-length arithmetic Rechnen *n* mit doppelter Stellenzahl, Doppelstellenzahlrechnen *n*

double length number, double precision number, dual length number, dual precision number, twin length number, twin precision number, two-length number, two-precision number [*In automatic data processing, a number having twice as many digits as are ordinarily used in a given computer*] doppelte Zahlenlänge, doppelte Zahlengenauigkeit *f*

double length word, dual length word, twin length word, two-length word doppelte Wortlänge, zweifache Wortlänge *f*

double level ((closed) loop) control → double step ((closed) loop) control

double level (sectional) radiator, dual level (sectional) radiator, twin level (sectional) radiator, two-level (sectional) radiator doppellagige Gliederheiz(ungs)körper, zweilagige Gliederheiz(ungs)-körper *m*

double liquid U-tube manometer, dual liquid U-tube manometer, twin liquid U-tube manometer, two-liquid U-tube manometer Doppelstoffmanometer, Zweistoffmanometer *n* [*Für kleine Drücke mittels U-Rohr wird in beiden Schenkeln eine spezifisch leichtere Flüssigkeit über eine spezifisch schwerere Flüssigkeit eingefüllt*]

double media filter, dual media filter, twin media filter, two-media filter Doppelschichtfilter, Zweischichtfilter *m, n*

double phase flow, dual phase flow, twin phase flow, two-phase flow [*Simultaneous flow of two phases of a fluid, usually gas-liquid flows*] Doppelphasenfließen, Zweiphasenfließen *n*

double pipe air-conditioning plant, dual pipe air-conditioning plant, twin pipe air-conditioning plant, two-pipe air-conditioning plant [*"Installation" may synonymously be used for "plant"*] Doppelrohrklimaanlage, Zweirohrklimaanlage *f* [*„Installation" kann synonym für „Anlage" verwendet werden*]

double pipe calorifier, dual pipe calorifier, twin pipe calorifier, two-pipe calorifier Doppelrohrwärmeübertrager, Zweirohrwärmeübertrager *m*

double pipe condenser, dual pipe condenser, twin pipe condenser, two-pipe condenser, tube-in-tube condenser [*Condenser constructed of concentric tubes in which the refrigerant circulates through the annular space and the cooling medium through the inner tube*] Doppelrohrverflüssiger, Doppelrohrkondensator, Zweirohrverflüssiger, Zweirohrkondensator *m*

double pipe drop system, twin pipe drop system, two-pipe drop system, dual pipe drop system [*central heating*] Zweirohrsystem mit oberer Verteilung, Doppelrohrsystem *n* mit oberer Verteilung [*Zentralheizung*]

double (pipe) elbow, dual pipe elbow, twin pipe elbow, (pipe) return bend, two-pipe elbow [*A symmetrical branch fitting in which two opposite branches are curved through 90° to join a main pipe*] (Rohr-)Doppelkrümmer, (Rohr)Doppelbogen, (Rohr)Etagenbogen, Doppelrohrbogen *m*, (Rohr-)Doppelknie *n*

double pipe forced hot water heating, dual pipe forced hot water heating, twin pipe forced hot water heating, two-pipe forced hot water heating Zweirohr-Pumpenwarmwasserheizung, Doppelrohr-Pumpenwarmwasserheizung *f*

double pipe heat exchanger, dual pipe heat exchanger, twin pipe heat exchanger, two-pipe heat exchanger [*Two pipes arranged concentrically, one within the other, and in which one fluid flows through the inner pipe and the other through the annulus between them. See remark under "heat exchanger"*] Doppelrohrwärme(aus)tauscher, Zweirohrwärme(aus)tauscher [*In der Industrie so genannt*] Doppelrohrwärmeübertrager, Zweirohrwärmeübertrager *m*

double pipe heating, dual pipe heating, twin pipe heating, two-pipe heating Doppelrohrheizung, Zweirohrheizung *f*

double pipe heating system, dual pipe heating system, twin pipe heating system, two-pipe heating system [*A steam or hot water heating system in which one pipe is used to supply the heating medium to the units and the other pipe is used to return the medium to the boiler*] Doppelrohrheiz(ungs)system, Zweirohrheiz(ungs)system *n*

double pipe hot water heating, double pipe warm water heating, dual pipe hot water heating, dual pipe warm water heating, twin pipe hot water heating, twin pipe warm water heating, two-pipe hot water heating, two-pipe warm water heating [*"Hydronic heating" and "wet heating" may synonymously be used for "hot water heating"*] Zweirohr-Warmwasserheizung, Zweirohr-WWH, Warmwasserheizung im Zweirohrsystem, Warmwasser-Zweirohrheizung *f* [*Bis 110 °C Vorlauftemperatur*]

double pipe hot water heating system, dual pipe hot water heating system, twin pipe hot water heating system, two-pipe hot water heating system [*See remark under "double pipe heating system". "Warm" may synonymously be used for "hot"*] Doppelrohrwarmwasserheiz(ungs)system, Zweirohrwarmwasserheiz(ungs)system *n*

double pipe induction air conditioning plant, dual pipe induction air conditioning, twin pipe induction air conditioning, two-pipe induction air conditioning [*"installation" may synonymously be used for "plant"*] Zweirohr-Induktionsklimaanlage, Zweirohr-Induktionsklimainstallation, Zweileiter-Induktionsklimaanlage, Zweileiter-Induktionsklimainstallation *f*

double pipe low-pressure steam heating, dual pipe low-pressure steam heating, twin pipe low-pressure steam heating, two-pipe low-pressure steam heating Zweirohrniederdruckdampfheizung, Doppelrohrniederdruckdampfheizung, Zweirohr-ND-Dampfheizung, Doppelrohr-ND-Dampfheizung *f*

double pipe plumbing, double stack plumbing, dual pipe plumbing, dual stack plumbing, twin pipe plumbing, twin stack plumbing, two-pipe plumbing, two-stack plumbing Doppelrohr-Grundstücksentwässerung, Zweirohr-Grundstücksentwässerung *f*

double pipe rising system, dual pipe rising system, two-pipe rising system, twin pipe rising system [*central heating*] Zweirohrsystem mit unterer Vertei-

D

lung, Doppelrohrsystem *n* mit unterer Verteilung [*Zentralheizung*]

double pipe steam heating, dual pipe steam heating, twin pipe steam heating, two-pipe steam heating Zweirohrdampfheizung, Doppelrohrdampfheizung *f*

double pipe steam heating system, dual pipe steam heating system, twin pipe steam heating system, two-pipe steam heating system Doppelrohrdampfheiz(ungs)system, Zweirohrdampfheiz(ungs)system *n*

double pipe system, dual pipe system, twin pipe system, two-pipe system [*Piping system in which the fluid withdraws from the supply passes through a heating or cooling unit to a separate return main*] Zweirohrsystem, Doppelrohrsystem *n*

double pipe vacuum heating, dual pipe vacuum heating, twin pipe vacuum heating, two-pipe vacuum heating Doppelrohr-Vakuumdampfheizung, Zweirohr-Vakuumdampfheizung, Zweirohr-Dampfvakuumheizung, Zweirohr-Niederdruckdampfheizung, Zweirohr-Dampfniederdruckheizung *f* [*DIN 4750*]

double point control, dual point control, twin point control, two-point control Doppelpunktreg(e)lung, Zweipunktreg(e)lung *f*

double point controlled, dual point controlled, twin point controlled, two-point controlled doppelpunktgeregelt, zweipunktgeregelt

double point suspension, dual point suspension, twin point suspension, two-point suspension [*escalator*] Doppelpunktauflagerung, Zweipunktauflagerung *f*

double pole on-off switch, dual pole on-off switch, twin pole on-off switch, two-pole on-off switch doppelpolige Ausschalter, zweipolige Ausschalter *m*

double pole plug, twin pole plug, dual pole plug, two-pole plug [*B. S. 546*] Doppelpolstecker, Zweipolstecker *m*

double pole pull(cord) (type) switch, dual pole pull(cord) (type) switch, twin pole pull(cord) (type) switch, two-pole pull(cord) (type) switch doppelpolige Zugschnurschalter, zweipolige Zugschnurschalter *m*

double pole single throw switch, DPST switch gekoppelte Schalter, Kopp(e)lungsschalter *m* [*z. B. 2 mal Ein/Aus*]

double slope method, dual slope method, twin slope method, two-slope method Doppelrampenverfahren, Zweirampenverfahren *n*, Doppelrampenmethode, Zweirampenmethode *f* [*integrierende Umsetzer*]

double pole terminal, dual pole terminal, twin pole terminal, two-pole terminal Doppelleiterklemme, Zweileiterklemme *f*

double port net(work), dual port net(work), twin port net(work), two-port net(work) Zweitor, Doppeltor *n* [*Bauelement der Baugruppe mit einem Eingang oder einem Ausgang bezogen auf den Signalfluss*]

double position controller, dual position controller, twin position controller, two-position controller Zweipunktregler, schaltende Regler *m* [*Er besteht aus Fühler, Schalter und Sollwertsteller*]

double precision number → double length number

double pressure controller, dual pressure controller, twin pressure controller [*It consists of two pressure bellows that operate a switch within an enclosed case. When applied to compressor control, one bellow functions on a change in suction pressure, the other on a change in discharge pressure*] Doppeldruckregler *m*

double (punch(ed)) card, twin (punch(ed)) card, dual (punch(ed)) card Verbund(loch)karte *f*

double punching, dual punching, twin punching Doppellochen *n*, Doppellochung *f*

double receiver, dual receiver, twin receiver Zwillingsempfänger, Doppelempfänger *m*

double refrigerant system, dual refrigerant system, twin refrigerant system, secondary refrigerant system [*System in which an evaporative refrigerant is used in a secondary circuit. Each system enclosing a separate body of an evaporative refrigerant is considered as a separate direct system*] Doppel-Kältemittelsystem *n*

double riser system, dual riser system, twin riser system, two-riser system [*System in which tow risers are used when a single riser sized for 1000 ft/minute at a minimum load will have an excessive pressure drop at maximum load. The smaller riser is sized for a minimum of 1000 ft/minute velocity at minimum load, while the larger is sized so the velocity through both risers will not be less than 1000 ft/minute at maximum load*] Doppelsteigrohrsystem, Doppelstandrohrsystem, Zweisteigrohrsystem, Zweistandrohrsystem *n*

double room fan, dual room fan, twin room fan, two-room fan Doppelraumlüfter, Doppelraumventilator, Zweiraumlüfter, Zweiraumventilator *m*

double room thermostat, day/night room thermostat, dual room thermostat, twin room thermostat [*It is arranged to control at a reduced temperature at night, and may be indexed (changed from day to night operation) at a remote point by hand or time clock, or from a time switch built into the thermostat itself*] Tag/Nacht-Thermostat *m*, Tag/Nacht-Temperaturregler *m*

double seal siphonic closet (fitment) → double trap full bore (type) siphonic closet (fitment)

double seated (slide) valve, dual seated (slide) valve, twin seated (slide) valve, balanced (slide) valve Doppelsitzschieber, Zweisitzschieber *m*

double seated valve, dual seated valve, twin seated valve Zweisitzventil, Doppelsitzventil *n*

double sided disk, double sided disc, dual sided disk, dual sided disc, twin sided disk, twin sided disc doppelseitige Diskette, zweiseitige Diskette *f*

double sink, dual sink, twin sink [*See remark under "sink"*] Doppelspüle *f*

double slope (analog(ue) digital) converter, dual slope (analog(ue) digital) converter, twin slope (analog(ue) digital) converter, two-slope (analog(ue) digital) converter [*"analog(ue)" digital converter" may be abbreviated by "ADC"*] A/D-Zweirampen-Umsetzer, A/D-Doppelrampenumsetzer, A/D-Z.-U., Zweirampen-A/D-Wandler, Zweirampen-Analog-Digital-Wandler *m* [*„...flanken..." kann synonym für „...rampen..." verwendet werden*]

double slope integration method, dual slope integration method, twin slope integration method, two-slope integration method Doppelrampen-Integrations-Methode, Zweirampen-Integrations-Methode *f*, Zweirampen-Integrations-Verfahren, Doppelrampen-Integrations-Verfahren *n*

double-slope-single-information method, dual-slope-single-information method, twin-slope-single-information method, two-slope-single-information method Integrationsverfahren *n* zur nochmaligen A/D-Wand(e)lung

double socket with earthing contact, twin socket with earthing contact, dual socket with earthing contact Doppel-Schutzkontakt(steck)dose *f*

double source mains infeed(ing), dual source mains infeed(ing), twin source mains infeed(ing), two-source mains infeed(ing) Doppeleinspeisung, zweifache Einspeisung *f* [*Einspeisung aus zwei voneinander unabhängigen jeweils für den Gesamtbedarf der Versorgung ausgelegten Energiequellen über getrennte Übertragungswege. V DIN 19 237*]

double speed master/slave, dual speed master/slave, twin speed master/slave, two-speed master/slave Doppelgeschwindigkeitslaufwerk, Zweigeschwindigkeitslaufwerk *n*

double stage burner, dual stage burner, twin stage burner, two-stage burner Zweistufenbrenner, Doppelstufenbrenner *m*

double stage changeover, two-stage changeover, dual stage changeover, twin stage changeover Doppelstufenumschaltung, Zweistufenumschaltung *f* [*z. B. Stadtgas – Erdgas*]

double stage starter, dual stage starter, twin stage starter, two-stage starter Doppelstufenanlasser, Zweistufenanlasser *m*

double state detector, dual state detector, twin state detector, two-state detector [*"detecting device" may synonymously be used for "detector" and "status" for "state"*] Zweipunktmelder *m*, Zweipunktmeldegerät *n*

double status point, dual status point, twin status point, two-status point [*A pair of signal inputs connected by separate ON and OFF outputs from electrical switchgear where it is vital to detect intermediate positions. "State" may synonymously be used for "status"*] Doppel-Betriebsmeldung *f*

double step action element, dual step action element, twin step action element, two-step action element Glied *n* mit Zweipunktverhalten, Element *n* mit Zweipunktverhalten

double step ((closed) loop) control, dual step ((closed) loop) control, twin step ((closed) loop) control, two-step ((closed) loop) control [*"two-level", "on-off" and „high-low" etc. may synonymously be used for "double step"*] Zweipunktreg(e)lung *f*

double strike, dual strike, twin strike Doppeldruck, Fettdruck *m*

double suction fan, dual suction fan, twin suction fan Doppel(an)saugventilator *m*

double swivel(ling) outlet, dual swivel(ling) outlet, twin swivel(ling) outlet Doppelschwenkauslauf, Doppelschwenkablauf, Doppelschwenkauslass, Doppelschwenkablass *m*

double temperature valve, dual temperature valve, twin temperature valve, two-temperature valve Doppeltemperaturventil *n*

double terminal, dual terminal, twin terminal Doppelklemme *f*

double thermistor sensing unit, dual thermistor sensing unit, twin thermistor sensing unit, two-thermistor sensing unit Doppelthermistor-Messschaltung, Messschaltung *f* mit zwei Thermistoren [*Thermomelder*]

double thermostat, dual thermostat, twin thermostat [*Two temperature controls in one enclosure*] Doppelthermostat *m* [*Kombination von Kesselthermostat und Überkochsicherung*]

double throw contact, dual throw contact, twin throw contact, two-throw contact [*Double throw means separate circuits are closed in both the energized and non-energized state*] Umschaltkontakt *m*

double throw contactor with neutral position, dual throw contactor with neutral position, twin throw contactor with neutral position, two-throw contactor with neutral position Wechselkontaktgeber *m*

D

D

double throw switch, twin throw switch, dual throw switch, two-throw switch Umschalter, Zweistellungsschalter, Schalter mit zwei Stellungen *m*

double tier terminal, dual tier terminal, twin tier terminal, two-tier terminal Doppelstockklemme *f*

double tone multi(ple) frequency, dual tone multi(ple) frequency, twin tone multi(ple) frequency, two-tone multi(ple) frequency Doppeltonwählverfahren *n*, Doppeltonwählmethode *f* [*Jede Zahl der Tastatur oder Wählscheibe wird durch die Summe zweier Töne repräsentiert. Dadurch wird auch ein Signalverfahren zur Unterscheidung von Telefon und Fernkopierer ermöglicht*]

double transistor, dual transistor, twin transistor Doppeltransistor *m*

double trap full bore (type) siphonic closet (fitment), double seal siphonic closet (fitment) Leersaug(e)abort mit Doppelgeruchverschluss, Absaug(e)abort mit Doppelgeruchverschluss *m*, Leersaug(e)toilette mit Doppelgeruchverschluss, Absaug(e)toilette mit Doppelgeruchverschluss *f*

double tube air conditioning installation, dual tube air conditioning installation, twin tube air conditioning installation, two-tube air conditioning installation Zweirohrklima(tisierungs)anlage, Doppelrohrklima(tisierungs)anlage *f*

double tube heat exchanger, twin tube heat exchanger, dual tube heat exchanger, two-tube heat exchanger Doppelrohrwärmeübertrager, Zweirohrwärmeübertrager *m* [*Siehe Anmerkung unter „Wärme(aus)tauscher“*]

double tumbler lock, twin tumbler lock, dual tumbler lock, two-tumbler lock Schloss mit zwei Zuhaltungen, Zweizuhaltungsschloss, Doppelzuhaltungsschloss *n*

double tuned amplifier, dual tuned amplifier, twin tuned amplifier Zweikreisverstärker, Doppelkreisverstärker *m*

double turn lock → twice-turning lock

double vanity, dual vanity, twin vanity Doppelwaschtisch *m*

double voltage coil, dual voltage coil, twin voltage coil Doppelspannungsspule, Zweispannungsspule *f*

double walled, dual walled, twin walled doppelwandig, zweiwandig

double wall(ed) hot water generator [*"twin-wall(ed)", "dual-wall(ed)", and "two-wall(ed)" may synonymously be used for "double wall(ed)"*] doppelwandige Warmwassererzeuger, zweiwandige Warmwassererzeuger, doppelwandige Warmwasserbereiter, zweiwandige Warmwasserbereiter, doppelwandige Boiler, zweiwandige Boiler *m*

double ward key → double-bit(ted) key

double ward lock → double-bit(ted) lock

double water supply system, dual water supply system, twin water supply system, double water distribution system, dual water distribution system, twin water distribution system Teilversorgungsnetzsystem, doppelte Wasser(versorgungs)netzsystem *n* [*Betriebswasser- und Trinkwasserversorgungssystem über getrennte Teilversorgungsnetze*]

double wattmeter method, dual wattmeter method, twin wattmeter method, two-phase method Doppelwattmetermethode, Zweiwattmetermethode *f*

double way alternate operation, twin way alternate operation, two-way alternate operation, dual way alternate operation Halb-Duplex-Betrieb, Halb-Duplex-Verkehr, Wechselbetrieb, Wechselverkehr *m*

double way heater valve, dual way heater valve, twin way heater valve, two-way heater valve Zweiwegeheiz(ungs)körperventil *n*

double way intercom(munication) unit, dual way intercom(munication) unit, twin way intercom(munication) unit, two-way intercom(munication) unit, duplex unit Gegensprechanlage *f*

double way mixing valve, dual way mixing valve, twin way mixing valve, two-way mixing valve Zweiwegemischventil *n*

double way (sectional) radiator valve, dual way (sectional) radiator valve, twin way (sectional) radiator valve, two-way (sectional) radiator valve Zweiwegeradiatorventil, Zweiwegegliederheiz(ungs)körperventil; Zweiwegeheiz(ungs)körperventil *n* [*Fehlbenennung*]

double way switch, dual way switch, twin way switch, two-way switch Wechselschalter *m*

double way thermostat, dual way thermostat, twin way thermostat, two-way thermostat [*One end of the thermostat controls the sensible cooler and the other the reheater*] Zweiwegethermostat, Zweiwegetemperaturregler *m*

double way universal mixing valve, dual way universal mixing valve, twin way universal mixing valve, two-way universal mixing valve Zweiwege-Universal-Mischventil *n* [*Wegen seiner drei Anschlüsse aus kommerziellen Gründen auch „Dreiwegeventil“ genannt*]

double way valve, dual way valve, twin way valve, two-way valve Zweiwegeventil *n* [*Es kann zwei Teilströme mischen und einen Strom trennen*]

double wire A.C. type, dual wire A.C. type, twin wire A.C. type, two-wire A.C. type Doppel-Draht-Wechselstromausführung, Zwei-Draht-Wechselstromausführung *f*

double wire bus, twin wire bus, dual wire bus, two-wire bus Zweidrahtbus, Doppeldrahtbus *m*

double wire circuit, dual wire circuit, twin wire circuit, two-wire circuit [*Metallic circuit formed by the adjacent conductors insulated from each other. It is possible to use the two conductors as a one-way transmission path, a half duplex path, or a full duplex path*] Zweidrahtschaltung, Doppeldrahtschaltung *f*

double wire control, dual wire control, twin wire control, two-wire control Doppelkontakt(gabe) *m*, (*f*)

double wire line, dual wire line, twin wire line, two-wire line Zweidrahtleitung, Doppeldrahtleitung *f*

double wire temperature sensor, dual wire temperature sensor, twin wire temperature sensor, two-wire temperature sensor Doppeldrahttemperaturfühler, Zweidrahttemperaturfühler *m*

double word, dual word, twin word Doppelwort *n* [*DIN 19 239*]

double word instruction, dual word instruction, twin word instruction, two-word instruction Doppelwortbefehl *m*

double zone dependency, dual zone dependency, twin zone dependency, two-zone dependency [*Two detectors in different zones have responded*] Doppelzonenabhängigkeit, Zweizonenabhängigkeit *f*

DO value → digital output value

dowel type Dübelart *f*

down nach unten, runter [*Schaltplan*]

down-blow air curtain Abwärtsluft(strom)schleier, Abwärtsluft(strom)vorhang *m*

downcomer → vertical cable

downcomer → downspout

downcounter Abwärtszähler, Rück(wärts)zähler *m*

downcounting Rück(wärts)zählen, Abwärtszählen *n*

downfeed(ing) Abwärtsvorschub *m*

downfeed(ing) (pipe) riser, downfeed(ing) (pipe) rising line Abstieg(rohr)leitung *f*

downflow central furnace [*Furnace designed with the airflow in an essentially vertical path, discharging air at or near the bottom of the furnace*] Abwärts-Zentralfeuerung, Zentral-Abwärtsfeuerung *f*

downflow-counterflow furnace [*Forced-warm-air furnace designed with the airflow through the furnace in an essentially vertical path downward, with the air discharged at or near the furnace bottom*] Abwärts-Gegenstromluftfeuerung, Gegenstromluft-Abwärts-Feuerung *f*

downflow furnace [*The circulating air flows downward, discharging at or near the bottom of the casing*] Abwärts-Warmluftfeuerung, Warmluft-Abwärtsfeuerung *f*

downjet Abwärtsstrahl *m*

downlighting, narrow-angle lighting Tiefstrahlung *f*

downlighting luminaire, narrow-angle luminaire Tiefstrahler *m*

to download fernladen, zentralladen

download(ing), down-line loading [*A particular type of file transfer that refers to the transfer of an executable program(me) or database to a remote device where it may be executed*] Fernladen, zentrale Laden, Zentralladen *n* [*Rechner-Rechner; Rechner-Endgerät*]

down milling Gleichlauffräsen *n*

downpipe → downspout

down service pipe Wasserfallrohr *n* [*In einem Haus oder Gebäude*]

down service pipes [*They connect the cistern (= storage tank) located at the top of the water installation of a house or building to the fittings*] Wasserfallrohre *f*

downspout, waste pipe, downcomer, downpipe, drop pipe, (rain) leader, fall pipe; conductor (US) [*A vertical pipe, often of sheet metal, used to conduct water from a roof drain or gutter to the ground or cistern*] (Regen)Fallrohr, Abfallrohr *n*

downstream nachgeordnet

downstream nachgeschaltet

downstream [*Electrically speaking, going away from the power source and toward the load*] lastseitig, lastwärts

down-swept flue Abwärts(ab)zug, Sturz(ab)zug, Abwärtsfeuerzug, Sturzfeuerzug *m*

down time, non-productive time Ausfallzeit, Brachzeit *f*

down-up counter → bidirectional counter

downward air distribution [*This distribution is commonly used for air conditioning where the inlet air is often cooler than that of the room; air is introduced at high level and extracted at low level*] Abwärtsluftverteilung *f*

downward arrow Abwärtspfeil, abwärts gerichtete Pfeil, Pfeil nach unten *m* [*Code*]

downward combustion abwärts gerichtete Verbrennung, Abwärtsverbrennung *f*

downward compatible abwärtsverträglich

downward travel → lowering

Dow oscillator, electron-coupled oscillator [*abbreviated: eco.*] Dow-Oszillator *m*

dowtherm boiler Kessel *m* für schweren Kohlenwasserstoff

D

DP → distributed periphery

DP → (data) processing

DP → dial pulse

DPC, destination point code Zielpunktkode, Zielpunktcode *m*, Zielkennung *f*

DPDM → digital pulse duration modulation

DPI → dots per inch

DPSK → differential phase shift(ing) keying

DPST switch → double pole single throw switch

DPU → door processing unit

DQDB, distributed queue dual bus Protokoll *n* für Hochgeschwindigkeitsübertragung mit Raten von 34 bis 155 Mbit/s

DR → digital readout

draft → (air) draft

draft air, draught air Zugluft *f*

draft check plate, flue damper, flue shutter, draught check plate, split firebridge Zugabsperrklappe für Flammrohre, Zugabsperrklappe für Feuerrohre *f*

draft exclusion, draught exclusion, exclusion of draught, draughtproofing, exclusion of draft, draftproofing Zug(ab)dichten, Dichtschließen *n*

draft fan, draught fan Zugventilator *m*

draft-free, draught-free zug(luft)frei

draft gage, draught gage [*It is used to measure air movement*] Zugmesser *m*

draft hood, draught hood, draft diverter, draught diverter [*Device placed in and made part of the vent connector from an appliance, or in the appliance itself, which is designed to ensure the ready escape of the combustion products in the event of no draft, backdraft, or stoppage beyond it; prevents a backdraft from entering the appliance and neutralizes the effect of stack action of a chimney flue on the operation of the appliance*] Zughaube *f*

draft indicator, draught indicator [*Instrument used to indicate or measure chimney draft and combustion gases*] Zuganzeiger *m*

draft inducer, draught inducer Zugerzeuger *m* [*z. B. ein Ventilator*]

drafting, drawing, draughting [*The process of using lines and numbers to depict the dimensions and specifications of something to be constructed, such as a building, device, equipment, machine or the like*] Zeichnen *n*

draft interruption, draught interruption Zugunterbrechung *f*

draft limiter, draught limiter, draft limit(ing) device, draught limit(ing) device Zugbegrenzer, Nebenluftbegrenzer, Zugregler, Nebenluftregler *m* [*DIN 4795*]

draft lobby, draught preventer, draught excluder, draught lobby, draft preventer, draft excluder Windfang *m*

draft meter, draught meter Zugmesser *m*

draft nuisance, draught nuisance Zug(luft)belästigung *f*

draftproof, draughtproof zugdicht

draft regulating register, draught regulating register, circular air grid Rosette *f*

draft regulator, draught regulator, damper [*This device is generally considered to be a valve, usually made of steel plates or sheet metal, and used to control the flow of air or gases in ducts and chimneys*] Zugregler *m*

draft seal(ing), draught seal(ing) Zug(luft)sperre *f*

draft sensation → sensation of draft

draftsman, draughtsman Zeichner *m*

draft standard, draught standard, standard draft, standard draught Normenentwurf *m*, Entwurfsnorm *f*

drafty (US), draughty (Brit.) zugig, windig [*Tür; Fenster*]

drag and drop Ziehen *n* und Ablegen [*Mausfunktion mit der Teile von Dokumenten kopiert oder verschoben werden*]

drag pointer, drag needle, trailing pointer, trailing needle Schleppzeiger *m* [*Messtechnik*]

drag washer Hakenscheibe, Schleppscheibe *f*

to drain entwässern, abwässern

drain [*A pipe, an open channel or underground conduit and the like for the conveyance of any water*] Drän *m*

drain Senke *f*

drain P-Pol *m* [*Feldeffekttransistor*]

drain Abfluss *m* [*Ein Feldeffekttransistoranschluss. Die Ladungsträger fließen hier aus dem Kanal heraus, vergleichbar mit dem Kollektor des Bipolartransistors*]

drain → current draw

drain → (underground) drain

drainage Entwässerung *f*

drain(age) channel, drain-off channel, draining channel [*Of a water system*] Entwässerungsrinne *f*

drain(age) cock, drain-off cock, draining cock [*It is placed at the lowest point of a water system or wet control heating installation for draining the system*] Entwässerungshahn, Abflusshahn, Ablaufhahn, Ausflusshahn, Auslaufhahn, Entleerungshahn *m*

drain(age) coil → ground(ing) reactor (US)

drain(age) connection, draining connection Entwässerungsanschluss *m*, Entwässerungsverbindung *f*

drain(age) device, drain(age) appliance, drain(age) piece, drain(age) item Entwässerungsgegenstand *m*, Entwässerungsvorrichtung *f*

drain(age) fixture, drain(age) object, draining fixture, draining object Entwässerungs(einbau)teil *m*, *n*

drainage gravity system → gravity drainage system

drainage hole, drainage opening, draining hole, draining opening Entwässerungsloch *n*, Entwässerungsöffnung *f*

drainage installation → estate drainage installation

drainage load, draining load Entwässerungslast, Entwässerungsmenge *f*

drain(age) loop, draining loop Entwässerungsschleife *f*

drain(age) pipe, draining pipe Entwässerungsrohr *n*

drainage (pipe) fitting Entwässerungs(rohr)fitting *m*, Entwässerungs(rohr)formstück, Entwässerungs(rohr)passstück *n*

drainage (pipe)line, draining (pipe)line [*"Piping" may synonymously be used for "(pipe)line"*] Entwässerungs(rohr)leitung *f*

drain(age) plug Ablassstopfen, Entwässerungsstopfen, Ablassstöpsel, Entwässerungsstöpsel *m*

drain(age) product, draining product Entwässerungsgegenstand *m*

drain(age) screw, draining screw Entwässerungsschraube *f*

drain(age) (slide) valve, draining (slide) valve Entwässerungsschieber, Entleerungsschieber *m*

drain(age) system, draining system Entwässerungssystem *n* [*Für Gebäude und Grundstücke. DIN 1986*]

drain cock Entleerungshahn, Entwässerungshahn *m*

drain current Dränstrom *m* [*Transistor. DIN 41 858*]

drain electrode Dränelektrode *f* [*Transistor. DIN 41 858*]

drainer [*These are available in stainless steel, pressed steel sheet, plastic, wood, alumin(i)um and asbestos cement. Other materials are available for special purposes, such as monel metal for its appearance, fireclay and procelain-enameled cast-iron*] (Ab)Tropffläche, (Ab)Tropfplatte *f* [*Teil eines Spültisches*]

draining Entwässern *n*

drain(ing) board [*sink*] Ablaufplatte *f* [*Spüle*]

drain(ing) cock → drain(age) cock

drain(ing) connection Entleer(ungs)anschluss *m* [*Vorratswasserheizer*]

drain manhole, drain manway [*A chamber into which branch drains are collected. It is used for inspection and cleaning purposes*] Dränschacht *m*, Dränmannloch *n*

to drain (off) water, to blow (off) water ablassen, entwässern [*Wasser aus Dampfkessel oder Rohrleitung*]

drain pan Entwässerungsschale *f*

drain pit [*A pit to receive nontoxic water for disposal, often to a dry well*] Entwässerungsgrube *f*

drain plug, drain stopper [*A device for temporarily sealing off a portion of a drain sometimes for testing purposes*] Dränpfropfen, Dränstopfen *m*

drain rod Stocherstange, Dränstocherstange *f*

drain terminal Dränanschluss *m* [*Transistor. DIN 41 858*]

drain test [*The testing of a drain for leakages*] Dränprüfung, Dränprobe *f*, Dränversuch *m*

(drain) trap → (bath)tub (drain) trap

DRAM → dynamic RAM

drastic product [*fuzzy logic*] drastische Produkt *n*, *t*-Norm *f*

drastic sum [*fuzzy logic*] drastische Summe, *t*-Konorm *f*

to draw entnehmen [*z. B. Strom aus dem Netz*]

DRAW, direct read after write Datenwiedergabe *f* direkt nach der Aufzeichnung

drawer Schublade *f*

drawer part → withdrawable part

drawer pull, drawer (hand) grip [*A metal, wood, or plastics grip for the front of a drawer. It is usually hollow underneath to receive the fingers*] Schubladengriff *m*

drawer roller Schubladenroller *m*

drawer (type) motor control centre (Brit.); drawer (type) motor control center (US) Einschubverteiler *m*

drawer unit section Einschubfeld *n*

drawing Zeichnung *f*

drawing → drafting

drawing exchange format → DxF

drawing not to scale → not-to-scale drawing

draw(ing) off → tapping

draw(ing) off tap → (water) tap

D

drawing paper Zeichenpapier, Zeichnungspapier *n*

drawing(s) set, set of drawings Zeichnungssatz, Plänesatz *m*

(drawing) title block (Zeichnungs)Schriftfeld *n*

to draw off → to tap

to draw off → to blow off

drench shower (rose), deluge shower (rose) [*A shower (rose) with a large output arranged to give the bather a rapid shower automatically in an emergency*] Schwallbrause *f*

drench (sprinkler) system, deluge (sprinkler) system, drencher system Überflutungsprinkleranlage, Sprühflutanlage *f* [*Die offenen Düsen versprühen beim Ansprechen der Anlage gleichmäßig das Löschwasser*]

dressing Abrichten *n*

dried sludge → dry sludge

drier, dryer [*Appliance that removes moisture*] Trockner *m*

drier coil, dryer coil Trocknerschlange, Trocknungsschlange *f*

drift [*Fine spray of water carried away from a cooling tower by the wind*] Wassernebel *m*

drift (Ab)drift, (Ab)Trift *f*, langsame Abwandern *n*

drift Nullpunktänderung *f* [*Unerwünscht bei Messgeräten*]

drift-corrected amplifier [*A type of amplifier that includes circuits designed to reduce gradual changes in output, used in analog(ue) computers*] abtriftberichtigte Verstärker, driftberichtigte Verstärker *m*

drift error [*An error arising in the use of an analog(ue) computer due to gradual changes in the output of circuits (such as amplifiers) in the computer*] (Ab)Triftfehler, (Ab)Driftfehler *m*

drill(ing) template Bohrschablone *f*

to drill to desired size → to redrill

drinking fountain Trinkbrunnen *m*

drinking water, potable water drinkable water Trinkwasser, Brauchwasser *n* [*Für den menschlichen Genuss und Gebrauch geeignetes Wasser mit Güteeigenschaften nach DIN 2000 (Leitsätze für die zentrale Trinkwasserversorgung) und nach DIN 2001 (Leitsätze für Einzeltrinkwasserversorgung*]

drinking water cooler, potable water cooler, drinkable water cooler Trinkwasserkühler, Brauchwasserkühler *m*

drinking water feed, potable water feed, drinkable water feed Trinkwasserzufuhr, Brauchwasserzufuhr *f*

drinking water jet, potable water jet, drinkable water jet Trinkwasserstrahl, Brauchwasserstrahl *m*

drinking water net(work), potable water net(-work), drinkable water net(work) Trinkwassernetz, Brauchwassernetz *n*

drinking water (pipe)line, potable water (pipe-)line Trinkwasser(rohr)leitung, Brauchwasser-(rohr)leitung *f*

drinking water pressure pipe, potable water pressure pipe, drinkable water pressure pipe Trinkwasser-Druckrohr, Brauchwasser-Druckrohr *n*

drinking water storage heater, potable water storage heater, storage drinking water heater, storage potable water heater Speicher-Trinkwasser(er)wärmer, Trinkwasser-Speicher(er)wärmer, Speicher-Brauchwasser(er)wärmer, Brauchwasser-Speicher(-er)wärmer *m*

drinking water supply, potable water supply, drinkable water supply Trinkwasserversorgung, Brauchwasserversorgung *f*

drinking water supply area, drinking water supply zone, potable water supply area, potable water supply zone, drinkable water supply area, drinkable water supply zone Trinkwasserversorgungsgebiet, Brauchwasserversorgungsgebiet *n*

drinking water supply pipe, potable water supply pipe, drinkable water supply pipe Trinkwasserrohr, Brauchwasserrohr *n*

drinks fridge Getränkeeisschrank, Eisgetränkeschrank *m*

to drip, to drop tropfen, abtropfen

drip-bagno, drop-bagno Schwitzbad *n*

drip(ping) eliminator → drip(ping) separator

dripping (off), dropping (off) [*A falling of liquid drop by drop*] (Ab)Tropfen *n*

drip(ping) (off) cock, drop(ping) (off) cock (Ab)Tropfhahn *m*

drip(ping) (off) sink, drop(ping) (off) sink [*It is fitted under a draw-off tap and usually fixed in a convenience closet (Schrank)*] (Ab)Tropfspüle *f*

drip(ping) (off) (water) tray, drip(ping) (off) (water) pan, drop(ping) (off) (water) tray, drop(ping) (off) (water) pan (Ab)Tropf(wasser-)wanne, (Ab)Tropf(wasser)schale *f*

drip(ping) separator, drip(ping) eliminator, drop(ping) separator, drop(ping) eliminator Tropfen(ab)scheider *m*

drip(ping) water, drop(ping) water Tropfwasser *n*

drip(ping) water connection, drop(ping) water connection Tropfwasseranschluss *m*

drip sink → **drip(ping) (off) sink**

drive Antrieb, Trieb *m*

drive Laufwerk *n* [*ist nicht Treiber*] [*z. B. einer Diskette, einer Festplatte oder einer optischen Speicherplatte*]

drive controller, drive regulator Laufwerkregler *m*, Laufwerkregelorgan *n*, Laufwerkreg(e)lungseinheit *f*

driver Treiber *m* [*Ab einer Auflösung von 480 Punkten in 256 Farben gibt es keine Norm zur Programmierung dieser Modi. Deshalb werden für diese Modi an Graphikkarten angepasste Treiber für die Anwendungsprogramme benötigt*]

driver Treiber *m* [*Ein Programm ohne Fenster, das Befehle und Aufträge vom Betriebssystem erhält und ausführt. Viele Einsteckkarten und CD-ROM-Laufwerke benötigen Treiber. Diese übersetzen die allgemeinen Aufträge in spezifische Befehlsfolgen, die nur das zugehörige Gerät versteht*]

driver [*The amplifier stage preceding the output stage in a receiver or transmitter*] Steuer(ungs)sender *m* und Verstärker

driver block Treiberbaustein *m*

driver diskette, driver felxible disk cartridge, driver floppy disk, driver floppy disk Treiberdiskette *f*

driver format Treiberformat *n*

drive rope, drive cable, driving rope, driving cable Antrieb(s)seil, Treibseil, Triebseil *m*

driver routine Treiberroutine *f*

driver spring Stiftfeder *f* [*Profilzylinder*]

driver stage Treiberstufe *f*

(drive) sheave → **sheave**

(drive) sheave elevator (US), (drive) sheave lift (Brit.) (Seil)Treibscheibenaufzug *m*

driving end sprocket Fahrstufenumsetzstation, Rollstufenumsetzstation *f*, Kopfteil *m*, *n* [*Obere Teil und untere Teil des Fahrtreppengerüstes*]

driving machine, drive machine [*Power unit which applies the energy necessary to raise and lower an elevator or dumbwaiter car or to drive an escalator, moving stairway or inclined lift*] Triebwerk *m*, Antriebsmaschine *f*

driving point impedance [*The complex ratio of applied alternating voltage to the resulting alternating current in an electron tube, network, or other transducer*] Eingangsscheinwiderstand *m*

driving system → **actuating system**

droop Statik *f*

droop compensation, reactive-current compensation Statikausgleich *m*

to drop → **to drip**

drop Tropfen *m*

drop cable Verbindungskabel *n* [*Es stellt die Verbindung , den Zugriff von und zu den Verteilerkabeln eines Netzwerkes her*]

drop cable (Ab)Zweigkabel, Nahbuskabel *n*

drop channel Abstaffelungskanal *m*

(drop-)dead halt [*A machine halt from which there is no recovery; such a halt may occur through a logical error in programming; examples in which a (drop-)dead halt could occur are division by zero and transfer to an instruction word*] absolute Stillstand *m*

drop door Falltür *f*

drop-in [*The accidental appearance of an unwanted bit, digit, or character on a magnetic recording surface or during reading from or writing to a magnetic storage device*] Stör(ungs)signal *n*

drop indicator [*Indicator for signal(l)ing, consisting of a hinged flap normally held up by a catch; the catch is released by an electromagnet, allowing the flap to drop when a signal is received*] Fallklappe *f*

drop indicator relay Fallklappenrelais *n*

dropless tropfenfrei, ohne Tropfen

droplet Tröpfchen *n*

droplet condensation [*Phenomenon encountered when the condensate does not wet the cold surface, but settles in the form of separate droplets*] Tröpfchenkondensation *f*

droplet separator, droplet interceptor Tröpfchen(ab)scheider *m*

drop (line) Fall(rohr)strang *m*

drop of a keyhole, keyhole drop Schlüssellochdeckel, Schlüssellochschild *m*

drop-on-demand, drop-on-requirement Tintenfluss *m* nach Bedarf [*Tintenstrahldrucker*]

dropout, scratch Kratzer *m* [*Er verläuft quer zur Spurrichtung der gespeicherten Daten*]

dropout [*Of a relay, the maximum current, voltage, power or such, at which it will release from its energized position*] Abfall(en) *m*, (*n*)

dropout Signalausfall *m*

dropout falsche Abspeicherung *f* [*Von Informationen durch Fehler im Informationsmedium (z. B. Band)*]

dropout falsche (Aus)Lesen *n* [*Von Informationen durch Fehler im Informationsmedium (z. B. Band)*]

dropout current Abfallstrom *m*

D

dropout error [*Loss of a recorded bit or any other error occurring in recorded magnetic tape due to foreign particles on or in the magnetic coating or to defects in the backing*] Lesespannungsausfall *m*, Beschichtungsloch *n*

dropout fuse, dropout fuze Ausfallsicherung, Trennsicherung *f*

dropout voltage [*The maximum voltage at which a relay or other magnetically operated device will release to its deenergized position*] Abfallspannung *f*

dropped ceiling → suspended ceiling

dropper bar Feldsammelschiene *f*

dropping resistor [*A resistor used in series with a load to decrease the volt-age applied to the load*] Vor(schalt)widerstand *m*

drop pipe → downspout

drop pocket Sortierfach, Ablegefach, Ablagefach *n*

drop-point slating → diagonal slating

drop repeater [*Microwave repeater that is provided with the necessary equipment for load termination of one or more circuits*] (Ab)Zweigverstärker *m*

droptight, dropproof tropf(en)sicher, tropf(en)fest

drop voltage, dropped voltage Abfallspannung *f*

drop wire [*Wire suitable for extending an open wire or cable pair from a pole or cable terminal to a building*] Einführungsdraht *m*

drum, rotary compartmented chamber Trommel, Crosley-Trommel *f* [*Nass-Gaszähler*]

drum cooler, cooling drum [*A rotating refrigerated cylinder whose surface is in contact with the liquid to be cooled*] Trommelkühler *m*, Kühl(ungs)trommel *f*

drum (line) printer Walzen(zeilen)drucker *m*

drum magnetic storage → magnetic drum storage

drum mark [*In automatic data processing, a character indicating the termination of a record on a magnetic drum*] Trommelmarke *f*

drum memory Trommelspeicher *m*

drum meter Trommelzähler *m*

drum plotter Trommelplotter *m*, Trommelzeichenmaschine *f* [*Siehe Anmerkung unter „Plotter"*]

drum runout Schlag *m*, Unwucht *f*

drum scanner Trommelabtaster *m*

drum (type) (gas) meter → wet (gas) meter

dry [*Free from or deficient in moisture*] trocken

to dry [*To separate or remove a liquid or vapo(u)r from another substance*] trocknen

dry air [*1. Air without entrained water vapo(u)r. 2. Air unmixed with or containing no water. Note: Composition of dry air is defined in ISO 2533-1975*], trockene Luft *f*

dry air blower Trockenluftgebläse *n*, Trockenluftbläser *m*

dry air cooler [*Cooler that removes sensible heat from dehydrated air*] Trockenluftkühler *m*

dry air density [*The weight of dry air, pounds per cubic foot*] Trockenluftdichte *f*

dry alarm valve station Trocken-Alarmventilstation *f* [*Sprinkleranlage*]

dry battery [*A battery made up of a series, parallel, or series-parallel arrangement of dry cells in a single housing to provide desired voltage and current values*] Trockenbatterie *f*

dry bulb [*air temperature measurement*] Trockenkugel *f*

dry bulb entering temperature → entering dry bulb temperature

dry-charged battery [*A storage battery in which the electrolyte is drained from the battery for storage, and which is filled with electrolyte and charged for a few minutes to prepare for use*] trockengeladene Batterie *f*

dry closet → earth closet

dry condensate (pipe)line trockene Kondens(at)-(rohr)leitung, trockene Kondenswasser(rohr)leitung *f*

dry condensate return Trockenkondensatrücklauf *m*

dry condenser trockene Kondensator, Fallrohrkondensator, Gefällekondensator *m*

dry cooling load trockene Kühl(ungs)last *f*

dry cooling tower, nonevaporative cooling tower trockene Kühl(ungs)turm *m*

dry drum filter Trockentrommelfilter, Trommeltrockenfilter *m*, *n*

dryer, drier [*Appliance that removes moisture*] Trockner *m*, Trocknungsgerät *n*

dryer coil → drier coil

dryer exit air duct, drier exit air duct Trocknerabluftkanal *m*

dry evaporation Trockenverdampfung *f*

dry evaporator [*Evaporator of the continuous tube type, where refrigerant from a pressure reducer is fed into one end and the suction line connects to the outlet end*] Trockenverdampfer *m*

dry expansion [*A method of refrigerant control in which one part of the cooler surface must be used to provide superheat to operate the regulator; it is therefore relatively ineffective*] Trockenausdehnung, Trockenexpansion *f*

(dry) expansion chiller, (dry) expansion refrigerating machine [*One of the two main types of evaporator for cooling water or brine, the other being the flood chiller. The (dry) expansion chiller cools water or brine by evaporating refrigerant inside the tubes of a heat exchanger similar in design to the horizontal shell and tube evaporator, but with the water or brine being on the outside of the tubes. Baffles are situated on the shell to ensure that the water or brine flows over the outside of the tubes and at right angles to the tubes in a cross baffled arrangement*] Trockenexpansionskältemaschine, Trockenausdehnungskältemaschine *f*

dry expansion evaporator Trockenausdehn(ungs-)verdunster, Trockenexpansionsverdunster *m*

dry (fire) extinguishing Trocken(feuer)löschen, Trockenbrandlöschen *n*

dry (fire) extinguishing powder Brandlöschpulver, (Feuer)Löschpulver *n*

dry (fire) extinguishing system Trocken(feuer)-löschsystem, Trockenbrandlöschsystem *n*

dry-fit battery wartungsfreie Batterie *f*

dry flashover voltage Funkenüberschlagspannung *f*

to dry galvanize trockenverzinken

dry (gas) meter, bellows (gas) meter, gas bellows meter [*A volumetric (gas) meter which measures the volume of gas by successive filling and emptying of bellows*] trockene (Gas)Zähler, Lederbalg(gas)zähler, Balgen(gas)zähler, (Gas)Zähler mit verform-baren Trennwänden, Trocken(gas)zähler *m*

dry glazing Trockenverglasung *f*

dry heating [*It utilizes air as a medium for conveying heat*] Luftheizung *f*

dry heating load trockene Heiz(ungs)last *f*

dry hydrant Trockenhydrant *m*

dry ice, solid CO_2, carbon dioxid(e) ice; solid carbon dioxid(e) [*proprietary term*] Kohlendioxydschnee, Kohlendioxidschnee; Kohlensäureschnee *m*, Trockeneis *n* [*Fehlbenennung*]

drying [*Removal of water from any substance*] Wasserentzug *m*, Trocknen *n*

drying by waste heat, waste heat drying Abwärmetrocknung, Trocknung mit Abwärme *f*

drying cabinet, drying cubicle Trocknungsschrank, Trockenschrank *m* [*Fehlbenennung*]

drying coil → drier coil

drying heat pump Trocknungswärmepumpe *f*

drying load Trocknungslast *f*

drying oven Trocknungsofen *m*

drying process Trocknungsvorgang *m*

dry joint kalte Lötstelle *f*

dry layer filter [*It has a dry filtering medium, as opposed to a viscous filter*] Trockenschichtfilter *m, n*

dry line Trockenleitung *f* [*Sie wird erst im Bedarfsfall mit Flüssigkeit gefüllt*]

dry (medium) (air) filter, strainer (type) (air) filter [*These filters usually consist of a frame and a dry replaceable medium of cellulose, glass fibres, specially treated paper, cotton batting, wool felt or synthetic material*] Trocken(luft)filter, Lufttrockenfilter *m, n*

dry meter → dry (gas) meter

dryout Austrocknen *n* [*Eine Form der Siedekrise in Abhängigkeit vom Dampfgehalt. Liegt auf Grund eines hohen Dampfgehaltes Ringströmung vor, bildet sich am Ende des Flüssigkeitsfilms eine trockene Stelle auf der Heizfläche. Dadurch ergibt sich ein Temperaturanstieg der zu einer Wärmeübergangskrise führt; der Flüssigkeitsfilm trocknet an der Wand aus*]

dry pendant sprinkler → suspended dry sprinkler

dry (pipe)line, dry piping Trocken(rohr)leitung *f* [*Brandbekämpfung*]

dry(-pipe) sprinkler system, dry(-pipe) automatic system [*Piping contains air under pressure which, when released by activated sprinkler heads, permits water to enter pipes automatically by action of dry(-pipe) valve and flow through sprinkler*] Trocken-(sprinkler)anlage *f*

dry(-pipe) valve Alarmventil *n* [*Sprinkleranlage*]

dry piston compressor [*Type of oil-free compressor in which no lubricating oil is used in the cylinder; the piston is usually equipped with down friction labyrinth rings*] Trockenkolbenverdichter, Trockenkolbenkompressor *m*

dry powder → ((fire) extinguishing) powder

dry powder (fire) extinguisher Trocken(feuer)-löscher, Trockenbrandlöscher *m*

dry recooling Trockenrückkühlung *f*

dry residue, dry rest, dry remainder Trockenrückstand, Trockenrest *m*

dry return (flow) Trockenrücklauf *m*

dry riser, dry rising (pipe)line [*A vertical pipe installed in a building for fire-fighting purposes being fitted with inlet connection at fire brigade access level*

D

D

and outlet valves at specific points in the building. The pipe is normally dry, but is capable of being charged with water, usually by pumping from the fire service appliance] Trocken-Steig(e)leitung *f*

dry (saturated) steam air humidification Trockendampfluftbefeuchtung, Trockendampfluftanfeuchtung *f*

dry (saturated) vapour; dry (saturated) vapor (US) [*Vapo(u)r at the saturation temperature corresponding to the existing pressure and without any liquid phase*] trockene Sattdampf, Trockensattdampf *m*

dry sensible heat, sensible dry heat trockene fühlbare Wärme, fühlbare trockene Wärme *f*

dry sludge, dried sludge Trockenschlamm, getrocknete Schlamm *m*

dry standpipe fire hose riser trockene (Feuer-)Löschschlauchschacht, trockene Brandlöschschlauchschacht *m*

dry substance, dry matter Trockensubstanz *f*

dry sump [*sewage disposal*] Pumpenkammer *f*

dry surface cooler Trocken-Oberflächenkühler *m*

dry suspended sprinkler → suspended dry sprinkler

dry ton(s) [*Sensible heat load expressed in tons*] Tonnen *f*

dry transformer luftgekühlte Transformator, Transformator mit Luftkühlung *m*

dry valve → dry(-pipe) valve

dry vane wheel (inferential) meter Trockenläufer, Trockenflügelradmesser *m*

dry wall hydrant Trockenwandhydrant *m*

dry waste trockene Abfall(stoff) *m* [*Kehricht, Küchenabfall, Asche, usw.*]

dry weather flow [*1. The foul water, together with infiltration, if any, flowing in a sewer or drain in dry weather. 2. The rate of flow of foul water, together with infiltration, if any, in a sewer or drain in dry weather. 3. The rainfall has not excluded one-tenth of an inch in 24 hours*] Trockenwetterabfluss *m*

dry well → seepage pit

DSA → directory system agent

DSA → data service adapter

DSE → data switching exchange

D (slide) valve, plain (slide) valve Muschel(flach)-schieber, D-Schieber *m*

DSM → (energy) demand-side-management

DSP → digital signal processor

DSP → digital signal processing

(D)SS → (dedicated) special system

DSS1 → digital subscriber (signal(l)ing) system (No. 1)

DST → daylight saving time

DSW, door switch Türschalter *m*

DTA → data-transfer-acknowledged

DTACK → data acknowledg(e)ment

DTB → data transfer bus

DTC → data-transfer-acknowledged

DTE → data terminal equipment

DTE backup Rufweiterschaltung *f*

DTL → diode transistor logic

DTP, desk(top) publishing Softwaresystem *n* zur Dokumentenbearbeitung

D-trap [*A trap like the letter D in shape. An obsolete form of intercepting trap for drains*] D-(Wasser-)(Geruch)Verschluss, (Wasser)(Geruch)Verschluss in D-Form *m*

DTU → data-transfer-unacknowledged

D(-type) flip-flop → D bistable element

dual...→ double...

dual-in-line → DIL

dual-in-line package, DIP, DIL package zweireihige Gehäuse, doppelreihige Gehäuse, Gehäuse mit parallelen Anschlüssen *n* [*Gehäuse mit zwei parallelen Reihen von Anschlussstiften zur Aufnahme elektronischer Schaltungen*]

dual slope Zwei-Flanken-Umsetzung *f* [*Auswertung von zwei Flanken einer Integrationsspannung*]

dubbing Überschreiten *n* [*Von bespielten Bandstellen*]

duckfoot bend, rest bend [*A right-angle fitting for a pipe with an integral seat which may be mounted on a support*] Fuß(rohr)bogen(stück) 90°, Fuß(rohr)-krümmer(stück) 90° *m*, (*n*) [*DIN 28 648*]

ducking Lautstärkeabsenkung *f* [*Einer Schallquelle durch eine andere*]

duct [*Tube, pipe, conduit, or continuous enclosed passageway used for conveying air, gases, or vapo(u)rs*] Kanal *m*

duct accessories Kanalzubehör *n*, *m*

duct air Kanalluft *f*

(duct) air velocity detector, (duct) air velocity sensor [*It is designed for duct mounting and uses a circuit which gives a 0-5 V dc output proportional to the measured air velocity. The measuring range is adjustable and can be set from 4 to 10 m/s air veloci-*

ty for the maximum output voltage of 5 V. The detector requires a 24 V ac supply] (Kanal)Luftgeschwindigkeitsfühler, (Kanal)Luftgeschwindigkeitsdetektor, (Kanal)Luftgeschwindigkeitssensor *m*, (Kanal)Luftgeschwindigkeitsdetektionsgerät *n*

duct connecting branch, duct connection branch Kanal(anschluss)stutzen *m*

duct construction Kanalbau *m*

duct conveyance → ducting

duct detector, duct detecting device, duct detection device Kanalmelder *m*, Kanalmeldegerät *n*

duct differential pressure sensor Kanal-Differenz(ial)druckfühler *m [z. B. Messblende]*

ducted [*warm air system*] kanalisiert

ducted heating Kanalheizung *f*

ducted hot air, ducted warm air Kanalwarmluft, Warmkanalluft *f*

duct elbow, duct el(l) Kanalknie *n*, Kanalwinkel(stück) *m*, (*n*)

(duct) enthalpy detector, (duct) enthalpy sensor [*Enthalpy is the sum of the sensible heat and the latent heat and represents the total heat content of the air being measured. It is measured in kJ/kg of dry air. The detector is for duct mounting only*] (Kanal-)Wärmeinhaltfühler, (Kanal)Wärmeinhaltsensor, (Kanal)Wärmeinhaltdetektor *m*, (Kanal)Wärmeinhaltdetektionsgerät *n*

duct fan Kanalventilator *m*

duct filter Kanalfilter *m*, *n*

duct fitting [*Duct system part used to join, adapt, or adjust other parts of the duct system*] Kanalfitting *m*

duct for services → services duct

duct friction Kanalreibung *f*

duct furnace [*1. Warm-air furnace installed in an air distribution duct to supply warm air for heating. This definition applies only to a warm-air heating appliance that depends for air circulation on a blower not furnished as part of the furnace. 2. Furnace designed for insertion or installation in a duct of an air distribution system to supply warm air for heating. The duct furnace depends for air circulation on a blower that is not a part of the furnace*] Kanalfeuerung, Kanalfeuerstätte *f*

duct gas Kanalgas *n*

duct heating Kanalheizung *f*

duct heat transfer, duct thermal transfer [*See remark under "heat transfer"*] Kanalwärmeübertragung *f*

duct humidity controller Kanalfeuchteregler, Kanalfeuchtigkeitsregler *m*

duct humidity detector, duct humidity sensor, duct humidistat, duct hygrostat Kanalfeuchtigkeitsfühler, Kanalfeuchtefühler *m* [*"Sensor", "Detektor" und "Detektionselement" können synonym für „Fühler" verwendet werden*]

ducting, duct conveyance, duct transport(ation) Kanalförderung *f*, Kanaltransport *m*

duct inlet velocity Kanaleingangsgeschwindigkeit *f*

duct installation Kanaleinbau *m [z. B. eines Luftfilters*]

duct leakage, duct leaking Kanalleckage *f*, Kanallecken *n*

ductless air-conditioning kanallose Klimatisierung, kanallose Klimareg(e)lung *f*

duct liner [*Acoustic liner to absorb sound*] Kanalschalldämpfer *m*

duct lining Kanalauskleidung, Kanalbekleidung, Kanalverkleidung *f*

duct muffler → duct sound attenuator

duct run, duct length Kanalstrecke *f*

duct sensor Kanalfühler, Kanalsensor *m* [*DIN 30 600, Reg.-Nr. 06155*]

duct shaft Kanalschacht *m*

duct sizing [*Calculation of dimensions of ducting for a given air distribution system*] Kanalbemessung *f*

duct sound attenuator, duct sound trap, duct sound damping device, duct muffler [*In ductwork, a device (usually prefabricated) especially designed to provide much greater sound attenuation than would be provided by an equal length of ductwork; the pressure drop through the device is greater than for an equal length of ductwork*] Kanal(schall)dämpfer *m*

duct sound damping, duct sound attenuation, duct sound abatement, duct sound deadening Kanalschalldämpfung *f*

duct system [*Continuous passageway for the transmission of air, which includes the ducts, duct fittings, dampers, plenums, fans, and accessory air-handling equipment*] Kanalsystem *n*

duct T(ee) Kanal-T(-Stück) *n*

duct temperature limiter, duct temperature limiting device Kanaltemperaturbegrenzer *m*

duct temperature measured value transmitter Kanaltemperaturmesswertgeber *m*

duct temperature regulator Kanaltemperaturregler *m*

duct temperature sensor Kanaltemperaturfühler, Kanaltemperatursensor *m*

duct thermal transfer → duct heat transfer

discharge duct Entleerungskanal *m*

duct thermostat Kanalthermostat *m*

duct transport(ation) → **ducting**

duct wall Kanalwand(ung) *f*

ductwork → **(air) ductwork**

ductwork, duct system, duct network Kanalsystem, Kanalnetz *n*

D

dull nickel-plated mattvernickelt

dull of system, system dull Systemflaute *f*

Dulong-Petit law [*The law that the product of the specific heat per gram and the atomic weight of any solid elements at room temperature have almost the same value, about 6.3 calories (264 joules) per degree Celsius*] Dulong-Petitsche Regel *f* [*1819 aufgestellt*]

dumbwaiter, dinner elevator (US); dinner lift (Brit.) Speisenaufzug *m*

dumbwaiter, service lift [*A special form of elevator, its height is limited to 1.2 m (4.0 ft) and its platform area to 0.8 m2 (9.0 ft2). It is used in medium-rise commercial or institutional buildings for mail handling and cart, package, or drawing deliveries*] Kleingeräteaufzug *m*

dummy Platzhalter *m* [*Für ein Programmteil, Modul oder Daten nach außen ein Original vortäuschend*]

dummy Füllzeichen, Leerzeichen *n* [*Aus formativen Gründen gespeicherter Wert. Auf einer Datei mit 80 Zeichen pro Zeile werden nicht beschriebene Positionen mit Leerzeichen gefüllt*]

dummy, useless imitation [*it has no operating features*] Attrappe, nutzlose Nachbildung *f*

dummy fuse Kurzschlusspatrone, Blindsicherung *f*

dummy instruction [*See remark under "instruction"*] belanglose Befehl, Blindbefehl *m*

dummy line identification Leeranschlusskennung *f*

dummy lock, flush-encased lock, rabbeted lock Einlassschloss, eingelassene Schloss *n*

dummy man → **eupatheoscope**

dummy state fiktive Systemzustand *m*

dummy tapping Blindanzapfung *f*

to dump umspeichern [*Information*]

dump, current dump Stromausfall, Netzausfall *m*

dump(ing) → **(bulk) dump(ing)**

dumping switch Entlastungsschalter *m*

dump level Datensicherungssteuerung *f*

dump request Datenblock-Anforderungsbefehl *m*

dump tape Speicherauszugband *n*

dunnage [*Strips placed between stored materials to permit air circulation between them*] Garnierung *f*

dunnage mat Garniermatte *f*

dunnage wood [*Wood strips placed between stored materials to permit air circulation between them*] Garnierholz *n*

dunning letter [*Slang*] → **prompt note**

duo cistern → **double flush(ing) cistern**

duplex → **either-way**

duplex Gegensprechbetrieb *m*, Gegensprechen *n*

duplex artificial line, duplex artificial circuit [*In electricity, a balancing network, simulating the impedance of the real line and distant terminal apparatus, which is employed in a duplex circuit for the purpose of making the receiving device unresponsive to outgoing signal currents*] Leitungsnachbildung *f*

duplex cable → **double cable**

duplex channel Duplexkanal, Doppelkanal *m*

duplex computer [*Two identical computers, either one of which can ensure continuous operation of the system when the other is shut down*] Duplexrechner, Doppelrechner *m*

duplex conference Gegensprechkonferenz *f*

duplexer, duplexing assembly [*radar*] Sende-Empfang-Schalter *m*

duplex mode Doppelbetriebsart *f* [*Gleichzeitig in beiden Richtungen einer Leitung*]

duplex outlet, duplex receptacle Doppel(netz)-steckdose *f*

duplex process, duplex practice, duplexing Duplexverfahren *n* Duplexmethode *f*

duplex star connection, star duplex connection Stern-Stern-Schaltung, Doppelsternschaltung, Sterndoppelschaltung *f*

duplex tag connector Zweifachflachstecker *m*

duplex transmission, full duplex [*Simultaneous independent transfer of data in two directions using software handshake mechanism*] Duplexübertragung, Zweitweg(e)übertragung *f*, Gegenbetrieb *m*

duplex unit → **double-way intercom(munication) unit**

duplicating register, repeat(ing)register Fernzählwerk, Fernzählgerät *n*

duplicator Vervielfältiger *m*

durability [*Capability of a structure, assembly, component or product to maintain serviceability for which it was designed and produced over at least a specified time; based on assumed and/or specified levels of use and maintenance*] Haltbarkeit *f*

duration of bounce, bounce duration Prellzeit *f*

duration of current → ON period

duress alarm Nötigungsalarm *m*

dusk lighting, dusk illumination Dämmer(ungs)-beleuchtung *f*

dust [*Solid particles, smaller than 100 micrometres, projected into air by natural forces, or by mechanical processes*] Staub *m*

(dust) air filter (Staub)Luftfilter *m, n*

dust attraction Staubanziehung *f*

dust collector, dust chamber Staubsammler *m*

dust counter, Kern counter [*A photoelectric apparatus which measures the size and number of dust particles per unit volume of air*] Staubzähler *m*

dust-counting chamber Staubzählkammer *f* [*Die Luftprobe wird in einer Kammer von etwa 5 cm³ Inhalt erfasst und die auf Objektträger sich setzenden Staubteilchen im Mikroskop gezählt*]

dust-counting microscope, microscope dust counter [*A microscope equipped for quantitative dust sample analysis; magnification is usually 100X*] Staubzählmikroskop *n*

dust detector, dust sensor Staublsensor, Staubfühler *m*

dust dry [*A stage in the drying of a finish, after which dust will not adhere to the surface*] staubtrocken

dust eliminator [*Device, usually applied at a dust generating source, that traps dust particles suspended in the air or gases passing through it*] Staubabscheider *m*

dust grain, dust particle Staubkorn, Staubteilchen *n*

dust ingress, ingress of dust Staubeintritt *m*

dust metering Staubmessung *f*

dustpan Kehrschaufel *f*

dust particle → dust grain

dust removal, dedusting Entstaubung *f*

dust repelling staubabweisend

dustspot test Verfärbungsprüfung, Verfärbungsprobe *f*, Verfärbungsversuch *m* [*ASHRAE Standard 52-68*]

dust storage capacity Staubspeichervermögen *n*, Staubspeicherfähigkeit *f*

dust-tight, dust-proof staubdicht, staubsicher

dust trap Staubfalle *f*

dust type Staubart *f*

Dutch oven furnace (US) → horizontal grate extended furnace

Dutch tile stove → cockle stove

duty → application

duty control system Einsatzleitsystem *n*

duty cycle, operation(al) cycle Arbeitszyklus *m*

duty cycling, operation(al) cycling [*A control method which alternates or cycles the sequence of plant use bringing standby plant regularly into use*] Arbeitstaktablauf *m*, zyklische Schalten *n*

duty roster → guard schedule

DVE, digital video effects digitale Videoeffekte *f*

DVM → digital voltmeter

DVST, direct view storage tube Speicherbildschirm *m*

DW → data word

dwarf steam boiler, automatic steam boiler Zwergdampfkessel, Dampfautomat *m*

dwelling Wohnung *f*

dwell(ing) Verweilen *n*

dwelling heating Wohn(ungs)heizung *f*

dwelling kitchen Wohn(ungs)küche *f*

dwell(ing) time Verweilzeit *f*

dwelling ventilation Wohnungslüftung, Wohnungsbe- und -entlüftung *f*

DxF, drawing exchange format Austauschformat *n* [*Format zum Austauschen rechnergestützter Zeichnungsdaten zwischen rechnergestützten Zeichnungssystemen auf Personalrechnerbasis*]

dying away → fading (out)

dynamic address allocation, dynamic address allotment, dynamic address assignment dynamische Adressenzuweisung, dynamische Adressenzuteilung, dynamische Adressenzuordnung *f*

dynamic allocation, dynamic allotment, dynamic assignment dynamische Zuweisung, dynamische Zuteilung, dynamische Zuordnung *f*

dynamically balanced dynamisch (aus)gewuchtet

dynamically type-coded, dynamically standardized dynamisch typisiert

dynamic behaviour (Brit.), dynamic behavior (US) dynamische Verhalten, Übertragungsverhalten, Zeitverhalten *n* [*Regler*]

dynamic (computer) memory dynamische (Rechner)Speicher *m*

dynamic digital boiler → sequential boiler

dynamic digital boiler circuit → sequential boiler circuit

dynamic digital circuit → **sequential circuit**

dynamic display dynamische Einblendung *f* [*Einblenden des aktuellen Zustandes oder Wertes von physikalischen, kommunikativen oder aus Verarbeitungfunktionen abgeleiteten Informationen in die zugehörige funktionale statische Darstellung*]

dynamic dump [*In automatic data processing, a dump performed during the execution of the program(me)*] Sofort-Speicherauszug, sofortige Speicherauszug *m*

dynamic graphics dynamische Anlagengraphik *f* [*Aktualisierte Anzeige von Anlagen oder Raumzuständen auf einem Graphikschirm mit Anlagenschema*]

dynamic image Bildvordergrund *m*

dynamic input [*IEC 117-15*] dynamische Eingang *m* [*DIN 40 700, Teil 14*]

dynamic link library, DLL dynamische Funktionsbibliothek *f* für Fensterprogramme

dynamic (loud)speaker, moving coil (loud)speaker, coil-driven (loud)speaker [*The moving diaphragm is attached to a coil that is conductively connected to the source of electric energy and placed in a constant magnetic field. The current through the coil interacts with the magnetic field, causing the coil and diaphragm to move back and forth in step with the current variations through the coil*] (elektro)dynamische Lautsprecher *m*

dynamic lowering circuit Senkbremsschaltung *f*

dynamic microprogramming dynamische Mikroprogrammierung *f*

dynamic noise limiting → **DNL**

dynamic plate resistance [*Opposition that the plate circuit of a vacuum tube offers to a small increment of plate voltage; it is the ratio of a small change in plate voltage to the resulting change in the plate current, other tube voltages remaining constant*] innere Röhrenwiderstand *m*

dynamic pressure [*The difference between pressure at rest and flowing pressure*] kinetische Druck, Staudruck *m* [*frühere Benennung: dynamische Druck*]

dynamic programming dynamische Optimierung, dynamische Programmierung *f*

dynamic RAM, DRAM, dynamic random access memory dynamische Speicher mit wahlfreiem Zugriff, dynamische Schreib-/Lesespeicher, dynamische Arbeitsspeicher *m* [*Ein Halbleiterspeicher dessen Inhalt im Gegensatz zu einem statischen Halbleiterspeicher periodisch aufgefrischt werden muss*]

dynamic random access memory dynamische Speicher *m* mit wahlfreiem Zugriff

dynamic range [*Difference in decibels between the overload level and the minimum acceptable signal level in a system or transducer*] Dynamikbereich *m* [*selten: n*]

dynamic sequential control [*Method of operation of a digital computer through which it can alter instructions as the computation proceeds, or the sequence in which instructions are executed, or both*] Programmablaufänderung *f*

dynamic stop [*A loop in a computer program(me) which is created by a branch instruction in the presence of an error condition, and which signifies the existence of this condition*] dynamische Halt *m*

dynamic subroutine [*Subroutine that involves parameters, such as decimal point position or item size, from which a relatively coded subroutine is derived by the computer itself*] dynamische Unterprogramm *n*

dynamic system → **((closed) loop) dynamic system**

dynamic track following dynamische Spurnachführung *f*

dynamic viscosity → **absolute viscosity**

dynamic voice allocation dynamische Stimmenzuordnung *f*

dynamic wattmeter, moving-coil wattmeter Drehspulen-Leistungsmesser *m*

dynamic working storage area freie Arbeitsspeicherbereich *m* [*selten: n*]

E

E → **stop bit**

early failure, early outage Frühausfall *m*, Frühversagen *n* [*DIN 40 042*]

early guard Frühwache *f*

early-make [*contact*] voreilend, früh

early opening → **advance(d) opening**

early recognition Früherkennung *f*

early warning → **advance(d) warning**

early warning fire detector, early warning fire detecting device, early warning fire detection device Frühwarnfeuermelder, Frühwarnbrandmelder, Frühwarnfeuermeldungsgeber, Frühwarnbrandmeldungsgeber, Frühwarnfeuerdetektor, Frühwarnbranddetektor *m*, Frühwarnfeuermeldegerät, Frühwarnbrandmeldegerät *n*

EAROM, electrically alterable ROM, electrically alterable read-only memory elektrisch umprogrammierbare Festwertspeicher, elektrisch veränderliche Festwertspeicher *m* [*Er kann elektrisch gelöscht und wieder programmiert werden*]

earth (Brit.) → ground (US)

to earth → to (connect to) earth

earth closet, dry closet Trockenabort *m*, Trockenklosett *n* [*ohne Spülung*]

earth conductor → lightning conductor

earth connection (Brit.); ground connection (US) Erdung *f*

earthed (Brit.); grounded (US) geerdet

earth fault Erdschluss *n*

earthing (Brit.) → grounding (US)

earth(ing) contact socket → socket with earth(ing) contact

earth-leakage detector Ableitstromanzeiger *m*

earth (mass) (Brit.) → ground (mass) (US)

(earth)quake alarm override, seismic alarm override Erdbebenalarmübersteuerung *f*

(earth)quake switch, seismic switch (Erd)Bebenschalter *m*

Eastern w.c. → squatting w.c.

easy key schlüssellose Zugang(s)kontrolle *f*, schlüssellose Zugang(s)system *n*

easy reach bequeme Reichweite *f*

easy to handle → handy

eavesdropping → listening

eavesdropping device → listening device

eave(s) gutter, eave(s) trough, eave(s) trow [*A shallow channel or conduit set below and along the eave(s) of a building to catch and carry off rainwater from the roof*] Trauf(rinn)e *f*

EB → executive block

EBAM → electron-beam(-accessed) memory

EBCDIC → extended binary-coded-decimal interchange code

EC → error correction

ECC → error check(ing) and correcting

eccentric compressor [*Sliding vane or crescent, in which longitudinal vanes slide radially in a rotor mounted eccentrically in a cylinder*] Radialverdichter, Radialkompressor *m*, Radialkompressionsmaschine *f*

eccentric valve Exzenterwelle *f*

eccentric worm pump, worm eccentric pump Exzenterschneckenpumpe *f*, Schneckenexzenterpumpe

Eccles-Jordan multivibrator → bistable multivibrator

ECL → emitter-coupled logic

ECMA, European Computer Manufactures Association ECMA [*Europäische Organisation für Standardisierung mit dem Ziel, Datenverarbeitungsrichtlinien zu entwickeln. ECMA ist eine uneigennützige Organisation, die als erste Gruppe ihrer Art das OSI (Transport Layer Protocol) definiert hat.*]

economic(al) boiler, multi(ple)-purpose boiler Mehrzweckkessel, Vielzweckkessel *m*

economical of space → space-saving

economic viability Wirtschaftlichkeit *f*

economizer, flue gas preheater [*An apparatus that uses warm flue gases to preheat feedwater entering a boiler, thus improving boiler efficiency and economy*] Rauchgas(röhren)vorwärmer, Röhrenvorwärmer mit Rauchgaus *m*

ECS → extended call station

ED → expedited data

EDC → error detection and correction

eddy Wirbel *m*

eddy flow → turbulent flow

eddy stress → Reynolds stress

eddy stress tensor → Reynolds stress tensor

EDFA → erbium doped fibre amplifier

edge → arris

edge offset, arris offset Kantenversatz *m*

edge-punched card, arris-punched card Karte *f* mit Kartenlochung

EDI → electronic data interchange

Edison battery, nickel-iron battery Nickel-Eisen-Batterie, Edisonbatterie *f*

to edit [*In automatic data processing, to modify the form or format of an output or input by inserting or deleting characters such as page numbers or decimal points*] editieren, druckreif machen, korrigieren, (druckfertig) aufbereiten, ändern, bearbeiten

edit buffer Pufferspeicher für edierte Programme, Pufferspeicher für korrigierte Programme *m*

edit(ing) Druckaufbereitung *f*, Editieren *n*

E

editing elektronische Schnitt *m*

editing clause [*automatic data processing*] Aufbereitungsklausel, Änderungsklausel, Bearbeitungsklausel, Editier(ungs)klausel *f*

editing instruction [*See remark under "instruction"*] Änderungsbefehl, Aufbereitungsbefehl, Bearbeitungsbefehl, Editier(ungs)befehl *m*

editing symbol [*automatic data processing*] Aufbereitungszeichen, Änderungszeichen, Bearbeitungszeichen, Editier(ungs)zeichen *n*

edit mode [*As opposed to a "performance mode"*] Edierbetriebsart *f* [*Alle Parameter können ediert werden*]

editor Parameteränderungsmenü *n*

editor program (US), editor programme (Brit.); editor [*A special program(me) by means of which a user can easily perform corrections, insertions, modifications or deletions in an existing program(me) or data file*] Dateibearbeitungs-Dienstleistungsprogramm, Änderungsprogramm, Aufbereitungsprogramm, Bearbeitungsprogramm, Textbearbeitungsprogramm, Editier(ungs)programm, Editorprogramm *n*

EDO, extended (data) output, extended out erweiterte Datenausgabe *f*

EDP → electronic data processing

EDP center → electronic data-processing center

EDR, equivalent direct radiation gleichwertige Direktstrahlung *f*

EDS → electronic data switching

EDS → exchangeable disk storage

EDT → equipped device table

educated language Hochsprache *f*

education Ausbildung *f*

educational computer Lernrechner, Ausbildungsrechner *m*

educational language, ELAN pädagogische Programmiersprache, Programmiersprache für pädagogische Arbeiten *f* [*Sie basiert auf Erfahrungen mit PASCAL und ALGOL und wurde Mitte der 1970er Jahre entwickelt*]

EEPLD → electrically erasable programmable logic device

EEPROM → electrically erasable programmable read only memory

EER → energy efficiency ratio

EEROM, electrically erasable ROM, electrically erasable read only memory elektrisch löschbare Festwertspeicher, elektrisch tilgbare Festwertspeicher *m*

effective address [*In automatic data processing, the address that is obtained by applying any specified indexing or indirect addressing rules to the specified address; the effective address is then used to identify the current operand*] wirksame Adresse *f*, effektive Adresse, Wirkadresse

effective area Wirkfläche, wirksame Fläche, effektive Fläche *f*

effective cross-section wirksame Querschnitt, effektive Querschnitt, Wirkquerschnitt *m*

effective direction Wirkrichtung, wirksame Richtung, effektive Richtung *f* [*DIN 6580*]

effective instruction → real instruction

effectiveness Wirksamkeit *f*

effectiveness level, average effectiveness level Wirksamkeitspegel *m*

effective radiated power, ERP effektive Strahlungsleistung, wirksame Strahlungsleistung *f* [*Kennzeichnendes Merkmal für das Fernfeld einer Halbwellendipolantenne*]

effective temperature, operative temperature [*Uniform temperature of a radiantly black enclosure in which an occupant would exchange the same amount of heat by radiation plus convection as in the actual nonuniform environment*] Effektivtemperatur, Empfindungstemperatur *f*

effective time [*The time during which computer equipment is in actual use and produces useful results*] Nutz(ungs)dauer, Nutz(ungs)zeit *f*

effect return → aux(iliary) return

efficiency Wirkungsgrad *m* [*Quotient aus der nutzbar abgegebenen Leistung und der zugeführten Leistung eines Systems im stationären Betriebszustand*]

efficiency drop, efficiency decrease Wirkungsgradabfall, Wirkungsgradrückgang *m*, Wirkungsgradabnahme *f*

efficiency loss cost(s) Wirkungsgradverlustkosten *f* [*Sie entstehen hauptsächlich als vermehrte Personal- und Energiekosten bei zumindest teilweise hoch laufenden aber unvollkommen funktionierenden Anlagen*]

efficiency test Wirkungsgradprüfung, Wirkungsgradprobe *f*, Wirkungsgradversuch *m*

effluent [*A fluid flowing away from a process, for example, the effluent of a septic tank*] behandelte Strömungsmedium *n*

EFL → error frequency limit

EFM, eight-to-fourteen modulation Acht-Vierzehn-Modulation, Von-8-auf-14-Modulation *f* [*Jeweils acht Datenbits werden in Form von 14 Datenbits abgespeichert, von denen aber nur 10 Datenbits genutzt wer-*

den. Dieser Code erlaubt hohe Bitraten bei geringer Aufnahmedichte und einfacher Synchronisation. Wird bei der digitalen Schallplatte verwendet]

EFS → error free seconds

EFT, electronic fund transfer elektronische Zahlungsweise *f*

egress of heat, heat egress Wärmeaustritt *m*

EIA, Electronic Industries Association Verband *m* der Rechner- und Elektronikhersteller der USA

EIB → European Installation Bus

EIBA → European Installation Bus Association

eigenvalue → characteristic value

eight-binary digit word → byte

eight-to-ten modulation, ETM Acht-Zehn-Modulation, Von-8-auf-10-Modulation *f* [*Bei der digitalen Aufzeichnung auf Magnetband in Vollformat verwendetes Kodierverfahren, jeweils 8 Datenbits werden als 10 Datenbits abgespeichert*]

eight-to-fourteen modulation → EFM

EI instruction → enable interrupt instruction

Einstein product Einstein-Produkt *n*, *t*-Norm *f*

Einstein sum Einstein-Summe, *t*-Norm *f*

either-way, duplex duplex, wechselseitig

ejected ausgeworfen

eject(ing) key Auswerftaste, Auswurftaste *f*

ejector Strahlsaugapparat *m* [*Turbinenanlage*]

ejector Ejektor *m*, absaugende Strahlpumpe *f*

ejector, sewage ejector (pump), (sewage) lift, f(a)eces ejector (pump) (Abwasser)Hebeanlage, Fäkalienhebeanlage *f*

el → elbow

ELAN → educational language

elapsed time Laufzeit *f*

elapsed time verflossene Zeit, vergangene Zeit *f*

elapsed time accumulating meter → accumulating elapsed time meter

elastic... → resilient...

elasto-plastic elastisch-plastisch

elbow, el(l) Knie *n*, Winkel(stück) *m*, (*n*)

elbug, electronic bug elektronische Morsetaste *f*

electively, optionally, alternat(iv)ely wahlweise, alternativ

electric [*Charged with or produced by electricity, e. g. electric light; electric shock; electric torch*] elektrisch

electric absorption chiller, electric absorption refrigerating machine Elektro-Absorptionskältemaschine *f*

electric act(uat)or, electric final controlling element Elektrostellglied *n* [*Siehe Anmerkung unter „Stellglied"*]

electric afterheating energy → afterheating electric energy

electric air filter Elektroluftfilter *m*, *n*

electric air heater Elektrolufterhitzer, Elektroluftheizer, Elektroluft(er)wärmer *m*, Elektroluftheiz(ungs)gerät *n*, Elektro-LH [*DIN 30 600*]

electric air heating, electric air warming Elektro-Luft(er)wärmung, Luft(er)wärmung durch Elektrizität *f*

electrical [*Relating to electricity, e. g. electrical appliance = Elektrogerät*] elektrisch

electrical accessories Elektrozubehör *n* [*selten*: *m*]

electrical appliance → electrical device

electrical-architectural plan → architectural-electrical plan

electrical article surveillance technique, electrical article security technique, electrical product surveillance technique, electrical product security technique elektrische Warensicherungstechnik, elektrische Artikelsicherungstechnik *f* [*Bei unbefugter Unterbrechung eines Stromkreises erfolgt Alarmgabe. Verwendung bei Waren, die der Kunde an- oder ausprobieren will, z. B. Kleidung, Spielwaren, Taschenrechner usw.*]

electrical auxiliary power elektrische Hilfsenergie *f*

electrical boiler, electrical water heater [*See remark under "water heater"*] Elektrowasserheizer, Elektroboiler *m*

electrical boiler Elektro(heizungs)kessel *m*

electrical booster heating Elektrozusatzheizung, Zusatzelektroheizung, elektrische Zusatzheizung *f*

electrical central storage heating, central electrical storage heating Elektrozentralspeicherheizung, Elektroblockspeicherheizung, Zentralelektrospeicherheizung, Blockelektrospeicherheizung *f*

electrical code [*A systematic body of rules governing the application and installation of electrically operated equipment and devices and electric wiring systems*] Elektrobestimmungen, Elektroordnung *f*

electrical commercial furnace → commercial electrical furnace

electrical condenser → condenser

(electrical) condenser (loud)speaker → electrostatic (loud)speaker

E

(electrical) conductance [*The real part of the admittance of a circuit; when the impedance contains no reactance, as in a direct-current circuit; it is the reciprocal of resistance, and is thus a measure of the ability of the circuit to conduct electricity. Designated G*] elektrische Wirkleitwert *m*

electrical conductivity, specific conductance [*The ratio of the electric current density to the electric field in a material*] (elektrische) Leitfähigkeit *f*, (elektrische) Leitvermögen *n*

E

electrical conduit → **conduit**

electrical conduit union → **conduit union**

electrical conduit with heavy protection → **conduit with heavy protection**

electrical conduit with light protection → **conduit with light protection**

electrical conduit with medium protection → **conduit with medium protection**

electrical continuous water heater Elektrodurchlauf(wasser)erhitzer *m*

electrical control → **electrical ((open) loop) control**

electrical controller elektrische Regler, Elektrikregler *m*

electrical correlator elektrische Zuordner *m*

(electric(al)) current line Stromleitung, Elektroleitung, elektrische Leitung *f*

electrical data Elektrodaten *f*

electrical device, electrical appliance Elektrogerät *n*, Elektrovorrichtung *f*

electrical dimmer Elektro-Helligkeitsregler, Elektro-Lichtsteller *m*

electrical domestic appliance → **electrical household appliance**

electrical elevator (US); electrical lift (Brit.) Elektroaufzug *m*

electrical energy elektrische Kraft, elektrische Energie *f*

electrical engineer Elektroingenieur *m*

electrical engineering Elektrotechnik *f*

electrical engineering within buildings Gebäude-Elektrotechnik, Elektro-Gebäudetechnik *f*

electrical equipment, electrical outfit [*See remark under "equipment"*] Elektroausrüstung, Elektroausstattung, Elektrobestückung *f*

electrical equipment room Elektrogeräteraum *m*

electrical eye → **photocell**

electrical fitter → **electrician**

electrical freight elevator (US) → **electrical goods elevator**

electrical furnace Elektroofen *m*

electrical goods elevator, electrical freight elevator (US); electrical goods lift, electrical freight lift (Brit.) Elektrogüteraufzug, Elektrotransportaufzug *m* [*Siehe Anmerkung unter „Güteraufzug"*]

electrical grid Elektronetz, elektrische Netz *n*

(electrical) ground → **zero voltage**

electrical heating Elektroheizung *f*

electrical hot water storage tank, electrical hot water storage vessel, electrical warm water storage tank, electrical warm water storage vessel Elektrowarmwasserspeicher, Elektro-WW-Speicher, Elektroboiler *m*, Elektrowarmwassergefäß, Elektro-WW-Gefäß *n* [*Siehe Anmerkung unter "Warmwasserspeicher"*]

electrical household appliance, electrical household device, electrical domestic appliance, electrical domestic device Elektrohaushalt(s)gerät *n*

electrical ignition elektrische Zündung *f*

electrical installation → **electrical plant**

(electrical) insulating material [*A material that is a very poor conductor of electricity*] Isolier(körper)-stoff, Isolatorstoff, Nichtleiterstoff *m*

(electrical) insulating oil Isolieröl *n*

(electrical) insulating putty Isolierkitt *m*

(electrical) insulating tape Isolierband *n*

(electrical) insulation Elektroisolierung, Isolierung *f*

(electrical) insulator [*A component or device made from material having great enough resistance to the flow of electric(al) current to be effectively considered as a nonconductor of current*] Isolator, Isolierkörper, Nichtleiter *m*

electrical intensity, electrical strength elektrische Feldstärke *f*

electrical interference → **interference**

electrical isolation, galvanic isolation, control-to-load isolation Potentialtrennung der Steuerkreise, Entkopp(e)lung der Steuerkreise, galvanische Trennung *f*

electrical life elektrische Lebensdauer *f*

electrical lift (Brit.) → **electrical elevator**

electrical location system → **paging (system)**

electrical locking Elektroverrieg(e)lung, elektrische Verrieg(e)lung *f*

electrically alterable elektrisch umprogrammierbar

electrically alterable ROM → EAROM

electrically conducting, electrically conductive elektrisch leitend

electrically energized → hot

electrically erasable elektrisch löschbar, elektrisch tilgbar [*Durch Anlegen einer definierten Spannung*]

electrically erasable programmable logic device, EEPLD, electrically erasable programmable logic appliance elektrisch löschbar programmierbare Logikgerät, elektrisch tilgbar programmierbare Lo-gikgerät *n*

electrically erasable programmable read only memory, EEPROM elektrisch löschbar programmierbare Nur-Lese-Speicher, elektrisch tilgbar programmierbare Nur-Lese-Speicher *m*

electrically erasable read-only memory → EEROM

electrically heated ceiling Elektroheiz(ungs)decke, elektrisch beheizte Decke *f*

electrically heated metal suspended ceiling, electrically heated metal hung ceiling Elektro-Metall-Heiz(ungs)decke, Metall-Elektro-Heiz(ungs)decke *f*

electrically heated tape → (pipe) heating tape

electrically-lit, electrically-illuminated elektrisch beleuchtet

electrically welded sleeve Elektroschweißmuffe *f*

electrical metrology elektrische Messtechnik *f*

electrical moment elektrische Moment *n*

electrical ((open) loop) control Elektrosteuerung *f*

electrical outfit → electrical equipment

electrical overload capacity elektrische Überlastbarkeit *f*

electrical paging (system) → paging (system)

electrical passenger elevator (US); electrical passenger lift (Brit.) Elektrofahrstuhl, Elektropersonenaufzug, elektrische Fahrstuhl, elektrische Personenaufzug *m*

electrical plant, electrical installation elektrische Anlage, elektrische Installation, Elektroanlage, Elektroinstallation *f*

electrical point Stromanschlussstelle *f*

electrical power elektrische Leistung *f*

electrical product surveillance technique → electrical article surveillance technique

electrical programmable and UV erasable ROM → erasable programmable read only memory

electrical protection Elektroschutz *m*

(electrical) resistance heating [*The generation of heat by electric conductors carrying current, degree of heating is proportional to the electrical resistance of the conductor; used in electrical home appliances, home or space heating, and heating ovens and furnaces*] (elektrische) Widerstandsheizung *f*

(electrical) resistance measurement [*The qualitative determination of that property of an electrically conductive material, component, or a circuit called electrical resistance*] (elektrische) Widerstandsmessung *f*

(electrical) resistance meter (elektrische) Widerstandsmesser *m*

(electrical) resistance thermometer, (electrical resistance pyrometer [*A thermometer in which the sensing element is a resistor whose resistance is an accurately known function of temperature*] (elektrische) Widerstandsthermometer *n*

(electrical) resistance wire [*Wire made from a metal or alloy having high resistance per unit length, such as Nichrome; used in wire-wound resistors and heating elements*] (elektrische) Widerstandsdraht *m*

electrical resistor → resistor

electrical staff locator → paging (system)

electrical storage (type) water heater → storage (type) electrical water heater

electrical storage water heating Elektrospeicherheizung mit Wasser, Elektrospeicherwasserheizung *f* [*Die durch Strom erzeugte Wärme wird in Wasser gespeichert*]

electrical storm → lightning storm

electrical strength → electrical intensity

electrical time-distribution system [*IEC 50(35)*] Zeitdienstsystem, Zentraluhrensystem *n*

electrical unit [*A standard in terms of which some electrical quantity is evaluated*] elektrische Einheit *f*

electrical unlocking Elektroentrieg(e)lung, elektrische Entrieg(e)lung *f*

electrical upgrade elektrische Leistungserhöhung *f*

electrical warm water storage tank → electrical hot water storage tank

electrical water heater → electrical boiler

electric arc lamp → arc lamp

(electric) asynchronous heating (elektrische) Asynchronheizung *f* mit Wärmespeicher

electric auto(matic) system Elektroautomatik *f*

electric base(board) convector, electric mopboard convector, electric washboard convector, electric scrub board convector, electric sanitary cove con-

E

vector (US); electric skirting (board) convector (Brit.); electric base plate convector [*Scotland*] Elektrosockelkonvektor, Elektrofußleistenkonvektor, Elektroscheuerleistenkonvektor, Elektroheizungsleistenkonvektor *m*

electric base(board) heating, electric scrub board heating, electric mopboard heating, electric washboard heating, electric sanitary cove heating (US); electric skirting (board) heating (Brit.); electric base plate heating (Scotland) Elektrosockelheizung, Elektrofußleistenheizung, Elektro(scheuer)leistenheizung *f*

electric bathroom heater Elektrobadezimmerheizer *m*

electric bell elektrische Klingel, Elektroklingel *f*

electric block storage heater, electric off-peak storage heater, electric night storage heater [*"heat emitter", "heating appliance", "heating apparatus" and "heating device" may synonymously be used for "heater"*] (Nachtstrom)Speicher(heizungs)block , (Nachtstrom)Block(heiz(ungs))-speicher *m*

electric block storgage heating, electric off-peak storage heating, electric night storage heating Nachtstrom(block)speicherheizung *f*

electric burner Elektrobrenner *m*

electric cable Elektrokabel, Stromkabel *n*

electric ceiling heating, ceiling electric heating Elektrodeckenheizung, Deckenelektroheizung *f*

(electric) charge → quantity of electricity

electric circuit → circuit

electric clock Elektrouhr *f*

electric closet Elektroschrank *m*

electric/coal bath stove Elektro-Kohle(n)-Badeofen *m*

electric compression heat pump, electric compressor heat pump Elektrokompressionswärmepumpe *f*

(electric) concrete floor heating (Elektro)Beton(fuß)bodenheizung *f*

(electric) condenser Elektrokondensator, elektrische Kondensator *m*

(electric) condenser bank → capacitor bank

(electric) condenser microphone → capacitor microphone

electric construction Elektrobau *m*

electric contact Elektrokontakt *m*

electric contactor → contact maker

electric continuous water heater Elektro-Durchflusswasser(er)wärmer *m*

electric convector Elektrokonvektor *m*

electric cord → power cord

electric cradle Elektro-(Fenster)Putzwagen *m*

electric defrosting [*Defrosting by means of electric heating*] Elektroentfrostung *f*

electric device [*In an electric system, a component that is intended to carry, but not to consume, electric energy, e. g. a switch*] Elektroenergieträger *m*

electric distribution board, electric distribution frame [*telephony*] Elektroverteiler *m*

electric domestic hot water heater, electric domestic warm water heater Elektro-Brauchwasser(er)wärmer, Elektro-Brauchwassererhitzer, Elektro-Brauchwasserheizer *m*

electric domestic hot water storage tank → domestic electric hot water storage tank

electric domestic refrigerator → domestic electric refrigerator

electric door release Elektrotüröffner *m*

electric door release system Elektrotüröffnersystem *n*

electric downflow wall heater Elektro-Abwärts-Wandheizer, Abwärts-Elektro-Wandheizer *m*

electric drinking fountain Elektro-Trinkbrunnen *m*

electric duct heater Elektrokanalheizer *m*

electric elevator (US) → electric (traction) lift

electric energy elektrische Energie, Elektroenergie *f*

electric (equipment) connection Elektro(geräte)anschluss *m*, Elektro(geräte)verbindung *f*

electric fan Elektroventilator, Elektrolüfter *m*

electric fan heater Elektroventilatorheizer, Elektrolüfterheizer *m*

electric fan high level heater [*It is mounted at high level on walls after the fashion of a unit heater. It is liable to cause a certain amount of noise and may be regarded as chiefly useful for industrial purposes*] Hoch-Elektroventilatorheizer, Hoch-Elektrolüfterheizer *m*

electric feeder, electric feeding line Elektrozuleitung, Elektrospeiseleitung *f*

electric field intensity, electric field strength, voltage gradient (elektrische) Feldstärke *f*

electric field intensity level, electric field strength level, voltage gradient level (elektrische) Feldstärkepegel *m* [*Logarithmiertes Verhältnis des Betrages des Vektors der elektrischen Feldstärke zum Bezugswert 1 μV/m*]

electric final controlling element → electric act(uat)or

electric fire → electric heating appliance

electric floor [*e. g. the Robertson Q-floor*] elektrotechnisch genutzte Decke *f*

electric floor heating, electric floor warming Elektro(fuß)bodenheizung, Elektro-FB-Heizung, (Fuß)Bodenelektroheizung, elektrische (Fuß)Bodenheizung *f* [*Die Warmluft wird durch einen Hohlraum unter den (Fuß)Boden geleitet. Es findet keine Wärmespeicherung statt*]

electric floor radiant heating, electric floor radiant warming, electric radiant floor heating, electric radiant floor warming elektrische (Fuß)Bodenstrahlungsheizung, Elektro(fuß)bodenstrahlungsheizung, Elektro-FB-Strahlungsheizung *f*

electric floor storage heating, electric floor storage warming elektrische (Fuß)Bodenspeicherheizung, Elektro-(Fuß)Bodenspeicherheizung *f*

electric flux elektrische Fluss, Elektrofluss *m*

electric flux density elektrische Kraftlinienzahl *f*

electric flux line elektrische Flusslinie, Elektroflusslinie *f*

electric flux pump Elektroflusspumpe *f*

electric food warmer Elektrospeisen(er)wärmer *m*

electric fryer Elektrobratpfanne *f*

electric heat Elektrowärme *f*

electric heater → electric heating appliance

electric heating [*It is by common practice accepted as local heating although in fact the basic heat source is coal or oil of the power station, the electricity being the distributing medium*] Elektroheizung *f*

electric heating appliance, electric heating apparatus, electric heating device, electric heater, electric fire Elektroheizung(ungs)gerät *n*, Elektroheiz(ungs)apparat, Elektroheizer, Elektroerhitzer, Elektro(er)wärmer *m*

electric heating bar Elektroheiz(ungs)stab *m*

electric heating coil Elektroheiz(ungs)schlange *f*

electric heating device → electric heating appliance

electric heating element Elektroheiz(ungs)element *n*

electric heating heat pump Elektroheiz(ungs)wärmepumpe *f*

electric heating insert elektrische Heiz(ungs)einsatz, Elektroheiz(ungs)einsatz *m*

electric heating panel Elektroheiz(ungs)platte *f*

electric heating tube Elektroheiz(ungs)röhre *f*

electric heating unit Elektroheiz(ungs)einheit *f*

electric heat pump Elektrowärmepumpe *f*

electric heat trace system → heat trace electric system

electric hospital lift (Brit.); electric hospital elevator (US) Kranken(haus)-Elektroaufzug, Elektro-Kranken(haus)aufzug, Krankenbahren-Elektroaufzug, Elektro-Krankenbahrenaufzug *m*

electrician, electrical fitter Elektriker, Elektroinstallateur, Elektromonteur *m*

electric ignition Elektrozündung *f*

electric illumination heat → electric lighting heat

electric incinerator Elektro(müll)verbrenner *m*

electric installation equipment elektrische Installationsgerät(e) *n*, (*f*)

electric interceptor → electric separator

electricity-measuring device Elektrizitätsmessgerät *n* [*z. B. Elektrizitätszähler*]

(electricity) meter (integrierende) (Elektro)Zähler *m*

electricity quantity → quantity of electricity

electricity utility EVU, Elektrizitätsversorgungsunternehmen *n*, Elektrizitätsgesellschaft *f*

electric kitchen Elektroküche *f*

electric (kitchen) range Elektro(küchen)herd, Küchenelektroherd *m*

electric lift (Brit.) → electric (traction) lift

electric lighting heat, electric illumination heat Elektrobeleuchtungswärme, Elektrolichtwärme *f*

electric locking Elektroverrieg(e)lung, Elektroabrieg(e)lung *f* [*Siehe Anmerkung unter „Verrieg(e)lung"*]

electric low-water alarm Elektro-Wassermangelsicherung *f*

(electric) main → power transmission line

electric main clock, electric principal clock Elektrohauptuhr, Hauptelektrouhr *f*

electric meter [*An electricity-measuring device that totalizes with time, such as a watt-hour meter or ampere-hour meter, in contrast to an electric instrument*] Elektrizitätszähler, Elektrizitätsmesser, Stromzähler, Strommesser, Elektrozähler, Elektromesser *m*

electric mopboard convector (US) → electric base(-board) convector

electric mopboard heating (US) → electric base(-board) heating

electric night storage heater → electric block storage heater

electric night storage heating → electric block storage heating

E

E

electric offpeak storage heater → **electric block storage heater**

electric offpeak storage heating → **electric block storage heating**

electric oil-filled radiator, oil-filled electric radiator Elektro-Öl-Radiator, Elektro-Öl-Gliederheizkörper *m*

electric on-off controller elektrische Zweipunktregler, unstetige Regler *m* [*Das Stellglied bei einer Wasserstandsreg(e)lung ist ein Magnetventil bei dem nur die beiden Stellungen „Auf" und „Zu" möglich sind*]

electric panel heater [*Output is a combination of low temperature radiation and convection*] Elektroplattenheizer *m*

electric panic release Elektropaniköffner *m*

electric passenger lift (Brit.); electric passenger elevator (US) Elektropersonenaufzug, Elektrofahrstuhl *m*

electric-pneumatic [*Not to be confused with "pneumatic-electric"*] elektropneumatisch, elektrisch-pneumatisch

electric preselector (gear) elektrische Gangvorwähler *m*

electric principal clock → **electric main clock**

electric radiant (ceiling) cable heater Elektro(decken)strahlungskabelheizer *m*

electric radiant floor heating → **electric floor radiant heating**

electric radiant heating, electric panel heating, radiant electric heating, panel electric heating [*Methods are electric(al) cables in ceiling or floor, prefab(ricated) ceiling panels, radiant glass panels, and electric(al) base(board). All four methods suitable to switch and thermostat control in each room*] Elektrostrahlungsheizung, Strahlungselektroheizung, elektrische Strahlungsheizung *f*

electric range → **electric (kitchen) range**

electric refrigerator Elektrokühlschrank *m*

electric resistance heating elektrische Widerstandsheizung, Elektrowiderstandsheizung *f*

electric reheat(ing) coil Elektronachwärmschlange *f*

electric resistance heater, electric resistance heating appliance, electric resistance heating apparatus, electric resistance heating device Elektrowiderstandsheizer, elektrische Widerstandsheizer *m*

electric room heating → **electric space heating**

electric sanitary cove convector (US) → **electric base(board) convector**

electric sanitary cove heating (US) → **electric base(board) heating**

electric scrub board convector (US) → **electric base(board) convector**

electric scrub board heating (US) → **electric base(board) heating**

electric separator, electric interceptor, electric trap [*See remark under "interceptor"*] Elektro(ab)-scheider *m* [*z. B. (Ab)Scheidung durch Ionentransport*]

electric service elevator (US); electric service lift (Brit.) Elektro-Speisenaufzug, Speisen-Elektroaufzug *m*

electric siren Elektrosirene *f*

electric skirting (board) convector (Brit.) → **electric base(board) convector**

electric skirting (board) heating (Brit.) → **electric base(board) heating**

electric space heating, electric room heating [*Permanently installed electric(al) heating as the principal source of space heating throughout an entire dwelling or business establishment*] elektrische Raumheizung, Elektroraumheizung *f* [*DIN 44 576*]

electric steam boiler elektrische Dampfkessel, Elektrodampfkessel *m*

electric storage heater, electric storage heat emitter, electric storage heating appliance, electric storage heating device, electric storage heating apparatus Elektrospeicherheizer, Speicherelektroheizer, Elektro-Warmwasseraufbereiter *m*

electric storage heating Elektrospeicherheizung, Speicherelektroheizung *f*

electric supply line Elektroversorgungsleitung, Stromversorgungsleitung *f*

electric supply main Elektro-Hauptleitung, Strom-Hauptleitung *f*

electric swing(ing) gate, electric slewing gate Elektroschwenktor, Schwenkelektrotor *n*

(electric) synchronous heating, synchronous electric heating (elektrische) Synchronheizung *f* [*Stromaufnahme und Heiz(ungs)energieabgabe fallen zeitlich zusammen*]

electric system [*The difference between electric and electronic systems lies in a somewhat esoteric refinement in equipment. In an electric (control) system, the thermostat sends its signal directly to the damper or valve motor. In an electronic (control) system, the thermostat sends a minute signal to an electronic relay, where the signal is measured and amplified. The relay in turn affects the control over the damper or valve, or over supplementary control circuits. The electronic (control) equipment may be used to activate any type of electric or pneumatic device, in any of numerous modes or cycles of operation*] Elektrosystem *n*

electric thermostat elektrische Thermostat, Elektrothermostat, elektrische automatische Temperaturregler *m*

electric towel rack, electric towel warmer Elektro-Handtuchgestell *n*, Elektro-Handtuchtrockner *m*

electric tracer heating elektrische Begleitheizung, Elektrobegleitheizung *f*

electric (traction) lift (Brit.); electric (traction) elevator (US) Elektroaufzug *m*

electric trap → electric separator

electric unit heater [*See remark under "unit heater"*] Elektrolufttheizer *m*

electric wall convector Elektrowandkonvektor, Wandelektrokonvektor *m*

electric wall heater Elektrowandheizer, Wandelektroheizer *m*

electric warm air central heating, electric hot air central heating Elektro-Warmluft-Zentralheizung, Warmluft-Elektro-Zentralheizung *f*

electric washboard convector (US) → electric base (board) convector

electric washboard heating (US) → electric base (board) heating (US)

electric water heater Elektrowarmwasserbereiter, Elektrowasserheizer, Elektrowasser(er)wärmer, Elektrowassererhitzer *m*

electric wave, Hertzian wave [*An electromagnetic wave, specially one whose wavelength is at least a few centimeters*] elektrische Welle, Elektrowelle *f*

electroacoustic(al) [*Pertaining to a device that involves both electric current and sound frequency pressures*] elektroakustisch, ela

electroacoustic(al) transducer elektroakustische Wandler, ela Wandler *m*

electroacoustics Elektroakustik *f*

electrochemical elektrochemisch

electrode boiler [*B. S. 1894*] Elektrodenkessel *m*

(electrode) dark current, DC [*The electrode current that flows when there is no radiant flux incident on the photocathode in a phototube or camera tube*] Dunkelstrom *m*

electrode level relay Elektrodenniveaurelais *n*

electrode potential, electrode voltage Elektrodenspannung *f*

electrode steam air humidifier Elektrodendampfluftbefeuchter, Elektrodendampfluftanfeuchter *m*

electrode steam generator Elektrodendampferzeuger *m*

electrode terminal Polklemme *f*

electrodynamic instrument, electrodynamic dynamometer elektrodynamische Instrument, elektrodynamische Dynamometer *n*

electrodynamic measuring element, electrodynamic measuring instrument, electrodynamic measuring mechanism [*"Measurement" may synonymously be used for "measuring"*] elektrodynamische Messwerk *n*

electrodynamic quotientmeter elektrodynamische Quotientenmessgerät, elektrodynamische Quotientenmesswerk *n*, elektrodynamsiche Quotientenmesser *m*

electro-gilding galvanische Vergoldung, Galvanovergoldung *f*

electro-hydraulic drive (unit) Elektrohydraulik(an)trieb *m*

electro-hydraulic elevator [*Direct-plunger elevator where liquid is pumped under pressure directly into the cylinder by a pump driven by an electric(al) motor*] Elektrohydraulikaufzug, Hydraulikelektroaufzug *m*

electroless plating stromlose Methode *f*, stromlose Verfahren *n* [*Zur Herstellung von Überzügen auf Metalloberflächen*]

electroluminescent panel, electroluminescent cell, electroluminescent lamp, luminescent cell [*A surface-area light source employing the principle of electroluminescence; consisting of a suitable phosphor placed between sheet-metal electrodes, one of which is essentially transparent, with an alternating current applied between the electrodes*] elektrolumineszierende Zelle, Elektrolumineszenzzelle, Elektrolumineszenzplatte *f*

electroluminor, electroluminescent phosphor elektrolumineszierende Phosphor *m*

electrolyte Elektrolyt *m*, elektrisch leitende Flüssigkeit *f*

electrolytic capacitor, polarized capacitor Elektrolytkondensator *m*

electrolytic tantalium capacitor → tantalium electrolytic capacitor

electromagnetic [*Having both magnetic and electric properties*] elektromagnetisch

electromagnetic coil → solenoid

electromagnetic compatibility, EMC elektromagnetische Verträglichkeit, EMV *f* [*VDE 0100*]

electromagnetic cylinder elektromagnetische Zylinder, EM-Zylinder *m* [*Tür*]

electromagnetic door activator elektromagnetische Türöffner *m*

E

electromagnetic (flow)meter [*A (flow)meter that offers no obstruction to liquid flow; two coils produce an electromagnetic field in the conductive moving fluid; the current induced in the liquid, detected by two electrodes, is directly proportional to the rate of flow*] elektromagnetische Strömungsmesser, elektromagnetische Durchflussmesser *m*

electromagnetic flush(ing) valve; electromagnetic flushometer (valve) (US) Magnet(klosett)(druck)-spüler *m*

E

electromagnetic interference, EMI [*Interference, generally at radio frequencies, that is generated inside systems, as contrasted to radio-frequency interference coming from sources outside a system*] elektromagnetische Störung *f*

electromagnetic radiation elektromagnetische Strahlung *f*

electromagnetic system Elektromagnetsystem *n*, Spannungsspule mit Eisenkern, Eisenkern-Spannungsspule *f*

electromagnetic valve → solenoid valve

electromagnetic wave [*A disturbance which propagates outward from any electric charge which oscillates or is accelerated; far from the charge it consists of vibrating electric and magnetic fields which move at the speed of light and are at right angles to each other and to the direction of motion*] elektromagnetische Welle *f*

electromechanical detector, electromechanical detecting device, electromechanical detection device elektromechanische Melder *m*, elektromechanische Detektionsmittel, elektromechanische Meldegerät *n*

electromechanical door locking elektromechanische Türverrieg(e)lung *f*

electromechanical lock elektromechanische Schloss *n*

electromechanical panic release elektromechanische Paniköffner *m*

electromechanical scanning elektromechanische Abtasten *n*

(electromechanical) short-circuit force Kurzschlusskraft *f*

electrometer Elektrometer *n* [*Hochempfindliches Messgerät zum Messen von Spannungen und Ladungen*]

electrometer tube [*A high-vacuum electron tube, having a high input impedance (low control-electrode conductance) to facilitate measurement of extremely small direct currents or voltages*] Elektrometerröhre *f*

electromotive actuator, electromotive actor elektromotorische Stellantrieb *m*

electromotive force, e.m.f., electromotance [*Force that causes electricity to flow when there is a difference of potential between two points. The unit of measurement is the volt*] elektromotorische Kraft, EMK, Urspannung *f*

electron beam, beam of electrons Elektronenstrahl *m*

electron-beam(-accessed) memory, EBAM [*A memory that uses a high-resolution electron beam to store information on a target in a vacuum tube*] Elektronenstrahlspeicher *m*

electron beam tube Elektronenstrahlröhre, Bildröhre *f*

electron-beam vacuum tube Elektronenstrahlvakuumröhre, Bildvakuumröhre *f*

electron-coupled oscillator → Dow oscillator

electronic access control system Elektronik-Zutrittsystem, Elektronik-Zugang(s)system *n*

electronic active element → active electronic element

electronic (air) filter Elektronik(luft)filter *m*, *n*

electronically readable elektronisch lesbar

electronic amplification of sound elektronische Schallverstärkung *f*

electronic anemometer [*Device sensitive to air movement past an electronic circuit to measure air velocity*] elektronische Anemometer *n*, elektronische Windmesser *m*

electronic article surveillance technique, electronic article security technique, electronic product surveillance technique, electronic product security technique elektronische Warensicherungstechnik, elektronische Artikelsicherungstechnik *f* [*Aufbringen von Klebe- oder Anheftetiketten, die mit einfachen Mitteln nicht entfernt werden können. Die Etiketten enthalten dünne, als passive Schwingungsgeber ausgebildete Metallstreifen. Eine am Ausgang installierte Überwachungsvorrichtung derart gekennzeichnete Ware führt zur Alarmgabe*]

electronic automatic machine → automatic electronic machine

electronic bell Elektronikklingel *f*

electronic bug → elbug

electronic building services engineering elektronische Gebäudetechnik, elektronische Haustechnik *f*

electronic calculator [*A calculator in which circuits perform calculations and show results on a digital display; the displays usually use either seven-segment light-emitting diodes or liquid crystals*] Elektronikrechenwerk *n*

(electronic) card lock (elektronische) Kartenschloss *n*

electronic climate controller, electronic climate regulator Elektronikklimaregler *m*

electronic ((closed) loop) control, electronic closed control elektronische Reg(e)lung, elektronische Regulierung, Elektronikreg(e)lung, Elektronikregulierung *f*, elektronische geschlossene Wirkungsablauf *m*

electronic closing system, electronic closure system elektronische Schließsystem *n*

electronic combination lock, electronic permutation lock elektronische Kombinationsschloss, elektronische Vexierschloss *n*

electronic component elektronische Bestandteil *m*

electronic concept Elektronikbegriff *m*

electronic control → electronic ((closed) loop) control

electronic control → electronic ((open) loop) control

electronic controller, electronic CTRL [*See remark under "controller"*] elektronische Regler, Elektronikregler *m*

electronic CTRL → electronic controller

electronic data elektronische Daten, Elektronikdaten *f*

electronic data interchange, EDI elektronische Datenaustausch, Elektronikdatenaustausch *m*

electronic data processing, EDP [*Processing data by using equipment that is predominately electronic in nature, such as an electronic digital computer*] elektronische Datenverarbeitung, EDV, Elektronikdatenverarbeitung *f*

electronic data-processing center, EDP center, computer center (US) [*British spelling is "centre"*] Rechenzentrum, Rechnerzentrum *n*

electronic data switching, EDS elektronische Datenvermittlung, Elektronikdatenvermittlung *f*

electronic dimmer [*See remark under "dimmer"*] Elektronik-Helligkeitsregler, Elektronik-Lichtsteller *m*

electronic display elektronische Anzeige *f*

electronic door lock Elektroniktürschloss *n*

electronic door opener elektronische Türöffner *m*, Elektroniktüröffner

(electronic) eavesdropping device → listening device

electronic engineering Elektroniktechnik *f*

electronic flame monitoring elektronische Flammenüberwachung *f*

electronic fund transfer → EFT

electronic heating, high-frequency heating, radio-frequency heating [*Heating by means of radio-frequency current produced by an electron-tube oscillator or an equivalent radio-frequency power source*] Elektronikheizung, Hochfrequenzheizung *f*

electronic ignition elektronische Zündung *f*

Electronic Industries Association → EIA

electronic ink elektronische Tinte *f*

electronic interference [*Any electrical or electromagnetic disturbance that causes undesirable response in electronic equipment*] elektronische Störung *f*

electronic life security elektronische Personensicherung *f*, elektronische Personenschutz, EPS *m*

electronic lime dissolver elektronische Kalk(auf)löser *m*

electronic line scanning elektronische Zeilenabtasten *n*

electronic lock Elektronikschloss *n*

electronic locking system elektronische Schließsystem, Elektronikschließsystem *n*

electronic mail, E-mail elektronische Post, Elektronikpost *f*

electronic mail search(ing) machine → search(ing) machine

electronic metrology elektronische Messtechnik, Elektronikmesstechnik *f*

electronic modular system, modular electronic system Elektronik-Baustein-System, Baustein-Elektronik-System, Elektronik-Modul-System, Modul-Elektronik-System *n*

electronic motion detector, electronic motion detecting device, electronic motion detection device elektronische Bewegungsmelder *m*, elektronische Bewegungsmeldegerät *n*

electronic muli(ple)-code lock mehrfachkodierte Elektronikschloss, vielfachkodierte Elektronikschloss *n*

electronic numerical integrator and calculator, ENIAC elektronische Rechner *m* [*von J.M. Brainard, J.P. Eckert, J.W. Mauchey und H.H. Goldstein, 1946*]

electronic ((open) loop) control elektronische Steuerung *f*, elektronische Steuern *n* [*Siehe Anmerkung unter "Steuerung"*]

electronic pen elektronische Stift, Lichtgriffel *m* [*Mit einem Rechner verbundenes schreibstiftähnliches Gerät zur Änderung der an einem Datensichtgerät erscheinenden Informationen*]

electronic permutation lock → electronic combination lock

E

E

electronic pressure controller, electronic pressure regulator Elektronikdruckregler *m*

electronic product surveillance technique → electronic article surveillance technique

electronic radiator controller, electronic radiator regulator elektronischer Heiz(ungs)körperregler *m*

electronic (raster) scanning [*Scanning in which an electron beam, controlled by electric or magnetic fields, is swept over the area of examination, in contrast to mechanical or electromechanical scanning*] elektronische (Raster)Abtasten *n*

electronic reader, electronic reading device Elektronikleser *m*, Elektroniklesegerät *n*

electronics [*Study, control and application of the conduction of electricity through gates or vacuum, or through semiconducting or conducting materials*] Elektronik *f*

electronic security elektronische Sicherung *f*

electronic security engineering elektronische Sicherungstechnik *f*

electronic security seal Elektronik-Sicherungsverschluss *m*

electronic security system elektronische Sicherungssystem *n*

electronics engineering Elektroniktechnik *f*

electronic sensor [*Installed at a shop exit to detect the unauthorized removal of merchandise that has been protected by security tags. If someone tries to leave the store without purchasing protected merchandise, an alarm sounds as the protected product passes through the sensors detection field*] elektronische Überwachungsvorrichtung *f*,, Sensor *m*

electronic shutter elektronische Verschluss *m*

electronics scrap Elektronikschrott *m*

electronic switching system elektronische Wählsystem, EWS, Elektronikwählsystem *n*

electronic system Elektroniksystem *n*

electronic terminal elektronische Klemme *f*

electronic timer elektronische Zeitglied, elektronische Zeitrelais *n*

electronic timetable Auskunftgeber *m* [*Er gibt sekundenschnell Auskunft über Zugverbindungen*]

electronic workbench Elektroniklabor(atorium) *n*

(electron) image tube → image converter

electron microscope Elektronenmikroskop *n*

electron microscopy Elektronenmikroskopie *f*

electron multiplier Elektronenvervielfacher *m*

electron radiation Bremsstrahlung *f*

electron ray tube → cathode-ray tube

electron track Elektronenspur *f*

electrooptical character recognition → optical character recognition

electro panic release Elektropaniköffner *m*

electroplated coat(ing) [*B. S. 1224*] elektrolytische Überzug *m*, elektrolytische Beschichtung *f*

electropneumatic controller, electropneumatic CTRL [*See remark under "controller"*] elektropneumatische Regler *m*

electro-precipitator → electrostatic precipitator

electrostatic actuator [*engineering acoustics*] elektrostatische Eichelektrode *f*

electrostatic attraction, Coulomb attraction [*The electrostatic force of attraction exerted by one charged particle to another charged particle of opposite sign*] elektrostatische Anziehung *f*

electrostatic charge elektrostatische (Auf)Ladung, (Auf)ladung statischer Elektrizität *f*

electrostatic discharge, ESD elektrostatische Entladung, Entladung statischer Elektrizität *f*

electrostatic field [*A time-independent electric field, such as that produced by stationary charges*] elektrostatische Feld *n*

electrostatic filter elektrostatische Filter *m*, *n*

electrostatic (loud)speaker, capacitor (loud-)speaker [*A (loud)speaker in which the mechanical forces are produced by the action of electrostatic fields; in one type the fields are produced between a thin metal diaphragm and rigid metal plate*] elektrostatische Lautsprecher, Kondensatorlautsprecher *m*

electrostatic memory elektrostatische Speicher *m*

electrostatic microphone → capacitor microphone

electrostatic plotter elektrostatische Strichzeichner *m*

electrostatic precipitator, electro-precipitator, precipitator (type) of filter elektrostatische (Ab-)Scheider, Elektro(ab)scheider *m*

electrostatic spraying, electrostatic coating [*Method of spraying in which an electrostatic potential is created between the work to be coated and the atomized paint particles. The charged particles of paint are attracted to the article being painted and are there deposited and discharged. The electrostatic potential is used in some processes to aid the atomization of the paint*] elektrostatische (Pulver)Spritzen, EPS *n*

electrostatic tweeter ((loud)speaker) [*A tweeter (loud)speaker in which a flat metal diaphragm is driven directly by a varying high voltage applied between the diaphragm and a fixed metal electrode*] elektrostatische Hochtonlautsprecher *m*

electrothermal elektrothermisch

electrothermal fuse elektrothermische Sicherung *f*

electrothermic cryogenic production, electrothermic production of low temperature elektrothermische Kälteerzeugung *f*

elegant design Gestaltung mit Stil, stilvolle Gestaltung *f*

element Bauelement, Element *n* [*Kleinstes für sich funktionsfähiges Bauteil einer RLT-Anlage oder eines freien Lüftungssystems*]

element Glied, Bauglied *n* [*Steuerungstechnik. Ein Objekt in einem Abschnitt eines Wirkungsweges, bei dem Eingangsgrößen in bestimmter Weise Ausgangsgrößen beeinflussen. Im allgemeinen Sinne kann ein Glied als System aufgefasst werden*]

element [*Portion of a part that cannot be renewed without destruction of the part; constituent of a microcircuit or integrated circuit that contributes directly to the operation*] Baustein *m*, Element *n*

element, constant Element *n*, Konstante *f*

elementary circuit, basic circuit [*See remark under "circuit"*] Grundschaltung, Basisschaltung *f*

elementary diagram → circuit diagram

elementary frequency [*IEC 50(551)*], **pulse frequency, basic frequency** Taktfrequenz *f* [*Leistungselektronik*]

elementary item [*An item considered to have no subordinate item in the COBOL language*] Grundfeld, Datenelement *n*

elementary particle, (fundamental) particle [*A particle which, in the present state of knowledge, cannot be described as compound, and is thus one of the fundamental constituents of all matter*] Elementarteilchen *n*

to elevate → to lift

elevated condensate tank, elevated condensation water tank, condensation water elevated tank Kondensathochbehälter, Kondenswasserhochbehälter *m*

elevated (water) reservoir → high service (water) reservoir

elevated zero range Bereich *m* [*selten: n*] mit angehobenem Nullpunkt

elevating door, lift(ing) door Hebetür, Hubtür *f*

elevation [*Scale drawing of vertical portions of a building*] Aufriss *m*

elevator (US); lift (Brit.) Aufzug *m*

elevator apartment building (US) → elevator residential building

(elevator) arrival lamp (US); (lift) arrival lamp (Brit.) (Aufzug)Ankunftslampe *f*, (Aufzug)Ankunftsmelder *m*

elevator attendant (US) → (passenger) lift attendant

elevator building (US); lift building (Brit.) Aufzuggebäude *n*

elevator cab (US); lift cab (Brit.) [*Enclosure consisting of walls and top built upon a car platform*] Aufzugkorbgerüst, Fahrkorbgerüst *n*

(elevator) car, elevator cage (US); (lift) car, (lift) cage (Brit.); passenger compartment [*The load-carrying unit of an elevator, including its platform, car frame, enclosure, and door or gate*] Fahrkorb, (Aufzug)Korb *m*, Fahrkabine, (Aufzug)Kabine *f*

(elevator) car access (US) → lift car access (Brit.)

(elevator) car annunciation, (elevator) cage annunciation (US); (lift) car annunciation, (lift) cage annunciation (Brit.) [*An electrical device in an elevator car which provides a visual indication of floor landings*] Geschossanzeiger, Etagenanzeiger, Stockwerkanzeiger *m*

(elevator) car annunciator (US) → lift car annunciator (Brit.)

(elevator) car buffer, (elevator) cage buffer (US); (lift) car buffer, (lift) cage buffer (Brit.) Fahrkorbaufsetzpuffer, Fahrkabinenaufsetzpuffer, (Aufzug)Korbaufsetzpuffer, (Aufzug)Kabinenaufsetzpuffer *m*

(elevator) car door, (elevator) cage door (US); (lift) car door, (lift) cage door (Brit.) Fahrkorbtür, Fahrkabinentür, (Aufzug)Korbtür, (Aufzug)Kabinentür *f*

(elevator) car frame, (elevator) cage frame (US); (lift) car frame, (lift) cage frame (Brit.) Fahrkorbrahmen, Fahrkabinenrahmen, (Aufzug)Korbrahmen, (Aufzug)Kabinenrahmen *m*

(elevator) car guide, (elevator) cage guide (US); (lift) car guide, (lift) cage guide (Brit.) Fahrkorbführung, Fahrkabinenführung, (Aufzug)Korbführung, (Aufzug)Kabinenführung *f*

(elevator) car guide rail (US) → lift car guide rail (Brit.)

(elevator) car illumination (US) → lift car illumination (Brit.)

E

(elevator) car keyboard, (elevator) cage keyboard (US); (lift) car keyboard, (lift) cage keyboard (Brit.) Fahrkorbtableau, Fahrkabinentableau, (Aufzug)Korbtableau, (Aufzug)Kabinentableau *n*

(elevator) car location indication (US) → lift car location indication (Brit.)

(elevator) car platform, (elevator) cage platform (US); (lift) car platform, (lift) cage platform (Brit.) Fahrkorbboden, (Aufzug)Korbboden, (Aufzug)Kabinenboden, Fahrkabinenboden *m*

(elevator) car (push)button, (elevator) cage (push)button (US); (lift) car (push)button, (lift) cage (push)button (Brit.) Fahrkorbdruckknopf, Fahrkabinendruckknopf, (Aufzug)Korbdruckknopf, (Aufzug)Kabinendruckknopf *m*

(elevator) car separating door (US) → lift car separating door (Brit.)

(elevator) car sliding door, (elevator) cage sliding door (US); (lift) car sliding door, (lift) cage sliding door (Brit.) Fahrkorbschiebetür, Fahrkabinenschiebetür, (Aufzug)Korbschiebetür, (Aufzug)Kabinenschiebetür *f*

(elevator) car switch, (elevator) cage switch (US); (lift) car switch, (lift) cage switch (Brit.) Fahrkorbschalter, (Aufzug)Korbschalter, (Aufzug)Kabinenschalter, Fahrkabinenschalter *m*

(elevator) car terminal box, (elevator) cage terminal box (US); (lift) car terminal box, (lift) cage terminal box (Brit.) Fahrkorbklemmenkasten, Fahrkabinenklemmenkasten, (Aufzug)Korbklemmenkasten, (Aufzug)Kabinenklemmenkasten *m*

elevator control (US) → lift control (Brit.)

elevator controller (US) → lift controller (Brit.)

elevator drive (US); lift drive (Brit.) Aufzugantrieb *m*

elevator dwelling building (US) → elevator residential building

elevator engineering (US); lift engineering (Brit.) Aufzugtechnik *f*

elevator entrance (US); lift entrance (Brit.) Aufzugeingang *m*

elevator for high-rise municipal flats (US); lift for high-rise municipal flats (Brit.) Wohnhochhausaufzug *m*

elevator governor (US); lift governor (Brit.) [*Safety device set to trip at excess speed*] Geschwindigkeitsauslöser, Geschwindigkeitswächter *m*

elevator guid(anc)e rail (US); lift guid(anc)e rail (Brit.) Aufzugführungsschiene *f*

elevator hoistway (US) → hoistway

elevator hoistway access, elevator shaft(way) access (US); lift well access (Brit.); lift channel access (South Africa); well access, hoistway access, shaft(way) access Fahrschachtzugang, Fahrkernzugang, Aufzugschachtzugang, Aufzugkernzugang *m*

(elevator) hoistway sliding door, (elevator) shaft(way) sliding door (US); (lift) well sliding door (Brit.); lift channel sliding door [*South Africa*] (Fahr)Schachtschiebetür, (Fahr)Kernschiebetür, Aufzugschachtschiebetür, Aufzugkernschiebetür *f*

elevatoring (US); lifting (Brit.) Aufzugbetrieb *m*

elevator installation (US); lift installation (Brit.) Aufzuganlage *f*

elevator interlock (US); lift interlock (Brit.) Aufzugverrieg(e)lung *f*

elevator landing (US); lift landing (Brit.); landing [*Portion of a floor, balcony, or platform used to receive and discharge passengers or freight*] (Aufzug-)Haltestelle *f*

(elevator) landing opening (US); lift landing opening (Brit.) Fahrschachtöffnung, Aufzugschachtöffnung *f* [*Durch eine Tür verschlossen*]

elevator machinery (US); lift machinery (Brit.) [*Machinery and its equipment used in raising and lowering the elevator car or platform*] Aufzugmaschinerie, Aufzugmaschinen *f*

elevator machine(ry) room (US); lift machine(ry) room (Brit.) Aufzugmaschinenraum *m*

elevator monitoring (US); lift monitoring (Brit.) Aufzugüberwachung *f*

elevator nonstop button (US); lift nonstop button (Brit.) Aufzugdurchfahrknopf *m*

elevator nonstop switch (US); lift nonstop switch (Brit.) [*Switch that when operated will prevent an elevator from making registered landing stops*] Aufzugdurchfahrschalter *m*

elevator ((open) loop) control (US); lift ((open) loop) control (Brit.) Aufzugsteuerung *f*

elevator pit (US) → (hatchway) pit

elevator potential switch (US); lift potential switch (Brit.) [*Magnetic switch that disconnects the power form the elevator apparatus when the supply voltage fails or decreases below a definite value and that is usually opened by various electrical safety devices*] Aufzugnullspannungsschalter *m*

elevator reader (US); lift reader (Brit.) Aufzugleser *m*

elevator residential building, elevator residence building, elevator dwelling building, elevator apartment building (US); lift residential building, lift residence building, lift dwelling building, lift apartment building (Brit.) Fahrstuhlwohnhaus, Aufzugwohnhaus *n*

elevator rope, elevator cable (US); lift rope, lift cable (Brit.) [*Wire rope used for supporting and moving the elevator cage or platform*] Aufzugseil *n*

elevator shaft(way) (US) → hoistway

elevator shaft(way) access (US) → lift well access (Brit.)

elevator sheave beam (US); lift sheave beam (Brit.) [*A steel beam, within an elevator machine(ry) room, which is beneath and supports elevator equipment; usually directly over the elevator hoistway*] Aufzugmaschinenträger *m*

elevator slack-rope switch, elevator slack-cable switch (US); lift slack-rope switch, lift slack-cable switch (Brit.) [*Device for automatically cutting off the power in case the hoisting cable becomes slack*] Aufzugschlaffseilschalter, Schlaffseilaufzugschalter *m*

elevator test(ing) tower (US); lift test(ing) tower (Brit.) Aufzugprüfturm, Prüfaufzugturm *m*

elevator type (US); lift type (Brit.) Aufzugart *f*

elevator(-type) car park (US); lift(-type) car park (Brit.) Aufzugparkgarage, Parkaufzuggarage *f*

elevator well (US) → hoistway

elevator winch (US) → lift winch (Brit.)

elevator with automatic push-button control (US); lift with automatic push-button control (Brit.) Selbstfahrer(aufzug) *m*

eligible area markierte Fläche *f* [*Kode*]

eliminator [*Device containing many stationary vanes or louvers designed to remove entrained water particles from an airstream*] Wasserabscheider *m*

el(l) → elbow

elongation maximale Auslenkung *f* [*Schwingung in der Audioelektronik*]

ELSI → extra large scale integration

elutriation [*The separation of coarse from fine material by the upward movement of circulating fluid*] Abschlämmen, (Aus)Schlämmen *n*

e.l.v. → extra low voltage

EM, end of medium, medium end Ende der Aufzeichnung, Aufzeichnungsende *n* [*Kode*]

EM → energy management

e-mail → electronic mail

e-mail search(ing) machine → search(ing) machine

emanation security [*The protection resulting from all measures designed to deny unauthorised persons information of value which might be derived from unintentional emission from other than telecommunications systems*] Geheimsicherung *f*

embedded eingebettet

ember [*The smouldering remains of a fire*] Glut *f*

EMC → electromagnetic compatibility

emergency → emergency situation

emergency... → escape...

emergency access Notzugang *m*

emergency actuation, emergency operation Notbetätigung, Notbedienung *f*, Notbetrieb *m*

emergency actuation level, emergency operation level Notbedien(ungs)ebene, Notbetätigungsebene *f*

emergency alarm Notalarm *m*

emergency button → emergency (stop) button

emergency call Notruf *m*

emergency call central station, central emergency call station Notrufzentrale *f*

emergency call key Notruftaste(r) *f*, (*m*)

emergency call system Notrufsystem *n*

emergency (case) → emergency situation

emergency chimney Notschornstein *m*

emergency cold Notkälte *f*

emergency connected load → connected emergency load

emergency de-energizing Notabschaltung *f*

emergency (electrical) paging (system), emergency (electrical) location system, emergency (electrical) staff locator (drahtlose) (Personen)Notsuchsystem, (Personen)Not(aus)rufsystem, Notrundfunksystem *n*, (Personen)Notsuchvorrichtung *f*

emergency electric service [*This service is installed for use when normal power fails*] Elektronotdienst *m*

emergency evacuation Noträumung *f*

emergency exit door Notausgangstür *f*

emergency heat Notwärme *f*

emergency key Nottaste *f*

emergency key Notschlüssel *m*

emergency key → (floor) master key

emergency light(ing), emergency illumination Notbeleuchtung *f*, Notlicht *n*

E

E

emergency light(ing) changeover switch, emergency illumination changeover switch Notbeleuchtungsumschalter, Notlicht(um)schalter *m*

emergency line Notleitung *f*

emergency mode (of operation), emergency operation(al) mode Notbetriebsart, Notbedienungsart, Notbetätigungsart *f* [*Siehe Anmerkung unter „Betriebsart"*]

emergency net(work) Notnetz *n*

emergency ((open) loop) control level Notsteuer(ungs)ebene *f*

emergency operation → emergency actuation

emergency operation(al) mode → emergency mode (of operation)

emergency operator communication level, emergency operator control level Notbedien(ungs)ebene *f* [*(In Europanorm = Servicebedieneinheit) DIN V 32734, VDI 3814 Bl.2 Manuelle Befehlsgabe und direkte Anzeige, mit fest zugeordneten Funktionen, Schalten, Stellen und Anzeigen für eingeschränkten Betrieb der TGA, unabhängig von Funktionen der Automations- oder Managementebene. Sicherheitsfunktionen der TGA dürfen durch die Notbedienung nicht eingeschränkt werden. Die Notbedieneinheiten werden gegen nicht autorisierten Gebrauch gesichert (Schloss, Schlüsselschalter). Unterscheide: Handbedienung*]

emergency panel Nottafel *f*, Nottableau *n*

emergency police call → police emergency call

emergency power, on(-)site power Notstrom *m*

emergency (power) generation, on(-)site (power) generation Notstromerzeugung *f*

emergency (power) generator, on(-)site (power) generator [*"Producer" may synonymously be used for "generator"*] Notstromgenerator, Notstromerzeuger *m*

emergency (power) load, on(-)site (power) load, standby power load Notstromlast *f*

emergency power plant, emergency power installation Notstromanlage *f*

emergency power supply, standby power supply, on(-)site power supply Notstromversorgung, Stromausfallsicherung, Ersatzstromversorgung *f*

emergency release [*Device to make the door or gate electric contacts or door interlocks inoperative in case of emergency*] Notauslösung *f*

emergency service, standby service Notdienst *m*

emergency set(ting) Not(ein)stellen *n*

emergency shower Branddusche, Notdusche, Feuerdusche *f*

emergency shutdown Notabschaltung *f*

emergency signal Notsignal *n*

emergency situation, emergency (case) Notfall, Notstand *m*, Notsituation *f*

emergency source, standby source [*It is used when normal electric power fails*] Notstromquelle *f*

emergency stop Nothalt *m*

emergency (stop) button Notausschalter *m*

emergency (stop) switch, safety switch Sicherheitsschalter, Not(halt)schalter *m* [*Aufzug*]

emergency system Notsystem *n*

emergency task force Krisenstab *m*

emergency (tele)phone Notruffernsprecher *m*, Notruftelefon, Notruftelephon *n*

emergency unlocking Notentrieg(e)lung *f*

emergency vehicle Notfahrzeug *n*

emergency warning Notwarnung *f*

emergency water, fire-fighting water Löschwasser, Feuerlöschwasser, Brandlöschwasser *n*

emergency water valve, fire-fighting water valve Brandlöschwasserschieber, Löschwasserschieber *m*

e.m.f. → electromotive force

E.M.F. controller, E.M.F. regulator EMK-Regler *m*

EMI → electromagnetic interference

emission, release of radiant energy, radiant energy release Emission *f*

emission loss [*radiator*] Leistungseinbuße *f*, Leistungsverlust *m*

emission monitoring Emissions(s)überwachung *f*

emissivity [*The radiating qualities of a source of heat*] Strahlungsvermögen *n*, Strahlungsfähigkeit *f*

emitter [*Electronics. A device that generates charged particles or waves of radiation*] Emitter *m*

emitter, emitter zone, emitter region [*In a bipolar transistor, the region that generates the charge carriers that proceed to the collector*] Emitterzone *f*

emitter, time pulse generator [*e.g. in a card punch*] Impulsgeber *m*

emitter amplifier Emitterverstärker *m*

emitter barrier [*Electronics. An area between the emitter and base regions of a transistor where rectification takes place*] Emitterschranke, Emittersperre *f*

emitter-base admittance (innere) Emitterbasisadmittanz *f* [*Admittanz zwischen innerem Emitterpunkt und innerem Basispunkt, wobei der innere Emitterpunkt oft dem Emitteranschein gleichgesetzt werden kann*]

emitter-base cutoff current Emitter(basis)reststrom I_{EBO} *m*

emitter-base voltage Emitter/Basis-Spannung *f*

emitter-coupled logic, ECL sättigungslose Rechnerschaltkreis *m*, emittergekoppelte Logik(schaltung) *f* [*Schaltkreisfamilie mit durch Transistoren und Widerständen realisierten Gattern*]

emitter current Emitterstrom *m*

emitter diffusion resistance Emitterdiffusionswiderstand *m* [*Kehrwert des Emitter-Diffusionsleitwertes*]

emitter follower [*A device in which amplification is achieved by applying the input signal to the base and taking the output signal from the emitter resistor*] Emitterwandler *m*

emitter region → emitter

emitter resistance [*In a bipolar transistor, the ratio of the emitter-to-base voltage to the emitter current*] Emitterwiderstand *m*

emitter yield Emissionswirkungsgrad, Injektionswirkungsgrad *m*, Emitterergiebigkeit *f* [*Verhältnis des Injektionsstromes zu dem über den Emitteranschluss fließenden Strom*]

emitter zone → emitter

emitting abgebend

employee theft Arbeitnehmerdiebstahl *m*

to empty → to discharge

empty casing, empty housing, hollow casing, hollow housing, hollow case, empty enclosure, empty case(work) Leergehäuse, Hohlgehäuse *n*

empty chamber, hollow chamber Leerkammer, Hohlkammer *f*

empty cube, hollow cube Hohlwürfel, Leerwürfel *m*

emptying → discharge

emptying cock, discharge cock Entleerungshahn, Ablasshahn, Abflusshahn, Ablaufhahn *m*

emptying noise → discharge noise

empty reel Leerspule *f*

EMS, expanded memory specification Standard *m* um einen Speicherplatz jenseits der 1-Mbyte-Grenze zu nutzen

emulation Emulation *f* [*Softwaremäßige Nachbildung (Simulation) eines Rechners oder Terminals, wobei nur das äußere Verhalten eines Systems nachgebildet wird. Das Mikroprogramm und die Befehlsausführung des einen Rechners werden von einem anderen Rechner nachgeahmt.*]

emulator Emulator *m* [*Ein Programm, das den internen Aufbau eines Rechners und dessen Mikroprogramm nachbildet. Dadurch können Programme des emulierten Rechners auf anderen Rechnern ablaufen*]

emulsion burner → (air/(fuel) oil) emulsion burner

emulsion flow → bubble flow

EN → enabling

to enable [*To turn on a device or place it in a state which will allow certain interrupts to occur in a processing unit. Interrupts are usually enabled by issuing a command or setting a switch or jumper*] freigeben

to enable [*In automatic data processing, to authorize an activity which would otherwise be suppressed, such as to write on a tape*] freigeben

enabled freigegeben [*Zustand einer Zentralaufbereitungsanlage, in dem sie nicht z. B. aufgrund eines direkten Speicherzugriffs gesperrt ist*]

enablement dependency, EN-dependency Freigabeabhängigkeit *f*

enablement input, EN-input Freigabeeinsatz *m*, Freigabeingabe *f*

enablement interrupt instruction, EI instruction Unterbrechungsfreigabebefehl *m*

enablement line, EN-line Freigabeleitung *f*

enablement → enabling

enablement module, EN-module, enable block, EN-block Freigabebaustein *m*, Freigabemodul *n*

enablement sequence, EN-sequence Freigabefolge *f*

enablement signal, EN-signal, enabling signal Freigabesignal *n* [*elektronische Steuerung*]

enablement switching command, EN-switching command Freigabeschaltbefehl *m*

enablement time, EN-time Freigabezeit *f*

enabling, EN, enablement Freigabe *f*, Freigeben *n* [*Das Zulassen der Weitergabe eines Signals oder der Ausführung eines Befehls, wenn die dazu erforderlichen Freigabesignale anstehen. E DIN 19 226T5, V DIN 19 237*]

enabling key, EN-key, enablement key Freigabetaste *f*

enabling pulse, EN-pulse, enablement pulse [*A pulse that prepares a circuit for some subsequent action*] Freigabeimpuls *m*

enabling voltage, EN-voltage, enablement voltage Freigabespannung *f* [*Stromkreis. DIN 19 237*]

EN-block → enablement module

to encapsulate, to enclose (ein)kapseln, allseitig schließen, ganz umschließen, voll umschließen

encapsulated, enclosed (ein)gekapselt, (allseitig) geschlossen, umschlossen

encapsulated arc lamp → enclosed arc lamp

encapsulated electronics, enclosed electronics (ein)gekapselte Elektronik, umschlossene Elektronik, (allseitig) geschlossene Elektronik *f*

encapsulated furnace → enclosed furnace

encapsulated fuse → cartridge fuse

encapsulated postscript, EPS Bilddateiformat *n*

encased (bath(ing)) tub → built-in (bath(ing)) tub

encased lock, build-in lock eingelassene Schloss, eingelegte Schloss *n*

to encipher, to encrypt [*To convert plain text into an unintelligible form by means of a cipher system*] chiffrieren, verschlüsseln

to enclose → to encapsulate

enclosed → encapsulated

enclosed arc lamp, encapsulated arc lamp Dauerbrennbogenlampe *f*

enclosed electronics → encapsulated electronics

enclosed furnace, encapsulated furnace [*Specific heating or heating and ventilating furnace incorporating an integral total enclosure and using only outside air for combustion*] geschlossene Feuerung, (ein-)gekapselte Feuerung, umschlossene Feuerung *f*

enclosed fuse (US) → cartridge fuse

enclosing masonry wall, space enclosing masonry wall Umfassungsmauer *f*

enclosing wall, space enclosing wall Umfassungswand *f*

enclosure, enclosing space umbaute Raum *m*

enclosure, cabin Kabine *f*

enclosure → casing

to encrypt → to encipher

end-around shift → cyclic(al) shift

end bit → stop bit

end consumer Endverbraucher *m*

end (elevator) landing (US) → terminal (elevator) landing

EN-dependency → enablement dependency

end feeder Endspeiser, Schlussspeiser *m*

ending angle Endwinkel, Schlusswinkel *m*

endless loop, continuous loop Endlosschleife *f*

endless paper printer → continuous paper printer

endless (rolled) towel, continuous (rolled) towel Endlos(rollen)handtuch *n*

end of block, block end Blockende, Satzende *n*

end-of-block character, block-end character Satzendezeichen, Blockendezeichen *n* [*Es bewirkt Lesehalt sobald ein Satz eingelesen ist*]

end of conversion, EOC, conversion end (Um)Wandlungsende, Umsetzungsende *n*

end-of-data mark, data mark end [*A character or word signal(l)ing the end of all data held in a storage unit*] Datenendemarke, Dateiendemarke *f*

end of file, EOF, file end [*Termination or point of completion of a quantity of data; end of file marks are used to indicate this print*] Dateiende *n*, Datenende, Dateischluss, Datenschluss *m*

end-of-file gap, EOF gap, file gap end [*A gap of precise dimension to indicate the end of a file on tape*] Dateiendelücke, Datenendelücke *f*

end-of-file mark, end-of-file indicator, file mark end, file indicator end [*A control character which signifies that the last record of a file has been read*] Dateiendemarke, Datenendemarke *f*

end-of-file routine, file end routine, EOF routine [*A program(me) which checks that the contents of a file read into the computer were correctly read; may also start the rewind procedure*] Dateiendeprogramm, Datenendeprogramm, Dateischlussprogramm, Datenschlussprogramm *n*

end-of-line amplifier, line end amplifier Zeilenend(e)verstärker, Zeilenschlussverstärker *m*

end-of-line character, line end character Zeilenend(e)zeichen, Zeilenschlusszeichen *n*

end-of-line resistor, line end resistor Endwiderstand(sgerät), Schlusswiderstand(sgerät) *m*, (*n*)

end-of-line unit, line end unit Endglied, Schlussglied *n*

end of medium, EM, medium end Ende der Aufzeichnung, Aufzeichnungsende *n* [*Kode*]

end of number → end of selection

end-of-perforated tape, perforated tape end Lochstreifenende *n* [*Zusatzfunktion; weitere Informationen können nicht mehr eingelesen werden*]

end of pipe, pipe end Rohrstirnfläche, Stirnfläche eines Rohres *f*

end of record, EOR, record end [*A gap of precise dimension (shorter than the end-of-file gap) which indicates the physical end of a record on a magnetic tape*] Satzende *n*, Satzschluss *m*

end-of-record word, EOR word, record word end [*The last word in a record, usually written in a special format that enables identification of the end of the record*] Satzendewort, Satzschlusswort *n*

end of selection, end of number, EON, selection end, number end Wahlende *n*, Wahlschluss *m*

end-of-selection signals, EOS, selection signal end Wahlendezeichen, Wahlschlusszeichen *n*

end-of-tape routine, tape routine end [*A program(me) which is brought into play when the end of tape is reached; may involve a series of validity checks and initiate the tape rewind*] Bandendeprogramm *n*

end of text, text end, EOT Textende *n*, Textschluss *m*

end-of-text character, text end character Textend(e)zeichen, Textschlusszeichen *n*

end of transmission, EOT, transmission end Übertragungsende *n*, Übertragungsschluss *m*

end-of-transmission block, ETB, transmission block end Übertragungsblockende *n*

end of volume, volume end, EOV (Magnet)Bandende *n*, (Magnet)Bandschluss *m*

endorser (circuit) [*In automatic data processing, a special feature available on most magnetic-ink character-recognition readers that imprints a band endorsement on successful document reading*] Indossiervorrichtung *f*

endothermic [*Pertaining to a reaction which occurs with the absorption of heat*] endotherm(isch)

end piece Endstück *n*

end position switch Endlagenschalter, Endstellungsschalter *m*

end setpoint → target setpoint

end stopper Verschlussstopfen *m*

end system Endsystem, Schlußystem *n*

end-to-end level measurement Pegelstreckenmessung *f*

end-to-end signalling (Brit.); end-to-end signaling (US) durchgehende Zeichengabe, durchgehende Signalisierung *f*

end unit [*COBOL*] Buchende *n*, Buchschluss *m*

end vector Ausgangsvektor, Endvektor *m* [*Alle Ausgangsgrößen eines betrachteten Systems bilden den Ausgangsvektor. DIN 19 226 Teil 1*]

end-window counter tube Fensterzählrohr *n*

endwise [*tank*] hochkant

energization fleeting contact, energizing fleeting contact Einschaltwischer, Einschaltwischkontakt *m*

energized → alive

energy [*Capacity for performing work. A particle or piece of matter may have energy because it is moving or because of its position in relation to other particles or pieces of matter*] Arbeitsfähigkeit *f*, Arbeitsvermögen *n*

energy → power

energy absorption [*The taking up of energy from radiation by the medium through which the radiation is passing*] Aufnahme, Energieaufnahme *f*

energy application, energy duty, energy use Energieanwendung, Energieverwendung, Energieeinsetzung *f*, Energieeinsatz, Energieverbrauch *m*

energy audit, energy billing Energie(ver)rechnung *f*

energy balance Energiebilanz *f*

energy barrier, energy holdoff, energy shutoff Energiesperre *f* [*Vollständige Einstellung der Energiezufuhr außerhalb der Gebäudenutzungszeit*]

energy billing → energy audit

energy column Energiesäule *f*

energy company Energie(versorgungs)unternehmen *n*

energy conditioning → power conditioning

energy-conscious energiebewusst

energy conservation → conservation of energy

energy conservation theorem → conservation of energy theorem

energy consultation, energy consulting Energieberatung *f*

energy consumption Energieverbrauch *m*

energy conversion Energieumwand(e)lung *f*

energy deficiency → energy shortage

energy demand, energy requirement Energienachfrage *f*, Energiebedarf *m*

energy drop Energieabfall, Energierückgang *m*

(energy) demand-side-management, DSM (Energie-)Nachfragemaßnahmen *f* [*Sie zielen darauf ab, das Nachfrageverhalten nach Energie zu beeinflussen*]

energy demand signal → energy requirement signal

energy duty → energy application

energy efficiency Energiewirkungsgrad *m*

energy efficiency ratio, EER Leistungszahl *f* im Kühl(ungs)betrieb [*Verhältnis der Kühl(ungs)leistung zur effektiven Leistungsaufnahme der Wärmepumpe*]

energy element Messwerk *n* [*Elektrizitätszähler*]

energy engineering Energietechnik *f*

energy expenditure → expenditure of energy

energy feed(ing) line Energiespeiseleitung *f*, Energiezuführ(ungs)leitung, Energiezuleitung

energy flow Energieströmung *f*, Energiestrom *m*

energy gap Energielücke *f*

energy hold-off → **energy barrier**

energy input [*mechanical efficiency*] zugeführte Energie *f*

E

energy loading Energiebelastung *f*

energy observation Energiebetrachtung *f*

energy management, EM Energiebewirtschaftung *f* [*Alle Maßnahmen mit dem Ziel, bei vorhandenen Gebäuden weniger Energie oder kostengünstigere Energie zu verbrauchen (z. B. durch Umstellung des Energieträgers) oder Energie unter Ausnutzung der Kraft-Wärme-Kopplung selbst bereitzustellen (z. B. durch Einbau eines BHKW), soweit dies ökologisch oder ökonomisch sinnvoll ist oder energieverbrauchsbedingte Emissionen zu mindern. Energiebewirtschaftung erstreckt sich auf alle Arten verbrauchter Energie in Gebäuden (für Heizung, Kühlung, Kraft, Licht usw.), aber auch auf die Verbraucher von Anlagen oder Einrichtungen, die dem Kerngeschäft eines Unternehmens dienen sowie für den gesamten Verbrauch von Wasser. Planungsleistungen bei Neubauten mit dem Ziel des Einsatzes energiesparender Baustoffe, Systeme oder Komponenten gelten nicht als Energiebewirtschaftung. Optimierungen im laufenden Betrieb von Anlagen sowie regelmäßige Dokumentation der Energieverbrauche gelten nicht als Energiebewirtschaftung, sondern gehören zur regulären Betriebsführung. Die Minderung von Energiekosten durch Anpassung von Energielieferverträgen gilt ebenfalls nicht als Energie-, sondern als Vertragsverwaltung.*]

energy management (and control) system → **facility management system**

energy medium Energieträger *m*

energy optimization Energieoptimierung *f* [*Verwendung eines Produkts oder Programms zur Steuerung oder Reg(e)lung einer Anlage bei minimalem Energieverbrauch ohne den Komfort zu beeinträchtigen*]

energy radiation Energiestrahlung *f*

energy recovering unit, energy recovery unit Energiewiedergewinnungseinheit, Energierückgewinnungseinheit *f*

energy recovery Energierückgewinnung, Energiewiedergewinnung, Leistungsrückspeisung *f*

energy recovery system Energierückgewinnungssystem, Energiewiedergewinnungssystem, Leistungsrückspeisesystem *n*

energy release Energiefreisetzung *f*

energy requirement signal, energy demand signal Energiebedarfssignal, Energienachfragesignal *n*

energy saving Energieeinsparen *n*

energy-saving, low-energy energiesparend

energy-saving button, low-energy button Energiesparknopf, Niedrigenergieknopf *m*

energy-saving function, low-energy function Energiesparfunktion, DTX-Funktion, Niedrigenergiefunktion *f*

energy-saving house, low-energy house, LEH Energiesparhaus, Niedrigenergiehaus *n*

energy-saving key Energiespartaste *f*

energy-saving office, low-energy office, LEO Energiesparbüro, Niedrigenergiebüro *n*

energy-saving switch, low-energy switch Energiesparschalter, Niedrigenergieschalter, Sparenergieschalter *m*

energy shortage, energy shortfall, energy deficiency Energiemangel *m*, Energieknappheit *f*

energy shutoff → **energy barrier**

energy source Energieträger *m*, Energiequelle *f* [*Stoffe und physikalische Erscheinungsformen der Energie, aus denen direkt oder durch Umwand(e)lung Nutz(ungs)energie gewonnen werden kann. Selten: Umwandlung*]

energy specific preparation → **power conditioning**

energy storage Energiespeicherung *f*

energy storage system Energiespeichersystem *n*

energy supply Energieversorgung, Energielieferung *f*

energy supply contract Energieliefervertrag, Energieversorgungsvertrag *m*

energy supply duct Energieversorgungskanal, Energielieferungskanal *m*

energy system, power system Energiesystem *n*

energy target [*The desired total energy use in a building or process for a given period*] Energieverbrauchsvorgabe *f*

energy tariff dependent switching energietarifabhängige Schalten *n* [*Mit der Veränderung des Energie-/Leistungspreises im Verlauf eines Tages werden bestimmte Anlagekomponenten zwecks Herabsetzung von Energiekosten ein- und ausgeschaltet*]

energy use → **energy application**

energy yield Energieausbeute, Energieergiebigkeit *f*

engine Benutzerprogramm *n* [*Es führt alle Aktionen des Benutzers, z. B. Suchen und Sortieren, aus*]

engineering Projektierung, Projektbearbeitung *f*

engineering [*The profession that creates all artefacts making up the modern world*] Technik *f*

engineering area Technikfläche *f* [*Gebäude*]

engineering drawing technische Zeichnung *f*

engineering firm Technikfirma *f*

engineering function Technikfunktion *f*

engineering industry technische Industrie *f*

engineering installations, engineering outfit technische Einbauten, technische Ausrüstung *f* [*In einem Gebäude*]

engineering layout [*A drawing of the design that is the basis for shop drawings*] Konstruktionszeichnung *f*

engineering management Anlagenbetreuung *f*

engineering metal Technikmetall *n*

engineering profession Technik(er)beruf *m*

engineering science → technology

engineering time [*The nonproductive time of a computer, reserved for maintenance and servicing*] Auszeit, technische Stillstandzeit *f*

engineering unit physikalische Einheit *f* (der Größe einer MSR-Messstelle) [*VDI/VDE 3695*]

to enhance verbessern, steigern

enhancement Anreicherung *f* [*Transistor*]

enhancement [*The process of increasing, improving, or intensifying specific uses*] Verbesserung, Steigerung *f*

enhancement factor Verbesserungsfaktor, Steigerungsfaktor *m*

enhancement mode (of operation), enhancement mode type, enhancement operation(al) mode [*Electronics. A characteristics of a field effect transistor's operation in which current flows through the source-drain channel only when a voltage has been applied to the gate, and increases as the voltage increases*] Anreicherungs(betriebs)art *f*, Anreicherungsarbeitsart [*Halbleiter(bauelement). DIN 41 858*]

ENIAC → electronic numerical integrator and calculator

EN-input → enablement input

EN-line → enablement line

EN-module → enablement module

EN-pulse → enablement pulse

enquiry, ENQ, inquiry Anfrage, Abfrage *f*

enriched angereichert

EN-sequence → enablement sequence

EN-signal → enablement signal

to enter → to input

entering air → inlet air

entering call, incoming call ankommende Ruf *m*

entering catch Einreiber *m*

entering catch closer Einreiberverschluss *m*

entering condenser water, incoming condenser water Eintrittskondensatorwasser, Eintrittsverflüssigerwasser *n* [*Absorptionskältemaschine*]

entering current aufgenommene Strom, Eingangsstrom *m*

entering dry bulb temperature, dry bulb entering temperature, incoming dry bulb temperature Trockenkugeleintrittstemperatur, Eintrittstrockenkugeltemperatur *f*

entering first selector, incoming first selector [*Connects incoming calls from outlying dial offices to local second selectors*] Eingangserstwähler, Ersteingangswähler *m*

entering pressure, initial pressure, incoming pressure [*Pressure of fluid entering the system*] Anfangsdruck, Eingangsdruck, Eintrittsdruck *m*

entering selector, incoming selector [*Selector associated with trunk circuits from another central office*] Eingangswähler *m*

entering supply, (incoming) supply, entry supply Zuleitung *f* [*Schaltplan*]

entering temperature, incoming temperature, entry temperature [*Temperature of a substance as it enters any equipment, plant, and the like*] Eintrittstemperatur, Eingangstemperatur *f*

entering terminal, incoming terminal Eintrittsklemme, Eingangsklemme *f*

entering terminal exchange, incoming terminal exchange Durchgangs-Auslandsvermittlungsstelle *f*

entering unit, incoming unit (Ein)Speisungseinheit *f*

entering utility service, incoming utility service, incoming utility line, entering utility line Anschlussversorgungsleitung *f*

entering water, incoming water Anschlusswasser *n*

enter key Eingabetaste *f*

enthalpy, (total) heat content, total heat [*Thermodynamic quality equal to the sum of the internal energy of a system plus the product of the pressure-volume work done on the system. h = E + pv, where h = enthalpy or total heat content, E = internal energy of the system, p = pressure, and v = volume*] Wärmeinhalt *m*, Enthalpie, Gesamtwärme *f*

enthalpy ((closed) loop) control, (total) heat content ((closed) loop) control, total heat ((closed) loop) control Enthalpiereg(e)lung, Wärme-Inhaltsreg(e)lung, Gesamtwärmereg(e)lung *f*

E

E

enthalpy control [*The process of optimizing the proportions of fresh and recirculated air in an air-handling unit for a space based on the enthalpy difference between indoor and outdoor air calculated from measurements of humidity and temperature, to minimize energy consumption*] Wärmeinhaltsoptimierung *f*

enthalpy controller, (total) heat content controller, total heat controller [*"Regulator" may synonymously be used for "controller"*] Enthalpieregler, Wärme-Inhaltsregler, Gesamtwärmeregler *m*

enthalpy recovery Enthalpierückgewinnung, Enthalpiewiedergewinnung *f*

enthalpy sensor → (duct) enthalpy detector

EN-time → enablement time

entitlement level Berechtigungsebene *f*

entity Instanz *f* [*Eine Funktionseinheit die Aufträge erteilt oder erhält, erhaltene Aufträge ablehnt oder annimmt und angenommene Aufträge ganz oder teilweise ausführt, weitergibt oder bei Unausführbarkeit zurückgibt. E DIN 44 300, Teil 1*]

entity Datenstrukturelement *n* [*z. B. graphisches Bildelement*]

entrained oil [*Oil droplets carried along by high-velocity refrigerant gas*] mitgerissene Öl *n*

entrained water [*Small water particles carried along by a rapidly moving airstream*] mitgerissene Wasser *n*

entrainment → induction

entrainment ratio → induction ratio

entrance → entry point

entrance box → (house) service box

entrance cable [*It brings power from an outside power line into a building*] Anschlusskabel, Eingangskabel *n*

entrance of dirt → dirt entrance

entrance ((open) loop) control Eingangssteuerung *f*

entrance switch Eingangsschalter *m*

(en)trapped air eingeschlossene Luft *f*

(en)trapped gull(e)y → trapped gull(e)y

entropic coding Kodierung *f* nach dem mittleren Nachrichteninhalt

entropy [*Thermodynamics. A property of a thermodynamic system, denoted by S, whose differential change is given by dS = dO/T, where dO refers to the differential amount of heat transfer introduced to the system in a reversible process, the T is the absolute temperature of the system*] Entropie *f*

entry → input (data)

entry Eintritt, Zutritt, Eingang, Zugang *m*

entry → (air) inlet

entry block [*In automatic data processing, the area of main memory reserved for the data which will be introduced at execution time*] Einführungsblock *m*

entry check(ing) Eingangskontrolle *f*

entry (data) → input (data)

entry/exit message Buchungsmitteilung, Buchungsnachricht *f*

entry field Eingangsfeld *n* [*Personalrechner; Programmierungsgerät*]

entry instruction [*In automatic data processing, the first instruction to be executed in a subroutine*] Einführungsbefehl *m*

entry item → input (data) item

entry opening Eingangsöffnung, Einstiegsöffnung *f*

entry point, entrance [*The location of a program(me) or subroutine at which execution is to start*] Eingangsstelle *f*

entry (pressure) loss [*Loss in pressure of a fluid stream flowing into a pipe, tube, duct, hood, or vessel*] Eingangsdruckverlust *m*

entry sorting [*Computer programming. A sorting method in which records or block of records are placed in a buffer and then sorted into the main list prior to receiving the next list*] Eingabesortierung *f*

entry temperature → entering temperature

envelope [*Communications. A curve drawn to pass through the peaks of a graph, such as that of a modulated radio-frequency carrier signal*] Umhüllende, Einhüllende, Hüll(kurv)e *f*

envelope delay time → group delay time

(envelope) delay distortion [*Electronics. A distortion in transmitted signals caused by different rates of change of phase shift over the frequency range of the equipment*] Verzögerungsverzerrung, Verzugsverzerrung *f* [*Gruppenlaufzeit*]

enveloping character Austauschzeichen *n*

environmental compatibility Umweltverträglichkeit *f*

environmental concern Umweltthema *n*

environmental consultant Umweltberater *m*

environmental energy Umweltenergie *f* [*Thermischer und mechanischer Energieinhalt der Biosphäre einschließlich der Strahlungsenergie der Sonne*]

environmental engineer Umweltingenieur *m*

environmental engineering Umwelttechnik *f*

environmental heat Umweltwärme *f* [*Wärmetechnik. Die aus der Umwelt gewonnene Wärme, vor allem die durch Sonnenstrahlung*]

environmentally acceptable, environmentally compatible umweltverträglich, umweltfreundlich

environmental noise Umweltgeräusch *n*

environmental radioactivity Umweltradioaktivität *f*

environmental room Umweltsimulierraum *m*

environmental security engineering Umweltsicherungstechnik *f*

environmental services [*All the plant and apparatus, including lighting, within a building intended to maintain the physical conditions of the internal environment*] Innenausrüstung, Innenausstattung, Innenanlagen, Innenbestückung *f*

environmental temperature [*A simplified equivalent to the comfort temperature, combining air temperature with mean radiant temperature. Not to be confused with "ambient temperature"*] Umwelttemperatur *f*

environment-induced umweltbedingt

EN-voltage → **enabling voltage**

EOC → **end of conversion**

EOF → **end of file**

EOF gap → **end-of-file gap**

EOF routine → **end-of-file routine**

EON → **end of selection**

EOR → **exclusive or**

EOR → **end-of-record**

EOR function → **exclusive OR function**

EOR word → **end-of-record word**

EOS → **end of selection signals**

EOT → **end of transmission**

EOT → **end of text**

EOV → **end of volume**

EPLD → **erasable programmable logic device**

EPS → **encapsulated postscript**

equal gleich

equality gate → **equivalence gate**

equalization, equalizing Entzerrung *f*, Entzerren *n*

equalizer Entzerrer *m*

equalizer, equalizing circuit, equalizing net(work) [*A net(work) designed to compensate for an undesired amplitude-frequency or phase-frequency response of a system or component; usually a combination of coils, capacitors, and resistors*] Entzerrerschaltung *f*, Entzerrer(kette) *m*, (*f*)

equalizer [*Pipe connection between two or more pieces of equipment made in such a way that the pressure in each piece is equal*] Druckausgleicher *m*

equalizer Ausgleicher *m*

equalizer (pipe)line [*It equalizes the gas or oil pressure in two or more pieces of equipment*] Druckausgleicher(rohr)leitung *f*

equalizing torque → **restoring torque**

equal loudness contour, equal loudness curve, Fletcher-Munson contour, Fletcher-Munson curve, audibility contour, audibility curve [*A curve on a graph of sound intensity in decibels versus frequency at each point along which sound appears to be equally loud to a listener*] Fletcher-Munson-Kurve *f*

equal percentage characteristic gleichprozentige Kennlinie *f*

equal-sized meshes Maschengleichheit *f*

equal-to sign Gleichheitszeichen *n*

equation [*A statement that each of two mathematical expressions is equal to the other*] Gleichung *f*

equation Gleichung *f* [*Die komprimierteste Form der Darstellung physikalischer Gesetzmäßigkeiten*]

equational format Gleichungsdarstellung *f* [*Impulsmessung*]

equation of continuity, continuity equation Stetigkeitsgleichung, Kontinuitätsgleichung *f* [*Sie besagt dass bei volumenbeständigen Strömungslinien (Fluiden) und bei stationären Strömungen durch jeden Querschnitt einer Stromröhre der gleiche Volumenstrom fließt*]

equilibrium ball valve, equilibrium floating valve [*See remark under "ball valve"*] Ausgleichsschwimmerventil, Ausgleichskugelventil *n*

equilibrium moisture Gleichgewichtsfeuchte, Gleichgewichtsfeuchtigkeit *f*

equilibrium position Gleichgewichtslage, Gleichgewichtsstellung *f*

equilibrium temperature Gleichgewichtstemperatur *f*

equilibrium valve Ausgleichsventil *n* [*Abortspülkastenvorrichtung bei der der steigende Schwimmer einen Kolben in eine Ventil(ab)dichtung drückt, um den Wasserzustrom anzuhalten. Das Ventil hat eine Vorrichtung, die den auf es ausgeübten Druck ausgleicht, so dass es auch bei großen Wasserdruckschwankungen funktioniert*]

to equip → **to outfit**

equipartition law [*In a classical ideal gas, the average kinetic energy, per molecule associated with any degree of freedom which occurs as a quadratic term in the expression for the mechanical energy, is equal to half of Boltzmann's constant times the absolute temperature*] Äquipartitionstheorem *n*, Gleichverteilungssatz *m* der kinetischen Energie

E

E

equipment → outfit

equipment → device

equipment battery Gerät(e)batterie *f*

equipment-bound gerät(e)gebunden

equipment c.b. → circuit-breaker for equipment

equipment characteristic (curve) Gerät(e)kennlinie *f*

equipment circuit-breaker → circuit-breaker for equipment

equipment connecting line, equipment connection line Gerät(e)anschlussleitung *f*

equipment connection fitting Gerät(e)anschlussarmatur *f*

equipment earth(ing) conductor (Brit.) → protection conductor

equipment ground(ing) conductor (US) → protection conductor

equipment interface Geräteschnittstelle *f*

equipment (item) → device

equipment monitoring Geräteüberwachung *f*

equipment object → outfit object

equipment piece → device

equipment protection Geräteschutz *m*

equipment-related gerätebezogen

equipment room Geräteraum *m*

equipment selection Geräte(aus)wahl *f*

equipment station → (central) machine(ry) station

equipment store Gerätelager, Gerätemagazin *n*

equipment system Gerätesystem *n*

equipment terminal Geräteklemme *f*

equipment wire, track(er) wire Schalt(ungs)draht *m*

(equipotential) bond(ing) strip → ground(ing) circuit (US)

equipped → outfitted

equipped device table, EDT, outfitted device table Geräteausstattungstabelle *f*

equipping → outfitting

to equip (with), to outfit (with) bestücken (mit), ausrüsten (mit), ausstatten (mit)

equivalence gate, biconditional gate, equality gate, exclusive-NOR gate, match gate [*"Element" may synonymously be used for "gate". A logic circuit that produces a binary output signal of 1 if its two binary input signals are the same, and an output signal of 0 if the input signals differ*] Äquivalenzvergleicher, Gleichwertigkeitsvergleicher *m*

equivalent gleichwertig, äquivalent

equivalent bit rate, data signalling rate; data signaling rate (US) (Daten)Übertragungsgeschwindigkeit *f*

equivalent bit rate selector → data signal(l)ing rate selector

equivalent circuit Ersatzschaltung, gleichwertige Schaltung *f*

equivalent circuit diagram Ersatzschalt(ungs)plan *m*, Ersatzschalt(ungs)bild *n* [*Besondere Ausführung eines erläuternden Schaltplanes für Analyse und Berechnung von Stromkreiseigenschaften. DIN 40 719, Blatt 1*]

equivalent dead time → delay dead time

equivalent delay timing → delay timing

equivalent direct radiation → EDR

equivalent friction loss [*Loss of pressure in a duct system due to changes in size or shape, bends, take-offs, or obstructions, expressed in terms of the length of straight pipe in which an equivalent friction loss would take place*] gleichwertige Reibungsverlust *m*

equivalent net(work) [*IEC 50(603)*] Ersatznetz(-werk) *n*

(equivalent sound pressure) level difference, sound reduction Schallpegeldifferenz D *f* [*Maß dB. Unterschied zwischen dem Schallpegel im Senderaum (L_1) und dem Schallpegel im Empfangsraum (L_2). $D = L_1 - L_2$. Diese Differenz wird bei der Prüfung der Luftschalldämmung von Wänden und Decken bestimmt. Sie ist auch vom Schallschluckvermögen des Empfangsraumes abhängig*]

equivalent temperature [*A method of measuring the combined effect of air temperature, mean radiant temperature and air movement in a room on bodily comfort. It takes no account of humidity and is measured by a eupatheoscope*] Äquivalenztemperatur, Gleichwertigkeitstemperatur *f*

ER → exception report

eradicator Virenbefreier *m* [*Befreiung von infizierten Daten*]

erasability Löschbarkeit, Tilgbarkeit *f*

erasable löschbar, tilgbar

erasable programmable logic device, EPLD löschbar programmierbare Logikgerät, tilgbar programmierbare Logikgerät *n*

erasable storage, flash storage [*Any storage medium which permits new data to be written in place of the old, such as magnetic disk or tape, but not punch(ed) card and punch(ed) tape*] Löschspeicher, Tilgungsspeicher *m*

to erase [*To clear the contents of a memory, e.g. that of an EPROM, by means of ultraviolet light*] löschen, tilgen

erase bit, erasing bit Kodeelement zum Datenlöschen, Codeelement zum Datenlöschen, Kodeelement zum Datentilgen, Codeelement zum Datentilgen *n*

erase counter, erasing counter Löschzähler, Tilgungszähler *m*

erase head, erasing head [*Electricity. A magnetic head used to erase material previously recorded on magnetic tape*] Löschkopf, Tilgungskopf *m*

erase lever, erasing lever Lösch(winkel)hebel, Tilgungs(winkel)hebel *m*, Riegelhebelplatte *f*

eraser [*Device for cleaning information from an EPROM*] Löschvorrichtung, Tilgungsvorrichtung *f*

eraser window [*Window through which ultraviolet rays are emitted to clear an EPROM cassette*] Löschfenster, Tilgungsfenster *n*

erase signal, erasing signal Löschsignal, Tilgungssignal *n*

erase speed, erasing speed [*Electricity. In charge storage tubes, the rate of erasing successive storage elements*] Löschgeschwindigkeit, Tilgungsgeschwindigkeit *f*

erase switch, erasing switch Löschschalter, Tilgungsschalter *m*

erasing [*Eliminating stored information in a charge-storage*] Löschen, Tilgen *n*

erasure Löschung, Tilgung *f*

erecting Aufstellen *n* [*Nicht verwechseln mit "Montieren"*]

erecting rated power → site rated power

erection Aufstellung *f* [*Nicht verwechseln mit "Montage"*]

erection room Aufstell(ungs)raum *m*

erasable programmable read-only memory, erasable PROM, EPROM, erasable programmable ROM, eprom, electrical programmable and UV erasable ROM selbstprogrammierbare Festwertspeicher, lösch- und programmierbare Festwertspeicher, tilg- und programmierbare Festwertspeicher *m* [*Der Inhalt kann vom Abnehmer mit speziellen Geräten selbst programmiert und gelöscht werden*]

ergonomy Arbeitsplatzgestaltung, Ergonomie *f*

erosion corrosion Erosionskorrosion *f*

ERP → effective radiated power

erroneous, false fehlerhaft, falsch

erroneous data → garbage

erroneous data collector → garbage collector

erroneous operation, false operation Fehlbedienung, Falschbedienung *f*

error, measurement error, measuring error, ERR Fehler, (Mess)Fehler *m*, (Mess)Abweichung *f* [*Messtechnik. Abweichung zwischen einem berechneten, beobachteten oder gemessenen Wert oder Bedingung und dem wahren, vorgegebenen oder theoretisch richtigen Wert oder Bedingung. In der Messtechnik: Messwert minus wahrer Wert. DIN 1319, Teile 1 und 3*]

E

error, ERR Fehler *m*

error, bug, fault Fehler *m*

error alarm, bug alarm, fault alarm Fehleralarm *m* [*Durch elektronische Datenverarbeitung festgehaltene oder von Hand eingegebene Daten über unzulässige Zustände, z. B. einer Zutritt(s)kontrollanlage; Beispiele: Tür zu lange offen, Tür ohne Karte geöffnet usw.*]

error burst [*The condition when more than 1 bit is in error in a given number of bits*] Fehlerbündel, Fehlerpaket *n*

error-check(ing) and correcting, ECC Fehlerkontrolle und -berichtigung *f*

error-correcting code, error correction code Fehlerberichtigungskode, Fehlerberichtigungscode, Fehlerkorrekturkode, Fehlerkorrekturcode *m* [*Fehlererkennungskode, bei dem eine Teilmenge der gestörten Zeichen aufgrund der Bildungsgesetze ohne Rückfrage korrigiert wird. DIN 44 300*]

error correction, error correcting, EC [*Automatically locating and correcting a machine error of dropping a bit or picking up an extraneous bit, without stopping the machine or having it go to a program(me) recovery routine*] Fehlerberichtigung, Fehlerkorrektur *f*

error correction routine [*A program(me) which corrects specific error conditions in another routine or subroutine*] Fehlerkorrekturprogramm, Fehlerberichtigungsprogramm *n*

error detecting code, error detection code, error recognition code Fehlererkennungskode, Fehlernachweiscode *m* [*Die Zeichen werden nach Gesetzen gebildet, die es ermöglichen, durch Störungen verursachte Abweichungen von diesen Gesetzen zu erkennen (DIN 44 300)*]

error detecting system, error recognition system, error detection system [*An automatic system which detects an error due to a lack of data, or erroneous data during transmission*] Fehlererkennungssystem, Fehlernachweissystem *n*

error detection, error recognition, error detecting Fehlererkennung *f*, Fehlernachweis *m* [*Codierungssystem, auf dessen Grundlage jedes Datensignal mit* *den spezifischen Richtlinien übereinstimmt, damit strukturelle Ausgangskriterien beim Empfang von Signalen automatisch erkannt werden können. Wird ein fehlerhafter Datenblock erkannt, wird dieser entweder aus dem zu sendenden Datenpaket gelöscht – mit oder ohne Hinweis auf einen durchgeführten Löschvorgang – oder das Datenpaket wird unter Hinweis auf einen fehlerhaften Datenblock übertragen.*]

E

error detection and correction, EDC, error detecting and correcting [*There are various methods of doing this, one of them being the parity check(ing)*] Fehlererkennung und -berichtigung, Fehlererkennung und -korrektur *f*, Fehlernachweis und -berichtigung, Fehlernachweis und -korrektur *m*

error diagnosis [*A computer print-out of an instruction or data statement, pinpointing an error in the instruction or statement and spelling out the type of error involved*] Fehlerdiagnose *f*

error elimination unit Fehlerschutzeinheit *f*

error feedback controller [*The process variable information is fed back to process being controlled*] Fehlerrückmeldung(s)regler *m*

error forecasting Fehlerprognose, Fehlervorhersage, Fehlervoraussage *f*

error free seconds, EFS fehlerfreie Sekunden, fehlerlose Sekunden *f*

error frequency limit, EFL [*The maximum number of single bit errors per unit of time that a computer will accept before a machine check interrupt is initiated*] Fehlerhäufigkeitsgrenze *f*

error interrupt(ion) [*The halt in execution of program(me) because of errors which the computer is not capable of correcting*] Fehlerunterbrechung *f*

error memory Fehlerspeicher *m*

error monitoring Fehlerüberwachen *n*, Fehlerüberwachung *f*

error probability Fehlerwahrscheinlichkeit *f*

error range [*A range of values such that an error condition will result if a specified data item falls within it*] Fehlerbereich *m* [*selten: n*]

error recovery (procedure), ER(P) [*A procedure invoked in response to a detected error that permits the information exchange to continue*] Fehlerbehebung, Fehlerbeseitigung, Fehlerkorrektur *f*

error register Fehlerregister *n*

error report [*A list produced by a computer showing the error conditions, such as overflows and errors resulting from incorrect or unwanted data, that are generated during program(me) execution*] Fehlerbericht *m*

error signal → ((closed) loop) error signal

error signal → ((open) loop) error signal

error tolerance Fehlertoleranz *f* [*Zulässige Fehlerabweichung vom gegebenen Maß*]

error tuning Fehlerabstimmung *f*

error type Fehlerart *f*

error variable [*See remark under "variable"*] Fehlergröße *f*

error voltage [*A voltage usually obtained from a selsyn, that is proportioned to the difference between the angular position of the input and output shafts of a servosystem; this voltage acts on the system to produce a motion that tends to reduce the error in position*] Fehlerspannung, Differenzspannung *f*

erythermal lamp Bestrahlungslampe, Höhensonne *f*

Esaki diode → tunnel diode

ESC → escape

escalator, moving staircase Rolltreppe, Fahrtreppe *f*

escalator cleaning machine Fahrtreppenreinigungsmaschine, Rolltreppenreinigungsmaschine *f*

escalator handrail Rolltreppenhandlauf, Fahrtreppenhandlauf *m*

escalator step Fahrstufe, Rollstufe *f*

escalator structure [*It contains the continuous band of steps plus its associated equipment*] Fahrtreppengerüst, Rolltreppengerüst *n*

escalator tracks Führungsschienen *f* [*Rolltreppe*]

(escalator) truss, support(ing) steelwork Rolltreppentragkörper, Fahrtreppentragkörper *m*

escalator tube Rolltreppenröhre, Fahrtreppenröhre *f*

(escalator) well, (escalator) shaft Fahrtreppenschacht, Rolltreppenschacht *m*

(escalator) well girder, (escalator) shaft girder Fahrtreppenschachtträger, Rolltreppenschachtträger *m*

escape, ESC Flucht *f*

escape, ESC Fluchtfunktion [*graphisches Kernsystem*], Codeumschaltung, Kodeumschaltung *f*

escape balcony, emergency balcony Fluchtbalkon, Rettungsbalkon, Notbalkon *m*

escape character, emergency character [*A character used to indicate that the succeeding character is or characters are expressed in a code different from the code currently in use*] Kodeumschalt(ungs)zeichen, Codeumschalt(ungs)zeichen *n*

escape door, emergency door Fluchttür, Rettungstür, Nottür *f*

escape door lock, emergency door lock Fluchttürschloss, Rettungstürschloss, Nottürschloss *n*

escape door ((open) loop) control, emergency door ((open) loop) control Fluchttürsteuerung, Nottürsteuerung, Rettungstürsteuerung *f*

escape door ((open) loop) control device, emergency door ((open) loop) control appliance Fluchttürsteuer(ungs)gerät, Nottürsteuer(ungs)gerät, Rettungstürsteuerungsgerät *n*, Fluchttürsteuer(ungs)-vorrichtung, Nottürsteuervorrichtung, Nottürsteuerungsvorrichtung, Rettungstürsteuerungsvorrichtung *f*

escape door release, escape door opener, emergency door release, emergency door opener Fluchttüröffner, Fluchttürauslöser, Rettungstüröffner, Rettungstürauslöser, Nottüröffner, Nottürauslöser *m*

escape door system, emergency door system Fluchttürsystem, Rettungstürsystem, Nottürsystem *n* [*Zur Zustandsüberwachung von Fluchttüren*]

escape exit, emergency exit Fluchtausgang, Notausgang, Rettungsausgang *m*

escape lighting, escape illumination, emergency lighting, emergency illumination Fluchtbeleuchtung, Notbeleuchtung, Rettungsbeleuchtung *f*

escape route, escape path, flight path, emergency route, emergency path, flight route Fluchtweg, Rettungsweg, Notweg *m*

escape route indicator, escape path indicator, emergency route indicator, emergency path indicator Fluchtweganzeiger, Rettungsweganzeiger, Notweganzeiger *m*

escape route lighting, escape route illumination, escape path lighting, escape path illumination, emergency route lighting, emergency route illumination, emergency path lighting, emergency path illumination Fluchtwegbeleuchtung, Rettungswegbeleuchtung, Notwegbeleuchtung *f*

(e)scutcheon, lock (e)scutcheon, key plate, key (e)scutcheon, lock plate [*Plate containing a keyhole*] Schlüssellochdeckel, Schlossdeckel *m*, Schlüssellochblech, (Schlüssel)Schild, Türschild, Schlossschild, Schlossblech *n*, Rosette *f*

ESD, electrostatic discharge Entladung statischer Elektrizität, elektrostatische Entladung *f*

espagnolette (bolt), Spanish sash bolt, cremorne bolt [*A fixing for tall windows and casement doors in which bolts are driven home simultaneously at the top and bottom of the window or door by turning a centrally positioned handle*] Espagnolett(en)verschluss, spanische Fensterdrehriegel, Drehstangenverschluss, Bajonettverschluss *m*, Espagnolett(en)-stange; Bajonettstange *f* [*Fehlbenennungen*]

estate → (real) property

estate management → (real) property management

esthetically pleasing, aesthetically pleasing, (a)estetically pleasant formschön, formvollendet, stilvoll

estimation überschlägliche Ermittlung, überschlägliche Bestimmung, Schätzung, überschlägige Ermittlung, überschlägige Bestimmung *f*

estimator Kalkulator *m* [*betriebliche Rechnungswesen*]

ETB → end-of-transmission block

ethane → refrigerant-170

Ethernet, carrier sense multiple access with collision detection, CSMA/CD Ethernet, CSMA/CD-Verfahren *n* [*Spezifikation eines Basisbandnetzwerkes. Es dient zum Zusammenschluss von Rechneranlagen unter Verwendung eines Koaxialkabels und eines Sende-Empfangs-Gerätes; z. B. SINEC H1*]

ethylene → refrigerant-1150

ETM → eight-to-ten modulation

ETX → data block and text end

EU → execution unit

eupatheoscope, dummy man Eupatheoskop *n* [*Temperatur, Strahlung und Lüftung berücksichtigendes Klimamessgerät*]

European article numbering europäische Artikelnummer(ier)ung *f*

European Computer Manufacturers Association → ECMA

European Installation Bus; EIB Europäische Installationsbus, EIB-Installationsbus, Installationsbus der Gebäudesystemtechnik *m* [*Der Europäische Installationsbus ist auf den Bereich des Elektrohandwerks zugeschnitten. Die Busteilnehmer werden über ein zweiadriges Kabel verbunden. Die Intelligenz des Systems ist dezentral bei den Komponenten bzw. den Busankopp(e)lern angesiedelt. Der EIB eignet sich insbesondere für die Feldebene und ist auch im Haus- und Wohnbereich einsetzbar.*]

European Installation Bus Association, EIBA Europäische Installations-Bus-Verband *m*

eutectic brine eutektische Sole *f*

to evacuate → to withdraw

to evacuate räumen

evacuated chamber → vacuum chamber

evacuated (thermal) insulation, vacuum insulation, insulating vacuum [*Insulation consisting of double balls, with the space between them, with or without packing, evacuated*] Vakuumisolierung, Vakuumisolation *f*

evacuated (thermal) porous insulation, vacuum porous insulation, porous insulating vacuum poröse Vakuumisolierung, poröse Vakuumisolation *f*

evacuated (thermal) powder insulation, vacuum powder insulation, powder insulating vacuum Pulver-Vakuumisolierung, Pulver-Vakuumisolation *f*

evacuation Vakuumziehen *n*

evacuation [*e.g. of a building in case of fire*] Räumung, Evakuierung *f*

evacuation alarm Räumungsalarm, Evakuierungsalarm *m*

evacuation order Räumungsbefehl, Evakuierungsbefehl *m*

to (e)valuate auswerten, bewerten

(e)valuating Auswerten, Bewerten *n* [*1. Das Auswerten von Messwerten bis zum Messergebnis wird dem Begriff „Messen" zugerechnet, dagegen gehört die weitere Verwertung von Messwerten und Messergebnissen nicht zum Messen. DIN 1319 Teil 1. 2. Beim Auswerten werden aus erfassten Größen durch Berechnen oder auch Sortieren Kenngrößen des Prozesses ermittelt. DIN 19 222*]

(e)valuating electronic module, (e)valuating electronic assembly (modulare) Auswerteelektronik, Auswerte-Elektronik-Baugruppe, Bewerte-Elektronik-Baugruppe, (modulare) Bewerteelektronik *f*

(e)valuation → (data) interpretation

(e)valuation, val., appraisal of value, appraisement (of value), assessment (of value) [*The determined or estimated value or price*] Wertermittlung, Bewertung, (Ab)Schätzung, Taxierung *f*

(e)valuation and intervention center (US); (e)valuation and intervention centre (Brit.) zentrale Auswert(ungs)- und Einsatzleitstelle *f*

(e)valuation electronics Auswerteelektronik, Bewerteelektronik, Auswertungselektronik, Bewertungselektronik *f*

(e)valuation point Auswerteort, Bewerteort, Auswertungsort, Bewertungsort *m*, Auswertestelle, Bewertestelle, Auswertungsstelle, Bewertungsstelle *f*

(e)valuation system → (data) (e)valuation system

(e)valuator, (e)valuating unit, (e)valuation unit Auswerteeinheit, Auswertungseinheit, Bewerteeinheit, Bewertungseinheit *f*, Auswertegerät, Auswertungsgerät, Bewertegerät, Bewertungsgerät *n*

to evaporate verdampfen

evaporating Verdampfen *n*

evaporation; vaporization (US); vapourization (Brit.) [*Conversion of liquid to the vapo(u)r state by the addition of latent heat*] Verdampfung *f*

evaporation burner → (oil) evaporation burner

evaporation condenser, evaporative condenser [*The removal of heat from the refrigerant is achieved by the evaporation of water from the exterior of the condensing surface, induced by the forced circulation of air and sensible cooling by the air*] Verdampfungskondensator, Verdampfungsverflüssiger *m*

evaporation fan Verdampfungsventilator, Verdampfungslüfter *m*

evaporation fan burner → fan evaporation burner

evaporation heat, evaporative heat Verdampfungswärme *f* [*Wärmemenge beim Übergang vom flüssigen in den dampfförmigen Zustand ohne Temperaturerhöhung*]

evaporation humidifier, evaporative humidifier Dampfbefeuchter, Dampfanfeuchter *m*

evaporation plant, evaporative plant, evaporation installation, evaporative installation Verdampfungsanlage *f*, Verdampfungsbetrieb *m*

evaporation pressure, evaporative pressure Verdampfungsdruck *m*

evaporation pressure controller, evaporative pressure controller Verdampfungsdruckregler *m*

evaporation temperature, evaporative temperature Verdampfungstemperatur *f*

evaporator Verdampfer *m* [*Wärme(aus)tauscher, mit dem der Wärmequelle von der Wärmepumpe Wärme entzogen wird*]

evaporator coil → expansion coil

evaporator-condenser cryogenic machine → cryogenic machine with evaporator and condenser

evaporator duty, evaporator performance, evaporator output Verdampferleistung *f*

evaporator pressure regulator, evaporator pressure controller Verdampferdruckregler *m*

evaporator tube Verdampferröhre *f*

evaporator water, evaporation water, evaporating water Verdampferwasser, Verdampfungswasser *n*

evapotranspiration Verdunstung *f* [*Übergang eines Stoffes (meist Wasser) vom flüssigen in den dampfförmigen Zustand, der sich nur an der Oberfläche des Stoffes und unterhalb der Verdampfungstemperatur ereignet. An der Oberfläche einer Flüssigkeit stehen Moleküle in der Mitte derselben, so dass eine Aggregatzustandsänderung unterhalb der Verdampfungstemperatur möglich ist*]

EVC (US) → lift control unit (Brit.)

even → level

evening peak (load) Abendspitze(nlast) *f* [*E-Werk*]

even(-numbered) g(e)radzahlig, g(e)rade

even parity g(e)rade Parität, g(e)rade Bitzahl *f* [*Die Anzahl der auf „1" befindlichen Bits (pro Wort oder Byte) wird mit dem Paritätsbit g(e)rade gemacht*]

event [*The moment of time at which a specified change of state occurs; usually marks the completion of an asynchronous input/output operation*] Ereignis *n*, Meldung bei Zustandsänderung, Zustandsänderungsmeldung *f* [*DIN 19 243, Teil 1*]

event analysis Ereignisanalyse *f*

event burst [*A group of events occurring together in time*] Ereignisschauer *m*

event check(ing) Ereigniskontrolle *f*

event checkpoint Ereigniskontrollstelle *f*

event class tag Ereignisklassenetikett *n*

event-controlled ereignisgeregelt

event-controlled ereignisgesteuert

event(-controlled) switching Ereignisschalten *n*

event(-controlled) switching program (US); event(-controlled) switching programme (Brit.) Ereignisschaltprogramm *n*

event counter [*A complete instrument for detecting, totalizing, and indicating a sequence of events*] Ereignis(folge(n))zähler *m*

event counting Ereignis(folge(n))zählung *f*, Ereignis(folge(n))zählen *n* [*Anhäufung von in einer Anlage auftretenden Ereignissen*]

event-depending, event-dependent ereignisabhängig

event display field Ereignismeldefeld, Ereignisanzeigefeld *n* [*Prozessmonitor*]

event (functional) sequence, flow Ereignisablauf *m*, Ereignisfolge *f*, Ereignishergang *m*

event initiated (computer) programme (Brit.); event initiated (computer) program (US) [*It is executed on the occurrence of an input/output operation or an alarm*] ereignisverursachte (Rechner)Programm *n*

event log Ereignisprotokoll *n*

event logging, eventlisting, event printing out Ereignisprotokollieren *n*

event long-term storage Ereignislangzeitspeicherung *f*

event occurrence → occurrence of an event

event ((open) loop) control Ereignissteuerung *f*

event path, event travel Ereignisweg, Ereignispfad *m*

event printer Ereignisdrucker *m*

event printout, event (log) protocol, event log printout [*Hard copy version of event log*] Ereignisprotokollausdruck *m*, Ereignisniederschrift *f*

event processing Ereignisverarbeitung *f*

event reaction Ereignisreaktion *f*

event recorder Ereignisschreiber *m*

event record(ing) Ereignisaufzeichnung *f* [*Aufzeichnungen aller Ereignisse oder programmierter Auswahl von Ereignissen, die während eines festgelegten Zeitraumes aufgetreten sind*]

event sequence → event (functional) sequence

event series → series of events

event signal Ereignissignal *n*

event signal burst Ereignissignalschauer *m*

event statistics Ereignisstatistik *f*

event storage [*It provides for the storage of the print address, state and information text, resulting from I/O and processing functions. Time and date stamps shall be stored with each information item*] Ereignisspeicherung *f*

event switching [*This function provides a logic output when a predefined event occurs*] Ereignisschaltung *f*

event switching programme; event switching program (US) Ereignisschalt(ungs)programm *n*

event table Ereignistabelle *f*

event text Ereignistext *m* [*Standard- oder Programmtext, der erscheint, wenn das entsprechende Ereignis auftritt.*]

event time Ereigniszeit(punkt) *f*, *(m)*

event trace diagram Ereignisverlaufsdiagramm *n*

event travel → event path

event treatment Ereignisbehandlung *f*

evidence gathering [*At a crime scene*] Spurenaufnahme, Spurenfeststellung *f*

evil smelling übelriechend, stinkend

examination → inspection

to examine sequentially → to scan

EXAPT, extended subset of automatically programmed machine tools Programmiersprache *f* für numerisch gesteuerte Werkzeugmaschinen, Weiterentwick(e)lung *f* der APT-Programmiersprache

to exceed überschreiten *n* [*Grenzwert*]

exceeding Überschreiten *n* [*Grenzwert*]

exceptionally sharp [*picture*] gestochen scharf, außerordentlich scharf

exception day Ausnahmetag *m*

E

E

exception report, ER Fehlsituationsbericht, Fehllagebericht *m*

excess air adjuster Luftüberschusseinsteller, Überschusslufteinsteller *m*

excess (charge) carrier Überschuss(ladungs)träger *m*

excess consumption Über(schuss)verbrauch *m*

excess(ive) (acoustic) noise Lärm *m* [*Unerwünschter, störender und gesundheitsschädlicher Schall. Alle Geräuschimmisionen die das körperliche, seelische und soziale Wohlbefinden beeinträchtigen. Lärm ist kein physikalischer, sondern ein sozial-psychologischer Begriff. Dies zeigt sich z. B. bei der Beurteilung von lauter Musik, die von einem als Vergnügen und Entspannung und vom anderen als belästigend empfunden wird.*]

excess(ive) air [*Air in excess of that theoretically required for complete combustion*] Luftüberschuss *m*, Überschussluft *f*

excess(ive) current → overcurrent

excess of heat, heat excess Wärmeüberschuss *m*, Überschusswärme *f*

to exchange → to interchange

exchange, public exchange Vermittlung, Fernsprechvermittlung *f*, Amt, Fernsprechamt *n*

exchange [*The exchange of the contents of memory locations with registers, accumulators etc.*] Austauschen *n*

exchangeable disk, exchangeable disc, cartridge disk, cartridge disc Wechsel(speicher)platte, Speicherwechselplatte *f*

exchangeable disk drive, exchangeable disc drive, cartridge disk drive, cartridge disc drive Wechsel(speicher)plattenlaufwerk, Speicherwechselplattenlaufwerk *n*

exchangeable disk storage, exchangeable disc storage, EDS [*It is used as a backing storage, in which the disks come in capsules, each containing several disks; the capsules can be replaced during operation of the computer and can be stored until needed*] Wechselplattenspeicher *m*

exchange baseband station, XBS Vermittlungsbasisbandstation *f*

exchange line → post-office line

exchange programmable (aus)tauschprogrammierbar, (aus)tauschprogrammierfähig

exchanger Tauscher, Austauscher *m*

exchange trunk line, public exchange trunk line Amtsfernleitung, Fernamt(s)leitung *f*

to excite → to actuate

exclamation mark Ausruf(ungs)zeichen *n*

exclusion of draft → draft exclusion

exclusive-NOR gate → equivalence gate

exclusive OR, EOR, nonequivalence Exklusiv-Oder(-Gatter) , XOR(-Gatter) *n*, Antiwertigkeit *f* [*Diese logische Funktion vergleicht zwei Eingangssignale, deren Ungleichheit am Ausgang angezeigt wird*]

exclusive or exor Exklusiv-Oder-Exor-Glied *n* [*Ein Grundfunktionsglied der Steuer(ungs)technik*]

exclusive OR function, EOR function, nonequivalence function Exklusiv-ODER-Funktion *f*

exclusive-OR gate → nonequivalence gate

exclusive-OR module → nonequivalence module

excreta → (human) excreta

to execute, to perform [*To carry out some action*] ausführen, durchführen

to execute, to perform [*Computer programming. 1. To run a program(me) on a computer. 2. To perform the actions specified by a program(me). 3. To execute instructions in a computer. 4. In COBOL, "to perform" is a verb that transfers control to a subroutine and executes it the specified numbers of time*] ausführen, durchführen

execution → performance

execution command, performance command Ausführungskommando *n* [*EXAPT*]

execution phase Ausführungsphase *f*

execution control program (US); execution control programme (Brit.) [*The program(me) delivered by the manufacturer which permits the computer to handle the program(me)s fed to it*] Organisationsprogramm *n*

execution instruction, performance instruction [*Instruction to determine how a specially written computer program(me) is to operate*] Ausführungsbefehl *m*

execution time, performance time [*The time during which actual work, such as addition or multiplication, is carried out in the execution of a computer instruction*] Ausführungszeit *f*

execution time Bearbeitungszeit *f*

execution unit, EU, performance unit ausführende Einheit, Ausführungseinheit *f*

executive Ablaufteil *m*, *n* [*Der größte Bestandteil des Organisationsprogrammes eines Betriebssystems*]

executive block, EB, executive module Organisationsbaustein, OB *m*, Organisationsmodul *n*

executive call routine organisatorische Aufruf *m*

executive operation organisatorische Operation *f* [*programmierbare Steuerung*]

executive routine, master routine, monitor routine [*A digital computer routine designed to process and control other routines*] Ablaufroutine *f*, Organisationsprogramm *n*

executive system → **operating system**

executive system supervisor → **operating system supervisor**

exhalation Ausatmung *f*

to exhale ausatmen

to exhaust air → **to extract air**

exhaust(ed) air, extract(ed) air [*It is discharged from a space to the outdoors as differentiated from air transferred from one space to an adjacent one*] Fortluft *f* [*Ins Freie abgeführte Luft*]

exhaust(ed) air blower, extract(ed) air blower Fortluftgebläse *n*, Fortluftbläser *m*

exhaust(ed) air damper, extract(ed) air damper Fortluft(stell)klappe *f*

exhaust(ed) (air) fan, extract(ed) (air) fan Fortluftventilator, Entlüfter, Entlüftungsventilator *m*

exhaust(ed) air heat, extract(ed) air heat Fortluftwärme *f*

exhaust(ed) (air) line, extract(ed) air line Fortluftleitung *f*

exhaust(ed) (air) shaft, extract(ed) air shaft, exhaust shaft Fortluftschacht *m*

exhaust(ed) (air) system, extract(ed) air system Fortluftsystem *n*

exhaust hood → **(kitchen) range hood**

exhaust(ed) heat, extract(ed) heat Fortwärme *f* [*Der Teil der Abwärme, der alle nicht durch Wärmerückgang genutzten Wärmewerte umfasst. Die Fortwärme wird ins Freie geführt.*]

exhaust ventilation Entlüftung *f*

exhaust (ventilation) system Entlüftungssystem *n*

exhibit Schaustück, Schauobjekt *n*, Schaugegenstand *m*

exhibition space, exhibition surface Ausstellungsraum *m*, Ausstellungsfläche *f*

EXIST. [*On drawings, abbreviation for "existing"*] bestehend, vorhanden

existing plant, existing installation bestehende Anlage, vorhandene Anlage, Altanlage, bestehende Installation, vorhandene Installation, Altinstallation *f*

existing temperature vorhandene Temperatur, Ist-Temperatur, Realtemperatur *f*

exit [*An instruction to leave the program(me)*] Anweisung *f* zum Verlassen, Beenden *n*

exit [*A place at which a repeated cycle of operations in a computer program(me) can be stopped*] Halt *m*

exit air [*It is ejected from a space and replaced by fresh air*] Abluft *f* [*Durch eine RLT-Anlage oder ein freies Lüftungssystem aus dem Raum abgezogene gesamte Luft. DIN 1946, Teil 1*]

exit air conditioning duct Abluftklimakanal *m*

exit air controller Abluftregler *m*

exit air damper Abluft(stell)klappe *f*

exit air disk valve, exit air disc valve Ablufttellerventil, Abluftscheibenventil *n*

exit air drier, exit air dryer Ablufttrockner *m*

exit air duct Abluftkanal *m*

exit air fan Abluftventilator *m*

exit air grid Abluftgitter *n*

exit air hood Ablufthaube *f*

exit air inlet Ablufteinlass *m*

exit air line Abluftleitung *f*

exit air luminaire, exit air light(ing) fixture, exit air light(ing) fitting Abluftleuchte *f*, Abluftbeleuchtungskörper *m*

exit air opening, exit air outlet Abluftöffnung *f*, Abluftdurchlass, Abluftauslass *m*

exit air pipe Abluftrohr *n*

exit air section Ablufttteil *m*, *n* [*Lochdecke*]

exit air shaft Abluftschacht *m*

exit air silencer, exit air muffler, exit air sound absorber Abluftschalldämpfer *m*

exit air temperature ((closed) loop) control Ablufttemperaturreg(e)lung *f*

exit air temperature ((open) loop) control Ablufttemperatursteuerung *f*

exit air unit Abluftgerät, AB-GR n [*DIN 1946, Teil 1*]

exit air valve Abluftventil *n*

exit air volume stream Abluftvolumenstrom *m*

exit delay Ausgangsverzögerung *f*

exit (signal(ling)) message Ausgangsmeldung *f*

expand fixing, expanded fastening, expanding fixing, expanding fastening Spreizbefestigungsmittel *n* [*z. B. Spreizdübel*]

expanded memory specification → **EMS**

expanded metal fence Streckmetallzaun *m*

expanded plastic fire → **foam plastic fire**

expanded polyeth(yl)ene Polyethylenschaum *m*

E

E

expander [*Transducer that for a given input amplitude range produces a large output range*] Erweiterungsschaltung *f*, Erweiterer *m*

expansibility Dehn(ungs)fähigkeit, Dehnbarkeit *f*, Dehn(ungs)-vermögen *n*

expansion Expansion, Ausdehnung *f*

expansion [*A process in which the effective gain of an amplifier is varied as a function of signal magnitude, the effective gain being greater for large signals than the small signals; the result is greater volume range in an audio amplifier and greater contrast range in facsimile*] Erweiterung *f*

expansion board Erweiterungsplatine *f*

expansion chiller → (dry) expansion chiller

expansion coil, evaporator coil [*Evaporator constructed of pipe or tubing*] Verdampfer(rohr)schlange *f*

expansion controller Regler ohne Hilfskraft, (Aus-)Dehn(ungs)regler, Expansionsregler *m* [*Er arbeitet nach dem Prinzip der Feuchtigkeitsausdehnung*]

expansion cooling [*Cooling of a substance by having it undergo adiabatic expansion*] Expansionskühlung, (Aus)Dehn(ungs)kühlung *f*

expansion cylinder (Aus)Dehn(ungs)zylinder *m*

expansion drive Dehn(ungs)antrieb *m* [*Armatur, DIN 30 600, Reg.-Nr. 06169*]

expansion element Expansionselement *n* [*Ventil, Kapillarrohr oder Drosselplatte im Kältekreislauf zwischen Verflüssiger und Verdampfer für die Absenkung vom Verflüssigungs- auf den Verdampfungsdruck*]

expansion fitting Kompensator, Ausgleicher *m*

expansion float valve → float expansion valve

expansion liquid (Aus)Dehn(ungs)flüssigkeit, Expansionsflüssigkeit *f*

expansion loop, compensating loop, compensation loop (Aus)Dehn-(ungs)bogen, Expansionsbogen, Lyrabogen *m* [*Rohrausgleicher*]

expansion piece, compensating piece, compensator Rohrausgleicher, Ausgleicher, Dehn(ungs)ausgleicher *m*, Ausgleichsstück *n*

expansion (pipe)line (Aus)Dehn(ungs)(rohr)leitung, Expansions(rohr)leitung *f*

expansion point Erweiterungspunkt *m* [*Mitteilungsübermitt(e)lung*]

expansion refrigerating machine → (dry) expansion chiller

expansion rack, ER Erweiterungsbaugruppenträger *m* [*Für die Ausrüstung einer speicherprogrammierbaren Steuerung mit Ein/Ausgabeeinheiten*]

expansion slot Steckplatz *m* [*Er dient zur Aufnahme von Steckkarten für die Systemerweiterung*]

expansion space (Aus)Dehn(ungs)raum, Expansion(s)raum *m*

expansion steam trap [*See remark under "steam trap"*] (Aus)Dehn(ungs)kondensatableiter, Dampfstauer, Expansionskondensatableiter *m*

expansion tank [*Partially filled tank, operated at atmospheric pressure, at the top of a water system for the accommodation of volume expansion and due to the contraction of water*] Expansionsbehälter, (Aus-)Dehn(ungs)behälter, Expansionstank, (Aus)Dehn-(ungs)tank *m*

expansion thermometer (Aus)Dehn(ungs)thermometer, Expansionsthermometer *n*

expansion trap [*hot water heater*] Expansionstopf, Ausdehn(ungs)topf *m*

expansion valve Expansionsventil *n*

expansion vessel Expansionsgefäß, (Aus)Dehngefäß, (Aus)Dehnungsgefäß *n* [*Eine das Ausdehnungsvolumen des flüssigen Wärmeträgers der zugehörigen Heiz(ungs)anlage speichernde Vorrichtung. DIN 4806*]

expected data erwartete Daten *f* [*Beschreibung der logischen Zustände am Ausgang eines digitalen Prüflings, wie sie aufgrund vorgegebener Prüfbitmuster erwartet werden*]

expedited data, ED Vorrangdaten *f*

expended energy aufgewendete Energie *f*

expenditure Aufwand *m*

expenditure of energy, energy expenditure Energieaufwand *m*

expert Sachverständige, Gutachter, Sachkundige, Fachmann *m*

expiry date Verfall(s)zeitpunkt *m*, Verfall(s)datum *n*

explanation component, explanatory component Erklärungskomponente *f*

explanation drawing, explanatory drawing Erklärungszeichnung *f*

explanatory diagram, explanatory chart erläuternde Schalt(ungs)plan *m* [*DIN 40 719, Blatt 1*]

explanatory table erläuternde Tabelle *f* [*DIN 40 719, Blatt 1*]

exploded view, exploded drawing Explosionszeichnung *f*

explorer Schaltzentrale *f* [*in Windows 95*]

explorer → probe

explosion hazard Explosionsgefährdung *f*

explosionproof, explosionsafe explosionssicher, explosionsgeschützt, explosionsfest

explosion protection Explosionsschutz *m*

exponent → power

export interface Exportschnittstelle *f*

exposed belichtet [*Photographie*]

exposed auf Putz, Aufputz...

exposed, naked unverkleidet, bloß, nackt

exposed (bath) mixer (valve), exposed (bath) mix(ing) valve, exposed (bath) blender, exposed (bath) blending valve Aufputz(bade)mischbatterie, Mischaufputz(bade)batterie *f*

exposed conductive part, conductive exposed part berührbare leitfähige Teil, leitfähige berührbare Teil, inaktive leitfähige Teil, leitfähige inaktive Teil *m, n* [*IEV 441-11-10. Leitfähiges Teil, welches ohne weiteres berührt werden kann und normalerweise nicht unter Spannung steht, welches aber im Störungsfall Spannung annehmen kann*]

exposed (sectional) radiator [*See remark under "radiator"*] unverkleidete Radiator, unverkleidete Gliederheiz(ungs)körper *m*

exposed wall freiliegende Wand *f*

exposure Belichtung *f* [*Photographie*]

exposure fire Übergriffbrand *m*, Übergrifffeuer *n*

exposure hazard → (fire) exposure hazard

expression [*A mathematical or logical statement written in a source language, consisting of a collection of operands connected by operations in a logical manner*] Ausdruck, Term *m*

extended erweitert

extended address field Erweiterungadress(en)feld *n*

extended binary coded decimal interchange code, EBCDIC [*A computer code that uses eight binary positions to represent a single character, giving a possible maximum of 256 characters*] erweiterte binärkodierte Dezimal-Austausch-Kode, erweiterte 8-Bit-Kode *m*

extended call station, ECS erweiterte Sprechstelle *f* [*Beschallungssystem*]

extended (data) output → EDO

extended furnace with horizontal grate → horizontal grate extended furnace

extended heat exchange surface, augmented heat exchange surface [*Increased area for heat flow by means of fins, discs, etc.*] erweiterte Wärmeaustausch(ober)fläche *f*

extended out → EDO

extended processing erweiterte Verarbeitung *f*

extended subset of automatically programmed machine tools → Exapt

extender Erweiterer *m*

extending ladder, telescopic ladder Ausziehleiter, Teleskopleiter *f*

extensibility Erweiterbarkeit *f*

extension → (system) extension

extension [*An example would be: data file name; DAT.nnn, where nnn is the extension*] Erweiterung *f*

extension Verlängerung *f*

extension Dateinamenteil *m, n* hinter dem Punkt

extension cable Verlängerungskabel *n*

extension frame Erweiterungsrahmen *m*

extension key, adjustable key, variable key, controllable key [*Key for sliding door locks, having a stem or shank adjustable as to its length to adapt the key to doors of various thicknesses*] Verstellschlüssel, Justierschlüssel, (Ein)Stellschlüssel *m*

extension line Nebenanschluss(leitung) *m, (f)*

extension pipe Verlängerungsrohr *n*

extension register Zusatzregister *n* [*Es wird zur Erweiterung (z.B. des Akkumulators bei Überlauf) verwendet*]

extension unit Nebenstelle *f* [*Zur Betätigung eines elektronischen Schalters*]

extensive building ausgedehnte Gebäude *n* [*z. B. Hochhaus*]

extensometer [*An instrument for measuring minute increases of length*] Feinlängenmesser *m*

external acoustic alarm → acoustic external alarm

external air, outside air, outdoor air, outer air, exterior air Außenluft *f* [*Die der Menge der Fortluft entsprechende, von außen angesaugte Luftmenge bis zum Zusammentreffen mit der Umluft*]

external air duct, outdoor air duct, outside air duct, outer air duct, exterior air duct Außenluftkanal *m*

external air duct connecting branch [*"outdoor", "outside", "outer" and "exterior" may synonymously be used for "external"*] Außenluftkanalstutzen *m*

external air inlet, external air intake, outdoor air inlet, outdoor air intake, outside air inlet, outside air intake, outer air inlet, outer air intake, exterior air inlet, exterior air intake Außenlufteinlass *m*

external air plant, external air installation [*"outdoor", "outside", "outer" and "exterior" may synonymously be used for "external"*] Außenluftanlage, Außenluftinstallation *f*

external air preheater, outer air preheater, outside air preheater, exterior air preheater, outdoor air preheater Außenluftvorwärmer, Außenluftvorheizer *m*

E

E

external air silencer, outside air silencer, outdoor air silencer, external air muffler, outside air muffler, outdoor air muffler [*"exterior" and "outer" may synonymously be used for "external"*] Außenluftschalldämpfer *m*

external (air) temperature sensor, outdoor (air) temperature sensor, outside (air) temperature sensor Außen(luft)temperatursensor, Außen(luft)-temperaturfühler *m*

external air volume stream, outside air volume stream, outdoor air volume stream Außenluftvolumenstrom *m*

external alarm Externalarm, örtliche Alarm *m* [*In der Intrusionsmeldetechnik nach DIN VDE*]

external bus [*A cable that acts as a means of data exchange. There are two main types, serial and parallel. In building services, it is most common to use serial types where data flows on just 2 cores of a cable*] äußere Bus, Außenbus *m*

external cabling, outside cabling, outdoor cabling, outer cabling, exterior cabling Außenverkabelung *f*

external casing, outer case, external case, exterior case, exterior casing, outer casing [*synonyms are: housing, enclosure, case(work)*] Außengehäuse *n*

external current Fremdstrom *m*

external defrosting, exterior defrosting [*Defrosting using heat from external source*] Außenentfrostung *f*

external design temperature → design external temperature

external door, exterior door [*Door in an entrance that provides access from the outside to the inside of a building*] Eingangstür *f*

external elevator, exterior elevator, outdoor elevator, outside elevator (US); external lift, exterior lift, outdoor lift, outside lift (Brit.) Außenaufzug, Fassadenaufzug *m*

external excitation, separate excitation Fremderregung *f*

external fire stairway, outside fire stairway, outside fire stair(case), outdoor fire stairway, outdoor fire stair(case), external fire stair(case), exterior fire stairway, exterior fire stair(case) Außenfluchttreppe, Fluchtaußentreppe, Außennottreppe *f*

external inclined grate furnace, exterior inclined grate furnace [*See remark under "furnace"*] Schrägrostvorfeuerung, Schrägrostvorfeuerstätte *f*

external jacket, outer jacket, exterior jacket Außenmantel *m*

external lift (Brit.) → external elevator (US)

external lighting, external illumination, outdoor lighting, outdoor illumination, outside lighting, outside illumination, exterior lighting, exterior illumination Außenbeleuchtung *f*

external measuring-circuit selector, measuring point selector Messstellenumschalter *m* [*DIN 43 782*]

external memory, peripheral memory, exterior memory, auxiliary memory, secondary storage (device), auxiliary storage (device) [*In automatic data processing, a storage (device) in addition to the main storage (device) of a computer; for example, magnetic tape, disk, or drum*] Externspeicher, Massenspeicher, Peripheriespeicher *m* [*Er gehört nicht zur Zentraleinheit*]

external optical alarm → optical external alarm

external passenger elevator, exterior passenger elevator (US); external passenger lift, exterior passenger lift (Brit.) Außenpersonenaufzug, Außenfahrstuhl *m*

external plumber (Bau)Klempner, (Bau)Spengler *m* [*Österreich; Süddeutschland; Schweiz*]

external plumbing (trade) (Bau)Klempnerei *f*, (Bau)Klempnerhandwerk *n*

external pushbutton box Außendruckknopfkasten *m* [*Aufzug*]

external radio act(uat)or Außenfunkbedienteil, Außenfunkbedienungsteil *m, n*

external-rotor split-pole motor, external-rotor shaded-pole motor Außenläufer-Spaltpolmotor, Spaltpol-Außenläufermotor *m*

external (source) feed → separate (source) feed

external stairway, outside stairway, exterior stairway, outdoor stairway Außentreppenhaus *n*

external standard temperature → standard external temperature

external steam, outside steam, exterior steam, outdoor steam Fremddampf *m*

external (tele)phone, exterior (tele)phone Auswärtsfernsprecher *m*, Auswärtstelefon *n* [*Im Gegensatz zum Hausfernsprecher*]

external unit, exterior unit, outdoor unit, outside unit Außengerät *n*

external wall gas heating oven Außenwand-Gasheiz(ungs)ofen *m*

to extinguish löschen

to extinguish → to become extinguished

external water feature, outside water feature, outdoor water feature, outer water feature, exterior water feature Außenbeckenattraktion *f* [*z. B. Springbrunnen*]

extinguisher → **(fire) extinguisher**

extinguishing → **(fire) extinguishing**

to extract air, to exhaust air mechanisch entlüften [*Raumluft durch Ventilatoren abführen*]

extract(ed)... → **exhaust(ed)...**

extract fan → **outlet fan**

extract(ion) Auszug, Abzug *m*

extract(ion) hood, extracting hood Entlüftungshaube *f*

extract(ion) line, extracting line Auszugsleitung, Abzugsleitung *f*

extract(ion) roof ventilation → **roof extract(ion) ventilation**

extract(ion) tube, extracting tube [*A tube for extracting vitiated air from rooms*] Entlüftungsröhre *f*

extract(ion) ventilation, extracting ventilation, extract(ion) venting, extracting venting Entlüftung *f*

extract(ion) ventilation valve, extract(ion) venting valve Entlüftungsventil *n*

extract(ion) vent(ilator), extracting vent(tilator) Entlüfter *m*

extraheavy, superheavy überschwer, superschwer, Schwerst...

extraheavy fuel oil [*3500 seconds*] Heizöl S, schwer(flüssig)e Heizöl, HS *n* [*DIN 51 603*]

extra large scale integration, ELSI [*merged transistor logic*] extragroße Integrationsdichte *f* [*Integrierte Schaltung mit 1001 bis 10 000 Transistorfunktionen auf einem Chip*]

extra low voltage, e.l.v. Kleinstpannung *f*

extraneous conductive part fremde leitfähige Teil, leitfähige fremde Teil *m*, *n* [*Ein leitfähiges Teil, das nicht Teil der elektrischen Anlage ist, jedoch ein elektrisches Potential, einschließlich des Erdpotentials, übertragen kann.*]

extraneous matter Fremdstoff, Störstoff *m*

extra strength (concrete) pipe hochfeste (Beton-) Rohr *n*

extraterrestrial [*Occurring or originating away from the earth*] außerirdisch

extremely durable → **highly durable**

external temperature compensated controller, outside temperature compensated controller, outside temperature compensated regulator [*It consists of the electronic controller and sensors and does not include the actuated equipment linked to the outputs*] außentemperaturabhängige Regler *m*

extreme value, record value Rekordwert, Extremwert *m*

extrinsic fibre optic sensor; extrinsic fiber optic sensor (US) faseroptische Fehlerstellenfühler *m* [*Die Faser dient zur möglichst störungsfreien Übertragung des Messwertes vom Messort zu einem entfernt gelegenen Auswerteort. Die Modulation erfolgt am Messort außerhalb der Faser, z.B. mittels integriert-optischer oder mikrooptischer Komponenten*]

extrinsic semiconductor [*A semiconductor whose electrical properties are dependent on impurities added to the semiconductor crystal, in contrast to an intrinsic semiconductor, whose properties are characteristic of an ideal pure crystal*] Fehlerstellenhalbleiter *m*

extrusion Extrudieren, Verschieben *n* [*Einer Kurve oder Fläche im Raum, um eine dreidimensionale Fläche oder ein dreidimensionales Objekt zu erzeugen*]

eye [*A general term for the centre of anything, as the eye of a volute*] Auge *n*

eye → **inspection hole**

eye Kausche *f* [*Seil*]

eye adaption Augenanpassung *f*

eye bolt Augenbolzen, Splintbolzen *m*

eye irritation(s) [*Condition caused by swimming pool water that is on the acid side of the pH range*] Augenbrennen *n*, Augenreizung(en) *f*

eye nut Ringmutter, Augenmutter *f*

eye-bolt and key Splintbolzen, Vorsteckerbolzen, Schließbolzen *m*

eye control Scharfeinstellung *f* [*Auf das betrachtete Objekt*]

eye screw [*A screw with an open loop head*] Augenschraube, Kopfschraube mit Augenkopf; Klappschraube *f* [*frühere Benennung*]

eye sketch flüchtige Skizze *f*

eyewash(ing) bath, eyewash(ing) shower [*A sanitary appliance installed in work places where there is risk of injury to eyes by solid particles or dangerous liquids. The appliance consists of a small bowl with two jets of water controlled by a hand or foot operated valve in such a way that the user can bathe the eyes without touching them*] Augenwaschanlage *f*

eyewash(ing) fountain Augenwaschbrunnen *m*

F

F → framing

fabric (heat) gain [*Through walls, windows, floors, doors and ceilings*] Rohbauwärmeabgabe *f*

fabric (heat) loss [*The heat passing through the fabric of a building or room by conduction, and being dissipated by radiation and convection. It is found by estimating the heat loss for each surface enclosing the space and summating the total*] Rohbauwärmeverlust *m*

facade design Fassadengestaltung *f*

facade elevator (US); facade lift (Brit.) Fassadenaufzug *m*

facade ladder Fassadenleiter *f*

facade system Fassadensystem *n*

face [*moviesetter*] Pinsel *m*

face berandete Freiformfläche *f* [*Sie schneidet eine Fläche an einer beliebigen Berandungskontur ab, d. h. die Flächenbeschreitung bleibt erhalten, die außerhalb der Berandung liegenden Flächenteile werden nicht dargestellt*]

FACE, field alterable control element anwenderprogrammierte Steuer(ungs)element *n*

face (air) velocity (Luft)Geschwindigkeit *f* im Stirnquerschnitt

face of joint, joint face (Fugen)(Ab)Dichtleiste, (Fugen)(Ab)Dichtungsleiste *f*

face protection Gesichtsschutz *m*

face protector Gesichtsschützer *m*

face towel Gesichtshandtuch *n*

face wash fountain Gesichtswaschbrunnen *m*

facilities, setups Einrichtungen *f*

facility → building

facility, setup Einrichtung *f*

facility Leistungsmerkmal *n*

facility automation system → building automation system

facility integrator → building (monitoring and) control system

facility management, FM, building (and energy) management (and control), B(E)M, BM(C) Gebäudeverwaltung, Gebäudeführung, Objektverwaltung, Objektführung, Betrieb(s)führung, Immobilienbewirtschaftung *f*, Gebäudebetrieb, Objektbetrieb *m* [*Technischer, kaufmännischer und infrastruktureller Gebäudebetrieb. Die Bennenung „facility management" kam 1987 als Marktbegriff des Büromöbelherstellers Herman Miller aus den USA nach Deutschland*]

facility management product → building management product

facility management system → building management system

facility management system information list → building management system information list

facility manager → building manager

facility planner, building planner Gebäudeplaner, Immobilienplaner, Objektplaner *m*

facing Auflage *f* [*Schaltstück*]

facsimile receiver [*The receiver used to translate the facsimile signal from a wire or radio communication channel into a facsimile record of the subject copy*] Fernkopiererempfänger *m*

facsimile unit Fernkopierer *m*

factor [*A quality by which a given quantity is multiplied or divided in order to indicate a difference in measurement*] Faktor *m*

factory adjustable werkseitig einstellbar, werkseitig justierbar, fabrikseitig einstellbar, fabrikseitig justierbar

factory-assembled werkmontiert, fabrikmontiert

factory-assembled heat pump, heat pump ready for connection anschlussfertige Wärmepumpe *f*

factory badge, factory card Werkausweis, Fabrikausweis *m*, Werkkarte, Fabrikkarte *f*

factory-built assembly fabrikfertige Schaltgerätekombination, werkfertige Schaltgerätekombination *f*

factory-coated werkbeschichtet, fabrikbeschichtet

factory ground Werkgelände, Fabrikgelände *n*

factory-insulated fabrikisoliert, werkisoliert

factory-packaged werkverpackt, fabrikverpackt

factory store Werklager, Fabriklager *n*

factory protection, industrial security service Werkschutz, Fabrikschutz *m*

factory-set werkeingestellt, fabrikeingestellt, eingestellt im Werk, eingestellt in der Fabrik

factory setting Werkeinstellung, Fabrikeinstellung *f* [*Einstellung eines Gerätes oder einer Anlage im Werk*]

factory visit Werkrundgang, Fabrikrundgang *m*

factory warranty Werkgarantie, Fabrikgarantie *f*

factory wiring Werkverdrahtung, Verdrahtung im Werk, Fabrikverdrahtung, Verdrahtung in der Fabrik *f* [*Gegensatz: Verdrahtung am Einbauort (field wiring)*]

fade-in, fading-in Einblenden *n* [*Ein Audio- oder Videosignal stufenlos von Null*]

to fade (out) → **to die away**

fader Ein/Ausblender *m* [*Schieberegler für sanftes Ein- und Ausblenden oder Mischen von Programmquellen (Bild; Ton)*]

fading (out), dying away, fade-out Schwund *m*, Schwinden, Abklingen *n*

f(a)ecal matter, night soil Fäkal(ien)schlamm *m* [*Der in Kleinkläranlagen, insbesondere solchen ohne Belüftung, anfallende Schlamm*]

f(a)ecal matter pit, midden Fäkal(ien)grube *f*

f(a)eces [*The waste material eliminated by the gastrointestinal tract*] Fäkalien *f*

f(a)eces ejector (pump) → **ejector**

FAI → **fresh air inlet**

to fail versagen, nicht funktionieren, ausfallen [*Gerät; Anlage; System*]

failbit Flankenformbit *n* [*Informationsaufteilung im Zeitschlitz*]

failed ausgefallen, versagt [*Gerät; Anlage; System*]

faillock(ing) verrieg(e)lungssicher [*Absperrung*]

(fail-)safe sicher, gefahrlos

(fail-)safe ausfallsicher

(fail-)safe störungssicher

(fail-)safe against earth faults erdschlusssicher, erdschlussfest, erdschlussbeständig

(fail-)safe against wire breakage drahtbruchsicher, drahtbruchfest

(fail-)safe state, (fail-)safe status gefahrlose Zustand, sichere Zustand *m*

(fail-)safe system [*A system designed to avoid the risk of danger in the event of a fault or faults*] gefahrlose System, sichere System *n*

fail-to-open circuit principle Arbeitsstromprinzip *n*

failure, outage cut Ausfall *m*, Versagen *n*

failure, defect Funktionsunfähigkeit *f*, Mangel *m* [*Einer Komponente in der Zuverlässigkeitstechnik*]

failure alarm, defect alarm Defektalarm, Mängelalarm, Mangelalarm *m* [*Durch technische Störung(en) der Brandmeldeanlage verursacht*]

failure analysis memory, outage analysis memory Ausfallanalysenspeicher, Versagensanalysenspeicher *m*

failure frequency, outage frequency Ausfallhäufigkeit, Versagenshäufigkeit *f*

failure indicator, defect indicator, failure indicating instrument, defect indicating instrument Mangelanzeiger, Mängelanzeiger *m*, Mangelanzeigeinstrument, Mängelanzeigeinstrument *n*

failure indicator, outage indicator, failure indicating instrument, outage indicating instrument Ausfallanzeiger, Versagensanzeiger *m*, Ausfallanzeigeinstrument, Versagensanzeigeinstrument *n*

failure instant, instant of failure, instant of outage, outage instant Ausfallzeitpunkt, Versagenszeitpunkt *m*

failure notification, outage notification Ausfallanzeige, Versagensanzeige *f*

failure notification → **defect notification**

failure of power (supply) → **power (supply) failure**

failure quota, outage quota, trouble quota Ausfallquote, Versagensquote *f*

failure shooting → **trouble shooting**

failure signal message, trouble signal message, outage signal message Ausfallmeldung, Versagensmeldung *f* [*Siehe Anmerkung unter „Meldung"*]

fall, slope [*Of a pipe, conduit, or channel usually expressed in inches per foot (or centimetre per meter) or in percent*] Gefälle *n*

fallback [*The system, electronic or manual, which is substituted for the computer system in case of breakdown*] Ersatzsystem *n*, Reserve *f*

to fall below unterschreiten [*Grenzwert*]

fall-forward Erhöhung *f* der Übertragungsgeschwindigkeit

falling below Unterschreiten *n* [*Grenzwert*]

falling film (surface) cooler → **Baudelot cooler**

falling latch Fallklinke *f*, Fallriegel *m*

falling latch spring lock → **spring lock with falling latch**

falling noise Fallgeräusch *n*

falling weight drive Fallgewichtsantrieb *m*

fall pipe → **downspout**

fall time Flankenabfallzeit *f* [*Halbleiter, DIN 41855*]

false → **erroneous**

false acceptance unberechtigte Freigabe, falsche Freigabe *f*

false firing Zünd(ungs)fehler *m*, Fehlzündung *f*

false ceiling → **suspended ceiling**

false key → **pick lock**

false air Falschluft *f* [*Luftmengen, die durch Undichtigkeit oder Rauchkanal und Schornstein eindringen und die Gasströmung stören*]

failure modes and effects and criticality analysis, FMECA Ausfallanalyse *f* [*Analysieren und Sichten der Ursachen und Folgen eines Ausfalls*]

false alarm Falschalarm, Fehlalarm, gefahrlose Alarm, gefahrfreie Alarm *m* [*DIN VDE 0833-2*]

false light, parasitic light, outside light Fremdlicht, Nebenlicht, Außenlicht *n*

F

false lock → blind lock

false operation → erroneous operation

false plug Blindstopfen, Blindstöpsel *m*

falsification → forging

family, series (Bau)Reihe, (Bau)Serie *f*

family (bath)tub, family bathing tub Familien(bade)wanne *f*

family characteristics, characteristics family Kennlinienfeld, Kennkurvenfeld *n*

FAMOS memory → float(ing) gate avalanche-injection metal-oxid(e) semiconductor memory

fan Ventilator *m*, VE, Lüfter *m* [*Strömungsmaschine zur Förderung von Luft und anderen Gasen bis zu einem Druck von 30 000 N/m² (Pa). Bei höheren Drücken heißen diese Ventilatoren „Gebläse". DIN 24163, Teil 1; DIN 24167, Teil 1; VDMA 24169, Teil 1*]

fan air-conditioning plant, fan air-conditioning installation [*The air is treated at a central plant and carried to and from the rooms by one or more fans and a system of ducts*] Ventilatorklima(tisierungs)-anlage *f*

fan air cooler Ventilatorluftkühler *m*

fan air density [*Density of air corresponding to the total pressure and total temperature at the fan inlet, when the fan is operating*] Ventilatorluftdichte *f*

fan air handling → forced ventilation

fan air heating Ventilatorluftheizung *f* [*Die Luft wird zentral erwärmt*]

fan-assisted ventilatorgestützt, ventilatorunterstützt

fan(-assisted) convection Ventilatorkonvektion *f*

fan(-assisted) heater Heizlüfter *m*

fan(-assisted) warm-air furnace → mechanical warm-air furnace

fan belt Ventilatorriemen *m*

fan belt guard Ventilatorriemenschutz *m*

fan blade Ventilatorblatt *n*, Ventilatorschaufel *f*

fan block → fan module

fan brake horsepower, fan bhp, fan BHP Ventilator-Brems-PS *f*

fan casing, fan case, fan housing, fan enclosure Ventilatorgehäuse *n*

fan chamber Ventilatorkammer *f*

fan characteristic (curve), fan curve Ventilatorkennlinie, Ventilatorkennkurve, Ventilatorcharakteristik *f*

fan circuit Ventilatorstromkreis, Ventilatorschaltkreis *m*, Ventilatorschaltung *f*

fan ((closed) loop) control Ventilatorreg(e)lung *f*

fan coil (unit) Ventilatorkonvektor, Gebläsekonvektor *m*, Klimatruhe *f* [*DIN 4704, Teil 5 (11.84)*]

fan coil (unit) for cooling only Kühl(ungs)ventilatorkonvektor *m*

fan coil (unit) for heating only Heiz(ungs)ventilatorkonvektor *m*

fan control (switch) [*Temperature-actuated switch controlling only the on-off operation of an air-circulating fan*] Lüfterschalter *m*

fan convector air conditioning plant, fan convector air conditioning installation Ventilator-Konvektor-Klimaanlage *f*

fan-cooled ventilatorgekühlt

fan cooling bewegte Kühlung, Ventilatorkühlung *f*

fan cooling tower Kühlturm mit Ventilator, Ventilatorkühlturm *m*

fancy hinge → ornated hinge

fan dimensions → fan size

fan discharge Ventilatorabführung *f*

fan draft → mechanical draft

fan drive Ventilatorantrieb *m*

fan efficiency Ventilatorwirkungsgrad *m*

fan enclosure → fan casing

fan engineering Ventilatortechnik *f*

fan evaporation burner, evaporation fan burner Verdampfungsbrenner mit Ventilator, Ventilatorverdampfungsbrenner *m*

fan evaporator unit Ventilator-Verdampfer-Einheit *f*

fan gas furnace Ventilator-Gasfeuerstätte, Ventilator-Gasfeuerung *f*

fan heater, fan heat emitter, fan heating device, fan heating appliance, fan heating apparatus Ventilatorerhitzer, Ventilator(er)wärmer, Ventilatorheizer, Ventilatorheiz(ungs)apparat *m*

fan hood → fan shroud

fan horse-power, fan h.p. Ventilator-PS *f*

fan housing → **fan casing**

fan-in Eingang(s)lastfaktor *m*, Eingängeanzahl, Eingang(s)fächerung, Eingang(s)signalverzweigung *f* [*Anzahl der Eingänge bei einer Türschaltung*]

fan inlet Ventilatoreintritt, Ventilatoreinlass *m*

fan jet Ventilatorstrahl *m*

fan module, fan block Ventilatorbaustein *m*, Ventilatormodul *n*

fan monitoring Ventilatorüberwachung *f*

fan motor Ventilatormotor *m*

fan motor controller, fan motor regulator Ventilatormotorregler *m*

fan/motor unit Ventilator/Motor-Einheit *f*

fanning action Ventilatorwirkung *f*

fan noise Ventilatorgeräusch *n*

fan ((open) loop) control Ventilatorsteuerung *f*

fan operation Ventilatorbetrieb *m*

fan-out Belastbarkeit, Ausgang(s)signalverzweigung *f*, Ausgang(s)lastfaktor, Ausgang(s)fächer(-ung) *m*, *(f)* [*Die mögliche Anzahl von Torschaltungen, die von einer (ursprünglichen) Torschaltung gesteuert werden können*]

fan outlet Ventilatoraustritt, Ventilatorablass, Ventilatorauslass, Ventilatorabgang *m*

fan-out module → **multi(ple)-transceiver**

fan pilot light Ventilatorkontrolllampe *f*

fan pressure Ventilatordruck *m*

fan-pressurized (US); fan-pressurised (Brit.) druckausgeglichen durch Ventilator

fan protection screen Ventilatorschutzgitter *n*

fan relay Ventilatorrelais *n*

fan return-air heating Ventilatorumkehrluftheizung, Umkehrventilatorluftheizung *f*

fan room Ventilatorraum *m*

fan rotations per minute → **fan speed**

fan r.p.m. → **fan speed**

fan running time Ventilatorlaufzeit *f*

fan run-on Ventilatornachlauf *m*

fan section Ventilatorteil *m, n*

fan shaft Ventilatorwelle *f*

fan(-shape(d)) nozzle Fächerdüse *f*

fan(-shape(d)) outlet Fächerauslass *m*

fan shroud, fan hood Schirmblech *n*, Ventilatorhaube *f*, Ventilatorstutzen, Ventilatorkragen *m*

fan size, fan dimensions Ventilatorgröße, Ventilatorabmessungen *f*

fan speed, fan r.p.m., fan rotations per minute Ventilatordrehzahl *f* (pro Minute)

fan speed reducer Ventilatordrehzahlminderer *m*

fan speed selector Ventilatordrehzahlwähler *m*

fan speed switch Ventilatordrehzahlschalter *m*

fan starter Ventilatoranlasser, Ventilatorstarter *m*

fan suspension Ventilatoraufhängung *f*

fan switch Ventilatorschalter *m*

fan system Ventilatorsystem *n*, Ventilatoranlage *f* [*Förderung der Raumluft durch Ventilatoren*]

fan-tail(ed) burner Rundstrahlbrenner *m*

fan throughput Ventilatorfördermenge, Luftfördermenge *f*

fan total pressure, fan static pressure, total fan pressure, static fan pressure Gesamtventilatordruck, Ventilatorgesamtdruck *m*

fan unit Ventilatoreinheit *f*

fan-ventilated ventilatorbelüftet

fan ventilation, fan vent(ilat)ing Ventilatorlüftung *f*

fan volume ((open) loop) control Ventilator-Volumensteuerung, Volumen-Ventilatorsteuerung *f*

fan wheel Ventilatorrad *n*

FAQs → **frequently asked questions**

far-end crosstalk Gegennebensprechen, Nebengegensprechen *n*

FAS → **building automation system**

FAS, full angle speaker Lautsprecher *m* nach Pfleiderer mit Punktstrahlerwirkung, Kugelstrahler *m* [*Er ist sehr breitbandig*]

fast... → **rapid...**

fastener, sash bolt Schubriegel, Schieb(e)riegel *m*

fastening Schließbeschlag *m*

fastening → **fixing**

fastenings Schließbeschläge *f*

FATG → **fixed azimuth tape guid(anc)e**

fast flaming fire lodernde Feuer *n*

fast Fourier transformation → **FFT**

fast freezer → **quick freezer**

fast-make-fast-break switch → **sensitive switch** [*IEC 163*]

fast mobility schnelle Mobilität *f* [*Mobilkommunikationssystem*]

fast traverse Eilgang, Schnellgang *m* [*Das Werkzeug fährt mit Maximalvorschub von einem Vorgang zum anderen*]

fast zero power, FZP schnelle Schaltungsauslegung *f*

F

fat gutter Fett(fang)rinne *f* [*Wasserhaube*]

father file → parent file

fat separator Fettabscheider *m* [*Wasserhaube*]

fat acid Fettsäure *f*

faucet (US) → (water) tap

fault, malfunction, breakdown, interference, error Fehler *m*, Störung *f*

fault alarm, malfunction alarm Stör(ungs)alarm, Fehleralarm *m*

F

fault alarm centre (Brit.); fault alarm center (US) [*"malfunction" may synonymously be used for "fault"*] Stör(ungs)alarmzentrale, Fehleralarmzentrale *f*

fault case → fault event

fault condition, fault state, fault status, malfunction condition, malfunction state, malfunction status Stör(ungs)zustand, Stör(ungs)status, Fehlerstatus, Fehlerzustand *m*

fault contact, malfunction contact Stör(ungs)kontakt, Fehlerkontakt *m*

fault counter, malfunction counter Stör(ungs)zähler, Fehlerzähler *m*

fault current, malfunction current Störstrom, Fehlerstrom *m*

fault current switch, malfunction current switch Stör(ungs)stromschutzschalter, Fehlerstromschutzschalter *m*

fault detector, malfunction detector, fault detecting device, malfunction detecting device, fault detection device, malfunction detection device Stör(ungs)melder, Stör(ungs)detektor, Fehlermelder, Fehlerdetektor, Stör(ungs)meldungsgeber, Fehlermeldungsgeber *m*, Stör(ungs)meldegerät, Fehlermeldegerät *n*

fault disconnection → shutdown on faults

fault elimination, malfunction elimination Störungsbeseitigung, Fehlerbeseitigung, Entstörung *f*

fault event, fault case, malfunction event, malfunction case Stör(ungs)ereignis *n*, Stör(ungs)fall *m*

fault forwarding, malfunction forwarding Störungsweitermeldung, Fehlerweitermeldung *f*

fault indicator, malfunction indicator Stör(ungs)-anzeiger, Fehleranzeiger *m*

fault lamp, malfunction lamp Stör(ungs)lampe, Fehlerlampe *f*

faultless, malfunctionless stör(ungs)frei, fehlerfrei

fault manager, malfunction manager Adresspunkt *m* für (System)Fehlermeldungen

fault monitoring, malfunction monitoring Störungsüberwachung, Fehlerüberwachung *f*

fault processing, malfunction processing Störungsbearbeitung, Fehlerbearbeitung *f*

fault protected switch, malfunction procedure switch elektronische Schalter mit Überspannungsschutz *m*

fault report, malfunction report Stör(ungs)bericht, Fehlerbericht *m*

fault shooting → trouble shooting

fault shutdown → shutdown on faults

fault signal, malfunction signal Stör(ungs)signal, Fehlersignal *n*

fault signal message, malfunction signal message Stör(ungs)meldung, Fehlermeldung *f* [*Siehe Anmerkung unter „Meldung"*]

fault signal message control panel, malfunction signal message control panel Stör(ungs)meldezentrale, Fehlermeldezentrale *f*

fault state → fault condition

fault switch(ing) off [*"Malfunction" may synonymously be used for "fault"*] Stör(ungs)abschaltung, Fehlerabschaltung *f*

fault switch(ing) over [*"Malfunction" may synonymously be used for "fault"*] Stör(ungs)umschaltung, Fehlerumschaltung *f*

fault to ground (US) → ground(ing) fault

fault tolerant, malfunction tolerant fehlertolerant, stör(ungs)tolerant

fault tree analysis, malfunction tree analysis Störungsbaumanalyse, Fehlerbaum-analyse *f*

fault variable, malfunction variable Störgröße, Störungsgröße, Fehlergröße *f*

fault voltage coil, malfunction voltage coil Störspannungsspule, Störungsspannungsspule, Fehlerspannungsspule *f*

fault voltage protection, malfunction voltage protection Störungsspannungsschutz, Fehlerspannungsschutz *m*

faulty insertion Fehleingabe *f* [*Elektronik*]

Faure storage battery, pasted-plate storage battery [*A storage battery in which the plates consists of lead-antimony supporting grids covered with a lead oxid(e) paste, immersed in weak sulfuric acid*] Pasten-Plattenbatterie *f*

fax polling Abruf *m* von Textinformationen

FCB → file control block

FC fan → forward-curved fan

fci, flux change per inch (magnetische) Flusswechsel *m* pro Zoll

FCS → frame check sequence

FD → floppy disk

FDC → floppy disk controller

FDDI → fibre distributed data interface

FDL → fieldbus data link

feasibility Durchführbarkeit, Machbarkeit *f*

feasibility study Durchführbarkeitsstudie, Machbarkeitsstudie *f*

feature, characteristic Merkmal *n*

feature connector Flachbandkabel *n* [*Zur internen Verbindung von Graphik- und Videokarten*]

feature fire(place), feature chimney offene Kamin, Attraktionskamin *m*

fecal matter → f(a)ecal matter

feces → f(a)eces

Federal Privacy Protection Law Bundesdatenschutzgesetz, BDSG *n* [*Bundesrepublik Deutschland*]

fee, rate, charge Gebühr *f*

fee acquisition, rate acquisition, charge acquisition Gebührenerfassung *f*

to feed → to charge

feedback Rückführung, Gegenkopp(e)lung, Rückkopp(e)lung *f* [*Übertragen eines Signals vom Ausgang zum Eingang zwecks Vergleich von Istwert und Sollwert*]

feedback ((closed) loop) control [*Closed loop control where the control activating (error) signal is derived from the setpoint and a signal indicating the current value of the controlled medium*] Rückführreg(e)lung, Rückführungsreg(e)lung *f*

feedback compensation → parallel compensation

feedback control circuit → ((closed) loop) control circuit

feedback loop Rückkopp(e)lungsschleife *f* [*Dieser Teil eines Regelkreises ermöglicht die Rückführung des gemessenen Istwertes zwecks Vergleich mit dem vorgegebenen Sollwertes*]

feedback optimizing Rückwärtsoptimierung *f*

feedback suppresser, feedback exterminator Rückkopplung(s)unterdrücker *m*

feedback value Istwert *m* [*Der Wert einer untersuchten Größe zum Messzeitpunkt*]

feedback variable Rückführ(ungs)größe *f* [*DIN 19 226, Teil 5*]

feed depth Vorschubtiefe *f*

feedforward ((closed) loop) control Reg(e)lung mit Störgrößenaufschaltung, Vorwärtsreg(e)lung *f*

feedforwarding Störgrößenaufschaltung *f* [*Eine Aufschaltung zur Verbesserung des Verhaltens der zu steuernden Größen, wobei als zusätzliche Eingangssignale messbare Störgrößen bewusst in geeigneter Weise in das System eingeführt werden*]

feed hole Führungsloch, Transportloch, Vorschubloch *n* [*Lochstreifen*]

feeding (Ein)Speisen, Zuführen, Zuleiten *n*, (Ein-)Speisung, Zuführung, Zuleitung *f*

feed(ing) Vorschub *m*, Vorschieben *n*

feeding Nachgleiten, Nachrutschen *n* [*Brennstoff bei der Kesselfeuerung*]

feed(ing) cable, feeder cable Speisekabel, Zuführungskabel, Zuleitungskabel *n*

feed(ing) cistern → (overhead) feed(ing) cistern

feed(ing) main (line) → principal feeder

feed(ing) cock, feeder cock Speisehahn, Zuführungshahn, Zuleitungshahn *m*

feed(ing) current circuit, feeder current circuit Speisestromkreis, Zuführungsstromkreis, Zuleitungsstromkreis *m*

feed(ing) device, feeder Speisegerät, Zuführungsgerät, Zuleit(ungs)gerät *n*

feed(ing) duct, feeder duct Speisekanal, Zuführungskanal, Zuleitungskanal *m*

feed(ing) line, feeder line Speiseleitung, Zuführungsleitung, Zuleitung *f*

feed(ing) path, feeder path Schaltweg *m* [*Elektrotechnik*]

feed(ing) pipe, feeder pipe Speiserohr, Zuführungsrohr, Zuleitungsrohr *n*

feed(ing) valve, feeder valve Speiseventil, Zuführungsventil, Zuleitungsventil *n*

feed(ing) voltage, feeder voltage Speisespannung, Zuführungsspannung, Zuleitungsspannung *f* [*Die an den Speiseanschlüssen der Versorgung notwendige Spannung. V DIN 19 237*]

feed table Anlegetisch *m*

feed terminal Speiseklemme *f*

feedthrough Übersprechen *n*

feedthrough, interface connection [*Electricity. A conductor that connects patterns on both sides of a printed circuit board*] Schnittstellenanschluss *m*, Schnittstellenverbindung, Anschaltung *f*

feedthrough capacitor [*A feedthrough terminal that provides a desired value of capacitance between the feedthrough conductor and the metal chassis or panel*

F

through which the conductor is passing; used chiefly for bypass purposes in ultra-high-frequency purposes] Durchführungskondensator *m*

feedthrough terminal, feedthrough insulator Durchgang(s)klemme *f*

(feed) track [*In automatic data processing, the longitudinal channel on a paper that contains the feed holes*] Spur *f*

feedwater, FW Speisewasser *n*

feedwater chemistry Speisewasserchemie *f*

F

feedwater heater Speisewassererhitzer, Speisewasser(er)wärmer, Speisewasserheizer *m*

feedwater preheater Speisewasservor(er)wärmer *m*

feedwater preheating Speisewasservor(er)wärmen *n*, Speisewasservor(er)wärmung *f*

feedwater pump → donkey pump

feedwater preparation plant, feedwater preparation installation Speisewasseraufbereitungsanlage *f*

feedwater regulator, feedwater controller Speisewasserregler *m*

feedwater suction line Speisewassersaugleitung *f*

feedwater tank → (overhead) feed(ing) cistern

feel Verhalten *n* [*Benutzerfreundlichkeit von Bedien(ungs)oberflächen*]

feeler, sensor, sensing element Sensor, Fühler *m*

feeling of comfort, comfort(able) feeling Behaglichkeitsgefühl, Wohlgefühl *n*

fee schedule for architects and engineers → (statutory) fee schedule for architects and engineers

feet (bath)tub Fuß(bade)wanne *f*

feet per minute → air cloth ratio

fel(d)spathic porcelain trap, hard (paste) porcelain trap [*See remark under "trap"*] Hartporzellangeruchverschluss *m*

felt (filter) insert Filz(filter)einsatz *m*

felting material [*heat insulation*] Verfilzungsstoff *m*, Verfilzungsmaterial *n*

FELV → functional extra low voltage

female bather Badende *f*

female coupling piece Aufschraubverschraubung *f*

female disconnect(ing) device Abschalt(ungs)-buchse *f*

female elbow coupling piece Knieaufschraubverschraubung *f*

female (pipe) fitting thread → cylindrical (pipe) fitting thread

female thread Innengewinde *n*

female thread cross Kreuz(stück) mit Innengewinde, Innengewindekreuz(stück) *n*

fence sensor, fence detector [*See remark under "sensor"*] Zaunsensor, Zaunfühler, Zaundetektor *m*, Zaundetektionselement *n*

fermentation Gärung *f*

fermentation heat Gärungswärme *f*

Fermi level Ferminniveau *n* [*Zur Besetzungswahrscheinlichkeit 0,5 für Ladungsträger gehörendes Energieniveau*]

ferrels → iron furniture

ferrimagnetic material [*It displays ferrimagnetism; the ferrites are the principal example*] ferrimagnetische (Werk)Stoff *m*, ferrimagnetische Material *n*

ferritic, containing iron, iron-containing ferritisch, eisenhaltig, eisenenthaltend

ferroelectric Ferroelektrikum *n*

ferromagnetic crystal, polar crystal [*A crystal of a ferromagnetic material*] ferromagnetische Kristall *m*

FET → field effect transistor

fetch-ahead Speichervorgriff, Vorgriff *m* auf Speicher

fetch machine cycle → process cycle

to fetch (out), to read out abrufen, (heraus)holen, auslesen, anzeigen [*Eines oder mehrerer Bytes (Befehle oder Daten) aus dem Speicher (memory). Geschieht im Ablaufzyklus (fetch machine cycle)*]

FF → bistable multivibrator

FF → form feed

FFT, fast Fourier transformation Fourierdarstellung, Fouriertransformation *f* [*Sie erzeugt mit schnellen Algorithmen aus einem zeitlichen Signal das zugehörige Spektrum*]

FI → (glass) fibre channel

fibre composite; fiber composite (US) Faserverbundstoff, Verbundfaserstoff *m*

fibre distributed data interface; fiber distributed data interface (US); FDDI Hochgeschwindigkeitsnetzwerk *n* in Ringstruktur mit gegenläufigen Glasfaserdoppelleitungen

fibre optic cable; fiber optic cable (US) Lichtwellenleiterkabel, LWL-Kabel *n*

fibre optic detector, fibre optic detection device, fibre optic detecting device [*US spelling is "fiber"*] faseroptische Melder, faseroptische Meldungsgeber, faseroptische Detektor *m*, faseroptische Meldegerät *n*

fibre optic gyroscope; fiber optic gyroscope (US); FOG faseroptische Gyroskop *n*

fibre optics; fiber optics (US) [*The technique of transmitting light through long, thin, flexible fibres of glass, plastic, or other transparent materials; bundles of parallel fibres can be used to transmit complete images*] Lichtwellenleitung *f*

fibre pen element; fiber pen element (US) Faserschreibeinsatz, Faserschreiber *m*

fibre sink → (glass) fibre sink

fibre waveguide → optical-fibre cable

fibre (wiring) conduit; fiber (wiring) conduit (US) [*These conduits consist of solid tubes composed of fiber or other bibulous substance, impregnated and coated with asphalt or coal-tar pitch or wax*] Faser-(Elt)Leerrohr, Faser-Elektroleerrohr *n*

fibrous filter, fibre(-pad) filter; fiber(-pad) filter (US) [*It has a fibrous medium of glass, cellulose, plastics, metal or the like*] Faserfilter *m, n*

fidelity (Klang)Wiedergabetreue *f*

fiducial value, reference value Bezug(s)wert, Referenzwert *m* [*In der Messtechnik ist Bezugswert der unter den gegebenen Bedingungen vorliegende (prinzipiell unbekannte) wahre Wert der Messgröße oder der (bekannte) konventionell richtige Wert der Messgröße. (Siehe DIN 1319 Teil 3/08.83, Abschnitte 2, 3 und 8.) Die Differenz zwischen dem Messwert und einem Bezugswert ist die Messabweichung. DIN 1319, Teil 1*]

field Sektor *m* [*Im Sinne von z. B. Heizungssektor*]

field [*In automatic data processing, a specified area, such as a group of card columns or a set of bit locations in a computer word, used for a particular category of data*] Feld *n*

field Feld *n*, Teilbereich *m* [*selten: n*] [*Bildschirm*]

field alterable control element → FACE

field appliance(s) → field device(s)

field backup fuse Feldvorsicherung *f*

field bus Feldbus, Zubringerbus *m* [*Serielles Übertragungssystem, das eine Vielzahl von Geräten der Automatisierungstechnik miteinander vernetzt. Unter „Feldbus" fallen Prozessbus und Sensor-/Aktorbus*]

field bus communication, field bus transmission Feldbuskommunikation, Feldbusübertragung, Zubringerkommunikation, Zubringerübertragung *f*

field bus data link, field bus DL, FDL Feldbuskoppelabschnitt, Zubringerbuskoppelabschnitt *m*, Feldbuskoppelstrecke, Zubringerbuskoppelstrecke *f*

field bus equipment Feldbusgerät(e), Zubringerbusgerät(e) *n, (f)*

field bus level Feldbusebene, Zubringerbusebene *f*

Field bus Message Specification → FMS

field bus module Feldbusmodul, Feldbusbaustein, Zubringerbusbaustein, Zubringerbusmodul *m*

field bus station Feldbusstation, Zubringerbusstation *f*

field change detector, field change detecting device, field change detection device Feldänderungsmelder *m*, Feldänderungsmeldegerät *n*

field code [*assembler*] Feldschlüssel *m*

field connection diagram [*Wiring schematic for input/output equipment*] Feldanschlussplan *m*

field control, (job)site control Baustellensteuerung *f*

field controller Feldregler *m* [*Er ist durch Gehäuseaufbau und Korrosionsschutz zur Aufstellung in einer Freiluftanlage (im Feld) geeignet. DIN 19 225*]

field current Feldstrom *m*

field device(s), field appliance(s) Feldgerät(e) *n, (f)*, Feldvorrichtung(en) *f* [*Gerät(e), das auf der Feldebene, wie z. B. anwendungsspezifische Steuer- und Regeleinheiten, Beleuchtungseinrichtungen und Schalter, Fühler, Antriebe usw.*]

field effect diode Feld(effekt)diode, FE-Diode *f*

field effect transistor, FET Feld(effekt)transistor, FET *m*

field effect triode transistor → triode field effect transistor

field effect varistor Feld(effekt)varistor *m*

field equipment cable Feldgerätekabel *n*

field excitation Felderregung *f*

field frequency → vertical frequency

field (il)luminance → adaption (il)luminance

field input device Geber mit Eingang(s)signal, Eingang(s)signalgeber *m*

field intensity → field strength

field intensity value → field strength value

field killing coil, killing field coil Gegenfeldspule, Feldgegenspule *f*

field leakage reactance Streureaktanz *f* der Erregerwick(e)lung

field length [*The number of columns, characters, or bits in a specified field*] Feldlänge *f*

field level Feldebene *f* [*Sensorik; Aktorik. Die Feldebene ist das Bindeglied zwischen den gebäudetechnischen Anlagen und den Automationsstationen. Ihre Aufgaben sind Messen, Melden, Schalten und Stellen*]

field (level) equipment Feld(ebenen)gerät(e) *n, (f)* [*Weitergabe der Befehle der Automationsebene an die gebäudetechnischen Anlagen*]

field (level) network, F(L)N Feld(ebenen)netz-(werk), FEN *n*

field line [*IEC 50(101)*] Feldlinie *f*

field management, (job)site management Baustellenbetrieb *m*

field mark end → end-of-field mark

field mesh Feldnetz *n* [*Netzelektrode*]

F

field of application, field of use Anwendungsfeld *n*, Anwendungsbereich *m* [*selten: n*]

field of vision, vision field Sehbereich, Sichtbereich *m* [*selten: n*], Sehfeld, Sichtfeld *n*

field output device Geber mit Ausgang(s)signal, Ausgang(s)signalgeber *m*

(field-)programmable gate array, (F)PGA frei programmierbare Gatterfeld *n*

(field-)programmable logic array, (F)PLA [*The internal connections of the logic gates can be programmed once in the field by passing high current through fusible links, by using avalanche-induced migration to short base emitter junction at desired interconnections, or by other means*] frei programmierbare logische Feld, logische frei programmierbare Feld *n*

field quantity, field variable Feldgröße *f* [*Dem Quadrat der Leistung proportionale physikalische Größe (z. B. elektrische Spannung)*]

field strength, field intensity Feldstärke *f*

field wiring, (job)site wiring Verdrahtung am Einbauort, Baustellenverdrahtung, Verdrahtung auf der Baustelle *f* [*Gegensatz: factory wiring (Werkverdrahtung)*]

FIFO → queue

FIFO memory, first-in-first-out memory Durchlaufspeicher *m*

figure, number figure Ziffer *f*

figure of merit, merit figure Gütemaß *n* [*Logarithmierte Verhältnis des Antennengewinnfaktors (bezogen auf Isotopenstrahler) zur Rauschtemperatur des Empfängereingangs einer Satellitenempfangsanlage*]

filament Leuchtdraht *m*

filamentary conductor Filamentleiter, Vielkernleiter, Mehrkernleiter *m*

filament lamp → incandescent lamp

filament lamp lighting, incandescent lamp lighting Glühlampenbeleuchtung *f*

filament light → incandescent light

filament (light) carbon → incandescent (light) carbon

filament voltage Fadenspannung, Heizspannung *f* [*Elektronenstrahlröhrenkat(h)ode*]

file Datei *f*, Datenbestand *m*

file conveyor plant, file conveyor installation, file conveying plant, file conveying installation Aktenförderanlage *f*

file end → end of file

file end routine → end-of-file routine

file entry Dateieintrag *m*

file gap end → end-of-file gap

file indicator end → end-of-file mark

file lift (Brit.); file elevator (US) Aktenaufzug *m*

file list Dateiliste *f*

file management Dateiverwaltung *f*

file management system Dateiverwaltungssystem *n*

file mark end → end-of-file mark

file printout [*Output from a computer printer consisting of a copy of the contents of a file held in some storage device, usually to assist in debugging a program(me)*] Dateiausdruck *m*

file processing [*Updating, sorting, or validating a data file*] Dateiaufbereitung, Dateibearbeitung *f*

file protection Dateischutz *m*

file reference [*An operation involving looking up and retrieving the information on file for a specified item or items*] Dateibezug *m*

file server Netzwerkrechner *m*, Rechner in einem Netzwerk [*Er speichert zentrale Dateien für andere Netzwerkrechner*]

file shredder Aktenvernichter, Aktenzerkleinerer, Akten(zer)schnitzler *m*

file transfer and management, FTAM Datenverwaltung *f* für verteilte Systeme

file transfer protocol, FTP Dateienübertragungsprotokoll *n* (zwischen zwei Rechnern)

filing → archiving

filing cabinet, files cabinet, filing cubicle, files cubicle Aktenschrank *m*

filled with smoke, smoke-filled verqualmt

fillet (Ecken)Abrundung *f*

fill(ing) branch Füll(ungs)stutzen *m*

fill(ing) character → idle character

fill(ing) cock Füll(ungs)hahn *m*

fill(ing) concrete → (in)fill(ing) concrete

fill(ing) hose Füll(ungs)schlauch *m*

fill(ing) level Füll(ungs)stand *m*, Füll(ungs)höhe *f*, Füll(ungs)niveau *n*

fill(ing) level alarm Füll(ungs)alarm, Füll(ungs)-stand(s)alarm, Füll(ungs)niveaualarm, Füll(ungs)-höhealarm *m*

fill(ing) level setpoint setter Füll(ungs)standsoll-wertsteller *m*

fill(ing) loss Füll(ungs)verlust *m*

fill(ing) loss pressure switch Füll(ungs)verlust-druckschalter *m*

fill(ing) material Füll(ungs)körper *m* [*Wasserrück-kühlwerk*]

fill(ing) noise Füll(ungs)geräusch *n*

fill(ing) (pipe)line Füll(ungs)(rohr)leitung *f*

fill(ing) pump Füll(ungs)pumpe *f*

fill thermal insulation → (loose-)fill thermal insulation

film coefficient → heat-transfer coefficient

film cooling tower [*Cooling tower with a packing over which the water spreads in a thin film*] Film-kühl(ungs)turm *m*

film forming condensation [*Phenomenon encountered when the condensate wets a cold surface to form a continuous film that separates this surface from the vapo(u)r*] Filmkondensation *f*

film handling noise Filmtransportgeräusch *n* [*Videokamera*]

film rust Flugrost *m*

filter [*In automatic data processing, a device or program(me) separating data or signals in accordance with specified criteria*] Filter *m, n*

filter, compensator [*Control system. An operating device or element used in a feedback control system to achieve stability and thus improve system performance*] Filter *m, n*

filter aid Filterhilfsmittel *n*, Filterhilfe *f* [*Zur Begünstigung der Filtration dem Abwasserschlamm zugemischtes Material um eine Koagulation der Feststoffe und leichtere Abgabe des Schlammwassers zu erreichen*]

filter base Filterunterlage *f* [*Poröse Unterlage für das Filtern von festen Stoffen aus Gasen und Flüssigkeiten*]

filter basket Filterkorb *m*

filter candle Filterkerze *f*

filter cassette Filterkassette *f*

filter ceiling Filterdecke *f*

filter cartridge, filter element Filterpatrone *f*, Filtereinsatz *m*, Filterelement *n*

filter cell Filterzelle *f*

filter cloth Filtertuch *n*

filter collector Filterkammer *f*

filter corner valve Filtereckventil *n*

filter drier, filter dryer Filtertrockner *m* [*Bauteil im Kühlkreislauf zur Aufnahme von Verunreinigungen und Feuchtigkeit*]

filter effluent → filtrate

filter fan Filterventilator *m*

filter(ing) efficiency, filtration efficiency Filterungswirkungsgrad, Filtrierwirkungsgrad, Filtrationswirkungsgrad *m*

filter insert Filtereinsatz *m*

filter interceptor → filter separator

filter layer Filterschicht, Filterlage *f*

filter medium [*Portion of a filtrating system that provides the liquid-solid separation, such as close-woven textiles, metal screens, papers, nonwoven fabrics, granular beds, or porous media*] Filtermedium *n*

filter/mixing box → (combination) box

filter/mixing section Filter/Misch-Teil *m, n*

filter monitoring Filterüberwachung *f*

filter pad Filtervlies *n*

filter pollution Filterverschmutzung *f*

(filter) separation (Filter)Abscheidung *f*

(filter) separation efficiency (Filter)Abscheidegrad, (Filter)Abscheidungsgrad *m*

filter separator, filter interceptor, filter trap [*See remark under "interceptor"*] Mechanik(ab)scheider *m* [*Filter mit Maschen, Spalten, Poren usw. zur Zurückhaltung von Teilchen*]

filter service, filter operation Filterbetrieb *m*

filter sieve Filtersieb *n*

filter system Filtersystem *n*

filter trap → filter separator

filter unit Filtereinheit *f*

filtrate, filter effluent Filtrat, Filtrationswasser *n* [*Nach der Filterung ablaufendes Wasser*]

filtration efficiency → filter(ing) efficiency

fin → (attached) (metal) fin

fin spacing [*Distance between two successive transverse fins on a tube*] Flossenabstand *m*

final apparent drying rate → apparent final drying rate

F

F

final basic cleaning, basic final cleaning Bauschluss(grund)reinigung *f*

final bit Endbit, Schlussbit *n*

final character Endzeichen, Schlusszeichen *n* [*Siehe Anmerkung unter „Zeichen"*]

final check(ing) Endkontrolle, Schlusskontrolle *f*

final circuit → **(building) final circuit**

final compression temperature → **compression final temperature**

final consumer Endverbraucher *m*

final controlling device, final controlling appliance Stellgerät *n*

final drying rate Endtrocknungsgeschwindigkeit *f*

final energy Endenergie *f* [*Zustand von Energieträgern beim Endverbrauch*]

final pressure [*Pressure of a fluid leaving a process*] Enddruck, Abgang(s)druck, Schlussdruck *m*

final position signal message; final position signalling message (Brit.); final position signaling message (US) Endstellungsmeldung, Schlüsselstellungsmeldung *f* [*Siehe Anmerkung unter „Meldung"*]

final route Kennzahlweg *m*

final sensitivity, final sensitiveness, final susceptibility Endempfindlichkeit *f* [*Bei Messgeräten, deren Empfindlichkeit längs der Skala nicht konstant ist, wird zwischen Anfangsempfindlichkeit und Endempfindlichkeit unterschieden*]

final setpoint → **target setpoint**

final stage → **output stage**

final tank [*sewage disposal*] Nachklärbecken *n*

final temperature Endtemperatur *f*

final user Endbenutzer *m*

final value Endwert *m*

final value theorem Endwertsatz *m*

final yield Endertrag *m*

finance management Finanz(en)verwaltung *f*

fin coil Flossenschlange *f*

fin condenser [*See remark under "(attached) (metal) fin"*] Lamellenkondensator, Flossenkondensator, Lamellenverflüssiger, Flossenverflüssiger *m*

fine adjustment, fine control Fein(ein)stellung, Feinverstellung, Feinjustierung *f*

fine dust filter, fine dust cleaner Feinstaubfilter, Staubfeinfilter *n, m* [*DIN 24185*]

fine dust mask Feinstaubmaske *f*

fine fitting, small fitting Feinarmatur *f*

fine leak(age) test Fein(ab)dichtigkeitsprüfung *f*

fine range → **incremental range**

fine rasp → **grated file**

fine sludge Feinschlamm *m*

finetime/protime Personal- und Projektzeit *f*

fine tuning Feinabstimmung *f* [*In Sende- und Empfangsgeräten*]

fine (wire) fuse, fine (wire) fuze Fein(draht)sicherung *f*

finger plate Fingerschild, Schonschild *m* [*Unter einem Türschloss um Verschmutzung durch Angreifen zu verhindern*]

fingerprint Fingerabdruck *m*

fingerprint comparison Fingerabdruckvergleich *m*

fingerprinting Fingerabdrucknahme *f*

fingerprint reader Fingerabdruckleser *m*

fingerproof fingersicher

finish Oberflächengestaltung *f*

finish(ing) (builders') hardware, builders' finish(ing) hardware, architectural (builders') hardware, architectural iron mongery [*Hardware which is seen and therefore given a good finish, such as door furniture, clothes hooks, and so on, as opposed to rough hardware. "Garnishment", "furniture" and "fittings" may synonymously be used for "hardware"*] sichtbare (Bau)Beschläge, Architektur(bau)beschläge *f*

finish(ing) signal message Fertigmeldung *f* [*Eine Meldung über die Beendigung eines gesteuerten Ablaufes*]

finite endlich

finite impulse response, FIR endliche Impulsreaktion *f*

finite impulse response filter, FIR filter endliche Impulsreaktionfilter *m, n*

finned berippt, lamelliert [*Röhre*]

fin(ned) heater Heizelement *n* [*Konvektor*]

fin(ned) tube Flossenröhre, Rippenröhre, Lamellenröhre *f*

fin(ned) tube heat exchanger, fin(ned) tube heat exchange unit, fin(ned) tube heat interchanger Rippenröhrenwärme(aus)tauscher (im engeren Sinne), Flossenröhrenwärme(aus)tauscher *m* (im engeren Sinne)

fin(ned) tube (sectional) radiator [*See remark under "radiator"*] Rippenröhrenradiator, Rippenröhrengliederheiz(ungs)körper; Rippenröhrenheiz-körper *m* [*Fehlbenennung*] [*„Flossen..." und „Lamellen..." können synonym für „Rippen..." verwendet werden*]

FIR → **finite impulse response**

fire → **heating appliance**

fire → **(air) heater**

fire [*A process of combustion characterised by heat or smoke or flame or any combination of these*] Feuer *n*, Brand *m*

fire acclerator Brandbeschleuniger, Feuerbeschleuniger *m*

fire alarm Brandalarm, Feueralarm *m*

fire alarm control panel → **fire signal message control panel**

fire alarm headquarters → **(community) fire alarm headquarters**

fire alarm interface Brandalarmschnittstelle, Feueralarmschnittstelle *f*

fire alarm override Feueralarmübersteuerung, Brandalarmübersteuerung *f*

fire alarm (push)button, fire alarm press button Brandalarmdruckknopf, Feueralarmdruckknopf *m*

fire algorithm Brandalgorithmus, Feueralgorithmus *m*

fire barrier Brand(schutz)abschluss, Feuer(schutz)-abschluss *m*, Brand(schutz)sperre, Feuer(schutz)-sperre *f*

fire behaviour (Brit.); fire behavior (US) Brandverhalten, Feuerverhalten *n* [*DIN 4102*]

firebox → **locomotive furnace**

fire/burglary detection system Brand-/Einbruchmeldesystem, Feuer-/Einbruchmeldesystem *n*

fire cause Brandursache, Feuerursache *f*

fire code Brandordnung, Feuerordnung *f*

fire compartment → **fire lobby**

fire construction [*A type of construction that uses incombustible materials within a structure to maximize protection from fire*] bauliche Brandschutz *m*

fire control, fire fighting, fire suppression Brandbekämpfung, Feuerbekämpfung *f*

fire damage → **fire loss**

fire damaged → **damaged by fire**

fire damper [*An approved automatic, or self-closing, noncombustible barrier designed to prevent the passage of air, gas, smoke or fire through an opening, a duct or plenum chamber*] Brand(schutz)klappe, Feuer-(schutz)klappe *f*

fire defense; fire defence (Brit.) abwehrende Brandschutz, abwehrende Feuerschutz *m*

fire department (US); fire brigade (Brit.) (öffentliche) Feuerwehr *f*

fire department access (US); fire brigade access (Brit.) Feuerwehrzufahrt *f*

fire department key box (US); fire brigade key box (Brit.) Feuerwehrschlüsselkasten *m*

fire department routing card (US); fire brigade routing card (Brit.) Feuerwehrlaufkarte *f*

fire department vehicles access → **access for fire department vehicles**

fire detecting system, fire detection system, fire recognition system Brandmeldesystem, Feuermeldesystem, Brandnachweissystem, Feuernachweissystem *n*

fire detection, fire recognition Branderkennung, Feuererkennung *f*, Brandnachweis, Feuernachweis *m*

fire detection plant, fire detection installation, fire recognition plant, fire recognition installation Brandmeldeanlage, Feuermeldeanlage, Brandnachweisanlage, Feuernachweisanlage *f*

fire detection sensor technique Brandmeldesensortechnik, Feuermeldesensortechnik *f*

fire detector, fire detection device, fire detection appliance Brandmelder, Feuermelder *m*

fire development Brandentwick(e)lung, Feuerentwick(e)lung *f*

fire door Brand(schutz)tür, Feuer(schutz)tür *f*

fire drencher → **hand (fire) extinguisher**

fire engineering Brandtechnik, Feuertechnik *f*

fire exhibition, fire show Brandschutzausstellung, Feuerschutzausstellung *f*

fire-exit device → **panic exit device**

(fire) exposure hazard [*The risk of fire spreading from a building, structure or other property to adjacent separate ones, or to another part of the same building, structure or other property, by radiated heat across the intervening space*] Brandübergriffgefährdung, Feuerübergriffgefährdung *f*

(fire) extinguisher Brandlöscher, (Feuer)Löscher *m*

(fire) extinguisher bracket (Feuer)Löscherhalterung, Brandlöscherhalterung *f*

(fire) extinguisher cabinet Brandlöscherschrank, (Feuer)Löscherschrank *m*

(fire) extinguishing Brandlöschen, (Feuer)Löschen *n*

(fire) extinguishing agent Brandlöschmittel, (Feuer)Löschmittel *n*

(fire) extinguishing engineering Brandlöschtechnik, (Feuer)Löschtechnik *f* [*DIN 14 011. Die Lehre über die zweckmäßige Anwendung der Löschverfahren und Bedienung der Löschgeräte*]

F

(fire) extinguishing liquid Brandlöschflüssigkeit, (Feuer)Löschflüssigkeit *f*

(fire) extinguishing nozzle Brandlöschdüse, (Feuer)Löschdüse *f*

(fire) extinguishing powder Brandlöschpulver, (Feuer)Löschpulver *n*

(fire) extinguishing system Brandlöschsystem, (Feuer)Löschsystem *n*

(fire) extinguishing valve Brandlöschventil, (Feuer)Löschventil *n*

F

fire-fighting → fire control

fire-fighting equipment box Brandlöschgeräteschrank, (Feuer)Löschgeräteschrank *m*

fire-fighting foam [*A colloidal solution of small gas bubbles produced by chemical reaction or mechanical agitation and used to extinguish hydrocarbon fire*] Brandlöschschaum, (Feuer)Löschschaum *m*

fire-fighting hose Brandlöschschlauch, (Feuer-)Löschschlauch *m*

fire-fighting hose cabinet Brandlöschschlauchschrank, (Feuer)Löschschlauchschrank *m*

fire-fighting hose connection, fire-fighting hose outlet Brandlöschschlauchanschluss, (Feuer)Löschschlauchanschluss *m*

fire-fighting hose drying tower Brandlöschschlauchturm, (Feuer)Löschschlauchturm *m*

fire-fighting hose riser Brandlöschschlauchschacht, (Feuer)Löschschlauchschacht *m*

fire-fighting installation, fire-fighting plant Brandlöschanlage, (Feuer)Löschanlage *f*

fire-fighting line [*A system of pipes and equipment used exclusively to supply water for extinguishing fires*] Brandlöschleitung, (Feuer)Löschleitung *f*

fire-fighting nozzle Brandlöschdüse, (Feuer)Löschdüse *f*

fire-fighting path, fire-fighting route Angriff(s)weg *m* [*Brandbekämpfung*]

fire-fighting point → fire point

fire-fighting powder Brandlöschpulver, (Feuer-)Löschpulver *n*

fire-fighting pump Brandlöschpumpe, (Feuer-)Löschpumpe *f*

fire-fighting drill Feuerwehrübung, Brandbekämpfungsübung *f*

fire-fighting equipment box Brandlöschgeräteschrank, (Feuer)Löschgeräteschrank *m*

fire-fighting foam [*A colloidal of small gas bubbles produced by chemical reaction or mechanical agitation and used to extinguish hydrocarbon fire*] Brandlöschschaum, (Feuer)Löschschaum *m*

fire-fighting squad, fire-fighting team, fire-fighting gang, fire-fighting party (Brand)Löschmannschaft, Feuerlöschmannschaft *f*

fire-fighting wall hydrant (Brandlösch)Wandhydrant, (Feuer)Löschwandhydrant *m*

fire-fighting water → emergency water

fire-fighting water (pipe)line Brandlöschwasser(rohr)leitung, (Feuer)Löschwasser(rohr)leitung *f* [*DIN 14 462*]

fire-fighting water (pipe)line net(work) (Feuer-)Löschwasser(rohr)netz(werk), Brandlöschwasser(rohr)netz(werk) *n*

fire-fighting water supply Brandlöschwasserversorgung, (Feuer)Löschwasserversorgung *f*

fire-fighting water system Brandlöschwasser, (Feuer)Löschwasser *n*

fire grading → fire(-resistance) grading

fire grading period → fire(-resistance) grading period

fire grate Brandrost, Feuerrost *m*

fire ground → fire location

fire hazard Brandgefährdung, Feuergefährdung *f*

fire-heated feuerbeheizt

fire hose rack (Feuer)Löschschlauchgestell, Brandlöschschlauchgestell *n*

(fire)hydrant, fire plug, water-post [*A device, by means of which water may be discharged from a pipe for extinguishing fire, provided with a valve for controlling the flow and a coupling for the attachment of a standpipe or hosepipe*] Brandlöschhydrant, (Feuer-)Löschhydrant *m*, Hydrant, (Feuer)Löschwasserständer *m* [*DIN 14 244*]

fire investigation Brandermittlung *f*

fire investigator Brandermittler *m*

fire ladder Feuerleiter, Brandleiter *f*

fire load(ing) Brandlast, Brandbelastung, Feuerlast, Feuerbelastung *f* [*Das auf die Grundfläche von Räumen oder Bauwerken bezogene Gesamtwärmepotential (die bei der Verbrennung frei werdende Wärmeenergie je Flächeneinheit). Die Brandlast ist von der Masse der vorhandenen brennbaren Stoffe und deren Heizwert abhängig. Sie ist für jeden Brandabschnitt bzw. Raum gesondert zu berechnen. Die Brandlast ist Bezugsgröße zur Ermittlung der Feuerwiderstandsklasse. DIN 18 230*]

fire lobby, fire compartment [*A space within a building designed to delay the passage or facilitate the control of fire*] Brandabschnitt, Feuerabschnitt *m* [*DIN 14 011*]

fire location, fire ground Brandort, Feuerort *m*

fire loss, fire damage Brandschaden, Brandverlust, Feuerschaden, Feuerverlust *m*

fireman → stoker

fireman's lift, firemen's lift (Brit.); fireman's elevator, firemen's elevator (US) Feuerwehraufzug *m*

fireman('s) switch, firemen('s) switch Feuerwehrschalter *m*

fire origin Brandherd, Feuerherd *m*

fire period Branddauer, Feuerdauer *f* [*DIN 18 230*]

fire pipe → flue (tube)

fire(place), fireside, (grate) chimney [*An open place built in wall, at the base of a chimney. Also called "hearth" in the sense of as the centre of family life*] (offene) Kamin *m*

fire plug → (fire) hydrant

fire point, fire-fighting point Brandlöschposten, (Feuer)Löschposten *m*

fire point Brennpunkt *m* [*brennbare Material*]

fire prevention Brandverhütung, Feuerverhütung *f*

fire product Brandschutzerzeugnis, Brandschutzprodukt, Feuerschutzerzeugnis, Feuerschutzprodukt *n*

fireproof → fire safe

fireproof damper, fire protection damper Brandschutzklappe, Feuerschutzklappe *f*

fireproof disk valve, fireproof disc valve, fire protection disk valve, fire protection disc valve Brandschutztellerventil, Feuerschutztellerventil, Brandschutzscheibenventil, Feuerschutzscheibenventil *n*

fireproof door, fire protection door Brandschutztür, Feuerschutztür *f*

fireproof duct, fire protection duct Brandschutzkanal, Feuerschutzkanal *m*

fireproof equipment, fire protection equipment Brandschutzgerät(e), Feuerschutzgerät(e) *n, (f)*

fireproof gate, fire protection gate Brandschutztor, Feuerschutztor *n*

fireproof glazing, fire protection glazing Brandschutzverglasung, Feuerschutzverglasung *f*

fireproof seal(ing), fire protection seal(ing) Abschottung *f*, Abschotten *n*

fireproof slab, fire protection slab Brandschutzplatte, Feuerschutzplatte *f*

fire propagation → fire spread(ing)

fire protection [*Measures and practices for preventing fire or for minimizing the probable loss of life or property resulting from a fire, by proper design and construction of buildings, by the use of detection and extinguishing systems, by the establishment of adequate fire fighting services and by the training of building occupants in the fire safety and evacuation procedures*] Brandschutz, Feuerschutz *m* [*DIN 4102*]

fire protection by technical systems anlagentechnische Feuerschutz, anlagentechnische Brandschutz *m* [*Durch technische Anlagen realisierbarer Brandschutz. Gesamtheit aller Maßnahmen, die durch Nutzung technischer Anlagen und Mittel präventiv (z. B. Brandmeldung) und operativ (z.B. Brandlöschung, Begrenzung und/oder Verminderung der Brandausbreitung) wirken*]

fire protection slab → fireproof slab

fire pump Löschwasserpumpe *f*

fire-raising → arson

fire range Brandbereich, Feuerbereich *m* (selten: *n*)

to fire rate brandklassieren, feuerklassieren

fire rated brandklassiert, feuerklassiert

fire recognition, fire detection Branderkennung, Feuererkennung *f*, Brandnachweis, Feuernachweis *m*

fire-related brandbezüglich, feuerbezüglich, brandbezogen, feuerbezogen

fire(-resistance) grading [*This concept has two-fold application. (a) It is applied to the classification or grading of the elements of structure of buildings in terms of their degree of resistance to fire. (b) With a broader meaning, it is applied to the classification of buildings according to the purpose for which they are used, that is, according to occupancy, and according to the fire resistance of the elements of which they are constructed.*] Brandklasseneinteilung, Brandklassierung, Feuerklasseneinteilung, Feuerklassierung *f*

fire(-resistance) grading period Brandklasse, Feuerklasse *f*

fire resistant, fire-resisting brandbeständig, feuerbeständig, brandsicher, feuersicher

fire-resistant compartmentation, (fire-resisting) compartmentation [*In the concept of fire-resistant compartmentation, a building is pictured of consisting of compartments all perfectly insulated from one another and the spread of fire as taking place by the successive failure of compartment boundaries. If the compartment boundaries are sufficiently fire resistant, it is argued, failure will not occur and fires will be confined to the compartment in which they start*] Brandabschnitteinteilung, Feuerabschnitteinteilung *f*

fire(-resisting) shutter, fire-resistant shutter [*A shutter which, together with its frame, is capable of satisfying for a stated period of time the criteria of fire resistance with respect to collapse and flame penetration*] Brandschutzabschluss, Feuerschutzabschluss *m* [*DIN 4102, Teil 5*]

F

fire-retardant brandhemmend, feuerhemmend

fire ring Brenner(rohr)konus *m*

fire riser → **fire standpipe**

fire safe, fireproof brandsicher, feuersicher

fire safety Brandsicherheit, Feuersicherheit *f*

fire safety engineering Brandsicherheitstechnik, Feuersicherheitstechnik *f*

fire security Brandsicherung, Feuersicherung *f*

fire setter → **arsonist**

F

fire severity Brandheftigkeit, Feuerheftigkeit *f*

fire show → **fire exhibition**

fireside → **fire(place)**

fire signal message Brandmeldung, Feuermeldung *f*

fire signal message control panel, fire alarm control panel Brandmelde(r)zentrale, Brandalarmzentrale, Feuermelde(r)zentrale, Feueralarmzentrale *f*

fire signal message system Brandmelde(r)system, BMS, Brandalarmsystem, Feuermelde(r)system, Feueralarmsystem *n*

fire smoke Brandrauch, Feuerrauch *m*

fire spread(ing), fire propagation Brandausbreitung, Brandfortpflanzung, Feuerausbreitung, Feuerfortpflanzung *f*, Brandübergriff, Feuerübergriff *m*, Brandübergreifen, Feuerübergreifen *n*,

(fire) sprinkler system Sprinkleranlage, Sprengeranlage *f*

(fire) sprinkler wet system → **wet(-pipe) (fire) sprinkler system**

fire standpipe, fire riser Brandlöschwasserstandrohr, (Feuer)Löschwasserstandrohr *n*

fire stress Brandbeanspruchung, Feuerbeanspruchung *f*

fire suppression → **fire control**

fire tank (Feuer)Löschwasserbehälter, Brandlöschwasserbehälter *m*

fire-tinned feuerverzinnt

fire test(ing) lab(oratory) Brandversuchsanstalt, Feuerversuchsanstalt, Brandprüf(ungs)anstalt, Feuerprüf(ungs)anstalt *f*

fire test(ing) room Brandversuchsraum, Brandprüfungsraum, Feuerversuchsraum, Feuerprüfungsraum *m*

fire trace Brandspur *f*

fire tube → **flue (tube)**

fire tube ferrule → **smoke tube ferrule**

fire tube (steam) boiler, smoke tube (steam) boiler, multi(ple)-tubular (steam) boiler, furnace flue (steam) boiler Heizrohrkessel, Feuerrohrkessel, Flammrohrkessel, Rauchrohrkessel *m*

fire type Brandart, Feuerart *f*

fire ventilator [*It opens automatically at a predetermined temperature, releasing smoke, heat and fumes, keeping the fire localized*] Lüfter *m* zum Absaugen von Rauchgasen und Dämpfen

fire venting [*The process of inducing heat and smoke to leave a building as quickly as possible*] Brandlüftung, Feuerlüftung *f*

fire vulnerable, vulnerable to fire brandanfällig, feueranfällig

fire wall Brandschutzwand, Feuerschutzwand *f*

firewall Rechnerschutzmechanismus *m* [*Er schützt die Rechner im Internet vor unberechtigtem Zugriff*]

fire warning Brandwarnung, Feuerwarnung *f*

"firewire", IEEE 1394 serielle Bus *m* [*Entwickelt von Apple, Texas Instruments und dem Standardisierungsgremium IEEE*]

fire zone Brandzone, Feuerzone *f*

FIR filter → **finite impulse response filter**

firing Feuerung *f*, Feuern *n*, Brand, Brenner *m* [*Ofen*]

firing controller Feuerungsregler *m*

firing device, firing appliance Feuerungsgerät *n*, Feuerungsvorrichtung *f*

firing equipment [*e.g. burner, stoker, etc.*] Brennausrüstung *f* [*Heiz(ungs)kessel*]

firing oil Brennöl *n*

firing pulse Zündimpuls *m*

firing pulse suppression Zündimpulsunterdrückung *f*

firmware [*Data and computer programs stored in a nonvolatile form such as hardwired or in a ROM/EPROM*] Firmware *f*

first... → **primary...**

first air [*In a clean room, air that issues directly from the HEPA filter before it passes over any work location. Not to be confused with "primary air"*] Reinraumluft *f*

first circuital law, Ampère(´s) law Durchflutungsgesetz *n*

first-class → **high-grade**

first-class steel → **high-grade steel**

first cost Anschaffungskosten *f*

first compact disk → **master**

first detector → **mixer**

first-in-first-out memory → **FIFO memory**

first of maxima defuzzification method, FOM defuzzification method Links-Max-Methode, Defussifizierungsmethode *f*, Links-Max-Verfahren, Defussifizierungsverfahren *n*

first order erste Ordnung, 1. Ordnung *f*

first order controlled system [*IEC 50(351)*], **first order directly controlled member** [*IEC 27-2A*], **first order process** Regelstrecke 1. Ordnung, Regelstrecke erster Ordnung *f*

first order delay element Verzögerungselement 1. Ordnung, Verzugselement 1. Ordnung *n*

first order hold element, hold first order element Halteglied 1. Ordnung, Halteglied erster Ordnung *n*

first order time delay element, first order lag element P-T_1-Glied, Verzögerungsglied erster Ordnung, Verzögerungsglied 1. Ordnung *n* [*E DIN 19 226 Teil 2. Das Übertragungsverhalten lautet: v(t) + Tv(t) = K u(t). T Zeitkonstante, K Beiwert.*]

first resonator → **buncher (resonator)**

first-up detection Erstwertmeldung, Erstwerterkennung *f*, Erstwertnachweis *m*

FIS → **fuzzy inference system**

fish fryer Fischbratpfanne *f*

fishing wire, shake Durchziehdraht, Durchzugsdraht *m* [*für Leiter*]

fishplate Lasche *f*

fission chamber Spaltkammer *f*

to fit passen

fit Passung, Einpassung *f*, Einpass, Pass *m*

fitment [*A decorative or functional item or component in a building which is fixed but not built in*] Einrichtung(sgegenstand) *f, (m)*

fitter → **installer**

fitter → **(workshop) fitter**

fitter, assembler Monteur, Schlosser *m* [*Nicht im Sinne eines Schlossers für Schlüssel und Schlösser*]

fitting [*System part used to join, adjust or adapt other parts as in piping or duct system*] Fitting *m*, Passstück *n*, Armatur *f*, Formteil *m, n* [*„Fitting" wird allgemein für „Armatur" verwendet, obwohl es ein „Passstück" ist*]

fitting → **pipe fitting**

five-conductor connection Fünf-Leiter-Anschluss *m*, Fünf-Leiter-Verbindung *f*

five days biochemical oxygen demand, five days B.O.D. [*This is found by determining the quantity of oxygen when sewage or sewage effluent is diluted with water containing oxygen and incubated at a constant temperature for five days*] fünftägige biochemische Sauerstoffbedarf *m*

five-electrode electron tube → **pentode**

fixation hole, fastening hole Befestigungsloch *n*

fixed automatic firefighting system → **automatic stationary firefighting system**

fixed azimuth tape guid(anc)e, FATG Azimutkontrollstifte *f* [*Bandführung in der digitalen Kompaktkassette*]

fixed bed filter Festbettfilter *m, n*

fixed circuit breaker [*CEE 19*] fest eingebaute Selbstschalter, fest angeschlossene Selbstschalter *m*

fixed command ((closed) loop) control → **fixed setpoint ((closed) loop) control**

fixed command controller → **fixed setpoint controller**

fixed connection fest geschaltete Verbindung *f*, fest geschaltete Anschluss *m*

fixed connector Gehäusestecker, feste Steckverbinder *m*

fixed cycle feste Zyklus *m* [*Festgespeichertes Unterprogramm für die Steuerung oder Reg(e)lung bestimmter Operationen*]

fixed (data) block [*Pertaining to an arrangement of data in which all the data blocks have the same number of words or characters, as determined by either the hardware equipment of the computer or the programmer*] feste (Daten)Satz *m*

fixed delay [*Contact element; IE (337-2B)*] nicht einstellbare Verzögerung *f* [*Schaltglied; VDE 0660, Teil 203*]

fixed disk, fixed disc Festplatte *f* [*Mit dem Laufwerk verbundene Magnetplatte. Sie kann nicht ausgewechselt werden*]

fixed disk drive, fixed disc drive Festplatten(an)trieb *m*

fixed disk memory, fixed disc memory Festplattenspeicher *m*

fixed equipment circuit [*This is a type of circuit in domestic installations. Apparatus with a rating of 3 kilowatts and above should be fused independently and requires separate circuits. This range includes immersion heaters, cookers and fixed electric fires*] Starkstromleitung, Kraftstromleitung *f* [*DIN 18 382*]

fixed field [*A field in computers, film selection devices, or punched cards, or a given number of holes along the edge of a marginal punched card, set aside, or "fixed", for the recording of a given type of characteristic*] ruhende Feld, stationäre Feld *n*

F

F

fixed fire (grate) bar, stationary fire (grate) bar feste (Feuer)Roststab *m*

fixed length record [*One of a file of records, each of which must have the same specified number of data units, such as blocks, words, characters, or digits*] Satz fester Länge, Festlängesatz *m*

fixed limit value feste Grenzwert *m*

fixed louver (US); fixed louvre (Brit.); fixed luffer [*A louver in which the blades do not move*] feststehende Luftraster, feste Luftraster *m*

fixed master method → (master-slave) polling

fixed memory → read-only memory

fixed memory chip → read-only memory chip

fixed nozzle Festdüse *f*

fixed outlet [*As opposed to a swivel(ling) outlet*] fest(stehend)e Auslauf, fest(stehend)e Ablauf, fest(stehend)e Auslass, fest(stehend)e Ablass *m*

fixed point Festpunkt *m*

fixed program computer (US); fixed programme computer [*It has a program(me) permanently wired-in*] Festprogrammrechner *m*

fixed-programmed ((open) loop) control festprogrammierte Steuerung *f* [*Verbindungsprogrammierte Steuerung, bei der Programmänderungen nicht vorgesehen sind. V DIN 19 237*]

fixed sequential (format) konstante Satzfolge *f*, feste Satzformat *n* [*Programmschreibweise, bei der Informationen in ihrer Reihenfolge festgelegt sind und die Bedeutung eines Wortes sich aus seiner Stellung im Satz ergibt. Alle möglichen Wörter vor dem letzten gewünschten Wort müssen vorhanden sein*]

fixed setpoint Festwert *m*

fixed setpoint ((closed) loop) control, fixed setpoint regulation, set value ((closed) loop) control, set value regulation, fixed command ((closed) loop) control, fixed command regulation Festwertreg(e)lung, Festwertregulierung *f*

fixed setpoint controller, set value controller, fixed command controller Festwertregler *m* [*Die Führungsgröße ist fest eingestellt bzw. innerhalb des Führungsbereiches einstellbar. DIN 19 225*]

fixed setpoint ((open) loop) control, fixed command ((open) loop) control, set value ((open) loop) control Festwertsteuerung *f* [*Automatische Steuerung, die die Werte der gesteuerten Größe konstanten Führungsgrößen angleicht*]

fixed setting Festeinstellung *f*

fixed (spray) shower feste Brause *f*

fixed storage [*A data storage not alterable by computer instructions, such as magnetic-core storage with a lockout feature*] Festspeicherung *f*

fixed-temperature method [*heat detection*] Festtemperaturverfahren *n*, Festtemperaturmethode *f*

fixed value Festwert *m*

fixed value capacitor Festwertkondensator *m*

fixed value ((closed) loop) control Festwertregelung, Festwertregelung *f*

fixed-wired → hard-wired

fixed word length [*The length of a computer machine word that always contains the same number of characters or digits*] Festwortlänge *f*

fixed zero Festnullpunkt *m*

fixed zero system Bezugsmesssystem *n* mit festem Nullpunkt *n* [*Die Koordinatenmaße werden auf einen Festnullpunkt bezogen*]

fixing, fastening Befestigung *f*, Befestigen *n*

fixing angle, fastening angle Befestigungswinkel *m*

fixing base, fastening base Befestigungsuntergrund *m*

fixing device, fastening device Befestigungsvorrichtung *f*

fixing hardware, fastening hardware [*e.g. clips, angles, inserts, bolts, etc.*] Befestigungszeug *n*

fixing system, fastening system Befestigungssystem *n*

fixture clearance [*Distance between fixtures; distance from a fixture to an obstruction; distance between a fixture and a wall*] Einbauteilabstand *m*

fixture, object affixed to a structure [*Any item which was once tangible personal property, but which by virtue of its affixation to real property is deemed to be permanently merged into it*] Einbauteil *m, n*

fixer (mason); (builder) mason [*Scotland and Cornwall*] [*A mason who sets prepared stones in walls, whether the stone be only facing or to the full wall thickness*] Maurer *m*

fixture trap → trap

FL → fluorescent lamp

flag → marker

flag bit, marker bit Merkerbit *m*

flag location, marker location Merkerzelle *f*

flake ice maker Flockeneismaschine, Eisflockenmaschine *f*

flame aperture → burner port

flame arc lamp Effektbogenlampe, Beck-Bogenlampe *f*

flame barrier Flammensperre *f*

flame detector, flame detecting device, flame detection device [*Detector that instantly senses infrared radiation emanating from flames, intended for areas of fast developing fires where ignition is almost instantaneous, high-ceiling areas or high-air-movement areas. To prevent false alarm, the unit is sensitive only to flame flickering sustained 3, 10 or 30 seconds, depending on the detector selected.*] Flammenmelder *m* [*DIN EN 54-10 (1991.09)*]

flame failure → lift-off

flame ignition Flammenzündung *f*

flame impingement [*Condition existing when the flame resulting from the combustion of the fuel comes into contact with any interior surface of the furnace*] Flammenaufprall *m*

flame lift-off → lift-off

flame mirror Flammen(beobachtungs)spiegel *m*

flame monitoring Flammenüberwachung *f*

flame monitoring device, flame monitoring appliance Flammenüberwachungsgerät *n*

flame monitoring wiring Flammenüberwachungsverdrahtung *f*

flame-on response time Anmeldezeit *f*

flame orifice → burner port

flame outage → lift-off

flame-out response time [*fire*] Abmeldezeit *f*

flame picture Flammenbild *n*

flame port → burner port

flameproof joint gap → gap of flameproof joint

flameproof treatment Flammenschutzbehandlung *f*

flame radiation Flammenstrahlung *f*

flame safeguard control [*System for sensing the presence or absence of flame and for indicating, alarming, or initiating control action*] Flammensicherheit(s)-vorrichtung *f*

flame sensor Flammensensor *m*

flame separation Flammenabheben *n*

flame sight hole, flame observation hole Flammenschauloch, Flammenbeobachtungsloch, Flammensichtloch *n*

flame signal Flammensignal *n*

flame soldering Flammenlöten *n*

flame spectrum [*An emission spectrum obtained by evaporating substances in a nonluminous flame*] Flammenspektrum *n*

flame speed [*The rate at which combustion moves through an explosive mix(ture)*] Flammengeschwindigkeit, Geschwindigkeit der Flammenfront *f*

flame spread, flame propagation Flammenübergriff *m*, Flammenausbreitung, Flammenfortpflanzung *f*

flame trap, (flame) flashback arrester (Flammen-) Rückschlagsicherung *f*

flammable refrigerant [*It is ignitable when mixed with air, e.g. ethyl chlorid(e), methyl chlorid(e), and hydrocarbon*] brennbare Kältemittel, entflammbare Kältemittel, feuergefährliche Kältemittel *n*

flanged geflanscht

flange(d) ball cock Flansch(en)kugelhahn *m*

flange(d) branch Flansch(en)stutzen *m*

flange(d) casing, flange(d) enclosure, flange(d) case, flange(d) housing Flansch(en)gehäuse *n* [*Siehe Anmerkung unter „Gehäuse"*]

flange(d) connection Flansch(en)anschluss *m*, Flansch(en)verbindung *f*

flange(d) motor Flansch(en)motor *m*

flange(d) parallel slide stop valve; flange(d) gate valve (US) Flansch(en)absperrschieber *m*

flange(d) pipe Flansch(en)rohr *n*

flange(d) return bend Flansch(en)doppelkrümmer *m*

flange(d) shutoff valve, flange(d) stop valve Flansch(en)absperrventil *n*

flange(d) taper(ed) pipe, flange(d) diminishing pipe, flange(d) reducer Flansch(en)-Übergang(s)-stück, Flansch(en)-Reduzierstück *n*

flange(d) three-way mix(ing) valve, flange(d) triple-way mix(ing) valve Dreiwege-Flansch(en)-Mischschieber, Flansch(en)-Dreiwege-Mischschieber *m*

flange plate Flanschblech *n*

flanger Phasenverschiebungseffekt *m*, Phasenverschiebungswirkung *f* [*Audio*]

flank clearance → backlash

flank(ing) sound Flankenübertragungsschall *m*

flap [*A flat, broad piece that is attached along one side to a larger body*] Klappe *f*

flank(ing) transmission (of sound) [*The transmission of sound from one room to an adjacent room, via common walls, floors or ceilings, flanking a partition between the rooms, when airborne sound is generated in the first room. Note. The term may also be used to describe any indirect path of sound transmission*] Flankenübertragung, Schalllängsleitung *f*

flap cover Klappdeckel *m*

flapper valve, cantilever valve [*Reel valve in which the flexible metal strip is attached at only one of its ends*] Klapp(en)ventil *n*

flap valve Drosselklappe, Ungezieferklappe *f* [*Rückschlagklappe an Wasseraustritten zur Vermeidung des Eindringens von Ungeziefer. Bei ins Grundwasser reichenden Becken auch als Auftriebssicherung verwendet weil sie sich bei leerem Boden auch dem Grundwasserdruck selbsttätig öffnet*]

flap valve [*A check valve with a hinged disk which opens when the flow is normal and is closed by gravity or by the flow when it tends to go backwards*] Rückschlagklappe, Armatur in Klappenform *f*

to flash blinken, aufleuchten

F

to flash aufflammen [*Feuer(ungs)technik*]

flashback, flame flashback Flammenrückschlag *m*

flashback arrester → flame trap

flash converter A/D-Umsetzer mit parallelen Komparatoren, parallele A/D-Umsetzer, Parallelumsetzer, Simultanumsetzer, A/D- Wandler mit parallelen Komparatoren, parallele A/D-Wandler, Parallelwandler, Simultanwandler *m*

flasher Blinker *m*

flasher lamp → flash lamp

flasher switch [*A switch, generally either motor-driven or using a combination heater element or bimetallic strip, that turns lamps on and off rapidly*] Blinkerschalter, Blinklichtschalter *m*

flash flue, flue without projections glatte Feuerzug *m*

flash gas [*Gas generated whenever pressure is reduced on a liquid held at boiling temperature*] entspannte Gas *n*

flashing blinkend, aufleuchtend

flashing Blinken, Aufleuchten *n*

flashing Verwahrung *f* [*Schornstein*]

flash(ing) frequency, flash(ing) rate, blinking frequency, blinking rate Blinkfrequenz *f*, Blinktakt *m*

flash(ing) lamp → flash lamp

flash(ing) light, strobe light Blitzlicht *n*

flash(ing) module Blinkmodul *n*

flash(ing) out Klarbrennen *n* [*luftleere Lampe*]

flash(ing)-over [*In rotating machinery, an unwanted arc that forms as a result of faulty insulation between commutator segments*] Lichtbogenüberschlag *m*

flash(ing) pulse Blinktakt *m*

flash(ing) tube Blinkröhre *f*

flash(ing) voltage Hochschaltspannung *f*

flash/interval ratio Blink-Pausen-Verhältnis *n*

flash lamp, stroboscopic lamp, flashing lamp, flasher lamp [*A gaseous discharge lamp used in a photoflash unit to produce flashes of light or short duration and high intensity for stroboscopic photography*] Stroboskoplampe, Blitzlampe *f*

flashlight, pocket lamp [*A portable, battery powered device that produces a beam of light*] Taschenlampe *f*

flashometer Lichtblitzanalysator *m*

flashover explosionsartige Verbrennung *f* von Rauchgas

flashover, spark discharge [*A disrupture, often luminous charge around or over the surface of an insulator, usually due to high voltage*] Überschlag, Übersprung *m*

flashover voltage, spark over voltage [*The voltage needed to cause a sudden discharge of elec-trical energy between electrodes or conductors that are separated by air or insulating material; frequently accompanied by light*] Überschlagspannung, Übersprungspannung *f*

flash point [*The minimum temperature of a combustible material at which their is sufficient vapo(u)rization to produce a combustible mix(ture) with air if ignited by a flame*] Flammpunkt, Entflammungspunkt *m*

flash steam entspannte Dampf *m*

flash storage → erasable storage

flash tank direkt gekühlte Zwischenkühler *m*

flash wheel Wurfrad *n* [*Fortschaffen von Wasser durch Werfen*]

flat Mietwohnung *f*

flat → level

flatbed plotter Flachbettzeichenmaschine, Tischzeichenmaschine *f* [*Stiftzeichenmaschine, bei der das Papier auf einen flachen Tisch aufgespannt ist*]

flatbed scanner Flachbettabtaster, Tischabtaster *m*

flat (blade) terminal Flachklemme *f*

flat bolt Flachriegel, platte Riegel, flache Riegel, Plattriegel *m*

flat bolt-nap, flat strike (plate), flat striking plate [*See remark under "bolt-nap"*] Flachschließblech *n*, Flachschließkappe *f*

flat-bottom(ed) flush(ing) cistern Flachbodenspülkasten *m*

flat-bottom(ed) toilet, flat-bottom(ed) closet Flachspültoilette, Ausspültoilette *f*, Flachspülklosett, Ausspülklosett *n*, Flachspülabtritt, Flachspülabort, Ausspülabtritt, Ausspülabort *m*

flat burner Flachbrenner *m*

flat conductor Flach(strom)leiter *m*

flat connector, flat-pin plug Flachsteck(verbind)er *m*

flat connector sleeve Flachsteck(verbind)erhülse *f*

flat convector Flachkonvektor *m*

flat cutting, flatting, ripping, ripsawing [*The resawing of timber parallel to one edge. British Standard 565*] Aufteilen *n*

flat duct Flachkanal *m* [*Lüftung*]

flat evaporator Flachverdampfer *m*

flat flame → laminar flame

flat flame burner, laminar flame burner Flachflammenbrenner, Flammenflachbrenner *m*

flat lath flache Latte, Flachlatte *f* [*eiserne Rolljalousie*]

flat (lock) striking plate, flat (lock) strike plate, flat (lock) strike(r) [*See remark under "(lock) striking plate"*] Flachschließblech, Lappenschließblech *n*

flatness [*Not to be confused with "levelness"*] Flachheit *f*

flat pack(ing) ring, flat sealing ring Flachdicht(ungs)ring *m* [*für lose Flansche*]

flat pack(ing) thyristor Scheibenthyristor *m*

flat pan WC, flat pan water closet Flachspülklosett *n* [*Bei diesem Klosett mit flacher Schüssel und geringem Wasserstand bleiben die Fäkalien sichtbar bis zum Spülen in der Schüssel liegen*]

flat-pin plug → flat connector

flat pin terminal Flach(stift)anschlussklemme *f*

flat plate heat exchanger Plattenwärme(aus)tauscher *m*

flat plate solar collector [*A nonconcentrating solar collector in which the absorbing surface is planar and approximately equal in area to the aperture*] Plattensonnenenergiekollektor *m*

flat rate Pauschaltarif *m*

flat roof breakthrough Flachdachdurchführung *f*

flat roof drainage Flachdachentwässerung *f*

flat roof inlet Flachdacheinlauf *m*

flat roof with air circulation → ventilation flat roof

flat screen Flach(bild)schirm *m*

flat sealing ring → flat pack(ing) ring

flat (sectional) radiator Flachradiator, Flachgliederheiz(ungs)körper *m*

flat slide valve Flachschieber *m*

flat shading Facettieren *n* [*Es verleiht jeder Fläche eines dreidimensionalen Objektes eine Farbe, daher wirkt das Objekt wie facettiert*]

flats lift (Brit.); flats elevator (US) Wohnhausaufzug *m*

flat square tube flache quadratische Bildröhre, quadratische flache Bildröhre *f*

flat steel Flachstahl, Bandstahl *m*

flat step [*escalator*] Horizontalstufe, Waag(e)rechtstufe *f*

flat strike (plate) → flat bolt-nap

flat(tened) (door) knob (Tür)Flachknopf *m*

flat terminal Flachklemme *f*

flatting → flat cutting

flat top Tisch *m* [*Teil eines großen Waschbeckens (= Waschtisch)*]

flat top response → bandpass response

(flat) twin engine Boxermotor *m*

flat web(bed) building wire → ribbon (type) web(bed) building wire

fleece Vlies, Vließ *n* [*Österreich*]

fleeting Wischen *n*

fleeting contact, momentary contact Wischkontakt, Wischer *m*

fleeting contact act(uat)or, momentary contact act(uat)or Taster *m* [*als Betätigungsglied*]

fleeting contact circuit, momentary contact circuit Wischerschaltung, Wischkontaktschaltung *f*

fleeting contact limit(ing) switch, momentary contact limit(ing) switch Grenztaster *m*

fleeting contact on energization, momentary contact on energization Einwischer, Einwischkontakt *m*

fleeting contact on de-energization → fleeting NC contact

fleeting contact time, contact fleeting time, momentary contact time, contact momentary time (Ein)Wisch(kontakt)zeit, Wischer(kontakt)zeit *f*

fleeting indication Kurzzeitangabe, (Ein)Wischerangabe *f*

fleeting NC contact, fleeting contact on de-energization, momentary contact on de-energization Kurzausschaltglied *n*, Ausschaltwischer, Wisch(er)ausschalter, Ausschaltwischkontakt *m*

fleeting NO contact Kurzeinschaltglied *n*, Einschaltwischer, Wisch(er)einschalter *m* [*DIN 40 713*]

fleeting pulse (Ein)Wischimpuls *m*

Fletcher-Munson contour → equal loudness contour

flexibility, bending property Biegsamkeit, Biegeeigenschaft, Biegefähigkeit *f*, Biegevermögen *n*

flexibility Flexibilität, Aus- und Umbaufähigkeit, Um- und Ausbaufähigkeit *f*, Aus- und Umbauvermögen, Um- und Ausbauvermögen *n*

flexible aus- und umbaufähig, um- und ausbaufähig, flexibel

flexible biegsam, flexibel

flexible busbars Band(bus)leitung *f*

flexible (electrical) conduit [*See remark under "conduit"*] biegsame Schutzrohr *n*

flexible limit(ing) value gleitende Grenzwert, Gleitgrenzwert *m*

F

flexible manufacturing cell, FMC [*Of CNC machines with tool and component storage*] flexible Fertigungszelle *f*

flexible multi(ple)-layer printed board, multi(-ple)-layer flexible printed board flexible Mehrlagenleiterplatte, flexible Viellagenleiterplatte *f*

flexible production cell, flexible work cell flexible Fertigungszelle, FFZ *f*

flexible production system, flexible work system flexible Fertigungssystem, FFS *n*

flexible switching gleitende Schalten, Gleitschalten *n*

flexitime, flexible time Gleitzeit, gleitende Zeit *f*

to flicker flimmern

flight path → escape route

flip chart Schaubild *n*

flip-flop (mechanische) Selbsthaltung *f* [*elektronische Türöffner*]

flip-flop (circuit) → bistable multivibrator

flip-flop element Kippglied, Kippelement *n* [*DIN 19 226, Teil 6*]

flipping of pages → page flipping

F(L)N → field (level) network

to float, to set versetzen [*Elektronik. Anschluss oder Leitung in den hochohmigen Zustand*]

float Schwimmer *m*

float casing, float housing, float case Schwimmergehäuse *n*

float charge Ladeerhaltungsspannung *f* [*Erforderliche Spannung zum Ausgleich der Selbstentladung einer Batterie bei einer bestimmten Temperatur*]

floated battery Notstrombatterie *f*

float expansion valve, expansion float valve [*A valve designed to maintain a constant liquid level in a flooded (refrigerant) evaporator*] Flüssigkeitsstandventil *n*

float housing → float casing

floating → not connected

floating, potential-free, voltage-free potentialfrei

floating ball Schwimm(er)kugel *f*

floating ball condensate discharging device, floating ball condensation water discharging device Schwimmer(kugel)-Kondens(at)ableiter, Schwimmer(kugel)-Kondenswasserableiter *m*

floating body Schwimmkörper *m*

floating controller Regler *m* mit regeldifferenzabhängiger Stell(ungs)geschwindigkeit

floating discharge valve → discharge floating valve

floating (electric) control [*It uses a force comparator similar to the two-position controller but does not use the snap-acting switch. It allows the common contact to "float" between the other two for some adjustable "null" or "deadband" range of sensed parameter. The two electrical circuits that can then be closed at different sensed values are not simply on ON command or OFF command*] Reg(e)lung *f* mit regeldifferenzabhängiger Stellgeschwindigkeit

floating electrode Schwimmelektrode *f*

floating gauge; floating gage (US) Schwimmerpegel *m*

floating gate Elektrode mit schwebendem Potenzial, isolierte Steuer(ungs)elektrode, schwebende Potenzialelektrode *f*

floating gate avalanche-injection metal-oxid(e)-semiconductor memory, FAMOS memory FAMOS-Speicher, Metall-Oxid-Halbleiterspeicher mit schwebendem Potenzial und Lawineninjektion *m*

floating ground erdfrei, massefrei, nicht geerdet, ungeerdet

floating indoor moisture state freie Zustand *m* der Innenraumfeuchte

floating low-water alarm → low-water alarm with float

floating magnetic switch → magnetic floating switch

floating measuring mechanism Schwimmermesswerk *n*

floating ((open) loop) control erdfreie Steuerung, erdfrei betriebene Steuerung *f*

floating outflow valve Schwimmerauslaufventil, Schwimmerausflussventil, Schwimmerablaufventil, Schwimmerabflussventil *n* [*Zur selbsttätigen Reglung des Wasserzuflusses oder -abflusses in Abhängigkeit vom Flüssigkeitsstand von Behältern*]

floating point Gleitkomma, Fließkomma *n* [*Vergrößerung des Zahlenbereichs eines Digitalrechners durch halblogarithmische Zahlendarstellung, d. h. die Zahlen werden in einem Zahlensystem zur Basis b in der Form m * b^e (m = Mantise, $0 \leq m < 1$, e = Exponent) dargestellt. Das Komma der Zahlen liegt nicht fest, sondern wird durch e bestimmt.*]

floating point number Gleitkommazahl *f*

floating point representation [*Computer programming. A number notation scheme in which a single number is represented by two numbers: mantissa field contains the significant digits and the exponent field the power of the radix indicating the magnitude. For example, the number 1,200,000 could be represented as the floating-point decimal 1.2,6. meaning $1{,}2x10^6$*] Gleitkommadarstellung *f*

floating solid Schwimm(fest)stoff *m* [*DIN 4045*]

float(ing) straight(-way) valve → straight(-way) float(ing) valve

float(ing) valve [*A valve on the supply pipe to a cistern, such as for a WC. The valve is connected by lever to floating ball which causes the supply to be turned off as the water level rises and turned on again when the water level falls*] Schwimmerventil *n*

float(ing) water level ((closed) loop) controller Schwimmerwasserstandregler *m*

floating zero, mov(e)able zero verstellbare Nullpunkt, bewegliche Nullpunkt, einstellbare Nullpunkt *m*

floating zero system, mov(e)able zero system System mit verstellbarer Nullpunktfestlegung, Kettenmaßsystem *n* [*Alle Bewegungen werden auf die letzte Position bezogen; sobald der Sollwert erreicht ist stellen sich die Zähler auf Null*]

float life Lebensdauer *f* [*Jahre, während derer eine Batterie eine Mindestkapazität im Ladeerhaltungsbetrieb realisieren kann*]

float(-operated) switch, floating switch Schwimmerschalter *m*

to flood (über)fluten, überschwemmen

flood Flut *f*

flood große Textmengen *f*

floodable (über)flutbar, überschwemmbar

flooded evaporation überflutete Verdampfung, überschwemmte Verdampfung *f*

flooding Überflutung, Überschwemmung *f*

floodlight Flutlicht *n*

floodlighting Anstrahlung, Flutlichtbeleuchtung *f*

flood-proofing of buildings Hochwasserschutz von Gebäuden, Gebäudehochwasserschutz *m*

flood(-type) hydrocooling Eiswassertauchkühlung *f*

floor, level, storey; story (US) Geschoss, Stockwerk *n*, Etage *f*

floor [*The surface within a room on which one walk(s). Abbreviated: flr.*] (Fuß)Boden *m*

floor, structural floor [*The horizontal structural part between two storeys as distinct from the flooring. Its main function is to provide support for the occupants, furniture and equipment of a room*] Decke, Tragdecke, tragende Decke *f*

floor air grill(e) → (air) floor grill(e)

floor air outlet, air floor outlet (Fuß)Bodenluftauslass, (Fuß)Bodenluftaustritt *m*

floor alarm panel, storey alarm panel; story alarm panel (US) Geschossalarmtafel, Etagenalarmtafel, Stockwerkalarmtafel *f*

floor bidet → floor(-mounted) bidet

floor blower → floor(-mounted) blower

floor bracket (Fuß)Bodenkonsole *f*

floor breakthrough, floor opening Deckendurchbruch *m*, Deckendurchführung *f* [*tragende Decke. Nicht verwechseln mit „(Fuß)Bodendurchbruch"*]

floor breakthrough, floor opening (Fuß)Bodendurchführung *f*, (Fuß)Bodendurchbruch *m*

floor bumper → floor (door) stop

floor cable heating → cable floor heating

floor care (Fuß)Bodenpflege *f*

floor care machine (Fuß)Bodenpflegemaschine *f*

floor catch → floor (door) stop

floor-ceiling bar Boden-Decken-Stange *f*

floor central distributor → floor manifold

floor channel (Fuß)Bodenrinne *f*

floor clearance [*The distance separating an object or part from a floor*] (Fuß)Bodenfreiheit *f*

floor closer → floor (door) spring

floor convector, floor convection heater [*See remark under "convector"*] (Fuß)Bodenkonvektor *m* [*In einem (Fuß)Bodenkanal eingebaut*]

floor (convenience) receptacle, floor convenience outlet, floor (power) outlet, floor electric outlet [*In an electric circuit, a receptacle which is mounted in an outlet box, set flush with the floor*] (Fuß)Bodensteckdose *f*

floor cooling (Fuß)Bodenkühlung *f*

floor diffuser, floor diffusor (Fuß)Bodendiffusor *m*

floor distributing frame, floor distribution frame, storey distributing frame, storey distribution frame; story distributing frame, story distribution frame (US) Geschossverteiler, Stockwerkverteiler, Etagenverteiler *m* [*Fernsprechwesen*]

F

floor (door) spring, floor (door) closer [*A spring housed into the floor and controlling the opening and shutting off of swing and revolving doors. The plate is level with the floor*] (Fuß)Boden(tür)schließer *m*

floor (door) stop, floor (door) bumper, floor (door) catch [*A door stop set in the floor to prevent a door swinging open too far*] (Fuß)Bodentürpuffer, (Fuß)Bodentüranschlag *m*

floor drain, floor inlet [*A fixture providing an opening in a floor to drain water into a plumbing system; in homes, usually fitted with a deep seal trap*] (Fuß-)Bodeneinlauf, (Fuß)Bodenablauf *m*

F

floor drainage (Fuß)Bodenentwässerung *f*

floor duct (Fuß)Bodenkanal *m*

floor duct heating (Fuß)Bodenkanalheizung *f*

floor electric outlet → floor (convenience) receptacle

floor finish → flooring

floor grating (Fuß)Bodenrost *m*

floor gravity water heating system, gravity floor water heating system [*"Stor(e)y" may synonymously be used for "floor"*] Stockwerk-Schwerkraft-Wasserheizungsanlage, Geschoss-Schwerkraft-Wasserheizungsanlage, Etagen-Schwerkraft-Wasserheizungsanlage, Schwerkraft-Stockwerk-Wasserheizungsanlage, Schwerkraft-Geschoss-Wasserheizungsanlage, Schwerkraft-Etagen-Wasserheizungsanlage *f*

floor grill(e) (Fuß)Bodengitter *n*

floor grill(e) → (air) floor grille

floor-heated (fuß)bodenbeheizt

floor heater, floor heat emitter, floor heating appliance, floor heating apparatus, floor heating device (Fuß)Bodenheizer, (Fuß)Boden(er)wärmer, (Fuß)Bodenerhitzer, (Fuß)Bodenheiz(ungs)apparat *m*

floor heating, floor warming; storey heating, storey warming (Brit.); story heating, story warming (US) Geschossheizung, Etagenheizung, Stockwerkheizung *f*

floor heating, floor warming (Fuß)Boden(flächen)heizung *f* [*Die Warmluft wird durch einen Hohlraum unter den (Fuß)Boden geleitet. Es findet keine Wärmespeicherung statt*]

floor heating cable, floor warming cable, storey warming cable, storey heating cable; story warming cable, story heating cable (US) Deckenheizungskabel *n*

floor heating coil, floor warming coil (Fuß)Bodenheiz(ungs)schlange *f*

floor heating manifold, floor warming manifold, floor heating central distributor, floor warming central distributor (Fuß)Bodenheiz(ungs)verteiler(rohr) *m, (n)*

floor heating tube, floor warming tube (Fuß)Bodenheiz(ungs)röhre *f [DIN 4726]*

floor hot water gravity heating, floor warm water gravity heating; storey hot water gravity heating, storey warm water gravity heating (Brit.); story hot water gravity heating, story warm water gravity heating (US) Geschoss-Schwerkraft-Warmwasserheizung, Etagen-Schwerkraft-Warmwasserheizung, Stockwerk-Schwerkraft-Warmwasserheizung *f*

floor hot water heating → floor warm water heating

floor hydronic heating → floor warm water heating

floor indicator, storey indicator; story indicator (US) Stockwerkanzeiger, Etagenanzeiger, Geschossanzeiger *m [Aufzug]*

flooring, floor finishing [*Any material used in laying a floor*] (Fuß)Bodenwerkstoff *m*, (Fuß)Bodenmaterial *n*

floor inlet → floor drain

floor installation [*e.g. of electric(al) heating cables*] (Fuß)Bodeneinbau *m*

floor key, storey key; story key (US) Geschosstaste, Etagentaste, Stockwerktaste *f [Aufzug]*

floor lamp, floor standard Stehlampe, Standlampe *f*

floor level (Fuß)Bodenhöhe, (Fuß)Bodenkante, (Fuß)Bodenkote *f*, (Fuß)Bodenniveau *n*,

floor main (line), storey main(line); story main(line) (US) Stockwerkleitung, Etagenleitung, Geschossleitung, (Ab)Zweigleitung, Abzweigung *f*

floor manifold, storey manifold; story manifold (US) [*"central distributor" may synonymously be used for "manifold"*] Geschossverteiler(rohr), Etagenverteiler(rohr), Stockwerk(s)verteiler(rohr) *m, (n)*

(floor) master key, emergency key [*For maid's use in a hotel. It unlocks the door under all conditions*] Etagenschlüssel, Geschossschlüssel, Stockwerkschlüssel, Notschlüssel *m*

floor mosaic (Fuß)Bodenmosaik *n*

floor-mounted, floor-standing (fuß)bodenmontiert, (fuß)bodenstehend

floor-mounted bidet [*See remark under "bidet"*] Standbidet, Standsitzwaschbecken, bodenstehende Bidet, bodenstehende Sitzwaschbecken *n*

floor mount(ing) (Fuß)Bodenaufstellung, (Fuß-)Bodenmontage *f*

floor opening → floor breakthrough

floor-operated → operated from floor

floor outlet → floor (convenience) receptacle

floor pad [*For the actuation of a door mechanism*] Mattenunterlagekontakt *m*

floor panel radiant heating, panel floor radiant heating (Fuß)Bodenstrahlungsplattenheizung, Strahlungs(fuß)bodenplattenheizung *f*

floor penetration Deckendurchführung *f* [*tragende Decke*]

floor penetration (Fuß)Bodendurchführung *f*

floor pipeline; storey pipeline (Brit.); story pipeline (US) Geschossrohrleitung, Stockwerkrohrleitung, Etagenrohrleitung *f*

floor plan; storey plan (Brit.); story plan (US) [*A drawing of a floor that shows the layout of the rooms and other details*] Geschossplan, Etagenplan, Stockwerkplan *m*

floor plug [*An electrical outlet whose face is level with or recessed into a floor*] (Fuß)Bodenstecker *m*

floor (power) outlet → floor (convenience) receptacle

floor radiant heating, radiant floor heating Strahlungsdeckenheizung, Deckenstrahlungsheizung *f* [*Eine Flächenheizung, bei der die Wärmeheiz(ungs)-rohre in der Decke angeordnet sind*]

floor (radiation) panel, radiant floor panel (Fuß-)Bodenstrahlungsheiz(ungs)platte *f*

floor range (Fuß)Bodenbereich *m* [*selten: n*]

floor reader Etagenleser, Geschossleser, Stockwerkleser *m* [*Aufzug*]

floor receptacle → floor (convenience) receptacle

floor register [*It is set flush with the floor*] (Fuß-)Bodenregister *n*

floor report; storey report (Brit.); story report (US) Geschossliste, Stockwerkliste, Etagenliste *f*

floor screed (Fuß)Bodenestrich *m*

floor spring → floor (door) spring

floor squeegee (Fuß)Bodenschwabber *m*

floor standard → floor lamp

floor-standing → floor-mounted

floor stop → floor (door) stop

floor storage heater, floor storage heat emitter, floor storage heating appliance, floor storage heating apparatus, floor storage heating device (Fuß-)Bodenspeichererhitzer, (Fuß)Bodenspeicher(er)-wärmer, (Fuß)Bodenspeicherheizer, (Fuß)Bodenspeicherheiz(ungs)apparat *m*

floor storage heating, floor storage warming, storage floor heating, storage floor warming (Fuß-)Bodenspeicherheizung *f* [*DIN 44 576*]

floor supply outlet [*See remark under "supply outlet"*] (Fuß)Bodenzuluftöffnung *f*

floor tube heating → tube floor heating

floor valve (Fuß)Bodenventil *n*

floor warming → floor heating

floor warm water gravity heating → floor hot water gravity heating

floor warm water heating, floor wet heating, floor hydronic heating, floor hot water heating, hot water floor heating [*"storey" (Brit.) and "story" (US) may synonymously be used for "floor"*] Stockwerk-Warmwasserheizung, Stockwerk-WWH, Etagen-Warmwasserheizung, Etagen-WWH, Geschoss-Warmwasserheizung, Geschoss-WWH, Warmwassergeschossheizung, Warmwasseretagenheizung, Warmwasserstockwerkheizung *f*

floor warm water heating, floor hot water heating, warm water floor heating, hot water floor heating Warmwasser(fuß)bodenheizung, (Fuß)Bodenwarmwasserheizung *f* [*DIN 4725*]

floppy disk, floppy disc, FD, diskette Diskette, flexible Speicherplatte, (flexible) Magnetplatte *f* [*Die magnetische Speicherschicht befindet sich auf einer biegsamen Kunstofffolie*]

floppy disk controller, floppy disc controller, FDC Laufwerksteuerung, (flexible) Magnetplattenlaufwerksteuerung *f*

floppy disk drive, floppy disc drive, diskette drive, FD drive Laufwerk, (flexible) Magnetplattenlaufwerk, (flexible) Speicherplattenlaufwerk, Diskettenlaufwerk *n* [*In das Laufwerk wird die Speicherplatte eingeschoben*]

florist cabinet, flower cabinet, florist cubicle, flower cubicle Blumenkühlschrank *m*

to flow fließen

to flow, to stream strömen

flow [*The forward continuous movement of a fluid, such as gases, vapo(u)rs, steams, or liquids, through open channels, pipes and tubes*] Durchfluss, Fluss *m*, Fließen, Durchfließen *n*

flow → sequence

flow, event (functional) sequence Ereignisfolge *f*, Ereignisablauf, Ereignishergang *m*

flow, stream Fluss *m*

flow → flow(ing)

flowability [*Capability of a fluid to (move by) flow*] Fließvermögen *n*, Fließfähigkeit, Fließbarkeit *f*

flow air → **flow(ing) air**

flow and return (flow) → **(forward) flow and return (flow)**

flow capacity (of a filter) [*Refrigerant gas flow that a filter will pass while maintaining a specified pressure drop across the clean filter. The flow capacity is expressed in 1lb/min (g/s) flow at a specified evaporator temperature and return gas temperature for the given refrigerant*] Fließkapazität, Kältemittelfließkapazität *f*

flow cell, flow chamber Durchlaufzelle *f*

flow chart Flussdiagramm, Ablaufdiagramm *n*, Flussplan, Ablaufplan *m*

flow check (test), flow checkout Fließkontrolle, Fließ(über)prüfung *f*

flow ((closed) loop) control [*Modulation of a fluid flow rate through a system of piping, channels, ductwork, or variable devices*] Fließreg(e)lung *f*

flow collector Vorlaufsammler *m*

flow computing → **(work) flow computing**

flow condition Fließbeschaffenheit *f*, Fließzustand *m*

flow condition Fließbedingung, Fließgegebenheit *f*

flow connection Strömungsanschluss *m*, Strömungsverbindung *f*

flow connection [*radiator*] Vorlaufanschluss *m*, Vorlaufverbindung *f*

flow controller, flow CTRL Fließregler *m*

flow control valve → **(automatic) flow control valve**

flow diagram [*A depiction of the physical layout of a process, including the locations of all activities and operations and the directions in which personnel or material travel between these locations*] Fließdiagramm *n*

flow distribution → **flow field**

flow divider Fließteiler *m*

flow elbow Messkrümmer *m*

flow engineering Fließtechnik *f*

flow equation, flowing equation [*Equation for the calculation of fluid flow through conduits or channels; consists of an interrelation of fluid properties, such as density or viscosity, enviromental conditions (such as temperature or pressure), and conduit or channel geometry and conditions, such as diameter, or surface roughness*] Fließgleichung *f*

flow equation → **flow(ing) equation**

flower cabinet → **florist cabinet**

flow factor, C_v-factor (Brit.); flow coefficient, C_v-coefficient (US) k_v-Wert, Fließkoeffizient *m*

flow field, flow distribution [*The velocity and the density of a fluid as functions of position and time*] Fließverteilung *f*

flow(-generated) noise [*Noise produced by the dynamic flow of fluid around an object moving through the fluid; it is caused by turbulence created by the boundary surface and eddy currents to the surface roughness and friction*] Fließrauschen *n*

to flow in einfließen

to flow in, to stream in einströmen

flow indicator (Durch)Flussanzeiger *m*

flow(ing) [*In engineering, a forward movement in a continuous stream or sequence of fluids or discrete objects or materials, as in a continuous chemical process or solids-conveying or production-line operation*] Fließen *n*

flow(ing) [*In physics, the movement of electric charges, gases, liquids, or other materials or quantities*] Fließen *n*

flow(ing) air, streaming air strömende Luft, Strömungsluft *f*

flow(ing) equation, streaming equation Strömungsgleichung *f*

flow(ing) limit Fließgrenze *f*

flow(ing) measurement, streaming measurement Strömungsmessung *f*

flow(ing)-out Ausfließen *n*

flow(ing)-out, streaming out Ausströmen *n*

flow(ing) phenomenon Fließerscheinung *f*

flow(ing) pressure, streaming pressure Strömungsdruck *m* [*Der statische Überdruck des strömenden Mediums*]

flow(ing) reversal, streaming reversal Strömungsumkehr *f*

flow(ing) transmitter Fließgeber *m*

flow(ing) velocity Fließgeschwindigkeit *f*

flow(ing) velocity, streaming velocity Strömungsgeschwindigkeit *f*

flow limit → **flow(ing) limit**

flow line Flusslinie, Ablauflinie, Wirkungslinie *f* [*Funktionsplan*]

flow line, streaming line [*The path traced by particular particles of a moving fluid, such as in pipe*] Stromlinie, Fließlinie *f*

flow measurement, fluid measurement Laufzeitmessung *f* [*Fluid*]

flowmeter, fluid meter Laufzeitmesser *m* [*Fluid*]

flowmeter [*For measuring workability (slump) and temperature of fresh concrete*] Fließmesser *m*

flow metrology Fließmesstechnik *f*

flow mixer, line mixer [*A mixing device that has nozzles and agitators, used to mix two or more liquids as they flow through*] Fließmischer, Flüssigkeit(s)-mischer *m*

flow mode, flow type, streaming mode, streaming type Strömungsart *f* [*Laminar oder turbulent*]

flow noise → flow(-generated) noise

flow of heat (energy) → heat (energy) flow

flow of persons Personenfluss, Personenstrom *m*

flow ((open) loop) control [*Modulation of a fluid flow rate through a system of piping, channels, duct-work, or variable devices*] Fließsteuerung *f*

flow(-operated) safety device [*burner*] nachgeschaltete Fließsicherung *f*

to flow out ausfließen

to flow out, to stream out ausströmen

to flow over überfließen

to flow over, to stream over überströmen

flow pattern [*Fluid mechanics. A pattern that is assumed by a two-phase (liquid-gas) combination flowing in a pipe, duct, or channel; relevant factors are the liquid-to-gas ratios, flow resistances, fluid velocities, fluid properties and pipe dimensions*] Fließmuster, Fließbild *n*

flow pressure Fließdruck *m*

flow pressure, streaming pressure Strömungsdruck *m*

flow probe, streaming probe Strömungssonde *f*

flow process [*A system in which fluids or solids are processed in continuous movement*] Fließverarbeitung *f*

flow range Strömungsbereich *m* [*selten: n*]

flow rate, rate of flow [*Weight of flowable material flowing per unit time*] Fließgewicht *n*

flow rate, rate of flow [*Time required for a given quantity of flowable material to flow a measured distance*] Fließzeit *f*

flow rate meter, rate of flow meter Fließzeitmesser *m*

flow rectifier Strömungsgleichrichter *m* [*DIN 30 600. Reg.-Nr. 06096*]

flow resistance [*Any factor within a conduit or fluid, such as surface roughness, viscous effects, bends, constrictions and expansions*] Fließwiderstand *m*

to flow-solder schwalllöten, einschwallen, aufschmelzlöten

flow soldering Schwalllöten, Einschwallen, Aufschmelzlöten *n*

flow straightener [*ventilation*] Gleichrichter *m*

flow stress, yield stress [*The extent of stress required to produce plastic flow*] Streckspannung *f*

flow switch, streaming switch Strömungsschalter *m*

flow temperature, streaming temperature Strömungstemperatur *f*

flow temperature Fließtemperatur *f*

flow temperature ((closed) loop) control, ((closed) loop) flow temperature control Vorlauftemperaturreg(e)lung *f*

flow thermostat Fließthermostat, Fließtemperaturregler *m*

flow thermostat, streaming thermostat Strömungsthermostat, Strömungstemperaturregler *m*

flow transmitter, streaming transmitter Strömungsgeber *m*

flow type → flow mode

flow(-type) heat exchanger → non-strorage heat exchanger

flow(-type) (water) heater, flow(-type) (water) heating appliance, instantaneous (water) heater, instantaneous (water) heating appliance, non-storage (water) heater, non-storage (water) heating appliance Durchlauf(wasser)heizer, Durchfluss(wasser)heizer, Durchlauf(wasser)(er)wärmer, Durchlauf(wasser)erhitzer, Durchfluss(wasser)erhitzer, Direkt(wasser)heizer, Direkt(wasser)(er)wärmer *m*, Therme *f* [*DIN 3368 und 3369*]

flow valve, self-closing valve, automatic closing valve [*A valve that closes itself when the flow of a fluid exceeds a particular value*] Selbstschlussventil, Automatikventil *n*

flow velocity → flow(ing) velocity

flow water Vorlaufwasser *n*

flue Schacht *m* [*Konvektortruhe*]

flue [*The chimney itself, if only one flue is enclosed*] Ein-Zug-Schornstein *m*

flue [*Passage leading from the combustion chamber to the stack*] Feuerzug, Abzug, Zug *m*

flue belt → flue ring

flue boiler Flammrohrkessel *m*

flue connection → principal flue

flue damper [*oil burner*] Drosselorgan *n*

flue damper → draft check plate

F

F

flue draught, flue draft Schornsteinzug *m*

flue dust → **fly ash**

flue effect [*deprecated*] → **chimney effect**

flue furnace Feuerzugfeuerung, Abzugfeuerung, Feuerzugfeuerstätte, Abzugfeuerstätte *f*

flue gas bypass Rauchgasumgehung *f*

flue gas chimney Rauchgasschornstein *m*

flue gas column [*in a chimney*] Rauchgassäule *f*

flue gas connection Rauchgasanschluss *m*, Rauchgasverbindung *f*

flue gas damper Rauchgasklappe *f*

flue gas dew point Rauchgastaupunkt *m*

flue gas examination Rauchgasuntersuchung *f*

flue boiler/fire tube boiler, flue boiler/smoke tube boiler Flammrohr-Rauchrohrkessel *m* [*Die Rauchrohre schließen sich an den vom Flammrohr gebildeten Feuerraum an. Sie liegen hinter, neben oder über dem Flammrohr und werden wie dieses vom Wasser oder Dampf umspült*]

flue gas, gaseous products of combustion Rauchgas *n* [*In einer Feuer(ungs)stätte erzeugt und über Schornstein abgeführt*]

flue gas cutoff device, flue gas shutoff device, flue gas cutoff appliance, flue gas shutoff appliance Rauchgasabsperrvorrichtung *f*

flue gas extraction Rauchgasabführung *f*, Rauchgasabzug *m*

flue gas flap, flue gas damper Rauchgasklappe *f*

flue gas flue Rauchgasfang, Rauchgasfänger *m*

flue gas pipe Rauchgasrohr *n*

flue gas preheater → **economizer**

flue gas propagation, flue gas distribution Rauchgasausbreitung, Rauchgasverbreitung *f*

flue gas purification Rauchgasreinigung *f*

flue gas recorder Rauchgasschreiber *m*, Rauchgasregistriergerät *n*

flue gas scrubber Rauchgaswäsche(r) *f*, (*m*)

flue gas side [*induced draft burner*] Rauchgasseite *f*

flue gas test section Rauchgasprüf(ungs)strecke *f*

flue gas throttling device, flue gas throttling appliance Rauchgasdrosselgerät *n*, Rauchgasdrosselvorrichtung *f*

flue gas tract, flue gas path [*heating furnace*] Rauchgasweg *m*

flueless (combustion) appliance [*Any combustion appliance for heating a room or for cooking which is not provided with a flue, the products of combustion being allowed to mix with the air of the room in which the appliance is situated*] Verbrennungsgerät ohne Feuerzug, feuerzuglose Verbrennungsgerät *n*

flue liner, flue lining Feuerzugfutter *n*

flue opening → **chimney tube**

flue pipe → **flue (tube)**

flue ring, flue belt, flue section Flammrohrschuss, Feuerrohrschuss *m*

flue thermostat Feuerzugthermostat *m*

flue tube, flue pipe, fire tube, fire pipe, furnace flue, smoke tube, smoke pipe [*It is vertical or nearly so, designed to convey all the products of combustion to the outside atmosphere*] Flammrohr, Feuerrohr, Rauchrohr *n*

flueway [*The clear space within a flue or flue lining for the passage of combustion gases*] Feuerzuglichtweite *f*

flue wall Zunge *f* [*Schornstein*]

flue without projections → **flash flue**

fluff separator Flusenabscheider *m*

fluff sieve Flusensieb *n*

fluid Fluid, Strömungsmedium, strömende Medium *n* [*Flüssigkeit; Gas; Dampf. DIN 5492*]

fluid-actuated, fluid-operated fluidbetätigt, fluidbetrieben

fluid computer [*A digital computer constructed entirely from air-powered fluid logic elements; it contains no moving parts and no electronic circuits; all logic functions are carried out by interaction between jets of air*] Fluidrechner, Luftstrahlrechner *m*

fluid-flow(-actuated) switch Strömungsmediumschalter *m*

fluid mechanics Strömungsmechanik *f* [*DIN 5492*]

fluid meter → **flowmeter**

fluid-operated → **fluid-actuated**

fluid science Strömungswissenschaft *f*

to fluorate fluorieren, anreichern mit Fluor

fluorescence Fluoreszenz *f* [*Aufleuchten mancher Stoffe während Belichtung, z. B. Flussspat*]

fluorescent [*Able to emit visible light only when exposed to the activating radiation*] fluoreszierend, (auf)leuchtend

fluorescent lamp, fluorescent tube, FL, cold cathode lamp, cold cathode tube [*A low-pressure electric-discharge lamp; ultraviolet-light radiation is generated by the passage of an arc through mercury vapo(u)r; the inner surface of the lamp tube is coated with a phosphor which absorbs the ultraviolet and* *converts some of it into visible light B.S. 1853*] Leuchtstofflampe, Leuchtstoffröhre , L-Lampe, L-Röhre *f* [*DIN 49 862*]

fluorescent lighting, fluorescent illumination, cold-cathode lamp lighting, cold-cathode lamp illumination Leuchtstoffbeleuchtung *f*

fluorescent light(ing) fitting, fluorescent luminaire [*A luminarie, usually complete with fluorescent lamps, sockets, ballast, reflector, and louver or diffusing medium*] Leuchtstoffleuchte *f*, Leuchtstoffbeleuchtungskörper *m*

fluorescent (light(ing)) strip, fluorescent illumination strip Leuchtstoffband *n*

fluorescent strip cove lighting, fluorescent strip coving lighting, fluorescent strip hollow cornice lighting [*See remark under "cove lighting"*] indirekte Leuchtstoffbandbeleuchtung *f* über Viertelstab

fluorescent tube → fluorescent lamp

fluorescent vacuum display → vacuum fluorescent display

fluorinated hydrocarbon Fluorkohlenwasserstoff, FKW *m*

to flush → to wash

flush Spülung *f*

flush → flush (with)

flush bolt [*A bolt for securing a door, let into the material so that it is level with the surface*] Kantenriegel *m*

flush closet → toilet

flush door [*It has a continuous smooth surface of plywood, coreboard or hardboard that covers a solid or hollow interior*] Flächentür, abgesperrte (Platten)Tür, Sperrtür *f* [*DIN 68 706*]

flush door bolt [*A door bolt so designed that when applied it is flush with the face or edge of the door*] bündige Schubriegel *m*

flush-encased (dead) bolt lock Einlassriegelschloss, eingelassene Riegelschloss *n*

flush-encased lock → dummy lock

flush fitting door release Stumpftürenöffner *m*

flush fixing, letting-in flush (on both meeting faces) Einstemmen *n* [*Einstemmband*]

flush handle [*This term is also used for the "Klappringmuschel"*] Klappringschild *n* [*Türbeschlag*]

flush handle [*This term is also used for the "Klappringschild"*] Klappringmuschel *f* [*Türbeschlag*]

flushing Abgleichen, Bündigmachen, Gleichmachen *n*

flushing Spülen *n*, Spülung *f*

flush(ing) basin → toilet bowl

flush(ing) bowl → toilet bowl

flush(ing) cistern → toilet cistern

flush(ing) closet → toilet

flush(ing) connection Spül(ungs)verbindung *f*, Spül(ungs)anschluss *m*

flush(ing) electronics [*See remark under "electronics"*] Spül(ungs)elektronik *f*

flush(ing) pan → toilet bowl

flush(ing) pipe → toilet flush(ing) pipe

flush(ing) point Spül(ungs)stelle *f*

flush(ing) rim Spül(ungs)rand *m*

flush(ing) tank → toilet cistern

flush(ing) tank [*A tank in which water or sewage is retained for period release through a sewer*] Rückhaltebehälter, Rückhaltetank *m*

flush(ing) (tank) closet → toilet

flush(ing) (tank) closet fitment → toilet fitment

flush(ing) (tank) toilet fitment → toilet fitment

flush(ing) trip Spül(ungs)auslöser *m* [*Toilette*]

flush(ing) trough [*A long water tank extending above and across a range of water closets and common to all. It is used particularly in institutional buildings, since it allows the flushing of several water closets at short intervals without waiting for a cistern to fill*] Spül(ungs)trog *m*

flush(ing) urinal cistern, urinal flush(ing) cistern Urinalspül(ungs)kasten *m*

flush(ing) valve; flushometer (valve) (US) [*In a flush cistern, a valve that automatically releases a flush when the water pressure in the cistern builds up*] (Druck)Spüler *m* [*DIN 3265*]

flush(ing) water → toilet flush(ing) water

flush left linksbündig

flush lever bolt Türkantriegel *m*

flush-mounted unter Putz montiert, bündig montiert

flush-mounted water meter Unterputzwasserzähler *m*

flush-mounted (water) mixing valve, flush-mounted (water) mixer valve [*See remark under "(water) mixing valve"*] Unterputzmischventil, Unterputzmischer(ventil) *m*, (*n*)

flush mounting, flush installation bündige Einbau, Unterputzeinbau *m*, bündige Montage, Unterputzmontage *f*

flush right rechtsbündig

flush toilet → toilet

F

F

flush wiring bündige Verdrahtung, Unterputzverdrahtung *f*

flush (with), flush to, level (to), dead level [*Pertaining to separate surfaces that are on the same level*] bündig (mit), niveaueben, flächenbündig, niveaugleich, flächengleich, niveaubündig

fluting, vertical groove [*radiator panel*] Profilierung, Riffelung *f*

flutter echo [*A radar echo consisting of a rapid succession of reflected pulses resulting from a single transmitted pulse*] Flatterecho *n*

flutter echo [*A rapid succession of echoes caused by sound reflection back and forth between two opposite walls; initiated by a single, sharp pulse of sound*] Flackerecho *n*

flux change per inch → fci

flux density Flussdichte *f*

flux gate Luftspalt *m*

flux-gate magnetometer [*A magnetometer in which the degree of saturation of the core by an external magnetic field is used as a measure of the strength of the Earth's magnetic field; the essential element is the flux gate*] Flussspaltmagnetometer *n*

flux jump(ing), Meissner(-Ochsenfeld) effect [*The expulsion of magnetic flux from the interior of a piece of superconducting material as the material undergoes the transition to the superconducting phase*] Flusssprung *m*

flux leakage [*Magnetic flux that does not pas through an air gap or other part of a magnetic circuit where it is required*] Flussstreuung *f*

fluxless flussmittelfrei

flux pump [*A cryogenic direct-current generator that converts a small alternating-current input to a large direct-current output when cooled to about 4 k; the output current builds up in a series of steps, much like the action of a pump*] Kryopumpe, Tieftemperaturpumpe *f*

fly ash, flue dust Flugasche *f*

flyback, retrace, return trace [*The time interval in which the electron beam of a cathode-ray tube returns to its starting point after scanning one line or one field of a television picture or after completing one trace in an oscilloscope*] Elektronenstrahlrücklauf *m*

flyback converter Sperrwandler *m* [*Schaltnetzteil*]

flyback transformer, horizontal output transformer, horizontal sweep transformer [*A transformer used in a television receiver to provide the horizontal deflection voltage, the high voltage for the second-anode power supply of the picture tube, and the filament voltage for the high-voltage rectifier tube*] Rücklauftransformator *m*

flying head fliegende Kopf *m* [*Lese- oder Schreibkopf an einem Magnettrommelspeicher der bei schneller Trommelrotation auf einer Luftschicht „schwimmt"*]

flying restart circuit Fangschaltung *f* [*Zur Zuschaltung eines Stromrichters auf eine laufende Maschine*]

flying scanner → optical scanner

flying scanning → optical scanning

flying-spot scanner tube Lichtpunktabtaströhre *f*

flyscreen Fliegengitter, Insektengitter *n*

FM → facility management

FM → frequency modulation

FMC → flexible manufacturing cell

FMECA → failure modes and effects and criticality analysis

FMS, Field bus Message Specification Anwenderdienste *f* für den Profibus [*DIN 19 245 Teil 2*]

FMS → building management system

FN → field net(work)

foam carryover, foamover Schaumübertrag *m*

foamed-in-place thermal insulation, foamed-in-situ thermal insulation Ortschaumwärmedämmung *f*

foam (fire) extinguisher Schaum(feuer)löscher, Schaumbrandlöscher *m*

foam (fire-fighting) installation, foam (fire-fighting) plant Schaum(feuer)löschinstallation, Schaum(feuer)löschanlage *f*

foam (fire) hydrant Schaum(feuer)hydrant *m*

foam (jet) gun Schaum(strahl)pistole *f*

foamover → foam carryover

foam plastic → cellular plastic

foam plastic fire, expanded plastic fire, cellular plastic fire Schaumstoffbrand *m*, Schaumstofffeuer *n*

foam soap dispenser → soap foam dispenser

focal length Brennweite *f*

focus Scharfeinstellung, Schärfeeinstellung *f*

focussing Scharfeinstellen, Schärfeeinstellen *n*

FOG → fibre optic gyroscope

FOG → (fuel) oil gauge

fog elimination, fog removal Entneb(e)lung, Nebelauflösung *f*

fog formation, fogging, fog generation Nebelbildung, Nebelentstehung *f*

fogger Versprühgerät *n* [Zur *Erzeugung eines Flüssigkeitsnebels*]

fogging Versprühen *n* [*Erzeugung eines Flüssigkeitsnebels*]

fogging Verschwinden *n* von Bildern im hellen Nebel

fog stream nozzle [*For extinguishing a fire*] Nebellöschdüse, Löschnebeldüse *f*

FOI → fibre optic interface

foil screen Folien(schutz)schirm *m*

foil strain gauge; foil strain gage (US) Foliendehn(ungs)messstreifen *m*

foldaway seat wegklappbare Sitz *m*

fold-back surface burner Klappkochplatte *f*

folded towel dispenser Falthandtuchspender *m*

folder [*User authentification method*] Ordner *m*

folding blind → folding shutter

folding paper dispenser Faltpapierspender *m*

folding shutter, folding blind Falladen *m* [*Er ist oben um Scharniere beweglich und wird durch Schnüre, Ketten oder Stutzen festgehalten*]

folding sliding gate, sliding folding gate Faltschiebetor, Schiebefalttor *n*

folding sliding gate wing → sliding folding gate wing

follower Folger *m*

follower regulator Folgeregler *m* [*Er arbeitet am Eingang mit parallel anliegenden, wertdiskreten, zeitkontinuierlichen, vorzugsweise binärkodierten Signalen. DIN 19 225*]

follow(ing) processing Weiterverarbeitung *f*

following unit → terminal interface

follow-up ((closed) loop) control, servo ((closed) loop) control Folgereg(e)lung, Nachlaufregelung, Servofolgereg(e)lung, Nachführungsregelung *f* [*Der Wert der Regelgröße folgt den veränderten Werten der Führungsgröße – E DIN 19 226, Teil 5*]

follow-up ((closed) loop) control system → servo ((closed) loop) control system

follow-up (fire) extinguishing (Feuer)Nachlöschen, Brandnachlöschen *n*

follow-up ((open) loop) control system → servo ((open) loop) control system

follow-up product Nachfolgeprodukt, Nachfolgeerzeugnis *n*

FOM defuzzification method → first of maxima defuzzification method

font, fount, character style [*A complete set of all the characters in one size of a typeface*] (Schrift)-Satz, Zeichensatz *m*, Schriftart, Schrift(zeichensatz)form, Schrifttype *f*

font [*The quantity of characters available to a computer*] Zeichenmenge *f*

fontitalic Kursivschrift *f*

fontwork Textgraphikeditor *m*

food preparation Speisenvorbereitung *f*

food warmer Speisenwärmer *m*

food waste(s), garbage, waste food [*Waste parts of food, as from a market or kitchen*] Nahrungsabfälle, organische Abfälle im engeren Sinne *f*

food waste(s) incinerator, waste food incinerator, garbage incinerator Nahrungsabfallverbrenner *m*, Nahrungsabfallverbrennungsanlage *f*

foot → base

foot actuation, foot operation Fußbedienung, Fußbetätigung *f*

foot bathtub [*A bath for washing the feet, and generally of ceramic ware, fixed at a low level and provided with hot and cold water taps and a waste outlet*] Fußwaschwanne, Fußbadewanne *f*, Fußbadebecken, Fußwaschbecken *n* [*DIN 13 214, 13215*]

footer Fußzeile *f*

foot of head [*The pressure exerted at the bottom of a column of water 1 ft high*] Wassersäulendruck *m*

foot-operated fußbedient, fußbetätigt

footprinter Ausleuchtzone *f* [*Die Fläche auf der Erde, die von einem Satelliten mit etwa gleicher Leistungsflussdichte bestrahlt wird. Hiervon hängt der erforderliche Durchmesser der Parabolantenne ab*]

foot scraper Fußabstreifer, Fußkratzer *m*

footstep detector, footstep detection device, footstep detecting device Tretmelder *m*, Tretmeldegerät *n* [*Er löst beim Betreten Alarm aus*]

foot valve Fußventil *n*

foot washer Fußwaschanlage *f*, Fußwäsche(r) *f*, (*m*)

forbidden zone Verbotszone *f*

to force a lock aufbrechen [*Schloss*]

force balance Kräfteausgleich *m*

to force [*To intervene manually in a computer routine and cause the computer to execute a jump instruction*] erzwingen

force, power [*Mechanics. Any external agent that causes a change in the motion of a free body, or that causes stress in a fixed body*] Kraft *f*

forced erzwungen

forced-air central furnace [*It is equipped with a fan or blower that provides the primary means for air circulation*] Zwangsluft-Zentralfeuerung, Zentral-Zwangsluftfeuerung *f*

forced-air circulation Zwangsluftumwälzung, Zwangsluftzirkulation *f*, Zwangsluftumlauf *m*

forced-air furnace [*It is equipped with a blower that provides the primary means for air circulation*] Zwangsluftfeuerung *f*

forced-air heater Zwangslufterhitzer, Zwangsluft(er)wärmer *m*

forced-air heating system [*Central warm-air heating system equipped with a fan or blower that provides the primary means for air circulation*] Zwangsluftheiz(ungs)system *n*

F

forced circulation Zwangsumlauf *m*, Zwangsumwälzung, Zwangszirkulation *f*

forced circulation air cooler Zwangsumlaufluftkühler *m*

forced circulation air cooling Zwangsumlaufluftkühlung *f*

forced circulation boiler Zwangsumlaufkessel *m*

forced circulation heat generator Zwangsumlaufwärmeerzeuger *m*

forced-circulation hot water (heating) system, forced-circulation warm water (heating) system Zwangsumlauf-Warmwasser-Heiz(ungs)system *n*

forced convection, forced thermal convection [*Thermodynamics. The heat transfer by a fluid resulting from forcing the fluid by some external agency*] erzwungene Konvektion, Zwangskonvektion *f*

forced convection (hot) air heating plant, forced convection (hot) air heating installation [*An assembly of heating element and distribution fan*] Ventilator-(Warm)Luftheiz(ungs)anlage *f* [*Die Luft wird mit Ventilator, also zwangsweise, in den zu beheizenden Raum gefördert*]

forced convector → fan-coil unit

forced entry gewaltsame Eintritt, gewaltsame Zutritt *m*, Eindringen *n*, Eindringung *f*

forced hot water heating, forced warm water heating, pumped hot water heating, pumped warm water heating Pumpenwarmwasserheizung, PWW-Heizung *f* [*bis 110 Grad C Vorlauftemperatur*]

forced hot water heating with lower distribution → pump hot water heating with lower distribution

forced key verdrehte Schlüssel *m*

forced lock verdrehte Schloss *n*

forced ventilation Zwangslüftung, verstärkte Lüftung *f*

forced warm air, warm forced air Zwangswarmluft, Warmzwangsluft, verstärkte Warmluft *f*

force-feed lubrication Zwangsschmierung *f*

forced-warm-air furnace [*Furnace equipped with blower to provide the primary means for circulating air*] Zwangswarmluftfeuerung, Zwangswarmluftfeuerstätte *f*

force warm water heating → forced hot water heating

force line → line of force

forcing configuration Grundfolge *f* [*Übertragung*]

fore-brim Stulp *m* [*Die Schlosskastenseite durch welche der Riegel ein- und ausgeht*]

forecasting Prognose, Vorhersage, Voraussage *f*

foreign make Fremdfabrikat *n*

forelock → bolt key

foreman glazier Glasermeister *m*

foreman plasterer (Ver)Putzermeister *m*

forged steel Schmiedestahl, geschmiedete Stahl *m*

forgery Schmiede *f*

forge scale → scale

forge scale formation → scale formation

forging, falsification, forgery Fälschung *f*

fork Aufspaltung, Verzweigung *f* [*Stelle im Programmablaufplan von der Zweige ausgehen; E DIN 44 300/10.85*]

fork Aufspaltung *f* [*Vorgang, bei dem eine Kopie eines Prozesses angelegt wird*]

forked fan → bifurcated fan

form Formular *n*

form → (screen) form

(formal) demand → prompt note

to format formatieren

format Format *n* [*Die Art der Anordnung von digitalen Daten in Speichern, auf Datenträgern oder in Nachrichtenkanälen. DIN 44 300 Teil 3*]

format character, layout character, FC [*A control character that determines the form in which the output data generated by a printer or display device are arranged*] Formatsteuer(ungs)zeichen *n* [*Code – E DIN 44 300/10.85*]; Formattrennfolge *f* [*Paketierer/Depaketierer*]

formation Bildung *f*

formatted formatiert, angeordnet [*Daten*]

formatted capacity Nettokapazität *f* [*Datenspeicher*]

formatter Formatierer *m* [*Programm zum Formatieren. Marken für Dateien, Spuren, Adressen, Prüfzeiten, Vorzeiten und Nachzeiten, z. B. auf einer Diskette*]

formatting Formatierung *f*

form feed(ing), FF [*The positioning of documents in order to move them past printing or sensing devices, either singly or in continuous rolls*] Formularvorschub *m*

form feed(ing) character [*A control character that determines when a printer or display device moves to the next page, form, or equivalent data unit*] Formularvorschubzeichen *n*

form stop [*A device which stops a machine when its supply of paper has run out*] Papiermangelhalt *m*

formula composer Formelsetzmaschine *f*

formula translation, Fortran Formelübersetzung, Fortran *f* [*Für die Verarbeitung mathematischer Probleme konstruierte Programmsprache*]

formwork oil → mould oil

for-statement [*ALGOL*] Laufanweisung *f*

forward-acting regulator, forward-acting controller [*Transmission regulator in which the adjustment made by the regulator does not affect the quantity which caused the adjustment*] Vorwärtsregler *m*

forward automatic closing valve → forward flow valve

forward/backward counter → bidirectional counter

forward converter (Durch)Flusswandler *m* [*Schaltnetzteil*]

forward counter Vorwärtszähler *m*

forward current [*It flows upon application of forward voltage*] Vorwärtsstrom, Strom in Leitrichtung *m*

forward flow Vorlauf *m*

(forward) flow and return (flow) Vorlauf *m* und Rücklauf

forward flow connection Vorlaufanschluss *m*, Vorlaufverbindung *f*

forward flow heating manifold → heating forward flow manifold

forward flow mixing valve Vorlaufmischventil, Mischvorlaufventil *n*

forward flow overtemperature Vorlaufübertemperatur *f*

forward flow (pipe)line Vorlauf(rohr)leitung *f*

forward flow pressure [*The pressure necessary to supply a fixture adequately*] Vorlaufdruck *m*

forward flow pump Vorlaufpumpe *f*

forward flow temperature Vorlauftemperatur *f*

forward flow temperature ((closed) loop) control, forward flow temperature regulation Vorlauftemperaturreg(e)lung *f*

forward flow temperature ((open) loop) control Vorlauftemperatursteuerung *f*

forward flow thermostat Vorlaufthermostat, automatische Vorlauftemperaturregler *m*

forward flow valve, forward self-closing valve, forward automatic closing valve [*A valve in a forward flow system that closes itself when the flow of a liquid exceeds a particular value*] Vorlaufselbstschlussventil *n*; Vorlaufselbstschlussschieber *m*

forward flow water Vorlaufwasser *n*

forward heating flow → heating forward flow

forwarding Weiterleiten *n*

forward metering Vorwärtsmessung *f*

forward plus inverse model → hybrid model

forward polling Vorwärtsabfrage *f*

forward reference Vorwärtsbezug *m* [*Bei einem Sprungbefehl die Angabe eines Sprungziel(e)s, das in einem nachfolgenden Programmteil liegt*]

forward safety flow → safety forward flow

(forward) scatter propagation → beyond-the-horizon communication

forward self-closing valve → forward flow valve

forward spacing Vorsetzen *n*

forward transfer signal, transfer forward signal Vorwärts-Transfer-Signal, Transfer-Vorwärts-Signal *n*

forward voltage Durchlassspannung, Vorwärtsspannung, positive Spannung *f*

FOS → fibre optic sensor

fossil fuel [*Oil, natural gas, coal and peat*] fossile Brennstoff *m*, fossile Brennmaterial *n*

foul air → vitiated air

foul air flue, vitiated air flue Schlechtluftabzug *m*

fouling (Flächen)Verschmutzung, Verunreinigungsablagerung *f* [*an wärmeübertragenden Flächen*]

foul water, waste(water) [*It is discharged into the drains of a building from baths, basins, sinks, showers and bidets, but not from WCs. "Waste(water)" as defined here must not be confused with "wastewater" which means "sewage" in the USA*] Schmutzwasser *n*

foundation earth Fundamenterder *m* [*VDE 0100, T. 200*]

foundation ground (connection) (US) → (concrete) foundation ground (connection)

fount → font

fountain Springbrunnen *m*, Fontäne *f*

four-address [*Pertaining to an instruction address which contains four address parts*] Viereradresse *f*

four arm handle Kreuzgriff *m*

four bits, four binary digits vier Bits, vier Binärstellen, Vierheit, Tetrade *f*

four-burner range Vierplattenherd, Vierbrennerherd *m*

four-chamber septic tank Vierkammerfaulgrube, Vierkammerkleinkläranlage *f [In Deutschland nur für Schmutzwasser zugelassen. DIN 4261]*

four-channel array Vierkanalfeld *n [Siehe Anmerkung unter „Feld"]*

four-channel data receiver Vierkanal-Datenempfänger *m*

F

four-compartment digestion pit Vierkammer(aus)-faulgrube, Vierkammergrube *f [Kleinkläranlage]*

four-conductor cable, four-core cable Vierleiterkabel *n*

four-conductor watt-hour meter Wirkverbrauch-Vierleiterzähler, Vierleiter-Wirkverbrauchzähler, Wattstunden-Vierleiterzähler, Vierleiter-Wattstundenzähler *m*

four-electrode electron tube → tetrode

four-hole bath(room) fitting Vier-Loch-Bad(ezimmer)armatur *f*

Fourier law of heat conduction, Fourier heat equation *[The law that the rate of heat flow through a substance is proportional to the area normal to the direction of flow and to the negative of the rate of change of temperature with distance along the direction of flow]* Fouriersche Gesetz, Fouriergesetz *n*

four-layer device Vierschichtenbauelement *n*, Vierschichtenzelle *f [Gemeinsame Bezeichnung für Vierschichtendiode und Vierschichten(schalt)triode]*

four-layer diode Vierschichtendiode *f [Eine Schaltdiode, die bei Überschreiten einer bestimmten Sperrspannung in der Richtung leitend wird, die das Symbol andeutet. Verwendung zum Schalten von Thyristoren und als Impulserzeuger]*

four-layer transistor Vierschichtentransistor *m [Transistor mit PNPN-Zonenfolge, dessen zwei mittlere Zonen sehr dünn sind. Die vierte, zu der Zonenfolge eines Dreischichttransistors hinzukommende Zone ist an dessen Kollektorzone angefügt. Ein solcher Transistor hat in Basisschaltung eine Stromverstärkung, die weit über 1 hinausgeht und die Größenordnung von 100 erreichen kann]*

four-pipe air conditioning plant, four-pipe air conditioning installation Vierrohrklimaanlage, Vierrohrklimainstallation, Vierleiterklimaanlage, Vierleiterklimainstallation *f*

four-pipe induction air conditioning plant, four-pipe induction air conditioning installation Vierrohr-Induktionsklimaanlage, Vierrohr-Induktionsklimainstallation, Vierleiter-Induktionsklimaanlage, Vierleiter-Induktionsklimainstallation *f*

four-point jack vierteilige Klinke *f [Telephonie]*

four-pole circuit Vierpolschaltung *f [Schaltung mit zwei Eingangsklemmen und zwei Ausgangsklemmen]*

four-pole equivalent circuit Vierpolersatzschaltung *f*

four-pole equivalent series circuit Vierpol-Reihen-Ersatzschaltung *f*

four-port Viertor, Viersatz *m [Bauelement oder Baugruppe mit einem Eingang und einem Ausgang bezogen auf den Signalfluss]*

four-port coupler Viertorkopp(e)ler, Viersatzkopp(e)ler *m*

four-position data connector Vierkontaktdatensteck(verbind)er *m*

four-position fan switch Vierstellungs-Ventilatorschalter *m [Aus, Max., Norm., Min.]*

four-quadrant multiplier, quarter-square multiplier *[A multiplier in an analog computer in which both the reference signal and the number represented by the input may be bipolar, and the multiplication rules for algebraic sign are obeyed]* Vierquadrantenmultiplikator *m*

four-rate counting mechanism, four-rate register Viertarifzählwerk *n*, Viertarifregler *m*

four-row coil vierreihige Schlange *f*

four-step(ped) characteristic Vierstufenkennlinie, vierstufige Kennlinie *f*

four-step(ped) core Vierstufenkern *m*

fourth power law → Stefan-Boltzmann law

four-tube air conditioning plant, four-tube air conditioning installation, four-pipe air conditioning plant, four-pipe air conditioning installation Vierrohrklimaanlage *f*

four-way cock Vierwegehahn, Vierwegekrahn *m [Norddeutschland] [DIN 30 600. Reg.-Nr. 00602]*

four-way mixer Vierwegemischer *m [Stellglied in einer Heiz(ungs)anlage]*

four-way switch *[A four-terminal switch (abcd) that operates ab-cd and ac-bd. Used to control outlets from three or more locations]* Vierwegeschalter *m*

four-way switching *[Outlet control from three or more locations]* Vierwegeschaltung *f*

four-way throttling valve Vierwege-Regelventil, Vierwege-Drosselventil *n [DIN 30 600, Reg.-Nr. 06163]*

four-way valve, recovering valve *[1. In a refrigerating system, a device which enables reversing the eva-*

porator and the condenser for hot gas defrosting. 2. In a heat pump system, a device to effect changeover between heating and cooling cycles] Vierwegeventil *n*, Umkehrventil

four-wire circuit [*Two-way circuit using two paths so arranged that the electric waves are transmitted in one direction only by one path and in the other direction only by the other path. Note: The transmission path may or may not employ four wires*] Vierdrahtschaltung *f*

four-wire technique Vierdrahttechnik *f*

(F)PGA → (field-)programmable gate array

(F)PLA → (field-)programmable logic array

fps → frames per second

fragmentary ice Stückeis *n*

fragmentation protection foil Splitterschutzfolie *f*

frame Rahmen *m*

fragmenting [*In automatic data processing, the breaking up of a document in its various components*] Fragmentieren *n*

Frahm frequency meter → (vibrating-)reed frequency meter

frame → central processing unit

frame Einzelbild *n* [*Innerhalb einer Videodatei*]

frame Wochenprogramm *n* [*Zeiterfassung*]

frame Bitsequenz *f*

frame Halb(fernseh)bild *n* [*Infolge des Zeilensprungverfahrens ergeben erst zwei nacheinander übertragene Halb(fernseh)bilder ein Voll(fernseh)bild*]

frame Lehreinheit *f*

frame Spalte *f* [*Lochstreifen*]

frame abortion Rahmenabbruch *m*

frame and socket (Elektro)Fassung *f*

frame buffer → bit map memory

frame burner Rahmenbrenner *m*

frame check sequence, FCS Zeichensequenz *f* für Übertragungsprüfung [*ISO-OSI Schichtenmodell*]

frame connector Steckerleiste *f*, Vielfachstecker, Mehrfachstecker *m*

frame flyback Bildrücklauf *m* [*Bildschirmgerät*]

frame frequency, picture frequency [*The number of times per second that the frame is completely scanned in television*] Bildfrequenz *f*

frame grabber Digitalisier-Videokarte *f* [*Zur Digitalisierung von Einzelbildern und Videofolgen*]

frame grabbing Einlesen *n* von Videobildern in Echtzeit, Echtzeiteinlesen von Videobildern

frame ground Schutzerde *f*

frame modus Vollbildmodus *m*

frame rating Modelleistung *f* [*Trafo*]

frame reject → FRMR

frames per second, fps Bilder *f* pro Sekunde

frame start indication Puls-Rahmentakt *m*

frame transfer, FT halbbildweise Auslesen *n* der Bildinformation [*Videokamera*]

framework Sammlung *f* wiederverwendbarer Komponenten

framing Bildstricheinstellung *f*

framing, F Start- und Endzeichen *n* eines Telegrammes [*ISO-OSI-Schichtenmodell*]

framing (Ein)Rahmung *f* [*Bei asynchroner Datenübertragung gehen jedem Zeichen ein oder mehrere Startbits voraus und werden von einem oder mehreren Stopbits abgeschlossen. Das Zeichen ist somit von Start- und Stopbits eingerahmt*]

framing pattern, frame pattern Rahmenmuster *n*

fraudulent insurance claim Versicherungsschwindel *m*

freak → (computer) freak

to free → to trip

free → anechoic

free → unobstructed

free air [*Air at atmospheric conditions at any specific location. Because altitude, barometric pressure, and temperature vary at different localities and at differential times, it follows that this term does not mean air under identical or standard conditions. Free air as a measure of volume may be applied either to displacement or capacity and in no way distinguishes between these two*] freie Luft, Freiluft *f*

free (air) jet Frei(luft)strahl *m* [*Er breitet sich im Raum allseitig aus wobei der Ausbreitungswinkel unabhängig von der Geschwindigkeit 23 Grad bis 25 Grad ist*]

free area [*The area of open space in a grille or diffuser through which air can pass without restriction or friction*] Freifläche *f*

free (available) chlorine frei wirksame Chlor, freie Chlor *n*

free-blowing freiblasend [*Ventilator*]

free-burning coal Magerkohle, Sandkohle *f*

free carbon freie Kohlenstoff *m*

free-carrying (sliding) gate freitragende Schiebetor *n*

free choice → user('s) choice

free convection, natural convection *[Thermodynamics. Heat transfer by fluid motion resulting from density gradients in the fluid]* freie Konvektion, natürliche Konvektion *f*

free cooling freie Kühlung *f* *[Direkte oder indirekte Kühlung mittels Außenluft, deren Temperatur unterhalb der Raumlufttemperatur liegt]*

free cross-section freie Querschnitt *m*

free delivery(-type) air cooler kanallose Luftkühler *m*

free discharge diffuser, free discharge diffusor Enddiffusor *m*

freedom of glare Blend(ungs)freiheit *f*

free energy → (Gibbs) free energy

free field *[In automatic data processing, a property of information retrieval devices which permits recording of information in the search medium without regard to preassigned or fixed fields]* Freifeld *n*

free field, anechoic field, nonreverberant field *[In acoustics, an area in which the sound makes spread-out without meeting any object]* (nach)hallfreie Fläche, (nach)hallwidrige Fläche, echofreie Fläche, echowidrige Fläche, reflexionsfreie Fläche, reflexionswidrige Fläche, reflektionsfreie Fläche, reflektionswidrige Fläche, schalltote Fläche *f*

free(field) room → anechoic room

free-field storage *[Data storage that allows recording of the data without regard to fixed or preassigned fields]* Freifeldspeicherung *f*

free-field termination → anechoic termination

free-flow oblique seat valve Schrägsitz-Freistromventil, Freistrom-Schrägsitzventil *n*

free-flow system Freiflusssystem, Freistromsystem *n*

free-flow valve Freiflussventil, Freistromventil *n*

free heat gain → incidental heat gain

free line Ruhezustand *m* *[dezentrale Zeichengabe]*

freely programmable frei programmierbar

free of allowance(s) zuschlagsfrei

free outflow freie Ausfluss *m* *[Der aus einer Öffnung gegen die Atmosphäre fließende Flüssigkeitsstrom]*

free outflowing freie Ausfließen *n*

free outlet (bath) mixer (valve), free outlet (bath) mix(ing) valve, free outlet (bath) blender, free outlet (bath) blending valve drucklose Mischbatterie, Mischbatterie mit freiem Auslauf, Freiauslauf-Mischbatterie *f*

free piston engine Freikolbenmotor *m*

free room → anechoic room

free outlet water heater *[See remark under "water heater"]* Wassererhitzer mit freiem Auslauf, Wasser(er)wärmer mit freiem Auslauf, Wasserheizer mit freiem Auslauf *m*

free-running freilaufend

free site frei Baustelle

free space → open space

free-standing freistehend

free-standing refrigerator Standkühlschrank, freistehende Kühlschrank *m* *[Standkühlschränke von 240 bis 290 Liter Nutzraum werden frei aufgestellt. Sie eignen sich für Speisekammer und Kühlräume]*

free-standing sensor, free-standing detector *[See remark under "sensor"]* freistehende Sensor, freistehende Detektor *m*, freistehende Detektionselement *n*

free termination → anechoic termination

free trial software lauffähige Software *f*

free tripping Freiauslösung *f*

free ventilation → natural ventilation

free ventilation by air density differences → natural ventilation by air density differences

freewheel(ing) diode Freilaufdiode, Rücklaufdiode, Blinddiode *f*, Stromverdopp(e)ler *m*

freezant *[Medium, usually a liquid cooled by the evaporator of a refrigerant system, used to freeze products]* Gefriermittel *n*, Gefrierstoff *m*

to freeze → to refrigerate

freeze digitale Speicherung *f* *[Videobild als Standbild]*

freeze concentration *[Concentrating a solution by partial freezing of water and removing ice crystals]* Gefrierkonzentration *f*

freeze danger, freezing danger Gefriergefahr *f*

freeze desalination *[Freezing saline or brackish water, so that salt-free ice crystals are formed, than removed and melted]* Gefrierentsalzung *f*

freeze dryer, freeze drier Gefriertrockner *m*

freeze drying, lyophilization *[Dehydration of a substance by freezing, and subsequent sublimation of ice]* Gefriertrocknung, Lyophilisation *f*

freeze-grinding → cryogrinding

freezer Gefrierapparat *m*, Gefriergerät *n*

freezer *[Device for freezing perishables]* Gefrierschrank *m*

freezer Gefrierfach *n* *[in einem Kühlschrank]*

freezer capacity *[Amount of produce which can be installed at one operation]* Gefrierschrankfassungsvermögen *n*

freezer-refrigerator unit → (combination) refrigerator-freezer unit

freezer room → frozen food storage room

freeze thaw resistance Gefrier-Auftau-Widerstand *m*

freeze-up, freezing-up Einfrieren *n*

freeze-up control (device), freezing-up control (device) [*Thermostatic device that senses the approach of an icing condition which would impede airflow and makes compensating adjustments*] Gefrierfühler *m*

freezing [*Process of changing a liquid substance or the liquid content of a food or other commodify to a solid state by removing heat*] Gefrieren *n*

freezing cabinets Gefriermöbel *f* [*Gefrierschränke und -truhen*]

freezing chest → chest freezer

freezing compartment → freezing section

freezing danger, freeze danger Gefriergefahr *f*

freezing desalination → desalination by freezing

freezing method Gefriermethode *f*, Gefrierverfahren *n*

freezing mix(ture) [*Physical chemistry. Any mix(ture) whose freezing point is lower than the freezing points of its individual components*] Gefriermischung *f*, Gefriergemisch *n*

freezing of unpacked products → bulk freezing

freezing plant, freezing installation [*It is designed to freeze products*] Gefrieranlage *f*

freezing plateau → heat arrest

freezing point, frost point [*For a particular pressure, the temperature at which a given substance will solidify or freeze upon removal of heat. Freezing point of water is 32 °F (0 °C)*] Gefrierpunkt, Eispunkt *m*

freezing process Gefriervorgang *m*

freezing protection → protection against freezing

freezing range Gefrierbereich *m* [*selten: n*]

freezing room [*Cold chamber usually kept at temperatures of −22° F (−30° C) to −31° F (−35° C), with high volume air circulation*] Gefrierraum *m*

freezing section, freezing compartment [*In a food processing plant, that section were foods are frozen*] Gefrierabteilung *f*

freezing tank [*Tank holding low-temperature fluid for process freezing*] Gefriertank *m*

freezing temperature → low temperature

freezing time, freezing period [*Time for any complete freezing process to take place*] Gefrierzeit, Gefrierdauer *f*

freezing tunnel → tunnel freezer

freezing-up → freeze-up

freight lift (Brit.) → goods elevator

freight-passenger elevator, goods-passenger elevator (US); freight-passenger lift, goods-passenger lift (Brit.) Lastenaufzug *m* [*Im Gegensatz zum Güteraufzug darf der Lastenaufzug Personen befördern*]

French boiler französische Kessel *m* mit äußeren SRiederöhren

French bolt, dormant bolt französische Riegel *m*, hebende Falle *f*

French espagnolette, French sash bolt französische (E)Spagnolettestange *f*

French key, full key, unbored key, undrilled key, preponderating key französische Schlüssel, Vorstecker *m*

French lock, stock lock Schloss *n* mit schießender Falle

French lock, tumbler lock französische Schloss, Zuhaltungsschloss, Schloss mit Zuhaltung *n* [*Erfunden von Johann Gottfried Freitag aus Gera*]

French lock with latch → spring lock with falling latch

French lock with spring bolt → stock lock with spring bolt

French sash bolt → French espagnolette

frequency [*The number of items occurring in a given category*] Häufigkeit *f*

frequency band Frequenzband *n*

frequency characteristic Frequenzgang *m*

frequency component Frequenzanteil *m*

frequency converter Frequenzumrichter, Frequenzwandler *m*

frequency divider, counting-down circuit [*A harmonic conversion transducer in which the frequency of the output signal is an integral submultiple of the input frequency*] Frequenzteiler *m*

frequency division multiplex, FDM Frequenz-Multiplex-System *n*

frequency division multiplexing, FDM Frequenzmultiplexing *n* [*Methode, bei der der für die Übertragung verfügbare Frequenzbereich in schwächere Bänder unterteilt wird, die jeweils als separate Kanäle dienen. Gleich der Breitbandtechnologie wird das Frequenzspektrum in Kanäle getrennt, sodass ein Anwender oder eine Anwendergruppe auf die Kanäle zugreifen kann*]

frequency diversity zeitversetzte Frequenzen *f*

frequency dividing network, crossover Frequenzweichen *f pl*

frequency domain → band

frequency group Frequenzgruppe *f*

frequency hopping, frequency jumping Frequenzspringen *n*

frequency modulation, single density, FM Frequenzmodulation, einfache Aufzeichnungsdichte *f* [*Methode der Datenaufzeichnung bei Disketten*]

frequency modulator Frequenzmodulator *m*

F

frequency multiplier Frequenzvervielfacher, Frequenzmultiplizierer *m*

frequency offset Frequenzabweichung *f*

frequency of operating cycles → operating frequency

frequency of operation → operating frequency

frequency of use, use frequency Benutzerhäufigkeit, Anwenderhäufigkeit *f* [*Beanspruchung von Schlössern. Zuordnung von Türgewicht und Schloss nach DIN 18 251*]

frequency pointer meter → pointer frequency meter

frequency range → band

frequency response Frequenzabhängigkeit *f*, Frequenzgang *m*

frequency response → ((closed) loop) frequency response

frequency response → ((open) loop) frequency response

frequency response characteristics → Bode diagram

frequency response locus Frequenzgangort(s)kurve *f*

frequency schedule Frequenzplan *m* [*Definition der Art und Weise, wie verschiedene Frequenzen eines Breitbandkabels für die Benutzung zugewiesen sind*]

frequency shift → Doppler shift

frequently asked questions, FAQs häufig gestellte Fragen *f pl*

fresh air Frischluft *f*

fresh air box Frischluftkasten *m*

fresh air content, fresh air proportion Frischluftanteil, Frischluftgehalt *m*

fresh air damper Frischluftklappe *f*

fresh air drying Frischlufttrocknung *f*

fresh air duct Frischluftkanal *m*

fresh air fan Frischluftventilator *m*

fresh air generator → air freshener

fresh air heating Frischluftheizung *f*

fresh air hood Frischlufthaube *f*

fresh air index → freshness index

fresh air inlet, fresh air intake, FAI [*A vent connection to a house drain, on the building side of the main drain trap*] Frischlufteinlass *m*

fresh air intake, FAI, fresh air inlet [*An opening through which outside air is brought into an air-conditioning system or into a boiler room*] Frischlufteinlass *m*

fresh air intake stack, fresh air inlet stack [*"Chimney" may synonymously be used for "stack"*] Frischluftschornstein, Frischluftkamin *m*

fresh air operation Frischluftbetrieb *m*

fresh air pipe Frischluftrohr *n*

fresh air preheating Frischluftvorwärmung *f*

fresh air proportion → fresh air content

fresh air shaft Außenluftschacht *m*

fresh air stream Frischluftstrom *m*

fresh air suction pipe Frischluft(an)saugrohr *n*

fresh air supply Frischluftversorgung *f*

fresh air thermostat Frischluftthermostat *m*

freshness index, fresh air index Frischluftindex *m*

freshwater Süßwasser, Frischwasser *n*

freshwater dispenser Süßwasserspender, Frischwasserspender *m*

freshwater feed(ing) Frischwasser(ein)speisung, Süßwasser(ein)speisung *f*

freshwater (pipe)line Frischwasser(rohr)leitung, Süßwasser(rohr)leitung *f*

freshwater softening plant, freshwater softening system Frischwasserenthärtungsanlage, Süßwasserenthärtungsanlage *f*, Frischwasserenthärtungssystem, Süßwasserenthärtungssystem *n*

Fresnel lens [*A thin lens constructed with stepped setbacks so as to have the optical properties of a much thicker lens*] Fresneloptik, Linsenfolie, Fresnellinse *f*

friction(al) gradient Reib(ungs)gefälle *n*

friction(al) loss Reib(ungs)verlust *m*

friction(al) pressure drop Reib(ungs)druckabfall *m*

friction(al) resistance Reib(ungs)widerstand *m*

friction physics → tribophysics

friendly fire [*It is confined to an intended place, for example, a furnace or fireplace*] Nutzfeuer *n*, Nutzbrand *m*

frigistor Kühl(ungs)element *n*

FRMR, frame reject Rahmenrückweisung *f*

front accessible frontzugänglich

front door bell Vordertürklingel, Fronttürklingel *f*

front door telephone → **intercom(munication) system**

front element Vorsatz *m*

front elevation Frontaufriss, Vorderaufriss *m*

front-end [*Of a microcomputer, under programmed instructions, performing data transfers and control operations to relieve a larger computer of these routines*] Vorrechner, Knotenrechner *m*

front entrance door Fronteingangstür *f*

front fixing, front mounting, front fastening Frontmontage, Frontbefestigung *f*

front load Endstelle *f* [*End-zu-End-Prüfung*]

front panel Fronttafel, Vordertafel *f*

front plate Frontplatte, Vorderplatte *f*

front (plate) Stulp *m* [*Einsteckschloss*]

front plug Frontstecker *m*

front plug lead Frontsteckerleitung *f*

front projection Aufprojektion *f*

front view Vorderansicht, Frontansicht *f*

frost [*Water from airborne droplets or vapo(u)r frozen on a surface*] Frost *m*

frost crack Frostriss *m*

frosted, coated with frost, frost-covered frostüberzogen, frostbedeckt, frostbeschichtet

frost formation [*Formation of frost on any cold part of a refrigerating circuit*] Frostbildung *f*

frost heave Frostaufbruch *m*, Frostbeule *f*

frost point → **freezing point**

frost (protection) control [*A special damage control function for switching all relevant parts of a plant into a defined state to avoid frost damage. It is triggered if the temperature drops below a limit detected by a specific sensor (frost thermostat) with latching, locking and acknowledg(e)ment within this function*] Frostschutzschaltung *f*

frost protection ((open) loop) control Frostschutzsteuerung *f*

frost protection security Frostschutzsicherung *f*

frost sensor, frost sensing device, frost pilot device, frost sensing element, frost (primary) detector, frost detecting element, frost detection element [*See remark under "sensor"*] Frostsensor, Frostfühler, Frostdetektor *m*, Frostdetektionselement *n* [*Siehe Anmerkung unter „Sensor"*]

frost thermostat Frostthermostat *m*

frozen food storage room, freezer room [*Room designed to receive and store frozen food*] Lebensmittelgefrierraum, Nahrungsmittelgefrierraum *m*

frozen freshwater gefrorene Süßwasser *n* [*Gegensatz: flüssige Süßwasser*]

frying plate Bratplatte *f*

FSK → **frequency shift keying**

FT → **frame transfer**

FTAM → **file transfer and management**

FTP → **file transfer protocol**

ft^3 per minute → **CFM**

fuel, burning substance, combustion substance Brennstoff *m*, Brennmaterial *n*

fuel-air mix(ture) Brennstoff-Luft-Gemisch *n*, Brennstoff-Luft-Mischung *f*

fuel cell Brennstoffzelle *f*

fuel degassifier Brennstoffentgaser *m*

fuel feeder, fuel feeding device Brennstoffbeschicker, Brennstoffaufgeber *m*

fuel feed(ing) [*On to the grate of a boiler*] Brennstoffbeschickung, Brennstoffaufgabe *f*

fuel feed(ing) interlock Brennstoffverrieg(e)lung *f*

fuel (feed) valve Brennstoffventil *n*

fuel(-fired) water heater [*See remark under "water heater"*] Brennstoffwasserheizer, Brennstoffwarmwasserbereiter *m*

fuel gas Rauchgas *n*

fuel hose Brennstoffschlauch *m*

fuel lifter → **fuel pump**

fuel oil [*Products of petroleum graded according to energy content and viscosity (No.1 to 6); No. 6 is heavy oil, with the highest energy content and viscosity rating. Oils with a flash point below 100°F (38°C) are excluded. Fuel oil includes heating oil, stove oil, furnace oil, and bunker fuel oil. Colloquially called "oil"*] Brennöl, Öl *n*

(fuel) oil gauge; (fuel) oil gage (US); FOG Brennölstandanzeiger *m*

fuel oil No. 2 [*USA*]; **light heating oil** Heizöl EL, leichte Heizöl *n*

fuel oil No. 6 [*USA*]; **heavy heating oil** Heizöl S, schwere Heizöl *n*

fuel oil No. 6 burner [*USA*] → **heavy heating oil burner**

fuel (pipe)line shut off Brennstoff(rohr)leitungs-Absperrung *n*

F

fuel pump, fuel lifter (Brennstoff-)Förderpumpe *f*

fuel store Brennstofflager *n*

fuel strainer Brennstoffseiher *m*

fuel type Brennstoffart *f*

fuel valve → fuel (feed) valve

full angle speaker → FAS

full automatic air conditioning installation, full automatic air conditioning plant vollautomatische Klimaanlage, vollautomatische Klimainstallation *f*

F

full circuit → complete circuit

full current Totalstrom *m*

full demineralization → complete demineralization

full duplex → duplex transmission

full duplex method Vollduplexverfahren *n*, Vollduplexmethode *f* [*Bei diesem Verfahren werden Daten gleichzeitig gesendet und empfangen*]

full heating, full warming Vollheizung *f*

full heat protection Vollwärmeschutz *m*

full key → French key

full load operation Volllastbetrieb *m*

full motor protection Motorvollschutz, Vollmotorschutz *m*

full screen display Ganzseitendarstellung *f* [*Die gesamte Nutzung der Bildschirmfläche vermeidet die schwarzen „Trauerränder"*]

full storage heating Vollspeicherheizung *f*

full stroke safety valve Vollhubsicherheitsventil *n*

full stroke travel time, travel time for full stroke [*valve; damper; piston*] Vollhubzeit *f*

full text [*As opposed to coded messages or mnemonic abbreviations*] Klartext *m*

full treatment → complete treatment

full warming → full heating

full-wave bridge module Vollwellenbrückenbaustein, Vollschwingung(s)brückenbaustein *m*, Vollwellenbrückenmodul, Vollschwingung(s)brückenmodul *n*

full-wave rectification Vollwellengleichrichtung *f*

full-way valve [*When fully opened, it does not impede the flow of water; used where water pressure is very low*] Durchgangsventil *n*

fully automatic hoistway gate → hoistway fully automatic door

fully desalinated water vollentsalzte Wasser, VE-Wasser, ganzentsalzte Wasser *n*

fully encapsulated electronics → complete encapsulated electronics

fully heated, completely heated vollbeheizt, ganzbeheizt

fully integrated → totally integrated

fully operating → operating at full capacity

fully recessed volleingelassen

fume Dunst *m*, Schwaden, Wrasen *f* [*Norddeutschland*]

fume hood Dunsthaube, Schwadenhaube, Wrasenhaube *f* [*Norddeutschland*]

function Funktion *f* [*Zusammenhang zwischen Ursache und Wirkung, zwischen Eingangsgröße und Ausgangsgröße oder Sensor und Aktor*]

functional nutzungsspezifisch, funktional, funktionell, zweckorientiert, zweckbezogen

function(al) area Funktionsfläche *f* [*Nicht verwechseln mit „Nutzungsbereich"*]

function(al) block → transmission block

function(al) check (test), function(al) checkout Funktionskontrolle, Funktions(über)prüfung *f*

function(al) coding Funktionskodierung, Funktionscodierung *f*

function(al) design funktionelle Gestaltung, zweckdienliche Gestaltung *f*

function(al) display, function(al) indication Funktionsanzeige *f*

function(al) division Funktions(auf)gliederung *f*

functional extra low voltage, FELV Funktionskleinstspannung, Kleinstfunktionsspannung *f*

function(al) failure, function(al) out(age) Funktionsausfall *m*, Funktionsversagen *n*

function(al) group Funktionsgruppe *f*

function(al) instruction [*See remark under "instruction"*] Funktionsbefehl *m*

function(al) interconnection Funktionsverbund *m*

function(al) interruption Funktionsunterbrechung *f*

function(al) key → functions key

functionally-linked funktionsverbunden

function(al) model Funktionsmodell *n*

function(al) monitoring Funktionsüberwachung *f*

function(al) number Funktionszahl *f*

functional ((open) loop) control Funktionssteuerung *f*

function(al) overvoltage Funktionsüberspannung *f*

function(al) procedure [*See remark under "procedure"*] Funktionsprozedur, Funktionsvorgehensweise *f*

function(al) punch → designation punch

function(al) reader Funktionsleser *m*

function(al) representation funktionelle Darstellung *f* [*MSR-Technik. DIN 19 227, Teil 1*]

function(al) safety Funktionssicherheit *f*

function(al) scope, function(al) extent Funktionsumfang *m*

function(al) security Funktionssicherung *f*

function(al) semantics Funktionssemantik *f*

function(al) sequence Funktionsablauf *m*, Funktionsfolge *f*

function(al) unit Funktionseinheit *f* [*Ein nach Aufgabe und/oder Wirkung abgrenzbares Gebilde*]

function-associated, function-related funktionsbezogen, funktionsorientiert

function block → ((closed) loop) function block

function call Funktionsaufruf *m* [*Weil nach Abarbeitung des Rumpfes ein Ergebniswert zurückgegeben wird, erfolgt der Aufruf in der Regel in einem Ausdruck, damit das Ergebnis weiterverarbeitet werden kann. Die Auswertung des Ausdrucks wird dabei so lange unterbrochen, bis der Funktionswert ermittelt ist*]

function code [*Special code which appears on a medium such as a paper tape and which controls machine functions such as a carriage return*] Funktionskode, Funktionscode *m*

function description Funktionsbeschreibung *f*

function key, functional key [*A special key on a keyboard to control a mechanical function, initiate a specific computer operation, or transmit a signal that would otherwise require multiple key strokes*] Funktionstaste *f*

function point Funktionsstelle *f*

function point diagram Funktionsstellenbild *n*

function-related → function-associated

function table [*Sets of computer information arranged so that an entry in one set selects one or more entries in the other set(s)*] Funktionstabelle, Funktionstafel, Arbeitstabelle, Arbeitstafel *f*

fundamental combination, basis combination Grundverknüpfung *f*

fundamental equation, basic equation Grundgleichung *f*

fundamental frequency [*Not to be confused with "basis frequency"*] Grundfrequenz *f*

fundamental heat circuit diagram Wärmegrundschaltplan, Grundschaltwärmeplan *m*

fundamental mode [*In physics, the normal mode of vibration having the lowest frequency*] Grundschwingung *f*

(fundamental) particle → elementary particle

fundamental points → basis

fundamental (tone) → pure tone

furnace [*Enclosed structure or chamber incorporating a burner, a heat exchanger, a jacket (or casing), a blower, a filter, and automatic control, in which heat is produced, as by burning fuel, or by converting electrical energy, to create a chemical or physical change upon a substance. An automatic humidifier is often included*] Feuerung, Feuerstätte *f* [*DIN 18 160*]

furnace → hot air furnace

furnace air ag(e)ing → circulating air ag(e)ing

furnace blower Feuerungsgebläse *n*, Feuerungsbläser *m*

furnace efficiency Umsetzungswirkungsgrad, Feuerungswirkungsgrad η (Eta F) *m* [*Das Verhältnis von nutzbar gemachter Wärmeenergie zu zugeführter Wärmeenergie*]

furnace engineering Feuerungstechnik *f*, Feuerstättentechnik [*DIN 18 160*]

furnace firebox, furnace combustion chamber Feuerungskammer *f*

furnace flue → flue (tube)

furnace flue (steam) boiler → fire-tube (steam) boiler

furnace front Feuerungsgeschränk *n*

furnace masonry Feuerungsausmauerung, Feuerstättenausmauerung *f*, Feuerungsmauerwerk, Feuerstättenmauerwerk *n*

furnace pressure controller Feuerungsdruckregler *m*

furnace room Feuerungsraum *m*

furnace stoker [*Device which automatically feeds solid fuel to a furnace. It is used mainly with coal*] automatische Feuerungsbeschicker, Feuerungsbeschickungsautomat *m*

furnace volume [*Space enclosed by the chamber floor, the refractory walls, the heat-absorbing water walls of a boiler firebox, and the crown sheet, shell, or water tubes of a boiler. If a target wall (bridgewall) is installed in the furnace, the furnace volume is reduced by the volume behind the bridgewall*] Feuerungsraum *m*

furnace with inclined grate and side hoppers (Kessel)Schüttfeuerung *f* mit seitlicher Brennstoffzufuhr

F

furnace with internal horizontal grate → locomotive furnace

furnace with step(ped) grate, step(ped) grate furnace Treppenrostfeuerung *f*

(to) furnish and (to) install liefern und einbauen

furniture → mountings

furniture connector Möbelstecker *m*

furniture item Möbelstück *n*

furniture vanity Möbelwaschtisch *m*

fuse, fuze [*Electricity. A protective device based on a wire or element that melts at low temperature. When the current through the fuse exceeds the fuse rating, the wire or element melts and opens the circuit*] (Schmelz)Sicherung *f*

fuse base, fuse mount Sicherungssockel *m*, Sicherungsunterteil *m, n*

fuse block Sicherungsleiste *f*

fuse board [*A board carrying fuses for a number of circuits in an electric installation*] Sicherungs(verteiler)tafel *f*

fuse carrier Sicherungseinsatzhalter, Sicherungseinsatzträger *m*

fuse compartment Sicherungsfach *n*

fused end abgesicherte Ausgang *m*, abgesicherte Ende *n* [*Ausgang eines Funktionsgliedes, bei dem der Ausgangsstrom im Kurzschlussfall durch eine Schmelzsicherung oder durch elektromechanische oder elektronische Maßnahmen unterbrochen wird. DIN 19 237*]

fuse enclosure insert Sicherungskasteneinsatz *m*

fuse failure relay Sicherungsausfallrelais *n*

fuseless sicherungslos, ohne Sicherung

fuse disconnect(ing) switch, fuse disconnector [*Electric switch designed to isolate part of the system when required. It is provided with fuses for the protection of the equipment. IEC 408*] schmelzbare Trennschalter, abgesicherte Trennschalter, Sicherungstrennschalter *m* [*VDE 0660, Teil 107*]

fuse holder [*Combination of a fuse base and its fuse carrier*] Sicherungshalter *m*, Sicherungsunterteil *m, n* mit Sicherungseinsatzträger [*IEC 50(441), 1974*]

fuse load isolator Sicherungslasttrenner *m* [*Elektrotechnik*]

fuse monitor Sicherungsüberwacher *m*

fuse monitoring Sicherungsüberwachung *f*

fuse mount → fuse base

fuse puller Sicherungsaufsteckgriff *m*

fuse tester Sicherungsprüfer *m*

fus(ibl)e contact → fus(ibl)e terminal

fus(ibl)e cutout Ausfallsicherung, Trennsicherung *f*

fus(ibl)e element Sicherungselement *n*, Schmelzleiter *m*

fus(ibl)e element sprinkler Schmelzleitersprinkler, Sicherungselementsprinkler, Schmelzleitersprenger, Sicherungselementsprenger *m* [*Er wird bei Wärmeeinwirkung durch das Schmelzen eutektischer Lote geöffnet*]

fus(ibl)e (hand) grip, fus(ibl)e handle, fus(ibl)e puller Sicherungs(aufsteck)griff *m*

fus(ibl)e indicator Schmelzanzeiger, Schmelzkörper *m*

fus(ibl)e interruptor → fus(ibl)e shunt

fus(ibl)e lead (cutout) Blei(schmelz)sicherung *f*

fus(ibl)e link [*Part of a fuse that carries the current of the circuit and all or part of which melts when the current exceeds a predetermined value*] Sicherungseinsatz, Schmelzeinsatz *m*, Durchschmelzverbindung *f*

fus(ibl)e (metal) strip Schmelz(metall)streifen *m*

fus(ibl)e plug, safety plug [*A protective device used on a heated pressure vessel (for example a steam boiler), and containing a fusible element that melts at a predetermined safe temperature to prevent the buildup of excessive pressure*] Schmelzpfropfen, Schmelzstöpsel *m*

fus(ibl)e plug low-water alarm → low-water alarm with fus(ibl)e plug

fus(ibl)e PROM, fus(ibl)e programmable read-only memory, fus(ibl)e programmable ROM [*A programmable read-only memory chip created by selectively blowing a series of fuses to create binary patterns on the chip*] ROM-Speicher mit Schmelzsicherung, programmierbare Festwertspeicher mit Schmelzsicherung, programmierte Nur-Lese-Speicher mit Schmelzsicherung *m*

fus(ibl)e puller → fus(ibl)e (hand) grip

fus(ibl)e shunt, fus(ibl)e interruptor Schmelzunterbrecher *m*

fus(ibl)e solder [*An alloy, usually containing bismuth, having a melting point below 361° F [183° C]*] Schmelzlot, Sicherungslot *n*

fus(ibl)e solder sprinkler Schmelzlotsprinkler, Schmelzlotsprenger *m*

fus(ibl)e switch [*A unit in which the fuses themselves move during the switching action. It is generally used to control heavy loads. IEC 408*] (Last)Schalter mit Sicherungen, Sicherungs(last)schalter *m* [*VDE 0660. Teil 107*]

fus(ibl)e switch-disconnector Sicherungs(last)trenn(schalt)er *m*

fus(ibl)e terminal, fus(ibl)e contact Sicherungsklemme *f*, Sicherungskontaktstück *n*

fus(ibl)e unit Schmelzeinsatz, Sicherungseinsatz *m*

fus(ibl)e wire Schmelzdraht, (Schmelz)Sicherungsdraht *m*

fusion heat → heat of fusion

future (placement) point Reserveeinbaustelle *f* [*Gebäudeinstallation*]

fuze → fuse

fuzzification Fuzzifizierung *f* [*Umsetzung von scharfen Signalwerten in Zugehörigkeitsgrade von linguistischen Werten*]

fuzzy AND Fuzzy-UND-Operator *m*, unscharfe UND-Verknüpfung *f*

fuzzy ((closed) loop) control unscharfe Reg(e)lung, Fuzzy-Reg(e)lung *f* [*Eine Art vorausschauende Regelung, die auf der Bewertung einer Anzahl subjektiver Eingänge basiert, mit dem Ziel, ein Ausgangssignal zu erzeugen, das die beste Regelgüte gewährleistet.*]

fuzzy controller, fuzzy regulator unscharfe Regler, Fuzzy-Regler *m*

fuzzy decision unscharfe Entscheidung, Fuzzy-Entscheidung *f*

fuzzy inference → fuzzy reasoning

fuzzy inference system, FIS, fuzzy reasoning system unscharfe Schlussfolgerungssystem *n*

fuzzy logic unscharfe Logik, Fuzzylogik *f* [*Direkte Umsetzung von empirischen Prozesswissen in Form von umgangssprachlichen Beschreibungen in ein technisches System*]

fuzzy number unscharfe Zahl, Fuzzyzahl *f*

fuzzy ((open) loop) control unscharfe Steuerung, Fuzzysteuerung *f*

fuzzy operator unscharfe Operator, Fuzzyoperator *m*

fuzzy-PID-controller PID-Regler *m*

fuzzy quantity → fuzzy set

fuzzy reasoning, fuzzy inference unscharfe Schließen *n*, unscharfe Schlussfolgerung *f*

fuzzy relation(ship) unscharfe Beziehung, Fuzzybeziehung *f*

fuzzy set, fuzzy quantity unscharfe Menge, Fuzzymenge *f*

FW, feedwater Speisewasser *n*

FZP → fast zero power

G

gaffer → gas fitter

gage (US) → gauge

gain, signal gain [*The increase in signal power that is produced by an amplifier; usually given as the ratio of output to input voltage, current, or power, expressed in decibels*] Verstärkung, Signalverstärkung *f*

gain → ((closed) loop) gain

gain → aerial gain

gain, reinforcement Verstärkung *f*

gain-crossover frequency [*Control system. The frequency at which the magnitude of a loop ratio is unity*] Durchtritt(s)kreisfrequenz *f*

gain margin Amplitudenreserve *f* [*Reg(e)lungstechnik*]

gain response → amplitude response

gallium arsenide Galliumarsenid, GaAs *n* [*Halbleitermaterial für photovoltaische Zellen*]

galvanic connection → direct connection

galvanic isolation → electrical isolation

galvanization, zinc coating Verzinkung *f*, Verzinken *n*

galvanized, zinc-coated verzinkt

gamma radiation Gammastrahlung *f*

GAN → global area network

ganged switch, multi(ple)-switch [*A switch in a group of switches arranged next to each other in ganged outlet boxes*] gekuppelte Schalter, Mehrfachschalter, Vielfachschalter *m*

Gantt chart see under **"bar graph"**

gap [*A uniformly magnetized area in a magnetic storage device (tape, disk), used to indicate the end of an area containing information*] Spalt *m*

gap, space Zwischenraum *m*, Lücke *f*

gap, discharge path Entladungsstrecke *f*

gap arrester Funkenableiter *m*

gap coding [*A process for conveying information by inserting gaps or periods of nontransmission in a system that normally transmits continuously*] Pausenkodierung, Lückenkodierung, Pausencodierung, Lückencodierung *f*

gap gauge; gap gage (US) Funkenstreckenlehre *f*

gapless lückenlos, ohne Lücke

gapless (magnetic) tape [*A magnetic tape upon which raw data is recorded in a continuous manner; the data are streamed onto the tape without the word gaps; the data still may contain signs and end-of-record marks in the gapless form*] Magnetband ohne Lücke, lückenlose Magnetband *n*

gap of flameproof joint, flameproof joint gap zünddurchschlagsichere Spaltweite *f*

gapped core Luftspaltkern, gescherte Kern, Kern mit Luftspalt, Scherkern *m*

G

gapping Scherung *f*, Scheren *n* [*Kern*]

gap range Unterbrechungsbereich, Lückenbereich *m* [*selten: n*] [*Messgerät*]

gap section Teilfunkenstrecke *f*

gap width [*magnetic head*] Spaltbreite *f*

garage door drive Garagentorantrieb *m*

garage door lock Garagentorschloss *n*

garage gate transmitter Garagentorsender *m*

garbage → food waste(s)

garbage, hash, gibberish, erroneous data, irrelevant data [*Data which are obviously meaningless, caused by human mistakes or computer malfunction*] Datenabfall *m*, Datenabfälle, unsinnige Rechnerergebnisse *f*

garbage collector, erroneous data collector, gibberish collector, irrelevant data collector, hash collector Speicher(be)reiniger, Speichersäuberer *m*

garbage incinerator → food waste(s) incinerator

Garden City latch (Brit.) → Canadian latch

garden gate Gartentor *n*

garden hydrant Gartenhydrant *m*

gargle basin Gurgelbecken *n* [*Zwischen beiden Becken eines Doppelwaschtisches*]

garish grell

garishness Grellheit *f*

garnet hinge Scheinhaken *m* [*Schlosserei*]

gas [*Vapo(u)r phase or state of a fluid*] Gas *n*

gas adsorption filter Gasadsorptionsfilter *m, n*

gas air heater [*It consists of a series of vertical tubes placed across the air stream, each tube having its gas burner at the base, and the top of the tubes connecting to a fume box which requires a flue to conduct the fumes out of the building, except in such places as foundries where it may be considered permissible to discharge it inside*] Gaslufterhitzer, Gasluft(er)wärmer *m*

gas air mix(ture), air gas mix(ture) Gas-Luft-Gemisch, Luft-Gas-Gemisch *n*, Gas-Luft-Mischung, Luft-Gas-Mischung *f*

gas alarm device, gas alarm appliance [*It is used in connection with a gas sensing detector*] Gasalarmgerät *n*, Gasalarmvorrichtung *f*

gas apparatus Gasapparat *m* [*Oberbenennung für Gasgerät und Gasfeuerstätte*]

gas appliance, gas device Gasgerät *n*, Gasvorrichtung *f*

gas application Gasanwendung *f*

gas assembler → gas fitter

gas at saturation temperature [*Pressure equilibrium. No superheat is present; if any heat is removed, some of the gas will condense to liquid*] Gas *n* bei Sättigungstemperatur

gas (base) cabinet range, gas undercounter range Gasunterbauherd, Unterbaugasherd *m*

gas bath stove Gasbadeofen *m*

gas bellows meter → dry (gas) meter

gas blower burner → blower (gas) burner

gas blower furnace → gas(-fired) blower furnace

gas boiler → gas (heating) boiler

gas bottle → service cylinder

gas bubble → bubble

gas burner Gasbrenner *m* [*DIN 4788*]

gas burner/oil burner boiler → oil burner/gas burner boiler

gas calorimeter Gaskalorimeter *n*, Gaswärmemesser *m*

gas cartridge Gaspatrone, Gaskartusche *f*

gas central distributor → gas manifold

gas central heating boiler, central gas heating boiler Gaszentralheiz(ungs)kessel, Zentralheiz(ungs)gaskessel *m*

gas chromatography [*Analytical chemistry. A technique for separating gas mix(tur)es, in which the gas is passed through a long column containing a fixed absorbent phase that separates the gas mix(ture) into its component parts*] Gaschromatographie *f*

gas circuit Gaskreislauf *m*

gas circulation water heater, gas circulation water heating appliance Gasumlaufwassererhitzer, Gasumlaufwasserheizer, Gasumlaufwasser(er)wärmer *m* [*Eine Gasanwendungsanlage für Pumpenwarmwasserheizung*]

gas ((closed) loop) control block, gas regulation block Gasregel(ungs)block, Gasreglungsblock *m*

gas closing cock → gas shut(ting)-off cock

gas clothes dryer, gas clothes drier Gaswäschetrockner *m*

gas-coal range Gas-Kohle-Herd *m*

gas cock Gashahn *m*

gas combustion chamber Gasverbrennungskammer *f*

gas compression horse-power, gas compression H.P. Gasverdichtungs-PS, Gaskompressions-PS *f*

gas compressor [*A machine which draws in a gas and delivers it at a higher pressure*] Gasverdichter, Gaskompressor *m*

gas condition, gas state Gasbeschaffenheit *f*, Gaszustand *m*

gas connection Gasanschluss *m*, Gasverbindung *f*

gas connection line Gasanschlussleitung, Gasverbindungsleitung *f*

gas consumption → gas input

gas consumption equipment, gas input equipment Gasverbrauch(s)einrichtung(en) *f* [*DIN 3362, 3364, 3369 und 3372*]

gas continuous water heater → gas no-storage water heater

gas control valve → gas ((open) loop) control valve

gas convector [*It is useful for warming halls, staircases, shops, stores, etc., where there is normally good enough ventilation to remove combustion products*] Gaskonvektor *m*

(gas) conversion burner, combined burner, combination gas-oil burner kombinierte (Gas-Öl-) Brenner *m* [*Er gestattet ein Umschalten von einem auf den anderen Brennstoff*]

gas cooker, gas cooking appliance, gas cooking unit Gaskocher *m*, Gaskochgerät *n*

gas cooling Gaskühlung *f* [*Kühlung mit Gas als Antriebsenergie*]

gas curtain [*Combustion products distributed across the doorway of a furnace to prevent air entering the working chamber*] Gasschleier, Gasvorhang *m*

gas cushion Gaspolster *n* [*Österreich: m*]

gas cutoff (device), gas switch(ing) off device, gas cutout (device), gas switching out device Gasabschalter, Gasausschalter *m*, Gasabschaltvorrichtung, Gasausschaltvorrichtung *f*

gas cylinder → service cylinder

gas damper [*See remark under "damper"*] Gasklappe *f*

gas detecting Gasmelden, Gaserkennen *n*

gas detection Gasmeldung, Gaserkennung *f*, Gasnachweis *m* [*Das Ergebnis des Gasmeldens*]

gas detector, gas detecting device, gas leakage detector, gas leakage detecting device, gas (leakage) detection device Gasmelder *m*, Gasmeldegerät *n*

gas device → gas appliance

(gas) discharge lamp, gas lamp Gasentladungslampe *f*

(gas) discharge valve, gas valve Gas(ablass)ventil *n*

gas dispatching zentrale Gasverteilung, zentrale Gasabfertigung *f*

gas distribution Gasverteilung *f*

gas (distribution) net(work) Gas(verteilungs)netz *n*

gas domestic boiler → domestic gas boiler

gas domestic hot water heater, gas domestic warm water heater Gas-Brauchwasser(er)wärmer, Gas-Brauchwassererhitzer, Gas-Brauchwasserheizer *m*

gas duct [*See remark under "duct"*] Gaskanal *m*

gas engine [*An internal combustion engine that uses gaseous fuel*] Gasmotor *m*

gas engine-driven gasmotorisch angetrieben, gasmotorgetrieben

gas engineering Gastechnik *f* [*Sie umfasst Gasgewinnung, Gasaufbereitung, Gasfortleitung, Gasspeicherung, Gasanwendung und Abgastechnik*]

gas engine heat pump Gasmotor-Wärmepumpe *f*

gas enrichment [*Increasing the heat content of a gas by mixing with a gas of higher heat content*] Gaserwärmung, Gaserhitzung *f*

gaseous gasförmig

gaseous fuel gasförmige Brennstoff *m*

gaseous products of combustion → flue gas

gaseous refrigerant gasförmige Kältemittel *n*

gas equipment Gasgerät(e) *n*, (*f*) [*Gasverbrauchseinrichtungen, deren Abgase nicht durch eine Abgasanlage ins Freie abgeführt werden, z. B. Gasherd*]

gas facility, gas setup Gaseinrichtung *f*

gas fill(ing) valve Gasfüll(ungs)ventil *n*

gas filter Gasfilter *m, n*

gas-fired gasbefeuert, gasgefeuert

gas-fired, gas-heated [*By the combustion of gas*] gasbeheizt, gasgeheizt

gas(-fired) absorption heat pump, gas-heated absorption heat pump Gasabsorptionswärmepumpe *f*

G

G

gas(-fired) blower furnace, gas-heated blower furnace, blower gas(-fired) furnace, blower gas-heated furnace Gebläsegasfeuerung, Gebläsegasfeuerstätte, Gasgebläsefeuerung, Gasgebläsefeuerstätte *f*

gas(-fired) boiler, gas-heated boiler Gaskessel *m*

gas(-fired) heat pump, gas-heated heat pump Gaswärmepumpe *f*

gas(-fired) hot air generator, gas(-fired) warm air generator, gas-heated hot air generator, gas-heated warm air generator Gaslufterhitzer, Gasluft(er)wärmer *m*

gas(-fired) hot water heating boiler, gas(-fired) warm water heating boiler, gas-heated hot water heating boiler, gas-heated warm water heating boiler Gaswarmwasserheiz(ungs)kessel *m*

gas(-fired) infrared heater Gas-Infrarotheizer, Gas-Infrarot(er)wärmer *m*

gas(-fired) radiant heater Gas-Strahlungsheizer *m*

gas(-fired) towel rail, gas-heated towel rail [*It is heated directly or indirectly by a gas burner*] gasbeheizte Handtuchschiene, gasgeheizte Handtuchschiene *f*

gas fitter, gaffer, gas assembler Gasmonteur, Gasschlosser *m*

gas float → **gas manifold**

gas flow(ing) Gasströmung *f*, Gasstrom *m*

gas furnace [*Enclosed chamber or structure in which heat is produced by burning gas*] Gasheiz(ungs)-ofen *m*, Gasfeuerung, Gasfeuerstätte, Gasfeuerungsanlage *f* [*DIN 4756*]

gas group [*The classification of gases according to their Wobble number*] Gasfamilie, Gasgruppe *f*

gas-heated → **gas-fired**

gas heater, gas heating appliance, gas heating device, gas heating apparatus [*A fluid appliance for heating one room mainly by emission of radiant heat*] Gasheizer, Gasheiz(ungs)apparat, Gaserhitzer, Gas(er)wärmer *m*, Gasheiz(ungs)gerät *n*

gas heater → **(recessed) (wall)) gas heater**

gas heating Gasheizung *f*

gas heating appliance → **gas heater**

gas (heating) boiler Gasheiz(ungs)kessel, Gaskessel *m*

gas heat pump Gaswärmepumpe *f*

gas heating range Gasheiz(ungs)herd *m* [*Er besteht aus Gasherd, Gasbackofen und Gasheiz(ungs)teil*]

gas heat value, gas thermal value Gaswärmewert *m* [*Sammelbenennung für „Brennwert" und „Heiz(ungs)wert"*]

gas high-pressure (pipe)line, high-pressure gas (pipe)line Gashochdruck(rohr)leitung, Hochdruckgas(rohr)leitung *f* [*DIN 2470*]

gas hose Gasschlauch *m* [*DIN 3383 Teil 1 und Teil 2*]

gas hot air generator, gas warm air generator Gas-Warmlufterzeuger *m*

gas hot water heater Gas-Warmwassererhitzer, Gas-Warmwasser(er)wärmer, Gas-Warmwasserheizer *m*

gas hot water heating boiler → **gas(-fired) hot water heating boiler**

gas house boiler → **domestic gas boiler**

gas house connection Gashausanschluss *m*, Gashausverbindung *f*

gas infrared radiant heater Gasinfrarotstrahlungsheizer *m*

gas infrared radiant heating plant, gas infrared radiant heating installation Gasinfrarotstrahlungsheiz(ungs)installation *f*

gas infrared radiator Gasinfrarotstrahler, Infrarotgasstrahler *m*

gas inlet, gas intake Gaseinlass, Gaseintritt *m*

gas input, gas consumption [*Amount of gas in cubic feet per hour or in British thermal units per hour that an appliance consumes*] Gasverbrauch *m*

gas input equipment → **gas consumption equipment**

gas insert burner, insert gas burner Gaseinsatzbrenner, Einsatzgasbrenner *m*

gas installation → **gas plant**

gasket, door gasket [*A continuous strip of resilient material attached to a door or doorframe to provide a tight seal(ing) between door and frame; acts as weather stripping, light seal(ing) and sound seal(ing)*] Tür(ab)dichtung *f*

gasket [*Any resilient material, used at a joint to prevent leakage*] Flach(ab)dichtung, (Ab)Dicht(ungs)scheibe *f*

gas kitchen range → **(kitchen) gas range**

gas lamp → **gas (discharge) lamp**

gas lance Gaslanze *f* [*Zum Anzünden festen Brennstoffes*]

gas leakage detector → **gas detector**

gas light Gaslicht *n*

gas lighting, gas illumination Gasbeleuchtung *f*

gas-like paint gasförmige Farbe *f* [*Farbdrucker*]

gas liquor Gasabwasser *n*

gas magneto valve Gasmagnetventil *n*

gas main Gastransportleitung *f*

gas manifold, gas central distributor, gas rail, gas float, burner rail [*The component of an appliance carrying gas to a number of separate burners*] Gasverteilerrohr, Gasverteilungsrohr, Gasverteilerstück, Gasverteilungsstück *n*

gas metering Gaszählung, Gasmessung *f*

gas monitor Gasüberwacher, Gasmonitor *m*

gas net(work) → gas (distribution) net(work)

gas meter [*An instrument for measuring, indicating and sometimes recording the volume of gas which passes through it, without interrupting the flow of gas*] Gaszähler, Gasmesser *m*

gas no-storage water heater, gas continuous water heater, no-storage gas water heater, continuous gas water heater Gasdurchlaufwasserheizer, Gasdurchlaufwarmwasserbereiter, Gasdurchlaufwasser(er)wärmer, Durchlaufgaswasser(er)wärmer *m*, Gas-Therme *f*

gas nozzle Gasdüse *f*

gas odour → gas smell

gas oil [*35 seconds*] Heizöl-EL, extra leichtflüssige Heizöl, HEL, EL-Heizöl *n*

gas(oline) (US); petrol (Brit.); motor spirit, naphtha Benzin *n*

gas(oline) (intercepting) trap (US) → petrol (intercepting) trap

gas(oline) interceptor (US) → petrol separator

gas ((open) loop) control valve Gassteuer(ungs)-ventil *n*

gas operated refrigerator, gas operated refrigerating unit, gas operated refrigeration unit [*See remark under "refrigerating unit"*] gasgetriebene Kälteaggregat *n*, gasgetriebene Kälteeinheit *f*

gas output, gas performance Gasleistung *f*

gas oven Gasofen *m*

gas piping, gas pipework Gas(rohr)leitung, Rohrgasleitung *f*

gas plant, gas installation Gasanlage *f*

gas plate Gasplatte *f*

gas pliers Gaszange *f*

gas port, gas orifice Gasöffnung *f*

gas pressure ((closed) loop) control, ((closed) loop) gas pressure control, gas pressure feedback control, gas pressure closed control, gas pressure regulation Gasdruckreg(e)lung, Gasdruckregulierung *f* [*Sie reduziert hohen Gasdruck ein- oder mehrstufig auf den Wert der nachfolgenden Anlage*]

gas pressure ((closed) loop) controller Gasdruckregler *m*

gas pressure differential [*wet (gas) meter*] (einseitige) Gasüberdruck *m*

gas pressure (fire) extinguisher → water (fire) extinguisher

gas pressure gauge; gas pressure gage (US) Gasmanometer *n*, Gasdruckmesser *m*

gas pressure monitor, gas pressure monitoring device Gasdruckwächter *m*, Gasdrucküberwachungsgerät *n*

gas pressure sensor Gasdrucksensor, Gasdruckfühler *m*

gasproof → gastight

gas radiant convector Gasstrahl(ungs)konvektor *m*

gas radiant panel Gasstrahl(ungs)platte *f*

gas radiation Gasstrahlung *f*

gas radiation heating, radiation gas heating Gasstrahl(ungs)heizung, Strahl(ungs)gasheizung *f*

gas radiator [*See remark under "radiator"*] Gasradiator, Gasgliederheiz(ungs)körper; Gasheiz(ungs)körper *m* [*Fehlbenennung*]

gas rail → gas manifold

gas range → (kitchen) gas range

gas range cock Gasherdhahn *m*

gas refrigerating system Gaskältesystem *n*

gas regulation block → gas ((closed) loop) control block

gas refrigerating plant, gas refrigerating installation, gas refrigeration plant, gas refrigeration installation Gaskälteanlage, Gaskälteinstallation *f*

gas regulator [*Device for controlling and maintaining a uniform gas supply pressure*] Gasregler *m*

gas release Gasfreigabe *f*

(gas) rest pressure (Gas)Ruhedruck *m* [*Der Druck des ruhenden Gases bei geschlossenen Hähnen*]

gas riser, gas rising (pipe)line [*A fuel gas supply pipe which extends vertically one story or more*] Gassteig(e)leitung, Steig(e)gasleitung *f*

gas room heater, room gas heater, gas room heating appliance, room gas heating appliance [*"space" may synonymously be used for "room"*] Gasraumheizer, Gasraumerhitzer, Gasraum(er)wärmer, Raumgasheizer, Raumgaserhitzer, Raumgas(er)wärmer *m*

gas safety controller, gas safety regulator [*It automatically opens an alternative gas supply into a main should the pressure of gas in that main fall*

G

below a predetermined value as a result of the failure or temporary inadequacy of the normal service of supply] Gassicherheitsregler *m*

gas safety valve Gassicherheitsventil *n*

gas saturation Gassättigung *f*

gas seal [*Any device other than a cock or valve used for preventing the passage of gas. In its most common form it employs water, liquor or tar as the sealing medium. Granular materials such as sand may also be used*] Gasverschluss *m*

gas security [*"security" should not be confused with "safety"*] Gassicherung *f* [*Siehe Anmerkung unter „Sicherung"*]

G

gas seepage Gaseinbruch *m*

gas service [*e.g. in buildings*] Gasbetrieb *m*

gas service (pipe)line [*It is used to convey unmeasured gas from the gas main to the premises. It extends underground to the inside face of the foundation wall*] Gasanschlussleitung *f*

gas setup → gas facility

gas shut(ting)-off cock, gas closing cock, gas closure cock, gas closeoff cock, gas stop cock Gas(ab)sperr(ungs)hahn, Gasschließhahn *m*

gas sieve Gassieb *n*

gassing current Gasungsstrom *m* [*Anteil des Ladestromes, der für die Erzeugung von Wasserstoff und Sauerstoff aus dem Elektrolyt verbraucht wird. Dieser Strom steigt mit zunehmender Spannung und Temperatur*]

gas single heating, single gas heating Gaseinzelheizung, Einzelgasheizung *f*

gas slide valve Gasschieber *m*

gas smell, gas odour, gas odor (US) Gasgeruch *m*

gas sniffer Gasschnüffler *m*, Gasspürgerät *n*

gas solenoid valve [*An electro-magnetically operated valve for opening and closing the gas supply*] Gasmagnetventil, Magnetgasventil *n*

gas space heater → gas room heater

gas special boiler Gasspezialheiz(ungs)kessel *m*

gas state → gas condition

gas stop cock → gas shut(ting)-off cock

gas storage Gasspeicherung, Gasbevorratung *f*

gas storage (water) heater, storage gas (water) heater [*See remark under "storage (water) heater"*] Gasvorratswasserheizer, Gasvorratswasser(er)wärmer, Gaswarmwasserspeicher, Gasspeicherwasserheizer, Gasspeicherwasser(er)wärmer, Speichergaswasserheizer *m*, Gasspeichergerät *n*

gas storing Gasspeichern, Gasbevorraten *n*

gas stream Gasstrom *m*

gas-supplied gasversorgt, gasbeliefert

gas supply Gaszuführung, Gaszufuhr *f* [*z. B. zu einem Brenner. Nicht verwechseln mit „Gasversorgung"*]

gas supply Gasversorgung *f*, Gasbelieferung

gas supply facility, gas supply setup Gasversorgungseinrichtung, Gasbelieferungseinrichtung *f*

gas supply plant, gas supply installation Gasversorgungsanlage, Gasbelieferungsanlage *f* [*DIN 3381*]

gas supply pressure Gasversorgungsdruck, Gasbelieferungsdruck *m*

gas surface burner Gaskochplatte, Gasheizplatte *f*

gas switch(ing) off device → gas cutoff (device)

gas takeover room Gasübergaberaum *m*

gas tank Gasbehälter, Gastank *m*

gas thermostat Gasthermostat, automatische Gas-Temperaturregler *m*

gas thread Gasgewinde *n*

gas throttle Gasdrossel(organ) *f*, (*n*)

gastight, gasproof gasdicht

gastightness Gasdichtheit, Gasdichtigkeit *f*

gas trace Gasspur *f*

gas transport(ation) Gasfortleitung, Gasförderung *f*, Gastransport *m*

gas turbine, simple-cycle (gas) turbine [*Prime mover in which gas, under pressure or formed by combustion, is directed against a series of turbine blades; the energy in the expanding gas is converted into mechanical energy supplying power at the shaft*] Gasturbine *f*

gas turbine electric-generating station, gas turbine electric-generating installation, gas turbine electric-generating plant, gas turbine power station, gas turbine power installation, gas turbine power plant Gasturbinen(-Elektro)kraftwerk, Verbrennungsturbinen(-Elektro)kraftwerk *n*

gas turbine heating power plant Gasturbinen-Heizkraftwerk *n*

gas type Gasart *f*

gas undercounter range → gas (base) cabinet range

gas use plant, gas use installation Gasanwendungsanlage *f*

gas utility → (public) gas utility

gas valve → (gas) discharge valve

gas vent Gasentlüftung *f* [*als Vorrichtung*]

gas wall(-hung) heating boiler, gas wall-suspended heating boiler Gaswandheiz(ungs)kessel *m*

gas warm air generator → gas hot air generator

gas water circulation heating → circulating gas water heating

gas water heater, gas water heating appliance [*See remark under "water heater"*] Gas(warm)wasserbereiter, Gaswassererhitzer, Gaswasserheizer, Gaswasser(er)wärmer, GWH *m*, Gas(heiz)therme *f*, Gaszentralheizgerät, Gas-WH *n*

gas water heating Gaswasserheizung *f*

gate [*The input electrode of a field effect transistor or thyristor device*] Torelektrode *f*

gate [*A mov(e)able or selective barrier that allows or regulates passage*] Tor *n*

gate Netz(werk)übergang *m* [*Übergang von einem Netz(werk) zu einem anderen*]

gate air curtain Torluftschleier, Torluftvorhang *m*

gate band Torband *n*

gate circuit [*Computer engineering. A basic circuit that controls the flow of binary information, each having a combination of input signals and a single output, such as an AND-gate, OR-gate, or NOR-gate*] Torschaltung *f*

gate cold air curtain Torkaltluftschleier, Torkaltluftvorhang, Kaltlufttorschleier, Kaltlufttorvorhang *m*

gate control → gate ((open) loop) control

gate-controlled delay time Zündverzögerung *f*, Zündverzug *m* [*Thyristor. DIN 41 786*]

gate-controlled rise time Durchschaltzeit *f* [*Thyristor. DIN 41 786*]

gate-controlled turnoff time Löschzeit *f* [*Abschaltthyristor. DIN 41 786*]

gate-controlled turnon time Zündzeit *f* [*Thyristor. DIN 41 786*]

gate drive Tor(an)trieb *m*

gate frame Torrahmen *m*

gate heating system Tor(be)heizungssystem *n*

gate hot air curtain → gate warm air curtain

gate infill(ing) Torfüllung *f*

gate lock Torschloss *n*

gate mounting, gate furniture item Torbeschlag *m*

gate non-trigger current höchste nichtzündende Steuerstrom *m* [*Thyristor. DIN 41 786*]

gate ((open) loop) control Torsteuerung *f*

gate pier Torpfeiler *m* [*Beton, Ziegel, Steine*]

gate post Torpfosten *m* [*Holz*]

gate travel, gate span Torspannweite *f*

gate trigger current Zündstrom *m* [*Thyristor*]

gate trigger voltage Zündspannung *f* [*Thyristor*]

gate turn-off current (Mindest)Abschaltstrom *m* [*Abschaltthyristor. DIN 41 786*]

gate turn-off thyristor, GTO abschaltbare Thyristor *m*

gate valve Schieber *m*

gate valve pit Schiebergrube *f*

gate warm air curtain, gate hot air curtain Torwarmluftschleier, Torwarmluftvorhang, Warmlufttorschleier, Warmluftturvorhang *m*

gateway Übersetzer, Umsetzer *m* [*Heterogene Systeme entstehen auch, wenn die beteiligten Hersteller sich auf ein Kommunikationsprotokoll verständigen und entsprechende Übersetzer (= gateways) zwischen einem neutralen Protokoll und ihrem Protokoll einsetzen*]

gateway Netzübergang, Übergang zwischen zwei Netzen *m* [*Streng genommen beschreibt ein "gateway" ein System, das Nachrichten zwischen verschiedenen Netzwerktypen überträgt, ein "voter" überträgt zwischen Netzwerken desselben Typs*]

gateway Netzwerkschnittstelle, Schnittstelle bei Netzwerken *f*

gateway exchange Knotenamt *n* [*Fernsprechtechnik*]

gathering → (data) collection

gating Ausblenden, Austasten, Ansteuern *n* [*Thyristor*]

gating [*Electrical engineering. The process of selecting those positions of a wave that exist during one or more selected time intervals or that have magnitudes between selected limits*] Anschnitt *m*

gating ((open) loop) control Anschnittsteuerung *f*

gating section Verknüpfungsfeld *n*

gauge; gage (US) [*Any of the wide variety of instruments that are used for measuring a parameter or characteristic of an object, such as its width, weight, volume, conductivity and pressure*] Messinstrument *n*

gauged mortar; gaged mortar (US); cement-lime mortar Zement-Kalk-Mörtel *n*

Gaussian minimum shift keying, GMSK Gaußsche Mindesttasten, Gaußsche Minimaltasten, Gauß-Minimumtasten *n* [*Die Übertragungsfunktion und die Impulsantwort des Tiefpasses haben die Form der Gaußschen Glockenkurve. Dieses Tasten führt zu einer wesentlichen und befriedigenden Begrenzung der HF-Bandbreite*]

Gay-Lussac('s) law, Charles' law [*At constant pressure, the volume of a fixed mass or quantity of gas varies directly with the absolute temperature; a close*

G

approximation] Gay-Lussacsche Gesetz, Gesetz *n* von Gay-Lussac [*1802*]

GD → ground

GDI → graphic device interface

geared machine → worm geared reduction unit

geared traction elevator (US); geared traction lift (Brit.) [*It uses a high-speed, low-torque electric(al) motor, with a worm-to-worm gear reduction between the motor shaft and the driving sheave of the hoist ropes. It is roped 1:1 between the car and the counterweight in the higher speed ranges and 2:1 in the lower speed ranges*] Getriebegegengewichtsaufzug *m*

G

gearing together → catching-in

gearless, without gear getriebelos, ohne Getriebe

gearless traction elevator (US); gearless traction lift (Brit.) [*It has a low-speed, high-torque electric motor with a drive sheave mounted directly on the motor armature shaft. Its speed ranges from 2.5 to 10 m/s (500 to 2000 ft/min). Its vertical rise is virtually unlimited. It is most effective with a roping of 2:1 between the car and the counterweight in the lower speed and lower travel ranges and 1:1 in the higher speed and higher travel ranges.*]

gear motion mortice lock → double bolt lock

to gear together → to catch in

gel(-type) battery Gel-Batterie *f* [*Blei-Säure-Batterie, bei welcher der Elektrolyt in eine Matrix aus Silikagel eingebunden ist*]

gemell [*old English*] **→ grimmer**

genemotor Motorengenerator, Einankerumformer *m*

general call, collective call Sammelruf *m*

general call key, collective call key Sammelrufschalter, Rufsammelschalter *m*

general diffuse lighting system, general diffuse illumination system, direct-indirect lighting system, direct-indirect illumination system direkte-indirekte Beleuchtungssystem *n*

general-diffuse luminaire, non-cutoff luminaire, general-diffuse light(ing) fitting, non-cutoff light(ing) fitting Freistrahler, freistrahlende Beleuchtungskörper *m*, freistrahlende Leuchte *f*,

generalized machine verallgemeinerte Maschine *f* [*Eine angenommene Maschine/Steuerungs-Kombination, aus deren Funktionen sich die Eingabedaten für die verallgemeinerte Software ergeben*]

(generalized) phase ((open) loop) control [*IEC 555-1*] (Phasen)Anschnittsteuerung *f* [*VDE 0838, Teil 1*] [*Leistungselektronik*]

generalized software verallgemeinerte Software *f* [*Anwendungssoftware, mit deren Hilfe ein Postprozessor definiert werden kann*]

general lighting, general illumination Allgemeinbeleuchtung *f*

generally recognized as safe unbedenklich sicher, „bombensicher"

general purpose branch circuit [*One that supplies a number of outlets for general lighting and convenience receptacles*] Abzweigstromkreis, Verbraucherstromkreis *m*

general purpose data register allgemeine Datenregister *n*

general purpose fuse Ganzbereich(s)sicherung, Vollbereich(s)sicherung *f*

general purpose interface adapter, GPIA Allzweckschnittstellenadapter *m*

general purpose interface bus → HPIB

generating station → powerhouse

generating station unit → power station unit

generator Generator *m*

generator Austreiber, Kocher, Generator *m* [*Ein Bauteil einer Sorptionsanlage, in dem unter Zufuhr von Wärme das Arbeitsmittel aus dem siedenden Arbeitstoffgemisch ausgetrieben (ausgedampft) wird*]

generator set, generating set Generatorsatz *m*

generator tube Austreiberröhre *f* [*Absorptionskälteanlage*]

genlock, generator lock Referenzsignalanbinder *m* [*Eine Videoquelle an einen vorgegebenen Takt, das Referenzsignal, anbindendes Gerät oder anbindende Baugruppe*]

genuine (spare) part Originalersatzteil *m, n*

geomagnetic field Erdmagnetfeld, Magneterdfeld *n*

geomagnetic noise [*Interference in radio communications arising from terrestrial magnetism*] Erdmagnetismusrauschen *n*

geometrical space acoustics geometrische Raumakustik *f*

geothermal district heating [*Use of geothermal energy passing through a heat exchanger to a secondary loop, from which the secondary fluid is circulated through the distribution system*] geothermische Fernheizung, geothermische Stadtheizung *f*

German lock, halfturning lock, semiturning lock deutsche Schloss, halbtourige Schloss, Halbtorschloss, Halbumdrehschloss *n* [*Es wird mit einem Schlüssel nur aufgeschlossen und mit einer am Schließhaken aufgebrachten Aufhaltung zugeschnappt*]

German (lock) bolt, halfturning (lock) bolt, semiturning (lock) bolt halbtourige (Schloss)Riegel, deutsche (Schloss)Riegel, Halbtourschlossriegel, Halbumdrehschlossriegel *m*

German rimlock deutsche Kastenschloss *n*

German spring lock deutsche Schloss, Schnappschloss *n*, Schnäpper *m*

GET → group execute trigger

get another display Bildwechsel *m* [*als Befehl*]

geyser → flow(-type) water heater

GF (US) → ground(ing) fault

GF(C)I → ground(ing) fault (circuit) interrupter

GFI → (glass) fibre channel

g.f.p. (US) → ground(ing) fault protector

Giaque-Debye method → adiabatic demagnetization

gibberish → garbage

gibberish collector → garbage collector

(Gibbs) free energy, Gibbs function freie Energie, Gibbs freie Energie, Gibbsfunktion *f*

Gibbs-Helmholtz equation [*An expression for the influence of temperature upon the equilibrium constant of a chemical reaction,* $(d \ln K^0/dT)_P=\Delta H^0/RT^2$, *where* K^0 *is the equilibrium constant,* ΔH^0 *the standard heat of the reaction at the absolute temperature T, and R the gas constant.* [*THERMO*] *1. Either of two thermodynamic relations that are useful in calculating the internal energy U or enthalpy H of a system; they may be written* $U=F-T(\delta F/\delta T)_V$ *and* $H=G-T(\delta G/\delta T)_P$, *where F is the free energy, G is the Gibbs free energy, T is the absolute temperature, V is the volume, and P is the pressure. 2. Any of the similar equations for changes in the thermodynamic potentials during an isothermal process.*] Gibbs-Helmholtz-Gleichung *f*

GIF, graphic interchange format Dateiformat *n* für pixelorientierte Graphiken

giga electron volts → BeV

girthed area abgewickelte Fläche *f*

glare Blendung *f* [*Störung des Sehvermögens durch allzu große Leuchtdichten (positive Blendung)*]

glaring Blenden *n*

glass balustrade Glasbalustrade *f* [*Rolltreppe*]

glass barrel sprinkler Glasfässchensprinkler, Glasfässchensprenger *m* [*Das Glasfässchen ist mit einer leichtsiedenden Flüssigkeit gefüllt. Bei Erwärmung auf die Öffnungstemperatur wird durch den Druckanstieg das Glas gesprengt und die Dichtung durch den Wasser- bzw. Luftdruck von der Düse geschleudert*]

glass block, glass brick Glas(bau)stein; Glasziegel *m* [*Fehlbenennung. DIN 18175*]

glass break(ing) detection Glasbruchmeldung, Glasbrucherkennung *f*, Glasbruchnachweis *m*

glass break(age) detector, glass break(age) detecting device, glass break(age) detection device Glasbruchmelder, GM *m*, Glasbruchmeldegerät *n*

glass break(ing) sensor Glasbruchsensor *m*

glass brick → glass block

(glass) bulb sprinkler Schirmsprinkler, Schirmsprenger *m*

glass cabin → glass enclosure

glass chimney Glasschornstein *m*

glass cloth jacket Glasgespinstmantel *m*

glass coil Glasschlange *f*

glass elevator (US); glass lift (Brit.) Panoramaaufzug, Glasaufzug, Panoramafahrstuhl, Glasfahrstuhl *m*

glass enclosure, glass cabin Glaskabine *f*

glasses washer, glasses washing machine, glasses washup machine Gläserspülmaschine *f*

(glass) fiber (US); (glass) fibre (Glas)Faser *f*

(glass) fiber blanket (US); (glass) fibre blanket (Glas)Fasermatte *f*

(glass) fiber blanket insulation (US); (glass) fibre blanket insulation (Glas)Fasermattendämmung *f*

(glass) fiber cable (US); (glass) fibre cable (Glas)Faserkabel, GFK *n*, Lichtwellenleiter, LWL *m*

(glass) fiber cord (US); (glass) fibre cord (Glas)Faserschnur, (Glas)Faserkordel *f*

(glass) fibre sink; (glass) fiber sink (US) [*See remark under "sink"*] Glasfaserspüle *f*

(glass) fibre channel, FI, GFI; (glass) fiber channel (US) (Glas)Faserkanal *m* [*In einem Faserkanalnetz können theoretisch 16 Millionen Teilnehmer verbunden sein*]

(glass) fibre switch; (glass) fiber switch (US) (Glas)Faserschalter *m*

glass gauge; glass gage (US) Wasserstandsglas *n*

glass insert Glaseinsatz *m*

glass lift (Brit.) → glass elevator (US)

glass monitoring Glasüberwachung *f* [*Glasbruchmelder; Folienüberwachung; Fadenzugkontakt*]

glass pad Glasvlies; Glasvließ *n* [*Österreich*] [*Filtermedium*]

glass panel radiant heating → radiant glass panel heating

glass radiant panel → radiant glass panel

glass semiconductor → amorphous semiconductor

glass shadow wings Glaslamellensonnenschutzsystem *n* [*Zur Beschattung verglaster Fassaden*]

glass shelf Glasablage(platte), Glasablegeplatte *f* [*Über einem Waschbecken*]

glass silk Glasseide *f*

glass vanity Glaswaschtisch *m*

glass wool Glaswolle *f*

G

glass wool disposable filter, disposable glass wool filter, disposable fibrous glass filter, fibrous glass disposable filter Glaswollewegwerffilter, Glaswolleeinmalfilter, Wegwerfglaswollefilter, Einmalglaswollefilter *m, n*

glass wool roll Glaswollebahn *f*

global area network, GAN weltweite Datenaustauschnetz, globale Datenaustauschnetz *n*

global broadcast [*A message addressed to all devices or nodes on all networks in a BACnet internet*] globale allgemeine Mitteilung, globale allgemeine Nachricht, globale allgemeine Information, allgemeiner globale Information, allgemeiner globale Nachricht, allgemeine globale Mitteilung *f*

to globalize globalisieren

global system for mobile communications →GSM

global variable globale Veränderliche *f*

global warming potential, GWP Treibhauspotential *n*

globe float valve → spherical float valve

globe knob → ball knob

globe lamp → ball lamp

globe lightning → ball lightning

globe mixer → ball mixer

globe valve, lift and lay valve, plug and set valve, set and plug valve, lay and lift valve, ball valve [*It has a spindle which moves up and down and uses a linear actuator. The overall height is therefore greater. This valve type can be used for low, medium or high pressure water and steam systems*] Kugelschieber *m*

globoid compressor Globoidverdichter *m*

glossary (push)button [*It allows access to an explanation of terms used in a program(me)*] Glossardruckknopf *m*

glossy paper Glanzpapier *n*

glove box Handschuhkasten *m*

glow-tube (type) fuse, globe-tube (type) fuse [*in lightning arresters*] Feinsicherung *f*

glyptal Sicherungslack *m* (für Einstellungen) [*Elektrotechnik*]

GMSK → Gaussian minimum shaft keying

Gmt → Greenwich mean time

GND (mass) → ground (mass) (US)

go-devil lock → pig lock

going masked Vermummung *f*

gold-bonded diode Golddrahtdiode *f*

golden board, golden device, known good board, known good device, KGD theoretisch fehlerfreie Baustein *m*, theoretisch fehlerfreie Leiterplattenbaugruppe *f*

gommet Tülle *f*

goods elevator, freight elevator (US); goods lift, freight lift (Brit.) Güteraufzug, Transportaufzug *m* [*Unterflurgüteraufzug; Kleingüteraufzug; Lagerhausaufzug; Behälteraufzug; vereinfachte Güteraufzug. Im Gegensatz zum Lastenaufzug darf der Güteraufzug keine Personen befördern*]

(goods) gravity conveyor (Güter)Schwerkraftförderer *m*

goods-passenger elevator (US) → freight-passenger elevator

goods/staff lift (Brit.) → staff/goods lift

gooseneck → bend offset

gooseneck microphone Schwanenhalsmikrophon *n*

gopher textbasierte Informationsdienst *m*

Go^s → quality of service

to go to local, GTL schalten auf Eigensteuerung

governor [*A device that automatically regulates the speed of an engine or machine by varying the supply of fuel according to the power demand*] Regler *m*

governor [*A device, especially one actuated by the centrifugal force of whirling weights opposed by gravity or by springs, used to provide automatic control of speed or power of a prime mover*] Regler *m*

governor gear Reglergetriebe *n*

GPIA → general purpose interface adapter

GPIB → HPIB

grab rail, grab bar Schienengriff *m*, Griffschiene, Greifstange, Griffstange, Haltestange, Halteschiene *f* [*z.B. an der Badewanne*]

graceful degradation [*A programming technique to prevent catastrophic system failure by allowing the machine to operate, though in a degraded mode, despite failure or malfunction of several integral units or subsystems*] herabgesetzte Arbeitsweise *f*

gradation Halbton *m*, Kontraststufe *f*

graded coke [*less favoured term*] → **sized coke**

graded index fibre; graded index fiber (US) Gradienten(profil)faser *f*

grade of service → **quality of service**

grading period → **(fire) grading period**

grading subgroup Zubringerteilgruppe *f* [*Fernmeldetechnik*]

gradual condensation, stepwise condensation Stufenkondensation, stufenweise Kondensation, Schrittkondensation, schrittweise Kondensation *f*

gradual nozzle steam trap [*See remark under "steam trap"*] (Stufen)Düsen(kondensat)abscheider *m*

to graduate graduieren, abstufen, messen [*In Grade einteilen*]

graduated cylinder [*Chemistry. A cylindrical container that is marked with divisions or units of measurements*] Messzylinder, Skalenzylinder, graduierte Zylinder, abgestufte Zylinder *m*

graduated key Skalenschlüssel, graduierte Schlüssel, Messschlüssel, abgestufte Schlüssel *m*

graduation → **scale (mark) spacing**

graduator Gradmesser *m*

grain fraction Kornanteil *m*, Kornfraktion *f*

graininess Körnung *f* [*Abtastschirm*]

graininess Körnigkeit *f*

grain sphericity Kornkugeligkeit *f*

grand master key, grand main key, grand principal key Generalhauptschlüssel, Hauptgeneralschlüssel *m*

granite vanity Granitwaschtisch *m*

granular carbon graupige Kohlenstoff, körnige Kohlenstoff, Kohle(n)grieß *m* [*Mikrophonkohle*]

graph graphische Darstellung *f* [*Ein Schaubild, z. B. von Bewegungsvorgängen, Betriebsvorgängen; Kräfteplan; Nomogramm; im engeren Sinne Darstellungen von Funktionen im rechtwinkligen Achsenkrenz veranschaulichen in einfacher Weise den Verlauf der abhängigen Größen über einen bestimmten Wertebereich. Positive Werte sind vom Nullpunkt aus nach rechts und nach oben, negative Werte nach links und nach unten abzutragen. Statt von einem Nullpunkt kann auch von einem anderen Bezugspunkt ausgegangen werden*]

graph follower → **curve follower**

graphic character Schriftzeichen *n*

graphic computer input graphische Rechnereingabe, Bild-Rechnereingabe *f*

graphic controller Graphikregler *m*

graphic curve → **line graph**

graphic data Bilddaten, graphische Daten *f*, Piktogramm *n*

graphic data processing graphische Datenverarbeitung, Bilddatenverarbeitung, Piktogrammverarbeitung *f*

graphic device interface, GDI graphische Geräteschnittstelle *f*

graphic display Graphikbild *n*

graphic editor Bildaufbereiter *m*

graphic hierarchy Bildhierarchie *f* [*Anzahl von schematischen Bildern auf einem Farbschirm, die von einem Übersichtsbild zu Bildern von Einzelheiten führen*]

graphic illuminated diagram [*IEC 113-2*], **graphic illuminated circuit diagram** Graphikleuchtschaltplan *m*

graphic interchange format → **GIF**

graphic level recorder Pegelschreiber *m* [*Elektroakustik*]

graphic memory Graphikspeicher *m*

graphic presentation graphische Darstellung *f*

graphic printer Graphikdrucker *m*

graphic programming graphische Programmierung *f*

graphic programming language graphische Programmiersprache *f*

graphic record graphische Aufzeichnung *f*, Schrieb *m*

graphic screen Graphikbildschirm *m*

Grashof number [*Dimensionless number used in the calculation of free convection of a fluid caused by a hot body, equal to the product of the fluid's coefficient of the thermal expansion, the temperature difference between the hot body and the fluid, the cube of a typical dimension of the body, and the square of the fluid's density, divided by the square of the fluid's dynamic viscosity. Also known as free-convection number. Symbol Gr, G_r or N_{Gr}.*] Grashofzahl *f*

grate bar, grating bar Feuer(ungs)roststab, Roststab *m*

(grate) chimney → **fire(place)**

grated file, fine rasp Feinraspel *f*

grate(d) gate → **barrier gate**

grate furnace, grating furnace Rostfeuerung, Rostfeuerstätte *f*

grate stoker, grating stoker, grate stoking device, grating stoking device Rostbeschicker *m* [*Kesselfeuerung*]

grating, grate [*A frame(work) of parallel or latticed bars for holding fuel in a fire(place), stove or furnace*] Feuer(ungs)rost, Rost *m*

grating → grid

gravitation(al) acceleration, acceleration due to gravity, gravity acceleration, acceleration of gravity Schwerkraftbeschleunigung, Schwerebeschleunigung, Gravitationsbeschleunigung, Erdbeschleunigung *f*

gravity air heating, air gravity heating Schwerkraftluftheizung, Auftriebsluftheizung *f*

gravity brake Schwerkraftbremse *f* [*z. B. in einem Pumpenschieber*]

gravity circulation Schwerkraftzirkulation, Zirkulation im Schwerkraftbetrieb, Schwerkraftumwälzung *f*, Schwerkraftumlauf *m*

G

gravity-closing schwerkraftschließend

gravity conveyor → (goods) gravity conveyor

gravity cooling Schwerkraftkühlung *f*

gravity drainage system, drainage gravity system Schwerkraftentwässerungssystem *n*

gravity elevator (US); gravity lift (Brit.) [*Gravity is the source of power and moves the car*] Schwerkraftaufzug *m*

gravity feed hopper (type) boiler Füll(ung(s))-schachtkessel *m*

gravity feed(ing) boiler Schwerkraftbeschickungskessel *m*

gravity floor heating Schwerkraftstockwerkheizung, Schwerkraftgeschossheizung, Schwerkraftetagenheizung *f*

gravity floor water heating system → floor gravity water heating system

gravity furnace [*Furnace that depends on the difference in density of warm and cool air circulation. The furnace may also use an integral fan or blower to overcome the internal furnace resistance to airflow*] Schwerkraftfeuerung, Schwerkraftfeuerstätte *f*

gravity heating system [*1. Central warm-air heating system through which air is circulated by gravity. 2. Comfort heating system consisting of gravity-type warm-air furnace, together with all air ducts or pipes and accessory apparatus installed in connection with it*] Schwerkraftwarmluftheiz(ungs)system, Warmluftschwerkraftheiz(ungs)system *n*

gravity hot-air furnace, gravity warm-air furnace [*It depends primarily on circulation of air through the furnace by gravity. This definition also includes any furnace approved with a booster-type fan, which does not materially restrict free circulation of air through the furnace when the fan is not in operation*] Schwerkraftwarmluftfeuerung, Warmluftschwerkraftfeuerung *f*

gravity hot water heating, gravity warm water heating Schwerkraftwarmwasserheizung, Warmwasserschwerkraftheizung *f*

gravity hot water system → hot water gravity system

gravity influence Schwerkrafteinfluss *m*

gravity lines method → lines of gravity method

gravity/pump hot water heating, gravity/pump warm water heating Schwerkraft-Pumpen-Warmwasserheizung, Pumpen-Schwerkraft-Warmwasserheizung *f*

gravity regulator Schwerkraftregler, Drehzahlregler mit Fliehkraftpendel *m*

gravity separator, gravity interceptor, gravity trap [*See remark under "interceptor"*] Schwerkraft(ab)scheider *m*

gravity smoke exhaust, gravity smoke removal, gravity smoke disposal Schwerkraftentrauchung, Schwerkraftrauchabführung, Schwerkraftrauchentsorgung *f*, Schwerkraftrauchabzug *m* [*Entrauchung durch Nutzung der Auftriebskräfte*]

gravity venting system, gravity ventilating system [*System that depends entirely on the heat from the fuel being used to provide the energy required to vent an appliance*] Schwerkraftlüftungssystem *n*

Gray Code Graykode, Graycode *m* [*Ein zyklisch permutierender Kode; ein einfacher Binärkode*]

gray scale, grey scale Graustufe *f*

gray scale discrimination, gray scale variance, grey scale discrimination, grey scale variance Graustufenunterscheidung, Graustufenunterschiedlichkeit *f*

gray scale range, grey scale range Graustufenbereich *m* [*selten: n*]

grease filter Fettfilter *m, n*

grease pan Fett(sammel)schale *f*

grease pump Fettpumpe *f*

grease removal → degreasing

grease separator, grease interceptor, grease trap [*See remark under "interceptor"*] Fett(ab)scheider, Fettfänger, Fettfang *m* [*DIN 4040*]

grease sludge Fettschlamm *m*

greasy fett(halt)ig

greek Datenstationsfilter *m, n*

green development grüne Bebauung *f*

greenhouse effect Treibhauswirkung *f*

greenhouse gas Treibhausgas *n*

green pricing grüne Tarif *m* [*Elektrizitätswirtschaft*]

Greenwich mean time, universal time, UT, GMT universelle Zeit *f*

grey ... → gray ...

grid → (current) grid

grid Bildschirmhilfsraster, Raster, Hilfsraster auf dem Bildschirm *m*

grid, grating [*It is made of crossed bars*] Lochgitter *n*

grid battery, C-battery [*The battery that supplies the steady bias voltage required by the control-grid electrodes of electron tubes in battery-operated equipment*] Gitter(vorspannungs)batterie *f*

grid ceiling Rasterdecke *f*

grid-connected netzgekoppelt, netzverbunden

grid-connected system Netzverbundsystem, Netzkoppelsystem *n*

grid disconnection switch → grid isolation switch

grid earth electrode, mesh earth electrode Maschenerder, Gittererder *m*

grid failure → power failure

grid failure alarm → power failure alarm

grid priming voltage → grid bias

grid system Rastersystem *n* [*Einbau elektronischer Ausrüstungen*]

gridiron pendulum [*A compensation pendulum in which the unequal thermal expansion of two metals is utilized to keep the length constant*] Rostpendel *n*

grid isolation switch, grid disconnection switch, network isolation switch, network disconnection switch Netzfreischalter *m*

grill(e) → (air) grill(e)

grilled pipe, rib(bed) pipe Rippenrohr *n*

grimmer, chymol; gemmel, jummel, gymow [*old English*] Türangel, Haspe *f*

grinding bench Schleiftisch *m*

grinding machine → mill

grip → handle

grip → (door) handle

grip mixer → (hand)grip mixer

gripping rail, gripping bar Griffstange, Haltestange, Greifstange *f*

grit spreading Streuen *n* gegen Glatteis

grommet, sleeve Tülle [*Leitungseinführung*]

groove(d) pulley → diverter pulley

groove(d) Rider (slide) valve → Rider (slide) valve with grooves

groove(d) sheave → diverter pulley

groove(d) treadway [*moving walk; moving ramp*] genutete Trittfläche *f*

groove(d) tube Rillenröhre *f*

groove for joint filler → joint filler groove

grooving [*Treadway surface of a moving walk grooved in a direction parallel to its travel for the purpose of meshing with combplates of the landing*] Längsnuten *f pl*

gross area Bruttofläche *f*

gross calorific value → higher heating value

gross capacity, gross performance, gross service, gross duty Bruttoleistung, Bruttokapazität *f*

gross cooling capacity, gross cooling duty, gross cooling performance, gross cooling service Bruttokühl(ungs)leistung, Bruttokühl(ungs)kapazität *f*

gross error grobe Fehler *m*

gross heating value → higher heating value

gross heat loss, gross thermal loss Bruttowärmeverlust, Wärmebruttoverlust *m*

gross specific energy → higher heating value

gross volume Bruttovolumen *n*, Bruttorauminhalt, BRI *m*

to ground (US) → to (connect to) earth

ground (US) → ground (mass)

ground, ground coat(ing) [*The first of several print coats*] Grundierlage, Grundierschicht *f*

ground Null-V-Potential *n* [*Spannungsmesswertsignal; Stell(ungs)signal*]

groundable (US); earthable (Brit.) erdungsmöglich

ground collector Erd(reich)kollektor *m*

ground connection (US) → grounding (US)

ground detector → ground sensor

ground driver Erdtreiber *m*

grounded, connected to ground, ground-connected (US); earthed, connected to earth, earth-connected (Brit.); at zero voltage, common geerdet

ground(ed) anode amplifier → cathode follower

ground(ed) conductor, conductor connected to ground (US); earth(ed) conductor, conductor connected to earth (Brit.) geerdete Leiter *m*

groundfloor elevator landing (US); groundfloor lift landing (Brit.) Erdgeschoss-Aufzughaltestelle *f*

ground heat Erd(reich)wärme *f*

ground heat pump Erd(reich)wärmepumpe *f*

ground hydrant [*As opposed to a wall hydrant*] Bodenhydrant *m*

G

grounding, ground connection, connection to ground (US); earthing, earth connection, connection to earth (Brit.) [*An electrical connection to earth through an earth electrode and suitable conductors to it*] Erdung, Erd(ungs)verbindung *f*, Erden *n*, Erder, Erd(ungs)schluss, Schutzleiterschluss *m*

ground(ing) arrangement (US) → ground(ing) system

ground(ing) blade (US); earth(ing) blade (Brit.) Erd(ungs)messer *m* [*am Schaltgerät*]

ground(ing) bolt (US); earth(ing) bolt (Brit.) Erd(ungs)bolzen *m*

G

ground(ing) bus(bar) (US); earth(ing) bus(bar) (Brit.) [*A bus(bar) in a panel or elsewhere, deliberately connected to ground*] Erd(ungs)sammelleitung, Erd(ungs)sammelschiene, Haupterd(ungs)schiene, Erd(ungs)hauptschiene *f*, Sammelerder *m* [*VDE 0100, Teil 200*]

ground(ing) cable (US); earth(ing) cable (Brit.) Erd(ungs)kabel *n*

ground(ing) circuit (connector) (US); earth(ing) circuit (connector) (Brit.); potential jumper, (equipotential) bond(ing) strip, bonding jumper Potentialausgleich(s)schiene *f*

ground(ing) clamp (US); earth(ing) clamp (Brit.) Erd(ungs)(rohr)schelle *f*

ground(ing) conductor (US); earth(ing) conductor (Brit.) [*Conductor run in an electrical system, which is deliberately connected to the ground electrode. Purpose is to provide a ground point throughout the system. Insulation colo(u)r is green*] Erd(ungs)leiter *m*, Erd(ungs)leitung, geerdete Schutzleiter *f* [*VDE 0100, Teil 200*]

ground(ing) connection (US); earth(ing) connection (Brit.) Erd(ungs)anschluss *m*

ground(ing) connector (US); earth(ing) connector (Brit.) Steckverbinder *m* mit Erd(ungs)anschluss

ground(ing) contact (US); earth(ing) contact (Brit.) Erd(ungs)kontakt(stück) *m*, (*n*)

ground(ing) contact plug (US); earth(ing) contact plug (Brit.) Schutzkontaktstecker, SCHUKO-Stecker *m* [*„SCHUKO" ist ein eingetragenes Warenzeichen*]

ground(ing) detector lamp (US); earth(ing) detector lamp (Brit.) Erd(ungs)schlussmeldelampe *f*

ground(ing) fault, ground(ing) leakage, GF, fault to ground, short-circuit to ground (US); earth(ing) fault, earth(ing) leakage, fault to earth, short-circuit to earth (Brit.) Erd(ungs)fehler, Erd(ungs)schluss *m*

ground(ing) electrode (US); earth(ing) electrode (Brit.) [*Conductor embedded in the earth, used for maintaining ground potential on conductors connected to it, and for dissipating into the earth, current conducted to it. It may be a rod, pole or strap driven or buried in the ground*] Erder *m*, Erd(ungs)elektrode *f*, Elektrodenerder [*VDE 0100, Teil 200*]

ground(ing) fault (circuit) interrupter, GF(C)I, g.f.c.i. (US); earth(ing) fault (circuit) interrupter (Brit.) [*A device that senses ground faults and reacts by opening the circuit*] Fehlerstromschutzschalter, FS-Schalter *m*

ground(ing) fault neutralizer (US); earth(ing) fault neutralizer (Brit.) Erd(ungs)fehlerlöscher *m*, Erd(ungs)fehlerlösch(er)spule, Lösch(er)spule, ESP *f* [*VDE 0532, Teil 20*]

ground(ing)-fault-neutralizer-grounded system (US); earth(ing)-fault-neutralizer-grounded system (Brit.) Netz mit über Lösch(er)spule geerdetem Sternpunkt, gelöschte Netz, induktiv geerdete Netz, Netz mit Erd(ungs)fehlerkompensation *n*

ground(ing) fault neutralizing (US); earth(ing) fault neutralizing (Brit.) Erd(ungs)fehlerlöschung *f*

ground(ing) fault protector, g.f.p. (US); earth(ing) fault protector (Brit.) Erd(ungs)fehlerschutzgerät *n*

ground(ing) grid (US); earth(ing) grid (Brit.) Erd(ungs)gitter *n*

ground(ing) jumper, ground(ing) strap (US); earth(ing) jumper, earth(ing) strap (Brit.) Erd(ungs)brücke, Erd(ungs)lasche, Laschenerde *f* Erd(ungs)verbund *m*

ground(ing) leakage (US) → ground(ing) fault

ground(ing) line (US); earth(ing) line (Brit.) Erd(ungs)leitung *f*

ground(ing) link (US); earth(ing) link (Brit.) Erd(ungs)brücke *f*

ground(ing) lug (US); earth(ing) lug (Brit.) Erd(ungs)nase *f*

ground(ing) loop (US); earth(ing) loop (Brit.) Erd(ungs)schleife *f*

ground(ing) metal sheet (US); earth(ing) metal sheet (Brit.) Erd(ungs)blech *n*

ground(ing) net(work) (US); earth(ing) net(work) (Brit.) Erd(ungs)netz *n*

ground(ing) pin (US); earth(ing) pin (Brit.) Erd(ungs)stift *m*

ground(ing) plate (US); earth(ing) plate (Brit.) Erd(ungs)platte *f*

ground(ing) potential, potential to ground (US); earth(ing) potential, potential to earth (Brit.) Erd(ungs)potential *n*

ground(ing) reactor (US); earth(ing) reactor (Brit.); discharge coil, drain(age) coil Erd(ungs)drossel *f*

ground(ing) return conductor (US); earth(ing) return conductor (Brit.) Erd(ungs)rückleiter *m*

ground(ing) rod (US); earth(ing) rod (Brit.) [*A copper rod driven into the earth to serve as the ground connection*] Erd(ungs)stab, Staberder *m*

ground(ing) screw (US); earth(ing) screw (Brit.) Erd(ungs)schraube *f*

ground(ing)(-type) strap (US) → ground(ing) jumper

ground(ing) strip (US); earth(ing) strip (Brit.); strip conductor electrode Erd(ungs)band *n*, Banderder *m*

ground(ing) switch (US); earth(ing) switch (Brit.) Erd(ungs)schalter *m* [*VDE 0670, Teil 2*]

ground(ing) terminal (US); earth(ing) terminal (Brit.) [*IEC 65*] Erd(ungs)klemme, Schutzleiterklemme *f*

ground(ing) system, ground(ing) arrangement (US); earth(ing) system, earth(ing) arrangement (Brit.) Erdungssystem *n* [*VDE 0100, T.200. Es besteht aus Erdungsleitungen und Erdern*]

ground(ing) transformer (US); earth(ing) transformer (Brit.) Erd(ungs)transformator, EdT, Nullpunktbildner, Löschtransformator *m*

ground(ing)(-type) receptacle → socket with earth(ing) contact

ground(ing) washer (US); earth(ing) washer (Brit.) Erd(ungs)unterlegscheibe *f*

ground(ing) wire (US); earth(ing) wire (Brit.) Erd(ungs)draht *m*

ground level Erdgleiche *f*

ground-level tank → low-level tank

ground (mass) (US); earth (mass) (Brit.); GND (mass) Masse(leitung), Erde, Betriebserde *f*, Bezugspotential, Nullniveau *n* [*nicht verwechseln mit „Schutzerde"*]

ground pipeline Grundrohrleitung *f* [*Sie kann erdverlegt oder nicht erdverlegt sein*]

ground plane (getrennte) Massefläche *f* [*Trennung von digitalem und analogem Schaltungsteil*]

ground probe Erd(reich)sonde *f*

ground protection law Bodenschutzrecht *n*

ground(s), terrain Gelände, Terrain *n*

ground search [*for buried objects*] Untergrundsuche *f*

ground sensor, ground detector [*See remark under "sensor"*] Bodensensor, Bodenmessfühler, Bodendetektor *m*

ground strip → earth strip

ground switch → lightning switch

ground/water heat pump installation, ground/water heat pump plant Erdreich/Wasser-Wärmepumpenanlage *f*

group (control) change Gruppenwechsel *m*

group controller Gruppenregler *m*

group (control) level Gruppenleitebene *f* [*Sie fasst alle Teile der Leitvorrichtung zusammen, die jeweils auf einen bestimmten Teilbereich der Einzelleitebene einwirken. V DIN 19 237*]

group delay time, envelope delay time Gruppenlaufzeit *f* [*Sie gibt an mit welcher Laufzeit die Umhüllende eines Signals mit seinem Frequenzband entlang einer Leitung läuft*]

group execute trigger, GET (Geräte)Gruppenauslöser *m* [*Programmierbare Messgerät. DIN IEC 625*]

group hunt(ing) [*automatic telephony*] Gruppenfreisuche(n) *f*, (*n*), Gruppenfreiwahl *f*

grouping → batching

group(ing) mark [*A character signal(l)ing the beginning or end of a group of data*] Gliederungsmittel *n* [*DIN 6763, Teil 1*], Gliederungszeichen *n*

group key Gruppenschlüssel *m*

group mixing valve Gruppenmischventil *n*

group of contacts → contact assembly

group of elevators (US); group of lifts (Brit.) Aufzuggruppe *f*

group ((open) loop) control Gruppensteuerung *f* [*Funktionseinheit zum Steuern eines zusammenhängenden Teilprozesses, die den dazugehörenden Einzel- oder Betriebssteuerungen übergeordnet ist. DIN 19 237*]

group ((open) loop) control level, open loop group control level Gruppensteuer(ungs)ebene *f* [*Gesamtheit aller Gruppensteuerungen. V DIN 19 237*]

group printing [*In automatic data processing, the printing of information summarizing the data on a group of cards or other records when a key change occurs*] Gruppendruck(en) *m*, (*n*)

group-related gruppenbezogen

group signal message Gruppenmeldung *f* [*Eine besondere Art der Sammelmeldung durch funktionsorientierte Zusammenfassung von Einzelmeldungen. Anmerkung: Die funktionsorientierte Zusammenfassung kann z. B. nach folgenden Gesichtspunkten durchgeführt werden: - nach gleichartigen Einzelmeldungen, eventuell aus verschiedenen Bereichen (z. B. Gruppenmeldung „Sicherungsfall"), – nach unterschiedlichen Einzelmeldungen einer Geräteeinheit*

G

(z. B. Gruppenmeldung, „Störung Antrieb XY"), – nach unterschiedlichen Einzelmeldungen eines zusammengehörenden Teilprozesses (z. B. Gruppenmeldung „Störung Kühlwasserversorgung").]

group supervisory system Gruppensammelsteuerung *f* [*Aufzug*]

group velocity gerichtete Gruppengeschwindigkeit *f*

growler Prüf(ungs)summer *m*

grub saw [*A handsaw used for cutting stone, such as marble, into slabs for shelves, mantelpieces, etc.*] Handsteinsäge, Steinhandsäge *f*

GSM, global system for mobile communications europäische ETSI-Standard *m* für mobile digitale Funktelefonnetze im 900-Mhz-Bereich

G

GTL → to go to local

GTO → gate turn-off thyristor

guard → protection device

guard → (security) guard

guard Sicherung *f*

guard against contact → contact guard

guard band Rasen *m*, Zwischenspuren *f*

guard circuit Schutzschaltung *f* [*Messtechnik*]

guard contact, protection contact Schutzkontakt *m*

guard digit Schutzziffer *f*

guarded → shielded

guarded luminaire, guarded light(ing) fitting Korbleuchte *f*, Korbbeleuchtungskörper *m*

guardhouse Eingangswache, Wache *f*, Wachhaus, Wachgebäude *n*

guardian Portier, Pförtner *m*

guard(ing) service → (manned) guard(ing) service

guard outfit, guard equipment Wachausrüstung, Wachausstattung, Wachbestückung *f*

(guard) patrol → (security) (guard) patrol

(guard) patrol company → (security) (guard) patrol company

(guard) patrol tour → (security) (guard) patrol tour

guard protection Wächterschutz *m*

guard rail(ing) [*For an automatically operated door, a rail(ing) used to separate and control traffic passing in opposite directions through the door*] Trenn(ungs)geländer *n*

guards → ward

guard schedule, duty roster (Wächter)Dienstplan, Wachdienstplan *m*

guard security Wächtersicherung *f*

guard service Wachdienst, Wächterdienst *m*

guard time Sicherheits(abstands)zeit *f* [*Informationsaufteilung im Zeitschlitz*]

gudgeon → hinge pin

guest room [*In a single-family or two-family dwelling, a room in the main or an accessory building, occupied or intended to be occupied by nonpaying guests*] Gästezimmer *n*, Gästeraum *m*

guest room [*In a multiple-family dwelling, a room occupied or intended to be occupied for hire*] Untermieterraum *m*

guest-room area, guest-room space [*This division includes the guest rooms together with guest bathrooms, clothes closets, and entrance vestibules, the sum of which is the area that the guest rents*] Gästezimmerbereich, Gästeraumbereich *m* [*selten: n*]

guest('s) key → room key

guest towel Gästehandtuch *n*

guest towel basket Gästehandtuchkorb *m*

guid(anc)e Führung *f*

guid(anc)e → (operator) prompting

guid(anc)e bush(ing) Führungsbuchse *f*

guid(anc)e drum Führungswalze *f* [*Elektrizitätszähler*]

guid(anc)e electrode Führungselektrode *f*

guid(anc)e line value Richtwert *m*

guid(anc)e plate Leitblech, Führungsblech *n*

guid(anc)e post Führungspfosten *m*

guid(anc)e rail Führungsschiene *f* [*ET; DIN 43 350*]

guid(anc)e rail → (car) guid(anc)e rail

guid(anc)e rail fastening, guid(anc)e rail fixing Führungsschienenbefestigung *f*

guid(anc)e rod Führungsstange *f*

guid(anc)e value Führungswert, Richtwert, Anhaltswert *m*

guid(anc)e vane Leitschaufel *f*

guid(anc)e vane pitch adjustment device Leitschaufel(ver)stellvorrichtung *f*

guid(anc)e track [*escalator*] Führungsbahn *f*

guided probe Pfadverfolgung *f* [*Fehlersuchverfahren im Zusammenhang mit der Funktionsprüfung von Digitalbaugruppen, bei dem fehlerhafte Signale bis zum Fehlerort verfolgt werden*]

gull(e)y [*A pit in the gutter by the side of a road. It is covered with a grating and water drains away from it through a trap, but silt can collect in the pit. The silt is periodically removed by a gull(e)y sucker*] Straßenablauf, Regeneinlauf, Straßeneinlauf, Regenablauf *m*

gull(e)y, yard gull(e)y, yard inlet [*A small grating and inlet to a drain to receive rainwater and waste water from sinks, baths, and basins*] Hofablauf, Hofeinlauf *m*

gull(e)y trap, yard trap [*A water trap provided in gull(e)ys to keep foul gases in the drains. It may get unsealed in very hot weather if the water evaporates. The remedy is to pour water into it*] Ablaufverschluss, Einlaufverschluss *m*

gum test Verharzungsprüfung, Verharzungsprobe *f*, Verharzungsversuch *m*

gun burner → (pressure) atomizing burner

gutter, roof gutter [*A channel for collecting rainwater from roofs*] Dach(regen)rinne, Regendachrinne *f*

gutter [*Colloquial. A channel for collecting surface water*] Gosse, Rinne *f*, Gerinne *n*

gutter hook [*A light metal strap used to secure or support a roof gutter*] Rinnenhaken *m*

gutter space [*The empty space around the sides, top and bottom of a panel-box, intended as a wiring space*] Sicherungskastenverdrahtungsraum *m*

GWP → global warming potential

gymnasium(-type) luminaire, gymnasium(-type) light(ing) fitting, unbreakable luminaire, unbreakable light(ing) fitting Turnhallenleuchte, ballwurfsichere Leuchte *f*, Turnhallenbeleuchtungskörper, ballwurfsichere Beleuchtungskörper *m*

gymow [*old English*] **→ grimmer**

gypsum plaster tile, gypsum board [*heated ceiling*] Gipsplatte *f*

gyro(scope) Kreisel *m*

gyro(scope) compass Kreiselkompass *m*

gyrostatic moment Kreiselmoment *n*

H

H → header

habitable bewohnbar

habitable room [*A space used for living, sleeping, eating, or combinations thereof, but not including bathrooms, closets, halls, storage rooms, utility and similar spaces*] Wohnraum *m*

Hagen-Poiseuille law [*In the case of laminar flow of fluid through a circular pipe, the loss of head due to the fluid friction is 32 times the product of the fluid's viscosity, the pipe length, and the fluid velocity, divided by the product of the acceleration of gravity, the fluid density, and the square of the pipe diameter*] Hagen-Poiseuille-Gesetz *n*

hairdresser basin Kopfwaschbecken *n*

hairline gauge block; hairline gage block (US) Strichendmaß *n*

half ... → semi ...

halfturning lock → German lock

halfturning (lock) bolt → German (lock) bolt

half-wave module Halbwellenbaustein, Halbschwingungsbaustein *m*, Halbwellenmodul, Halbschwingungsmodul *n*

halide [*Chemical compound of the type MX, where X is fluorine, chlorine, iodine, bromine, or astaline, and M is another element or organic radical*] Halid *n*

hall Simulation *f* von Räumlichkeit [*Audio*]

hall button [*Signal(ling) device at a landing outside the hoistway enclosure for calling an elevator car*] Aufzugrufknopf *m*

Hall effect Halleffekt *m*, Hallwirkung *f*

Hall-effect element Hall-Element *n* [*Fühler zum Nachweis und zur Messung magnetischer Felder über die Hall-Spannung*]

Hall generator Hall-Generator *m* [*Bauelement zur Halleffektmessung*]

hall heating Hallenheizung, Großraumheizung *f*

(hall) lantern [*Signal light at a landing that indicates the arrival of an elevator car and its direction of travel*] Ankunft(s)lampe *f*

halogenated hydrocarbon, halocarbon Halogenkohlenwasserstoff *m*, Halon *n*

halogenated hydrocarbon refrigerant, halocarbon refrigerant [*Freon; Arcton; Isecon; etc.*] halogenierte Kohlenwasserstoffkältemittel, Halonkältemittel *n*

halogen lamp Halogenlampe *f*

halogen lamp dimmer [*See remark under "dimmer"*] Halogenlampe(n)helligkeitsregler, Halogenlampe(n)lichtsteller *m*

halogen-quenched counter tube Halogenzählröhre *f*

halon cylinder Halonflasche *f*

halon fire extinguisher → BCF fire extinguisher

halt instruction → program stop (US)

halt instruction Haltbefehl *m*

hammered plate gehämmerte Blech, Hammerblech *n*

H

hammer scale → scale

hammer scale formation → scale formation

hamper space [*e.g. for soiled linen*] Korbplatz *m* [*z. B. für Schmutzwäsche*]

ham radio → amateur radio

hand → pointer

hand action, manual action Handbetätigung *f*

hand ac(tua)tor, manual ac(tua)tor Handauslöser *m*

hand adjustment, manual adjustment Handjustierung, Handverstellung *f*

hand/automatic switch, manual/automatic switch Hand/Automatikschalter *m*

hand backflow(ing) shutter, manual backflow(ing) shutter Handrückstauverschluss *m*, Handrückflussverschluss [*Vorrichtung zum manuellen Abschluss einer anschließenden Leitung(sstrecke) bei Rückstau*]

hand bypass, manual bypass Handumführung *f*

handcharge Vorarbeiter, Postengeselle *m*

hand ((closed) loop) control, manual ((closed) loop) control, hand feedback control, manual feedback control, hand closed control, manual feedback control, hand regulation, manual regulation Handreg(e)lung, Handregulierung *f*

hand closing, manual closing Handeinschaltung *f*

hand control → hand ((closed) loop) control

hand control → hand ((open) loop) control

hand controller, manual controller [*See remark under "controller"*] Handregler *m*

hand data system, manual data system [*See remark under "data system"*] Handdatensystem *n*

hand deaerator, manual deaerator [*Hand-operated apparatus for removing air from some other substance or object*] Handentlüfter *m*

hand defrosting, manual defrosting [*Defrosting by natural or manual means with manual indication and termination of overall defrost operation*] Handentfrostung *f*

hand diverter (assembly), manual(ly-operated) diverter (assembly) Handumsteller *m* [*Bad(e)batterie*]

hand dryer → (hot air) hand dryer

hand dumb waiter (US) → hand service lift

hand elevator, manual elevator (US); hand lift, manual lift (Brit.) [*It is moved by manual energy*] Handaufzug *m*

hand-fed, manually fed handgespeist, handbeschickt [*z. B. eine Feuerung*]

hand feedback control → hand ((closed) loop) control

hand(-feed) punch, manual punch [*A device which punches holes in punch cards and moves the cards as they are punched as a direct result of pressure on the keys of a keyboard, and which requires feeding and removal of the cards by hand, one at a time*] Handlocher *m*

hand fire alarm, manual fire alarm Druckknopffeuermelder *m* [*(Glas)Scheibe einschlagen; Knopf drücken*]

hand-fired boiler, manually-fired boiler handbefeuerte Kessel, handgefeuerte Kessel *m*

hand fire detector, manual fire detector, hand fire detecting device, manual fire detecting device, hand fire detection device, manual fire detection device Handbrandmelder, Handbranddetektor, Handfeuermelder, Handfeuerdetektor *m*, Handfeuermeldegerät, Handbrandmeldegerät *n*[*„manuell" kann synonym für „Hand… " verwendet werden*]

hand (fire) extinguisher, manual (fire) extinguisher, portable (fire) extinguisher, fire drencher Hand(feuer)löscher *m*

hand firing, hand stoking, manual firing, manual stoking Handfeuern, Feuern mit Handbeschickung *n*

(hand)grip → handle

(hand)grip mixer, handle mixer Griffmischer, Handgriffmischer *m* [*Bad*]

hand-held, manually held handgeführt

hand(-held) programming unit, manual programming unit Handprogrammiergerät *n*

hand(-held) scanner, manual scanner handgeführte Abtaster *m*

hand(-held) transmitter, manual transmitter Handsender *m* [*IR-Fernbedienungsgerät*]

handhole [*Small exterior concrete box. intended as a pulling or splicing point for underground cables*] Handloch *n*

handing over → handover

hand input, manual input, input by hand Handeinsatz *m*, Handeingabe *f*

hand intervention, manual intervention Handeingriff *m*

hand lavatory (basin) → hand (wash)basin

handle → heft

handle, handgrip, grip Griff, Handgriff *m*

handle → (door) handle

handle drehbare Riegel *m*

handle bar, (hand)grip bar Griffstange *f*

handle bolt → bent bolt

handle cap Griffkappe *f*

handle knob Griffknopf *m*

handle mixer → (hand)grip mixer

handler Verwalter *m* [*Bussystem*]

handler Verwaltungsprogramm *n* [*z. B. für eine Datenbank*]

handles Handhabungsbeschläge *f*

hand lift (Brit.) → hand elevator (US)

handling → (air) mixing

handling → (air) circulation

handling → manipulation

hand mode, manual mode Handbetriebsart, Betriebsart Hand *f* [*Die Steuer(ungs)vorrichtung arbeitet durch Eingriff des Bedienenden nur in Abhängigkeit von Verrieg(e)lungen. V DIN 19 237*]

hand (motor) starter, manual (motor) starter Handanlasser *m*

hand oil burner, manual oil burner handbediente Ölbrenner *m* [*DIN 4787, Teil 1. Ein Brenner, der sich vom teilautomatischen Brenner dadurch unterscheidet, dass er nicht mit selbsttätig wirkenden Zünd- und Steuereinrichtungen ausgerüstet ist. Die einzelnen Schritte des Zündvorganges werden entweder unmittelbar an dem Brenner vor Ort oder durch Fernbetätigung von Hand durch das Bedienungspersonal durchgeführt. Bei Fernbetätigung werden die einzelnen Schritte durch Signalisierung am Bedienungsort erkennbar.*]

hand ((open) loop) control, manual ((open) loop) control [*See remark under "((open) loop) control"*] Handsteuerung *f*

hand operation, manual operation Handbedienung *f* [*DIN 19 222, DIN V 32 734, Teil von „Leiten" nach DIN 19 222 (Leittechnik, Begriffe), jedoch keiner Ebene zuordenbar. Unterscheidung in fest zugeordnete Funktionen und dialogorientierte Bedienung. Unter Handbedienung ist die Einwirkung auf den Prozess unter Mitwirkung der Automationsstation zu verstehen. Die Handeingriffe sind in der Anwendungssoftware unter der gewählten Betriebsart vorgesehen und werden geführt. Entsprechende Schalt- und Stellbefehle gibt die Automationsstation aus. Unterschied: Notbedienebene*]

handover, handing over Übergabe *f* [*Zeitpunkt, an welchem ein Gebäude mit allen Einrichtungen fertig gestellt, die Inbetriebsetzung erfolgt, und der beratende Ingenieur und/oder Endkunde mit dem Betrieb des Gebäudes zufrieden ist.*]

handover unterbrechungsfreie Umschalten *n* [*Verlässt ein mobiles Gerät während einer Verbindung eine Funkzelle und trifft in einer benachbarte Funkzelle ein, so darf die Verbindung nicht abbrechen*]

hand override, manual override Handübersteuerung, Übersteuerung von Hand *f*

hand-portable tragbar

handrail Handlauf *m*

hand override switch, (manual) override switch Handübersteuer(ungs)schalter, Schalter für manuellen Eingriff, Schalter für Handeingriff, (Hand)Korrekturschalter *m*

hand-power elevator (US); hand-power lift (Brit.) [*Elevator utilizing manual energy to move the car*] Handaufzug *m*

handrail drive [*escalator*] Handlaufantrieb *m*

handrail driving wheel [*escalator*] Handlaufantriebsrad *n*

handrail inlet guide [*escalator*] Handlaufeinführung *f*

hand reception, manual reception bediente Empfang *m*

hand regulating cap, manual regulating cap Handregulierkappe *f*

hand regulation → hand ((closed) loop) control

hand release, manual release Handauslösung *f* [*Zum Auslösen einer Vorrichtung von Hand*]

hand restart, manual restart Handwiederanlauf *m*

hand service lift, manual service lift (Brit.); hand dumb waiter, manual dumb waiter (US) Handspeise(n)aufzug *m*

handshake Hand-in-Hand-Arbeit *f* [*Schnittstelle. Es werden bestimmte Vorgänge zur Kommunikationskontrolle benutzt*]

H

handshaking Steuer(ungs)signale *f* mit dem Partnergerät

handshaking Einleitung(svorgang) *f*, (*m*), Beginnabgleich *m* [*Datenkommunikation*]

hand shower mixer, hand shower blender [*"Manual" may synonymously be used for "hand"*] Handbrausemischer *m*

hand slide-in reader, manual slide-in reader, hand slide-in reading device, manual slide-in reading device Handeinschubleser, Einschubhandleser *m*

hand (spray) shower, manual (spray) shower, mov(e)able (spray) shower Handbrause, Schlauchbrause *f*

H

hand-started single-phase motor, manually started single-phase motor Einphasenwechselstromanwurfmotor, Einphasenmotor ohne Hilfswick(e)lung *m*

hand starter, manual starter Handanlasser, Handstarter *m*

hand station, manual station Handstation *f* [*Geräte oder Gerät werden oder wird von Hand bedient*]

hand stoking → hand firing

hand switch, manual switch Handschalter *m*

hand switchroom, manual switchroom Handamtsraum *m* [*Fernsprechzentrale*]

hand thermostat, hand temperature controller, manual thermostat, manual temperature controller Handthermostat, Handtemperaturregler *m*

hand throwover, manual throwover Handkippen, Handumschlagen *n*

hand valve, manual valve Handventil *n*

hand vice, manual vice [*locksmith*] Feilkloben, Handkloben *m*

hand (wash)basin, hand lavatory (basin) Hand(wasch)becken *n*

hand(water) mixing valve, non-thermostatic (water) mixing valve [*"mixer (valve)", "showermixer", "blender" and "blending mixer" may synonymously be used for "mixing valve"*] Bad(e)-Handbatterie, Misch-Handbatterie *f*

handwheel Handrad *n*

handwoven tapestry → arras

handy handlich

hangar door Hallentür *f*

hangar gate Hallentor *n*

hanger → (pipe) hanger

hanger distance, hanger interval Schellenabstand *m*

hanger strap Schellenlasche *f*

hanger with steel rod → (pipe) hanger with steel rod

hanging → suspended

hangup Arbeitsablaufblockierung, Blockierung *f* [*Elektronik. Zustand, der das System aus eigener Kraft nicht mehr verlassen kann, z. B. infolge einer Endlosschleife*]

hard copy [*Data in a permanent and tangible form such as printed, punched or handwritten*] Papierkopie *f*

hard disk, hard disc, rigid magnetic disk, rigid magnetic disc Festplatte, Magnetplatte *f* [*Ihre magnetische, gewöhnlich nicht austauschbare, Schicht ist auf ein festes Trägermaterial aufgebracht (z. B. Aluminium oder Keramik)*]

hardening, quenching [*Rapid cooling from specific temperatures of metallic products, particularly steel, in order to increase hardness*] Abschrecken *n*

hard lead pipe → antimonial lead pipe

hard lead pressure pipe → antimonial lead pressure pipe

hard (paste) porcelain trap → fel(d)spathic porcelain trap

hard (protection) label, hard (protection) tag [*shoplifting prevention*] Hart(sicherungs)etikett *n*

hard shadow Hartschatten *m*

to hard solder → to braze

hard solder Hartlot *n* [*DIN 8513*]

hard soldering → brazing

hardware → (builders') iron supplies

hardware, HW Bausteine *f*, elektronische Bauteile *f* und Geräte *f* einer DV-Anlage [*Rechner; Speicher; Drucker; Bildschirm; Kabel usw.*]

hardware interface Bausteineschnittstelle *f*

hardware programme (Brit.); hardware program (US) Arbeitsanweisungen *f* an die Zentraleinheit

hard-wired, fixed-wired [*Connected permanently by means of electrically conducting wires*] festverdrahtet

hardy, rugged kräftig, robust

harmonic current Stromharmonie *f*, harmonische Strom *m*

harmonic detector, harmonic detecting device, harmonic detection device Klirrfaktormessbrücke *f*

harmonic distortion, nonlinear distortion [*Of a system or transducer characterized by the appearance in the output of harmonics other than the fundamental component when the input wave is sinusoidal*] harmonische Verzerrung, nichtlineare Verzerrung *f*

harmonic induction torque asynchrone Zusatzdrehmoment *n*

harmonic oscillation harmonische Schwingung, Sinusschwingung *f* [*In der musikalischen Akustik wird eine Sinusschwingung als Ton bezeichnet*]

(harmonic) response (characteristic) method → frequency response (characteristic) method

harmonic tone harmonische Ton, Oberton *m*

harmonic tone scale Obertonskala, harmonische Tonskala *f*

harmonic wave Oberwelle, harmonische Welle *f*

harmonic wave limitation Oberwellenbegrenzung, harmonische Wellenbegrenzung *f*

Harry-sound Mischung *f* und Überblendung von Audiosignalen

hash Rauschstörung *f* [*Zerhackerbetrieb*]

hash → garbage

hash collector → garbage collector

hash-filter capacitor Symmetriekondensator *m*

hash total [*In automatic data processing, a sum obtained by adding together numbers having different meanings; the whole purpose is to ensure that the correct number of data have been read by the computer*] Kontrollsumme *f*

hasp → (door) crook

hasp and staple, clasp and clamp Haspe und Krampe, Kette und Haspe *f*, Anwurf, Überwurf, Überfall und Krampe, Kettel *m*, Schließband *n* [*Zum Befestigen eines Vorlegeschlosses*]

hasp lock Überfallschloss *n*

hatch Luke *f*

hatchway → hoistway

hatchway air removal → hoistway air removal

(hatchway) pit; elevator pit (US); lift pit (Brit.) [*Portion of the hatchway extending below the level of the bottom landing to provide for bottom overtravel and clearance and for parts that refuse place below the bottom limit of car travel*] Aufzuggrube, Schachtgrube *f*

hazard [*Not to be confused with "risk"*] Gefährdung *f*

hazard analysis Gefährdungsanalyse *f*

hazard degree Gefährdungsgrad *m*

hazard location Gefährdungsstelle *f*

hazardous gefährdend

hazard reduction Gefährdungsminderung *f*

hazard to life, life hazard Personengefährdung, Lebensgefährdung *f*

hazard zone Gefährdungszone *f*

haze Dunst *m* [*Kaum sichtbarer Rauch*]

hazy image → out-of-focus

hazy picture → out-of-focus

HBES → home and building electronic system

HC → semicarry

HCR, hypothetical reference circuit Bezugsleitung, Bezugsverbindung *f*

H.C.V. → higher heating value

HDD → heating degree day(s)

HDLC → high-level data link control

HDSS → holographic data storage system

HDTV → high definition television

HDX interface → semiduplex interface

head [*The terms "head" and "pressure" are often mistakenly used interchangeably; however, head is the height of a fluid column supported by fluid flow, while pressure is the normal force per unit area. For liquids, it is convenient to measure the head in terms of the flowing liquid. With gas or air, however, it is customary to measure pressure on a column of liquid*] Höhe *f*

head [*A device that reads, records or erases data on a storage medium such as drum or tape; examples are a small electromagnet or a sensing or punching device*] Kopf *m*

head amplifier Vorverstärker *m*

headcrash Kopfaufschlag, Festplattenabsturz *m* [*Schreibkopf; Lesekopf*]

head disk assembly, head disc assembly Schreib-/Lesekopf *m*

headend Netzwerkkabelkopf, Kabelkopf *m*

headend equipment Endgerät(e) *n,(f)*

headend eqipment interface Endgerät(e)-Schnittstelle *f*

headend line Endstellenleitung *f*

header Kennsatz *m*

header, distribution header, manifold, central distributor [*Pipe or tube (extruded, cast, or prefabricated) to which a number of other pipes or tubes are connected*] Sammelrohr, Verteiler(rohr), Verteil(ungs)-rohr, Verteil(er)stück, Verteilungsstück *n*, Sammler *m*, Verteilerkopfleitung, Kopfleitung, Verteil(er)batterie, Verteilungsbatterie *f*

header (US) → principal duct

header, H (Schicht)Steuer(ungs)zeichen *n* [*ISO-Osi-Schichtenmodell*]

header card [*A card that contains supplemental information related to the data on the succeeding cards*] Kopfkarte *f*

H

header data file Definitionsdatei *f*

header (line), heading Kopfzeile, Überschrift *f*

head hole Kopfloch *n*

heading Anstauchen *n*

heading start → **start of heading**

heading up [*e.g. sewage in a drainage system*] Rückstau(en) *m,(n)*

head level, head height Kopfhöhe *f*

head line Kopfzeile, Titelzeile *f*

head line → **pressure line**

head loss Druckhöhenverlust *m*

head of water seal(ing), water seal(ing) head Sperwasserhöhe *f* [*Wirksame Höhe der Wassersäule in einem Geruchsverschluss in mm. DIN 4045*]

head-on von vorn(e)

head pressure ((open) loop) control Förderdrucksteuerung *f*

head rest Kopfstütze *f*

head room lichte Höhe, Lichthöhe *f*

head setting time Kopfberuhigungszeit *f* [*Speicherlaufwerk*]

head (spray) shower Kopfbrause *f*

to head up [*e.g. sewage in a drainage system*] rückstauen

health engineering Gesundheitstechnik *f*

hearing hazard Hörgefährdung *f*

hearth [*The stone or brick floor of a fire(place), often extending out into the room*] Kaminboden *m*

hearth → **fire(place)**

hearth fire Herdfeuer *n*, Herdbrand *m*

to heat heizen, (er)wärmen

to heat beheizen [*Raum*]

heat, thermal energy Wärme *f*

heat [*The quality of being hot*] Hitze *f*

heat absorber, thermal absorber Wärmeabsorber, Wärmeaufnehmer *m*

heat absorption, thermal absorption Wärmeaufnahme, Wärmeabsorption *f*

heat absorption coefficient, thermal absorption coefficient Wärmeabsorptionskoeffizient, Wärmeaufnahmekoeffizient *m*

heat accumulation → **heat storage**

heat accumulation capacity → **heat storage capacity**

heat accumulation device → **heat storage device**

heat accumulation management → **heat storage management**

heat accumulator, thermal accumulator Wärmespeicher *m* [*Vorrichtung zum Ausgleich von Zeit- und Leistungsunterschieden zwischen Wärmeangebot und Wärmeabnahme*]

heat action, heat effect, thermal action, thermal effect Wärmewirkung *f*

heat actuated valve, thermally actuated valve [*Automatic valve which uses the heat generated by the resistance of an electrical component in opening or closing itself*] elektrothermische Ventil *n*

heat act(uat)or, thermal act(uat)or thermische Stellantrieb *m*

heat addition, thermal addition [*Carnot cycle*] Wärmezufuhr, zugeführte Wärmemenge *f*

heat admission → **admission of heat**

heat air treatment method, thermal air treatment method thermische Luftbehandlungsverfahren *n*, thermische Luftbehandlungsmethode *f* [*Diese Verfahren sind Heizen, Kühlen, Befeuchten und Entfeuchten*]

heat and mass transfer by convection konvektive Wärme- und Stoffaustausch *m*

heat and moisture flow through openings by convection konvektive Wärme- und Feuchteaustausch *m* durch Öffnungen

heat arrest, freezing plateau, thermal arrest [*During the freezing of produce, that part of the graph of temperature against time, in which the temperature remains virtually constant. It is induced by the time required to remove latent heat.*] Gefrierplateau *n* konstanter Temperatur, Gefrierabschnitt *m* konstanter Temperatur, Gefrierkurve *f* konstanter Temperatur

heat balance, thermal balance [*Statement that shows the changes in a system from heat and work input to output losses*] Wärmebilanz *f*, Wärmehaushalt *m*

heat barrier, heat break, thermal barrier, thermal break Wärmesperre, Thermosperre *f*

heat boundary resistance, thermal boundary resistance, heat contact resistance, thermal contact resistance [*Ratio of temperature difference to heat flux across the boundary between two distinct media – solid/solid or solid/fluid*] thermische Grenzflächenwiderstand *m*

heat bridge Wärmebrücke *f* [*Ein örtlich begrenzter Bereich in der Konstruktion, der gegenüber der benachbarten Stoffschicht eine erhöhte Wärmestromdichte und verminderte innere Oberflächentemperatur aufweist*]

heat build-up, thermal build-up Wärme(auf)stau *m*

heat call(ing), call(ing) for heat Wärme(an)forderung *f*

heat capacity, thermal capacity Wärmekapazität *f*

heat circulation, thermal circulation Wärmeumwälzung, Wärmezirkulation *f*, Wärmeumlauf *m*

heat ((closed) loop) control Wärmereg(e)lung *f*

heat ((closed) loop) control relay Wärmregel(ungs)relais *n*

heat coil [*Protective device which uses a mechanical element which is allowed to move when the fusible substance that holds it in place is heated above a predetermined temperature by current in the circuit*] Feinsicherung(spatrone), Hitzdrahtspulensicherung *f*

heat comfort, thermal comfort Wärmebehaglichkeit *f*

(heat) conductance, thermal conductance, C [*The amount of heat, expressed in British Thermal Units (Btu), transmitted in 1 hour from surface to surface of 1ft2 of material or a combinations of materials for each degree of temperature difference between the two surfaces. It should be noted that this value is not expressed in terms per inch of thickness, but of transmission from surface to surface*] Wärmedurchlasszahl *f* in britischen Wärmeeinheiten [*Nicht gleichzusetzen mit der deutschen Wärmedurchlasszahl (Ë)*]

heat conduction, thermal conduction, conductive heat transfer [*Thermodynamics. The flow of thermal energy through a substance from a higher to a lower temperature region*] Wärmeleitung *f*

heat conduction transfer Wärmeübertragung *f* durch Wärmeleitung

heat conductivity, thermal conductivity [*The heat flow across a surface per unit area per unit time, divided by the negative of the rate of change of temperature with distance in a direction perpendicular to the surface*] Wärmeleitfähigkeit *f*, Wärmeleitvermögen *n*

heat conductivity cell, catharometer, thermal conductivity cell, katharometer [*An instrument for detecting the presence of small quantities of gases in air by measuring the resulting change in thermal conductivity of the air*] Katharometer *n*

heat conductor, thermal conductor, conductor od heat, caloric conductor [*A substance with a high thermal conductivity*] Wärmeleiter *m*

heat conservation shutter, thermal conservation shutter [*fan*] wärmehaltende Lamelle *f*

heat consumer, heat user Wärmeverbraucher, Wärmenützer, Wärmeabnehmer *m*

heat consumption, heat use Wärmeverbrauch *m*, Wärmenützung, Wärmeabnahme *f*

heat content → enthalpy

heat content ((closed) loop) control → enthalpy ((closed) loop) control

heat content controller → enthalpy controller

heat control → heat ((closed) loop) control

heat control → heat ((open) loop) control

heat control relay → heat ((open) loop) control relay

heat convection, thermal convection Wärmekonvektion *f*

heat cracking, heat crack formation Wärmerissbildung *f*

heat curtain, thermal curtain Wärmeschleier, Wärmevorhang *m*

heat demand, heat requirement, heat need Wärmebedarf *m*

heat density Wärmedichte *f*

heat detector → bolometer

heat diffusion, thermal diffusion [*Phenomena in which a temperature gradient in a mix(ture) of gases gives rise to a flow of one constituent relative to the whole mix(ture)*] Thermodiffusion *f*

heat diffusivity, thermal diffusivity [*Physical quantity that determines the rate of heat propagation in transient-state processes. Thermal conductivity divided by the product of density and specific heat. Unit: m^2/s*] thermische Diffusivität, Temperaturleitzahl *f*

heat dilatation, thermal dilatation Wärmeausdehnung *f*

heat discharge installation, heat discharge plant Wärmeabführ(ungs)anlage, Wärmeabzuganlage, Wärmeableit(ungs)anlage *f* [*VdS, DIN 14 011*]

heat dissipating wärmeabgebend, wärmeabführend

heat dissipation Wärmeabgabe, Wärmeabfuhr *f*

heat distribution, distribution of heat, thermal distribution Wärmeverteilung *f*

heat distribution network Wärmeverteilungsnetz(werk) *n*

heat drive, thermal drive thermische Antrieb *m* [*Thermostatkopf*]

heat drying, thermal drying thermische Trocknung, Trocknung im engeren Sinne *f*

heat-drying, thermal-drying wärmetrocknend

heat ecomomizing, heat-saving wärmesparend

heat economy, thermal economy Wärmewirtschaft *f*

heated erwärmt, gewärmt

heated beheizt

heated → (air) heated

H

heated acoustic ceiling, acoustic heated ceiling Akustikheiz(unter)decke, Heizakustik(unter)decke *f*

heated air → warm air

heated all-concrete floor beheizte Vollbetondecke *f*

heated ceiling beheizte (Unter)Decke, beheizte nichtragende Decke, Heiz(ungs)(unter)decke *f*

heated floor, heating floor beheizte (Fuß)Boden, Heiz(ungs)(fuß)boden *m*

heated hollow (building) block floor beheizte Hohlblock(stein)decke, beheizte Hohlsteindecke, Heiz(ungs)hohlsteindecke, Heiz(ungs)hohlblock(stein)decke *f*

heated slab [*Concrete slab-on-ground floor, heated by integral wires, cables, pipes, or ducts, that radiates heat to the space*] Beton(fuß)bodenheiz(ungs)platte *f*

H

heated space, heated room [*Space within a building which is provided with a positive heat supply. Finished living space within a basement, or registers or heating devices to supply heat to a basement space automatically define that space as heated space*] beheizte Raum *m*

heated (structural) floor beheizte Decke, Heizungsdecke *f*

heated towel rack beheizte Handtuchständer *m*

heated towel rail, heated towel holder, heated towel bar beheizte Handtuchschiene, beheizte Handtuchstange *f*, beheizte Handtuchhalter *m*

heat effect → heat action

heat efficiency, thermal efficiency Wärmewirkungsgrad, thermische Wirkungsgrad *m*

heat egress → egress of heat

heat (electrical) noise power, thermal (electrical) noise power Wärmerauschleistung, Wärmestörleistung *f*

heat elimination, heat removal, heat rejection Wärmeabfuhr, Wärmeabführung *f*, Wärmeentzug *m*

heat emitter, thermal emitter Wärmeabgeber *m*

heat emitting wärmeabgebend

heat energy, thermal energy Wärmeenergie *f*

heat (energy) flow, flow of heat (energy), flow of thermal energy, thermal energy flow [*Passage of heat from one point to another or one space to another by conduction, convection, refrigeration, or radiation*] Wärmefluss *m*

heat energy storage, thermal energy storage Wärmeenergiespeicherung *f*

heat engine, thermal engine Wärmekraftmaschine *f*

heat engineering, thermal engineering Wärmetechnik *f*

heat enviroment, thermal enviroment Wärmeumwelt *f*

heater Heiz(ungs)körper *m* [*Sammelbenennung für Radiator, Konvektor und Flachheiz(ungs)körper*]

heater → heating appliance

heater → (floor(-mounted)) (room) heater

heater battery Heizerbatterie *f*

heater bracket Heiz(ungs)körperkonsole *f*

heater characteristic (curve) Heiz(ungs)körperkennlinie, Heiz(ungs)körperkennkurve *f*

heater coil → zig-zag heating tube

heater furnace → (air) heater furnace

heater inlet Heiz(ungs)körpereinlauf *m*

heater mat [*Electrically heated wire mesh or heated liquid network embedded in a cold store floor to prevent freezing of water in the coil below*] Heiz(ungs)matte *f*

heater mounting Heiz(ungs)körperaufstellung *f*

heater outlet Heiz(ungs)körperauslauf *m*

heater pump → (air) heater pump

heater recess Heiz(ungs)körpernische *f*

heater-storage (combination) Heizer-Speicher(-Kombination) *m, (f)*

heater switch Heiz(ungs)körperschalter *m*

heater tube Heiz(ungs)körperröhre *f*

heater valve Erhitzerventil, Erwärmerventil *n*

heat evolution, thermal evolution Wärmeentwick(e)lung *f*

heat excess → excess of heat

heat exchange [*Process of heat transfer in which heat flows from one substance to another*] Wärme(aus)tausch *m* [*Siehe Anmerkung unter „Wärme(aus)tauscher"*]

heat exchange area Wärme(aus)tauschfläche *f*

heat exchange of man with his environment Wärmeaustausch *m* Mensch/Umgebung

heat exchanger, heat exchange unit, heat interchanger Wärme(aus)tauscher (im engeren Sinne), Umformer *m* [*Apparat, bei dem Wärme zwischen zwei Medien ausgetauscht wird.*]

heat exchanger, heat exchange unit, heat interchanger Wärme(aus)tauscher *m* (im weiteren Sinne), Wärmeübertrager [*In der technischen Praxis haben sich – bedauerlicherweise – viele falsche Benennungen eingeschlichen . So ist zum Beispiel „Wärme(aus)tauscher" schlichtweg falsch, da ein „Austausch" ein gegenseitiges Geben und Nehmen beinhaltet, die Wärmeübertragung jedoch einen recht* *einseitigen Vorgang darstellt: Das Medium mit dem höheren Temperaturniveau gibt Wärme ab, ohne dafür etwas zurückzuerhalten. Demzufolge wäre für Bauteile mit Wärmeübertragungsfunktionen ausschließlich die Benennung „Wärmeübertrager" zu verwenden*]

heat expansion, thermal expansion [*Increase in one or more of the dimensions of a solid body or a liquid volume, caused by a temperature rise*] Wärmeexpansion, Wärmeausdehnung *f*

heat expansion coefficient, thermal expansion coefficient Wärmeexpansionskoeffizient, Wärmeausdehnungskoeffizient *m*

heat extraction Wärmeabzug, Wärmeauszug *m*

heat flow, thermal flow [*Thermodynamics. 1. The transfer of energy from one substance to another as a result of temperature difference. 2. The amount of such transfer for a unit time*] Wärmestrom *m*

heat flow density, thermal flow density Wärmestromdichte *f*

heat fluid, thermal fluid Wärmefluid, Wärmeströmungsmedium *n* [*Flüssigkeit; Gas; Dampf*]

heat gain Wärmegewinn *m*

heat generated by illumination → illumination (generated) heat

heat generated by people, people-generated heat Personenwärme *f*

heat generating → heat producing

heat generation, heat production Wärmeerzeugung *f* [*DIN 3440*]

heat generation installation, heat generation plant, heat generating installation, heat generating plant [*"Production" may synonymously be used for "generation"*] Wärmeerzeugungsanlage *f* [*DIN 4751 und DIN 4759*]

heat generator, heat producer, hear-generating device, heat-producing device Wärmeerzeuger, WE *m* [*Vorrichtung, mit der zugeführte Energie in nutzbare Wärme umgeformt wird*]

heat-humidity recovery device, heat-humidity recovering device Wärme-Feuchtigkeits-Rückgewinner, Wärme-Feuchte-Rückgewinner, WFRG *m* [*DIN 1946, Teil 1*]

heat impact, thermal impact Wärmestoß *m*

heat inertia, thermal inertia Wärmeträgheit, thermische Trägheit *f*

heating heizend

heating (Er)Wärmen, Erhitzen *n*, Erwärmung *f*

heating [*Process of adding heat energy causing a rise in temperature, or a transfer of sensible heat into latent heat*] Heizung *f*, Heizen *n*

heating Beheizen *n*, Beheizung *f*

heating ... → (air) heating ...

heating accessories Heiz(ungs)zubehör *n,m*

heating air Heiz(ungs)luft *f*

heating air-conditioning unit [*Specific air-treating combination consisting of means for ventilation, air circulation, air cleaning, and heat transfer, with control means for heating*] Heiz(ungs)klimaeinheit *f*

heating air outlet Heiz(ungs)luftauslass, Heiz(ungs)luftaustritt *m*

heating and cooling unit → (combination) heating and cooling unit

heating appliance → (air) heating device

heating area Heiz(ungs)fläche *f*

heating assembler → heating fitter

heating ball cock Heiz(ungs)kugelhahn *m*

heating bar Heiz(ungs)stab *m*

heating basement Heiz(ungs)kellergeschoss, Heizungskellerstockwerk, Heizkellerstockwerk *n*, Heizkelleretage, Heizungskelleretage *f*

heating battery [*The heating surface of a calorifier*] Heiz(ungs)batterie *f*

heating bill Heiz(ungs)rechnung *f*

heating boiler Heiz(ungs)kessel *m*

heating boiler plant, heating boiler installation Heiz(ungs)kesselanlage, Heiz(ungs)kesselinstallation *f*

heating booster → booster heater

heating builder Heizungsbauer *m*

heating by town gas, town gas heating Stadtgasheizung *f*

heating by waste heat, waste heat heating Abwärmeheizung *f* [*Durch Zwischenschaltung von Wärmespeichern wird der fast wertlose Abdampf von Fördermaschinen, Dampfhämmern usw. für Heizzwecke genutzt. Auch Auspuffgase sind verwendbar*]

heating cabinet, heating cubicle Wärmeschrank *m*

heating cable Heiz(ungs)kabel *n*

heating calculation Heiz(ungs)berechnung *f*

heating cartridge Heiz(ungs)patrone *f*

heating casing → heating jacket

heating ceiling Heiz(ungs)(unter)decke *f*

heating cellar Heiz(ungs)keller *m*

heating central distributor → heating manifold

heating central station, central heating station Heiz(ungs)zentrale *f*

H

heating chamber Heiz(ungs)kammer *f*

heating characteristic (curve), heating curve Heiz(ungs)kennkurve, Heiz(ungs)kennlinie *f*

heating circuit Heiz(ungs)schaltung *f*

heating circuit diagram Heiz(ungs)schaltschema *n*

heating circuit distribution cabinet, heating circuit distribution cubicle Heiz(ungs)verteilerschrank *m*

heating circuit manifold, heating circuit central distributor Heiz(ungs)verteiler(rohr) *m,(n)*, Heiz(ungs)verteilerstück *n*

heating circulation Heiz(ungs)umlauf *m*

heating circulation pump → circulating heating pump

H

heating climate Heiz(ungs)klima *n*

heating ((closed) loop) control, heating regulation, heating closed control [*See remark under "((closed) loop) control"*] Heiz(ungs)reg(e)lung, Heiz(ungs)regulierung *f*

heating coil → heating (pipe) coil

heating comfort Heiz(ungs)komfort *m*, Heiz(ungs)wohlbehagen *n*

heating computer Heiz(ungs)rechner *m*

heating concrete floor → concrete heating floor

heating concrete slab → concrete heating slab

heating contactor Heiz(ungs)stromschütz *n*

heating contractor Heizungsunternehmer *m*

heating control → heating ((open) loop) control

heating controller Heiz(ungs)regler *m*

heating/cooling thermostat, summer/winter thermostat [*It can have its action reversed and, where required, its set point raised or lowered by the indexing method. This type of thermostat is used to actuate controlled devices, such as valves or dampers, that may regulate a heating medium at one time and a cooling medium at another*] Heizung(s)/ Kühlung(s)-Thermostat, Sommer/Winter-Thermostat *m*

heating cost(s) settling Heiz(ungs)kostenabrechnung *f*

heating cubicle → heating cabinet

heating current, heating electricity Heiz(ungs)strom *m*, Heiz(ungs)elektrizität *f*

heating curve → (controller characteristic) heating curve

heating cycle Heiz(ungs)kreislauf *m*

heating degree-day, HDD [*Degree-day for departure below the base*] Heiz(ungs)gradtag *m*

heating designer Heizungskonstrukteur *m*

heating device → (air) heating device

heating device → heating appliance

heating duct Heiz(ungs)kanal *m*

heating effect Heiz(ungs)wirkung *f*

heating element Heiz(ungs)element, Heiz(ungs)glied *n*

heating energy Heiz(ungs)energie *f*

heating energy demand, heating energy requirement Heiz(ungs)energiebedarf *m*

heating engineer Heiz(ungs)ingenieur *m*

heating engineering Heiz(ungs)technik *f*

heating equipment [*Two or more asemblies capable of performing a complete heating function*] Heizgerät(e) *n, (f)*

heating equipment room Heiz(ungs)gerät(e)raum *m*

heating expert Heizungssachverständige, Heizungsgutachter *m*

heating facade Heizungsfassade *f*

heating fitter, heating assembler Heizungsmonteur, Heizungsschlosser *m*

heating flame Heizflamme *f*

heating flange Heizflansch *m*

heating floor, heated floor Heiz(ungs)(fuß)boden, beheizte (Fuß)Boden *m*

heating flue Heiz(ungs)zug *m*

heating (forward) flow, forward heating flow Heiz(ungs)vorlauf *m*

heating forward flow manifold, forward flow heating manifold Heiz(ungs)vorlaufverteiler(rohr), Vorlaufheiz(ungs)verteiler(rohr) *m,(n)*

heating gas path Heiz(ungs)gasweg *m*

heating heat Heiz(ungs)wärme *f*

heating heat flow Heiz(ungs)wärmestrom *m*

heating heat generator, hating heat producer Heiz(ungs)wärmegenerator, Heiz(ungs)wärmeerzeuger *m*

heating heat pump [*Refrigerating system designed primarily to utilize the heat rejection from the system for a delivered heating function*] Heiz(ungs)wärmepumpe *f*

heating hot water, heating warm water Heizungswarmwasser, Heizwarmwasser *n*

heating insert Heiz(ungs)einsatz *m* [*z. B. für Kachelöfen*]

heating installation, heating plant Heiz(ungs)anlage *f*

heating interruption Heiz(ungs)unterbrechung *f*

heating jacket, heating casing Heiz(ungs)mantel *m*, Heiz(ungs)ummantelung *f*

heating line → heating (pipe)line

heating lining Heiz(ungs)verkleidung *f*

heating load Heiz(ungs)last *f* [*Die Heiz(ungs)last wird bestimmt durch die Nutzeranforderungen, die klimatischen Bedingungen und die konstruktive und die bautechnische Ausführung der Raumumschließungsflächen (Wände, Decken, Fußböden, Fenster, Türen und Öffnungen aller Art) Sie tritt auf, wenn die vom Raum an die Umgebung abgegebene Wärmemenge größer ist als die dem Raum zufließende und in ihm entstehende Wärmemenge*]

heating load adap(ta)tion Heiz(ungs)lastanpassung *f*

heating load curve Heiz(ungs)lastkurve *f*

heating load factor, HLF Heiz(ungs)lastfaktor *m*

heating log Heiz(ungs)tagebuch *n*

heating main Heiz(ungs)haupt(rohr)leitung, Hauptheiz(ungs)(rohr)leitung *f*

heating manifold, heating central distributor Heiz(ungs)verteiler(rohr) *m, (n)*, Heiz(ungs)verteilungsrohr, Heiz(ungs)verteilerstück, Heiz(ungs)-kreisverteiler(rohr), Heiz(ungs)verteilungsstück *n*, Heiz(ungs)verteilerbatterie, Heiz(ungs)verteilungsbatterie *f*

heating medium Heiz(ungs)mittel *n*, Heiz(ungs)träger *m* [*Im Heizkessel oder in einem anderen Wärmeübertragungsgerät (z. B. Gegenstromapparat, Lufterhitzer und anderen) erwärmtes Medium, das durch Rohrleitungen oder Kanäle zum Verwendungsort strömt um dort – meist über Heiz(ungs)flächen – Wärme abzugeben*]

heating medium flow Heiz(ungs)mittelstrom, Heiz(ungs)trägerstrom *m*

heating medium return Heiz(ungs)trägerrücklauf, Heiz(ungs)mittelrücklauf *m*

heating medium valve Heiz(ungs)trägerventil, Heiz(ungs)mittelventil *n*

heating mixer Heiz(ungs)mischer *m*

heating network Heiz(ungs)netz(werk) *n*

heating oil [*See remark under "fuel oil"*] Heizöl *n* [*DIN 51603. Es verbrennt nicht als Flüssigkeit, sondern wird vor der Verbrennung in einem Brenner durch Verdampfen (Ölofen) oder Zerstäuben (Heizkessel) aufbereitet*]

heating oil barrier Heizölsperre *f* [*DIN 4043*]

heating oil burner Heizölbrenner *m*

heating oil hose Heizölschlauch *m*

heating oil line Heizölleitung *f*

heating oil meter Heizölzähler, Heizölmesser *m*

heating oil preheater Heizölvor(er)wärmer, Heizölvorerhitzer *m*

heating oil room Heizöllagerraum *m*

heating oil separator, heating oil trap, heating oil interceptor [*See remark under "interceptor"*] Heizöl(ab)scheider, Heizölfang, Heizölfänger *m* [*DIN 1999*]

heating oil storage Heizöllagerung *f*

heating oil (storage) tank Heizöl(lager)behälter, Heizöl(lager)tank *m*

heating oil supply Heizölversorgung *f*

heating oil supply plant, heating oil supply installation Heizölversorgungsanlage, Heizölversorgungsinstallation *f*

heating-only heat pump Wärmepumpe *f* nach DIN 8900

heating ((open) loop) control Heiz(ungs)steuerung *f*

heating operation(s) Heiz(ungs)betrieb *m*

heating output, heating performance Heiz(ungs)-leistung *f*

heating oven, heating stove Heiz(ungs)ofen *m*

heating pad Heiz(ungs)kissen, Wärmekissen *n*

(heating) performance coefficient → coefficient of (heating) performance

heating pipe Heiz(ungs)rohr *n*

heating (pipe) coil Heiz(ungs)(rohr)schlange *f*

heating (pipe) network Heiz(ungs)rohrnetz *n*

heating plant, heating installation [*The entire system for heating enclosed spaces or an enclosed space including either a boiler, piping, and radiators or a furnace, ducts, and air outlets*] Heiz(ungs)anlage, Heiz(ungs)installation *f*

heating power plant, power heating plant Heizungskraftwerk, HKW *n*

heating programme switch; heating program switch (US) Heiz(ungs)programmschalter *m*

heating program selector (US); heating programme selector Heiz(ungs)programmwahlschalter *m*

heating pump Heiz(ungs)pumpe; Heiz(ungs)wärmepumpe *f* [*Fehlbenennung*] [*DIN 8900, Teile 2–4*]

heating register Heiz(ungs)register *n*

heating regulation → heating ((open) loop) control

heating return Heiz(ungs)rücklauf *m*

heating ribbon Heiz(ungs)band *n*

H

heating riser, heating rising (pipe)line Steig(e)-rohrleitung für Heizung, Heiz(ungs)steig(e)rohrleitung *f*

heating room Heiz(ungs)raum *m* [*Raum, in dem eine Feuerstätte oder mehrere Feuerstätten für feste, flüssige oder gasförmige Brennstoffe zur zentralen Beheizung, zur Warmwasserbereitung oder zur Betriebs- und Wirtschaftswärmeerzeugung mit einer Gesamtheiz(ungs)leistung von 48 kW aufgestellt ist oder sind*]

heating room drainage Heiz(ungs)raumentwässerung *f*

heating rules Heiz(ungs)regeln *f* [*DIN 4701*]

heating screed Heiz(ungs)estrich *m* [*DIN 18 560 Teil 2*]

heating season Heiz(ungs)saison *f*

heating (self-contained) air conditioner, heating (self-contained) air conditioning unit [*See remark under "air conditioner"*] Vollklimagerät *n* [*Zusätzlich zur Klimatisierung auch Beheizung durch Anschluss an eine Zentralheizungsanlage, Gasheizung oder elektrische Heizung. Es kann auch am Ölbrenner eingebaut werden*]

heating setting Heiz(ungs)einstellung *f*

heating socket Heiz(ungs)steckdose *f* [*Ein Verbindungselement zum Anschluss mobiler Heiz(ungs)körper*]

heating specialist Heiz(ungs)fachmann *m*

heating station Heiz(ungs)station *f*

heating steam, steam for heating Heiz(ungs)-dampf *m*

heating steam pressure Heiz(ungs)dampfdruck *m*

heating stove, heating oven Heiz(ungs)ofen *m*

heating strip Heiz(ungs)leiste *f*

heating strip ceiling → strip heating ceiling

heating substation Heiz(ungs)unterstation *f*

heating surface Heiz(ungs)oberfläche *f*

heating switch Heiz(ungs)schalter *m*

heating system [*One in which heat is transferred from a source of energy through a distribution network to spaces to be heated*] Heiz(ungs)system *n*

heating technican Heiz(ungs)techniker *m*

heating time optimizer Heiz(ungs)zeitoptimierer *m*

heating tube Heiz(ungs)röhre *f*

heating tube cluster Heiz(ungs)röhrenbündel *n*

heating tubes concrete slab, concrete slab with heating tubes Betonplatte *f* mit Heiz(ungs)röhren

heating-up Anheizen, Aufheizen, Aufladen [*z. B. bei der Nachtstromspeicherheizung*]

heating-up period, heating-up time Aufladezeit, Aufheizzeit, Anheizzeit *f* [*z. B. bei der Nachtstromspeicherheizung*]

heating-up process Aufheiz(ungs)vorgang *m*

heating value [*The constant quantity of heat generated by complete combustion of a unit of specific fuel*] Brennwert *m* (kJ/kg oder kJ/m_n^3) [*DIN 5499.1.72*]

heating/ventilating/air-conditioning, HVAC Heizung/Lüftung/Klimareg(e)lung *f*

heating/ventilating/air-conditioning engineering → HVAC engineering

heating/ventilation/air conditioning/sanitary/-electric engineering Versorgungstechnik, H-L-K-S-E-Technik *f*

heating/ventilating/air conditioning services → HVAC services

heating ventilator Heiz(ungs)lüfter *m* [*Siehe Anmerkung unter „Lüfter"*]

heating voltage Heiz(ungs)spannung *f*

heating warm water → heating hot water

heating water Heiz(ungs)wasser *n* [*Wasser als Medium zur Versorgung von Luft(er)wärmern*]

heating water chamber Heiz(ungs)wasserraum *m*

heating water circulation pump Heiz(ungs)wasser-Umwälz(ungs)pumpe *f*

heating water filter Heiz(ungs)wasserfilter *m, n*

heating water flow Heiz(ungs)wasserstrom *m*

heating water forward flow Heiz(ungs)wasservorlauf *m*

heating water heat Heiz(ungs)wasserwärme *f*

heating water pipe Heiz(ungs)wasserrohr *n*

heating water preparation Heiz(ungs)wasser(auf)-bereitung *f*

heating water return flow Heiz(ungs)wasserrücklauf *m*

heating work Heiz(ungs)arbeiten *f*

heat input Wärmeeingabe *f*

heat insulation, isulation of heat, thermal insulation [*The ability of materials to conduct heat which closely follows their respective densities. Materials with low densities are often called heat insulators*] Wärmedämmung *f*

heat interchanger → heat exchanger

heat lag, thermal lag [*Time elasping between the initiation of a thermal phenomenon and the appearance of its effect*] Wärmeverzug *m*

heat lamp Heiz(ungs)lampe *f*

heat level → level of heat

heat load, thermal load Wärmelast *f*

heat loss calculation Wärmeverlustberechnung *f*

heat management Wärmebewirtschaftung *f*

heat meter Wärmezähler *m*

heat need, heat requirement, heat demand Wärmebedarf *m* [*DIN 4701*]

heat need calculation → heat requirement calculation

heating medium forward flow Heiz(ungs)trägervorlauf, Heiz(ungs)mittelvorlauf *m* [*Die in der Zeiteinheit durch einen Heizkörper strömende Flüssigkeits- oder Dampfmenge, d. h. Quotient aus Menge und Zeit. Einheit kg/h oder ltr/h*]

heat network, thermal network Wärmenetz *n*

heat noise, thermal noise thermische Rauschen *n*

heat nominal load → nominal heat load

heat nominal output → nominal heat output

heating of compression, compression heat [*Heat generated when air is compressed*] Verdichtungswärme *f*

heat of condensation, condensation heat [*The increase in enthalpy accompanying the conversion of 1 mole vapo(u)r into liquid at constant pressure and temperature*] Kondensationswärme, Verflüssigungswärme *f* [*Physikalisch ist „Verflüssigungsenthalpie"*]

heat of convection → convection heat

heat of cooling, cooling heat [*Increase in enthalpy during cooling of a system at constant pressure, resulting from an internal change such as a allotropic transformation*] Kühl(ungs)wärme *f*

heat of fusion, fusion heat, heat of melting, melting heat [*1. Latent heat involved in changing between the solid and the liquid states. 2. Heat energy required to cause a change of state from solid to liquid at constant temperature. For ice-to-water: 143 Btu/lb (333.8 kJ/kg)*] Schmelzwärme *f*

heat-of-light system [*Heat due to lighting intensity is utilized for heating spaces in buildings*] Beleuchtungswärmeheizung *f*

heat of liquid Flüssigkeitswärme *f*

heat of reaction, reaction heat [*In the sorption process*] Reaktionswärme *f*

heat of vaporization (US); heat of vapourization (Brit.) Verdampfungswärme *f*

heat ((open) loop) control Wärmesteuerung *f*

heat ((open) loop) control relay Wärmesteuerungsrelais *n*

heat output, thermal output Wärmeabgabe *f* [*Nutzbare Wärme, die vom Wärmeerzeuger an den Wärmeträger abgegeben wird*]

heat peak, thermal peak Wärmespitze *f*

heat performance, thermal performance Wärmeleistung *f*

heat pipe Wärmerohr *n* [*Ein an beiden Enden geschlossenes Rohr, mit im allgemeinen durch Rippen vergrößerter Oberfläche, das in der Lage ist, durch Verdampfung und nachfolgende Verflüssigung eines Stoffes, oft eines fluorierten Kältemittels, Wärme axial zu transportieren. Ersatzmöglichkeiten in der Klimatechnik als rekuperative Wärmeübertrager für die Wärmerückgewinnung*]

heat pipe heat exchanger, heat pipe heat exchange unit, heat pipe heat interchanger Wärmerohrwärme(aus)tauscher *m* [*In der Industrie so genannt*], Wärmerohrwärmeübertrager [*Siehe Anmerkung unter „Wärme(aus)tauscher"*]

heat pipe heat recovery unit Wärmerohrwärmerückgewinner *m*

heat printer, heat writer, thermal printer, thermal writer Thermaldrucker, Wärmedrucker, Thermodrucker *m*

heat producer → heat generator

heat producing, heat generating, calorific wärmeerzeugend

heat-producing device → heat generator

heat production → heat generation

heat production installation → heat generation installation

heat protection, thermal protection Wärmeschutz *m* [*DIN 4108*]

heat protection against spurious response → heat safeguard against spurious response

heat protection in building construction, thermal protection in building construction Wärmeschutz im Hochbau, Hochbau-Wärmeschutz *m*

heat protection ordinance, thermal protection ordinance Wärmeschutzverordnung *f*

heat pump Wärmepumpe *f* [*Sie nimmt einen Wärmestrom mit niedriger Temperatur auf und gibt ihn samt dem zum Betrieb erforderlichen Antrieb(s)energiestrom mit höherer Temperatur als Nutzwärmestrom ab. DIN 8900; DIN 8901; DIN 255 Teil 1*]

heat pump air conditioning installation, heat pump air conditioning plant Wärmepumpe-Klima(tisierungs)anlage, Klima(tisierungs)anlage mit Wärmepumpe(nschaltung) *f*

heat pump boiler, heat pump water heater, heat pump hot water generator, heat pump warm water generator [*See remark under "water heater"*] Wärmepumpenboiler, Wärmepumpenwasserheizer, Wärmepumpenwasserbereiter, Wärmepumpenwarmwassererzeuger, Wärmepumpenspeicher *m*

heat pump chiller, heat pump refrigerating system [*See remark under "chiller"*] Wärmepumpen-Kälteanlage *f*

heat pump circuit Wärmepumpenschaltung *f*

heat pump ((closed) loop) control Wärmepumpenreg(e)lung *f*

heat pump compressor Wärmepumpenverdichter *m*

heat pump dryer, heat pump drier Wärmepumpentrockner *m*

heat pump heating Wärmepumpenheizung *f*

heat pump heating line Wärmepumpenheizungsleitung *f*

heat pump heating plant, heat pump heating installation Wärmepumpenheiz(ungs)anlage, WPHA *f* [*Sie besteht aus Wärmepumpenanlage und Wärmenutzungsanlage*]

heat pump ((open) loop) control Wärmepumpensteuerung *f*

heat pump operator Wärmepumpenbeteiber *m*

heat pump plant, heat pump installation Wärmepumpenanlage *f* [*Sie besteht aus Wärmepumpe und Wärmequellenanlage*]

heat pump ready for connection → factory-assembled heat pump

heat pump refrigerant Wärmepumpenkältemittel *n*

heat pump system [*The system usually consists of a series of water source heat pump units connected by a two-pipe water circuit. The units get heat from the circuit on the heating cycle and give it to the water circuit on the cooling cycle. Heat transfer occurs when some areas need cooling. The heat in the water resulting from the cooling process in one area is absorbed from the water and used in another area. The common water circuit becomes the medium for the heat exchange.*] Wärmepumpensystem *n*

heat pump unit anschlussfertige Wärmepumpe *f*

heat pump water heater Wärmepumpenwasser(er)wärmer, Wärmepumpenwassererhitzer *m*

heat pump water return, HPWR Wärmepumpenwasserrücklauf *m*

heat pump water supply, HPWS Wärmepumpenwasserversorgung *f*

heat quantity, quantity of heat Wärmemenge *f*

heat quantity count(ing) Wärmemengenzählung *f*

heat quantity metering Wärmemengenmessung *f*

heat radiation, thermal radiation Wärmestrahlung *f*

heat radiator, thermal radiator Wärmestrahler *m*

heat ray, thermal ray Wärmestrahl *m*

heat reclaim chiller Wärmewiedergewinnungs-Kältemaschine *f*

heat recovery Wärmerückgewinnung, WRG, Wärmewiedergewinnung *f* [*Maßnahme zur Rückführung von Abwärme in dasselbe System*]

heat recovery coil of runaround type, runaround (type) heat recovery coil geschlossene Wärmerückgewinnungsschlange *f*

heat recovery device, heat recovery unit Wärmerückgewinner, WRG *m*

heat recovery heat pump Wärmepumpe *f* zur Wärmerückgewinnung

heat recovery liquid cooler Wärmerückgewinnungs-Flüssigkeitskühler *m*

heat recovery plant, heat recovery installation Wärmerückgewinnungsanlage *f*

heat recovery system Wärmerückgewinnungssystem *n* [*Verknüpfung eines oder mehrerer, auch verschiedener Wärmerückgewinner einschließlich ihrer anwendungsbezogenen Zusatzvorrichtungen mit einer raumlufttechnischen Anlage*]

heat recovery unit, heat recovery device Wärmerückgewinner, WRG *m*

heat recovery ventilator, HRV Wärmerückgewinnungslüfter *m*

heat reflective wärmereflektierend

heat rejection → heat elimination

heat release, thermal release Wärmeabgabe *f*

heat removal → heat elimination

heat replica, thermal replica thermische Abbild, Wärmeabbild *n*

heat requirement → heat need

heat requirement calculation, heat need calculation Wärmebedarfsermittlung, Wärmebedarfsberechnung *f*

heat resistance, thermal resistance Wärmeübergangswiderstand *m*

heat resistivity, thermal resistivity Wärmeleitwiderstand *m*

heat-responsive temperature Wärme-Ansprechtemperatur *f*

heat safeguard against spurious response, heat protection against spurious response [*gas burner*] Fremdlichtschutz *m*

heat-saving, heat-economizing wärmesparend

heat screen, thermal screen Wärmeblende *f*

heat sensing device → bolometer

heat-sensitive wärmeempfindlich

heat-sensitive paper → temperature-sensitive paper

heat-sensitive (type) (automatic) fire detector automatische Hitzebrandmelder, automatische Hitzefeuermelder *m*

heat sensor → bolometer

heat shock, thermal shock Wärmestoß *m*

heat sink, dissipator [*Electricity. A mass of metal that is added to a device for the purpose of absorbing and dissipating heat; used with power transistors and many types of metallic rectifiers*] Wärmeableiter m

heat sink, cold source [*Substance or environment into which unwanted heat is directed*] Wärmesenke *f*

heat sink Kühl(ungs)körper *m* [*Halbleiter*]

to heat-soak durchwärmen

heat source, thermal source [*Substance or environment from which heat is taken*] Wärmequelle *f*

heat source plant, heat source installation, thermal source plant, thermal source installation Wärmequelleanlage, WQA *f* [*Sie besteht aus Wärmequelle und den zum Transport der Wärmeträger erforderlichen Einrichtungen*]

heat stability, thermal stability thermische Stabilität *f*

heat storage, thermal storage, heat accumulation, thermal accumulation [*1. Temporary storage of high- or low-temperature energy for later use. 2. Accumulation of energy in a body or system in the form of sensible heat (temperature rise) or latent heat (change of phase). 3. Technology and systems of accumulating heating capacity for subsequent use*] Wärmespeicherung *f*

heat storage capacity, heat accumulation capacity Wärmespeicherfähigkeit *f*, Wärmespeichervermögen *n* [*Die Eigenschaft eines Stoffes bei Temperaturänderungen Wärme aufzunehmen oder abzugeben*]

heat storage device [*A container plus all the contents of the container used for storing thermal energy. The transfer fluid and accessories, such as heat exchangers, agitators, circulating pumps, flow-switching devices, valves, and baffles which are integral with the thermal storage container are considered a part of the storage device*] Wärmespeicher *m*

heat storage effect Wärmespeicherwirkung *f*

heat storage management, heat accumulation management Speicherwärmebewirtschaftung *f* [*Von Bauteilen wie Betondecken, Fußböden und Wänden mit Wasserkreisläufen als Wärmeträger*]

heat stress, thermal stress Wärmespannung *f*

heat supply [*To a room*] Wärmezufuhr, Wärmezuführung, Wärmeversorgung, Wärmeleitung *f*, Wärmenachschub *m*

heat supply central station → central heat supply station

heat supply from cogeneration Wärmeauskopp(e)lung *f* [*Fernheizung*]

heat switch, thermal switch, thermostatic switch, temperature-controlled switch Thermoschalter *m*

heat trace electric system, electric heat trace system, trace electric heat system Elektrobegleitheizung *f*

heat trace steam system, steam heat trace system, trace steam heat system Dampfbegleitheizung *f*

heat trace system, trace heat system Begleitheizung *f*

heat-transfer coefficient, film coefficient, thermal-transfer coefficient [*Thermodynamics. The amount of heat exchanged across a unit area in a unit amount of time through a temperature difference of 1 °C between the system boundary and the surrounding fluid*] Wärmeübertragungskoeffizient, Wärmeübergabekoeffizient *m*

heat transfer device Wärmeübertrager *m*, Wärmeübergabevorirchtung *f*

heat trap, thermal trap Wärmefalle *f*

heat unit, thermal unit Wärmeeinheit *f*

heat use → heat consumption

heat use plant, heat use installation, heat consumption plant, heat consumption installation Wärmenutzungsanlage *f* [*Anlage zum Wärmeträgertransport vom Wärmeerzeuger zum Wärmeverbraucher und der Wärmeverteilung einschließlich aller Zusatzeinrichtungen*]

heat user's point → heat consumption point

heat value, thermal value [*The energy released in combustion, usually expressed in units of calories per gram or BTUs per pound*] Wärmewert *m*

heat-water-recycling system Wärme-Wasser-Wiederverwendungs-System *n* [*Dusch- und Badewasser wird nach Füllung, Zwischenspeicherung und Chlorung zur Toilettenspülung verwendet*]

heat writer → heat printer

heavy current Starkstrom, Hochstrom *m*

heavy-duty hochleistungsfähig

heavy-duty alloy hochfeste Legierung *f*

heavy-duty contact Starkstromkontakt *m*

heavy-duty hook Schwer(aufhänge)haken *m*

H

heavy-duty in-line fuse switch-disconnector schaltbare Sicherungslasttrennleiste *f*

(heavy-gauge) steel conduit, high-strength conduit [*US spelling is "gage"*] Stahlpanzer(schutz)-rohr, Stapa(schutz)rohr *n*

heavy heating oil → fuel oil No. 6 [*USA*]

heavy heating oil burner; fuel oil No. 6 burner [*USA*] Schwerölbrenner *m*

heavy noise → objectionable noise

heavy oil → residual fuel oil

heavy pilot duty hohe Schaltleistung *f*

heck [*local term*] **→ door latch**

heft, handle Heft *n*, Griff *m*

H

height Bauhöhe *f* [*z. B. bei einem Radiator*]

height, level [*The vertical dimension above a horizontal reference datum*] Höhe *f*, Niveau *n*, Stand *m*

(Heizmann('s)) plate superheater Plattenüberhitzer, Heizmannsche Überhitzer *m*

helical fin → spiral fin

helical fin tube → spiral fin tube

helical rotary (refrigerant) compressor → screw (refrigerant) compressor

helical scan method Schrägspurverfahren *n*, Schrägspurmethode *f* [*Aufzeichnung von Videosignalen*]

Helmholtz resonator [*An enclosure having a small opening consisting of a straight tube of such dimensions that the enclosure resonates at a single frequency determined by the geometry of the resonator*] Lochresonator, Helmholtzresonator *m*

help, aid Hilfe *f*

help call(ing) → call(ing) for help

helper application Hilfsprogramm *n*, Hilfsapplikation *f* [*Ein selbständig lauffähiges Programm, das der Anwender nachträglich in seinen Browser einbindet*]

HEPA filter → absolute filter

hermetic(ally-sealed) hermetisch (abgeschlossen), luftdicht gekapselt

hermetic motorcompressor hermetische Motorverdichter, Kapselmotorverdichter *m*

herpolhode cone, space cone [*Mechanics. A cone, fixed in space, swept by the processing angular-velocity vector of a rotating free body executing Poinsot motion*] Rastpolkegel *m*

HES → home electronic system

heterodyne oscillator → beat-frequency oscillator

heterodyne modulator → mixer

heterogeneous building (monitoring and) control system heterogene Gebäudeleitsystem *n* [*Es ist aus Komponenten verschiedener Hersteller aufgebaut*]

HF → high-frequency

HFC heating → high-frequency capacitance heating

HF modem → high-frequency modem

Hick('s) double-ported (slide) valve, Hick('s) dual-ported (slide) valve, Hick('s) twin-ported (slide) valve Hicksche (Doppel)Schieber *m*

HICO, high coercive hochkoerzitiv [*Magnetstreifen mit einer hohen Sicherheit gegen Entmagnetisierung, z. B. bei Scheckkarten*]

hidamet → high damping metal

hidden → concealed

hidden base, concealed base Unterputzsockel *m*

hidden battery, concealed battery Unterputzbatterie *f*

hidden (convenience) outlet → hidden socket outlet

hidden design, hidden type, hidden version, concealed design, concealed type, concealed version Unterputzausführung, Unterputzbauweise, Unterputzbauart, UP-Ausführung, UP- Bauweise, UP-Bauart *f*

hidden fitting, concealed fitting Unterputzarmatur *f*

hidden housing, concealed housing, concealed casing, hidden casing Unterputzgehäuse, Gehäuse UP *n*

hidden light, concealed light, indirect light indirekte Licht, mittelbare Licht *n*

hidden lighting (system), concealed lighting (system), concealed illumination (system), indirect lighting (system), indirect illumination (system), hidden illumination (system) [*It is characterized by relatively low source brightness, good diffusion, freedom from shadows, and low moderate utilization*] indirkte Beleuchtung(sanlage) *f*

hidden line removal, concealed line removal Ausblenden verdeckter Karten, verdeckte Kartenausblenden *n*

hidden socket outlet, hidden receptacle outlet (US); hidden receptacle, hidden (convenience) outlet [*"Concealed" may synonymously be used for "hidden"*] Unterputzsteckdose *f*

hidden wiring, concealed wiring Unterputzverdrahtung *f*

HID lamp → high-intensity discharge lamp

hierarchical übergeordnet, hierarchisch

hierarchical computer Leitrechner, übergeordnete Rechner, hierarchische Rechner *m*

hierarchical (computer) memory structure hierarchische (Rechner)Speicheraufbau, übergeordnete (Rechner)Speicheraufbau *m*

hierarchical level Hierarchieebene *f*

hierarchical object-orien(ta)ted picture system, HOOPS 3D-Grafiksystem, HOOPS *n*

hifi → high fidelity

high hoch [*Zustand mit hohem Pegel (logisch H). Bei positiver Speisespannung entspricht „H" dem Zustand „EIN"*]

"high" hoch, hohe [*Anzeige*]

High logisch „Eins" *f* [*Pegelfeststellung für alle Daten- und Steuer(ungs)signale*]

high air velocity system Hochgeschwindigkeitssystem *n* [*System mit Luftgeschwindigkeiten über 10 m/s in den Verteil- und Sammelleitungen*]

high ash coal asche(n)reiche Kohle *f*

high-boiling hochsiedend

high boost → high-frequency compensation

high-boy upflow furnace [*The blower is located below the heat exchanger*] hohe Aufwärts-Warmluftfeuerung, hohe Warmluft-Aufwärtsfeuerung *f*

high cabinet, high cubicle Hochschrank *m* [*im Badezimmer*]

high-capacity → high-performance

high-capacity steel heating boiler, high-performance steel heating boiler, high-output steam heating boiler, high-duty steel heating boiler Hochleistungs-Stahl(heiz(ungs))kessel *m*

high-challenge fire Großbrand *m*, Großfeuer *n*

high coercive → HICO

high cold(ness) Großkälte *f* [*Kältemittel*]

high cut Frequenzbegrenzung *f* [*Begrenzung höherer Frequenzen zur Rauschminderung*]

high damping metal, hidamet Metall *n* mit großem Verlustfaktor d [*z. B. Gusseisen mit viel Lamellengraphit*]

high definition television, HDTV hochauflösende Fernsehen *n*

high-duty → high-performance

high-duty ... → high-capacity ...

high efficiency Hochwirkungsgrad *m*

high efficiency boiler Hochwirkungsgradkessel *m*

high efficiency particulate air filter → absolute filter

high-end housing, luxury housing, de-luxe housing [*As opposed to affordable housing*] Luxuswohnung *f*

high-end product Höchstgüteerzeugnis, Höchstgüteprodukt *n*

higher calorific value → higher heating value

higher heating value, gross heating value, total heating value, higher calorific value, gross calorific value, gross specific energy, H.C.V. [*It includes the latent heat of vapo(u)rization and is determined when water vapo(u)r in the fuel combustion products is condensed*] Brennwert *m*, Verbrennungswärme, V *f* [*früher: obere Heizwert* H_o]

higher level übergeordnete Ebene, höhere Ebene *f*

higher level function übergeordnete Funktion, Funktion auf höherer Ebene, höherer Funktion *f*

higher level language → high-level language

higher order controlled system [*IEC 50(351)*], **higher order directly controlled member** [*IEC 27-2A*], **higher order process** Regelstrecke höherer Ordnung, höhere Regelstrecke *f*

higher order language → high-level language

higher programming language höhere Programmiersprache *f*

high fidelity, hifi Tonwiedergabe *f* mit hoher Klanggüte, hohe Wiedergabetreue *f* [*DIN 45 500*]

high-frequency, HF, radio-frequency, RF Hochfrequenz, HF *f* [*Übertragung elektromagnetischer Wellen im Megahertzbereich; 3 bis 30 MHz*]

high-frequency cable, radio-frequency cable Hochfrequenzkabel *n*

high-frequency capacitance heating, high-frequency capacitive heating, HFC heating kapazitive Hochfrequenzheizung *f*

high-frequency compensation, high boost, radio-frequency compensation [*Increasing the amplification at high frequencies with respect to that at low and middle frequencies in a given band, such as in a video band or an audio band*] Hochfrequenzausgleich *m*

high-frequency energy, radio-frequency energy Hochfrequenzenergie *f*

high-frequency heating → electronic heating

high-frequency modem, HF modem, radio-frequency modem, RF modem Hochfrequenz-Modulator/Demodulator, HF-Modulator/Demodulator *m*, Hochfrequenz-Modem, HF-Modem *n* [*Ein zur Umsetzung von digitalen Datensignalen in analoge Signale (und umgekehrt) verwendetes Gerät, das diese von und in die zugewiesenen Frequenzen moduliert/demoduliert*]

high-frequency reader, radio-frequency reader Hochfrequenzleser *m*

high-frequency thawing, radio-frequency thawing Hochfrequenzauftauen *n*

high-gloss chrome-plated hochglanzverchromt

high-grade, first-class hochwertig, erstklassig

high-grade steel, first-class steel Edelstahl *m*

high hazard location Hochgefährdungsstelle *f*

high-impedance hochohmig

high-insulating hochisolierend

high-insulating hochdämmend

high-intensity discharge lamp, HID lamp Entladungslampe großer Lichtstärke, lichtstarke Entladungslampe *f*

H

high-(leakage-)reactance transformer Streufeldtransformator *m*

high-level hochangeordnet, hochliegend

high-level data link control, HDLC bitorientierte Datenträgerprotokoll, Binärzeichenprotokoll *n* [*Protokoll an der Übertragung von Informationen durch Folgen von Binärzeichen orientiert*]

high-level (flushing) cistern → high-level toilet cistern

high-level language, HLL, higher level language, higher order language [*A computer programming vocabulary that can be used to create instructions to enable a CPU to manipulate data*] höhere Programmiersprache, Hochsprache *f*

high-level tank → elevated tank [*water supply*]

high-level toilet cistern, high-level (closet) cistern, high-level (flush(ing)) cistern, high-level W.C. (flush(ing)) cistern hochhängende (Abort-) Spülkasten, Hoch(spül)kasten, hochhängende (Abort)Spültank, Hoch(spül)tank *m*

high-level water closet, high-level flush toilet, high-level (type of) W.C., high-level suite, high-level washdown water closet [*A room in which high-level water closet pans are installed, but only one pan may also be installed*] Spülabort mit Hochspülkasten, Wasserabort mit Hochspülkasten, Hochspülabort *m*, Spülklosett mit Hochspülkasten, Wasserklosett mit Hochspülkasten, Hochspülklosett *n*

high-level (water) reservoir → high service (water) reservoir

high limit(ing) control (device) [*Device responsive to change in temperature and used for the purpose of interrupting the fuel supply when plenum or bonnet temperatures exceed predetermined settings*] Brennstoffunterbrecher *m*

high limit(ing) value, upper limit(ing) value obere Grenzwert *m*

high line Überstreichung *f* [*Kode*]

high-low ((closed) loop) control → double level ((closed) loop) control

high-low signal selector Extremwertauswahleinheit *f*

highly durable, extremely durable strapazierbar, strapazierfähig

highly purified air hochgereinigte Luft *f*

highly-restricted zone → high-security zone

highly stressed concrete → concrete under severe condition(s)

highly toxic hochgiftig

high-mounted luminaire, high-mounting luminaire Hochleuchte, Obenleuchte *f*

high-order bit werthohe Bit *n*

high outlet Hochauslass, Obenauslass *m*

high output hohe Aussteuerbarkeit *f* [*Tonband*]

high-output → high-performance

high-output ... → high-capacity ...

high-pass filter [*Filter having a single transmission band extending from some cutoff frequency, not zero, up to infinite frequency*] Hochpass(filter) *m*, *(m, n)*

high-performance, high-duty, high-output, high-capacity leistungsstark

high-performance ... → high-capacity ...

high-polished hochglanzpoliert

high power Starkstrom *m*

high-power diode laser Hochleistungsdiodenlaser *m*

high-power NOR gate verstärkte NOR-Stufe *f*

high-pressure air-atomizing burner [*See remark under "air-atomizing burner"*] Hochdruck-Ölzerstäuberbrenner, Öl-Hochdruckzerstäuberbrenner *m*

high-pressure air conditioning induction equipment Hochdruck-Klima-Induktionsgerät(e) *n*, *(f)*

high-pressure air-conditioning plant, high-velocity air-conditioning plant, high-pressure air-conditioning installation, high-velocity air-conditioning installation Hochdruckklimaanlage *f* [*DIN 1946*]

high-pressure air conditioning plant with induction units, high-velocity air conditioning plant with induction units, high-pressure air conditioning installation with induction units Hochdruck-Induktions(klima)anlage, Hochdruck-(Klima-) Anlage mit Induktionsgeräten, Hochdruck-Klimainduktionsanlage, Hochgeschwindigkeits-Induktions-(klima)anlage *f*

high-pressure air distribution, high-velocity air distribution Hochdruckluftverteilung, Hochgeschwindigkeitsluftverteilung *f*

high-pressure air heating, high-velocity air conditioning Hochdruckluftheizung *f*

high-pressure air outlet, high-velocity air outlet Hochdruckluftauslass, Hochgeschwindigkeitsluftauslass *m*

high-pressure atomization Hochdruckzertstäubung *f*

high-pressure atomizing burner Hochdruckzerstäuberbrenner, Hochdruckzerstäubungsbrenner *m*

high-pressure belt Hochdruckgürtel *m*

high-pressure boiler Hochdruckkessel *m*

high-pressure cast(-iron) (convector) radiator [*See remark under "radiator"*] Guss(eisen)radiator in Hochdruckausführung, Hochdruckguss(eisen)radiator *m*

high-pressure (central heating) system, central high-pressure heating system [*It is designed for high-pressure operation in order to value the diameter of the circulation pipes required*] Hochdruckzentralheiz(ungs)system *n*

high-pressure ((closed) loop) control, high-pressure feedback control, high-pressure regulation Hochdruckreg(e)lung, Hochdruckregulierung *f*

high-pressure cold water, cold high-pressure water Hochdruckkaltwasser, Kalthochdruckwasser *n*

high-pressure compressor Hochdruckkompressor, Hochdruckverdichter *m*

high-pressure condenser cutout → condenser high-pressure cutout

high-pressure control → high-pressure ((closed) loop) control

high-pressure (convector) radiator, high-pressure convector heater [*See remark under "radiator"*] Hochdruckradiator, Hochdruckgliederheiz(ungs)körper, Hochdruckheiz(ungs)körper *m*

high-pressure discharge lamp Hochdruckentladungslampe *f*

high-pressure district water heating, high-pressure long-distance water heating Heißwasserfernheizung *f*

high-pressure double duct air conditioning plant, high-pressure double duct air conditioning installation Hochdruck-Zweikanal-Klimaanlage, Zweikanal-Hochdruck-Klimaanlage *f*

high-pressure duct net(work) Hochdruckkanalnetz *n*

high-pressure fan Hochdruckventilator, Hochdrucklüfter *m*

high-pressure feedback control → high-pressure ((closed) loop) control

high-pressure gas [*Town gas compressed to a pressure above that at which it is normally applied*] Hochdruckgas *n*

high-pressure gas burner Hochdruckgasbrenner *m*

high-pressure gas (pipe)line → gas high-pressure (pipe)line

high-pressure gauge; high-pressure gage (US) Hochdruckmanometer *n*

high-pressure heating Hochdruckheizung *f*

high-pressure heating system [*See remark under "system"*] Hochdruckheiz(ungs)system *n*

high-pressure hot water absorption chiller, high-pressure hot water absorption refrigerating system [*See remark under "chiller". H.P.H.W. may synonymously be used for "high-pressure hot water"*] Heißwasser-Absorptionskältesystem *n*

high-pressure hot water heating, H.P.H.W. heating, hot high-pressure water heating, high-temperature water heating Heißwasserheizung *f*

high-pressure induction air conditioning plant, high-pressure induction air conditioning installation, induction high-pressure air conditioning plant, induction high-pressure air conditioning installation Hochdruck-Induktionsklima(tisierungs)anlage, Induktions-Hochdruckklima(tisierungs)anlage, HD-Induktionsklima(tisierungs)anlage, Induktions-HD-Klima(tisierungs)anlage *f*

high-pressure induction unit [*See remark under "induction unit"*] Hochdruckinduktionsgerät, HD-Induktionsgerät *n*

high-pressure lamp Hochdrucklampe *f*

high-pressure long-distance water heating → high-pressure district water heating

high-pressure ((open) loop) control Hochdrucksteuerung, HD-Steuerung *f*

high-pressure pipe, solid pipe Hochdruckrohr, robuste Rohr, HD-Rohr *n*

high-pressure refrigerant Hochdruck-Kältemittel *n*

high-pressure regulation → high-pressure ((closed) loop) control

high-pressure (sectional) radiator [*See remark under "radiator"*] Hochdruckradiator, Hochdruckgliederheiz(ungs)körper *m*

high-pressure side Hochdruckseite *f*

high-pressure single-lever fitting, high-pressure one-lever fitting Hochdruck-Ein-Hebelarmatur *f*

high-pressure steam Hochdruckdampf, HD-Dampf *m*

high-pressure steam boiler Hochdruckdampfkessel, Dampfhochdruckkessel *m*

high-pressure steam generator Hochdruckdampferzeuger, Dampfhochdruckerzeuger *m*

high-pressure steam heat generating plant, high-pressure steam heat generating installation Hochdruckdampf-Wärmeerzeugungsanlage *f*

high-pressure steam heating Hochdruckdampfheizung, HD-Dampfheizung *f* [*Der Druck liegt mehr als 0,07 Mpa über Umgebungsdruck*]

high-pressure steam heating net(work) Hochdruckdampfheiz(ungs)netz(werk) *n*

high-pressure steam (pipe)line, steam high-pressure (pipe)line Dampfhochdruckleitung, Hochdruckdampfleitung, HD-Dampfleitung *f*

H

high-pressure steam system [*See remark under "system"*] Hochdruckdampfsystem, Dampfhochdrucksystem *n*, Hochdruckdampfanlage, Dampfhochdruckanlage *f*

high-pressure superheated water three-line system, high-pressure superheated water triple-line system Hochdruckheißwasser-Dreileitersystem *n* [*Es hat zwei getrennte Vorlaufleitungen und eine gemeinsame Rücklaufleitung*]

high-pressure switch Hochdruckschalter *m*

high-pressure system [*See remark under "system"*] Hochdrucksystem *n*

high-pressure two-duct air-conditioning plant, high-pressure two-duct air-conditioning installation Hochdruckzweikanalklimaanlage, Zweikanalhochdruckklimaanlage *f*

high-pressure vent(ilation) plant, high-pressure vent(ilation) installation, high-pressure vent(ilat)ing plant, high-pressure vent(ilat)ing installation Hochgeschwindigkeitslüftungsanlage *f* [*Diese Anlage hat keine Zuluftkanäle. Die Zuluft wird mit hoher Geschwindigkeit dem Raum so zugeführt, dass die Lüftung der Aufenthaltszone durch den Rückstrom erfolgt*]

high rate washer Hochgeschwindigkeitswäsche(r) *f,(m)*

high-reactance transformer → high-(leakage-) reactance transformer

high resolution hohe Auflösung *f*

high resolution printing Drucken *n* mit hoher Auflösung

high-rupturing capacity contactor → HRC contactor

high rupturing capacity disconnector changeover switch → HRC disconnector changeover switch

high rupturing capacity disconnector switch → HRC disconnector switch

high safety Hochsicherheit *f*

high-safety lock Hochsicherheitsschloss *n*

high-safety system Hochsicherheitsystem *n*

high security Hochsicherung *f*

high security camera Hochsicherungskamera *f*, Hochsicherungsphotoapparat *m*

high security (designed) überlistungssicher

high security lock Hochsicherungsschloss *n*

high security zone, highly restricted zone Hochsicherungszone *f*

high service (water) reservoir, high service (water) tank, high-level (water) reservoir, high-level (water) tank, elevated (water) reservoir, elevated (water) tank Hoch(wasser)behälter, Hoch(wasser)tank *m* [*Er hängt hoch*]

high-solving hochauflösend

high-speed air outlet Hochgeschwindigkeitsluftauslass *m*

high-speed burner hochtourige Brenner *m*

high-speed dubbing Schnellüberspielen *n* [*Kassette*]

high-speed elevator, express elevator (US); high-speed lift, express lift (Brit.) Schnellaufzug *m*

high-speed freight lift. high-speed goods lift, high-speed load lift, high-speed truck lift, express freight lift, express goods lift, express load lift, express truck lift (Brit.) [*"lift" is called "elevator" in the USA*] Schnellgüteraufzug, Schnellfrachtaufzug, Schnelllastenaufzug *m*

high-speed passenger elevator, express passenger elevator (US); high-speed passenger lift, express passenger lift (Brit.) Schnellfahrstuhl, Schnellpersonenaufzug *m*

high-speed printer, HSP, express printer Schnelldrucker *m*

high-speed reader, express reader [*The fastest input device existing at a particular time in the state of the engineering*] Schnellleser *m*

high-speed run, express run Schnellfahrt *f* [*Aufzug*]

high-speed sliding folding gate, high-speed folding sliding gate, express sliding folding gate, express folding sliding gate Schnelllauftor *n* [*Schiebefalttor mit hoher Öffnungs- und Schließgeschwindigkeit*]

high-strength steel conduit → (heavy-gauge) steel conduit

high-stroke spring safety valve Hochhub-Feder-Sicherheitsventil *n*

high-temperature cooling Hochtemperaturkühlung *f*

high-temperature gasket, high-temperature joint ring Hochtemperatur(ab)dicht(ungs)ring *m*, Hochtemperatur(flach)(ab)dichtung *f* [*z. B. für Radiatoren*]

high-temperature hot water → high-pressure hot water

high-temperature liquid Heißflüssigkeit *f*

high-temperature waste heat → waste superheat

high-temperature water → high-pressure hot water

high-temperature water generation Heißwassererzeugung, Heißwasserbildung, Heißwasserbereitung *f*

high-temperature water generator Heißwassererzeuger, Heißwasserbereiter *m*

high-temperature water ground storage tank Heißwasser-Erdspeicher(tank), Heißwasser-Erdvorratstank, Heißwasser-Erdlagertank *m*

high-temperature water heating → high-pressure hot water heating

high-temperature water heating plant, high-temperature water heating installation Heißwasserheiz(ungs)anlage, Heißwasserheiz(ungs)installation *f* [*Sie wird als geschlossene Anlage gebaut. Die Abgrenzung zu den geschlossenen Warmwasserheizungsanlagen ist verschwommen und von Land zu Land verschieden*]

high-temperature water takeoff Heißwasserentnahme *f*

high tension → high voltage

high-tension arc → high-voltage arc

high tension service → primary service

high-tension switching station → high-voltage switching station

high-threshold logic Logik *f* mit hoher Störschwelle

high-usage route Regel(ungs)weg, Reglungsweg *m* [*Fernmeldetechnik*]

high-vacuum breakdown Hochvakuumdurchschlag *m*

high-vacuum cock Hochvakuumhahn *m*

high-vacuum pump Hochvakuumpumpe *f*

high-vacuum thermal insulation Hochvakuumwärmedämmung *f*

high value room [*e.g. a computer room*] hochwertige Raum *m*

high-velocity air ... → high-pressure air ...

high-velocity duct Hochgeschwindigkeitskanal *m*

high voltage, high tension Hochspannung *f*

high-voltage arc, high-tension arc Hochspannungsbogen *m*

high-voltage busbar, high-tension busbar [*See remark under "busbar"*] Hochspannungssammelschiene, Hochspannungsstromschiene, Hochspannungsverteilerschiene *f*

high-voltage discharge luminaire, high-voltage discharge light(ing) fitting, high-tension discharge luminaire, high-tension discharge light(ing) fitting Hochspannungsentladungsleuchte *f*, Hochspannungsentladungsbeleuchtungskörper *m*

high-voltage engineering, high-tension engineering Hochspannungstechnik *f*

high-voltage ignition, high-tension ignition Hochspannungszündung *f*

high-voltage service → primary service

high-voltage supply, high-tension supply Hochspannungsversorgung *f*

high-voltage switching station, high-tension switching station Hochspannungsschaltanlage *f*

highway → (data) highway

highway Vielfachleitung, Multiplexleitung *f*, Vielfachweg, Multiplexweg *m* [*Nachrichtentechnik*]

hinge → (door) hinge

(hinge and) loop → butt (hinge)

hinge box, box of a hinge Pfanne *f* [*Türangel*]

hinge(d) lid, spring flap lid Klappdeckel *m* [*Steckdose*]

hinge(d) lid Scharnierdeckel *m*

hinge(d) mirror Klappspiegel *m*

hinge joint → turning joint

hinge pin, gudgeon Scharnierstift *m*

hinge plate [*locksmith*] Hakenband *n*

hinge stocks Scharnierkluppe *f*

hinge strap Schippenband *n*, Bandlappen *m*

hip (bath)tub, hip bath(ing tub) [*A (bath) tub in which one bathes in a sitting position*] Sitz(bade)wanne *f*

hiss [*Audio-frequency noise having subjective characteristics*] Zischen *n*

histogram Säulendiagramm, Histogramm *n*

historical database Vergangenheitsdatenbank, historische Datenbank *f* [*Eine Datenbank mit vorwiegend älteren Daten, die jedoch nützlich sein können, um z. B. Vergleiche anzustellen oder zukünftige Daten voraussagen zu können.*]

H

history Ablaufspeicherung *f* [*Speicherung des Ablaufes der durch den Benutzer eingegebenen Befehlsfolge*]

history buffer Ablaufzwischenspeicher *m*

history data Langzeitdaten, historische Daten *f*

hit and miss ((open) loop) control fehlermögliche Steuerung *f*

hit rate [*The ratio of the number of records accessed in a run to the number of records in the file, expressed as a percentage*] Trefferprozent(e) *n,(f)*, Trefferprozentsatz *m*

HLF → **heating load factor**

HLL → **high-level language**

HLR → **home location register**

H

hoarding (Brit.) → **billboard (US)**

hodman Maurerhelfer, Maurergehilfe *m*

hog fuel Holzrest(e)brennstoff *m*

hohlraum → **blackbody**

hohlraum radiation → **blackbody radiation**

to hoist → **to lift**

hoist(ing) installation → **lift(ing) installation**

hoist(ing) machine, shaftway machine, lift(ing) machine [*Power unit that applies the energy necessary to raise and lower an elevator*] Aufzug(an)triebwerk, Förder(an)triebwerk *n*

hoist(ing) platform → **lift(ing) platform**

hoist(ing) rope, hoist(ing) cable, lift(ing) rope, lift(ing) cable Aufzugseil, Förderseil *n*

hoisting speed, lifting speed Hubgeschwindigkeit, Hebegeschwindigkeit *f*

hoistway, hatchway, shaft(way), well; lift hoistway, lift shaft(way), lift well (Brit.); elevator hoistway, elevator shaft(way), elevator well (US); lift channel [*South Africa*] [*"Core" and "tower" may synonymously be used for "well", "hoistway", and "shaft(way)"*] Fahrschacht, Aufzug(fahr)schacht, Fahrkern, Aufzugkern, Schacht, Kern *m*

hoistway access → **elevator hoistway access**

hoistway biparting door, shaft(way) biparting door; hoistway centre-opening door, shaft(way) centre-opening door (Brit.); hoistway center-opening door, shaft(way) center-opening door (US) [*The sections or parts open any simultaneously from each other*] zentralöffnende und zentralschließende Fahrschachtschiebetür *f*

hoistway door, hoistway gate, landing door, landing gate, shaft(way) door, shaft(way) gate; lift well door, lift well gate (Brit.); elevator well gate, elevator well door (US); lift channel door [*South Africa*] [*It completely fills the door opening giving access to the elevator or dumbwaiter car at any landing. It is of solid construction, with or without vision panels, regardless of design or method of operation*] Fahrschachttür, Aufzugschachttür, Aufzugkerntür, Fahrkerntür *f*

hoistway door interlock, hoistway gate interlock Fahrschachttürverrieg(e)lung, Aufzugschachttürverrieg(e)lung, Aufzugkerntürverrieg(e)lung, Fahrkerntürverrieg(e)lung *f*

hoistway door locking device, hoistway gate locking device [*It secures a hoistway door or gate in the closed position and prevents it from being opened from the landing side except under certain specified conditions*] Fahrschachttürsicherung, Aufzugschachttürsicherung, Aufzugkerntürsicherung, Fahrkerntürsicherung *f*

hoistway enclosure, shaft(way) enclosure [*Fixed structure, consisting of vertical walls or partitions, that isolates the hoistway from all other parts of the building or from an adjacent hoistway and in which the hoistway doors and door assemblies are installed*] Fahrschachtwände, Aufzugschachtwände *f*

hoistway fully automatic door, hoistway fully automatic gate, fully automatic hoistway door, fully automatic hoistway gate vollautomatische Fahrschachttür, vollautomatische Aufzugtür, vollautomatische Fahrkerntür, vollautomatische Aufzugkerntür *f*

hoistway gate interlock → **hoistway door interlock**

hoistway manually operated door, hoistway manually operated gate handbetriebene (Fahr)Schachttür, handbetriebene Aufzugschachttür, handbetriebene (Fahr)Kerntür, handbetriebene Aufzugkerntür *f*

hoistway power-operated door, hoistway power-operated gate [*Door opened and closed by power other than by hand, gravity, springs, or the movement of the elevator car*] kraftbetriebene (Fahr)Schachttür, kraftbetriebene Aufzugschachttür, kraftbetriebene (Fahr)Kerntür, kraftbetriebene Aufzugkerntür *f*

hoistway pit, shaft(way) pit Fahrschachtgrube, Aufzugschachtgrube *f*

hoistway switch (US) → **(lift) well switch (Brit.)**

hoistway wall (US); (lift) wall (Brit.) Fahrschachtwand *f*

to hold [*To retain information in a computer storage device for further use after it has been initially utilized*] aufbewahren

holder → **(lamp)holder**

holder Halter(ung) *m,(f)*

hold first order element → **first order hold element**

hold(ing) Belegung *f* [*Fernmeldetechnik*]

hold(ing) Anhalten *n* [*Eine Gehäuseanschlussbezeichnung beim Mikroprozessor 8080. Der Mikroprozessor wird vorübergehend angehalten und die Busse werden hochohmig gemacht*]

hold(ing) acknowledge Anhaltequittung *f* [*Eine Gehäuseanschlussbezeichnung beim Mikroprozessor 8080. Der Mikroprozessor liefert so lange ein Signal, solange er sich im Haltezustand befindet*]

hold(ing) bar Haltestange *f*

hold(ing) coil [*Part of a magnetic starter or relay that causes the device to operate when energized*] Haltespule *f*

hold(ing) down dowel Niederhaltedübel *m*

hold(ing) element Halt(e)glied, Halt(e)element *n*

hold(ing) mode, interruption mode [*The state of an analog computer in which its operation is interrupted without altering the values of the variable, it is handling, so that computation can continue when the interruption is over*] Halt-Betriebsart, Betriebsart „Halt" *f*

hold(ing) step Spannungsspitze *f* beim Umschaltvorgang

hold(ing) time Belegungsdauer, Belegungszeit *f*

hold(ing) zeroth order element → **zeroth order hold(ing) element**

holdoff → **barrier**

hold tag Sperrvermerk *m* [*Gütesicherung*]

hold-up Überfall *m*

hold-up alarm Überfallalarm *m*

hold-up alarm system Überfallalarmsystem *n*, Überfallalarmanlage *f*

hold-up detection Überfallmeldung *f* [*Der von einem Melder einer Überfallmeldeanlage ausgelöste Alarm*]

hold-up detection central installation, central hold-up detection installation Überfallmeldezentrale *f* [*Sie erfasst Informationen aus der Überfallmeldeanlage, z. B. Auslösung von Geldscheinkontakten, Handmeldern, Fußkontaktmeldern, und setzt sie in eine Überfallmeldung um. Diese Zentrale ist ständig scharf geschaltet und muss nach höchstens 180 Sekunden eine weitere Meldung geben können*]

hold-up detection installation Überfallmeldeanlage, ÜMA *f* [*Sie dient Personen zum Hilferuf bei Überfällen*]

hold-up detector, hold-up detecting device, hold-up detection device Überfalltaster, Überfallmelder, Überfalldetektor, Überfallmeldungsgeber *m*, Überfallmeldegerät *n* [*Er ist ständig scharf geschaltet*]

hold-up monitoring Überfallüberwachung *f*

hold-up protection Überfallschutz *m*

hold-up video camera Videoüberfallkamera, Überfallvideokamera *f*

hole [*Not to be confused with "opening"*] Loch *n* [*Nicht verwechseln mit „Öffnung"*]

hole → **cut-out**

hole Loch *n*, positive Leitungsträger *m*

hole circle Lochkreis *m*

hole disk, hole disc Lochscheibe *f*

hole disk siren, hole disc siren Lochsirene *f* [*Die in einem Kreis angeordneten Öffnungen in einem unter Luftdruck stehenden Gehäuse werden durch eine sich drehende Lochscheibe abwechselnd geöffnet und geschlossen. Dadurch entsteht ein unterbrochener Ton*]

hole electron pair Loch-Elektron-Paar, Defektelektron-Elektron-Paar *n*

hole mask Lochmaske *f*

holistic ganzheitlich, allumfassend

Hollerith code [*A code used to represent letters, numbers, or special symbols to be punched in standard 80-column punch cards*] Hollerithkode, Hollerithcode *m*

Hollerith string [*A sequence of characters preceded by an H and a character count in FORTRAN, as 4HJTOP*] Hollerith(zeichen)kette *f*

hollow (building) block floor heating Hohlblock(stein)deckenheizung, Hohlsteindeckenheizung *f*

hollow cast-iron section, cast-iron hollow section Gusseisenglied *n* [*gusseiserne Gliederheizkörper*]

hollow cube → **empty cube**

hollow section Glied, Heizkörperglied *n*

hollow section Hohlprofil *n*

(hollow) wall tie, tie [*A rigid, corrosion-resistant metal tie fitted into the bed joints across the cavity of a cavity wall*] (Wand)Anker, Hohlwandanker *m*

hologram (interferometer) [*The special photographic plate used in holography; when this negative is developed and illuminated from behind by a coherent gas-laser beam, it produces a three-dimensional image*] Hologramm *n*, dreidimensionale Aufnahme *f*

hologram (interferometer) coding Hologrammkodierung, Hologrammcodierung *f* [*Die Echtheitsmerkmale (unsichtbar) sind in Form eines Hologramms aufgebracht und können über ein spezielles Lesegerät ausgelesen (verglichen) werden*]

holographic (computer) memory holographische (Rechner)Speicher *m*

holographic data storage system, HDSS holographische Datenspeicher(ungs)system *n*

H

home address, original address Ausgangsadresse *f*

home and building electronic system, HBES Elektroniksystem *n* für Wohn- und Nichtwohnbauten

homebanking → remote banking

home computer Heimrechner *m*

home contact Normalstellungskontakt *m* [*Schalter*]

home electronic system, HES, house electronic system Hauselektroniksystem *n* [*Ein elektronische Geräte im Haus miteinander verbindendes System*]

home freezer Heimgefriermöbel *n*

home key Eigentaste *f* [*Fernmeldetechnik*]

home location register, HLR Heimregister *n* [*Datensatz eines Mobilfunkteilnetzes*]

H

home position Grundstellung *f* [*Vorgesehener Signalzustand der speichernden Glieder einer Steuerungseinrichtung als Voraussetzung für die Problemverwirklichung. DIN 19 237*]

home record Mitlesestreifen, Kontrollstreifen *m* [*Nachrichtentechnik*]

home thermostat Heimthermostat *m*

home (wiring) run [*The run between the panel and the first outlet in the branch circuit. Looking upstream, it is the wiring run between the last outlet and the panel*] Schlussverdrahtungszug, Endverdrahtungszug *m*

home woodworking shop → woodworking home shop

home workshop Heimwerkstatt *f*

homogeneous building (monitoring and) control system homogene Gebäudeleitsystem *n* [*Es ist aus Komponenten eines einzigen Herstellers aufgebaut*]

homopolar generator → acyclic machine

honeycomb(ed) humidifier, honeycomb(ed) humidificator Wabenbefeuchter, Wabenanfeuchter *m*

honeycomb(ed) rectifier Wabengleichrichter *m*

hood, bonnet [*Air-intake device connected to a mechanical exhaust system for collecting vapo(u)rs, fumes, smoke, dust, steam, heat or odo(u)rs, from , at or near the equipment, place, or area where generated, produced, or released*] Haube *f*

hood [*A screen placed over a cathode-ray used to decrease amount of emitted light*] Kappe, Haube *f*

hood [*A usually dome- or funnel-shape(d) canopy or overhang that acts as a ventilator*] Lufthaube *f*, Haubenlüfter *m*

hood air → bonnet air

hook → (door) crook

hook latch bolt Hakenfalle *f* [*Schloss*]

hook pipe Hakenrohr *n*

hook tube Hakenröhre *f*

hook-up wire → jumper

hoop in the wards [*locksmith*] Schlossreif *m*

HOOPS → hierarchical object-orien(ta)ted picture system

hooter Hupe *f*

hopper → access hopper

hopper → (card) hopper

hopper closet, hopper toilet Trichterklosett *n*, Trichterabort *m*, Trichtertoilette *f*

horizontal air heating equipment horizontale Luftheiz(ungs)gerät(e), waag(e)rechte Luftheiz(ungs)gerät(e), liegende Luftheiz(ungs)gerät(e) *n*, *(f)*

horizontal arrangement liegende Anordnung, horizontale Anordnung, waag(e)rechte Anordnung *f*

horizontal autowalk, horizontal passenger conveyor, horizontal passenger conveyer Horizontalfahrsteig, Horizontalrollsteig, Waag(e)rechtfahrsteig, Waag(e)rechtrollsteig *m*

horizontal bar graph, horizontal bar chart [*See remark under "bar graph"*] waag(e)rechte Balkendiagramm, horizontale Balkendiagramm *n*, waag(e)rechte graphische Balkendarstellung, horizontale graphische Balkendarstellung *f*

horizontal central furnace, central horizontal furnace [*It is designed for low headroom installation with the airflow through the appliance in a horizontal path*] Warmluftfeuerung *f* mit waag(e)rechter Luftführung

horizontal definition → horizontal resolution

horizontal design horizontale Ausführung, liegende Ausführung, waag(e)rechte Ausführung *f* [*z. B. Kessel*]

horizontal (drain) trap, horizontal U-bend, horizontal stench trap, horizontal siphon trap, horizontal fixture trap [*See remark under "trap"*] liegende (Wasser)(Geruch)Verschluss, Schlangensiphon *m*

horizontal flyback → line flyback

horizontal frequency, line frequency Zeilenfrequenz, Horizontalfrequenz *f* [*Sie wird in Kilohertz gemessen und gibt an, wie viele Zeilen der Elektronenstrahl eines Monitors in einer Sekunde abtastet*]

horizontal furnace [*Furnace in which the air flows horizontally with air inlet and discharge openings at opposite ends of the casing*] Horizontalfeuerung, Waag(e)rechtfeuerung, Horizontalfeuerstätte, Waag(e)rechtfeuerstätte *f* [*DIN 18 160*]

horizontal grate extended furnace, extended furnace with horizontal grate; Dutch oven furnace (US) [*See remark under "furnace"*] Planrostvorfeuerung, Planrostvorfeuerstätte *f* [*DIN 18160*]

horizontal grate furnace [*See remark under "furnace"*] Planrostfeuerung, Planrostfeuerstätte *f* [*DIN 18160*]

horizontal line liegende Leitung, horizontale Leitung, waag(e)rechte Leitung *f*

horizontal(ly) moving latch schießende Falle, waag(e)recht bewegliche Falle, horizontal bewegliche Falle *f* [*Kastenschloss*]

horizontal output transformer → flyback transformer

horizontal passenger conveyor → horizontal autowalk

horizontal (pipe)line waag(e)rechte (Rohr)Leitung, liegende (Rohr)Leitung, horizontale (Rohr)Leitung *f*

horizontal reciprocating compressor, horizontal piston compressor [*The cylinder or cylinders are arranged horizontally*] Waag(e)rechtkolbenverdichter, Waag(e)rechtkolbenkompressor, Horizontalkolbenverdichter, Horizontalkolbenkompressor *m*

horizontal reference datum Bezug(s)waagerechte, Bezug(s)horizontale *f*

horizontal resolution, horizontal definition [*The number of individual picture elements or dots that can be distinguished in a horizontal scanning line of a television or facsimile image*] horizontale Aufklärung, waag(e)rechte Aufklärung *f*

horizontal (roof) gutter Liege(dach)rinne *f*

horizontal section Verziehung *f* [*Eine waag(e)rechte Leitung zwischen zwei senkrechten*]

horizontal shell-and-tube condenser [*It consists of a cylindrical shell in which the refrigerant condenses on a number of tubes through which the water flows in one or more passes*] Horizontalröhrenverflüssiger, Horizontalröhrenkondensator *m* [*„waag(e)recht" kann synonym für „Horizontal" verwendet werden*]

horizontal siphon trap → horizontal (drain) trap

horizontal sliding door Horizontalschiebetür, Waag(e)rechtschiebetür *f*

horizontal sliding window Horizontalschiebefenster, Waag(e)rechtschiebefenster *n*

horizontal sloping slat (Lüftungs)Lamelle, (Lüftungs)Kieme *f*

horizontal solution waag(e)rechte Auflösung, horizontale Auflösung *f*

horizontal spacing → character density

horizontal stench trap → horizontal (drain) trap

horizontal sweep transformer → flyback transformer

horizontal travel [*autowalk*] Horizontalfahrt, Waag(e)rechtfahrt *f*

horizontal U-bend → horizontal (drain) trap

horn [*Tubular or rectangular enclosure for radiating or reducing acoustic waves*] Schalltrichter *m*

horn Horn, Hupe *n*

horn (spark) gap Hörner(funken)strecke, Horn(funken)strecke *f*

horsepower [*The effort necessary to raise 33,000 lb a distance of 1 ft in 1 minute*] PS *f*

hose Schlauch *m*

hose basket Schlauchkorb *m*

hose braiding → braiding

hose cabinet, hose cubicle Schlauchschrank *m*

hose connection Schlauchanschluss *m*, Schlauchverbindung *f*

hose connector valve, hose connection valve Schlauchanschlussventil, Schlauchverbindungsventil *n*

hose coupling Schlauchkupp(e)lung *f*

hose down (Ab)Spülen *n* mit Schlauch

hose fitting Schlaucharmatur *f*

hose line Schlauchleitung *f*

hose rack Schlauchgestell *n*

hose reel Schlauchtrommel *f*, Schlauchhaspel *f, m*

hose reel cabinet, hose reel cubicle Schlauchtrommelschrank, Schlauchhaspelschrank *m*

hose screw(ed) connection → screw(ed) hose connection

hose screwing Schlauchverschraubung *f*

hose (spray) shower Schlauchbrause *f*

hose threaded valve Schlauchgewindeventil, Gewindeschlauchventil *n*

hose water Schlauchwasser *n*

hospital (bath)tub, hospital bathing tub Krankenhaus(bade)wanne, Hospital(bade)wanne *f*

hospital heating plant, hospital heating installation Krankenhausheiz(ungs)anlage, Hospitalheiz(ungs)anlage *f*

hospital information system [*The collection, evaluation, verification, storage, and retrieval of information about a patient*] Krankendatensystem *n*

hospital lift, infirmary lift (Brit.); hospital elevator, infirmary elevator (US) Kranken(haus)aufzug, Krankenbettenaufzug, Hospitalaufzug *m*

H

hospital radiator [*Each section of the radiator is a single smooth unit to facilitate cleaning*] Krankenhausradiator, Krankenhausgliederheiz(ungs)körper; Hospitalradiator, Hospitalgliederheiz(ungs)körper *m*

hospital sink [*See remark under "sink"*] Krankenhausspüle, Hospitalspüle *f*

hospital slop sink Krankenhausausguss, Hospitalausguss *m*

host (computer), non-system computer eingetragene Benutzerrechner, Dienstleistungsrechner, Fremdrechner *m*

host elevator Gästefahrstuhl, Gästeaufzug *m*

hostile fire [*It occurs where it is not intended to be*] Schadenfeuer *n*, Schadenbrand *m*

H

host processor zentrale Systemeinheit, Zentralsystemeinheit *f*

hot → alive

hot, very warm [*Having much heat, characterized by a relatively or abnormally high temperature*] heiß

hot, warm warm

hot air, very warm air Heißluft *f*

hot air, warm air [*The American Society of Heating, Refrigerating and Air-conditioning Engineers, Inc. makes no distinction between both terms, nor does presumbly anyone else*] Warmluft *f*

hot air blower, warm air blower Warmluftgebläse *n*, Warmluftbläser *m*

hot air central heating system, warm air central heating system, central hot air heating system, central warm air heating system [*It consists of a heat exchanger with an outer casing or jacket; electrical heating unit together with its supply and return duct system*] Warmluftzentralheiz(ungs)system, Zentralwarmluftheiz(ungs)system *n*

hot air circulation, warm air circulation Warmluftumwälzung, Warmluftzirkulation *f*, Warmluftumlauf *m*

hot air convector, warm air convector Warmluftkonvektor *m*

hot air curtain, warm air curtain Warmluftschleier, Warmluftvorhang *m*, Warmluftschürze *f* [*In die Tür- oder Tor- oder Portalöffnung wird von unten, oben oder von der Seite (den Seiten) warme Luft geblasen*]

hot air discharge → hot air outlet

hot air distribution, warm air distribution Warmluftverteilung *f*

hot air drying, warm air drying Warmlufttrocknung *f*

hot air duct, warm air duct Warmluftkanal *m*

hot air duct riser → hot air rising duct

hot air fan(-assisted) furnace → mechanical warm air furnace

hot air floor heating, warm air floor heating Warmluftdeckenheizung *f*

hot air flow → warm air flow

hot air furnace, (warm air) furnace [*Self-contained, direct-fired, or electrically heated furnace designed to supply heated air through ducts to spaces that require it*] Warmluftfeuerung, Warmluftfeuerstätte *f*

hot air generator, warm air generator Warmlufterzeuger, direkt beheizte Luft(er)wärmer *m* [*Raumheizgerät mit eingebautem Wärmeerzeuger zur Erwärmung und Förderung von Luft. DIN 4794*]

(hot air) hand dryer, (hot air) hand drier, warm air hand dryer, warm air hand drier (Warmluft-) Händetrockner *m*

hot air heater, warm air heater, warm air heating appliance, hot air heating appliance Warmluftheiz(ungs)gerät *n*, Warmluftheizer *m*

hot air heating, warm air heating Warmluftheizung *f*

hot air large space heating → large space (warm) air heating

hot air main duct, hot air principal duct, warm air main duct, warm air principal duct Warmlufthauptkanal, Hauptwarmluftkanal *m*

hot air mechanical furnace → mechanical warm air furnace

hot air outlet, warm air outlet, hot air discharge, warm air discharge Warmluftaustritt, Warmluftauslass *m*

hot air output, warm air output Warmluftleistung *f*

hot air plate duct, warm air plate duct, warm air sheet duct, hot air sheet duct Warmluftblechkanal, Blechwarmluftkanal *m*

hot air riser (pipe), warm air riser (pipe), warm air rising pipe, hot air rising pipe Warmluftsteigrohr, Steigwarmluftrohr *n*

hot air rising duct, warm air rising duct, warm air duct riser, hot air duct riser Warmluftsteigkanal, Steigwarmluftkanal *m*

hot air room heater, warm air room heater Warmluftraumheizer, Warmluftzimmerheizer *m*

hot air supply, warm air supply Warmluftversorgung *f*

hot air system, warm air system [*A system which circulates warm air; hence, it could include anything from a blast coil heated by steam to a parlor stove. More strictly, however such a system is defined as one containing a direct-fired furnace over which air is circulated. When the air circulation is by natural gravity* *the system is referred to as a gravity warm-air system, a type which has been superseded by the forced warm-air system in which the air circulation is by means of a fan*] Warmluftsystem *n*

hot boiler water, warm boiler water Kesselwarmwasser *n*

hot-carrier diode → Schottky (barrier) diode

hot-cathode fluorescent lamp Glühkat(h)odenleuchtstofflampe *f*

hot-cold resistant, cold-hot resistant temperaturwechselfest, temperaturwechselbeständig

hot curve Auslösekennlinie *f* mit Vorlast [*Überlastrelais, DIN IEC 255, T. 8*]

hotel lock Hotelschloss *n*

hotel security Hotelsicherung *f*

hotel sink [*See remark under "sink"*] Hotelspüle *f*

hot forced water → forced hot water

to hot galvanize feuerverzinken

hot gas [*Refrigerant gas in the high-pressure side of the system*] Heißgas *n*

hot gas line Heißgasleitung *f* [*Verbindungsleitung im Kältekreislauf vom Verdichter zum Verflüssiger*]

hot heating water, warm heating water Heiz(ungs)warmwasser, Warmheiz(ungs)wasser *n*

hot high-pressure water heating → high-pressure hot water heating

hot key Tastenkombination *f* [*Mit ihr lässt sich ein bestimmter Befehl noch schneller aufrufen*]

hotline telefonische Hilfs- und Auskunftsdienst *m*

hot plate [*kitchen range*] Kochplatte, Heizplatte *f*

hot plugging Stecken *n* während des Betriebs

hot potable water, hot drinking water, hot drinkable water, warm potable water, warm drinking water, warm drinkable water Brauchwarmwasser, Trinkwarmwasser, warme sanitäre Brauchwasser, sanitäre warme Brauchwasser, Warmbrauchwasser, Warmtrinkwasser *n*

hot pressure Betriebsdruck *m* [*Lampe*]

hot pump water, warm pump water, pump hot water, pump warm water Pumpenwarmwasser, PWW *n*

hot service counter Warm(ausgabe)tresen *m*

hot side, warm side [*heat pump*] warme Seite *f*

hot spot heiße Stelle *f*

hot swap Systemkomponentenaustausch *m* [*bei laufendem Betrieb*]

hot water [*Temperature over 120° F, 50°C*] Warmwasser, WW *n* [*Nicht verwechseln mit „Heißwasser"*]

hot water air heater Warmwasserlufterhitzer, Warmwasserluftheizer, WW-Lufterhitzer, WW-Luftheizer *m*

hot water air heating Warmwasserluftheizung, WW-Luftheizung *f* [*Luftheizung mit Warmwasser als Zwischenmedium*]

hot water base (board) convector [*See remark under "base (board) convector"*] Warmwassersockel(leisten)konvektor, Warmwasserfußleistenkonvektor, Warmwasserscheuerleistenkonvektor *m* [*„WW" kann synonym für „Warmwasser" verwendet werden*]

hot water base (board) heating Warmwasser-Sockel(leisten)heizung, Warmwasser-Fußleistenheizung, Sockel(leisten)warmwasserheizung, Fußleistenwarmwasserheizung *f* [*„WW" kann synonym für „Warmwasser" verwendet werden*]

hot water base (board) heating unit Warmwasser-Sockel(leisten)heiz(ungs)körper, Warmwasser-Fußleistenheiz(ungs)körper, Warmwasser-Scheuerleistenheiz(ungs)körper *m*, Warmwasser-Heiz(ungs)leiste *f* [*„WW" kann synonym für „Warmwasser" verwendet werden*]

hot water boiler Warmwasserkessel, WW-Kessel *m*

hot water cavity ceilling heating → hot water void ceiling heating

hot water central distributor → hot water manifold

hot water central heating → central hot water heating

hot water central (preparation) installation → central hot water (preparation) installation

hot water cock Warmwasserhahn, WW-Hahn *m*

hot water coil Warmwasserschlange, WW-Schlange *f*

hot water compact heat pump Warmwasserkompaktwärmepumpe, WW-Kompaktwärmepumpe *f* [*Wärmepumpe und Wasserspeicher sind eine Einheit, also nicht getrennt aufgestellt*]

hot water connection Warmwasseranschluss, WW-Anschluss *m*, Warmwasserverbindung, WW-Verbindung *f*

hot water consumption (pipe)line Warmwasserverbrauchs(rohr)leitung, WW-Verbrauchs(rohr)leitung *f*

hot water continous heater, continous hot water heater Warmwasserdurchlauferhitzer, WW-Durchlauferhitzer Warmwasserdurchlauf(er)wärmer *m*

hot water convector Warmwasserkonvektor, WW-Konvektor *m*

hot water (convector) radiator [*See remark under "radiator"*] WW-Radiator, WW-Gliederheiz(ungs)körper, Warmwassergliederheiz(ungs)körper; Warmwasserheiz(ungs)körper *m* [*Fehlbenennung*]

hot water cylinder [*A cylinder for storing hot water*] Warmwasser(speicher)zylinder, WW-(Speicher)Zylinder *m*

hot water demand, hot water requirement, hot water need, warm water need Warmwasserbedarf, WW-Bedarf *m*, Warmwassernachfrage, WW-Nachfrage *f*

hot water dispenser Warmwasserspender, WW-Spender *m*

hot water distribution Warmwasserverteilung, WW-Verteilung *f*

hot water distribution line Warmwasserverteilungs(rohr)leitung, WWVL *f*

hot water distribution plant, hot water distribution installation Warmwasserverteilungsanlage, WWV-Anlage *f*

H

hot water distributor Warmwasserverteiler, WW-Verteiler *m*

hot water district heating (system) → district (hot) water heating (system)

hot water draw-off point Warmwasserentnahmestelle, WW-Entnahmestelle *f*

hot water fitting Warmwasserarmatur, WW-Armatur *f*

hot water floor heating → floor warm water heating

hot water forced circulation system [*It has a booster or circulating pump installed to circulate the water in the system mechanically*] Warmwasser-Zwangsumlaufsystem, WW-Zwangsumlaufsystem *n*

hot water forward flow pipe [*A pipe taking hot water from a boiler to a storage cylinder or heating panel*] Warmwasservorlaufrohr, WW-Vorlaufrohr *n*

hot water generator → water heater

hot water gravity system, gravity hot water system, hot water gravity heating, gravity hot water heating [*Gravity circulating system that depends on the difference in the weight or in the density between the hot water and the cold water to create circulation within the piping system*] Schwerkraftwarmwasserheizung, Warmwasserschwerkraftheizung, Warmwasserheizung mit Schwerkraftumtrieb, WW-Schwerkraftheizung *f* [*bis 110 °C Vorlauftemperatur*]

hot water heat Warmwasserwärme, WW-Wärme; Brauchwasserwärme *f* [*veraltete Benennung*]

hot water heater Warmwasserbereiter, WWB *m*

hot water heating, hydronic heating, wet heating Warmwasserheizung, WWH *f* [*Sie arbeitet mit Heißwasser bis zu einer Temperatur von 120 °C bzw. mit Warmwasser bis zu einer Temperatur von 100 °C als Wärmeträger. Diese Temperaturen beziehen sich nach dem deutschem Recht auf die Einstellung des Sicherheitstemperaturbegrenzers (STB)*]

hot water inlet, hot water intake Warmwassereinlauf, Warmwasserzufluss, WW-Einlauf, WW-Zufluss *m*

hot water line Warmwasserleitung, WW-Leitung *f*

hot water manifold, hot water central distributor Warmwasserverteiler(rohr), WW-Verteiler(rohr) *m, (n)*

hot water meter Warmwasserzähler, WW-Zähler *m*

hot water need → hot water demand

hot water outlet Warmwasserauslass, WW-Auslass *m*

hot water panel heating, hot water radiant heating Warmwasserflächenheizung, Warmwasserstrahlungsheizung, Strahlungswarmwasserheizung, Flächenwarmwasserheizung, WW-Flächenheizung, WW-Strahlungsheizung *f*

hot water pipe Warmwasserrohr, WW-Rohr *n*

hot water (pipe)line Warmwasser(rohr)leitung, WW-(Rohr)Leitung *f*

hot water plant, warm water plant, hot water installation, warm water installation Warmwasseranlage, WW-Anlage *f*

hot water preparation Warmwasserbereitung, WWB, Warmwassererzeugung *f*

hot water preparation plant, hot water preparation installation Warmwasserbereitungsanlage, WWB-A, Warmwassererzeugungsanlage *f*

hot water priority circuit Warmwasservorrangschaltung, WW-Vorrangschaltung *f* [*Der Heizkessel läuft ausschließlich für die Warmwasserbereitung. Während der Warmwasserbereitung wird der Heizkreis nicht mit Wärme versorgt*]

hot water pump Warmwasserpumpe, WW-Pumpe *f*

hot water pump heating Warmwasserpumpenheizung, WW-Pumpenheizung *f*

hot water radiator Warmwasserradiator, Warmwassergliederheizkörper, WW-Radiator, WW-Gliederheizkörper *m*

hot water requirement → hot water demand

hot water solar collector system Warmwasser-Solarkollektoranlage, WW-Solarkollektoranlage *f*

hot water split heat pump Warmwassersplitwärmepumpe, WW-Splitwärmepumpe *f* [*Wärmepumpe, deren Kältemittelkreislauf beim Anliefern in Einheiten getrennt ist. Bei der Installation wird sie mit einem Wasserspeicher verbunden wobei sie vom Wasserspeicher getrennt aufgestellt werden kann*]

hot water (storage) tank, hot water (storage) vessel, HWS tank, HWS vessel Warmwasserspeicher, WWS, WW-Speicher, Boiler *m*, Warmwassergefäß, WW-Gefäß *n* [*Das englische Wort „boiler" für Kessel* *wird im Deutschen fälschlicherweise für Warmwasserspeicher verwendet; falsch deshalb weil im Speicher nicht gewärmt wird. DIN 4801 bis 4804. DIN 4805 (10.80) gilt für elektrische Heizeinsatzanschlüsse*]

hot water storey heating (Brit.) → **hot water floor heating**

hot water supply Warmwasserversorgung, WW-Versorgung, Warmwasserbelieferung *f* [*Umfasst Wasser(er)wärmer mit Zuleitung und Verteileranlage bis zu den Entnahmestellen*]

hot water supply boiler [*Boiler completely filled with water that furnishes hot water to be used externally to itself at pressures exceeding 160 psig or at temperatures not exceeding 250° F at or near the boiler outlet*] Warmwasserversorgungskessel, Warmwasserbelieferungskessel, WW-Versorgungskessel, WW-Belieferungskessel *m*

hot water tank → **hot water (storage) tank**

hot (water) tap Warmwasserzapfhahn, WW-Zapfhahn *m*

hot (water) tap(ping) line Warmwasserzapfleitung *f*

hot water tracing heating, hot water trace heating Warmwasserbegleitheizung, Begleitwarmwasserheizung, WW-Begleitheizung *f*

hot water (under)floor heating WW-(Fuß)Bodenheizung, Warmwasser-FB-Heizung, (Fuß)Bodenwarmwasserheizung *f* [*Eine (Fuß)Bodenheizung mit im (Fuß)Boden verlegten Rohren oder anderen Hohlprofilen, die von Wasser als Heiz(ungs)medium durchströmt werden. DIN 4725*]

hot water valve Warmwasserventil, WW-Ventil *n*

hot water vessel → **hot water (storage) tank**

hot water void ceiling heating, hot water cavity ceiling heating Warmwasser-Hohlraumdeckenheizung, Hohlraumdecken-Warmwasserheizung, WW-Hohlraumdeckenheizung, Hohlraumdecken-WW-Heizung *f*

hot wire [*A resistive wire in an electric relay that expands when heated and contracts when cooled*] Hitzdraht *m*

hot wire anemometer, hot wire wind speed meter Hitzdrahtanemometer *n*, Hitzdrahtwindgeschwindigkeitsmesser *m*

hour glass Stundenglas *n*, Sanduhr *f*

hourly air (ex)changes → **AC(P)H**

hourly demand, hourly requirement, hourly need Stundenbedarf *m*, Stundennachfrage *f*

hourly work rate Regiestundensatz *m*

hour-of-day dial Tagesscheibe *f*

house [*A building for human habitation*] Haus, Wohnhaus *n*

house ... → **domestic ...**

house air-raid shelter → **(air-raid) house shelter**

house cable Hausanschlusskabel, Hausverbindungskabel *n*

house central station Hauszentrale *f*

house chimney Hausschornstein, Kamin *m* [*DIN 18160, Teil 1*]

house coal Haus(brand)kohle *f*

house connection Hausanschluss *m*, Hausverbindung *f*

house connection box Hausanschlusskasten, Hausverbindungskasten *m*

house connection line Hausanschlussleitung, Hausverbindungsleitung *f*

house connection room Hausanschlussraum, Hausverbindungsraum *m*

house distribution station Hausverteilerstation *f*

house door lock Haustürschluss *n*

house drain Hausdrän *m*

house drainage Hausentwässerung *f*

house engineering → **house (service) engineering**

(house) entrance box → **(house) service box**

house evacuation Hausräumung *f*

house gas boiler → **domestic gas boiler**

house gas piping [*The system of piping within a structure or a building, either exposed or concealed, which conveys gas from the outlet of the service meter or line to appliances at various places throughout the building or structure. Any piping underground which contains measured gas is also house gas piping*] Hausgasleitung *f*

house heating Hausheizung *f*

house heating plant, house heating installation Hausheiz(ungs)anlage *f*

housing heating pump Hauswärmepumpe *f*

household consumer → **household user**

household cooking appliance, household cooking device Haushalt(s)kochgerät *n*

household device, household appliance Haushalt(s)gerät *n*

household dishwasher, household dishwashing machine, household washup machine Haushaltsgeschirrspüler *m*, Haushaltsgeschirrspülmaschine *f*

household freezer Haushaltsgefrierschrank *m*

household kitchen Haushaltsküche, Hausfrauenküche *f*

household meter Haushaltsmesser, Haushaltszähler *m*

H

household rate → residential rate

household refrigerator Haushaltskühlschrank *m*

household user, household consumer Haushaltsabnehmer, Haushaltsverbraucher *m*

household washing machine, household washer, clothes washing machine, clothes washer Haushaltswaschmaschine, Waschmaschine *f*

household water purifier Haushaltswasserreiniger *m*

house interceptor → house trap

housekeeper → janitor

housekeeping [*Those operations or routines which do not contribute directly to the solution of a computer program(me), but rather to the organization of the program(me)*] Programmorganisation, Programmverwaltung *f*

H

housekeeping digit, service bit/digit Ziffer *f* zur inneren Systemorganisation [*Übertragung*]

housekeeping run [*In automatic data processing, the performance of a program(me) or routine to maintain the structure of files, such as sorting, merging, addition of new records, deletion or modification of existing records*] Programmorganisationsbetrieb, Programmverwaltungsbetrieb *m*

house laundry Hauswäscherei, Waschküche *f*

house lock Hausschloss *n*

housemaid(s) sink → bucket sink

house mail box Hausbriefkasten *m*

house management Hausbewirtschaftung, Hausführung, Hausverwaltung *f*, Hausbetrieb *m*

house manager Hausverwalter, Hausbewirtschafter *m*

house network Hausnetz *n*

house (pipe)line Hausanschluss(rohr)leitung, Hausverbindungs(rohr)leitung *f*

house pressure controller Hausdruckregler *m* [*Gasgerät*]

house presure water filter Hausdruckwasserfilter *m,n*

house range → (kitchen) range

house rubbish Hausmüll *m*

house separator → house trap

(house) service box, (house) entrance box (Haus-) Anschlusskasten *m*

house (services) engineering Haustechnik *f*

house sewage; house wastewater (US) Hausabwasser *n*

house sewer Hausabwasserleitung *f*

house shelter → (air-raid) house shelter

house slow combustion range → domestic slow combustion range

house staff Hauspersonal *n*, Hausbelegschaft *f*

house (sub)station Hausanschluss(unter)station *f*

house supply Hausversorgung *f*

house telephone → residence telephone

house trap, house interceptor, house separator [*A trap between the house drain and the house sewer*] Haus(ab)scheider *m*

house wastes → wastes

house water meter Hauswasserzähler *m*

house wiring Hausverdrahtung *f*

housing, casing, enclosure, case(work) Gehäuse *n*

housing (bath)tub Wohn(ungs)(bade)wanne *f*

h parameter → hybrid parameter

H.P.H.W. → high-pressure hot water

H.P.H.W. heating → high-pressure hot water heating

HPIB, GPIB, general purpose interface bus Schnittstellen-Bussystem *n* von Hewlett-Packard [*Verwendung für programmierbare Messsysteme*]

HPWR → heat pump water return

HPWS → heat pump water supply

HRC contactor, high rupturing capacity contactor Hochleistungsschütz *n*

HRC disconnector changeover switch, high-rupturing capacity disconnector changeover switch Hochleistungstrennumschalter *m*

HRC disconnector switch, high rupturing capacity disconnector switch Hochleistungs-Trenn(schalt)er *m*

HRV → heat recovery ventilator

HSI → human/system interface

HSP → high-speed printer

HTML, Hyper Text Markup Language Befehlssprache *f* für Dokumente im World Wide Web (WWW)

HTTP, Hyper Text Transport Protocol Untermenge *f* [*Von TCP/IP, die für das World Wide Web (WWW) benutzt wird*]

hub Anschlussverteiler *m*

hub Nabe *f*

hub Drückernuss *f* [*Zylinderkastenschloss*]

hub → bell

hub pipe → bell pipe

hub ratio Nabenverhältnis *n*

HUD → U.S. Department of Housing and Urban Development

hue Farbton *m*

human body waste, (human) excreta, soil (sewerage) (menschliche) Auswurfstoffe, Exkremente *f*

(human) comfort → comfort

human engineering Anthropotechnik *f*

humanly occupied space [*Space normally frequented or occupied by people but excluding machinery rooms and walk-in coolers used primarily for refrigerated storage*] menschenbelegte Raum, personenbelegte Raum *m*

human occupancy Personenbelegung, Menschenbelegung *f*

human operator, attendant [*lift*] Führer, Bediener *m*

human /system interface, HSI [*It describes the interaction between an operator and a system, e.g. device, equipment, plant and the like*] Bedien(ungs)-oberfläche, Mensch-System-Schnittstelle, MSS *f*

humid air cooler Feuchtluftkühler *m*

humid air cooling Feuchtluftkühlung *f*

humid cooling load feuchte Kühl(ungs)last *f*

humid heat [*Ratio of increase of enthalpy per unit mass of air, with its associated moisture, to the rise of temperature under conditions of constant pressure and specific humidity*] feuchte Wärme *f*

humidification Befeuchtung, Anfeuchtung *f*

humidification performance, humidification output, humidifying output, humidifying performance Befeuchtungsleistung, Anfeuchtungsleistung *f*

humidification water, humidifying water Befeuchtungswasser, Anfeuchtungswasser *n*

humidifier, humidificator Anfeuchter, Befeuchter *m*

humidifier pump, humidifactor pump Befeuchterpumpe, Anfeuchterpumpe *f*

humidifier water, humidifactor water Befeuchterwasser, Anfeuchterwasser *n*

to humidify [*To add water vapo(u)r or moisture to any moisture-absorbing material, including the atmosphere*] befeuchten, anfeuchten

humidifying Befeuchten, Anfeuchten *n*

humidifying cooling, humidification cooling Befeuchtungskühlung, Anfeuchtungskühlung *f*

humidifying effect, humidification effect [*Product of the mass of water evaporated times the latent heat at the evaporating temperature*] Befeuchtungseffekt, Anfeuchtungseffekt *m*, Befeuchtungswirkung, Anfeuchtungswirkung *f*

humidifying efficiency, humidification efficiency Befeuchtungswirkungsgrad, Anfeuchtungswirkungsgrad *m*

humidifying load, humidification load Befeuchtungslast, Anfeuchtungslast *f* [*Dampfmassestrom, der dem Raum zugeführt werden muss, um einen angestrebten Raumluftzustand aufrecht zu erhalten. DIN 1946, Teil 1*]

humidifying nozzle, humidification nozzle Befeuchtungsdüse, Anfeuchtungsdüse *f*

humidifying run, humidification run Befeuchtungsstrecke, Anfeuchtungsstrecke *f*

humidifying tank, humidification tank Befeuchtungs-Tank, Anfeuchtungs-Tank *m*

humidity, moisture [*Meteorology. Atmospheric water vapo(u)r content expressed in several measures, such as relative humidity*] Luftfeuchte, Luftfeuchtigkeit *f*

humidity balance Feuchtigkeitsausgleich, Feuchteausgleich *m*

humidity ((closed) loop) control Feuchtereg(e)lung, Feuchtigkeitsreg(e)lung *f*

humidity content, humidity percentage Feuchtegehalt, Feuchteanteil, Feuchteprozentsatz, Feuchtigkeitsgehalt, Feuchtigkeitsanteil, Feuchtigkeitsprozentsatz *m*

humidity control → hygrostat

humidity detector, humidity sensor Feuchtigkeitsfühler, Feuchtigkeitssensor, Feuchtigkeitsdetektor, Feuchtefühler, Feuchtedetektor, Feuchtesensor *m*, Feuchtigkeitsdetektionsgerät, Feuchtedetektionsgerät *n*

humidity measurement Feuchtemessung, Feuchtigkeitsmessung *f*

humidity regulation → ((closed) loop) humidity control

humidity release Feuchtigkeitsabgabe, Feuchteabgabe *f*

humidity sensor [*See remark under "sensor"*] Feuchtesensor, Feuchtigkeitssensor, Feuchtefühler, Feuchtigkeitsfühler *m*

humid latent heat, latent humid heat feuchte latente Wärme, latente feuchte Wärme *f* [*Wasserdampf aus der Luft wird an den kalten Flächen des Kühlers niedergeschlagen, wobei die Wasserverdampfungswärme frei wird und auf das Kühlwasser in den Rohren übergeht*]

humid room Feuchtraum *m*

hum(ming) Brummen *n*

hung → suspended

hung ... → suspended ...

hunt group, multi(ple)-terminals with the same data number Sammelrufnummer *f*

hunting [*A condition where a control function is unable to achieve stability and is constantly adjusting its output. An undesirable condition*] Instabilität *f*

hunting [*video*] Bildverschiebung *f*, Bildverschub *m*

hunting [*computer*] Nachlauf *m*, Selbstreg(e)lung *f*

hunting [*automatic telephony*] Freiwahl *f*

hunting, cycling [*It is a periodic change in the controlled variable from one value to another*] Pendeln, Schwingen, Einpendeln *n*

hunting [*Fluctuation caused by the controls attempting to establish an equilibrium against difficult conditions*] Einpendeln, Einschwingen, Pendeln *n*

HVAC engineering, heating/ventilating/air conditioning engineering Heizung-Lüftung-Klima-Technik *f*

H

HVAC services, HVAC process, heating/ventilating/air conditioning services [*The mechanical apparatus used to heat or cool the air or water to provide comfort to the occupant. The process could also be mixing of two air streams as in a fan system mixed air economizer. A water to air heating coil is used to explain the impact of the process on the control loop behavio(u)r*] HLK-Ausrüstung, Heizung-Lüftung-Klima-Ausrüstung *f*

H-vector H-Vektor *m*, magnetische Feldstärke *f*

h.v. service → primary service

HW → hardware

HWS tank → hot water storage tank

h,x-directed ((closed) loop) control [*An engineered scheme that controls an air-conditioned space, including temperature, humidity and enthalpy*] H,x-geführte Reg(e)lung *f*

hybrid circuit Hybridschaltung, Mischschaltung *f*

hybrid computer → analogue(-to)-digital computer

hybrid connector Mischleiste(nsteck(verbind)er) *f, (m)*

hybrid hardware ((open) loop) control [*The control of and communication between the various parts of a hybrid computer*] Hybridrechnersteuerung, Analog-Digital-Rechnersteuerung *f*

hybrid input/output [*The routine required to handle inputs to and outputs from a computer system comprising digital and analog computers*] Hybrid-Eingang-Ausgang *m*

hybrid integrated circuit [*A circuit in which one or more discrete components are used in combination with integrated-circuit construction*] integrierte Hybridschaltung, Hybridintegrationsschaltung, Integrationshybridschaltung *f*

hybrid junction, bridge hybrid [*A transformer, resistor, or wave-guide circuit or device that has four pairs of terminals so arranged that signal entering at one terminal pair divides and emerges from the two adjacent terminal pairs, but is unable to reach the opposite terminal pair*] Hybride *f*

hybrid microcircuit [*Microcircuit in which thin-film, thick-film, or diffusion techniques are combined with separately attached semiconductor chips to form the circuit*] Hybridmikroschaltung, Mikrohybridschaltung, Mikromischschaltung, Mischmikroschaltung *f*

hybrid model, forward plus inverse model Hybridmodell *n*, Mischmodell

hybrid net(work), mixed structure net(work) Hybridnetz, Mischnetz, hybride Netz, gemischte Netz *n*

hybrid parameter, h parameter [*One of a set of four transistor equivalent circuit parameters that conveniently specify transistor performance for small voltages and currents in a particular circuit*] Hybridparameter *n*

hybrid plug connector Misch(leisten)-Messerleiste *f*

hybrid programming [*Hybrid system routines that handle timing, function generation, and simulation*] Hybridprogrammierung *f*

hybrid relay [*A relay in which solid-state elements are combined with moving contacts*] Hybridrelais *n*

hybrid simulation [*The use of a hybrid computer for purposes of simulation*] Hybridsimulation *f*

hybrid socket connector Misch(leisten)-Federleiste *f*

hybrid tube Hybridröhre *f*

hydrant → (fire) hydrant

hydrant [*A form of connection incorporated in a water main to enable a hose to be attached and a continuous supply of water obtained*] Hydrant *m*

hydrant duckfoot bend, hydrant rest bend [*See remark under "duckfoot bend"*] Hydrant(en)-Fußbogen 90° *m* [*DIN 28 648*]

hydration water Hydratwasser *n* [*In Feststoffteilen eingeschlossenes Wasser*]

hydraulic act(uat)or [*See remark under "act(uat)or"*] Hydraulikstellgerät *n*

hydraulic auxiliary power hydraulische Hilfsenergie *f*

hydraulic circuit, hydraulic switching [*A circuit that operates by hydraulic control of water or fluid currents rather than by electric currents*] hydraulische Schaltung, Hydraulikschaltung *f* [*In Heiz- und Raumluftanlagen. VDI 2073*]

hydraulic computer [*A computer in which electric current and gates are replaced by fluid and valves*] Hydraulikrechner *m*

hydraulic diameter hydraulische Durchmesser *m*

hydraulic driving machine, hydraulic drive machine Hydrauliktriebwerk *n* [*Aufzug*]

hydraulic elevator (US); hydraulic lift (Brit.) [*The energy is applied by means of a liquid under pressure in a cylinder equipped with a plunger or ram*] Hydraulikaufzug, hydraulische Aufzug, Aufzug mit hydraulischem Antrieb *m*

hydraulic goods/staff lift (Brit.) → **hydraulic staff/goods lift**

hydraulic head → **water head**

hydraulic lift (Brit.) → **hydraulic elevator**

hydraulic lime, greystone lime hydraulische Kalk *m*

hydraulic ((open) loop) control hydraulische Steuerung, Hydrauliksteuerung *f*

hydraulic passenger elevator (US); hydraulic passenger lift (Brit.) Hydraulikfahrstuhl, Hydraulikpersonenaufzug *m*

(hydraulic) plunger, (hydraulic) piston, (hydraulic) jack, (hydraulic) ram [*Supporting member of hydraulic elevator that is attached to the bottom of the elevator car*] Hubkolben, Hubstempel *m*

hydraulic resistance hydraulische Widerstand *m*

hydraulic shock, water hammer [*An internal pressure stress imposed in piping systems by a sudden change in liquid velocity, as by the sudden stopping of flow*] Wasserschlag, Wasserstoß, Druckschlag, Druckstoß *m*

hydraulic staff/goods lift, hydraulic goods/staff lift (Brit.); hydraulic staff/goods elevator, hydraulic goods/staff elevator (US) Hydraulik-(Personen)Lastenaufzug *m*

hydraulic switch Hydraulikschalter *m*

hydraulic switching → **hydraulic circuit**

hydraulic window opening device hydraulische Fensteröffner *m*

hydrocarbon Kohlenwasserstoff, KW *m*

hydrochloric acid Salzsäure *m*

hydrocooling Kaltwasserkühlung *f*

hydrodynamic strömungsgünstig, hydrodynamisch

hydroelectric elevator (US); hydroelectric lift (Brit.) [*Direct-plunger elevator where liquid is pumped under pressure directly into the cylinder by a pump driven by an electric motor*] Kolbenaufzug *m*

hydrometer [*A direct-reading instrument for indicating the density, specific gravity, flow velocity, or some other characteristic of liquids*] Hydrometer *n*

hydrometry Hydrometrie *f*

hydronic heating → **hot water heating**

hydrostatic modulus → **modulus of volume elasticity**

hydrostatic pressure hydrostatische Druck *m*

hygrometer Luftfeuchtemesser, Luftfeuchtigkeitsmesser *m*, Hygrometer *n*

hygrostat, moisture sensor Hygrostat *m*, Feuchtefühler *m* mit Schaltelement

hyperbolic cooling tower [*Cooling tower of hyperbolic shape which depends on natural draft for air movement through the tower. It can be either crossflow or counterflow*] hyperbolische Kühlturm *m*

hyperlink Verweis *m* zu weiteren Daten

Hyper Text Markup language → **HTML**

Hyper Text Transport Protocol → **HTTP**

hypothetical reference circuit → **HCR**

hypothetical reference connection hypothetische Bezugsverbindung, angenommene Bezugsverbindung *f*

hysteresis Hysterese, Hysteresis *f* [*Physik. Fortdauer der Wirkung nach Aufhören der Ursache*]

hysteresis error, hysteresis deviation Hysteresisfehler, Hysteresefehler *m*, Hysteresisabweichung, Hystereseabweichung *f* [*DIN IEC 770*]

hysteresis loop Hysteresisschleife, Hystereseschleife *f*

I

IAE → **integral of absolute value of error**

IAQ → **room air quality**

IAQ procedure → **room air quality procedure**

IAS → **immediate access storage**

IB → **interface bus**

IBMS → **integrated broadband mobile system**

I.B.P → **initial boiling point**

IC → **integrated circuit**

IC → **interlock code**

I

I & C → **control and instrumentation technique**

ICB → **incoming calls barred**

I²C-bus, interintegrated circuit bus „Kabelbaum" *m* zur Datenübertragung [*Zwischen integrierten Schaltungen und Rechnern*]

IC card Chipkarte *f* [*ISO 7816*]

ICCS → **integrated communications cabling system**

ICE → **in-circuit emulator**

ice bank evaporator [*Evaporator immersed in water, on which an ice bank is formed*] Eiserzeugungsverdampfer *m*

ice maker, ice machine Eismaschine *f*

ice maker heat pump, ice machine heat pump Eismaschinenwärmepumpe *f*

ice making, ice producing Eiserzeugung, Eisherstellung *f*

ice sensor Eisfühler *m*

ice storage plant, ice storage installation Eisspeicheranlage, Eisspeicherinstallation *f*

ice storage system, ice chiller [*Thermal storage system, used for chilling processes or for comfort cooling, that uses the phase change of ice to water. Ice is formed during periods of low refrigerating demand, for delivery of cooling during periods of high refrigerating demand*] Eisspeichersystem *n*

ice storage vessel, ice storage tank Eisspeicher(behälter) *m*

ice store basement → **(block) ice store basement**

ice water [*Water chilled, as with ice*] Eiswasser *n*

ice water, meltage ice, melting ice Eiswasser, geschmolzene Eis *n*

icing → **(contact) icing**

I ((closed) loop) control parameter I-Reglerkennwert *m*

IC memory → **integrated-circuit memory**

icon Bildsymbol, Piktogramm *n*

ICR → **intelligent character recognition**

ID → **indentification**

ID, inside diameter, internal diameter Innendurchmesser *m*

ID, internal dimension, inside dimension Innenmaß *n*, Innenabmessung *f*

ID card ... → **badge (card) ...**

IDE, integrated drive electronics, AT-Bus Festplatte *f* mit integriertem Adapter

ideal combustion → **perfect combustion**

ideal displacement flow, perfect displacement flow ideale Verdrängungsströmung *f* [*Idealisierte Form der Raumluftströmung, bei der eine gleichmäßige Ausbreitung ohne Mischen erfolgt*]

ideal fluid, perfect fluid ideale Fluid, ideale Strömungsmedium *n* [*Es bewegt sich reibungslos durch ein Rohr*]

ideal gas, perfect gas Idealgas, perfekte Gas, vollkommene Gas, ideale Gas *n*

ideal liquid, incompressible liquid, perfect liquid [*In practice to liquid it is completely incompressible*] Idealflüssigkeit, ideale Flüssigkeit, vollkommene Flüssigkeit *f* [*Sie ist reibungsfrei und raumbeständig (= inkompressibel) gedacht*]

ideal mixing flow, ideal room air flow, perfect mixing flow, perfect room room air flow ideale Mischströmung *f* [*Idealisierte Form der Raumluftströmung, bei der infolge vollständiger Durchmischung der Raumluftzustand an jeder Stelle gleich ist*]

ideal radiator → **blackbody**

ideal radiator radiation → **blackbody radiation**

identification, ID Identifizierung, Kenn(zeichn)ung, Gleichsetzung *f*

identification card ... → **badge (card) ...**

identification device Identifizierungsgerät, Kenn(zeichn)ungsgerät *n*

identification product Identifizierungserzeugnis, Kenn(zeichn)ungserzeugnis, Identifizierungsprodukt, Kenn(zeichn)ungsprodukt *n*

identification rose Identifizierungsrosette, Kenn(zeichn)ungsrosette *f* [*Über den Schließzylinder geschoben*]

identification tag → **tag**

identifier Identifikator *m* [*Eine Zeichenkette zur Identifizierung oder Benennung eines Programmelements*]

identifier, data identifier [*A symbol whose purpose is to specify a body of data*] (Daten)Kennzeichen *n*, (Daten)Kennzeichner *m*

identifier Stichwort *n*

to identify identifizieren

identity Identität, völlige Gleichheit *f*

identity card → **badge (card)**

identity confidentiality Identitätsvertraulichkeit *f*

identity photo, identification photo Passbild, Passfoto *n*

identity photo comparison, identification photo comparison Passbildvergleich, Passfotovergleich *m*

identity photo comparison method, identification photo comparison method Passbildvergleichsverfahren *n*, Passbildvergleichsmethode *f*

IDE (PC/AT) AT-Bus *m* [*Dieser Schnittstellenstandard wurde von der Firma Conher als Schnittstelle für den OEM-Markt entwickelt*]

idioelelectirc selbstelektrisch

idle character, fill(ing) character Leerzeichen, Füllzeichen *n* [*Es hat keine Bedeutung, sondern füllt Felder zwischen gespeicherten Daten aus*]

idle current → reactive current

idle running, idling Leerlauf *m*

idle state Ruhezustand *m*

idle wait(ing) state Ruhewartezustand *m* [*programmierbare Messgeräte*]

IDN, integrated digital network Integration *f* erster Art, Integration 1.Art [*Kombination digitaler Vermittlungstechnik und digitaler Übertragungstechnik*]

IEEE 1394 → "firewire"

IEEE-488 (US) IEC-(615-)Bus *m* [*Übertragung der Daten byteweise auf je einen Takt hin*]

I-element, integrating element I-Glied, integrierende Glied, I-Element, integrierende Element *n*

IFC → interface clear

IF rejection → intermediate frequency rejection

IGES → initial graphics exchange standard

IG-FET, insulated gate-FET, insulated-gate field-effect transistor Isolierschicht-Feldeffekttransistor, Oberflächen-Feldeffekttransistor, IG-FET, Feldeffekttransistor mit isolierter Steuer(ungs)elektrode *m*

to ignite zünden

igniter fuse → ignition fuse

ignition Zündung *f*

ignition aid Zündhilfe *f*

ignition burner [*A remov(e)able burner incorporated as an integral part of a solid fuel appliance, for igniting the fuel*] Zündbrenner *m*

ignition chamber [*Portion of the furnace directly above the grates and hearth*] Zündkammer *f*

ignition delay → delayed ignition

ignition device, ignition appliance Zündorgan *n*, Zündvorrichtung *f*

ignition electrode Zündelektrode *f*

ignition energy Zündenergie *f*

ignition flame burner Zündflammenbrenner *m*

ignition fuse, ignition fuze, igniter fuse, igniter fuze Zündsicherung *f*

ignition gas Zündgas *n*

ignition gas opening Zündgasöffnung *f*

ignition gas pushbutton Zündgasdruckknopf *m*

ignition gas valve Zündgasventil *n*

ignition limit Zündgrenze *f*

ignition line Zündleitung *f*

ignition ((open) loop) control Zündsteuerung *f*

ignition protective system Zündschutzart *f*

ignition rate Zündgeschwindigkeit *f*

ignition safety valve Zündsicherheitsventil *n*

ignition source Zündquelle *f*, Zündinitial *n* [*Energieträger, der (bei mehrfacher Energieumwandlung als letzter) die für die Zündung eines brennbaren Systems erforderliche Temperatur hervorruft*]

ignition spark Zündfunken *m*

ignition temperature, ignition point, kindling point, kindling temperature [*Temperature at which a combustible material will unite with oxygen in the atmosphere, and combustion will commence*] Zündtemperatur *f*

ignition transformer Zündtransformator, Zündtrafo *m*

ignition valve Zündventil *n*

ignitor → pilot electrode

IIC → interface integrated circuit

IIR filter → infinite impulse response filter

IIS → Internet information server

illicit (radio) device → listening device

ill-lit, ill-illuminated schlecht beleuchtet

illuminated, lit beleuchtet

illuminated diagram [*IEC 113-2*], **illuminated circuit diagram** Leuchtschaltplan *m*, Leuchtschaltschema *n*

illuminated indicator → indicator light

illuminated pushbutton Leuchtdruckknopf *m*

illuminating engineering Beleuchtungstechnik, Lichttechnik *f*

illuminating gas, light(ing) gas Leuchtgas *n*

illumination, lighting Beleuchtung, künstliche Beleuchtung *f*, Kunstlicht *n* [*Fehlbenennung, weil Pleonasmus*]

illumination alarm, light(ing) alarm Beleuchtungsalarm, Lichtalarm *m*

illumination booth, light(ing) booth Beleuchterkabine, Beleuchtungskabine *f*

illumination cable, light(ing) cable Licht(strom)-kabel, Beleuchtungskabel *n*

I

illumination calculation, light(ing) calculation Beleuchtungsberechnung, Lichtberechnung *f*

illumination circuit, light(ing) circuit Beleuchtungsstromkreis, Lichtstromkreis *m*

illumination column, (light(ing)) column (Brit.); (illumination) mast, light(ing) mast (US) (Licht)Mast, Beleuchtungsmast, Leuchtenmast *m*

illumination component, light(ing) component Beleuchtungsanteil, Lichtanteil *m*

illumination current, light(ing) current Beleuchtungsstrom, Lichtstrom *m*

illumination designer, light(ing) designer Beleuchtungsgestalter, Lichtgestalter *m*

illumination diffuser, diffusing unit, light-diffusing unit, light(ing) diffuser Lichtstreuelement *n*

illumination element, light(ing) element Beleuchtungselement, Lichtelement *n*

I

illumination engineer, light(ing) engineer Beleuchtungsingenieur, Lichtingenieur *m*

illumination engineering, light(ing) engineering Beleuchtungstechnik, Lichttechnik *f*

illumination gallery, light(ing) gallery Beleuchtungsbrücke, Beleuchterbrücke, Lichtbrücke *f*

illumination(-generated) heat, heat generated by illumination, light(ing)(-generated) heat, heat generated by lighting Beleuchtungswärme, Lichtwärme *f*

illumination intensity, light(ing) intensity Beleuchtungsstärke, Lichtstärke *f*

illumination line, light(ing) line Lichtleitung, Beleuchtungsleitung *f*

illumination load, light(ing) load Beleuchtungslast, Lichtlast *f*

illumination loss, light(ing) loss Lichtverlust, Beleuchtungsverlust *m*

illumination mast, light(ing) mast Beleuchtungsmast, Leuchtenmast, Lichtmast *m* [*DIN 49 778 und DIN 49 779*]

illumination mast with arm, illumination column with bracket (US) [*"Light(ing)" may synonymously be used for "illumination"*] Auslegerbeleuchtungsarm, Auslegerleuchtenmast, Ausleger(licht)mast *m*

illumination ((open) loop) control, light(ing) ((open) loop) control Beleuchtungssteuerung, Lichtsteuerung *f*

illumination outlet, light(ing) outlet Lichtsteckdose, Beleuchtungssteckdose *f*

illumination panel, light(ing) panel Lichtschalttafel, Beleuchtungsschalttafel *f*

illumination point, light(ing) point Lichtstelle, Beleuchtungsstelle *f*

illumination source, light(ing) source Beleuchtungsquelle, Lichtquelle *f*

illumination switch → light(ing) switch

illumination system, light(ing) system Lichtinstallation, Beleuchtungsinstallation *f*, Lichtsystem, Beleuchtungssystem *n*

illumination tariff, light(ing) tariff Lichttarif, Beleuchtungstarif *m*

illumination trunking, light(ing) trunking [*In large offices and factories it is customary to run trunking down the building and attach the lighting fittings to them*] Beleuchtungsinstallationskanal, Lichtinstallationskanal, Lichtsystemkanal, Beleuchtungssystemkanal *m*

illumination tube, light(ing) tube Beleuchtungsröhre, Lichtröhre, Leuchtröhre *f*

ILT → inter-line transfer

IM → intermodulation

IM → interface module

IM → initialization mode

image [*In automatic data processing, a copy of the information combined in a punched card or other data record recorded on a different data medium*] Bild, Abbild *n*, Wiedergabe *f*

image, picture [*The scene reproduced by a television or facsimile receiver*] Bild *n*

image amplifier Bildverstärker *m* [*Ein Bildwandler, der die Bildintensität innerhalb eines Spektralbereichs verstärkt (v_1=v_2)*]

image blanking signal Bildaustastsignal, BA-Signal *n*

image blemish Störfleck *m* [*Elektronenstrahlröhre*]

image caption Bildtitel *m*, Bildüberschrift, Bildunterschrift *f*

image clarity Bildreinheit, Bildklarheit *f*

image comparison Bildvergleich *m*

image composing elektronische Photomontage, elektronische Fotomontage *f*

image content Bildinhalt *m*

image conversion Bildwandlung *f*

image converter, (electron) image tube Bildwandlerröhre *f* [*Sie transformiert zweidimensionale elektromagnetische Wellen (Bilder) mit oder ohne Verstärkung von einem Spektralbereich in einen anderen. So werden Bilder aus einem dem Auge nicht zugänglichen Frequenzbereich v_1 (z. B. Infrarotstrahlung, Ultraviolettstrahlung oder Röntgenstrahlung) sichtbar gemacht (Frequenzbereich v_2)*]

image enhancement Bildreinigung, Bildverbesserung *f*

image force [*The electrostatic force on a charge in the neighbo(u)rhood of a conductor, which may be thought of as the attraction to the charge's electric image*] Bildkraft *f*

image geometry Bildgeometrie *f* [*Bildschirmgerät*]

image impedance [*Impedance that, together with other impedances, will simultaneously terminate all inputs and outputs to a transducer in such a way that at each of its inputs and outputs the impedance in both directions will be equal*] Wellenwiderstand *m* [*4-Pol*]

image intensifier [*Any electronic device which, when actuated by a light image, reproduces a similar image of enhanced brightness, and which is capable of operating at very low light levels without introducing spurious brightness variations (noise) into the reproduced image*] Bildverstärker *m*

image management Bildverwaltung *f*

image memory, image storage Bildspeicher *m*

image plane [*Optics. A plane that lies perpendicular to the axis of an optical system in which an image is formed*] Bildebene *f*

image processing [*Processing pictorial inspection information by computer*] Bildverarbeitung *f*

image recording Bildaufzeichnung *f*

image refresh rate (Bild)Wiederholrate *f*, (Bild)Frequenz *f* [*Sie gibt an wie oft in einer Sekunde das Bild auf dem Monitor durch die Graphikkarte neu aufgebaut wird*]

image rejection Spiegelfrequenzabstand *m*, Spiegelfrequenzunterdrückung, Spiegelfrequenzfestigkeit, Spiegelfrequenzsicherheit *f* [*Unterschied zwischen dem Pegel des auf die Zwischenfrequenz umgesetzten Signals mit Empfangsfrequenz und dem auf die Zwischenfrequenz umgesetzten Signals mit Spiegelfrequenz*]

image sensor, image sensing device, image pilot device, image sensing element, image (primary) detector, image detector element, image detection element [*See remark under "sensor"*] Bildsensor, bildverarbeitende Sensor, Bildfühler *m*, Bilddetektionselement *n* [*Sensor zur elektronischen Aufnahme von Bildern für eine visuelle Beobachtung oder eine Bildverarbeitung. Kameraröhren (z. B. Vidikon) und Halbleiterbildsensoren (CCD-Sensor; CCD-Bildwandler; CID-Element*]

image storage → image memory

image transmission system, ITS Bildübertragungssystem *n*

image tube → image converter

image viewing Bildbetrachtung *f*

imaging system Bilderkennungssystem *n*

IM distortion → intermodulation distortion

immediate access storage, IAS Speicher *m* mit unmittelbarem Zugriff [*Die Zeit für den Zugriff auf eine beliebige Speicherstelle ist unabhängig vom vorherigen Zugriff und üblicherweise von derselben Größe wie die Zykluszeit des Prozessors*]

immediate addressing unmittelbare Adressierung *f* [*Der Operand eines Maschinenbefehls steht direkt hinter dem Befehl. Weil ein Befehl nur Konstanten beinhalten kann, ist diese Adressierung für Variable nicht möglich*]

immediate alarm Sofortalarm *m*

immediate (data) processing → demand (data) processing

immediate (memory) access Sofort(speicher)zugriff *m*

immersible → submersible

immersion collar, immersed collar, submersible collar Tauchhülse *f*, Eintauchhülse

immersion condenser, submerged condenser, immersed condenser, submersible condenser [*Condenser in which the piping is submerged in a vessel containing cooling water*] Tauchrohrverflüssiger, Tauchrohrkondensator, Eintauchrohrverflüssiger, Eintauchrohrkondensator *m*

immersion cooling Tauchkühlung *f*

immersion electrode humidifier, submersible electrode humidifier, submerged electrode humidifier, immersed electrode humidifier (Ein)Tauchelektrodenbefeuchter, (Ein)Tauchelektrodenanfeuchter *m*

immersion heater, submerged heater (Ein)Tauchheizer, (Ein)Taucherhitzer, (Ein)Tauch(er)wärmer *m*

immersion intake, submerged intake Taucheinlauf, Eintaucheinlauf *m*

immersion pipe, immersed pipe, submersible pipe, submerged pipe Tauchrohr, Eintauchrohr *n*

immersion pump, immersed pump, submersible pump, submerged pump, immerged pump Tauchpumpe, Nassläuferpumpe, Eintauchpumpe *f*

immersion rotary (water) meter, immersed rotary (water) meter, submersible rotary (water) meter, submerged rotary (water) meter Nassläufer *m*

immersion tank → dip tank

immersion temperature sensor, submersible temperature sensor, immersed temperature sensor, submerged temperature sensor Tauchtemperatursensor, Eintauchtemperatursensor *m*

immersion thermostat, immersed thermostat, immersion temperature controller, immersed temperature controller, temperature submersible controller, submersible thermostat, submersible temperature controller, temperature immersion controller, submerged thermostat, submerged temperature controller [*It is used for the control of high pressure fluids or gases. "Regulator" may synonymously be used for "controller"*] Tauchthermostat, Tauchtemperaturregler, Temperatureintauchregler, Eintauchthermostat, Eintauchtemperaturregler *m* [*Siehe Anmerkung unter „Thermostat"*]

immersion tube, immersed tube, submersible tube, submerged tube Tauchröhre, Eintauchröhre *f*

immersion water detector, immersion water sensor [*It is designed for mounting in pockets inside water pipes thus providing a means for the element to be removed for servicing without the need for the system to be drained down. It works on the same principle as the room detector and has no local setting adjustment*] (Ein)Tauch-Wasserdetektor, (Ein)Tauch-Wasserfühler *m*, (Ein)Tauch-Wasserrohr, (Ein-)Tauch-Wasserdetektionselement *n*

impact Stoß, Aufprall *m*

impact avalanche transit time diode → IMPATT diode

impact load [*A suddenly applied load, as from reciprocating machinery, rolling loads, etc.*] Stoßlast *f*

impact matrix printer → matrix impact printer

impact printer, impact writer anschlagende Drucker, Anschlagdrucker *m*

impact sound, structure-borne sound, contact sound Körperschall *m* [*Schall, der sich in festen Stoffen fortpflanzt (Erschütterungen ergeben Körperschall)*]

impact sound detector, impact sound detecting device, contact sound detector, contact sound detecting device, structure-borne sound detector, structure-borne sound detecting device Körperschallmelder, KSM *m*, Körperschallmeldegerät *n* [*Er wird an den Fenstersender und/oder Türsender angeschlossen und reagiert auf Durchbruch und Öffnung von Fenstern und Türen*]

impact sound insulation index I [*ISO/R 717*] bewertete Norm-Trittschallpegel *m*

impact sound intensity, structure-borne sound intensity, contact noise intensity Körperschallstärke *f*

impact sound transmission, structure-borne sound transmission, contact noise transmission Körperschallübertragung *f*

impact sound (transmission) level, contact noise (transmission) level, structure-borne sound (transmission) level [*In a given octave band, between two rooms. The average octave band pressure level, throughout one room, produced by impacts delivered to the structure of the other room by a standard tapping machine*] Körperschallpegel *m*

imparity bit, imparity binary digit Imparitätsbit *n* [*Stets eine ungerade Anzahl von binären Einsen zu erzeugen hat den Vorteil, dass niemals alle Bits in einem Wort oder Byte gleichzeitig den Wert 0 annehmen. Diese Lösung heißt „Imparitätsbit"*]

IMPATT diode, impact avalanche transit time diode Lawinenlaufzeitdiode *f*

impedance [*Literally the opposition that a circuit offers to the flow of alternating current. Primarily used in the sound field as an impedance characteristic of amplifiers, microphones and accessories*] Impedanz *f*, (elektrische) Scheinwiderstand *m*

impedance bridge [*A device similar to a Wheatstone bridge, used to compare impedance which may contain inductance, capacitance, and resistance*] Impedanzmessbrücke, Scheinwiderstandsmessbrücke *f*

impedance converter Impedanzumsetzer, Impedanzumwandler, Scheinwiderstandsumsetzer, Scheinwiderstandsumwandler, Impedanzkonverter, Scheinwiderstandskonverter *m*

impedance coupling [*Coupling of two signal circuits with an impedance*] Impedanzkopp(e)lung, Scheinwiderstandskopp(e)lung *f*

impedance drop [*The total voltage drop across a component or conductor of an alternating current circuit, equal to the sum of the resistant drop and the reactance drop*] Wechselstromspannungsabfall, innere Spannungsabfall *m*

impedance matching network, line-building-out network [*A network of two or more resistors, coils, or capacitors used to couple two circuits in such a manner that the impedance of each circuit will be equal to the impedance into which it locks*] Impedanzanpassungsnetz, Scheinwiderstandsanpassungsnetz *n*

impedance of actual value input Istwerteingangswiderstand *m*

impedance voltage Nennkurzschlussspannung, Kurzschlussnennspannung *f*

impeller Laufrad *n*

imperative statement [*A statement in a symbolic program(me) which is translated into actual machine-language instructions by assembly routine*] unbedingte Anweisung *f*

impingement Aufprall *m*

implementation Ausführung, Durchführung *f*

implication Wenn-Dann-Zusammenhang *m*, Implikation *f*

implied demand, implied requirement vorausgesetzte Forderung *f*, vorausgesetzte Erfordernis *n*

to impress → **to apply**

impressed current anode Fremdstromanode *f* [*Korrosionsschutz*]

impulse Impuls, *I m*, Bewegungsgröße *f* [*kg.s*] [*Das Produkt aus Masse und Geschwindigkeit eines Massepunktes, also das Zeitintegral der Kraft. DIN 4044*]

impulse-duration modulation, PDM, impulse-width modulation Impulsdauermodulation *f*

impulse excitation [*Excitation produced by a voltage or current variation of relatively short duration; used to initiate oscillation in the resonant circuit of an oscillator*] Impulserregung *f*

impulse generator, pulse generator [*An apparatus which produces very short surges of high-voltage or high-current power by discharging capacitors in parallel or in series*] Impulsgenerator, Impulserzeuger *m*

impulse modulation [*Modulation of a signal in which it is replaced by a series of impulses, equally spaced in time, whose strengths (integrals over time) are proportional to the amplitude of the signal at the time of the impulse*] Impulsmodulation *f*

impulse response [*The response of a system to an impulse which differs from zero for an infinitesimal time, but whose integral over time is unity; this impulse may be represented mathematically by a Dirac delta function*] Impulsantwort *f*

impulse separator, sync separator [*In a television receiver, the circuit that separates the horizontal synchronizing impulses in the received signal from the vertical synchronizing impulses*] Impulssieb *n*

impulse signalling ; impulse signaling (US) [*Conveying information by means of on-off conditions transmitted down a line or over free space*] Impulssignalgebung *f*

impulse sound level Impulsschallpegel *m*

impulse sound level meter Impulsschallpegelmesser *m*

impulse tachometer [*A tachometer in which each rotation of a shaft generates an electric(al) pulse and the time rate of pulses is then measured; designed as capacitor-current, inductory, or interrupted direct-current tachometer*] Impulstachometer *n* [*Umgangssprachlich: m*]

impulse train [*An input consisting of an infinitive series of unit impulses, equally separated in time*] Impulsreihe *f*, Impulszug *m*

impulse transmission [*Form of signal(l)ing which employs impulses of either or both polarities for transmission to indicate the occurrence of transition in the signals, used principally to reduce the effects of low-frequency interference; the impulses are generally formed by suppressing the low-frequency components, including direct currents of the signals*] Impulsübertragung *f*

impulse value Impulswert *m*

impulse-width modulation → **impulse-duration modulation**

impulse withstand voltage, impulse withstand tension Stehstoßspannung, Haltestoßspannung, Impulsspannungsfestigkeit *f*

impurity Verunreinigung *f*

inaccessible unzugänglich

(inadvertent) wrong operation → **maloperation**

(in) all poles, all-pole allpolig, ganzpolig

in-band signalling (Brit.); in-band signaling (US) bandinterne Signalgebung, bandinterne Signalisierung *f* [*Übertragung von Informationen über bestimmte Frequenzen, die in einem Trägerkanal liegen oder über Frequenzen, die normalerweise der Übertragung von Nachrichten dienen*]

in-built → **built-in**

in-built cupboard → **closet**

incandescence [*The emission of visible radiation by a hot body*] Weißglühen *n*, Weißglut *f*

incandescent weißglühend

incandescent bulb, filament bulb Glühbirne, Weißbirne *f*

incandescent illumination, incandescent lighting, filament illumination, filament lighting Glühlampenbeleuchtung, Weißlampenbeleuchtung *f*

incandescent lamp, filament lamp [*An electric lamp that produces light when metal(lic) filament is heated white-hot in a vacuum by passing an electric current through the filament*] Glühlampe, Weißlampe *f* [*DIN EN 60 061-1*]

incandescent lamp lighting → **filament lamp lighting**

incandescent light, filament light Glühlicht, Weißlicht *n*

incandescent (light) carbon, filament (light) carbon Glühlichtkohle, Weißlichtkohle *f*

incandescent luminaire, incandescent light(ing) fitting, filament luminaire, filament light(ing) fitting Glühleuchte, Weißleuchte *f*, Glühbeleuchtungskörper, Weißbeleuchtungskörper *m*

incendiarism → **arson**

incendiary → **arsonist**

inching mode → **jog(ging) mode**

inch-plank Zollbrett *n*

incidence of light, light incidence Lichteinfall *m*

incidental heat gain, free heat gain [*The total heat acquired by a building from solar radiation and any other external source of heat (e.g. lighting, occupants), that does not form part of the installed heating system of the building. Incidental heat gain can contribute to fuel savings only if the temperature-control system of the building is designed to take account of such heat gain*] freie Wärmeabgabe *f*

incident angle → **acceptance angle**

incident history Störungsablauf *m*

incident light illumination Auflichtbeleuchtung *f*

incident light plane Auflichtebene *f*

I

incinerator [*For the reduction of waste material, garbage, refuse, or the like by incineration*] Verbrennungsanlage *f*

incipient fire Entstehungsfeuer *n*, Entstehungsbrand *m*

in circuit → **alive**

in-circuit emulator, ICE systemeigene Prüfemulator, Emulations(- und Prüf)Adapter *m* [*System zur Simulation des Verhaltens eines Mikroprozessors sowie zur Aufzeichnung von Signalen und Befehlen. Diese Emulatoren werden meist in Verbindung mit Entwicklungssystemen angeboten; es gibt jedoch bereits Produkte, die über V.24-Schnittstellen an Großrechner angeschlossen werden können*]

in-circuit test Prüfung *f* von Elektronikbaugruppen [*Die Bauelemente werden nacheinander geprüft*]

inclination → **slope**

incline angle → **angle of incline**

inclined autowalk, inclined passenger conveyor geneigte Fahrsteig, geneigte Rollsteig *m* [*Neigungswinkel maximal 12°*]

(inclined) (building) (service) chute [*An inclined tube or shaft that conveys and controls the sliding of objects such as mail, laundry, and garbage to a lower level*] Rutsche *f*

inclined elevator (US); inclined lift (Brit.) Schrägaufzug *m*

inclined grate furnace [*See remark under "furnace"*] Schrägrostfeuerung, Schrägrostfeuerstätte *f*

inclined internal grate furnace → **internal inclined grate furnace**

(inclined) laundry chute Wäscherutsche *f* [*Eine senkrechte Wäscherutsche wird „Wäscheschlucker" genannt*]

(inclined) letter chute → **letter chute**

inclined lift Schrägaufzug *m*

inclined-seat (slide) valve, slant(ed) seat (slide) valve Schrägsitzschieber *m*

inclined-seat valve, slant(ed)-seat valve Schrägsitzventil *m*

inclined travel, inclined run [*escalator*] Neigungsfahrt, Schrägfahrt *f*

incombustible [*less correct term*] → **noncombustible**

income value → **capitalized value**

incoming ... → **entering ...**

incoming (air) ... → **inlet (air) ...**

incoming calls barred, ICB ankommende Anrufe verhindert, eingehende Anrufe verhindert

incoming line Zulaufleitung *f*

in commission, in operation in Betrieb

incompressible liquid → **ideal liquid**

to incorporate, to place in storage, to store (ein)speichern, (ein)lagern

incorporated eingeschlossen, hinzugenommen

incorporating, placing in storage, storing, storage, incorporation (Ein)Lagern, (Ein)Speichern *n*, (Ein)Lagerung, (Ein)Speicherung, Lagerhaltung *f*

in correct time order → **chronological**

increase → **rise**

increased safety erhöhte Sicherheit *f*

increased security erhöhte Sicherung *f*

incremental compiler, incremental compiling routine inkrementelle (Sprach)Übersetzer *m* [*Siehe Anmerkung unter „Sprachübersetzer"*]

incremental power gain inkrementelle Leistungsverstärkung *f*

incremental range, fine range Feinbereich *m* [*selten: n*][*Messgerät*]

incrementer Aufwärtszähler, Hochzähler *m*

incrementing timer Vorwärtszeitglied *n*

incrustation, crust formation Verkrustung *f*, Krustenbildung, Ansatz *m*

in-degree Eingang(s)grad *m*

in-demand energy lastgemäße Energie *f*

to indent, to notch (ein)kerben

to indent (aus)zahnen

to indent einrücken

indented connector, notch(ed) connector (Ein-) Kerbverbindung *f*

indented (door) falling latch (aus)gezahnte (Tür)Fallklinke *f*

independent → offline

independent → stand-alone

independent conformity Kennlinienübereinstimmung *f* bei Kleinstwerteinstellung [*DIN IEC 770*]

independent heat pump installation, independent heat pump plant, self-sufficient heat pump installation, self-sufficient heat pump plant unabhängige Wärmepumpenanlage *f* [*Sie beheizt ein Gebäude ohne zusätzliche(n) Wärmeerzeuger*]

independent manual operation, snap(-action) operation unabhängige Handbetätigung [*VDE 0660, Teil 101*], bedienungsunabhängige Schaltung, Sprungschaltung, Schnappschaltung, Kippschaltung *f*

independent mechanical system unverriegelte Mechanik *f* [*Drucktaste*]

independent variable unabhängige Größe *f*

to index [*To produce a machine-orderable set of record surrogates, as indexing a book*] indizieren

to index [*To compute a machine location by indexing as is done by index registers*] indizieren

index [*A list of record surrogates arranged in order of some attribute expressible in machine-orderable form*] Index *m*

index [*The portion of a computer instruction which indicates what index register (if any) is to be used to modify the address of an instruction*] Index *m*

index register, base register, B box, B line, B register, B store, modifier register [*A hardware element which holds a number that can be added to (or, in some cases, subtracted from) the address portion of a computer instruction to form an effective address*] Indexregister, Basisregister *n*

index-sequential access (method), ISA(M) [*Computer programming. A technique for rapidly accessing arbitrary records by means of index markers*] indexsequentielle Zugriff(smethode) *m,(f)*

India-rubber wire (cable) Gummiaderleitung *f*

indicating fuse, indicator fuse [*Also written "fuze"*] Sicherung *f* mit Unterbrechungsanzeige(r)

indicating gap [*surge arrester*] Anzeigefunkenstrecke *f*

indicating instrument → indicator

indicating limit(ing) monitor, limit(ing) indicating monitor anzeigende Grenzwertüberwacher *m*

indicating maximum-demand mechanism anzeigende Maximumwerk *n*, anzeigende Maximummechanismus *m*

(indicating) pointers Zählwerk *n*, Zählmechanismus *m* [*Flügelradwasserzähler*]

indicating target Schauzeichen *n*

indication [*In ultrasonic testing, determination of the presence of a flaw by detection of a reflected ultrasonic beam*] Anrisserkennung *f*

indication, measuring instrument indication, measuring value [*IEC 50301*] Messwert *m*

indication Anzeige *f*

indication error Anzeigefehler *m*

indication-free anzeige(n)frei, ohne Anzeige

indicative data [*Data which describe a specific item*] Beschreibungsdaten *f*

indicator, indicating instrument [*An instrument for obtaining a diagram of the pressure-volume changes in a running positive-displacement engine, compressor, or pump cylinder during the working cycles*] (Diagramm)Anzeiger *m*, (Diagramm)Anzeigeinstrument *n*

indicator and ((open) loop) control panel [*"Board" may synonymously be used for "panel"*] Anzeige- und Steuer(ungs)tafel *f*, Anzeige- und Steuer(ungs)tableau *n*

indicator and operating panel, indicator and operating board Anzeige- und Bedien(ungs)tafel *f*, Anzeige- und Bedien(ungs)tableau *n*

indicator annunciator → annunciator

indicator box Anzeigekasten *m*

indicator diagram [*A pressure-volume diagram representing and measuring the work done by or on a fluid while performing the work cycle in a reciprocating engine, pump, or compressor cylinder*] Tangentialdruckdiagramm, Indikatordiagramm *n*

indicator dial Zeigerscheibe *f*

indicator element [*Electronics. The component within a system that displays the greatest variation in a given parameter*] Anzeigeelement *n*

indicator function → characteristic function

indicator fuse, indicator fuze Anzeigesicherung, Sicherung mit Unterbrechungsanzeige *f*

indicator lamp [*A neon lamp whose on-off condition is used to convey information*] Anzeigelampe *f*

indicator light [*IEC 337-2C*], **illuminated indicator, pilot light** Leuchtmelder *m* [*VDE 0660, Teil 205*], Meldeleuchte *f*

indicator light with voltage reducer [*IEC 337-2C*] Leuchtmelder *m* mit Spannungsreduzierung [*VDE 0660, Teil 205*] Meldeleuchte *f* mit Spannungsreduzierung

indicator panel → annunciator

indicator register Anzeigeregister *n*

I

indicator tube [*An electron-beam tube in which useful information is conveyed by the variation in cross section of the beam at a luminescent target*] Anzeigeröhre *f*

indirect absorption (refrigeration) unit, indirect absorption refrigerating unit, indirect absorption chiller [*Unit in which the refrigerant evaporator is not in contact with the air to be conditioned*] indirekte Absorptionskälteaggregat, mittelbare Absorptionskälteaggregat *n*, indirekte Absorptionskältesatz, mittelbare Absorptionskältesatz *m*

indirect-acting electrical measuring instrument selbstabgleichende elektrische Messgerät, elektrische selbstabgleichende Messgerät *n*

indirect-acting mechanical balance (measuring) instrument Kompensationsmessgerät *n* mit mechanischem Nullabgleich

I

indirect address, multi(ple-)level address [*An address in a computer instruction that indicates a location where address of the referenced operand is to be found*] indirekte Adresse, mittelbare Adresse *f*

indirect addressing [*A programming device whereby the address part of an instruction is not the address of the operand but rather the location in storage where the address of the operand may be found*] indirekte Adressierung, mittelbare Adressierung *f*

indirect air, forced air [*A fan or other device induces a stream of air over the surface to be cooled*] Zwangsluft *f*

indirect building (transfer) station, indirect transfer station indirekte Übergabestation, indirekte Anschlussstation, indirekte Gebäudeübergabestation, indirekte Gebäudeanschlussstation, indirekte Hausübergabestation, indirekte Hausanschlussstation *f* [*„Mittelbare" kann synonym für „indirekte" verwendet werden*]

indirect cooling [*Cooling in which a liquid such as brine or water, cooled by the refrigerant, is circulated to the material or space to be cooled, or is used to cool air so circulated*] indirekte Kühlung, mittelbare Kühlung *f*, indirekte Kühlen, mittelbare Kühlen *n*

indirect domestic water heating [*The water in the boiler heats a coil or annular heater in the storage cylinder, the water in the boiler re-circulates on a closed circuit. This has the advantage of avoiding scale or fur deposits in the boiler and piping*] indirekte Hauswasserheizung *f*

indirect-drive machine, indirect-driving machine [*Electric driving machine, the motor of which is connected indirectly to the drive sheave, drum, or shaft by means of a belt or chain through intermediate gears*] indirekte (An)Triebmaschine, mittelbare (An)Triebmaschine *f*

indirect earthing (Brit.); indirect grounding (US) mittelbare Erdung, indirekte Erdung *f*

indirect functional earthing (Brit.); indirect functional grounding (US) mittelbare Betriebserdung, indirekte Betriebserdung *f* [*VDE 0100, T.100*]

indirect glare indirekte Blendung, Umfeldblendung, mittelbare Blendung *f*

indirect heating indirekte Heizung, mittelbare Heizung *f*, indirekte Heizen, mittelbare Heizen *n* [*Heizen mit Zwischenschalten eines Mediums zwischen Wärmequelle und zu heizendem Medium. DIN 1946 Teil 1*]

indirect (hot water) (storage) cylinder, indirect (hot water) storage tank [*The stored water is heated by contact with a heat exchanger rather than directly heated in the boiler. The arrangement prevents furring by hard water. "Warm water" may synonymously be used for "hot water"*] Warmwasserspeicher *m* mit indirekter Wasser(er)wärmung

indirect hot water system, indirect warm water system indirekte Warmwassersystem, mittelbare Warmwassersystem *n*

indirect lighting (system) → hidden lighting (system)

indirect luminaire Indirektleuchte *f*, Indirektbeleuchtungskörper *m*

indirect magnetic (registration) water meter magnetisch-induktive Wassermesser *m*

indirect material Hilfs(werk)stoff *m*, Hilfsmaterial *n*

indirect overcurrent release, indirect overcurrent trip [*IEC 157-1*] indirekte Überstromauslöser [*VDE 0660, Teil 101*], Sekundärstromauslöser, Wandlerstromauslöser, mittelbare Überstromauslöser *m*

indirect overcurrent releasing, indirect overcurrent tripping indirekte Überstromauslösung, Sekundärstromauslösung, Wandlerstromauslösung, mittelbare Überstromauslösung *f*

indirect-pressure tunnel terminal Buchsenklemme mit Druckstück, Druckstückbuchsenklemme *f*

indirect rectifier Zwischenkreisgleichrichter *m*

indirect release, indirect trip indirekte Auslöser, mittelbare Auslöser, Sekundärauslöser *m*

indirect subventilation indirekte Nebenlüftung, mittelbare Nebenlüftung *f* [*Nicht verwechseln mit „indirekte Sekundärlüftung"*]

indirect transfer station → indirect building (transfer) station

indirect trip → indirect release

indirect warm water system → indirect hot water system

individual branch circuit, individual branch line [*One that supplies only a single piece of electrical equipment*] Einzelabzweig(leitung) *m*, *(f)*, Abzweigstromkreis, Ringleitungsabzweig *m*, Stichleitung *f*

individual chimney → single chimney

individual commissioning Einzelinbetriebnahme *f*

individual control Einzelregulierung, Einzelreg(e)lung *f*

individual disconnection [*The opposite to disconnection on all poles*] Einzelabschaltung, Einzelausschaltung, Einzel(ab)trennung *f*, Einzel(ab)trennen, Einzelabschalten, Einzelausschalten *n*

individual drinking water supply, individual drinkable water supply, individual potable water supply Einzeltrinkwasserversorgung, Einzelbrauchwasserversorgung *f* [*DIN 2001*]

individual enclosure, individual housing, individual casing, individual case Einzelgehäuse *n*

individual-local-remote selection Einzel-Ort-Fernwahl *f*

individual lock (mechanism) Einzelschließmechanismus *m*, Einzelschließung *f*

individually anodized, piecewise anodized stückeloxiert

individually screened cable einzeln (ab)geschirmte Kabel *n*

individual ((open) loop) control Einzelsteuerung, Antriebssteuerung *f* [*Funktionseinheit zum Steuern eines Stellgliedes. V DIN 19 237*]

individual ((open) loop) control level Einzelsteuer(ungs)ebene, Antriebssteuer(ungs)ebene *f* [*Gesamtheit aller Antrieb(s)steuerungen*]

individual reset(ing) Einzelrückstellung *f*

individual room ((closed) loop) control Einzelraumreg(e)lung *f*

individual room controller Einzelraumregler *m*

individual room heater Einzelraumheizer *m*

individual room heating Einzelraumheizung *f*

individual room ((open) loop) control Einzelraumsteuerung *f*

individual security patrol(l)ing Separatbewachung, Separatsicherung *f*

individual service Einzeldienst *m*

individual shower Einzelbrause *f*

individual springs Einzelfederung *f*

individual ventilation system Einzellüftungsanlage *f*

individual vitiated air plant, individual vitiated air installation Einzelschlechtluftanlage, Schlechtlufteinzelanlage *f*

individual water supply Einzelwasserversorgung *f*

individual water supply installation, individual water supply plant Einzelwasserversorgungsanlage *f*

indoor... → room...

indoor and outdoor lighting protection system → outdoor and indoor lighting protection system

indoor climate Innenklima *n*

indoor dew point Innentaupunkt *m*

indoor erection → indoor installation

indoor fire-extinguishing system Gebäudefeuerlöschsystem, Hausfeuerlöschsystem *n*

indoor ground(ing) switch (US); indoor earth(ing) switch (Brit.) [*IEC 129*] Innenerdungsschalter *m* [*VDE 0670, Teil 2*]

indoor hydrant Innenhydrant *m*

indoor installation, indoor mounting, indoor erection Inneneinbau *m*, Inneninstallation, Innenmontage, Innenaufstellung *f* [*nicht verwechseln mit „Montage"*]

indoor installation assembly [*IEC 439-1*] Innen-Schaltgerätekombination, Schaltgerätekombination für Innenaufstellung *f* [*VDE 0660, Teil 500*]

indoor isolating switch, indoor isolator, indoor disconnect(ing) switch, indoor disconnector [*IEC 129*] Innentrenn(schalt)er, Innentrennschutzschalter, Innen-TS *m* [*VDE 0670, Teil 2*]

indoor lighting protection Innenblitzschutz, innere Blitzschutz *m* [*Zur Vermeidung von Gewitterüberspannungsschäden*]

indoor location Innenstandort *m*

indoor monitoring Innenüberwachung *f*

indoor mounting → indoor installation

indoor power circuit-breaker Innenleistungsschalter *m*

indoor shot, indoor photograph Innenaufnahme *f*, Innenphoto(graphie) *n*, *(f)*

indoor (swimming) pool Innen(schwimm)becken *n*

indoor switch [*IEC 265*] Innenraumlastschalter *m* [*VDE 0670, Teil 3*]

indoor switch(ing) station Innenschaltanlage *f*

indoor traffic installation, indoor traffic plant Innenverkehrsanlage *f* [*Personen- und Güterverkehrsanlage in einem Gebäude*]

indoor (utility) substation Innenunterstation *f*

indoor zone Innenzone *f* [*Nicht an eine Fassade oder Fassaden angrenzender Raum oder angrenzende Räume*]

induced draft burner, induced draught burner Saugzugbrenner *m*

I

induced draft fan, induced draught fan Saugzugventilator *m*

induction Induktion *f* [*Mitnehmen von Raumluft (Sekundärluft) durch einen Luftstrahl (Primärluft)*]

induction air conditioning plant, induction air conditioning installation Induktionsklimaanlage, Induktionsklimainstallation *f* [*Bei jedem Luftauslass ist ein Wärme(aus)tauscher für Sekundärluft vorhanden, der je nach Jahreszeit von Warmwasser oder Kaltwasser durchflossen wird. Manchmal werden auch zwei getrennte Wärme(aus)tauscher eingebaut, so dass die Primärluft nur nach dem Außenluftbedarf dimensioniert wird und die Sekundärwärme(aus)tauscher die verbleibende Heiz- und Kühllast abführen*]

induction generator Asynchrongenerator, Induktionsgenerator *m*

induction high-pressure air conditioning plant → high-pressure induction air conditioning plant

induction loop Induktionsschleife *f*

induction luminaire Induktionsleuchte *f*

induction machine Asynchronmaschine, Induktionsmaschine *f*

induction measuring element Induktionsmesswerk, Ferrarismesswerk *n*

induction microphone [*A moving-coil microphone that uses an induction coil to produce an electro-acoustical signal*] Induktionsmikrophon *n*

induction nozzle Induktionsdüse *f*

induction ratio, air diffusion entrainment ratio, entrainment ratio air diffusion [*The ratio of total (primary plus secondary) air to primary air. This definition is not restricted to the secondary air only as the German definition does*] Induktionsverhältnis *n* [*Sekundärluft-Volumenstrom bezogen auf Primärluft-Volumenstrom. Diese Definition bezieht sich also nicht auf den Gesamtluftvolumenstrom (total air) wie die englische. DIN 1946, Teil 1*]

induction regulator, phase transformer [*A transformer with a primary winding in shunt and secondary winding in series with a circuit; it gradually adjusts the voltage or phase relation of the circuit by changing the relative magnetic coupling of the primary and secondary windings*] Drehtransformator *m*

induction relay Induktionsrelais, Ferrarisrelais, Motorrelais *n*

induction unit [*An air terminal device that delivers a small quantity of conditioned (= primary) air through high-velocity jets, to induce a large quantity of room (= secondary) air into the supply airstream. A heating coil may be located in the primary or secondary airstream. Primary air often is predominantely outside air. Units generally discharge directly through a grill(e) into the space due to limited downstream static pressure capability*] Induktions(klima)gerät *n*, Klimakonvektor *m*

induction (voltage) regulator Drehumspanner *m*

inductive ballast induktive Vorschaltgerät *n*

inductive breaking capacity induktive Schaltvermögen *n*, induktive Schaltfähigkeit *f*

inductive circuit [*Electricity. A circuit that combines inductive reactance and resistance*] induktive Stromkreis *m*

inductive identification induktive Kennung *f*

inductive impedance induktive Scheinwiderstand *m*, induktive Impedanz *f*

inductive load, lagging load [*A load that can be represented as an inductor or a combination of an inductor and a resistor*] induktive Last *f*

inductive proximity switch, inductive non-contact switch, inductive contactless switch induktive Näherungsschalter, induktive berührungslose Schalter, berührungslose induktive Schalter *m*

inductive reactance [*The magnitude of the impedance of an inductor, which varies proportionally with frequency and is measured in ohms*] induktive Blindwiderstand *m*, Induktanz, Induktivität *f*

inductive residual voltage [*IEC 147-0C*] induktive Restspannung *f* [*DIN 41 863*]

inductive susceptance induktive Blindleitwert *m*

inductor [*Electricity. A wire coil that is often wound on magnetic material*] Induktor *m*, Induktionsspule, Drosselspule, Drahtspule *f*

industrial air conditioning [*See remark under "air conditioning"*] Industrieklimareg(e)lung, Industrieklimatisierung *f*

industrial air curtain industrielle Luftschleier, Industrieluftschleier, industrielle Luftvorhang, Industrieluftvorhang *m*

industrial blower heater → centrifugal fan unit heater

industrial boiler Industriekessel *m*

industrial contaminant Industrieverseuchungsstoff, Industrieverseuchungsmittel *n*

industrial conveyance equipment Industriefördergerät(e) *n*,(*f*)

industrial cooling Industriekühlung *f*

industrial design Industrieausführung, Industriekonstruktion *f*, Industrieentwurf *m*

industrial electronics Industrieelektronik *f*

industrial elevator (US) → industrial lift (Brit.)

industrial engineering Industrietechnik *f*

industrial espionage Industriespionage *f*

industrial exhaust(ed) air plant, industrial extracted air plant, industrial exhaust(ed) air installation, industrial extracted air installation Industrieabluftanlage *f*

industrial exhaust system industrielle Absaug(ungs)anlage *f*

industrial facility, industrial plant, industrial installation Industrieanlage *f*

industrial floor heating Industrie(fuß)bodenheizung *f*

industrial gas boiler Industriegaskessel *m*

industrial heat Industriewärme *f*

industrial heating Industrieheizung *f*

industrial heating ventilator Industrieheiz(ungs)lüfter *m* [*Siehe Anmerkung unter „Lüfter"*]

industrial heat pump Industriewärmepumpe *f*

industrial incinerator [*Incinerator for the reduction of industrial waste by incineration*] Industriemüllverbrennungsanlage *f*

industrial installation → **industrial facility**

industrial key Industrieschlüssel *m*

industrial lift (Brit.); industrial elevator (US) Industrieaufzug *m*

industrial plant → **industrial facility**

industrial premise Industrieanwesen *n*

industrial radiant heating panel Industrie-Strahl(ungs)plattenheiz(ungs)körper, Industrie-Strahl(ungs)plattenheizer *m*

industrial refuse Industriemüll *m*

industrial security Industriesicherung *f*

industrial security service → **factory protection**

industrial sewage; industrial wastewater (US) Industrieabwasser *n*

industrial ventilation, industrial vent(ilat)ing Industrielüftung *f*

industrial ventilation plant, industrial ventilation installation, industrial vent(ilat)ing plant, industrial vent(ilat)ing installation Industrielüftungsanlage, Industrielüftungsinstallation *f*

industrial ventilation system, industrial vent(ilat)ing system Industrielüftungssystem *n*

industrial waste Industrieabfall(stoff), Industriemüll *m*

industrial waste heat Industrieabwärme *f*

inert gas explosionsfreie Gas, träge Gas, Edelgas *n*

inertia base Trägheitsgrundplatte, Beharrungsgrundplatte *f*

infant mortality Frühausfall *m* [*Der Frühausfall einer Komponente tritt infolge nicht entdeckter Material- und/oder Fertigungsmängel auf*]

inference Schlussfolgerung *f*

(in)fill(ing) concrete (Aus)Füll(ungs)beton *m*

infiltration [*The uncontrolled entry of unconditioned outside air directly into the building, results from natural forces, e.g. wind and buoyancy due to the temperature difference between inside and outside*] Infiltration *f*

infiltration degree day Infiltrationsgradtag *m*

infiltration heat gain [*It is caused by the introduction of warm outdoor air through cracks around doors and windows into the conditioned spaces*] Infiltrationswärmegewinn *m*

infiltration (of air through the crackage of doors and windows) Selbstbelüftung, Fugenbelüftung *f*

infinite, boundless [*Having no limits whatsoever*] unendlich, grenzenlos

infinite impulse response filter, IIR filter unendliche Impulsanwortfilter *m,n*

infinitely stufenlos, gradlos

infinitely variable, infinitely adjustable stufenlos (ein)stellbar, stufenlos verstellbar, gradlos (ein)stellbar, gradlos verstellbar

infinitely variable speed, infinitely adjustable speed Drehzahl 100 % (ein)stellbar, stufenlos verstellbare Drehzahl, stufenlos (ein)stellbare Drehzahl, Drehzahl 100 % verstellbar, gradlos verstellbare Drehzahl, gradlos (ein)stellbare Drehzahl *f*

infinitesimal winzig

infinity plug Dosenstecker *m*

infirmary lift (Brit.) → **hospital lift**

to inflame entflammen

inflame furnace → **direct-fired furnace**

inflatable aufblasbar

inflow Zufluss, Zulauf, Einlauf *m*

inflow line Zuflussleitung, Zulaufleitung, Einlaufleitung *f*

influent filter water [*Water that flows into a filter*] Filtereinlaufwasser, Filterzulaufwasser, Filterzuflusswasser *n*

information, data item information, datum information Information, Datuminformation *f* [*Ein Datum, welches zusätzlich einen Bedeutungsinhalt für den Menschen darstellt. Beispiel: Die Datenfolge 5987432107 wird dann zur Information, wenn man weiß, dass diese Ziffernfolge z. B. eine Fernsprechnummer ist*]

I

information Information *f* [*Im Sinne der Gemeinsprache: Kenntnisse über Sachverhalte und Vorgänge. Nicht verwechseln mit dem Informationsbegriff nach C. Shannon. E DIN 44 300, Teil 1*]

information center (US); information centre [*A facility whose purpose is to store, process, and retrieve information according to user's needs*] Informationszentrum *n*

information(al) content Informationsgehalt *m*

information bit, information binary digit [*Bit that is generated by the data source but is not used by the data-transmission system*] Informationsbit, Informationsbinärzeichen, Informationsdualzeichen *n*, Informationsbinärstelle, Informationsdualstelle, binäre Informationsziffer *f*

information broker Informationsvermittler *m*

information carrier → information medium

information channel [*A facility used to transmit information between data-processing terminals separated by large distances*] Informationskanal *m*

information content, negentropy [*A numerical measure of the information generated in selecting a specific symbol (or message), equal to the negative of the logarithm of the probability of the symbol (or message) selected*] Negentropie *f*, negative Entropie, Informationsgehalt *m*

information (content) unit, negentropy unit [*A unit of information content, equal to a bit, nit, or hartley, according to whether logarithms are taken to base* 2,*e*, *or* 10] Informationseinheit *f*

information cycle Informationskreislauf *m*

information exchange, information interchange [*Telecommunications. The exchange of information between two communication entities*] Informationsaustausch *m*

information flow [*The graphic representation of data collection, data processing, and report distribution throughout an organization*] Informationsfluss *m*

information input Informationseingabe *f*

information larency Informationsdiebstahl *m*

information management [*Telecommunications. A body of definitions, standards, and techniques for information processing and transmissions among communication systems*] Informationsleitung, Informationsführung *f*

information medium, information carrier Informationsträger *m*

information node Informationsknoten *m*

information parameter Informationsparameter *m* [*Diejenige Kenngröße eines Signals, deren Wert oder Werteverlauf aufgrund bekannten oder unterstellten Abmachungen Information darstellt. E DIN 19 226, Teil 5*]

information processing [*The manipulation of data so that new data (implicit in the original) appear in a useful form*] Informationsverarbeitung *f*

information production Informationsbeschaffung *f*

information provider Informationserbringer, Informationsbereitsteller *m*

information question [*Any of actual or anticipated questions which may be posed to an information retrieval system*] Informationsfrage *f*

information request → request for information

information retrieval Informationsauffindung, Informationswiedergewinnung *f*

information securing Informationssichern *n*

information security Informationssicherung *f*

information separator [*A character that separates items or fields of information in a record, especially a variable-length record*] Informationstrennzeichen *n* [*E DIN 44 300, Teil 2*]

information sorting Meldungsverzweigung *f* [*Fernwirktechnik*]

information source, information origin [*A system which produces informations by making successive selections from a group of symbols*] Informationsquelle, Informationsherkunft *f*, Informationsursprung *m*

information supply Informationsversorgung *f*

information system [*Any means for communicating knowledge from one person to another, such as by simple verbal communication, punched-card systems, optical coincidence systems based on coordinate indexing, and completely computerized methods of storing, searching, and retrieving of information*] Informationssystem *n*

information theory [*A branch of communications theory which is devoted to problems in coding, and which provides criteria for comparing different communication systems on the basis of signaling rate, using a numerical measure of the amount of information gained when the content of a message is learned*] Informationstheorie *f*

information theory [*In mathematics, the branch of probability theory concerned with the likelihood of the transmission of messages, accurate to within specified limits, when the bits of information composing the message are subject to possible distortion*] Informationstheroie *f*

information word → data word

infraacoustic(al) telegraphy Unterlagerungstelegraphie, Unterlagerungstelegrafie *f*

infrared Infrarot, Ultrarot *n*

infrared air heater, infrared air heating device, infrared air heating appliance, infrared air heating apparatus Infrarotluft(er)wärmer, Ultrarotluft(er)wärmer, Infrarotlufterhitzer, Ultrarotlufterhitzer, Infrarotheizer, Ultrarotheizer, Infrarotheizungsapparat, Ultrarotheiz(ungs)apparat, Infrarotheizapparat *m*

infrared analyser Infrarot-Analysegerät *n* [*Es benutzt die verschiedene Absoptionsfähigkeit von mehratomischen Gasen für infrarote Strahlen. Jedes Gas absorbiert einen charakteristischen Wellenbereich*]

infrared band elimination filter → infrared stop filter

infrared barrier → infrared (light) barrier

infrared bolometer [*A bolometer adapted to detecting infrared radiation, as opposed to microwave radiation*] Infrarotstrahl(ungs)melder, Infrarotstrahl(ungs)detektor, Ultrarotstrahl(ungs)melder, Ultrarotstrahl(ungs)detektor *m*

infrared CO_2 meter, infrared carbon dioxid(e) meter [*Instrument for determination of carbon dioxid(e) content, based on absorption of infrared radiation*] Infrarot-Kohlendioxidmesser, Infrarot-Kohlensäureanhydridmesser, Ultrarot-Kohlendioxidmesser, Ultrarot-Kohlensäureanhydridmesser *m*

infrared communications receiver, Nancy communications receiver [*A device that intercepts or demodulates infrared radiation that may carry intelligence*] Infrarot-Nachrichtenempfänger, Ultrarot-Nachrichtenempfänger *m*

infrared data transmission Datenübertragung per Infrarot, IrDA, Datenübertragung per Ultrarot, Infrarot-Datenübertragung, Ultrarot-Datenübertragung *f*

infrared detector, infrared detecting device, infrared detection device Infrarotmelder, Ultrarotmelder *m*, Infrarotmeldegerät, Infrarotdetektionsmittel, Ultrarotmeldegerät, Ultrarotdetektionsmittel *n* [*Nicht alle Infrarotmelder sind Flammenmelder, sie können auch auf heiße Raumgase reagieren*]

infrared emitting diode, IRED Infrarotdiode, Ultrarotdiode *f*

infrared (en)coding Infrarotkodierung, Infrarotcodierung, Ultrarotkodierung, Ultrarotcodierung *f*

infrared flame detector, infrared flame detecting device, infrared flame detection device [*See remark under "flame detector"*] Infrarotflammenmelder, Ultrarotflammenmelder *m*, Infrarotflammenmeldegerät, Infrarotflammendetektionsmittel, Ultrarotflammenmeldegerät, Ultrarotflammendetektionsmittel *n*

infrared gas heating device, infrared gas heating appliance Infrarotgasheiz(ungs)gerät *n*, Infrarotgasheiz(ungs)vorrichtung *f*

infrared gas radiator Infrarotstrahler, Ultrarotstrahler, Gasheiz(ungs)strahler; Hochtemperaturstrahler *m* [*Diese Benennung ist den anderen vorzuziehen*]

infrared heating system Infrarotheiz(ungs)system *n*

infrared image converter, infrared image tube, infrared imaging converter, infrared imaging tube, infrared imaging device, infrared image device [*A device that transforms invisible infrared images into visible ones through a system of light-sensitive components*] Infrarot-Bildwandlerröhre *f* [*Siehe Anmerkung unter „Bildwandlerröhre"*]

infrared interface Infrarotschnittstelle, Ultrarotschnittstelle *f*

infrared light Infrarotlicht, Ultrarotlicht *n*

infrared (light) barrier Infrarot(licht)schranke, Ultrarot(licht)schranke, Infrarot(licht)sperre, Ultrarot(licht)sperre *f*

infrared (light) ray Infrarot(licht)strahl, Ultrarot(licht)strahl *m*

infrared motion detector, infrared motion detecting device, infrared motion detection device Infrarotbewegungsmelder, Infrarotbewegungsdetektor, Infrarotbewegunsmeldungsgeber, Ultrarotbewegungsmelder, Ultrarotbewegungsdetektor, Ultrarotbewegunsmeldungsgeber *m*, Infrarotbewegungsmeldegerät, Ultrarotbewegungsmeldegerät *n*

infrared (open) controlled infrarotgsteuert, ultrarotgesteuert

infrared passband filter Infrarotdurchlassfilter, Ultrarotdurchlassfilter *m, n*

infrared perimeter barrier → perimeter infrared barrier

infrared radiant heater Infrarotstrahl(ungsheiz)er, Ultrarotstrahl(ungsheiz)er *m*

infrared radiation, IR [*Electromagnetic radiation lying in the infrared spectrum, with wavelengths between 750 nanometers and 1 millimeter*] Infrarotstrahlung, infrarote Strahlung, Ultrarotstrahlung, ultrarote Strahlung *f*

infrared radiation heating, radiation infrared heating, IR heating Infrarotstrahlungsheizung, Ultrarotstrahlungsheizung *f*

infrared ray → infrared (light) ray

infrared receiver [*A device that intercepts infrared radiation thought to be carrying intelligence*] Infrarotempfänger, Ultrarotempfänger *m*

I

infrared remote ((open) loop) control, remote infrared ((open) loop) control Infrarotfernsteuerung, Ultrarotfernsteuerung *f*

infrared remote operation, remote infrared operation, infrared teleoperation Infrarotfernbedienung, Ferninfrarotbedienung, Ultrarotfernbedienung, Fernultrarotbedienung *f*

infrared scanner [*An infrared detector mounted on a motor-driven platform which causes it to scan a field of view line by line, much as in television*] Infrarotabtaster *m*, Ultrarotabtaster

infrared stop filter, infrared rejection filter, infrared band elimination filter Infrarotsperrfilter, Ultrarotsperrfilter *m,n*

infrared transmitter Infrarotsender, Ultrarotsender *m*

I

infrasonic; subsonic [*deprecated*] [*Pertaining to signals, equipment ,or phenomena involving frequencies below the range of human hearing, hence below about 15 (or sometimes 20) hertz*] unhörbar tief

infrasound [*Vibrations of the air at frequencies too low to be perceived as sound by the human ear, of about 20 hertz or less*] Infraschall *m*

infrasound frequency, infrasonic frequency; subsonic frequency [*deprecated*] [*Frequency lying below the audio-frequency range*] unhörbar tiefe Frequenz, tiefe unhörbare Frequenz, Infraschallfrequenz *f*

ingress → (air) inlet

ingress of dust → dust ingress

ingress protection, IP Schutzart *f* [*Beschreibt mit zwei Kennziffern den Schutz gegen mechanische Beanspruchung und Eindringen von Wasser*]

in-ground cable → burial cable

in-ground installation → burial installation

to inhale, to breathe in einatmen

inherent delay Eigenverzögerung *f*, Eigenverzug *m*

inherent error, inherited error [*An error resulting from incorrect ini-tial values caused by uncertain measurement(s), inadequate precision and/or human mistakes*] Eigenfehler *m*

inherent explosion [*A result from the normal process of a risk on the premises in which explosion occurred, and not brought on by external causes, for example, a dust explosion in a grain mill*] Eigenexplosion *f*

inherently short-circuit-proof kurzschlusssicher, kurzschlussfest [*VDE 0100, Teil 200*]

inherent moment of inertia Eigenträgheitsmoment, Trägheitseigenmoment *n*

inherent resistance Eigenwiderstand *m*

inherent short-circuit capacity Eigenkurzschlussfestigkeit, Eigenkurzschlusssicherheit *f*

inherited error, inherent error [*An error resulting from incorrect initial values caused by uncertain measurement(s), inadequate precision and/or human mistakes*] Eigenfehler *m*

to inhibit [*To hinder, restrain, obstruct, or prohibit from doing something*] sperren, blockieren

inhibited gesperrt, blockiert

inhibited oil Wirkstofföl, legierte Öl *n*

inhibit(ing) input [*A terminal to which an inhibit(ing) signal is applied*] Sperrsignaleingabestelle *f*

inhibit(ing) signal, inhibit(ing) input [*A signal that halts the operation of a circuit in a digital computer*] Sperrsignal *n*

inhouse firmenintern

inhouse monitoring Eigenüberwachung *f*

inhouse outsourcing Abspaltung in einem eigenen Betrieb, firmeninterne Abspaltung *f* [*Die Ausgliederung auf ein Unternehmen, welches von Teilen aus der bisherigen Belegschaft gegründet wurde, um Arbeitsplätze zu sichern*]

initial address Anfangsadresse *f*

initial boiling point, I.B.P. Siedebeginn, Siedeanfang *m*

initial condition [*Computer programming. A specific condition that must be adhered to in order to a computer program(me) or subroutine to execute correctly, such as the location of inputs or outputs, relationship between inputs, and clock signals*] Eingangsbedingung, Eingangsgegebenheit *f*

initial condition → initial state

initial corona pulse Koronazündimpuls, Zündkoronaimpuls *m*

initial current Anfangsstrom *m*

initial efficiency Anfangswirkungsgrad *m*

initial graphics exchange standard, IGES Anfangsnorm *f* für Graphikaustausch [*Sie entstand 1980 in den USA und wurde 1981 in der Version 1 als ANSI-Norm übernommen*]

initialization [*The process of establishing a known state, usually from a power up condition. Initialization may require re-establishment of a hole's logical or physical address*] Initialisierung, Bereitstellung, Vorbereitung, Einleitung *f*

initialization mode, IM Initialisierungsbetrieb, Bereitstellungsbetrieb, Vorbereitungsbetrieb, Einleitungsbetrieb *m*

initial letter Großbuchstabe *m*

initial overpressure [*See remark under "overpressure"*] Anfangsüberdruck *m*

initial position Anfangsstellung, Anfangslage *f*

initial pressure → **entering pressure**

initial reach Anfangshaltung *f* [*DIN 1045. Erste Haltung in Fließrichtung*]

initial sensitivity , initial sensitiveness, initial susceptibility Anfangsempfindlichkeit *f* [*Bei Messgeräten, deren Empfindlichkeit längs der Skala nicht konstant ist, wird zwischen Anfangsempfindlichkeit und Endempfindlichkeit unterschieden*]

initial setting mode → **setting-up mode**

initial short-circuit alternating current, initial alternating short-circuit current Anfangskurzschlusswechselstrom, Kurzschlussanfangswechselstrom *m*

initial state, initial condition Anfangszustand *m*, Anfangsbeschaffenheit *f*

initial value [*The value of a variable prior to any computation or processing*] Ausgangswert, Anfangswert *m*

initial value theorem Anfangswertsatz, Ausgangswertsatz *m*

to initiate → **to trigger**

initiation action → **triggering**

initiation control alarm → **alarm for initiation control**

injection burner Einspritzbrenner, Injektionsbrenner *m*

injection gun Injektionsstrahlerzeuger, Einspritzstrahlerzeuger *m*

injection laser Injektionslaser, Einspritzlaser *m* [*Er verwandelt im Unterschied zum Festkörper- und Gaslaser einen Strom direkt in Laserstrahlen*]

injection steam cooler, steam (injection) cooler Injektionsdampfkühler, Einspritzdampfkühler *m*

injection switching Injektionsschaltung, Einspritzschaltung *f*

injector, inspirator, mixer [*An apparatus in which gas issuing from a jet entrains some or all of the air required for combustion. Alternatively, air issues from the jet and entrains the gas, but in this case the terms "inspirator" and "mixer" are more generally used. After entrainment of air or gas the mix(ture) usually passes to the burner mixing tube where mixing is completed*] Injektor(mischer) *m*, Strahlpumpe *f*

injector Impfstück *n* [*Dosier(ungs)anlage*]

injector [*Device for forcing water into a boiler against the pressure of the boiler by means of a steam jet*] Dampfstrahlpumpe *f*, Injektor *m*

injurious to health → **insanitary**

ink Paste *f* [*Dickschichttechnik für hybrid integrierte Schaltkreise*]

ink Tinte *f*

ink bleed(ing) [*In character recognition, the capillary extension of ink beyond the original edges of a printed or handwritten character*] Tintenabsonderung *f*, Tintenabstoßen, Tintenabsondern *n*

ink cartridge Tintenpatrone *f*

ink(-filled) pen Tintengriffel *m*

ink(-filled) pen recording Tintengriffelaufzeichnung *f*

(ink-)jet, bubble jet Tintenstrahl *m*

(ink-)jet matrix printer, matrix ink(-jet) printer, bubble jet matrix printer Matrixtintenstrahldrucker, Tintenstrahlmatrixdrucker *m*

(ink-)jet method, bubble jet method Tintenstrahl(schreib)verfahren, Schreibverfahren mit Tintenstrahl *n*, Tintenstrahl(schreib)methode, Schreibmethode mit Tintenstrahl *f*

(ink-)jet paper, bubble jet paper Tintenstrahlpapier *n*

(ink-)jet printer, bubble jet printer, ink printer [*"Writer" may synonymously be used for "printer"*] Tinten(spritz)drucker, Spritztintendrucker, Tintenstrahldrucker, Strahltintendrucker *m*

ink reading head Tintenlesekopf *m*

ink recorder, ink recording instrument [*A recorder that employs an ink-filled pen or capillary tube to produce the graphic record*] Tintenschreiber *m*

ink smudge [*In character recognition, the overflow of ink beyond the original edges of a printed or handwritten character*] Tintenverschmierung *f*

(ink-)spray plotter, bubble sprayer Spritztintenstrichzeichner, Tintenstrichspritzzeichner *m*

ink squeezeout [*In character recognition, the overflow of ink from the stroke centerline to the edges of a printed or handwritten character*] Tintenverlauf *m*

ink-vapor recording (US); ink-vapour recording (Brit.); ink-mist recording [*A type of recording in which vaporized ink particles are directly deposited on the record sheet*] Tintennebelaufzeichnung *f*

inlet, intake Eintritt, Einlass, Eingang *m*

inlet air, ingoing air, incoming air, intake air, supply air, entering air [*Air entering a space from an air conditioning, heating, or ventilating apparatus*] Zuluft *f*

inlet air ceiling, ingoing air ceiling, incoming air ceiling, intake air ceiling, entering air ceiling, supply air ceiling Zuluftdecke *f*

inlet air collar Zuluftring *m* [*Ventilator*]

I

inlet air disk valve, ingoing air disk valve, intake air disk valve, entering air disk valve, supply air disk valve [*Also spelled "disc"*] Zulufttellerventil, Zuluftscheibenventil, Tellerzuluftventil, Scheibenzuluftventil *n*

inlet air duct, ingoing air duct, incoming air duct, intake air duct, supply air duct, entering air duct Zuluftkanal *m*

inlet air-exit air unit, ingoing air-exit air unit, incoming air-exit air unit, intake air-exit air unit, entering air-exit air unit, supply air-exit air unit Zuluft-Abluft-Gerät, ZU-AB-GR *n* [*DIN 1946, Teil 1*]

inlet air fan, incoming air fan, ingoing air fan, intake air fan, supply air fan, entering air fan Zuluftventilator *m*

inlet air grill(e), ingoing air grill(e), incoming air grill(e), intake air grill(e), entering air grill(e), supply air grill(e) Zuluftgitter *n*

I

inlet air hole, ingoing air hole, incoming air hole, intake air hole, entering air hole, supply air hole Zuluftloch *n*

inlet air jet, ingoing air jet, intake air jet, entering air jet, supply air jet Zuluftstrahl *m*

inlet air luminaire, ingoing air luminaire, intake air luminaire, entering air luminaire, supply air luminaire [*"Light(ing) fitting" and "light(ing) fixture" may synonymously be used for "luminaire"*] Zuluftleuchte *f*, Zuluftbeleuchtungskörper *m*

inlet air pressure, supply air pressure Zuluftdruck *m*

inlet air sensor, ingoing air sensor, incoming air sensor, intake air sensor, entering air sensor, supply air sensor Zuluftfühler *m*

inlet air shaft, ingoing air shaft, incoming air shaft, intake air shaft, entering air shaft, supply air shaft Zuluftschacht *m*

inlet air temperature, ingoing air temperature, incoming air temperature, intake air temperature, supply air temperature Zulufttemperatur *f*

inlet air unit, ingoing air unit, incoming air unit, intake air unit, entering air unit, supply air unit Zuluftgerät, ZU-GR *n* [*DIN 1946, Teil 1*]

inlet branch Eintrittsstutzen, Einlassstutzen *m*

inlet core area → (air) inlet core area

inlet duct Eintrittskanal, Einlasskanal *m*

inlet end Eintrittsende, Einlassende *n*

inlet nozzle Eintrittsdüse, Einlassdüse *f*

inlet pipe Eintrittsrohr, Einlassrohr *n*

inlet pressure [*Absolute total pressure at the inlet flange of the compressor*] Eintrittsdruck, Einlassdruck *m*

inlet spin [*See remark under "spin"*] Eintrittsdrall, Einlassdrall *m*

inlet system → (air) inlet system

inlet temperature [*Total temperature at the inlet flange of the compressor*] Eintrittstemperatur, Einlass-Temperatur *f*

inlet valve Eintrittsventil, Einlassventil *n*

inlet ventilation Belüftung *f*

in line im Netz, in der Leitung

in line nebeneinander

in-line freezing kontinuierliche Gefrieren, stetige Gefrieren *n*

in-line furnace [*Furnace in which the air inlet and discharge openings are in the top of the casing*] Feuerung *f* mit obenliegenden Luftöffnungen

in-line master switch Zwischenhauptschalter, Hauptzwischenschalter *m*

in-line pump, pipeline-installed pump Rohr(leitung(s))einbaupumpe, Streckenpumpe *f*

inner armature-current ((closed) loop) control unterlagerte Ankerstromreg(e)lung *f*

inner hearth [*fire place*] Kaminherd, Kaminboden *m*, Herdplatte *f*

inner jacket Innenmantel *m*

inner lead Elektrode *f* [*Glühlampe*]

innovation Neuerung *f*

inode Informationsknoten *m*

in operation → in commission

inoperative funktionsunfähig, funktionsuntüchtig

in-order back end, retirement Befehl(s)überprüfung *f* [*Überprüfung welche Befehle vollständig ausgeführt sind*]

in-out ((closed) loop) control → bang bang ((closed) loop) control

in-phase current, restrictive current [*The current component that is parallel to the voltage phase*] Wirkstrom *m*

in-phase rejection → common-mode rejection

in-phase signal → common-mode signal

in-phase voltage → common-mode voltage

in-phase voltage ((closed) loop) control → common-mode voltage ((closed) loop) control

in-phase voltage range → common-mode voltage range

in-process tester, in-process inspector Fertigungsprüfer, Herstellungsprüfer *m*

to input, to enter, to put in [*1. To convey or transfer energy or data to a device, process or system. 2. To introduce data from auxiliary storage or an input device into main memory*] eingeben

input → input power

input [*The point at which energy enters a device*] Eingang *m*

input [*Something that is entered*] Eingabe *f*, Eingegebene *n*

input → input (data)

input Verbrauch *m* [*Verflechtungsbilanzierung*]

input [*The terminals of an electrical or electronic device to which the power or signal is applied*] Klemmen *f*

input [*The process of delivering data to a data-processing machine*] Eingabe, Dateneingabe *f*, Eingeben, Dateneingeben *n*, Einsatz, Dateneinsatz *m*

input amplifier Eingangsverstärker *m*

input area, input block, input storage, input section [*A section of internal storage reserved for storage of data or instructions received from an input unit such as card or tape*] Eingabe(speicher)bereich *m* [*selten: n*]

input by hand → hand input

input card Eingabekarte *f*

input (cavity) resonator → buncher (resonator)

input configuration, input circuit Eingangsschaltung *f*

input connector Eingangsstecker *m*

input current, primary current Eingangsstrom, Primärstrom, aufgenommene Strom *m*

input (data) [*In automatic data processing, the data that is delivered to a data-processing device from the external world*] Eingabe(daten), Primärdaten, Ausgangsdaten *f*

input (data) block, entry block [*A block of data read or transferred into a computer*] Eingabe(daten)-block *m* [*DIN 66 201, Teil 1*]

input (data) item, input datum, entry (data) item, entry item [*See remark under "data item"*] Eingabedatum *n*

input device → input unit

input element → input unit

input (equipment) [*The equipment used for transferring data and instructing into an automatic data-processing system*] (Daten)Eingabegerät(e) *n,(f)* [*E DIN 44 300/10.85*]

input field Eingabefeld *n*

input file Eingabedatei *f* [*Siehe Anmerkung unter „Datei"*]

input format Eingabe(datum)format *n*

input frequency, primary frequency Eingangsfrequenz, Primärfrequenz *f*

input gap [*An interaction gap used to initiate a variation in an electron stream; in a velocity-modulated tube it is in the buncher (resonator)*] Eingangsspalt *m*

input impedance [*The impedance across the input terminals of a four-terminal network when the output terminals are short-circuited*] Eingangsimpedanz *f*, elektrische Eingangsscheinwiderstand *m*

input instruction Eingabebefehl *m*

input (kVA) → input power

input level Eingangspegel *m* [*Logarithmiertes Verhältnis des Spannungswertes oder Leistungswertes am Eingang eines Eintors (=Zweipol) oder Zweitors (= Vierpol) zu einem anderen Spannungswert bzw. Leistungswert*]

input-limited [*Pertaining to a system or operation whose speed or efficiency depends mainly on the speed of input into the machine rather than the speed of the machine itself*] eingabebegrenzt

input line Eingangsleitung *f*

(input) magazine [*A part of a card-handling device which supplies the cards to the processing portion of the machine*] (Eingabe)Magazin *n*

input matrix Eingangsmatrix *f*

input mode Eingabemodus *m*

input network Eingangsnetz(werk) *n*

input noise Eingangsrauschen *n*

input noise voltage Eingangsrauschspannung *f*

input offset current, offset input current Eingangsfehlstrom, Fehleingangsstrom *m*

input offset current drift, offset input current drift Temperaturkoeffizient *m* des Eingangsfehlstromes [*Operationsverstärker*]

input offset voltage, offset input voltage Eingangsfehlspannung, Fehleingangsspannung *f* [*Operationsverstärker. Angelegte Spannung um den Ausgang auf Null zu bringen*]

input offset voltage drift, offset input voltage drift Temperaturkoeffizient *m* der Eingangsfehlspannung [*Operationsverstärker*]

input/output, I/O [*The transfer of data between a computer and a peripheral device*] Eingabe/Ausgabe *f*, E/A

input/output adapter → channel adapter

input/output buffer, I/O buffer [*An area of computer memory used to temporarily store data and instructions transferred into and out of a computer, permitting several such transfers to take place simultaneously with processing of data*] Eingabe/Ausgabe(zwischen)speicher, Eingabe/Ausgabepuffer *m*

input/output bound, I/O bound, input/output limited, I/O limited [*Pertaining to a system or condition in which the time for input and output operations exceeds other operations*] eingabe/ausgabegebunden, eingabe/ausgabe-begrenzt

input/output bus, I/O bus Eingabe/Ausgabe-Bus *m*

input/output channel, I/O channel [*The physical link connecting the computer to an input or to an output device*] Eingabe/Ausgabe-Kanal *m*

input/output controller, I/O controller [*An independent processor which provides the data paths between input and output devices and main memory*] Eingabe/Ausgabe-Regler *m*

I

input/output control system, IOCS [*A set of flexible routines that supervise the input and output operations of a computer at the detailed machine-language level*] Eingabe-Ausgabe-Überwachungssystem *n*

input/output control unit, I/O control unit [*The piece of hardware which controls the operation of one or more of a type of devices such as tape drives or disk drives; this unit is frequently an integral part of the input/output device itself*] Eingabe/Ausgabe-Steuer(ungs)werk *n*, Eingabe/Ausgabe-Steuer(ungs)einheit, Eingabe/Ausgabe-Leitwerk, Eingabe/Ausgabe-Leiteinheit *f*

input/output device, input/output appliance, I/O device, I/O appliance [*A unit that accepts new data, sends it into the computer for processing, receives the results, and translates them into a usable medium*] Eingabe/Ausgabe-Gerät *n*, Eingabe/Ausgabe-Vorrichtung *f*

input/output interface, I/O interface Eingabe/-Ausgabe-Schnittstelle *f*

input/output interrupt, I/O interrupt [*A technique by which the central processor needs only initiate an input/output operation and then handle other matters, while other units within the system carry out the rest of the operation*] Eingabe/Ausgabe-Unterbrechung *f*

input/output interrupt identification, I/O interrupt identification [*The ascertainment of the device and channel taking part in the transfer of information into or out of a computer that causes a particular input/output interrupt, and of the status of the device and channel*] Eingabe/Ausgabe-Unterbrechungs-Bestimmung *f*

input/output interrupt indicator, I/O interrupt indicator [*A device which registers an input/output interrupt associated with a particular input/output channel; it can be used in input/output interrupt identification*] Eingabe/Ausgabe-Unterbrechungsanzeiger *m*

input/output limited → input/output bound

input/output list, I/O list [*A list of variables specifying the storage locations: 1. From which data is to be read. 2. Into which data is to be written*] Eingabe/Ausgabe-Liste *f*

input/output medium, unit record Datenträger, Informationsträger *m*

input/output module, I/O module, input/output block, I/O block Eingabe/Ausgabe-Modul *n*, Eingabe/Ausgabe-Baustein *m*

input/output periphery, I/O periphery Ein/Ausgabe-Peripherie *f*

input/output register, I/O register [*Computer register that provides the transfer of information from inputs to the central computer, or from it to output equipment*] Eingabe/Ausgaberegister *n*

input/output regulator → input/output controller

input/output request, IORQ Eingabe/Ausgabe-Forderung *f* [*Sie zeigt an, dass die niederwertigen Bits des Adressbusses die Adresse einer Ein/Ausgabeeinheit und nicht die ganze eines Speichers darstellen*]

input/output switching, I/O switching [*A technique in which a number of channels can connect input and output devices to a central processing unit; each device may be assigned to any available channel, so that several different channels may receive a particular device during the execution of a program*] Eingabe-Ausgabeschaltung *f*

to input parameters → to parameterize

input power, (power) input, input (kVA), absorbed real power, real power absorbed [*The power fed into an electrical or electronic device*] Eingangsleistung, Primärleistung, Aufnahme(leistung), aufgenommene (Wirk)Leistung, Leistungsaufnahme *f*

input processor Eingabeprozessor *m*

input request Eingabeaufforderung *f*, Eingabewunsch *m*

input resonator → buncher (resonator)

input section [*The part of a program(me) which controls the reading of data into a computer memory from external devices*] Eingabeteil *m,n*

input section → input area

input signal [*The signal fed into an electrical or electronic device*] Eingangssignal, Eingabesignal *n*

input signal conditioning Eingangssignalanpassung, Eingabesignalanpassung *f*

input (state) vector Eingangsvektor *m* [*Sämtliche Eingangsgrößen eines betrachteten Systems bilden den Eingangsvektor. E DIN 19 226, Teil 1*]

input station [*A terminal is an in-plant communications system at which data can be entered into the system directly as events take place, enabling files to be immediately updated*] Eingabestation, Eingangsstation *f*

input storage area → **input area**

input time interval Eingabezeitspanne *f*

input unit, input element, input device Eingabeeinheit *f*, Eingabegerät *n*

input variable [*See remark under "variable"*] Eingangsgröße *f*

input vector → **input (state) vector**

input voltage Eingangsspannung *f* [*Spannung am Eingang eines Eintors (= Zweipol) oder Zweitors (= Vierpol)*]

inquiry → **enquiry**

in reader Innenleser *m*

inrush current Anzugsstrom *m* [*Kurzzeitiger Strom einer Magnetspule beim Einschalten*]

inrush current → **(lamp) inrush current**

inrush load [*VA required to start a device, e.g. a transformer*] Einschaltlast *f*

insanitary, injurious to health, detrimental to health [*Contrary to sanitary principles*] gesundheitsschädlich, ungesund

inscription Inschrift *f*

to insert einfügen, einsetzen

to insert einblenden [*z. B. Teilbilder*]

insert Einsatz *m*, Einfügung *f*

inserted light(ing) fitting, inserted luminaire [*See remark under "light(ing) fitting"*] Einbauleuchte *f*, Einbaubeleuchtungskörper *m*

insert gas burner → **gas insert burner**

insertion [*Ratio of the power delivered to the part of a transmission system following a transducer to the power delivered to the same part of the system before the insertion of the transducer*] Einfügungsverstärkung *f*

insertion burner Einschubbrenner *m*

insertion label Einlegeschild *n* [*Mit Bildzeichen oder Beschriftung*]

insertion reader Einsteckleser *m*

insertion thermostat [*It is mounted on a duct with the control element in the duct*] Einsteckthermostat, Einstecktemperaturregler *m*

insert point Einschleifpunkt *m* [*Dynamikprozessor*]

inset box → **(wall) inset box**

inside blind, indoor blind Innenblende *f*

inside diameter → **ID**

inside dimension → **ID**

in-situ equipment system Einzelgerätesystem *n* [*Die Förderung oder teilweise Förderung behandelter oder unbehandelter Luft für den gesamten Versorgungsbereich mit Geräten an den Versorgungsstellen*]

in-situ polyurethane foam → **polyurethane in-situ foam**

insolation Sonneneinstrahlung, Besonnung *f*

inspection, examination Abnahmekontrolle *f*

inspection, examination Inspektion *f* [*Erfassung und Beurteilung des Istzustandes*]

inspection and repair and maintenance Instandhaltung *f* [*Den Zustand des Gegenstandes über der Schadensgrenze halten. Es ist allerdings nicht nötig und wirtschaftlich nicht tragbar, den Istzustand auf voller Höhe des Sollzustandes zu halten*]

inspection cock, sampling cock Probe(n)(ent)nahmehahn *m*

inspection date Inspektionstermin *m*

inspection directorate Zentralprüfamt *n*

inspection door → **cleanout door**

inspection hole, inspection opening, eye Schauloch, Inspektionsloch *n*, Schauöffnung, Inspektionsöffnung *f*

inspection gully → **cleaning gully**

inspection manhole, inspection chamber Revisionsschacht, Inspektionsschacht *m*, Revisionsmannloch, Inspektionsmannloch *n*

inspection plate [*Handhole cover which, when properly removed, makes possible the inspection of the internal parts of a machine or the like*] Schaudeckel *m*

inspection socket outlet, inspection receptacle outlet (US), inspection (convenience) outlet Revisionssteckdose *f*

inspirator → **injector**

inspirator of burner, burner inspirator Brennermischer *m*

instability [*The controlled variable oscillates between two values and also not reaches a steady state condition*] Labilität *f*

to install, to mount, to erect installieren, montieren, aufbauen, einbauen

I

installation → building services

installation → (pipe) installation

installation, erection, mounting, evecting, installing Aufbau *m*, Montage, Installation *f*, Installieren, Montieren, Einbauen *n*

installation, plant, physical plant Anlage *f*, Betrieb *m* [*Eine Gesamtheit der Einrichtungen technischer und organisatorischer Art zur Vorbereitung, Durchführung und Ergebnisprüfung von technischen Prozessen, die in einem unmittelbaren Zusammenhang stehen. DIN 66 201, Teil 1*]

installation bus Installationsbus *m*

installed capacity, nameplate capacity installierte Leistung, Namensschildleistung *f*

installed load → connected load

installer, fitter Einbauer, Monteur *m*

I

installing → installation

instant Zeitpunkt *m*

instantaneous contactor relay [*IEC 337-1*], **instantaneous control relay** [*deprecated*] unverzögerte Hilfsschütz(e) *m,(f)*

instantaneous demand, instantaneous requirement, instantaneous need Sofortbedarf, Stoßbedarf *m*, Sofortnachfrage, Stoßnachfrage *f*

instantaneous gas water heater → flow(-type) gas water heater

instantaneous value → actual value

instant(aneous) (water) heater → flow(-type) (water) heater

instant of failure → failure instant

instant-start fluorescent lamp → slim-line lamp

in steps → step by step

institutional business Großverbrauchergeschäft *n*

instruction Anweisung *f*

instruction [*In automatic data processing, a pattern of digits which signifies to a computer that a particular operation is to be performed and which may also indicate the operands (or the locations of operands) to be operated on*] Instruktion *f*, (programmierbare) Befehl *m* [*„Befehl" wird meistens bei maschinenorientierten Sprachen verwendet. Bei anderen Programmiersprachen heißt Befehl „elementare Anweisung" f*]

instruction address [*In automatic data processing, the address of the storage location in which a given instruction is stored*] Befehlsadresse *f*

instruction address register [*A special storage location, forming part of the program(me) controller, in which addresses of instructions are stored in order to control their sequential retrieval from memory during the execution of a program(me)*] Befehlsadressenregister *n*

instruction area [*In automatic data processing, a section of storage used for storing program(me) instructions*] Befehlsabschnitt *m*

instruction block Befehlsblock *m*

instruction code [*In automatic data processing, that part of an instruction which distinguishes it from all other instructions and specifies the action to be performed*] Befehlskode, Befehlscode, Befehlsschlüssel *m*

instruction command Textbefehl [*Fernwirktechnik*], Standardtextbefehl *m*

instruction complement Befehlsvorrat *m*

instruction constant [*A dummy instruction of the type K = 1, where K is irrelevant to the program(me)*] Befehlskonstante *f*

instruction counter, control counter, instruction location counter, control location counter, programme counter (Brit.); program counter (US) Befehlszähler *m*, Befehlsregister *n* [*E DIN 44 300/10.85*]

instruction cycle [*Computer programming. The time required to fetch an instruction from storage, and decode and execute it*] Befehlszyklus *m*

instruction deck [*Set of cards punched to contain a symbolic coded program(me) to be read into a computer*] (Loch)Befehlskartensatz *m*

instruction decoder Befehlsentschlüss(e)ler *m*

instruction execution Befehlsausführung *f* [*Datenverarbeitung*]

instruction execution check(ing) Befehlsausführungskontrolle *f* [*Datenverarbeitung*]

instruction fetch Befehlsabruf *m*

instruction file Befehlsdatei *f*, Befehlsdatenbestand *m*

instruction flow chart Befehls(ablauf)diagramm *n*

instruction format Befehlsformat *n*, Befehlsaufbau *m* [*Ein Befehl in einer Maschinensprache besteht aus einem Operationsteil und einem Adressteil. Der Operationsteil gibt an, was zu tun ist, der Adressteil wo sich die betreffenden Operanden befinden. Das genaue Format richtet sich nach der Art der zur Verfügung stehenden Rechenanlage*]

instruction frame „Tastenzeichen" *n* [*Baudzahl*]

instruction initiation Befehlsgabe *f* [*Sie wird grundsätzlich unterschieden in Dauer(befehls)- und Impulskontaktgabe*]

instruction input Befehlseingabe *f*

instruction length [*The number of bits that define or represent a single instruction (sometimes called a word) in the main memory of a computer. The actual length of a valid instruction varies with the type of computer, and a given computer may have multiple instruction lengths*] Befehlslänge *f*

instruction line Befehlsleitung *f*

instruction list Befehlsliste *f* [*Die Anweisungsliste einer speicherprogrammierten Steuerung ist die zeilenweise Auflistung aller Steuer(ungs)anweisungen mit entsprechenden, den Steuer(ungs)ablauf erläuternden Klartextkommentaren*]

instruction modification [*In automatic data processing, a change, carried out by the program(me), in an instruction so that, upon being repeated, this instruction will perform a different operation*] Befehlsänderung *f*

instruction module Befehlsmodul *n*, Befehlsbaustein *m*

instruction output Befehls(aus)gabe *f*

instruction processing unit Befehlsverarbeitungseinheit *f*

instruction register; programme register (Brit.); program register (US) Befehlsregister, BR *n* [*Es enthält im Laufe der Abarbeitung eines Programmes stets den augenblicklich bearbeiteten Befehl, der aus dem Speicher in das Befehlsregister gebracht wurde. Hier wird der Befehl entschlüsselt*]

instruction set, instruction repertory Befehlssatz, Befehlsvorrat, Befehlskatalog *m*, Befehlsspektrum *n* [*DIN 44 300. Der Satz von Befehlen in Maschinensprache, der von einem Rechner ausgeführt werden kann. Wichtige Befehlsarten sind: 1) Arithmetische und logische Befehle, 2)Befehle zur Steuerung des Programmablaufs, 3) Befehle zum Transfer von Daten*]

instruction stream Befehlsstrom *m*

instruction text Befehlstext *m*

instruction time [*In automatic data processing, the time required to carry out an instruction having a specified number of addresses in a particular computer*] Befehlszeit *f*

instruction transfer [*In automatic data processing, an instruction which transfers control to one or another subprogram(me), depending upon the value of some operation*] Übergabebefehl, Übermitt(e)-lungsbefehl, Weitergabebefehl *m*

instruction type Befehlsart *f*

instruction word [*A computer word or fundamental unit of data storage, containing an instruction*] Befehlswort *n*

to instrument instrumentieren

instrumentation [*Designing, manufacturing, and utilizing physical instruments or instrument systems for detection, observation, measurement, automatic computation, communication, or data processing*] Mess- und Steuer- und Regelungstechnik, MSR-Technik *f*

instrumentation Instrumentierung, Instrumentenbestückung, Instrumentation *f*

instrumentation and control technique → **control and instrumentation technique**

instrument autotransformer Sparschalt(ungs)-messwandler, Messwandler in Sparschaltung, Messsparwandler *m*

instrument casing → **instrument housing**

instrument housing, instrument casing, instrument case, instrument shelter Instrumentengehäuse *n*

instrument panel [*Any panel on which various instruments and controls are mounted*] Instrumententafel *f*

instrument resistor [*For certain measuring instruments such as the voltmeter, a multiplier resistance used to increase the range of the instrument*] Instrument(en)widerstand *m*

instrument shelter → **instrument housing**

instrument transformer Messwandler, Messtransformator *m* [*Ein Wandler mit erhöhter Genauigkeit. E DIN 19 226, Teil 5*]

insufficient definition → **lack of focus**

insufficient focus → **lack of focus**

insulance → **insulation resistance**

to insulate isolieren

to insulate dämmen

insulated against structure-borne sound körperschallgedämmt

insulated base, insulating base, insulation base Isolier(stoff)sockel *m*

insulated divider Schienenstütze *f*

insulated gate-FET → **IG-FET**

insulated heater wire isolierte Heizleiter *m* aus Widerstandsdraht [*Elektro(fuß)bodenheizung*]

insulating Dämmen *n* [*Schutz gegen Wärme- und Schalldurchgang*]

insulating Isolieren *n*

insulating ... → **(electrical) insulating ...**

insulating vacuum → **evacuated (thermal) insulation**

insulation Isolierung, Isolation *f*

I

insulation Dämmung *f* [*Schutz gegen Wärme- und Schallldurchgang*]

insulation → (electrical) insulation

insulation adhesive Dämmkleber, Dämmkleb(e)-stoff *m* [*Schutz gegen Wärme- und Schalldurchgang*]

insulation from earth, insulation from ground, insulation against earth, insulation against ground Isolation gegen Erde, Erde-Isolation, Isolierung gegen Erde, Erde-Isolierung *f*

insulation of heat → heat insulation

insulation rating [*Electricity. A measurement that rates insulating materials according to factors such as dielectric strength, tensile strength, and leakage resistance*] Isolationsbemessungsdaten *f*

insulation resistance, insulance [*The resistance to the flow of current through an insulating material resulting form an impressed direct voltage; usually expressed in ohms*] Isolationswiderstand *m*

I

insulator → (electrical) insulator

in-system systemintern

intake → inlet

integer, whole number [*It may be positive, negative or zero. The terms exclude fractions, square roots, and imaginary numbers. An element of the set {..., -2, -1, 0, 1, 2, ...}*] ganze Zahl, dezimalstellenlose Zahl, Ganzzahl, Festkommazahl, natürliche Zahl *f*

integral action coefficient [*IEC 546*] Integrierbeiwert, I-Beiwert *m*

integral action factor [*ANSI C 85.1*] Integrierfaktor, I-Faktor *m*

integral action limiter Integrieranteilbegrenzer, I-Anteilbegrenzer *m*

integral action time [*ANSI C 85.1*], **integral action time constant** [*IEC 50(351)*] Integrierzeit, I-Zeit *f* [*Für I-Regler wird häufig unter Bezug der Eingangs- und Ausgangsgrößen auf ihre Bereiche (Normierung) der reziproke Wert von K_E als Integrierzeit T_E angegeben. Diese ist kein Kennwert des I-Gliedes, da sie von der Normierungsfestlegung abhängig ist. E DIN 19 226, Teil 5*]

integral compensation [*Control systems. Compensation in which the output is directly proportional to the input*] Integrierausgleich, I-Ausgleich *m*, Integrierkompensation, I-Kompensation *f*

integral component Integrieranteil, I-Anteil *m*

integral controller Integrierregler, I-Regler, integral wirkende Regler *m* [*Er kann im Gegensatz zum P-Regler den Regelunterschied voll ausregeln; er arbeitet aber ziemlich träge und neigt zum Pendeln*]

integral control (mode) [*It contributes to the output of the controller an amount equal to the integral of the error signal over time, multiplied by the integral gain*] Integrierreg(e)lung(sart), I-Regelung(sart), I-Reglung(sart) *f*

integral gain Integrierverstärkung, I-Verstärkung *f* [*Reg(e)lung*]

integrally-fused circuit breaker [*IEC 157-1*] Leistungsschalter *m* mit integralen Sicherungen

integral modem, integral modulator/demodulator [*A modem built into a system or device, enabling it to communicate over a telephone line*] Integriermodem, Integriermodulator/Demodulator *m*

integral of absolute value of error, IAE Betragsregel(ungs)fläche *f*

integral of squared error, ISE quadratische Regel(ungs)fläche *f*

integral of squared time multiplied by squared error, ISTSE zeitquadratisch gewichtete quadratische Regel(ungs)fläche *f*

integral of time multiplied by absolute value of error, ITAE zeitgewichtete Betragsregel(ungs)fläche *f*

integral of time multiplied by squared error, ITSE zeitlinear gewichtete quadratische Regel(ungs)fläche *f*

integral of time-weighted absolute error Integral *n* des zeitlichen absoluten Fehlers

integral switching device Einbauschaltgerät *n*, Schaltersteckdose *f*

to integrate integrieren, zusammenschließen

integrated integriert, ganzheitlich, vollheitlich, zusammengeschlossen

integrated broadband mobile system, IBMS integrierte breitbandige Mobilkommunikationssystem *n*

integrated circuit, IC [*Microelectronic circuits embedded on a small piece of silicia. Often referred to as a chip*] Festkörperschaltung, (monolithisch) integrierte Schaltung, integrale Schaltung, Verbundschaltung, ganzheitliche Schaltung, vollheitliche Schaltung, IS *f* [*„Schaltkreis" kann synonym für „Schaltung" verwendet werden*]

integrated-circuit memory, IC memory integrierte Speicherschaltung *f*

integrated-circuit regulator, integrated-circuit controller integrierte Reglerschaltung *f*

integrated communications cabling system, ICCS integrierte Kommunikationsverkabelungssystem *n* [*Ein Konzept von Siemens zur leistungsfähigen Verkabelung von Systemen und Geräten der Steuer(ungs)-, Informations- und Kommunikationstechnik unterschiedlicher Hersteller für Gebäude und Standorte beliebiger Größe. Es sind weniger Kabel erforderlich*]

integrated-demand memory Zählspeicher *m*

integrated digital circuit → digital integrated circuit

integrated digital network → IDN

integrated drive electronics → IDE

integrated façade integrierte Fassade *f* [*Gebäudeaußenwand mit wassergefüllten Stahlstützen*]

integrated interface circuit → interface integrated circuit

integrated recording head, IRH integrierte Aufnahmekopf *m* [*DDC*]

integrated room control, IRC Einzelraumreglungssystem *n*

integrated room controller, integrated room regulator Einzelraumregler *m*

integrated service(s) digital network, ISDN Netz(werk) zusammengeführter digitaler Dienste, diensteintegrierende Fernmeldenetz(werk), digitale Universalnetz(werk), dienstintegrierende Digitalnetzwerk, digitale Fernmeldenetz(werk) *n* [*Mehrere Dienste sind unter einer Nummer erreichbar (Telefon, Telefax, Modem, Datex/J [BtX], Dateitransfer und Remote Computing). Das entsprechende Endgerät wird durch die Diensterkennung im ISDN-Protokoll angesprochen. Die Übertragung erfolgt auf digitalem Wege. Die Grundlage dieses Netzwerkes ist das vorhandene Telefonnetz*]

integrated system vernetzte System, integrierte System *n*

integrating analog(ue)(-to)-digital converter integrierende Analog-Digital-Wandler *m* [*Er bildet über das Messintervall den Mittelwert der Eingangsspannung*]

integrating element → I-element

integrating of trades, trade integrating gewerkeübergreifend

integration amplifier Integrationsverstärker, ladungsempfindliche Verstärker *m*

integrator circuit Integrationsschaltung *f*

integrity Integrität, Unversehrtheit *f*

intelligent building hochtechnisierte Gebäude *n* [*Gebäude mit flexibler, sowie nutzer- und betreiberfreundlicher Infrastruktur. Ausgerüstet mit digitaler Gebäudeautomation, Sicherheitstechnik, Kommunikations- und Informationstechnik. („intelligent" hat im Englischen eine andere Bedeutung, im Deutschen können Sachen nicht „intelligent" sein!*]

intelligent character recognition, ICR intelligente Zeichenerklärung *f*

intelligent (data) station programmierbare (Daten)Station *f*

intelligent (peripheral) I/O module, smart (peripheral) I/O card intelligente Peripheriebaugruppe, IP, intelligente E/A-Baugruppe, signalverarbeitende Baugruppe *f*

intelligent sensor intelligente Sensor *m* [*Ein Sensor, bei dem eine Anzahl verschiedener oder gleichartiger physikalischer Messgrößen über einen oder mehrere Messköpfe erfasst, korrigiert und vorverarbeitet wird*]

intelligent terminal Bildschirmgerät *n* mit begrenzter Rechnerleistung

intelligibility Verständlichkeit *f*

intended life [*IEC 505*] geforderte Lebensdauer *f* [*Isoliersystem, VDE 0302, Teil 1*]

intense air(borne) noise, air(borne) intense noise Luftlärm *m* [*Er breitet sich in der Luft aus*]

intensifier electrode Nachbeschleunigungselektrode, Verstärkerelektrode *f*

intensity → level

intensity, level, volume, loudness [*These are interchangeable terms when used loosely; however, loudness is a subjective assessment of the intensity of a sound. As intensity, in watts per square meter (W/m²), is difficult to measure directly, it is estimated from the pressure variations in the air resulting from the passage of the sound. The root mean square (rms) value of the pressure fluctuations, expressed in pascals (Pa) or micropascals (µPa), is proportional to the intensity when there is no significant reflected sound and where the source is not close*] Schallstärke, Lautstärke, Lautheit, Schallintensität, Lautintensität *f*

intensity of radiation, radiation intensity Strahlungsintensität, Strahlungsstärke *f* [*Wärmestromdichte je Wellenlängeneinheit*]

interaction Eingreifen, Einwirken *n* [*Des Menschen auf die Leiteinrichtung oder Stellglieder eines Prozesses. DIN 19 222*]

interaction Wechselwirkung *f*, gegenseitige Beeinflussung

interaction gap Wechselwirkungsspalt *m* [*Elektronenstrahlröhre*]

interactive [*A mode of operation of a computer system in which a sequence of alternating entries and responses between a user and the system takes place in a manner similar to a dialogue between two persons*] interaktiv, gegenseitig wirkend

interactive area [*screen*] Dialogfeld *n* [*Bildschirm*]

interactive component Dialogkomponente *f*

interactive form Dialogmaske *f*

interactive mode, conversational mode Dialog(-betrieb), Dialogverkehr *m*, Dialogverarbeitung, interaktive Verarbeitung *f* [*Betriebsart oder Program-*

I

miersystem, mit Dialog zwischen einem speicherprogrammierbaren Automatisierungsgerät und Bediener. Zum Beispiel wird nach der Eingabe „Start" vom System zurückgefragt, ob der Start erfolgen soll]

interactive multi(ple-)media interaktive Nutzung *f* (durch Anwender) [*Strukturierung der Information*]

interactive point selection [*Method of accessing data points via address tree/menue structure*] interaktive Bedienerführung *f*

INTERBUS-C Produktlinie *f* verträglich mit Bitbus-Standard [*Hergestellt von Phoenix Contact*]

INTERBUS-S Sensor/Aktorbus *m* [*Entwickelt von Phoenix Contact*]

intercarrier Differenzträger *m*

intercarrier method Differenzträgerverfahren *n*, Differenzträgermethode *f*

I

intercepting trap, disconnect(ing) trap, interceptor [*A trap fixed on a drain to prevent the passage of sewer gas into the drain*] Kanalgassperre *f*

interception of calls, call interception Fernsprechhinweisdienst *m*

interceptor, separator, trap [*A device to trap, remove, or separate deleterious, hazardous, or undesirable matter (such as oil, grease, gas(oline) (US)/petrol (Brit.), sand and sediment) from normal waste conveyed through it, permitting normal sewage or liquid waste to discharge into the disposal terminal by gravity*] Abscheider *m*

interceptor → collector

interchangeability Austauschbarkeit, Auswechselbarkeit *f*

interchangeable austauschbar, auswechselbar

interchangeable anti-siphonage device, interchangeable anti-syphonage device Aufsatz-Rohrbelüfter *m*

interchange(able) circuit Schnittstellenleitung [*DIN 44 302*], Anpassungsschaltung *f*

interchange(able) ((closed) loop) power control Übergabeleistungsreg(e)lung *f*

interchange(able) (data) file Austauschdatei, Auswechseldatei *f*

interchange(able) power Übergabeleistung, Austauschleistung *f* [*Netz*]

interchange(able) power control → interchange(able) ((closed) loop) power control

interchange(able) reactive power Übergabeblindleistung *f*

inter-character timeout Auszeit zwischen Sendesignalen, Sendesignalauszeit *f*

intercom(munication) system, interphone, front door telephone [*The installation in houses to facilitate answering the door*] Eigenverständigungsanlage, Gegensprechanlage, (Haus-)Türsprechanlage, (Wechsel)Sprechanlage, Akustikanlage *f*, (Haus)Türtelefon *n*

intercondenser Zwischenkondensator, Zwischenverflüssiger *m*

interconnected arrangement side by side Batterieschaltung *f* [*z. B. von Druckausdehnungsgefäßen*]

interconnected computer system Bildschirmtext-Rechner-Verbund *m*

interconnected energy system Energieverbund *m*

interconnect(ing) server → bridge

interconnect(ing) wire Schaltdraht, Verbindungsdraht *m*

interconnection diagram Verbindungsplan *m* [*Nicht verwechseln mit „Anschlussplan"*] [*DIN 40 719, Blatt 1*]

interconnection mask, mask interconnection Maskenverdrahtung, Verdrahtungsmaske *f* [*integrierte Schaltung*]

interconnection of blocks, blocks interconnection Verschalten *n* von Bausteinen [*programmierbares Steuer(ungs)gerät*]

interconnection (switching) diagram [*IEC 113-1*] Verbindungs(schalt)plan *m* [*DIN 40 719*]

intercooler, interstage cooler, intermediate cooler [*Apparatus for cooling compressed gas or vapo(u)r between the compression stages*] Zwischenkühler *m*

intercooling, interstage cooling Zwischenkühlung *f*, Zwischenkühlen *n*

interface Anpassungsglied, Anpassglied *n*, Anpassungsschaltung, Anpassschaltung, Nahtstelle, Schnittstelle *f* [*Elektronische Schaltung zur Anpassung verschiedener Geräte oder Bausteine, um die Verbindung untereinander herzustellen*]

interface adapter, interface adaptor SSA, Schnittstellenanpasser *m*

interface allocation Schnittstellenzuordnung *f*

interface allocator Schnittstellenzuordner, Schnittstellenzuweiser, Schnittstellenzuteiler, Schnittstellenbeleger *m*

interface board, interface panel Schnittstellentafel *f*

interface bus, IB Schnittstellenbus *m*

interface circuit Schnittstellenschaltung *f*

interface clear, IFC Schnittstellenfunktion *f* (zu-)rücksetzen [*Programmierbare Messgerät. DIN IEC 625*]

interface ((closed) loop) control Schnittstellenreg(e)lung *f*

interface condition, interface state Schnittstellenzustand *m*

interface connection → feedthrough

interface control → interface ((open) loop) control

interface device Schnittstellengerät *n*

interface flag Schnittstellenmerker *m*

interface integrated circuit, IIC, integrated interface circuit integrierte Schnittstellenschaltung *f*

interface management bus Schnittstellensteuer(ungs)bus *m*

interface manifold Schnittstellenverteiler *m*

interface message Schnittstellennachricht, Schnittstellenmitteilung *f* [*programmierbares Messgerät*]

interface module, IM Schnittstellenbaugruppe *f*, Schnittstellenbaustein *m*, Schnittstellenmodul *n*

interface multiplier Schnittstellenvervielfacher *m*

interface ((open) loop) control Schnittstellensteuerung *f*

interface panel → interface board

interface printout, interface writeout Schnittstellenausdruck *m*, Schnittstellendruckausgabe *f*

interface protection Schnittstellenschutz *m*

interface runtime Schnittstellenbelegungszeit *f*

interface section Schnittstellenebene *f*

interface signal line Schnittstellensignalleitung *f*

interface state → interface condition

interface system Schnittstellensystem *n*, Prozessperipherie *f*

interface technique Schnittstellentechnik *f*

interface terminal Schnittstellenklemme *f*

interface tester, interface testing device Schnittstellenprüfer *m*

to interface (to) schneiden [*Schnittstelle*]

interface unit Schnittstelleneinheit *f*

interface writeout → interface printout

interfacing (An)Kopp(e)lung *f*

interference → disturbance

interference, intervention Eingriff *m*

interference, electrical interference, disturbance [*Communication. Any undesired energy that tends to interfere with the reception of desired signals*] (Empfangs)Störung *f*

interference, wave interference [*The variation with distance or time of the amplitude of a wave which results from the superposition (algebraic or vector addition) of two or more waves having the same, or nearly the same, frequency*] Schwingung(s)weitenstörung *f*

interference analyzer, interference analyser [*An instrument that discloses the frequency and amplitude of unwanted input*] ungewollte Eingabeanalysator *m*

interference fading [*Fading of the signal produced by different wave components travel(l)ing slightly different paths*] Stör(ungs)schwund *m*, Schwundstörung *f*

interference fit Festsitz *m*, Übermaßpassung *f*

interference immunity (Funk)Stör(spannungs)festigkeit *f*

interference pattern Stör(ungs)muster *n*

interference prediction [*An estimate of the level of interference a given device might experience in an electromagnetic environment*] Stör(ungs)vorhersage *f*

interference reduction, interference suppression, interference rejection [*Electronics. A process by which interference in an electrical circuit from outside factors, such as radio transmitters and lightning, is minimized*] Entstörung, Stör(ung(s))unterdrückung *f*

interference voltage, noise voltage, intervention voltage Fremdspannung, Rauschspannung, Störspannung *f* [*Neben der Nutzsignalspannung (Spannung des vorgesehenen Signals) auftretende störende Spannung*]

interfering → intervening

inter-integrated circuit bus → I²C-bus

interior decorator, designer Innenarchitekt *m*

interior environment Innenumwelt *f*

interior motion detector, interior motion detecting device, interior motion detection device [*burglar alarm*] Innenbewegungsmelder, Innenbewegungsdetektor *m*, Innenbewegungsmeldegerät *n*

interior-reflected lamp innenverspiegelte Lampe *f*

interior temperature sensor Raumtemperaturfühler *m*

interior water feature, indoor water feature, inside water feature Innenwasserbeckenattraktion *f* [*z. B. Springbrunnen*]

interlace Bildaufbau *m* [*Es werden zuerst alle ung(e)raden Bildzeichen und dann alle g(e)raden ausgegeben*]

interlaced monitor Zeilensprungverfahrensmonitor *m*

I

interlacing Halbbildverfahren *n*, Halbbildmethode *f* [*Rasterbildschirm*]

interlacing method Zeilensprungverfahren *n*, Zeilensprungmethode *f* [*graphische Datenverarbeitung*]

interleaved [*e.g. paper hand towels*] ineinander gelagert

interleaving [*book*] Durchschießen *n*

interleaving geschachtelte Adressierung, Überlagerung, verschachtelte Adressierung, Verwürfelung, Verschachtelung, Kodespreizung, Codespreizung *f* [*Sie gestattet bei Mehrrechnersystemen eine Verminderung von Zugriffskonflikten und erleichtert den Zugriff*]

interleaving abwechselnde Ansprechen *n* [*Speicherbank*]

inter-line transfer, ILT zeilenweise Auslesen *n* der Bildinformation [*Videokamera*]

I

to interlink verketten

interlock [*A safety switch or system that prevents the operation of equipment when a hazard exists, such as the mechanism that shuts down high voltage circuits when protective doors to the circuit are opened*] Verrieg(e)lung *f*

interlock [*A hardware or software mechanism that coordinates processes so that they function independently without interferring with each other, but achieve proper coordination and sharing of resources*] Verblockung *f*

interlock code, IC Verknüpfungskodekennzeichen, Kennzeichen für „Verknüpfungscode" *n*

interlock deactivating key, defeater key Entrieg(e)lungsschlüssel *m*

(interlock) deactivation, unlocking Entriegeln *n*, Entrieg(e)lung *f* [*Schaltgerät*]

(interlock) deactivation by (means of) key, key defeating Schlüsselentrieg(e)lung, Schlüsselentriegeln *f* [*Schaltgerät*]

interlocked doors → double door interlock

interlocked round doors → round interlock doors

interlocked with contactor, contactor-interlocked schütz(en)verriegelt

interlock(ing) Verriegelung *f*

interlocking (door) latch zwangläufige (Tür)-Schloss *n*

interlocking joint Verfalzung *f*

interlocking switch socket [*IEC 117-8*] verriegelte Steckdose *f* [*DIN 40 717*]

interlocking system [*Arrangement of interlocks to ensure that all interrelated parts of a system are operating properly together*] Verbund *m*

interlock level Verrieg(e)lungsebene *f*

interlock lid switch → lid interlock

interlock monitoring device, interlock monitoring appliance, interlock monitor [*See remark under "monitoring device"*] Verrieg(e)lungsüberwachungsgerät *n*, Verrieg(e)lungsüberwachungsvorrichtung *f*

interlock relay [*A relay having two or more armatures that are linked mechanically or interconnected electrically in such a way that the position of any one of the armatures causes or prevents the motion of another one*] Verrieg(e)lungsrelais *n*

interlock switch [*It is usually attached to a cabinet door or housing, that removes power form a circuit when the door is opened, thus providing protection against high-voltage electric shock*] Verrieg(e)lungsschalter *m*

intermateable connection (zusammen)steckbare Verbinder *m*

intermediate alarm memory Alarmzwischenspeicher *m*

intermediate balustrade Zwischenbalustrade *f* [*Bei nebeneinanderliegenden Fahrtreppen oder Fahrsteigen*]

intermediate control level [*Middle level in a 3-level hierarchical structure*] Zwischenleitebene *f*

intermediate distributor Zwischenverteiler *m*

intermediate flange gate valve, intermediate flange pipeline valve Zwischenflanschschieber *m*

intermediate frequency rejection, IF-rejection Zwischenfrequenzabstand, ZF-Abstand *m*, ZF-Unterdrückung, ZF-Festigkeit, ZF-Sicherheit *f* [*Unterschied zwischen dem Pegel des auf die Zwischenfrequenz umgesetzten Signals mit Empfangsfrequenz und dem Signal welches auf der Zwischenfrequenz unmittelbar (also ohne Umsetzung) einwirkt*]

intermediate marker, intermediate flag Zwischenmerker *m* [*programmierbare Steuer(ungs)gerät*]

intermediate memory, intermediate storage [*The portion of the computer storage facilities that usually stores information in the processing stage*] Zwischenspeicher *m*

intermediate pressure, interstage pressure [*Pressure prevailing between stages of multistage compression*] Zwischendruck *m*

intermediate-pressure stage, pressure interstage [*That part of the installation between pressure regulators fitted in series*] Zwischendruckstufe, Druckzwischenstufe *f*

intermediate support Zwischenauflager *n* [*Fahrtreppe; Fahrsteig*]

to intermesh ineinandergreifen

intermeshing Vermaschung *f* [*Gütersicherung. DIN 40 042*]

intermittent ((closed) loop) control → discontinuous ((closed) loop) control

intermittent direct current → discontinuous direct current

intermittent ((open) loop) control → discontinuous ((open) loop) control

intermittent operation → discontinuous operation

intermittent periodic duty [*IEC 34-1*] Aussetzbetrieb *m* [*Kurzbezeichnungen: S3, S4, S5*]

intermittent setting → discontinuous setting

intermodulation, IM [*Modulation of the components of a complex wave by each other, producing new waves whose frequencies are equal to the sums and differences of integral multiples of the component frequencies of the original complex wave*] Intermodulation, Differenztonbildung, Zwischenmodulation *f*

intermodulation distortion, IM distortion [*Nonlinear distortion characterized by the appearance of output frequencies equal to the sums and differences of integral multiples of the input frequency components; harmonic components also present in the output and usually not included as part of the intermodulation distortion*] Intermodulation(s)verzerrung, Zwischenmodulation(s)verzerrung, Differenztonverzerrung *f*

internal alarm, local alarm Internalarm, örtliche Alarm *m* [*In der Brandmeldetechnik nach allgemeinem Sprachgebrauch und VdS-Richtlinien*]

internal bus [*The internal bus connection of computer components (CPU, controllers), usually 8, 16 or 32 bits broad*] innere Bus, Innenbus *m*

internal cooling load innere Kühl(ungs)last *f*

internal designer, internal architect Innenarchitekt *m*

internal diameter → ID

internal dimension → ID

internal draft, internal draught innere Schräge *f* von 7° [*Toleranzangabe beim Gesenkschmieden*]

internal dry cooling load innere trockene Kühlungslast *f*

internal handle [*door*] Innengriff *m*

internal humid cooling load innere feuchte Kühl(ungs)last *f*

internal inclined grate furnace, inclined internal grate furnace [*See remark under "furnace"*] Schrägrostinnenfeuerung, Innenschrägrostfeuerung, Schrägrostinnenfeuerstätte, Innenschrägsrostfeuerstätte *f*

internal pipe diameter Rohr(licht)weite *f*, Rohrinnendurchmesser *m*

internal piping Innenrohrleitung *f*

internal plumber Installateur, Installationshandwerker *m*

internal plumbing (trade) Installationshandwerk, Installateurhandwerk *n*

internal pressure Innendruck *m*

internal thermal resistance, thermal internal resistance innere Wärmewiderstand *m* [*Halbleiter(bauelement); DIN 41 858*]

internal tolerance innere Toleranz *f* von + .03 - .06 [*Toleranzangabe beim Gesenkschmieden*]

international call Ausland(s)(an)ruf *m* [*Durchgangsverbindung über internationale Leitung(en)*]

international candle → candela

Internet information server, IIS, Internet information service provider Internetinformations(-dienst)anbieter *m*

Internet Relay Chat → IRC

internet service provider, ISP Internet(dienst)anbieter *m*

internetwork(ing) [*A set of two or more networks interconnected by routers. In a BACnet internetwork, there exists exactly one message path between any two needs*] Kommunikationsverbund *m*, Datenvernetzung, Vernetzung (von Datennetzen), Kopplung (von Datennetzen), Koppelung (von Datennetzen) *f*

interoperability Zusammenwirken *n* [*Gewerke*]

interorganizational organisationsübergreifend

interphase reactor → balance coil

interphase transformer [*Autotransformer or a set of mutually coupled reactors used in conjunction with tree-phase rectifier transformers to modify current relations in the rectifier system to increase the number of anodes of different phase relations which carry current at any instant*] Saugdrossel(spule) *f*

interphone → intercom(munication) system

interpolation point Stützpunkt *m* [*Zur Bildung nichtlinearer Kennlinien*]

interpole → commutating pole

interpole field → commutating (pole) field

interposing relay → coupling relay

interpretation → (data) interpretation

interpretation system → (data) evaluation system

interpreted data ausgewertete Daten, ausgelegte Daten *f*

interpreter, conversational compiler Interpretierer *m* [*Eine Funktionseinheit in einem digitalen Rechensystem, die eine Anweisung analysiert und deren Ausführung auslöst, bevor sie die nächste Anweisung behandelt (= interpretiert). E DIN 44 300, Teil 4*]

interpreter Dolmetscher *m*

interpreter plant, interpreter installation Dolmetscheranlage, Simultananlage *f*

interrogation input → **query input**

to interrupt unterbrechen, trennen, ausschalten

interrupted acknowledg(e)ment Quittung für Programmunterbrechung, Unterbrechungsquittung, Programmunterbrechungsquittung *f*

interrupted condition-code mask Unterbrechungsanzeigemaske, Anzeigemaske zur Programmunterbrechung, Programmunterbrechungsanzeigemaske *f*

I

interrupted operation → **discontinuous operation**

interrupt-enable Unterbrechungsfreigabe, Programmunterbrechungsfreigabe, Freigabe der Programmunterbrechung *f* [*Eine Gehäuseanschlussbezeichnung beim Mikroprozessor 8080. Ein Signal 1 am Ausgang bestätigt der Eingabeeinrichtung, dass der Mikroprozessor Akkumulator und Register in den Kellerspeicher übertragen und dessen Adressbereich im Stapelspeicher gespeichert hat. Jetzt werden die Informationen der Eingabeeinrichtung übernommen*]

interrupter Unterbrecher *m*

interrupter switch → **switch disconnector**

interruptible use Abschaltnutzung, unterbrechbare Nutzung, abschaltbare Nutzung *f*

interrupting contact rating → **contact interrupting rating**

interruption → **cut-out**

interruption cable → **jumper cable**

interruption cost(s), break cost(s) Unterbrechungskosten *f* [*Sie entstehen bei Funktionsausfällen, das heißt bei ungeplanten Stillsetzungen der Anlagen. Die gesamten fixen und die meisten variablen Kosten aller von den ungeplanten Stillsetzungen betroffenen Unternehmensbereiche tragen hierzu bei*]

interruption key Alarmtaste, Unterbrechungstaste *f* [*programmierbare Steuer(ungs)gerät*]

interruption level Unterbrechungsebene *f* [*Datenverarbeitung*]

interruption mode → **hold(ing) mode**

interruption of occupation → **break of occupation**

interruption of use → **break of use**

interruption point → **break point**

interruption request, IRQ Programmunterbrechungsanfrage, Unterbrechungsanfrage *f* [*Eine Gehäuseanschlussbezeichnung beim Mirkoprozessor 8080. Ein Signal 1 am Eingang informiert den Mikroprozessor darüber, dass von außen Daten oder Befehle eingegeben werden sollen. Sie können aber erst dann auf den Bus gegeben werden, wenn der Ausgang INTE ein Signal abgibt*]

interruption service routine; interruption service programme (Brit.); interruption service program (US); ISR Unterbrechungsdienstprogramm, Alarmprogramm *n* [*Unterprogramm, zu dem bei Eintreffen einer Unterbrechungsanforderung gesprungen wird*]

interruption setpoint → **break setpoint**

interruption setting → **break setting**

interruption signal → **break signal**

interrupt mode → **hold(ing) mode**

interruptor switch → **(switch) disconnector**

interruption word overflow [*PC*] Alarmwortüberlauf *m*

interrupt-request, IRQ Programmunterbrechungs-Anfrage, Unterbrechungsanfrage *f* [*Eine Gehäuseanschlussbezeichnung beim Mikroprozessor 8080. Ein Signal 1 am Eingang informiert den Mikroprozessor darüber, dass von außen Daten oder Befehle eingegeben werden sollen. Sie können aber erst dann auf den Bus gegeben werden, wenn der Ausgang INTE ein Signal abgibt*]

intersection operation Durchschnittsoperation bei Mengen, UND-Operation, Mengen-Durchschnittsoperation *f*

interstage cooler → **intercooler**

interstage cooling → **intercooling**

interstage pressure → **intermediate pressure**

insterstage transformer Zwischentransformator *m*

interval, spacing Abstand *m* [*Betrag des Unterschiedes zwischen den Pegeln zweier verschiedener Signale an derselben Stelle in einem System*]

interval (time-)delay relay, interval (time-)lag relay, (time-)delay interval relay, (time-)lag interval relay Wischrelais *n*

intervening, interfering Eingreifen *n* [*Steuern; Regeln; Auswerten; Überwachen; Optimieren; Schützen; Schalten*]

intervention, interference Eingriff *m*

interzonal air flow Luftströmung zwischen den Zonen, Zonenluftströmung *f*

intranetworking Kommunikation *f* innerhalb eines Netz(werk)es

intra-plaster installation Imputzinstallation *f*

intrinsically(-)safe eigensicher

intrinsic conduction Eigenleitung *f*

intrinsic fibre optic sensor; intrinsic fiber optic sensor (US) faseroptische Eigensensor *m* [*Die Wandlung der Messgröße in ein optisches Signal (Modulation) erfolgt in der Faser*]

intrinsic semiconductor, *i*-type semiconductor [*A semiconductor in which the concentration of charge carriers is characteristic of the mature itself rather than of the content of impurities and structural defects of the crystal*] Eigenhalbleiter *m*

intrinsic sensor [*See remark under "sensor"*] Eigensensor *m*

intruder Eindringling *m*

intruder detection, intrusion detection Eindring(ungs)meldung, Eindringling(s)meldung, Intrusionsmeldung *f*

intruder detector, intrusion detector, intruder detecting device, intrusion detecting device, intruder detection device, intrusion detection device Eindring(ungs)melder, Eindringling(s)melder, Intrusionsmelder *m*, Eindring(ungs)meldegerät, Eindringling(s)meldegerät, Intrusionsmeldegerät *n*

intruder monitoring, intrusion monitoring Eindring(ungs)überwachung, Eindringling(s)überwachung, Intrusionsüberwachung *f*

intruder-proof, intruder-resistant, intrusion-proof, intrusion-resistant eindring(ungs)sicher, eindringling(s)sicher, intrusionssicher

intruder protection, intrusion protection Eindring(ungs)schutz, Eindringling(s)schutz, Intrusionsschutz, Schutz vor (widerrechtlichem) Eindringen *m*

intruder recognition system, intrusion recognition system Eindring(ungs)erkennungssystem, Eindringling(s)erkennungssystem, Intrusionserkennungssystem *n*

intruder safety → safety against intruders

intruder security, security against intruders Sicherung gegen Eindringlinge, Eindring(ungs)sicherung *f*

intruder trap, intrusion trap Eindring(ungs)falle, Intrusionsfalle, Eindringling(s)falle *f*

intrusion Aufschaltung *f* [*Telefon*]

intrusion Intrusion *f*, (widerrechtliche) Eindringen *n*

intrusion alarm → burglar(y) alarm

intrusion detection → intruder detection

in use → busy

inventory → on-hand quantity

inverse boundary layer approach umgekehrte Grenzschichtmethode *f*, umgekehrte Grenzschichtverfahren *n* [*Diffusor*]

inverse direction of operation, inverse operation(al) direction [*junction transistor*] umgekehrte Betrieb *m* [*Sperrschichttransistor*]

inverse-time zeitverzögert

inversion alternate mark code → alternate mark inversion code

inversion flame Umkehrflamme *f* [*Die Rauchgase strömen entlang der Flamme zurück und dann in Rippenkammern, Nachtschaltheizflächen und ähnlichem*]

inverted binary Einerkomplement *n* [*Es entspricht dem Binärcode, wobei jedoch Null und Eins ausgetauscht sind. Die Umformung von „normalem" Binärcode und Code im Einerkomplement erfolgt durch eine einfache Inversion der Bits. Beispiel: Die Zahl 182 wird als 01001001 dargestellt*]

inverted siphon → dip pipe

inverter, dc-to-ac inverter, power inverter, dc-to-ac converter, inverted rectifier [*A device for converting direct current into alternating current; it may be electromechanical, as in a vibrator or synchronous inverter, or electronic, as in a thyratron inverter circuit*] Umkehrer, Wechselstromrichter, Umrichter, Inverter *m*, Umkehrschaltung, Umkehrstufe *f*, NICHT-Element, NICHT-Glied *n*

inverter, frequency transformer Frequenzwandler *m*

inverter circuit → NOT circuit

inverting amplifier, sign-reversing amplifier Umkehrverstärker, invertierende Verstärker *m* [*Ein gegengekoppelter Operationsverstärker, der am Ausgang stets eine Spannung mit entgegengesetzter Polarität zur Eingangsspannung führt*]

invisible radiation unsichtbare Strahlung *f*

to invoke aufrufen [*im Kommunikationssystem*]

in-wall installation Inwandinstallation *f* [*Gegensatz: Vorwandinstallation*]

in-wall valve Inwandventil *n* [*Gegensatz von Wandventil*]

in word mode wortweise

I/O → input/output

I/O allocation table [*It controls input and output data relative to channel number and addresses index position*] E/A-Zuweisungsliste, E/A-Rangierliste *f*

I/O bus → peripheral bus

IOCS → input/output central system

IOCS → input/output contral system

iodate jodsaure Salz *n*

I/O interface → **input/output interface**

ionic (loud)speaker Ionenlautsprecher *m*

ion implantation Inonenimplantation, Ioneneinpflanzung *f*

ionization alarm (signal) [*See remark under "alarm (signal)"*] Ionisationsalarm(signal) *m, (n)*, Ionisationsstörungssignal *n*

ionization chamber Ionisationskammer *f*

ionization fire detector, ionization fire detecting device, ionization fire detection device Ionisations(brand)melder, Ionisationsfeuermelder, I-Melder *m*, Ionisations(brand)meldegerät, Ionisationsfeuermeldegerät, I-Meldegerät *n*

ionization ignition fuse, ionization ignition fuze Ionisationszündsicherung *f* [*Die Flamme leitet einen Gleichstrom von der Elektrode zur Masse. Bei Verlöschen der Flamme wird der Strom unterbrochen*]

I

ionization inception voltage Ionisationseinsetzspannung *f*

ionization line Ionisationslinie, I-Linie *f* [*Linie zum Anschluss von Brandmeldern nach dem Ionisationskammerprinzip*]

ionization monitoring Iionisationsüberwachung *f*

ionization sensor, ionization sensing device [*See remark under "sensor"*] Ionisationsfühler, Ionisationssensor, Ionisationsdetektor *m*, Ionisationsdetektionselement *n*

ionization smoke detector, ionization smoke detecting device, ionization smoke detection device Ionisationsrauchmelder, I-Rauchmelder *m*, Ionisationsrauchmeldegerät, I-Rauchmeldegerät *n*

to ionize [*To make conductive by the emission from a radioactive source, usually americanium 243*] ionisieren

ionizing radiation [*Any radiation displaying electrons from atoms or molecules, thereby precluding ions. Examples: alpha, beta, gamma radiation; short-wave ultra-violet light. This radiation may produce severe stain or tissue damage*] ionisierende Strahlung *f*

ion migration Ionenwanderung *f*

IORQ → **input/output request**

ip → **ingress protection**

IR → **instruction register**

IR → **infrared radiation**

IRC, Internet Relay Chat Konferenzraum *m* im Internet, Internet-Konferenzraum

IRC → **integrated room control**

IRED → **infrared emitting diode**

IRH → **integrated recording head**

IR heating → **infrared radiation heating**

iron angle → **angle iron**

iron (builders') mounting, builders' iron mounting Eisen(bau)beschlag, Baueisenbeschlag *m*

iron chimney connector → **iron smoke pipe**

iron-containing → **ferritic**

ironer Bügelapparat *m*

iron furniture, ferrels Eisen(bau)beschläge *f*

ironing board Bügelbrett *n*

ironmongery → **(builders') iron supplies**

iron (pipe) hanger, pipe iron hanger (Rohr)Schelleisen *n*

iron powder [*Cast iron or pig iron in powder form*] Eisenpulver *n*

iron smoke pipe, iron stove pipe, iron chimney connector Eisen-Rauchrohr, Eisen-Ofenrohr *n* [*DIN 1298*]

iron stove Kanonenofen, Eisenofen, eiserne Ofen *m*

iron supplies → **(builders') iron supplies**

IRQ → **interruppt-request**

irrelevant data → **garbage**

irrelevant data collector → **garbage collector**

irreversible thermodynamics → **nonequilibrium thermodynamics**

irritant refrigerant, irritating refrigerant [*Refrigerant that has an irritating effect on the eyes, nose, throat, and lungs*] Reizkältemittel, Kältereizmittel *n*, Reizkältestoff, Kältereizstoff *m*

irritation chemical (agent), irritant chemical (agent) Reizstoff *m*, Reizmittel *n*

ISA(M) → **index-sequential access (method)**

ISDN → **integrated services digital network**

ISE → **integral of squared error**

Isecon → **halogenated hydrocarbon refrigerant**

isenthalpic process isenthalpische Prozess *m* [*z. B. der Joule-Thomson-Prozess*]

isentropic [*Having constant entropy; at constant entropy*] isentrop

isentropic compression [*It occurs without any change in entropy*] isentrope Verdichtung *f*

isentropic expansion [*It occurs without any change in entropy*] isentrope Ausdehnung *f*

isentropic exponent Isentropenexponent *m*

isentropic process, reversible and adiabatic process [*In thermodynamics, a change that takes place without any increase or decrease in entropy such as a* *process which is both reversible and adiabatic*] reversible adiabatische Prozess *m*, adiabatische reversible Prozess, isentrope Prozess

I signal → **common-mode signal**

islanding → **network splitting**

isobar Isobare *f* [*Verbindungslinie zwischen Orten gleichen Luftdrucks*]

isobutane → **refrigerant-601**

isolatable (ab)trennbar

to isolate [*To separate two portions of a process system by means of line blanks; used as safety measure during maintenance or repair, or to redirect process flows*] (ab)trennen

isolated building [*As opposed to row building*] alleinstehende Gebäude, freistehende Gebäude *n*

isolating capacitor Trennkondensator *m*

isolating link Trennlasche, Trennbrücke *f*

isolating resistor Trennwiderstand *m*

isolating switch → **(switch) disconnector**

isolating terminal Trennklemme, schaltbare Klemme *f*

isolating transformer Trennwandler *m*

isolation (Ab)Trennung *f*

isolation amplifier [*It is used to minimize the effects of following circuit on the preceding circuit*] (Ab)Trenn(ungs)verstärker *m*

isolation appliance → **(safety) isolation device**

isolation device → **(safety) isolation device**

isolation diode [*A diode used in circuit to allow signals to pass in only one direction*] Richtungsdiode, Einwegdiode *f*

isolation location [*A location in a computer memory which is protected by some hardware device so that it cannot be addressed by a computer program(me) and its contents cannot be accidentally altered*] Trennstelle *f*

isolation net(work) [*A net(work) inserted in a circuit or transmission line to prevent interaction between circuits on each side of the insertion point*] Trennschaltung *f*, Puffernetz(werk) *n*

isolation switch, disconnection switch Freischalter *m*

isolation transformer [*Transformer inserted in a system to separate one section of the system from undesired influences of other sections*] Isolierwandler *m*

isolator [*A passive attenuator in which the loss in one direction is much greater than that in the other direction; a ferrite isolator for waveguides is an example*] Richtungsleitung, Einwegleitung *f*

isolator [*Any device that absorbs vibration or noise, or prevents its transmission*] Dämpfer *m*

isotherm Isotherme, Linie gleicher Temperatur, Gleichtemperaturlinie *f*

isothermal (free) (air) jet [*It has the same temperature as the surrounding air*] isotherme freie (Luft)Strahl, freie isotherme (Luft)Strahl *m*

ISP → **internet service provider**

ISR → **interruption service routine**

to issue [*from an opening*] austreten [*aus einer Öffnung*]

ISTSE → **integral of squared time multiplied by squared error**

ITAE → **integral of time multiplied by absolute value of error**

item laufende Nummer *f* [*Beschriftung im Zeichnungsschriftfeld*]

item Position *f* [*Leistungsverzeichnis*]

item of equipment → **device**

iterative network, recurrent network, lattice network Kettenleiternetz(werk) *n*

ITS → **image transmission system**

ITSE → **integral of time multiplied by squared error**

***i*-type semiconductor** → **intrinsic semiconductor**

J

jabber control [*LAN*] Behebung *f* der Dauerbelegung

jack → **(hydraulic) plunger**

jack (Anfrage)Klinke *f* [*Fernsprecher*]

jack [*In electric(al) engineering, a connecting device into which a plug can be inserted to make circuit connections; may also have contacts that open or close to perform switching functions when the plug is inserted or removed*] Buchse *f*

jack Klinkenstecker *m*

jacket, casing, case, sheath(ing) Mantel *m*, Ummantelung *f*

jacket condensation water, jacket condensate Mantelkondensat, Mantelkondenswasser *n*

jacket cooling [*Cooling in which produce is placed either in a room surrounded by a jacket of cold air, or in a tank surrounded by a jacket of cold liquid*] Mantelkühlung *f*

jacketed cold room, jacketed cold chamber [*It is entirely surrounded by insulated cavities or corridors in which cold air is circulated*] Mantelkälteraum *m*, Mantelkältekammer *f*

jacket pipe, casing pipe Mantelrohr *n*

jacket tube → casing tube

jacket vessel, casing vessel Mantelgefäß *n*

jacket water [*In a compressor or engine, the water used in cooling the cylinder head and/or walls*] Mantelwasser *n*

J

jack input Buchseneingang *m*

jack lamp, plug-in lamp Stecklampe *f*

jack panel Buchsenfeld *n*

jalousie [*A window having a series of horizontal glass slats, each hinged at the top and opening out at the bottom*] Glasjalousiefenster, Glasladenfenster *n*

jalousie [*A window blind or shutter made of series of horizontal slats, each hinged at the top and opening out at the bottom, letting in light and air but not direct sunlight and rain*] Jalousie *f*, Rollladen *m*

jam → backing

jammer → (noise) jammer

jamming Störung *f* durch Störsender

jamming modulation Störmodulation *f*

jamming transmitter Störsender *m*

jam sync Repositionierung *f* [*Sie erfordert einen lückenlosen Zeitkode auf dem Endband und einen Rückkanal für den Zeitkode des Recorders zum Schnittpunkt. Dann kann die Schnittbearbeitung auch nach einer Unterbrechung bildgenau fortgesetzt werden*]

janitor, housekeeper Hausmeister, Hauswart *m*

jerrycan Benzinkanister *m*, Benzinkanne *f*

jet → (ink-)jet

jet [*A strong, well-defined stream of compressible fluid, either gas or liquid, issuing from an orifice or nozzle or moving in a contracted duct*] Strahl *m*

jet burner Strahlbrenner *m*

jet condenser Einspritzkondensator *m*

jet cooling [*Quick-chilling process using air at very high speed forced around the products, usually while those products are continuously moving on a conveyor belt*] Strahlkühlung *f*

jet effect [*e.g. of air on entering a room*] Strahlwirkung *f*

jet fan Strahlventilator *m*

jet freezing, blast freezing [*Quick freezing process using cold air at very high speed forced onto the produce*] Strahlgefrieren *n*

jet liquid pump → liquid jet pump

jet method → (ink-)jet method

jet outlet tap Hahn *m* mit Umstellung Strahldusche/Brause

jet pipe Strahlrohr *n*

jet plotter → (ink-)jet plotter

jet printer → (ink-)jet printer

jet pump [*An ejector or injector (pump)*] Strahlpumpe *f*

jet series → series of jets

(jet) shower Strahldusche, Dusche *f* [*Das Wasser wird im geschlossenen Strahl gegen den Körper geleitet. Gegenteil: Brause, bei der das Wasser in feiner Verteilung gegen den Körper geleitet wird*]

(jet) shower cubicle, (jet) shower cab(in), (jet) shower cell, (jet) shower recess, (jet) shower enclosure, (jet) shower compartment, (jet) shower stall Strahlduschkabine, Duschkabine, Strahlduschzelle, Duschzelle, Strahlduschnische, Duschnische *f*, Strahlduschstand, Duschstand *m*

jet straightener Strahlregler *m*

jet throw, throw of jet Strahlweite *f*

jet tube Strahlröhre *f*

jet velocity Strahlgeschwindigkeit *f*

jet ventilation → mixed ventilation

jet water Strahlwasser *n*

JFET, junction-field-effect transistor, jugfet, junction-gate-FET Sperrschichtfeldeffekttransistor *m*, Sperrschichtgitter-FET *n*

jitter Zittern *n* [*Bei einem Oszillogramm das unerwünschte Zittern des Bildes oder eines Teiles davon in waag(e)rechter Richtung*]

jitter, phase jitter Zittern, Phasenzittern *n*, Synchronisationsstörung *f* [*Es wird hervorgerufen durch Geschwindigkeitsschwankungen beim Abtasten und ergibt einen ungleichmäßigen Datenfluss, der aber im Prozessor ausgeglichen wird. Allgemeine zeitliche Instabilität der Nulldurchgänge von Digitalsignalen kann zu Bitfehlern führen, wenn Toleranzgrenzen überschritten werden*]

JK flipflop (circuit), JK bistable multivibrator *[See remark under "bistable multivibrator"]* bistabile JK-Multivibrator *m*, bistabile JK-Kippstufe, JK-Triggerschaltung *f*

job Auftrag *m [Anforderung an eine Rechenanlage, eine Datenverarbeitungsleistung zu erbringen]*

job control Auftragssteuerung *f*

job-control language → command language

job engineering, work engineering Arbeitstechnik *f*

(job)site control → field control

(job)site management → field management

(job)site wiring → field wiring

jockey compressor Kleinkompressor, Kleinverdichter *m*

jockey pump *[A small pump to boost output of a system]* kleine Druckverstärkerpumpe *f*

jog shuttle Suchdrehscheibe *f [An Video- und Schnittgeräten. Diese Scheibe ermöglicht bildgenaues Positionieren auf dem Videoband]*

jog(ging) mode, inching mode Betriebsart *f* Tippen *[Betriebsart bei Ablaufsteuerungen, in der das Weiterschalten der Ablaufkette auf den nächsten Schritt abhängig vom Bedienereingriff auf einen für die Ablaufkette nur einmal für diesen Zweck vorhandene Signalgeber erfolgt. V DIN 19 237]*

joining Anbindung *f*

joint distribution Verbundverteilung *f*

jointed flush, flush-jointed, jointed flut, flut-jointed vollfugig, ganzfugig

joint editing gemeinsame Editieren *n [Von Dokumenten im Notizblock]*

joint face → face of joint

joint filler groove, groove for joint filler Fugendichtung(s)hohlkehle *f*

joint hinge Scharnierband *n*

joint hinge eye Scharnier(band)öse *f*

jointless floor(ing) fugenlose (Fuß)Boden *m*

joint permeability Fugendurchlässigkeit *f*

joint section Rollenlappen *m*, Teilung *f [Scharnier]*

joint ventilation Fugenlüftung *f [Sie wird durch Undichtigkeiten von Fenstern, Türen und anderen Öffnungen eines Raumes verursacht]*

joint viewing gleichzeitige Darstellung *f* von Anwendungsfenstern (auf den beteiligten Systemen)

(Joule-)Thomson effect *[Ratio of temperature change to pressure change dt/dp of an actual gas in a process of throttling or expansion without doing work or interchanging heat]* (Joule-)Thomson-Effekt *m*, (Joule-)Thomson-Wirkung *f*

journal "activity" Kommunikationsdatenbericht *m [Bericht über wichtige Kommunikationsdaten, wie z. B. Empfänger, Sendezeit und Datum]*

joystick, joy stick Eingabegerät *n [Zur Steuerung der Fadenkreuzposition auf dem Bildschirm]*

joystick, joy stick *[A lever that pivots in all directions, used widely in computer games as a locator device because its action is faster than that of the directional arrow keys; it also serves as an input device in computer-aided design systems]* Markier(ungs)-stift *m*

jugfet → JFET

juke-box Schallplattenwechselautomat *m*

jummel *[old English]* **→ grimmer**

jump (unbedingte) Sprung *m*

jump address *[The address to which the program(me) will proceed upon receipt of a jump signal]* Sprungadresse *f*, Sprungziel *n [Elektronik]*

jump address correction → correction of jump address

jump address modification → modification of jump address

jump element Sprungelement *n*

to jumper überbrücken *[durch Schaltdraht]*

jumper Steckbrücke, Steckverbindung *f*

jumper Erdungsbügel, Kurzschlussbügel *m*

jumper, jumping wire, hook-up wire Überbrückungsdraht, Schaltdraht *m*, Drahtbrücke *f*

jumper Verbindungsleiter *m*, Verbindungsleitung *f [mit 2 Steckern]*

jumper Strombrücke *f*

jumper cable Überbrückungskabel *n*

jumpering card Rangierkarte *f*

jumper module, jumper block Brückenbaustein, Brückenblock *m*, Brückenmodul *n [elektronische Steuerung]*

jump(-forced) response Sprungantwort *f [DIN 19 226. Die zeitliche Änderung der Stellgröße bei einer sprungweisen Änderung der Regelgröße]*

jump instruction → branching instruction

jump label Sprungmarke *f [programmierbare Steuer(ungs)gerät]*

jump signal *[The signal which causes the program(me) to bypass a section of the program(me), and proceed to the jump address]* Sprungsignal *n*

junction → connection

junction Sperrschicht *f*

J

junction box [*Metal box in which tap to circuit conductors is made. It is not an outlet since no load is fed from it directly*] Anschlussdose, Abzweigdose, Verbindungsdose *f* [*DIN IEC 23.F*]

junction box receiver Anschlussdosenempfänger, Abzweigdosenempfänger, Verbindungsdosenempfänger *m*

junction breakdown Sperrschichtdurchschlag *m*

junction diode Flächendiode, FD *f*

junction-field-effect transistor → JFET

junction gate-FET → JFET

junction-gate field effect transistor, PN FET Feldeffekttransistor mit PN-Übergang, Sperrschichtfeldeffekttransistor *m*

junction transistor Sperrschichttransistor, Flächentransistor *m*

justification, margin justification Randausgleich *m* [*Rechts- und linksbündige Begrenzung der Textzeilen*]

justified block, right justified format Blocksatz *m* [*Textverarbeitung*]

just-in-time zeitgerecht, rechtzeitig

just revert-reverse value Wiederansprechistwert *m* [*Relais*]

just value Istwert *m* [*Relaisprüfung*]

K

K, temperature on the Kelvin scale Kelvintemperatur *f*

Kata thermometer Kata-Thermometer *n* [*Von Hill 1914 eingeführt. Es bestimmt die Kühlstärke der Luft und misst geringe Luftgeschwindigkeiten*]

Kata thermometer value Katawert, Kühlwert *m*

katharometer → heat conductivity cell

keep → (door) keep

keep-alive arc Hilfsentladung *f*

to keep closed zuhalten

keeper (plate) → (lock) striking plate

keeping Aufbewahren *n*

Kelvin ((absolute) temperature) scale [*Thermodynamics. It is based on the absolute zero temperature (0K) and the ice point of water (273.16 K). The scale is defined so that the ratio of the temperatures of two heat reservoirs is equal to the ratio of heat transfer from the hotter reservoir by a heat engine to the amount of heat transfer to the colder reservoir during a Carnot cycle*] Kelvinskala, absolute Skala, thermodynamische Skala *f*

Kelvin bridge Kelvinbrücke *f*

kerf, saw-cut Sägeschnitt *m* [*In Holz, Stein Kunststoff, Kohle usw.*]

kerf Schnittfuge *f*, Schnittspalt *m*

Kern counter → dust counter

kernel Kern *m* [*Der im Hauptspeicher befindliche Teil des Betriebssystems*]

kerning Unterschneiden *n* [*Verringern eines etwas zu großen Abstandes zwischen bestimmten Zeichen*]

kettle, open vessel Kessel *m*, offene Gefäß *n*

kettle furnace Kesselofen *m*

kevel, cavil [*A strong piece of wood used for securing a rope's end*] Seilpoller *m*

to key tasten

key, roughness [*This texture enables plaster, mortar, glue, etc., to grip it with a mechanical bond*] Rauh(igk)eit *f*

key Taste(r) *f, (m)*

key [*A sequence of symbols that controls the operations of encipherment and decipherment*] Schlüssel *m*

key Schlüssel *m* [*Bezeichner von digitalen Daten*]

key Schlüssel *m* [*zum Auf- oder Zuschließen*]

key → key(ing) element

key-activated switch → keylock switch

key-actuated → key-operated

key address [*data processing*] Schlüsseladresse *f*

key bearer, key holder, key possessor Schlüsselinhaber, Schlüsselträger *m*

key-bit(t) → bit(t)

key blank Schlüsselrohling *m*

keyboard, keypad Tastatur *f*, Tastenfeld, Tastenwerk *n*

keyboard overlay Tastaturfolie *f*, Tastaturaschablone, Tastenfeldfolie, Tastenfeldschablone, Tastenwerkschablone, Tastenwerkfolie *f*

keyboard transmitter, keypad transmitter Tastatursender, Tastenfeldsender, Tastenwerksender *m*

keyboard unit, keypad unit Tastgerät *n*

keyboard with membrane-switch arrays, keypad with membrane switch arrays Tastatur mit Folienschaltern, Folienschaltertastatur *f* [*E DIN 42 115*]

key bow → bow of a key

key box Schlüsselkasten *m*

key bunch, bunch of keys Schlüsselbund *n* [*Österreich: m*]

key cabinet [*Cabinet that incorporates a number of hinged leaves with hooks for storing keys. A door pocket holds records*] Schlüsselschrank *m*

key change (Schlüssel)Schließungsverschiedenheit *f*

key check(ing) Schlüsselkontrolle *f*

key coding Schlüsselkodierung, Schlüsselcodierung *f*

key controller system Energiesparsystem *n* mit Zimmerschlüssel [*Wenn der Gast seinen Schlüssel am Empfang abholt wird durch das Abnehmen des Schlüssels vom Halter die Heizung im betreffenden Zimmer elektrisch eingeschaltet und beim Einhängen des Schlüssels auf den Halter abgeschaltet*]

key debouncing Tastenentprellung *f*

key defeating → (interlock) deactivation by (means of) key

key dialling Tastenwahl *f*

key duplication safety → safety against key duplication

keyed verpolsicher

key (e)scutcheon → (e)scutcheon

key field [*data processing*] Schlüsselfeld *n*

key group [*computer*] Schlüsselgruppe *f*

key holder Schlüsselhalter *m* [*als Person*]

key holder identity Schlüsselhalteridentität *f*

keyhole Schlüsselloch *n*

keyhole cover Schlüssellochblech, Schlüssellochschild *n*

key hole drop → drop of a key hole

key identification Schlüsselidentifizierung, Schlüsselkenn(zeichn)ung *f*

key (identification) tag [*Metal or fibre identification tag attached to keys*] Schlüsselanhänger *m*

to key in eintasten, eingeben

keying [*The forming of signals, such as for telegraph transmission, by modulating a direct-current or other carrier between discrete values of some characteristic*] Tasten *n*

key(ing element) [*In automatic data processing, a data item that serves to uniquely identify a data record*] Kennzeichen *n*

keying error ratio [*Communication. The ratio of the number of characters incorrectly transmitted to the total number of characters in a message*] Tastfehlerverhältnis *n*

keying frequency [*Communication. In facsimile, the maximum number of times a second that a black-line signal occurs when scanning the subject copy*] Tastfrequenz *f*

keying-in, bonding [*e.g. a brick wall to another*] Einbinden *n*, Einbindung *f*

keying-in Eintasten, Eingeben *n*

key-in-knob lock → cylinder lock

key instruction Tastenbefehl *m*

key interlock Schlüsselsperre *f*

key label Tastenschild *n* [*Mit Bildzeichen oder Beschriftung*]

keyless lock schlüssellose Schloss *n*

key lock Schlüsselschloss *n*

key lock Tastsperre *f*

keylock switch, key(-operated) switch, lockswitch, key-actuated switch [*A switch that can be operated only by inserting and turning a key such as that used in ordinary locks*] Schlüsselschalter *m*; Kontaktschloss *n* [*Fehlbenennung*]

key mnemonic Tastenkürzel *n*

key of cock, cock key Hahnschlüssel *m*

key-operated, key-actuated tastenbetätigt, tastenbedient

key(-operated) button, key-actuated button [*IEC 337-2*] Schlüsseltaste *f*

key(-operated) switch → keylock switch

key(-operated) telephone set, key-actuated telephone set Tastenfernsprecher *m*, Tastentelefon *n*

keypad → keyboard

keypad transmitter → keyboard transmitter

keypad unit → keyboard unit

key pipe of the lock → (lock) key pipe

key plate → (e)scutcheon

key pressure Tastendruck *m*

key rack Schlüsselbrett *n*

key ring Schlüsselring *m*

key rose Schlüsselrosette *f*

key row Tastenreihe *f*

key security Schlüsselsicherung *f*

key security Tastensicherung *f*

key selector Schlüsselwahlschalter, Wahlschlüsselschalter *m*

key service Schlüsseldienst *m*

keyset Tastensatz *m*

key set Schlüsselsatz *m*

key shank → **(lock) key pipe**

key slot Schlüsseleinschnitt, Schlüsselschlitz *m*

(key) step → **bit(t)**

key switch [*In automatic data processing, a switch that is operated by depressing a key on the keyboard of a data entry terminal*] Tastenschalter *m*

key symbol Tastensymbol, Tastersymbol *n*

keytable Tastaturtabelle *f*

key tag → **key (identification) tag**

key turn Schlüsseldrehung *f*

key ward Schlüsselbart *m*

keyword [*It is recognized by the given computer program(me) as a reference for a given object, action, or state. The vocabulary of a computer program(me) is made up of key words*] Schlüsselwort, Kennwort *n*

K

KGD → **golden board**

kick(ing) plate [*A metal plate placed along the lower edge of a door to prevent the marring of the finish by shoe marks*] Trittplatte, Stoßplatte, Sockelplatte, Abweisplatte *f*, Sockelblech, Stoßblech, Abweisblech, Trittblech *n*

kick(ing) strip [*A metal strip placed along the lower edge of a door to prevent the marring of the finish by shoe marks*] Trittleiste, Stoßleiste, Sockelleiste, Abweisleiste *f*

to kill abbrechen [*Prozess*]

killing field coil → **field killing coil**

kilocalorie, large calorie Kilokalorie, kcal *f* [*War die Einheit der Wärmemenge. Sie war festgelegt durch die Wärmemenge, die 1 kg Wasser bei 760 Torr Luftdruck von 14,5 auf 15,5 °C erwärmt. Sie ist ersetzt durch Joule (J) oder Ws bzw. kWh. 1 kcal ≈ 4,2 kJ ≈ 1,2 Wh.*]

kilowatt-hour meter [*B.S. 37*] Kilowattstundenzähler, Energiezähler, KWh-Zähler, Wirkverbrauch(s)-zähler *m*

kilowatt-hour rate Arbeitspreis *m* [*Stromtarif*]

kindling temperature → **ignition temperature**

kinematic viscosity [*The absolute viscosity of a fluid divided by its density*] kinematische Zähigkeit, kinematische Viskosität *f*

kinetic energy [*Energy caused by the velocity of the molecules*] Bewegungsenergie, kinetische Energie, Geschwindigkeitsenergie; lebende Kraft *f* [*frühere Benennung*]

king valve → **principal valve**

kink Stufe *f* [*optische Halbleiterlaserkennlinie*]

kink Knick *m*

kiosk substation Kompaktstation *f*

kit → **assembly**

kitchen cabinet, kitchen cubicle Küchenschrank *m*

kitchen closet, kitchen built-in cupboard, kitchen in-built cupboard Kücheneinbauschrank, Küchenwandschrank, Einbau-Küchenschrank, Wand-Küchenschrank *m*

kitchen court, kitchen yard Küchenhof *m*

kitchen connection Küchenanschluss *m*, Küchenverbindung *f*

kitchen designer, kitchen decorator Küchenarchitekt *m*

kitchen discharge air valve Küchenabluftventil *n*

kitchen dishwasher, kitchen dishwashing machine Küchengeschirrspüler *m*, Küchengeschirrspülmaschine *f*

kitchen engineering Küchentechnik *f*

kitchener → **(kitchen) range**

kitchener hood → **(kitchen) range hood**

kitchenette Kleinküche *f*

kitchen garbage, kitchen food waste, kitchen waste food Küchen(speise)abfall *m*, Küchen(speise)abfälle *f*

(kitchen) gas range, gas kitchen range, gas kitchener Küchengasherd, Gasküchenherd *m*

kitchen line Küchenzeile, Küchenreihe *f*

kitchen mantle Rauchfang, Schurz *m* [*Küchen-herd*]

kitchen oven Küchenofen *m*

(kitchen) range, domestic range, house range, kitchener (Küchen)Herd, Kochherd, Hausherd *m*

(kitchen) range (base) cabinet, domestic range (base) cabinet, house range (base) cabinet [*"cubicle" may synonymously be used for "cabinet"*] (Küchen)Herd(unter)schrank, Hausherd(unter)-schrank, Kochherd(unter)schrank *m*

(kitchen) range burner, domestic range burner, house range burner, kitchener burner (Küchen-) Herdplatte, Koch(herd)platte, Hausherdplatte *f*

(kitchen) range furnace, domestic range furnace, house range furnace, kitchener furnace (Küchen-) Herdfeuerung, Kochherdfeuerung, Hausherdfeuerung, (Küchen)Herdfeuerstätte, Kochherdfeuerstätte, Hausherdfeuerstätte *f*

(kitchen) range hood, domestic range hood, house range hood, kitchener hood, (cooking) exhaust hood (Küchen)Herdhaube, Hausherdhaube, Kochherdhaube, Dunst(abzug(s))haube, Entlüftungshaube, Wrasenhaube *f* [*Niederdeutsch*], Dunstabzug *m*

(kitchen) range with oven, domestic range with oven, house range with oven (Küchen)Herd mit Backofen, Hausherd mit Backofen, Kochherd mit Backofen *m*

kitchen sink [*See remark under "sink"*] Küchenspüle *f*

kitchen ventilation Küchenlüftung *f*

kitchen waste food → **kitchen garbage**

(kitchen) (waste) shredder (Küchen)Abfallzerkleinerer *m*

kitchen waste water → **dishwashing waste water**

kitchen with eating facility Essküche *f*

kitchen without eating facility Arbeitsküche *f*

kitchen yard → **kitchen court**

kit construction method → **assemblage construction method**

kit fault switch(ing) off → **assembly fault switch(ing) off**

kit fault switch(ing) over → **assembly fault switch(ing) over**

klaxon → **claxon**

klystron [*An evacuated electron-beam tube in which an initial velocity modulation imparted to electrons in the beam results subsequently in density modulation of the beam; used as an amplifier in the microwave region or as an oscillator*] Klystron *n*

knapsack sprayer Rucksackspritze *f*

knee Winkel *m*, Knie *n* [*Formstück*]

knee frequency → **corner frequency**

knee pipe Knierohr, Winkelrohr *n*

knob Olive *f*, Knauf *m* [*Schloss*]

knob-a-channel-system [*television*] Mehrkanalmischer, Vielkanalmischer *m* [*Jedem Kanal ist ein Handeinsteller zugeordnet, mit dem der Verstärkungsgrad des jeweiligen Kanals verändert werden kann*]

knob cylinder Knaufzylinder *m*

knocker → **(door) knocker**

knockout ausbrechbare Vorpressung *f* [*Isolierpressstoffleuchte*]

knot → **node**

knowhow, special knowledge Fachwissen *n*, Fachkenntnisse *f*

knowledge base Wissensbasis, Wissensgrundlage *f*

knowledge community Wissensgemeinschaft *f*

knowledge engineering Wissenstechnik *f* [*Die Überführung menschlichen Wissens in eine formale für das Expertensystem adäquate Form*]

knowledge management Wissensbetreuung *f*

knowledge processing Wissensverarbeitung *f*

known good device → **golden board**

knuckle length Gewerbelänge, Länge der Scharnierrolle, Rollenlänge, Scharnierrollenlänge *f*

knurled knob Rändelknauf, Rändelknopf *m*

knurling Rändelung *f*

Korndorfer starting method Dreischaltermethode *f*, Dreischalterverfahren *n*

KSR → **keyboard send/receiver**

L

LA → **logical analyzer**

label [*In automatic data processing, a data item that serves to identify a data record much in the same way as a key is used, or a symbolic name used in a program(me) to mark the location of a particular instruction or routine*] Marke , Sprungmarke *f* [*VDI 2880, Blatt 4*]

label Etikett *n* [*Schweiz; Österreich; sonst veraltet*], Etikette *f*

label, term, value Wertname *m*

label Kennsatz *m*

label Kennzeichen *n*

label printer, label writer Etikettendrucker *m*

label strip Etikettenstreifen *m*

label (type) microphone Knopflochmikrophon, Ansteckmikrophon *n*

lab(oratory) bench Labor(atoriums)tisch *m*

lab(oratory) fitting Labor(atoriums)armatur *f* [*Siehe Anmerkung unter „Armatur"*]

lab(oratory) resistor Messwiderstand *m* [*DIN IEC 477*]

lab(oratory) sewage; lab(oratory) wastewater (US) Labor(atoriums)abwasser *n*

labyrinth steam trap [*See remark under "steam trap"*] Labyrinthkondensatableiter *m*

L

lack of focus, lack of definition, insufficient focus, insufficient definition, blur(ring), out-of-focus condition Unschärfe, Verschwommenheit *f*

ladder diagram, LAD Kontaktplan, Koppelplan, KOP *m* [*Eine an der Relaisschaltungstechnik orientierte, graphische Beschreibungsform für Verknüpfungs- und Ablaufsteuerungen*]

ladder hook Leiterhaken *m*

ladder net(work) Serienschaltung von zwei L-Toren, Doppel-Tore-Serienschaltung, Kettenschaltung, Abzweigschaltung von L-Zweitoren *f*, Leiternetzwerk *n*

ladder resistor net(work) Widerstandsleiternetz(-werk) *n*

ladder web Leiterkordel *f* [*Jalousie*]

ladies' toilet Damentoilette *f*

to lag nacheilen

to lag [*To prevent heat loss(es)*] wärmeisolieren

lag → delay

LAG → listener address group

lag angle, delay angle Nacheilwinkel, Verzögerungswinkel, Verzugswinkel *m*

L

lagging nacheilend

lagging load → inductive load

lagging power factor correction circuit, lagging p.f. correction circuit induktive Schaltung *f* [*Leuchtstofflampe*]

lag interval relay → interval (time-delayed) relay

lag module Verzögerungsbaustein *m*, Verzögerungsmodul *n* [*Siehe Anmerkung unter „Baustein"*]

lag time → dead time

lag (time) block → dead (time) block

lag (time) element → dead (time) element

laid-up core verseilte Ader *f*

lambda-locked-loop method, LLL method Durchflussmessung *f* mit Frequenznachführung [*Verfahren zur Durchflussmessung, bei dem die zur Kompensation der Wellenlängenunterschiede zwischen stromaufwärts und stromabwärts gesendeten Ultraschallsignalen erforderliche Frequenznachführung ein Maß für die Strömungsgeschwindigkeit ist*]

Lambertian source Lambertstrahler *m*

Lambert('s) law [*The law that the luminous intensity in a given direction radiated or reflected by a perfectly diffusing plane surface varies as the cosine of the angle between that direction and the normal to the surface*] Lambertsche Richtungsgesetz, Richtungsgesetz von Lambert *n*

Lambeth saddle, pipe saddle Rohrsattel *m*

lamella Lamelle *f*

lamella block Lamellenblock *m*

lamella burner Lamellenbrenner *m*

lamella field wirbelfreie Feld, Lamellenfeld *n*

lamella rib(bed) pipe → rib(bed) lamella pipe

lamella tube ceiling heating Lamellenrohrdeckenheizung, Deckenlamellenrohrheizung *f*

laminar boundary layer laminare Grenzschicht *f*

laminar flame, flat flame Flachflamme *f*

laminar flame burner → flat flame burner

laminar flow, streamline (flow), LF [*Fluid flow in which all the particles move in substantially parallel paths*] laminare Strömung, geordnete Strömung, Schicht(en)strömung, turbulenzarme Verdrängungsströmung, Laminarströmung, Bandströmung *f* [*Fehlbenennung: „Verdrängungsströmung"*]

laminar premixe(ed) flame → premix(ed) laminar flame

laminar (water) jet Laminar(wasser)strahl *m*

laminated glass [*A transparent safety glass in which two or more glass sheets are bonded together with intervening layer(s) of plastic materials or cellulose compounds so that, when broken, the glass will tend to adhere to the plastic rather than fly*] Verbundsicherheitsglas, VSG, Mehrscheibensicherheitsglas, MSG, Vielscheibensicherheitsglas *n*

(laminated) stator core Ständerblechpaket *n*

laminated ventilation, laminated vent(ilat)ing Schichtlüftung, Schichtventilation *f*

lamp bulb → light bulb

lamp complement of luminaire, number of lamps per luminaire Leuchtenbestückung, Beleuchtungskörperbestückung *f*

lamp heat Lampenwärme *f*

lampholder, lamp socket Lampenhalter *m*, (Lampen)Fassung *f*

lampholder adapter, lampholder adaptor, lampholder plug, lamp socket adapter, lamp socket adaptor, lamp socket plug (Lampen)Fassungssteckverbinder *m*

(lamp) inrush current [*The initial surge of current when an incandescent filament lamp is turned on; may be as much as 50 times the rated current and may last several tenths of a second for high-wattage lamps*] Einschaltstrom *m* [*Trafo und DIN 41 745*]

lamp socket → lampholder

lamp socket plug → lampholder adapter

lamp switch Lampenschalter *m*

lamp test(ing) device Lampenprüf(ungs)vorrichtung *f*

lamp wattage Lampenleistung *f*

lamp wire connector Lüsterklemme *f*

lamson tube [*A pneumatic tube for conveying documents in containers*] Druckluftpostrohr, Pressluftpostrohr, pneumatische Postrohr *n*

LAN → local area net(work)

land Steg(ansatz) *m*, Anschlussfläche *f* [*gedruckte Schaltung*]

landing Podest *m* [*Österreich: n*], Treppenabsatz, Flur *m*

landing → elevator landing

landing [*Stationary area at the entrance to or exit from a moving walk or moving walk system*] Plattform *f*

landing angle [*escalator*] Antrittswinkel *m*

landing call [*lift*] Haltestellenruf, Stockwerkruf, Geschossruf, Etagenruf *m*

landing call button Haltestellendruckknopf, Stockwerkdruckknopf, Geschossdruckknopf, Etagendruckknopf *m*

landing door → hoistway door

landing gate → hoistway door

landing hydrant (Treppen)Podesthydrant, Treppenabsatzhydrant *m*

landing opening → (elevator) landing opening

landing platform [*escalator*] Antrittsplateau *n*, Antrittsplattform *f*

landing zone, parting zone schreibfreie Spur *f* [*Festplatte. Auf diese Spur werden keine Daten geschrieben*]

landless hole lötaugenlose Loch *n*

land radio network Landfunknetz *n*

land(s) → (real) property

landscape office Großraumbüro *n*

land(s) management → (real) property management

language Sprache *f*

language, text [*An organized system of data presentation*] Text *m*

language cabling Verkabelung für Sprachen, Sprachenverkabelung *f*

language-coded sprach(en)kodiert, sprach(en)codiert, sprach(en)verschlüsselt

language indication, language announcement Meldung der Sprache, Anzeige der Sprache, Sprachanzeige, Sprachmeldung *f* [*Mitteilungsübermitt(e)lung*]

language layer Sprachebene, Sprachschicht *f*

language processing Sprachverarbeitung *f*

language processing computer Sprachverarbeitungsrechner *m*

language recognition Spracherkennung *f*

lantern → (hall) lantern

lantern [*EN 40*] Mastleuchte *f*

lantern fixing, lantern fastening [*EN 40*] Mastleuchtenanschluss *m*

LAP, link access procedure Übertragungssteuer(ungs)verfahren, Leitungszugang(s)verfahren *n*

lap jointed window Stulpenfenster *n*

lap-riveted [*Said of two plates overlapped and then joined by riveting*] überlappt genietet

laptop computer tragbare Kompaktrechner *m*

larder sink → pantry sink

large calorie → kilocalorie

large domestic incinerator [*It has a combustion area not over 12 ft³ or a capacity not exceeding a 75–76 hours burning rate*] Haushalt(s)-Großverbrennungsanlage *f*

large fire Großbrand *m*, Großfeuer *n*

large fitting, rough fitting Grobarmatur *f*

large heat pump Großwärmepumpe *f*

large-scale-integrated circuit, LSIc großintegrierte Schaltung, Großschaltung *f* [*Schaltkreis mit Packungsdichten von etwa 500 bis 10 000 Thermistorfunktionen pro Chip*]

large(-scale) integration → LSI

large(-scale) kitchen Großküche *f*

large space (warm) air heating, (warm) air large space heating, hot air large space heating, large space hot air heating (Warm)Luftgroßraumheizung, Großraum(warm)luftheizung *f*

Lascar closet [*deprecated*] **→ squatting w.c.**

laser copying device Laserkopierer *m*

laser disk, laser disc Bildplatte *f* [*Auf ihr befinden sich analog abgespeicherte Videodaten*]

laser plotter Laserstrichzeichner *m*

laser print Laserdruck *m*

laser scanner Laserabtaster *m*, Laserlesegerät *n*

laserwriter, laserprinter Laserdrucker *m*

laserwriter driver, laserprinter driver Laserdruckertreiber *m*

last in first out → LIFO

last of maximal defuzzification, LOM defuzzification Rechts-Max-Methode *f*, Rechts-Max-Verfahren *n*

last-up detection Letztwertmeldung, Letztwerterkennung *f*, Letztwertnachweis *m*

latch, upper latch obere Pegelbereich *m* [*selten: n*] [*9 Bit. 16-Bit-A/D Wandler*]

L

latch, lower latch untere Pegelbereich *m* [*selten: n*] [*7 Bit. 16-Bit-A/D Wandler*]

latch Klinke *f*

latch-down and bolt shot riegelfest

latch French lock → **spring lock with falling latch**

latching mechanism, breaker mechanism Schaltschloss *n* [*Leistungsschalter*]

latch lock [*This type of lock has a spring bolt only and is operated by a handle on both sides. It is essentially a fastening for the easy and frequent closing and opening of a door without providing any security and so is generally only used on internal doors*] Klinkenschloss, Drückerschloss, Schnäpperschloss *n*

late-break contact Spätöffner, nacheilende Öffner *m*

latency [*In automatic data processing, the waiting time between the order to read/write some information from/to a specified place and the beginning of the data-read/write operation*] Wartezeit *f* [*DIN 19 237*]

latent air conditioner capacity → **(useful) latent air conditioner capacity**

latent cooling load latente Kühl(ungs)last *f* [*Raumlufttechnik. Erforderliche Wärmestrom, um einen Dampfmassenstrom bei Lufttemperatur zu kondensieren, so dass bei konstanter Lufttemperatur eine bestimmte absolute Luftfeuchte im Raum aufrechterhalten wird*]

latent heat [*Change of enthalpy during a change of state*] Latentwärme, latente Wärme *f*

latent heat(ing) load latente Heiz(ungs)last, latente Wärmelast *f* [*Raumlufttechnik. Erforderliche Wärmestrom um bei konstanter Raumlufttemperatur die Wassermasse zu verdunsten, die den Raum zum Aufrechterhalten einer bestimmten absoluten Luftfeuchte zugeführt werden muss*]

latent heat of condensation, condensation latent heat, latent condensation heat [*Quantity of heat released on change of unit mass of vapo(u)r to liquid with no change in temperature*] latente Kondensationswärme *f*

latent heat of fusion [*The amount of heat that must be added to the unit mass of a solid substance to change it to a liquid without any change in temperature*] latente Schmelzwärme *f*

latent heat of vapourization, vapourization latent heat (Brit.); latent heat of vaporization, vaporization latent heat (US); latent heat of volatilization, votatilization latent heat [*Heat energy required to cause a change of state from a saturated liquid to a saturated vapo(u)r, measured in Btu/lb$_m$ (J/kg)*] latente Verdampfungswärme *f*

latent heat storage, latent thermal storage [*Use of phase change of a medium for storing heating capacity*] Latentwärmespeicherung, latente Wärmespeicherung *f*

latent heat storage device, latent thermal storage device Latentwärmespeicher, latente Wärmespeicher *m*

latent humid heat → **humid latent heat**

latent root → **characteristic value**

lateral ... → **side ...**

late warning Spätwarnung *f*

lathe Drehbank *f*

lather dispenser Seifenschaumspender, Schaumseifenspender *m*

lathing Lattung *f*

lathing Rotieren *n* [*Gestaltung eines dreidimensionalen Objekts*]

latrine → **(pit) latrine**

lattice gate Gittertor *n*

lattice grill(e) Gitterrost *m*

lattice network → **iterative network**

laundry Wäsche *f*

laundry Wäscherei *f*

laundry cart Wäschereiwagen *m*

laundry centre; laundry center (US) Waschsalon *m*

laundry chute → **(inclined) laundry chute**

laundry dryer, laundry drier Wäschetrockner *m*

(laundry) lint extraction (Wäsche)Flusenabsaugung *f*

(laundry) lint screen (Wäsche)Flusensieb *n*

laundry steam Wäschereidampf *m*

lavalliere [*French*] Krawattenmikrophon *n*

lavatory, washroom [*A room equipped with basin(s) for washing hands and face*] Waschraum *m*

lavatory [*a euphemism*] → **toilet**

lavatory (basin) → **(wash)basin**

lavatory trough → **trough lavatory**

lavatory-vanity unit Waschbecken-Waschtisch-Einheit *f*

layer [*CD*] Informationsschicht *f*

layer asynchrone Fenster *n* [*Bei Datensichtstationen mit Bildfensterdarstellung*]

layer Schicht, Beschreibungsebene, Ebene *f*

layer 1 → **physical layer**

layer 2 → **data link layer**

layer 3 → **network layer**

layer 4 → **transport layer**

layer 5 → session layer

layer 6 → presentation layer

layer 7 → application layer

layer cryoton Schichtkryoton, Schichttieftemperaturschaltelement *n*

layering Gliederung *f* [*z. B. von Daten*]

layers Protokoll *n* [*Zwischen einem Leitrechner und einer Datensichtstation mit Bildfensterdarstellung*]

layer technique, level technique Ebenentechnik *f*

laying → (pipe) laying

laying up → shutdown

to lay out → to arrange

layout, arrangment, disposition Anordnung *f*

layout [*First drawings of a system showing the relation of the system components. Usually followed by dimensions and notes*] Auslegung *f*

to lay up, to shut down stillegen, stillsetzen

LCB, line ((open) loop) control block Leitungssteuer(ungs)block *m*

LCD → liquid crystal display

LC-OFC-cable, linear oxygen free copper cable Lautsprecherkabel *n* aus sauerstoffreiem Kupfer

LCR → least cost routing

LDR → light dependent resistor

leachability of lead, lead leachability Bleiabgabe *f*

lead Blei *n*

lead-acid battery Bleisäurebatterie *f*

lead bath(tub) trap [*See remark under "trap"*] Blei-Badewannen-(Wasser)(Geruch)Verschluss *m*

lead block, lead module Vorhalt(e)baustein *m*, Vorhalt(e)modul *n*

lead draining pipe, lead drain(age) pipe, lead discharge pipe, lead waste pipe Bleiabflussrohr, Bleiablaufrohr, Bleientwässerungsrohr, Bleidränrohr *n* [*Fehlbenennungen: Bleiabfluss m, Bleiablauf m*]

leader → downspout

lead-in Durchführung *f* [*z. B. durch ein Fenster*]

lead-in bell, weather head Übergang(s)kopf *m* [*Leitungseinführung*]

lead-in connection piece Einführungsanschlussstück, Einführungsverbindungsstück *n*

leading auxiliary switch voreilende Hilfsschalter, VHi *m*

leading depth → total depth

leading dimension → total dimension

leading-edge method Direkt-Laufzeitmessung von Ultraschallimpulsen, Laufzeit-Direktmessung von Ultraschallimpulsen *f*

leading make contact Frühschließer *m*

lead jacket → lead sheath

lead/lag module, lead/lag block Vorhalt(e)verzögerungsbaustein *m*, Vorhalt(e)verzögerungsmodul *n*

lead leachability → leachability of lead

lead-out terminal → bushing terminal

lead module → lead block

lead pipe Bleirohr *n*

lead (pipe)line Blei(rohr)leitung *f*

to lead-seal plombieren

lead-seal Plombe *f*

lead-seal borehole Plombierbohrung *f*, Plombierbohrloch *n*

lead-seal screw Plombierschraube *f*

lead sheath, lead jacket Bleimantel *m*

lead trap [*See remark under "trap"*] Blei-(Wasser-)(Geruch)Verschluss *m*

lead washer Bleiunterlegscheibe *f*

lead waste pipe → lead draining pipe

lead wire → supply conductor

leak Leck *n*, undichte Stelle *f*

leakage, leaking Lecken *n*, Leckage *f*

leakage air, leaking air Leck(age)luft *f*

leakage air flow Leck(age)luftstrom *m*

leakage conductance, leakance, leakage permeance Streuleitwert, Kehrwert des Isolationswiderstandes *m*

leakage current → stray current

leakage detection Leck(age)meldung, Leck(age)erkennung *f*, Leck(age)nachweis *m*

leakage detection cable Leck(age)meldekabel *n*

leakage detecting system, leakage detection system Leck(age)meldesystem *n*

leakage detector, leakage detecting element, leak-age detection element Leck(age)melder, Leck(age)detektor, Leck(age)meldungsgeber, Leck(age)sucher *m*, Leck(age)meldegerät, Leck(age)-suchgerät *n*

leakage determination Leck(age)feststellung *f*

leakage distribution Leckageverteilung *f*

leakage inductance Streuinduktivität *f*

leakage loss Streuverlust *m*

leakage path Streuweg, Streupfad *m*, Streuspur *f*

leakage permeance → leakage conductance

L

leakage power Verlustleistung (durch Ableitströmung), Leck(age)leistung *f*

leakage reactance Streureaktanz *f*, Streublindwiderstand *m*

leakage suppresion winding Schubwick(e)lung *f*

leakage tester, leakage testing device, leakage testing appliance Leck(age)sucher *m*, Leck(age)-suchgerät *n*

leakage tightness, leakproofness Leck(age)freiheit *f*

leakage water detector, leakage water detecting device, leakage water detection device Leck(age)-wassermelder, Leck(age)wasserdetektor *m*, Leck-(age)wassermeldegerät *n*

leakance → leakage conductance

leaking → leakage

leakless, leakproof lecksicher, leckfest, leckfrei

leak-off (pipe)line Leck(rohr)leitung *f*

leak plugging Leckstellen(ab)dichtung *f*

leakproofness → leakage tightness

leaky undicht, leck

L

lean current Magerstrom, Armstrom *m*

lean flame Magerflamme, Armflamme *f*

lean gas [*Gas of relatively low calorific value*] Armgas, Magergas *n*

leased-circuit data network Direktrufnetz *n*

least cost routing, LCR gebührenoptimale Nutzung, kostenoptimale Nutzung *f* [*Telefonnetzverbund*]

least significant bit, LSB, least significant binary digit Bit mit dem niedrigsten Stellenwert, niedrigstwertige Bit, niedrigwertigste Bit, Niedrigstwertbit, geringwertigste Bit, geringstwertige Bit *n*, Binärstelle mit dem niedrigsten Wert, niedrigwertigste Binärstelle, niedrigstwertige Binärstelle *f*

least significant digit, LSD Ziffer mit dem niedrigsten (Stellen)Wert, niedrigstwertige Ziffer, niedrigwertigste Ziffer, Niedrigstwertziffer *f*

least significant number, LSN Zahl mit dem niedrigsten (Stellen)Wert, niedrigstwertige Zahl, niedrigste Wertigkeitszahl, niedrigwertigste Zahl, Niedrigstwertzahl *f*

leather bellow Lederbalg *m*

leather gasket [*See remark under "gasket"*] Leder-(ab)dicht(ungs)scheibe *f*

leaving abgehend, ausgehend, austretend

leaving temperature Ausgang(s)temperatur, Abgang(s)temperatur *f*

LED → light emitting diode

LED display, light admitting diode display, light emitter diode display Leuchtdiodenanzeige *f*

ledge-back lavatory (basin), ledge-back (wash) basin Waschbecken mit Rückwand, Waschbecken mit Spritzwand *n*, Waschtisch mit Rückwand, Waschtisch mit Spritzwand *m*

ledged door → batten door

left half *s*-plane, LHP linke *s*-Halbebene *f*

left-hand bowl sink Linksbeckenspüle *f*

left-hand installation Linkseinbau *m*

left side view Seitenansicht von links, linke Seitenansicht *f*

left-thread nipple Linksgewindenippel *m*

LEH → energy-saving house

Leidenfrost's phenomenon [*A phenomenon in which a liquid dropped on a surface that is above a critical temperature becomes insulated from the surface by a vapo(u)r layer, and does not wet the surface as a result*] Leidenfrostsche Phänomen *n*

leisure centre (Brit.); leisure center (US) Freizeitzentrum *n*

LEMP → lightning electromagnetic impulse

length [*One of two horizontal dimensions, normally the greater*] Länge *f*

lenghtening Verlängerung *f*, Ansatz *m*

lengthening piece Verlängerungsstück, Ansatzstück *n*

length limiter, length limiting device Längenbegrenzer *m*

length overall → total length

lens-fault optische Abbildungsfehler *m* [*Die optischen Abbildungsfehler eines Kameraobjektivs werden zur Steigerung des Wirklichkeitseindrucks in eine Rechnergraphik eingerechnet*]

lens floodlight Linsenscheinwerfer *m*

Lenz's law [*Electromagnetism. A law stating that an induced electromotive force in a conductor is always polarized in a direction so as to oppose the change that causes the induced electromotive force*] Lenzsche Regel, Behinderungsregel *f*

LEO → energy-saving office

let-in flap Einstemmlappen, Fi(t)schen(band)lappen *m*

to let off ablassen

let-out area Ausgabestauraum *m*

letter-associated, letter-related buchstabenbezogen

letter bomb, postal bomb Briefbombe *f*

letter box, postal box Briefkasten *m*

letter-box plate with bell-push and card-holder, letter(-drop) plate with bell-push and card holder Briefeinwurf *m* mit Klingelknopf und Namensschildausschnitt

letter chute, mail chute, inclined mail chute, inclined letter chute Briefrutsche, Postrutsche *f*

letter demanding payment → prompt note

lettering Beschriftung *f*

letter keyboard Buchstabentastatur *f*, Buchstabentastenfeld *n*

letter(-keyed) lock, wheel lock Buchstabenschloss, Alphabetschloss *n*

letter plate, letter-box plate, letter-drop plate [*A front-door fitting to receive letters, consisting of a slotted plate fixed to the outside of the door. B.S. 2911*] Briefeinwurf *m*

letter quality, LQ Buchstabengüte, Briefqualität *f* [*Fehlbenennung*]

letter slot → mail slot

letter symbol [*IEC 27*], **code letter, distinctive letter, classification letter** Kennbuchstabe *m*, Buchstabenkennzeichen *n*

letter(-type) combination lock, letter(-type) dial lock Buchstabenschloss *n*

letting-in flush (on both meeting faces), flush fixing Einstemmen *n* [*Einstemmband*]

to level egalisieren, abgleichen, nivellieren [*Nicht im Sinne von „vermessen"*]

level → flush (with)

level → channel

level [*automatic telephony*] Gruppenschritt *m*

level [*automatic telephony*] Höhenschritt *m* [*Hebedrehwähler*]

level Ebene *f* [*Hilfsmittel zur Strukturierung von Bildinhalten*]

level, intensity, strength Stärke, Intensität *f*

level → intensity

level Pegel *m* [*Logarithmierte Verhältnis der Beträge zweier Leistungen oder Feldgrößen*]

level, flat, even, plane eben, flach

level → floor

level change value [*automatic control*] Sprungwert *m*

level connection Ebenenverbindung *f*

level control → ((closed) loop) level control

level control → ((open) loop) level control

level controller Niveauregler, (Höhen)Standregler, Pegelregler, Höhenregler *m*

level converter Pegelumsetzer *m* [*Siehe Anmerkung unter „Pegel"*]

level crossing distribution Klassendurchgang(s)verteilung *f*

level-dependent niveauabhängig, (höhen)standabhängig, pegelabhängig, höhenabhängig

level detector, level detecting device, level detection device Niveaumelder, Höhenmelder, (Höhen-)Standmelder *m*, Niveaumeldegerät, (Höhen)Standmeldegerät, Höhenmeldegerät, Pegelmeldegerät *n*

level difference → (equivalent sound pressure) level difference

level gauge; level gage (US) Niveaumesslehre, Pegelmesslehre, (Höhen)Standmesslehre *f*

level indicator Standanzeiger *m*

level limiter, level limiting device Pegelbegrenzer, Höhenbegrenzer, Niveaubegrenzer, (Höhen)Standbegrenzer *m*

level line Niveaulinie, Höhen(stand)linie, Standlinie, Pegellinie *f*, Meterriss *m*

level measurement (Höhen)Standmessung, Höhenmessung, Niveaumessung, Pegelmessung *f*

levelness [*No to be confused with "flatness"*] Ebenheit *f*

level of authority, authority level Kompetenzstufe *f*

level of deviation → deviation level

level of heat, heat level Wärmestand, Wärmepegel *m*

level of security, security level Sicherungsstufe *f*

level of training, training level [*operating staff*] Ausbildungsstand *m*

level ((open) loop) control Niveausteuerung *f*

level probe Niveausonde, (Höhen)Standsonde, Höhensonde, Pegelsonde *f*

level recorder controller Registrier-Niveauregler, schreibende Niveauregler *m*

level reduction Pegelminderung *f*

level relay Niveaurelais, (Höhen)Standrelais, Pegelrelais, Höhenrelais *n*

level 1 relay → repeater

level 3 relay → router

level sensor Niveausensor, (Höhen)Standsensor, Pegelsensor, Höhensensor *m*

level switch Niveauschalter, Pegelschalter, (Höhen-)Standschalter, Höhenschalter *m*

level switching Niveauschaltung, Pegelschaltung, (Höhen)Standschaltung, Höhenschaltung *f*

level technique → layer technique

level (to) → flush (with)

level value Pegelwert, Niveauwert, (Höhen)Standwert, Höhenwert *m*

lever(-actuated) switch, lever-operated switch Hebelschalter, Griffschalter, Greifschalter *m*

lever bolt Treibriegel *m*

lever (door) handle, door lever handle [*A door handle in the form of a pivoted horizontal bar, which is pushed downwards to withdraw the latch*] Drücker, Türdrücker *m*

lever handle Hebelgriff *m*

lever mechanism Hebelmechanismus *m*

lever mixer Hebelmischer *m*

lever mixing tap Hebelmischer, Hebelmischhahn *m*

lever-operated, lever-actuated hebelbedient, hebelbetätigt

lever(-operated) limit switch, lever-actuated limit switch Hebel(grenz)schalter, Griff(grenz)schalter, Greif(grenz)schalter *m*

lever(-operated) linkage mechanism, lever-actuated linkage mechanism Hebelgestänge(an)trieb *m* [*Schaltgerät*]

lever(-operated) mixer (valve), lever(-operated) mix(ing) valve, lever(-operated) blender, mechanical mixer, lever(-operated) blending valve, single-lever mixer (valve), one-lever mixer (valve) Hebelmischbatterie, Mischhebelbatterie *f*, Einhebelmischer *m*

L

lever scale Hebelwaage *f*

(lever) tumbler Zuhaltung *f*

(lever) tumbler lock Zuhaltungsschloss *n*

Lewis key Tangentialpassfeder, Pass-Tangentialfeder *f*

LF → laminar flow

LF → line feed

LHP → left-half *s*-plane

liability defect Gewährleistungsmangel *m*

library Bibliothek *f* [*In der Elektronik eine Sammlung von Unterprogrammen, die sich an beliebiger Stelle in das Hauptprogramm einfügen lassen*]

Library of Recommended Architecture, LORA Bibliothek *f* der empfohlenen Architektur des 20. Jahrhunderts [*Sie erlaubt als digitales Nachschlagewerk ein Projekt zur Aufnahme vorzuschlagen und mit Texten und Bildern zu erläutern*]

library routine Bibliotheks(verwaltungs)programm *n*

licence provider Lizenzgeber *m*

lid interlock switch, interlock lid switch Deckelverrieg(e)lungsschalter, Deckelabrieg(e)lungsschalter *m*

life (cycle), lifetime, life span, service life Nutzungsdauer, Lebensdauer *f*

life (cycle) activity Nutzungsdauertätigkeit, Lebensdauertätigkeit *f* [*Die Summe aller Tätigkeiten einer gebäudetechnischen Anlage während ihrer Lebensdauer*]

life (cycle) cost(s) Lebensdauerkosten, Nutzungsdauerkosten *f*

life (cycle) expectancy Nutzungsdauererwartung, Lebensdauererwartung *f*

life (cycle) monitoring → service life monitoring

life hazard → hazard to life

life safety Personensicherheit *f*

life security Personensicherung *f*

life-zero lebende Nullpunkt *m* [*In Einheitssignalen mit lebendem Nullpunkt ist dem Messanfang nicht der Signalwert Null, sondern ein höherer Wert zugeordnet*]

LIFO, last in first out Lesen *n* der zuletzt gespeicherten Informationen

to lift, to elevate, to hoist heben

lift (Brit.) → elevator (US)

lift → ejector [*sewage*]

lift and lay valve → globe valve

lift attendant (Brit.) → (passenger) lift attendant

(lift) car (Brit.); (elevator) car (US) Aufzugkorb, Fahrkorb *m*

lift car access (Brit.); (elevator) car access (US) Aufzugkorbzugang, Fahrkorbzugang *m*

lift car annunciator (Brit.); (elevator) car annunciator (US) Aufzuggeschossanzeiger, Aufzugstockwerk(s)anzeiger, Aufzugetagenanzeiger *m*

lift car door (Brit.); (elevator) car door (US) Aufzugkorbtür, Fahrkorbtür *f*

lift car guide rail (Brit.); (elevator) car guide rail (US) Aufzugkorbführungsschiene, Fahrkorbführungsschiene *f*

lift car illumination (Brit.); (elevator) car illumination (US) Aufzugkorbbeleuchtung, Fahrkorbbeleuchtung *f*

lift car location indication (Brit.); (elevator) car location indication (US) Aufzugkorbstandortanzeige, Fahrkorbstandortanzeige *f*

lift car separating door (Brit.); (elevator) car separating door (US) Aufzugkorbtrenntür, Fahrkorbtrenntür *f*

lift channel [*South Africa*] **→ hoistway**

lift control (Brit.); elevator control (US) Aufzugsteuerung *f*

lift controller (Brit.); elevator controller (US) Aufzugsteuer(ungs)schalter *m*

lift engineering (Brit.); elevator engineering (US) Aufzugtechnik *f*

lift(ing) door → elevating door

lift(ing) gate, elevating gate Hubtor, Hebetor *n*

lift(ing) installation, hoist(ing) installation, elevating installation Hebeanlage, Hubanlage *f* [*Abführung von dauernd zufließendem Wasser unterhalb der Rückstauebene*]

lift(ing) machine → hoist(ing) machine

lift(ing) platform, hoist(ing) platform, elevating platform Hebebühne, Hubbühne, Hebeplattform, Hubplattform *f*

lift(ing) pump, hoist(ing) pump, elevating pump Hebepumpe, Hubpumpe *f*

lift(ing) rope → hoist(ing) rope

lift(ing) sliding door Hebe-Schiebe-Tür, Hub-Schiebe-Tür *f* [*Der Flügel wird in geschlossenem Zustand über den Beschlag abgesenkt*]

lift(ing) sliding tilt(ing) door Hebe-Schiebe-Kipptür, Hub-Schiebe-Kipptür *f* [*Der Flügel wird in geschlossenem Zustand über den Beschlag abgesenkt und die Tür kann gekippt werden*]

lift(ing) speed → hoisting speed

lift latch (US) → Canadian latch

lift-off, blow-off, flame failure, flame outage, flame lift-off Abreißen, Ausfallen, Versagen, Abheben *n* [*Flamme*]

lift potential switch (Brit.); elevator potential switch (US) Aufzugnullspannungsschalter *m*

lift slack cable switch (Brit.); elevator slack cable switch (US) Aufzugschlaffseilschalter *m*

(lift) wall (Brit.) → hoistway wall (US)

lift winch (Brit.); elevator winch (US) Aufzugwinde *f*

light [*Radiating energy evaluated according to its capacity to produce visual sensation*] Licht *n*

light ... → illuminated ...

light absorption smoke detector, light absorption smoke detecting device, light absorption smoke detection device Durchlichtrauchmelder, Durchlichtrauchdetektor, Durchlichtrauchmeldungsgeber *m*, Durchlichtrauchmeldegerät *n*

light admission → admission of light

light amount → amount of light

light barrier Lichtschranke, Lichtsperre *f*

light barrier electronics Lichtschrankenelektronik, Lichtsperrenelektronik *f*

light barrier ((open) loop) control Lichtschrankensteuerung, Lichtsperrensteuerung *f*

light beam Lichtstrahl *m*

light beam curv(atur)e Lichtstrahlkrümmung *f*

light beam reach Lichtstrahlreichweite *f*

light brightness ((closed) loop) control (Licht)Helligkeit(s)reg(e)lung *f*

light bulb, lamp bulb [*The glass envelope enclosing the luminous element or material of an electric lamp; usually made of glass, quartz, or similar material; its shape usually is designed by a letter (e.g. T – tubular, G - globe, etc.), followed by a number which indicates the maximum diameter of the bulb in lengths of an inch*] Birne *f*

light carrier Lichtträger *m*

light ceiling Lichtdecke *f*

light ((closed) loop) control Lichtreg(e)lung *f*

light colored (US); light coloured (Brit.) hellfarbig

light colour (Brit.); light color (US) Hellfarbe *f*

light control → light ((open) loop) control

light control element → light ((open) loop) control element

light controller Lichtregler *m*

light current Lichtstrom *m*

light current, weak current Schwachstrom *m*

light current (distribution) network → weak current (distribution) network

light density Lichtdichte *f*

light-dependent resistor, LDR, light-sensitive resistor lichtabhängige Widerstand, Photowiderstand, Fotowiderstand *m*

light designer Lichtgestalter *m*

light diffusion Lichststreuung *f*

light-directing lichtlenkend, lichtrichtend

light distribution Lichtverteilung *f*

light emitting diode, LED, light emitter diode lichtemittierende Diode, Leuchtdiode, Lumineszenzdiode, Halbleiterdiode *f*

light emitting diode display → LED display

light engineering Lichttechnik *f*

lightening method gleichzeitige Darstellung *f* der Komponenten Signale Y und U und V [*elektronische Bildübertragung*]

light evaluation Lichtmessebene *f* [*Video-Alarm-System*]

light filtration Lichtfiltration *f*

light frame Lichtrahmen *m*

light fuel oil [*200 seconds*] Heizöl-L, leichtflüssige Heizöl, HL *n*

L

light guide → **optical fibre**

light heating load Schwachheiz(ungs)last *f*

light heating oil → **fuel oil No. 2**

light incidence → **incidence of light**

light incidence angle [*See remark under "angle of incidence"*] Lichteinfall(s)winkel *m*

light(ing) ... → **illumination ...**

light(ing) fitting → **luminaire**

light(ing) fixture → **luminaire**

lighting installation, lighting plant Beleuchtungsanlage *f*

lighting torch Zündbrenner *m*

light in (push) button → **(push) button light**

light liquid Leichtflüssigkeit *f* [*Ihre Dichte ist kleiner als die Wasserdichte*]

light liquid separator Leichtflüssigkeit(s)(ab)scheider *m*

light loss factor, maintenance factor Verminderungsfaktor *m* [*Beleuchtungstechnik*]

light measured-value → **light pickup**

L

light-metal baseboard heating element Leichtmetallsockelheizkörper, Leichtmetallfußleistenheizkörper *m*, Leichtmetallheiz(ungs)leiste *f*

light microscope → **optical microscope**

light mirror Lichtspiegel *m*

lightning arrester [*A protective device that allows voltages higher than a certain value to be discharged to ground*] Blitzableiter, Überspannungsableiter *m*

lightning channel [*The superheated, irregular, ionized pathway in the atmosphere created by the passage of a lightning bolt*] Blitzkanal *m*

lightning conductor [*Any pathway that lightning can follow to ground, such as the trunk of a tree, a power line, or a communication antenna*] Blitzfortleiter *m*

lightning counter Blitzzähler *m*, Blitzzählgerät *n*

lightning current → **lightning (stroke) current**

lightning damage Blitzschaden *m*

lightning danger Blitzgefahr *f*

lightning discharge [*The neutralization of a large electric(al) potential in the atmosphere*] Blitzentladung *f*

lightning electromagnetic impulse, LEMP elektromagnetische Blitzstoß *m*

lightning [*A luminous, high-current, electric discharge with a long path that flows between a cloud and the ground, between two clouds, or between two parts of the same cloud*] Blitz *m*

lightning generator [*A generator that produces surge voltage similar to lightning, for testing insulators and other high-voltage components*] Kurzschlussgenerator, Stoß(leistungs)generator *m*

lightning impulse protection level Blitzstoßspannungspegel *m*

lightning overvoltage Blitzüberspannung *f*

lightning protection, protection (of structures) against lightning Blitzschutz *m*

lightning protection capacitor Blitzschutzkondensator *m*

lightning protection grounding Blitzschutzerdung *f*

lightning protection plant, lightning protection installation, lightning protection system Blitzschutzanlage *f*, Blitzschutzsystem *n*

lightning protection potential equalization Blitzschutz-Potentialausgleich *m* [*Das Verbinden der Blitzschutzanlage mit Rohrleitungen und anderen metallenen Installationen, mit Erdungsanlagen und geerdeten Teilen von Starkstrom- und Fernmeldeanlagen*]

lightning storm, electrical storm Blitzgewitter *n*

lightning stroke Blitzschlag *m*

lightning (stroke) current Blitz(schlag)strom *m*

lightning surge [*A transient electric disturbance in an electric circuit produced by lightning*] Blitzstoß *m*

lightning surge flashover voltage Überschlag-Blitzstoßspannung *f*

lightning surge protection ratio Blitzstoßüberspannungsverhältnis *n*

lightning surge sparkover voltage Ansprechblitzstoßspannung *f*

lightning surge sparkover voltage/time curve Ansprechkennlinie *f* der Blitzstoßspannungen

lightning surge voltage Blitzstoßspannung *f*

lightning surge (voltage) strength Blitzstoßspannungsfestigkeit *f*

lightning switch; earthing switch [*IEC 129*]; **grounding switch (US)** [*A manually operat-ed switch that is employed during electrical storms to connect a radio antenna to ground rather than to the radio receiver*] Erdungsschalter *m* [*VDE 0670, Teil 2*]

light oil → **distillate fuel oil**

light-oil burner Leichtölbrenner *m*

light ((open) loop) control Lichtsteuerung *f*

light ((open) loop) control element Lichtsteuer(ungs)element *n*

light organ Lichtorgel *f*

light pen [*A peripheral used to position a cursor on a VDU by means of pointing at the screen*] Lichtgriffel, Lichtstift, elektronische Zeichenstift *m*

light penetration Lichteintritt *m*

light-pen programmer Lichtgriffel-Programmiergerät *n*

light pickup, light measured-value Lichtmesswertgebertransmitter *m*

light planning Lichtplanung *f*

light point Lichtanschlussstelle *f*

light PVC-sheathed flexible cord leichte PVC-Schlauchleitung *f* [*DIN VDE 0281 Teil 401*]

light quantum → photon

light receptor Lichtempfänger *m*

light reflection value Lichtreflektionswert *m*

light response characteristic Lichtempfindlichkeits-Charakteristik *f*

light response curve Lichtempfindlichkeitskurve *f*

light screen Lichtblende *f*

light-sensitive resistor → light-dependent resistor

light sensor Lichtsensor *m*

light shaft → air shaft

light shaft grate → air shaft grate

light shaft security → air shaft security

light shelf Lichtregal *n*

light signal Lichtsignal *n* [*z. B. eine Blitzleuchte*]

light speed → speed of light

light switch Lichtschalter *m*

lighttight lock Lichtschleuse *f*

light time switch Lichtzeitschalter, Zeitlichtschalter *m*

light transmitter, opto-transmitter Lichtsender *m*

light trap Lichtfalle *f*

light wave Lichtwelle *f*

light well → air shaft

light well grate → air shaft grate

light well security → air shaft security

lime-carbonic acid equilibrium Kalk-Kohlensäure-Gleichgewicht *n*

lime control, lime suppression Kalkbekämpfung *f*

lime dissolver Kalk(auf)löser *m*

lime incrustation, lime crust formation Kalkverkrustung, Kalkkrustenbildung *f*

lime scale, lime crust, lime deposit Kalkkruste, Kalkablagerung *f*, Kalkansatz, Kalkbeleg *m*

lime spot Kalkfleck *m*

liming-up Verkalken *n*, Verkalkung *f*

limitation Begrenzung *f*

limited distance modem → baseband modem

limited refrigerant charge [*Charge that does not exceed the design pressure at complete evaporation*] begrenzte Kältemittelfüllung *f*

limiter, limit(ing) device Begrenzer *m*, Begrenzungsvorrichtung *f* [*Vorrichtung, die bei Erreichen des eingestellten Grenzwertes den Betrieb unterbricht und nicht wieder einschaltet*]

limiter diode → breakdown diode

limiting Begrenzen *n*

limit(ing) ((closed) loop) control Begrenzungsreg(e)lung, Grenzreg(e)lung *f*

limit(ing) contact maker, limit(ing) contactor, limit(ing) contact mechanism Begrenzungskontaktgeber *m*

limit(ing) continuous current, continuous limit(ing) current [*relay; output circuit. IEC 50 (446)*] Grenzdauerstrom, Begrenzungsdauerstrom *m*

limit(ing) control → limit(ing) ((open) loop) control

limit(ing) controller, limit(ing) CTRL [*See remark under "controller"*] Begrenzungsregler, Grenzregler *m*, Begrenzungsorgan, Grenzorgan *n*, Begrenzungsregel(ungs)einheit, Grenzregel(ungs)einheit *f*

limit(ing) current Begrenzungsstrom, Grenzstrom *m*

limit(ing) current calibration Begrenzungsstromeichung, Grenzstromeichung *f*

limit(ing) deviation, deviation limit(ing) Begrenzungsabmaß, Grenzabmaß *n*

limit(ing) element, limiter element Begrenzungsglied, Grenzglied *n*

limit(ing) frequency Grenzfrequenz, Begrenzungsfrequenz *f*

limit(ing) function Begrenzungsfunktion, Grenzfunktion *f*

limit(ing) indicating monitor → indicating limit(ing) monitor

limit(ing) monitor Grenzwertüberwacher *m*

limit(ing) ((open) loop) control [*A safety device on a boiler, refrigerator, or air-conditioning system which shuts off the system and actuates alarms when unsafe conditions are detected*] Begrenzungssteuerung, Grenzsteuerung *f*

limit(ing) pulse Grenzimpuls, Begrenzungsimpuls *m*

limit(ing) quality level, LQL rückzuweisende Qualitätsgrenz(lag)e, Ausschussgrenz(lag)e *f*

limit(ing) signal Grenzsignal, Begrenzungssignal *n*

limit(ing) signal transmission Grenzsignalübertragung, Begrenzungssignalübertragung *f*

limit(ing) switch Grenzschalter, Endlagenschalter, Begrenzungsschalter, Endstellungsschalter *m*

limit(ing) thermostat Begrenzungsthermostat, Grenzthermostat, Begrenzungstemperaturregler, Grenztemperaturregler *m*

limit(ing) value Begrenzungswert, Randwert, Grenzwert *m* [*DIN 40 200, Oktober 1981*]

limit(ing) value monitoring Grenzwertüberwachung, Begrenzungswertüberwachung *f*

limit(ing) value of mean on-state current Dauergrenzstrom, Dauerbegrenzungsstrom *m* [*Thyristor. DIN 41 786*]

limit signal Grenz(wert)signal *n*

limit violation → **off-limit condition**

line Linie *f*

line Zeile *f*

line → **channel**

line → **communication path**

line → **(pipe)line**

L

line Strecke *f*

line adapter → **(pipe)line adapter**

linear absorption heat detector, linear absorption heat detecting device, linear absorption heat detection device linienförmige Wärmemelder, Extinktionswärmemelder *m*, linienförmige Wärmemeldegerät, Extinktionswärmemeldegerät *n* [*1. Er wertet eine durch Wärme hervorgerufene Veränderung des Isolationswiderstandes zwischen zwei Leitern eines Kabels oder die Veränderung des Leitwertes aus. 2. Sein Fühler besteht aus einem dünnen Kupferrohr, dessen in ihm befindliche Luft sich bei Erwärmung ausdehnt*]

linear absorption smoke detector, linear absorption smoke detecting device, linear absorption smoke detection device linienförmige Rauchmelder, Extinktionsrauchmelder *m*, linienförmige Rauchmeldegerät, Extinktionsrauchmeldegerät *n*

linear act(uat)or, linear final controlling element Linearstellglied *n* [*Siehe Anmerkung unter „Stellglied"*]

linear adaptive element → **adaptive linear element**

linear (air) diffuser → **slot(ted) (air) diffuser**

linear controller Linearregler, Schieberegler *m*

linear crystal oxygen free copper cable → **LC-OFC-cable**

linear displacement lineare Verdrängung *f*

line(ar electrical) parameter → **transmission-line constant**

linear final controlling element → **linear act(uat)or**

linear interpolation Geradzwischenpolation *f* [*Gerechnete Verbindung programmierter Punkte zu einer Geraden, d. h. die die Punkte verbindenden Segmente sind gerade Linien*]

linearity [*The relationship that exists between two quantities when a change in one of them produces a directly proportional change in the other*] Linearität *f*

linearity Linearität *f* [*Genauer: Nichtlinearität. Die maximale Abweichung der Übertragungsfunktion eines Analog-Digital-Umsetzers oder Digital-Analog-Umsetzers von der linearen Idealkurve*]

linearity error Linearitätsfehler *m*

linearity fluctuation Linearitätsschwankung *f*

linearization Linearisierung *f*

linear language lineare Sprache *f*

linear measurement method → **direct measurement method**

linear movement electric act(uat)or with electric motor Schubstellglied *n*, Schub(an)trieb *m*

linear output Hall effect transducer, LOHET integrierte Halleffektsensor *m* [*Nach dem Halleffekt arbeitender Sensor mit Hallelementen auf Halbleiterbasis, meist mit Konstantstromquelle und Verstärker*]

line arrangement, lineup Ausrichtung *f*

linear resistor lineare Widerstand *m* [*Widerstand mit linearer I-O-Kennlinie*]

linear smoke detector, linear smoke detecting device, linear smoke detection device lineare Rauchmelder *m*, lineare Rauchmeldegerät *n*

linear-time invariant ((closed) loop) control lineare zeitvariante Regel(ungs)system, zeitvariante lineare Regel(ungs)system *n*

linear variable-differential transformer → **differential transformer**

line-balance converter → **balun**

line-bound leitungsgebunden

line breakage monitoring, line fracture monitoring Leitungsbruchüberwachung *f*

line-building-out network → **impedance matching network**

line compensation Leitungsabgleich *m* [*Messgerät*]

line compensation resistance Leitungsabgleichwiderstand *m* [*Messgerät*]

line component Leitungs(bestand)teil *m, n*

line conditioner Leitungskühlmittel *n*

line connection Leitungsanschluss *m*, Leitungsverbindung *f*

line control Leitungssteuerung *f*

line control block → **LCB**

line cord → **power cord**

line core; line insulated conductor (US) Linienader *f*

line cross section Leitungsquerschnitt *m*

line coupler Linienkopp(e)ler *m* [*Komponente im Installationsbussystem, um Linien miteinander zu verbinden. Telegramme werden über einen Linienkoppeler weitergegeben oder auf die Linie begrenzt*]

lined verkleidet

line deflection Zeilenablenkung *f*

line deletion Zeilenlöschung, Zeilentilgung *f*

line disturbance → **line trouble**

line ditch Leitungsgraben *m* [*V-förmig*]

line-dotted strichpunktiert, punktgestrichelt

line driver Zeilentreiber *m*

line duct Leitungskanal *m*

line end amplifier → **end-of-line amplifier**

line end character → **end-of-line character**

line end resistor → **end-of-line resistor**

line feed, LF Zeilenvorschub, Zeilenschritt *m* [*Formatssteuer(ungs)funktion*]

line feed inseration Einfügen von Zeilenvorschub, Zeilenvorschubeinfügen *n* [*Leitungsmerkmal*]

line feed padding Füllzeichen *n* nach Zeilenvorschub [*PAD-Parameter; Leitungsmerkmal*]

line fire detector, line fire detecting device, line fire detection device Linienbrandmelder, Linienfeuermelder *m*, Linienbrandmeldegerät, Linienfeuermeldegerät *n*

line flyback, horizontal flyback Zeilenrücklauf *m* [*Bildschirmgerät*]

line fracture monitoring → **line breakage monitoring**

line frequency → **horizontal frequency**

line fuse, line fuze Leitungssicherung *f*

line graph, graphic curve graphische Darstellung *f* (durch Schaulinie(n)) [*DIN 461*]

line heat detector, line heat detecting device, line heat detection device linienförmige Wärmemelder *m*, linienförmige Wärmemeldegerät *n* [*Er arbeitet mit einer Auswerteeinheit und einem Detektionskabel oder Rohr*]

line holder Leitungshalter *m*

line home position Zeilenanfang *m* [*Kode*]

line in/ line out Übertragung der reinen Audiosignale, reine Audiosignalübertragung *f*

line installation → **line laying**

line insulated conductor (US) → **line core**

line interruption Leitungsunterbrechung *f*

line interval → **line spacing**

line jump Zeilensprung *m*

line key Leitungstaste(r) *f, (m)*

line laying, line installation Leitungsverlegung *f*, Leitungseinbau *m*

line length Leitungslänge *f*

line length Zeilenlänge *f*

line level [*Communication. Signal level in decibels at a particular position on a transmission line*] Leitungsschalldruckpegel *m*

line lock Netzsynchronisierung *f* [*Kamerafunktion*]

line loss → **(pipe)line loss**

line manipulated variable, section manipulated variable [*ANSI C 85.1*] Streckenstellgröße *f* [*DIN 19 226*]

line mixer → **flow mixer**

line monitoring Leitungsüberwachung *f*

line multiplier Linienmultiplikator *m*

linen chute Wäscherutsche *f*

linen cupboard Wäscheschrank *m*

linen dryer, linen drier Wäschetrockner *m*

linen heater Wäsche(er)wärmer *m*

line node Leitungsknoten *m*

line noise → **(pipe)line noise**

line number [*A number used to identify a line of text in a file processed by a text editor*] Zeilennummer *f*

line of equal dryness [*Mollier diagram*] Linie *f* gleicher spezifischer Dampfmenge

line of force, force line Kraft(fluss)linie *f*

line ((open) loop) control Leitungssteuerung *f*

line ((open) loop) control block → **LCB**

line output transformer Zeilenausgangs-Transformator *m*

line parameter → **transmission-line constant**

line penetration → **service penetration**

line pipe Leitungsrohr *n*

line plotter Strichzeichner *m*

line printer, LP Zeilendrucker, Paralleldrucker *m*

line project Leitungsbauvorhaben *n*

line reaction → **(manipulated) line reaction**

line resolution Zeilenauflösung *f*

line routing Leitungsführung, Leitungstrassierung *f*

L

line scanning Zeilenabtasten *n*

line section Leitungsstrecke *f*, Leitungsabschnitt *m*

line segment Polylinie *f*

line shaft Leitungsschacht *m*

line side [*The side of a device electrically closest to the current source*] Netzseite *f*

line size Spalten pro Zeile, Spalten je Zeile *f*

line smoke detector, line smoke detecting device, line smoke detection device linienförmige Rauchmelder *m*, linienförmige Rauchmeldegerät *n*

lines of gravity method, gravity lines method [*ventilation*] Schwerlinienverfahren *n*, Schwerlinienmethode *f*

line spacing, line interval Zeilenabstand *m*

lines per minute → LPM

line system Leitungsanlage *f*

line system → (pipe)line system

line-to-earth voltage, phase-to-earth voltage (Brit.); line-to-ground voltage, phase-to-ground voltage (US) Leitungsspannung *f* gegen Erde

line topology Linientopologie *f* [*Siehe Anmerkung unter „logische Topologie"*]

L

line trench Leitungsgraben *m* [*rechteckig*]

line trouble, line disturbance Leitungsstörung *f*

line type Leitungsart *f*

line unit Leitungseinheit *f*

line up, line arrangement Ausrichtung *f*

line voltage thermostat [*A thermostat that is connected directly to the line. Full power is fed through it to the controlled heater or air conditioner*] Netzspannungsthermostat, Leiterspannungsthermostat, Speisespannungsthermostat *m*

to line with hemp einhanfen, verhanfen

to link (ver)binden, verketten [*getrennte Systeme oder Programmteile*]

link Brücke, Überbrückung *f*

link → channel

link Verweis *m*

link access procedure → LAP

linkage Gelenk *n*

linkage editor → linker

linked list, chained list verkettete Liste, Kettenliste *f* [*Die Einträge sind nicht notwendigerweise sequentiell gespeichert*]

linked switches [*Two or more electric switches which are mechanically connected by operating arms or levers, so as to operate at the same time or in a desired sequence*] mechanisch verbundene Schalter *f* [*Mehrzahl*]

linker, linkage editor Binder, Verbinder *m* [*Wenn der Kompilierer den Quelltext ohne Fehlermeldung übersetzt hat, muss der entsprechende Maschinenkode meist noch mit den von anderen benötigten Programmteilen zusammengebunden werden. Diese Programmarbeit führt der Binder durch*]

linker (programme) (Brit.); linker (program) (US) [*It reads one or more object files created by language processors and produces an executable image file, a shareable image file, or a system image file*] Binderprogramm, Verbinderprogramm *n*

link fuse, link fuze [*An exposed fuse which is mounted on electrically insulated supports*] Schmelzsicherung *f*

linking [*The resolution of external references between object modules used to create an image, the acquisition of referenced library routines, service entry points, and data for the image and the assignment of virtual addresses to components of an image. Could be done during the development process or under windows, is most often done during loadtime using DLLs*] Verknüpfung, Kopp(e)lung *f*

linking loader, linkage loader Bindelader *m* [*Ein Ladeprogramm zum Zusammenfügen mehrerer unabhängig voneinander erstellter Programmteile zu einem Gesamtprogramm*]

linking (together) → chaining

link layer Schicht 2, Ebene 2, Sicherungsschicht *f* [*ISO/OSI- Schichtenmodell. Aufbau einer (möglichst) fehlerfreien Verbindung zwischen Endsystem und Netzzugang. Zusammenfassung der Daten in Blöcke; Blocksynchronisation, Fehlererkennung und Korrektur*]

link loading Kennzeichenabschnitt(s)last *f*

link protocol Übermitt(e)lungsvorschrift *f*

link-up Kopp(e)lung *f* [*Rechnerzentralgerät oder Rechnerentwicklungssystem*]

linoleum nosing, nosing for linoleum Linoleum(schutz)schiene, Linoleum(schutz)leiste *f*

linseed-oil paint Leinöl(anstrich)farbe *f*

lint extraction → (laundry) lint extraction

lint screen → (laundry) lint screen

lip guard [*drinking fountain*] Lippenschutz *m*

lip seal(ing) Lippen(ab)dichtung *f*

liquefaction; liquifaction [*obsolete*] Verflüssigung *f*

liquefaction and separation of gases engineering [*"Liquifaction" is obsolete*] Tieftemperaturverfahrenstechnik *f*

liquefaction heat Verflüssigungswärme *f* [*Wärmemenge beim Übergang vom dampfförmigen in den flüssigen Zustand*]

liquefied petroleum gas, LPG, liquid petroleum gas Flüssiggas *n* [*Die handelsüblichen Qualitäten der C3- und C4-Kohlenwasserstoffe Propan, Propen, Butan, Buten und deren Gemische*]

liquid-borne sound Flüssigkeitsschall *m*

liquid bubble → bubble

liquid butane Flüssigbutan *n*

liquid charge, liquid fill(ing) Flüssigkeitsfüllung *f*

liquid chilling system → chiller

liquid chlorine Flüssigchlor *n*

liquid collector Flüssigkeitssammler *m*

liquid column Flüssigkeitssäule *f*

liquid-controlled flüssigkeitsgesteuert

liquid-controlled, liquid-regulated flüssigkeitsgeregelt

liquid coolant, cooling liquid Kühl(ungs)flüssigkeit *f*

liquid cooler [*This may be either a brine or water cooler*] Flüssigkeitskühler *m*

liquid cooling plant, liquid cooling installation Flüssigkeitskühl(ungs)anlage *f*

liquid cooling set Flüssigkeitskühl(ungs)satz *m*

liquid cryogen, cryogenic liquid Kryoflüssigkeit, Kälteflüssigkeit *f*

liquid crystal display, LCD [*A screen in which a liquid crystal material sandwiched between two sheets of glass exhibits computer data when a voltage appears across leads that are connected to a character forming segment etched on to its inner glass surface*] Flüssigkristallanzeige *f*

liquid crystal oscilloscope Flüssigkeitskristalloszilloskop *n*

liquid dosing system Flüssigkeitsdosiersystem *n*

liquid expansion sensor Flüssigkeitsausdehnungssensor *m*

liquid expansion thermometer Flüssigkeitsausdehnungsthermometer *n*

liquid fire Flüssigkeitsbrand *m*

liquid (fire) extinguishing agent Flüssiglöschmittel, flüssige Löschmittel *n*

liquid flow Flüssigkeitsströmung *f*, Flüssigkeitsstrom *m*

liquid freshwater flüssige Süßwasser *n* [*Gegensatz: gefrorene Süßwasser*]

liquid fuel Flüssigbrennstoff *m* [*DIN 51 603 Teil 1*]

liquid fuel boiler Flüssigbrennstoffkessel *m*

liquid gas Flüssiggas *n* [*DIN 51 622*]

liquid gas boiler Flüssiggaskessel *m*

liquid gas heating appliance, liquid gas heater Flüssiggasheiz(ungs)gerät *n*, Flüssiggasheizer *m*

liquid gas tank Flüssiggasbehälter *m*

liquid gold Flüssiggold *n*

liquid hammer Flüssigkeitsschlag *m*

liquid-heated flüssigkeitsbeheizt

liquid hydrogen Flüssigwasserstoff *m*

liquid-immersed transformer [*It has its core and coils submerged in an insulating liquid such as oil*] flüssigkeitsgetauchte Transformator *m*

liquid indicator [*A device, frequently combined with a strainer, located in the liquid line of a refrigeration system and having a sight port by which the liquid flow may be observed for presence of bubbles*] Blasenanzeiger *m*

liquid jet jump, jet liquid pump Strahlflüssigkeitspumpe, Flüssigkeitsstrahlpumpe *f*

liquid level controller, liquid level regulator Flüssigkeitsniveauregler, Flüssigkeits(höhen)standregler *m*

liquid level indicator Flüssigkeitsstandanzeiger *m*

liquid level meter Flüssigkeitsstandmesser *m*

liquid line [*A tube or pipe carrying the refrigerant liquid from the condenser or receiver of a refrigerating system to a pressure-reducing device*] Kälteflüssigkeitsleitung, Flüssigkältemittelleitung *f*

liquid meter Flüssigkeitsmesser *m*

liquid metering Flüssigkeitsmessung *f*

liquid nitrogen Flüssigkeitsstickstoff *m*

liquid-only system Nur-Flüssigkeitssystem *n* [*Energiezufuhr an den Versorgungsstellen nur durch Flüssigkeit. DIN 1946, Teil 1*]

liquid particle Flüssigkeitsteilchen *n*

liquid phase Flüssigkeitsphase *f*

liquid (pipe)line Flüssigkeits(rohr)leitung *f*

liquid (pipe)line solenoid valve Flüssigkeits(rohr)leitung(s)magnetventil *n*

liquid pressure gauge; liquid pressure gage (US) Flüssigkeitsdruckmesser *m*, Flüssigkeitsmanometer *n*

liquid refrigerant → refrigerant liquid

liquid refrigerant evaporation pressure Flüssigkältemittel-Verdampfungsdruck *m*

liquid refrigerant evaporation temperature Flüssigkältemittel-Verdampfungstemperatur *f*

liquid refrigerant line Flüssigkältemittelleitung *f*

liquid refrigerant pressure reducer → refrigerant liquid pressure reducer

L

liquid refrigerant vapor (US) → **refrigerant liquid vapor**

liquid ring compressor Flüssigkeitsringverdichter, Flüssigkeitsringkompressor *m*

liquid sensor, liquid feeler Flüssigkeitsfühler, Flüssigkeitssensor, Hydrofühler, Hydrosensor *m*

liquid separation Flüssigkeitsabscheidung *f*

liquid separator Flüssigkeitsabscheider *m*

liquid shutoff valve Flüssigkeits(ab)sperrventil *n*

liquid sight glass Flüssigkeitsschauglas *n*

liquid soap dispenser Flüssigseifenspender *m*

liquid suction line Flüssigkeitssaugleitung *f*

(liquid) tank (Flüssigkeits)Tank, (Flüssigkeits-)Behälter *m*

liquid vacuum pump Flüssigkeitsringvakuumpumpe *f*

liquid waste Flüssigmüll, Flüssigabfall(stoff) *m*

liquid waste appliance → **ablutionary water appliance**

liquid water [*As opposed to ice, snow and hail*] Flüssigwasser *n*

liquifying [*obsolete*] → **condensing**

L

liquor [*Solution used in absorption refrigeration*] Lösungsmittel *n*, Löser *m*

to list auflisten

to list → **to scan**

to listen hören

listener Datenempfänger, Empfänger *m* [*An einen IEC-Bus angeschlossen*]

listener Hörer *m*

listener address group, LAG Höreradressengruppe *f*

listener's threshold Hörschwelle *f*

listening (Zu)Hören *n*

listening, bugging, eavesdropping Abhören *n*

listening device, bug(ging device), (electronic) eavesdropping device, illicit (radio) device Abhörvorrichtung, „Wanze" *f*

lister Empfänger *m* [*Ein an einen Bus angeschlossenes Datenaufnahmegerät*]

listing Kommentarliste *f* [*Sie ordnet die abgespeicherten Kommentare direkt den Anschlussbezeichnungen des Steuergerätes zu (z. B. Eingang/Ausgang, Merker, Zeit, Zähler)*]

listing → **scanning**

listing → **logging**

list item Listenelement *n*

list processor → **LISP**

lithium bromide Lithiumbromid *n*

lithium chlorid(e) cell Lithiumchloridzelle, LiCl-Zelle *f*

lithium chlorid(e) hygrometer Lithiumchloridfeuchtemesser, Lithiumchloridfeuchtigkeitsmesser *m*

lithium chromate inhibitor Lithiumchromathemmstoff *m*

lithium ion battery Lithiumionenbatterie *f*

litmus paper Lackmuspapier, Anzeigepapier *n* [*Für Säuren und Laugen*]

little bell-rope Klingelschnur, Schellenschnur *f*, Schellenband, Klingelband *n*

live → **alive**

live part, active part aktive Teil, spannungsführende Teil *m, n* [*IEV 826-03.01. Jeder Leiter oder jedes leitfähige Teil, das dazu bestimmt ist, bei ungestörtem Betrieb unter Spannung zu stehen, einschließlich des Neutralleiters, aber vereinbarungsgemäß nicht der PEN-Leiter*]

live picture Echtzeitbild, Realzeitbild *n*

live potted plant Topfpflanze *f*

live room [*A room characterized by a usually small amount of sound absorption*] Hallraum *m*

live steam Frischdampf *m*

living area Wohnbereich *m* [*selten: n*], Wohnfläche *f*

living floor; living storey (Brit.); living story (US) Wohnetage *f*, Wohnstockwerk, Wohngeschoss *n*

living kitchen Wohnküche *f*

living space Aufenthaltszone *f* [*In DIN 44 576 Teil 1 „Verweilfläche" genannt*]

living space Wohnraum *m*

LL → **logical link**

LLC → **logical link ((open) loop) control**

LLC → **low layer compatibility**

LLL method → **lambda-locked-loop method**

LOA → **length overall**

to load, to charge belasten

LOAD Ladebefehl *m* [*Die Anweisung für das „Laden" einer Zahl ist eine im Speicher (memory) befindliche Information. Die Adresse, unter der diese Information zu finden sein muss, wird vom Programmzähler festgelegt. Der Befehl heißt "LOAD"*]

load → **charge**

loadability Belastbarkeit *f*

load adap(ta)tion → **load matching**

load block → **load module**

load-break switch [*deprecated*] → **switch disconnector**

load calculation, load computation Lastberechnung *f*

load capacity in intermittent operation Belastbarkeit *f* bei Aussetzbetrieb

load card [*data processing*] Ladekarte *f*

load change Lastwechsel *m*

load circuit, load switching Lastschaltung *f*

load cycling [*A control method where management of plant output is achieved by means of variable on/off periods of operation*] Aussetzbetrieb *m*

load-dependent lastabhängig

load disconnection Lastabwurf *m*, Lasttrennung *f*

load disconnector Lasttrennschalter, Lastabwurfschalter *m*

load diversity Lastungleichheit, Lastungleichmäßigkeit *f*

to load down herunterladen

load elevator (US) → freight elevator

loader Lader *m* [*allgemeine Benennung*]

loader Lader *m* [*Er transportiert ausführbare Maschinenprogramme von einem Hintergrundspeicher in den Hauptspeicher*]

loader initial program (US); loader initial programme (Brit.) Umlader *m* [*Programm zum Starten eines Rechners. Das Systemprogramm steht immer am Anfangsteil eines Betriebssystems, so dass das Programm anschließend gestartet werden kann*]

loader routine, loading routine; loader program, loading program (US); loading programme, loader programme (Brit.) [*data processing*] Ladeprogramm *n*

load-free ... → no-load ...

load impedance Lastscheinwiderstand *m*

load-independent current eingeprägte Strom *m*

load-independent voltage eingeprägte Spannung *f* [*Sie behält ihren Wert auch bei stärkster Belastung der Quelle*]

loading error [*Error due to the adverse effect of a load on the transducer or signal source driving it*] Belastungsfehler *m*

loading error [*Error in transferring data from a data source to a computer memory*] Übertragungsfehler *m*

loading limit Belastungsgrenze *f*

loading peak Belastungsspitze *f*

loading programme (Brit.); loading program (US) [*data processing*] Ladeprogramm *n*

load interconnection Lastverbund *m*

load interrupter → (switch) disconnector

load isolator Lasttrenner *m* [*Elektrotechnik*]

load key [*data processing*] Ladetaste *f*

load limit Lastgrenze *f*

load limitation Lastbegrenzung *f*

load matching, load adap(ta)tion Lastanpassung *f*, Lastanpassen *n*

load module, load block Lademodul *n*, Ladebaustein *m*

load output → load rating

load peak Ladespitze *f*

load point [*data processing*] Ladepunkt *m*

load range Lastbereich *m* [*selten: n*]

load rating, load output Nennlast *f*

load reduction factor Lastabschlagsfaktor *m*

load shedding [*The function of switching off electrical equipment if the predicted instantaneous current exceeds a limit. This function therefore avoids the payment of maximum demand tariffs.*] Lastabwurf *m*, Lastabwerfen *n*

load shedding and indication contact Lastabwurf- und Anzeigekontakt *m*

load shedding circuit Lastabwurfschaltung, Entlastungsschaltung *f*

load shedding contact Lastabwurfkontakt, Entlastungskontakt *m*

load shedding relay Lastabwurfrelais, Entlastungsrelais *n*

load side [*The side of a device electrically furthest from the current source*] Lastseite *f*

load switch Lastschalter *m*

load transfer switch Lastumschalter *m*

load variation Lastschwankung *f*

local air exchange indicator lokale Luftaustauschkennzeichen, örtliche Luftaustauschkennzeichen *n*

local alarm → internal alarm

local area net(work), LAN Netz(werk) für örtliche Bereiche, lokale Datenaustauschnetz(werk), örtliche Datenaustauschnetz(werk) *n* [*Datenkommunikationssystem, welches unabhängigen Vorrichtungen der Datenverarbeitung eine Kommunikation hoher Datenrate auf begrenztem geographischen Gebiet ermöglicht*]

local broadcast [*A message addressed to all devices or nodes on the same network as the originator*] lokale allgemeine Mitteilung, lokale allgemeine Nachricht, lokale allgemeine Information, örtliche allgemeine Mitteilung, örtliche allgemeine Nachricht, örtliche allgemeine Information *f*

local computer net(work) lokale Rechnernetz *n*

locale Örtlichkeit *f*

local heat Nahwärme *f* [*Gegensatz: Fernwärme (long distance heat)*]

local heater → room heater

local heating capacity → room heating capacity

localized corrosion → pitting

local line → access line

local loop → access line

locally administered address [*See remark under "burned-in address"*] überlagerte Token-Ring-Adresse *f*

locally assembled → assembled on site

local net(work) Ort(s)netz *n*

local operating network → LON

local operator terminal Vorortbedienung *f* [*Einheit für Bedienereingriffe an der Automationsstation*]

local reading, direct reading Direktablesung *f*

local schedule Belastungsplan *m* [*Elektrizität*]

local silicon oxydation örtliche Siliziumoxidation, gezielte Siliziumoxidation, lokale Siliziumoxidation *f* [*auch „Oxydation" geschrieben*]

local supply örtliche Versorgung, lokale Versorgung *f*

local switching Vor-Ort-Schaltung *f*

local warning örtliche Warnung, lokale Warnung *f*

L

location Standort *m*

location Speicherplatz *m*, Speicherstelle *f* [*Durch eine Adresse eindeutig gekennzeichnete Speicherstelle eines Wortes*]

location [*equipment*] **→ placement**

location call → paging call

location diagram Anordnungs(schalt)plan *m* [*DIN 40 719, Blatt 1*]

location server Rechner *m* im Netzwerkbetrieb [*Er hat die Aufgabe der zentralen Verwaltung der Anwendungsprogramme, der Anwender und der Geräte*]

location service → paging service

location system → paging system

location update procedure Aktualisierung *f* [*Verwaltungsdaten in einem Mobilfunknetz nach einer Gesprächsweitergabe an eine Basisstation, die einer anderen Hauptstation untergeordnet ist*]

to lock sperren, absperren; zusperren, verriegeln, abriegeln, verschließen, abschließen, (zu)schließen

lock [*A device which fastens in position, as a door, gate, window, etc.; may be operated by a key or a dead bolt*] Schloss *n*

lock Schleuse *f*

lockable (ab)schließbar, verschließbar

lockable verriegelbar, abriegelbar

lockable thumb grip knob Sperrgriff, Daumensperrgriff *m*

lock bar → bar

(lock) bar handle → bar handle

(lock) bar head → bar head

lock barrel Schließzylinder, Schlosszylinder *m*

lock block [*A wood(en) block in a flush door, into which the lock is set*] Schlossblock *m*

lock border Umschweif *m*

lock bottom Schlossboden *m*

(lock) box, (lock) case, (lock) casing Kasten, Schlosskasten *m*

lock broach → broach

lock catch Klinkhaken, Schließhaken *m*

lock chain Hemmkette *f*

locked verriegelt, abgeriegelt

locked-rotor current [*IEC 34-1*]**; breakway starting current** Anzugsstrom *m* [*Elektromotor. VDE 0530, Teil 1*]

locker Spind *m, n*

locker room Spindraum *m*

lock (e)scutcheon → (e)scutcheon

lock for armo(u)red security vehicles Geldtransportfahrzeugschleuse *f*

lock hardware Schlossbeschläge *f*

locking Sperren, Absperren *n*, Sperrung, Absperrung *f*

locking Arretieren *n*, Arretierung *f*

locking Verriegeln, Abriegeln *n*, Verrieg(e)lung, Abrieg(e)lung *f* [*Ein Verfahren, mit welchem gemeinsam genutzte Daten vor dem gleichzeitigen Zugriff anderer Rechner geschützt werden*]

locking Schließen, Abschließen *n*

locking → (door) locking

lock(ing) bolt, lock(ing) fastener (Schloss)Riegel, Schließriegel *m* [*In einem Schloss der Riegel der in das Schließblech, in die Schließkappe oder den Schließhaken einfällt und dadurch die Tür zuhält*]

locking bracket Schließwinkel *m*

locking cam Schließbart *m*

locking contact Schließkontakt *m*

locking cylinder core Schließzylinderkern *m*

locking device Schließvorrichtung *f*

locking disc, locking disk Schließscheibe *f*

locking element Schließelement *n*

lock(ing) fastener → lock(ing) bolt

locking level Schließebene, Trennungsebene *f* [*Stiftzuhaltung*]

locking mechanism Schließmechanismus *m*

lock(ing) mounting Schlossbeschlag *m*

lock(ing) mountings, lock(ing) furniture Schlossbeschläge *f*

locking plate Schlossblech *n*

locking plate contact Schlossblechkontakt *m*

locking release Schlossöffner *m*

lockings, door lockings (Tür)Verschlussbeschläge *f*

locking system Schließsystem *n*

(lock) key pipe, key pipe of the lock, (lock) key shank Schlüsselrohr (des Schlosses), Schlossschlüsselrohr, Schlüsselrohr im Schloss *n*, Führungshülse im Schloss, Schlossführungshülse *f*, Schlüsselschaft *m*

lock locked from both sides Schloss *n* von zwei Seiten zu schließen

lockmaking, lock manufacture Schlossbau *m*, Schlossfertigung, Schlossherstellung *f*

lock nail Schlossnagel *m*

lock nut Kontermutter *f*

lock of Bramah → Bramah('s) (detector) lock

lock of Chubb → Chubb('s) (detector) lock

lockout [*e.g. of a burner*] Ausfall *m*, Versagen *n*

lockout [*The shut-down condition of the control system such that restart cannot be accomplished without manual reset*] Sperre, Verrieg(e)lung, Einschaltsperre, Einschaltverrieg(e)lung *f*

lockout handle for padlock, padlocking feature Vorhängeschlosssperre, Vorhängeschlossverrieg(e)lung *f*

lockout release [*burner*] Entrieg(e)lung *f*

lockout time Sperrzeit *f*

lock pin Schlossstift *m*

lock plate → (e)scutcheon

lock plate → (lock) striking plate

(lock) pushbutton [*Button in the knob of a cylindrical or bored(-type) lock used to lock the opposite knob in a fixed position*] Schlossdruckknopf *m*

lock screw Schlossschraube *f*

lock security Schlosssicherung *f*

lock set Schlossgarnitur *f*

lock signal Verrieg(e)lungssignal *n*

locksmith Schlosser *m*

locksmith('s) work Schlosserarbeiten *f*

lock spring Schlossfeder *f*

(lock) striking plate, (lock) strike(r), (lock) strike plate, lock plate, keeper (plate) [*A metal plate which is set in a doorjamb and is either pierced or recessed to receive the bolt or latch of a lock, fixed on a door*] Schließblech, Schlossblech *n*

lockswitch → keylock switch

lock system Schleusensystem *n*

lock type, type of lock Schlossart *f*

lock with a cross-bolt, cross-bolt lock Doppelriegelschloss *n*

lock with double turn → twice-turning lock

lock with one turn and a half anderthalbtourige Schloss, eineinhalbtourige Schloss *n*

lock with three bolts, three bolts lock, triple bolts lock Kastenschloss *n* mit Falle und Nachtriegel

LOCO, low coercive niedrigkoerziv [*Magnetstreifen mit einer geringen Sicherheit gegen Entmagnetisierung*]

locomotive furnace, firebox, furnace with internal horizontal grate Planrostinnenfeuerung, Innenplanrostfeuerung, Feuer(ungs)büchse *f*

lofting → skinning

to log aufzeichnen, protokollieren [*Ereignisse*]

log [*A chronical record of a series of events. Each event is identified by time, day, value and physical reference*] (chronologische) Aufzeichnung *f*, Protokoll *n*, Bericht *m*

logatom intelligibility Silbenverständlichkeit *f*

log display Anzeige *f* von (chronologischen) Aufzeichnungen

log file, protocol file, log data, protocol data Protokolldatei, Protokolldaten *f*

logger Mitschreiber *m* [*Rechner*]

logging, listing, printing out Protokollieren *n* [*Das spontane, zyklische oder auf Abruf erfolgende Wiedergeben einer Aufzeichnung in für den Menschen lesbarer Form. DIN 19 222*]

logging capacity [*The number of data elements that can be stored by an outstation when undertaking a log*] Protokollkapazität *f*

logging family Protokollfamilie *f*

logging message Protokollnachricht, Protokollmitteilung *f*

logging printer, logger Protokoll(ausgabe)drucker *m*

logic Logik *f* [*Rechenelementesystem auf den Grundschaltungen Entweder-Oder, Und-Nicht usw.*]

logic [*General term for the various types of gates, flip-flops, and other on/off circuits used to perform problem-solving functions in a digital computer*] elektronische Logik *f*, logische Schaltung, Verknüpfung *f*

logic [*The basic principles and applications of truth tables, interconnections of on/off circuit elements, and other factors involved in mathematical computation in a computer*] mathematische Logik *f* [*Die Gesetze, For-*

meln und Symbole der mathematischen Logik werden in der Elektronik angewendet (z. B. Schaltungsalgebra, Boolsche Algebra, Logikschaltungen). Dabei ist eine Zuordnung zwischen logischen Zuständen und physikalischen Größen der Schaltung (z. B. Spannungspegel erforderlich]

logic(al) algebra, Boolean algebra, switching algebra, circuit algebra zweiwertige Kombinatorik, Schalt(ungs)algebra, Logikalgebra, Boolsche Algebra der Logik, symbolische Logik, Kreisalgebra *f* [*entwickelt vom englischen Mathematiker Georg Boole (1815-1864)*]

logic(al) analyzer, LA Logikanalysator, logische Analysator, Verknüpfungsanalysator, LA *m* [*Ein Oszilloskop, bei dem mittels eines elektronischen Umschalters 8 oder 16 Kanäle mit logischen Signalen gleichzeitig auf einem Bildschirm dargestellt werden. Zur Untersuchung digitaler (logischer) Schaltungen verwendet*]

logic(al) and circuit UND-Schaltung *f*

logic(al) array → logic(al) circuit

logic(al)block addressing Logikblockadressierung, LBA, logische Blockadressierung, Verknüpfungsblockaddressierung *f*

L

logic(al) block number [*A number used to identify a block on a mass storage device. The number is a volume-relative address rather than the physical (device-oriented) address or its virtual (file-relative) address*] Logiksatznummer, logische Satznummer, Verknüpfungssatznummer *f*

logic(al) board Logiktafel, Logikleiterplatte, logische Tafel, logische Leiterplatte, Verknüpfungstafel, Verknüpfungsleiterplatte *f*

logic(al) character logische Zeichen, Logikzeichen *n*

logic(al) circuit, combinatorial logic(al) system, logic(al) array [*A computer circuit that provides the action of a logic function or logic operation*] logische Schaltung, Logikschaltung, Verknüpfungsschaltung *f*

logic(al) ((closed) loop) controller Verknüpfungsregler, Logikregler *m*

logic(al) condition Logikbedingung, Verknüpfungsbedingung, logische Bedingung *f*

logic(al) control → logic(al) ((open) loop) control

logic(al) converter Logikwandler, logische Wandler, Verknüpfungswandler *m*

logic(al) data set logische Datensatz, Logikdatensatz, Verknüpfungsdatensatz *m*

logic(al) decision [*Choice or ability to choose between alternatives. This amounts to answer yes or no to a question of equality or relative magnitude*] Logikentscheidung, Verknüpfungsentscheidung, logische Entscheidung *f*

logic(al) design, logic(al) type, logic(al) version [*The design of a computer at the level which considers the operation of each functional block and the relationship between those blocks*] Logikentwurf, logische Entwurf, Verknüpfungsentwurf *m*

logic(al) device Logikbaustein, Verknüpfungsbaustein, logische Baustein *m*

logic(al) diagram [*A graphical representation of the logic design or a portion thereof; displays the existence of functional elements and the paths by which they interact with one another*] Funktionsplan, logische Schaltplan, Logikschaltplan, Verknüpfungsschaltplan *m*

logic(al) element → logic(al) gate

logic(al) expression Logikausdruck, Logikterminus, logische Ausdruck, logische Terminus, Verknüpfungsausdruck, Verknüpfungsterminus *m*

logic(al) flowchart [*Detailed statement of the work order in terms of logic, or built-in operations and characteristics, of a specified machine. Concise symbolic notation is used to represent the information and describe the input, output, standard set of block symbols. A coding process normally follows a logical flowchart*] Verknüpfungsablaufplan, Logikablaufplan, logische Ablaufplan *m*

logic(al) function logische Funktion, Schaltfunktion, binäre Verknüpfungsfunktion *f*

logic(al) gate, logic(al) element, combinative element [*Electronics. A device or circuit that has no output until it is triggered into operation by one or more enabling signals, or until an input signal exceeds a prederminded threshold amplitude*] Verknüpfungsglied, Schaltglied, Logikelement, Logikglied *n*

logic(al) grid Logikraster, Verknüpfungsraster, logische Raster *m*

logic(al) ground (US); logic(al) earth (Brit.) Logikmasse, Logikerde, logische Masse, logische Erde, Verknüpfungsmasse, Verknüpfungserde *f* [*IEC 625*]

logic(al) ground return (US); logic(al) earth return (Brit.) logische Masserückleitung, Logikmasserückleitung, Verknüpfungsmasserückleitung, Logikerderückleitung, logische Erderückleitung, Verknüpfungserderückleitung *f*

logic(al) identity element Äquivalenzelement, binäre Schaltelement, Gleichwertigkeitselement *n*

logic(al) instruction [*A digital computer instruction which forms a logical combination (on a bit-by-bit basis) of its operands and leaves the result in a known location*] Verknüpfungsbefehl, logische Befehl, Logikbefehl *m*

logic(al) interlock(ing) logische Verrieg(e)lung *f*

logic(al) level [*One of the two voltages whose values have been arbitrarily chosen to represent the binary digits 1 and 0 in a particular data-processing system*] Logikpegel, logische Pegel, Verknüpfungspegel *m*

logic(al) link, LL Logikverbindung, logische Verbindung, Verknüpfungsverbindung *f*

logic(al) link ((open) loop) control, LLC [*ISO 8802-2*] Steuer(ungs)verfahren *n* im Übermitt(e)lungsabschnitt

logic(al) module, logic(al) block Logikmodul, logische Modul, Verknüpfungsmodul *n*, Logikbaustein, logische Baustein, Verknüpfungsbaustein *m*

logic(al) multiplication [*The multiplicate binary operation of a Boolean algebra*] Logikmultiplikation, logische Multiplikation, Verknüpfungsmultiplikation *f*

logic(al) (nesting) depth Verknüpfungstiefe *f*

logic(al) one logisch Eins *f*

logic(al) ((open) loop) control Logiksteuerung, logische Steuerung, Verknüpfungssteuerung; Führungssteuerung *f* [*Fehlbenennung. V DIN 19 237*]

logic(al) operation [*A nonarithmetical operation in a computer, such as comparing, selecting, making reference, matching, sorting and merging, where logical yes-or-no quantities are involved*] Logikverknüpfung, logische Verknüpfung, Verknüpfung(soperation) *f*

logic(al) operations sequence, logic(al) operations chain logische Verknüpfungsfolge, logische Verknüpfungskette *f*

logic(al) polarity Logikpolarität, Verknüpfungspolarität, logische Polarität *f*

logic(al) record [*A group of adjacent, logically related data items*] Datensatz *m*

logic(al) sum [*A computer addition in which the result is 1 when either one or both input variables are 1, and the result is 0 when the input variables are both 0*] logische Summe, Verknüpfungssumme, Logiksumme *f*

logic(al) system Logiksystem, logische System, Verknüpfungssystem *n* [*Durch Stecken oder Löten von Logikbausteinen hergestellt*]

logic(al) tester Logikprüfer, logische Prüfer, Verknüpfungsprüfer *m*

logic(al) type → **logic(al) design**

logic(al) unit logische Einheit, Verknüpfungseinheit, Logikeinheit *f*

logic(al) version → **logic(al) design**

login Einwählen *n* [*In ein System*]

log-in Anmelden *n*, Identifikation *f* [*In einem Netz, das dem Zustandekommen der Verbindung folgt*]

logistic closed control loop operative Regel(ungs)-kreis *m*

logoff Ausloggen, Abmelden *n*

logoff, logout Abmelden *n* [*Eines Benutzers bei einem Datenverarbeitungssystem*]

log printer Protokolldrucker *m*

LOHET → **linear output Hall effect transducer**

LOM defuzzification → **last of maximal defuzzification**

LON, local operating network lokal arbeitende Netz(werk), örtlich arbeitende Netz(werk) *n* [*Ein Produkt der Echelon Corp., USA. Vorbild für diesen Standard sind die LANS in Rechnernetzwerken. Mit LON sinkt die Zahl der benötigten Komponenten im Netzwerk, jedoch basiert das Konzept auf speziellen Chips (zunächst Neuron-Chips), in denen das Protokoll der LONtalk-Kommunikation fest hinterlegt ist. Die Komponenten sind lizenzpflichtig. 1996 wurde das Protokoll LONtalk freigegeben, d.h. es kann durch andere Chiphersteller auf beliebigen Prozessoren implementiert werden. LON eignet sich für die Feldebene.*]

long bend → **long (sweep) bend**

long cross garnet, long tee hinge Langband, Haspenband *n* [*Es ist länger als 1/3 der Türflügelbreite*]

long-distance cold supply plant, long-distance cold supply installation Fernkälteversorgungsanlage *f*

long-distance cold water supply → **cold water long-distance supply**

long-distance fuel oil (pipe)line Fernheiz(ungs)-öl(rohr)leitung *f*

long-distance gas Ferngas *n*

long-distance gas (pipe)line Gasfernleitung, Ferngasleitung *f*

long-distance heat Fernwärme *f*

long-distance heating Fernheizung *f*

long-distance heat utility Fernheizungsunternehmen *n*

long-distance net(work) Fernnetz(werk) *n*

long-distance steam net(work) Ferndampfnetz-(werk) *n*

long-haul net(work) Langstreckennetz(werk) *n*

longitudinal measure Längenmaß *n*

longitudinal redundance check(ing), LRC Längsparitätskontrolle *f* [*Magnetbandspeicher. „Kreuzsicherung" genannt, wenn spalten- und zeilenweise bei einem Datenblock angewendet*]

longitudinal uniformity Längsgleichmäßigkeit *f*

long shackle padlock Hangschloss mit hohem Bügel, Vorhängeschloss mit hohem Bügel *n*

L

(long) slot burner Schlitzbrenner, Langlochbrenner *m*

long (sweep) bend Knierohr, Knie(stück) *n*

long tee hinge, long cross garnet Langband, Haspenband *n* [*Es ist länger als 1/3 der Türflügelbreite*]

long-term langfristig

long-term event storage (function) [*It provides for the storage of the point address, state and information text, resulting from input/output and processing functions. Time and data stamps shall be stored with each item of information. A means of retrieving the data must be available*] Langzeitereignisspeicherung *f*

long-term material Langzeit(werk)stoff *m*, Langzeitmaterial *n*

long-term monitoring Langzeitüberwachung *f*

long-term stability error Langzeitstabilitätsfehler *m* [*Messgerät*]

long-term storage → archiving of historic(al) data

long thread Langgewinde *n*

long-throw nozzle Langwurfdüse, Weitwurfdüse *f*

long-time heat storage device, long-time thermal storage device Langzeit(wärme)speicher, Saison(wärme)speicher, Jahreszeit(wärme)speicher *m* [*(Wärme)Speicherung während Wochen oder Monaten von der warmen zur kalten Jahreszeit*]

L

longword, L WORD [*Four contiguous bytes (32bits) starting on an addressable byte boundary. Bits are numbered from right to left*] Langwort *n*

look Durchsuchung *f* [*(Sortierte) Textdateien nach Zeichen mit übereinstimmenden Zeichenketten durchsuchen*]

look Erscheinung *f* [*Benutzerfreundlichkeit von Bedien(ungs)oberflächen*]

look-ahead Vorgriff *m* [*Während der Ausführung das Holen der aktuellen Ausführung auf dem Hauptspeicher*]

lookall Gesamtdurchsuchung *f* [*Sämtliche (zugänglichen) Textdateien eines Dateisystems durchsuchen*]

loop → (butt) hinge

loop Schleife, Ringleitung *f*

loopback Prüf(ungs)schleife *f* [*Datenübertragung*]

loopback Zurückkoppeln *n* [*Kommunikationsystem*]

loop control (device) → control (device)

loop control process → control process

loop(ing)-in Einschleifung *f*

loop of oscillation → antinode

loop process → control process

loop variable → (closed) loop variable

loop variable → (open) loop variable

(loose-)fill thermal insulation Schüttwärmedämmung, Wärmeschüttdämmung *f*

loose flange [*It is not fixed to a fitting or pipe but free to be rotated or moved longitudinally*] lose Flansch, Losflansch *m*

Lorenz relation → Wiedmann-Franz law

loss coefficient Widerstandszahl *f*, Verlustbeiwert *m* $\zeta = \frac{2\text{Ø}_{fie}}{\bar{u}_1^2}$ [*eine Diffusionskennzahl*]

loss heat Verlustwärme *f*

loss heat flow Verlustwärmestrom *m*

loss-of-service duration, loss-of-service time, service-off duration, service outage duration, service outage time, service off time Dienstunterbrechungsdauer *f*

lost motion → backlash

lost return → prevential yield

lost yield → prevential yield

lot tolerance percent defective, LTPD rückzuweisende Qualitätsgrenzlage *f*

loudness → intensity

(loud)speaker [*A device that converts electrical signal energy into acoustical energy, which it radiates into a bounded space or into an outdoor space*] Lautsprecher *m*

(loud)speaker dividing network → crossover network

(loud)speaker horn Lautsprechertrichter, Beschallungstrichter *m*

(loud)speaker line Lautsprecherleitung, Beschallungsleitung *f*

(loud)speaker monitoring Lautsprecherüberwachung, Beschallungsüberwachung *f*

(loud)speaker system Lautsprecheranlage, Beschallungsanlage *f*

loud-speaking telephone Lautfernsprecher *m*, Lauttelefon *n*

louvre(d) ceiling Rasterdecke *f*

louvre(d) luminaire, louvre(d) light(ing) fitting Rasterleuchte *f*, Rasterbeleuchtungskörper *m*

louvre(d) recessed luminaire, louvre(d) recessed light(ing) fitting Rastereinbauleuchte *f*, Rastereinbaubeleuchtungskörper *m*

louvres → slatted blind, jalousie

Low logisch »Null« *f* [*Pegelfestlegung für alle Daten- und Steuer(ungs)signale*]

"lo(w)" niedrig, gering [*Anzeige*]

low air velocity system Niedriggeschwindigkeitssystem *n* [*System mit Luftgeschwindigkeiten bis 10 m/s in den Verteil- und Sammelleitungen*]

low-angle sunlight flach einfallende Sonnenlicht *n*

low-bay reflector Tiefstrahler *m*

low-boiling niedrigsiedend

low-boiling naphtha Leichtbenzin *n*

low-boy upflow furnace [*The blower is located adjacent to the heat exchanger*] niedrige Aufwärts-Warmluftfeuerung, niedrige Warmluft-Aufwärtsfeuerung *f*

low byte niedrig(wertige) Byte *n*

low capacity disconnector Leertrenn(schalt)er *m*

low coercive → **LOCO**

low-delay verzögerungsarm, verzugsarm

low density niedrige Dichte, geringe Dichte *f*

low-energy → **energy-saving**

lower die → **bottom swage**

lower energy efficiency, bottom energy efficiency untere Wirkungsgrad *m* [*eine Diffusionskennzahl*]

lower heating value, net heating value [*It is obtained when the latent heat of vapo(u)rization is not included. When the heating value of a fuel is specified without designation higher or lower, it generally means the higher heating value in the USA. Lower heating value is mainly used for internal combustion engine fuels*] Heizwert, H_u [*früher: untere Heizwert*]

lowering, downward travel, downward run Abwärtsfahrt *f*, Absenken *n* [*Aufzug*]

lowering gate Senktor *n*

lowering temperature Absenktemperatur *f*

lowering temperature setpoint (value) Absenktemperatursollwert *m* [*Siehe Anmerkung unter „Sollwert"*]

lower latch → **latch**

lower-level unterlagert

lower plate → **bottom plate**

lower power device → **LPD**

low flow niedrige Durchfluss *m* [*im Sonnenkollektorkreis*]

low frequency → **voice frequency**

low-gas pressure alarm Gasdruckmangelsicherung *f*

low-gas quantity alarm Gasmangelsicherung *f*

low Joule value biogas, LBG niedrigkalorige Biogas *n*

low-lag verzögerungsarm, verzugsarm

low layer compatibility, LLC Verträglichkeit auf niederer Ebene, niedere Verträglichkeit *f*

low level Niedrigstand, Tiefstand *m*

low-level tief angebracht, niedrig angebracht, tiefliegend, niedrigliegend

low-level computer language, low-level machine language, low-level computer code, low-level machine code niedere Maschinensprache, niedere Rechnerprogrammiersprache *f*

low-level cut-out Wassermangelsicherung *f*

low-level screen [*These screens are used mainly to subdivide space in offices. Since these partitions are not secured at the top, it is necessary to provide cross walls or right-angle junctions to achieve stability*] Blende, Niedrigtrennwand *f*

low-level signal Niedrigpegelsignal, L-Signal, Kleinsignal *n*

low-level syphonic (water) closet, low-level siphonic (water) closet Tiefspülsiphonklosett *n* [*Siehe Anmerkung unter „Siphonklosett"*]

low-level tank, ground-level tank Tiefbehälter *m*

low-level toilet cistern, low-level (closet) cistern, low-level flush(ing) cistern [*"Tank" may synonymously be used for "cistern". A flushing cistern that is mounted on or just above the WC pan, rather than mounted at a height on the wall*] tiefhängende (Abort)Spülkasten, tiefliegende (Abort)Spülkasten, tiefhängende Toilettenspülkasten, tiefliegende Toilettenspülkasten, tief angebrachte (Abort)Spülkasten, tief angebrachte Toilettenspülkasten, tiefhängende (Abort)Spülapparat, tiefliegende (Abort)Spülapparat, tief angebrachte (Abort)Spülapparat *m*

low-level water closet, low-level flush toilet, low-level (type of) W.C., low-level suite, low-level washdown water closet Spülabort mit Tiefspülkasten, Wasserabort mit Tiefspülkasten, Tiefspülabort *m*, Spülklosett mit Tiefspülkasten, Wasserklosett mit Tiefspülkasten, Tiefspülklosett *n*

low light level camera Restlichtkamera *f*, Restlichtphotoapparat, Restlichtfotoapparat *m*

low limit(ing) value, bottom limit(ing) value untere Grenzwert *m*

low-loss verlustarm

low(-mounted) vanity Friseurwaschtisch *m*

low-noise geräuscharm

low-noise rauscharm [*Tonband*]

low-noise fan Flüsterventilator, geräuscharme Ventilator *m*

low-noise level Geräuscharmut *f*

low note attenuation Bassanhebung *f*

low-order bit wertniedrige Bit *m*

low-pass filter [*Filter having a single transmission band extending from zero to some cutoff frequency, not infinite*] Tiefpass(filter) *m*, (*m*,*n*)

L

low point *[pipe system]* Tiefpunkt *m*, Tiefstelle *f*

low power Kleinleistung *f* *[z. B. bei einer Schaltung]*

low-power Schottky transistor-transistor logic, low-power Schottky TTL, LS-TTL schwache Schottky Transistor-Transistor-Logik *f* *[Die Schottky-Dioden liegen parallel zu den Basis-Kollektor-Strecken, die Arbeitsgeschwindigkeit wird dadurch gesteigert]*

low pressure Niederdruck, ND *m*

low-pressure air-conditioning installation with induction unit, low-pressure air-conditioning plant with induction unit Niederdruck-Induktions(klima)-anlage, Niederdruck-(Klima)Anlage mit Induktionsgerät, Niederdruck-Klimainduktionsanlage *f*

low-pressure air-conditioning plant, low-pressure air-conditioning installation Niederdruckklimaanlage *f*

low-pressure (bath) mixer (valve) Niederdruckbad(e)batterie, Niederdruckmischbatterie *f*

low-pressure boiler *[Electric, gas, or oil-burning appliance designed to supply low-pressure steam or hot water for space heating application. A low-pressure steam boiler operates at or below 15 psig steam pressure; a hot water boiler operates at or below 160 psig water pressure and 250° F water temperature]* Niederdruckkessel *m*

low-pressure chemical vapour deposition (Brit.); low-pressure chemical vapor deposition (US); LPCVD Niederdruckschichtabscheidung *f* *[aus der Gasphase]*

low-pressure condensate return pump, low-pressure condensation water return pump Niederdruckkondens(at)rücklaufpumpe, Niederdruckkondenswasserrücklaufpumpe *f*

low-pressure cutoff switch *[An electric switch operated by the pressure of the gas supply which stops the gas compressor or other appliance when the gas pressure falls below a prederminated minimum]* Niederdruckabschalter, Niederdruckausschalter *m*

low-pressure discharge lamp Niederdruckentladungslampe *f*

low-pressure distributor druckarme Verteiler *m*, hydraulische Weiche *f*

low-pressure fan Niederdruckventilator, Niederdrucklüfter *m*

low-pressure float(ing) valve Niederdruckschwimmerventil, Schwimmerniederdruckventil *n*

low-pressure gas *[Town gas at the pressure at which it is normally supplied usually between 2 in. (5 cm) and 6in. (15 cm) water ga(u)ge, but sometimes up to 20 in. (51 cm) water ga(u)ge]* Niederdruckgas *n*

low-pressure gas plant, low-pressure gas installation Niederdruckgasanlage, Niederdruckgasinstallation *f*

low-pressure gas (supply) (pipe)line Niederdruckgas(versorgungs)(rohr)leitung *f*

low-pressure gas tank Niederdruckgasbehälter, ND-Gasbehälter *m*

low-pressure gravity return (flow) Niederdruck-Schwerkraft-Rücklauf *m*

low-pressure heating Niederdruckheizung, ND-Heizung *f*

low-pressure high-temperature water Niederdruckheißwasser *n*

low-pressure hot water, low-pressure warm water Niederdruckwarmwasser, ND-Warmwasser *n*

low-pressure hot water boiler, low-pressure warm water boiler *[It operates at or below 160 psig water pressure and 250° F water temperature]* Niederdruckwarmwasserkessel, ND-(Warmwasser)Kessel *m*

low-pressure hot water storage tank Niederdruckwarmwasserspeicher, ND-(Warmwasser)Speicher *m*

low-pressure induction unit *[See remark under "induction unit"]* Niederdruckinduktionsgerät, ND-Induktionsgerät *n*

low-pressure plant, low-pressure installation Niederdruckanlage, Niederdruckinstallation, ND-Installation, ND-Anlage *f*

low-pressure (refrigerant) receiver, accumulator *[This vessel acts as the separator for the mix(ture) of vapo(u)r and liquid returning from the evaporator. A constant refrigerant level is usually maintained by conventional control devices]* Niederdruckkältemittelsammler, ND-Kältemittelsammler *m*

low pressure side Niederdruckseite, ND-Seite *f*

low-pressure steam Niederdruckdampf, ND-Dampf *m*

low-pressure steam boiler *[It operates at or below 15 psig steam pressure]* Niederdruckdampfkessel, ND-Dampfkessel *m*

low-pressure steam boiler plant, low-pressure steam boiler installation Niederdruckdampfkesselanlage, ND-Dampfkesselanlage *f*

low-pressure steam generator Niederdruckdampferzeuger, ND-Dampferzeuger *m* *[DIN 4750 für sicherheitstechnische Anforderungen]*

low-pressure steam heating *[It uses steam at pressure between 0 and 15 psig]* Niederdruckdampfheizung, ND-Dampfheizung *f*

low-pressure steam pipe Niederdruckdampfrohr, ND-Dampfrohr *n*

L

low-pressure steam radiator Niederdruck-Dampfgliederheizkörper, Niederdruck-Dampfradiator; Niederdruck-Dampfheizkörper *m* [*Fehlbenennung*]. [*„ND“ kann synonym für „Niederdruck“ verwendet werden*]

low-pressure steam sectional boiler Niederdruckdampfgliederkessel, ND-Dampfgliederkessel *m*

low-pressure superheated water generation plant, low-pressure superheated water generation installation Niederdruck-Heißwassererzeugungsanlage, ND-Heißwassererzeugungsanlage *f*

low-pressure supply Niederdruckversorgung, ND-Versorgung *f*

low-pressure switch Niederdruckschalter, ND-Schalter *m*

low-pressure vacuum steam heating → **vacuum low-pressure steam heating**

low-pressure warm water → **low-pressure hot water**

low profile flache Bauweise, niedrige Bauweise *f* [*Gerät; Apparat; Instrument*]

low profile design flache Ausführung, niedrige Ausführung *f*

low profile keyboard, low profile keypad Flachtastatur *f*, Flachtastenfeld *n*

low-quantity alarm Mangelsicherung *f*

low-radiating strahlungsarm

low rank coal Armkohle *f*

low-resistance niederohmig

low resolution niedrige Auflösung *f*

low speed, low rotations per minute niedrige Drehzahl *f*, niedrige Umdrehungen *fpl* pro Minute

low-speed motor Langsamläufer *m*

low temperature (US) → **sub-zero temperature**

low temperature Niedrigtemperatur, Niedertemperatur *f*

low temperature, freezing temperature Gefriertemperatur *f*

low-temperature air heater Niedrigtemperaturluftheizer, NT-Luftheizer, Niedrigtemperaturlufterhitzer, NT-Lufterhitzer, Niedertemperaturluftheizer, Niedertemperaturlufterhitzer *m*

low-temperature boiler Niedertemperaturkessel, Niedrigtemperaturkessel, NT-Kessel *m* [*DIN 4702*]

low-temperature chimney [*One handling the products of combustion with a temperature at the point of leaving equipment, of 600° F (315 °C) or less under normal operating conditions*] Niedrigtemperaturschornstein, NT-Schornstein, Niedertemperaturschornstein *m*

low-temperature convector, low-temperature convection heater [*See remark under "convector"*] Niedertemperaturkonvektor(heizkörper) *m*

low-temperature corrosion Tieftemperaturkorrosion *f*

low-temperature cut out Niedrigtemperaturausschalter, Niedertemperaturausschalter *m*

low-temperature district heating [*Use of fluids less than 50 °C (200 °F) (in round numbers) for district heating supply; permitting use of energy sources formerly discarded*] Niedertemperaturfernheizung, Niedertemperaturstadtheizung, Niedrigtemperaturfernheizung, Niedrigtemperaturstadtheizung, NT-Fernheizung, NT-Stadtheizung *f*

low-temperature engineering Niedertemperaturtechnik, Niedrigtemperaturtechnik *f*

low-temperature floor heating Niedrigtemperatur-(Fuß)Bodenheizung, NT-(Fuß)Bodenheizung, Niedertemperatur-(Fuß)Bodenheizung *f*

low-temperature heat Niedrigtemperaturwärme, Niedertemperaturwärme *f*

low-temperature heater Niedertemperaturheiz(ungs)körper, Niedrigtemperaturheiz(ungs)körper, NT-Heiz(ungs)körper *m* [*Siehe Anmerkung unter „Heiz(ungs)körper“*]

low-temperature heat generating installation Niedrigtemperatur-Wärmeerzeugungsanlage, NT-Wärmeerzeugungsanlage, Niedertemperatur-Wärmeerzeugungsanlage *f*

low-temperature heating Niedrigtemperaturheizung, NT-Heizung, Niedertemperaturheizung *f*

low-temperature heating system Niedrigtemperaturheiz(ungs)system, NT-Heiz(ungs)system, Niedertemperaturheiz(ungs)system *n*

low-temperature heating water Niedrigtemperatur-Heiz(ungs)wasser, NT-Heiz(ungs)wasser *n*

low-temperature heat supply Niedertemperatur-Wärmeversorgung, NT-Wärmeversorgung *f*

low-temperature hot water, low-temperature warm water Niedrigtemperaturwarmwasser, Niedertemperaturwarmwasser, NT-Warmwasser *n*

low-temperature ignition Niedrigtemperaturzündung, NT-Zündung, Niedertemperaturzündung *f*

low-temperature incinerator Niedrigtemperaturverbrenner *m*, Niedrigtemperaturverbrennungsanlage *f*

low-temperature oil/gas heating boiler Tieftemperatur-Öl/Gas-Heiz(ungs)kessel, Niedrigtemperatur-Öl/Gas-Heiz(ungs)kessel *m*

low-temperature radiant heating, radiant low-temperature heating Niedrigtemperaturstrahl(ungs)heizung, NT-Strahl(ungs)heizung, Niedertemperaturstrahl(ungs)heizung *f*

L

low-temperature radiation Niedrigtemperaturstrahlung, NT-Strahlung, Niedertemperaturstrahlung *f*

low-turbulence turbulenzarm

low velocity air inlet Niedriggeschwindigkeit(s)lufteinlass *m*

low velocity duct net Niedriggeschwindigkeitskanalnetz *n* [*Ventilatorluftheizung mit kleiner räumlicher Ausdehnung*]

low voltage Niederspannung, Unterspannung *f*

low-voltage connector pair, l.v. connector pair Niedervoltsteckverbindung *f*, Niedervoltsteckverbinder *m*

low-voltage differential signal(l)ing, l.v. differential signal(l)ing Niederspannungs-Differenz-Signalübertragung *f*

low-voltage feed(ing) line Niederspannungsspeiseleitung *f*

low-voltage fitting Niederspannungsarmatur, Schwachstromarmatur *f*

low-voltage halogen lamp Niedervolthalogenlampe *f*

low-voltage illumination, low-voltage lighting, l.v. illumination, l.v. lighting Niedervoltbeleuchtung *f*

L

low-voltage logic Niederspannungslogik *f*

low-voltage power supply unit, low-voltage power pack, l.v. power pack, l.v. power supply unit Niederspannungsnetzteil *m,n*

low-voltage service → secondary service

low-voltage switchgear, low-voltage switch(ing) station, l.v. switchgear, l.v. switch(ing) station Niederspannungsschaltanlage *f*

low-voltage switching [*A system of outlet control by low voltage switches and relays*] Niederspannungsschalten *n*

low-voltage thermostat, l.v. thermostat Niederspannungs-Thermostat *m*

low water Wassermangel *m*

low-water alarm Wassermangelsicherung *f*

low-water alarm whistle Wassermangelpfeife *f*

low-water alarm with float, floating low-water alarm Wassermangelsicherung mit Schwimmer, Schwimmer-Wassermangelsicherung *f*

low-water alarm with fus(ibl)e plug, fus(ibl)e plug low-water alarm Wassermangelsicherung *f* mit Schmelzstöpsel

low-water level Niedrigwasserstand *m*

low-water valve Wassermangelventil *n*

low-wiring verdrahtungsarm

lozenge(-shaped) handle → diamond(-shaped) handle

LP → line printer

LPCVD → low pressure chemical vapour deposition

LPD, lower power device Funkanlage *f* geringer Leistung

LPG → liquefied petroleum gas

LPM, lines per minute Zeilen *f* pro Minute [*Drucker*]

LQ → limit(ing) quality

LQ → letter quality

LQL → limit(ing) quality level

LRC → longitudinal redundancy check(ing)

LSB → least significant bit

LSD → least significant digit

L-shaped kitchen L-förmige Küche *f* [*Eine Küche, deren Großgeräte und Schränke an zwei benachbarten Wänden angeordnet sind*]

LSI, large-scale integration hohe Integration *f* [*Integration mit 100 bis 500 Gatter pro Chip bei integralen Schaltkreisen*]

LSIC → large-scale-integrated circuit

LSN → least significant number

LS-TTL → low-power Schottky transistor-transistor logic

LTPD → lot tolerance percent defective

lubrication Schmierung *f*

lubrication fitting Schmiernippel *m*

lug Lasche, Fahne *f*

lug-in-ready einsteckfertig

lumen-hour Lumenstunde *f*

lumen per watt Lumen *n* je Watt

luminaire, light(ing) fitting, light(ing) fixture [*A complete lighting unit consisting of a lamp with the parts designed to distribute the light, to position and protect the lamp, and to connect the unit to a power supply*] Leuchte *f*, Beleuchtungskörper *m*

luminaire mirror Leuchtenspiegel *m*

luminance; brightness [*obsolete*] [*The ratio of the luminous intensity in a given direction of an infinitesimal element of a surface containing the point under consideration, to the orthogonally projected area of the element on a plane perpendicular to the given direction*] Helligkeit, Leuchtdichte *f*

luminance carrier → picture carrier

luminance signal Leuchtdichtesignal *n*

luminescence [*The phenomenon in which a substance absorbs primary radiation which gives rise to a later emission of light by the substance*] Lumineszenz *f*, Stoffleuchten, kalte Leuchten *n*

luminescent cell → **electroluminescent panel**

luminous flame Leuchtflamme *f*

luminous flame burner Leuchtflamme(n)brenner *m*

luminous tube Leuchtröhre, Lichtröhre *f*

lux, lx Lux, lx *n* [*Einheit der Beleuchtungsstärke, 1 lx ist vorhanden, wenn der Lichtstrom 1 Lumen auf die Fläche von 1 m² strahlt*]

luxury (bath)tub, deluxe (bath)tub Luxus(bade)-wanne *f* [*in Feuerton oder Marmor*]**luxury fluorescent tube** → **deluxe fluorescent tube**

luxury shower → **deluxe shower**

LVDT → **differential transformer**

L WORD → **longword**

lye Lauge *f*

lyophilization → **freeze drying**

Lyot filter → **bifringent filter**

M

MAC → **medium access ((open) loop) control**

to machine maschinell bearbeiten

machine address → **absolute address**

machine code, machine language, computer code, computer language Maschinensprache *f*, Rechnerprogrammiersprache, Programmiersprache eines Rechners, Maschinenkode, Rechnerkode *m* [*Die Maschinensprache besteht aus Befehls-Bitmustern, die der Rechner direkt dekodieren und ausführen kann. Programme in anderen Sprachen müssen in Maschinensprache übersetzt werden*]

machined maschinell bearbeitet

machine data acquisition Maschinendatenerfassung, MDE *f*

machine language Maschinensprache *f*

machine-orient(ta)ted language maschinenorientierte (Programmier)Sprache *f*

machine readable maschinenlesbar [*Daten*]

machine rundown → **rundown**

machine(ry) construction, machine(ry) building Maschinenbau *m*

machine(ry) level → **mechanical (equipment) floor**

machine(ry) room, mechanical (equipment) room [*Room in a building where major items of equipment are grouped together*] Anlagenraum, Technikraum, Betriebsraum, Maschinenraum, Triebwerk(s)raum *m*

machine(ry) sequential circuit Maschinenfolgeschaltung *f*

machine(ry) storey → **mechanical (equipment) floor**

machine switching system → **automatic exchange**

machine word Maschinenwort *n*

machining sequence maschinelle Bearbeitungsablauf *m*

machining tolerance maschinelle Bearbeitungstoleranz *f* [*Maximal zulässige Bearbeitungstoleranz aufgrund der Werkzeugmaschinendynamik*]

macro [*A statement that requests a language processor to generate a set of instructions*] Makro *n*

macrofunction Makrofunktion *f*

macro-instruction Makrobefehl *m* [*Programmteil zur Lösung einer Teilaufgabe*]

magamp [*abbreviated*] → **magnetic amplifier**

magazine → **(input) magazine**

magnet automatic closing valve → **automatic magnet closing valve**

magnetic amplifier, transducer [*A device that employs saturable reactors to modulate the flow of alternating-current electric(al) power to a load in response to a lower-energy-level direct-current input signal. Abbreviated: magamp*] Magnetverstärker, Transduktor *m*

(magnetic) bubble (computer) memory, (magnetic) bubble (computer) storage, MBM Magnetblasen(rechner)speicher *m*

magnetic card Magnetkarte *f*

magnetic card (computer) memory, magnetic card (computer) storage Magnetkarten(rechner)-speicher *m*

magnetic card reader Magnetkartenleser *m*, Magnetkartenlesegerät *n*

magnetic catch Magnetschnäpper *m*

magnetic checkpoint magnetische Kontrollstelle *f*

magnetic clamp Haftmagnet *m*

magnetic cooling → **adiabatic demagnetization**

M

magnetic core Magnetkern, Kern, Ferritkern *m*, Joch *n*

(magnetic) core memory, (magnetic) core storage (Magnet)Kernspeicher, Ferritkernspeicher, Jochspeicher *m*

magnetic data medium, magnetic data carrier, magnetic storage device, magnetic storage medium, magnetic output magnetische Datenträger *m*, magnetische Datenaufbewahrungsmittel *n*

magnetic defrosting valve Magnetabtauventil *n*

magnetic disk, magnetic disc Magnetscheibe *f*

magnetic disk (computer) memory, magnetic disk (computer) storage, magnetic (computer) memory, magnetic (computer) storage Magnetband(rechner)speicher *m*

(magnetic) disk storage, (magnetic) disc storage [*An external computer storage device consisting of one or more disks spaced an a common shaft, and magnetic heads mounted on arms that reach between the disks to read and record information on them*] Plattenspeicher, Magnetplattenspeicher *m*, (Magnet)Platteneinheit *f*

magnetic displacement → **magnetic induction**

magnetic drum Magnettrommel *f*

M

magnetic drum (computer) memory, magnetic drum (computer) storage Magnettrommel(rechner)-speicher *m*

magnetic drum storage, drum magnetic storage Magnettrommelspeicher, Trommelmagnetspeicher *m*

magnetic field Magnetfeld *m*

magnetic float(ing) switch, float(ing) magnetic switch Magnetschwimmerschalter *m*

magnetic flow magnetische Fluss *m*

magnetic flow meter Magnetflussmesser *m*

magnetic head Magnetkopf *m*

magnetic induction, magnetic displacement, magnetic flux density, magnetic vector [*A vector quantity that is used as a quantitative measure of magnetic field; the force on a charged particle moving in the field is equal to the particle's charge times the cross product of the particle's velocity with the magnetic induction (mks units)*] magnetische Induktion, (magnetische) Flussdichte, Kraftliniendichte *f*

magnetic interceptor → **magnetic separator**

magnetic iron ore filter → **magnetite filter**

magnetic lime dissolver magnetische Kalk(auf)-löser *m*

magnetic lock Magnetschloss *n*

magnetic north Magnetischnord *n*

magnetic recording Magnetaufzeichnen *n*, Magnetaufzeichnung, MAZ *f*

(magnetic) remote switch (magnetische) Fernschalter *m*

magnetic rod Magnetstab *m*

magnetic separator, magnetic interceptor, magnetic trap [*See remark under "interceptor"*] Magnet(ab)-scheider *m*

magnetic soap holder Magnetseifenhalter *m*

magnetic soundtrack [*Any soundtrack that is recorded onto a magnetic tape, as in an audio cassette recording or the audio portion of a motion picture*] Magnettonspur *f*

magnetic storage medium magnetische Speicherschicht *f* [*Diskette*]

magnetic stray field → **stray magnetic field**

magnetic strip Magnetstreifen *m*

magnetic strip badge, magnetic strip card Magnetstreifenausweis(karte) *m,(f)*

magnetic strip reader Magnetstreifenleser *m*

magnetic tape (computer) memory, magnetic tape (computer) storage Magnetband(rechner)speicher *m*

(magnetic) tape drive, (magnetic) tape station, (magnetic) tape transport, (magnetic) tape unit [*A device designed to control the movement of magnetic tape past the heads, which enables the tape to be read, recorded (written upon), or rewound*] (Magnet-) Bandlaufwerk *n*

(magnetic) tape unit Magnetbandeinheit *f*

magnetic temperature [*As measured by a magnetic thermometer*] magnetische Thermometertemperatur *f*

magnetic thermometer [*It indicates temperature from the measurement of magnetic properties (e.g. susceptibility) of any suitable substance; used for very low temperatures*] Magnetthermometer *n*

magnetic trap → **magnetic separator**

magnetic vector → **magnetic induction**

magnetite filter, magnetic iron ore filter, octahedral iron ore filter Magnetitfilter *m,n*

magnetite heating water filter Magnetitheiz(ungs)-wasserfilter *m,n*

magneto ignition Magnetzündung *f*

magnetomotive force, MMf, mmf, m.m.f. magnetomotorische Kraft, MMK *f*

magnetomotive memory magnetomotorische Speicher *m*

magneto optical disk, magneto optical disc, rewritable optical disk, rewritable optical disc, ROD, MOD Aufnahme- und Lösch-Digitalplatte, magnetooptische

Platte *f* [*Ein mehrfach beschreibbares Massenspeichermedium*]

magneto resistive head, MRH Wiedergabekopfspalte *f* [*digitale Kompaktkassette*]

magnetoresistor Feldplatte *f*, magnetische Widerstand *m* [*Dieses Element wandelt eine magnetische Induktionsänderung*]

magnetoresistor potentiometer Feldplattenpotentiometer, magnetische Widerstandspotentiometer *n*

magnet wire Dynamodraht *m*

magnification mirror Vergrößerungsspiegel *m*

magnifying power → power

magnitude vector → vector with magnitude

mailbox [*A software data structure that is treated as a record-orien(ta)ted device for general interprocess communication. Communication using mailbox is similar to other forms of device-independent I/O. Senders perform a write to a mailbox, the receiver performs a read from that mailbox*] elektronische Briefkasten, Briefkastendienst *m*

mail chute → letter chute

mailing list Adressenverteilerliste *f*

mailing service Mitteilungsdienst *m* [*Ein Multimedia-Kommunikationsdienst*]

mail slot, letter slot Briefschlitz *m*

main → power transmission line

main... → principal...

mainboard Rechnerhauptplatine *f*

main distribution board → distribution center

main distribution frame → trunk distribution frame (US)

main electric service [*It enters the building below ground level*] Elektrozuleitung *f*

main frame → central processing unit

main fuse → company's fuse

main gas service [*It enters the building below ground level*] Gaszuleitung *f*

main piped water supply [*It usually runs in the street or footpath, if any*] Haupt(rohr)wasserversorgung *f*

mains cable Netzkabel *n*

main service → (plot) main service

mains infeed(ing) of controller, mains infeed(ing) of regulator (Ein)Speisung *f* einer Steuer(ungs)einrichtung [*Bereitstellung der von der Versorgung benötigten Energie. V DIN 19 237*]

mains-operated circulating pump Netzumwälzpumpe *f*

mains plug Netzstecker *m*

mains recovery program (US); mains recovery programme (Brit.) Netzwiederkehrprogramm *n* [*Eine programmierte Folge die festlegt, welche Maßnahmen vom System zu ergreifen sind, wenn die Netzversorgung zurückkehrt*]

mains voltage Netzspannung *f*

mains water, network water, tap water, water for general use Leitungswasser, Brauchwasser *n*

maintainability [*Specific time period within which a system or facility can operate in an economical, fully restored condition*] Wartbarkeit *f*

maintained contact Stetigkontakt *m* [*Er kommt von Schützen oder Stellschaltern wie Nocken- und Wahlschaltern*]

maintained-contact control, maintained-contact operation Stetigkontaktgabe *f*

maintained signal → continuous signal

maintenance, upkeep Wartung, Bewahrung des Sollzustandes, Unterhaltung *f* [*DIN 31 051*]

maintenance factor → light loss factor

maintenance function, upkeep function Unterhaltungswahrscheinlichkeit, Wartungswahrscheinlichkeit *f*

maintenance signal message, upkeep signal message Wartungsmeldung, Unterhaltungsmeldung *f* [*Siehe Anmerkung unter „Meldung"*]

main water service [*It enters the building below ground level*] Wasserzuleitung *f* [*Sie bringt das Wasser vom Straßenhauptrohr in das Gebäude*]

major consumer Großverbraucher, Großabnehmer *m*

major fire, serious fire Großbrand *m*, Großfeuer *n*

make Fabrikat *n*

to make active → to activate

make break-time Ein-Ausschaltzeit *f*

make contact, a-contact, normally open contact, NO contact, make contact element [*IEC 337-1*] Schließer; Schließkontakt, Arbeitskontakt *m*, Einschalt(ungs)glied *n* [*frühere Benennung*] [*Beim Betätigen schließendes und im unbetätigten Zustand offenes Schaltstück eines elektrischen Kontaktes. VDE 0660, Teil 200*]

make contact opening time Schließerrückfallzeit *f* [*Monostabile Relais. DIN IEC 255, Teil 1-00*]

to make good [*e.g. a chasing*] einputzen

maker, manufacturer, producer Hersteller, Fabrikant *m*

make time Einschaltzeit *f*

to make up nachholen [*z. B. Kesselwasser*]

makeup, replacement Nachspeisung *f* [*z. B. von verdunstetem Wasser*]

M

makeup air, replacement air [*1. Air brought into a building from the outside to replace that exhausted. 2. In a clean room, air introduced to the secondary air system for ventilation, pressurisation, and replacement of exhaust air*] Zuluft *f*

makeup flow, replacement flow Nachströmung *f* [*z. B. Zuluft*]

makeup mirror Schminkspiegel *m*

makeup (pipe)line, replacement (pipe)line Nachhol(rohr)leitung, Nachfüll(ungs)(rohr)leitung *f*

makeup water, replacement water [*It is added to any vessel to maintain a predetermined liquid device*] Nachholwasser, Ergänzungswasser, Nachfüll(ungs)-wasser *n*

to (make) warm → to warm

making [*contact*] Schließen *n*, Schließung *f*

making/break(ing) unit [*IEC 56-1*], **unit** Schaltelement *n* [*LS, VDE 0670, Teil 101*]

making current Einschaltstrom *m*

making current release Einschaltstromauslöser *m*

making duty → switching capacity

making fleeting contact → fleeting contact on making

making performance → switching capacity

making service → switching capacity

maladjustment Falschjustierung, Fehljustierung, Falschverstellung, Fehlverstellung *f*

male disconnect(ing) device Abschalt(ungs)stecker *m*

male (pipe) fitting thread → conical (pipe) fitting thread

male plug Einschraubstopfen *m*

male thread Außengewinde *n*

malfunction → fault

malleable cast iron, MCI schmiedbare Guss(eisen), Temperguss(eisen) *m, (n)*

malleable cast iron (pipe) fitting, MCI (pipe) fitting Temperguss(eisen)fitting *m*, Temperguss-(eisen)(rohr)formstück, Tempergussrohrverbindungsstück *n* [*DIN. 2950-2973. Siehe Anmerkung unter "Fitting"*]

maloperation, (inadvertent) wrong operation Fehlbedienung *f*

MAMDANI controller rationale Fuzzy-Regler *m*

MAN, metropolitan area network Metropolenregionnetz(werk) *n*

to manage bewirtschaften

to manage leiten, führen

to manage betreuen

management Bewirtschaftung *f*

management [*The attainment of organisational goals by means of planning, organisation and efficient use of resources, especially human resource(s)*] Leitung, Führung *f*

management Betreuung *f*

management-buy-out Beschäftigungsgesellschaft *f*

management by objectives, objective management [*A management technique based on setting measurable objectives for employees, departments, or projects, and then using the objectives to direct and evaluate performance*] sachliche Leitung, sachliche Führung *f*

management information system [*Telecommunications. A communication system in which data are processed for use in decision-making by individuals in management*] Leitungsinformationssystem, Führungsinformationssytem *n*

management level, ML [*The level within the communications hierarchy that consists of monitoring and operator units (MOVs) and gateways' data interface unit (DIUs) to other systems. The management functions take place at this level*] Leitebene, Führungsebene *f* VDI 3814-2 Anwendungsrichtlinie; GAEB StLB 071 LV-Text CEN; TC 247 WG 3 Teil 2 Entwurf

management level network, MLN [*The net(work) used for the interconnection of management level devices. The communications infrastructure within the management level*] Leitebene(n)netz(werk), Führungsebene(n)netz(werk) *n*

management science [*the application of quantitative techniques to the solution of managerial problems*] Leitungswissenschaft, Führungswissenschaft *f*

management system Leit(ungs)system, Führungssystem *n*

manager → building manager

managing Betreuen *n*

managing Bewirtschaften *n*

managing[*See remark under "management"*] Führen, Leiten *n*

Mandani-controller relationale Fuzzyregler *m*

Mandani implication Mandaniimplikation *f*

Mandani inference Max-Min-Inferenz, unscharfe Schlussfolgerungsmethode *f*, unscharfe Schlussfolgerungsverfahren *n*

mandatory obligatorisch, verbindlich, verpflichtend

mandrel → (pipe) mandrel

manganese removal Entmanganung, Manganbeseitigung *f*

M

manhole, manway; (access) chamber [*deprecated*] [*A chamber constructed on a drain or sewer so as to provide access thereto for inspecting, testing or the clearance of abstruction*] Mannloch, Einstiegloch, Einsteigloch *n*, Einstiegschacht, Einsteigschacht *m*

manhole cover, manway cover [*The remov(e)able cover of a manhole*] Schachtdeckel, Mannlochdeckel, Einstiegsschachtdeckel *m* [*Er ist abnehmbar*]

manhole frame Mannlochrahmen, Mannlochaufsatz, Schachtrahmen, Schachtaufsatz *m*

manhole pit, manway pit Domschacht *m*

manhole ventilation Sammelschachtlüftung, Sammelmannlochlüftung *f*

manhole with tumbling bay → backdrop manhole

manifold → header

Manila paper Manilapapier *n* [*z. B. zur Dichtung bei Warmwasserradiatoren*]

(manipulated) line reaction Stellverhalten der Strecke, Streckenstellverhalten *n*

manipulated variable [*ANSI C 85.1. In control systems, a quantity or condition that is changed by the computer, based on system feedback, to cause a shift in the value of some regulated conditions*] Stellgröße *f* [*DIN 19 226*]

manipulating capacity Stellvermögen *n*, Stellfähigkeit *f* [*DIN 1926, Teil 5*]

manipulating point Stellort, Stellpunkt *m*

manipulating range Stellbereich *m* [*selten: n*] [*DIN 19 226, Teil 5*]

manipulating speed Stellgeschwindigkeit *f* [*DIN 19 226, Teil 5*]

manipulating time Stellzeit *f*

manipulation Führung *f*

manipulation Stellen *n*

manipulation, handling Handhabung *f*, Umgang *m*

manipulation (bath)tub, patient (bath)tub Patientenbadewanne *f*

man/machine interface, man-to-machine interface Mensch-Maschine-Schnittstelle *f*

(manned) guard(ing) service Werkschutz *m*

Mannheim gold Mannheimer Gold, Prinz-Rupperts-Metall *n*

manometer, differential pressure gauge [*US spelling is "gage". "Meter" may synonymously be used for "gauge"*] Manometer *n*, Differenzdruckmesser *m*

mantisse Basis *f* [*Hinter dem Komma stehende Ziffer der Logarithmen*]

man-to-machine interface → man/machine interface

man-to-system interface, man-system interface, man/system interface Mensch-System-Schnittstelle *f*

manual, handbook Handbuch *n*

manual ... → hand ...

manually accessible → accessible to hand

manually held → hand-held

manually lighted gas boiler [*Boiler in which gas to the main burners is turned on only by hand*] handgezündete Gaskessel *m*

manually operated releasing signal, manually operated trip(ping) signal Handauslösesignal *n*

manual position [*automatic telephony*] Vermittlungsplatz *m*, Vermittlungsstelle *f*

manufactured product Anfertigungserzeugnis, Anfertigungsprodukt, Herstellungserzeugnis, Herstellungsprodukt *n*

manufacturer → maker

manufacturer independence Anfertigungsunabhängigkeit *f* [*Zusammenschließen von Geräten verschiedener Hersteller in einem Netz; häufig werden verschiedene Protokolle verwendet*]

manway → manhole

manway cover → manhole cover

MAU, multi(ple)-station access unit Token-Ringleitungsverteiler *m*

map [*Matrix format for presenting logic states*] Karte *f*

MAP, manufacturing automation protocol Kommunikationsnetzwerk, Breitbandnetzwerk mit offener Struktur, Netzwerkkonzept für Fertigungseinrichtungen *n* [*Ein herstellerunabhängiges, offenes Netzwerk. Ein Kommunikationssystem für den Datenaustausch zwischen Rechnern, Steuerungen und Maschinen. Es baut auf Koaxialkabel, Bus-Topologie und Token-Zugriff(s)verfahren auf*]

map control „Totmannschaltung" *f* für mobiles Sicherheitspersonal [*Zur Absicherung von Mitarbeitern, die sich per Funk regelmäßig mit einem Kurzdatentelegramm in der Zentrale melden. Das Programm im Leitrechner verwaltet die Rufe und kontrolliert ob die vereinbarten Zeitintervalle eingehalten werden*]

map manager Operator *m* Sicherheitsdienste [*Er ermöglicht die rechnergesteuerte Planung, Disposition und Überwachung von Wach- und Schutzpersonal im Außendienst*]

marble (bath)tub Marmor(bade)wanne *f*

marble tile(d) stove, marble Dutch stove Marmorkachelofen *m* [*DIN 1294*]

marble vanity Marmorwaschtisch *m*

margin justification → justification

M

Mariotte(s) law → Boyle('s) law

marker, flag Merker *m* [*Binäres Speicherglied im Programmspeicher der Steuer(ungs)vorrichtung zur Speicherung ausgewählter binärer Daten. DIN 19 226, Teil 5*]

marker bit → flag bit

marker location → flag location

market garbage, market food waste, market waste food Marktabfall *m*, Marktabfälle *f* [*Bezieht sich nur auf Nahrungsmittelabfälle, nicht auf Verpackungsmaterial*]

market leader Marktführer *m*

market niche Marktlücke *f*

mark sensing card Zeichenlochkarte *f*

markup language Strukturierungssprache *f*

MASER → microwave amplifier by stimulated emission of radiation

mask Maske *f* [*Bitmuster zur gezielten Ausblendung von Binärstellen*]

mask → template

masking pattern universal subband integrated coding and multiplexing Rundfunk-Datenreduktionsmethode *f*

masking set Blendrahmensatz *m*

M

mask interconnection → interconnection mask

mason → fixer (mason)

masonry (air) duct Mauerwerk(luft)kanal, Mauerwerklüftungskanal *m*

masonry furnace Mauerwerkfeuerung, Mauerwerkfeuerstätte *f*

masonry rehab(ilitation) Mauerwerksanierung *f*

masonry wall box Mauerkasten *m*

masonry wall breakthrough, masonry wall penetration Mauerdurchbruch *m*, Mauerdurchführung *f*

mass [*Quantity of matter in a body; measured in terms of resistance to acceleration by a force; i.e., its inertia. In SI (Système International d'unités) the standard unit of mass is the kilogram (kg); In I-P (Inch-pound) it is the slug*] Masse *f*

mass conservation → conservation of mass

mass diffusivity (ratio), diffusion coefficient [*Ratio of the mass flow of a substance diffusing through a surface of unit area to the rate of variation in the concentration of this substance normal to this surface*] Diffusionskoeffizient, Streukoeffizient *m*

mass-flow sensor, mass-flow feeler Strömungssensor, durchströmte Sensor, Strömungsfühler, durchströmte Fühler *m*

mass occupancy Massenbelegung *f*

mass storage [*Mass memory is a misnomer*] Massenspeicher *m* [*Er nimmt große Datenmengen auf. Es sind Magnetplatten, optische Speicher und Magnetbänder*]

mass transfer, mass transport [*Transfer of one component of a mix(ture), relative to the motion of the mix(ture). It is the result of a concentration gradient*] Massentransport, Massentransfer *m*, Massenübertragung *f*

mass transport(ation), public transport(ation), passenger-carrying, (passenger) bulk movement Personen(be)förderung *f*, Personentransport *m*

mass transport(ation) capacity, passenger-carrying capacity, public transport(ation) capacity, (passenger) bulk movement capacity Personen(be)förder(ungs)kapazität, Personen(be)förder(ungs)leistung, Personentransportkapazität, Personentransportleistung *f*

mass velocity [*Ratio of the mass flow rate of a fluid in a pipe to the cross-sectional area of the pipe*] Massengeschwindigkeit *f*

master → (bus) master

master → master (system)

master Grundschaltung *f*

master aerial → central aerial

master board Grundleitplatte *f*

master controller, master switch Hauptschalter *m*

master (data) file, data master file Hauptdatei, Generaldatei *f*

master distributor Erstverteiler *m*

master fitter Meistermonteur, Monteurmeister, Obermonteur *m*

master gas fitter Meistergasmonteur, Gasmeistermonteur *m*

master key → change key

master key → (floor) master key

master-keyed lock → change-keyed lock

master memory → principal memory

master plumber Sanitärschlossermeister *m*

master radio aerial → central radio aerial

master routine → executive routine

(master-slave) polling, master-slave method, fixed master method zentrale Bussystemsteuerung, zentrale Buszuteilung *f*

master standard Hauptnormal *n*

master-submaster controller (US) → reset (action) controller (Brit.)

master (system) [*It gains control of the communication process and indicates a transaction*] Hauptsystem *n*

master T.V. aerial → **central T.V. aerial**

matchboarding Profilbretter *f*

match gate → **equivalence gate**

matching Anpassen *n*, Anpassung *f*

matching resistor, terminating resistor Abschlusswiderstand *m*

material handling elevator Güteraufzug, Materialaufzug *m*

material(s) bill → **BOM**

material(s) collection point (Werk)Stoffsammelstelle, Materialsammelstelle *f*

material(s) damage Sachschaden *m*

material(s) lift (Brit.); material(s) elevator (US) [*A special form of dumbwaiter or small-sized elevator with an automated transfer device for the loading and unloading of boxes or carts*] Kastenaufzug *m*

material(s) lock Materialschleuse *f*

material(s) testing institute → **board of agrément**

mating piece, counter piece Gegenstück *n*

matrix [*In automatic data processing, a latticework of input and output leads with logic elements connected at some of their intersections*] Matrix *f*

matrix [*In mathematics, a rectangular array of numbers or scalars from a vector space*] Matrix *f*

matrix [*A binding agent used to make an agglomerate mass*] Binder *m*

matrix Leiterbett *n*

matrix [*A recessed mo(u)ld in which something is formed or cast*] Matrize *f*

matrix calculus Matrizenrechnung *f*

matrix connection Matrixschaltung *f*

matrix diode Programmierdiode *f*

matrix document Textschablone *f*, Formtext, Haupttext *m*

matrix impact printer, impact matrix printer Matrixnadeldrucker, Nadelmatrixdrucker *m*

matrix ink-jet printer → **ink-jet matrix printer**

matrix inversion Matrixumkehr *f*

matrix module, matrix block Matrixbaustein *m*, Matrixmodul *n*

matrix printer, stylus printer, wire printer, dot-matrix-printer [*A printer in which each character is printed as a pattern of dots with needles*] Matrixdrucker, Mosaikdrucker *m*

matrix print head, stylus print head, wire print head, dot-matrix print head Nadeldruckkopf *m*

matrix printing, stylus printing, wire printing [*In automatic data processing, a high-speed printing in which characterlike configurations of dots are printed through the proper selection of wire ends (needles) from a matrix of wire ends*] Matrixdrucken, Mosaikdrucken *n*

matrix storage → **coordinate storage**

matrix theory [*The algebraic study of matrices and their use in evaluating linear processes*] Matrizentheorie *f*

matrix unit, matrix switcher [*video*] Matrizenschalter *m*

matter → **substance**

matter extraneous to air luftfremde Stoff, Luftfremdstoff, Luftstörstoff, luftstörende Stoff *m*

matter in suspension → **suspended solid**

matter phase → **state**

matting coal, caking coal, close burning coal, binding coal Backkohle, backende Kohle, kuchenbildende Kohle *f*

mattock, back Breithacke, Lettenhaue, Radehaue, Radeharke, Lettenhacke, Radehacke, Radehaue, Breithaue, Breitenharke, Lettenharke *f*

mattress array → **billboard array**

mature(d) ausgereift, reif

mat wiring Mattenverdrahtung *f*

MAU → **medium attachment unit**

mavar → **parametric amplifier**

maverick Ausreißer *m* [*Gütesicherung*]

MAXDLyT → **maximum delay time**

max-dot → **max-prod**

maxibar Getränkeautomat *m*

maximum arc voltage, peak arc voltage Höchstlichtbogenspannung, Spitzenlichtbogenspannung *f*

maximum blocking voltage, peak blocking voltage Höchstsperrspannung, Spitzensperrspannung *f*

maximum bolometer, maximum thermal detector, maximum thermal detecting device, maximum thermal defection device (Thermo)Maximalmelder *m*, (Thermo)Maximalmeldegerät *n* [*Er löst bei Erreichen einer bestimmten maximalen Temperatur (oft rund 12 °C) Alarm aus*]

maximum consumption, peak consumption Spitzenverbrauch, Höchstverbrauch *m*

maximum crowd, peak crowd Personenspitzenmenge, Personenhöchstmenge *f*

maximum current, peak current Spitzenstrom, Höchststrom *m*

M

maximum delay time, MAXDlyT, peak delay time Spitzenverzögerungszeit, Höchstverzögerungszeit *f*

maximum demand, maximum need, peak demand, peak need Spitzenbedarf *m*

maximum demand heating plant, maximum demand heating installation, peak demand heating plant, peak demand heating installation Spitzenheizwerk *m*

maximum demand limitation, peak demand limitation Spitzenbedarfbegrenzung *f*

maximum detector, maximum detecting device, maximum detection device Maximalmelder *m*, Maximalmeldegerät *n* [*Er spricht bei Überschreitung einer bestimmten Größe für genügend lange Zeit an. DIN EN 54, Teil 1, 04.77*]

maximum distribution, peak distribution Spitzenwertverteilung *f*

maximum(-duty) operation, maximum load operation, peak(-duty) operation, peak load operation Spitzen(last)betrieb, Höchst(last)betrieb *m*

maximum efficiency, peak efficiency Spitzenwirkungsgrad, Höchstwirkungsgrad *m*

maximum envelope power, peak envelope power, PEP Spitzenhüllkurvenleistung, Höchsthüllkurvenleistung *f*

M

maximum function Maximum-Operation *f*, *t*-Konorm *f*

maximum heat demand, peak heat demand, peak heat requirement, maximum heat requirement Spitzenwärmebedarf *m*

maximum heat detector, maximum heat detecting device, maximum heat detection device Wärmemaximalmelder, Maximalwärmemelder *m*, Wärmemaximalmeldegerät, Maximalwärmemeldegerät n [*Es wird das Erreichen oder Überschreiten einer festgelegten Temperatur für eine genügend lange Zeit als Alarmsignal ausgewertet*]

maximum heating, peak heating Spitzenheizung, Höchstheizung *f* [*Gegensatz: Grundheizung*]

maximum installation pressure → maximum plant pressure

maximum level, peak level Spitzenpegel, Höchstpegel *m* [*Schallpegel bei schwankenden Geräuschen, der nur während sehr kurzer Zeit erreicht bzw. überschritten wird. Nicht verwechseln mit „Scheitelpegel"*]

maximum level meter, peak level meter Spitzenwertmesser, Höchstwertmesser *m*

maximum limitation, peak limitation Spitzenbegrenzung, Höchstbegrenzung *f*

maximum load, peak load Spitzenlast, Höchstlast *f*

maximum load connection, peak load connection Spitzen(last)zuschaltung, Höchst(last)zuschaltung *f*

maximum load consumption, peak load consumption Spitzen(last)verbrauch, Höchst(last)verbrauch *m*

maximum load controller, maximum load regulator, peak load controller, peak load regulator Spitzen(last)regler, Höchst(last)regler *m*

maximum load limitation, peak load limitation Spitzen(last)begrenzung, Höchst(last)begrenzung *f* [*Die Begrenzung der elektrischen Last eines Gebäudes auf einen vorgegebenen Maximalwert*]

maximum load optimization, peak load optimization, PLO Höchst(last)optimierung, HLO, Spitzen(last)optimierung, SLO *f*,

maximum load shedding, peak load shedding Spitzen(last)abwurf, Höchst(last)abwurf *m*

maximum operation → maximum(-duty) operation

maximum output rate of change, peak output rate of change Ausregel(ungs)steilheit *f* [*DIN 41 745*]

maximum performance meter, peak performance meter, maximum output meter, peak output meter Spitzenleistungsmesser, Höchstleistungsmesser *m*

maximum plant pressure, maximum installation pressure, peak plant pressure, peak installation pressure Anlagenhöchstdruck, Anlagenspitzendruck *m*

maximum position, peak position Spitzenstellung, Höchststellung *f*

maximum power point, MPP höchste Leistungspunkt, maximale Leistungspunkt *m* [*Auf einer Strom-Spannungs-Kennlinie. Für eine typische Siliziumzelle liegt dieser Punkt ungefähr bei 0,45 V*]

maximum pressure, peak pressure Spitzendruck, Höchstdruck *m*

maximum pressure line, peak pressure line Spitzendrucklinie, Höchstdrucklinie *f*

maximum-quality product, peak-quality product Spitzenerzeugnis, Spitzenprodukt *n*

maximum reverse recovery current, peak reverse recovery current Rückstromspitze *f* [*Thyristor; Diode*]

maximum tariff, peak tariff Spitzentarif, Höchsttarif *m*

maximum tariff current, peak tariff current Spitzentarifstrom, Höchsttarifstrom *m*

maximum-temperature detector, maximum-temperature detecting device, maximum-temperature detection device Thermomaximalmelder, Maximalthermomelder *m*, Thermomaximalmeldegerät, Maximalthermomeldegerät *n*

maximum-temperature radiant heating, radiant maximum-temperature heating Spitzentemperaturstrahlungsheizung, Höchsttemperaturstrahlungsheizung *f*

maximum thermal detector → maximum bolometer

maximum time, peak time t_{max}-Zeit *f* [*Regel(ungs)-technik*]

maximum usable read number, peak usable read number höchste nutzbare Abfragehäufigkeit, maximale nutzbare Abfragehäufigkeit *f*

maximum value, peak value Scheitelwert, Spitzenwert, Höchstwert, Größtwert, Maximumwert, Maximalwert *m*

maximum withstand current, peak withstand current Haltespitzenstrom, Haltehöchststrom *m*

Maxwell(ian) distribution [*A function giving the number of molecules of a gas in thermal equilibrium whose velocities lie within a given, infinitesimal range of values, assuming that the molecules obey classical mechanics and do not interact*] Maxwellsche Geschwindigkeitsverteilung *f*

MBG → medium Joule biogas

MBM → (magnetic) bubble (computer) memory

MC → microcomputer

MCA layer → medium access control layer

m.c.b. → miniature circuit breaker

MCC → motor control center

MCC → management control center (US)

MCI → malleable cast iron

MCI (pipe) fitting → malleable cast iron (pipe) fitting

MCS → principal call station

mcts → multi(ple)-channel transceiving system

MDF → trunk distribution frame (US)

MDR → (computer) memory data register

mean accuracy → accuracy of the mean

mean failure detection time, average failure detection time mittlere Ausfallerkennungszeit *f*

mean height → mean level

mean information content mittlere Informationsgehalt, mittlere Informationsinhalt *m*

mean level, average level mittlere Pegel *m* [*Siehe Anmerkung unter „Pegel"*]

mean level, mean height, average level, average height mittlere Stand *m*, mittlere Höhe *f*, mittlere Niveau *n*

mean noise factor, average noise factor Bandrauschzahl *f*

mean of maximum defuzzification, MOM defuzzification, middle of maximum defuzzification Maximum-Mittelwert-Verfahren *n*, Maximum-Mittelwert-Methode *f*

mean outgoing quality, average outgoing quality, AOQ Durchschlupf *m*

mean outgoing quality limit, average outgoing quality limit, AOQL maximale Durchschlupf *m*, Durchschlupfgrenze *f*

mean peak noise, average peak noise mittlere Geräuschspitze, durchschnittliche Geräuschspitze *f*

mean radiant temperature, MRT, average radiant temperature [*Uniform surface temperature of a radiantly black enclosure in which an occupant would exchange the same amount of radiant heat as in the actual nonuniform environment*] mittlere Strahlungstemperatur, durchschnittliche Strahlungstemperatur *f*

mean sample number, average sample number, ASN durchschnittliche Stichprobenumfang, mittlere Stichprobenumfang *m*

mean temperature, average temperature [*Arithmetic mean of temperatures measured at a given point during a given lapse of time, or at a given time in various points of a space or body*] mittlere Temperatur, Durchschnittstemperatur *f*

meat keeper [*Enclosed compartment or container which is designed for the storage of fresh meat at or near 32° F(0° C)*] Fleischfach *n*

mean effectiveness level → average effectiveness level

mean throughflow, average throughflow mittlere Durchfluss *m* [*zeitlicher Mittelwert des Durchflusses*]

mean time between errors, MTBE, average time between errors mittlere Ausfallzeit *f* [*Mittlere Zeit zwischen zwei Fehlerausfällen*]

mean time between failures, MTBF, mean failure detection time, MFDT, average time between failures, average failure detection time mittlere Zeit zwischen zwei Ausfällen, MZZA, mittlere Fehlererkennungszeit, mittlere Betriebszeit *f* [*Sie liegt bei den meisten sporadisch eintretenden Schäden zwischen 10 bis 1 000 Bauteilbetriebsjahren. Wenn zum Beispiel bestimmte Schaltgeräte eine MZZA von 100 Jahren besäßen, würde in einer großen elektrischen Verteilung mit 100 Schaltgeräten ein Gerät durchschnittlich jährlich ausfallen*]

mean time between maintenance → MTBM

mean time between repairs → MTBR

mean time between replacements → MTBR

mean time between stops → MTBS

mean time to failure, MTTF, average time to failure mittlere Ausfallzeit *f* [*Mittlere Zeit bis zum Ausfall eines Gerätes oder Bauteils*]

mean time to first failure, MTTFF, average time to first failure mittlere Lebensdauer *f*

M

mean time to repair, MTTR, average time to repair mittlere Stör(ungs)dauer *f* [*Durchschnittliche Zeitspanne zwischen dem erstmaligen Eintreten einer Störung und der Wiederherstellung der Funktion*]

mean transfer rate, average transfer rate mittlere Übertragungsgeschwindigkeit, durchschnittliche Übertragungsgeschwindigkeit *f* [*Mittelwert der Übertragungsgeschwindigkeit von Daten zur Charakterisierung des Übertragungskanals. Programmbedingte Unterbrechungen im Datenfluss werden nicht berücksichtigt*]

mean value sensor, average value sensor Mittelwertfühler, Mittelwertsensor, Durchschnittswertfühler, Durchschnittswertsensor *m*

mean-ward Blechreif, Mittelbruch *m* [*Im Eingerichte des Schlosses. Auf ihm dreht sich der Barteinschnitt*]

measurable variable [*See remark under "variable"*] Messgröße, messbare Größe *f*

measurand Messwert *m* [*Der zu messende Wert*]

to measure → **to meter**

measure Maß *n*

measure [*A plan or course of action intended to attain some object*] Maßnahme, Vorkehrung *f*

measured gas [*It has passed through and the volume of which has been recorded by a meter*] gemessene (Gas)Volumen *n*

measured value Messwert, gemessene Wert *m* [*Produkt aus Maßzahl und Einheit*]

measured-value indicator, measured-value indicating instrument Messwertanzeiger *m* [*Gerät, das die Maßgröße mit der vorgegebenen Einheit als Messwert analog oder digital anzeigt*]

measured-value log Messwertprotokoll *n*

measured-value processing Messwertverarbeitung *f*

measured-value report Messwertbericht *m*

measured-value scan Messwertabfrage *f*

measured-value transmitter → **pickup**

measurement difference Messabweichung *f* [*Der Unterschied zwischen dem Messwert und seinem Bezugswert. DIN 1319, Teil 1*]

measurement error → **error**

measurement point Messort *m*, Messstelle *f*

measurement(s) of heat (capacities), calorimetry, calorimetric measurement(s) Kalorimetrie *f*, Wärmemessung(en) *f(pl)*

measurement variable, measuring variable Messgröße *f* [*Physikalische Größe, die gemessen wird*]

measure value → **actual value**

measuring Messen *n* [*Der experimentelle Vorgang, durch den ein spezieller Wert einer physikalischen Größe als Vielfaches einer Einheit oder eines Bezugswertes ermittelt wird. DIN 1319, Teil 1 1/11.71. DIN 19 222*]

measuring amplifier Messverstärker *m* [*Ein Verstärker mit erhöhter Genauigkeit. E DIN 19 226, Teil 5*]

measuring ampule Messampulle *f*

measuring and control engineering MSR-Technik, Mess-, Steuer- und Regeltechnik *f*

measuring center (US) → **measuring station**

measuring chamber [*deprecated*] → **diaphragm chamber**

measuring chamber Messkammer *f*

measuring circuit, measuring arrangement Messschaltung *f*

(measuring) compartment Kammer *f* [*Nass-Gaszähler*]

measuring converter Messumsetzer *m*

measuring current Messstrom *m*

measuring data bus Messdatenbus *m*

measuring data processing Messdatenverarbeitung *f*

measuring equipment for gas furnaces Messgerät(e) für Gasfeuerung, Messgerät(e) für Gasfeuerstätten n, (*f*) [*Jede Gasfeuerstätte muss einreguliert werden. Man benötigt hierfür:*
a) eine Stoppuhr zur Bestimmung der Belastung (Gaseinstellung),
b) ein Abgasthermometer zur Messung der Abgastemperatur vor der Strömungssicherung bzw. am Kesselabgasstutzen (Messbereich bis ca. 400 °C) sowie am Schornsteinkopf (Messbereich bis ca. 120 °C),
c) ein U-Rohr-Manometer zur Messung des Anschluss- und Brennerdrucks (Messbereich bis 50 mbar (500 mm WS),
d) ein CO_2-Messgerät (z. B. Brigon-CO_2-Prüfer),
e) ein CO-Gasspürgerät (z. B. Dräger-Gasspürgerät mit Messröhrchen 0-0,3% CO).

measuring error Messfehler *m*

measuring gas Messgas *n*

measuring hose Messschlauch *m*

measuring instrument indication → **indication**

measuring instrument with circuit control devices [*IEC 50(301)*] Messgerät mit Signalgeber, Signalgebermessgerät *n*

measuring light ray Messlichtstrahl *m*

measuring line Messleitung *f*

measuring liquid Messflüssigkeit *f*

measuring matrix Messmatrix *f*

measuring/open control/closed control switchgear cubicle → switchgear cubicle for measuring and open control and closed control

measuring point, measuring place Messort *m*, Messstelle *f* [*Der Ort des Regelkreises, an dem der Wert der Regelgröße erfasst wird*]

measuring point selector → external measuring-circuit selector

measuring probe, measuring explorer Messsonde *f*

measuring pump Messpumpe *f* [*Mit dieser Pumpe wird über die Pumpenkennlinie die umgewälzte Wassermenge ermittelt*]

measuring range [*IEC 688-1*] Messbereich, MB *m* [*selten: n*] [*Innerhalb dieses Bereiches wird die Messgröße erfasst*]

measuring relay Messrelais *n*

measuring resistor Messwiderstand *m*

measuring scale Messskala *f*

measuring scatterband Messstreubreite *f*

measuring signal Messsignal *n*

measuring signal processing Messsignalverarbeitung *f*

measuring socket Messstutzen *m*

measuring station; measuring center (US); measuring centre (Brit.) Messwarte, Messstation *f*

measuring step Messschritt *m*

measuring transducer, (signal) transducer Messumformer, Signalumformer, Messumsetzer, Signalumsetzer, Signalwandler, Messwandler *m* [*Ein Gerät, welches ein Eingangssignal, gegebenenfalls unter Verwendung einer Hilfsenergie, möglichst eindeutig in ein damit zusammenhängendes Ausgangssignal umformt. Die physikalische Größe und der Wertebereich des Ausgangssignals müssen den Erfordernissen des nachgeschalteten Baugliedes entsprechen (z. B. Einheitssignal)*]

measuring value → indication

measuring valve, measurement valve Messventil *n*

measuring variable → measurement variable

measuring vector Messvektor *m*

measuring voltage transformer Messspannungswandler, Messspannungsübertrager, Spannungswandler für Messzwecke, Spannungsübertrager für Messzwecke *m*

meat preparation room Fleischvorbereitungsraum *m*

mecatronic [*Japanese term*] **→ precision engineering**

mechanical (air) extract(ion) mechanische (Luft-)Auszug, mechanische (Luft)Abzug *m*

mechanical (air) humidifier mechanische (Luft-)Befeuchter, mechanische (Luft)Anfeuchter *m*

mechanical (air) inlet mechanische (Luft)Einlass *m*

mechanical and electrical and plumbing connection → MEP connection

mechanical and electrical and plumbing equipment → MEP equipment

mechanical and electrical and plumbing equipment room → MEP room

mechanical and electrical plants, mechanical and electrical installations betriebstechnische Anlagen *fpl*, betriebstechnische Installationen *fpl*

mechanical article surveillance technique, mechanical article security technique, mechanical product surveillance technique, mechanical product security technique mechanische Warensicherungstechnik, mechanische Artikelsicherungstechnik *f* [*Sicherung durch Ketten, Seile oder Präsentation der Waren in verschlossenen Vitrinen*]

mechanical atomizer → pressure atomizer

mechanical (boiler) stoking mechanische (Kessel)Beschickung *f*

mechanical building service mechanische Gebäudedienst *m*

mechanical calculator mechanische Rechenwerk *n*

mechanical chiller, mechanical refrigerating machine [*It uses mechanical energy input to generate chilled water*] mechanische Kältemaschine *f*

mechanical clock mechanische Uhr *f*

mechanical company → building services company

mechanical compressor mechanische Verdichter, mechanische Kompressor *m*

mechanical computer mechanische Rechner *m*, Rechenmaschine *f* [*Der erste mechanische Rechner Z1 und die algorithmische Programmsprache „plant Kalkül" wurden 1936 von Konrad Zuse (gestorben 18.12.1995) erfunden. Das berühmteste Modell von Zuse ist der 1941 entstandene mechanische Rechner Z3. Dieser programmgesteuerte Rechner hatte eine Speicherkapazität von 64 Wörtern*]

mechanical core → service core

mechanical deaeration mechanische Entlüftung *f*

mechanical draft, mechanical draught, fan draft, fan draught [*There are two main types: forced draft (Druckzug) and induced draft (Saugzug)*] Ventilatorzug *m*

mechanical draft water cooling tower, mechanical draught water cooling tower [*A water cooling tower through which air movement is effected by one or more fans. There are two main types: forced draft with fans located at the air inlet and induced draft with fans located at the air exhaust*] Ventilator-Wasserkühlturm *m*

mechanical drive mechanische Antrieb *m*

M

mechanical efficiency [*Ratio of the compression energy or work of a compressor to the energy or work input*] mechanische Wirkungsgrad *m*

mechanical energy, mechanical power mechanische Energie, mechanische Kraft *f*

mechanical engineer → building services engineer

mechanical engineering → building (services) engineering

mechanical engineering research → (building) services engineering research

mechanical (equipment) floor, machine(ry) floor [*"Level", "storey" and "story" (US) may synonymously be used for "floor"*] Betriebsgeschoss, Betriebsstockwerk, Technikgeschoss, Technikstockwerk, Maschinengeschoss, Maschinenstockwerk, Anlagengeschoss, Anlagenstockwerk *n*, Betriebsetage, Techniketage, Maschinenetage *f*

mechanical (equipment) room → machine(ry) room

mechanical exhaust(ed) air system, mechanical extracted air system, mechanical exhaust system mechanische Fortluftsystem *n*

mechanical exhaust ventilation mechanische Entlüftung *f*

mechanical extract system → mechanical outlet system

M

mechanical interlock(ing) Verrieg(e)lung, Abrieg(e)lung *f*

mechanical level → mechanical (equipment) floor

mechanical life mechanische Lebensdauer *f*

mechanical loading mechanische Belastung *f*

mechanical lock konventionelle Schloss *n*

mechanically delayed release Auslöser mit mechanischem Hemmwerk, mechanische Hemmwerkauslöser *m*

(mechanically) short-circuit-proof, resistant to short circuit(s) kurzschlussfest, kurzschlusssicher [*VDE 0100. Teil 200*]

mechanical measurement mechanische Messung *f*

mechanical mixer (valve) → lever(-operated) mixer (valve)

mechanical oscillation mechanische Schwingung *f*

mechanical outlet system, mechanical extract system mechanische Ausgang(s)system *n*

mechanical plant, mechanical installation mechanische Anlage, mechanische Installation *f*

mechanical power → mechanical energy

mechanical product surveillance technique → mechanical article surveillance technique

mechanical refrigerating machine → mechanical chiller

mechanical refrigerating system, mechanical refrigeration system [*This system uses mechanical compression to remove the refrigerant from the low-pressure side and to deliver it to the high-pressure side of the system*] mechanische Kältesystem *n*

mechanical safety engineering mechanische Sicherheitstechnik *f*

mechanical scanning mechanische Abtasten *n*

mechanical security mechanische Sicherung *f*

mechanical security engineering mechanische Sicherungstechnik *f*

mechanical shutter mechanische Verschluss *m*

mechanical smoke exhaust, mechanical smoke removal, mechanical smoke disposal maschinelle Entrauchung, maschinelle Rauchabführung, maschinelle Rauchentsorgung, Entrauchung mit maschineller Unterstützung *f*, maschinelle Rauchabzug *m*,

mechanical station → (central) machine(ry) station

mechanical storey → mechanical (equipment) floor

mechanical supply air mechanische Zuluft *f*

mechanical temperature controller, mechanical temperature regulator mechanische Temperaturregler *m*, mechanische Temperaturregelorgan *n*

mechanical trade [*e.g. plumbing trade*] technische Gewerk *n*

mechanical variable [*See remark under "variable"*] mechanische Größe *f*

mechanical ventilation, power ventilation, mechanical vent(ilat)ing, power vent(ilat)ing [*It depends on a mechanical device to provide a positive draft*] mechanische Lüftung, mechanische Ventilation, mechanische Be- und Entlüftung *f*

mechanical warm air furnace, fan(-assisted) warm air furnace, warm air mechanical furnace, warm air fan(-assisted) furnace [*"Hot air" may synonymously be used for "warm air"*] Ventilator-Warmluftheizung, Warmluft-Ventilatorheizung *f*

mechanical waste gas damper mechanisch befestigte Abgasklappe, mechanisch befestigte Abgassperrvorrichtung *f* [*Sie wird durch eine Hilfsenergie betätigt und sperrt im Stillstand der Feuerung den Abgasweg ab*]

mechanical water atomization mechanische Wasserzerstäubung *f*

mechanism Mechanismus *m*, Werk *n*

media access ((open) loop) control, MAC Zugriff(s)steuerung *f* auf das Übertragungsmedium

medical bathroom medizinische Baderaum *m*

medicine chest Medikamentenschränkchen *n*, Arzneimittelschränkchen, Hausapotheke *f*

medium [*The physical transmission entity. Typical media are, amongst others, twisted-pair wire, tube optic cable, and coaxial cable*] Träger *m*

medium access Trägerzugriff *m* [*Siehe Anmerkung unter „Zugriff"*]

medium access control layer, MCA layer (Daten-) Sicherungsschicht, Schicht 2 *f* [*OSI-Schichtenmodell*]

medium attachment unit, MAU [*LAN*] Mediumanschlusseinheit *f*

medium cold(ness) Mittelkälte *f* ($\dot{Q}_0 \approx 85\,000 \ldots 400\,000$ kJ/h) [*Kältemittel*]

medium complexity mittlere Verknüpfungstiefe *f*

medium end → EM

medium fuel oil Heizöl-M, mittelflüssige Heizöl, HM *n*

medium heating plant, medium heating installation mittlere Heiz(ungs)anlage *f*

medium Joule biogas, MBG mittelkalorige Biogas *n*

medium-level (flushing) cistern, medium-level (flushing) tank [*A flushing cistern that is mounted at a medium-level on the wall*] halbhochhängende Spülkasten, halbhochhängende Spültank *m*

medium-load method Mittellastverfahren *n*, Mittellastmethode *f*

medium pressure Mitteldruck *m*

medium-pressure ball valve, medium-pressure globe valve [*This is for about 100 lb per sq in. with an inlet hole about 1/4 in. diameter, suitable for ground-floor W.C.s in houses. See remark under "ball valve"*] Mitteldruckschwimmerventil *n*

medium-pressure fan Mitteldruckventilator, Mitteldrucklüfter *m*

medium-pressure radial fan, radial medium-pressure fan Mitteldruckradialventilator, Radialmitteldruckventilator *m*

medium-pressure steam, MPS Mitteldruckdampf *m*

medium-pressure steam heating, MPS heating Mitteldruckdampfheizung *f*

medium-pressure warm water, medium-pressure hot water Mitteldruckwarmwasser *n*

medium-pressure water heating installation, medium-pressure water heating plant [*Up to 120° C*] Mitteldruck-Wasserheiz(ungs)anlage *f*

medium resolution mittlere Auflösung *f*

medium scale integration, MSI [*merged transistor logic*] Integration mit 12 bis 100 Gatter pro Chip, mittlere integrierte Schaltung, integrierte mittlere Schaltung *f*

medium specific heat mittlere spezifische Wärme, spezifische mittlere Wärme *f*

medium-temperature chimney [*One handling the products of combustion, at the point of leaving equipment, with a temperature between 600 to 1000° F (315 to 538 ° C) under normal operating conditions*] Mitteltemperaturschornstein *m*

medium-temperature equipment Mitteltemperaturgerät(e) *n*, *(f)*

medium-temperature incinerator Mitteltemperatur-Verbrennungsanlage *f*

medium-temperature water system [*It operates with temperatures below 350 degrees F and permits design to a pressure rating of 125 to 150 psig*] Mitteltemperatur-Wassersystem *n*

medium-voltage switchgear Mittelspannungsschaltanlage *f*

ME(GA)FLOP → million floating-point operating per second

Meissner(-Ochsenfeld) effect → flux jump(ing)

meltage ice → ice water

melting heat Schmelzwärme *f*

melting stock Schmelzgut *n*

meltback transistor Schmelzperlentransistor *m*

member company → company member

membership Zugehörigkeit *f* [*Fuzzy-Logik*]

membership card reader Mitglied(s)kartenleser *m*

membership degree → degree of membership

membership function core → core of membership function

membership function support → support of membership function

membership verification Mitgliedschaftsnachprüfung *f*

membrane → diaphragm

memory → (computer) memory

memory address register → address register

memory capacity, storage capacity Speicherfähigkeit, Speicherkapazität *f*, Speichervermögen *n*

memory cycle → cycle time

memory effect Gedächtniswirkung *f* [*Akku*]

memory-programmable → (computer) memory-programmable

memory protection, storage protection Speicherschutz *m*

men-machine interface, MMI Mensch-Maschine-Schnittstelle *f*

men's dormitory Männerschlafsaal *m*

M

MEP connection, mechanical and electrical and plumbing connection haustechnische Anschluss, gebäudetechnische Anschluss *m*

MEP equipment, mechanical and electrical and plumbing equipment Gebäudetechnikgerät(e), Haustechnikgerät(e) *n*, *(f)*

MEP room, mechanical and electrical and plumbing equipment room haustechnische Raum, gebäudetechnische Raum *m*

merchandiser, showcase Vitrine, Schautruhe *f*, Schaukasten *m*

merchandise security Warensicherung *f*

merchandise security system Warensicherungssystem *n* [*Es erkennt nicht bezahlte Waren vor dem Ausgang*]

mercury-arc rectifier → mercury-vapor rectifier

mercury differential pressure ga(u)ge → mercury U(-shape(d)) tube manometer

mercury (discharge) lamp Quecksilber(entladungs)-lampe *f*

mercury-in-steel thermometer Quecksilber-Stahl-Thermometer *n*

mercury pressure gauge; mercury pressure gage (US) Quecksilbermanometer *n*, Quecksilberdruckmesser *m*

M

mercury salt Quecksilbersalz *n*

mercury thermometer [*Bulb and attached glass capillary tube containing mercury which expands or contracts with changes in temperature; so marked that the end of the mercury column indicates the ambient temperature*] Quecksilber(glas)thermometer *n*

mercury-tungsten lamp → mixed-light lamp

mercury U(-shape(d)) tube manometer, mercury differential pressure ga(u)ge Quecksilber-U-Rohr-Manometer *n*, Quecksilber-U-Rohr-Differenzdruckmesser *m*

mercury-vapour lamp (Brit.); mercury-vapor lamp (US) Quecksilberdampflampe *f*

mercury-vapour mixed-light lamp (Brit.); mercury-vapor mixed-light lamp (US) Quecksilberdampfmischlichtlampe, Mischlicht-Quecksilberdampflampe *f*

mercury-vapor rectifier (US); mercury-arc rectifier, mercury vapour rectifier [*A gas tube rectifier in which mercury vapor is the conducting medium. The cathode may be either a hot cathode or a cool cathode*] Quecksilberdampfgleichrichter *m*

to merge mischen, zusammenfügen, einbinden [*Von zwei geordneten Dateien, so dass die Ordnung bestehen bleibt. E DIN 44 300/10.85*]

merged transistor logic, MTL integrierte Schaltung *f*

merit figure → figure of merit

mesh Maschennetz *n*

mesh earth electrode → grid earth electrode

meshed vermascht, gemascht

mesh filter Maschenfilter *m*, *n*

mesh generator Netzgenerator *m* [*Ein komfortables Programm zur FE-Modellerstellung*]

mesh (network) Maschennetz *n*

mesh voltage Dreiecksspannung, kleinste verkettete Spannung, verkettete kleinste Spannung *f*

mesh welded fence → welded mesh fence

mesh wire fence, wire mesh fence Maschendrahtzaun, Drahtmaschenzaun *m*

message delay Mitteilungsverzögerung, Nachrichtenverzögerung *f*, Mitteilungsverzug, Nachrichtenverzug *m* [*Das Zurückhalten einer Nachricht während einer vorausbestimmten Zeitspanne, so z. B. das Verzögern einer Nachricht mit niedriger Priorität über das Fernsprechnetz, bis die Gebühren tiefer liegen (z. B. Nutzung des Nachttarifs*]

message distributor Mitteilungsverteilung, Nachrichtenverteilung *f*

message handling system, MHS Mitteilungsumgangssystem, Nachrichtenumgangssystem, Mitteilungübermittlungssystem, Nachrichtenübermittlungssystem *n* [*Das internationale Gesamtsystem der Electronic Mail über X.400*]

message identification Mitteilungskenn(zeichn)ung, Nachrichtenkenn(zeichn)ung *f*

message indicator [*Communication. Element placed within a message to serve as a guide to the selection or derivation and application of the correct key to facilitate the prompt description of the message*] Nachrichtenanzeiger, Mitteilungsanzeiger *m*

message interception Lauschangriff *m*

message interface Nachrichtenschnittstelle, Mitteilungsschnittstelle *f*

message processing Mitteilungsaufarbeitung, Nachrichtenaufarbeitung, Textmeldungsaufarbeitung *f*

message queue [*The stacking of messages according to some priority rule as the messages await processing*] Nachrichtenwarteschlange, Mitteilungswarteschlange *f*

message route → communication path [*IEC 625. Programmable measuring apparatus*]

message switching Mitteilungsumschaltung, Nachrichtenumschaltung *f*

messaging Mitteilen, Benachrichtigen *n*

messenger calling system Botenrufanlage *f*

metal-cased conductor Panzeraderleitung *f*

metal-ceramic surge protector Metallkeramikableiter *m*

(metal) fin → (attached) (metal) fin

metallically separated galvanisch getrennt

metal(lic) ceiling Metall(unter)decke *f*

metal(lic) expansion thermometer Metall-Ausdehnungs-Thermometer *n* [*Zur Temperaturmessung wird der Unterschied der Ausdehnung zweier Metalle mit verschiedenen Ausdehnungszahlen benutzt; nicht verwechseln mit Bimetallthermometer*]

metal(lic) film resistor Metallschichtwiderstand *m*

metal(lic) (hose) braiding Metall(schlauch)umspannung *f*

metal(lic) knob Metallknopf *m*

metal(lic) louver (US); metal(lic) louvre (Brit.) Metall-Lüftungsklappe, Strahlungsmetallplatte *f*

metal(lic) pad Metallvlies; Metallvließ *n* [*Österreich*] [*Filtermedium*]

metal(lic) pressure gauge; metal(lic) pressure gage (US) Metallmanometer *n*, Metalldruckmesser *m*

metal(lic) radiant panel, radiant metal(lic) panel Metallstrahlungsplatte, Strahlungsmetallplatte *f*

metal(lic) rectifier Trockengleichrichter *m*

metal organic chemical vapour deposition (Brit.); metal organic chemical vapor deposition (US); NOCVD Metallschichtabscheidung *f*

metal oxid(e) semiconductor → MOS

metal-plastics capacitor Metall-Kunststoff-Kondensator, MK-Kondensator *m*

metal-semiconductor contact, Schottky contact Halbleiter-Metall-Kontakt, Metall-Halbleiter-Kontakt, Schottkykontakt *m*

metal sheet waste gas pipe → sheet metal waste gas pipe

metal vapour; metal vapor (US) Metalldampf *m*

metal vapour discharge luminaire, metal vapour discharge light(ing) fitting (Brit.) [*US spelling is "vapor"*] Metalldampfentladungsleuchte *f*, Metalldampfentladungsbeleuchtungskörper *m*

metal window casement turn, metal window casement fastener Metallfenstervorreiber *m*

to meter, to measure [*An object, event, or process*] messen

meter [*An instrument giving direct indication of the quantity of liquid flowing, without calling for calculation or skilled manipulation*] Zähler, Messer *m*

meter [*Device installed by the utility company for measuring the quantity of a medium used by the consumer*] Verbrauchszähler, Zähler *m*

meter → (electricity) meter

meter board Zählertafel *f*

meter cabinet, meter cubicle Zählerschrank *m*, Zählerkasten *m*

metered flow [*Design engineering. A flow that is controlled to allow a specific quantity to flow in a specified time or in harmony with an event*] gemessene (Durch)Fließen *n*, gemessene (Durch)Fluss *m*

metering, building metering Gesamtverbrauchsmessung, Verbrauchsmessung eines Gebäudes *f* [*Sie wird von Wasser- und Energieversorgern durchgeführt*]

meter protective switch, meter protecting switch, meter protection switch Zählervorschalter *m*

meter readability Zählerablesbarkeit *f*

meter reader → reading gauge

meter reading Zählerablesung *f*

meter selector switch Zählerwählschalter *m*

meter terminal Zählerklemme *f*

meter unit Ableseeinheit, Zählereinheit, Ablesungseinheit *f*

method [*The means, procedure, or technique used to carry out some process*] Methode *f*, Verfahren *n*

methylchlorid(e) → refrigerant-40

methylene chlorid(e) → refrigerant-30

metric sized metrisch dimensioniert, metrisch bemessen

metrology [*The science of measurement*] Metrologie, Maßkunde *f*

metropolitan area network → MAN

MFBT → mean time between failures

m.-g. set → motor-generator set

MHS → message handling system

microblower Mikrogebläse *n*, Mikrobläser *m* [*Vom Klosettspülrohr abgehende Entlüftungsleitung; saugt die Luft unmittelbar aus dem Klosettbecken ab und bläst sie ins Freie*]

microbore (heating) system [*A very small-bore, wet central heating system in which there is a manifold closely situated to a group of radiators, to which the flow and return pipes from the boiler are installed and which serves the radiators by very small (6 mm dia.) flow and return pipes*] Mikrorohrheiz(ungs)system *n*

microcomputer, MC [*A digital computer whose central processing unit (CPU) resides on a single semi-conductor integrated circuit chip*] Mikrorechner *m*

microcomputer programming → PL/M

microcontroller, micro CTRL [*See remark under "controller"*] Mikroregler *m*, Programmschaltwerk, Mikroregel(ungs)organ *n*, Mikroregel(ungs)einheit *f*

M

micro current Mikrostrom *m*

micro-disconnection, micro-cutout Mikroabschaltung, Mikrotrennung, Mikrounterbrechung *f*

microenvironment [*The environmental conditions that exist in a small, localised habitat*] Mikroumwelt, Mikroumgebung, unmittelbare Umgebungsbedingungen *f* [*VDE 0109*]

micro-fin(ned) tube Mikroflossenröhre *f*

microinstruction [*Computer programming. A low-level control mechanism stored in a control word in a CPU that causes a single data-register state transition; examples include "add", "shift", and "store". In a microcoded machine, one machine instruction is executed by a sequence of microinstructions*] Mikrobefehl *m*

micromodule, microblock [*Cube-shaped, plug-in, miniature circuit composed of microelements; each microelement can consist of a resistor, diode, capacitor, transistor, or a combination of these*] Mikromodul *n*, Mikrobaustein *m*

micromodule circuit, microblock circuit Mikromodulschaltung, Mikrobausteinschaltung *f* [*In der Mikromodultechnik werden Widerstände, Dioden, Transistoren und Kondensatoren mit kleinen Kapazitäten zu einer Schaltung vereinigt und mit Kunststoff umpresst. Es gibt Dünnfilmschaltung und Dickschichttschaltung*]

microphone call station Mikrophonsprechstelle, Mikrofonsprechstelle *f*

microphone (loud)speaker unit Mikrophon-Lautsprecher-Einheit, MLE, Mikrofon-Lautsprecher-Einheit *f*

miroprobe [*Spectroscopy. A device used to irradiate a sample by focus(s)ing a beam of electrons on a very small area in order to spectroscopically analyse the resulting emitted X-rays*] Elektronensonde *f*

microprocessing unit, MPU [*A microprocessor with external memory, input/output interface devices, buffer, clock and driver circuits*] Mikroprozessoreinheit *f*

microprocessor, MP Mikroprozessor *m*, MP [*Die Zentraleinheit eines Mikrorechners μC. Die auf einem Chip realisierten Funktionen des Mikroprozessors sind in der Regel das Rechnerwerk, auch Arithmetikeinheit genannt, die Arbeitsregister und die Auslaufsteuerung*]

microprocessor ((closed) loop) control Mikroprozessorreg(e)lung *f*

microprocessor ((open) loop) control Mikroprozessorsteuerung *f*

microscopic dust counter → **dust counting microscope**

microswitch Kontakt *m* [*WC-Spülkasten*]

microtechnique [*French term*] → **precision engineering**

microwave amplifier by stimulated emission of radiation, MASER MASER *m*

microwave barrier Mikrowellenschranke *f*

microwave beam Mikrowellenstrahl *m*

microwave circuit Mikrowellenschaltung *f*

microwave detector, microwave detecting device, microwave detection device Mikrowellenmelder, Mikrowellendetektor, Mikrowellenmeldungsgeber *m*, Mikrowellenmeldegerät *n*

microwave (frequency) thawing Mikrowellenauftauen *n*

microwave motion detector, microwave motion detecting device, microwave motion detection device Mikrowellenbewegungsmelder *m*, Mikrowellenbewegungsmeldegerät *n*

microwave oven Mikrowellenofen *m*

microwave range Mikrowellenkochgerät *n*, Mikrowellenherd *m*

midden → **f(a)ecal matter pit**

middle conductor Mp-Leiter, Mittel(punkt(s))leiter, Nullleiter *m*

middle of maximum defuzzification → **mean of maximum defuzzification**

mid frequency Mittenfrequenz, Abstimm(ungs)frequenz *f*

midget central station, small central station Kleinzentrale *f*

midpoint Nullphase *f*

mid-season Übergangsjahreszeit *f*

mildew resistant schimmelwidrig, schimmelbeständig, schimmelfest, schimmelsicher

mild heat → **warmth to touch**

mild steel section Flussstahlprofil *n*

mill [*A building containing machinery, as a sawmill*] Werk *n*

mill, grinding machine Mühle *f*

milled pattern Fräsmuster *n*

Miller generator → **bootstrap integrator**

milling out speed → **speed of milling out**

million floating-point operations per second, ME(GA) FLOP 1 Million Gleitkommaoperationen *f* pro Sekunde

million operations per second, MOPS Million Rechenoperationen *f* je Sekunde

MIME, multi(ple)-purpose internet mail extension Dateianhang *m* für E-Mail mit umgewandeltem Textformat

M

mimic diagram Funktionsabbild [*IEC 50(603)*] Blindschaltbild, Anlagenbild, A-Bild *n*

mineral fiber felt (US); mineral fibre felt (Brit.) Mineralfaserfilz *m*

mineral fiber thermal insulation (US); mineral fibre thermal insulation Mineralfaserwärmedämmung *f*

mineral material Mineral(werk)stoff *m*

miniature circuit breaker, m.c.b. (Leitungs)Schutzschalter, LS-Schalter, Kleinselbstschalter, Sicherungsautomat *m*

miniaturized controller → compact controller

minibore central heating, central minibore heating Zentralheizung mit Kleinstrohrweiten; Sammelheizung mit Kleinstrohrweiten *f* [*Fehlbenennung*]

miniburner Kleinbrenner *m*

minicryogenic machine Kleinkältemaschine, Kleintieftemperaturmaschine, Kleintiefkühl(ungs)maschine *f*

mini-gap switch → microswitch

minimum clearance [*The minimum distance separating two objects or parts*] Mindestfreiheit *f*, Kleinstspiel, Mindestspiel *n*, Mindestabstand, Kleinstabstand *m*

minimum conductivity Mindestleitfähigkeit *f*, Mindestleitvermögen *n* [*Unterer Grenzwert der Leitfähigkeit des Fluids bei gegebenen Fehlergrenzen eines magnetisch-induktiven Durchflussmessers*]

minimum flow pressure Mindestfließdruck *m*

minimum function Mindestfunktion, *t*-Norm *f*

minimum-fuse, minimum-fuze sicherungsarm

minimum limit Mindestgrenze *f*

minimum-phase element Mindestphasenelement *n*

minimum position Mindeststellung *f*

minimum shift keying, MSK Mindesttastung *f* [*Modulation digitaler Signale*]

minimum space requirement, minimum space need Mindestraumbedarf *m*

minimum value limiter Mindestwertbegrenzer *m*

min-max composition Min-Max-Zusammensetzung *f*

minor loop [*In a feedback control system, a small continuous network that contains both forward elements and feedback elements*] Nebenschleife *f*

mirror, specular glass reflector Spiegel *m*

mirror cabinet, mirror cubicle Spiegelschrank *m* [*im Badezimmer*]

mirror(-coat)ed lamp verspiegelte Lampe *f*

mirror(-coat)ing [*Optics. A very thin layer of metal, such as silver, that is evaporated onto a polished surface, such as glass, to create a reflective surface or two layers of dielectric materials evaporated onto a glass surface in a vacuum to form a reflector of narrow band of wavelengths*] Verspieg(e)lung *f*

mirror(ed) door Spiegeltür *f*

mirror-equipped spiegelbestückt, bespiegelt [*Mit Spiegeln ausgestattet*]

mirror image → reflected image

mirroring [*The effect that causes ordinary glass to reflect like a mirror*] Spieg(e)lung *f*

mirror luminaire Spiegelleuchte *f*

mirror reflection Spiegelreflexion *f*

mis-aligned; mis-alined (US) fehl ausgerichtet, fehl ausgefluchtet, falsch ausgerichtet, falsch ausgefluchtet

misframing fehlerhafte Bildstricheinstellung *f*

misfunction Fehlfunktion *f*

mismatch(ing) Fehlanpassung *f*, Fehlanspassen *n*

MIS-MOS-IG-FET Isolierschicht-Feldeffekttransistor, Oberflächen-Feldeffekttransistor *m*

mis-recognition Fehlerkennung *f*

missing Fehlen *n*

missing Inkonsistenz *f* eines Datensystems

missing code fehlende Kode, fehlende Code, Fehlkode, Fehlcode, Ausfallkode, Ausfallcode *m*

mist (Raum)Tropfennebel, (Raum)Sprühnebel *m* [*Gemisch aus feuchter Luft und dispers verteiltem Wasser. Der Nebel bildet sich durch Auskondensieren von Wasser aus übersättigter Luft*]

mist droplet [*Meteorology. A mist particle whose size is between those of a haze droplet and a fog droplet*] Sprühnebeltröpfchen, Nebeltröpfchen *n*

misting (Raum)Tropfenverneb(e)lung, (Raum)Sprühverneb(e)lung *f*

misting [*mirror*] Beschlagen *n*

mist(ing) fan Sprühnebelventilator, Trofennebelventilator, Nebelventilator *m*

misuse, tamper Missbrauch, schädigende Eingriff *m*, Fehlanwendung *f*

misuse-proof, tamper-proof missbrauchsicher, sicher gegen schädigenden Eingriff

misuse switch → tamper switch

mixdown Abmischung *f*, Abmischen *n* [*Elektronisches Zusammenmischen verschiedener Kanäle am Mischpult*]

mixed air Mischluft *f* [*Primärluft plus Sekundärluft*]

M

mixed air and exhaust air system, mixed air and extracted air system Misch- und Fortluftsystem *n* [*Es werden nur Außenluft, Umluft und Fortluft maschinell gefördert*]

mixed air box → (air) mixing box

mixed air conditioning plant, mixed air conditioning installation Mischluftklimaanlage *f*

mixed air controller Mischluftregler *m*

mixed air curtain Mischluftschleier, Mischluftvorhang *m*

mixed air heating Mischluftheizung *f*

mixed air heating plant, mixed air heating installation Mischluftheiz(ungs)anlage, Mischluftheizinstallation, Mischluftheiz(ungs)installation *f*

mixed air operation Mischluftbetrieb *m*

mixed air plant, mixed air installation Mischluftanlage, Mischluftinstallation *f* [*Es werden nur Außenluft und Umluft maschinell gefördert*]

mixed air plenum → (air) mixing box

mixed condensation Mischkondensation *f*

mixed diagram gemischte Schalt(ungs)plan *m* [*DIN 40 719, Blatt 1*]

mixed flow fan [*Its characteristics combine both centrifugal and axial airflow*] halbaxiale Ventilator, meridianbeschleunigte Ventilator *m*

mixed gas [*A mix(ture) of combustible gases individually produced by two or more different processes*] Mischgas *n*

mixed-light lamp, blended lamp, self-ballasted mercury lamp, mercury-tungsten lamp Mischlichtlampe, Verbundlampe *f*

mixed mode analysis gleichzeitige Analyse *f* digitaler und analoger Schaltungskomponenten

mixed pair verdrillte Telefonkabel *n*

mixed structure network → hybrid network

mixed ventilation, mixed vent(ilat)ing, jet ventilation, jet vent(ilat)ing Verdünnungslüftung, Mischlüftung, Strahllüftung *f* [*„Be- und Entlüftung" und „Ventilation" können synonym für „Lüftung" verwendet werden. Bei dieser Lüftung wird Induktion angestrebt. Folgt die Luft zunächst tangential einer Wand oder Decke wird eine Strömungswalze gebildet. Dies geschieht meist bei Induktionsgeräten oder Auslässen mit Coandawirkung*]

mixed water Mischwasser *n* [*Das im Gebäude und auf dem Grundstück anfallende und in die öffentliche Kanalisierung abgeleitete Wasser, zusammengesetzt aus Regen- und Schmutzwasser. DIN 4045*]

mixer [*general term*] Mischer *m*

mixer [*A device having two or more inputs, usually adjustable, and a common output; used to combine separate audio or video signals linerarly in desired proportions to produce an output signal*] Mischer *m*

mixer, first detector, heterodyne modulator, mixer-first detector [*The stage in a superheterodyne receiver in which the incoming modulated radio-frequency signal is combined with the signal of a local radio-frequency oscillator to produce a modulated intermediate-frequency signal*] Überlagerungsmischer *m*

mixer → injector

mixer head [*Portion of an injection(-type) (Bunsen) burner, usually enlarged, into which primary air flows to mix with the gas stream*] Misch(er)kopf *m*

MIXFLA → stationary laminar premixed flat one-dimensional flame

mixing [*general term*] Mischen *n*

mixing → (air) mixing

mixing battery Mischbatterie *f*

mixing box → (air) mixing box

mixing chamber [*Space or passage connecting the ignition and combustion chambers*] Mischkammer *f*

mixing cock Mischhahn *m* [*Für Warm- und Kaltwasser*]

mixing damper → (air) mixing damper

mixing pipe Mischrohr *n*

mixing plenum → (air) mixing box

mixing preheater Mischvorerhitzer, Mischvor(er)wärmer *m*

mixing section [*In a duct system a section for mixing airstreams at differing temperatures or humidities, or both*] Mischstrecke *f*

mixing valve Mischventil *n*

ML → management level

MLN → management level network

MMF → magneto motive force

MMI → man-machine interface

mnemonic [*A symbol chosen to assist the human. Used to express complicated words, names or phrases usually by reference to the first letter or syllable of the original*] Mnemonik, Gedächtnishilfe *f*, Kürzel *n*, Pseudokode, Pseudocode *m*

mobile (air) heating device, mobile (air) heating appliance [*Approved, unvented (air) heating appliance designed for human comfort and not secured or attached to a building by any means other than fuel piping or electrical wiring*] bewegliche Luftheiz(ungs)gerät *n*

mobile operator terminal [*A keyboard and display unit suitable for local connection to a facility manage-*

M

ment system. Often used to initial(l)ing setting up an outstation] bewegliche Bedienerplatz, mobile Bedienerplatz *m*

mobile (radio) station, MS [*A radio station that is used while in motion or during stops at unparticipated locations*] mobile Funkstation, bewegliche Funkstation *f*

mobile switching centre (Brit.); mobile switching center (US); MSC Steuer(ungs)einheit *f* [*Hochkomplexe Steuer(ungs)einheit eines internationalen digitalen Kommunikationsstandardnetzes für Funktelefone, in dem die Teilnehmerdaten eines Bereiches verwaltet werden*]

mock fireplace Kaminattrappe *f*, Attrappenkamin *m*

mock lock → blind lock

MOD → magneto-optical disk

modal ((closed) loop) control, modal regulation modale Reg(e)lung *f*

mode [*A manner, method, or state of acting or doing*] Modus *m*, Art *f*

mode-hopping Intensitätsverschiebung *f* [*Bei indexgeführten Lasern von einem Modus zum anderen*]

model-building (work)shop Modell(bau)werkstatt *f*

model correction Modellkorrektur *f* [*Parameteränderung nach verbesserten Informationen aus dem Modellprozess*]

model reference adaptive system → MRAS

model series Modellreihe, Modellserie *f*

(model) updating (Modell)Nachführung *f*

modem, modulator/demodulator [*Carrier frequency engineering. A combination modulator and demodulator at each end of a telephone line to convert binary digital information to audio tone signals suitable for transmission over the line and vice versa*] Modulator/Demodulator, Modem *m*

modem interface Modemschnittstelle *f*

modem port Modemanschluss *m*

modification of jump address, jump address modification [*See remark under "jump address"*] Sprungzieländerung, Sprungadressänderung *f*

modifier register → index register

modular [*Composed of standardized units or sections*] bausteinförmig, modular

modular 6/4 → modular plug

modular air-conditioning plant, modular air-conditioning installation [*On-site assembly of prefabricated components, each with a functional role in air conditioning (air circulation, air filtration, cooling, heating, humidification, etc.)*] modulare Klima(tisierungs)-anlage *f*

modular basic unit, basic modular unit modulare Grundeinheit, Zentrale *f*

modular computer modulare Rechner *m*

modular construction Modul(ar)bau, Baukastenbau *m*

modular controller Modul(ar)regler *m*

modular electronic system → electronic modular system

modular equipment Modul(ar)gerät(e) *n*, (*f*)

modularity [*Computer programming. The property of a program(me) that is composed of a collection of well-defined, logically self-contained segments or subroutines*] Modularität *f*

modular panelboard, modular distribution board Bausteinverteiler(tafel) *m*, (*f*)

modular pillar terminal, modular tunnel terminal Anreihbuchsenklemme *f*

modular plug, modular 6/4, remote jack 11, RJ 11 Westernstecker *m* [*„Western" stammt von der amerikanischen Telekommunikationsgesellschaft Western Union ab, aus der Zeit, als noch Telegramme depeschiert wurden*]

modular size, modular dimensions Modulgröße, Modulabmessungen *f*

modular space grid modulare Raumraster *m*

modular system, unitized system Bausteinsystem, Modulsystem *n*

modular system(s) engineering modulare Systemtechnik *f*

modular terminal Anreihklemme *f*

modular terminal block Reihenklemme *f*

modular tunnel terminal Anreihbuchsenklemme *f*

modulating, throttling [*The action of a control from its maximum to minimum position in either predetermined steps or increments of movements as caused by its actuating medium*] Drosseln *n*

modulating control → ((closed) loop) control

modulating control amplifier → ((closed) loop) control amplifier

modulating signal, modulating wave [*Telecommunications. A wave that varies some characteristics of the carrier, such as the frequency, phase or amplitude*] Modulationssignal *n*

modulating thermostat [*Temperature controller employing a potentiometer winding inserted of switch contacts*] Modulationsthermostat *m*

modulation [*The process or the result of it whereby a message is changed into information-bearing signals that not only unambiguously represent the message but also are suitable for propagation over the transmitting medium to the receiver*] Modulation *f*

M

modulation output, modulation performance Modulationsleistung *f*

modulation rate Schrittgeschwindigkeit (in Baud), Baudschrittgeschwindigkeit *f* [*Datenübertragung. DIN 44 302*]

modulator Signalumsetzer, Modulator, Signalumwandler *m* [*Er setzt analoge Impulse in digitale Signale um*]

modulator/demodulator → modem

module [*A portion of a program(me) or program(me) library, as in a source module, object module or image module*] Modul *m*

module [*A board, usually made of plastic covered with an electrical conductor, on which logic devices (such as transistors, resistors, and memory clips) are mounted, and circuits connection these devices, are etched, as in a logic module*] Baugruppe *f*, Baustein *m*, Modul *n*

module Modul *n*, Karte *f* [*Ein getrennt steckbarer Baustein*]

module controller Bausteinregler, Modulregler *m* [*Er besteht aus miteinander leicht kombinierbaren Bausteinen, die zu unterschiedlichem Regelverhalten zusammengeschaltet werden. DIN 19 225*]

module expansion slot Modulsteckplatz, Bausteinsteckplatz *m*

M

module-external flag bausteinexterne Merker, modulexterne Merker *m*

module frame → card rack

module frame tier → card rack tier

modulus Modul *m*, (technische) Verhältniszahl *f* [*Mehrzahl: Module*]

modulus of volume elasticity, bulk modulus, bulk modulus of elasticity, hydrostatic modulus, compression modulus, modulus of compression [*The ratio for the change in pressure applied on a body to the corresponding fractional change in volume that this pressure produces; usually denoted by B or K*] Kompressionsmodul *m*

moist, damp, slightly wet [*These terms imply less wetness than "wet"*] feucht, leicht nass, angenässt

to moist-clean feuchtreinigen

moisture [*Climatology: The quantity of precipitation or the precipitation effectiveness. Meteorology: The water vapo(ur)r content of the atmosphere (also called humidity), or the total water substance (gaseous, liquid, and solid) present in a given volume of air. Physical chemistry: Water that is dispersed through a gas in the form of water vapo(u)r or small droplets, dispersed through a solid, or condensed on the surface of a solid*] Feuchte, Feuchtigkeit *f*

moisture concentration, moisture accumulation Feuchtigkeitsansammlung, Feuchteansammlung *f*

moisture content Feuchtegehalt, Feuchtigkeitsgehalt *m*, absolute Feuchte, absolute Feuchtigkeit *f*

moisture controller, moisture regulator Feuchteregler, Feuchtigkeitsregler *m*

moisture eliminator Feuchte(ab)scheider, Feuchtigkeits(ab)scheider *m*

moisture gain (to a building) Feuchteabgabe, Feuchtigkeitsabgabe *f*

moisture laden path Feuchtigkeitsweg, Feuchteweg *m*

moisture migration, moisture movement Feuchtigkeitswanderung, Feuchtewanderung *f*

moisture recovery Feuchtigkeitsrückgewinnung, Feuchterückgewinnung *f* [*Maßnahmen zur Wiederbenutzung des Feuchtegehaltes der einen Versorgungsbereich verlassenen Luft. DIN 1946, Teil 1*]

moisture removal → dehydration

moisture sensor → hygrostat

molar heat capacity, molecular heat capacity [*The amount of heat required to raise 1 mole of a substance 1 °C in temperature*] Atomwärme *f*, Mol(ekular)wärme, molekulare Wärmekapazität C_m *f*

molar mass [*Physical chemistry. The mass in grams of one mole of a given substance; the practical unit of measurement in g mol^{-1}*] molare Masse, Molmasse *f*

molecular beam [*A narrow beam of neutral molecules, typically found in a vacuum*] Molekularstrahl *m*

molecular sieve Molekularsieb *n*

molecular volume [*The volume occupied by one mole of a substance in gaseous form at standard temperature and pressure; it equals molecular weight divided by density*] Molvolumen *n*

mole volume, molar volume, molal volume [*The volume occupied by one mole of a substance in the form of a solid, liquid, or gas*] Molvolumen *n*

Mollier diagram, Mollier chart [*Graph of enthalpy versus entropy of a vapo(u)r on which isobars, isothermals, and lines of equal dryness are plotted*] h-x-Diagramm nach Mollier, Mollier-h-x-Diagramm *n*

molten lead, casting lead Gussblei *n*

MOM defuzzification → mean of maximum defuzzification

momentary contact → fleeting contact

momentary switch Tastschalter *m*

moment of momentum → angular momentum

moment of mentum operator → angular momentum operator

momentum moment → angular momentum

to monitor [*To measure a condition in a system by means of meters or instruments*] überwachen

to monitor [*Electronics. To check by means of a receiver the operation of a telephone, radio, television, or similar transmitter in order to ascertain certain characteristics, such as to a prescribed frequency band*] überwachen

to monitor [*To check or evaluate something on a constant or regular basis*] überwachen

to monitor, to oversee continuously [*Program(me) during execution*] laufend überwachen

monitor [*Any instrument that periodically measures or regulates any condition in a system that must be maintained within prescribed limits*] Betrieb(s)systemkern *m*

monitor [*The person who watches a monitor*] Überwacher *m*

monitor, video display unit Monitor *m* [*Bildschirm einschließlich Steuer(ungs)elektronik*]

monitor board [*In telephony, a console from which a supervisory operator can intercept calls that are being handled by other operators*] Überwachungstafel *f*

monitor check(ing) routine Überwachungsprogramm *n*

monitor control dump [*Computer programming. A memory printout that is produced regularly after a program(me) has been executed*] Überwachungs-Speicherausdruck *m*

monitor display [*The capability of some processes to temporarily halt the central processing unit and display information from main storage and internal registers*] Überwachungsdarstellung *f*

monitoring Überwachen *n*, Überwachung *f* [*Das Überprüfen ausgewählter Größen auf Einhaltung vorgegebener Werte, Wertebereiche oder Schaltzustände. Die zu überwachenden Größen können direkt gemessen oder das Ergebnis einer Auswertung sein. DIN 19 222*]

monitoring amplifier [*A power amplifier used primarily for evaluation and supervision of a program(me)*] Überwachungsverstärker *m*

monitoring and control, control and monitoring Leiten *n* [*Leiten nach DIN 19 222 (Begriffe der Leittechnik) von betriebstechnischen Anlagen (BTA). Die Gesamtheit aller Maßnahmen, die, vorwiegend unter Mitwirkung des Menschen, den erwünschten Ablauf des Prozesses bewirken; im wesentlichen bei Produktionsprozessen*]

monitoring and control engineering, control and monitoring engineering Leittechnik *f*

monitoring and control system, control and monitoring system Leitsystem *n*

monitoring and intervention centre (Brit.); monitoring and intervention center (US) zentrale Überwachungsstelle und Einsatz(leit)stelle, Überwachungszentrale und Einsatz(leit)zentrale *f*

monitoring and operator station, operator and monitoring station Leitstation *f*

monitoring and operator unit, MOU, operator and monitoring unit Überwachungs- und Bedien(ungs)gerät *n* [*Gerät unterschiedlicher Größe, von kleinen Anzeigegeräten, die in der Anlage selber installiert sind bis zu großen rechnergestützten Geräten mit Bedienoberfläche, die dem Bediener erlauben mit dem Gebäudeautomatisationssystem eine Verbindung herzustellen*]

monitoring and operator unit station, MOU Überwachungs- und Bedien(ungs)einheit *f*

monitoring camera Überwachungskamera *f*, Überwachungsphotoapparat, Überwachungsfotoapparat *m*

monitoring contact Überwachungskontakt *m*

monitoring desk Überwachungspult *n*, Überwachungstisch *m*

monitoring device, monitor, monitoring appliance [*Computer engineering. A hardware device that measures electrical events such as pulses or voltage levels in a digital computer*] Überwachungsgerät *n*, Überwachungsvorrichtung *f*

monitoring electrode Überwachungselektrode *f*

monitoring function Überwachungsfunktion *f*

monitoring jack Mithörklinke *f*

monitoring key [*A device that allows an operator to listen on a telephone circuit without impairing the transmission*] Mithörtaste *f*

monitoring loudspeaker Überwachungslautsprecher *m*

monitoring panel Überwachungstableau *n*, Überwachungstafel *f*

monitoring point Überwachungspunkt *m*

monitoring program (US); monitoring programme (Brit.) Kontrollprogramm *n* [*Es zeigt den Stand der Abarbeitung des in einem automatischen Datenverarbeitungssystem laufenden Programmes an*]

monitoring slide-in module Überwachungseinschubmodul *n*

monitoring switchboard, monitoring switchpanel Überwachungsschalttafel *f*

monitoring system [*Design engineering. A system that collects the output sensors and human impact to construct a history of events of a process, machine or equipment*] Überwachungssystem *n*

monitor interface module Monitoranschaltbaugruppe *f*

monitor loudspeaker [*A small loudspeaker mounted in a console or cabinet rack (or adjacent to these) which provides an audible indication of the sound level* *of the system and which permits preliminary adjustment of the amplifier output before distribution to remote loudspeakers]* Überwachungslautsprecher *m*

monitor (program) (US); monitor (programm) (Brit.) Monitor *m*, Kontrollprogramm *n [Ein Programm, das durch Mitteilungen über den augenblicklichen Stand der Abarbeitung eines Arbeitsprogrammes im Rechenautomat informiert]*

to monitor remotely fernüberwachen

monitor routine → executive routine

monobloc blender → single-hole mixer (valve)

monobloc fitting Monoblockarmatur *f [Siehe Anmerkung unter „Armatur"]*

monobloc mixer (valve) → single-hole mixer (valve)

monoblock (wash)basin mixer, monoblock lavatory mixer Monoblockwaschbeckenmischer *m*

monoboard computer Einplatinenrechner *m*

monochromatic filter → bifringent filter

monochrome television, black-and-white television Schwarz-Weiß-Fernsehen *n*

monochrome view(ing) screen, black-and-white view(ing) screen Schwarz-Weiß-Bildschirm *m*

M

mono-edge triggering Einflankenansteuerung *f*

monoflop, single shot monostablile Kippschaltung *f*

monolith Monolith *m [Festkörperschaltung aus einem Kristallblock]*

monolithic monolithisch *[Anordnung sämtlicher Elemente einer integrierten Schaltung in einem Halbleiterkristallstück]*

monolithic circuit Monolithschaltung *f*

monolithic display monolithische Anzeigeeinheit *f*

monolithic image sensor Festkörperbildsensor *m*

monomode fibre; monomode fiber (US) Monomodelichtwellenleiter *m*

mono sound Monoklang *m*

monostable multivibrator → sinlge-shot multivibrator

monoton(icit)y *[Electronics. In an analog-to-digital converter, a condition in which the output increases as the input voltage increases]* Monotonie *f*

Monte Carlo method *[A technique which obtains a probabilistic approximation to the solution of a problem by using statistical sampling techniques]* Monte-Carlo-Verfahren *n*, Monte-Carlo-Methode *f*

monthly average Monatsmittel *n*, Monatsdurchschnitt *m*

month timer Monatsschaltuhr *f*

mood music → background music

moonlight camera Restlichtkamera *f [Videokamera als Signalquelle für die Videoüberwachung]*

mopboard (US) → base (board)

mopboard convector (US) → base (board) convector

mopboard heater (US) → base (board) heater

MOPS → million operations per second

mop sink *[It has a deep basin, usually used by janitors]* Tiefbeckenausguss *m*

morning boost(ing) Schnellaufheizung *f [Eine Sollwertanhebung nach einer Heizungsabschaltung über Nacht oder Wochenende]*

Morse lamp, blinker lamp Blinklampe *f*

mortice (Brit.); mortise (US); tenon hole, pocket Zapfenloch, (Ein)Steckloch, Ankerloch *n*, kleine Aussparung *f [nicht durchgehend]*

mortice deadlock (Brit.); mortise deadlock (US); tenon hole deadlock, pocket deadlock (Ein)Steckschubriegelschloss ohne Feder, federlose Schubriegel(ein)steckschloss, federlose (Ein)Steckschubriegelschloss *n*

mortice installation (Brit.); mortise installation (US); tenon hole installation, pocket installation (Ein)Steckmontage *f [Einsteckschloss]*

mortice lock, mortice latch (Brit.); mortise lock, mortise latch (US) *[1. Lock designed to be installed in a mortise rather than applied to the door(s) surface. 2. Lock to be mortised into the edge of a door rather than applied to its surface]* (Ein)Steckschloss *n*

mortice lockset (Brit.); mortise lockset (US); tenon hole lockset, pocket lockset (Ein)Steckschlossgarnitur *f*

mortice sliding door lock (Brit.); mortise sliding door lock (US); tenon hole sliding door lock, pocket sliding door lock (Ein)Steckschiebetürschloss, Schiebetür(ein)steckschloss *n*

MOS, metal oxid(e) semiconductor Metalloxidhalbleiter *m [Schaltkreistechnik, welche einen einfachen und platzsparenden technologischen Aufbau bei hoher Packungsdichte ermöglicht. Es gibt N-Kanal-MOS und P-Kanal-MOS]*

mosaic character set Mosaikzeichensatz *m [Videotext]*

most significant bit, MSB, most significant binary digit Bit mit dem höchsten (Stellen)Wert, höchstwertige Bit, Höchstwertbit, hochwertigste Bit *n*, Binärstelle mit dem höchsten Wert, hochwertigste Binärstelle, höchstwertige Binärstelle *f*

most significant digit, MSD Ziffer mit dem höchsten (Stellen)Wert, höchstwertige Ziffer, hochwertigste Ziffer, Höchstwertziffer *f*

most significant number, MSN Zahl mit demhöchsten (Stellen)Wert, höchstwertige Zahl, hochwertigste Zahl, Höchstwertzahl *f*

motion blur Bewegungsunschärfe *f*

motion detection Bewegungsmeldung, Bewegungserkennung *f*, Bewegungsnachweis *m*

motion detector, motion detecting device, motion detection device Bewegungsmelder *m*, Bewegungsmeldegerät *n* [*Er kann zwischen Einbrecher und Bewohnern nicht unterscheiden und deshalb wird er im privaten Bereich häufig nur als zusätzliche Raumfalle eingesetzt. Typische Einsatzorte in privaten Wohnhäusern sind Garagen, Anbauten, Schuppen, Dachböden, usw.*]

motion estimation Bewegungsabschätzung *f*

motionless air → still air

motion safety system Bewegungssicherheitssystem *n*

motion vector Bewegungsvektor *m*

motor act(uat)or → positioning motor

motorboating [*Undesired oscillation in an amplifying system or transducer, usually of a pulse type, occurring at a subaudio or low audio frequency*] Blubbern *n*

motor bracket Motorkonsole *f* [*Elektromotor*]

motor brake horsepower, motor bhp, motor BHP Motor-Brems-PS *f*

motor chamber, motor room Motorraum *m* [*Aufzug*]

motor-compressor unit, motor-compressor set, motor-compressor assembly, motor-compressor pack(age) Motorkompressorsatz *m*, Motorkompressoreinheit, Motorkompressorkombination, Motorkompressorgruppe *f*, Motorkompressoraggregat *n* [*„Verdichter" kann synonym für „Kompressor" verwendet werden*]

motor control center (US); motor control centre (Brit.); MCC [*A single metal-enclosed assembly containing a number of motor controllers and possibly other devices such as switches and control devices*] (fabrikfertige) Motorschaltschrank, (fabrikfertige) Motorsteuer(ungs)schrank, (fabrikfertige) Motorverteiler *m*

motor control contact Motorschaltkontakt *m*

motor cradle Motorwippe *f*

motor-fan compartment, motor-fan case, motor-fan casing, motor-fan housing Motor-Ventilator-Gehäuse *n*

motor flying restart circuit Motorfangschaltung *f* [*Zur Zuschaltung eines Stromrichters auf einen laufenden Motor*]

motor-generator set, m.-g. set [*Combination unit that provides variable voltage from ac service for operation of a dc motor driving unit*] Umformersatz, Maschinenumformer, Motorgenerator, (rotierende) Umformer *m*

motor(ized) barometric draught regulator, motor(ized) barometric draft regulator, motor(ized) (barometric) damper Motor-Schornsteinzugregler, Motor-Kaminzugregler *m*

motor(ized) bolt Motorriegel *m*

motor(ized) fan unit Ventilatoraggregat *n*

motor(ized) control Motorsteuerung *f*

motor(ized) power bolt Elektromotorriegel *m*

motor(ized) protection switch Motorschutzschalter *m*

motor(ized) protector relay Motorschutzrelais *n*

motor(ized) pull-in reader Motoreinzug(s)leser *m*

motor(ized) revolving door Motordrehtür *f*

motor(ized) siren Motorsirene *f* [*Ein Elektromotor treibt die Lochscheibe oder den Lochzylinder und den Verdichter (= Kompressor) zur Tonerzeugung an*]

motor-operated, motorized motorbetätigt, motorisiert

motor spirit → gas(oline) (US)

motorstair(way) → moving stair(way)

motor starter Motorstarter *m* [*Der europäische Motorstarter fasst alle Geräte zum Schalten, Steuern und Schützen eines Motors in einem Gehäuse zusammen*]

motor vehicle, automotive vehicle [*It has rubber tyres and does not run on rails*] Auto(mobil), Motorfahrzeug *n*

motor vehicle fleet management Fuhrparkverwaltung *f*

MOU → monitoring and operator unit station

moulded knockout; molded knockout (US) Vorprägung *f*

moulded polycarbonate [*CD-ROM*] Polykarbonatschicht *f*

mould fungus; mold fungus (US); aspergus negri [*Latin*] Schimmelpilz *m*

mould oil, formwork oil, shuttering oil; mold oil (US) Schal(ungs)öl *n*

to mount → to assemble

to mount → to install

to mount einhängen [*Dateisystem*]

mount [*Structure supporting any apparatus, as a gun, searchlight, telescope, surveying instrument, etc.*] Unterbau *m*

mounted on site → assembled on site

mounting → installation

mounting Aufstellung *f*

M

mounting, builders' mounting *[B.S. 1391]* (Bau-)Beschlag *m*

mounting accessories, installation accessories Montagezubehör(teile), Installationszubehör(teile), Einbauzubehör(teile) *n, m, (f)*

mounting aid, installation aid Installationshilfsmittel, Einbauhilfsmittel, Montagehilfsmittel *n*

mounting base, installation base Gehäuseboden *m* *[Halbleiter]*

mounting certificate, installation certificate Einbauzeugnis, Installationszeugnis, Montagezeugnis *n*

mounting depth, installation depth Einbautiefe, Installationstiefe, Montagetiefe *f [Abstand von Vorder- zur Rückseite eines eingebauten Gerätes]*

mounting drawing → assembly drawing

mounting frame, installation frame Einbaurahmen, Installationsrahmen, Montagerahmen *m*

mounting height, installation height Einbauhöhe, Installationshöhe, Montagehöhe *f [Abstand von Unterkante zur Oberkante eines eingebauten Gerätes]*

mounting instruction, installation instruction Einbauanleitung, Installationsanleitung, Montageanleitung *f*

mounting level, installation level Montagehöhe, Einbauhöhe, Installationshöhe *f*

mounting opening, installation opening Einbauöffnung, Einbringöffnung *f [Fahrtreppe; Fahrsteig]*

mounting plate, base plate, bed plate, installation plate Grundplatte, Montageplatte, Aufstell(ungs)-platte *f*

mounting plinth, installation plinth Montage-(stand)sockel, Einbau(stand)sockel, Installations-(stand)sockel *m*

mount(ing) rack → card rack

mount(ing) rack, rack mount(ing) Gestelleinschub-Baugruppenträger, Gestelleinschub-BGT *m*

mount(ing) rack tier → card rack tier

mount(ing)rail Tragschiene *f*

mountings, furniture, builders' mountings, builders' furniture (Bau)Beschläge *f*

mounting sequence, installation sequence Montagereihenfolge, Einbaureihenfolge, Installationsreihenfolge *f*

mounting site, installation site Einbauort, Installationsort, Montageort *m*

mounting unit, installation unit, Montageeinheit, Installationseinheit, Einbaueinheit *f*

mouse Steuerkugel, Rollkugel, (mechanische) Maus *f [Eine in zwei Achsen bewegliche Kugel mit welcher Positionsanzeiger und sonstige Markierungshilfen innerhalb der Bildschirmdarstellungsfläche bewegt werden. Diese Positionierungshilfe dient der gezielten graphischen Eingabe und der Dialogunterstützung zwischen Anwender und Abbildung]*

mouse cursor, mouse mov(e)able visible mark Mausvorlaufzeiger, Mauskursor *m*, Mausbildmarke *f [Siehe Anmerkung unter „Vorlaufzeiger"]*

(mouth)rinsing basin (Mund)Spülbecken *n*

mov(e)able handrail → running handrail

mov(e)able mobility drahtlose Mobilität *f [Innerhalb eines Gebäudes mit hochbitratigen Netzwerken mit gleichzeitiger Integration von Telekommunikationsdiensten mittlerer und niedriger Dienstrate bei höherer Mobilität]*

mov(e)able (spray) shower → hand (spray) shower

mov(e)able visible mark → cursor

mov(e)able zero → floating zero

mov(e)able zero system → floating zero system

moved air bewegte Luft *f*

moved character übertragene Zeichen *n*

moved picture retrieval Bewegtbildabruf *m*

movement detector, movement detection device, movement detecting device Bewegungsmelder *m*

movement monitoring Bewegungsüberwachung *f*

move with offset Übertragen *n* mit Stellenversetzung

moving coil (loud)speaker → dynamic (loud)speaker

moving coil wattmeter → dynamic wattmeter

moving contact Schaltkontakt *m*, bewegliche Schaltstück *n*

moving curtain filter → roll filter

moving fire (grate) bar, mov(e)able fire (grate) bar, rocking fire (grate) bar bewegliche (Feuer-)Roststab *m*

moving ramp *[It has a slope or angle of inclination exceeding 3° with the horizontal]* Fahrrampe, Rollrampe *f*

moving staircase → escalator

moving telecontact, remote moving contact Fernschaltkontakt *m*, bewegliche Fernschaltstück *n*

moving walk *[A passenger-carrying device on which passengers stand or walk, and in which the passenger-carrying surface remains parallel to its direction of motion and is uninterrupted]* Fahrsteig, Rollsteig *m*, Personenförderband *n*

moving walk system Fahrsteiggruppe, Rollsteiggruppe, Personenförderbandgruppe *f*

MP → microprocessor

MPP → maximum power point

MPR → multi(ple)-protocol routing

MPS → medium-pressure steam

MPS heating → medium-pressure steam heating

MPU → microprocessing unit

MPX → multiplexer

MRAS, model reference adaptive system Modellreferenzsystem *n*

MRH →magneto resistive head

MRT → mean radiant temperature

MS → mobile (radio) station

MSB → most significant bit

MSC → mobile switching centre (Brit.)

MSD → most significant digit

MsF method → multi(ple)-stage(d) flash method

MSI → medium scale integration

MSK → minimum shift keying

MSN → multi(ple)-subscriber number

MSN → most significant number

MTA → my talker address

MTBE → mean time between errors

MTBF → mean time between failures

MTBM, mean time between maintenance mittlere Unterhaltungsabstand *m*, mittlere Unterhaltungsintervall *n* [*Mittlere Zeit zwischen zwei Unterhaltungen*]

MTBR, mean time between replacements mittlere Austauschzeit, mittlere Auswechselzeit *f* [*Mittlere Zeit zwischen zwei Austauschen*]

MTBR, mean time between repairs mittlere Reparaturzeit *f* [*Mittlere Zeit zwischen zwei Reparaturen*]

MTBS, mean time between stops mittlere Unterbrechungszeit *f* [*Mittlere Zeit zwischen zwei Unterbrechungen*]

MTL → merged transistor logic

MTTF → mean time to failure

MTTFF → mean time to first failure

MTTR → mean time to repair

MU → (computer) memory unit

mud accumulation Verschlammung *f*

mud bucket Schlammeimer *m*

mud cock Schlammhahn, Abschlämmhahn *m*

mud collector Schlammfang, Schlammfänger, Schlammsammler *m*

mud drum [*water-tube boiler*] Schlammtrommel *f*

mud (pipe)line Schlamm(rohr)leitung *f* [*Heizkessel*]

mud valve [*A type of boiler fitting*] Schlammventil, Abschlämmventil *n*

muffled [*sound*] dumpf, gedämpft

muffler → acoustic absorber

multilingual operation mehrsprachige Bedienung, vielsprachige Bedienung *f* [*Die Kommunikation zwischen Bediener und System kann in einer von mehreren möglichen Sprachen erfolgen*]

multipipe heat exchanger → multitubular heat exchanger

multiple → shunt

multi(ple)-emitter Mehrfachemitter, Vielfachemitter *m* [*Transistor*]

multi(ple)-blade damper Mehrfachklappenzugregler, Vielfachklappenzugregler *m*

multi(ple)-box(-type) assembly Mehrfachkastenbauform, Vielfachkastenbauform *f*

multi(ple-)byte instruction Vielbitgruppe-Hinweis, Mehrbitgruppe-Hinweis *m*, Vielbitgruppe-Anweisung, Mehrbitgruppe-Anweisung *f* [*Eine Anweisung deren Bitzahl größer als die Wortlänge des Rechners ist. Diese Anweisung muss deswegen in zwei oder mehr Speicherplätzen abgelegt werden*]

multi(ple)-cabinet(-type) assembly Mehrfach-Schrankbauform, Vielfach-Schrankbauform *f*

multi(ple)-cell system Mehrzellensystem, Vielzellensystem *n*

multi(ple)-chamber (air) lock Mehrkammer(luft)-schleuse, Vielkammer(luft)schleuse *f*

multi(ple)-channel system Mehrkanalsystem, Vielkanalsystem *n*, Mehrkanalanlage, Vielkanalanlage *f*

multi(ple)-channel transceiving system, mcts Mehrkanal-Sende- und Empfangssystem, Vielkanal-Sende- und Empfangssystem *n*, Mehrkanal-Sende- und Empfangsanlage, Vielkanal-Sende- und Empfangsanlage *f*

multi(ple)-channel transmitter Mehrkanalsender, Vielkanalsender *m*

multi(ple)-circuit single-pipe (heating) system, multi(ple)-circuit single-pipe (heating) system Mehrkreis-Einrohrsystem, Vielkreis-Einrohrsystem *n*

multi(ple)-circuit switch Vielfachschalter, Mehrfachschalter *m*

multi(ple)-compartment digestion pit Mehrkammer(faul)grube, Mehrkammerausfaulgrube, Vielkammer(faul)grube, Vielkammerausfaulgrube *f* [*Kleinkläranlage. DIN 4261*]

multi(ple)-criteria detector, multi(ple)-criteria detecting device, multi(ple)-criteria dectection device Mehrkriterienmelder, Vielkriterienmelder *m*, Mehrkriterienmeldegerät, Vielkriterienmeldegerät *n*

M

multi(ple)-frequency telephone Mehrfrequenztelefon, Vielfrequenztelefon *n*, Mehrfrequenzfernsprecher, Vielfrequenzfernsprecher *m*

multi(ple)-fuel burner Mehrstoffbrenner, Vielstoffbrenner *m* [*In ihm werden flüssige und andere Brennstoffe gleichzeitig oder wechselweise verfeuert*]

multi(ple)-function controller Mehrfunktionsregler, Vielfunktionsregler *m*

multi(ple)-function drive Mehrfunktionsantrieb, Vielfunktionsantrieb *m*

multi(ple)-function telephone Mehrdienstendvorrichtung, Vieldienstendvorrichtung *f*, multifunktionale Fernsprecher *m*, multifunktionale Telefon *n*

multi(ple)-function terminal Mehrfunktionsklemme, Vielfunktionsklemme *f*

multi(ple)-gas boiler Mehrgaskessel, Vielgaskessel *m*

multi(ple)-gas burner Mehrgasbrenner, Vielgasbrenner *m*

multi(ple)-hoistway Mehrfach-Fahrschacht, Vielfach-Fahrschacht, Mehrfach-Aufzugschacht, Vielfach-Aufzugschacht *m*

multi(ple)-hole mixer fitting Mehrloch(misch)batterie, Vielloch(misch)batterie *f*

multi(ple)-image print Vielfachbilddruck, Mehrfachbilddruck *m*

M

multi(ple)-interface Mehrfachschnittstelle, Vielfachschnittstelle, Mehrfachanpassungsschaltung(s)-stelle, Vielfachanpassungsschaltung(s)stelle *f*

multi(ple)-jet burner [*Conversion burner consisting of a number of individual burner tubes or jets mounted on one or more manifolds. Each jet in itself is a complete burner with an orifice, primary air openings, and flame ports*] Mehrfachdüsenbrenner, Vielfachdüsenbrenner *m*

multi(ple)-jet meter Vielstrahlzähler, Mehrstrahlzähler *m*

multi(ple)-layer board vanity Mehrschicht(en)platte-Waschtisch, Vielschicht(en)platte-Waschtisch *m*

multi(ple)-layer filter Mehrschichtenfilter, Vielschichtenfilter *m*, *n*

multi(ple)-layer flexible printed board → flexible multi(ple)-layer printed board

multi(ple) layer thermal insulation Mehrschichtenwärmedämmung, Vielschichtenwärmedämmung *f*

multi(ple)-level address → indirect address

multi(ple)-level cabinet Mehrere-Tage-Truhe *f* [*Tiefkühlmöbel*]

multi(ple)-level interruption Vielpegelunterbrechung, Mehrpegelunterbrechung *f* [*Unterbrechung mit hoher Priorität kann eine andere mit niedriger Priorität auf mehreren Ebenen unterbrechen*]

multi(ple)-line representation mehrpolige Darstellung, vielpolige Darstellung *f* [*DIN 40 719, Blatt 1*]

multi(ple)-loop automatic control system, automatic control multi(ple)-loop system vermaschte Regelkreis *m*

multi(ple)-media computing Mehrmedia-Informationsverarbeitung, Vielmedia-Informationsverarbeitung *f*

multi(ple)-media mail mehrmediale Informationstransfer, vielmediale Informationstransfer *m*

multi(ple)-meter Vielfachmessgerät, Mehrfachmessgerät *n*, Vielfachmesser, Mehrfachmesser *m*

multiple of power Potenzvielfalt *f*

multi(ple)-paired [*cable*] hochpaarig, vielpaarig

multi(ple) pass mehrmalige Durchlauf, vielmalige Durchlauf *m*

multi(ple)-pass welding Mehrraupenschweißen, Vielraupenschweißen *n*

multi(ple)-path signal, shadow Mehrwegsignal, Vielwegsignal *n*

multi(ple)-point regulation, multi(ple)-point ((closed) loop) control, multi(ple)-point feedback control, multi(ple)-point closed control [*See remark under "regulation"*] Mehrpunktreg(e)lung, Vielpunktreg(e)lung, Mehrpunktregulierung, Vielpunktregulierung *f*

multi(ple)-point water heater [*See remark under "water heater"*] Wassererhitzer für Gruppenversorgung, Wasser(er)wärmer für Gruppenversorgung, Wasserheizer für Gruppenversorgung *m*

multi(ple)-pole mehrpolig, vielpolig

multi(ple)-pole circuit breaker mehrpolige Leistungs(schutz)schalter, vielpolige Leistungs(schutz)-schalter *m*

multi(ple)-pole terminal mehrpolige Klemme, vielpolige Klemme *f*

multi(ple)-position controller, multi(ple)-position regulator, multi(ple)-step controller, multi(ple)-step regulator Mehrpunktregler, Vielpunktregler *m* [*Er arbeitet mit mehr als zwei Schaltlagen*]

multi(ple)-position mode Mehrplatzbetrieb, Vielplatzbetrieb *m*

multi(ple)-programming, multi(ple)-tasking Mehrfachprogrammieren, Vielfachprogrammieren *n* [*Form der (scheinbar) gleichzeitigen Bearbeitung mehrerer Programme durch ein Rechnersystem*]

multi(ple)-protective earthing → protection multiple earthing

multi(ple)-protocol routing, MPR Weiterleitung *f* ohne Hardware, Rangieren *n* ohne Hardware

multi(ple)-purpose boiler → **economic(al) boiler**

multi(ple)-purpose controller, multi(ple)-purpose regulator Mehrzweckregler, Vielzweckregler *m* [*Er ist in Verbindung mit geeignetem Messumformer und Stellglied für verschiedene Regelgrößen verwendbar*]

multi(ple)-purpose internet mail extension → **MIME**

multi(ple)-purpose shower (bath)tub Mehrzweckbrause(bade)wanne, Vielzweckbrause(bade)wanne *f*

multi(ple)-room building Vielraumgebäude, Mehrraumgebäude *n*

multi(ple)-room fan Mehrraumventilator, Vielraumventilator *m*

multi(ple)-room heating Mehrraumheizung, Vielraumheizung *f*

multi(ple)-room tile stove heating Mehrraum-Kachelofenheizung, Vielraum-Kachelofenheizung *f*

multi(ple)-sensor detector, multi(ple)-sensor detecting device, multi(ple)-sensor detection device Mehrsensormelder, Vielsensormelder *m*, Vielsensormeldegerät, Mehrsensormeldegerät *n*

multi(ple)-server system Mehr-Diensteanbieter-System, Viel-Diensteanbieter-System *n*

multi(ple)-shell condenser → **closed shell-and-tube condenser**

multi(ple)-sounder Mehrfachhörmelder, Vielfachhörmelder *m*

multi(ple)-speed contactor Polumschaltschütz *n*

multi(ple)-speed motor polumschaltbare Motor *m*

multi(ple)-stage absorption chiller, multi(ple)-stage absorption refrigerating machine mehrstufige Absorptionskältemaschine, vielstufige Absorptionskältemaschine *f*

multi(ple)-stage(d) mehrstufig, vielstufig

multi(ple)-stage(d) air conditioning plant, multi(ple)-stage(d) air conditioning installation mehrstufige Klima(tisierungs)anlage, vielstufige Klima(tisierungs)anlage *f*

multi(ple)-stage(d) compressor mehrstufige Verdichter, mehrstufige Kompressor, vielstufige Verdichter, vielstufige Kompressor *m*

multi(ple)-stage(d) expansion [*Allowing refrigerant to pass through two or more expansion valves in series with each other*] mehrstufige Expansion, vielstufige Expansion, mehrstufige Ausdehnung, vielstufige Ausdehnung *f*

multi(ple)-stage(d) flash method, MsF method vielstufige Entspannungsverdampfungsmethode *f*, mehrstufige Entspannungsverdampfungsverfahren *n* [*Meer(es)wasserentsalzung*]

multi(ple)-state detector, multi(ple)-state detecting device, multi(ple)-state detection device Mehrpunktmelder, Vielpunktmelder *m*, Mehrpunktmeldegerät, Vielpunktmeldegerät *n*

multi(ple)-station access unit → **MAU**

multi(ple)-step controller → **multi(ple)-position controller**

multi(ple)-subscriber number, MSN Mehrfachrufnummer, Vielfachrufnummer *f*

multi(ple)-switch → **ganged switch**

multi(ple)-tasking → **multi(ple)-programming**

multi(ple)-teletotalizer, multi(ple) remote totalizer Mehrfachferntotalisator, Vielfachferntotalisator *m*

multi(ple)-terminal with the same data number → **hunt group**

multi(ple)-transceiver, fan-out module, fan-out unit Schnittstellenvervielfacher, SSV *m*

multi(ple) tubes, tube bundle [*condenser*] Röhrenbündel *n*

multi(ple)-tubular (steam) boiler → **fire-tube (steam) boiler**

multi(ple)-unit Mehrfacheinheit, Vielfacheinheit *f*

multi(ple)-use gleichzeitige Nutzung *f* [*durch mehrere Rechner*]

multi(ple)-use of air Luftmehrfachnutzung, Luftvielfachnutzung *f*

multi(ple)-user Mehrbenutzer, Vielbenutzer, Mehranwender, Vielanwender *m*

multi(ple)-user operating system Ansteuer(ungs)-system *n* für mehrere Bedien(ungs)plätze

multi(ple)-user system Mehrbenutzer(rechner)system, Vielbenutzer(rechner)system *n* [*Es hat nur eine Zentraleinheit an die mehrere DV-Arbeitsplätze angeschlossen sind*]

multi(ple)-variable zusammengesetzte Größe, Mehrgröße, Vielgröße *f*

multi(ple)-variable system Mehrgrößensystem, Vielgrößensystem *n*

multi(ple)-(wash)basin installation → **battery lavatories (US)**

multi(ple)-windowing Mehrfenstertechnik, Vielfenstertechnik *f*

multiplexer, MPX Multiplexer, Mehrfachumschalter, Vielfachumschalter *m* [*z. B. Messstellenumschalter, Steuer(ungs)wertumschalter. Mehrere anstehende Messsignale werden nacheinander auf einen Verstärker mit nachfolgendem schnellen Analog-Digital-Umsetzer geschaltet*]

M

multiplex operation Multiplexbetrieb *m* [*Eine Betriebsart einer Funktionseinheit, bei der mehrere Aufträge abwechselnd in Zeitabschnitten verzahnt abgewickelt werden. E DIN 44 300 T 9*]

multi(ple)-zone air conditioning Mehrzonenklimatisierung, Mehrzonenklimareg(e)lung, Vielzonenklimatisierung, Vielzonenklimareg(e)lung *f*

multiplier → range switch

multiplier, multiplying element Multiplizier(er)glied, Multiplizierer, Vervielfacher(glied) *m*, (*n*) [*Ein Multiplizier(er)glied (Funktionseinheit) ist ein Glied mit 2 Eingängen und einem Ausgang. Der Augenblickswert des Ausgangssignals ist gleich dem Produkt aus den Augenblickswerten der Eingangssignale und einem festen Bewertungsfaktor nach der Beziehung:* $v(t) = k_0 \cdot (k_1 \cdot u_1(t)) \cdot (k_2 \cdot u^2(t))$ *Die Eingangssignale dürfen die zulässigen Wertebereiche nicht überschreiten. Die entsprechende Baueinheit heißt Multiplizierer. E DIN 19 226, Teil 5*]

multiplier chart Vervielfachschaubild *n*

multiprocessing Mehrprozessorbetrieb *m* durch Rechnerkopp(e)lung

multiprogramming Mehrprogrammbetrieb, Vielprogrammbetrieb *m*

Multipurpose Internet Mail Extension → MIME

multitasking gleichzeitige Abarbeitung *f*, gleichzeitige Programmablauf *m* [*Auf einem PC laufen mehrere Programme zur gleichen Zeit ab*]

multitubular heat exchanger, multipipe heat exchanger [*It consists of an assembly of tubes or pipes which convey one of the heat exchange fluids. See remark under "heat exchanger"*] Vielfachrohrwärme(aus)tauscher, Mehrfachrohrwärme(aus)tauscher [*In der Industrie so genannt*]; Vielfachröhrenwärmeübertrager, Mehrfachröhrenwärmeübertrager *m*

multiusing Mehrfachverwendung, Vielfachverwendung *f*

multivariable zusammengesetzte Größe *f*

multizone [*Spatial division of a building having different air-conditioning loads*] Multizone *f*

multizone unit [*Air-conditioning unit capable of handling variable loads from different sections of a building simultaneously*] Multizonengerät *n*

municipal sewage; municipal wastewater (US) kommunale Abwasser, Gemeindeabwasser *n*

municipal solid waste incinerator Gemeindemüllverbrennungsanlage, kommunale Müllverbrennungsanlage *f*

municipal water kommunale Wasser, Gemeindewasser *n*

mushroom flame Pilzflamme *f*

mushroom lamp Pilzlampe *f*

musical scale Tonskala *f*

musical signal Musiksignal *n*

musical sound Musikton *m*

music practice room Musikzimmer *n*

mute switch Stummschalter *m*

muting Stumm(ab)schaltung, Rauschunterdrückung, Abschwächung *f* [*Bei Empfängern*]

muting signal Stummsignal *n*

mutual exclusion ausschließliche Zugriff *m*

MUX → multiplexer

MV → multivibrator

my talker address, MTA eigene Sprecheradresse *f*

N

N → negative

N. A. → numerical aperture

nab [*A metal plate or box which is set in a doorjamb and is either pierced or recessed to receive the bolt or latch of a lock, fixed on a door*] Schließblech *n* [*DIN 18 251*]

nailed lock angenagelte Schloss *n*, angeschlagene Schloss

nailhead bonding → ball point bonding

NAK → negative acknowledg(e)ment

name file Namen(s)datei *f*

name max maximale Zeichenanzahl *f* [*Dateiname*]

nameplate Bezeichnungsschild *n*, Namensschild

nameplate capacity → installed capacity

nameserver Adressenumsetz(ungs)rechner *m* [*Er setzt symbolische Namen in IP Adressen um. IP = intelligente Peripheriebaugruppe*]

Nancy communications receiver → infrared communications receiver

N

NAND, sheffer stroke, NOT-AND [*A logic operator having the characteristic that if P, Q, R, ... are statements, then the NAND of P, Q, R, ... is true if at least one statement is false, false if all statements are true. Derived from NOT-AND*] NAND-Verknüpfung, Nicht-UND, UND mit Inversion am Ausgang, UND-NICHT, Scheffer-Funktion *f*, invertierte UND-Gatter, NAND-Gatter *n*

NAND element, NOT-and-AND element UND-Element *n* mit negativem Ausgang

NAND module, NAND block, NOT-and-AND module, NOT-and-AND block UND-Baustein *m*, UND-Modul *n*

naphtha → gas(oline) (US)

to narrow → to contract

narrow-angle lighting → downlighting

narrow-angle luminaire → downlighting luminaire

natural air circulation [*Air circulation caused by differences of density induced by differences of temperature*] natürliche Luftzirkulation, natürliche Luftumwälzung *f*

natural air conditioning [*See remark under "air conditioning"*] natürliche Luftbehandlung, natürliche Klimareg(e)lung, natürliche Luftaufbereitung, natürliche Klimatisierung, natürliche Luftveredelung *f* [*Eine Technik, die mit bauphysikalischen Methoden eine Er-wärmung, Abkühlung und Lüftung von Gebäuden ohne mechanische Anlagen erreicht*]

natural binary binäre Darstellung *f* [*Sie eignet sich nur für die Repräsentation positiver Eingangssignale. Ein binäres Wort bei dem alle Stellen gleich null sind, entspricht dem Wort null, ein binäres Wort mit allen Bits gleich eins entspricht dem vollen Spannungsbereich des Wandlers. Beispiel: Die Zahl 182 wird als 10110110 dargestellt*]

natural circulation [*Movement of heated or cooled air, water or other fluid unassisted by mechanical means*] natürliche Umwälzung *f*

natural convection → free convection

natural convection air cooler [*One depending on natural convection for air circulation*] natürliche Konvektionsluftkühler, freie Konvektionsluftkühler *m*

natural convection condenser [*Condenser cooled by natural circulation of atmospheric air*] natürliche Konvektionskondensator, natürliche Konvektionsverflüssiger, freie Konvektionskondensator, freie Konvektionsverflüssiger *m*

natural convection cooling [*1. Cooling a fluid by natural convection over the cooling surface. 2. Cooling a space by means of air cooled by natural convection*] natürliche Konvektionskühlung, freie Konvektionskühlung *f*

natural cooling freie Kühlung, natürliche Kühlung *f* [*Kühlung mit Außenluft, deren Temperatur unterhalb der Raumlufttemperatur liegt*]

natural damped angular frequency → damped natural angular frequency

natural draft, natural draught Naturzug, Normalzug *m*

natural-draft boiler, natural-draught boiler Naturzugkessel, Normalzugkessel *m*

natural-draft burner, natural-draught burner [*Burner which depends primarily on the natural draft created in the chimney or venting system to induce the air required for combustion into the burner*] Naturzugbrenner, Normalzugbrenner *m*

natural draft cooling tower → atmospheric (cooling) tower

natural exhaust ventilation natürliche Entlüftung *f*

natural feed Zulauf unter natürlichem Gefälle, natürliche Zulauf *m*

natural frequency Einschwingfrequenz *f* (in Hochspannungsnetzen)

natural frequency, resonance frequency [*Frequency at which resonance exists*] Eigenfrequenz *f*

natural fresh air Naturfrischluft, natürliche Frischluft *f*

natural fuel [*Such as wood or peat*] natürliche Brennstoff, Naturbrennstoff *m*

natural gas engine Naturgasmotor, Erdgasmotor *m*

natural gas (heating) boiler Erdgas(heiz)kessel, Naturgas(heiz)kessel *m*

natural history museum Naturkundemuseum, Naturgeschichtemuseum *n*

natural (hot) air heating plant, natural (hot) air heating installation natürliche (Warm)Luftheiz(ungs)anlage, natürliche (Warm)Luftheiz(ungs)installation *f* [*Die Luft wird auf natürlichem Wege in den zu beheizenden Raum gefördert*]

natural language natürliche Sprache *f*

naturally ventilated natürlich gelüftet

natural period Eigenschwingungszeit *f* [*Akustik*]

natural refrigerant natürliche Kältemittel *n*

natural ventilation, free ventilation [*For natural ventilation there are two causes of air motion through the building: the wind, and the differences in density between the inside and outside air. The two forces may act independently or in combination*] freie Lüftung, natürliche Lüftung *f*

natural ventilation by air density differences, free ventilation by air density differences Auftrieb(s)lüftung *f* [*Freie Lüftung durch Temperaturunterschied zwischen Außenluft und Innenluft*]

N

natural ventilation system, standard vent(ilat)ing system freie Lüftungssystem *n*; Lüftungseinrichtung *f* [*Funktionseinheit zur freien Lüftung ohne maschinelle Luftförderung*]

natural warmth natürliche Wärme *f*

natural water natürliche Wasser, nicht aufbereitete Wasser, unaufbereitete Wasser *n*

NAUN, nearest active upstream neighbo(u)r unmittelbare Nachbar *m* stromaufwärts [*Tokenring. Bitstromempfang einer Station oder eines Knotens*]

NB → negative big

***n*-butane → refrigerant-600**

NC → network congestion

NC → normally closed

N/C → numerical control (system)

ncc → normally closed contact

NC contact → break contact element

***n*-conducting → *n*(-type)-conducting**

NDAc → not data accepted

NDIS, network device interface specification Standard *m* für die Anbindung von Netzwerkkarten (Hardware) an Netzwerkprotokolle (Software)

near critical fluid Fluid in der Nähe des kritischen Punktes, fastkritische Fluid *n*

nearest active upstream neighbo(u)r → NAUN

near-floor level (Fuß)Bodennähe *f*

near heat Nahwärme *f* [*Gegensatz: Fernwärme*]

near-letter quality, NLQ annähernde Buchstabengüte *f*

to need, to require benötigen, brauchen

need → demand

need assignment → DA

need characteristic number → characteristic requirement number

need-guided → requirement-guided

needle → pointer

needle bearing Nadellager *n*

needle jet Nadelstrahl *m*

needle mechanism → pointer mechanism

needle position → pointer position

needle pulse Nadelimpuls *m*

need period → requirement period

to negate negieren, verneinen

negate element Negationselement, Negationsglied, Negierungsglied, Negierungselement, Verneinungsglied, Verneinungselement *n*

negation Negierung, Verneinung *f*

negation indicator amplifier → amplifier with negation indicator

negative, N negativ, NE

negative acknowledg(e)ment, NAK negative Quittung *f* [*Verneinende Empfängerantwort und Aufforderung zur wiederholten Übertragung eines fehlerhaften Blocks*]

negative big, NB negativ-groß, NG

negative deviation → system deviation

negative check-back signal message negative Rückmeldung, NAK *f*

negative delayed feedback → delayed negative feedback

negative feedback Rückführung, Gegenkopp(e)lung *f*

negative logic negative Logik *f* [*H-Pegel = Wert 0; L-Pegel = Wert 1*]

negative medium, NM negativ-mittel, NM

negative modulation, negative transmission [*1. The transmission of a television signal in which an increase in brightness causes a corresponding reduction of power. 2. The transmission of a facsimile in which an increase in brightness causes a reduction frequency*] Übertragung mit negativer Modulation, negative Modulationsübertragung *f*

negative overpressure negative Überdruck, Unterdruck *m*

negative positioning negative Positionierung *f*, negative Instellungbringen *n* [*Bewegung auf der negativen Seite der Koordinatenachse*]

negative pressure [*Less than atmospheric pressure*] Unterdruck *m*

negative pressure region Unterdruckbereich *m* [*selten: n*]

negative small, NS negativ-klein, NK

negative temperature coefficient, NTC [*A condition that is indicated by a decrease in some physical quantity as the temperature is increased*] negative Temperaturkoeffizient *m*

negative temperature coefficient resistor, NTC resistor Heißleiter, Thermistor, Thermenid, NTC-Widerstand *m*

negative temperature coefficient thermistor → NTC thermistor

negative temperature coefficient thermistor resistor → NTC thermistor resistor

negative transmission → negative modulation

negentropy → information content

negentropy unit → information content unit

N

negligible vernachlässigbar

neon lamp, neon bulb, neon tube [*A tube that operates a bright red glow (or, if heated with mercury, bright blue) when the neon gas inside it is ionized by an electric current, commonly used in outdoor signs and as an indicator light*] Neonlampe, Neonröhre *f*

neon lamp [*misnomer*] → **cold cathode lamp**

NEP → **noise-equivalent power**

Nernst theorem [*A theorem stating that for a homogeneous thermodynamic system, as the temperature approaches absolute zero, the rate of change of the entropy approaches zero and the entropy itself converges to a minimum value*] Nernstsche Wärmetheorem *n*, Nernstsche Wärmelehrssatz *m*

nested [*program(me)*] verschachtelt, geschachtelt

nested level (Ver)schachtelungsebene *f*

nested tables, stacking tables [*Series of small tables, graduating in size, so that one can be set inside the other*] Stapeltische *f*

nesting [*The act of calling up a further subroutine from a subroutine already called up, e. g. in an interrupt program(me)*] Schachtelung, Verschachtelung *f*, Schachteln, Verschachteln *n*

nesting store Kellerspeicher *m*

net... → **(useful)...**

net data rate Nettodatenrate *f* [*Das Verhältnis von übertragenen Nutzdatenbits zu den insgesamt übertragenen Bits. Die Nettodatenrate liegt im Bereich von 4 % – 8 % (z. B. bei Bitbus und Profibus) bis zu ungefähr 95 % (bei Interbus)*]

net (heating) load Netto(heiz)last *f*

net heating value → **lower heating value**

net heat loss, net thermal loss Nettowärmeverlust, Wärmenettoverlust *m*

netnews → **USENET**

net output, net performance Nettoleistung *f*

net positive suction head, NPSH Haltedruckhöhe *f* [*Netto-Energiehöhe im Eintrittsquerschnitt der Pumpe*]

net refrigerating capacity, net refrigeration capacity [*Remaining rate of heat removal from all sources by the evaporator of a refrigerating system, at stated conditions, after deducting internal and external heat transfers to the evaporator that occur before distribution of the refrigerating medium and after its return*] Nettokälteleistung *f*

Netscape Navigator Bookmarks → **hotlist**

net thermal loss → **net heat loss**

nettle wrapping Nesselbandage *f*

net volume Nettovolumen *n*, Nettorauminhalt, NRI *m*

net weight Nettogewicht, Reingewicht *n*

to network vernetzen

network → **(computer) network**

network Netz(werk) *n* [*Eine Reihe von Knoten, die durch Kommunikationskanäle miteinander verbunden sind, wobei Kontroll-Fehlermelde- oder Reorganisationsfunktionen durch Rechner oder Bedienfunktionen realisiert werden können.*]

network architecture [*The logical organization of hardware and software elements that make up switching nodes, transmission links, terminal nodes and operating and administrative systems, along with their interfaces and communications protocols that together define a telecommunications network*] Kommunikationsarchitektur *f*

network capability Netz(werk)fähigkeit *f*, Netz(werk)vermögen *n*

network characteristic (curve) Netz(werk)kennlinie *f*

network communication Netz(werk)kommunikation *f*

network congestion, NC Netz(werk)überlastung *f*, Netz(werk)stau *m*

network control center (US); network control centre (Brit.) Netz(werk)warte *f*

network controller Netz(werk)regler *m*

network-dependent netz(werk)abhängig

network device, network appliance Netz(werk)gerät *n*, Netz(werk)vorrichtung *f*

network device interface specification → **NDIS**

network diagram Netzplan *m* [*DIN 40 719, Blatt 1*]

network disconnection switch → **grid isolation switch**

network drive Netz(werk)laufwerk *n*

network driver netz(werk)spezifische Treiber *m*

networked netz(werk)betrieben

network file system Netz(werk)dateisystem *n*

network for fixed connections Direktrufnetz(werk) *n*

network identification signal Netz(werk)kennung *f* [*dezentrale Zeichengabe*]

networking Vernetzen *n*, Vernetzung *f*

network input Netz(werk)eingang *m*

network integration Netz(werk)integration *f*

network interface Netz(werk)schnittstelle *f*

network interface controller, NIC [*A communication device which enables data processing equipment to be combined to form a network*] Netz(werk)schnittstellenregler *m*

network interface unit, NIU Netz(werk)schnittstelleneinheit *f*

N

network isolation switch → **grid isolation switch**

network layer, layer 3 Netz(werk)schicht, dritte Schicht, Netz(werk)ebene, Vermittlungsschicht, Vermittlungsebene, Schicht 3, Ebene 3 *f*

network management Netz(werk)verwaltung, Netz(werk)bewirtschaftung *f* [*Ein speziell zur Durchführung von verwaltungstechnischen Aufgaben eingerichteter Dienst wie Netzwerk-Topologie und Software-Konfiguration, Laden der Software auf Ferngeräte, Überprüfung der Netzwerk-Leistung, Überwachung des Netzwerk-Betriebes, Fehlerdiagnose und –behebung*]

network master relay [*ANSI C37 100*] Maschennetz(werk)relais, Rückwattrelais, Schutzrelais für Maschennetz(werk)schalter *n*

network node Netz(werk)knoten *m*

network operating system, NOS Netz(werk)betriebssystem *n*

network operator Netz(werk)betreiber *m*

network performance, network output Netz(werk)leistung *f*

network pressure Netz(werk)druck *m*

network printer, network writer Netz(werk)drucker *m*

network protector Netz(werk)schalter *m*

network provider, network supplier Netz(werk)erbringer, Netz(werk)bereitsteller, Netz(werk)versorger *m*

N

network pump Netz(werk)pumpe *f*

network relay [*A voltage or power type relay designed to protect and control a low-voltage AC network*] Maschennetz(werk)relais *n*

network resource [*Any physical or logical entity which may be accessed via a communication medium*] Netz(werk)betriebsmittel *n*

network security Netz(werk)sicherung *f*

network service Netz(werk)dienst *m* [*Aus einem Netzwerk verfügbare Anwendungen, z. B. Datenübertragung*]

network service server, NSS Netz(werk)dienstgeber *m*

network splitting, islanding Inselbildung *f*

network table Netz(werk)tabelle *f*

network termination, NT Netz(werk)abschluss *m*

network terminator, NT(BA) ISDN-Netz(werk)abschluss *m* (am Basisanschluss)

network terminator primary multiplex, NTPM Netz(werk)abschluss *m* des Primärmultiplexanschlusses

network timing Netz(werk)synchronisation *f*

(network) user Netz(werk)benutzer, Netz(werk)teilnehmer *m*

(network) user identification, NUI Netz(werk)benutzerkennung *f* [*für Telefonzugänge*]

network water → **mains water**

neutral bar Mittelleiterschiene *f*

neutralization → **protection multiple earthing**

neutralizing sewage Neutralisationsabwasser, aggressive Abwasser *n*

neutral pole N(eutral)-Pol *m*

neutral relay, nonpolarized relay [*A relay with a motion characterstic of its armature and independent of the current flow direction through its coil*] neutrale Relais, unpolarisierte Relais, nichtpolarisierte Relais *n*

neutral system → **system neutral**

neutral zone after echo → **dead zone after echo**

neutron dosimetry Neutronendosismessung *f*

newcomer Einsteiger, Neuling *m*

newform Formatänderung *f* [*Mit dem Kommando "newform" wird das Format einer Textdatei geändert*]

News → **USENET**

newsclip Nachrichtentelegramm *n*

newsletter Rundschreiben *n*

next-channel-system [*television*] AB-Mischer *m* [*Je zwei Eingangssignale können hierbei über eine Kreuzschiene angewählt und überblendet werden*]

nibble Halbbitgruppe *f*, Halbwort, Vier-Bit, Halbbyte, Vier-Bit-Wort *n* [*Die Hälfte einer Bitgruppe*]

NIC → **network interface controller**

niche, recess Nische *f*

nickel-cadmium battery, nicad battery Nickel-Kadmium-Batterie *f*

nickel-iron battery → **Edison battery**

night bell Nachtklingel *f*

night block storage heater, block (night) storage heater, block off-peak storage heater, off-peak block storage heater Nachtstromheiz(ungs)speicherblock *m*

night bolt Nachtriegel *m*

night charge Nachtladung *f*

night charging, off-peak charging Nachtaufheizung *f*

night charging period, off-peak charging period Nachtaufheizzeit *f*

night charging with supplementary day heating, off-peak charging with supplementary day heating Nachtaufheizen *n* mit Tagesnachheizen

night cooling Nachtkühlung *f*

night cooling operation Nachtkühl(ung(s))betrieb *m*

night current, off-peak current Nachtstrom *m*

night current operation, off-peak current operation Nachtstrombetrieb *m*

night electric current, night electricity, off-peak electric current, off-peak electricity Nachtenergie *f*, Nachtstrom *m*

night electric supply, off-peak electric supply Nachtstromversorgung *f*

night hot water heating, off-peak hot water heating, night (warm) water heating, off-peak (warm) water heating Nachtstrom(warm)wasserheizung *f*

night illumination, night lighting, off-peak illumination, off-peak lighting Nachtbeleuchtung *f*

night latch; night latch bolt (US) Nachtfalle *f* [*Schloss*]

night latch lock [*It has a spring bolt which is operated by a key from the outside and a knob inside*] Nachtklinkenschloss *n*

night lock [*A spring latch is operated by a sliding knob on the inside and by a key on the outside*] Nachtschloss *n*

night lowering Nachtabsenkung *f* [*Heizung*]

night(-only) tariff, off-peak tariff Nachtstromtarif *m*

night operation(s) Nachtbetrieb *m*

night purge [*Use of external cold air to cool a building overnight*] Nachtluftkühlung *f*

night safe Nachttresor *m*

night setback, night lowering [*See remark under "setback"*] Nachtabsenkung, Nachtrückstellung *f* [*Absenkung der Raumbedingungen um einen bestimmten Wert während der Nacht*]

night setback heating resistor, night lowering heating resistor, off-peak setback heating resistor, off-peak lowering heating resistor Nachtabsenkung(s)heizwiderstand, Nachtrückstellung(s)heizwiderstand *m*

night setback mode, night lowering mode, unoccupied mode, off-peak (setback) mode Nachtabsenkung(s)betriebsart, Nachtrückstellung(s)betriebsart *f*

night setting Nachteinstellung *f*

night soil → f(a)ecal matter

night storage heater, off-peak storage heater Nachtspeicherheizer, Speichernachtheizer *m*

night storage (hot) water heater, night storage (hot) water heating appliance, off-peak storage (hot) water heater, off-peak storage (hot) water heating appliance Nachtstrom-Wassererhitzer, Nachtstromwasserheizer *m*

night storage unit, off-peak storage unit Nachtspeichereinheit, Speichernachteinheit *f*

night switching Nachtschaltung *f*

night switching off Nachtabschaltung *f*

night ventilation Nachtlüftung *f*

night (warm) water heating → night hot water heating

night water heating, off-peak water heating Nacht-Wasserheizung, Wasser-Nachtheizung *f*

nil Nullzeichen *n*

nipple [*Short length of externally threaded pipe for connecting two internally threaded fittings*] Nippel, Gewindenippel *m*, Anschlussstück *n*

NIPTS → permanent threshold shift

nitrogen chemistry Stickstoffchemie *f*

nitrogen-cooled stickstoffgekühlt [*z. B. ein Sensor*]

nitrogen cushion Stickstoffpolster *m* [*Österreich n*]

nitrogen dioxid(e) Stickstoffdioxid *n*

nitrogenous, containing nitrogen stickstoffhaltig

NIU → network interface unit

NLQ → near-letter quality

NM → negative medium

NMR → noise to mask ratio

NO → normally open

Nobel Laureate Nobelpreisträger *m*

No contact → make contact

NOCVD → metal organic chemical vapour deposition

node [*Telecommunications. The location at which transmitting or receiving equipment is connected to a communications network*] Knoten, Verknüpfungspunkt *m*

node, data entry point, data entrance point Dateneintritt(s)stelle *f*

no flow state, no flow status, no flow condition Nichtfließzustand, fließlose Zustand *m*

noise Rauschen *n* [*zufällige Geräuschstörung*]

noise Geräusch *n*

noise Stör(ungs)größe *f* [*Von außen wirkende Größe in Steuerungen und Reg(e)lungen, die zu Fehlwirkungen führt*]

noise analyser [*Electronics. An instrument that measures, plots, displays, and/or tabulates noise as a function of frequencies*] Rauschanalysator *m*

noise analysis [*Electricity. The study of the sources of noise in a signal processing system, typically involving the determination of the frequency components responsable for the noise*] Rauschanalyse *f*

N

noise behaviour (Brit.); noise behavior (US) Geräuschverhalten *n* [*Von Armaturen, Geräten usw.*]

noise cancelling → noise suppression

noise control [*Acoustical engineering. The prevention or reduction of noise, either at its source using such equipment as balancing machinery at the listener's ear using methods such as shielding*] Geräuschbekämpfung, Geräuschabwehr *f*

noise (distortion), signal disturbance Rauschen *n*

noise emission Geräuschemission *f*

noise-equivalent power, electrical noise-equivalent power, NEP rauschäquivalente Leistung, äquivalente Rauschleistung *f*

noise (e)valuation → noise rating

noise factor [*Electronics. The ratio of the total noise power per unit bandwidth at the output of a system to the portion of the noise power that is due to the input termination, at the standard noise temperature of 290 K*] Rauschverhältnis *n*

noise filter [*Acoustical engineering. A device added to a radio receiver to prevent spurious signals from entering*] Geräuschsperre *f*, Geräuschfilter *m, n*

noise filter, anti-interference filter [*Electronics. A device that prevents spurious signals form passing through an AC power line*] Rauschfilter, Stör(schutz)-filter *m, n*

noise generation Geräuschbildung, Geräuscherzeugung *f*

noise generator [*Acoustical engineering. A device that produces noise; sometimes used for calibrating sound measurement devices*] Geräuscherzeuger, Geräuschbildner *m*

noise-induced permanent threshold shift → permanent threshold shift

noise intensity Geräuschstärke *f*

(noise) jammer Störer, Störsender *m*

noise killer, noise suicide circuit [*Telecommunications. A device added to a communications circuit to prevent it from interfering with the operations of nearby circuits*] Rauschtöter, Rauschvernichter *m*

noiseless, clean geräuschfrei

noiseless, clean rauschfrei

noiseless channel, clean channel [*Telecommunications. A communication channel in which the effects of unwanted signals are negligible*] rauschfreie Kanal *m*

noise level [*The intensity of unwanted sound averaged over a specified frequency range and time interval, and weighted with frequency in a specified manner, usually expressed in decibels relative to a specified reference*] Rauschpegel *m*

noise limiter, noise silencer, noise suppressor, electrical noise limiter, electrical noise silencer, electrical noise suppressor [*A limiter circuit that cuts off all noise peaks that are stronger than the highest peak in the desired signal being reached, thereby reducing the effects of atmospheric or man-made interference*] Rauschunterdrücker, Entstörer *m*

noise limiting → noise suppression

noise measurement [*Electricity. Any of a wide range of measurements of random and nonrandom electrical noise, but usually noise-power measurement*] Rauschmessen *n*, Rauschmessung *f*

noise measurement, acoustic noise measurement [*The process of quantitatively determining one or more properties of acoustic noise*] Geräuschmessung *f*, Geräuschmessen *n*

noise-modulated rauschmoduliert

noise pollution [*Acoustics. Unwanted or harmful sounds in the environment*] Geräuschbelästigung *f*

noise pollution [*For a given locale, the average of the measured noise level over a period of time*] Geräuschdurchschnitt *m*

noise power ratio, NPR [*Electrical engineering. The ratio of the noise power at the input of a circuit to the noise power in the noise source*] Rauschverhältnis *n*

noise (pressure) level Geräuschniveau *n*

noise prevention Geräuschverhütung *f*

noise production Geräuschbildung, Geräuscherzeugung *f*

noise propagation Geräuschfortpflanzung *f*

noise quadripole, electrical noise quadripole Rauschvierpol *m*

noise radiating body Geräusch(ab)strahler *m*

noise radiation Geräusch(ab)strahlung *f*

noise rating, noise (e)valuation [*Acoustics. The (e-)valuation of noise level in terms of loudness, perceived level, a noise-criterion curve, or other reference criteria*] Geräuschbewertung *f*

noise rating curve Geräuschbewertungskurve, Geräuschbewertungslinie *f*

noise rating number [*Acoustics. A numerical representation of the level of noise that can be tolerated in a given situation*] Geräuschbewertungszahl *f*

noise ratio Stör(ungs)abstand, Signalstör(ungs)-abstand *m*

noise reduction [*Acoustical engineering. The use of electronic circuity or mechanical means to reduce audible rise that is recorded or reproduced by a sound system, without affecting the desired sounds*] Geräuschminderung *f*

noise reduction coefficient [*Acoustical engineering. The noise reduction capability, specified at 500 hertz or more, of an acoustic material in a room lined with this material, determined by calculating an average of the sabin absorption coefficients at 250, 500, 1000 and 2000 hertz*] Geräuschminderungskoeffizient *m*

noise resistance Rauschwiderstand *m*

noise shaping Rauschgestaltung *f*

noise silencer → **noise limiter**

noise silencing → **noise suppression**

noise standard Rauschnormal *n*

noise suicide circuit → **noise killer**

noise suppression [*Acoustical engineering. A method for minimizing surface noise when an audio tape or phonograph record is being played*] Geräuschunterdrückung, Geräuschentstörung *f*

noise suppression, noise cancelling, noise silencing, noise limiting [*Electrical engineering. A method for minimizing or elimination to effects of spurious electric signals*] Rauschunterdrückung, Rauschentstörung *f*

noise temperature [*Electronics. Given an ideal resistor whose resistance is that of the terminal pair in a system, the temperature at which the ideal resistor would generate the same noise power as that present at the system terminal*] Rauschtemperatur *f*

noise temperature ratio Rauschtemperaturverhältnis *n*

noise testing, electrical noise testing [*The measurement of the power dissipated in a resistance termination of given value joined to one end of a telephone or telegraph circuit when no test power is applied to the circuit*] Rauschprüfen *n*

noise to mask ratio, NMR Rausch/Maskierungs-Verhältnis *n*

noise transmission Geräuschübertragung *f*

noise tube [*Electronics. A type of gas-filled tube that generates white noise*] Rauschröhre *f*

noise window Rauschfenster *n*

no-load current, unloaded current, load-free current [*Electricity. The flow of current into a network with an open-circuited output*] Leerlaufstrom *m*

no-load loss, unloaded loss, load-free loss [*Electricity. The power lost in a device such as a motor or transformer operating at rated voltage and frequency, but with the load disconnected. IEC 76-1*] Leerlaufverlust *m* [*VDE 0532, Teil 1*]

no-load start(ing), unloaded start(ing), load-free start(ing) [*Practice of starting a compressor after equalizing pressures in high- and low-side pressures*] unbelastete Start(en), lastfreie Start(en) *m*, (*n*)

nominal capacitor current → **capacitor nominal current**

nominal capacity Nennkapazität *f*

nominal condensing capacity, condensing nominal capacity Kondensatornennleistung, Verflüssigernennleistung *f*

nominal heat load, nominal thermal load, heat nominal load, thermal nominal load Nennwärmelast, Nennwärmebelastung *f* [*Die vom Hersteller auf dem Geräteschild angegebene höchstzulässige Wärmebelastung in Watt (W) oder Kilowatt (kW), bezogen auf den Heizwert H_u die bei der Geräteeinstellung nicht überschritten werden darf.*]

nominal heat output, nominal thermal output, heat nominal output, thermal nominal output Nennwärmeleistung *f* [*Die bei der Nennwärmeberechnung von einer Verbrauchsvorrichtung für den Betriebszweck nutzbar gemachte Wärmemenge in Watt (W) oder Kilowatt (kW)*]

nominal linewidth [*In FAX transmission, the average separation between the centres of adjacent recording or scanning lines*] Nennzeilenbreite *f*

nominal power Nennleistung *f*

nominal refrigerating output, rated refrigerating output Nennkälteleistung *f*

non-actuating → **non-tripping**

non-air-entrained concrete Nicht-Luftporenbeton *m*

non-appearance → **non-initiating**

non-arithmetic shift → **cyclic(al) shift**

non-attendant service Selbstfahrbetrieb *m* [*Aufzug*]

non-automatic detector, non-automatic detecting device nichtautomatische Melder *m*, nichtautomatische Meldegerät *n* [*Er wird direkt oder indirekt durch Personen ausgelöst, z. B. Druckknopfmelder, Fußkontaktmelder*]

non-automatic fire detector, non-automatic fire detecting device nichtautomatische Brandmelder, nichtautomatische Feuermelder *m*, nichtautomatische Brandmeldegerät, nichtautomatische Feuermeldegerät *n* [*Er wird von Personen mittelbar oder unmittelbar betätigt. DIN VDE 0833*]

non-blocking nichtsperrend

non-blocking access [*In telephony, the ability to connect any incoming line within the switching center to any outgoing line, provided that the outgoing line is unoccupied*] nichtsperrende Zugang *m*

non change over Nichtumschaltung *f*

non change over system Zweirohr-Kaltwasser-Klimaanlage, Kaltwasser-Zweirohr-Klimaanlage, Zweirohr-Klimaanlage ohne Umschaltung *f* [*Den Induktionsgeräten wird nicht mehr abwechselnd Warm- oder Kaltwasser, sondern nur noch Kaltwasser zugeführt*]

N

non-chokable, unchokable verstopfungsfrei

noncircular, nonround unrund, nicht rund

non-combustibility Nichtbrennbarkeit *f*

non-combustible; incombustible [*Less current term*] [*Terms applied to a material which is incapable of supporting combustion, such as plaster, mortar, concrete, metal, etc.*] unbrennbar, nichtbrennbar

non-company computer betriebsfremde Rechner, Fremdrechner *m*

non-compliance Nichtübereinstimmung *f* [*z. B. eines Messgerätes mit den Anforderungen*]

nonconcentrating solar collector [*A solar collector in which the absorber heat flux is not greater than the solar irradiance across the aperture area. It may or may not contain optical elements to direct the radiant flux onto the absorber*] nichtkonzentrierte Sonnenkollektor, unkonzentrierte Sonnenkollektor *m*

noncondensable gas nichtkondensierbare Gas *n*

non-condensing nichtkondensierend

non-condensing gas Hilfsgas *n* [*Absorptionskälteanlage*]

non-conditioned → non-processed

non-conditioned return air, non-processed return air, non-reprocessed air Umluft *f*

non-conditioned return air bypass duct, non-processed return air bypass duct, non-reprocessed air bypass duct Umluft-Umführungskanal *m*

non-conditioned return air chamber, non-processed return air chamber, non-reprocessed air chamber Umluftkammer *f*

non-conditioned return air duct, non-processed return air duct, non-reprocessed air duct Umluftkanal *m*

non-conditioned return air suction, non-reprocessed air suction, non-processed return air suction Umluftansaugung *f*

non-contact detector → proximity detector

non-contact effect → proximity effect

non-contact sensor → proximity sensor

non-contact switch → proximity switch

non-cooking kitchen Garküche *f*

non-cutoff luminaire → general-diffuse luminaire

non-data mode Nichtdatenbetrieb *m*

nondemand, nonrequirement, nonneed Nichtbedarf *m*

nondepletable unerschöpflich

nondepletable energy source [*Source of energy (excluding minerals) derived from incoming solar radiation, including photosynthetic processes, from phenomena including wind, waves and tides, lake or pond thermal differences, and energy derived from the internal heat of the earth, including nocturnal thermal exchanges*] unerschöpfliche Energiequelle *f*

nondestructive zerstörungsfrei

nondestructive read(ing) nichtlöschende Lesen *n*

nondimensioned parameter → dimensionless number

non-document file Programmdatei *f* [*Textverarbeitung*]

non-drinking water → non-potable water

non-energy use nichtenergetische Nutzung *f*

nonequivalence → exclusive OR

nonequivalence function → exclusive OR function

nonequivalence gate, exclusive-OR gate, nonequivalence element, exclusive-OR element Antivalenzglied, Exklusiv-ODER-Gatter *n*

nonequivalence module, nonequivalence block, exclusive-OR module, exclusive-OR block Antivalenzbaustein *m*, Antivalenzmodul *n*

nonerasable storage [*Any storage device or medium whose contents cannot be erased during processing, such as punched cards or microfilm*] nichtlöschbare Speicherung *f*

nonevaporative cooling tower → dry cooling tower

nonfaradaic path [*A route followed by current travel(l)ing through an electrolytic solution, moved along by a continuous charging and discharging of stored energy (capacitance)*] nichtfaradische Stromweg *m*

non-fire rated brandklassenfrei

non-freeze gefrierfrei

non-fuel incinerator, refuse incinerator [*Incinerator that uses no fuel for combustion other than the refuse to be incinerated, except that a gas flame (without blower) or other approved means may be used to accomplish primary ignition. The term "refuse" includes garbage, rubbish and trade waste*] Müllverbrenner *m*, Müllverbrennungsanlage *f*

nonimpact printer [*Any of a class of printers that produce output without the printing element actually strik-ing the paper, including electrophotographical, thermo-graphic, ink-jet, and laser printers*] (an)schlagfreie Drucker *m*

non-initiating, non-appearance, non-initiation Ausbleiben *n* [*Flamme*]

noninterlaced monitor Monitor *m* ohne Zeilensprungverfahren

noninteracting wechselwirkungsfrei, nichtwechselwirkend, entkoppelnd

N

noninteracting ((closed) loop) control entkoppelte Reg(e)lung, nichtwechselwirkende Reg(e)lung, wechselwirkungsfreie Reg(e)lung *f*

noninverting amplifier nichtinvertierende Verstärker *m [Ein gegengekoppelter Operationsverstärker, der im Betrieb am Ausgang eine Spannung führt, die dasselbe Vorzeichen wie die Eingangsspannung hat. Bei Wechselspannungssignalen verlaufen Eingangs- und Ausgangsspannung in Phasen.]*

nonisothermal (air) jet nichtisotherme (Luft)Strahl *m*

nonlinear nichtlinear [*Siehe Anmerkung unter „Linearität"*]

nonlinear amplifier [*A device in which a change in the input signal does not result in a proportional change in the output signal*] nichtlineare Verstärker *m*

nonlinear controller nichtlineare Regler *m*

nonlinear distortion → harmonic distortion

nonlinearized value Rohwert, nichtlinearisierte Wert *m*

nonline data reduction [*The processing of information as rapidly as it is received by the computing system*] schritthaltende Datenreduktion, schritthaltende Datenverdichtung *f*

non-locking nichtverriegelnd

non-maskable interruption nichtmaskierbare Unterbrechung, unmaskierbare Unterbrechung *f*

nonmetal(lic) conduit Nicht-Metall(installations)-rohr, Nicht-Metallschutzrohr *n* [*Elektrotechnik*]

nonmetal(lic) pipe Nichtmetallrohr *n*

non-moving stair stationäre Treppe, ortsfeste Treppe *f* [*Im Gegensatz zur Rolltreppe*]

nonoccupancy, nonoccupation Nichtbelegung *f*

nonpolarized relay → neutral relay

non-potable water, non-drinking water Nichttrinkwasser, Betriebswasser *n* [*Brauchwasser für das keine Trinkwassergüte verlangt wird*]

nonpreemptive allocation, nonpreemptive assignment, nonpreemptive allotment nichtunterbrechende Zuordnung, nichtunterbrechende Zuteilung, nichtunterbrechende Zuweisung *f*

non-processed, non-conditioned nicht aufbereitet, nicht veredelt, unveredelt, unaufbereitet [*Luft*]

non-processed... → non-conditioned...

non-public potable water supply, non-public drinking water supply, well potable water supply, well drinking water supply Eigentrinkwasserversorgung, örtliche Trinkwasserversorgung, Brunnen-Trinkwasserversorgung *f* [*Der Brunnen befindet sich auf dem Grundstück*]

nonredundant system, simplex structure [*A computer system designed with only the minimum number and variety of hardware components needed to accomplish its purpose*] nichtüberzählige System *n*

nonrequirement → nondemand

non-retentive marker spannungsabhängige Merker *m*

non-return angle(-type) valve → angle(-type) non-return valve

non-return ball valve, ball non-return valve, check(ing) ball valve, ball check(ing) valve Kugelrückschlagventil, Rückschlagkugelventil *n*

non-return pointer → slave pointer

non-return straight(way) valve → straight(way) non-return valve

non-return-to-zero recording → charge-on-ones recording

non-return valve → check(ing) valve

nonreverberant → anechoic

nonreverberant field → free field

nonreverberant room → anechoic room

nonreverberant termination → anechoic termination

non-rising stem, non-rising spindle steigungslose Spindel, nichtsteigende Spindel *f* [*Schieber*]

non-rotation lock, non-rotative lock Verdreh(ungs)-sicherung *f*

nonround → noncircular

non-safety lock gewöhnliche Schloss, ordinäre Schloss *n* [*früher: Schloss*]

non-self-regulating controlled system [*IEC 50(351)*], **non-self-regulating directly controlled member, non-self-regulating process** Regelstrecke ohne Ausgleich, ausgleichlose Regelstrecke *f* [*Sie strebt einem unendlichen Wert zu*]

non-sharp switching Unscharfschaltung *f*

nonsmoking occupancy Nichtraucherbelegung *f*

nonstable flip-flop, mono(flip-)flop monostabile Kippglied, Zeitglied *n*, monostabile Kipper, Verzögerungsmultivibrator, monostabile Multivibrator *m*, monostabile Kippschaltung, Verzögerungsschaltung *f* [*Schaltungsanordnung mit nur einem stabilen Schaltzustand*]

non-standard sequence Sonderschaltung *f*

nonsteady state running, nonsteady condition running instationäre Betrieb *m*

non-storage heat exchanger, non-storage heat exchange unit, non-storage heat interchanger, flow(-type) heat exchanger, flow(-type) heat exchange unit, flow(-type) heat interchanger

N

Durchlaufwärme(aus)tauscher (im engeren Sinne), Durchlaufumformer *m* [*Die Wärme wird zwischen zwei Medien ausgetauscht*]

non-storage (type) gas water heater → **flow (-type) gas water heater**

non-storage (water) heater → **flow(-type) water heater**

non switched/permanent circuit Standleitung, Punkt-zu-Punkt-Verbindung *f* [*Eine Verbindungsart von Kommunikationsteilnehmern*]

nonsynchronous [*In electricity, not related in frequency or speed to other quantities within a system*] nichtsynchron

nonsynchronous transmission [*In electronics, a process that does not use a clock to control the spacing of signals that are transmitting data*] (Signal)Übertragung ohne Uhr, uhrlose (Signal)Übertragung *f*

non-system computer → **host (computer)**

non-technical nichttechnisch, untechnisch

non-thermostatic (water) mixing valve → **hand (water) mixing valve**

non-time delay fuse flinke Sicherung *f*

non-tripping, non-actuating Nichtauslösung *f*, Nichtauslösen *n*

non-turbulent wirbelfrei

nonuniform connection teilgesteuerte Schaltung, gemischte Schaltung *f* [*Leistungselektronik*]

nonuniform flow ungleichförmige Fließen *n*

nonuniform rational basic splines → **NURBS**

non-use time Nichtnutzungszeit, ungenutzte Zeit, verlorene Zeit, verschwendete Zeit *f*

non-vented condensate return, closed condensate return geschlossene Kondensatrücklauf *m*

non-ventilated (flat) roof, warm roof [*A roof, where the layer of thermal insulation is installed near the outside rather than near the ceiling. The air in the roof space is therefore similar to the warmer temperatures of the building interior and the roof is less likely to suffer from condensation than a cold roof*] einschalige Flachdach, Warmdach, nicht durchlüftete Flachdach *n*

nonvolatile, not volatile nullspannungssicher

nonvolatile, nonerasable, not volatile, not erasable unverlierbar, nicht flüchtig, nicht verlierbar [*Speicherinhalt*]

nonvolatile RAM → **NVRAM**

non-von Neumann architecture → **dataflow system**

non-walk-in lift (Brit.); non-walk-in elevator (US) nicht begehbare Aufzug, Kleinlastenaufzug, (Klein)Güteraufzug *m* [*Aktenaufzug; Speiseaufzug; Bücheraufzug*]

non-wearing verschleißfrei

no-operation instruction, NOP instruction, skip instruction, blank instruction Nulloperationsbefehl, Leerbefehl, Überspringbefehl *m* [*Befehl, der den Rechner veranlasst nichts zu tun, außer mit der Abarbeitung des nächsten Befehls fortzufahren*]

no parity keine Parität *f*

NOR element → **NOT-and-OR element**

Norfolk latch → **Canadian latch**

nor logic (operator), Peirce logic (operator) Nicht-Oder-Verknüpfung *f*

normal burner Normalbrenner *m*

normal form Normalform, kanonische Form *f*

normal gap switch Schalter *m* mit normaler Kontaktöffnung

normalized fuzzy set normalisierte unscharfe Menge, unscharfe normalisierte Menge *f*

normally closed, NC stromlos geschlossen [*elektrothermische Stell(an)trieb*]

normally closed, NC normal geschlossen [*Relaiskontakt*]

normally closed contact → **break contact element**

normally open, NO normalerweise offen, stromlos geöffnet, normalerweise geöffnet, stromlos offen [*Elektrotechnik*]

normally open contact → **make contact**

normally open gate Einschalttor *n*

normally open switch, NO switch Schließer *m* [*Schalter*]

normal network Normalnetz *n*

normal position Ruhelage *f*

northlight (roof) gutter [*A valley gutter having a flat sole and sides angled to suit the unequal pitches of a northlight roof*] Sheddachrinne *f*

NOS → **network operating system**

no scale → **not to scale**

nosing for linoleum → **linoleum nosing**

no smoking Rauchverbot *n*

no-storage gas water heater, continuous gas water heater, gas no-storage water heater Durchlauf-Gas-Wasserheizer, Durchlauf-Gas-Wasser(er)wärmer, Durchlauf-Gas-Wassererhitzer, Gas-Durchlauf-Wasserheizer, Gas-Durchlauf-Wassererhitzer, Gas-Durchlauf-Wasser(er)wärmer *m*

no-storage water heater, continuous water heater Durchlaufwasserheizer, Durchlaufwarmwasserbereiter, Durchlaufwasser(er)wärmer, Durchflusswasserheizer, Durchflusswasser(er)wärmer, Durchflusswarmwasserbereiter *m*, Therme *f*

NO switch → **normally open switch**

NOT-AND → **NAND**

NOT-and-AND element → **NAND element**

NOT-and-AND module → **NAND module**

NOT-and-OR element, NOR element ODER-Element *n* mit negativem Ausgang

notation Schreibweise *f*

to notch → **to indent**

notch(ed) connector → **indented connector**

notch of the bolt → **bolt notch**

NOT circuit, NOT gate, inverter circuit [*Electronics. A logic circuit in which the output is always the opposite of the input, so that if voltage applied to the gate is high the voltage across the output will be low, and vice versa*] Negationsschaltung *f*, Negator, Inverter *m*

not conditional jump nichtbedingte Sprung *m*

not connected, floating nichtangeschlossen, potenzialfrei

not data accepted, NDAc Daten nicht übernommen *f*

notebook (elektronische) Notizbuch *n*

NOT element Nichtglied, Negativglied *n*

notice of defects, notification of defects, customer's complaint(s) Mängelrüge, Baumängelbenachrichtigung *f* [*Schreiben für die Nachfristsetzung zur Mängelbeseitigung*]

notification of defects → **notice of defects**

not intrinsically(-)safe nichteigensicher

NOT logic (operator) NICHT-Verknüpfung, Negativverknüpfung *f*

not lossy method Komprimier(ungs)verfahren *n* ohne Informationsverlust , Komprimier(ungs)methode *f* ohne Informationsverlust

NOT module, NOT block NICHT-Modul, Negativmodul *n*, NICHT-Baustein, Negativbaustein *m*

not-or gate → **nor gate**

not ready for data, NRFD nicht bereit für Daten [*Eine Funktion einer Schnittstellenleitung*]

not stored instruction nicht gespeicherte Befehl *m*

not to scale, n.t.s., no scale unmaßstäblich, nichtmaßstab(s)gerecht, nichtmaßstäblich

not-to-scale drawing, drawing not to scale Skizze *f*

not volatile → **nonvolatile**

N-output sequence N-Ausgangssequenz *f* [*Sequenzfunktion mit N Ausgangsstufen*]

no-volt proof nullspannungssicher, nullspannungsfest

no-volt release [*IEC 157-1*] Unterspannungsauslöser *m* [*Ansprechspannung 35%-10% der Netzspannung*]

nozzle [*A short tube with a taper or constriction used to spread up or direct a flow of fluid*] Düse *f*

nozzle Schlüsselführungsrohr *n*

nozzle atomization Düsenzerstäubung *f*

nozzle body Düsenkörper *m*

nozzle chamber Düsenkammer *f*

nozzle convector Düsenkonvektor *m*

nozzle core Düsenkern *m*

nozzle filter Düsenfilter *m, n*

nozzle humidifier Düsenbefeuchter, Düsenanfeuchter *m* [*Das Wasser wird durch Zerstäubungsdüsen zu einem dichten Nebel von Wassertropfen zerstäubt, durch den die Luft hindurchgefördert wird*]

nozzle induction outlet Düseninduktionsauslass *m*

nozzle jet Düsenstrahl *m*

nozzle-mixing burner [*Separate streams of air or oxygen and combustible gas are mixed at the burner port*] Düsenmischbrenner, Mischdüsenbrenner *m*

nozzle pipe Düsenrohr *n*

nozzle pipe net(work) Düsenrohrnetz *n*

nozzle pressure Düsendruck *m*

nozzle sieve Düsensieb *n*

nozzle ventilation Düsenlüftung *f*

nozzle water fire extinguishing plant, nozzle water fire extinguishing installation Sprühwasserlöschanlage *f* [*Das Löschwasser wird nach dem Auslösen über Rohrleitungen an die ständig offenen Düsen herangeführt. Die dem offenen Düsenrohrnetz vorgeschaltete Ventilstation hält im Bereitschaftszustand den Löschwasserfluss geschlossen. Nicht verwechseln mit „Sprinkleranlage"*]

NPR → **noise power ratio**

NPSH → **net positive suction head**

NRFD → **not ready for data**

NRZ recording → **charge-on-ones recording**

NS → **negative small**

NS contact → **break contact (element)** [*IEC 337-1*]

NSS → **network service server**

NT(BA) → **network terminator**

NTC → **negative temperature coefficient**

NTC resistor → **negative temperature coefficient resistor**

NTC thermistor, negative temperature coefficient thermistor Heißleiter *m*

NTC thermistor resistor, negative temperature coefficient thermistor resistor Heißleiterwiderstand *m*

2N-terminal Mehrpol, Vielpol *m* [*Netzwerk*]

NTPM → **network terminator primary multiplex**

N-trap Stickstoffstörstelle *f* [*lichtemittierende Diode*]

n.t.s. → **not to scale**

n(-type)-conducting n-leitend

n(-type)-conductor n-Leiter *m*

nuclear force microscope → **atomic force microscope**

nuclear waste → **atomic waste**

NUI → **(network) user identification**

nuisance, annoyance Belästigung, Lästigkeit *f*

nuisance alarm, annoyance alarm Täuschungsalarm *m*

number checking unit Nummernprüfer *m* [*Tabelliermaschine*]

number cruncher intensive Rechner *m*

number end → **end of selection**

numberer Nummernwerk *n* [*Rechentechnik*]

number indicating system Nummerngeber *m* [*Fernmeldetechnik*]

numbering machine Nummeriermaschine *f*, Laufnummerngeber *m*

number of degree-days, degree-days number Gradtagzahl, Heizgradtagzahl *f*

number of lamps per luminaire → **lamp complement of luminaire**

N

number of people Personenaufkommen *n*, Personen(an)zahl *f*

number of stages Stufen(an)zahl *f* [*z. B. bei einer Verdichtung*]

number representation Zahlendarstellung *f* [*Im Rechner werden Zahlen im Dualsystem oder im BCD-Kode verarbeitet. Zur Ein- oder Ausgabe werden drei oder vier Bits im Oktal- oder Hexadezimalsystem zusammengefasst*]

number string → **string of numbers**

number system Zahlensystem *n*

number transfer bus Hauptleitung vom Speicher zum Rechner, Speicher-Rechner-Hauptleitung *f*

number unobtainable tone Besetztzeichen *n*

numeral → **digit**

numeral system, numeration system [*A method of representing numbers by numerals in which each numeral is associated with a unique number*] Zahlensystem, Ziffernsystem *n*

numeral(-type) dial lock, numeral(-type) combination lock Ziffernschloss, Zahlenschloss *n*

numerator polynominal Zählerpolynom *n*

numeric [*In computers, pertaining to data composed wholly or partly of digits, as distinct from alphabetic*] numerisch, zahlenmäßig

numerical [*In mathematics, pertaining to numbers*] numerisch, zahlenmäßig

numerical aperture, N.A. [*In optics, a measure of the resolving power of a microscope objective, equal to the product of the refractive index of the medium in front of the objective and the sine of the angle between the outmost ray entering the objective and the optical axis*] numerische Apertur, zahlenmäßige Apertur *f* [*Lichtwellenleiter*]

numerical combination lock Zahlenkombinationsschloss, Kombinationszahlenschloss *n*

numerical control (system), NC, N/C [*A control system for machine tools, and same industrial processes, in which numerical values corresponding to desired positions of tools or controls are recorded on punched paper tapes, punched cards, or magnetic tapes so that they can be used to control the operation automatically*] numerische Wekzeugmaschinensteuerung *f*

numerical display numerische Anzeige *f* [*Sichtanzeige der Soll- und Istpositionen in Form von Dezimalzahlen*]

numerical display unit numerische Anzeigeneinheit, Zahlenanzeigeneinheit *f*

numerical drawing numerische Zeichnung *f*

numerical index of vector group Schaltgruppenziffer *f* [*Trafo*]

numerical keyboard, numerical keypad numerische Tastenfeld *n*, numerische Tastatur *f*

numerical memory Zahlenspeicher *m*

numerical number numerische Nummer *f*

numerical section Zifferntell *m, n*

numerical string [*A contiguous sequence of bytes representing digits and possibly a sign*] numerische Reihung *f*

numerical tape [*In automatic data processing, the tape required by a computer operating a machine tool*] numerische Band *n*

numeric coding, digit coding [*Coding in which only digits are used, usually binary or octal*] numerische Kodierung, numerische Codierung, Zahlenkodierung, Zahlencodierung, Ziffernkodierung, Zifferncodierung *f*

numeric control [*In automatic data processing, the action of program(me)s written for specialized computers which operate machine tools*] Rechner-Werkzeugmaschinensystem, Werkzeugmaschinen-Rechnersystem *n*

numeric data [*Data consisting of digits and not letters of the alphabet or special characters*] numerische Daten *f*

numeric printer [*In automatic data processing, old type of printer which positioned its keys to print a field in one operation, rather than one digit at a time*] numerische Drucker *m*

NURBS, non-uniform rational basic splines mathematische Form *f* von Freiformflächen und -kurven

nurse(s') call (intercom(munication)) system (Kranken)Schwester(n)rufanlage *f*

Nusselt equation Nusseltgleichung *f*

Nusselt number [*In thermodynamics, a dimensionless number used in the study of forced convection which gives a measure of the ratio of the total heat transfer to conductive heat transfer, and is equal to the heat transfer coefficient times a characteristic length divided by the thermal conductivity. Symbolized* N_{nu}] Nusselt(-Kenn)zahl *f*, dimensionslose Wärmeübergangskoeffizient *m*

nut Mutter *f*

nutating disk flow meter, nutating disc flow meter Taumelscheibenzähler *m*

nylon cord Nylonschnur *f*

Nyquist criterion Nyquist-Kriterium *n*

Nyquist plot Nyquist-Ortskurve *f*

O

object affixed to a structure → fixture

object code [*Instructions which can be directly interpreted by the computer*] Maschinencode, Maschinenkode, Objektcode, Objektkode *m*, Maschinensprache, Objektsprache *f*

object colour (Brit.); object color (US) gebundene Farbe *f*

objectionable matter Fremdstoff *m*

objectionable noise, heavy noise Lärm *m*

objective life Lebensdauerrichtwert *m* [*DIN IEC 434*] Richtlebensdauer *f*

objective management → management by objectives

object linking and embedding → OLE

object management Objektbewirtschaftung *f*

object master data Objektstammdaten *f*

object model(l)ing technique Objektmodellierungstechnik *f*

object parameter window Objektparameterfenster *n* [*Bedienen und Beobachten aller Parameter*]

object program (US); object programme (Brit.) Maschinenprogramm *n* [*Es wird vom Programmschlüssler oder vom Sprachübersetzer erzeugt und kann direkt geladen werden*]

object representation Objektdarstellung *f*

object schedule Objektverzeichnis, OV *n*

object security → security of objects

obligatory call relay Pflichtrufrelais *n*

oblique cut Schrägschnitt *m*

oblique flow filter Schrägstromfilter *m, n*

obliquely inclined, obliquely sloped schräggeneigt

oblique saddle junction piece Einlass mit schrägem Flansch, Schrägflanscheinlass *m* [*Kanalisationszeug*]

oblique seat(ing), angle seat(ing) Schrägsitz(form) *m, (f)* [*Schieber; Ventil*]

oblong hole punching Langlochstanzen *n*

obscuration Verdunk(e)lung, Strahlen(ab)deckung *f* [*Rauchmeldung*]

observability Beobachtbarkeit *f*

observability matrix Beobachtbarkeitsmatrix *f*

observable optisch merklich, beobachtbar, bemerkbar

observable canonical form Beobachtungsnormalform *f*

observation... → sight...

observing matrix Beobachtungsmatrix *f*

to obtain [*Conditions in a locality*] (vor)herrschen

occupance [*Within a space or room*] Insassen *f*

occupancy [*building*] Belegung, Präsenz *f*

occupancy cycle Belegungszyklus, Präsenzzyklus *m*

occupancy detector, occupancy detection device, occupancy detecting device Anwesenheitsmelder *m*, Anwesenheitsmeldegerät *n*

occupancy number Belegungs(an)zahl, Präsenz(an)zahl *f* [*Gebäude*]

occupancy sensor Präsenzsensor, Belegungssensor *m*

occupant → (building) occupant

occupant → (room) occupant

occupation Belegung *f*

occupation period Belegungszeitraum *m*

occupied mode → **day mode**

occupation setpoint Belegungssollwert, Belegungsvorgabewert *m*

occupied make contact belegte Schließer *m*

occurrence of an event, event occurrence Auftreten eines Ereignisses, Erscheinen eines Ereignisses *n*, Ereignisauftritt *m*, Ereigniserscheinung *f*

OC curve → **operating characteristic curve**

OCR → **optical character recognition**

octahedral iron ore filter → **magnetite filter**

octal oktal [*Zahlendarstellung im Zahlensystem mit der Basis 8 (Oktalsystem)*]

octal system [*A counting system with a base of eight. Used for internal representation of data in computer systems*] Oktalsystem *n*

octave band [*Electronics. A band of frequencies in which the limits have the ratio of 2 to 1; an octave band extends from a lower frequency to twice the lower frequency*] Oktavband *n*

octave band midfrequency Oktavmittenfrequenz, Mittenoktavfrequenz *f*

octave filter Oktavfilter *m, n*

octet string Oktettreihung, Oktettkette *f*

ODBC → **open database connectivity**

odd ung(e)rade(zahlig)

odd corner section → **corner odd section**

odd-even bit → **parity bit**

odour control (Brit.); odor control (US) Geruch(s)-bekämpfung *f*

odourization (Brit.); odorization (US) [*The addition of a chemical substance to gas in order to meet the legal requirement that it can be identified by colo(u)r or smell*] Odorierung *f*

odour-laden (Brit.); odor-laden (US) geruchsbelastet, geruchsbeladen

odour nuisance (Brit.); odor nuisance (US) Geruchsbelästigung *f*

odour threshold (Brit.); odor threshold (US) Geruchsschwelle *f*

OE → **output enable**

OEIC → **optoelectronic integrated circuit**

"off" aus [*Anzeige*]

off, open geöffnet, offen [*Schalter*]

off-balance mass Unwuchtmasse *f*

off cycle [*Period when an equipment or a system is not in operation*] Stillstandzeit *f*

off-delayed rückfallverzögert

offering Aufschalten *n* [*Telephon, z. B. zum Mithören*]

off-hook abgehoben [*Telephonhörer*]

office building lift (Brit.); office building elevator (US) Bürogebäudeaufzug *m*

office computer → **business computer**

office illumination, office lighting Bürobeleuchtung *f*

office landscape Bürolandschaft *f*

office landscape screen [*A fixed or mov(e)able, free-standing, rearrangeable interior space divider; may incorporate acoustical properties*] Bürolandschaftsblende *f*

office printer, office writer Bürodrucker *m*

officer Beauftragte *m*

office security Bürosicherung *f*

office service Bürodienst *m*

official periodical offizielle Zeitschrift *f*

off-limit condition, limit violation Grenzwertüberschreitung *f*

off-line Stapelverarbeitung *f*

off-line, independent leitungsgehemmt, nicht auf Leitung (gestellt), außerhalb der Leitung, indirekt, unabhängig, nicht prozessgekoppelt, nicht verbunden [*Arbeitsprozesse; Maschineneinheiten*]

off-line Datenträgerzwischenverbindung *f* [*Verbindung zum Rechner über Datenzwischenträger*]

off-line equipment → **auxiliary equipment**

off-line operation unabhängige Betrieb *m* [*Verarbeitung auf peripheren Geräten unabhängig von der Zentraleinheit*]

off-line storage → **archiving**

off-loading operator Abräumstation *f* [*Geschirrspülmaschine*]

off-load method Freilastverfahren *n*, Freilastmethode *f*

off-modus niedrigste Strom(ein)spar(ungs)taste *f*

OFF period, OFF time Ausschalt(ungs)dauer, Ausschalt(ungs)zeit *f*

off position → **position of rest**

off-position interlock, zero-position interlock Nullstellungszwang, Ausstellungszwang *m*

off push-button Ausdruckknopf *m*, Austaste(r) *f, (m)*

offset → **displacement**

offset → **steady-state deviation from desired value**

offset Nullpunktverlegung *f* [*A/D- und D/A-Wandler*]

offset absence Kennliniensteigerung *f* gleich Null [*Reg(e)lung*]

offset address Distanzadresse *f*

offset (bend) → **bend offset**

offset conduit nipple Sprungnippel *m*

offset current Fehlstrom *m*

offset hinge → **cleaning hinge**

offset single-compartment sink top, offset one-compartment sink top versetzte Einbeckenspültisch *m* [*Das Spülbecken ist nicht mittig angeordnet*]

offset terminal gekröpfte Anschlussklemme *f*

off switch with spring return Aus-Tastschalter, Tast-Ausschalter *m*

offset switch Korrekturschalter *m*

OFF time → **OFF period**

ogee (roof) gutter, ogee rainwater gutter, ogee R.W. gutter, O.G. (roof) gutter, O.G. rainwater gutter, O.G.R.W. gutter [*An eaves gutter having an upright back and a combined sole and front of ogee shape*] Sima(regen)rinne, Simadachrinne *f* [*Nicht verwechseln mit „(Dach)Rinne mit Sima"*]

OHS, oval(-headed) screw Ovalkopfschraube, Kopfovalschraube *f*

oil [*colloquially*] → **fuel oil**

oil air heater Öl-Lufterhitzer, Öl-Luft(er)wärmer *m*

oil air heating installation, oil air heating plant Öl-Luftheiz(ungs)anlage *f* [*Siehe Anmerkung unter „Luftheiz(ungs)anlage"*]

oil atomization burner Ölzerstäubungsbrenner *m*

oil atomizer; oil atomiser (Brit.) Ölzerstäuber *m*

oil atomizing burner → **atomizing oil burner**

oil bath(-type) filter Ölbadfilter *m,n*

oil blower burner Ölgebläsebrenner, Gebläseölbrenner *m*

oil blower furnace [*See remark under "furnace"*] Öl-Gebläsefeuer(ungs)stätte, Öl-Gebläsefeuerung(sanlage), Gebläse-Ölfeuer(ungs)stätte, Gebläse-Ölfeuerung(sanlage) *f*

oil boiler → **oil(-fired) boiler**

oil burner Ölbrenner *m* [*DIN 4787, Teil 1*]

oil burner/gas burner boiler, gas burner/oil burner boiler Wechselbrandkessel, Brandwechselkessel *m* [*Durch Ein- und Ausschwenken des Öl- oder Gasbrenners kann der Betreiber den Brennstoff wechseln, kein Umbau der Kessel- und Feuerungsteile wie beim Umstellbrandkessel*]

oil burner pump Ölbrennerpumpe *f*

oil-burning → **oil-fired**

oil burner nozzle [*Nozzle for mechanically atomizing fuel oil in a burner*] Öl(brenn)düse, Ölzerstäuberdüse, Druckzerstäuberöldüse, Luftzerstäuberöldüse *f*

oil-burning... → **oil(-fired)...**

oil catcher Ölfang, Ölfänger *m*

oil central distributor → **oil manifold**

oil circulating pump Ölumwälzpumpe, Ölzirkulationspumpe, Ölumlaufpumpe *f*

oil consumption meter Ölverbrauchszähler, Ölverbrauchsmesser *m*

oil control, oil closed-loop control Ölreg(e)lung *f*

oil cooler Ölkühler *m*

oil cooling [*Transfer of heat from oil in pipe or tubing to a refrigerant or brine*] Ölkühlung *f*

oil dashpot Ölpuffer *m*

oil derivative Ölderivat *n*, Ölabkömmling *m*

oil district heating → **district oil heating**

oil domestic hot water heater, oil domestic warm water heater Öl-Brauchwasser(er)wärmer, Öl-Brauchwassererhitzer, Öl-Brauchwasserheizer *m*

oil drip(ping) tray, oil (drain) tray, oil drip(ping) pan, oil (drain) pan Öl(auffang)wanne, Öl(auffang)schale, Öl(ab)tropfwanne , Öl(ab)tropfschale, Ölentwässerungswanne, Ölentwässerungsschale *f*

oil failure switch → **oil pressure cutout**

oil-filled electric radiator → **electric oil-filled radiator**

oil-fired, oil-burning ölbefeuert, ölgefeuert

oil(-fired) absorption chiller, oil(-fired) absorption refrigerating machine, oil-burning absorption chiller, oil-burning absorption refrigerating machine ölgefeuerte Absorptionskältemaschine, ölbefeuerte Absorptionskältemaschine *f*

oil(-fired) absorption heat pump, oil-burning absorption heat pump ölgefeuerte Absorptionswärmepumpe, ölgefeuerte AWP, ölbefeuerte Absorptionswärmepumpe, ölbefeuerte AWP *f*

oil(-fired) heating, oil-burning heating Ölheizung *f*

oil(-fired) hot air heater, oil(-fired) warm air heater, oil-burning hot air heater, oil-burning warm air heater ölbeheizte Warmluftheizer, ölbeheizte Warmluft(er)wärmer *m*

oil(-fired)-only boiler, fuel oil-only boiler, oil-burning-only boiler Spezialölkessel *m* [*DIN 4702, Blatt 1 (Januar 1967). Kessel nur für Öl, also weder zum Brennstoffwechsel noch zur Umstellung auf einen anderen Brennstoff*]

oil(-fired) space heater, oil(-fired) unit heater, oil-burning space heater, oil-burning unit heater Ölraumheizer *m*

O

oil foam burner Ölschaumbrenner *m*

oil(-fired) storage (water) heater, oil-burning storage (water) heater [*See remark under "storage (water) heater"*] Öl-Vorratswasserheizer, ölgefeuerte Vorratswasserheizer *m*

oil(-fired) vapourising burner (Brit.); oil(-fired) vaporizing burner (US) [*"Oil-burning" may synonymously be used for "oil(-fired")*] Öl-Verdampfungsbrenner , ölgefeuerte Verdampfungsbrenner *m*

oil furnace [*See remark under "furnace"*] Ölfeuer(ungs)stätte, Ölfeuerung(sanlage) *f*

oil gasification burner Ölvergasungsbrenner *m*

oil gauge → (fuel) oil gauge

oil heat Ölwärme *f*

oil-heated central heating, central oil-heated heating ölbeheizte Zentralheizung, Öl-Zentralheizung *f*

oil heater Ölheizer *m*

oil heating Ölheizung *f*

oil heating appliance, oil heating device, oil heating apparatus, oil heater, oil fire Ölheiz(ungs)gerät *n*, Ölheizer, Ölheiz(ungs)apparat, Ölerhitzer, Öl(er)wärmer *m*

oil-hydraulic elevator (US); oil-hydraulic lift (Brit.) [*The energy is applied by means of oil under pressure in a cylinder equipped with a plunger or piston*] ölhydraulische Aufzug *m*

oil ignition Ölzündung *f*

oil interceptor → oil separator

oil leak(age) detector, oil leak(age) detecting device, oil leak(age) detection device Ölleck(age)melder *m* , Ölleck(age)meldegerät *n*

oil-less compressor → oil-free compressor

oil level Ölstand *m*

oil level sight glass Ölstandschauglas, Ölstandsichtglas *n*

oil loop [*Loop placed at the bottom of a riser to force oil to travel up the riser*] Ölschleife *f*

oil manifold, oil central distributor Ölverteiler(rohr) *m*, *(n)*

oil mist, atomized oil Ölnebel, Öltröpfchennebel *m*, zerstäubte Öl *n*

oil mist separator, atomized oil separator Ölnebel(ab)scheider *m*

oil pan → oil drip(ping) tray

oil path Ölweg *m*

oil (pipe)line preheater Öl(rohr)leitungsvorwärmer *m*

oil preheater Ölvorwärmer *m*

oil firing Ölfeuerung *f*, Ölfeuern *n*

oil-free compressor, oil-less compressor [*No lubricating oil is used in the cylinder*] öllose Verdichter, öllose Kompressor *m*

oil preheating Ölvorwärmen *n*, Ölvorwärmung *f*

oil pressure burner → pressure oil burner

oil pressure controller Öldruckregler *m*

oil pressure monitoring Öldrucküberwachung *f*

oil pressure regulating valve Öldruckregulierventil *n*

oil pressure switch Öldruckschalter *m*

oil return flow valve Ölrücklaufventil *n*

oil return line Ölrücklaufleitung *f*

oil safety switch Ölsicherheitsschalter *m*

oil scraper ring Ölabstreifring *m*

oil (sectional) radiator [*See remark under "radiator"*] Ölradiator, Ölgliederheiz(ungs)körper *m*

oil pressure cutout, oil failure switch, differential oil pressure switch [*Safety device used to stop the compressor when the oil pressure reaches a preset, abnormally low value*] Öldruckausschalter *m*

oil pressure/steam pressure atomizer; oil pressure/steam pressure atomiser (Brit.) Öldruck/Dampfdruck-Zerstäuber *m*

oil return flow Ölrücklauf *m*, Ölrückführung *f* [*z. B. aus dem Verdampfer zum Verdichter*]

oil return flow (pipe)line Ölrücklauf(rohr)leitung *f*

oil return flow pressure atomizer; oil return flow pressure atomiser (Brit.) Ölrücklaufdruckzerstäuber *m* [*Der Düsenknopf hat ein Ölrücklaufventil*]

oil separator, oil interceptor, oil trap [*See remark under "interceptor"*] Ölabscheider, Ölfänger, Ölfang, Heizölabscheider, Heizölfänger, Heizölfang *m* [*DIN 4043*]

oil service tank Ölbetriebsbehälter, Ölbetriebstank *m*

oil sight glass Ölschauglas *n*

oil solenoid valve Ölmagnetventil *n*

oil storage tank Öllagerbehälter, Öllagertank *m*

oil suction Ölansaugung *f*

oil sump Ölsumpf *m*

oil supply Ölversorgung *f*

oil tank [*It may be a storage tank or a service tank*] Ölbehälter, Öltank *m*

oil thread test Ölfadenprobe, Ölfadenprüfung *f*, Ölfadenversuch *m*

oil throughput Öldurchsatz *m* [*Meistens in kg/Stunde gemessen*]

oil trace Ölspur *f*

oil trap → **oil separator**

oil tray → **oil drip(ping) tray**

oil vapour loss (Brit.); oil vapor loss (US) Öl-Wasserdampf-Verlust *m*

oil water heater, oil water heating appliance[*See remark under "water heater"*] Ölwassererhitzer, Ölwasser(er)wärmer, Ölwasserheizer, Öl(warm)wasserbereiter *m*, Öl(heiz)therme *f*, Ölzentralheizgerät *n*

oil-wetted ölbenetzt

OIS, optical image stabilizer Antiwackelvorrichtung, Gegenwackelvorrichtung *f*

OLE, object linking and embedding Objektaufnahme *f* in Dokumenten

omnidirectional rund(her)um [*für alle Richtungen*]

OMUP, operations and maintenance user part Betriebs- und Wartungsbenutzerteil *m, n* [*Signalisierungssystem Nr. 7*]

on-and-off thermostat [*Thermostat designed to open or close an electric circuit in response to temperature change*] Ein-Aus-Thermostat *m*

on-delay Ansprechverzögerung *f*, Ansprechverzug *m*

on-delayed ansprechverzögert

ondulation Wellung *f*

ON duration → **ON period**

one-level code → **absolute programming**

one-to-one function → **bifunction**

on-hand building quantity → **building stock**

on-hand quantity, inventory, stock (Lager)Bestand, Magazinbestand *m*

online direkte Verbindung, Anbindung, unmittelbare Verbindung *f*, direkte Anschluss *m* [*Arbeitsprozesse; Maschineneinheiten*]

online lebendig, speicherfrei

online [*Pertaining to equipment capable of interacting with a computer*] rechnerangeschlossen, rechnerverbunden, rechnergekoppelt

online leitungsverbunden, leitungsangeschlossen, leitungsgekoppelt

online direkt (gekoppelt), direkt verbunden, direkt angebunden, unmittelbar verbunden, unmittelbar angebunden, unmittelbar (gekoppelt) [*Arbeitsprozesse; Maschineneinheiten*]

online equipment, online device [*The equipment or devices in a system whose operation is under control of the central processing unit, and in which information reflecting current activity is introduced into the data-processing system as soon as it occurs*] fest angeschlossene Gerät *n*, fest angeschlossene Vorrichtung *f*

online library rechnerangeschlossene Bibliothek *f*

online mode [*In automatic data processing, a mode of operation in which all devices are responsive to the central processor*] Ansprechbetriebsart *f*, Ansprechbetriebsmodus *m*

on-line ((open) loop) operation, direct digital ((open) loop) operation offene Prozesskopp(e)lung *f*

on-off ((closed) loop) control → **double step ((closed) loop) control**

on-off code Ja-Nein-Kode, Ja-Nein-Code *m*

on-off controller → **switching controller**

on-off differential hysteresis → **differential travel "H"**

on-off switch → **power switch**

on-off switching Ein-Aus-Schaltung *f*, Ein-Aus-Schalten *n*

on-off switch with spring return Tast-Ein-Aus-Schalter *m*, Ein-Aus-Tastschalter

on-line [*The state in which a piece of equipment or a subsystem is connected and powered to deliver its proper output to the system*] Systemintegration *f*

on-line connection Direktanschluss *m*, Direktverbindung *f* [*Über eine Drahtleitung zum Rechner*]

on-line disk file, on-line disc file [*A magnetic disk directly connected to the central processing unit (CPU), thereby increasing the memory capacity of the computer*] fest angeschlossene Plattendatei *f*

on-line operation [*Computer operation in which input data are fed into the computer directly from observing instruments or other input equipment, and computer results are obtained during the progress of the event*] schritthaltende Betrieb, prozessgekoppelte Betrieb, prozessverbundene Betrieb, prozessungebundene Betrieb *m*

on-line process interfacing, on-line process interface connection direkte Prozesskopp(e)lung *f* [*Eingabe- oder Ausgabedaten werden ohne Handgriff übertragen. DIN 66 201, Teil 1*]

on-off ((open) loop) control Ein/Aus-Steuerung *f*

on-off operation An-Aus-Betrieb, Ein-Aus-Betrieb *m* [*Betrieb ohne Zwischenstellungen*]

on-off signaling (US); on-off signalling (Brit.) Amplitudenmodulation *f*

ON period, ON duration, period of current, current period, current duration Einschalt(ungs)dauer *f* [*Stromkreis*]

on-position Ein-Stellung *f*

(on-)screen monitoring (Bild)Schirmüberwachung *f*

on(-)site incineration An-Ort-Müllverbrennung *f*

on-size error Überlauf *m* [*COBOL*]

O

on-state current Durchlassstrom *m* [*Thyristor. DIN 41786*]

on-state slope resistance Ersatzwiderstand *m* [*Thyristor. DIN 41786*]

on-state voltage Durchlassspannung *f* [*Thyristor. DIN 41786*]

on-stream desludging of cooling systems Reinigung *f* von Kühlanlagen während des Betriebs, Entschlammen *n* von Kühlanlagen während des Betriebs

on-switch with spring return Eintastschalter, Tasteinschalter *m*

on-tap, ready for tapping, ready for being tapped zapfbereit, zapffertig

on-the-fly printer [*In automatic data processing, a high-speed line printer using continuously rotating print wheels and fast-acting hammers to print the letters contained on one line of text so rapidly that the characters appear to be printed simultaneously*] Drucker mit drehender Typenwalze, Typenwalzedrucker *m*

on the suction side → suction-side

ON time, time of current, current time Einschaltungszeit *f* [*Stromkreis*]

OP → operator panel

OPA → operation(al) amplifier

opal-flashed → flashed-opal

opcode, operation code [*The pattern of bites within an instruction that specify the operations to be performed*] Operationskode, Operationscode *m* [*E DIN 44300, 10.85*]

O

open → off

open drucklos, offen [*z. B. Wasser(er)wärmer; Badeofen; Behälter*]

openable öffnungsbar, öffnungsfähig

open action flow offene Wirkungsablauf *m*

open action path offene Wirkungsweg *m*

open air Freiluft *f*

open air mounting, open air erection Aufstellung im Freien, Freiaufstellung *f*

open air storage Freiluftlagerung, oberirdische Lagerung im Freien *f*

open air switchgear Freiluftschaltanlage *f*

open area, open ground Freigelände *n*

open area design Freiflächenentwurf *m*, Freiflächengestaltung *f*

open area monitoring Freigeländeüberwachung, Außenüberwachung *f* [*Geländeüberwachung außerhalb allseits umschlossener Bauten*]

open area monitoring system, open ground monitoring system Freigeländeüberwachungssystem *m*

open area protection, open ground protection Freilandschutz *m*

open area security, open ground security Freigeländesicherung *f*

open cabinet, open cubicle offener Schrank *m*

open circuit Unterbrechung *f*

open circuit voltage Leerlaufspannung *f* [*Einer beleuchteten photovoltaischen Zelle oder eines Moduls, wenn kein Strom fließt. Die Leerlaufspannung ist die größtmögliche Spannung der Zelle oder des Moduls*]

open circuit water cooling Mischwasserkühlung *f*

open collector driver offene Kollektortreiber, Treiber mit offenem Kollektor *m*

open collector output offene Kollektorausgang, Ausgangs-Transistor mit offenem Kollektor *m*

open combustion chamber offene Verbrennungskammer *f*

open communication [*ISO 7498*] offene Kommunikation *f*

open (communication) system [*ISO 7498*] offene Kommunikationssystem *n*

open condensate return → vented condensate return

open compressor offene Verdichter, offene Kompressor *m* [*Die bewegten Teile (Kurbelwellenende oder Kolbenstange) befinden sich außen und müssen sicher abgedichtet werden, um Kältemittelverluste und Eindringen von Feuchtluft zu verhindern*]

(open) control → ((open) loop) control

open (control) loop [*A signal path without feedback*] offene Regelkreis *m*, Steuerung *f*

open control signal → ((open) loop) control signal

open database connectivity, ODBC Zugriffsschnittstelle *f* [*Sie erleichtert den Zugriff aus einem Anwendungsprogramm auf unterschiedliche Datenbankensysteme*]

open discharge safety valve offene Sicherheitsventil, Sicherheitsventil ohne Dampfabblasestutzen *n*

"open door" monitoring „Tür offen"-Überwachung *f* [*Überwachung der über einen einstellbaren Zeitraum hinaus geöffneten Tür(en)*]

open-end(ed) coil offene Spule *f*

open-end(ed)line leerlaufende Leitung *f*

open-end(ed) system [*In character recognition, a system in which the input data to be read are derived from sources other than the computer with which the character reader is associated*] offene Zeichenerkennungssystem *n*

opener Öffner *m*

open expansion vessel offene Ausdehnungsgefäß *n* [*Sonnenheiz(ungs)anlage. Ein mit dem Kollektorkreislauf und der Atmosphäre unabsperrbar verbundener Behälter. Er nimmt die beim Aufheizen entstehende Volumenänderung des Wärmeträgers auf. DIN 30 600, Reg.-Nr. 06172*]

open fire offene Brand *m*, offene Feuer *n*

open-flame furnace → direct-fired furnace

open flow channel Freispiegel(ge)rinne *f, (n)*

open gravity water heating offene Schwerkraftwasserheizung *f*

open ground → open area

open ground monitoring system → open area monitoring system

open ground security → open area security

opening coil Ausschaltspule, Abschaltspule *f*

opening contact Öffnungskontakt *m*

opening handle Öffnungsgriff *m*

opening indicator Öffnungsanzeiger *m*

open heating [*A heating piping system in which the circulating water return main is connected to an open-vented elevated tank which serves as a reservoir to accommodate expansion and contraction of the fluid, and as an inspection point for the condition of the fluid*] offene Heizung *f*

opening monitoring Öffnungsüberwachung *f*

open(ing) position Offenstellung *f* [*Schlüssel*]

opening rods Öffnungsgestänge *n* [*hochliegende Kippfenster*]

open hot water heating, open wet heating, open hydronic heating offene Warmwasserheizung, offen WWH *f* [*Das Ausdehnungsgefäß steht mit der Atmosphäre in Verbindung, so dass eine unzulässige Drucksteigerung im Rohrnetz nicht auftreten kann*]

opening detector, opening detecting device, opening detection device Öffnungsmelder, Öffnungsdetektor *m*, Öffnungsmeldegerät *n*

opening direction, direction of opening Öffnungsrichtung, Auf-Richtung *f*

opening spring Öffnungsfeder *f*

opening temperature → trip(ping) temperature

opening threshold Öffnungsschwelle *f*

opening time → trip(ping) time

opening travel, opening path Öffnungsweg, Öffnungspfad *m* [*Kontakt*]

opening valve Öfnungsventil *n*

open interface offene Schnittstelle, offene Anpassungsschaltung *f* [*Siehe Anmerkung unter „Schnittstelle"*]

open line Leitungsunterbrechung *f*

open loop → open (control) loop

(open loop and closed loop) control circuits Steuer(ungs)- und Regelkreise, MSR *f* [*VDI/VDE 3695*]

((open) loop) boiler control → boiler ((open) loop) control

((open) loop) control Steuerung *f*, Steuern *n* [*Ein Vorgang in einem System, bei dem eine oder mehrere Größen als Eingangsgrößen andere Größen als Ausgangsgrößen aufgrund einer vorgegebenen Gesetzmäßigkeit des Systems beeinflussen. Der hierzu erforderliche Wirkungsablauf vollzieht sich über eine Steuer(ungs)kette oder ein Übertragungsglied*]

((open) loop) control algorithm Steuer(ungs)algorithmus *m*

((open) loop) control block Steuer(ungs)block *m* [*Er beinhaltet die kodierten numerischen Werte, die von der Maschine über die Steuerung in Schalt- und Bewegungsvorgänge umgewandelt werden*]

((open) loop) control bus Steuer(ungs)bus *m* [*Mikrorechner. Über ihn teilt zum Beispiel die Zentral-einheit den Speichern mit, ob gelesen oder geschrieben werden soll. Zum Steuer(ungs)bus gehören auch die Unterbrechungs- und Taktleitungen*]

((open) loop) control bit, ((open) loop) control binary digit Steuer(ungs)binärzeichen, binäre Steuer(ungs)zeichen, Steuer(ungs)bit, SB *n*, Steuer(ungs)binärstelle, Steu-er(ungs)dualstelle *f* [*Siehe Anmerkung unter „Binärzeichen"*]

((open) loop) control chain Steuer(ungs)kette *f*

((open) loop) control circuit Steuer(ungs)kreis *m*

((open) loop) control clock Steuer(ungs)uhr *f*

((open) loop) control command → control ((open) loop) command

((open) loop) control design engineer Steuer(ungs)konstrukteur, Steuer(ungs)bauer *m*

((open) loop) control device Steuer(ungs)vorrichtung *f* [*Gesamtheit der Bauglieder, welche die aufgabenmäßige Beeinflussung der Steuer(ungs)strecke bewirkt*]

((open) loop) control element Steuer(ungs)glied *n*

((open) loop) control engineering Steuer(ungs)technik *f*

((open) loop) control equipment Steuer(ungs)gerät(e) *n, (f)*

((open) loop) control frame Steuer(ungs)block *m* [*Er enthält keine Nutzdaten, sondern dient der Übertragungssteuerung*]

((open) loop) control function [*A term used to describe the form of control, e. g. compensation, night setback, etc.*] Steuer(ungs)funktion *f*

O

((open) loop) control gas valve Steuer(ungs)gasventil *n*

((open) loop) control identifier Steuer(ungs)kennzeichen, Identitätskennzeichen *n*

((open) loop) control information Steuer(ungs)information *f*

((open) loop) control instruction Steuer(ungs)befehl *m*

((open) loop) control interconnection Steuer(ungs)verbund *m*

((open) loop) controllability Steuer(ungs)fähigkeit, Steuerbarkeit *f*, Steuer(ungs)vermögen *n*

((open) loop) controllable steuerbar

((open) loop) controlled gesteuert

((open) loop) controlled mercury-arc rectifier gesteuerte Quecksilberdampfgleichrichter *m*

((open) loop) controlled system → plant

((open) loop) controlled temperature profile gesteuerte Temperaturprofil *n*

((open) loop) controlled variable [*Any variable that is controlled in an open-loop system*] Steuer(ungs)größe *f*

((open) loop) controller Steuer(ungs)gerät *n*

((open) loop) controlling Steuern *n*

((open) loop) control mechanism Steuer(ungs)mechanismus *m*

((open) loop) control mode Steuer(ungs)betrieb *m*, Steuer(ungs)art *f*

O

((open) loop) control network Steuer(ungs)netz(werk) *n*

((open) loop) control output Steuer(ungs)ausgangsgröße *f*

((open) loop) control parameter Steuer(ungs)kennwert, Steuer(ungs)parameter *m*

(open) loop) control process → control process

((open) loop) control program (US); ((open) loop control programme (Brit.) Programm einer Steuerung, Steuer(ungs)programm *n* [*Die Gesamtheit aller Anweisungen und Vereinbarungen für die Signalverarbeitung, durch die eine zu steuernde Anlage (Prozess) aufgabengemäß beeinflusst wird. E DIN 19 227, Teil 5 und V DIN 19 237*]

((open) loop) control protocol Steuer(ungs)protokoll *n*

(open) loop) control range Steuer(ungs)bereich, Nenneinstell(ung)sbereich *m* [*selten: n*] [*DIN 41 745*]

(open) loop) control relay Steuer(ungs)relais *n*

((open) loop) control section Steuer(ungs)bereich *m*, [*Ein Funktionsbereich eines Vermittlungsmoduls*]

((open) loop) control sequence Steuer(ungs)ablauf, Ablauf von Steuer(ungs)vorgängen *m*

((open) loop) control signal, open control signal, CS [*A set of (im)pulses used to identify the channels to be followed by transferred data*] Steuer(ungs)signal *n*

((open) loop) control signal message Steuer(ungs)meldung *f*

((open) loop) control store Steuer(ungs)speicher *m*

(open) loop) control system Steuerungssystem *n* mit offenem Wirkungsablauf, Steuerkette *f* [*Die Ausgabe – der Istwert – kann mit der Eingabe – dem Sollwert – nicht verglichen werden*]

((open) loop) control panel, ((open) loop) control board Steuer(ungs)tafel, Zentrale *f*, Steuer(ungs)tableau *n*

((open) loop) control terminal Steuer(ungs)terminal *n*

((open) loop) control valve Steuer(ungs)ventil *n*

((open) loop) control variable Steuer(ungs)größe *f*

(open) loop dead time Steuer(ungs)totzeit, Totsteuer(ungs)zeit *f*

((open) loop) error signal Steuer(ungs)differenz *f*, Steuer(ungs)unterschied *m*

((open) loop) frequency response Frequenzgang *m* des offenen Regel(ungs)kreises

open-loop gain offene Regelkreisverstärkung *f*

((open) loop) group control level → group open loop control level

((open) loop) level control Niveausteuerung, Standsteuerung *f* [*Flüssigkeit*]

open loop operation offene Betrieb *m* [*Die in einem Prozess entnommenen Daten werden erst nach Entscheidung zu Eingriffen in den Betrieb verwendet*]

((open) loop) output control, output (open) loop control Ausgabesteuerung *f*

open loop path offene Wirkungsweg *m*

((open) loop) position control, (open) loop positioning Lagersteuer(ungs)kreis *m*

((open) loop) project control Projektsteuerung *f*

((open) loop) power control, power (open) loop control Leistungssteuerung *f*

((open) loop) pump control Pumpensteuerung *f*

((open) loop) speed control, speed (open) loop control Drehzahlsteuerung *f*

((open) loop) telecontrol, open telecontrol Fernsteuerung *f*, offene Fernwirkungsablauf *m*

(open) loop time Steuer(ungs)zeit *f*

((open) loop) timer control → timer ((open) loop) control

((open) loop) timing-pulse control, timing-pulse ((open) loop) control gesteuerte Taktgebung, Taktsteuerung *f*

((open) loop) valve-controlled ventilgesteuert

(open) loop variable [*See remark under "variable"*] Steuer(ungs)größe *f*

open-outlet water heater Kochendwassergerät *n*

open position Offenstellung, geöffnete Stellung, Ausschaltstellung, offene Stellung *f*

open pump heating offene Pumpenheizung *f*

open shell-and-tube condenser [*Condenser in which the water passes in a film over the inner surfaces of the tubes, which are open to the atmosphere*] offene Rohrbündelverflüssiger, offene Rohrbündelkondensator, offene Bündelrohrverflüssiger, offene Bündelrohrkondensator *m*

open shop [*A data-processing central organization in which individuals from outside the data-processing community are permitted to implement their own solutions to problems*] Kunden-Datenzentrum *n*

open solar heating plant, open solar heating installation, open sun heating plant, open sun heating installation offene Sonnenheiz(ungs)anlage, offene Solarheiz(ungs)anlage *f* [*Sie hat eine unabsperrbare Verbindung zur Atmosphäre, so dass der zulässige Gesamtüberdruck auch ohne Sicherheitsvorrichtung nicht überschritten wird*]

open space, free space Freiraum *m*

open sprinkler offene Sprinkler, offene Sprenger *m* [*Sprühwasserlöschanlage*]

Open Systems Interconnection → OSI

open telecontrol → ((open) loop) telecontrol

open thermodynamic, thermodynamic open system [*In thermodynamics, a system across whose boundaries both matter and energy may pass*] offene thermodynamische System, thermodynamische offene System *n*

open throughout durchgehend offen

open vessel → kettle

open voltage Auslösespannung *f*

open water heater, open water heating appliance offene Wassererhitzer, offene Wasser(er)wärmer, offene Wasserheizer, offene Warmwasserbereiter *m* [*Auslauf stets offen*]

open water heating offene Wasserheizung *f* [*Sie steht über einer unverschließbaren Öffnung an der höchsten Stelle der Anlage in direkter Verbindung mit der Atmosphäre*]

open window unit → sabin

to open with the picklock → to pick with the picklock

operability Betriebsfähigkeit *f*

operable betriebsfähig

operand Operand *m* [*Information oder Rechnergröße. Bei speicherprogrammierbaren Steuerungen: Ausgang, Eingang, Merker, Zeitglied innerhalb einer Anweisung.*]

operand address Operandenadresse *f*

operand(-precision) register [*In automatic data processing, a special register found in some minicomputers which can be programmed from 8- to 32-bit precision*] Operandregister *n*

to operate → to run

to operate → to attend

operate delay, response delay Ansprechverzögerung *f*, Ansprechverzug *m* [*Relais*]

operated from floor, floor-operated (fuß)bodenbedient, (fuß)bodenbetätigt

operated in the switching mode → clocked

to operate in the switching mode → to clock

operate time [*Electricity. Total elapsed time from application of energizing current to a relay coil until the time the contacts have opened or closed*] Ansprechzeit, Reaktionszeit, Impulszeit *f*

operate time [*In automatic data processing, the phase of computer operation when an instruction is being carried out*] Befehls(ausführungs)zeit *f*

operate voltage, operate tension Ansprechspannung, Auslösespannung, Funktionsspannung *f* [*RSA-Empfänger*]

operating Bedienen *n*

operating amplitude → response amplitude

operating at full capacity, fully operating voll ausgelastet, ganz ausgelastet

operating characteristic → operating feature

operating characteristic curve, OC curve, operating feature curve [*In hypothesis testing, a plot of the probability of accepting the hypothesis against the true state of nature*] Annahmekennlinie, Annahmekennkurve *f*

operating cycle [*In automatic data processing, the portion of a memory cycle required to perform an operation; division and multiplication usually require more than one memory cycle to be completed*] (Betriebs)Gang *m*, (Betriebs)Takt, (Betriebs)Zyklus *m*, (Betriebs)Spiel *n*

operating data Betriebsdaten *f* [*Daten zur Darstellung der für die Betriebsleitung relevanten technischen und organisatorischen Größen eines Betriebes. DIN 66201, Teil 1*]

operating delay [*Computer time lost because of mistakes or inefficiency of operating personnel or users of the system, excluding time lost because of defects in program(me)s or data*] Betriebsverzögerung *f*, Betriebsverzug *m*

operating distance "S", sensing distance Schaltabstand „S" *m* [*Näherungsschalter*]

operating environment Bedien(ungs)oberfläche *f*

operating equipment Bedien(ungs)gerät(e) *n, (f)*

operating factor relative Einschalt(ungs)dauer *f* [*Relais*]

operating feature, operating characteristic Betriebsmerkmal *n*

operating gangway [*IEC 439-1*] Bedien(ungs)gang *m* [*VDE 0660, Teil 500*]

operating hierarchy Bedien(ungs)hierarchie *f*

operating I^2+ Abschalt-I^2+ *n* [*Schmelzsicherung*]

operating instruction [*A detailed description of the actions that must be carried out by a computer operator in running a program(me) or group of interrelated program(me)s, usually included in the documentation of a program(me) supplied by a programmer or system analyst, along with the source program(me) and flow charts*] Betriebsbefehl, Bedienungsbefehl *m*

operating interval Einschaltzeit *f* [*Schaltbetrieb*]

operating manual Betriebshandbuch, Bedienungshandbuch *n*

operating outdoor temperature → outdoor operating temperature

operating part → operator

operating point [*Point on a family of characteristic curves of a vacuum tube or transistor where the coordinates of the point represent the instantaneous values of the electrode voltages and currents for the operating conditions under study or consideration*] Betriebspunkt *m*

operating position [*Terminal of a communication channel which is attended by an operator; usually the term refers to a single operator, such as radio operator's position or a telephone operator's position; however, certain terminals may require more than one operating position*] Endgerät(e)arbeitsplatz, Bedienungsplatz *m*

operating pressure Betriebsdruck *m*

operating principle Betriebsgrundsatz *m*

operating ratio, computer ratio [*In automatic data processing, the time during which computer hardware operates and gives reliable results divided by the total time scheduled for computer operation*] Rechnerverhältnis *n*

operating region, operating range [*plant*] Betriebsbereich *m* [*selten: n*]

operating reliability Betriebszuverlässigkeit *f*

operating sequence [*Sequence of specified operations and time intervals*] Schaltreihe, Schaltfolge *f*

operating sound pressure level Betriebsschalldruckpegel *m*

operating state [*e.g. "ON"*] Betriebszustand *m*

operating stroke Gleitstrecke *f*

operating switch, service switch Betriebsschalter *m*

operating system, OS, executive system Betriebssystem, (Rechner)Grundprogramm, BS *n*, Systemsoftware *f* [*Programmpaket, das die Bearbeitung von Programmen durch den Rechner ermöglicht und die Abwicklung von Programmen steuert und überwacht. DIN 44 300*]

operating system supervisor, executive system supervisor [*The control program(me) of a set of program(me)s which guide a computer in the performance of its tasks and which assist the program(me) with certain supporting functions*] Betriebssystemsteuer(ungs)programm *n*

operating temperature Betriebstemperatur, Auslösetemperatur *f* [*Sprinkler*]

operating theatre [*hospital*] Operationsraum *m*

operating threshold → response threshold

operation, run, service Betrieb *m*

operation [*data processing*] Arbeitsgang, Rechengang, Betriebsgang *m*, Operation, Rechenoperation *f*

operation Bedienung *f*

operation Schaltspiel *n* [*Rundsteuer(ungs)anlageempfänger, VDE 0420*]

operation(al) amplifier, OPA, Op-Amp [*An amplifier having high direct-current stability and high immunity to oscillation, generally achieved by using a large amount of negative feedback; used to perform analog computer function such as summing and integrating*] Operationsverstärker, OPV *m*

operation(al) authorization Betriebsberechtigung *f*

operation(al) calculus, operation(al) analysis [*A technique by which problems in analysis, in particular differential equations, are transformed into algebraic problems, usually the problem of solving in polynomial equation*] Operatorenrechnung *f*

operation(al) characteristic (curve) Betriebskennlinie, Arbeitskennlinie *f*

operation(al) code Operationsschlüssel *m* [*Darstellung des mit dem Befehl jeweils geforderten Verarbeitungsschrittes*]

operation(al) collective display → **collective operation(al) display**

operation(al) counter [*IEC 214*] Zählwerk *n*, Zählmechanismus *m*

operation(al) decoder [*In automatic data processing, a device that examines the operation contained in an instruction of a computer program(me) and sends signals to the circuits required to carry out the operation*] Arbeitsentkodierer, Betriebsentkodierer *m*

operation(al) enabling Bedien(ungs)freigabe *f*

operation(al) engineering Betriebstechnik *f*

operation(al) fault, operation(al) malfunction Betriebsfehler, Arbeitsfehler *m*, Betriebsstörung, Arbeitsstörung *f*

operation(al) fire protection measures betriebliche Brandschutzmaßnahmen, betriebliche Feuerschutzmaßnahmen *f*

operation(al) gain → **((closed) loop) gain**

operation(al) interruption Betriebsunterbrechung, Arbeitsunterbrechung *f*

operation(al) keyboard, operation(al) keypad Bedien(er)tastatur, Bedienungstastatur *f*

operation(al) label [*In automatic data processing, a combination of letters and digits at the beginning of the tape which uniquely identify the tape required by the system*] Bandkennung *f*

operationally reliable betriebszuverlässig, arbeitszuverlässig

operation(al) platform Bedien(ungs)bühne *f*

operation(al) rated current → **rated operation(al) current**

operation(al) readiness Betriebsbereitschaft, Arbeitsbereitschaft *f*

operation(al) reliability Betriebszuverlässigkeit, Arbeitszuverlässigkeit *f*

operation(al) report Betriebsbericht, Arbeitsbericht *m*

operation(al) research Betriebsforschung, Arbeitsforschung *f*

operation(al) safety Betriebssicherheit, Arbeitssicherheit *f*

operation(al) security Betriebssicherung, Arbeitssicherung *f*

(operation(al)) serviceability Betriebsfähigkeit *f*, Betriebsvermögen *n*

operation(al) signal message Betriebsmeldung *f* [*Siehe Anmerkung unter „Meldung"*]

operation(al) standby program (US); operation(al) standby programme (Brit.) [*The program(me) operating in the standby computer when in the duplex mode of operation*] Betriebsreserveprogramm, Reservebetriebsprogramm, Arbeitsreserveprogramm, Reservearbeitsprogramm *n*

operation(al) system Betriebssystem, betriebstechnische System, Arbeitssystem, arbeitstechnische System *n*

operation(al) voltage → **running voltage**

operation amplifier Operationsverstärker, Opamp *m*

operation and maintenance user part → **OMUP**

operation code → **opcode**

operation console [*Equipment which provides for manual intervention and monitoring computer operation*] Bedien(er)pult, Bedienungspult *n*

operation data detection Arbeitsdatenerfassung, Arbeitsdatenerkennung, Betriebsdatenerfassung, Betriebsdatenerkennung *f*, Betriebsdatennachweis, Arbeitsdatennachweis *m* [*Anwendung eines Prozessrechensystems zur Durchführung der erforderlichen Maßnahmen, um Betriebsdaten in maschinell verarbeitungsfähige Formen am Ort ihrer Verarbeitung bereitzustellen. DIN 66 201, Teil 1*]

operation department Betriebsabteilung, Arbeitsabteilung *f*

operation hour → **running hour**

operation hours counting → **running hours counting**

operation in parallel with bus Sammelschienenparallellauf, Parallesammelschienenlauf *m*

operation interruption Betriebsunterbrechung, Arbeitsunterbrechung *f*

operation monitoring Betriebsüberwachung, Arbeitsüberwachung, Operationsüberwachung *f*

operation number [*Number designating the portion of an operation, or its equivalent subroutine, in the sequence of operations composing a routine*] Betriebsgangnummer, Arbeitsgangnummer, Operationszahl *f*

operation part [*In automatic data processing, that portion of a digital computer instruction which is reserved for the operation code*] Betriebsteil, Arbeitsteil *m, n*

operation path Operationspfad *m*, logische Kette *f*

operation position Betriebsstellung, Arbeitsstellung *f*

operation quietness → **quietness of operation**

operation readiness Betriebsbereitschaft *f*

operation register [*In automatic data processing, a register used to store and decode the operation code for the next instruction to be carried out by a computer*] Betriebsregister, Arbeitsregister *n*

operation sequence → **sequence of operations**

operations manual, operations handbook Betriebshandbuch, Arbeitshandbuch *n*

O

operation(s)-modifying adapter, adapter to modify operation(s) [*Also spelled "adaptor"*] Kodierteil *m,n* für Funktionsänderung

operations science Betriebswissenschaft, Arbeitswissenschaft *f*

operations sequence chart Schaltfolgeprogramm *n* [*DIN 40 719, Teil 11*]

operation state, operation status Betriebsstatus, Betriebszustand, Arbeitsstatus , Arbeitszustand *m*

operation time, work(ing) time Betriebszeit, Arbeitszeit *f*

operative temperature → effective temperature

operator, operating part Bedien(ungs)teil *m, n*

operation time [*In automatic data processing, the time elapsed during the interpretation and execution of an arithmetic or logic operation by a computer*] Bearbeitungszeit *f*

operator Operator *m* [*Symbol, das die arithmetische oder logische Operation darstellt*]

operator [*A person whose duties include operation, adjustment, and maintenance of a piece of equipment*] Bediener, Operateur *m*

operator, owner Betreiber *m* [*Die für den Betrieb einer Anlage verantwortliche Verwaltung bzw. die für den Betrieb fachkundige und verantwortliche Person*]

operator → (building) operator

operator access privilege criterium Bedienerzugriffskriterium *n*

O

operator action, operator activity, operator intervention Eingriff *m*, Eingreifen *n* [*Das Einwirken des Bedieners auf die Leiteinrichtung oder Stellglieder des Prozesses. DIN 19 222*]

operator activity logbook Bedien(er)tätigkeitsliste *f*, Protokoll der Bedienerhandlungen *n* [*Aufzeichnungen der Handlungen des Bedieners mit einem GAS (Gebäudeautomationssystem). Im Minimum sind dies die An- und Abmeldezeiten*]

operator and monitoring station → monitoring and operator station

operator and monitoring unit → monitoring and operator unit

operator authentication [*The corroboration that the operator logging on to a device is as claimed*] Bedien(ungs)teilechtheit *f*

operator board → operator panel

operator communication level, operator control level Bedien(ungs)ebene *f*

operator-conscious bedienerbewusst, bedienungsbewusst

operator elevator (US); operator lift (Brit.) Bedieneraufzug, Führeraufzug *m*

operator-entry-completed bit Bedien(ungs)ausführbit *n* [*programmierbares Steuer(ungs)gerät*]

operator fatigue Bedienerermüdung *f*

operator-friendly bedienerfreundlich, bedienungsfreundlich

(operator) guid(anc)e → (operator) prompting

operator intervention → operator action

operator position user part → OPUP

operator ID, operator identity Bedienerkennung *f* [*Im Zusammenhang mit Passwort benutzt zur Identifizierung eines bestimmten Bedieners. In Abhängigkeit der Bedienerkennung können bestimmte Dienste für einen Bediener freigegeben oder gesperrt werden*]

operator interruption [*In automatic data processing, a step whereby control is passed to the monitor, and a message, usually requiring a typed answer, is printed on the console typewriter*] Bedienungsunterbrechung *f*

operator keyboard, operator keypad Bedien(er)tastatur, Bedienungs-Tastatur *f*, Bedien(er)tastenfeld, Bedienungs-Tastenfeld *n*

operator not working → automatic switch to night service

operator panel, OP, operator board, operator field Bedien(er)feld, Bedienungsfeld *n*

(operator) prompting, (operator) guid(anc)e Bedien(er)führung, Bedienungsführung *f* [*Neben einer direkten Eingabemöglichkeit alphanumerischer Adressen kann die Auswahl von Adressen mittels Frage- und Antwortspiel erfolgen. Nach Eingabe des ersten Adresskriteriums werden hierzu alle Möglichkeiten mit einem Codezusatz (Ziffer oder Buchstabe) angezeigt. Nach Eingabe einer dieser Ziffern (oder Buchstaben) wird diese und alle Möglichkeiten des nächsten Adresskriteriums mit Codezusatz angezeigt. Entsprechend der Zahl der Anwahlkriterien (z. B. Ort-Gewerk-Funktion-Information) kann so in wenigen Eingabeschritten eine Information angewählt werden. In der gleichen Art können auch Parametrierungen vorgenommen werden, d. h. nach Aufruf einer Adresse werden die einzelnen Parametereingabemöglichkeiten angezeigt. mittels eines blinkenden Zeichens werden die Stellen gekennzeichnet, an denen Parameter einzugeben sind. Nach jeder Eingabe springt das blinkende Zeichen selbsttätig zur nächsten Eingabemöglichkeit. Bei nicht zulässigen Eingaben erfolgt ein Hinweis auf dem Bildschirm. VDI 3814 Blatt 1 (neu)*]

operator station, OS Bedien(ungs)- und Beobachtungssytem (= B + B) *n* mit graphischer Darstellung

operator terminal, service terminal Bedien(er)terminal, Bedienungsterminal *n*

operator unit Bedien(er)einheit, Bedienungseinheit *f*

operator (work)station Bedien(er)station, Bedienungsstation, Leitstation *f*, Leitplatz, Bedien(er)-platz, Bedienungsplatz *m*

opp., opposite entgegengesetzt [*Beschriftung im Zeichnungsschriftfeld*]

opposite side [*The side of a building opposite of the wind pressure side*] Abströmseite, Winddruckgegenseite *f*

optical alarm device optische Alarmgeber, optische Signalgeber *m*, optische Alarmier(ungs)vorrichtung *f* [*Blitzleuchten, Lampen, Rundumkennleuchten, usw.*]

optical cable Lichtleiterkabel, Lichtwellenkabel *n*

optical card optische Speicherkarte *f*

optical character reader Klarschriftleser *m*

optical character recognition, electrooptical character recognition, OCR [*In automatic data processing, that branch of character recognition concerned with the automatic identification of hand-written or printed characters by any of various photoelectric methods*] optische (Schrift)Zeichenerkennung *f*

optical communication [*The use of electromagnetic waves in the region of the spectrum near visible light, for the transmission of signals representing speech, pictures, data pulses, or other information, usually in the form of a laser beam modulated by the information signal*] optische Kommunikation *f*

optical comparator Optiklehre, Lehre mit Optik *f*

optical computer [*It uses varies combinations of holography, lasers, and mass-storage memories for such applications as ultra-high speed signal processing, image deblurring, and character recognition*] optische Rechner *m*

optical coupler → **optocoupler**

optical decoder kodierte Drehgeber, codierte Drehgeber *m* [*Gerät zur Umwand(e)lung analoger Daten, die einen bestimmten Winkel angeben, in kodierte, digitale Daten*]

optical density [*The logarithm to the base 10 of the reciprocal of the transmission factor*] optische Dichte *f*

optical detector, optical detecting device, optical detection device optische Melder *m*, optische Meldegerät *n*

optical disk, optical disc optische Speicherplatte *f*, Rundscheibendatenträger *m* [*Datenträger als runde Scheibe auf die physikalisch unterschiedliche Informationen mit Laserstrahl aufgezeichnet und abgelesen werden*]

optical distance → **optical path**

optical drive optische Laufwerk *n*

optical efficiency optische Wirkungsgrad *m*

optical electronic integrated circuit → **optoelectronic integrated circuit**

optical electronic semiconductor → **optoelectronic semiconductor**

optical encoder optische Kodierer, optische Codierer, Kodierer mit photoelektrischer Abtastung, Codierer mit photoelektrischer Abtastung *m*

optical exchangeable disk, optical exchangeable disc, optical cartridge disk, optical cartridge disc optische (Speicher)Wechselplatte *f*

optical external alarm, external optical alarm optische Externalarm, optische örtliche Alarm, örtliche optische Alarm *m*

optical fibre, light guide; optical fiber (US) [*A long thin thread of fused silicia, or other transparent substance, used to transmit light*] Lichtwellenleiter, LWL *m*

optical-fibre cable, fibre waveguide, optical waveguide [*US spelling is "fiber". A waveguide in which a light-transmitting material such as a glass or plastic fibre is used for transmitting information from point to point at wavelengths somewhere in the ultraviolet, visible-light, or infrared portions of the spectrum*] Lichtwellenleiterkabel, LWL-Kabel *n*

(optical) glass fibre cable; (optical) glass fiber cable (US) Glasfaserkabel *n*

optical identification optische Kennung *f*

optical image stabilizer → **OIS**

optical index Lichtmarke *f* [*Messgerät*]

optical isolator → **optocoupler**

optical length → **optical path**

optically absorptive lichtschluckend

optical memory [*A computer memory that uses optical techniques which generally involve an addressable laser beam, a storage medium which responds to the beam for writing and sometimes for erasing, and a detector which reacts to the altered character of the medium when it uses the beam to read out stored data*] optische Speicher *m*

optical-link pilot wire system Schutzsystem mit Lichtwellenleiter, Lichtwellenleiter-Schutzsystem *n*

optical microscope, light microscope, photon microscope Lichtmikroskop *n*

optical path, optical distance, optical length [*For a ray of light travel(l)ing along a path between two points, the optical path is the integral, over elements of length along the path, of the refractive index*] Brech(ungs)indexintegral *n*

optical pattern [*photography*] Lichtband *n*

optical pointer Lichtzeiger *m*

O

optical reader [*A computer data-entry machine that converts pointed characters, bar or line codes, and pencil-shaded areas into a computer-input code format*] optische Leser *m*

optical scanner, flying scanner, spot scanner optische Abtaster *m*

optical scanning, flying scanning, spot scanning optische Abtastung *f*

optical signal converter optische Signalgeber, optische Signalwandler, optische Signalumsetzer *m* [*Alarmgabe über Rundumkennleuchte oder elektronische Blitzleuchte. DIN 6136 Teil 1*]

optical signal message display optische Meldungsanzeige *f*

optical signal transmitter optische Signalgeber *m*

optical smoke detector, optical smoke detecting device, optical smoke detection device optische Rauchmelder, Streulichtrauchdetektor, Streulichtrauchmelder, optische Rauchdetektor, optische Rauchmeldungsgeber, Streulichtrauchmeldungsgeber *m*, optische Rauchmeldegerät, Streulichtrauchmeldegerät *n* [*Er nutzt den Tyndalleffekt aus, gemäß dem Lichtstrahlen durch Reflexion an der Oberfläche kleiner (Rauch)Teilchen diffus gestreut werden*]

optical sound, photographic sound optische Schall *m*

optical sound recorder, photographic sound recorder optische Schallaufzeichnungsgerät *n*

optical sound reproducer, photographic sound reproducer optische Schallwiedergabegerät *n*

O

optical storage [*Storage of large amounts of data in permanent form on photographic film or its equivalent, for nondestructive readout by means of a light source and photodetector*] optische Speicherung *f*

optical time-domain reflectometer, OTDR LWL-Reflektometer, Lichtwellenleiter-Reflektometer, Rückstreumessplatz *m*, optisches Impulsreflektometer *n*

optical video disk, optical video disc, optical storage disk, optical storage disc optische Speicherplatte *f*

optical waveguide → optical-fibre cable

optimization, system optimization [*Systems Engineering. 1. Broadly, the efforts and processes of making a decision, a design, or a system as perfect, effective, or functional as possible. 2. Narrowly, the specific methodology, techniques, and procedures used to decide on the one specific solution in a defined set of possible alternatives that will best satisfy a selected criterion*] (System)Optimierung *f*, (System)Optimieren *n*

optimization function Optimierungsfunktion *f* [*Eine Funktion, die zur Änderung eines Parameters im GAS (Gebäudeautomationssystem) führt, die ihrerseits die Effizienz des Systems erhöht*]

to optimize [*Automatic data processing. To rearrange the instructions or data in storage so that a minimum number of time-consuming jumps or transfers are required in the running of a program(me)*] optimieren

optimizer [*A control function which provides economical switch on/off of plant in response to its measured inputs*] Optimierer *m*

optimum ((closed) loop) control [*A system to maintain the controlled variable in the optimum condition evaluated from the viewpoint of both minimization of energy consumption and the creation of a comfortable indoor environment*] Optimumreg(e)lung *f*

optimum code[*In automatic data processing, a computer code which is particularly efficient with regard to a particular aspect; for example, minimum time of execution, minimum or efficient use of storage space, and minimum coding time*] Optimierungskode, Optimierungscode *m*

optimum off reduzierter Betrieb *m* zum frühesten Zeitpunkt [*Funktionsarbeit einer Heiz(ungs)optimierung*]

optimum on mögliche Wiedereinschaltung *f* der Heizung zum spätest zulässigen Zeitpunkt [*Funktionsarbeit einer Heiz(ungs)optimierung*]

optimum output power optimale Ausgangsleistung *f* [*Elektronenstrahlröhre*]

optimum refrigerant charge [*Charge achieving maximum possible refrigerant effect within design limitations*] optimale Kältemittelfüllung *f*

optional, at option wahlweise

optional file Wahldatei *f*

optional key Wahltaste *f*

optimum start optimale Einschalten *n* [*Ein Programm, das eine Anlage zum spätest möglichen Zeitpunkt vor einer Belegung einschaltet*]

optimum start/stop optimale (Ein/Aus)Schalten *n* [*Von Anlagen für Heiz- und Kühlbetrieb im Sinne spätestmöglich EIN, frühestmöglich AUS*]

optimum stop [*A control system or algorithm which stops plant operation at the earliest possible time such that the residual capacity of the space and the system will prevent conditions in the space deteriorating beyond pre-set limits by the end of the occupancy period*] Optimumsperre *f*

optionally → electively

optional skip wahlweise Überlesen, wahlweise Übergehen, wahlweise Überspringen *n* [*Siehe Anmerkung unter „Überspringen"*]

optional stop instruction bedingter Haltbefehl *m*

optional word Wahlwort *n*

optocoupler, optical coupler, optical isolator Optokopp(e)ler, optische Kopp(e)ler, optische Isolator *m*, optoelektronische Koppelelement *n* [*Bauelement zur galvanischen Trennung*]

optoelectronic optoelektronisch

optoelectronic fuse monitoring optoelektronische Sicherungsüberwachung *f*

optoelectronic integrated circuit, OEIC, optical electronic integrated circuit monolithisch integrierte optoelektronische Komponente, optoelektronische monolithisch integrierte Komponente *f*

optoelectronic semiconductor, optical electronic semiconductor Optohalbleiter, optische Halbleiter *m*

opto-isolator, optical-isolator [*A device consisting of a close-coupled light emitter and photo-sensor which, when connected in a communication line , will provide electrical isolation, but allows signal transmission*] optische Isolator, Optokopp(e)ler *m*

OPUP, operator position user part Vermittlungsplätze-Benutzerteil *m, n* [*Signalisierungssystem Nr.7*]

OR ODER (-Glied), inklusive ODER *n* [*einfache Verknüpfung*]

OR-before-AND logic ODER-vor-UND-Verknüpfung *f*

orbital angular momentum, orbital moment momentum, orbital momentum moment Bahndrall, Bahndrehimpuls *m*

or-block → or-module

orchestra lift (Brit.); orchestra elevator (US) Orchesteraufzug *m*

order, sequence Reihenfolge *f*

order → (supply) order

order(ing) details Bestell(ungs)angaben, Bestellungseinzelheiten, Bestelleinzelheiten *f*

ordering text Bestell(ungs)text *m*

order management Auftragsverwaltung, Bestell(ungs)verwaltung *f*

order number Bestell(ungs)nummer, Auftragsnummer *f*

order of priority, priority order Vorrangfolge *f*

order-related auftragsbezogen

order without user's inspection [*ISO-R 404*] Bestellung *f* ohne Ablieferungsprüfung durch den Besteller

order with user's inspection [*ISO-R 404*] Bestellung *f* mit Ablieferungsprüfung durch den Besteller

ordinary tough rubber sheated flexible cord Gummischlauchleitung *f* [*DIN VDE 0282 Teil 804*]

organic nitrogen organische Stickstoff *m*

organic Rankine cycles, bottoming cycles ORC-Prozess *m* [*Erzeugung mechanischer Energie aus Abwärme*]

organizer Ausrichter, Organisator, Veranstalter *m* [*Kongresse; Messen; Tagungen usw.*]

(organ) pipe (Orgel)Pfeife *f*

Or-gate ODER-Gatter(schaltung), ODER-Torschaltung *n (f)*

orifice → port

orifice gauge; orifice gage (US) Messblende *f*

orifice plate Drosselblech *n*

original address → home address

origin(ation) Ursprung *m*

origin of an electrical installation Speisepunkt *m* einer elektrischen Anlage

O-ring Rund(ab)dicht(ungs)ring *m*

Or instruction ODER-Befehl *m*

OR logic (operator), OR logic function ODER-Verknüpfung *f*

or-module, or-block Oder-Baustein *m*, Oder-Modul *m*

ornamentally styled → decoratively styled

ornated hinge, ornative hinge, fancy hinge Dekor(ations)scharnierband, Ornamentscharnierband, Zierscharnierband, Schmuckscharnierband *n* [*Für Fenster, Türen und Tore*]

OROM → optical read only memory

Orsat analyzer [*Analytical chemistry. An apparatus used to determine the composition of a mix(ture) of gases by passing the mix(ture) through a series of solvents; each of which absorbs only one of the gases*] Orsatgerät *n*

orthogonal (computer) memory orthogonale (Rechner)Speicher *m*

OS → operator station

OS → operating system

oscillating compressor Schwing(ungs)verdichter *m*

oscillating piston (liquid) meter Hubkolbenzähler *m*

oscillation Schwingung *f*

OTDR → optical time-domain reflectometer

osmotic pressure osmotische Druck, Osmosedruck *m*

other alle anderen [*Eine Benutzerklasse*]

other listener address fremde Höreradresse *f*

other talker address fremde Sprecheradresse *f*

outage → failure

outboard drive Außenantrieb *m*

outdegree Ausgangsgrad *m*

outdoor advertising Außenwerbung, Außenreklame *f*

outdoor aerial (Brit.); outdoor antenna (US) Außenantenne, Freiluftantenne *f*

outdoor (air) temperature setpoint control Sollwertführung *f* nach Außentemperatur

outdoor and indoor lightning protection system, indoor and outdoor lightning protection system Gesamtblitzschutzsystem *n* [*Äußere plus innere Blitzschutz*]

outdoor daily temperature range Außentagestemperaturbereich *m [selten: n]*

outdoor dew point Außentaupunkt *m*

outdoor erection Außenaufstellung *f*

outdoor heating Freiflächenheizung f [*Heizung um Flächen im Freien eis-, schnee- und frostfrei zu halten*]

outdoor installation, outdoor plant Freiluftanlage *f*

outdoor installation assembly, outdoor plant assembly [*IEC 439-1*] Freiluft-Schaltgerätekombination, Schaltgerätekombination für Freiluftaufstellung *f* [*VDE 0660, Teil 500*]

outdoor lightning protection Außenblitzschutz, äußere Blitzschutz *m* [*Faradaykäfig aus Draht mit einer Erdung*]

outdoor monitoring Außenüberwachung *f*

outdoor operating temperature, operating outdoor temperature Außenbetriebstemperatur, Betriebsaußentemperatur *f*

outdoors Freie *n*

outdoor section [*heat pump*] Außenteil *m, n*

outdoor steam → external steam

outdoor temperature sensor → external (air) temperature sensor

outdoor wall installation Außenwandeinbau *m*

outer... → external...

outer protection window → protection outer window

to outfit, to equip (with) bestücken (mit), ausrüsten (mit), ausstatten (mit)

outfit, equipment Ausstattung, Ausrüstung, Bestückung *f*

outfit object, equipment object Austattungsgegenstand, Ausrüstungsgegenstand, Bestückungsgegenstand *m* [*„Objekt" kann synonym für „Gegenstand" verwendet werden*]

outfitted, equipped bestückt, ausgestattet, ausgerüstet

outfitted device table → equipped device table

outfitting, equipping Ausstatten, Ausrüsten, Bestücken *n*

outflow Ausfluss, Auslauf *m* [*Der aus einer Öffnung fließende Flüssigkeitsstrom*]

outflow float(ing) valve Auslaufschwimmerventil, Ablaufschwimmerventil, Ausflussschwimmerventil, Ausfließschwimmerventil, Abflussschwimmerventil *n*

outflow valve Auslaufventil, Ausfließventil, Ablaufventil, Abflussventil, Ausflussventil *n* [*Zur Wasserentnahme aus einer Kaltwasser, Warmwasser oder Mischwasser führenden Zuflussleitung. Erzeugung von Mischwasser in dieser Armatur ist nicht möglich*]

outflow vertical valve → vertical outflow valve

(outgoing) cable feeder → cable branch (line)

outgoing cable unit, cable outgoing unit Kabelabgang(seinheit) *m, (f)*

outgoing call abgehende Ruf *m*

outgoing effluent behandelte Abwasser *n*

outgoing guard delay abgehende Überwachungszeit, Sperrzeit *f* [*dezentrale Zeichengabe*]

outgoing signal message gehende Meldung, ausgehende Meldung, abgehende Meldung *f* [*Eine Meldung, abgeleitet vom Wechsel des Eingabesignals aus dem zu meldenden Betriebszustand*]

outgoing switch Abgangsschalter *m*

outlet Auslauf, Ablauf, Ausfluss *m* [*Als Gegenstand und als Vorgang (= Austritt, Auslass)*]

outlet (box) [*A metal box containing wires from a branch circuit and connection to wires from electric load*] Steckdose *f*

outlet connecting branch, outlet connection branch Auslaufstutzen, Entnahmestutzen, Ausflussstutzen *m*

outlet connection Ablaufanschluss, Auslaufanschluss *m*, Ablaufverbindung, Auslaufverbindung *f* [*z. B. von Wasch- und Geschirrspülmaschinen*]

outlet fan, extract fan Ausgangsventilator *m*

outlet pipe Auslaufrohr, Ausflussrohr, Entnahmerohr *n*

outlet vent Dunstrohr *n*

outline font Laserdruckerzeichensatz *m* [*Die Zeichen sind durch geometrische Formen (Kurven; Linien) definiert*]

outline illumination, outline lighting Umrissbeleuchtung, Kontour(en)beleuchtung *f*

outline-like konturengleich, umrissgleich

out-of-band signalling; out-of band signaling (US) Kennzeichengabe *f* mit getrenntem Zeichenkanal, Signalisierung *f* mit getrenntem Zeichenkanal

out(-of-)door unit → open air unit

out-of-focus image, out-of-focus picture, blurred image, blurred picture, hazy image, hazy picture unscharfe Bild, ungenau fokussierte Bild, verschwommene Bild *n*

out-of-ground hydrant → above-ground hydrant

O

out of order → DER

out-of-order signal Antwortkennzeichen „Außer Betrieb“, Antwortkennzeichen „gestört“ *n*

out-of-parity fehlerhafte Parität, fehlerhafte Gleichstellung *f*

outpatient clinic Ambulanzklinik *f*

to output, to put out ausgeben, abgeben, vergeben [*Rechnerinformation an die Peripherie, z. B. Drucker, Festplatten, Diskettenlaufwerke, Strichzeichner usw.*]

output Aufkommen *n* [*Verflechtungsbilanzierung*]

output → (data) output

output [*The activity of transmitting the generated information(s)*] Informationsübertragung, Mitteilungsübertragung, Nachrichtenübertragung *f*

output → data carrier

output [*The product of a system*] Ausstoß *m*

output → output (unit)

output [*The current, voltage, power, driving force, which a circuit or device delivers*] Ausgangsleistung, (abgegebene) Leistung, Leistungsabgabe, Abgabeleistung *f*

output [*The readable storage medium upon which generated data are written, as in hard-copy output*] Datenaufzeichnung(s)speicher *m*

output [*The terminal from which energy or information is emitted by a system or device*] Ausgang *m*

output Ausstoß *m*, Ausgabe, Gabe *f*

output acknowledg(e)ment Ausgangsbestätigung, Ausgangsquittierung *f*

output amplifier → output (channel) amplifier

output attenuator Ausgang(s)abschwächer, Ausgang(s)spannungsteiler *m*

output balance stationäre Zustand *m* [*Regler*]

output block, output area [*A portion of the internal storage of a computer that is reserved for receiving, processing, and transmitting data to be transferred out*] Ausgabedatenspeicher *m*

output buffer Ausgabepuffer *m* [*Er übernimmt die Rechenwerkergebnisse und gibt sie an Ausgabegeräte*]

output bus driver [*A device that power-amplifies output signals from a computer to allow them to drive heavy circuit loads*] Ausgangsbus-Treiber *m*

output capacitance; output capacity [*obsolete*] [*Of a n-terminal electron tube, the short-circuit transfer capacitance between the output (terminal) and all other terminals, except the input terminal, connected together*] Ausgangskapazität *f*

output card Ausgangskarte *f*

output carry over Ausgangsübertrag *m*

output (channel) amplifier Ausgangs(kanal)verstärker *m*

output characteristic (curve) Ausgangskennlinie, Ausgangskennkurve *f* [*Zusammenhang zwischen Ausgangsstrom und Ausgangsspannung eines aktiven Bauelements für eine konstante Steuer(ungs)spannung bzw., einen konstanten Steuer(ungs)strom*]

output characteristics (display) field Ausgangskennlinienfeld *n* [*In ihm sind die Ausgangskennlinien eines aktiven Bauelements für mehrere konstante Steuer(ungs)spannungen oder Steuer(ungs)ströme aufgetragen*]

output circuit, secondary circuit Ausgangskreis, Ausgangs(be)schaltung *f*, Sekundärkreis *m*

output closed loop control → performance closed loop control

output conductance Ausgang(s)leitwert *m* [*Eingang kurzgeschlossen*]

output contact Ausgang(s)kontakt *m*

output control → ((open) loop) output control

output converter lastseitige Stromrichter *m*

output current Ausgangsstrom, Sekundärstrom *m*

output cut-off frequency Ausgangsgrenzfrequenz *f*

output data, secondary data [*The data actively transmitted from within the computer to an external device, or onto a permanent recording medium (paper, microfilm)*] Ausgabedaten, Sekundärdaten, übertragende Daten *f*

output deck → output stack

output device → output (unit)

output disable Ausgang *m* gesperrt

output disable time, output disabling time Ausgang(s)sperrzeit *f*

output displacement factor lastseitige Verschiebungsfaktor *m*

output drop → performance drop

output element Ausgabeglied *n* [*Funktionsglied zur Aufbereitung und Ausgabe von Ausgabesignalen bzw. Ausgabedaten. V DIN 19 237*]

output enable, OE Ausgangsfreigabe *f*

output enable input Ausgangsfreigabeeingang *m*

output equipment Ausgabegerät(e) *n,(f)* [*1. Gerätetechnische Verwirklichung von Ausgabegliedern und/-oder einer Ausgabeeinheit. DIN 19 237. 2. In einer Ausgabeeinheit eine Baueinheit, durch die Daten aus einer Rechenanlage ausgegeben werden. E DIN 44 300. Teil 5*]

output file Ausgabedatei *f*

output flag Ausgangsmerker *m*

O

output frequency Ausgangsfrequenz *f*

output gap [*An interaction gap by means of which usable power can be abstracted from an electron stream in a microwave tube*] Auskoppel(ungs)spalt, Auskopplungsspalt *m*

output heat Abgabewärme, Leistungswärme *f* [*Sie wird vom Wärmeerzeuger an den Wärmeträger abgegeben*]

output impedance [*The impedance presented by a source to a load*] Ausgangsimpedanz *f*, Ausgang(s)-scheinwiderstand, Innenwiderstand *m* (eines Messsenders)

output indicator [*A meter or other device that is connected to a radio receiver to indicate variations in output signal strength for alignment and other purposes, without indicating the exact value of the output*] Schaltzustand(s)anzeige(r), Ausgangsleistungsanzeige(r) *f, (m)* [*Er hat geringe Genauigkeit*]

output instruction Ausgangsbefehl *m*

output jack Auskoppel(ungs)buchse, Ausgangsbuchse, Auskopplungsbuchse *f*

output-limited [*Restricted by the need to await completion of an output operation, as in process control or data processing*] notwendigerweise begrenzt, notwendigerweise beschränkt, notwendigerweise eingeschränkt

output listing [*data processing*] Programmprotokoll *n*

output load Ausgangslast *f*

output loading capability [*IEC 147-OD*] Ausgang(s)-belastbarkeit *f*

O

output magazine [*In automatic data processing, a mechanism that accumulates cards after they have passed through a machine*] Kartenmagazin *n*

output matrix Ausgangsmatrix *f*

output medium Ausgabedatenträger *m*

output message [*See remark under "message"*] Ausgabemitteilung *f*, Ausgabenachricht *f*, abgegebene Mitteilung, abgegebene Nachricht

output need → **output requirement**

output noise Ausgangsrauschen *n*

output noise ratio Ausgangsrauschverhältnis *n*

output ((open) loop) control → **((open) loop) output control**

output point Ausgabepunkt *m*

output power → **power output**

output power amplifier Ausgangsleistungsverstärker *m*

output pressure Ausgangsdruck *m* [*Druckregler*]

output procedure Ausgabevorgang, Ausgabeablauf, Ausgabeprozess *m*

output program (US); output programme (Brit.) Ausgabeprogramm *n*

output pulse Ausgangsimpuls *m*

output pulse priority Ausgangsimpulsvorrang *m*

output pulse valency Ausgangsimpulswertigkeit *f*

output punch [*A device that transcribes data generated by a computer onto paper tape or punch cards*] Umschreiber *m*, Umschreibgerät *n*

output (Q) Ausgang (A) *m*

output quantity Ausgangsgröße *f* [*Eine beeinflusste und erfassbare Größe eines Systems. DIN 19 226, Teil 1*]

output rate Ausgaberate, Ausgangsrate *f* [*Die maximale Anzahl der digitalen Daten je Zeiteinheit im Hauptspeicher, die im Dauerbetrieb am Ausgang der Analog- bzw. Digitalausgabe als Werte der Signalparameter bereitgestellt werden können. DIN 66 201. Teil 1*]

output redirection Umleitung der Ausgabe, Ausgabeumleitung *f*

output referred to a potential, potential output potentialbezogene Ausgang *m* [*DIN 41 855*]

output relay Ausgangsrelais, Ausgaberelais, Endstufenrelais *n*

output request Ausgabeabfrage, Ausgabeanfrage *f*

output requirement, performance requirement, output need, performance need Leistungsbedarf *m*

output signal Ausgabesignal, Ausgangssignal *n* [*Ein von der Signalverarbeitung zur Ausgabe bereitgestelltes Signal. DIN 19 226 Teil 5; DIN 19 237*]

output spooling Ausspulen *n*

output spooling file Ausspuldatei *f*

output stack, output deck Ausgabestapel, Datenausgabestapel *m*

output stage, final stage [*electronic equipment*] Ausgangsstufe, Endstufe, Auskoppel(ungs)stufe, Auskopplungsstufe *f*

output statement Ausgabeanweisung *f*

output storage area Ausgabespeicher(ungs)bereich *m [selten: n]*

output time Ausgabezeit *f* [*Die Zeit, die benötigt wird, um im Hauptspeicher vorhandene digitale Daten am Ausgang der Analog- oder Digitalausgabe als Werte des Signalparameters bereitzustellen. DIN 66 201 Teil 1*]

output transformer [*The iron-core audio-frequency transformer used to match the output stage of a radio receiver or an amplifier to its loudspeaker or other load*] Ausgangstransformator, Ausgangsübertrager, AÜ, Nachübertrager, NÜ, Ausgangstrafo, Nachtrafo, Nachtransformator *m*

output (unit), output device [*In computers, a unit which delivers information from the computer to an external device or from internal to external storage*] Ausgabeteil *m, n*, Ausgabeeinheit *f*, Ausgabegerät *n* im Rechner eingebaut [*DIN 19 237*]

output value Ausgabewert, Ausgang(s)wert *m*

output variable Ausgang(s)größe *f*

output vector Ausgangsvektor *m*

output voltage, output tension Ausgangsspannung, Abgabespannung *f*

output voltage string, output tension string Ausgangsspannungsbereich, Abgabespannungsbereich *m* [*selten: n*] [*Operationsverstärker*]

output word [*Any running word into which an input word is to be translated*] Ausgangswort *n*

output work queue Warteschlange *f* der Ausgangsarbeiten

out reader Außenleser *m*

outside light → **false light**

outsider Außenstehende *m,f*

outside steam → **external steam**

outside (temperature) detector, outside (temperature) sensor Außentemperatursensor, Außentemperaturdetektor, Außentemperaturfühler *m*, Außentemperaturdetektionsgerät *n*

outside temperature measured value transmitter Außentemperatur-Messwertgeber *m*

outside wall fan Außenwandventilator *m*

outside wall gas heating furnace Gas-Außenwandheiz(ungs)ofen, Außenwand-Gasheiz(ungs)ofen *m* [*DIN 3364*]

outside wall-mounted furnace Außenwandfeuerung, Außenwandfeuerstätte *f*

outside wall-mounted gas room heater Außenwand-Gasraumheizer *m*

outside wall-mounted room heater Außenwandraumheizer *m*

outsourcing Dienst(leistungs)vergabe, Ausgliederung aus dem Kerngeschäft, Fremdvergabe, Vergabe an externe(n) Dienstleister *f*

outstation, universal network controller, UNC [*The field device to which all the systems to be connected is connected. It comprises two parts; the first is the central processing unit (CPU) which monitors and controls the building services, and the second is the communication card which handles all the data transfers between the CPU and any other connected devices on the system*] Außenstation, Außenstelle, Unterstation, Unterstelle *f*

oval (door) knob Oval(tür)knopf *m*, Türovalknopf

oval(-headed) screw, OHS Ovalkopfschraube, Kopfovalschraube *f*

'O' value → **total thermal transmittance value for building fabrics**

oval wheel meter Ovalradzähler *m*

oven [*A heating enclosure for baking, heating or drying*] Ofen *m*

oven furnace [*deprecated*] → **direct-fired furnace**

overcoming time factor Überwindungszeitfaktor *m*

overcoupling Überkoppeln *n*

overcurrent, excess(ive) currrent Überstrom *m*

overcurrent (blocking) device, eccess(ive) current (blocking) device [*A device, such as a fuse or a circuit breaker, designed to protect a circuit against excessive current by opening the circuit. IEC 214*] Überlast(ungs)schutzgerät *n*

overcurrent discrimination, excess(ive) current discrimination Überstromselektivität *f*

overcurrent protection, excess(ive) current protection Überstromschutz *m*

overcurrent protective breaker, excess(ive) current protective breaker Überstrom-Schutzschalter *m*

overcurrent protective device, excess(ive) current protective device [*In electrical work, a device, either a fuse or circuit breaker, provided to open (interrupt) an electric circuit to prevent damage to an electric conductor in the event of an excessive current*] Überstromschutzvorrichtung *f*

overcurrent protective fuse, excess(ive) current protective fuse Überstromsicherung *f*

overcurrent relay, excess(ive) current relay Überstromrelais *n*

overcurrent release, overcurrent trip, excess(ive) current release, excess(ive) current trip Überstromauslöser *m* [*VDE 0660, T.101*]

overdamped system PT_2-System *n* mit $D > 1$ [*Kriechzahl*]

overdimensioned → **oversized**

overdimensioning → **oversizing**

overdoor heater [*It comprises a heat exchanger, single or multiple centrifugal or tangential fan/motor assembly and a discharge slot grill(e) in a casing with fitted or remote controls*] Übertürheizer, Übertürerhitzer *m*

overdoor heating Übertürheizung *f*

to overdrive übersteuern [*Verstärker*]

overflow Überlauf *m* [*Bereichsüberschreitung bei einer arithmetischen Operation nach oben*]

overflow → **spill(-over)**

O

overflow connection → **trap outlet**

overflow footing report group Überlauffuß *m* [*COBOL*]

overflow grill(e) Überlaufgitter, Überströmgitter *n*

overflow heading report group Überlaufkopf *m* [*COBOL*]

overflow hole Überströmloch, Überlaufloch *n* [*Waschbecken; Badewanne*]

overflowing Überlaufen, Überströmen *n*

overflow(ing) (error) Überlauf(en) *m*,*(n)*, Überschreiten *n*, Überschreitung *f* [*Speicher; Zählvorrichtung*]

overflow(ing) pipe Überlaufrohr *n*

overflow(ing) water Überlaufwasser, Überströmwasser *n*

overflow level → **spill(-over) level**

overflow mixing battery Überlaufmischbatterie, Überströmmischbatterie *f* [*DIN 44 897*]

overflow outlet Überlaufaustritt, Überlaufablass, Überlaufauslass, Überlaufabgang, Überlaufablauf, Überströmaustritt, Überströmablass, Überströmauslass, Überströmabgang, Überströmablauf *m*

overflow pipe [*A pipe connected to a vessel, sanitary appliance, sewer or chamber to discharge excess water*] Überlaufrohr, Überströmrohr *n*

overflow slot Überlaufschlitz, Überströmschlitz *m*

overflow storage heater Überlaufspeicherheizer, Überströmspeicherheizer *m*

O

overhead Zusatzdaten *f* [*Sie werden bei der Übermitt(e)lung von Informationen zusätzlich meist zu Steuer(ungs)zwecken übertragen*]

overhead ... → **ceiling(-mounted) ...**

overhead door closer, overhead door closing device, overhead door closing mechanism, overhead device for door closing Oben(-Tür)schließvorrichtung, Oben(-Tür)schließer *f*

(overhead) feed(ing) cistern, (overhead) feedwater cistern, (overhead) feed(ing) tank, (overhead) feedwater tank [*A cold water storage cistern, which is supplied from the service mains through a rising main and ball cock, and is used to provide cold water to the hot water system and, usually, to all other cold water points in the building, except those specifically for drinking purposes*] Speisewassertank, Speisewasserbehälter *m*, Speisewasserzisterne *f*

overhead lighting, overhead illumination Überkopfbeleuchtung, Obenbeleuchtung *f*

overhead mounting, overhead installation Obenmontage, Überkopfmontage, Obeninstallation, Überkopfinstallation *f*, Obeneinbau, Überkopfeinbau *m*

overhead projector Überkopfprojektor, Obenprojektor *m*

overhead radiant heater Überkopfstrahl(ungs)heizer, Obenstrahl(ungs)heizer *m*

overhead service Freileitungs(haus)anschluss *m*

overhead service entry Freileitungseinführung *f* [*elektrische Hausanlage*]

overhead shower, fixed shower Überkopfdusche, Hochdusche, Obendusche *f*

overhead spray shower, fixed spray shower Überkopfbrause, Hochbrause, Obenbrause *f*

overhead thermal radiation Überkopfwärmestrahlung, Obenwärmestrahlung *f*

overhead track → **(trolley) track**

overheater Überhitzer *m*

overheating Überhitzung *f*

overheating heat Überhitzungswärme *f*

overhouse aerial, roof aerial (Brit.); overhouse antenna, roof antenna (US) Dachantenne *f*

to overhumidify über(an)feuchten, überbefeuchten [*Luft*]

over-igniting Überzünden *n*

to overlap [*To perform some or all of an operation concurrently with one or more other operations*] überlappen

overlap(ping) Überlappung *f*

to overlay einblenden [*Bildschirmgerät*]

overlay Einblendung *f* [*Bildschirmgerät*]

overlay [*In automatic data processing, a technique for bringing routines into high-speed storage from some other form of storage during processing, so that several routines will occupy the same storage locations at different times; overlay is used when the total storage requirements for instructions exceed the available main storage*] Überlagerung *f*

overload [*A condition of excess current, more current flowing than the circuit was designed to carry*] Überlast *f*

overloading Über(be)lastung *f*, Über(be)lasten *n*

overloading → **congestion**

overload current Überstrom *m* [*Thyristor. DIN 41 786*]

overload-proof überlastfest, überlastsicher

overload-proof end überlastfeste Ausgang, überlastsichere Ausgang *m*, überlastfeste Ende, überlastsichere Ende *n* [*Ausgang eines Funktionsgliedes, der gegen Zerstörung durch zeitlich unbegrenzt anstehende beliebige passive Last geschützt ist. DIN 19 237*]

overload relay Überlastrelais *n*

overload shedding Überlastabwurf *m*

overmax operator Über-MAX-Operator *m* [*Alle t-konormen außer MAX*]

overplot Zeichnungsüberlagerung *f* [*Das Legen einer Kontur über eine Zeichnung*]

overpotential ... → overvoltage ...

overpressure, above-atmospheric pressure atmosphärische Druckdifferenz *f*, Überdruck *m* [*Die Differenz zwischen einem absoluten Druck P_{abs} und dem jeweiligen (absoluten) Atmosphärendruck P_{amb} ist die atmosphärische Druckdifferenz P_e; sie wird Überdruck genannt: $P_e = P_{abs} - P_{amb}$ Der Überdruck P_e nimmt positive Werte an, wenn der absolute Druck größer als der Atmosphärendruck ist; er nimmt negative Werte an, wenn der absolute Druck kleiner als der Atmosphärendruck ist.*
Anmerkung 1: Bisher wurde von Überdruck nur gesprochen, wenn der absolute Druck größer als der Atmosphärendruck war; war er kleiner, wurde die durch die Differenz $P_{amb} - P_{abs}$ definierte Größe „Unterdruck" verwendet. Den Unterdruckbereich kennzeichnen nunmehr negative Werte des Überdruckes.
Das Wort „Unterdruck" darf nicht mehr als Benennung einer Größe, sondern nur noch für die qualitative Bezeichnung eines Zustandes verwendet werden. Beispiele: „Unterdruckkammer"; „Im Saugrohr herrscht Unterdruck".
Anmerkung 2: Der Bereich der Drücke unterhalb des Atmosphärendruckes wird auch Vakuumbereich genannt (siehe DIN 28 400 Teil 1). In der Vakuumtechnik wird stets der absolute Druck angegeben.]

overpressure blower Überdruckgebläse *n*, Überdruckbläser *m*

overpressure safety switch Überdrucksicherheitsschalter *m*

overpressure ventilation, overpressure vent(ilat)ing Überdrucklüftung *f*

overpressure whistle Überdruckpfeife *f*

overpressurization Überdruckbildung *f*

to overpunch überlochen

overpunch Überloch *n*

overrange Bereichsüberschreitung *f*, Überbereich *m* [*selten: n*] [*Messgerät*]

to overreach übergreifen

to override [*e.g. to override a value temporarily*] übergehen, unbeachtet lassen, nicht beachten

to override übersteuern

override alarm (control) → alarm override control

override (control), overriding Übersteuerung *f*, Übersteuern *n*

override switch → hand override switch

overriding relay Übersteuerungsrelais, Maximalrelais *n*

oversampling Vervielfachung *f* der Abtastrate, Über(ab)tasten *n*

oversaturated übersättigt

to oversee continuously → to monitor

overshoot → overtravel

overshoot (amplitude) Überschwing(ungs)weite *f*

overshoot angle Überschwing(ungs)winkel *m* [*Schrittmotor*]

overshoot factor Überschwing(ungs)faktor *m* [*Verstärker*]

overshoot time → overtravel time

oversize, overdimension Übergröße *f*

oversized, overdimensioned überdimensioniert, überbemessen, übergroß

oversizing, overdimensioning Überbemessung, Überdimensionierung *f*

overspeed, over r.p.m, over revolutions per minute Überdrehzahl *f*

overspeed governor Geschwindigkeitsbegrenzer *m* [*Aufzug*]

overstroke Überhub *m*

overtemperature Übertemperatur *f*

overtemperature alarm Übertemperaturalarm *m*

overtemperature protection Übertemperaturschutz *m*

overtemperature protector light Übertemperaturschutzlampe *f*

overtemperature security Übertemperatursicherung *f*

over-the-horizon propagation → beyond-the-horizon communication

overtravel Überhub *m* [*Ventil*]

overtravel, overrun, overshoot Auslauf *m*, Überfahren *n*, Durchfahrt, Überfahrt *f* [*Aufzug*]

overtravel, overshoot Ausreg(e)lung *f*, Nachlauf *m* [*Betätigungselement; Schaltglied; Relais*]

overtravel time [*ANSI C 37.90*]; **overshoot time** Nachlaufzeit, Überschwing(ungs)zeit, Ausregel(ungs)zeit *f* [*Relais; Hilfsstromschalter*]

overturning Überdrehen *n*

overview Übersicht *f*

overview field Übersichtsfeld *n* [*Prozessmonitor*]

overview panel Übersichtstafel *f*, Übersichtstableau *n*

overvoltage, overpotential Überspannung *f*, Durchschlag *m*

overvoltage blocking device, overpotential blocking device Überspannungssperre, Durchschlagsperre *f*

O

overvoltage protection, overpotential protection Überspannungsschutz, Durchschlagschutz *m*

overvoltage protector, overpotential protector Überspannungssicherung, Durchschlagsicherung *f*

overvoltage release, overpotential release Überspannungsauslöser, Durchschlagauslöser *m*

overwarming Über(er)wärmen *n*, Über(er)wärmung *f*

to overwrite überschreiben [*Einschreiben von Daten in einen Speicher, wobei die vorherigen zerstört werden*]

owl [*DOP test*] optische Messgerät *n* [*Zur Kontrolle der Ölnebelteilchen von 0,3 µm auf Grund des unter einem Winkel von 29 ° abgestrahlten Streulichtmaximums*]

owner [*The person who has the exclusive right to modify data*] Modifikationsberechtigte *m, f*

owner → operator

owner Bauherr(schaft) *m,(f)*

owner provided, owner supplied bauseits (geliefert)

OW unit → sabin

oxy(gen)-depletion Sauerstoffmangel *m*, Sauerstoffverarmung *f*

oxy(gen)-depletion device [*A safety device shutting off the gas supply in the event of reduced oxygen levels in a room*] Sauerstoffmangelsicherung *f*

oxy(gen)-distribution system Sauerstoffverteilungsystem *n*

oxygen tent Sauerstoffzelt *n*

ozone depletion Ozonverarmung *f*, Ozonmangel *m*

P

P → positive

Pa → pascal

pacing Nachrichtenmengendosierung *f*, Synchronisation *f*

to pack [*To reduce the amount of storage required to hold information by changing the method of encoding the data*] verringern

package Paket *n*

package(d) bath(room) Fertigbad, Kompaktbad *n*

package(d) boiler [*A boiler plus a number of its controls and accessories*] Fertigkessel, Kompaktkessel *m*

package(d) integrated circuit integrierte Schaltkreis *m* in Normengehäuse

package directory Paketverzeichnis *n*

package(d) terminal air-conditioning plant, package(d) terminal air-conditioning installation, ptac [*Self-contained air-conditioning plant for through-the wall installation. Note: A ptac often uses a wall sleeve*] Fertigklimaanlage, Kompaktklimaanlage *f*

package(d) unit [*A unit plus a number of its accessories*] Fertigeinheit, Kompakteinheit *f*

packer Dateiverkleinerungs-Hilfsprogramm *n*

packet assembler/disassembler → PAD

packet assembly/disassembly facility, PAD Paketier(ungs)-Depaketier(ungs)-Einrichtung *f*

packet length election Paketlängen(aus)wahl *f*

packet level ready Paketschicht *f* betriebsbereit [*Protokoll*]

packet-mode operation Paketbetrieb, Paketmodus *m*

packet-orient(ta)ted, packet-related paketorientiert, paketbezogen

packet send sequence number Paketsendelaufnummer *f*

packet sequency Paketreihung, Paketfolge *f*

packing (Ab)Dichtung, Packung, Liderung *f*

packing density [*The amount of information per unit of storage medium, as characters per inch or tape, bits per inch or drum, or bits per square inch in photographic storage*] Packungsdichte, Speicherdichte *f*

packing grooves flange Dichtungsrillenflansch *m*

pack(ing)less, without pack(ing), stuffingless, without stuffing stopfbuchsenlos, packungslos

packing routine [*A subprogram(me) which compresses data so as to eliminate blanks and reduce the storage needed for a file*] Packungsteilprogramm *n*, Packungsroutine *f*

pad Lötauge *n* [*Platinenfläche*]

pad Vlies; Vließ *n* [*Österreich*] [*Filtermedium*]

pad Tablett *n*

PAD, packet assembler/disassembler, packet assembly/disassembly facility Paketierer/Entpaketierer *m*, Paketierungs-Entpaketierungsvorrichtung *f* [*Eine Steuer(ungs)einheit, die Daten zu oder von einem in Zeichenbetriebsart arbeitendem Gerät puffert und diese in die für den X.25-Betrieb erforderlichen Pakete zusammenstellt. Diese Einheit ist eine Erweiterung des CCITT X.25*]

padding [*The adding of meaningless data (usually blanks) to a unit of data to bring it up to some fixed size*] Stopfen *n*

padding bit Stopfbit *n*

padlock → portable lock

padlockable absperrbar mit Vorhängeschloss

padlock hasp, portable lock hasp Überfall *m* [*Hangschloss*]

padlocking feature → lockout handle for padlocks

padlock staple, staple of a padlock, portable lock staple Haspe(n) *f,(m)*

to page ausrufen [*Eine Person über Lautsprecher*]

to page blättern

page addressing Seitenadressierung *f*, Abschnittadressierung [*Durch Aufteilung des Speichers in Abschnitte (Seiten) kann man den Aufwand der Adressierung innerhalb der gleichen Seiten verringern, indem kürzere Adressen verwendet werden. Der Nachteil ist, dass eine Adressierung außerhalb der Seite nur mit höherem Aufwand möglich ist.*]

page boundary, page address [*The address of the first (lowest) word or byte within a page of memory*] Seitenadresse, Abschnittsadresse *f*

page description language Seitenbeschreibungssprache *f*

to page down, to page backward rückwärtsblättern, zurückblättern

page flipping, flipping of pages Durchblättern *n* von Bild(er)folgen

page frame Seitenrahmen *m*, Speicherbereich *m [selten: n]* für eine Seite, Kachel *f*

page home line Seitenanfang *m* [*Code*]

page image format, PIF Seitendarstellung *f* [*Kode*]

page make-up Umbruch *m*

page mode seitenweise Betrieb *m*

page numbering Seitennummerierung *f*

to page out → to roll out

page printer [*A high-speed printer used to transpose messages received on paper tape to full-page format, by printing characters one at a time*] Seitendrucker *m*

pager, beeper, radio paging device [*Telecommunications. An electronic device that receives or sends a signal in order to notify its user about messages*] Personenfunkgerät *n*

page reader [*In character recognition, a character reader capable of processing cut-form documents of varying sizes; sometimes capable of reading information in reel forms*] Seitenleser *m*

page read mode seitenweise Lesen *n*

pagesize Zeichen(anzahl) *f* pro Seite

page terminator, PT Seitenende *n* [*Kode*]

to page up vorwärtsblättern

page write mode seitenweise Schreiben *n*

paging (Seiten)Auslagerung *f*, Seitenwechsel *m* [*Es wird nicht der gesamte Prozess ausgelagert, sondern nur (die wenig benutzten) Programmteile (Seiten)*]

paging area Auslagerungsspeicher, Hintergrundspeicher, Seitenwechselspeicher, Seitenauslagerungsspeicher *m* [*Auf ihm werden Programmteile (Seiten) ausgelagert*]

paging call, location call Suchruf, Ausruf, Funkruf *m*

paging device, paging appliance Seiten(wechsel)-speicher, externe (Wechsel)Speicher *m* [*Reicht der Platz im reellen Speicher für Daten und Programme nicht aus, so werden Teile auf den externen Speicher gebracht und wieder in den Arbeitspeicher übertragen, wenn sie zur Verarbeitung benötigt werden. Diese auswechselbaren Teile (2048 bzw. 4096 Bytes) werden Seiten und externe Speicher Seitenspeicher genannt*]

paging service, location service Rufdienst, Personenrufdienst, Suchrufdienst, Ausrufdienst, Funkrufdienst, Ortsrufdienst *m*

paging system, location system, staff locator, electrical paging (system), electrical location system, electrical staff locator (drahtlose) (Personen)Suchsystem, Ausrufsystem, Personen(aus)rufsystem, Rundrufsystem, Funkrufsystem *n*

pain threshold Schmerzschwelle *f*

paint mist filter Farb(en)nebelfilter *m,n*

paint remover Farbenentferner *m*

pair of adapter plates, pair of adaptor plates Zwischenplattenpaar, Adapterplattenpaar *n*

pair of conductors, conductor pair Leiterpaar *n*

P-algorithm P-Algorithmus *m* [*Proportionaler Regelalgorithmus. Das Ausgangssignal des Reglers ist direkt proportional zu Abweichung des Sollwerts vom Istwert.*]

pallet lift (Brit.); pallet elevator (US) Palettenaufzug *m*

pallet(-type) moving walk [*Series of connected and power-driven pallets that together constitute the treadway*] Palettenfahrsteig, Fahrsteig mit Paletten, Fahrsteig mit Fahrplatten, Fahrplattenfahrsteig *m*

palm tree frond, compound palm leaf [*Used for heat-insulation*] Verbundpalmenblatt *n*

PAL system → phase alteration line system

PAM → pulse amplitude modulation

pan Schale *f*

P

pan → **(door) pan**

pan (Bildausschnitt)Schwenk(en) *m,(n)*

pan → **toilet bowl**

pancake coil Scheibenspule *f*

panel [*radio, telephone, etc.*] Wanne *f*, Feld *n*, Einschub *m*

panel → **annunciator**

panel → **console**

panel air ceiling → **radiant cooling ceiling**

panel air heating → **air panel heating**

panel air system → **radiant cooling system**

panel bus Nahbus *m* [*20-100 m*]

panel ceiling heating → **ceiling panel heating**

panel ceiling radiant heating → **ceiling panel radiant heating**

panel cooler [*Refrigerated flat surface, cooling mainly by radiation*] Plattenkühler *m*

panel cutout Tafelausschnitt *m*

panel floor radiant heating → **floor panel radiant heating**

panel heater [*A radiant heater often set flush with the wall or ceiling surface and having a flat front plate heated either by a concealed coil of hot-water pipes or by an electric element*] Paneelheizer *m*

panel heating [*Heat is transmitted by both radiation and convection from panel surface to both air and surrounding surfaces*] Paneelheizung *f*

panel jack [*radio, etc.*] Einbaubuchse *f*

panel office [*telephony*] Stangenwähleramt *n*

P

panel radiant heating, radiant panel heating [*The heating medium may be hot water in pipes buried in floors, walls, or ceilings, or electric resistance heating elements buried in walls or ceilings in panels mounted on the walls or ceilings*] Strahl(ungs)plattenheizung *f*

panel radiator [*Heating unit placed on flush with a flat wall surface and intended to function essentially as a radiator*] Flachradiator, Plattenheiz(ungs)körper, Plattenradiator, Flachheiz(ungs)körper, Paneelheiz(ungs)körper *m*

panel socket outlet Tafelsteckdose *f*

panel switch Tafelschalter *m*

panel radiant heater Plattenstrahlungsheizer, Plattenstrahlungserhitzer *m*

panel wall radiant heating → **wall panel radiant heating**

panic announcement Panikdurchsage *f*

panic bolt [*A patent bolt fitted to the exit of a public building and operated by pressure at the middle of the door*] Panikriegel *m*

panic button Panik(druck)knopf, Schlagtaster *m*

panic exit device, fire-exit device, panic hardware [*A door locking device used on exit doors; the door latch releases when a bar across the inside of a door is pushed*] Panikbeschlag, Stangenbeschlag *m*

panic gate Paniktor *n*

panic lock Panikschloss, Schloss mit Panikbeschlag *n*

panic release Paniköffner *m*

panic switch Panikschalter *m*

panning, roaming Schwenken, Verschieben *n* [*Eine Grundfunktion der Bildschnitt(s)änderung. Kamera; Scheinwerfer*]

panoramic lift (Brit.); panoramic elevator (US) Panoramaaufzug, Sichtaufzug, Panoramafahrstuhl, Sichtfahrstuhl *m*

pantograph Storchenschnabel, Scherenstromabnehmer *m*

pantry sink, larder sink Speisekammerausguss *m*

paper capacitor [*A fixed capacitor consisting of two strips of metal foil that are separated by insulating material and rolled together into a compact tube*] Papierkondensator mit Metallfolien, gewöhnlicher Papierkondensator, Wickelkondensator *m*

paper cup dispenser Pappbecherspender *m*

paper coat(ing) Papierbeschichtung *f*, Papierüberzug, Papierbelag *m*

paper (disposal) chute, paper (disposal) shaft, paper disposer Papierabwurfanlage, Abwurfanlage für Papier *f*, Papier(abwurf)schacht, Abwurfschacht für Papier, Papierschlucker, Schlucker für Papier *m*

paper feed(ing) Papierzufuhr, Papierzuführung *f*, Papiervorschub *m*

paper hanger Tapezierer *m*

paper incinerator Papierverbrenner *m*

paper-insulated cable [*B.S. 480*] papierisolierte Kabel, Papierkabel *n*

paper-insulated lead-covered cable Papierbleikabel, Bleipapierkabel *n*

paper roll towel dispenser Papierrollenhandtuchspender *m*

paper sheet Papierbogen *m*

(paper) tape, punch(ed) tape [*A paper ribbon in which data may be represented by means of partially or completely punched holes*] Papierlochstreifen, Loch(papier)streifen *m*

(paper) tape punch(er) (Papier)Lochstreifenstanzer, Papierstreifenlocher *m*

(paper) tape reader, punch(ed) tape reader [*A device which reads the holes in computer tape, and translates them into alphanumeric characters on a typewrit-*

er, i.e. normal transcript. A card reader does the same with computer cards] (Papier)Lochstreifenleser, Lochpapierstreifenleser *m*

(paper) tape unit [*The mechanism which handles punched paper tape and usually consists of a paper tape transport, sensing and recording or perforating heads, and associated electrical and electronic equipment*] Lochpapierstreifengerät, Papierlochstreifengerät *n*

paper towel dispenser Papierhandtuchspender *m*

paper waste, waste paper Abfallpapier, Altpapier *n*, Papierabfall *m*

paraffin oil [*Petroleum lubricating oil fraction in which straight- and/or branched-chain hydrocarbon structures predominate*] Paraffinöl *n*

paraffin scale Paraffingatsch *m*

paragraph sign → pilcrow

parallel → shunt

parallel, simultaneous [*Transmission of, storage of, or logical operations of the part of a word, character, or other subdivision of a word in a computer, using separate facilities for the various parts*] parallel, gleichzeitig, simultan

parallel assisted heat pump installation → assisted parallel heat pump installation

parallel circuit [*One where all the elements are connected across the voltage source. Therefore, the voltage on each element is the same but the current through each may be different*] Parallelschaltung *f*

parallel compensation, feedback compensation [*Improvement of the response of a feedback control system by placing a compensator in the feedback path, in contact to cascade compensation*] Querkompensation *f*

parallel computer [*A computer that can carry out more than one logic or arithmetic operation at one time*] Parallelrechner *m*

parallel connection → connection-in-parallel

parallel counter synchrone Zähler *m* [*Seine durch parallele Übertragungssignale ausgelösten Zählglieder schalten gleichzeitig. V DIN 19 237*]

parallel data transmission [*A transmission method where the bits which compose words (i.e. some multiple of eight bits) e.g. bytes are transmitted simultaneously. Usually applied to connect peripherals such as printers to computer systems*] Paralleldatenübertragung *f*

parallel-flow heat exchanger [*Heat exchanger in which fluids flow approximately parallel to each other and in the same direction. See remark under "heat exchanger"*] Parallelwärme(aus)tauscher [*In der Industrie so genannt*]; Parallelwärmeübertrager *m*

paralleling bridge Parallelverbinder *m*

parallel operation Parallelbetrieb *m* [*Wärmepumpenanlage mit zweitem Wärmeerzeuger bei der beide Wärmeerzeuger gleichzeitig in Betrieb sein können und gemeinsam die Beheizung durchführen*]

parallel operation [*Performance of several actions, usually of a similar nature, by a computer system, simultaneously through provision of individual similar or identical devices*] Parallelbetrieb *m*

parallel poll Parallel(ab)frage *f*

parallel poll disable, PPD Ausschließen *n* von der Beantwortung der Parallel(ab)frage [*programmierbare Messgerät; DIN IEC 625*]

parallel poll enable, PPE Einstellung *f* zur Antwort auf die Parallel(ab)frage

parallel poll standby state → PPSS

parallel poll unconfiguring Zurücksetzen *n* der Parallel(ab)frage

parallel poll idle state, PPIS Parallel(ab)frageruhezustand *m* [*programmierbare Messgerät; DIN IEC 625*]

parallel poll(ing) Parallel(ab)frage, parallele (Ab)Frage *f* [*IEC-Bus*]

parallel poll(ing) unconfigure, PPU Zurücksetzen *n* der Parallel(ab)frage

parallel programming [*A method for performing simultaneously the normally sequential steps of a computer program (US)/ programme (Brit.), using two or more processors*] Parallelprogrammieren *n*

parallel representation Paralleldarstellung *f*

parallel resonant circuit → antiresonant circuit

parallel (roof) gutter → box(ed) (roof) gutter

parallel-serial converter Parallel-Serien-Umsetzer, Parallel-Serien-Umwandler *m* [*Er setzt parallel dargestellte Daten in zeitlich sequentiell dargestellte Daten um. DIN 19 226, Teil 5*]

parallel-serial-parallel converter Parallel-Serien-Parallel-Wandler, Parallel-Serien-Parallel-Umsetzer *m*

parallel transfer, simultaneous transfer [*Of all bits in a storage location constituting a character or word. Not to be confused with "parallel transmission"*] bitparallele Übertragung, gleichzeitige Übertragung *f*

(para)magnetic cooling → adiabatic demagnetization

parameter Parameter, Einstell(ungs)wert, Kennwert *m* [*Bei festliegender Struktur eines Systems bestimmen seine Parameter in welchem Ausmaß die Ge-bilde des Systems aufeinander wirken. E DIN 19 226, Teil 1*]

parameter card [*A punched card containing the values of parameters needed for a program(me)*] Parameter(loch)karte, Parametersteuer(ungs)karte *f*

P

parameterization, parameter setting Parametrierung *f*, Parametrieren *n* [*Vorgabe der anlagenspezifischen Einstellwerte (Parameter), die für die gewünschten Programmfunktionen erforderlich sind. Parameter können sein: Sollwerte, Grenzwerte, Kennlinien, Zeiten usw. Die Parameter können während des Betriebs eingegeben, angezeigt und verändert werden. VDI 3814 Blatt 1 (neu)*]

to parameterize, to assign parameters, to input parameters parametrieren [*Den Bausteinen eines Reg(e)lungs- und Steuerungssystems Werte ihrer Kenngrößen zuweisen, die ein gewünschtes Verhalten bewirken. E DIN 19 226 T. 5*]

parameter measurement unit, PMU Parametermesseinheit *f* [*Teil einer analogen Messvorrichtung eines Prüfautomaten zur Messung der Parameter von Ein- und Ausgangssignalen (Spannungspegel; Ströme; Flankensteilheiten)*]

parameter of a conic section, conic section parameter Parameter eines Kegelschnittes, Kegelschnittparameter *m* [*Die im Brennpunkt auf der Achse senkrechte Sehne*]

parameter of D-controller D-Regler-Kennwert *m*

parameter of I-controller I-Regler-Kennwert *m*

parameter of P-controller P-Regler-Kennwert *m*

parameter-RAM, PRAM kleine Speicherbaustein *m* [*Er speichert mit Hilfe einer Batterie dauerhaft Systemeinstellungen wie Mausgeschwindigkeiten, Monitor(en)-einstellungen usw.*]

parameter separator Trennzeichen *n* [*Kode*]

parameter value Parameterwert *m*

P

parameter word [*A word in a computer storage containing one or more parameters that specify the action of a routine or subroutine*] Parameterwort *n*

parametric amplifier, mavar, paramp, reactance amplifier, variable-reactance amplifier parametrische Verstärker *m*

parametrized parametr(is)iert

parapet induction unit → **spandrel (wall) induction unit**

parapet (roof) gutter → **box(ed) (roof) gutter**

parasitic light → **false light**

parasitic signal → **unwanted signal**

parcel service Paketdienst *m*

parent file, father file Ausgangsdatei *f*

parent process identification, PPID Elternprozess-Identifikation *f*

pare-soleil [*French*] → **brise-soleil**

parity [*The use of a self-checking code in a computer employing binary digits in which the total number of 1's or 0's in each permissible code expression is always even or always odd*] Parität, Gleichstellung *f*

parity bit, PB, parity check, odd-even bit, odd-even check [*An additional nondata bit that is attached to a set of data bits to check their validity; it is set so that the sum of one-bits in the augmented set is always even or always odd*] Paritätsbit, Gleichstellungsbit, Kontrollbit, Prüf(ungs)bit, PB *n*

parity error, PE [*A machine error in which an odd number of bits are accidentally changed, so that the error can be detected by a parity check*] Paritätsfehler, Gleichstellungsfehler *m*

parity test Übertragungsprüfung, Paritätsprüfung, zeichenweise Informationssicherung, Gleichstellungsprüfung *f*

parking lot Parkplatz *m*

part, piece Stück *n*, Teil *n, m*

part (Werk)Stück *n*

partial air conditioning plant, partial air conditioning installation Teilklimaanlage *f*

partial air heating system Teilluftheiz(ungs)anlage *f* [*Sie deckt nur einen Teil des Wärmebedarfs des Raumes*]

partial airstream, partial airflow Teilluftstrom *m*, Teilluftströmung *f*

partial circulating air conditioning plant, partial circulating air conditioning installation Umluft-Teilklimaanlage *f*

partial condenser → **dephlegmator**

partial failure, partial outage Teilausfall, Teilfehler *m*, Teilversagen *n*

partial (gas) pressure [*Portion of total (gas) pressure of a mix(ture) attributable to one component (Dalton's law of gases)*] Teil(gas)druck, Gas-Teildruck *m*

part(ial) heating load Teilheiz(ungs)last, Heiz(ungs)teillast *f*

partial line down, PLD Teilzeile *f* abwärts [*Kode*]

partial line up, PLU Teilzeile *f* aufwärts [*Kode*]

part(ial) load Teillast *f*

part(ial)-load nozzle Teillastdüse *f* [*Durch sie strömt Gas zum Brenner*]

part(ial)-load operation Teillastbetrieb *m*

part(ial)-load value, PLV [*Single number figure-of-merit expressing part-load efficiency for equipment or the basis of weighted operation at various partial-load capacities for the equipment; expressed in kilowatts per hour of refrigeration*] Teillastwert *m*

partially automatic oil burner teilautomatische Ölbrenner *m* [*DIN 4787, Teil 1. Ein Brenner, der sich vom automatischen Brenner nur dadurch unterscheidet, dass die Inbetriebnahme des Brenners von Hand durch Bedienungspersonal eingeleitet wird, und dass nach einer Brennerabschaltung keine automatische Wiederinbetriebnahme erfolgt.*]

partially glazed teilverglast

partially separate sewerage (system), partially separate system (of sewers) [*The greater part of the rainwater is passed to surface water sewers but the run-off from back roofs and yards is permitted to be discharged to the soil sewers*] Teil-Trennkanalisation *f*

partially wired teilverdrahted

part(ial) miscibility Teilaustauschbarkeit *f*

partial monitoring Teilüberwachung *f*

partial outage → **partial failure**

partial overall monitoring → **total partial monitoring**

partial (pipe) run, partial (pipe) section Teil(rohr)-strang *m*

partial register stall Registerleseverzögerung *f* [*Verzögerung, die sich durch das Lesen eines Registers ergibt, nachdem ein Teilbereich desselben Registers beschrieben wurde. Ein Beispiel ist das Lesen eines Registers in seiner „32-Bit-Form", nachdem es durch eine 8- oder 16-Bit-Operation beschrieben wurde*]

partial space heating Teilraum(be)heizung *f*

partial storage heating Teilspeicher(be)heizung *f*

partial total monitoring → **total partial monitoring**

partial warming Teil(er)wärmen *n*, Teil(er)wärmung *f*

particle → **elementary particle**

particle velocity [*Acoustics. The instantaneous value of the distance travel(l)ed by a particle per unit time in a medium that is displaced from its equilibrium state by the passage of a sound wave*] Schnelle *f*

parting zone → **landing zone**

partitioning Unterteilung *f* [*Ein Grundkonzept der funktionalen Modellierung*]

partitioning Aufteilung *f* [*Aufteilung eines Systems in Einheiten, die so beschaffen sind, dass sie integriert werden können*]

partitioning Festlegung *f* der Zugehörigkeitsfunktionen für die Fuzzifizierung

partition noise Stromverteilungsrauschen *n*

part model Bauteilmodell *n*

part model structure lokalisierte Bauteilmodell *n*

partner key Partnerschlüssel *m*

part number Teil(e)nummer *f*

parts list, pieces list Stückliste, Teileliste *f*

pascal, torr, Pa [*A unit of pressure equal to the presssure resulting from a force of 1 newton acting uniformly over an area of 1 square meter (1 Pa= 1N/m²*] Pascal *n* [*DIN 1301*]

Pascal's law [*The law that a confined fluid transmits externally applied pressure uniformly in all directions, without change in magnitude*] Pascalsche Gesetz *n*

pass, tube bundle pass, tube nest pass (Röhren-) Gang *m*

pass [*A complete cycle of reading, processing, and writing in a computer*] Durchlauf *m*

pass Durchlauf, Verarbeitungsschritt *m* [*Einmaliges Verarbeiten eines Quellenprogramms durch Assemblierer bzw. Kompilierer beim Übersetzungsvorgang*]

pass(age), passage way Durchgang *m*

passage of current → **current passage**

passage (compressed-)air tube installation, passage pneumatic tube installation Durchlaufrohrpost(-anlage), Rohrpostdurchlaufanlage *f*

passband filter Durchlassfilter *m, n*

passenger Fahrgast, Passagier *m*

(passenger) bulk movement → **mass transport(ation)**

(passenger) bulk movement capacity → **mass transport(ation) capacity**

passenger-carrying → **mass transport(ation)**

passenger-carrying capacity → **mass transport(ation) capacity**

passenger-carrying surface → **continuous bent treadway**

passenger compartment → **(lift) car**

passenger conveying system Personentransportsystem *n*

passenger conveyor → **autowalk**

passenger conveyor with upper transition curve → **autowalk with upper transition curve**

passenger elevator [*Elevator used primarily to carry persons other than the operator and persons necessary for loading and unloading*] Fahrstuhl, Personenaufzug *m*

passenger flow Personenstrom, Personenfluss *m*

passenger lift (Brit.); passenger elevator (US) Fahrstuhl *m*, Personenaufzug *m* [*DIN 15 301 und DIN 15 302*]

(passenger) lift attendant (Brit.); (passenger) elevator attendant (US) Fahrstuhlbediener, Fahrstuhlführer, Personenaufzugbediener, Personenaufzugführer *m*

P

passenger (lift) cage, passenger (lift) car (Brit.); passenger (elevator) cage, passenger (elevator) car (US) Fahrstuhlkabine, Personenaufzugkabine *f*, Fahrstuhlkorb, Personenaufzugkorb *m*

passenger service Personenbetrieb *m* [*z. B. bei einem Aufzug*]

passenger traffic Personenverkehr *m*

passer-by Vorübergehende *m, f*

passing Vorbeigehen *n* [*Personen*]; Vorbeifahren *n* [*Fahrzeug*]

passive air lock passive Luftschleuse *f* [*Luftschleuse ohne Anschluss an eine RLT-Anlage*]

passive detector, passive detection device, passive detecting device passive Melder, passive Meldungsgeber, passiver Detektor *m*, passives Meldegerät *n* [*Er überwacht seinen Bereich auf eine bestimmte physikalische Größe und gibt bei deren Änderung eine Meldung ab*]

passive device passive Bauelement *n* [*Ein Bauelement ohne Verstärker- oder Gleichrichtereigenschaften, z. B. Kondensator, Spule oder Widerstand*]

passive infrared detector, passive infrared detecting device, passive infrared detection device Passiv-Infrarot-Melder, Passiv-Infrarot-Meldungsgeber *m*, Passiv-Infrarot-Meldegerät *n*

passive station Wartestation *f*

passive infrared motion detector, passive infrared motion detecting device, passive infrared motion detection device Passiv-Infrarot-Bewegungsmelder, Passiv-Infrarot-Bewegungsmeldungsgeber *m*, Passiv-Infrarot-Bewegungsmeldegerät *n*

P

passive injector source passive Injektionsquelle *f* [*Messung des Luftaustausches zwischen Zonen mit mehreren Spurengasen*]

passive transducer [*Transducer that has no power source other than the input signal(s), and whose output signal power cannot exceed that of the input*] passive Wandler, passive Messwertumformer, passive Messwertgeber, passive Signalumformer, passive Signalgeber *m* [*DIN 19 237*]

passkey → change key

passkey(ed) lock → change key(ed) lock

passkey system Generalschließanlage, Hauptschließanlage *f*

pass-through window Durchreiche(fenster) *f, (n)*

password [*A unique string of characters that a programmer, computer operator or user must supply to meet security requirements before gaining access to a computer's operating system, computer program or data*] Passwort, Schlüsselwort, Kennwort, Kodewort, Codewort *n*, Parole *f*

password authorization Schlüsselwortberechtigung, Passwortberechtigung, Codewortberechtigung, Kennwortberechtigung, Kodewortberechtigung *f*

password check(ing) Passwortkontrolle, Kodewortkontrolle, Codewortkontrolle, Kennwortkontrolle, Schlüsselwortkontrolle *f*

passwording Passwortprogramm, Kennwortprogramm, Codewortprogramm, Kodewortprogramm, Schlüsselwortprogramm *n*

password level Passwortebene, Kennwortebene, Kodewortebene, Codewortebene, Schlüsselwortebene *f*

password protection Passwortschutz, Codewortschutz, Schlüsselwortschutz, Kennwortschutz, Kodewortschutz *m*

paste buffer [*blob*] Zwischenablage *f*

pasted-plate storage battery → Faure storage battery

P.A. system → (public-address) (loud)speaker system

patch Teilfläche *f*, Freiformflächenelement *n* [*Rechnerintern in Vierecke zerlegte Freiformfläche*]

patch Korrektur *f* [*Ändern von Programmteilen ohne deren Länge zu ändern*]

paternoster (lift) (Brit.) → continuous lift

path → travel

path → channel

path [*The logical sequence of instructions followed by a computer in carrying out a routine*] Befehlsreihenfolge, Strecke *f*

path Pfad *m*

path attenuation, path damping, path abatement, path deadening [*Power attenuation between transmitter and receiver, due to any cause*] Streckendämpfung, Streckenabschwächung, Wegdämpfung, Wegabschwächung *f*

path ((open) loop) control → PC

patient (bath)tub → manipulation (bath)tub

patrol Revierdienst *m*

patrol → (security) (guard) patrol

patrol company → (security) (guard) patrol company

patrol tour → (security) (guard) patrol tour

pattern Muster *n*

pattern Punktbild *n*

pattern(ed) ceiling strukturierte Decke, gemusterte Decke, Musterdecke *f* [*z. B. eine Rasterdecke*]

pattern matching Musteranpassen *n*, Musteranpassung *f*

pause time maximal erlaubte Zeit *f* [*Speicherprüfung*]

pay station, public (switched) (tele)phone station, call box öffentliche Telefon *n*, öffentliche Fernsprecher *m*

pay television, pay TV, video on demand Bezahl(t)-fernsehen *n*

PB → parity bit

PB → positive-big

PC, pulsating current pulsierende Strom *m*

PC, photoconductor Photoleiter *m*

PC → workstation (computer)

PC → photoresistor

PC, personal computer Personalrechner *m* [*Im engeren Sinne ein IBM-verträglicher Personalrechner*]

PC, path ((open) loop) control, travel ((open) loop) control Pfadsteuerung, Weg(e)steuerung *f*

P.C.E. → pyrometric cone equivalent

P-channel MOS transistor, P-MOS transistor P(-Kanal)-MOS-Transistor *m*

PCL, printer command language Druckerkommandosprache *f*

P ((closed) loop) control → proportional ((closed) loop) control

P controller, PCTRL, proportional(action) controller [*See remark under "controller"*] P-Regler *m*, P-Regelorgan *n*, P-Regel(ungs)einheit *f* [*Regler nur mit P-Anteil im Regelalgorithmus. DIN 19 225*]

PCS → print contrast signal

PCTRL → P controller

PD → pressure-dependent

PDAU, physical delivery access unit Zugriffseinheit *f* für elektronische Briefübermitt(e)lung

PD controller, PD regulator, proportional-plus-derivative controller, proportional-plus-derivative regulator, rate(-action) controller, rate(-action) regulator PD-Regler *m* [*PD-Regler mit zusätzlicher D-Aufschaltung. DIN 19 225*]

PD element, proportional-plus-derivative-action element PD-Glied *n* [*Die Parallelschaltung von P-Glied und D-Glied*

$$v\,(t) = K_P\,u(t) + K_D\,\frac{d\,u(t)}{dt} = K_P\left[u(t) + T_V\,\frac{d\,u(t)}{dt}\right]$$

$$T_V = \frac{K_D}{K_P}\ \textit{Vorhaltzeit}$$

Übertragungsfunktion $G(s) = K_P\,(1 + T_V s)$
Bezogene Anstiegsantwort mit $u(t) = ut$
$v(t) = K_P\,(u * t * T_V u)$
$t = T_V;\ v = 0.$
E DIN 19 226 Teil 2]

PDN, public data net(work) öffentliche Datennetz *n* [*Ein auf Paket- oder Verbindungsgrundlage arbeitendes Netzwerk, das einem großen Anwenderkreis zur Verfügung steht. Es bietet viele wertvolle Dienste und ist kostengünstig, da Kommunikationsquellen gemeinsam genutzt werden. Durch integrierte Redundanz gewöhnlich sehr zuverlässig.*]

PE → phase-encoding

PE → parity error

PE (Brit.) → protection ground

peak... → maximum...

pe(a)rlite thermal insulation, pe(a)rlite heat insulation Perlitwärmedämmung *f*

pearl lamp [*A lamp "frosted" on the inside surface of the glass*] innenmattierte Lampe *f*

pecker [*The part of a paper tape reader that senses holes in the tape*] Lochfühler *m*

Peclet number [*Chemical engineering. A dimensionless parameter that is the product of the Reynolds and Prandtl numbers, used in forced convection for internal flow.*] Pecletzahl *f*

PECVD → plasma enhanced chemical vapour deposition

pedal actuation, pedal operation Fußhebelbedienung, Fußhebelbetätigung, Pedalbedienung, Pedalbetätigung *f*

pedestal Säule *f*

pedestal ashtray Säulenaschenbecher *m*

pedestal drinking fountain Säulenbrunnen *m*, Brunnensäule *f*

pedestal (wash)basin Säulenwaschbecken *n*

pedestal W.C., pedestal water closet Säulenabort *m*

pedestal urinal Säulenurinal *n*, Urinalsäule *f*

pedestrian access ((open) loop) control Fußgängerzugang(s)steuerung *f*

pedestrian gate Fußgängertor, Gehtor *n*

pedestrian traffic Gehverkehr, Fußgängerverkehr *m*

peeler, peeling machine Schälmaschine *f*

peeling chamber, (peeling) hopper Schältopf *m* [*Schälmaschine*]

peer device, peer appliance Partnergerät *n*

peer entity [*Entity within the same layer*] Partnerinstanz *f* [*E DIN ISO 7498*]

peer entity authentication [*The corroboration that a peer entity in an association is the one claimed*] Partnerinstanzechtheit *f*

peer-to-peer communication Buskommunikation jeder mit jedem, Kopp(e)lung von Automatisierungsstationen ohne Leitzentrale, Partner-zu-Partner-

Kommunikation, Querkommunikation *f* [*Die Kommunikation zwischen beteiligten Automationsstationen, bei der es erlaubt ist alle Informationen spontan oder auf Anforderung von anderen Stationen zu übermitteln. Der Informationsaustausch kann in einer oder in beiden Richtungen individuell mit einer oder mit mehreren oder mit allen Kommunikationspartnern er-folgen. IEC 870-1-4:1994*]

peer-to-peer LAN, peer-to-peer local area network Partner-zu-Partner-Netz(werk) *n* für örtliche Bereiche

peer-to-peer network installation, peer-to-peer network plant Partner-zu-Partner-Netzanlage *f*

peg hinge → pin hinge

Peirce logic (operator) → nor logic (operator)

P element, proportional action element P-Glied, Proportionalglied *n* [*Es hat das Übertragungsverhalten* $v(t) = K_P\,u(t)$. *Dabei heißt die Konstante* K_P *Proportionalbeiwert (P-Beiwert). Die Übergangsfunktion hat den Verlauf* $h(t) = K_P \cdot \sigma\,(t)$. *E DIN 19 226 T2, E DIN 19 226 T5*]

pelmet Vorhangleiste, Blende *f* [*Fenster*]

pel path Weg *m* der photographischen Elemente

Peltier cooler Elektroluftkühler, Peltierkühler *m*

Peltier effect [*Evolution or absorption of heat produced by an electric current passing across junctions of no suitable dissimilar metals, alloys, or semiconductors*] Peltiereffekt *m* [$dQ_p = p \cdot dI.$] [*1834 von Peltier entdeckt*]

PELV → protection extra low voltage

pen (Zeichen)Stift *m*

PEN (Brit.) → protection ground neutral (US)

pen computer (Zeichen)Stiftrechner *m*

pendulum detector, pendulum detection device, pendulum detecting device Pendelmelder *m*, Pendelmeldegerät *n*

pendulum luminaire, pendulum light(ing) fitting Pendelleuchte *f*, Pendelbeleuchtungskörper *m*

pen eraser Radierstift *m*

penetration Durchführung *f* [*Durch eine Decke, Wand, Mauer oder (Fuß)Boden*]

penetration, penetrating Durchdringen *n*, Durchdringung *f*

penetration layer Darstellungsschicht, Schicht 6 *f* [*OSI-Schichtenmodell*]

penetration monitoring Durchdringungsüberwachung *f*

Penn (slide) valve Pennsche Schieber *m*

pen plotter Stiftzeichenmaschine *f*, Stiftzeichner *m*

penruler digitale Kurvenlineal *n*

pentagrid converter Fünfgittermischröhre *f*

penthouse Überfahrt *f* [*Aufzug*]

penthouse machine room Überfahrtmaschinenraum *m* [*Aufzug*]

pentode, five-electrode electron tube [*It contains an anode, a cathode, a control electrode, and two electrodes that are ordinarily grids*] Pentode, Fünfpolröhre *f*

people-generated heat → heat generated by people

people load Personenlast *f* [*In einem Raum*]

people lock Personenschleuse *f*

people-mover plant, people-mover installation Personen(be)förderungsanlage *f*

people outside air rate Personen(-Außen)luftrate *f* [*Außenluftvolumenstrom, bezogen auf eine Person*]

PEP → maximum envelope power

perceived air quality empfundene Luftgüte *f*

perceived noise level, PNL Lärmigkeit *f* [*Maß für die Störminderung von Lärm*]

percentage → proportion

percentage duty cycle, proportion duty cycle Einschalt(ungs)dauer *f* [*Ignitron*]

percentage humidity [*Ratio of the weight of water vapo(u)r associated with a pound of dry air for the weight of water vapo(u)r associated with a pound of dry air saturated at the same temperature*] Feuchteprozentsatz, Feuchtigkeitsprozentsatz *m*

perceptibility Wahrnehmbarkeit *f*

perceptibility Wahrnehmungsvermögen *n*, Wahrnehmungsfähigkeit *f*

perception Wahrnehmung *f*

perfect combustion, ideal combustion vollkommene Verbrennung, ideale Verbrennung *f*

perfect fluid → ideal fluid

perfect gas → ideal gas

perfect liquid → ideal liquid

perfect mixing flow vollkommene Mischströmung *f* [*Eine Raumströmungsform*]

perforated ceiling [*One composed of perforated panels used to distribute the air uniformly throughout the ceiling or portion thereof*] Loch(platten)(unter)decke, Lochtafel(unter)decke *f*

perforated ceiling panel Loch(unter)deckentafel, Deckenlochtafel *f*

perforated ceiling ventilation Loch(unter)deckenlüftung *f*

perforated grill(e) Lochgitter *n*

perforated plate ceiling Lochblech(unter)decke *f*

perforated rail Lochschiene *f*

perforated slab Lochplatte *f*

perforated soap tray → self-draining soap tray

(perforated) tape transmitter (Loch)Streifengeber *m*

to perform → to execute

performance, execution [*In compound work, the term "performance" may also mean in German "Leistung" as, for instance, "performance data" (Leistungsdaten)*] Ausführung *f*

performance closed loop control, output closed loop control Leistungsreg(e)lung *f*

performance coefficient → coefficient of (heating) performance

performance contracting Verträge *f* auf der Grundlage von Erträgen durch Energieeinsparung

performance data, power data [*The information assessing the characteristics and effectiveness of a device, instrument, machine, or substance during use*] Leistungsdaten *f*

performance drop, output drop Leistungsabfall, Leistungsrückgang *m*, Leistungsminderung *f*

performance factor [*Ratio of the useful output capacity of a system to the input required to obtain it. The units of capacity and input need not be consistant*] Leistungsfaktor *m*

performance increase Leistungssteigerung, Leistungszunahme, Leistungserhöhung *f*

performance need → output requirement

performance requirement → output requirement

performance time → execution time

performance unit → execution unit

performer Leistungsgerät *n*

perform statement Laufanweisung *f* [*Datenverarbeitung*]

perimeter [*The total length of a closed curve; for example, the perimeter of a polygon is the total length of its sides*] Umfang *m*, Perimeter *n* [*Schweiz: m*]

perimeter air conditioning, perimeter air treatment [*Air-conditioning by a system which feeds air through registers located along the outer walls, supplied through ducts from a plenum chamber*] Perimeterklimareg(e)lung, Perimeterklimatisierung, Perimeterluftbehandlung, Perimeterluftaufbereitung, Perimeterluftvered(e)lung *f*

perimeter barrier Perimeterschranke, Perimetersperre *f*

perimeter duct Perimeterkanal *m*

perimeter floor outlet Perimeter(fuß)bodenauslass *m*

perimeter glass Außenhautglas *n* [*Gebäude*]

perimeter hot-air heating, perimeter warm-air heating [*Warm-air ducts embedded in the concrete slab of a basementless house, around the perimeter, receive heated air from a furnace and deliver it to the space through registers placed in or near the floor. Air is returned to the furnace from registers near the ceiling*] Perimeter(warmluft)heizung, Warmluftperimeterheizung *f*

perimeter infrared barrier, infrared perimeter barrier Perimeterinfrarotschranke, Infrarotperimeterschranke, Perimeterinfrarotsperre, Infrarotperimetersperre, Perimeterultrarotschranke, Ultrarotperimeterschranke, Ultrarotperimetersperre *f*

perimeter loop Perimeterschleife *f*

perimeter-radial (duct) system, perimeter-radial distribution system [*For crawl space or basement*] Perimeterradialkanalsystem *n*

perimeter register system Perimeterregistersystem *n*

perimeter riser Perimetersteig(e)rohr *n*

perimeter security Perimetersicherung *f*

perimeter side-wall outlet Perimeterseitenwandauslass *m*

periodical recording on mass memory zyklische Langzeitspeicherung, Speicherung für Analysefunktionen *f*

period of current → ON period

peripheral bus, I/O bus, input-output bus Peripheriebus, Übergabebus, E/A-Bus, Ein-/Ausgabebus *m*

peripheral cable grounding (US); peripheral cable earthing (Brit.) periphere Kabelerdung *f*

peripheral component interconnect bus PCI-Bus *m*, [*Vom Prozessorhersteller Intel eingeführt*]

peripheral (device) Peripheriegerät *n* [*Nicht zur Zentraleinheit gehörende und vom Rechner ansprechbare Speicher- und Ein-/Ausgabeeinheit*]

peripheral interchange program (US); peripheral interchange programme (Brit.); PIP Peripheriedatenaustauschprogramm *n*

peripheral interface adapter, peripheral interface adaptor, PIA periphere Schnittstellenvorsatz *m*

peripheral memory → external memory

peripheral message specification → PMS

peripheral monitoring Außenhautüberwachung *f*, organisatorisch/personelle Maßnahmen *f pl* [*Überwachung aller Einstiegsmöglichkeiten. Regel ist die Öffnungs- und Flächenüberwachung. Definition nach Pflichtenkatalog der deutschen Kriminalpolizei*]

P

peripheral protection Außenhautschutz, Peripherieschutz *m* [*Alle Sicherheitsmaßnahmen an der Gebäudehülle, um das Objekt oder die Objekte im Gebäude zu schützen. Er besteht aus mechanischen/baulichen Maßnahmen (Außenhautsicherung), aus elektronischen Maßnahmen (= Detektionsmaßnahmen) und/ oder organisatorischen/personellen Maßnahmen (Außenhautüberwachung)*]

peripherals Peripheriegeräte *f* [*Sämtliche an die Zentraleinheit angeschlossene Hardware*]

peripheral section Peripheriebereich *m* [*selten: n*] [*Ein Funktionsbereich eines Vermittlungsmoduls*]

peripheral security Außenhautsicherung, mechanisch/bauliche Maßnahmen *f pl*

peripheral unit Peripherieeinheit *f* [*E DIN 44 300/10.85*]

periphery [*The bounding curve of a surface*] Peripherie *f*

periphery [*The surface of a solid*] Festkörperoberfläche *f*

permanent → **continuous**

permanent basic load Dauergrundlast, Dauerbasislast *f*

permanent communication link Dauerkommunikationsstrecke, Kommunikationsdauerstrecke *f*

permanent coupling nichtschaltbare Kupp(e)lung *f*

permanent enable, permanent EN Dauerfreigabe *f*

permanent instruction Dauerbefehl *m*

permanent locking Dauerverrieg(e)lung *f*

(permanent-magnet) moving-coil element Drehspul(en)messwerk *n*

P

permanent operation → **continuous operation**

permanent pressure (fire) extinguisher Dauer(druck)(feuer)löscher, Dauer(druck)brandlöscher *m*

permanent sound level Dauerschallpegel *m*

permanent steam output → **continuous steam output**

permanent stress → **continuous stress**

permanent threshold shift, PTS, noise-induced permanent threshold shift, NIPTS irreparable Hörschwellenverschiebung *f* [*durch starken Lärm*]

permanent unlocking Dauerentrieg(e)lung *f*

permanent ventilation Dauerbelüftung, Dauerbe- und -entlüftung *f*

permanent (water) hardness permanente (Wasser)-Härte, Dauerwasserhärte, Nichtkarbonathärte, NKH *f*

permission drawing Genehmigungszeichnung, Bewilligungszeichnung, Erlaubniszeichnung *f* [*Sie ist der Baugenehmigungsbehörde vorzulegen*]

permutation lock → **combination lock**

perpetuating testimony method Beweissicherungsverfahren, selbständige Beweis(aufnahme)verfahren, (vorsorgliche) Beweis(aufnahme)verfahren *n*

persistent command Dauerbefehl *m* [*Fernwirktechnik*]

personal computer → **PC**

personal detachment → **privacy**

personal identification number, personal identity number, PIN Personennummer, persönliche Identifikationsnummer, personenbezogene Identifikationsnummer *f* [*DIN 6763, Teil 1*]

personal laser writer, personal laser printer Einzelplatzlaserdrucker, Lasereinzelplatzdrucker *m*

personal radio call system Personenfunkrufsystem *n*

person identification, person ID Personenidentifizierung, Personenkenn(zeichn)ung *f*

person/machine interface Personen-Maschinen-Schnittstelle *f*

person's file Personendatei *f*

person's lock Personenschleuse *f*

person's warmth Personenwärme *f*

PERT → **project evaluation and review technique**

pertains to gehört zu [*Beschriftung im Zeichnungsschriftfeld*]

pertinent einschlägig [*z. B. Vorschrift*]

perturbation [*Physics. An influence or a system that only disturbs the system slightly*] geringfügige Störung *f*

Petersen coil → **arc-suppression coil**

petrol (Brit.) → **gas(oline) (US)**

petroleum lubricating oil Erdölschmieröl *n*

petroleum product Mineralölerzeugnis *n* [*ISO 6245 – 1982*]

petrol (intercepting) trap, petrol separator, petrol interceptor (Brit.); gas(oline) (intercepting) trap, gas(oline) interceptor, gas(oline)separator, gasoline interceptor (US) [*A chamber, the outlet of which is fitted with a trapping band, or a series of such chambers, in which any petrol (Brit.) / gas(oline) (US) present in the drains collects on the top of the contained water and evaporates through a vent pipe carried to higher level*] Benzin(ab)scheider, Benzinfang, Benzinfänger *m* [*DIN 1999*]

'P' (fixture) trap, 'P' stench trap, 'P' plumbing trap [*A trap with the inlet vertical and the outlet inclined slightly below the horizontal*] P-Geruchsverschluss *m*

PGA → **programmable gain amplifier**

P-gate thyristor kat(h)odenseitig steuerbare Thyristor *m*

PH [*Chemistry. A symbol for the logarithm of the reciprocal of the hydrogen-ion concentration of an aqueous solution, used to express its acidity alkalinity. At 25 Grad C (77 °F), a neutral solution, such as pure waters, has a pH of 7; a pH under 7 indicates that the solution is acidic, and a pH over 7 indicates that the solution is alkaline*] Wasserstoffexponent pH *m*

phase advancer Blindleistungsmaschine *f*

phase alteration line system, PAL system [*A colo(u)r television system used in Europe, in which the phase of the colo(u)r subcarrier is changed from scanning line to scanning line, requiring transmission of a line switching signal as well as a colo(u)r burst*] PAL(-System) *n* [*Es arbeitet mit 625 Zeilen, 833 Pixeln pro Zeile und 50 Hz (50 Halbbilder pro Sekunde)*]

phase angle [*Electricity. The angle of a steady-state electrical quantity, such as voltage, current, or impedance, described as a phasor*] Phasenwinkel *m*

phase angle, phase difference [*Physics. The relative angular displacement between a periodic quantity and a reference angle or between two sinusoidally varying quantities of identical frequencies*] Phasenwinkel, Phasenunterschied *m*

phase approach Phasenanschnitt *m*

phase change [*Shift of a material or system from one phase to another, such as liquid to gas, liquid to solid, solid to gas*] Phasenwechsel *m*

phase change method Phasenwechselverfahren *n*, Phasenwechselmethode *f*

phase change of ice to water Abschmelzen *n*, Abschmelzung *f*, [*Eisspeicher(anlage)*]

phase-changing coefficient → phase constant

phase comparator Phasenvergleicher *m*

phase-comparison relaying [*A method of detecting faults in an electric power system in which signals are transmitted from each of two terminals every half cycle so that a continuous signal is received at an intermediate point if there is no fault between the terminals, while a periodic signal is received if there is a fault*] Phasenvergleich(s)relaissystem *n*

phase conductor [*In a polyphase circuit, any conductor other than the natural conductor*] Außenleiter, Phasenleiter, Hauptleiter *m*

phase constant, phace-changing coefficient, wavelength constant Phasenkonstante *f*

phase control → (generalized) phase (open) (loop) control

phase control, phase shift control [*A control that changes the phase angle at which the alternating current-line voltage fires a thyratron, ignition, or other controllable gas tube*] Zündeinsatzsteuerung *f*

phase control Phasensteuerung *f*

phase converter [*It changes the number of phases in an alternating-current power source without changing the frequency*] Phasenwandler, Phasenumformer, Phasenumrichter *m*

phase coupler Phasenkopp(e)ler *m*

phase crossover [*A point on the plot of the loop ratio at which it has a phase angle of 180 Grad*] Phasenschnitt *m*

phase crossover frequency Phasenschnittfrequenz *f* [*Bodediagramm*]

phase delay Phasenverzögerung *f*, Phasenverzug *m*

phase difference → phase angle

phase displacement Phasenverschiebung *f*

phase-earth-phase fault zweiphasige Erdschluss, doppelphasige Erdschluss, zweipolige Kurzschluss mit Erdberührung, doppelpolige Kurzschluss mit Erdberührung, zweipolige Erdschluss, doppelpolige Erdschluss, Zweipolerdschluss, Doppel(pol)erdschluss *m*

phase-encoding, PE Richtungstaktschrift, Zwei-Phasen-Schrift *f*

phase factor → power factor

phase-failure monitoring Phasenausfallüberwachung *f*

phase-failure relay Phasenausfallrelais, Unterspannungsrelais *n*

phase gating Phasenanschnitt *m*

phase gating (open) (loop) control Phasenanschnittsteuerung *f*

phase jitter → jitter

phase lamp Phasenlampe *f*

phase-locked (control) loop, PL(C)L Nachlaufsynchronisation, phasensynchronisierte (Regel)Schleife, Phasenregelschleife *f* [*VCO und Phasendetektor*]

phase modulation → PM

phase of matter → state

phase ((open) loop) control → (generalized) phase ((open) loop) control

phase-out Ausstieg *m*, Anwendungsende *n*

phase selection Umschalten *n*, Phasen(aus)wahl *f*

phase separation Phasentrennung *f*

phase sequence Phasen(reihen)folge, Leiterfolge, Klemmenfolge, Drehfeldrichtung *f*

phase-sequence monitoring Phasen(reihen)folgeüberwachung, Drehfeldüberwachung *f*

phase-shift control → phase control

phase shifter [*A device used to change the phase relation between two alternating-current values*] Phasenschieber, Phasendreher *m*

phase-shift keying, PSK Phasenumtastung *f* [*Eine der drei prinzipiellen Modulationsarten für digitale Signale. Sie existiert in reiner Form als Umtastung der Phase des HF-Trägers um 45 Grad, 90 Grad oder 180 Grad*]

phase tapping-off, phase location Phasenabgriff *m*

phase-to-earth voltage (Brit.) → line-to-earth voltage

phase to phase Phase *f* gegen Phase

phase transformer → induction regulator

phasor [*Electricity. A polar (magnitude and angle) representation of a complex (real and imaginary) number; generally used to represent steady-state electrical quantities, such as voltage, current, and impedance at a single frequency*] Zeiger *m*

PHIGS, Programmer's Hierarchical Interactive Graphics Standard US-Standard *m* einer Programmschnittstelle für graphikfähige Geräte

phonetic transcription Lautschrift *f*

photocathode, photokathode [*A photosensitive surface that emits electrons when expressed to light or other suitable radiation; used in phototubes, television camera tubes, and other light-sensitive devices*] Photokat(h)ode, Fotokat(h)ode *f*

photocell, electrical eye, photoelectrical cell, photosensor [*A solid-state photosensitive electron device whose current-voltage characteristic is a function of incident radiation*] elektrisches Auge *n*, Photozelle, Fotozelle *f*, Lichtfänger, Lichtfang, Lichtsensor *m*

photoconductivity Photoleitfähigkeit, Photowiderstandsfähigkeit *f*, Photowiderstandsvermögen *n*

photodiode, photoconductor diode Photodiode, Fotodiode *f* [*Sie ändert bei Belichtung ihren Widerstand*]

photodiode image converter, photodiode (electron) image tube Photodiodenbildwandler(röhre) *m*, (*f*)

photoeffect lichtelektrische Effekt, photoelektrische Effekt, Photoeffekt, Fotoeffekt *m*

photoelectrical cell → photocell

photoelectric lighting controller, photoelectric lighting switch, sun relay, solar relay Dämmer(ungs)schalter *m*

photoelectric semiconductor photoelektrische Halbleiter *m*

photoelement Photoelement, Fotoelement *n*

photograph → shot

photometry Licht(stärke)messung *f*

photo microscope → optical microscope

photomultiplier cell [*Electronics. A device combining a photodetector to sense light and an electron multiplier to amplify the resultant current*] Photovervielfacher, Fotovervielfacher *m*

photon, light quantum [*A massless particle, the quantum of the electromagnetic field, carrying energy momentum and angular momentum*] Lichtquantum, Photon *n*

photon counting Photon(e)zählung *f* [*Messverfahren zum Nachweis geringer optischer Signalleistung bei vorgegebener Lichtmeterlänge*]

photo punching pliers Photozange *f* zum (Aus)Stanzen

photoresistor Photowiderstand, Photoleiter *m* [*Ein lichtabhängiger Widerstand, bei dem der innere fotoelektrische Effekt im Halbleitermaterial ausgenutzt wird*]

photosemiconductor Photohalbleiter *m*

phototransistor [*A junction transistor that may have only collector and emitter leads or also a base lead, with the base exposed to light through a tiny lens in the housing; collector current increases with light intensity, as a result of amplification of base current by the transistor structure*] Phototransistor *m*

photovoltaic cell, phototronic photocell photovoltaische Zelle *f*

photovoltaic facade Photovoltaikfassade *f*

photovoltaic generator, photovoltaic array photovoltaische Generator *m*

photovoltaics Photovoltaik *f*

photovoltaics plant, photovoltaics installation, photovoltaics system Photovoltaikanlage, Photovoltaikinstallation *f* [*Sie erzeugt Strom aus der Sonnenstrahlung*]

pH range of colo(u)r change Umschlag(s)intervall *m*

ph-value; potentia hydrogenii [*Latein*] Wasserstoffexponent *m* [*Wasser leitet elektrischen Strom, wenn an zwei Elektroden, die in das Wasser eintauchen, eine elektrische Spannung angelegt wird. Auch chemisch reines Wasser ist in geringem Maße stromleitend. Das bedeutet, dass Ladungsträger, sogenannte Ionen, vorhanden sein müssen. Bei chemisch reinem Wasser sind dies positiv geladene H^+-Ionen und negativ geladene OH^--Ionen. Genaue Untersuchungen haben ergeben, dass sich das positiv geladene Wasserstoff-Ion an ein ungespaltenes Wassermolekül anlagert. Es kommt zur Bildung von H_3O^+-Ionen. Messungen für chemisch reines Wasser mit einer Temperatur von 25 °C ergeben folgende Ionen-Konzentrationen: In einem Liter Wasser sind ca. 1/10 000 000 Gramm H^+-Ionen und gleichviel OH^--Ionen vorhanden. Mathematisch mit Zehnerpoten-*

P

zen ausgedrückt, sind also 10^{-7} Gramm H^+-Ionen und 10^{-7} Gramm OH^--Ionen vorhanden. Da diese Konzentration der Wasserstoff-Ionen in einem Liter chemisch reinem Wasser sehr klein ist, wird zur Vereinfachung nur der negative Exponent der Basis 10 – das ist die Ziffer 7 – zur Kennzeichnung der Wasserstoff-Ionen-Konzentration herangezogen und als pH-Wert bezeichnet. Wenn die Konzentration der H^+-Ionen gleich der Konzentration der OH^--Ionen ist, so handelt es sich um neutrales Wasser (pH = 7). Gibt man dem Wasser Säure zu, so steigt die Konzentration der H^+-Ionen, z. B. auf 10^{-4} an (pH = 4). In gleichem Maße nimmt die Konzentration der OH^--Ionen auf 10^{-10} ab. Gibt man neutralem Wasser eine Lauge zu, nimmt die Konzentration der H^+-Ionen ab, z. B. auf 10^{-9}. Das Wasser hat dann einen pH-Wert von 9 (pH = 9). Die Konzentration der negative OH^--Ionen steigt dabei auf 10^{-5}. Die Summe aus positiv und negativ geladenen Ionen ergibt stets den konstanten Wert 10^{-14}.]

physical wirklich, sachlich, richtig, physikalisch

physical address [*The code used by a computer system to identify a particular peripheral in a network or a location in memory*] physikalische Adresse *f*

physical (barrier) security [*It involves the protection of an asset through the erection of physical barriers, such as locks, bars, and grilles*] Sperrensicherung *f*

physical data point physikalische Datenpunkt *m* [*Punkt in einer Automationsstation mit physikalischer Verbindung zu einem Eingang oder Ausgang*]

physical delivery access unit → PDAU

physical dimensions → physical size

physical input/output control system → PIOCS

physical interface physikalische Schnittstelle *f*

physical layer, layer 1 physikalische Schicht, erste Schicht, physikalische Ebene, erste Ebene, Bitübertragungsschicht, Bitübertragungsebene, Schicht 1, Ebene 1 *f* [*Für die Kommunikation in digitalen Netzwerken*]

physical line [*data processing*] metallische Leiter *m*

physical memory physikalische Speicher *m* [*Der in Form von Speicherbausteinen vorhandene Speicher*]

physical plant → installation

physical point address technische Adresse *f* [*Die vom System vorgesehene technische Adresse einer physikalischen Information, z. B. Außentemperatur*]

physical position Stellung *f* [*z. B. eines Ventils*]

physical protocol Bitübertragungsprotokoll *n* [*E DIN ISO 7498*]

physical record → block

physical service Bitübertragungsdienst *m* [*E DIN ISO 7498*]

physical size, physical dimensions mechanische Größe, mechanische Maße, mechanische Abmessungen *f*

physical unit physikalische Einheit, Baueinheit *f* [*E DIN 44 300/10.85*]

physical variable [*Pressure, length, voltage, power, enthalpy, temperature, speed, velocity etc.*] physikalische Größe *f*

physical water treatment physikalische Wasserbehandlung *f*

picker (knife) Kartenmesser *n*

picking schlüssellose Öffnen, Aufsperren *n* [*Der Versuch mit systemfremden Sperrwerkzeugen den Schlosszylinder zu öffnen*]

picking-off, picking-up, tapping Abgreifen, Anzapfen *n* [*Herstellen eines elektrischen Kontaktes zu zwei bestimmten Punkten einer Schaltung oder eines Gerätes, um die dazwischen bestehende Spannung messen oder an ein anderes Schaltelement anlegen zu können*]

picking security → security against picking

picking tool Aufsperrwerkzeug *n*

picking window Sichtfenster *n* [*Photozange*]

pick-lock, skeleton-key, lock-picker, false key Diet((h)e)rich, Diebeschlüssel, Sperrhaken, Aufsperrhaken *m*

pickoff, tap, pickup Abgriff *m*, Anzapfung *f* [*Ein zwischen den Anschlussklemmen eines Schaltelements befindlicher fester oder verschiebbarer Kontakt (z. B. an eine Spulenwindung angelöteter Draht*]

pickup → pickoff

pickup [*A device that converts a sound, scene, measurable quantity, or other form of intelligence into corresponding electric signals, as in a microphone, phonograph pickup, or television camera*] Messfühler, Sensor, (Messwert)Aufnehmer *m*

to pickup aufnehmen

to pickup [*relay*] ansprechen

pickup → (call) pickup

pickup, measured-value transmitter Messwertgeber *m*

pickup current Anzug(s)strom *m* [*Relais*]

pickup delay Einschaltverzögerung *f*, Einschaltverzug *m*

pickup microphone Richtmikrophon, Richtmikrofon *n*

pickup time Anzug(s)zeit *f* [*Rechner*]

pickup value Anzug(s)wert *m* [*Rechner*]

to pick with the picklock, to open with the picklock, to pick with the lockpicker, to open with the lockpicker aufsperren [*Schloss mit Dietrich*]

P

PI control → proportional-plus-integral ((closed) loop) control

PI controller → proportional-plus-integral ((closed) loop) controller

picture, image [*The scene reproduced by a television or facsimile receiver*] Bild *n*

picture, image [*Produced by a photographic process*] Bild, Photo, Foto *n*, Aufnahme *f*

picture [*Automatic data processing. In COBOL. A symbolic description of each data element or item according to specified rules concerning numerals, alphanumerics, location of decimal points, and length*] Bild *n*

picture → template

picture carrier, luminance carrier [*In communications, a carrier frequency located 1. 25 megahertz above the lower frequency limit of a standard National Television Systems Committee (USA) television signal; in colo(u)r television, it is used for transmitting colo(u)r information*] Leuchtdichteträger *m*

picture element, pixel, point Bildpunkt *m* auf Rasterbildschirm

picture enhancement → image enhancement

picture frame → frame

picture frequency → frame frequency

picture monitoring system Bild(er)überwachungssystem *n*

picture postcard Bildpostkarte *f*

picture tube Bildröhre, Aufnahmeröhre *f*

PID algorithm → proportional-plus-integral-plus-derivative algorithm

PID control → proportional-plus-integral-plus-derivative control

PID controller → proportional-plus-integral-plus-derivative controller

piece → part

piece of equipment → device

pieces list → parts list

piecewise anodized → individually anodized

pie chart Tortendiagramm *n*

to pierce, to punch lochen, stanzen, ausstanzen

pier table [*joinery*] Pfeilertisch *m*

piezoelectric acceleration pickup piezoelektrische Beschleunigungsaufnahme *f*

piezoelectricity [*Solid-state physics. A voltage or electric field in a dielectric crystal produced by the piezoelectric effect; this can occur in certain nonmetallic materials, such as quartz crystals, ceramics or plastics*] Piezoelektrizität *f*

piezoelectric metrology piezoelektrische Messtechnik *f*

piezoelectric microphone, crystal microphone piezoelektrische Mikrophon, Kristallmikrophon *n*

piezoelectric transducer Piezowandler *m*

piezo sounder [*Colloquially called "piezo alarm"*] Piezohörmelder *m*, Piezohörmeldegerät *n*

PIF → program information file (US)

PIF → page image format

pig lock, scraper lock, go-devil lock Molchschleuse *f*

pilcrow, paragraph sign Abschnitt(s)zeichen *n* [*Kode*]

pillar (discharge) valve → upright (discharge) valve

pillar hydrant → above-ground hydrant

pillar mixer (valve), pillar mix(ing) valve, pillar blender, pillar blending valve Standmischbatterie, Mischstandbatterie *f*

pillar tap [*A tap, suitable for mounting on a horizontal surface, having a vertical inlet and a nozzle bent to discharge in a downward direction*] Standhahn *m*

pillar valve Standventil *n*

pilot contactor Vorstufenschütz *n*

pilot-controlled valve → servo-operated valve

pilot device → sensor

pilot duty Schaltleistung *f*

pilot electrode, ignitor [*An element that triggers and sustains a discharge in a switching tube*] Vorelektrode *f*

pilot flame [*Flame utilized to ignite the gas at the main burner or burners*] Zündflamme *f*

pilot flame ignition Zündflammenzündung *f*

pilot gas Zündgas *n*

pilot generator → tachometer generator

pilot lamp [*A small incandescent or neon light indicating that a device is operating*] Meldelampe *f*

pilot light [*A small permanent flame used to ignite a gas burner*] Zündflamme *f*

pilot light → indicator light

pilot motor → positioning motor

pilot-operated (slide) valve [*It ultilizes the pressure difference between upstream and downstream sides to act upon a diaphragm or piston to move the valve. Usually single-seated, for two position applications only, and used where large forces are required for valve operations*] Pilotschieber *m*

pilot setting Piloteinstellung *f*

pilot switch Hilfsstromschalter als Begrenzer/Regler, Begrenzer/Regler-Hilfsstromschalter *m*

P

pilot valve [*A small valve where the opening and closing directly influences a larger valve, as in a servo system*] Pilotventil *n* [*Nicht verwechseln mit „pilotgesteuertem Ventil"*]

pin Zapfen *m*

pin Stift *m*

pin Kloben *m*, Angel, Haspe, Hespe *f* [*Beschlag*]

PIN → personal identification number

pin and driver Stiftpaar *n* [*Zylinderkastenschloss*]

pin-and-socket hinge Zapfen *m* und Pfanne *f*, Zapfenband *n*

pin base Stecksockel *m* [*Relais*]

pinch point kleinste Temperaturabstand *m*

pin connection Steckverbindung *f*

pin drill Stiftsenker *m*

pin flat bed writer, pin flat bed printer Nadelflachbettdrucker, Flachbettnadeldrucker *m*

ping-pong Übertragung *f* nach dem Zeitgetrenntlagerverfahren [*Telekommunikation*]

pin grid array Anschlussstifte *f* in einem quadratischen Feld

pin hinge Scharnierband *n* mit Vorstecker

pinhole burner [*A burner assembly of which the burner parts are a number of very small holes*] Lochbrenner, Rohrbrenner *m*

pin lock, pipe-key(ed) lock, thorn lock Dornschloss *n*

pin-tong Bandzange *f* [*Schlosserwerkzeug*]

pin tumbler Stiftzuhaltung *f* [*Schloss*]

pin (type) insulator Stützisolator *m*

pin writer, pin printer Nadeldrucker *m*

PIOCS, physical input/output control system [*An extension of the hardware, constituting an interface between programs and data channels; opposed to LIOCS, logical input/output control system*] wirkliche Eingabe/Ausgabe-Regelsystem *n*

PIP → peripheral interchange program (US)

pipe Rohr *n* [*1. Beim deutschen Schlüssel der hohle Teil. 2. Die das Schlüsselloch einfassende Rille. 3. Die kleine Krampe zum Aneinanderbefestigen der Schlossteile*]

pipe, tube, tubing [*The terms pipe, tubing and tube can be used interchangeably. However, copper is usually referred to as tube or tubing, and other materials are referred to as pipe*] Rohr *n* [*Zum Transport von pulverförmigen, gasförmigen und flüssigen Stoffen*]

pipe → (organ) pipe

pipe aerator Rohrbelüfter *m*

pipe bend Rohrknie *n*

pipe bracket Rohrkonsole *f*

pipe branch Rohr(ab)zweig *m*

pipe branching Rohrverzweigung, Rohrabzweigung *f*

pipe bundle Rohrbündel *n*

pipe burner Rohrbrenner *m*

pipe burst Rohrbruch *m*

pipe burst detector, pipe burst detecting device, pipe burst detection device Rohrbruchmelder *m*, Rohrbruchmeldegerät *n*

pipe burst security Rohrbruchsicherung *f*

pipe casing → pipe jacket

pipe ceiling heating Rohrdeckenheizung, Deckenrohrheizung *f* [*Nahtlose Rohre von 3/8", 1/2" oder 1/4" werden in der Decke verlegt, entweder im Deckenbeton oder im Deckenputz*]

pipe clip Rohrschelle *f*

pipe coil → coil

pipe component Rohrbauteil *m*, *n*

pipe connection → pipe joint

pipe convector Rohrkonvektor *m* [*Er besteht aus Heiz(ungs)rohr mit Warmluft-Blechummantelung*]

pipe (convector) radiator Rohrradiator *m*

pipe corrugations Rohrwellung *f*

(pipe) curved fitting → bend

pipe cutting Rohrschneiden *n*

pipe dimensioning table → pipe sizing table

pipe dimensions → pipe size

pipe disk, pipe disc Scheiben(rohr)rippe *f*

pipe (drain) trap, pipe stench trap, pipe U-bend, pipe fixture trap, pipe siphon trap, pipe water trap, pipe water stop, pipe water barrier [*See remark under "trap"*] Rohr(wasser)(geruch)verschluss *m*, Rohrwasservorlage *f*

pipe duct Rohrkanal *m*

pipe earth(ing) (Brit.); pipe ground(ing) (US) Rohrerdung, Rohrerde *f* [*Blitzschutz*]

pipe elbow, pipe el(l) [*Pipe fitting to make an angle between connected pipes. If angle is not specified, 90° angle is implied*] Rohrknie *n*, Rohrwinkel(stück) *m*, (*n*)

pipe end, end of pipe Rohrstirnfläche, Stirnfläche eines Rohres *f*

pipe fastener Rohrhalter *m*

pipe fitter Rohrschlosser, Rohrmonteur *m*

pipe fitting [*The process of installing a pipe system*] Rohrmontage *f*

P

pipe fitting [*See remark under "fitting"*] (Rohr)Fitting, (Rohr)Formstück, (Rohr)Passstück, Rohrverbindungsstück *n*, Rohrarmatur *f* [*Muffe; Winkel; T-Stück; Verschraubung; Bogen; Nippel usw. „Fitting" wird allgemein für „Armatur" verwendet obwohl „Fitting" ein „Passstück" ist*]

pipe floor heating Rohr(fuß)bodenheizung, (Fuß)Bodenrohrheizung *f*

pipe flow(ing) Rohrströmung *f*, Rohrströmen *n*

pipe flushing Rohrspülung *f*

pipe friction Rohrreibung *f*

pipe friction(al) resistance Rohrreibungswiderstand *m*

pipe friction number Rohrreibungszahl *f*

pipe grid Rohrregister *n* [*Strahlungsheizung*]

(pipe) hanger (Rohr)Schelle *f*

(pipe) hanger with steel rod Hänge(rohr)schelle *f*

pipe heating area Rohrheiz(ungs)fläche *f* [*Rohrschlange oder Rohrregister*]

(pipe) heating tape, electrically heated tape [*Such tapes with a low wattage loading are used for wrapping pipework in chemical and oil plants to maintain suitable viscosity in external pipework in cold weather*] (Elektro)Heizband(age), (Elektro)Begleitheizband(age) *n*,(*f*)

pipe heat loss Rohrwärmeverlust *m*

pipe identification Rohrkennzeichnung *f*

pipe identification ring Rohrkennzeichnungsring *m*

(pipe) installation, (pipe) laying (Rohr)Verlegung *f*, (Rohr)Einbau *m*

pipe interior Rohrinnere *n*

pipe isolator Rohrtrenner *m*

pipe item Rohrposition *f* [*In einem Leistungsverzeichnis*]

pipe jacket, pipe casing, pipe case Rohrmantel *m*, Rohrummantelung *f*

pipe joint, pipe connection Rohrverbindung *f*, Rohranschluss *m*

pipe-key(ed) lock → pin lock

pipeless furnace [*Gravity furnace in which the entire heat output is delivered through one opening directly above the combustion chamber. Return air enters the unit through grille work around the perimeter of the outlet grille*] rohrlose Feuerung, rohrlose Feuerstätte *f*

(pipe)line, piping (Rohr)Leitung *f*

(pipe)line accessories, piping accessories (Rohr)-Leitungszubehör *m*, *n*

(pipe)line adapter, (pipe)line adaptor, piping adapter, piping adaptor Ausbaustück *n* [*Ausbaubares Rohrleitungsstück um einen Schieber, eine Messvorrichtung usw. einzubauen*]

(pipe)line arrangement, (pipe)line run, piping arrangement, piping run (Rohr)Leitungsanordnung, (Rohr)Leitungsführung, Rohrführung, Rohranordnung *f*

(pipe)line characteristic (curve), piping characteristic (curve) (Rohr)Leitungskennlinie, (Rohr)Leitungskennkurve *f*

pipeline(d) burst cache, piping burst cache Höchstgeschwindigkeitszwischenspeicher *m*

(pipe)line flange, piping flange (Rohr)Leitungsflansch *m*

(pipe)line heat load, piping (heat) load [*Unusable heat loss or gain from piping*] (Rohr)Leitungslast *f*

(pipe)line installation, piping installation (Rohr)-Leitungsanlage *f*

pipeline-installed pump → in-line pump

(pipe)line length, piping length (Rohr)Leitungslänge *f*

(pipe)line loss, piping loss (Rohr)Leitungsverlust *m*

(pipe)line manhole, piping manhole (Rohr)Leitungsschacht *m*

(pipe)line noise, piping noise (Rohr)Leitungsgeräusch *n*

(pipe)line pressure, piping pressure (Rohr)Leitungsdruck *m*

(pipe)line run, piping run (Rohr)Leitungsführung *f*

(pipe)line section, piping section (Rohr)Leitungsabschnitt *m*

(pipe)line system, piping system (Rohr)Leitungsanlage *f* [*Oberbennenung für alle innerhalb oder außerhalb von Gebäuden verlegten Rohrleitungen nach der Hauptabsperrvorrichtung*]

(pipe)line valve → gate valve

(pipe)line valve pit → gate valve pit

pipelining Fließbandverarbeitung *f* [*Verarbeitung von Programmen, durch die sich die Gesamtverarbeitungszeit verkürzen lässt*]

pipe main [*A pipe for distributing to or collecting flowing fluid from various branches*] Hauprohrleitung, Rohrhauptleitung *f*

pipe manufacturer, pipe maker Rohrhersteller *m*, Rohrfabrikant *m*

pipe net(work) design Rohrnetzauslegung *f*

pipe net(work) engineering Rohrnetztechnik *f*

pipe net(work) monitoring Rohrnetzüberwachung *f*

P

pipe net(work) resistance Rohrnetzwiderstand *m*

pipe packing Rohr(ab)dichtung, Rohrpackung, Rohrliderung *f*

pipe penetration Rohrdurchführung *f* [*Durch eine Decke, Wand, Mauer oder (Fuß)Boden*]

pipe plug Rohrstopfen, Rohrstöpsel *m*

pipe puller Rohrzieher *m*

pipe radiator → pipe (convector) radiator

pipe reflection, pipe reflexion Abspieg(e)lung, Rohrabspieg(e)lung *f*

pipe resistance Rohrwiderstand *m*

(pipe) return bend → twin (pipe) elbow

pipe run [*Portion of a fitting having its end in line or nearly so, as distinct from the branch or side opening as of a tee*] Rohrfittingstück *n*

pipe run [*Assembly of more than one piece of pipe*] Rohrstrecke *f*

pipe saddle → Lambeth saddle

pipe sawing Rohrsägen *n*

pipe schedule [*Pipe size system listing outside diameters and wall thicknesses for a variety of pressures*] Rohrbemessungssystem *n*

pipe shaft Rohrschacht *m*

pipe siphon trap → pipe (drain) trap

pipe size, pipe dimensions Rohrgröße, Rohrabmessungen, Rohrmaße *f*

pipe sizing Rohrbemessung, Rohrbemaßung *f*

pipe sizing table, pipe dimensioning table Rohrbemessungstabelle, Rohrbemaßungstabelle *f*

pipe stench trap → pipe (drain) trap

pipe support Rohrhalterung, Rohrstütze *f*

pipe-support(ing) roller Rohrtragrolle *f*

pipe system → pipework

pipe tapping Entnahmestutzen *m* [*Messblende*]

pipe thread Rohrgewinde *n*

(pipe) tracer heating (Rohr)Begleitheizung *f*

pipe trap → pipe (drain) trap

pipette Pipette *f*, Tropfglas *n*

pipe U-bend → pipe (drain) trap

pipe vertex Rohrscheitel *m*

pipe wall Rohrwand(ung) *f*

pipe water trap → pipe (drain) trap

pipe well Rohrbrunnen *m*

pipework, piping, pipe system Rohrsystem *n*

pipe wrapping Rohrbandagierung, Rohrumwick(e)lung *f*

PI/PID algorithm PI/PID-Algorithmus *m*

piping [*Either pipe or tubing, or both*] Rohrstrang *m*

piping → pipework

piston cold water set Kolbenkaltwassergarnitur *f*

piston compressor machine Kolbenkompressormaschine *f*

piston displacement → (compressor) piston displacement

piston pressure pump → pressure piston pump

piston seat Kolbensitz *m*

piston suction pump → suction piston pump

pit → (hatchway) pit

pit Vertiefung *f*

pitch [*That attribute of auditory sensation in terms of which sound may be ordered on a scale related primarily to frequency*] Tonhöhe *f* [*Zuordnung eines Tones oder Tongemisches zu einer bestimmten Skala zwischen tiefen und hohen Tönen. Anmerkung: Das Wort Tonhöhe entstammt der Musik. Allgemein ist die Tonhöhe ein psychologisches Merkmal des Hörereignisses. In der physikalischen Akustik wird die Tonhöhe durch die Frequenz gekennzeichnet*]

pitch angle [*adjustable fan blade*] Einstell(ungs)winkel *m*

pitch interval Tonhöhenintervall *n* [*Akustik*]

(pit) latrine Latrine *f*, Grubenabort *m*

Pitot tube Staudruckmesser *m*

pitting, localized corrosion Lochfraß *m*

pit valve Grubenventil *n*

to pivot zapfenlagern [*in Zapfen lagern*]

pivot Körner *m* [*Körnerpfanne*]

pivot Lagerzapfen, Drehzapfen *m*

pivoted zapfengelagert, gezapft

pixel → picture element

(pixel) resolution (Bildpunkt)Auflösung *f* [*Anzahl der darstellbaren Bildpunkte auf dem Bildschirm oder dem (Raster)Plotter je Flächeneinheit*]

PL → programming language

place → digit

place Ort *m*

to place in storage → to incorporate

place-and-finish mortar → swift setting mortar

placement [*Of a room or rooms in relation to another room or other rooms*] Anordnung *f*

placement, location [*of an equipment*] Einbaustelle *f*, Einbauort *m*

P

place of assembly, assembly place Versammlungsort *m*, Versammlungsstelle *f*

placing in storage → incorporating

plain flange, blind flange Blindflansch *m*

plain language, plain text Klartext *m*

plain link Brückenlasche, Überbrückungslasche *f*

plain scale Breitenmaßstab *m*

plain (slide) valve → single (slide) valve

plain (slide) valve → D (slide) valve

plain text Klartext *m*

plain text log Klartextprotokoll *n*

plain tube [*As opposed to a fin(ned) tube*] Glattröhre *f*

plain writing Klarschrift *f*

plaiting Geflecht *n*

planar diode Planardiode *f*

planar transistor Planartransistor *m* [*Er enthält alle Anschlüsse auf einer Ebene des Halbleiterkristalls und erlaubt die einfache Verbindung von mehreren Transistoren durch dieselben Techniken wie bei der Transistorproduktion*]

Planck radiation formula, Planck's law, Planck distribution law [*A formula for the intensity of radiation emitted by a blackbody within a narrow band of frequencies (or wavelengths), as a function of frequency, and of the blackbody's temperature*] Plancksche Strahlungsformel *f*

plan drawing Planzeichnung *f*

plane → level

plane diffuser, plane diffusor glatte Wanne *f* [*Leuchte*]

plane model Ebenenmodell *n* [*Automatisierung*]

plane projecting diffuser, plane projecting diffusor glatte überstehende Wanne, überstehende glatte Wanne *f* [*Leuchte*]

plane scanner Flachbettabtaster *m*

plane selection Ebenenauswahl *f* [*Wegbedingung*]

planing Abrichten, Pritschen *n* [*gehämmerte Blech*]

plank [*Usually a comparatively wide board more than 1" thick which is laid with its wide dimension uppermost and used as a walkway or similar purposes*] Bohle, Planke *f*

plank bottom, plank cover Bohlenbelag, Plankenbelag *m*

plank walkway, plank footway Bohlengehweg, Bohlenfußweg, Plankengehweg, Plankenfußweg *m*

planned maintenance Wartung *f*

planned stop wahlweise Halt *m* [*Zusatzfunktion*]

planning aid Planungshilfe *f*

planning approach Planungsansatz *m*

planning document Planungsunterlage, Planungsvorlage *f*, Planungsdokument *n*

planning fundamental Planungsgrundlage *f*

planning manual, planning handbook Planungshandbuch *n*

planning principle Planungsgrundsatz *m*

planning tool Planungsmittel *n*

plant; directly close-controlled member [*IEC 27-2A*] **close-controlled system** [*IEC 50(351)*] Regel(ungs)strecke *f* [*Siehe Anmerkung unter „Strecke"*]

plant Betrieb *m*, Werk *n*

plant → installation

plant builder, installation builder Anlagenbauer, Anlagenerrichter *m*

plant building, plant construction, installation building, installation construction Anlagenbau *m*, Anlagenerrichtung, Anlageninstallation *f* [*Fehlbennenung*]

plant building engineer, plant construction engineer, installation building engineer, installation construction engineer Anlagenbauer, Anlagenbauingenieur *m*

plant ((closed) loop) control, installation ((closed) loop) control Anlagenreg(e)lung *f*

plant-conditioned, installation-conditioned anlagenbedingt

plant data, installation data Anlagedaten *f* [*DIN 66 201*]

plant description, installation description Anlagenbeschreibung *f*

plant display, installation display Anlagenbild *n*

plant engineer, installation engineer Betriebsingenieur, Werkingenieur *m*

plant engineer, installation engineer Anlageningenieur *m*

plant engineering, installation engineering Anlagentechnik *f*

plant engineering Betrieb(s)technik, Werktechnik *f*

plant equipment, installation equipment, plant outfit, installation outfit Anlagenausstattung, Anlagenausrüstung, Anlagenbestückung *f*

plant fire-fighting squad, plant fire-fighting team, plant fire-fighting gang, plant fire-fighting party Werk(feuer)löschmannschaft, Werkbrandlöschmannschaft *f*, Werk(feuer)löschtrupp, Werkbrandlöschtrupp *m*

P

plant logistics system, installation logistics system Betriebsdatenerfassung, Werkdatenerfassung *f*

plant machinery, installation machinery Anlagenmaschinen *f*

plant management, installation management Anlagenbewirtschaftung *f*

plant mode, installation mode Anlagenbetriebsart *f*

plant mode strategy module, installation mode strategy module Strategiebaustein für die Anlagenbetriebsart, Anlagenbetriebsart-Strategiemodul *m*

plant ((open) loop) control, installation ((open) loop) control Anlagensteuerung *f*

plant operation monitoring, installation operation monitoring Anlagenbetriebsüberwachung *f*

plant operator, installation operator Anlagenbediener, Anlagenfahrer *m*

plant operator, installation operator Anlagenbetreiber *m*

plant outfit → plant equipment

plant performance, installation performance Anlagenleistung *f*

plant production level, installation product level Produktionsleitebene, Fertigungsleitebene *f*

plant reliability, installation reliability Anlagenzuverlässigkeit *f*

plant schematic, installation schematic Anlagenbild *n*

plant security, installation security Betriebsschutz, Werkschutz *m*, Betiebssicherung, Werksicherung *f*

plant security, installation security Anlagenschutz *m*, Anlagensicherung *f*

plant time constant, installation time constant Streckenzeitkonstante *f*

plant water, installation water Anlagenwasser *n*

plaque Platten-Luftverteiler, Luft-Plattenverteiler *m*

plasma enhanced chemical vapour deposition (Brit.); plasma enhanced chemical vapor deposition (US); PECVD plasmaunterstützte Schichtabscheidung *f* [*Plasma wird durch eine Glimmentladung im Druckbereich von ungefähr (0,013-1,3)102 Pa erzeugt*]

plaster Gips(stein) *m*

plaster gebrannte Gips, Gipsmörtel, Stuck *m*

plaster Putz *m*

plaster base Putzträger *m* [*z. B. Streckmetall*]

plaster ceiling Putzdecke *f*

plasterer's mixing board Putzermischbrett *n*

plaster panel Putztafel *f*

plaster surface mounting Aufputzmontage *f*

plaster-surrounded tubes ceiling heating, plaster-surrounded pipes ceiling heating Putz-Rohrdeckenheizung, Rohrdeckenheizung mit Heizrohren im Deckenputz *f* [*Nahtlose Rohre von $^3/_8$", $^1/_2$" oder $^1/_4$" im Deckenputz verlegt*]

plastic (air) duct Kunststoff(luft)kanal, Kunststofflüftungskanal *m*

plastic-air waste gas system Kunststoff-Luft-Abgassystem *n* [*Brennwerttechnik*]

plastic ball valve, plastic floating valve [*See remark under "ball valve"*] Kunststoffschwimmerventil, Kunststoffkugelventil *n*

plastic (bath)tub Kunststoff(bade)wanne *f*

plastic capacitor Kunststoffkondensator, Folienkondensator *m*

plastic coated kunststoffbeschichtet

plastic (convector) radiator Kunststoffradiator, Kunststoffgliederheiz(ungs)körper, Kunststoffheiz(ungs)körper *m* [*Fehlbenennung*]

plastic cracking plastische Rissbildung *f*

plastic deformation plastische Verformung, plastische Formänderung *f*

plastic drainer [*See remark under "drainer"*] Kunststoff(ab)tropffläche, Kunststoff(ab)tropfplatte *f*

plastic fabric tape Kunststoff-Gewebe-Band *n*

plastic fill material Kunststoff-Füllkörper *m* [*Kühlturm*]

plastic fire Kunststoffbrand *m*, Kunststofffeuer *n*

plastic floating valve → plastic ball valve

plastic jacket Kunststoffmantel *m*

plastic leadless chip carrier, PLCC IS-Anschlüsse auf (Anschluss)Fahnen, integrierte Schaltungsanschlüsse auf (Anschluss)Fahnen *f*

plastic mounting channel Kunststoff-Montageschiene *f* [*für Lichtbandanordnung*]

plastic nozzle Kunststoffdüse *f*

plastic pad Kunststoffvlies; Kunststoffvließ *n* [*Österreich*] [*Filtermedium*]

plastic pipe Kunststoffrohr *n*

plastic pipe (under)floor warming, plastic pipe (under)floor heating Kunststoffrohr(fuß)bodenheizung, Kunststoffrohr-FB-Heizung *f*

plastic radiator → plastic convector radiator

plastic tube Kunststoffröhre *f*

plastic welding Kunststoffschweißen *n*

to plate plattieren

plate → anode

plate Dickblech, dicke Blech *n* [> *5 mm*]

plate Platte *f*

plate (air) duct Blech(luft)kanal, Blechlüftungskanal *m*

plate and knob Knopf-Stützringschild *n* [*Türbeschlag*]

plate apron Blechschürze *f*

plate baffle, plate deflector, plate deflecting device Blechschikane, Blechführ(ungs)wand *f*, Blechumlenker, Blechablenker, Blechstrombrecher *m*, Blechpralleil *m*, *n*

plate duct Blechkanal *m*

(plated) wire memory Drahtspeicher *m*

plate electrode Plattenelektrode *f*

plate enclosure, plate casing, plate case, plate housing Blechgehäuse *n*

plate fin [*A strip of thin sheet metal having a punched hole into which a pipe fits*] Blechlamelle, Blechflosse *f*

plate freezing [*Heat removal by direct contact of the packaged product with refrigerated plates*] (Kontakt)-Plattengefrieren *n*

plate heat exchanger, plate heat exchange unit, plate heat interchanger [*See remark under "heat exchanger"*] Plattenwärmeübertrager; Plattenwärme(aus)tauscher *m* [*In der Industrie so genannt. Siehe Anmerkung unter „Wärme(aus)tauscher"*]

plate jacket, plate casing, plate case Blechmantel *m*, Blechummantelung *f*

plate rack, plate stand Tafelgeschirrgestell *n*, Tafelgeschirrständer *m*

plate spring Blattfeder *f*

plate superheater → **(Heizmann's) plate superheater**

plate trimming → **trimming the plates**

plate (vent(ilation)) duct Blech(lüftungs)kanal, Lüftungsblechkanal *m*

platinum-coated platin(um)beschichtet, platin(um)-überzogen

platinum resistance thermometer, Callendar's thermometer [*The basis of the International Practical Temperature Scale of 1968 from 259. 35° to 630.74°C; used in industrial thermometers in range to 650°C; capable of high accuracy because platinum is noncorrosive, ductile, and nonvolatile, and can be obtained in a very pure state*] Platinwiderstand(s)thermometer *n*

platinum-rhodium thermocouple Platinrhodium-thermoelement *n*

play, clearance [*The distance separating two objects or parts*] Freiheit *f*, Spiel(raum) *n*, (*m*), Abstand *m*

playback Wiedergabe *f*

player Zuspielgerät *n*

playroom Spielzimmer *n*

PLC → **programmable logic ((open) loop) control**

PLCC → **plastic leadless chip carrier**

PL(C)L → **phase-locked (control) loop**

PLD → **programmable logic device**

PLD → **partial line down**

pliers kleine Zange *f*

plinth Standsockel, Sockel *m*

PLL → **phase locked loop**

PL/M, programming for microcomputer, microcomputer programming Mikrorechnerprogrammsprache, Programmsprache für Mikrorechner *f* [*Höhere Universalprogrammsprache, basierend auf PL/1 (programming language = höhere Programmsprache)*]

PLO → **maximum load optimization**

plot → **(real) property**

plotter Plotter, Koordinatenschreiber, Planzeichner, Kurvenschreiber, X-Y-Schreiber, Zeichenautomat *m*, Zeichenmaschine *f*, automatische Zeichengerät, (selbsttätige) Zeichengerät *n* [*Ein impulsgesteuertes Gerät zur Darstellung von Linien, Kurven und Punkten*]

plough bit Nuthobeleisen *n*

PLSS → **private listening sound system**

PLU → **partial line up**

to plug → **to plug (in(to))**

plug Stopfen, Stöpsel *m*

plug, taper(ed) plug, tapering plug drehbare Kegel *m*, (Hahn)Küken *n*

plug Dorn, Scharnierstift *m* [*Schloss*]

plug → **socket inlet**

plug and play Einstecken *n* und Loslegen [*Die Fähigkeit von Windows 95 Einsteckkarten und externe Geräte automatisch zu erkennen und die richtigen Treiber dafür selbstständig zu installieren*]

plug and set valve → **globe valve**

plug-and-socket connection, plug-in connection, pluggable connection Steckverbindung, elektrische Steckverbindung *f* [*Lösbare elektrische Verbindung zwischen Brenner und Wärmeerzeuger. DIN 4791*]

plugboard, wiring board [*An array of jacks or sockets in which wires (or other elements) may be plugged to control the action of an electromechanical device in a data processing system such as a printer*] Stecktafel, Schalttafel, Bedien(ungs)tafel, Schaltplatte, Steckplatte, Bedien(ungs)platte *f*

plug gauge; plug gage (US) Messdorn, Lehrdorn *m*

plugging (into), plugging-in (Ein)Stecken *n*

plugging space [*cable termination*] Füllraum *m* [*Kabelendverschluss*]

plug-in einsteckbar

plug-in act(uat)or [*See remark under "act(uat)or"*] Einsteckaktor *m*

plug-in bus duct [*Bus duct with built-in power tap-off points. Tap-off is made with a plug-in switch, circuit breaker, or other fitting*] Einsteck(sammel)schienenkanal, Einsteckstromschienenkanal *m*

plug-in cap Stecksockel *m* [*Lampe*]

plug-in card Steckkarte *f*

plug-in connection → plug-and-socket connection

plug-in connector Steckverbinder *m*, Steckkupp(e)lung *f*

plug-in contact Steckkontakt, Einfahrkontakt *m*

plug-in interface Einsteckschnittstelle *f*

plug-in lamp → jack lamp

plug-in link Steckbrücke *f*

plug-in package Steckblock *m* [*DIN 43 350*]

plug-in power supply unit Steckernetzteil *m, n*

plug-in-ready (ein)steckfertig

plug-in scanner, plug-in reader Einsteckleser, Steckleser *m* [*Ausweis(karte) wird in den Leser eingesteckt*]

to plug (in(to)) (ein)stecken

plug-in tube Steckersicherung *f*

plug-in unit [*IEC 547*] Einsteckkassette *f*

plug-on riser bus [*microcomputer control*] Steckschiene *f*

plug Steckerstift *m*

plug screw → screw plug

plumber [*A tradesman who shapes and fixes flexible metal roofing, cuts bends, and joins water pipes, instals soil pipes, waste pipes, and water systems, sometimes gas pipes also. The usual metals he works in are copper, lead, iron, zinc and mild steel, through the term plumber originally meant lead worker. In the north of England, plumbers are usually glaziers*] Klempner; Installateur *m*

plumber(s) labo(u)rer, plumber(s) mate Klempnergehilfe *m*

plumbing [*Transferring a point at one level to a point vertically below or above it, usually with a plumb bob or plumb rule*] (Ab)Loten, (Ab)Senkeln *n*

plumbing [*The practice, materials, and fixtures used in the installation, maintenance, extension, and alteration of all piping, fixture, plumbing appliances, and plumbing appurtenances in connection with any of the following: Sanitary drainage or storm drainage facilities and venting systems and the public or private water supply systems, within any building, structure, or conveyance. This definition covers the extension of such lines described above to a point not exceeding five (5) feet beyond the foundation walls of the building or structure. Not included in this definition are installations of gas piping; chilled water piping in connection with refrigeration, process and comfort cooling; hot water piping in connection with building heating; and piping for fire sprinklers and standpipes.*] Klempnerei *f*, Installateurwesen, Installationswesen *n*

plumbing installation, sanitary installation sanitäre Installation *f*

plumbing trade → (internal) plumbing trade

plumbing trade → (external) plumbing trade

plunger Stößel *m*

plunger → (hydraulic) plunger

plunger, solid piston [*As of a force pump*] Vollkolben *m*

plunger coil, plunger electromagnet Tauchspule *f*

plunger elevator, ram elevator (US); plunger lift, ram lift (Brit.) [*Hydraulic elevator, having a ram or plunger directly attached to the underside of the car platform*] Hydraulikaufzug *m* mit Hubkolben unter dem Fahrkorb

plunger with roller → roller plunger

PLV → part(ial)-load value

plywood parquet, plywood squares Sperrholzparkett *n*

PM → pulse modulation

PM, phase modulation Phasenmodulation *f*, Richtungsschritt *m*

PM → positive-medium

PMOS P-Kanal-MOS *f* [*MOS-Technologie. Hat gegenüber NMOS größeren Kristallflächen- und Leistungsbedarf sowie höhere Schaltgeschwindigkeiten*]

P-MOS transistor → P-channel MOS transistor

PMS, peripheral message specification Dienststruktur *f* des INTERBUS-S verträglich zu FMS und MMS

PMU → parameter measuring unit

PN → rated pressure

pneumatic ... → (compressed-)air ...

pneumatically outfitted, pneumatically equipped pneumatisch bestückt, pneumatisch ausgestattet, pneumatisch ausgerüstet

pneumatic conveyor → airtube conveyor

pneumatic-electric [*Not to be confused with "electric-pneumatic"*] pneumatisch-elektrisch

pneumatic tube → carrier tube

PN FET → **junction-gate field-effect transistor**

pn junction pn-Übergang *m*

PNL → **perceived noise level**

pocket computer Taschenrechner *m*

pocket lamp → **flashlight**

pocket scanner Taschenabtaster *m*

POCSAG, post-office code standardisation advisory group Funkrufstandard *m* für Adressierung und Übertragung kodierter Daten mit 512 bit/s.

POD → **processor-orient(at)ed device**

to point ausfugen

point [*Mathematics. The mathematical symbol, usually denoted by a period (= full stop), that separates the integral part of a quantity from the fractional part of the quantity represented by a positional notation; e.g. decimal point in a base ten notation, binary point in a base two notation*] Komma *n*

point [*A physical source or destination for data in the form of analog(ue) or digital signals*] Punkt *m*

point [*A location in space*] Punkt *m*

point [*An element of a geometry that indicates position but not distance, direction or size*] Punkt *m*, Stelle *f*

point → **data point**

point → **picture element**

point address Punktadresse *f*

point-contact transistor [*A transistor having a base electrode and two or more point contacts located near each other on the surface of an n-type semiconductor*] Punkt-Kontakt-Transistor *m* [*W. Shockley, J. Bardeen und W.H. Bratais haben dafür 1956 den Nobelpreis für Physik erhalten*]

point control criterion [*Criterion or parameter controlling event reporting from and operator access to points*] Datenpunkt-Sperrkriterium *n*

point descriptor [*The information containing the characteristics of a control or monitoring point*] Punktbeschreiber *m*

point detector, point detecting device, point detection device Punktmelder *m*, Punktmeldegerät *n*

point(ed) bolt Spitzbolzen, Scharfbolzen *m*

pointer Adressenverweis *m* [*Ein Speicherplatz, der eine Adresse enthält. Bei häufiger Verwendung wird hierfür ein Register benutzt*]

pointer Anschläger *m* [*Schloss*]

pointer, needle, hand Zeiger *m* [*z. B. bei einer Uhr*]

pointer frequency meter, frequency pointer meter Zeigerfrequenzenmesser, Frequenzzeigermesser *m*

pointer mechanism, needle mechanism Zeigermechanismus *m*, Zeigerwerk *n*

pointer mercury thermometer → **dial mercury thermometer**

pointer position, needle position Zeigerstellung, Zeigerlage *f*

pointers → **(indicating) pointers**

pointer thermometer → **dial thermometer**

pointer zero positioning device Zeigernullstellungsvorrichtung *f* [*Berichtigung des mechanischen Nullpunktes bei Zeigerinstrumenten*]

point heat → **concentrated heat**

point identification text, point name Datenpunkt-Identifikationstext *m*

pointing device Eingabegerät, Zeigegerät *n* [*graphische Benutzeroberfläche*]

point input → **data point input**

point in time Zeitpunkt *m*

point number Kommazahl *f*

point of presence, POP örtliche Internetanbieter *m*

point of sale event monitoring, POS/EM elektronische Kassenkontrolle *f*, Kassenüberwachungssystem *n*

point scale Punktskala *f*

point selection Anwahl *f*

points file Geometriedatei *f*

point source of light, point light source Punktleuchtquelle, Punktlichtquelle *f*

point-to-point connection Punkt-zu-Punkt-Verbindung *f*, Punkt-zu-Punkt-Anschluss *m* [*Von zwei Knoten ohne auf der Sendestrecke liegende Knoten zu passieren*]

point-to-point protocol → **PPP**

point-to-point transfer Punkt-zu-Punkt-(Daten-) Übertragung *f* [*Zwischen zwei (Daten)Endgeräten, d. h. einem Sender und einem Empfänger*]

point-to-point wiring freizügige Verdrahtung, X-Verdrahtung, Wildverdrahtung *f*

point-to-point with straight cut Punkt- und Streckensteuerung *f* [*Punktsteuerung mit der Möglichkeit eines gesteuerten Vorschubs längs der Koordinatenachsen oder in einem festen Winkel zu ihnen*]

point(-type) fire detector, point(-type) fire detecting device, point(-type) fire detection device Punkt-Brandmelder, Punkt-Feuermelder *m*, Punkt-Brandmeldegerät, Punkt-Feuermeldegerät *n*

poison label Giftetikett *n*

poison(ous) chemical Giftchemikalie *f*

polar crystal → **ferromagnetic crystal**

polar current Doppelstrom *m*

polarity [*The direction of current flow in a d-c circuit. By convention, current flows from +10 to -, Electron flow is usually in the opposite direction*] Polarität *f*, Polung *f*, Leiterfolge *f*

polarity inversion, reversal of polarity, pole changing, pole reversal Umpolung, Dahlanderschaltung, Pol(aritäts)umkehr, Verpolung *f*, Pol(aritäts)wechsel *m*

polarity tester Polaritätsprüfer *m*

polarization piece Schlüsselstück *n* [*Steckvorrichtung*]

polarized gepolt, polarisiert, polunverwechselbar, verpolsicher

polarized capacitor → electrolytic capacitor

to pole polen

pole [*An electrical connection point. In a panel, the point of connection. On a device, the terminal that connects to the power*] Pol *m*

pole placement Polvorgabe *f*

pole power → power per pole

police emergency call detector, police emergency call detecting device, police emergency call detection device Polizeinotrufmelder *m*, Polizeinotrufmeldegerät *n*

police engineering Polizeitechnik *f*

polished plate glass mirror Kristallspiegel *m*

poll Abfrage *f*

polling Sendeaufruf *m*, (kontinuierliche) Abfrage, zentrale Zugriff(s)steuerung *f* [*Methode zur Kontrolle der Übertragungssequenzen durch Geräte an einer Mehrpunkt-Leitung; jedes Gerät wird zum Warten aufgefordert, bis der Prüfprozessor die Übertragung freigibt.*]

polling Faxabruf *m* auf eigene Kosten [*Von einem anderen Faxgerät*]

polling → (master-slave) polling

polling mode Abrufbetrieb *m* [*DIN 44 300, Teil 9*]

polling system Abfragesystem *n* [*Fernwirktechnik*]

polling timeout delay Abfragezeit *f*

pollutant, contaminator, contaminant [*A thing that contaminates*] Verunreinigungsstoff, Verunreiniger *m*

pollutant concentration, contaminator concentration, contaminant concentration Verunreinigerstärke, Verunreinigungsstoffstärke *f*

polluted household water, contaminated household water Haushalt(s)schmutzwasser *n*

polluter-pay principle Verursachergrundsatz *m*

pollution → contamination

pollution [*Destruction or impairment of the purity of the environment*] Verunreinigung *f* [*ist nicht Verschmutzung*]

pollution flashover Fremdschichtüberschlag *m*

pollution load Verunreinigungslast *f*

pollution severity Fremdschichtgrad *m* [*Isolator*]

pollution severity level Fremdschichtklasse *f* [*Isolator*]

pollution signal Fremdsignal *n*

polygonal (pipe) bend, polygonal bend fitting, polygonal (pipe) curved fitting [*See remark under "bend"*] Polygon(-Rohr)-Krümmer(stück) *m*, (*n*), Polygonbogenrohr, Polygonkrümmerrohr, Polygon(-Rohr)-Bogen(stück), Mehrfachknie(stück) *n*

polytropic polytrop [*Zustandsänderung eines Gases ohne Änderung von Druck, Volumen und Temperatur*]

polyurethane fire PU(R)-Schaumstoffbrand *m*, Polyurethanbrand, PU(R)-Schaumstofffeuer, Polyurethanfeuer *n*

polyurethane insitu foam, insitu polyurethane foam PU(R)-Ortschaum, Polyurethan-Ortschaum *m*

POM → printer output microfilm

pony motor → starting motor

pool → (swimming) pool

pool care system → (water) pool care system

pooled test Ringprobe, Ringprüfung *f*, Ringversuch *m*

pool surrounding area → (swimming) pool surrounding area

POP → point of presence

pop-a-plug system Ausdehnungskolben *m* [*Ein hydraulischer Kolben von 3000 kg oder 18 t Gewicht, der einen zweiteiligen mechanischen Verschluss im Innern eines undichten Wärmetauschers ausdehnt*]

to pop up [*data processing*] Hervorholen *n* [*Stapel*]

pop-up menu Aufblend(ungs)-Bildschirmmenu *n*

pop-up waste Ablauf mit Hebepfropfen, Hebe-pfropfenablauf *m*

porous bearing Sinterlager *n*

porous insulating vacuum → evacuated (thermal) porous insulation

port, opening, orifice, aperture [*Not to be confused with "hole"*] Öffnung *f* [*Nicht verwechseln mit „Loch"*]

port Schnittstelle *f* [*Zwischen einem Rechnerkanal und einem Peripheriegerät*]

port Anschluss *m* [*Ein physikalischer Zugriff(s)punkt auf einem Rechner, Multiplexer oder Netz(werk), wo Signale zugestellt oder weggenommen werden*]

P

port E/A-Kanal *m*, Ein/Ausgabekanal [*komplexer Baustein*]

portability Übertragbarkeit *f*, Verpflanzbarkeit [*Die Eigenschaft von Programmen auf unterschiedlichen Rechnern ohne oder nur mit geringfügigen Änderung eingesetzt werden zu können*]

portable [*As opposed to built-in*] frei, nicht eingebaut

portable [*equipment*] ortsveränderlich

portable data format plattenunabhängige Dateiformat *n*

portable (fire) extinguisher → hand (fire) extinguisher

portable lock, padlock; catenatum [*Latin*] (Vor-)Hangschloss, Vorlegeschloss, Anlegeschloss, Anwerfschloss, (Vor)Hängeschloss *n* [*DIN 7465*]

portable server → serving cart (on casters)

port driven input/output, port driven I/O Ein-/Ausgabe mit logischem E/A-Kanal, E/A mit logischem E/A-Kanal *f*

porter's service Pförtnerdienst *m*

port 0 Ein/Ausgang(s)kanal 0 *m* [*Mikrorechner. Zur bitparallelen Ein- oder Ausgabe mit einer Wortlänge von 8 Bit und Vergleichereingängen. Die Ansprechschwelle wird mit der Spannung U_{ref} an Pin 4 gewählt*]

POS/EM → point of sale event monitoring

to position positionieren, in Stellung bringen, in Lage bringen

position (Speicher)Stelle *f*

positional notation, positional representation Stellungs(wert)schreibweise, Positions(wert)schreibweise, Lage(wert)schreibweise *f*

P

position checkback signal message; position checkback signalling message (Brit.); position checkback signaling message (US) Stellungsrückmeldung, Positionsrückmeldung, Lagerückmeldung *f*

position control → ((closed) loop) position control

position control → ((open) loop) position control

position detector, position detecting device, position detection device Stell(ungs)melder, Positionsmelder, Lagemelder *m*, Stell(ungs)meldegerät, Positionsmeldegerät, Lagemeldegerät *n*

position indicator Stell(ungs)anzeiger, Positionsanzeiger, Verfahr(ungs)anzeiger *m*

positioning Stellen, Instellungbringen, Positionieren, Verfahren *n*

positioning → ((closed) loop) position control

positioning act(uat)or Stellenantrieb *m* [*Er kann nicht nur in die Auf- und Zu-Stellung gefahren werden und so kann z. B. ein Steuersignal von 0 – 10 V den Stellantrieb innerhalb der gegebenen Auflösung von 0 bis 100 % seines Hubs steuern*]

positioning command Stell(ungs)kommando, Positionskommando, Verfahr(ungs)kommando *n*

positioning command with floating signal Stell(ungs)kommando mit Dreipunktsignal, Positionskommando mit Dreipunktsignal, Verfahr(ungs)kommando mit Dreipunktsignal *n*

positioning command with analog(ue) signal Stell(ungs)kommando mit stetigem Signal, Positionskommando mit stetigem Signal, Verfahr(ungs)kommando mit stetigem Signal *n*

position(ing) controller Lageregler, Stellungsregler, Positionsregler *m*

positioning data Wegdaten *f*

positioning motor, motor act(uat)or, servo-motor, pilot motor Stell(ungs)motor, Servomotor, Positionsmotor *m* [*Der Antrieb des Stellgliedes sind Membran, Rückholfeder und Spindel. Der Stellmotor gehört zum Regler*]

positioning table Positionstisch, numerisch gesteuerte Tisch, Stell(ungs)tisch *m*

positioning signal Stell(ungs)signal, Positionssignal *n*

positioning time Stell(ungs)zeit, Positionszeit *f*

position monitoring Stell(ungs)überwachung, Positionsüberwachung, Lageüberwachung *f*

position of rest, rest position, off position, position at rest, zero position [*IEC 337-2A*] Ruhestellung, Ruhelage, AUS-Stellung, Nullstellung, Ruheposition, Raststellung, Rastposition, AUS-Lage, Nulllage, Rastlage *f* [*Hilfsstromschalter, VDE 0660, 202*]

position parameter Stell(ungs)parameter, Positionsparameter, Lageparameter *m*

position pulse → commutator pulse

position repeater Stell(ungs)rückmelder, Positionsrückmelder *m*

position setpoint, position desired value Stell(ungs)sollwert, Stell(ungs)vorgabewert, Positionssollwert, Positionsvorgabewert, Lagesollwert, Lagevorgabewert *m*

position slide switch, position sliding-dolly switch Stell(ungs)schiebeschalter, Positionsschiebeschalter *m*

position switch [*IEC 337-1*] Endschalter, Positionsschalter, Stell(ungs)schalter, Lageschalter *m* [*VDE 0660 Teil 200*]

position switch [*telephony*] Platzumschalter *m*

position vector, radius vector [*Mathematics. Given a point in P in Euclidean n-space, the vector with length equal to the distance from the origin to P, and orientation in the direction from the origin to P*] Ortsvektor *m*

positive zwangsläufig

positive, P positiv, PO

positive-big, PB positiv-groß, PG

positive column, positive glow [*The luminous glow, often striated, that occurs between the Faraday dark space and the anode in a glow-discharge tube*] Anodenglimmlicht *n*

positive contact protection, automatic contact protection zwangsläufige Berührungsschutz *m*

positive displacement compressor, volumetric compressor Volumenverdichter, Volumenkompressor *m*

positive displacement (gas) meter, volumetric (gas) meter [*It measures directly the volume of gas which passes through it*] Volumen(gas)zähler, Volumen(gas)messer *m*

positive drive Zwangsläufigkeit, nichtbegrenzte Einwirkung, unbegrenzte Einwirkung *f*, formschlüssige Antrieb, zwangsläufige Antrieb *m* [*Hilfsstreckenschalter. Verbindung zwischen Bedienteil und Schaltglied*]

positive feedback positive Rückkopp(e)lung, Mitkopp(e)lung *f*

positively driven zwangs(an)getrieben

positively interlocked zwangsverriegelt

positively opening zwangsöffnend

positively ((open) loop) controlled zwangsgesteuert

positive-medium, PM positive-mittel, PM

positive meter → displacement meter

positive operating pressure Betriebsüberdruck *m*

positive overpressure positive Überdruck *m*

positive positioning positive Positionierung *f*, positive Instellungbringen *n* [*Bewegung auf der positiven Seite der Koordinatenachse*]

positive pressure burner Überdruckbrenner *m*

positive pressure region [*burner*] Überdruckbereich *m* [*selten: n*]

positive quantity meter → positive total flow meter

positive-small, PS positiv-klein, PK

positive temperature coefficient resistor Kaltleiter, PTC-Widerstand *m*

positive temperature coefficient thermistor → PTC thermistor

positive temperature coefficient thermistor resistor → PTC thermistor resistor

positive total flow meter, positive quantity meter, displacement total flow meter, displacement quantity meter Verdrängungsvolumenmesser, Verdrängungsdurchflussmesser *m*

post Pfosten *m*

postal box → letter box

postal knocker [*A letter plate which includes a door knocker*] Briefeinwurf *m* mit (Tür)Klopfer

post-alloy-diffused diffundiert nach dem Legieren

post-arc current Nach(lichtbogen)strom *m* [*Schaltgerät*]

postcooling Nachkühlen *n*

post copy processor → post printer

postheating Nach(er)wärmen, Nacherhitzen, Nachheizen *n*

post hydrant → above-ground hydrant

post-mortem dump Speicherabzug *m*

post-office code standardisation advisory group → POCSAG

post-office line, (direct) exchange line, subscriber's line Amtsleitung *f* [*Fernsprechanlage*]

post printer, post writer, post copy processor tastengesteuerte Drucker, Tastendrucker *m*

postprocessor nachgeschaltete Programmumsetzer *m* [*Umsetzen von Daten in ein anderes Format*]

postscript printer definition data, postscript writer definition data, PPD data Definitionsdatei *f* [*Papierformat*]

post-treatment water installation Haus-(Wasser-) Aufbereitungsanlage, Nachaufbereitungsanlage *f*

post writer → post printer

potable hot water → domestic hot water

potable water → drinking water

potatoe chipper, potatoe chipping machine Kartoffelschnitzler *m*, Kartoffelschnitzelmaschine *f*

potatoe peeler, potatoe peeling machine Kartoffelschälmaschine *f*, Kartoffelschäler *m*

potatoe starch interceptor, potatoe starch separator, potatoe starch trap Kartoffelstärke(ab)scheider *m*

potentia hydrogenii [*Latin*] **→ pH-value**

potential wirbelfrei [*Bewegung*]

potential Potential *n*

potential circuit → voltage circuit

potential-dependent potentialbehaftet, potentialabhängig

potential equality Potentialgleichheit *f*

potential flow [*Flow in which the velocity of flow is the gradient of a scalar function, known as the velocity potential*] Potentialströmung, dehnungsfreie Strömung *f*

potential-free → floating

potential jumper → ground(ing) circuit (US)

potentially separated potentialgetrennt

potential output → output referred to a potential

potential relay [*Voltage-controlled switch used to disconnect the starting windings of a compressor motor*] Potentialrelais *n*

potential to ground (US) → ground(ing) potential

potential transformer Spannungstransformator, Spannungswandler *m*

pourpoint Fließpunkt *m* [*Die niedrigste Temperatur, bei welcher Öl unter Bedingungen nach ISO 3016 noch fließt*]

pourpoint Fließgrenze *f* [*DIN 51597. Sie ist vom Hersteller des Wärmeträgers anzugeben; die Fließgrenze muss mit den Angaben über die Pumpfähigkeit des Wärmeträgers korrespondieren*]

powder-coated pulverbeschichtet, pulverüberzogen

powder(ed) soap dispenser Pulverseifenspender, Seifenpulverspender *m*

powder (fire) extinguisher Pulver(feuer)löscher, Pulverbrandlöscher *m*

powder insulating vacuum → evacuated (thermal) powder insulation

power [*The power of a point P in the plane of a circle with respect to the circle is the value of the product PA times PB, where A and B are the points of intersection with the circle of any secant through P, and the length PA and PB are directed*] Potenz *f*

power, magnifying power [*telescope; microscope*] Vergrößerungskraft, Vergrößerungsleistung *f*

power, ability, capability, capacity Vermögen *n*, Fähigkeit *f*

power, intensity Stärke *f*

power → force

power, performance [*The time rate at which work is performed (energy is transformed) by a system*] Leistung *f*

power, exponent [*For instance, the expression x^b can be read as "x raised to power b" or "x to the bth power"*] Exponent *m*, Hochzahl *f*

power, energy [*1. Any type of physical energy, as of electricity, water, and so on. 2. Energy produced by a machine, as distinguished form hand labo(u)r*] Kraft, Energie *f*

power acceptance, accepted power Leistungsaufnahme *f* [*(Schmelz)Sicherung*]

power amplification, power gain [*The ratio of the increased output power delivered by a transducer to the power absorbed by the input circuit of the transducer*] (Wirk)Leistungsverstärkung *f*, (Wirk)Leistungsgewinn *m*

power arm Leistungszweig *m* [*Leistungselektronik*]

power attenuation, power loss [*The ratio of the power absorbed by the input circuit of a transducer to the power delivered to a specified load; usually dissipated as heat and expressed in decibels*] Leistungsverlust *m*, Verlustleistung *f*

power bell → power (current) bell

power boiler Kraftwerkkessel *m*

power bus Hauptsammelschiene *f*

power cabinet, power cubicle Verteilerschrank, Elektroverteilerschrank *m*

power capacitor Leistungskondensator *m*

power characteristic number → characteristic power number

power ((closed) loop) control → ((closed) loop) power control

power conditioning, energy conditioning, power specific preparation, energy specific preparation Energiekonditionierung *f*

power consumption, current consumption Leistungsaufnahme *f*, Stromverbrauch *m* [*z. B. gemessen in kWh*]

power contactor Leistungsschütz(e), Schaltschütz(e) *m*, (*f*), Leistungsschützen, Schaltschützen *m*

power controller Leistungsregler *m*

power converter → converter

power cord, line cord, electric cord [*A two-wire cord terminating in a two-prong plug at one end and connected permanently to a radio receiver or other appliance at the other end; used to make connections to a source of power*] Netzanschlussschnur, Anschlussschnur *f*

power current Starkstrom *m*

power data → performance data

power demand, power requirement Leistungsbedarf *m*

power density [*Electromagnetism. The amount of power per unit area in a radiated microwave or other electromagnetic field, usually expressed in watts per square centimeter*] Leistungsdichte *f*

power divider [*Electromagnetism. A device used to produce a desired distribution of power at a branch point in a waveguide system*] Leistungsteiler *m*

power-down mode Netzausschaltmodus, Netzabschaltmodus *m*

power-down sequence Netzausschalt(ab)lauf, Netzabschalt(ab)lauf *m*, Netzausschaltfolge, Netzabschaltfolge *f*

power drive Kraft(an)trieb, Energie(an)trieb *m*

P

power-driven, power-operated kraft(an)getrieben, energie(an)getrieben, kraftbetrieben, energiebetrieben

power drop → power lowering

power electronics Leistungselektronik *f*

power elevator (US); power lift (Brit.) [*Elevator utilizing means other than gravity or manual energy to move the car*] Maschinenaufzug *m*

power engineering Energietechnik *f*

power factor, phase factor [*Electricity. 1. The ratio of active power to apparent power. For sinusoidal voltage and current, the power factor is cos (È) when (È) is the phase difference between the voltage and current. 2. The ratio of total watts to the root-mean-square (RMS) volt-amperes of an AC circuit*] Leistungsfaktor *m*

power failure, grid failure, power fault, grid fault, power malfunction, grid malfunction Netzausfall, Stromausfall, Netzfehler, Stromfehler *m*, Netzpanne, Strompanne, Netzunterbrechung, Stromunterbrechung, Netzstörung, Stromstörung *f*

power failure alarm, grid failure alarm, power fault alarm, grid fault alarm, power malfunction alarm, grid malfunction alarm Netzausfallalarm, Stromausfallalarm, Netzpannenalarm, Strompannenalarm, Netzstörungsalarm, Stromstörungsalarm, Netzfehleralarm, Stromfehleralarm, Netzunterbrechungsalarm, Stromunterbrechungsalarm *m*

power feeding Strom(ein)speisung *f*

power flow [*Electromagnetism. The rate of electromagnetic energy transport across a specified unit area by an electromagnetic field*] Kraftfluss *m*

power fluctuation, power variation Leistungsschwankung *f*

power-for-size ratio Ausnutzungsverhältnis *n* [*elektrische Maschine*]

power frequency [*The frequency at which electric power is generated and distributed*] Netzfrequenz, Betriebsfrequenz *f*

power fuse Grobsicherung *f*

power gain → power amplification

power gate Leistungsgatter *n*

power grid, power current grid Starkstromnetz *n*

power heating plant → heating power plant

power-heat link(-up), power-heat coupling Kraft-Wärme-Kopp(e)lung *f*

powerhouse, generating station, power station Kraftwerk *n*

powerhouse unit → power station unit

power indicator Betriebskontrolllampe *f*

(power) input → input power

power interconnection Leistungsverbund *m*

power interruptor Stromunterbrecher, Stromtrenner *m*

power inverter → inverter

powerless → power-off

power level [*Electricity. The amount of power that is transmitted beyond a point in an electric system, usually expressed in decibels, indicating the ratio to a given reference power (such as one watt or one milliwatt)*] Feldstärke *f*

power lift (Brit.) → power elevator (US)

power limiter, power limiting device Leistungsbegrenzer *m*

powerline carrier [*Electricity. A carrier-frequency, generally below 600 kHz, that is used to transmit control signals between stations over power lines without interfering with line's normal transmission*] Trägerfrequenz *f* auf Hochspannungsleitungen

powerline interference [*Telecommunications. Unwanted signals that are caused by radiation from electric(al) power transmission lines*] Störsignale von Starkstromleitungen, Störsignale von Kraftstromleitungen *f*

powerline monitor [*Electronics. A device that continuously tracks the voltage level across a power line*] Spannungswächter an einer Starkstromleitung, Spannungswächter an einer Kraftstromleitung *m*

powerline transposition [*Electricity. An ordered interchange of position within an open wire-line and its phases which is implemented so that effects of mutial capitance and inductance with consequent interference are minimized or balanced without material distortion of the normal wire configuration at a distance comparable to the wire separation*] Starkstromleitungskreuzung, Kraftstromleitungskreuzung *f*

power loop control → ((closed) loop) power control

power loop control → ((open) loop) power control

power loss → power attenuation

power (loud)speaker Groß(flächen)lautsprecher *m*

power lowering, power drop Leistungsabnahme, Leistungssenkung *f* [*Bei reduziertem Bedarf*]

power malfunction → power failure

power measuring relay Leistungsmessrelais *n*

power monitor Speisespannungsüberwacher *m* [*elektrische Steuerung*]

power noise Leistungsrauschen *n*

P

power-off, powerless stromlos, ohne Strom

power ((open) loop) control → ((open) loop) power control

power-operated → power-driven

power output, output power [*Electronics. The energy that is delivered by a power amplifier to another component, such as an audio speaker; measured in watts*] abgegebene Leistung, Leistungsabgabe, Ausgangsleistung *f*, Leistungsausgang *m*

power pack [*Electronics. A unit that converts the voltage of a power line or battery into energy at voltage levels suitable for a given device, as in a battery-operated radio, toy, or the like, Not to be confused with "power package"*] Netzteil *m*, *n*, Netz(anschluss)gerät, Stromversorgungsgerät *n* [*„...einheit" kann synonym für „...teil" und „...gerät" verwendet werden*]

power panel Verteilertafel, Elektroverteilertafel *f*

power peak Leistungsspitze *f*

power per pole, pole power Polleistung *f*

power per unit surface, unit surface power Flächenleistung *f*

power plug Netzstecker *m*

power point Kraftanschlussstelle *f*

power protection Leistungsschutz *m*

power ratio Leistungsverhältnis *n*

power rectifier Leistungsgleichrichter, Netzgleichrichter *m*

power relay Leistungsrelais, Hauptstromrelais *n*

power requirement, power demand Leistungsbedarf *m*

P

power reserve → running reserve

power reserve Leistungsreserve *f*

power resistor [*Electricity. A heavy-duty resistor designed to carry large currents without overheating; may be cooled by air convection, air blasts, or water*] Leistungswiderstand *m*

power return Netzwiederkehr *f*

power reversal Leistungsumkehr *f*

power ringing Ruf *m* mit Maschinen- oder Polwechslerstrom

power section Leistungsteil *m*, *n*

power semiconductor (device) [*It is capable of dissipating appreciable power (generally over 1 watt) in normal operation; may handle currents of thousands of amperes or voltages up into thousands of volts, at frequencies up to 10 kilohertz*] Leistungshalbleiter *m*

power service Elektrohausanschluss *m*

power service protector Sicherungsleistungstrenner *m*

power skirting Elektrosockelleiste *f*

power socket outlet and plug Starkstromsteckvorrichtung, Kraftstromsteckvorrichtung *f*

power/space ratio Ausnutzungsgrad *m* [*elektrische Maschine*]

power specific preparation → power conditioning

power stage Leistungsstufe *f*

power station → powerhouse

power station unit, generating station unit, powerhouse unit Kraftwerk(s)block *m*

power supply Netzteil *m*, *n*

power supply Stromversorgung *f*

power supply center (US); power supply centre (Brit.) Energieversorgungszentrale, Stromversorgungszentrale *f*

power (supply) failure, failure of power (supply) Stromausfall *m*

power (supply) removal → removal of power (supply)

power switch, on-off switch, on/off switch Leistungsschalter, Ein-Aus-Schalter *m*

power switchboard [*Electricity. The part of a switch box that contains one or more panels within switches mounted for the purpose of control, circuit-protection measurement, and equipment regulation; it can also carry main power switching and interruption devices with associated connections*] Leistungsschalttafel, Hauptstromschalttafel *f*

power switching [*Electricity. A transference from one high-level power supply circuit output to another*] Leistungsschalten *n*

power switching section Leistungsschaltfeld *n*

power system → energy system

power system control Netzleittechnik *f*

power system control centre (Brit.); power system control center (US) Netzleitstelle, Netzleitstation *f*

power system management Netz(betriebs)führung *f*

power system protection Netz(betriebs)schutz *m*

power system stabilizer, PSS Schlupfstabilisator, Pendeldämpfer, Schwankungsdämpfer *m* [*Dämpfung von Wirkleistungsschwankungen im Netz*]

power transformer [*An iron-core transformer having a primary winding that is connected to an alternating-current power line and one or more secondary windings that provide different alternating voltage values*] Leistungstransformator, Leistungsübertrager, LT *m*

power transistor [*It delivers and dissipates large amounts of power without malfunctioning; commonly used in audio and switching circuits*] Leistungstransistor, End(stufen)transistor *m*

power transmission → **power (current) transmission**

power transmission line, (electric) main [*The facility in an electric power system used to transfer large amounts of power from one location to a distant location; distinguished from a subtransmission or distribution line by higher voltage, greater power capability, and greater length*] Überlandleitung *f*

power tube (US) → **power (amplifier) tube**

power unit Netzsatz, Leistungssatz *m*, Netzgerät *n*, Netzteil, Leistungsteil *m*, *n*

power unit Kraftwerk(s)block *m*

power-up/power-down sequence Netzeinschalt-Netzausschalt(ab)lauf, Netzeinschalt-Netzabschalt(ab)lauf *m*, Netzeinschalt-Netzausschaltfolge, Netzeinschalt-Netzausschaltfolge *f*

power-up sequence Netzeinschaltfolge *f*, Netzeinschalt(ab)lauf *m*

power variation → **power fluctuation**

power ventilation → **mechanical ventilation**

power winding [*Electricity. The output winding through which a controlled current flows in a magnetic amplifier or a saturable reactor*] Leistungswick(e)lung *f*

PPD → **parallel poll disable**

PPD → **predicted percentage of dissatisfied persons**

PPD data → **postscript printer definition data**

PPE → **parallel poll enable**

PPI → **programmable peripheral interface**

PPID → **parent process identification**

PPIO unit → **programmable parallel I/O unit**

PPIS → **parallel poll idle state**

PPL → **phase locked loop**

"P" plumbing trap → **"P" (fixture) trap**

PPM → **pulse position modulation**

PPM, pulse phase modulation Impulsphasenmodulation *f*

PPP, point-to-point protocol Teilnahmeprotokoll *n* [*Mit diesem Protokoll kann ein Rechner auch über eine analoge (Modem) oder digitale (ISDN) Telefonleitung am Internet teilnehmen*]

PPSS, parallel poll standby state Parallelabfragebereitschaft *f* [*programmierbare Messgerät; DIN IEC 625*]

PPU → **parallel poll(ing) unconfigure**

P pulse → **commutator pulse**

practice exercise Praxisübung *f*

practitioner Praktiker *m*

PRAM → **parameter-RAM**

Prandtl number [*Fluid mechanics. A dimensionless quantity used in momentum and heat transfer calculations, given by the kinematic viscosity of a fluid divided by the molecular diffusivity*] Prandtl-Zahl *f*

preaction sprinkler plant, preaction sprinkler installation vorgesteuerte Sprinkleranlage, vorgesteuerte Sprengeranlage *f*

pre-aerated burner [*A burner for which the gas and primary air under pressure are premixed*] Vormischbrenner *m*

pre-alarm Voralarm *m* [*Gegensatz: Hauptalarm*]

preamble Präambel *f* [*Eine von einem Block übertragene Bitfolge, um Übereinstimmung des Zeittaktes und anderer Schaltkriterien im Kanal sicherzustellen. In der Ethemet-Spezifikation beträgt die Präambel 64 bits*]

preamplifier Eingang(s)verstärker, Vorverstärker *m*

to pre-assemble vorfertigen, konfektionieren

pre-assembled cable set vorgefertigte Kabelsatz, konfektionierte Kabelsatz *m*

pre-assembly Vorfertigung, Konfektionierung *f*

to pre-assign, to pre-allot, to pre-allocate vorbelegen, vorzuweisen, vorzuordnen, vorzuteilen

pre-carbonation (treatment) Vorkarbonatisieren *n*

precast concrete flue block [*B.S. 1289*] Betonfeuerzugblock(stein), Betonfeuerzugstein *m*

precaution measure Vorsicht(s)maßnahme *f*

preceiver Kanalwähler-Vorverstärker-Kombination *f* [*Zur Wiedergabe werden noch Leistungsverstärker und Lautsprecher benötigt*]

precipitate Niederschlag, Bodensatz *m*

precipitation Niederschlag *m*

precipitation water Niederschlag(s)wasser *n*

precipitator, precipitation apparatus [*Device to remove fine ash, tars, dusts, or smoke particles from flue gases or other gaseous streams*] Fäller, Fällapparat *m*

precipitator (type) of filter → **electrostatic precipitator**

precision [*Closeness of agreement among repeated measurements of the same characteristic by the same method under the same conditions*] Präzision *f*

precision engineering; mecatronic [*Japanese term*]; **microtechnique** [*French term*] Feinwirktechnik *f*

precompiler, precompiling routine Vor(sprach)übersetzer, Vorbefehlsübersetzer, Vorkompilierer *m*, Vorübersetzungsprogramm, vorkompilierende Programm *n* [*Siehe Anmerkung unter „Sprachübersetzer"*]

preconditioned, pretreated [*air*] vorbehandelt, vorklimatisiert

P

precooler Vorkühler *m*

precooling Vorkühlen *n*, Vorkühlung *f*

pre-defined condition bestimmte Voraussetzung *f*

predetermined vorbestimmt, vorgegeben

predicted mean vote PWV-Index *m* [*Thermische Behaglichkeit. Die mittlere Beurteilung des Raumklimas, wie sie durch eine große Anzahl von Personen vorgenommen würde*]

predicted percentage of dissatisfied persons, PPD persons Anzahl *f* der voraussichtlichen unzufriedenen Personen [*Behaglichkeitsgleichung. DIN EN ISO 7730*]

predictive maintenance, PDM zustandsorientierte und vorausschauende Unterhaltung *f*

predictive ((open) loop) control [*A control method on an algorithm which attempts to predict future states*] vorhersagende Steuerung *f*

preemphasis Höhenanhebung *f* [*Videoaufzeichnung*]

preemptive allocation, preemptive assignment, preemptive allotment unterbrechende Zuordnung, unterbrechende Zuteilung, unterbrechende Zuweisung *f*

prefab(ricated) vorgefertigt, konfektioniert

prefab(ricated) bathroom Badezimmerzelle *f*

prefab(ricated) connection wire, prefab(ricated) connecting wire konfektionierte Steckdraht, vorgefertigte Steckdraht *m*

prefab(ricated) program (US); prefab(ricated) programme Fertigprogramm *n*

prefabrication Vorfertigung, Konfektionierung *f*

to prefill vorfüllen

prefilter [*In an air-conditioning system, a filter before the main filter to remove the larger particles; usually has a lower efficiency than the main filter and has a low pressure drop characteristic*] Vorfilter *m, n*

prefiltering Vorfiltern *n*

prefiltration Vorfilterung *f*

pre-handover Vorübergabe *f*

to preheat vorerhitzen, vor(er)wärmen, vorheizen

preheater → **(air) preheater**

preheating → **(air) preheating**

preheat(ing) coil Vorwärmschlange, Vorheiz(ungs)-schlange *f*

preheat(ing) lamp Glühstartlampe, Warmstartlampe *f*

preheat(ing) period → **boost(ing) period**

prehumidifier [*It may be used before cooling unit*] Vorentfeuchter *m*

preignition Vorzündung *f*

preinstallation Voreinbau *m*, Vorinstallation *f*

preinstalled vorinstalliert, voreingebaut

preliminary air scavenging, preliminary air purging Luftvorspülen *n*

prelubrication Vorschmierung *f*

premise WENN-Teil *m/n*, Prämisse, Bedingung, Voraussetzung *f*

premix(ed) combustion Vormischverbrennung, vorgemischte Verbrennung *f*

premix(ed) flame Vormischflamme, vorgemischte Flamme *f*

premix(ed) flat one-dimensional flame eindimensionale flache Vormischflamme, flache eindimensionale Vormischflamme *f*

premix(ed) laminar flame, laminar premix(ed) flame laminare Vormischflamme, vorgemischte Flachflamme *f*

premixing Vormischen *n*

premix(ing) air burner → **air-premix(ing) burner**

premix(ing) radiation burner Vormisch-Strahlungsbrenner *m*

prenumbered vornummeriert

preparation to start up Einrüsten *n* im Systemanlauf

preparatory shelf Werkstattsammellager *n*

prepayment gas meter Münzgaszähler, Münzgasmesser *m*

prepayment meter [*B.S. 37*] Münzzähler, Münzmesser *m*

prepiped vorverrohrt, vorberohrt

preponderating key → **French key**

preprocessing → **(data) preprocessing**

preprossessor vorgeschaltete Umsetzprogramm *n* [*Zum Umsetzen von Daten aus einem anderen Format*]

prerotating swirl Vordreh(ungs)wirbel *m*

prescreening Bestückungsprüfung, Bestückungsprobe *f*, Bestückungsversuch *m* [*Von elektronischen Baugruppen zur Auffindung von Fertigungsfehlern*]

prescaler Vorteiler *m* [*Frequenzteiler*]

to preselect vorwählen

preselection Vorwahl *f*

preselection command Vorwahlkommando *n*

preselection switch Vorwahlschalter *m*

preselector Vorwähler *m*

presentation Darstellung *f*

presentation image Begriffsvorrat *m* [*E DIN ISO 7498*]

presentation layer, layer 6 Darstellungsschicht, Darstellungsebene, sechste Schicht, sechste Ebene, Schicht 6, Ebene 6 *f* [*OSI. Auf dieser Ebene wird festgelegt welche Datenkodierung vereinbart werden soll*]

to preset voreinstellen

preset voreingestellt

preset circulating current Kreisstromvorgabe *f*

preset counter Vorwahlzähler *m* [*Er summiert Eingangssignale auf und gibt bei Erreichen einer oder mehrerer vorgewählter Zahlen je ein Signal ab*]

preset rupture point Sollbruchstelle *f*

presetter Voreinsteller *m*

presetting element Voreinstellglied, Voreinstellelement *n*

presetting of coordinates, coordinates presetting Koordinatenspeicherung *f*

pre-shutoff Vorabsperrung *f*

pre-shutoff valve Vorabsperrventil *n*

presignal Vorsignal *n*

prespin Vordrall *m*

to press drücken [*Taste*]

press-button → (push)button

pressed steel Pressstahl *m*

pressed steel (bath)tub Pressstahl(bade)wanne *f*

pressure Druck *m* [*Die physikalische Größe Druck p ist der Quotient aus der Normalkraft F_N, die auf eine Fläche wirkt, und dieser Fläche A.*

$$p = \frac{F_N}{A}.$$

In der Technik werden verschiedene Druckgrößen benutzt, überwiegend Differenzen zweier Drücke, die im Sprachgebrauch der Technik ebenfalls Druck genannt werden.]

pressure air supply [*See remark under "air supply"*] Druckbelüften *n*, Druckbelüftung *f*

pressure atomizer, mechanical atomizer Druckzerstäuber, mechanische Zerstäuber *m* [*Die Zerstäubung erfolgt durch Brennstoffdruck*]

(pressure) atomizing burner; (pressure) atomising burner (Brit.); gun burner [*A liquid-fuel burner that atomises the unignited fuel into a fine spray as it enters the combustion chamber*] Zerstäuberbrenner, Zerstäubungsbrenner *m*

pressure at rest → at-rest pressure

pressure balance, pressure equalization Druckausgleich *m*

pressure boiler Druckkessel *m*

pressure booster station, pressure boosting station, pressure rising station Druckerhöhungsstation *f*

pressure build-up Druckaufbau *m*

pressure calibration device, pressure calibrating device Druckkalibriergerät *n*

pressure central distributor → pressure manifold

pressure characteristic (curve) Druckkennlinie, Druckkennkurve *f*

pressure ((closed) loop) control → pressure regulation

pressure ((closed) loop) controller Druckregler *m*

pressure compensator [*A device in a governor which exerts a force equal and opposite to the differential pressure across the governor valve which, unless so balanced, would tend to alter the position of the valve*] Druckausgleicher *m*

pressure control → pressure regulation

pressure control → pressure ((open) loop) control

pressure-controlled [*((closed) loop) control*] druckgeregelt

pressure-controlled [*((open) loop) control*] druckgesteuert

pressure cooling Druckkühlung *f*

pressure cut-out switch Druckausschalter *m*

pressure-dependent, PD [*Varying flow rate through a flow control device in response to changes in system pressure*] druckabhängig

pressure detector, pressure detecting device, pressure detection device Druckmelder *m*, Druckmeldegerät *n* [*z. B. Trittmatte*]

pressure dew point Drucktaupunkt *m*

pressure difference → differential pressure

pressure drop, pressure decrease, pressure reduction [*Loss of pressure due to friction or lift*] Druckabfall, Druckrückgang *m*, Druckabnahme, Druckminderung, Druck(ab)senkung *f*

pressure elimination Druckvernichtung *f*

pressure energy Druckenergie *f*

pressure engineering Drucktechnik *f*

pressure equalization → pressure balance

pressure equalizing duct, pressure balancing duct Druckausgleichskanal *m*

pressure equalizing tank, pressure balancing tank Druckausgleichsbehälter *m*

pressure equation Druckgleichung *f*

pressure equilibrium Druckgleichgewicht *n*

P

pressure exerting part Druckübertragungsteil *m, n* [*Klemme*]

pressure expansion vessel Druckausdehn(ungs)gefäß *n* [*DIN 30 600, Reg.-Nr. 06171*]

pressure fan Druckventilator *m*

pressure gauge; pressure gage (US) Manometer *n*, Druckmesser *m* [*Messen der Spannkraft von Gas und Dampf*]

pressure gauge cock; pressure gage cock (US) Manometerhahn, Druckmesserhahn *m*

pressure gauge shut-off valve; pressure gage shut-off valve (US) Manometerabsperrventil, Druckmesserabsperrventil *n*

pressure gauging; pressure gaging (US) Druckmessung *f*, Druckmessen *n*

pressure gradient Druckgefälle *n*

pressure increase, pressure rise Druckerhöhung *f*, Förderdruck, theoretische Gesamtdruckunterschied *m* [*Ventilator*]

pressure interrupt(ion) Druckunterbrechung *f*

pressure interstage → **intermediate-pressure stage**

pressure jet Druckstrahl *m*

pressure jet boiler Druckstrahlkessel *m*

pressure jet burner Druckstrahlbrenner *m*

pressure jet miniburner Druckstrahlkleinbrenner *m*

pressure jet (oil) burner Druckstrahl(öl)brenner *m*

pressureless → **depressurized**

pressureless powder tank (fire) extinguisher Auflade(feuer)löscher, Aufladebrandlöscher [*Löscher mit drucklosem Pulverbehälter und an ihm außen oder innen angebrachter Treibgasflasche*]

pressure level Druckniveau *n*, Druck(höhen)stand, Druckpegel *m*

pressure limit Druckgrenze *f*

pressure limit(ing) device, pressure limiter Druckbegrenzer *m*

pressure line, head line Druckleitung, Förderleitung *f*

pressure liquid Druckflüssigkeit *f*

pressure loadability Druckbelastbarkeit *f*

pressure loop Druckschleife *f*

pressure loss Druckverlust *m*

pressure loss percentage Druckverlustanteil, Druckverlustprozentsatz *m*, Autorität, Ventilautorität *f* [*Ventilbemessung in der Heiz(ungs)technik. Dieser Anteil ist auf den Gesamtdruckabfall im Netz bezogen*]

pressure maintenance, pressure maintaining Druckhaltung *f* [*Haltung vorgegebener Drücke in Räumen, Rohrleitungen und Kanälen*]

pressure maintenance pump, pressure-maintaining pump Druckhaltungspumpe, Druckhaltepumpe *f*

pressure maintenance valve, pressure-maintaining valve [*Valve which maintains a uniform pressure on its outlet side irrespective of pressure variation on its inlet side*] Druckhaltungsventil, Druckhalteventil *n*

pressure manifold, pressure central distributor Druckverteiler(rohr) *m*, (*n*)

pressure mat, contact mat Kontaktmatte, Druckmatte *f*

pressure oil burner, oil pressure burner Druckölbrenner, Öldruckbrenner *m*

pressure ((open) loop) control Drucksteuerung *f*

pressure(-operated) switch Druckschalter, Druckwächter *m*

pressure pickup, pressure measured value transmitter Druckmesswertgeber, Druckaufnehmer *m*

pressure pipe Druckrohr *n*

pressure pipe flange Druckrohrflansch *m*

pressure pipe flow Druckrohrströmung *f*

pressure (pipe)line Druck(rohr)leitung *f*

pressure piston pump, piston pressure pump Druckkolbenpumpe, Kolbendruckpumpe *f*

pressure preheater Druckvor(er)wärmer, Druckvorerhitzer *m*

pressure probe Drucksonde *f*

pressure propagation Druckfortpflanzung *f*

pressure range Druckbereich *m* [*selten: n*]

pressure ratio, compression ratio, CR [*Ratio of the absolute discharge pressure to the absolute inlet pressure*] Druckverhältnis *n*

pressure recorder Druckschreiber *m*

pressure recovery coefficient Druckrückgewinnungskoeffizient *m* [*Eine Diffusionskennzahl*]

pressure reducer Druckminderer *m* [*Sicherheitsarmatur für eine vorgegebene oder einstellbare Druckminderung innerhalb eines Drucksystems*]

pressure reducing → **pressure drop**

pressure reducing angle valve → **angle pressure reducing valve**

pressure reducing straight(-way) valve → **straight(-way) pressure reducing valve**

pressure reduction → **pressure drop**

pressure reflection coefficient → **acoustic (pressure) reflection coefficient**

P

pressure regulating valve Druckregulier(ungs)ventil, Druckregel(ungs)ventil *n*

pressure regulation, pressure ((closed) loop) control Druckreg(e)lung *f*

pressure regulator Druckregler *m* [*Druckregler im allgemeinen Sinn(e) haben die Aufgabe den Druck in einem System konstant zu halten. Sie bestehen grundsätzlich aus einem Ventil, einer Membran und einer Steuerleitung ohne Hilfsenergie*]

pressure relief Druckentlastung *f*

pressure relief device [*Pressure-actuated valve or rupture member designed to relieve excessive pressure automatically*] Druckentlastungsgerät *n*, Druckentlastungsvorrichtung *f*

pressure relief valve [*Pressure-actuated valve held closed by a spring or other means and designed to relieve pressure in excess of its setting automatically*] Druckentlastungsventil *n*

pressure-relieved druckentlastet

pressure resistance Druckwiderstand *m*

pressure retention Druckhaltung *f*

pressure rise → **pressure increase**

pressure rising station → **pressure booster station**

pressure scale Druckskala *f*

pressure selector Druckwähler *m*

pressure-sensitive druckempfindlich

pressure sensor Druckfühler, Druckdetektor, Drucksensor *m*

pressure sequential key Druckfolgetaste(r) *f*, (*m*)

pressure sequential switch Druckfolgeschalter *m*

pressure setpoint Drucksollwert *m*

pressure shock Druckstoß *m*

pressure shutoff valve Druckabsperrventil *n*

pressure side [*compressor*] → **delivery side**

pressure-side druckseitig

pressure spring thermometer Druckfederthermometer *n*

pressure steam air humidifier Druckdampf-Luftbefeuchter, Druckdampf-Luftanfeuchter *m*

pressure-sustaining valve [*Valve providing maintenance of designated pressure levels within a system*] Druck(er)haltungsventil *n*

pressure switch [*A switch that automatically turns on or off in response to changes of pressure in a system*] Druckwächter *m*

pressure tapping Druckentnahme *f* [*Messblende*]

pressure test(ing) Abdrücken *n*, Druckprüfung, Druckprobe *f*, Druckversuch *m*

pressure tube [*Small line carrying pressure to the element of a pressure controller*] Druckröhre *f*

pressure (type) water heater, unvent(ilat)ed water heater [*See remark under "water heater"*] geschlossene Wasserheizer, geschlossene Wasser(er)wärmer, geschlossene Wassererhitzer, druckfeste Wasserheizer, druckfeste Wassererhitzer, druckfeste Wasser(er)wärmer *m*

pressure tank plant, pressure vessel plant, pressure tank installation, pressure vessel installation Druckbehälteranlage, Druckgefäßanlage, Druckspeicheranlage, Drucktankanlage *f* [*Wasserversorgung*]

pressure transmitter Druckfühler, Drucktransmitter *m* [*Zur Erfassung von Absolut- und Differenzdrücken in Flüssigkeiten und Gasen*]

pressure vent Druckdurchlass *m*

pressure vessel, pressure tank Druckgefäß *n*, Druckspeicher, Druckbehälter, Drucktank *m* [*Wasserversorgung*]

pressure water filter Druckwasserfilter *m, n*

pressure wire Messader *f* [*Kabel*]

pressure zone isolation Druckzonentrennung *f* [*Rohrleitungssystem in einem Gebäude*]

pressure zone microphone [*Trademark of Messr. Crown*], **boundary layer microphone, BLM** Grenzlagermikrophon *n*

pressurization, pressurizing (US); pressurisation, pressurising (Brit.) Druckhaltung *f*

pressurized hot water Heißwasser *n*

pressurized hot water heating Heißwasserheizung *f* [*Sie arbeitet mit Temperaturen über 120 Grad C*]

prestop Vorhalt, Endschalterhaltung *m* [*Automatische Vorabschaltung zur Vermeidung eines Überfahrens*]

prestressed concrete pressure vessel Spannbetondruckgefäß *n*

presumptive address → **address constant**

presumptive instruction → **basic instruction**

pretreated → **preconditioned**

pretrimmed wallpaper [*Selvages are removed at the factory so that the wallpaper is ready to hang*] Fertigtapete *f*

prevention Verhütung, Vorsorge *f*

prevention of contamination Reinhaltung *f*

prevention of harmful emissions Abstrahlverhütung, Abstrahlvorsorge *f*

preventive vorbeugend, vorsorglich

preventive check(ing) Präventivkontrolle, vorbeugende Kontrolle, vorsorgende Kontrolle *f*

P

preventive fire protection vorbeugende Feuerschutz, vorsorgende Feuerschutz, vorbeugende Brandschutz, vorsorgende Brandschutz *m* [*Durch Absperrvorrichtungen gegen Brandübertragung*]

preventive maintenance → **care**

pre-warning Vorwarnung *f*

prewashing stage Vorabräumzone *f* [*Geschirrspülmaschine*]

to prewire vorverdrahten

prewired [*The only wiring required on the site is that between the terminal (strip) and the actuating device*] vorverdrahted

prewiring Vorverdrahtung *f*

PRF → **pulse repetition rate**

price file [*It allows to make the right estimating, buying and invoicing decisions*] Kalkulationsdatei *f*

prick-through (type) lampholder Schnellanzapfstecker *m* für Lampen

primary air, first air Primärluft, Erstluft *f* [*Luft, die als Erste in einen Raum strömt*]

primary air, first air [*Any air which is mixed with fuel at or in a burner prior to burning*] Primärluft, Erstluft *f*

primary central air conditioning plant → **combined air conditioning plant**

primary air conditioning plant, first air conditioning plant, primary air conditioning installation, first air conditioning installation Primärluftklima(tisierungs)anlage, Erstluftklima(tisierungs)anlage *f*

primary air nozzle, first air nozzle Primärluftdüse, Erstluftdüse *f*

P

primary air supply Primärluftzufuhr, Erstluftzufuhr *f*

primary air temperature Primärlufttemperatur, Erstluftterrnperatur *f*

primary call, first call Erst(an)ruf, Primär(an)ruf *m*

primary call, first call Erstgespräch, Primärgespräch *n*

primary circuit, first circuit [*The flow and return circuit through which water circulates between the water heater and the cylinder, tank or calorifier*] Primär(umwälz(ungs))kreis, Erst(umwälz(ungs))kreis *m*

primary circulation, first circulation [*The circulations of water in the primary circuit of a hot water system*] Primärumwälzung, Primärzirkulation, Erstumwälzung, Erstzirkulation *f*

primary commissioning, first commissioning Erstinbetriebnahme, Primärinbetriebnahme *f*

primary condenser, primary cooler, first condenser, first cooler [*Apparatus for condensing tar and water from raw gas and for cooling gas to a temperature suitable for the first process to which it is to be subjected*] Primärkondensator *m*

primary conductor, first conductor Primär(strom)leiter, Erst(strom)leiter *m*

primary control unit, central control unit Leitzentrale, zentrale Leitstelle *f*

primary current → **input current**

primary damage, first damage, damage caused by natural hazards Elementarschaden *m*

(primary) detector → **sensor**

primary energy, first energy Primärenergie, Erstenergie *f* [*Energieinhalt von noch nicht umgewandelten Energieträgern*]

primary headend equipment, first headend equipment Erstendeinrichtung, Primärendeinrichtung *f*

primary heating water, first heating water Primärheiz(ungs)wasser, Erstheiz(ungs)wasser *n*

primary light source, first light source Primärlichtquelle, Erstlichtquelle *f*, Selbstleuchter *m*

primary memory → **principal memory**

primary mounting, first mounting [*As opposed to "replacement mounting"*] Erstmontage, Primärmontage *f*

primary protection, first protection Primärschutz, Erstschutz *m*

primary return, first return [*The pipe conveying water from the cylinder or tank to the heater or boiler*] Primärrücklaufrohr, Erstrücklaufrohr *n*

primary service, high-voltage service, high-tension service [*above 600 V.*] Hochspannungsbetrieb *m*

primary (supply) line, first (supply) line, primary (supply) service, first (supply) service Primär(versorgungs)leitung, Erst(versorgungs)leitung *f*

primary water [*The water in contact with the hot surface of the boiler*] Primärwasser, Erstwasser *n*

primary water net, first water net Primärwassernetz, Erstwassernetz *n*

primary (water) supply, first (water) supply [*It is instantly available*] Wasserprimäreinspeisung, Wasserersteinspeisung *f* [*z. B. in eine Sprinkleranlage*]

prime coat(ing) Grundierung, Grund(ier)schicht, Grund(ier)beschichtung *f*

primed grundiert

primer [*Any material that is used to produce initial coat(ing) for a surface in preparation for a finishing coat(ing)*] Grundier(ungs)mittel *n*

primer electrode Starterelektrode *f*

primitive → **(service-)primitive**

primus stove (Brit.) Brennspirituskocher *m*

principal air treatment, main air treatment Lufthauptbehandlung *f* [*Teilbehandlung, der Luft, die nach der Zeit der Luftbehandlungsfunktionen oder nach der Leistung die wesentlichste ist*]

principal alarm, main alarm, master alarm Hauptalarm *m* [*Gegensatz: Voralarm*]

principal axis of inertia, main axis of inertia Hauptträgheitsachse *f*

principal cable, master cable, main cable Stammkabel *n*

principal call station, master call station, MCS, main call station Leitsprechstelle *f*

principal clock, master clock, main clock Hauptuhr, Mutteruhr *f*

principal cock, main cock, master cock Haupthahn *m*

principal computer, master computer, main computer Leitrechner *m*

principal (computer) memory, main (computer) memory, master (computer) memory Haupt(rechner)speicher, Rechnerhauptspeicher *m*

principal contact, main contact, master contact Hauptkontakt *m*, Hauptschaltglied, Hauptschaltstück, Hauptkontaktstück *n*

principal controller, principal regulator, master controller, master regulator, main controller, main regulator Führungsregler, Hauptregler *m* [*Er gibt die Führungsgröße für einen oder mehrere Folgeregler vor. DIN 19 225*]

principal control panel, principal control board, master control panel, master control board, main control panel, main control board Hauptzentrale, Hauptbedien(er)tafel, Hauptbedienungstafel *f*, Hauptbedien(er)feld, Hauptbedienungsfeld *n*

principal current breaker, main current breaker Haupt(strom)unterbrecher *m*

principal data, master data, main data Stammdaten *f*

principal dimmer, master dimmer, main dimmer Hauptlichtsteller, Haupthelligkeitsregler, Summenlichtsteller, Summenhelligkeitsregler *m*

principal display, master display, main display Leitmarke *f*

principal distribution board; distribution centre (US); distribution centre (Brit.); main distribution board, master distribution board (Spannungs-) Hauptverteilung *f*, Hauptverteilerkasten *m*

principal distribution box, master distribution box, main distribution box Hauptverteilung(skasten) *f*, (*m*)

principal duct, master duct, main duct; header (US) [*A duct for distributing to or collecting flowing fluid from various branches*] Hauptkanal *m*

principal feeder, master feeder, feed(ing) main (line), main feeder Hauptspeiseleitung, Speisehauptleitung, Hauptzuleitung *f*

principal flue, main flue, uptake, master flue, flue connection; breeching (US) [*Casing at the (end of a boiler) opening of a device carrying products of combustion where a flue is connected as part of a system to convey the products of combustion (from the boiler) to a chimney or vent*] Fuchs(kanal) *m*

principal forward flow, main forward flow, master forward flow Hauptvorlauf *m*

principal forward flow pipe, main forward flow pipe, master forward flow pipe Hauptvorlaufrohr, Vorlaufhauptrohr *n*

principal frame computer, main frame computer, master frame computer Großrechneranlage, Großrechenanlage *f*

principal fuse → company's fuse

principal gas valve, main gas valve, master gas valve Hauptgasventil, Gashauptventil *n*

principal gate valve, main gate valve, master gate valve (US); sluice junction valve, parallel slide junction valve Hauptschieber *m* [*Dampfabsperrorgan*]

principal group key, main group key, master group key Hauptgruppenschlüssel *m*

principal incoming terminal, main incoming terminal, master incoming terminal Installationsklemme *f*

principal index, master index, main index Anlagenkennung *f*

principal input, main input, master input Hauptanschluss *m* [*Schaltplan*]

principal leakage, main leakage Luftspaltstreuung *f*

principal light switch, master light switch, main light switch Hauptlichtschalter *m*

principal (line) pipe, main (line) pipe, master (line) pipe Haupt(leitungs)rohr, Leitungshauptrohr *n*

principal memory, principal storage, main memory, main storage, master memory, master storage, user memory, active memory, primary memory, core, central memory, central storage Arbeitsspeicher, ASP, Hauptspeicher, Zentralspeicher *m*, Speicherwerk *n*

principal meter, master meter, main meter Hauptzähler *m*

principal return air duct, main return air duct, master return air duct Hauptrückluftkanal, Rückluft-hauptkanal *m*

principal setpoint, master setpoint, main setpoint Führungssollwert *m*

principal sewer, main sewer, master sewer [*See remark under "sewer"*] Hauptkanal *m*, Hauptleitung *f*

P

principal spindle controller, main spindle controller, master spindle controller, net(work) spindle controller Hauptspindelregler *m*

principal station, master station, main station [*A station that can select and transmit a message to a slave station*] Hauptstation *f*

principal storage → principal memory

principal supply network, main supply network, master supply network Hauptnetz *n*

principal switch, main switch, master switch Hauptschalter *m*

principal switch assembly kit, main switch assembly kit, master switch assembly kit Hauptschaltereinbausatz *m*

principal transmitter, master transmitter, main transmitter Hauptsender *m*

principal valve, main valve, master valve, king valve [*1. Stop valve between receiver an liquid main. 2. Stop valve in boiler head*] Haupt(ab)sperrschieber, Hauptschieber *m*

principle Grundsatz *m*

principle benefit(s) grundsätzliche Funktionalität *f*

printable, writeable bedruckbar, druckbar

print contrast signal, PCS Druckkontrastsignal, Kontrastdrucksignal *n*

printed circuit gedruckte Schaltung *f*

printed(-circuit) board, PCB, track Leiterplatte, Printplatte, Flachbaugruppe, Leiterbahn *f* [*EN 50020*]

printed(-circuit) board assembly, PCB assembly bestückte Leiterplatte, bestückte Printplatte, bestückte Flachbaugruppe, bestückte Leiterbahn *f*

P

printed(-circuit) board layer, PCB layer Leiterplattenlage, Printplattenlage, Leiterplattenschicht, Printplattenschicht, Flachbaugruppenlage, Flachbaugruppenschicht, Leiterbahnlage, Leiterbahnschicht *f*

printed(-circuit) bus card, PCB bus card, printed (-circuit) bus plate Bus-Leiterplatte, Bus-Printplatte, Bus-Flachbaugruppe, Bus-Leiterbahn *f*

printed letter Druckbuchstabe *m*

printer, writer, copy processor Drucker *m*

printer changeover switch, writer changeover switch, printer transfer switch, writer transfer switch, printer double-throw switch, writer double-throw switch, printer throwover switch, writer throwover switch [*"Copy processor" may synonymously be used for "printer" and "writer"*] Druckumschalter *m*

printer command language → PCL

printer connection, writer connection Druckeranschluss *m*, Druckerverbindung *f*

printer data, writer data Druckerdaten *f*

printer description, writer description Druckerbeschreibung *f*

printer double-throw switch → printer changeover switch

printer drive, writer drive Druckerlaufwerk *n*

printer ink, writer ink Druckertinte *f*

printer interface, writer interface Druckerschnittstelle *f*

printer module, writer module, printer block, writer block Druckerbaustein, Druckermodul *m*

printer output microfilm, POM Druckerausgabemikrofilm *m*

printer spooling, writer spooling Druckerspulbetrieb *m*

printer throwover switch → printer changeover switch

printer transfer switch → printer changeover switch

print(ing) cycle, writing cycle Druckzyklus *m*

print(ing) error search, writing error search Druckfehlersuche *f*

print(ing) head, writing head Druckkopf *m*

printing manager, writing manager Druckverwalter *m* [*Er bestimmt, an welchen Drucker ein Auftrag geschickt wird. Außerdem speichert er Aufträge zwischen. Somit kann weitergearbeitet werden, bevor ein Ausdruck beendet ist*]

print(ing) mechanism, writing mechanism Druckwerk *n*

printing out → logging

print(ing) speed, print(ing) rate, writing rate, writing speed Druckgeschwindigkeit *f*

to print out, to write out ausdrucken

printout, writeout Ausdruck *m*, Druckausgabe *f*

printserver separate Rechner *m* mit Druckanschluss

to prioritize vorziehen, bevorzugen

priority Priorität, Bevorzugung *f*, Vorrang, Vorzug *m*

priority circuit Vorrangschaltung, Prioritätsschaltung, Vorzugsschaltung *f*

priority degree → degree of priority

priority level Vorrangstufe, Prioritätsstufe, Vorzugsstufe *f*

priority order → order of priority

priority scanning Vorrangabtastung, Prioritätsabtastung, Vorzugsabtastung *f*

priority selector (switch) Vorrangwähler, Vorrangwahlschalter, Prioritätswähler, Prioritätswahlschalter, Vorzugswähler, Vorzugswahlschalter *m*

prison(-type) lock Gefängnis(zellen)schloss *n*

privacy, personal detachment Abgeschiedenheit, Abgeschlossenheit *f*

privacy protection Datenschutz *m* [*Sachlage, bei der die schutzwürdigen Belange Betroffener vor Beeinträchtigung, die von der Verarbeitung von Daten ausgeht, bewahrt sind. Betroffene können natürliche oder juristische Personen oder Personenvereinigungen sein, aber nur insoweit, als Daten über sie verarbeitet werden oder durch Verarbeitung von Daten auf ihre Identität geschlossen werden kann. E DIN 44 300 T 1*]

private individual, private person Privatperson *f*

private listening sound system, PLSS störungsarme Hörwiedergabesystem *n*

private sewage treatment; private wastewater treatment (US) [*Sewage treatment other than in central city treatment plants*] Privatabwasserbehandlung *f*

private (swimming) pool → residential (swimming) pool

private value Bodenwert, (unbebaute) Grund(stücks)wert, Liegenschaftswert, Landwert *m*

privilege Befugnis *f*, Vorrecht, Sonderrecht *n*

probability factor Wahrscheinlichkeitsfaktor *m*

probe, explorer Sonde *f*

probe cable, explorer cable Sondenkabel *n*

probe card, explorer card Sondenkarte *f* [*Sie trägt Prüfspitzen zur Kontaktierung eines Chips auf einer Halbleiterscheibe*]

probe index (Ultra)Schallaustrittspunkt *m* [*DIN 54 119*]

probe ((open) loop) control, explorer ((open) loop) control Sondensteuerung *f*

problem statement Fragestellung, Problemstellung *f*

procedural language problemorientierte (Programmier)Sprache *f*

procedural semantics [*Artificial intelligence. Representation of the meaning of words by procedures that construct the meanings, as distinguished from a declarative, dictionarylike representation*] prozedurale Semantik *f*

procedure → process

procedure Prozedur *f* [*DIN 44 300, Teil 4. Ein Programmbaustein, der eine zur Lösung einer Aufgabe vollständige Anweisung, aber nicht notwendig alle erforderlichen Vereinbarungen, umfasst. Die beim Aufruf anzugebenden Ein- und Ausgabegrößen müssen jedoch innerhalb der Prozedur vereinbart werden; sie werden formale Prozedurparameter, die ihnen entsprechenden Größen in der Rufanweisung oder im rufenden Ausdruck aktuelle Prozedurparameter genannt.*]

procedure, sequence of programming instructions [*Computer engineering. It collectively accomplishes some task, may be referenced at multiple points in a program(me), and may have one or more input and output parameters; generally used in reference to high-level languages*] Programmier(ungs)befehlsfolge *f*

procedure statement Prozeduranweisung *f* [*Anweisung zum Abruf einer Prozedur wobei die Prozedurparameter definiert werden müssen*]

to proceed [*In some process or course of action*] vorgehen

to process [*Computer engineering. To perform operations on data in a computer*] aufbereiten

to process [*to perform a series of activities*] tätigen

process, (systematic) procedure, unit operation Prozess *m*, Vorgang *m* [*In einem Rechensystem die Gesamtheit der Vorgänge, die an der jeweiligen Ausführung eines Programms beteiligt sind und von einer Instanz gesteuert werden. DIN 44 300, Teil 1*]

process Prozess *m* [*Eine Gesamtheit von aufeinander einwirkenden Vorgängen in einem System, durch die Maschine, Energie, oder Information umgeformt, transportiert oder gespeichert werden. DIN 19 226, Teil 1*]

process → controlled system [*IEC 50(351)*]

process analog(ue) detector, process analog(ue) detecting device, process analog(ue) detection device Prozessanalogmelder *m*, Prozessanalogmeldegerät *n*

process ((closed) loop) control Prozessreg(e)lung *f*

process control [*1. Control for automatic regulation of operations or processes, or both. 2. Closed loop control as applied to processes or operations, or both. Note: HVAC systems may be considered as processes*] Prozessleitung, Prozessführung *f*

process (control) computer Prozessrechner *m* [*Digitale Rechenanlage zur Erfassung von Prozessdaten (Erfassungseinheit) und gegebenenfalls über solche zur Abgabe von Steuer(ungs)werten (Ausgabeeinheit) im Echtzeitbetrieb*]

process control facility Prozessleiteinrichtung, Prozessführungseinrichtung *f* [*Sie umfasst alle für das Leiten erforderlichen Geräte und Programme sowie im weiteren Sinne auch Anweisungen und Vorschriften. DIN 19 222*]

process-controlled → computer-controlled

process controller [*A device that measures an input and than tries to maintain it at a desired value by adjusting an output device*] Prozessregler *m*

process control level, process management level Prozessleitebene, Prozessführungsebene *f* [*Sie fasst alle Teile der Leitvorrichtung zusammen, die auf die Gruppenleitebene einwirken. DIN 19 222*]

P

process cooler Prozesskühler *m*

process cooling Prozesskühlung *f*

process cycle, fetch machine cycle [*See remark under "to fetch"*] Abrufzyklus *m*

process data processing Prozessdatenverarbeitung *f*

process diagnostic detector, process diagnostic detecting device, process diagnostic detection device Prozessdiagnosemelder *m*, Prozessdiagnosemeldegerät *n*

process field bus Profibus *m* [*Feldbus, in DIN 19 245 Teil 1 und 2 genormt*]

process heat Prozesswärme *f*

process industry Verfahrensindustrie *f*

processing Verarbeitung *f*

processing → (data) processing

process(ing) engineering Verfahrenstechnik *f*

process(ing) machine → computer

process(ing) module, process(ing) block Verarbeitungsbaustein *m*, Verarbeitungsmodul *n*

process(ing) plant, process(ing) installation verfahrenstechnische Anlage *f*

process interface Prozessschnittstelle *f*

process (I/O) image Prozessabbild, PA *n* [*Digitale Darstellung eines Prozesses*]

process management level → process control level

process manipulation Prozessführung *f*

process medium Prozessträger *m*

process monitoring Prozessüberwachung *f*

process ((open) loop) control Prozesssteuerung *f*

process ((open) loop) control level Prozesssteuer(ungs)ebene *f* [*DIN 19 226, Teil 5*]

processor Prozessor *m* [*Steuer(ungs)einheit bei einem System; bei Rechnersystemen stets Zentraleinheit (CPU = central processing unit)*]

processor → (data) processor

processor field bus, profibus Prozessfeldbus, Prozesszubringerbus *m*

processor-orient(at)ed device, POD prozessororientierte Gerät *n* [*Hardwarezusatz in einem Prüfautomaten mit dem sich prozessortypische Signale (Steuer(ungs)signale; Adress- und Datenbuszustände) in der Regel passend jeweils zu einem bestimmten Prozessor nachbilden lassen*]

process-orient(at)ed sequential control prozessabhängige Ablaufsteuerung, prozessgeführte Ablaufsteuerung *f* [*1.) Eine Ablaufsteuerung, deren Weiterschaltbedingungen vorwiegend von Signalen der gesteuerten Anlage (Prozess) abhängig sind, wird prozessabhängige Ablaufsteuerung genannt. E DIN 19 226 I 5 2.) Weiterschaltbedingungen nur von Signalen. V DIN 19 237*]

processor loading [*The amount of input/output (I/O) that a processor (CPU) will be responsible to support*] Prozessoraufladung *f*

processor module, processor block Prozessormodul, Prozessorbaustein *m*

processor power Prozessorleistung *f*

process plan Ablaufplan *m*

process signal Prozesssignal *n* [*Träger von Informationen über den jeweiligen sich verändernden Prozesszustand*]

process variable Prozessgröße *f*

producer → maker

producer gas, air gas, CO [*A highly flammable, toxic gas obtained by burning coal or coke with a restricted supply of air, or by passing air or steam through a bed of incandescent fuel so that the carbon dioxid(e) formed is converted into carbon monoxid(e); a cheap gas with low caloric content that is used where transportation is not required*] Aerogas, Luftgas, Generatorgas *n*

product directory Erzeugungsverzeichnis *n*

production Erzeugung, Herstellung, Fertigung *f*

production cell, work cell Fertigungszelle, Herstellungszelle, Erzeugungszelle *f*

production of low temperatures → cryogenie production

production system, work system Fertigungssystem, Herstellungssystem, Erzeugungssystem *f*

professional Fachmann *m*

professional course Berufslehrgang *m*

profibus → processor field bus

profile, shape, section, unit, trim Profil *n* [*Bauprodukt*]

profile cylinder Profilzylinder *m*

profiled steel plate heating panel, profiled steel plate heater panel, profiled steel plate panel heater Flachheiz(ungs)körper, Flachradiator *m* [*Raumheiz(körp)er mit geringer Bautiefe und profilierter Ansichtsfläche für Wasserheizungen. Es gibt sechs Hauptgruppen: Flachrohrheiz(körp)er, Stahlplatten, Plattenheiz(körp)er aus zwei Blechen, gusseiserne Flachheiz(körp)er, Stahlplattenheiz(körp)er mit Konvektionsschächten auf der Rückseite und zwei- oder dreilagige Plattenheiz(körp)er, die aber nur im weiteren Sinne zu den Flachheiz(körp)ern gerechnet werden*]

P

profile for interior work, shape for interior work, trim for interior work, section for interior work, unit for interior work Ausbauprofil *n* [*Verglasungsleiste; Türdichtung; Fensterdichtung; Einbau-Vorhangschiene; Badewannen-Wanddichtung usw.*]

profile selection Profil(aus)wahl *f* [*PAD-Befehl*]

profitability calculation Rentabilitäts(be)rechnung *f*

profit center (US); profit centre (Brit.) Umsatzbereich, Verantwortungsbereich für Leistung und Erfolg *m, n* [*Horizontale Gliederung einer Firma*]

profit security Gewinnsicherung *f*

program (US); programme (Brit.) [*A sequence of audio signals alone, or audio and video signals, transmitted for entertainment or information*] Programm *n*

program (US); programme (Brit.) [*A detailed and explicit set of instructions for accomplishing some purpose, the set being expressed in some language suitable for input into a computer, or in machine language*] Programm *n*

program compatibility (US); programme compatibility (Brit.) [*The type of compatibility shared by two computers that can process the identical program(me) or program(me)s written in the same source language or machine language*] Programmverträglichkeit *f*

program counter (US) → instruction counter

program development (US); programme development (Brit.) Programmentwick(e)lung *f*

program development time (US); programme development time (Brit.) [*The total time taken on a computer to produce operating program(me)s, including the time taken to compile, test, and debug program(me)s, plus the time taken to develop and test new procedures and techniques*] Programmentwick(e)lungszeit *f*

program element (US); programme element (Brit.) [*Part of a central computer system that carries out the instruction sequence scheduled by the programmer*] Programmelement *n*

program flow (US); programme flow (Brit.) Programmablauf *m* [*E DIN 44 300, Teil 4*]

program flowchart (US); programme flowchart (Brit.) Programmablaufplan *m*

program information file (US); programme information file (Brit.); PIF Programm-Information-Datei *f*

program invocation management (US); programme invocation management (Brit.) Zusammenstellen *n* und Starten und Anhalten von Programmen

program jump (US); programme jump (Brit.) Programmverzweigung *f* [*Sie wird in Abhängigkeit von einer Bedingung durch einen Sprungbefehl geschaffen*]

program library (US); programme library (Brit.) [*An organized set of computer routines and program(me)s*] Programmarchiv *n*, Programmbibliothek *f*

program load and bind (US); programme load and bind (Brit.) [*data processing*] Programmladen *n* und -verknüpfen

programmable calculator [*An electronic calculator that has some provision for changing its internal program(me), usually by inserting a new magnetic card on which the desired calculating program(me) has been stored*] programmierbare Rechenwerk *n*

programmable function programmierbare Funktion *f*

programmable gain amplifier, PGA programmierbare Verstärker *m*

programmable gate array → (field-)programmable gate array

programmable logic array → (field-)programmable logic array

programmable logic device, PLD programmierbare Logikbaustein *m*

programmable logic ((open) loop) control, PLC programmierbare Verknüpfungssteuerung, programmierbare logische Steuerung *f*

programmable parallel I/O unit, PPIO unit programmierbare parallele E/A-Einheit, parallel programmierbare E/A-Einheit *f*

programmable peripheral interface, PPI programmierbare Ein/Ausgabebaugruppe *f*

programmable point Einstell(ungs)punkt *m* [*Ein Punkt für den ein Benutzer ein Überwachungs- oder Steuer-(ungs)programm einstellen kann*]

programmable read-only memory, PROM, programmable ROM [*A large-scale integrated circuit chip for storing digital data; it can be erased with ultra-violet light and reprogrammed, or can be programmed only once either at the factory or in the field*] ROM-Speicher, programmierbare Festwertspeicher, programmierbare ROM-Speicher, programmierbare Nur-Lesespeicher, nichtlöschbare Programmspeicher *m*

program maintenance (US); programme maintenance (Brit.) [*The updating of computer program(me)s both by error correction and by alteration of program(me)s to meet changing needs*] Programmunterhaltung *f*

programme (Brit.) → program (US)

programme code bug (Brit.); program code bug (US) Programmfehler *m*

programmed logic array, PLA [*Array of AND/OR logic gates that provides logic functions for a given set of inputs programmed during manufacture and serves as a read-only memory*] programmierte logische Feld, logische programmierte Feld *n*

P

programmed ((open) loop) control system Programmsteuerung *f* [*Aufzug*]

programmed stop, programmed halt [*A halt that occurs deliberately as the result of an instruction in the program(me)*] programmierte Halt *m*, programmierte Unterbrechung *f*

programme flow (Brit.); program flow (US) Programmablauf *m*

programme flow chart, programme flow diagramm (Brit.); program flow chart, program flow diagram (US) Programmablaufdiagramm, Programmfließdiagramm, Programmrechendiagramm *n*, Programmablaufplan, Programmfließplan, Programmrechenplan, PAP *m*

programme invocation management Zusammenstellen *n* und Starten und Anhalten von Programmen

programme package (US) [*A program(me) that is written for a general and widely used application in such a way that its usefulness is not impaired by the problems of data or organization of a particular use*] Programmpaket *n*

programmer [*A person who prepares sequences of instructions for a computer, without necessarily converting them into the detailed codes*] Programmierer *m*

programmer and eraser Programmier- und Löschgerät *n*

programmer's Hierarchical Interactive Graphics Standard → PHIGS

programming, computer programming Programmieren, Rechnerprogrammieren *n*

programming card Programmierkarte *f*

P

programming environment Programmierungsumgebung *f*

programming for microcomputer → PL/M

programming interface Programmierschnittstelle *f*

programming language, PL [*In automatic data processing, the language used by a programmer to write a program(me) for a computer*] Programmierersprache *f* [*ist nicht Maschinensprache*]

programming ((open) loop) control system [*Control (system) that provides for performing various operations in predetermined sequences related to time or other variables*] Programmiersteuer(ungs)system *n*

programming unit, PU Programmiergerät *n*, Programmiereinheit *f*

program register (US) → instruction register

program run (US); programme run (Brit.) Programmlauf *m* [*Er setzt ein Absolutkodeprogramm voraus*]

program set station (US); programme set station (Brit.); PSS Programmgeber, Zeitplangeber *m*

program status word (US); programme status word (Brit.); PSW [*An internal register to the central processing unit denoting the computer state at a moment in time*] Programmzustandswort *n*

program step (US); programme step (Brit.) [*A single operation in a computer program, typically one instruction*] Programmschritt *m*

program stop (US); programme stop (Brit.); halt instruction, stop instruction [*An instruction built into a computer program(me) that will automatically stop the machine under certain conditions, or upon reading the end of processing or completing the solution of a program(me)*] (programmierte) Haltbefehl *m*

progression condition → step enabling condition

progressive plane wave fortschreitende ebene Welle, ebene fortschreitende Welle *f* [*Schalltechnik*]

project, scheme Projekt *n*

project control → ((closed) loop) project control

project control → ((open) loop) project control

project engineer, scheme engineer Projektingenieur *m*

project engineering, scheme engineering technische Projektbearbeitung, technische Projektierung und Inbetriebnahme *f*

project evaluation and review technique, PERT Technik *f* zur Planung, Kontrolle und Berichterstattung von Projekten

project implementation, scheme implementation Projektdurchführung, Projektausführung *f*

projecting diffuser überstehende Wanne *f* [*Leuchte*]

projection Ausladung, Vorkragung *f*, Überstand *m*

projection line Maßhilfslinie *f*

projection plane Projektionsebene *f*

projection thermography Projektionsthermographie *f*

project management, scheme management Projektbewirtschaftung *f*

projector lamp Lichtwurflampe *f*

project regulation → ((closed) loop) project control

project manager, scheme manager Projektbetreuer, Projektleiter *m*

prompting → (operator) prompting

prompting mode Abfragemodus *m*

prompting time Abfragezeit *f*

prompt note, prompt notification, demand note, demand notification, reminder (of debt due), letter demanding payment, (formal) demand [*Slang: dunning letter*] Mahnbrief *m*, Mahnschreiben *n*, Mahnung *f*

prompt sequence Eingabeaufforderungsfolge *f*

proof against surge current(s), proof against transient current(s) stoßstromsicher, stoßstromfest

proofreading Korrekturlesen *n*

proof of touch → contact-protected

propagation → acoustic propagation

propagation constant → acoustic propagation constant

propagation delay Signalwechselzeit *f* [*Zeit zwischen einem Signalwechsel am Eingang eines Prüflings und dem Signalwechsel an einem bestimmten inneren Knoten oder einem bestimmten Ausgang*]

propagation speed Fortpflanzungsgeschwindigkeit *f*

propagation velocity Ausbreitungsgeschwindigkeit *f* [*Geschwindigkeit, mit der sich die von einer Antenne ausgestrahlten elektromagnetischen Feldlinien im Raum ausbreiten. Sie ist gleich der Lichtgeschwindigkeit*]

propane → refrigerant-290

propane → refrigerant-1270

propeller blade Propellerschaufel, Schraubenschaufel *f*

propeller fan Propellerventilator, Schraubenventilator *m*

propeller fan (type) heater Propellerventilatorheizer, Schraubenventilatorheizer *m*

propeller stirrer, propeller agitator Propellerrührer, Schraubenrührer *m*

propeller (water) meter → rotary (water) meter

property [*A particular characteristic of an object type*] Eigenschaft *f*

property → (real) property

property management → (real) property management

property value management, value of property management Vermögenswertbewirtschaftung *f*

proportion, percentage Anteil, Prozentsatz *m*

proportion action; proportion behavior (US);

proportion behaviour (Brit.) P-Verhalten *m* [*Regler*]

proportional (action) controller → P controller

proportional action element → P element

proportional amplifier Proportionalverstärker *m*

proportional band [*The deviation necessary to produce the full range of control action. It can be expressed in terms of a physical quantity or as a percentage of the controller scale range*] Proportionalbereich *m* [*selten: n*] [*Reg., DIN 19 226*]

proportional ((closed) loop) control, proportional regulation [*A regime where the controlling element has an output signal proportional to its input signal*] Proportionalreg(e)lung, P-Reg(e)lung *f*

proportional controller, proportional regulator [*A controller whose output is proportional to the error signal*] Proportionalregler, P-Regler *m*

proportional counting tube, proportional counter tube Proportionalzählröhre *f*, Proportionalzählrohr *n* [*Fehlbenennung*]

proportional degree [*Of a (closed) loop control*] P-Grad, Proportionalgrad *m*

proportional differential controller, proportional differential regulator Proportional-Differential-Regler, PD-Regler *m*

proportional element Proportionalglied, Proportionalelement *n*

proportional heating cost(s) meter Heiz(ungs)kostenverteiler *m*

proportional integral controller, proportional integral regulator Proportional-Integral-Regler, PI-Regler *m*

proportionality law Proportionalitätsgesetz *n*

proportional-plus-derivative action PD-Verhalten, Proportional-Differential-Verhalten *n*

proportional-plus-derivative-action ((closed) loop) controller PD-Regler, proportional-differential-wirkende Regler *m*

proportional-plus-derivative-action element → PD-element

proportional-plus-derivative controller → PD controller

proportional-plus-integral action PI-Verhalten, Proportional-Integral-Verhalten *n*

proportional-plus-integral amplifier Proportional-Integral-Verstärker, PI-Verstäker *m*

proportional-plus-integral ((closed) loop) control, PI control PI-Reg(e)lung, Proportional-Integral-Reg(e)lung *f*

proportional-plus-integral ((closed) loop) controller, PI controller PI-Regler, Proportional-plus-Integral-Regler, proportional-integral-wirkende Regler *m*

proportional-plus-integral control with state feedback PI-Zustandsreg(e)lung *f*

proportional-plus-integral-plus-derivative action Proportional-Integral-Differentialverhalten, PID-Verhalten *n*

proportional-plus-integral-plus-derivative algorithm, PID algorithm [*It has the same operation as the PI algorithm with the addition of an influence on the output determined by the rate of change of deriva-*

tion between set(ting) point and input signal] Proportional-Integral-Differential-Algorithmus, PID-Algorithmus *m*

proportional-plus-integral-plus-derivative ((closed) loop) controller PID-Regler, Proportional-Integral-Differential-Regler *m*

proportional-plus-integral-plus-derivative control, PID control Proportional-Integral-Differential-Reg(e)lung, PID-Reg(e)lung *f*

proportional-plus-integral-plus-derivative controller, PID controller Proportional-Integral-Differential-Regler, PID-Regler, PI-Regler mit Vorhalt, Vorhalt-PI-Regler, proportional-integral-differential-wirkende Regler *m*

proportional regulation → proportional ((closed) loop) control

proportional step change, P step change Proportionalsprung *m*

proprietary record [*numerical control*] reservierte Satz *m*

prop(yl)ene → refrigerant-1270

prop(yl)ene glycol [*A clear, colo(u)rless liquid used to depress the freezing point of water for use in a secondary coolant in HVAC&R systems. Note: Inhibitors are required to control corrosion caused by glycol solutions*] Propylenglykol *n*

prospective making current unbeeinflusste Einschaltstrom *m*

prospective peak current unbeeinflusste Spitzenstrom, unbeeinflusste Starkstrom *m*

protected against finger contact fingersicher, fingergeschützt

protected area, protection area Schutzfläche, geschützte Fläche *f*

protected location Schutzspeicherstelle, geschützte Speicherstelle *f* [*Auf sie kann nur von einem berechtigten Benutzer zugegriffen werden*]

protected object Schutzziel, Schutzobjekt, geschützte Ziel, geschützte Objekt *n*, Schutzgegenstand, geschützte Gegenstand *m*

protection Schutz *m*

protection Ersatzschaltung *f* [*Ringnetz*]

protection against contact → contact guard

protection against drilling, drilling protection Bohrschutz *m*

protection against freezing, freezing protection, freeze protection Gefrierschutz, Einfrierschutz *m*

protection against sound, sound protection Schallschutz *m* [*DIN 4109 und VDI 2715*]

protection area → protected area

protection circuit, protecting circuit, protective circuit Schutzschaltung *f*

protection conductor, protecting conductor, protective conductor; equipment ground(ing) conductor (US); equipment earth(ing) conductor (Brit.) Schutzleiter *m*, SL [*VDE 0100, T. 200*]

protection conductor terminal, protective conductor terminal, protecting conductor terminal; equipment ground(ing) conductor terminal (US); equipment earth(ing) conductor terminal (Brit.) Schutzleiterklemme, SL-Klemme *f* [*VDE 0100, T. 200*]

protection controller, protection regulator, protective controller, protective regulator, protecting regulator, protecting controller Schutzregler *m*

protection current, protecting current, protective current Schutzstrom *m*

protection device, protecting device, protective device, guard Schutzvorrichtung *f*

protection earth (Brit.) → protection ground

protection extra low voltage, protective extra low voltage, PELV, protecting extra low voltage Schutzkleinspannung, Kleinschutzspannung *f*

protection from fire Brandschutz, Feuerschutz *m*

protection grill(e), protective grill(e), protecting grill(e) Schutzgitter *n*

protection ground, protective ground, protecting ground (US); protection earth, protective earth, protecting earth, PE (Brit.) Schutzerde *f*

protection ground neutral, protective ground neutral, protecting ground neutral (US); protection earth neutral, protective earth neutral, protecting earth neutral, PEN (Brit.) Neutralschutzerde, Mittelschutzerde, Schutzerde neutral *f*

protection hood, protecting hood, protective hood Schutzhaube *f*

protection label, protecting label, protective label Schutzetikett *n*

protection measure, protecting measure, protective measure Schutzmaßnahme, Schutzvorkehrung *f*

protection multiple earthing, protective multiple earthing, protecting multiple earthing, TIV system, multi(ple)-protective earthing, neutralization [*"Grounding" is used for "earthing" in the USA*] Nullung *f*

protection of data privacy, data privacy protection Datenschutz *m*

protection (of structures) against lightning → lightning protection

protection outer window, protecting outer window, protective outer window, outer protection window,

outer protective window, outer protecting window Vorsatzfenster *n*

protection type, protective type, protecting type Schutzart *f*

protocol, communication protocol Protokoll, Kommunikationsprotokoll, Übermitt(e)lungsprotokoll *n*, Niederschrift, Kommunikationsniederschrift, Übermitt(e)lungsniederschrift *f* [*Umfasst alle für eine Kommunikation notwendigen Vereinbarungen, die Festlegung über die technischen, zeitlichen und inhaltlichen Einzelheiten des Informationsaustausches*]

protocol analyzer Protokollanalysator *m*

protocol converter Protokollumwandler, Protokollumsetzer *m*

protocol discriminator Protokollunterscheidungskennzeichen *n*

protocol error Protokollfehler *m*

protocol file → log file

protocol layer Protokollschicht *f*

protocol level Protokollebene *f*

protocol output Protokollausgabe *f*

protocol printer, protocol writer Protokolldrucker, Protokollschreiber *m*

protractor Winkelmesser *m*

to provide and fix, to supply and fix liefern und einbauen

provided by the Owner bauseitig (beigestellt), bauseits (beigestellt)

provider, supplier Erbringer, Bereitsteller, Versorger, Lieferant *m*

provision of water, water supply Wasserversorgung *f*

proximity detector, non-contact detector, contactless detector [*"Detecting device" and "detection device" may synonymously be used for "detector". It produces an electric signal when approached by an object or approaching an object*] Näherungsmelder, Näherungsdetektor, Näherungsmeldungsgeber *m*, Näherungsmeldegerät *n*

proximity effect, non-contact effect, contactless effect [*Redistribution of current in a conductor brought about by another conductor*] Näherungswirkung *f*

proximity sensor, proximity feeler, non-contact sensor, non-contact feeler, contactless sensor, contactless feeler Näherungssensor, Näherungsfühler, berührungslose Sensor, berührungslose Fühler *m*

proximity switch, non-contact switch, contactless switch Näherungschalter, berührungslose Schalter *m*

PRR → pulse repetition rate

PS → positive-small

pseudocode → abstract code

pseudoinstruction Pseudobefehl *m* [*Ein Befehl an den Programmschlüssler der nicht übersetzt wird*]

pseudo-random sequence, PRS Zufallsfolge *f*

"P" (siphon) trap, "P" strench trap [*See remark under "(siphon) trap"*] P-Geruchsverschluss, P-Siphon, P-Traps *m*

PSK → phase-shift keying

PSS → program set station

PSS → power system stabilizer

P step change → proportional step change

P(S)TN → public (switched) (tele)phone net(work)

PSW → program status word (US)

psychrometer, wet and dry bulb thermometer [*A device comprising two thermometers, one a dry bulb, the other a wet or wick-covered bulb, used in determining the moisture content or relative humidity of air and other gases*] Psychrometer *n*, Luftfeuchtigkeitsmessser, Luftfeuchtemesser *m*

psychrometric formula [*The semi-empirical relation giving the vapo(u)r pressure in terms of the barometer and psychrometer reading*] (Sprungsche) Psychrometerformel *f* [*Entwickelt von Sprung 1888*]

psychrometric table, psychrometric chart [*A graphical representation of certain thermodynamic relations. The dry-bulb temperature is the abscissa and the water content of the atmosphere (in pounds of water per pound of dry air, or as vapour pressure in millibars) is the ordinate. The relative humidity is then plotted as a series of curves, and the wet-bulb temperature as a series of diagonal lines. The enthalpy is also plotted as a series of diagonal lines. The chart can therefore be used to determine the amount of heat or cooling power required to change air at one temperature and humidity to that at another temperature and humidity by air conditioning*] Psychrometertabelle, Prychrometertafel, Thermohygrometertabelle, Thermohygrometertafel *f*

PT → page terminator

ptac → packaged terminal air-conditioning system

PTC thermistor, positive temperature coefficient thermistor Kaltleiter *m*

PTC thermistor resistor, positive temperature coefficient thermistor resistor Kaltleiterwiderstand *m*

PTN → public (switched) (tele)phone net(work)

PTS → permanent threshold shift

PU → programming unit

P

(public-address) (loud)speaker system, public address system, P.A. system, sound-amplification system, sound reinforcement system Lautsprechersystem, Beschallungssystem *n*

publication Veröffentlichung *f*

public building öffentliche (zugängige) Gebäude, zugängige öffentliche Gebäude *n*

public (building) monitoring, public construction monitoring öffentliche Bauaufsicht, öffentliche Bauüberwachung *f*

public channel öffentliche Kanal *m*

public convenience öffentliche Toilettenanlage, Bedürfnisanstalt *f*

public current grid, public power grid Stromnetz, Elektrizitätsnetz, E-Netz, öffentliche Stromnetz, öffentliche Elektrizitätsnetz, öffentliche E-Netz *n*

public data net(work) → PDN

public drainage öffentliche Entwässerung *f*

public exchange → exchange

public exchange trunk line → exchange trunk line

(public) gas utility (öffentliche) Gasversorgungsunternehmen *n*, (öffentliche) Gasanstalt *f*

public health engineer → sanitation engineer

public health engineering → sanitation

public key öffentliche Schlüssel *m* [*Verschlüsselung*]

public land mobile radio network bewegliche öffentliche Landfunknetz, öffentliche bewegliche Landfunknetz *n*

public network öffentliche Netz *n*

public power grid → public current grid

public power supply öffentliche Stromversorgung *f*

public room Publikumsraum, öffentliche Raum *m*

public service → (public) utility (company)

public sewer öffentliche Abwasserleitung *f*

public stairway öffentliche Treppenhaus *n*

public supply öffentliche Versorgung *f*

public supply (pipe)line öffentliche Versorgungs(rohr)leitung *f*

public (switched) (tele)phone coin box, coin-box (tele)phone Münzfernsprecher *m*, Münztelefon *n*

public (switched) (tele)phone net(work), P(S)TN öffentliche Fernsprechnetz, öffentliche Telefonnetz *n*

public (switched) (tele)phone station → pay station

public transport(ation) → mass transport(ation)

public transport(ation) capacity → mass transport(ation) capacity

(public) utility (company), public service (öffentliche) Versorgungsunternehmen *n*, (öffentliche) Versorgungsträger, (öffentliche) Versorgungsbetrieb *m*

(public) utility shaft, (public) service shaft (öffentliche) Versorgungsschacht, (öffentliche) Installationsschacht *m*

public water closet, public W.C. öffentliche Toilette *f*, öffentliche Klosett *n*, öffentliche Abort *m*

public water supply öffentliche Wasserversorgung *f*

to publish veröffentlichen, herausgeben, bekanntmachen

puff pipe → anti-siphonage pipe

puff pipe with leakage fitting → anti-siphonage pipe with leakage fitting

to pull [*e.g. in a relay*] synchron werden, in Tritt fallen

pull-and-let-go chain Ziehkette *f* [*Spülabort*]

pull box → (cable) pull box

(pull-)cord switch → suspended switch

to pull down herunterziehen

pull-down menu Bildschirmmenü *n* [*Bildschirmmenü, bei dem durch Identifizieren eines Menüelementes ein Untermenü auf dem Bildschirm aufgeblendet wird, aus dem durch Herunterziehen des Kursors ein Menüfeld ausgewählt wird*]

pull handle → (door) pull handle

pull(ing) wire Zugdraht, Ziehdraht *m*

pull-in scanner Einzugsabtaster *m* [*Ausweis(karte) wird in einen Schlitz gesteckt und vom System transportiert, sie kann gegebenenfalls auch einbehalten werden*]

pull knob Zugknopf, Ziehknopf *m*

pull-knob pop-up waste Ziehknopf-Ablaufgarnitur, Zugknopf-Ablaufgarnitur *f* [*Waschbecken; Waschtisch; Bidet*]

pull ring → (door) pull ring

pull switch → hung switch

pull-through converter Fädelwandler *m*

to pull up hochziehen

pulsating burner, pulsation burner Pulsationsbrenner *m*

pulsating current → PC

pulsating oil flow gepulste Ölzufuhr *f* [*Ölbrenner*]

pulsating room temperature pulsierende Raumtemperatur *f* [*Periodische Temperaturschwankung um die Empfindungstemperatur. Die Schwankungsamplitude soll nicht größer als 1 K sein*]

pulse Stromstoß, Impuls *m*

P

pulse acceptance Impulsverarbeitung *f*

pulse amplifier [*An amplifier designed specifically to amplify electric(al) pulses without appreciably changing their wave forms*] Impulsverstärker *m*

pulse amplitude Impulshöhe, Impulsamplitude *f*

pulse-amplitude modulation, PAM [*Amplitude modulation of a pulse carrier*] Impulsamplitudenmodulation, IAM, Impulshöhenmodulation *f*

pulse analyzer Impulsanalysator *m*

pulse bandwidth Impulsbandbreite *f*

pulse bus bar Impulssammelschiene *f*

pulse contact Impulskontakt *m*

pulse counter Impulszähler *m*

pulse decay time [*The interval of time required for the trailing edge to decay from 90% to 10% of the peak-pulse amplitude*] Impulsabklingzeit *f*

pulse(d) input [*The representation of a value by a series of abrupt and relatively short cyclic changes in a signal*] Impulseingang *m*

pulse gas burner Impulsgasbrenner *m*

pulse generator → impulse generator

pulse generator [*In electronics, a generator that produces repetitive pulses or signal-initiated pulses*] Impulserzeuger, Impulsgenerator *m*

pulse group → pulse train

pulse-height discriminator → amplitude discriminator

pulse-height selector, amplitude selector [*A circuit that produces a specified output pulse only when it receives an input pulse whose amplitude lies between two assigned values*] Diffraktions-Amplitudenwähler *m*

pulse induction circuit Impulsinduktionsschaltung *f*

pulse interleaving, pulse interlacing Impulsverflechtung, Impulsverschachtelung *f*

pulse interval → pulse spacing

pulse maintained-contact control, pulse maintained-contact regulation Impulskontaktgabe *f*

pulse modulation, PM Impulsmodulation *f*

pulse motor → stepping motor

pulse overshoot Impulsspitze *f*

pulse phase demodulation Impulsphasendemodulation *f*

pulse phase modulation → PPM

pulse position modulation, PPM [*Modulation of a pulse carrier wherein the value of each instantaneous sample of a modulating wave varies the position in time of a pulse relative to its unmodulated time of occurrence*] Impulslagenmodulation, PLM *f*

pulse recurrence time, pulse repetition time Impulsfolgezeit *f*

pulse regeneration [*The process of restoring pulses to their original relative timings, forms, and magnitudes*] Impulsverbesserung, Impulsregenerierung *f*

pulse repetition rate, pulse recurrence rate, PRR, pulse repetition frequency, pulse recurrence frequency, PRF [*The number of times per second that a pulse is transmitted*] Impuls(folge)frequenz *f*

pulse rise time [*The interval of time required for the leading edge of a pulse to rise from 10% to 90% of the peak pulse amplitude*] Impulsanstiegszeit, Impulszunahmezeit, Impulserhöhungszeit *f*

pulse rounding Impulsverschleifen *n*

pulse run count Impulsgruppenzahl *f*

pulse sequence Impulsfolge *f*

pulse shaper [*A transducer used for changing one or more characteristics of a pulse, such as a pulse regenerator or pulse stretcher*] Impulswandler *m*

pulse shaping Impulsformung *f*

pulse shaping element Impulsformglied, Impulsformelement *n*

pulse signal Impulssignal *n*

pulse signalling technique (Brit.); pulse signaling technique (US) Impulsmeldetechnik *f* [*Der jeweils empfangene Impuls (z. B. 1.Impuls entspricht Melder 1; 2.Impuls entspricht Melder 2 usw.) ist die Meldeadresse*]

pulse spacing, pulse interval Impulszeitabstand *m*

pulse-spacing modulation, pulse-interval modulation [*A form of pulse-time modulation in which the pulse spacing is varied*] Impulsabstand(s)modulation *f*

pulse summation device Impulssummiergerät *n*

pulse switching command Impulsschaltkommando *n* [*Feinwirktechnik*]

pulse tone Impulston *m*

pulse train, pulse group, impulse train, impulse group Impulsfolge, Impulsgruppe, Impulsreihe *f* [*Periodischer Vorgang, der aus einer Folge von gleichen Impulsen besteht*]

pulse transformer Impulsübertrager *m*

pulse transmitter [*A pulse-modulated transmitter whose peak-power-output capabilities are usually large with respect to the average-power-output rating*] Impulssender *m* [*Fernmessen*]

pulse-triggered flipflop → DC flipflop

pulse valency Impulswertigkeit *f*

pulverized coal burner Staubkohle(n)brenner, Pulverkohle(n)brenner *m*

pulverized coal fuel Pulverkohle(n)brennstoff, Staubkohle(n)brennstoff *m*

pumpable pumpbar, pumpfähig

pumpable concrete Pumpbeton *m*

pumpage Pumpenfördermenge *f*

pump assembly, pump set, pump kit Pumpen(bau)satz *m*, Pumpenaggregat *n*, Pumpenkombination, Pumpengruppe, Pumpengarnitur *f*

pump(-assisted) heating Pumpenheizung *f*

pump booster Pumpendruckerhöhungsanlage *f*

pump characteristic (curve) Pumpenkennlinie, Q-H-Linie *f* [*Das Betriebsverhalten einer Heizungsumwälz(ungs)pumpe wird auf dem Prüfstand ermittelt. Dabei wird für verschiedene Druckhöhen die zugehörige Wassermenge festgestellt. Die Verbindungslinie dieser vielen möglichen Betriebszustände heißt Pumpenkennlinie oder Q-H-Linie*]

pump circulation Pumpenzirkulation, Pumpenumwälzung *f*, Pumpenumlauf *m*

pump cold water → **cold pump water**

pump control → **((closed) loop) pump control**

pump control → **((open) loop) pump control**

pump cover, pump bonnet Pumpenhaube *f*

pump cycle Pumpenkreislauf *m*

pump efficiency Pumpenwirkungsgrad *m* [*DIN 4046*]

pump hot water → **hot pump water**

pump hot water heating → **forced hot water heating**

pump hot water heating with lower distribution, forced hot water heating with lower distribution Pumpenwarmwasserheizung mit unterer Verteilung, Flächenheizung *f* [*nicht area heating*]

pump(ing) basin Pumpenvorlage *f*, Pumpenbecken *n* [*Becken aus dem eine Flüssigkeit durch Pumpen abgeführt wird*]

pumping down, pumping off [*Removing a liquid from something by pumping*] Abpumpen, Wegpumpen *n*

pump(ing) (pressure) head Pumpen(druck)höhe, Pumpenförderhöhe *f*

pump(ing) station Pumpstation *f*, Pumpwerk *n*

pump kit → **pump assembly**

pump management Pumpenführung *f*

to pump off, to pump down abpumpen, wegpumpen

pumpout, pumping out Auspumpen *n*

pump pressure controller, pump pressure regulator Pumpendruckregler *m*

pump regulation → **((closed) loop) pump control**

pump run Pumpenlauf *m*

pump run-on Pumpennachlauf *m*

pump set → **pump assembly**

pump (slide) valve Pumpenschieber *m*

pump suction level, pump suction head Pumpensaughöhe *f*

pump sump → **wet sump**

pump warm water → **hot pump water**

to punch → **to pierce**

punch(ed) card, tabulating card Lochkarte *f*

punch(ed) card bed [*In automatic data processing, the metal plate along which the card travels to the punching and reading station*] Lochkartenplatte *f*

punch(ed) card buffer [*A temporary device into which the output of a computer is transmitted before it is entered on a punch card by a card punch, and in which the data remain if for any reason the card punch cannot operate*] Lochkartenpuffer *m*

punch(ed) card check(ing) [*In automatic data processing, the verification carried out by the computer to ensure that all the data keypunched on a card has been correctly read into the memory*] Lochkartenkontrolle *f*

punch(ed) card code [*In automatic data processing, the representation of characters on a punched card by means of punching one or more holes per column*] Lochkartenkode, Lochkartencode *m*

punch(ed) card image [*A one-to-one representation of the contents of a punched card, such as a matrix in which a 1 represents a hole and a 0 represents the absence of a hole*] Lochkartenbild *n*

punch(ed) card jam, punch(ed) card wreck [*A condition in which one or more punch cards become jammed along the card bed of a machine that is processing them*] Lochkartenstau *m*

punch(ed) card machine [*May be a segregating unit or a punched-card sorter*] Lochkartentrenner, Lochkartensortierer *m*

punch(ed) card reader [*A mechanism that senses information punched on cards, using wire brushes, metal feelers, or a photoelectric system*] Lochkartenleser *m*

punch(ed) card recorder Lochkartenstanzer *m*

punch(ed) card reproducer [*A machine that makes a duplicate of a punched card*] Lochkartenvervielfältiger *m*

punch(ed) card row [*In automatic data processing, a row of punching positions parallel to the long edge of a punched card*] Lochkartenzeile *f*

punch(ed) card segregation → **segregation**

P

punch(ed) card sorter [*A machine used to arrange punched cards into an appropriate sequence for further processing*] Lochkartensortierer *m*

punch(ed) card storage Lochkartenspeicher *m*

punch(ed) card-to-card transceiving [*A system that makes possible instantaneous and accurate duplication of punched cards over telephone and telegraph network between locations separated by either just a few miles or thousands of miles*] Datenübertragung *f* von Lochkarten auf Lochkarten

punch(ed) card verifier [*In automatic data processing, an electromechanical device which allows the operator to check that a card has been properly keypunched*] Lochkartenprüfer *m*

punch(ed) paper tape Papierlochstreifen, Lochpapierstreifen *m*

punch(ed) tag Identifizierungslochkarte, Lochidentifizierungskarte *f*

punch(ed) tape → paper tape

punch(ed) tape reader → paper tape reader

punching wedge, wedge for punches Lochungskeil *m*

punch-in/out Ein-/Aussteiger *m* [*aus einer Aufnahme*]

punctuation Zeichensetzung *f*

punctuation mark Setzzeichen *n*

pup [*A short make-up of pipe*] Rohrpassstück, Passrohrstück *n*

pure air → clean air

pure binary code, binary pure code reine binäre Kode, reine binäre Code, binäre reine Kode, binäre reine Code *m*

pure tone, fundamental (tone), simple tone [*A sound in which the sound pressure varies sinusoidally with time. The waveform may be represented by sin*] (einfache) Ton, reine Ton, Grundton *m* [*Sinusförmige Schallschwingung im Hörbereich. Anmerkung: Die in der Musik übliche Bestimmung des Begriffes Ton weicht im allgemeinen von der in der Akustik üblichen ab. Der Musiker nennt das Schallergebnis, das einer einzelnen Note entspricht, Ton, während es sich nach der Terminologie der Akustik in den meisten Fällen – wegen der Obertöne – um einen Klang handelt*]

purge [*Removal of unburned gases from a combustion chamber*] Gas(ab)scheiden *n*

purge [*Device to remove noncondensable gases from a refrigerating system*] Gas(ab)scheider *m*

purge → blowing through

purge... → blow off...

purge valve, purging valve, blowdown valve [*Device to allow fluid to flow out of a system, particularly noncondensable gases*] Ablassventil, Auslassventil *n*

purging [*ventilation*] Spülen *n*

purging → blowing through

push-and-latch feature Steck-Rast-Technik *f*

push button → (lock) push button

push button, push knob, press button [*Small mov(e)able knob or button, within a socket, the movement or which actuates a bell, electrical or otherwise*] Druckknopf *m*

push button alarm, press button alarm Druckknopfalarm *m*

push button box, press button box Druckknopfkasten *n*

push button bracket, press-button bracket Druckknopfwinkel *m*

push button ((closed) loop) control, press-button ((closed) loop) control Druckknopfregler *m*

pus hbutton detector, press-button detector, push-button detection device, press-button detection device, push button detecting device, press-button detecting device Druckknopfmelder, Druckknopfdetektor, Druckknopfmeldungsgeber *m*, Druckknopfmeldegerät *n*

push button dialling, press-button dialling [*automatic telephone, American spelling: dialing*] Druckknopfwahl *f*

push button (fire) detector, (push)button (fire) detecting device, press-button (fire) detector, press-button (fire) detecting device Druckknopf(feuer)melder, Druckknopfbrandmelder, Druckknopf(feuer)detektor, Druckknopfbranddetektor, Druckknopf(feuer)meldungsgeber, Druckknopfbrandmeldungsgeber *m*, Druckknopf(feuer)meldegerät, Druckknopfbrandmeldegerät *n* [*DIN 14 655; ÖNorm F 3045*]

push button light, press button light, light in pushbutton, light in press-button Druckknopflicht *n*

push button ((open) loop) control, press button ((open) loop) control Druckknopfsteuerung *f*

push button station, press button station [*A switching or controlling device equipped with marked buttons. Depression of the buttons opens or closes contacts in electrical circuits*] Druckknopffeld *n*

(push)button switch, press button switch Druckknopfschalter *m*

pushdown stack register Stapelspeicher *m*

pushdown store Keller, Nachrückspeicher *m* [*Arbeitsspeicher, in dem die jeweils neu aufgenommen Datengrößen einen obersten Platz besetzen und alle vorher gespeicherten Daten um einen Speicherplatz nach unten rutschen*]

push(ing) handle Stoßgriff, Drückgriff *m*

P

push(ing) rose(tte) Schubrosette *f* [*z. B. bei einem Auslaufventil*]

push-pull Gegentakt *m*

push-pull amplifier [*Amplifier in which there are two identical signal branches connected so as to operate in phase opposition and with input and output connections each balanced to ground*] Gegentaktverstärker *m*

push-pull chopper Gegentaktzerhacker *m*

push-pull circuit Gegentaktschaltung *f*

push-pull communication Gegentaktverkehr *m*

push-pull heat exchanger Kapillargebläse, Kirchmeier-Kapillargebläse *n*

to push shut zudrücken [*Deckel*]

push-through badge (card), push-through identity card Druchzugsausweis(karte), Durchziehausweis(karte) *m, (f)*

push-through reader → swipe reader

push-through scanner Durchzugsabtaster, Durchziehabtaster *m* [*Die Ausweiskarte muss entlang einer Lesekopfschiene geführt werden*]

to put in → to input

to put out → to output

puzzle lock → combination lock

PV-module photovoltaische Modul *m*

pyroelectric(al) pyroelektrisch

pyroelectricity Pyroelektrizität *f*

pyrolysis smoke Pyrolyserauch *m*

Q

QBE → quality building environment

Qo^{s} → quality of service

quadrant switch Quadrantenschalter *m*

quadraspeed Vierfachgeschwindigkeit *f*

quadrature demodulation Quadraturdemodulation *f*

quadrature modulation [*Modulation of two carrier components 90° apart in phase by separate modulating function*] Quadraturmodulation *f*

quad(ruple) vierfach

quad(ruple) input NAND NAND-Gatter *n* mit vier Eingängen

quad(ruple) split screen vierfach geteilte (Bild)-Schirm *m*

qualified active area → active qualified area

qualifier [*A portion of a command string that modifies a command verb or command parameter by selecting one of several options. A qualifier, if present, follows the command verb or parameter to which it applies and is in the formal "/qualifier option". For example, in the command string "xcopy filename/s/e/v", the /s/e/v qualifier indicates that the user wants the subdirectories, even if empty and verifies the copy process*] Kennzeichner *m*

qualifying date Stichtag *m*

qualifying date reading Stichtagablesung *f*

quality approval Gütesiegel, Qualitätssiegel *n*

quality assurance Qualitätssicherung, Gütesicherung *f*

quality building environment, QVE Gütegebäudeumfeld, Qualitätsgebäudeumfeld *n*

quality control Qualitätslenkung, Gütelenkung *f* [*Die vorbeugenden, überwachenden und korrigierenden Tätigkeiten bei der Realisierung der Einheit mit dem Ziel, die Qualitätsanforderung zu erfüllen. DIN 55 350, Teil 11*]

quality control report → report based on quality control

quality engineering Qualitätstechnik, Gütetechnik *f*

quality level Qualitätsstand, Gütestand *m*, Qualitätsklasse, Qualitätsstufe, Güteklasse, Gütestufe *f*

quality loss Güteverlust *m*, Verschlechterung, Qualitätsverlust *f*

quality mark Gütezeichen, Qualitätszeichen *n*

quality of service, Qo^{s}, service quality, grade of service, Go^{s}, service grade Dienstgüte, Dienstqualität *f*

quality-proofed gütegesichert, qualitätsgesichert

quality reduction → reduced quality

quality test → soundness test

quality verification Güteverifikation, Qualitätsverifikation, Verifikation der Güte, Verifikation der Qualität *f*, Gütenachweis, Nachweis der Güte, Qualitätsnachweis, Nachweis der Qualität *m*

Quan. per Ass'y → quantity per assembly

quantifiable mengenmäßig bestimmbar

quantifiable variable Mengengröße, mengenmäßig bestimmbare Größe *f*

quantitative water management Wassermengenwirtschaft, Mengenwasserwirtschaft *f*

quantity [*In computers, a positive or negative real number in the mathematical sense; the term quantity is preferred to the term number in referring to numerical data; the term number is used in the sense of natural number and reserved for "the number of digits", the "number of operations", and so forth.*] Anzahl *f*

quantity [*In mathematics, any expression which is concerned with value rather than relation*] Menge *f*

quantity meter → total flow meter

quantity meter Mengenzähler *m*

quantity metering Mengenzählung *f*

quantity of electricity, electricity quantity, (electric) charge (elektrische) Ladung *f*

quantity of heat → heat quantity

quantity per assembly, Quan. per Ass'y Anzahl pro Zusammenstellung, Zusammenstellungsanzahl *f* [*Beschriften im Zeichnungsschriftfeld*]

quantity pulse Mengenimpuls *m*

quantized Hall effect quantisierte Halleffekt *m*, quantisierte Hallwirkung *f*

quantizing errror Quantisierungsfehler *m*

quantum coupled device Quantentransistor *m*

quarter-bend Krümmer *m* *[90°]*

quarterly (journal) Vierteljahreszeitschrift *f*

quarter of brick → closer

quarter-square multiplier → four-quadrant multiplier

quarter (turn) valve Viertelventil *n*

quartic vierten Grades, viertgradig

quartz clock Quarzuhr *f*

quartz time switch, quartz clock relay, quartz timer Quarzschaltuhr, Schaltquarzuhr *f*

quenching → hardening

query Anfrage, Abfrage, Frage *f*, Abruf *m* [*einer in einer Datenbank gespeicherten Information*]

query input, interrogation input Abfrageeingang, Frageeingang, Anfrageeingang, Abrufeingang *m* [*Speicher*]

query instruction Anfragebefehl, Abfragebefehl, Fragebefehl, Abrufbefehl *m*

query language Abfragesprache, Anfragesprache, Fragesprache, Abrufsprache *f*

quick... → rapid...

quick-break switch → switch disconnector

to quicken, to accelerate beschleunigen

quick freezer, rapid freezer, fast freezer, quick freezing apparatus, rapid freezing apparatus, fast freezing apparatus Schnellgefrierapparat, Schnellfrostapparat *m*

quickly accessible → readily accessible

quick-make-quick-break switch → sensitive switch [*IEC 163*]

quiescent current [*electron tube*] Ruhestrom *m*

quiescent sedimentation Ruhesedimentation, Sedimentation im Ruhebetrieb *f*

quietness Geräuscharmut, Ruhe *f*

quiet-running geräuscharm laufend, leise laufend, ruhig laufend

quintet Fünfbiteinheit *f*

quotientmeter Quotientenmessgerät, Quotientenmesswerk *n*, Quotientenmesser *m*

R

R → refrigerant

R-10 → refrigerant-10

R-11 → refrigerant-11

R-12 → dichlorodifluoromethane

R-13 → refrigerant-13

R-14 → refrigerant-14

R-21 → refrigerant-21

R-22 → monochlorodifluoromethane

R-30 → refrigerant-30

R-40 → refrigerant-40

R-50 → refrigerant-50

R-113 → refrigerant-113

R-114 → refrigerant-114

R-170 → refrigerant-170

R-290 → refrigerant-290

R-600 → refrigerant-600

R-601 → refrigerant-601

R-717 → (anhydrous) ammonia refrigerant

R-1150 → refrigerant-1150

R-1270 → refrigerant-1270

rabbet → rebate

rabbet door release → rebate door release

rabbeted lock → dummy lock

rabbit-ear faucet (US) Löffelhahn *m* [*In Toiletten, anstelle der Zugkette für die Wasserspülung*]

RAC → remote access control(ling)

raceway [*Any support system, open or closed, for carrying electric wires, cables, or big bars*] Leitungskanal *m*

rack [*An open frame(work) to receive articles, as a place of storage*] (Lager)Gestell *n*

rack Baueinheit *f* [*System mit einer bestimmten Funktionalität*]

racking up zeilenweise Rollen *n* [*Bildschirmgerät*]

rack position Einbauplatz *m* (im Baugruppenträger), Steckplatz

rad Einheitszeichen *n* der SI-Einheit für den ebenen Winkel

rad, radiation absorbed dose Energiedosis bei ionisierender Strahlung, ionisierende Strahlungsenergiedosis *f*

RAD → rapid application development

radar, radio detecting and ranging [*These terms are erroneously used for monitoring by microwave motion detectors*] Funkmessmethode, Funkortungsmethode *f*, Funkmessverfahren, Funkortungsverfahren *n* [*Verfahren zur Entfernungsmessung und Ortung von Objekten mit Hilfe gebündelter von einem Sender ausgehender elektromagnetischer Wellen*]

radar motion detector, radar motion detecting device, radar motion detective device Radar-Bewegungsmelder, Radar-Bewegungsmeldungsgeber *m*, Radar-Bewegungsmeldegerät *n* [*Er arbeitet nach dem Dopplereffekt*]

R

radial blower Radialgebläse *n*, Radialbläser *m*

radial centrifugal fan, straight-blade(d) centrifugal fan Radialventilator, Radiallüfter *m*

radial duct Radialkanal *m*

radial duct fan Radialkanalventilator, Radialkanallüfter, Kanalradialventilator, Kanalradiallüfter *m*

radial fan Radialventilator, Radiallüfter *m*

radial (flow) jet Radialstrahl *m*

radial medium-pressure fan → medium-pressure radial fan

radial pipe fan Radialrohrventilator, Radialrohrlüfter, Rohrradiallüfter, Rohrradialventilator *m*

radial reciprocating compressor, radial piston compressor [*Three or more cylinders are disposed around a common crank*] Radialkolbenverdichter, Radialkolbenkompressor, Kolbenradialverdichter, Kolbenradialkompressor *m*

radial roof fan, roof radial fan Dachradialventilator, Radialdachventilator, Dachradiallüfter, Radialdachlüfter *m*

radiance, steradiance spezifische Ausstrahlung, Strahlungsdichte *f*

radiant absorption [*Transformation of radiant energy to a different form of energy by interaction with matter*] Strahlungsabsorption *f*

radiant air heating → air panel heating

radiant cable → (ceiling) radiant cable

radiant ceiling heating → ceiling radiant heating

radiant ceiling panel → ceiling radiation panel

radiant coil Strahlungsschlange *f*

radiant-convective heat exchange Strahlungs-Konvektions-Wärme(aus)tausch *m*

radiant/convector electric heater → (combined) radiant/convector electric heater

radiant/convector gas fire; convector fire [*deprecated*] [*A gas fire so designed that the radiant heat is supplemented by convected heat*] Strahlungs-Konvektions-Gasofen, Konvektions-Strahlungs-Gasofen *m*

radiant cooling ceiling, panel air ceiling Strahlungskühl(ungs)decke *f*

radiant cooling system, panel air system [*System using panel coolers to absorb radiant heat*] Strahlungskühl(ungs)system *n*

radiant drying Strahlungstrocknung *f*

radiant electric heating → electric radiant heating

radiant energy release → emission

radiant field Strahlungsfeld *n*

radiant floor heating → floor radiant heating

radiant glass panel, glass radiant panel [*An alumin(i)um grid forms resistance element for heating on the back of the panel*] Glasstrahlungsheiz(ungs)platte, Strahlungsglasheiz(ungs)platte *f*

radiant glass panel heating, glass panel radiant heating Strahlungsglastafelheizung, Glastafelstrahlungsheizung *f*

radiant heat, radiated heat [*1. Heat transmitted through an object which may not itself be not. 2. Heat transmitted by a hot object*] Strahlungswärme *f*

radiant-heated strahlungsbeheizt

radiant heater, radiant heat emitter, radiant heating appliance, radiant heating apparatus, radiant heating device Heiz(ungs)strahler, Strahlungserhitzer, Strahlungs(er)wärmer, Strahlungsheizer, Strahlungsheiz(ungs)apparat, Wärmestrahler *m*

radiant heat gain Strahlungswärmeabgabe *f*

radiant heating, radiation heating [*1. Heating system that radiates heat from a surface to its surroundings. 2. Any of various means of heating objects or persons by radiation without heating the intervening air*] Strahlungsheizung *f*

radiant heating device, radiant heating appliance, radiation heating device, radiation heating appliance Strahlungsheiz(ungs)gerät *n*

radiant heating panel Strahlungplattenheiz(ungs)-körper *m*

radiant heating pipe, radiation heating pipe Strahlungsheiz(ungs)rohr *n*

radiant heating system, radiation heating system Strahlungsheiz(ungs)system *n*

radiant intensity [*Electromagnetism. The electromagnetic radiation power per unit solid angle about the direction under consideration; expressed in units of watts per steradian*] Strahlungsstärke *f*

radiant low temperature heating → low temperature radiant heating

radiant maximum-temperature heating → maximum-temperature radiant heating

radiant metal(lic) panel → metal(lic) radiant panel

radiant panel heating → panel radiant heating

radiant power [*Electromagnetism. 1. The time-rate of electromagnetic energy passage across the unit surface. 2. The rate at which electromagnetic energy is emitted from a source*] Strahlungsleistung, abgestrahlte Leistung *f*, Strahlungsfluss, Energiefluss *m*

radiantproof strahlungsundurchlässig

radiant reflectance [*Electromagnetism. The ratio of electromagnetic power reflected from an object to that which is incident of the object*] Strahlungsreflektionsgrad *m*

radiant room heater Strahlungs(raum)heizer *m*

radiant (space) heat gain Strahlungs(raum)wärmegewinn *m*

radiant storage heat Strahlungsspeicherwärme, Speicherstrahlungswärme *f*

radiant-strip (type) heater → strip (type) radiant heater

radiant superheater [*A part of a steam-generating unit in which steam is heated above its saturation temperature by the transfer of radiant heat from combustion products to the steam*] Strahlungsüberhitzer *m*

radiant surface burner → surface radiant burner

radiant temperature Strahlungstemperatur *f*

radiant thermometer Strahlungsthermometer *n*

radiant transmittance [*The ratio of electromagnetic power transmitted through an object to that which is incident on the object*] Strahlungsdurchlässigkeit(-sgrad) *f*, (*m*), Strahlungsdurchlass(grad) *m*

radiant(-type) boiler [*A water-tube boiler in which the tubes are arranged around the outer surface of the combustion area of the furnace*] Strahlungskessel *m*

radiant wall heating → wall radiant heating

to radiate [*Physics. To give off rays, as of light or heat; to emit radiation*] strahlen

radiated heat → radiant heat

radiated heat transfer → radiation heat transfer

radiating body → radiator

radiation [*Physics. 1. The emission of waves or particles from a source and the propagation of these waves or particles through a medium; when used without a modifier, the term usually refers to electromagnetic radiation. 2. By extension, a stream of particles such as electrons, neutrons, protons, or alpha particles. 3. The complete process by which waves or particles are emitted, pass through a medium, and are absorbed by another body. 4. The energy transferred by such processes (Strahlungsenergie = radiant energy)*] Strahlung *f*

radiation absorbed dose, rad absorbierte Strahlungsdosis *f*

radiation angle [*Electromagnetism. An angle measured with respect to a horizontal or radial reference indicating the direction of electromagnetic energy propagation from a transmitting antenna*] Strahlungswinkel *m*

radiation appliance, radiation device Strahler *m*

radiation background Strahlungshintergrund *m*

radiation burner Strahlungsbrenner *m*

radiation ceiling Strahlungsdecke *f*

radiation component Strahlungskomponente *f*

radiation constant Strahlungskonstante *f*

radiation cooling [*Electronics. Any process of losing heat other than by conduction or convection*] Strahlungskühlung *f*

radiation cooling ceiling Strahlungskühl(ungs)-(unter)decke *f*

radiation cooling screen Strahlungskühl(ungs)-schirm *m*

radiation cooling surface Strahlungskühl(ungs)fläche *f*

radiation detector, radiation detection device, radiation detecting device Strahlungsmelder, Strahlungsdetektor, Strahlungsmeldungsgeber *m*, Strahlungsmeldegerät *n*

R

radiation device → **radiation appliance**

radiation gas heating → **gas radiation heating**

radiation heat Strahlungswärme *f*

radiation heat flow Strahlungswärmestrom *m*

radiation heating, radiant heating Strahlungsheizung *f*

radiation heating device → **radiant heating device**

radiation heat transfer, radiated heat transfer Wärme(aus)tausch durch Strahlung, Strahlungswärme(aus)tausch *m*

radiation infrared heating → **infrared radiation heating**

radiation intensity → **intensity of radiation**

radiation load Strahlungslast *f*

radiation loss [*The amount of heat that escapes from a hole into the atmosphere due to conduction, radiation, and convection*] Strahlungsverlust *m*

radiation sensor Strahlungssensor, Strahlungsfühler *m*

radiation shield [*Separate panel interposed between heating surfaces and jackets to reduce heat losses through radiation*] Strahlungsschild *m*

radiation sound pressure → **acoustic radiation pressure**

radiation source Strahlungsquelle *f*

radiator, radiating body [*A body that emits radiant energy*] Strahler, strahlende Körper *m*

radiator [*Heat source that transmits heat by radiation and convection*] Wärmestrahler, strahlende Wärmekörper *m*

radiator, sectional radiator [*A heating unit usually exposed to view within the room or space to be heated; transfers heat by radiation to objects within visible range, and by conduction to the surrounding air, which in turn is circulated by natural convection*] Radiator, Gliederheiz(ungs)körper; Heiz(ungs)körper *m* [*Fehlbenennung*] [*Er besteht aus aneinandergereihten Gliedern gleicher Größe*]

radiator bracket, sectional radiator bracket Radiatorkonsole, Gliederheiz(ungs)körperkonsole; Heiz(ungs)körperkonsole *f* [*Fehlbenennung*]

radiator controller, sectional radiator controller Radiatorregler, Gliederheiz(ungs)körperregler; Heiz(ungs)körperregler *m* [*Fehlbenennung*]

radiator heating, sectional radiator heating Radiator(en)heizung, Gliederheiz(ungs)körperheizung; Heiz(ungs)körperheizung *f* [*Fehlbenennung*]

radiator inlet, sectional radiator inlet Radiatoreinlauf, Gliederheiz(ungs)körpereinlauf; Heiz(ungs)körpereinlauf *m* [*Fehlbenennung*]

radiator lining, sectional radiator lining Radiatorverkleidung, Gliederheiz(ungs)körperverkleidung; Heiz(ungs)körperverkleidung *f* [*Fehlbenennung*]

radiator mounting, sectional radiator mounting Radiatoraufstellung, Gliederheiz(ungs)körperaufstellung; Heiz(ungs)körperaufstellung *f* [*Fehlbenennung*]

radiator outlet, sectional radiator outlet Radiatorauslauf, Gliederheiz(ungs)körperauslauf; Heiz(ungs)körperauslauf *m* [*Fehlbenennung*]

radiator recess, sectional radiator recess Radiatornische, Gliederheiz(ungs)körpernische; Heiz(ungs)körpernische *f* [*Fehlbenennung*]

radiator section, sectional radiator section Glied, Radiatorglied; Heiz(ungs)körperglied *n* [*Fehlbenennung*]

radiator valve, sectional radiator valve Radiatorventil, Gliederheiz(ungs)körperventil; Heiz(ungs)körperventil *n* [*Fehlbenennung*]

radiator with prefab(ricated) connection Fertigheiz(ungs)körper; Heiz(ungs)körper mit vorgefertigtem Anschluss *m* [*Fehlbenennung*]

radio Hörfunk *m*

radioactivity radioaktive Strahlung *f*

radio act(uat)or, wireless act(uat)or Funkbedienteil *m, n*

radio alarm, wireless alarm Funkalarm *m*

radioautography → **autoradiography**

radio broadcast, wireless broadcast Rundfunksendung, Radiosendung *f*

radio bus, wireless bus Funkbus *m*

radio call, wireless call Funkruf *m*

radio call receiver, wireless call receiver Funkrufempfänger *m*

radio call system, wireless call system Funkrufsystem *n*

radio central station → **central radio station**

radio channel, wireless channel Funkkanal *m*

radio clock, wireless clock Funkuhr *f*

radio communication, wireless communication Funkkommunikation *f*

radio contact transmitter, wireless contact transmitter Funkkontaktsender *m*

radio coverage → **radio range**

radio detector, radio detecting device, radio detection device, wireless detector, wireless detecting device, wireless detection device Funkmelder, Funkmeldungsgeber *m*, Funkmeldegerät *n*

radio detecting and ranging → **radar**

R

radio data system, RDS, wireless data system verbesserte Verkehrsfunksystem *n* [*Die Informationen werden über UKW-Sender übertragen*]

radio disturbance field strength, radio interference field strength Funkstör(ungs)feldstärke *f*

radio disturbance, radio interference Funkstörung *f*

radio disturbance limit (value), radio interference limit (value) Funkstör(ungs)grenzwert *m*

radio disturbance power, radio interference power Funkstör(ungs)leistung *f*

radio disturbance suppression, radio interference suppression Funkentstörung *f*

radio disturbance suppression capacitor, radio interference suppression capacitor Funkentstör(ungs)kondensator *m*

radio disturbance suppression choke, radio interference suppression choke Funkentstör(ungs)drossel *f*

radio engineering, wireless engineering Funktechnik *f*

radio-frequency → high-frequency

radio-frequency ground → rf ground

radio-frequency interference reactor → RFI reactor

radio-frequency power, RF power Hochfrequenzleistung, HF-Leistung *f*

radio-frequency preheating → rf preheating

radio glass break detector, radio glass break detection device, radio glass break detecting device [*"Wireless" may synonymously be used for "radio"*] Funkglasbruchmelder, Funkglasbruchmeldungsgeber *m*, Funkglasbruchmeldegerät *n*

radio installation, radio plant, wireless installation, wireless plant Funkanlage, Funkinstallation *f*

radio interference, wireless interference Funkstörung *f*

radio interference suppression → disturbance suppression

radio link, wireless link Funkverbindung *f*

radio linked alarm system, wireless linked alarm system drahtlose Alarm(ierungs)system, Funkalarm(ierungs)system *n*

radio linked burglar(y) signal message system, wireless linked burglar(y) signal message system Einbruchmeldeanlage mit Funkübertragung, Einbruchmeldungsanlage mit Funkübertragung, drahtlose Einbruchmeldeanlage, drahtlose Einbruchmeldungsanlage, Einbruchfunkalarmanlage *f* [*Im Gegensatz zu einer drahtgebundenen Einbruchmeldeanlage benutzt eine Funkalarmanlage für den Transport der Meldungen zwischen den Komponenten elektromagnetische Funkwellen*]

radio local loop, wireless local loop, WLL drahtlose Teilnehmeranschluss *m*, drahtlose Teilnehmersystem *n*

radio location, wireless location Funkortung *f*

radio message system, radio message service, wireless message system, wireless message service Funkrufdienst *m*, Funkrufsystem *n*

radio motion detector, radio motion detecting device, radio motion detection device [*"Wireless" may synonymously be used for "radio"*] Funkbewegungsmelder, Funkbewegungsdetektor, Funkbewegungsmeldungsgeber *m*, Funkbewegungsmeldegerät *n* [*Ein Raumüberwachungsmelder, der die Temperaturstrahlung eines Menschen registriert und per Funk eine Meldung an die Alarmzentrale gibt*]

radio net(work), wireless net(work) Funknetz *n*

radio net(work) termination, wireless net(work) termination, WNT drahtlose Netzabschluss *m*

radio noise, wireless noise Funkrauschen *n*

radio ((open) loop) control Funksteuerung *f*

radio pollution signal, wireless pollution signal Funkfremdsignal, Fremdfunksignal *n*

radio paging device → pager

radio range, radio coverage, wireless range, wireless coverage Funkreichweite *f*

radio relay, wireless relay Richtfunk *m*

radio relay signal, wireless relay signal Richtfunksignal *n*

radio remote ((open) loop) control, wireless remote ((open) loop) control Funkfernsteuerung, Fernfunksteuerung *f*

radio (remote) reading, wireless (remote) reading (Fern)Ablesung per Funk, Funk(fern)ablesung *f*

radio security system, wireless security system drahtlose Sicherungssystem, DSS *n*

radio shielding, wireless shielding Funkentstörung *f*

radio signal, wireless signal Funksignal *n*

radio smoke detector, radio smoke detecting device, radio smoke detection device [*"Wireless" may synonymously be used for "radio"*] Funkrauchmelder, Funkrauchdetektor, Funkrauchmeldungsgeber *m*, Funkrauchmeldegerät *n*

radio space temperature ((closed) loop) control, ((closed) loop) radio space temperature control Funkraumtemperaturreg(e)lung *f*

radio space temperature ((closed) loop) controller, ((closed) loop) radio space temperature controller Funkraumtemperaturregler *m*

radio station terminal, radio terminal station, RST, RTS Endfunkstation, Funkendstation *f*

radio suppressed, wireless suppressed funkentstört

radio telecontrol, wireless telecontrol Funkfernsteuerung, Fernfunksteuerung *f*

radio telephony, wireless telefony Sprechfunk *m*

radio teleprinter, wireless teleprinter Funkfernschreiber *m*

radio teletype method, RTTY method, wireless teletype method Funktelexverfahren, Funkfernschreibverfahren *n*, Funktelexmethode, Funkfernschreibmethode *f*

radio teletyper, RTTY, wireless teletyper Funkfernschreiber *m*

RAM → random-access (computer) memory

ram elevator (US) → plunger elevator

ramp-down time Sollwertrücklauf *m*, "Rampe" *f*

ramp(-forced) response Anstieg(s)antwort *f* [*Die Ausgangsfunktion v(+) eines Übertragungsgliedes, an dessen Eingang eine Anstiegsfunktion oder Rampe gelegt wird*]

ramp function, accelerating function Anstieg(s)-funktion, Beschleunigungsfunktion *f*, Hochlauf *m*

ramp function generator, accelerating function generator Hochlaufgeber, Sollwertgeber, Hochlaufintegrator, Sollwertintegrator, Anstiegsgeber, Beschleunigungsgeber *m*

RAMPS, resource allocation and multi(ple)project scheduling Verteilungsverfahren *n*, Verteilungsmethode *f* [*Verteilung der Kapazitäten und Zeitplanung für Probleme, bei denen mehrere Projekte zu planen sind*]

ramp time Flanke *f*

random wahlfrei, beliebig

random access, direct access wahlfreie (Speicher)-Zugriff, beliebige (Speicher)Zugriff, direkte (Speicher)Zugriff *m*

random-access (computer) memory, ram, RAM löschbare Datenspeicher, Schreib-Lese-Speicher, Lese-Schreib-Speicher, Direktzugriffsspeicher, Arbeitsspeicher, (Rechner)Speicher mit wahlfreiem Zugriff, (Rechner)Speicher mit beliebigem Zugriff, Direkteingriffsspeicher *m* [*E DIN 44 300/10.85. Ein Halbleiterspeicher, der beliebig oft gelesen oder beschrieben werden kann und den Inhalt nach Ausschalten verliert*]

random-access stored-programme machine; random-access stored-program machine (US) Speicherprogrammaschine mit wahlfreiem Zugriff, Speicherprogrammaschine mit beliebigem Zugriff *f*

random card Besucherkarte *f*

random (check) generator Zufallsgenerator *m*

random check(ing) → spot check(ing)

random-controlled zufallsgesteuert

random (data) processing asynchrone (Daten)Verarbeitung *f*

random logic Zufallslogik *f*

random padding Auffüllen *n* mit Zufallsbits [*Einer zu unterschreibenden Nachricht*]

random processing wahlfreie Verarbeitung *f*

random sample Stichprobe *f*

random sampling, random sample taking Stichprobe(n)nahme *f*

random sequence Zufallsfolge *f*

random storage Direktzugriffsspeicher *m*

range Reihe(nanlage) *f*, Stand *m*

range [*The region between the limits within which a quantity is measured, received, or transmitted, expressed by stating the lower and the upper range value. For instance, a temperature sensor is defined by its range of, say, 0-200 °F and the accuracy of ±1.0 °F at full range*] Bereich *m* [*selten: n*]

range → kitchener

rangeability → (valve) rangeability

range alarm Bereich(s)alarm *m*

range connector Bereich(s)stecker *m*

range finder, RF, ranger Entfernungsmesser, Abstandsmesser *m*

range monitoring Bereich(s)überwachung *f*

range of audibility, audibility range Hör(barkeit(s))-bereich *m* [*selten: n*], Hörweite *f*

range of frequencies → band

range of shading, shading range Graunuancenskala *f*

range of sizes Größenbereich *m* [*selten: n*]

range of urinal stalls → urinal stall range

range ((open) loop) control panel Bereich(s)zentrale, Bereich(s)steuer(ungs)tafel *f*

range switch, multiplier Bereich(s)schalter, Voltmeterschalter *m*

Rankine cycle Rankinekreislauf *m*

rapid-access storage, fast-access storage, quick-access storage [*The section of a computer storage, from which data may be obtained most readily*] Schnellzugriffsspeicherung *f*

rapid-acting, quick-acting, fast-acting flink [*Sicherung*]

rapid air heater, quick air heater, fast air heater Schnellufterhitzer, Schnelluft(er)wärmer *m*

rapid application development, RAD, quick application development, fast application development schnelle Anwendungsentwicklung *f*

R

rapid auto(matic)-reclosure → **auto(matic)-reclosure**

rapid-chilling, quick-chilling, rapid-cooling, quick-cooling, fast chilling, fast cooling Schnellkühlen *n*, Schnellkühlung *f*

rapid-closing outlet valve, quick-closing outlet valve, fast-closing outlet valve Schnellschlussablaufventil, Schnellschlussauslaufventil, Schnellschließablaufventil, Schnellschließauslaufventil *n*

rapid-closing valve, fast-closing valve, quick-closing valve Schnellschlussventil, Auslaufventil mit Hebelgriff, Hebelgriff(auslauf)ventil *n*

rapid-filling cistern, fast-filling cistern, quick-filling cistern Schnellfüll(ungs)spülkasten *m*

rapid exit, quick exit, fast exit [*in emergency situations*] Flucht *f*

rapid flashing, quick flashing, fast flashing schnelle Blinken, Schnellblinken *n*

rapid-freeze locker, quick-freeze locker, fast-freeze locker Schnellgefrierschrank, Tiefgefrierschrank *m*

rapid freezer → **quick freezer**

rapid-freezing, deep-freezing, quick-freezing, fast-freezing [*Freezing of a food or other commodity at a rapid rate of temperature reduction to produce a desired crystalline structure in the frozen product*] Tiefgefrieren, Schnellgefrieren *n*

rapid heater, fast heater, quick heater Schnellheizer *m*

rapid heating up, quick heating up, fast heating up Schnellaufheizung *f*

rapidly accessible → **readily accessible**

rapid-make-rapid-break switch → **sensitive switch** [*IEC 163*]

rapid release fastener, fast release fastener, quick release fastener Schnellverschluss *m*

rapid sand filter, fast sand filter Schnell-Sandfilter *m, n*

rapid-start lamp, quick-start lamp, fast-start lamp Sofortstartlampe, Schnellstartlampe, sofortzündende Lampe *f*

rapid steam generator, fast steam generator, quick steam generator Schnelldampferzeuger *m*

rapid stop valve, quick stop valve, fast stop valve Schnell(ab)sperrventil *n*

rapid TN scheme, fast TN scheme, quick TN scheme schnelle Nullung *f*

rapper → **(door) knocker**

raster display Rasteranzeige *f*

raster plotter Rasterzeichenmaschine *f*

raster scan zeilenweise Bildaufbau *m*

raster (scan(ning)) screen Abtastbildschirm, Rasterbildschirm *m* [*Der Elektronenstrahl tastet die Schirmoberfläche zeilenweise ab*]

RAT → **register alias table**

rate → **fee**

rate acquisition → **fee acquisition**

rate action, derivative action, D-action Vorhalt *m*, D-Aufschaltung, Differentialaufschaltung *f* [*Aufschaltung einer Größe, die der Änderungsgeschwindigkeit der Regelgröße proportional ist, auf den Reglereingang*]

rate(-action) controller → **PD controller**

rate-1/2-convolutional-encoder Faltungscoder *m*, Faltungskoder [*Er fügt redundante Informationen so zum laufenden Datenstrom hinzu, dass man diese auf der Empfangsseite zur Bitfehlerkorrektur verwenden kann. Der Faltungscoder verdoppelt zunächst die Bitrate auf 384 kbit/s und gibt sie in Form von zwei Datenstömen C0 und C1 mit je 192 kbit/s ab*]

rated input Nennaufnahme *f*

rated operation(al) current, operation(al) rated current Nennbetriebsstrom, Betriebsnennstrom *m*

rated pressure, PN Nenndruck *m* [*DIN 2401, Teil 1*]

rated refrigerating output → **nominal refrigerating output**

rated size Nennabmessungen, Nenngröße *f*

rated speed [*The maximum speed or data rate of a device or facility, without allowance for delaying functions such as checking or tabbing*] Nenngeschwindigkeit *f*

rate effect Geschwindigkeitseffekt *m*, Geschwindigkeitswirkung *f*

rate-growth method Stufenziehverfahren *n*, Stufenziehmethode *f* [*Herstellung eines npn-Transistors*]

rate of airflow → **air rate**

rate of change and turn → **churnrate**

rate of flow → **flowrate**

rate of heatflow → **heat transfer**

rate-of-rise Anstieg(s)geschwindigkeit *f*

(rate-of-)rise detector, (rate-of-)rise detecting device Differentialmeldegerät *n*, Differentialdetektor, Differentialmelder, Differentialmeldungsgeber *m* [*Er spricht an, wenn die Änderungsgeschwindigkeit der gemessenen Kenngröße einen bestimmten Wert für eine genügend lange Zeit überschreitet. DIN EN 54 T1 04.77*]

(rate-of-)rise method [*heat detection*] Anstieg(s)verfahren *n*, Anstieg(s)methode *f*

R

rate of transmission, transmission rate Übertragungsgeschwindigkeit *f*

rate range Wertebereich *m [selten: n]*

rate time, derivative(-action) time Vorhalt(e)zeit, Differenzierzeit *f [Symbol* T_V*. Zeit, die ein P-Regler bei konstanter Änderungsgeschwindigkeit der Regelgröße benötigt, um die gleiche Änderung der Stellgröße zu bewirken, die ein PD-Regler infolge seiner D-Aufschaltung sofort bewirkt]*

rating Einstufung *f*

ratio, relationship Verhältnis *n*

ratio equation, relationship equation Verhältnisgleichung *f*

ratio pyrometer → double colour pyrometer

ratio unit, relationship unit Verhältnismaß *n*, Verhältniseinheit *f*

rattling Klappern *n*

raw air → unconditioned air

raw industrial waste rohe Industrieabfall *m*

raw mode zeichenweise Eingabeverarbeitung *f [Datensichtstation]*

raw sewage → crude sewage

raytracing, backward raytracing Zurückverfolgen, Strahlverfolgen, Lichtstrahlverfolgen *n [Ein photorealistisches Bild wird über das Zurückverfolgen eines Lichtstrahls vom Betrachter aus zur Lichtquelle hin für alle Bildelemente (= Bildpunkte) einer Szene berechnet]*

razor blade slot *[A slot for the disposal of razor blades in a hollow wall interior]* Rasierklingenschlitz *m*

razor socket Rasierersteckdose *f*

RCCB → residual-current circuit breaker

r.c.d. → residual current device

R

R 14 cell; c-cell (US) Babyzelle *f*

RC switch → remote control switch

RCTL, resistor-capacitor-transistor-logic Rechnerschaltkreis *m* bestehend aus einer Widerstands-Kondensator-Transistor-Kombination

RD → refrigerant discharge

RDD → replaceable database driver

RDS → radio data system

RE → resolution enhancement

reach(able space) Reichweite *f*

reach(ing) height Reichhöhe *f*

reach of screw, screw reach Einschraublänge *f*

reactance amplifier → parametric amplifier

reaction boiler Überdruckkessel, Hochleistungskessel *m*

reaction heat → heat of reaction

reactive current, wattless current, idle current, blind current Blindstrom *m*

reactive-power controller, VAr controller, blind-power controller Blindleistungsregler *m [Er erfasst die Blindleistung und schaltet die benötigten Kompensationskondensatoren automatisch nach Bedarf zu und ab]*

to read back (from memory) *[In automatic data processing, to find and select specific information]* abrufen (vom Speicher)

reader, reading device, reading apparatus *[In automatic data processing, a device that converts information from one form to another, as from punched paper tape to magnetic tape]* Leser, Leseapparat *m*, Lesegerät *n*

reader electronics Leseelektronik *f*

reader housing, reader casing, reader case Lesegehäuse *n*

reader interface Leseschnittstelle *f*

reader interface device Leseschnittstellengerät *n*

reader-interpreter *[In automatic data processing, a service routine that reads an input string, stores programs and data on random-access storage for later processing, identifies the control information contained in the input string, and stores this control information separately in the appropriate control lists]* Leseinterpret(ier)er *m*

reader plug Lesestecker *m*

reader-punch equipment *[In automatic data processing, an input/output unit which can punch computer results on cards and read card data into the computer]* Einlese-Stanz-Gerät *n*

reader terminal Leseterminal *n*

read head *[In automatic data processing, a device that converts digital information stored on a magnetic tape, drum, or disk into electrical signals usable by the computer arithmetic unit]* Lesekopf *m*

read hopper Vorratsfach *n [Lesegerät]*

readily accessible, quickly accessible, rapidly accessible *[Capable of being reached quickly for operation, renewal, or inspections, without requiring those to whom ready access is requisite to climb over or remove obstacles or to resort to portable ladders, chairs, scaffolds, etc.]* schnell zugänglich

to read-in *[In automatic data processing, to sense information contained in source and transmit this information to an internal storage]* eingeben, einlesen

readiness Bereitschaft *f*

readiness review [*In automatic date processing, an on-site examination of the adequacy of preparations for effective utilization upon installation of a computer, and to identify any necessary corrective actions*] (Betriebs)Bereitschaftsprüfung *f*

reading, read operation Lesen *n*, Lesung *f*

reading access Lesezugriff, Lesungszugriff *m*

reading access time Lesezugriffszeit, Lesungszugriffszeit *f*

reading accuracy Lesegenauigkeit, Lesungsgenauigkeit *f*

reading current Lesestrom *m*

reading device → reader

reading distance Leseentfernung *f*

reading error Lesefehler *m*

reading field Lesefeld *n*

reading gauge; reading gage (US); meter reader Ablesegerät *n*

reading gun Lesepistole *f*, Lesestrahlerzeuger *m* [*Oszillograph*]

reading (im)pulse Leseimpuls *m*

reading lamp Leselampe *f*

reading mode Lesebetriebsart *f*

reading pen Lesestift *m*

reading range Lesebereich *m* [*selten: n*]

reading rate [*In automatic data processing, number of characters, words, fields, blocks, or cards sensed by an input sensing device per unit of time*] Lesezeichenanzahl *f*

reading rule → (reference) reading rule

reading screen [*In optical character recognition (OR), the transparent component part of most character readers through which appears the input document to be recognized]* Leseschirm *m*

reading station, sensing station [*In automatic data processing, the position in a punched-card machine at which the data on the card are read, by sensing the positions of the holes, and converted into electrical impulses*] Lesestation *f*

reading time (interval) [*In automatic data processing, the time (interval) between the instant at which information is called for from storage and the instant at which delivery is completed in a computer*] Lesezeit(spanne) *f*

reading wand Lesestift *m* [*Strichkode*]

reading/write head, reading/writing head Lese-/Schreib-Kopf *m*

read-in program (US); read-in programme (Brit.) [*In automatic data processing, a computer program that can be put into a computer in a simple binary form and allows other programs to be read into the computer in more complex forms*] Einleseprogramm *n*

to reajust, to reset nachstellen

read mostly memory, RMM Vorwiegend-Lese-Speicher *m*

to read off, to take a reading ablesen

read only schreibgeschützt [*Attribut in einer Datei*]

read-only memory, read-only storage, ROM, fixed memory, fixed storage Fest(wert)speicher, (Nur-)Lesespeicher *m* [*E DIN 4430/10.85*]

read-only memory chip, read-only storage chip, ROM chip, fixed memory chip, fixed storage chip Fest(wert)speicherbaustein, (Nur)Lesespeicherbaustein *m*

read operation → reading

to read out → to fetch (out)

readout, display [*In automatic data processing, the presentation of output information by means of lights, printed or punched tapes or cards, or other methods*] Anzeige, Sichtanzeige *f*

read-punch unit [*In automatic data processing, an input/output unit of a computing system which punches computed results into cards, reads input informations into the system and segregates output cards; the read-punch unit generally consists of a card feed, a read station, a punch station, another read station, and two output card stackers*] Lese-Stanz-Einheit, Stanz-Lese-Einheit *f*

read signal → (computer) memory read signal

read stacker Kartenleserablage *f*, Kartenlesermagazin *n*

read time → access time

read-while-writing, read/write mode [*In automatic data processing, the reading of a record or group of records into storage from tape at the same time another record or group of records is written from storage to tapes*] Lesen-Schreiben *n*

read/write channel [*In automatic data processing, a path along which information is transmitted between the central processing unit of a computer and an input, output, or storage unit under the control of the computer*] Lese-Schreib-Kanal *m*

read/write check(ing) indicator [*In automatic data processing, a device incorporated in certain computers to indicate upon interrogations whether or not an error was made in reading or writing; the machine can be made to stop, retry the operation, or follow a special subroutine, depending upon the result of the interrogation*] Lese-Schreib-Kontrollanzeiger *m*

read/write cycle Lese-Schreib-Zyklus *m*

read/write head, combined head [*In automatic data processing, a magnetic head that both senses and records data*] Lese-Schreib-Kopf *m*

read/write head access aperture, combined head access aperture Lese-Schreib-Kopfaussparung *f*

ready fertig, bereit [*Eine Gehäuseanschlussbezeichnung beim Mikroprozessor 8080. Dieses Signal zeigt dem Mikroprozessor, dass auf dem Bus Daten zur Verfügung stehen und übernommen werden können*]

ready → clear

ready assembled in the factory werkseitig bestückt, werkbestückt, fabrikseitig bestückt, fabrikbestückt

ready data abfragebereite Daten, abfragefertige Daten *f*

ready for being connected anschließfertig, anschlussbereit, anschlussfertig, anschließbereit

read/write register Lese-Schreib-Register, Schreib-Lese-Register *n*

read/write mode → read-while-writing

read/write memory, RWM Lese-Schreib-Speicher, Schreib-Lese-Speicher *m*

ready for next message, RFNM bereit für das nächste Datenpaket, fertig für das nächste Datenpaket [*Meldung innerhalb eines paketsortierten Datenübertragungsprotokolls*]

ready for operation betriebsfertig, betriebsbereit

ready for sharp switching scharfschaltbereit, scharfschaltfertig [*Gefahrenmeldeanlage; Sicherungsbereich*]

ready for tapping → on-tap

ready-to-be connected anschlussfertig, verbindungsfertig, anschlussbereit, verbindungsbereit

ready-to-fit, ready-to-install einbaufertig, einbaubereit

ready-to-be plugged in steckerfertig, steckerbereit

ready-to-operate betriebsbereit, betriebsfertig

ready to send → RTS

real → actual

real capacity, actual capacity Istfassungsvermögen, tatsächliche Fassungsvermögen, wirkliche Fassungsvermögent, Realfassungsvermögen *n*, Istfassungsfähigkeit, tatsächliche Fassungsfähigkeit, wirkliche Fassungsfähigkei, Realfassungsfähigkeit *f*

real clearance, actual clearance Istspiel, tatsächliche Spiel, wirkliche Spiel, Realspiel *n*

real condition → real state

real controller, actual controller nichtideale Regler *m*

real dimension, actual dimension Istmaß, Maß 1:1 *n*, Istabmessung, Abmessung 1:1 *f*

real dimensions → real size

real displacement flow, actual displacement flow reale Verdrängungsströmung, wirkliche Verdrängungsströmung *f* [*Reale Form der Raumluftströmung, bei der der Raum nahezu ohne Mischen gleichzeitig durchströmt wird*]

real efficiency, actual efficiency Istwirkungsgrad, tatsächliche Wirkungsgrad, wirkliche Wirkungsgrad, reale Wirkungsgrad *m*

(real) estate → (real) property

(real) estate management → (real) property management

real fluid, actual fluid reale Fluid, wirkliche Fluid, tatsächliche Fluid *n*

real instruction, effective instruction, actual instruction [*The computer instruction that results from changing a basic instruction during program(me) modification*] wirksame Befehl, Wirkbefehl, effektive Befehl, wirkliche Befehl, tatsächliche Befehl *m*

realization Durchführung *f*

real-life solution wirklichkeitsnahe Lösung *f*

real liquid, actual liquid reale Flüssigkeit, wirkliche Flüssigkeit, tatsächliche Flüssigkeit *f*

real loader, absolute loader, actual loader Absolut(auf)lader, absolute (Auf)Lader *m* [*Elektronik. Er lädt das Absolutprogramm in den Speicher*]

real memory, actual memory Real(rechner)speicher *m*

real mixing flow, real room air flow reelle Mischströmung, tatsächliche Mischströmung, wirkliche Mischströmung *f* [*Reale Form der Raumluftströmung, bei der der Raumluftzustand infolge starker Durchmischung im Aufenthaltsbereich an jeder Stelle nahezu gleich ist*]

real open (communication) system [*A real system which complies with the requirements of OSI standards in its communication with other real systems*] echte offene Kommunikationssystem, offene echte Kommunikationssystem *n*

real power absorbed → input power

(real) property, realty, (real) estate, land(s), plot Liegenschaft *f*, Land *n*, Boden *m*, Grund(stück) *m,(n)*

(real) property drainage installation, estate drainage installation, (plot) drainage installation, realty drainage installation Grundstück(s)entwässerungsanlage *f* [*DIN 1986*]

(real) property main service, (plot) main service [*It enters the building below ground level*] Zuleitung, Grundstück(s)zuleitung *f*

R

(real) property management, realty management, (real) estate management, land(s) management, plot management [*A management, usually computer based, designed for the administration of an estate of property. Such a system is intended as a tool for carrying out such estate management functions as leasing, plant and building maintenance and optimizing use. "Administration" may synonymously be used for "management"*] Liegenschaftsverwaltung, Landverwaltung, Immobilienverwaltung, Immobiliarverwaltung, Grund(stücks)verwaltung, Bodenverwaltung *f*

real sign, actual sign [*COBOL*] Druckvorzeichen *n*

real size, real dimensions, actual size, actual dimensions Istgröße, Istabmessungen, Istmaße, Größe 1:1 , Abmessungen 1:1, Maße 1:1 *f*

real speed, actual speed Istdrehzahl, tatsächliche Drehzahl, wirkliche Drehzahl, reale Drehzahl, Realdrehzahl *f*

real state, actual state, real status, actual status, real condition, actual condition Istzustand, Iststatus, tatsächliche Zustand, wirkliche Zustand, tatsächliche Status, wirkliche Status, reale Zustand, reale Status *m*

real system, actual system [*A set of one or more computers, the associated software, peripherals, terminals, human operators, physical processes, information transfer means, etc., that forms an autonomous whole capable of performing information processing and/or information transfer. ISO 7498*] Realsystem, Echtsystem, Istsystem, tatsächliche System, wirkliche System, reale System *n*

real time, actual time Echtzeit, Realzeit, Istzeit, Uhrzeit, tatsächliche Zeit, wirkliche Zeit *f*

real time clock, RTC, actual time clock Echtzeituhr, Realzeituhr, Istzeituhr *f*

real time computer, actual time computer Echtzeitrechner, Realzeitrechner, Istzeitrechner, Uhrzeitrechner *m*

real time (en)coding, real time encryption, actual time (en)coding, actual time encryption Echtzeiteinkodieren, Uhrzeiteinkodieren, Realzeiteinkodieren, Istzeiteinkodieren, Echtzeiteincodieren, Uhrzeiteincodieren, Realzeiteincodieren, Istzeiteincodieren *n*

real time format, actual time format Echtzeitdarstellung, Realzeitdarstellung, Istzeitdarstellung, Uhrzeitdarstellung *f*

real time mode, actual time mode Echtzeitaufzeichnungsmodus, Echtzeitaufzeichnungsbetrieb, Uhrzeitaufzeichnungsmodus, Uhrzeitaufzeichnungsbetrieb, Realzeitaufzeichnungsmodus, Realzeitaufzeichnungsbetrieb, Istzeitaufzeichnungsmodus, Istzeitaufzeichnungsbetrieb *m*

real time operating system, real time OS Echtzeitbetriebssystem, Uhrzeitbetriebssystem *n*

real time processing, real time management, actual time processing, actual time management [*Expressions used to refer to a system in which the processing of input data to obtain a result occurs in an acceptable short time for process control*] Echtzeitverarbeitung, Uhrzeitverarbeitung, Realzeitverarbeitung, Istzeitverarbeitung, Dialogzeitverarbeitung, Sofortzeitverarbeitung, Direktzeitverarbeitung, schritthaltende Verarbeitung *f*

real time simulation, actual time simulation Echtzeitsimulation, Realzeitsimulation, Uhrzeitsimulation, Istzeitsimulation *f*

realty → (real) property

realty management → (real) property management

real value, actual value Sachwert, Substanzwert, Realwert *m* [*Er setzt sich zusammen aus dem Wert des Grund(stück)es wenn es unbebaut wäre (= Bodenwert) und dem Wert der baulichen Anlage (= Bauwert)*]

real value → actual value

real value memory, actual value memory Sachwertspeicher, Realwertspeicher, Substanzwertspeicher *m*

real value security, actual value security Sicherung gegen Angriffe auf Sachwerte, Sicherung gegen Angriffe auf Substanzwerte, Sicherung gegen Angriffe auf Realwerte *f*

real zero, actual zero [*charge transfer device*] Nullladung *f*

reasonable value limit Plausibilitätsgrenze *f*

reasoning Schließen, Folgern *n*

rebate, rabbet [*A shoulder or access cut into the edge of a piece of wood; used to receive the edge of a similar piece*] Falz *m*

rebate door release, rabbet door release Falztüröffner *m*

to recalculate, to recompute nachrechnen

to recall rückrufen, wieder (herbei)rufen

recall Wieder(herbei)ruf, Rückruf *m* [*Vermitt(e)lung*]

received backward channel data Hilfskanalempfangsdaten *f* [*Schnittstellenleitung*]

received character timing empfangsseitige Abtast(ungs)markierung *f*

received data Empfangsdaten *f*

received line signal detector Empfangssignalpegel *m*

received message Empfangsmeldung, Empfangsnachricht, Empfangsmitteilung *f*

received signal strength indicator, RSSI Empfangsfeldstärkeindikator, Empfangsfeldstärkeanzeiger *m*

R

received voice answer empfangene Sprachantwort *f*

receive frequency Empfangsfrequenz *f*

receive mode Empfangsbetriebsart *f*

receive-only device, receive-only appliance Nur-Empfangs-Gerät *n*, Nur-Empfangs-Vorrichtung *f*

receive-only display Bildschirm *m* mit Nur-Rechner-Empfang

receive-only printer Hartkopiergerät *n*

receiver Aufnehmer *m*, Zwischenkammer *f* [*Dampf*]

receiver → (refrigerant) receiver

receiver Empfänger *m* [*Bei Logikfamilien die empfangende Stelle*]

receiver → (telephone) receiver

receiver buffer Empfangspuffer *m*

receiver condenser [*Water-cooled condenser comprising a shell with the bottom located below the heat-exchanging elements to form a receiver; in principle designed to hold the entire refrigerant charge of the installation*] Kältemittelgefäßkondensator, Kältemittelkastenkondensator *m*

receiver diode Empfängerdiode *f*

receiver primary → display primary

receiver signal element timing Empfangsschritttakt *m*

receiving amplifier Empfangsverstärker *m*

receiving fluctuation Empfangsschwankung *f*

receiving headend Empfangsendstelle *f*

receiving property Empfangseigenschaft *f*

receiving room, sound receiving room (Schall)Aufnahmeraum, (Schall)Empfangsraum *m*

receiving station Empfangsstation *f* [*Rohrpost(anlage)*]

receiving station [*European Standard 54*] Empfangszentrale *f*

receiving wire, receiver wire Empfangsdraht *m*

receptacle (US) → socket outlet

recess Aussparung *f*, Rücksprung, Absatz *m*

recess → niche

recessed [*of an item of equipment and the like*] eingelassen, Einbau...

recessed aussparend, rückspringend

recessed ceiling floodlight, ceiling recessed floodlight Einbau-(Unter)Deckenflutlicht *n*, Einbau-(Unter)Deckenfluter *m*

recessed ceiling reflector lamp, ceiling recessed reflector lamp Einbau(Unter)Deckenstrahler, (Unter)Deckeneinbaustrahler *m*

recessed convector Einbaukonvektor *m*

(recessed) (wall)) gas heater, (recessed) (wall)) gas-fired heating unit [*Completely self-contained unit, recessed in a wall and located entirely above the floor of the space it is intended to heat*] Einbau-Gasheizer *m*

rechargeable wieder aufladbar [*Akku*]

recipe manipulation Rezeptänderung *f* [*Daten; Prozeduren*]

reciprocating chiller, piston chiller, reciprocating refrigerating system, piston refrigerating system Kolben(kompressor)kälteanlage, Kolbenverdichterkälteanlage *f* [*Siehe Anmerkung unter „Kompressorkälteanlage"*]

reciprocating clearance, piston clearance [*Space between the top face of piston and the valve assembly when the piston is on the top of its stroke*] Kolbenfreiheit *f*

reciprocating compressor, piston compressor [*Positive-displacement compressor in which the change in internal volume of the compression chamber(s) is accomplished by the reciprocating motion of one or more pistons*] Kolbenverdichter, Kolbenkompressor *m*

reciprocating compressor assembly, piston compressor assembly [*"Set", "pack(age)" and "unit" may synonymously be used for "assembly"*] Kolbenverdichtersatz, Kolbenkompressorsatz *m* [*„Kombination", „Gruppe", „Aggregat", „Einheit" und „Block" können synonym für „Satz" verwendet werden*]

reciprocating compressor refrigerating plant, piston compressor refrigerating plant Kolben-Kompressionskälteanlage *f*, Kolben-Kompressionskältesystem *n*

reciprocating cooler, piston cooler Kolbenkühler *m*

reciprocating meter, piston(-type area) meter [*A variable area, constant-head fluid-flow meter in which the position of the piston, moved by the buoyent force of the liquid, indicates the flow rate*] Kolbenzähler *m*, Kolbenmesser *n*

reciprocating slide valve, piston slide valve (Ring)-Kolbenschieber *m*

reciprocating stroke, piston stroke Kolbenhub *m*

reciprocating water meter, piston water meter Kolbenwasserzähler, Kolbenwassermesser *m*

to recirculate wieder umwälzen, erneut umwälzen

recirculated air [*Air taken from a space and returned to that space after being passed through a conditioning system*] Umluft *f* [*DIN 1946 Teil 2*]

recirculated air conditioning Umluftklimatisierung *f*

recirculated air conditioning plant, recirculated air conditioning installation Umluftklimaanlage, Umluftklimainstallation, HKBE-UM *f*

R

recirculated air(flow) damper, recirculated airstream damper Umluftklappe *f* [*Drosselung und/oder (Ab)Sperrung durch senkrecht zum Luftstrom drehbare Flächen*]

recirculated air unit Umluftgerät, UM-GR *n* [*DIN 1946, Teil 1*]

recirculation Wiederumwälzung *f*

reclaimable heat → **recoverable heat**

reclosing Wiederschließen *n*

reclosure Wiederschließung *f*

recoding Umkodierung, Umcodierung, Umschlüsselung *f* [*Übertragung der in einem Kode dargestellten Informationen in einen anderen Kode*]

recognition → **detection**

recognition differential, detection threshold [*The minimum detectable signal-to-noise ratio, giving a measure of a sonar systems effectiveness*] Erkennungsschwelle, Nachweisschwelle *f*

recognition sensitivity, detection sensitivity, detection sensitiveness, detection susceptibility, recognition sensitiveness, recognition susceptibility Erkennungsempfindlichkeit, Nachweisempfindlichkeit *f*

recognition system, detection system Erkennungssystem, Nachweissystem *n*

recognizability Erkennbarkeit, Nachweisbarkeit *f*

recognized private operating agency → **RPOA**

recommended value, counselled value, advised value Richtwert *m*

recommissioning Wiederinbetriebnahme *f*

reconfiguration, reconfiguring Nachprojektieren, Nachkonfigurieren *n*

reconnaissance squad check(ing), attendance check(ing), attendance management Anwesenheit(s)kontrolle, Präsenzkontrolle *f*

recooled rückgekühlt

recooling Rückkühlung *f*

recooling plant, recooling installation Rückkühl(ungs)anlage *f*

to record aufnehmen, aufzeichnen

record Aufnahme, Aufzeichnung *f*

record density → **character density**

record drawing → **as-built drawing**

recorded aufgezeichnet, aufgenommen

record key gespeicherter Schlüssel, Vergleich(s)schlüssel *m* [*Hinzufügung eines Satzes zu einer Datei*]

record(ed) sound Aufnahmeschall, aufgenommene Schall *m*

record end → **end of record**

recorder, recording instrument Aufnahmegerät *n*, Aufnehmer *m*

recorder buffer, ROB Gültigmacher *m* [*Er stellt sicher, dass in der Reihenfolge, in der die Befehle abgesetzt worden sind, diese auch wieder gültig gesetzt (retired) werden*]

recording Aufnehmen, Registrieren, Aufzeichnen *n* [*Das Festhalten von Größen zum Weiterverarbeiten oder Dokumentieren. Das fortlaufende Aufzeichnen des Werteverlaufs von Größen in einer für den Menschen leicht auswertbaren Form heißt Registrieren. DIN 19 222*]

recording barometer → **barograph**

recording calorimeter [*See remark under "calorimeter"*] Schreibkalorimeter, Registrierkalorimeter *n*

recording density Aufnahmedichte, Speicherdichte *f* [*Angegeben in Bit pro Fläche*]

recording device → **data recorder**

recording head Aufnahmekopf *m*

recording hygrometer Feuchteschreiber, Feuchtigkeitsschreiber, Hygrograph *m*

recording instrument → **recorder**

recording on mass memory Langzeitspeicherung *f*

recording period Aufzeichnungszeitraum, Aufnahmezeitraum *m*

recording pressure gauge; recording pressure gage (US) Registriermanometer, Schreibmanometer *n*, Registrierdruckmesser, Schreibdruckmesser *m*

recording section Meldeschrank *m* [*Telefonie*]

recording thermometer → **thermograph**

record input/output Übertragung von Sätzen, Satzübertragung, Sätzeübertragung *f*

record locking Datensatzebenenverrieg(e)lung, Satzsperre *f* [*Siehe Anmerkung unter „Verrieg(e)lung"*]

record structure → **(data) record structure**

record subtype Satzunterart *f*

record type Satzart *f*

record value → **extreme value**

record word end → **end-of-record word**

recoverable heat, reclaimable heat [*Portion of thermal input to a prime mover that is not converted to mechanical power and can be reclaimed for utilization*] wiedergewinnbare Wärme, regenerierbare Wärme *f*

recovered energy [*Energy used, which would otherwise be wasted*] wiedergewonnene Energie, rückgewonnene Energie *f*

recovered heat Rückgewinnungswärme *f* [*Teil der Abwärme, der durch Wärmerückgewinnung in dasselbe System zurückgeführt wird*]

R

recovering valve → **four-way valve**

recovery Rückgewinnung, Verwertung *f*

recovery Wiederbereitschaft *f*

recovery loss Rückgewinnungsverlust, Verwertungsverlust *m*

recovery unit, recovery device Rückgewinner, Verwerter *m*

recovery voltage Wiederkehrspannung *f*

recreation room Freizeitraum, Erholungsraum *m*

rectangular (air) duct Rechteck(luft)kanal, U-Luftkanal *m*

rectangular airtube conveyor, rectangular airtube system, rectangular pneumatic conveyor, rectangular pneumatic system Zettelrohrpost *f* [*Keine Rohrbüchsen, sondern flache rechteckige Rohre*]

rectangular (bath)tub, rectangular bathing tub Rechteck(bade)wanne, U-Badewanne *f*

rectangular duct, U duct Rechteckkanal, U-Kanal *m*

rectangular expansion piece, rectangular compensating piece Rohrschenkelausgleicher *m* [*Rohrleitung*]

rectangular slating englische (Dach)Schiefer(ab)deckung *f*

rectification [*chemistry*] Rektifikation *f*, [*Reinigung durch wiederholte Destillation*]

recuperator Rekuperator, Wärme(aus)tauscher ohne Speicherung *m*

recurrent network → **iterative network**

recycling Stoffkreislauf, (Material)Kreislauf *m*, Wiederverwertung *f* [*Wiedergewinnung und Wiederverwendung von Abfallstoffen zwecks Verringerung der Abfallmengen und Rückgewinnung von Rohstoffen*]

recycling management (Material)Kreislaufbewirtschaftung, Stoffkreislaufbewirtschaftung *f*

R

recycling plant, recycling installation (Material)-Kreislaufanlage, Stoffkreislaufanlage *f*

red signal → **danger signal**

to redrill, to drill to desired size ausbohren

to redesign umkonstruieren

reduced instruction set computer, RISC Rechner *m* mit reduziertem Befehlssatz [*30 – 50 Befehle*]

reduced operation herabgesetzte Betrieb *m*

reduced quality, quality reduction Güteeinbuße, Qualitätseinbuße *f*, Abstrich *m*

reduced-voltage starter → **autotransformer starter**

reduced-voltage start(ing) → **soft start(ing)**

reducer Minderer, Verringerer *m*

reducer → **taper(ed) pipe**

reductance magnetische Widerstand *m*

reduction Minderung, Verringerung *f*

redundancy Redundanz, Überladung *f*, Überfluss *m* [*Nicht notwendiger Teil einer Information*]

redundancy Redundanz, Überladung *f*, Überfluss *m* [*Maß für den Ordnungszustand eines endlichen Schemas (Zeichenfolge; Nachricht)*]

redundancy Mehrfachspeicherung gleicher Daten, Vielfachspeicherung gleicher Daten, Redundanz, Überbestimmung *f* [*Eine Information besteht aus mehr Zeichen als zur eindeutigen Bestimmung des Inhalts notwendig ist. Dadurch erhöhte Sicherheit bei der Signalerkennung, auch wenn kleine Teile des Signals verloren gehen, ist dann die eindeutige Entschlüsselung der Information immer noch gesichert. Unter „redundanten Systemen" versteht man auch Doppel- und Dreifachsysteme, welche sich gegenseitig überwachen und bei Ausfall eines Systems immer noch richtige Informationen liefern.*]

redundancy function Redundanzfunktion, Überbestimmungsfunktion *f* [*Mehrfachausführung bestimmter Funktionen, die so konfiguriert sind, dass die gerade laufende Gebäudebetriebssystemfunktion zu Ende geführt wird, sollte eine der Mehrfachausführungen ausfallen*]

reed switch, magnetic reed switch [*A switch that has contacts mounted on ferromagnetic reeds sealed in a glass tube, designed for actuation by an external magnetic field*] Reedschalter, Blattfederschalter, Zungenschalter *m*

reed relay [*A relay having contacts mounted on magnetic reeds sealed into a length of small glass tubing; an actuating coil is wound around the tubing or mounted on an auxiliary ferrite-core structure, to provide the magnetic field required for relay operation*] Schutzgaskontaktrelais *n*

reed pipe Zungenpfeife *f*

reed frequency meter → **(vibrating-)reed frequency meter**

reeded stair tile gerillte Trittplatte *f* [*Treppe*]

reel Spule *f*

reengineering Wiederprojektierung, Wiederprojektbearbeitung, Nachprojektierung, Projektnachbearbeitung *f*

to re-enter wieder eingeben, erneut eingeben

refeeding Nachspeisen *n*, Nachspeisung *f*

reference Verweis, Hinweis, Bezug *m*, Referenz *f*

reference address → **address constant**

reference burst → **colour burst (Brit.)**

reference call → **call by reference**

reference element Bezug(s)element, Referenzelement *n* [*Elektronisches Bauelement für eine hochgenaue Bezugsspannung. Beispiel: Z-Diode in Reihe geschaltet mit einer Gleichrichterdiode*]

reference file Bezug(s)datei, Referenzdatei *f* [*Sie enthält die Beschreibung aller Datenpunkte in firmenneutraler Form*]

reference input variable [*IEC 27-2A*], **reference variable** [*IEC 50(351)*], **command variable** Führungsgröße *f* [*Größe einer Reg(e)lung oder Steuerung, der die Regel(ungs)größe oder Steuer(ungs)größe in vorgegebener Abhängigkeit folgen soll. Ist bei der Reg(e)lung diese Führungsgröße zeitlich konstant, so heißt sie „Sollwert der Reg(e)lung"*]

reference level Bezug(s)pegel, Referenzpegel *m*

reference line Symbollinie *f* [*Liniensystem für Schweißangaben nach ISO-Entwurf*]

reference point Referenzpunkt, Bezug(s)punkt *m*

reference potential Bezug(s)potential, Referenzpotential *n*

reference pressure Bezug(s)druck, Referenzdruck *m*

reference range Bezug(s)bereich, Referenzbereich *m* [*selten: n*]

reference reaction of the control loop Führungsverhalten des Regelkreises, Regelkreis-Führungsverhalten *n*

(reference) reading rule Ableselineal, Ablesungslineal *n* [*Schreiber*]

reference signal Bezug(s)signal, Referenzsignal *n*

reference signal generator Führungssignalbildner *m* [*Ein Glied, das bei Anlegen des Signals einer Zielgröße am Eingang ein daraus abgeleitetes Führungssignal am Ausgang in der Weise bildet, dass kritische Grenzwerte des Führungssignals oder seiner Ableitungen oder auch anderer Prozesssignale nicht überschritten werden. E DIN 19 226*]

reference sound Bezug(s)schall, Referenzschall *m*

reference sound power Bezug(s)schallleistung, Referenzschallleistung *f*

reference sound pressure Bezug(s)schalldruck *m*, Referenzschalldruck

reference sound source Bezug(s)schallquelle, Referenzschallquelle *f*

reference tape Bezug(s)band, Referenzband *n*

reference tone Bezug(s)ton, Referenzton *m*

reference value → **fiducial value**

reference variable [*IEC 50(351)*] → **reference input variable**

reference voltage Sollwertspannung, Bezug(s)spannung, Richtspannung, Referenzspannung *f*

reference voltage setting Sollwertspannungsvorgabe, Bezug(s)spanungsvorgabe, Richtspannungsvorgabe, Referenzspannungsvorgabe *f*

to refill nachfüllen, wiederbefüllen

refillable nachfüllbar, wiederbefüllbar

refilling Wiederbefüllung, Nachfüllung *f*

refill(ing) ink Nachfüll(ungs)tinte, Wiederbefüll(ungs)tinte *f*

refiltration Nachfilterung *f*

refitting table Umrüsttabelle *f*

reflected glare Rückstrahlblendung *f*

reflected image, mirror image Spiegelbild *n*

reflecting (type) water level indicator Reflexions-Wasserstandsanzeiger *m*

reflector [*Element of luminaire designed to direct light in desired direction*] Reflektor *m*

reflector (electric(al)) heater; reflector (electric(al)) fire (Brit.) Reflektor-Elektroheizkörper *m*

reflow Schmelzen *n* [*der Zinnschicht bei der Platinenherstellung durch Erhitzen*]

reflux valve → **check(ing) valve**

refractive index Brech(ungs)index *m*, Brech(ungs)zahl *f*

to refresh auffrischen

refresh display → **(vector) refresh display**

refresh display screen Bildwiederhol(ungs)schirm, Wiederhol(ungs)bildschirm *m* [*Der Elektronenstrahl schreibt die Graphik als Vektoren bis zu 20 mal pro Sekunde auf die Bildschirmfläche*]

refreshing [*Redrawing constantly the image on the screen*] (Bild)Wiederholung *f*

refresh(ing) memory (Bild)Wiederhol(ungs)speicher *m*

refresh(ing) rate Bildaufbaurate *f* [*Monitor*]

refresh(ing) screen Wiederhol(ungs)bildschirm *m*

refresh(ing) tube → **(vector) refresh(ing) tube**

refrigerant, R, refrigerating medium, refrigerating fluid [*In a refrigerating system, the medium of heat transfer which picks up heat by evaporating at a low temperature and pressure, and gives up heat on condensing at higher temperature and pressure*] Kältemittel, Arbeitsmittel *n*

refrigerant-10, R-10, carbon tetrachlorid(e), CCl_4 Tetrachlorkohlenstoff *m*, Tetrachlormethan, Kohlenstofftetrachlorid *n*

refrigerant-11, R-11, trichloro(mono)fluormethane, CCl_3F Kältemittel 11, R 11, Freon 11 *n*

R

refrigerant-12, R-12, dichlorodifluormethane, CCl_2F_2 Kältemittel 12, R 12, Freon 12, Difluordichlormethan *n*

refrigerant-13, R-13, chlorotrifluormethane, CC_1F_3 Kältemittel 13, R 13, Freon 13 *n*

refrigerant-14, R-14, carbon tetrafluorid(e), CF4 Tetrafluorkohlenstoff *m*, Kohlenstofftetrafluorid, Tetrafluormethan *n*

refrigerant-21, R-21, dichloro(mono)fluoromethane, $CHCl_2F$ Dichlorofluormethan, Freon 21 *n*

refrigerant-22, R-22, Chlorodifluormethane, $CHClF_2$ Kältemittel 22, R 22, Freon 22, Difluormonochlormethan *n*

refrigerant-30, R-30, methylene chlorid(e), dichloromethane, CH_2Cl_2 Kältemittel 30, R 30, Methylenchlorid, Dichlormethan *n*

refrigerant-40, R-40, methylenchlorid(e), chloromethane, CH_3Cl Kältemittel 40, R 40, Freon 40 *n*

refrigerant-50, R-50 Kältemittel 50, R 50 *n*

refrigerant-113, R-113, Freon 113, trichlorotrifluoroethane, $C_2F_3Cl_3$ Kältemittel 113, R 113, Freon 113 *n*

refrigerant-114, R-114, dichlorotetrafluoromethane, $C_2F_4Cl_2$ Kältemittel 114, R 114, Freon 114 *n*

refrigerant-170, R-170, ethane, CH_3CH_3 Kältemittel 170, R 170, Ethan; Äthan *n [frühere Benennung]*

refrigerant-290, R-290, propane, $CH_3CH_2CH_3$ Propan *n*

refrigerant-600, R-600, *n*-butane, $CH_3CH_2CH_2CH_3$ Butan, Normalbutan, *n*-Butan *n*

refrigerant-601, R-601, isobutane, $CH(CH_3)_3$ Isobutan *n*

refrigerant-1150, R-1150, ethylene, CH_2=CH_2 Eth(yl)en; Äth(yl)en *n [frühere Benennung]*

refrigerant-1270, R-1270, prop(yl)ene, CH_2 Prop(yl)en *n*

R

refrigerant absorber Kältemittelansorber *m* [*Siehe Anmerkung unter „Absorber"*]

refrigerant/air heat exchanger, refrigerant/air heat exchange unit, refrigerant/air heat interchanger Kältemittel/Luft-Wärme(aus)tauscher *m* [*Siehe Anmerkung unter „Wärme(aus)tauscher"*]

refrigerant blend, refrigerant mix(ture) Kältemittelgemisch *n*, Kältemittelmischung *f*

refrigerant bottle → refrigerant cylinder

refrigerant charge Kältemittelladung *f*

refrigerant (charging) valve [*Valve, located on the liquid line, through which refrigerant is charged into the system*] Kältemittel(lade)ventil *n*

refrigerant chemistry Kältemittelchemie *f*

refrigerant circulation pump Kältemittelzirkulationspumpe, Kältemittelumwälzpumpe, Kältemittelumlaufpumpe *f*

refrigerant coil Kältemittelschlange *f*

refrigerant collector Kältemittelsammler *m*

refrigerant column Kältemittelsäule *f*

(refrigerant) compressor [*Component of a refrigerating system which increases the pressure of a compressible refrigerant fluid and simultaneously reduces its volume, while moving the fluid through the device*] (Kältemittel)Verdichter, (Kältemittel)Kompressor *m*

refrigerant condensate Kältemittelkondensat *n*

refrigerant condenser → condenser

refrigerant control valve → refrigerant expansion valve

refrigerant-cooled kältemittelgekühlt

refrigerant cycle [*A process in which refrigerant repeatedly changes state and transfers heat from one place to another. Heat is absorbed from the surrounding by evaporation of the refrigerant and heat is released during liquefaction. The cycle may be used to operate a refrigerator or a heat pump*] Kältemittelkreislauf *m*

refrigerant cylinder, refrigerant bottle [*Cylindrical steel pressure vessel with operable closure for storage and transport of refrigerant*] Kältemittelflasche *f*

refrigerant density Kältemitteldichte *f*

refrigerant discharge, RD, refrigerant emptying Kältemittelentleerung *f*

refrigerant discharge gas → discharge (refrigerant) gas

refrigerant displacement compressor → displacement (refrigerant) compressor

refrigerant disposal Kältemittelentsorgung *f*

refrigerant distribution Kältemittelverteilung *f*

refrigerant distributor Kältemittelverteiler *m*

refrigerant drier, refrigerant dryer Kältemitteltrockner *m*

refrigerant drying Kältemitteltrocknung *f*

refrigerant expansion valve, refrigerant control valve [*Valve that meters refrigerant and maintains pressure-difference between the high-pressure side and the low-pressure side of a mechanical refrigerating system while the unit is operating*] Kältemittelexpansionsventil, Kältemittelentspannungsventil *n*

refrigerant fill(ing) valve Kältemittelfüll(ungs)ventil *n*

refrigerant filter dryer, refrigerant filter drier Kältemittelfiltertrockner *m*

refrigerant flow(ing) Kältemittelfließen *n*, Kältemittelfluss *m*

refrigerant gas Kältemittelgas *n*

refrigerant heat exchanger, refrigerant heat exchange unit, refrigerant heat interchanger Kältemittelaustauscher (im engeren Sinne), Kältemittelumformer *m*

refrigerant heat Kältemittelwärme *f*

refrigerant heating Heizung mit Kältemittel, Kältemittelheizung *f* [*Vorwiegend bei luftgekühlten Kälteanlagen werden im Kältekreislauf vor Eintritt in den luftgekühlten Verflüssiger ein (oder auch mehrere) wassergekühlte Verflüssiger eingebaut, die direkt Brauchwasser und Heiz(ungs)wasser erwärmen. Auch direkte Luft(er)wärmung, analog zur direkten Luftkühlung, wird ausgeführt, vorwiegend für Verkaufsräume in Supermärkten*]

refrigerant heat transfer Kältemittelwärmeübergang *m*, Kältemittelwärmeübertragung *f*

refrigerant hose Kältemittelschlauch *m*

refrigerant hydrolysis [*Splitting of refrigerant compounds by reaction with water; e.g., reaction of dichlorodifluormethane or methylchlorid(e) with water, forming acid products*] Kältemittelhydrolyse *f*

refrigerant injection Kältemitteleinspritzung *f*

refrigerant line Kältemittelleitung *f*

refrigerant liquid, refrigerating liquid, liquid refrigerant Kälteflüssigkeit *f*, Flüssigkältemittel *n*

refrigerant liquid pressure reducer, refrigerating liquid pressure reducer, liquid refrigerant pressure reducer [*Device in a refrigerating system, in which the pressure of the liquid is reduced from that of condensed liquid to that of the evaporator*] Flüssigkältemitteldruckminderer, Kälteflüssigkeit(s)druckminderer *m*

refrigerant liquid vapor, (liquid) refrigerant vapor (US); refrigerant liquid vapour, (liquid) refrigerant vapour Kältemitteldampf *m*

refrigerant (mass-)flow rate [*Of a refrigerant pumped or moved through a system at specified temperature and pressure*] Kältemittelfließgeschwindigkeit *f*

refrigerant meter(ing device) [*It controls the flow of liquid refrigerant to an evaporator*] Kältemittelmesser *m*

refrigerant mix(ture) → **refrigerant blend**

refrigerant-oil separator Kältemittel-Öl-(Ab)Scheider *m*

refrigerant operating pressure Kältemittelbetriebsdruck *m*

refrigerant (pipe)line, refrigeration (pipe)line, refrigerant piping, refrigeration piping Kältemittel(rohr)leitung *f*

refrigerant power, refrigerant performance, refrigerant capacity Kältemittelleistung, Kältemittelkapazität *f*

refrigerant pressure-imposing device [*Device or portion of equipment used for the purpose of increasing refrigerant pressure*] Kältemitteldruckerhöhungsapparat *m*

refrigerant processing Kältemittelaufbereitung *f*

refrigerant pump Kältemittelpumpe *f*

refrigerant pump-out system [*Dedicated apparatus for transfer of refrigerant from chiller or other refrigerating system to a separate and distinct storage vessel*] Kältemittelübertragungssystem *m*

(refrigerant) receiver [*Vessel for storing the refrigerant liquefied by the condenser*] Kältemittelsammler *m*

refrigerant recovery [*Removing a refrigerant in any condition from a system and storing it in an external container*] Kältemittelwiedergewinnung, Kühlmittelrückgewinnung *f*

refrigerant recycling [*Reducing contaminants in used refrigerants by separating oil, removing noncondensables, and using devices such as filterdriers to reduce acidity moisture and particulate water*] Kältemittelkreislauf *m*

refrigerant screw compressor → **screw (refrigerant) compressor**

refrigerant side Kältemittelseite *f*

refrigerant sight glass, refrigerant view(ing) glass, refrigerant observation glass Kältemittelschauglas, Kältemittelbeobachtungsglas *n*

refrigerant solution Kältemittellösung *f* [*bei Absorptionsanlagen*]

refrigerant stream Kältemittelstrom *m*

refrigerant subcooling Kältemittelunterkühlung *f*

refrigerant subcooler Kältemittelunterkühler *m*

refrigerant suction line Kältemittelsaugleitung *f*

(refrigerant) suction vapor (US); (refrigerant) suction vapour (Brit.) (Kältemittel)Saugdampf *m*

refrigerant sump Kältemittelsumpf *m*

refrigerant supply Kältemittelzufuhr, Kältemittelversorgung, Kältemittelzuführung *f*

refrigerant system Kältemittelsystem *n*

refrigerant system chemistry Kältemittelsystemchemie *f*

refrigerant user Kältemittelanwender, Kältemittelbenutzer *m*

refrigerant valve → **refrigerant (charging) valve**

R

refrigerant vapour; refrigerant vapor (US) Kältemitteldampf *m* [*Er wird in Kolben-, Turbo- oder Rotationsverdichtern verdichtet. Das Kältemittel ist kein Wasser*]

refrigerant vapour jet pump (Brit.); refrigerant vapor jet pump (US) Dampfstrahlpumpe, Kältemitteldampfstrahlpumpe *f*

refrigerant vapour jet pump plant, refrigerant vapour jet pump system, refrigerant vapour jet pump installation (Brit.) [*US spelling is "vapor"*] Dampfstrahlkälteanlage *f* [*Das Kältemittel wird mit Hilfe einer Dampfstrahlpumpe, die Kältemitteldampf aus dem Verdampfer absaugt, auf den Kondensatordruck verdichtet*]

refrigerant volume flow Kältemittelvolumenstrom *m*

refrigerant weight [*Weight of refrigerant required for proper functioning of a closed system*] Kältemittelmasse *f*; Kältemittelgewicht *n* [*frühere Benennung*]

refrigerated... → chilled...

refrigerating → refrigeration

refrigerating absorption system, refrigeration absorption system Kälteabsorptionsanlage, Absorptionskälteanlage *f*

refrigerating air, refrigeration air Kälteluft *f*

refrigerating apparatus, refrigeration apparatus Kälteapparat *m*

refrigerating capacity, refrigerating duty, refrigeration capacity, refrigeration duty [*Rate at which a system can remove heat, usually stated in tons or British thermal units (Btu) per hour*] Kälte *f*

refrigerating compressor, refrigeration compressor Kältekompressor, Kälteverdichter *m*

refrigerating engineering, refrigeration engineering [*Design, manufacture, application, repair and operation of refrigerating machinery and its primary equipment*] Kältemaschinentechnik *f*

R

refrigerating equipment, refrigeration outfit, refrigeration equipment, refrigerating outfit [*It includes any or all of the following: compressor, condenser, generator, absorber, (liquid) receiver, connecting pipe, evaporator, or complete unit system*] Kälteausrüstung, Kälteausstattung, Kältebestückung *f*

refrigerating fluid, refrigeration fluid [*Fluid used to transfer heat between cold refrigerant and the substance or bodies to be cooled, by circulation of the fluid without change of state, or by evaporation-condensation process at essentially equal pressures*] Kältefluid, Kälteströmungsmedium *n*

refrigerating liquid → refrigerant liquid

refrigerating load, refrigeration load [*Amount of heat per unit time, imposed on a refrigerating system for the required rate of heat removed. The heat is "imposed" by whatever requires cooling. In other words, if a warm container of food is placed in a refrigerator, it imposes a "load" on the system; it has imposed heat on the system. The refrigerator will, over a certain period of time, remove the heat until the container of food reaches a certain preset temperature*] Kühl(ungs)last *f*

refrigerating medium → refrigerant

refrigerating oil, refrigeration oil [*Stable fluid which is compatible with system components, will form a friction-reducing film between rubbing surfaces, seal critical clearances, and has low-temperature properties suitable to the application*] Kältemaschinenöl *n*

refrigerating output, refrigerating performance, refrigeration output, refrigeration performance Kälteleistung *f*

refrigerating plant, refrigeration plant, refrigerating installation, refrigeration installation Kälteanlage *f*

refrigerating plant for cooling, refrigerating installation for cooling, refrigeration plant for cooling, refrigeration installation for cooling Kühl(ungs)-anlage *f*

refrigerating system → chiller

refrigerating (system) machinery, refrigerating (system) machines, refrigeration (system) machinery, refrigeration (system) machines [*These machines form a part of the refrigeration system including any or all of the following: compressor, condenser, generator, absorber (adsorber), liquid receiver, connecting piping, and evaporator*] Kältemaschinen *f*

refrigerating unit, refrigeration unit, refrigerator [*It comprises a compressor, evaporator, condenser, and expansion device, used for refrigerating and for extracting heat to cool an enclosure*] Kälteaggregat *n*, Kälteeinheit *f*

refrigeration, refrigerating [*The removal of heat from air, water, brine or any other substance which it is desired to cool, and its transference to a suitable cooling medium usually water, sometimes the atmosphere. In the usual engineering sense of the term, refrigeration involves reducing the temperature of the air or whatever it may be, below that of the cooling medium, and, as the Second Law of Thermodynamics states, this cannot be done without the expenditure of energy*] Kälteerzeugung *f*

refrigeration central station → central refrigeration station

refrigeration equipment, refrigerating equipment Kältegerät(e) *n*, (*f*)

refrigeration (pipe)line → refrigerant (pipe)line

refrigeration ton → ton of refrigeration

refrigerator → cold chamber

refrigerator [*An apparatus for making (block) ice*] (Block)Eisapparat *m*

refrigerator [*A plant for making (block) ice*] (Block)-Eisanlage *f*

refrigerator → cold store

refrigerator [*Cabinet for keeping food or other items cool*] Kühlschrank *m*

refrigerator → refrigerating unit

refrigerator compression Kühlschrankkompression *f*

refrigerator-freezer (unit) → (combination) refrigerator-freezer (unit)

refusal list → (access) refusal list

refuse [*Term that includes garbage, rubbish and trade waste*] Müll *m*

refuse chute Müllrutsche *f*

refuse coal, waste coal Abfallkohle *f*, Kohle(n)abfall *m*

refused connection abgelehnte Verbindung, verweigerte Verbindung *f*, abgelehnte Anschluss, verweigerte Anschluss *m*

refuse incinerator → non-fuel incinerator

regedit Registrier(ungs)editor *m*

regenerative air preheater Regenerativluftvorwärmer *m*

regenerative braking Nutzbremsen *n*, Nutzbremsung *f*

regenerative cooling [*Process of utilizing heat which must be rejected or absorbed in one part of the cycle to perform a useful function in another part of the cycle*] Regenerativkühlung *f*

regenerative cycle gas turbine [*A simple-cycle gas turbine to which has been added a heat exchanger to warm incoming combustion air with heat from exhaust gases to increase thermal efficiency*] geschlossene Gasturbine, Heißluftturbine *f*

regenerative heat exchanger [*It is constructed from a porous material through which the warm and the cold fluids are ducted alternately. See remark under "heat exchanger"*] Regenerativwärmeübertrager *m*

regenerative heating [*Heat rejected in one part of a system used to perform a useful function in another part*] Regenerativheizung, Kreislaufverbundheizung *f*

regenerative heat recovery device Regenerativwärmerückgewinner *m*

regenerator [*An apparatus for effecting the transference of heat from one gas to another, the heat being absorbed from the hotter gas by a mass of refractory or metallic filling during one period and given up by the filling to the colder gas during the subsequent period. It is incorrect to refer to a recuperator as a regenerator, though this is sometimes done in respect of recuperators used with retort setting*] Regenerator, Wärme(aus)tauscher mit Speicherung, Speicherungswärme(aus)tauscher *m*

register, registering mechanism, counting mechanism [*electricity meter*] Zählwerk *n* [*Elektrizitätszähler*]

register [*It is commonly considered as an outlet which discharges the air in a confined jet*] Strahl(en)auslass *m*

register Register *n* [*Sehr schneller Speicher mit relativ geringer Kapazität, der anders als durch Maschinenadressen aufrufbar ist*]

register → (air) register

register Register *n* [*Ein vorübergehend kleine Einheiten digitaler Informationen speicherndes Gerät. Diese werden mit kurzer Zugriff(s)zeit abgegeben. Auf die Bits wird meist parallel zugegriffen*]

register [*A damper to control a chimney draught*] Zugklappe *f*

register alias table, RAT Registerumbenennung *f* [*(Logische) Register-Referenzen, die durch das auszuführende Programm definiert wurden, werden aufgelöst und in physikalische umgewandelt*]

register block → register module

register changeover device, registering mechanism changeover device, counting changeover device Zählwerkumschaltvorrichtung *f*

register constant, register mechanism constant, counting mechanism constant Zählwerkkonstante *f*

(registered) trade-mark (eingetragene) Warenzeichen *n*

registering mechanism → register

registering mechanism changeover device → register changeover device

registering mechanism constant → register constant

registering mechanism module → register module

register module, register block, registering mechanism module, registering mechanism block, counting module, counting block Zählwerkbaustein, Zählwerkblock *m*

register outlet Registerauslass *m*

register-register instruction Register-Register-Befehl *m* [*Ein Befehl, bei dem sich der Operand in einem Register befindet und das Ergebnis wieder in einem Register abgelegt wird.*]

register renaming Zuordnung *f* eines physikalischen Registers

registry Registrierdatenbank *f*

R

regular reflection → specular reflection

regulated, controlled reguliert

regulated central air → central regulated air

regulated variable → ((closed) loop) controlled variable

regulating → ((closed) loop) controlling

regulating circuit, regulating arrangement, tap(ping) circuit, tap(ping) arrangement Regelkreis *m* [*Trafo; Stellwick(e)lung*]

regulating flap valve Regulierdrehklappe *f* [*Luftzug*]

regulating power → control power

regulating signal Regelsignal *n* [*Physikalische Träger, z. B. elektrischer Strom oder Druckluft, der durch Regelstrecke und Regler hindurchlaufenden Information*]

regulation → ((closed) loop) control

regulator Regler *m*

rehab(ilitation) Sanierung *f*

reheater, reheating appliance, reheating apparatus, reheating device Nach(er)wärmer, Wieder(er)wärmer, Nacherhitzer, Wiedererhitzer, Nachheizer, Wiederheizer *m*

reheating Nach(er)wärmen, Wieder(er)wärmen, Nachheizen, Wiederheizen *n*

reheat(ing) coil Wieder(er)wärmungsschlange, Nach(er)wärmungsschlange, Nachheiz(ungs)schlange *f*

Reid vapour pressure, vapour pressure Reid (Brit.); Reid vapor pressure, vapor pressure Reid (US) Reid-Dampfdruck, Dampfdruck nach Reid *m* [*DIN 51754*]

re-ignition Wiederzündung *f*

reinforced concrete upper floor slab Stahlbetondeckenplatte, Deckenstahlbetonplatte *f*

reinforcement → gain

reinforcement training Weiterausbildung, Weiterschulung *f*

reinforcing fibre, reinforcement fibre; reinforcing fiber, reinforcement fiber (US) Bewehrungsfaser, Armierungsfaser *f*

re-insulation Nachdämmung *f*

rejection Ablehnung, Zurückweisung *f*

related ramp response bezogene Anstieg(s)antwort *f*

related to plunger travel stößelbezogen

relationship, ratio Verhältnis *n*

relative address relative Adresse, bezogene Adresse *f*

relative addressing relative Adressierung, bezogene Adressierung *f* [*Verbesserte Form der Seitenadressierung. Es wird eine Abweichung von der vorherigen Adresse von ± $^1/_2$ Seite angegeben. Dadurch wird der erhöhte Aufwand bei den Seitenübergängen vermieden, der jedoch bei Adressierungen außerhalb des relativen Seitenbereichs wieder voll auftritt.*]

relative admissible humidity → admissible relative humidity

relative (air) humidity [*1. Amount of moisture in the air stated in terms of percentage of total saturation at the existing dry-bulb temperature. 2. Ratio of the weight of water vapour usually present in a unit volume of air to the weight that would be present if the air were saturated with vapour at its actual temperature*] relative Luftfeuchtigkeit, relative Luftfeuchte *f*

relative auxiliary switch Relativhilfsschalter, RHi, Ausgelöstmelder *m* [*Er schaltet nur, wenn eine Auslösung des Schalters durch Überstromabschaltung, Kurzschlussabschaltung oder Spannungsauslöser erfolgt. Es tritt kein Öffnungswischer auf.*]

relative density [*The ratio of the weight of unit volume of dry gas to that of unit volume of dry air under the same conditions of temperature and pressure*] (Brenngas)Dichteverhältnis *n*

relative dielectric constant → dielectric constant

relative efficiency relative Wirkungsgrad *m*

relative glare Relativblendung *f*

100% relative humidity, dew point, saturation point Taupunkt *m*

relative ozone depletion potential, RODP Ozonabbaupotenzial *n* [*Der kennzeichnende Wert für die Gefährdung der Ozonschicht*]

relative permittivity → dielectric constant

relative thawing-off period relative Abtaudauer *f* [*Verhältnis der Zeitspannen, in der die Abtauvorrichtungen in Betrieb sind, zu der Zeitspanne zwischen zwei Abtauvorgängen*]

relay Zweipunktglied *n* [*Systemtechnik*]

relay Relais *n*

relay amplifier Relaisverstärker *m*

relay board Relaistafel *f*

relay ((closed) loop) control Relaisreg(e)lung *f*

relay ((closed) loop) controller Relaisregler *m*

relay coil Relaisspule *f*

relay computer Relaisrechner *m* [*Er wurde vom Bauingenieur Konrad Zuse von 1935 bis 1941 entwickelt*]

relay connector board Relaiskoppeltafel *f*

relay contact Relaiskontakt *m*

relay driver Relaistreiber *m*

relay drop(ping) time Relaisabfallzeit *f*

relay element Relaiselement *n*

R

relay flash(ing) unit Relaisblinkeinheit *f*

relay for a.c.-d.c. operation, a.c.-d.c. relay, universal relay Allstromrelais, Wechselstrom-Gleichstrom-Relais, Universalrelais *n*

relay hum → relay noise

relay interface Relaisschnittstelle *f*

relay noise, relay hum Relaisgeräusch *n*

relay ((open) loop) control Relaissteuerung *f*

relay ((open) loop) controller Relaissteuer(ungs)-gerät *n*

relay tester, relay testing device Relaisprüfer *m*, Relaisprüfungsgerät *n*

to release → to trip

to release, to start, to actuate [*To set a piece of equipment into operation; to put into motion*] starten

release, releasing, starting, actuation, actuating, putting into motion [*Setting a piece of equipment into operation*] Starten *n*

release → triggering

release key → tripping key

release of radiant energy → emission

release on request, ROR Freigabe *f* mit Anforderung [*Mikroprozessorsystembus*]

release trap Offenfalle *f*

releasing → release

releasing, tripping Auslösen, Entsperren, Freigeben *n*

relief opening Entlastungsöffnung *f*

relief pipe → blowoff pipe

relief (pipe)line → blowoff (pipe)line

relief piston Entlastungskolben *m*

relief valve [*Valve of the spring-loaded type, without disk guides on its pressure side, used on a hot-water heating boiler, a hot-water supply boiler, or a water storage tank to relieve excess pressure*] Druckentlastungsventil, Entlastungsventil, Überdruckventil, Entspannungsventil, Druckentspannungsventil *n*

relining Einziehen *n* von Rohren in defekte Rohrleitungen

to relinquish, to give up aufgeben, freigeben

reliquefaction Wiederverflüssigung *f*

to reliquefy wiederverflüssigen

relocatability Verschieblichkeit, Verschiebbarkeit *f*

relocatable verschiebbar [*Nicht an eine absolute Adresse gebunden*]

to relocate → to roll out

relocating loader verschiebbare Programmlader, Lader für verschiebbare Programme *m* [*Von diesem Lader kann ein verschiebbares Objektprogramm an einen beliebigen Platz im Programmspeicher für die Ausführung gebracht werden*]

relubrication Nachschmierung *f*

remaining bleibend, verbleibend, zurückbleibend

remaining desired value deviation bleibende Sollwertabweichung *f* [*DIN 19 226, Teil 5*]

remaining error Restfehler, bleibende Fehler *m*

remaining error signal Restregeldifferenz, bleibende Regeldifferenz *f* [*DIN 19 226, Teil 5*]

remanence relay, retentive(-type) relay Remanenzrelais, Restmagnetismusrelais *n*

remedy Abhilfe *f*

reminder (of debt due) → prompt note

remote [*Pertaining to devices or nodes on a different network than the referenced device or node*] fern

remote access, teleaccess Fernzugriff *m* [*auf eine bestimmte Speicherstelle und einen bestimmten Speicherbereich*]

remote access control(ling), remote access check(ing), RAC, remote access management [*"tele..." may synonymously be used for "remote"*] Fernzutrittskontrolle, Ferneintrittskontrolle; Fernzugangskontrolle *f* [*frühere Benennung*]

remote adjustment, teleadjustment Fernverstellung, Fernjustierung, Fern(ein)stellung *f*

remote alarm, telealarm Fernalarm *m*

remote alarming, remote alarm release, remote actuation, remote generation Fernalarmierung, Fernalarmgabe, Fernalarmauslösung *f*, Fernalarmieren *n*

remote alarm transmission → alarm remote transmission

remote banking, telebanking, homebanking Fernkontenführung, Kontenfernführung *f*

remote batch computing [*The running of program(me)s, usually during nonprime hours, or whenever the demands of real-time or time-sharing computing slacken sufficiently to allow less pressing program(me)s to be run*] Fernstapelrechnen, Stapelfernrechnen *n*

remote batch processing [*Batch processing in which an input device is located at a distance from the main installation and has access to a computer through a communication link*] Fernstapelverarbeitung, Stapelfernverarbeitung *f*, Fernstapelbetrieb, Stapelfernbetrieb *m*

remote bridge entfernte Brücke, entlegene Brücke, abgelegene Brücke *f*

R

remote BSC ortsunterschiedliche Positionierung *f* von Basissteuer(ungs)einheit und Hauptstation [*Festnetzverwaltung*]

remote computing system, remote computer system, remote data processing system, remote (information) processing system, data teleprocessing system [*A data processing system that has terminals distant from the central processing unit, from which users can communicate with the central processing unit and compile, debug, test, and execute program(me)s*] Fernrechnersystem, Rechnerfernsystem *n*

remote computing system language, remote computer system language [*It is used for communications between the central processing unit and remote consoles for a remote computer system, generally incorporating a procedure-orient(at)ed language such as FORTRAN, but also containing operating statements, such as instructions to debug or execute program(me)s*] Fernprogrammiersprache, Fernrechnersystemsprache *f*

remote condition → remote state

remote console [*A terminal in a remote computing system that has facilities for communicating with, and exerting control over, the central processing unit, and which may have devices for reading and punching cards or paper tape, any of various types of display units, printers, and a keyboard device for direct communication with the control processing unit*] Fernrechnersystemkonsole, Rechnerfernsystemkonsole *f*

remote control [*Control of a quantity which is separated by an appreciable distance from the controlling quantity; examples include master-slave manipulators, telemetering, telephone, and television*] Fernsteuerung, Fernführung *f* [*DIN 41 745*]

remote controller Fernregler *m*

remote control switch, RC switch [*A magnetically operated mechanically held switch, normally used for remote switching of power blocks*] Fernschalter *m*

remote cryogenic supply Fernkryoversorgung, Fernkälteversorgung *f*

remote data processing → data teleprocessing

remote data station, remote data terminal [*A terminal in a data-processing system at which data can be sent to or received from a central computer over telephone or telegraph circuits, but which exerts no direct operating control over the central computer*] Datenfernstation, Ferndatenstation *f*

remote data transmission → data remote transmission

remote debugging → remote testing

remote debugging method → remote testing method

remote de-energization Fernabschaltung *f*

remote diagnosis Ferndiagnose *f*

remote display → remote readout

remote drive Fernantrieb *m*

remote enable Fernsteuer(ungs)freigabe *f* [*PMG, DIN IEC 625*]

remote enable signal, REN Fernsteuer(ungs)signal *n*

remote error sensing Fehlerfernerfassung, Fernfehlererfassung *f*

remote-fed ferngespeist

remote generation → remote alarming

remote gyrostat, remote humidistat, remote humidity control [*See remark under "humidity control"*] Fernfeuchtigkeit(s)sensor, Fernfeuchtesensor *m*

remote hand operation → remote manual operation

remote heating Fernheizung *f*

remote heat supply Fernwärmeversorgung *f*

remote ignition Fernzündung *f*

remote indication → remote readout

remote indicator, remote indicating instrument, remote indicating device [*An indicator located at a distance from the data-gathering sensing element, with data being transmitted to the indicator mechanically, electrically or by wires, or by means of light, radio, or sound wave*] Fernanzeiger *m*, Fernanzeigeinstrument *n*, Fernanzeigevorrichtung *f*

remote (information) processing system → remote computing system

remote infrared ((open) loop) control → infrared remote ((open) loop) control

remote infrared operation → infrared remote operation

remote input/output unit, rio unit ferngesteuerte Ein/Ausgabe-Einheit *f*

remote inquiry [*Interrogation of the content of an automatic data-processing equipment storage unit from a device remotely displaced from the storage unit site*] Fernabfrage *f*

remote jack 11 → modular plug

remote job entry, RJE Stapelfernverarbeitung *f*

remote line Fernleitung *f*

remote line concentrator Wählfernschalter *m* [*Fernmeldetechnik*]

remote line duct Fernleitungskanal *m*

remote/local ... Fern/Eigen ...

remote/local function [*IEC 625*] Fern/Eigen-Umschaltfunktion *f* [*DIN IEC 625*]

remotely controlled switch, remotely actuated switch Fernschalter, fernbetätigte Schalter, fernbediente Schalter *m*

remotely interchangeable fernbedient auswechselbar, fernbedient austauschbar, fernauswechselbar, fernaustauschbar

remotely non-switchable nicht fernschaltbar

remotely regulated station ferngeregelte Fernmeldeamt *n*

remotely resettable fernrückstellbar

remotely switchable fernschaltbar

remote management, telemanagement Fernleitung, Fernführung *f*

remote manual operation, remote hand operation Handfernbedienung, Fernhandbedienung *f*

remote measuring device, remote measurement device Fernmessgerät *n*, Fernmesser *m*

remote message Nachricht *f* an externer Stelle

remote metering, telemetering, telemetry, remote measuring [*Transmitting the readings of instruments to a remote location by means of wires, radio waves, or other means*] Fernmesstechnik, Fernmessungstechnik *f*

remote monitor, telemonitor Netzwerküberwachungsprogramm *n*

remote motor ((closed) loop) control Motorfernreg(e)lung, Fernmotorreg(e)lung *f*

remote motor ((open) loop) control Motorfernsteuerung, Fernmotorsteuerung *f*

remote moving contact → moving telecontact

remote multi(ple-)media detectable access Fernzugriff *m* auf Datenbanken mit multimedialen Informationsbestandteilen

remote multiplexer (terminal) Fernmultiplexer *m*

remote network, telenetwork Fernnetz(werk) *n*

remote nodal station, remote station nodal, RSN abgesetzte Knotenfunkstation, Fernknotenfunkstation *f*

remote ((open) loop) control Fernsteuerung *f*

remote operation, teleoperation Fernbedienung, Fernbetätigung *f*

remote operation transmitter, teleoperation transmitter Fernbedienungssender, Fernbetätigungssender *m*

remote operator('s) panel Fernbedien(ungs)feld *n*

remote pick(ing) up [*Pick(ing) up a radio or television program(me) at a remote location and transmitting it to the studio or transmitter over wire lines or a shortwave radio link*] Fernaufnehmen *n*

remote point entlegene Stelle *f*

remote position indicator Fernstellungsanzeiger, Stellungsfernanzeiger *m*

remote primary detector → remote sensor

remote procedure error, RPE, teleprocedure error ferne Ablauffehler, Ablauffehler der Gegenstation *m* [*Netzmeldung*]

remote processing → teleprocessing

remote processing system → remote computing system

remote reading gauge; remote reading gage (US); remote meter reader Fern(ab)leser *m*

remote readout, remote display, remote indication [*See remark under "readout"*] Fernanzeige *f*

remote recording thermometer → remote thermograph

remote refrigeration system, remote refrigerating system [*1. It has the condensing unit outside and the evaporator inside the area to be cooled. 2. The compressor or generator is located in a space other than the cabinet or fixture containing the evaporator*] Fern-Kälteumwälz(ungs)system *n*

remote reset(ting) Fernrücksetzen (eines Ausgangs), Fernrückstellen (eines Ausgangs) *n*

remote selection, teleselection Fernanwahl *f*

remote selector, teleselector Fernanwähler *m*

remote sensing [*Sensing, by a power supply, of voltage directly at the load, so that variations in the load lead drop do not affect load regulation*] (Spannungs)-Istwert-Fernerfassen *n*

remote sensor, remote sensing element, remote primary detector, telesensor [*See remark under "sensor"*] Fernsensor, Fernfühler *m* [*Siehe Anmerkung unter "Sensor"*]

remote service, teleservice Fernwartung *f*

remote setting, telesetting Ferneinstellung *f*

remote shopping, teleshopping Ferneinkauf *m*

remote signalling; remote signaling (US) Fernsignalisierung *f*

remote state, REMS, remote condition Fernsteuer(ungs)zustand *m* [*Programmierbare Messgerät; DIN IEC 625*]

remote state indication panel Fernzustandsanzeigetafel *f*

remote station nodal → remote nodal station

remote subscriber unit, RSU abgesetzte Teilnehmereinheit, Fernteilnehmereinheit *f*

remote supervisory control → supervisory remote control

to remote-switch, to teleswitch fernschalten

R

remote switch → **(magnetic) remote switch**

remote switching, teleswitching Fernschaltung *f*, Fernschalten *n*

remote switch(ing) unit, teleswitch(ing) unit Fernschalt(ungs)einheit *f*

remote system → **stand-alone PV system**

remote tap indicator Stellungsfernanzeiger, Fernstellungsanzeiger *m* [*Trafo*]

remote temperature measuring installation, remote temperature measuring plant Ferntemperaturmessanlage, Temperaturfernmessanlage *f*

remote testing, remote debugging [*The testing and correction of computer program(me)s at a remote console of a remote computing system*] Fernentstörung *f*, Fernentstören *n*

remote testing method, remote debugging method [*A method of testing and correcting computer program(me)s; programmers do not go to the computer centre, but provide detailed instructions to be carried out by computer operators along with the program(me) and associated test data*] Fernentstörungsverfahren *n*, Fernentstörungsmethode *f*

remote thermograph, remote recording thermometer Fernthermograph, schreibende Ferntemperaturmesser *m*, schreibende Fernthermometer *n*

remote thermometer Fernthermometer *n*

remote transmission Fernübertragung *f*

remote transmission request, RTR Fernübertragungsaufforderung *f*

remote video transmission → **video remote transmission**

remote water level indicator Fernwasserstandsanzeiger *m*

removal Beseitigung, Entfernung *f*

removal of (boiler) scale → **descaling**

removal of power (supply), power (supply) removal Stromabschaltung *f*

removal of sewage (from buildings) → **sanitary drainage**

removable, detachable abnehmbar [*Geräteteil*]

to remove ash(es) → **to deash**

to remove boiler scale → **to descale**

removed section (Profil)Schnitt *m* außerhalb der Ansicht

to remove gas → **to degas**

to remove smoke, to desmoke entrauchen

REMS → **remote state**

REN → **remote enable signal**

to rename umbenennen

renaming Umbenennen *n*

rendering, rendition Wiedergabe *f*

rendering photorealistische Darstellung *f* [*Die Gestaltung, Berechnung und Darstellung einer naturgetreuen Oberfläche eines Volumenmodells (Beleuchtungs- und Schattierungsmodell)*]

rendering factor, rendition factor Wiedergabefaktor *m* [*Beleuchtungstechnik*]

renewable erneuerungsfähig, erneuerbar

renown Ruf *m* [*z. B. internationale Ruf = international renown*]

rent(al) management Mietverwaltung *f*

rent collector Kassiergerät *n* [*Elektrizitätszähler*]

repaint schnelle Graphikneuaufbau *m* [*Auf dem Bildschirm ohne Bildausschnittveränderung*]

repair Instandsetzung, Reparatur *f*, Wiederherstellung *n*

repair service Reparaturdienst, Instandsetzungsdienst *m*

repair switch Reparaturschalter, Instandsetzungschalter *m*

repair trade Reparaturgewerbe, Instandsetzungsgewerbe *n*

repeatability [*1. Closeness of agreement among repeated measurements of the same variable under the same conditions. 2. Closeness of agreement among consecutive measurements of the output for the same value*] Wiederholbarkeit *f*

repeater, level 1 relay Regenerator *m* [*Eine Komponente im LAN-Netz, die digitale Signale regeneriert. Der Regenerator wird benutzt, wenn die Übertragungsstrecke die maximale Segmentlänge übersteigt*]

repeater (Zwischen)Verstärker *m*

repeater Rückmelder *m*

repeater → **(signal) repeater**

repeat(ing) Wiederholen *n* [*Aufzeichnung*]

repeat(ing) coil Übertrager *m* [*Fernsprecher*]

repeat(ing) counter Wiederholungszähler *m*

repeat(ing) cranking Wiederholungskurbelanlassen *n*

repeat(ing) loop Wiederholungsschleife *f*

repeat(ing) register → **duplicating register**

repeat(ing) signal message; repeat(ing) signalling message (Brit.); repeat(ing) signaling message (US) Stellungsrückmeldung *f*

repetition accuracy Wiederholungsgenauigkeit *f*

to replace, to substitute, to change [*One object for another*] (aus)wechseln, ersetzen, (aus)tauschen

R

replaceable → changeable

replaceable database driver, RDD, changeable database driver Austauschdatenbanktreiber *m*

replaceable part, changeable part Austauschstück *n*

replacement → makeup

replacement, change Austausch *m*, Auswechs(e)lung *f*

replacement (lock) striking plate, change (lock) striking plate, replacement (lock) strike(r), change (lock) strike(r), replacement (lock) strike plate, change (lock) strike plate Austauschschlossblech *n*

replacement mounting [*As opposed to "first mounting"*] Austauschmontage, Auswechs(e)lungsmontage *f*

replacement refrigerant → alternate refrigerant

replacement sampling → sampling with replacement

replicated photopolymer [*dual-layer disk*] Photopolymerschicht *f*

report Bericht *m*

report [*data processing*] Liste *f*

report based on quality control, quality control report Werkzeugnis *n* [*Euronorm 21-62*]

report generator Listengenerator *m* [*Datenverarbeitung*]

report printer, report writer Berichtdrucker *m*

report program generator (US); report programme generator (Brit.); RPG (problemorientierte) Programmgenerator *m*

representation Darstellung *f*

representation Schreibweise *f*

representation unit Ziffernschritt *m* [*Kleinste Zu- oder Abnahme zwischen zwei Ausgangswerten. DIN 4472*], Auflösung *f*

to reproduce vervielfältigen

to reproduce → to scan

reproducibility Vervielfältigbarkeit *f*

reproducing → scanning

reproducing Vervielfältigen *n*

reproduction Vervielfältigung *f*

reprogrammable wiederprogrammierbar, umprogrammierbar

reprogrammable read-only memory, reprogrammable PROM, R-PROM wiederprogrammierbare Festwertspeicher, umprogrammierbare Festwertspeicher *m*

reprogramming Umprogrammieren, Wiederprogrammieren *n*

repulsion induction motor [*Motor with commutated bars in the rotor and a centrifugal mechanism that, when the motor approaches full running speed, short-circuits the commutator*] Repulsionsinduktionsmotor *m*

to repump umpumpen

request Anforderung *f*, Anforderungspaket *n* [*Token-Zugriff(s)verfahren*]

requested connection angeforderte Verbindung *f*, angeforderte Anschluss *m*

request for information, information request Anlegen der Adresse, Adressenanlegen *n* [*Zur Angabe von Daten eines Speichers*]

request for information, information request Aufruf *m*

request signal message Hinweismeldung *f* [*Eine Aufforderung zur Durchführung bestimmter Maßnahmen*]

required distance between buildings, required distance between structures brandschutztechnisch erforderliche Abstand, feuerschutztechnisch erforderliche Abstand *m*

required ventilation, required vent(ilat)ing Lüftung nach Bedarf, Bedarfslüftung, Be- und Entlüftung nach Bedarf, Bedarfsbe- und Entlüftung *f*

requirement → demand

requirement characteristic number → characteristic requirement number

requirement (closed-loop) controlled → demand (closed-loop) controlled

requirement control → demand control

requirement-guided, need-guided, demand-guided bedarfsausgeführt

requirement period, need period, demand period Bedarfsperiode *f*

requiring acknowled(e)gement quittier(ungs)pflichtig

rescue [*Not to be confused with "escape"*] Rettung *f*

research finding, research result Forschungsergebnis *n*

reservation station Belegungsstation *f*

reserve busbar Reservesammelschiene, Reservestromschiene *f*

reserve circuit Reserveleitung *f* [*Fernmeldetechnik*]

reserve conduit Leerrohr *n* [*Elektroanlage*]

reserve controller director Bereitschaftswächter *m* [*Im nicht unterbrechbaren Regelkreis*]

reserve(d) security keying Sperrschließung *f*, Sperrschlüssel *m*

reserve power Gangreserve *f* [*Uhr; Schreiber*]

R

reserve power Reserveleistung *f* [*Elektronetz*]

reserve symbol Ausweichzeichen *n* [*Schaltzeichen*]

reserve wire Reserveader *f* [*Fernmeldetechnik*]

to reset zurückgehen [*Relais*]

reset → residue

reset(-action) controller (Brit.); master-submaster controller (US) [*It is a combination of the integral controller and a proportional reset. This gives a form of controller with proportional and integral action but it is important to note that this does not constitute what is properly termed a P + I (proportional plus integral) controller. It is normally used as a supply air controller with proportional adjustment from space temperature. It is commonly used in large spaces such as auditoriums where it takes a long time for the system to change the temperature*] Rückstellregler *m*

reset-set flip-flop (circuit) → RS flip-flop (circuit)

resettability Rückstellbarkeit, Rücksetzbarkeit *f*

reset(ting) Rückstellen (eines Ausgangs), Rücksetzen *n* (eines Ausgangs) [*In der Regel auf Nullpegel*]

reset(ing) key Rückstelltaste, Rücksetztaste *f*

reset(ing) ratio Rückstellverhältnis, Rücksetzverhältnis *n*

reset(ing) time Rückstellzeit, Rücksetzzeit *f*

to reset to zero → to clear

residence telephone, house telephone Hausanschluss, Wohnungsanschluss, Privatanschluss *m*

residence time Verweilzeit *f*

residential building [*house; hospital; hotel; etc.*] Personengebäude *n*

residential lockset Wohnungsschlossgarnitur *f*

residential rate, household rate Haushalt(s)tarif *m*

R

residual-current circuit breaker, RCCB [*It monitors the current in a circuit and breaks the circuit if a leak to earth is detected*] Differnzstromschalter, DI-Schalter *m*

residual current device, r. c. d., (residual) current-operated device Fehlerstromschutzvorrichtung *f*

residual fuel oil, heavy oil, fuel oil No. 6 schwere Heizöl, Heizöl S *n*

residual heat Restwärme *f*

residual heat sensor Restwärmesensor, Restwärmefühler *m*

residual modulation → carrier noise

residual pressure [*In freeze drying, the pressure of the gas (usually air) at the trap*] Restdruck *m*

residual relay Nullstromrelais *n*

residue, rest Rest, Rückstand *m*

resilient feedback, elastic feedback elastische Rückführung, elastische Gegenkopp(e)lung, federnde Rückführung, federnde Gegenkopp(e)lung *f* [*Rückführung, bei welcher ein mit der Zeit verschwindender Anteil der Ausgangsgröße auf den Eingang zurückgeführt wird, wodurch die wirksame Verstärkung im Laufe der Zeit größer wird*]

resilient floor cover(ing), resilient flooring, resilient floor (finish) [*"Elastic" may synonymously be used for "resilient"*] elastische (Fuß)Boden(belag), federnde (Fuß)Boden(belag) *m*

resilient support, elastic support elastische Auflage, federnde Auflage, elastische Unterstützung, federnde Unterstützung *f*

resilient suspension, elastic suspension elastische Aufhängung, federnde Aufhängung *f*

resin-bonded harzgebunden

resinous electricity negative Reibungselektrizität *f*

resinous powder Farbpulver *n* [*Trockenschrift*]

resin (water) softener → (water) softening resin

resistance... → (electrical) resistance...

resistance characteristic, resistance feature Widerstandsmerkmal *n*

resistance grounding (US); resistance earthing (Brit.) [*Electrical grounding in which lines are connected to ground be a resistive (totally dissipative) impedance*] Widerstandserdung *f*

resistance lamp [*Electric lamp used to prevent the current in a circuit from exceeding a desired limit*] Widerstandslampe *f*

resistance loss [*Power loss due to the current flowing through resistance; its value in watts is to the resistance in ohms multiplied by the square of the current in amperes*] Widerstandsverlust *m*

resistance soldering Widerstandslöten *n*

resistance thermometer Widerstandsthermometer *n*

resistant to short circuit(s) → (mechanically) short-circuit-proof

resistive attenuator → absorptive attenuator

resistive load Widerstandslast, Ohmsche Last, Wirk(strom)last *f*

resistor, electrical resistor [*A device designed to have a definite amount of resistance; used in circuits to limit current flow or to provide a voltage drop*] (elektrische) Widerstandsgerät *n*

resistor bank Widerstandsbank, Widerstandsgruppe, Widerstandsbatterie *f*, Widerstandssatz *m*

resistor-capacitor-transistor-logic → RCTL

resistor core [*Insulating support in which a resistor element is wound or otherwise placed*] Widerstandskern *m*

resistor monitoring Widerstandsüberwachung *f*

resistor network [*An electrical network consisting entirely of resistors*] Widerstandsnetz(werk) *n*

resistor-transistor-logic → **RTL**

resolution [*The degree to which equal values of a quantity can be discriminated in a device; the smallest change in the input signal that will result in a measurable change in the output signal*] Feinheit *f*

resolution Auflösung *f*

resolution → **(pixel) resolution**

resolution enhancement, RE Beschleunigung *f* der Auflösung durch chemische Reaktion

resolution error [*An error of an analog computing unit that results from its inability to respond to changes of less than a given magnitude*] Auflösungsfehler *m*

resonance absorber Resonanzabsorber *m*

resonance frequency → **natural frequency**

resonance selectivity → **selectivity of resonance**

resonance sharpness → **sharpness of resonance**

resource Betriebsmittel *n*

resource allocation and multi(ple)-project scheduling → **RAMPS**

to respire → **to breathe**

respiration → **breathing**

respiratory heat, breathing heat Atmungswärme, Atemwärme *f*

respiratory heat exchange, breathing heat exchange Atmungswärmeaustausch, Atemwärmeaustausch *m*

respondent Ansprechpartner *m*

response → **reply**

response Frequenzbereich *m* [*selten: n*] [*Mikrophon*]

response Ansprechen *n* [*z. B. Melder*]

response Reaktion *f*

response amplitude, operating amplitude Ansprechamplitude *f*

response behaviour (Brit.); response behavior (US) Ansprechverhalten *n*

response characteristic [*The response as a function of an independent variable, such as direction or frequency, often presented in graphical form*] Ansprechcharakteristik *f*

response curve Ansprechkurve, Ansprechlinie *f* [*Näherungsschalter*]

response delay → **operate delay**

response element Ansprechelement *n*

response grade [*heat detector*] Ansprechklasse *f* [*Wärmemelder*]

response pressure Ansprechdruck *m*

response property Ansprecheigenschaft *f*

response rate Ansprechgeschwindigkeit *f*

response susceptibility, response sensitiveness, response sensitivity Ansprechempfindlichkeit, Ansprechanfälligkeit *f*

response telegram Antwort *f* der Inselzentrale auf einen Leitzentralenbefehl

response temperature Ansprechtemperatur *f*

response threshold, operating threshold Ansprechschwelle *f* [*Kleinste für die Auslösung eines Vorgangs oder Signals ausreichende Amplitude des Steuer(ungs)-signals*]

response time [*analog/digital converter*] Messzeit *f*

response time [*pointer*] Einschwingzeit *f*

response time [*In automatic data processing, the delay experienced in time sharing between request and answer; a delay which increases when the number of users or the system increases*] Antwortzeit, Beantwortungszeit, Rückmeldungszeit, Rückmeldezeit *f*

response time [*In a control system, the time required for the output of a control system or element to reach a specific fraction of its new value after application of a step input or disturbance*] Ansprechzeit *f*

response value Ansprechwert *m*

response voltage Ansprechspannung *f*

restart Wiederanlauf *m*, Repetition *f*

restaurant floor drive (system) [*tower restaurant*] Restaurant-Drehantrieb *m*

rest contact, normally closed contact Ruhekontakt *m* [*Schaltgerät*]

resting frequency → **carrier frequency**

restorable change rückführbare Änderung *f* [*DIN 40 042*]

to restore wiederherstellen, instandsetzen

restore → **(data) backup**

to restore wiederspeichern, wiedereinbringen, umspeichern [*Daten in ein System*]

to restore a load wiedereinschalten eines Verbrauchers

restored energy time Wiederaufladezeit *f* [*unterbrechungsfreie Stromversorgung*]

restored voltage wiederkehrende Spannung *f*

restorer diode Klemmdiode *f*

restoring element [*IEC 211*] Rückstellvorrichtung *f* [*EZ; VDE 0418,1*]

re-storing pulse Umspeicher(ungs)impuls *m*

restoring torque, equalizing torque, equalization torque Richt(dreh)moment, Ausgeichs(dreh)moment, Rückstell(ungs)(dreh)moment, rücktreibende (Dreh)Moment *n*

rest position → position of rest

rest pressure → (gas) rest pressure

restricted area Sperrgebiet *n*

restricting point, restriction point, restrictive point Drosselstelle *f* [*Durchflussmessung*]

restriction Drosselung *f* [*Durchflussmessung*]

restriction Gesprächsbeschränkung *f*

restrictive current → in-phase current

restrictor, restricting device, restricting appliance Drossel(ungs)gerät *n* [*Durchflussmessung*]

to restrict the flow → to throttle

retail security equipment Einzelhandelssicherungsgerät(e) *n*, (*f*)

retail shop Einzelhandelsgeschäft *n*, Einzelhandelsladen *m*

retainer → holder

retarder Verzögerer *m*

retention, retenting Zurückhaltung *f*, Rückhalt *m*

retention flame, retenting flame Sparflamme *f* [*Gaskocher*]

retentive(-type) relay → remanence relay

retinal image Gesichtsempfindung *f*

retina-pattern access control method Augenhintergrunderkennung, Netzhauterkennung *f* [*Methode zur Identifizierung einer Person durch Vergleich des Blutgefäßmusters auf der Netzhaut*]

to retire zurückschreiben [*Ein als gültig ausgewiesenes Ergebnis vom physikalischen in das logische Register zurückschreiben*]

to retire gültig setzen, übernehmen [*Befehl*]

retired übernommen, gültig gesetzt [*Befehl*]

retired [*See German remark under "to retire"*] zurückgeschrieben

retirement → in-order back end

retrace → flyback

retraining Nachsynchronisation(svorgang), Resynchronisation(svorgang) *f*,(*m*) [*Siehe Anmerkung unter „Synchronisation(svorgang)"*]

retransmission, retransfer Wiederübertragung *f* [*Erneute Übertragung eines Paketes, das bei der ersten Übertragung nicht angekommen ist, oder beim Empfang eines fehlerhaften Paketes oder wenn eine Kollision erkannt wurde*]

to retrofit nachausrüsten

return → (common) return

return Rücklauf *m*

return Rückleiter *m*

return address Rückkehradresse *f* [*Die Adresse, zu der nach Beendigung des Unterprogramms zurückgekehrt wird*]

return air [*Air extracted from a space, and totally or partially returned to an air conditioner, furnace or other heat source*] Rückluft *f*

return air cabinet, return air cubicle Rückluftschrank *m*

return air damper Rückluftklappe *f*, Rückluftschieber *m*

return air duct Rückluftkanal *m*

return (air) fan Rückluftventilator *m*

return (air) filter Rückluftfilter *m*, *n*

return air grill(e), return air grating, return air grate Rückluftgitter *n*, Rückluftrost *m*

return air heater [*A device that heats the return air for reuse in an air-conditioning system*] Rücklufterhitzer, Rückluft(er)wärmer, Rückluftheizer *m*

return air inlet Rücklufteinlass *m*

return air plenum (chamber) [*See remark under "plenum (chamber)"*] Rückluftüberdruckkammer *f*

return axial valve Rücklaufventil *n*

return bend → double bend

return chamber [*manifold*] Rücklaufkammer *f*

return collector Rücklaufsammler *m*

return connection Rücklaufanschluss *m*, Rücklaufverbindung *f*

return corner valve, corner return valve Rücklaufeckventil, Eckrücklaufventil *n*

return distribution Rücklaufverteilung *f*

return fan Rückluftventilator *m*

return flow Rückstrom *m*

return flow Rücklauf *m*

return flow blending Rücklaufbeimischung *f*

return flow collector Rücklaufsammler *m*

return flow compressor [*Compressor in which the suction valve is located in the cylinder head of the compressor*] Rücklaufverdichter, Rücklaufkompressor *m*

R

return flow connection Rücklaufanschluss *m*, Rücklaufverbindung *f*

return flow duct Rücklaufkanal *m*

return flow duration Rückflussdauer *f*

return flow(ing) Rückfluss *m*, Rückfließen, Rückströmen *n*

return flow limitation Rücklaufbegrenzung *f*

return flow main Rücklaufhauptleitung, Hauptrücklaufleitung *f*

return flow overtemperature Rücklaufübertemperatur *f*

return flow pipe Rücklaufrohr *n*

return flow (pipe)line, return flow piping Rücklauf(rohr)leitung *f*

return flow pressure Rücklaufdruck *m*

return flow preventer Rückflussverhinderer *m*

return flow pump Rücklaufpumpe *f*

return flow riser pipe, return flow rising pipe; return flow riser main, return flow rising main [*deprecated*] Rücklaufsteig(e)rohr, Rücklaufstandrohr *n*

return flow screwing Rücklaufverschraubung *f*

return flow sensor Rücklaufsensor *m*

return flow side [*heat exchanger*] Rücklaufseite *f*

return flow socket Rücklaufstutzen *m*

return flow straight(-way) valve, return flow straight-through valve Rücklaufdurchgangsventil *n*

return flow stream Rücklaufstrom *m*

return flow T(ee) Rücklauf-T(-Stück) *n*

return flow temperature Rücklauftemperatur *f*

return flow temperature limitation Rücklauftemperaturbegrenzung *f*

return flow temperature limiter, return flow temperature limiting device Rücklauftemperaturbegrenzer *m*

return flow temperature sensor Rücklauftemperatursensor *m*

return flow thermometer Rücklaufthermometer *n*

return flow thermostat Rücklaufthermostat *m*

return flow valve Rücklaufventil *n*

return flow water Rücklaufwasser *n*

return flow water (pipe)line Rücklaufwasser(rohr)leitung *f*

return flow water temperature Rücklaufwassertemperatur *f*

return (gas) flue, return smoke flue zurückkehrende (Feuer)Zug, zurückkehrende Rauch(gas)kanal, zurückkehrende Abzug, zurückkehrende Rauch(gas)abzug *m*

return key Zeilenschalter *m*

return line Rücklaufleitung *f*

return message [*See remark under "message"*] Rückmeldung *f*

return oil, spill oil Rücklauföl *n*

return safety (pipe)line → safety return (pipe)line

return spring Rückholfeder *f*

return temperature limiter, return temperature limiting device, return temperature limiting appliance Rücklauftemperaturbegrenzer *m*

return to non-data mode Datenbetriebsablösung *f*, Datenbetriebsablösen *n*

return to opertional readiness Wiedereinschaltbereitschaft *f*

return to output Wiederleistungsbereitschaft *f*

return trace → flyback

return voltage Rückspannung *f*

reuse Wiedergebrauch *m*, Wiederverwendung *f*

reverberant nachhallig, nachhallend

reverberation [*The subtle prolongation of sound in a room caused by continued multiple reflections after the sound source has stopped*] Nachhall *m*

reverberation room, diffuse-field room Nachhallraum *m* [ISO 3741]

reverberation time, RT [*The time taken for a sound to decay to 1 million of its original intensity (i.e. by 60 dB)*] Nachhallzeit T, Nachhalldauer T *f*

reversal of polarity → polarity inversion

reverse-acting thermostat [*It decreases the control pressure upon an increase in temperature*] umkehrbare Thermostat *m*

reverse breakdown voltage Rückwärtsdurchbruchspannung, Rückwärtsdurchschlagspannung *f* [*rückwärtssperrende Thyristor. DIN 41 786*]

reverse counter Rückwärtszähler *m*

reverse current Rückstrom *m*

reverse current cleaning → anodic cleaning

reverse current relay Rückstromrelais *n*

reverse current release [*IEC 157-1*] Rückstromauslöser *m* [*VDE 0660 Teil 101*]

reversed Carnot cycle umgekehrte Carnot Kreisprozess *m*, links laufende Carnot Kreisprozess [*1824*]

reversed oblique → backlash

R

reversed return flow Umkehrrücklauf *m*

reverse feedback Gegenkopp(e)lung *f*

reverse function Umkehr(ungs)funktion *f*

reverse osmosis, RO Umkehrosmose, umgekehrte Osmose, Molekularfiltermethode *f*, Molekularfilterverfahren *n*

reverse playback Umkehrwiedergabe *f*

reverse polarity protection Verpolungsschutz *m*

reverse polling Rückwärtsabfragen *n [Fernwirktechnik]*

Reverse-Return-System Tichelmannsystem *n* [*Heiz(ungs)wasserverteilung*]

reverse solidus Schrägstrich nach links, Linksschrägstrich *m* [*Kode*]

reverse video Inversdarstellung, Reversdarstellung *f* [*Invertierte Wiedergabegerät von Text und Graphik auf einem Sichtgerät*]

reverse voltage Sperrspannung *f* [*Diode; DIN 41781*]

reverse voltage protection Verpolungsschutz *m* [*DIN 41745*]

reversible and adiabatic process → insentropic process

reversible card Umkehrkarte, Wendekarte *f*

reversible cell → (storage) battery

reversible counter → bidirectional counter

reversible escalator Umkehrrolltreppe, Zweirichtungsrolltreppe *f*

reversible-flue furnace [*Furnace in which the upward direction of the flue gas flow is reversed before reacting the flue collar*] Umkehr-Feuerzugfeuerung *f*

reversible heat pump, reverse cycle heat pump Umschaltungswärmepumpe, umschaltbare Wärmepumpe *f*

reversible key Wendeschlüssel *m*

reversible motor Umkehrmotor, drehrichtungsumschaltbare Motor, Motor für zwei Drehrichtungen *m*

reversible oil pump Umkehrölpumpe *f*

reversible system Umkehrsystem, Wendesystem *n*

reversing Umschalten, Wenden, Reversieren *n*

reversing ((closed) loop) control Umkehrreg(e)lung, Wendereg(e)lung *f*

reversing contactor Wende(schalt)schütz(e), Umkehr(schalt)schütz(e) *m, (f)*, Wende(schalt)schützen, Umkehr(schalt)schützen *m*

reversing controller, reversing regulator Umkehrregler *m*, Umkehrregelorgan *n*, Umkehrregel(ungs)einheit *f*

reversing on-off switch with spring return Wende-Ein-Aus-Schalter, Ein-Aus-Wendeschalter *m*

reversing ((open) loop) control Umkehrsteuerung *f*

reversing operation Wendebetrieb, Umkehrbetrieb *m*

reversing single-phase motor Einphasenmotor mit zwei Drehrichtungen, Einphasenreversiermotor, Einphasenumkehrmotor, umsteuerbare Einphasenmotor *m*

reversing span element Umkehrspannenelement *n*

reversing switch Wendeschalter, Umkehrschalter, Drehrichtungswendeschalter, Drehrichtungsumkehrschalter *m*

reversing switch with spring return Wendetastschalter, Tastwendeschalter, Umkehrtastschalter, Tastumkehrschalter *m*

reversing valve Umkehrventil, Wendeventil *n*

reversing valve relay Umkehrventilrelais, Wendeventilrelais *n*

revised revidiert, überarbeitet

revision Änderung *f* [*Beschriftung im Zeichnungsschriftfeld*]

revision memory Revisionsspeicher *m*

revolving-armature machine, rotary armature machine, rotating armature machine, rotation armature machine, rotative armature machine Außenpolmaschine *f*

revolving barrier Drehsperre *f*

revolving damper → swivel damper

revolving door Drehtür *f*

revolving light Rundumkennlicht, Drehkennlicht *n*

revolving machine Innenpolmaschine *f*

revolving vane anemometer [*Device consisting of rotating propeller-type vanes; the air velocity is indicated from the rotational speed of the vanes*] Flügelradanemometer *n*, Flügelradwindmesser *m*

Rewarming Nach(er)wärmung *f*, Nach(er)wärmen *n*

Rewiring Nachverdrahtung *f*, Nachverdrahten *n*

rewritable optical disk → magneto optical disk

Reynold's number, Damköhler number V, DaV [*A dimensionless number which is significant in the dimension of a model of any system in which the effect of velocity is important in controlling the velocities or to flow pattern of a fluid; equal to the density of a fluid, times its velocity, times a characteristic length, divided by the fluid viscosity. Symbolized* NR_e] Reynoldssche Zahl *f*

Reynold's stress, eddy stress [*The net transfer of momentum across a surface in a turbulent fluid because of fluctuations in fluid velocity*] Reynoldsspannung *f*

R

Reynold's stress tensor, eddy stress tensor [*A tensor whose components are the components of the Reynold stress across three mutually perpendicular surfaces*] Reynoldsspannungstensor *m*

RF → **range finder**

RF → **high-frequency**

RF ground, radio-frequency ground Hochfrequenzerder *m*

RFI reactor, radio-frequency interface reactor Funkentstör(ungs)drossel *f*

RF modem → **high-frequency modem**

RFNM → **ready for next message**

RF power → **radio-frequency power**

rf preheating, radio-frequency preheating [*Preheating of plastics-mo(u)lding materials by radio frequencies of 10 to 100 megahertz per second to facilitate the mo(u)lding operation or to reduce the mo(u)lding-cycle time*] Hochfrequenz(er)wärmer, Hochfrequenzerhitzer *m*

rheostat [*A variable resistor having a resistance element and a continuously adjustable contact arm*] Rheostat *m*

rheostatic ((closed) loop) controller Widerstandsregler *m*, Widerstandsregelorgan *n*, Widerstandsregel(ungs)einheit *f*

rheostatic ((open) loop) control [*A direct current system of control by varying restistance and reactance in the field circuit of the hoisting motor. Alternating current liftmotors start across the line*] Rheostatsteuerung, Widerstandssteuerung *f*

rheostatic rotor starter [*IEC 292-3*] Widerstandsläuferanlasser *m* [*VDE 0660 Teil 301*]

rheostatic starter [*IEC 292-3*] Widerstandsanlasser, Widerstandsstarter, Metallanlasser, Metallstarter *m*

rheostatic voltage controller, rheostatic voltage regulator Widerstandsspannungsregler *m*, Widerstandsspannungsregel(ungs)organ *n*, Widerstandsspannungsregel(ungs)einheit *f*

RHP → **right half** ***s*****-plane**

rhythmic light Taktfeuer *n*

ribbed gerippt, riefig

rib(bed) cast-iron tube central heating unit, rib(bed) cast-iron tube central heater Gusseisenrippenrohrheiz(ungs)körper *m*

rib(bed) coil, rib(bed) attemperator Rippenheiz(ungs)schlange, Rippenschlange *f*

rib(bed) cooling pipe, cooling rib(bed) pipe Rippenkühlrohr, Kühlrippenrohr *n*

rib(bed) heating element Rippenheiz(ungs)element, Rippenheiz(ungs)glied *n*

rib(bed) heating tube Rippenheiz(ungs)rohr *n*

rib(bed) lamella pipe, lamella rib(bed) pipe Lamellenrippenrohr, Rippenlamellenrohr *n*

rib(bed) pipe → **gilled pipe**

rib(bed) plate heat exchanger berippte Plattenwärmeübertrager, Rippen-Plattenwärmeübertrager *m* [*Siehe Anmerkung unter „Wärme(aus)tauscher"*]

rib(bed) (sectional) radiator Rippenradiator; Rippenheiz(ungs)körper *m* [*Fehlbenennung*]

rib(bed) spiral pipe → **spiral rib(bed) pipe**

rib(bed) tube Rippenrohr *n*

rib(bed) tube heat exchanger Rippenrohrwärmeübertrager *m* [*Siehe Anmerkung unter „Wärme(aus)-tauscher"*]

rib(bed) tube section Rippenrohrglied *n*

ribbon cable (Flach)Bandkabel *n*

ribbon cartridge Farbbandkassette *f*

ribbon(-type) ceiling (radiant) panel, ribbon(-type) radiant panel Bandstrahler *m*, Bandstrahlplatte *f* [*Stahlplattenheizer*]

ribbon (type) web(bed) building wire, flat web(bed) building wire, flat web(bed) cable Stegleitung, NYIF *f* [*Diese Leitungen sind in einem Kunststoffmantel flach nebeneinander angeordnet und werden auf die unverputzte Wand oder Decke genagelt*]

rib efficiency Rippenwirkungsgrad *m*

rib type Rippenart *f*

ride number [*lift*] Fahrten(an)zahl *f*

Rider flat (slide) valve Riderflachschieber *m*

Rider (slide) valve with grooves, groove(d) Rider (slide) valve Riderschieber mit Einschnitten, Einschnitte-Riderschieber *m*

Rider (slide) valve with slots, slot(ed) Rider (slide) valve Riderschieber mit Spalten, Spalten-Riderschieber *m*

ride smoothness → **smoothness of ride**

Riefler clock [*A clock employing a pin escapement and a mercurial pendulum; used as a standard*] Rieflersche Kompensationspendel *n*

right half ***s*****-plane, RHP** rechte *s*-Halbebene *f*

right(-hand) bowl sink Rechtsbeckenspüle *f*

right(-hand) hinge Rechtsband *n* [*Beschlag für Fenster und Türen*]

right(-hand) installation Rechtseinbau *m*

right(-hand) lock rechte Schloss, Schloss DIN rechts, Rechtsschloss *n*

right-justified format Blocksatz *m*

right parenthesis (runde) Klammer zu

right rotation Rechtsrotation *f* [*Eine Bit-Manipulation*]

right side view Seitenansicht von rechts, rechte Seitenansicht *f*

right-thread nipple Rechtsgewindenippel *m*

right welding Nach-Rechts-Schweißen *n*

rigid (electrical) conduit (biege)steife (Installations)Rohr, starre (Installations)Rohr, Trägerrohr, starre Schutzrohr, Stangenrohr, (biege)steife Schutzrohr *n* [*Elektrotechnik*]

rigid foam insulation Hartschaumdämmung *f*

rigid magnetic disk → **hard disk**

rigid metallic (electrical) conduit (biege)steife Metallschutzrohr, starre Metallschutzrohr *n*

rigid nonmetallic (electrical) conduit (biege)steife nichtmetallische Schutzrohr, starre nichtmetallische Schutzrohr *n*

riller rail, runner rail Laufschiene *f* [*Schiebetür*]

rim Wulstrand *m*

rim Umschweif *m* [*Schloss*]

rim capacitor Wulstrandkondensator *m*

rim lock, straight lock Kastenschloss, Aufschraubschloss, aufliegende Schloss *n*

rim lock door release Kastenschlosstüröffner *m*

rim pin Umschweifstift *m* [*Zwischen Schlossblech und Schlossdeckel innerhalb des Schlosskastens angebracht*]

rim strike Schließkasten *m* [*Zylinderkastenriegelschloss*]

ring air space → **air annulus space**

ring bus detector, ring bus detecting device, ring bus detection device Ringbusmelder *m*, Ringbusmeldegerät *n*

ring earth electrode, annulus earth electrode Ringerder *m*

ringer [*telephone*] Rufsatz *m*

ringer level, ringing level Rufsatzpegel *m*

ringer volume, ringing volume Rufsatzlautstärke *f*

ring-gap arcing chamber → **annular-gap arcing chamber**

ring ground → **annular ground**

ringing ... → **call(ing) ...**

ringing apparatus Geläut(e) *n*, Läutevorrichtung *f*

ringing wire Klingeldraht *m*

ring lock Ringschloss *n*

ring-main unit Ringnetzstation, Ringkabelstation, Lastschaltanlage *f* [*für Ringkabelnetz*]

ring (pipe) hanger Ring(rohr)schelle *f*

ring proximity switch ringförmige (An)Näherungsschalter *m*

ring punching einteilige Blechring *m*, Blechronde *f*

ring shift → **cyclic(al) shift**

ring slide valve → **annular slide valve**

ring topology Ringtopologie *f* [*Siehe Anmerkung unter „logische Topologie"*]

ring-valve plate Ringplattenventil *n*

rinsing Abspülen *n*

rinsing basin → **(mouth) rinsing basin**

rinsing sink [*A metal sink, in which water can be heated either electrically or by gas, for the immersion of culinary utensils and table-ware at a temperature high enough to achieve freedom from all harmful bacteria in an active form*] beheizte Spülbecken *n*

rinsing water → **(dishes) rinsing water**

rio unit → **remote input/output unit**

RIP → **routing information protocol**

ripping → **flat cutting**

ripple carry counter synchrone Zähler *m*

ripple counter→ **asynchronous counter**

rip-rap noise knallartige Geräusch *n*

ripsawing → **flat cutting**

RISC → **reduced instruction set computer**

rise, travel [*Of an elevator, escalator, etc., the vertical distance between the bottom terminal landing and the top terminal landing*] Förderhöhe *f*

rise, increase Anstieg *m*, Erhöhung, Zunahme *f*

rise, raising, upward travel, upward run Aufwärtsfahrt *f* [*Aufzug*]

rise detector → **(rate-of-)rise detector**

rise method → **(rate-of-)rise method**

riser, standpipe Standrohr, Steig(e)rohr *n*

riser pipe collector, rising pipe collector Standrohrsammler *m*

riser (pipe) valve Steig(e)rohrventil, Standrohrventil *n*

rise time Anregelzeit *f* [*In der Reg(e)lungsstechnik die Zeitspanne, die beginnt, wenn der Wert der Regelgröße nach einem Sprung der Stör- oder Führungsgröße den Toleranzbereich verlässt, und die endet, wenn der Wert der Regelgröße wieder in diesen Toleranzbereich eintritt. Gegensatz: Ausregelzeit*]

rise time Anstiegszeit *f* [*Die erforderliche Zeit, um einen Stromkreis von einem Zustand in einen anderen zu schalten*]

rising arm barrier Hub(arm)schranke, Hub(arm)-sperre, Hebe(arm)schranke, Hebe(arm)sperre *f*

rising main, riser, ascending main [*A main water, gas or electric service which, having entered the building below ground level, rises to a main metering or distribution point, or passes through one or more floors of the building, or from one property to another positioned above*] Steig(e)leitung, Standleitung *f*

rising out of synchronism Außertrittziehen *n*

rising pipe water meter, rise pipe water meter Steigrohrwasserzähler *m*

rising run, riser run, ascending run Steig(e)strang, Standstrang *m*

rising warded mechanism, warded rising mechanism steigende Riegelgesperre *n* [*Buntbartschloss*]

risk [*Not to be confused with "hazard"*] Risiko *n*

risk zone Risikozone *f*

rivet bearing → **clenching**

riveting on site, site riveting Baustellennieten *n*

RJ 11 → **modular plug**

RJE → **remote job entry**

RMM, read mostly memory Vorwiegend-Lese-Speicher *m*

RMS, root mean square quadratische Mittelwert, Effektivwert *m*

r.m.s. value of the sound pressure, root-mean-square volume of the sound pressure quadratische energetische Schalldruckmittelwert, energetische quadratische Schalldruckmittelwert *m*

RO → **reverse osmosis**

roaming Wechseln *n* in ausländische Netze

roaming → **panning**

ROB → **recorder buffer**

robbery Diebstahl *m*

robe hook, clothes hook Kleiderhaken *m*

robot Roboter, Handhabungsautomat *m*

robotics Robotik, Robotertechnik *f* [*Untersuchung und/oder Anwendung von Automaten*]

rocker (dolly), dolly, rocker button (Schalter)Wippe *f*

rocker (dolly) switch, dolly switch, rocker button switch [*CEE 24*] Wipp(en)schalter *m* [*VDE 0632*]

rocker(-dolly) switch with pilot lamp, rocker (-dolly) switch with target light indicator Wipp(en)-kontrollschalter *m*

rocker key Wipptaste(r) *f, (m)*

rocking chair Schaukelstuhl *m*

rocking fire (grate) bar → **moving fire (grate) bar**

ROD → **magneto optical disk**

rodding ... → **cleaning ...**

rodding door → **cleanout door**

rod door spring → **tail door spring**

RODP → **relative ozone depletion potential**

To rod through stochern

Roentgenlucent → **radio(trans)parent**

rollback lock Rücklaufsperre, Rückdrehsperre *f*

rolled flange Walzflansch *m*

rolled section Walzprofil *n*

rolled steel product Walzstahlerzeugnis *n*

roller → **(sliding door) roller**

roller curtain (type) filter → **continuous roll (type) filter**

roller guid(anc)e Rollenführung *f*

roller head Anfahrrolle, Startrolle *f*

roller lever Rollenhebel, Schwenkhebel *m*

roller plunger, plunger with roller Rollenstößel *m* [*Positionsschalter*]

roller shutter Rolladen *m*

roller(-type) moving walk [*It is belt supported by a succession of rollers with their axes at right angles to the direction of the treadway motion*] Fahrsteig mit Bandunterstützung durch Querwalzen, Querwalzen-fahrsteig *m*

roll filter, moving curtain filter [*Filter with a continuous belt on movable rolls to bring clean filter air into the airstream, either manually or automatically*] (Roll)Band(luft)filter *m, n*

roll forward Nachvollziehen *n* von Operationen einer Datenbank aufgrund eines Logfilters

rolling curtain Rollvorhang *m*

rolling gate Rolltor *n*

rolling piston (refrigerant) compressor [*A special type of small rotary compressor eccentrically within the stator, used in domestic refrigerators*] Rollkolben-(kälte)verdichter, Rollkolben(kälte)kompressor *m*

rolling shutter Rolladen *m*

roll in/roll out Auslagerung *f* [*Vorübergehende Speicherung eines Teils des Arbeitsspeicherinhaltes auf einen peripheren Speicher um den Arbeitsspeicher oder einen Teil davon für einen anderen Prozess frei zu machen*]

roll off Filterflanke *f* [*Optimizer*]

R

to roll out, to relocate, to page out auslagern [*Datenträger*]

rollover Mehrfachbetätigung, Vielfachbetätigung *f*

roll paper dispenser Rollenpapierspender *m*

roll (tape) filter Roll(band)filter *m,n*

roll (type) disposable media filter → continuous roll (type) filter

ROM chip → read-only memory chip

roof aerial (Brit.) → overhouse aerial

roof air supply fan Dachzuluftventilator *m*

roof car [*Structure for the suspension of work(ing) platform, providing for its horizontal movement to working positions*] Dachwagen *m*

roof-car-powered (work(ing)) platform Dachwagen(arbeits)plattform *f*

roof cavity air removal Dachhohlraumentlüftung *f*

roof cavity ventilation Dachhohlraumlüftung *f*

roof central heating installation, roof central heating plant Dachheiz(ungs)zentrale *f*

roof (central) station, central roof station Dachzentrale *f*

roof cowl Dachaufsatz *m*

roof cowl ventilation Dachaufsatzlüftung *f*

roof cradle, roof boat, wall lift Fassadenreinigungswagen, Fassadenputzwagen *m*

roof cradle machine, roof boat machine, wall lift machine mechanische Fassadenreinigungswagen, mechanische Fassadenputzwagen *m*

roof curtain [*deprecated*] **→ roof screen**

roof design Dachkonstruktion, Dachauslegung *f*, Dachentwurf *m*

roof disposal water, disposal roof water Dachablaufwasser, Ablaufdachwasser *n*

R

roof drainage Dachentwässerung, Dachabwässerung *f*

roof extract(ion) fan Dachventilator *m*

roof extract(ion) hood → cowled roof extractor

roof extract(ion) ventilation, extract(ion) roof ventilation Dachentlüftung *f*

roof fan Dachventilator *m*

roof gutter → gutter

(roof) gutter heating (Dach)Rinnen(be)heizung *f*

(roof) gutter iron (Dach)Rinneneisen *n*

(roof) gutter with ogee (mo(u)lding) (Dach)Rinne mit Sima, Regenrinne mit Sima, Sima(regen)rinne, Simadachrinne *f* [*Die Rinne hat eine Blechverkleidung, die Sima, die in schwungvoller Form die vordere Rinnenoberkante mit dem Gesims scheinbar zu einem Ganzen zusammenfügt"*]

roof heating station Dachheiz(ungs)zentrale *f*

roof heat pump Dachwärmepumpe *f*

roof insulation Dachdämmung *f*

roof(-mounted) window washing system dachmontierte Fensterwaschanlage, dachmontierte Fensterreinigungsanlage *f*

roof plant room, roof installation room Dach-Anlagenraum *m*

roof radial fan → radial roof fan

roof screen; roof curtain [*deprecated*] [*A vertical screen fitted internally to the roof of a building to divide the roof into bays, so that smoke and hot gases from a fire are contained within the bay of origin*] Dachbrandschott, Branddachschott *n*

roof spire ventilation, roof spire vent(ilat)ing Dachreiterlüftung *f* [*Luftwechsel durch Auftriebs- und Windwirkung von Dachreitern mit geeigneten Öffnungen*]

roof spray cooling system Dachspritzkühl(ungs)anlage *f*

roof station → roof (central) station

roof swimming pool Dachschwimmbecken *n*

roof tiler Ziegel(dach)decker *m*

roof-top air conditioner [*Packaged air conditioner mounted on a roof, the conditioned air being discharged directly into the rooms below or through a duct system*] Dachklimaanlage, Dachklimazentrale *f*

roof ventilation, roof vent(ilat)ing Dachaufsatzlüftung, Dachreiterlüftung *f* [*Die freie Lüftung, die sich durch Aufsätze, kurze Schächte oder Entlüftungsöffnungen im Dach einstellt. Sie beruht hauptsächlich auf dem thermischen Auftrieb, der sich durch den Temperaturunterschied zwischen außen und innen ergibt*]

roof ventilator [*Not to be confused with "roof fan"*] Dachlüfter *m* [*Nicht verwechseln mit „Dachventilator"*]

roof water Dachwasser *n*

room, space [*In a building, a particular portion, an enclosure or division separated from other divisions by walls or partitions*] Raum *m* [*Auch „Zimmer" genannt, wenn er Wohnzwecken dient*]

room absorption, space absorption, indoor absorption [*An expression of the total absorption influence within a room*] Raumabsorption, Innenabsorption *f*

room acoustics, space acoustics, indoor acoustics Raumakustik, Innenakustik *f*

room air, space air, indoor air Raumluft, Innenluft *f*

room air circulation, space air circulation, indoor air circulation Raumluftumwälzung, Innenluftumwälzung *f*

room air conditioner, room air conditioning unit, indoor air conditioner, indoor air conditioning unit, space air conditioner, space air conditioning unit [*See remark under "air conditioner"*] Raumklimagerät *n*, Raumklimaaggregat, Wohnungsklimagerät, Wohnungsklimaaggregat, Innenklimagerät, Innenklimaaggregat *n*

room air cooler, indoor air cooler, space air cooler Innenluftkühler, Raumluftkühler *m*

room (air) cooling, space (air) cooling, indoor (air) cooling Raum(luft)kühlung, Innenluftkühlung *f*

room air diffusion, indoor air diffusion, space air diffusion Raumluftdiffusion, Innenluftdiffusion *f*

room air engineering, indoor air engineering, space air engineering Raumlufttechnik, RLT, Innenlufttechnik *f*

room air exhaust, space air exhaust, indoor air exhaust Raumluftabsaugung, Innenluftabsaugung *f*

room (air) fan, indoor (air) fan, space fan Raumventilator, Innenventilator *m*

room air humidifier, space air humidifier, indoor humidifier Raum(luft)befeuchter, Raum(luft)-anfeuchter, Innen(luft)befeuchter, Innen(luft)anfeuchter *m*

room air humidity, space air humidity, indoor air humidity Raumluftfeuchte, Raumluftfeuchtigkeit, Innenluftfeuchte, Innenluftfeuchtigkeit *f*

Room air induction unit, space air induction unit, indoor air induction unit [*One consisting of a cooling coil (or cooling and heating coil) and means for delivery of preconditioned air received under pressure from an external source mixed with recirculated air by air induction process to the space being conditioned. This device is normally designed for free delivery of air into rooms*] Vier-Leiter-Induktionsgerät *n*

room air intake, room air inlet, space air intake, space air inlet, indoor air intake, indoor air inlet Raumlufteinlass, Innenlufteinlass *m*

room air quality, IAQ, indoor air quality, space air quality Raumluftgüte, Innenluftgüte *f*

room air quality procedure, space air quality procedure, IAC procedure (direkte) Messverfahren *n* der Raumluftgüte

room air return, space air return, indoor air return Raumluftrückführung, Innenluftrückführung *f*

room air space, indoor air space Innenluftraum, Raumluftraum *m*

room air stream, space air stream, indoor air stream, room air flow, space air flow, indoor air flow Raumluftströmung, Innenluftströmung *f*

room (air) temperature, space (air) temperature, indoor (air) temperature Raumlufttemperatur, Innenlufttemperatur *f*

room air velocity, space air velocity, indoor air velocity Raumluftgeschwindigkeit, Innenluftgeschwindigkeit *f*

room allocation, room allotment, room assignment, space allocation, space allotment, space assignment, indoor allocation, indoor allotment, indoor assignment, Raumzuordnung, Raumzuteilung, Raumzuweisung *f*

room blower, indoor blower, space blower Raumgebläse, Innengebläse *n*, Raumbläser, Innenbläser *m*

room book, space book Raumbuch *n* [*Flächen; Ausstattung; Möbel*]

room bus [*See remark under "bus"*] Raumbus *m*

room bus controller, room bus regulator Raumbusregler *m*, Raumbusregelorgan *n*, Raumbusregel(ungs)einheit *f*

room bus interface Raumbusschnittstelle *f*

room bus module Raumbusmodul, Raumbusbaustein *m*

room climate, space climate, indoor climate Innenklima, Raumklima *n*

room ((closed) loop) control, space ((closed) loop) control, indoor ((closed) loop) control [*The control of the environment in a room*] Raumreg(e)lung, Innenreg(e)lung *f*

room ((closed) loop) controller, space ((closed) loop) controller, indoor ((closed) loop) controller Raumregler, Innenregler *m*

room coil, space coil, indoor coil [*As opposed to an outdoor coil*] Raumschlange, Innenschlange *f*

room conditions, indoor conditions, space conditions Raumverhältnisse, Innenverhältnisse *f*

room cooler → room (air) cooler

room cooling load, space cooling load, indoor cooling load Raumkühl(ungs)last, Innenkühl(ungs)last *f*

room cooling system, space cooling system, indoor cooling system Raumkühl(ungs)system, Innenkühl(ungs)system *n*

room detector → room sensor

room dew point, indoor dew point, space dew point [*Temperature in a room at which the water vapo(u)r reaches saturation*] Raumtaupunkt, Innentaupunkt *m*

room door lock, space door lock, indoor door lock Raumtürschloss, Innentürschloss *n*

room door reader, indoor door reader, space door reader Raumtürleser, Innentürleser *m*

R

room dryer, room drier, space dryer, space drier, indoor dryer, indoor drier Raumtrockner, Innentrockner *m*

room enclosure surface temperature, space enclosure surface temperature Raumumschließungs-Oberflächentemperatur *f* [*Oberflächentemperatur der Wände, der Decke und des (Fuß)Bodens*]

room environment, indoor environment, space environment Raumumwelt, Innenumwelt *f*

room evaporator coil, indoor evaporator coil, space evaporator coil Raumverdampferschlange *f*

room fan, indoor fan, space fan Raumventilator, Innenventilator *m*

room fan heater, space fan heater, indoor fan heater, room fan heating unit, space fan heating unit, indoor fan heating unit Ventilator(-Raum)heiz(ungs)körper *m* [*Siehe Anmerkung unter „Heiz(ungs)körper"*]

room gas heater → gas room heater

room gas (pipe)line, indoor gas (pipe)line, space gas (pipe)line Gasinnenleitung, Innengasleitung *f*

room heat, space heat, indoor heat Raumwärme, Innenwärme *f*

(room) heater → (floor(-mounted)) (room) heater

room heater, room heating appliance, room heating apparatus, room heating device, room fire, room heating unit [*"Space" and "local" may synonymously be used for "room"*] Raumheizer, Raum(er)wärmer, Raumerhitzer, Raumheiz(ungs)apparat *m*, Raumheiz(ungs)gerät *n* [*„Innen" kann synonym für „Raum" verwendet werden*]

room heating Zimmerheizung, Zimmererwärmung *f*

room heating, space heating, indoor heating, local heating Raumheizung, Raum(er)wärmung *f*

room heating capacity, space heating capacity, indoor heating capacity, local heating capacity Raumheiz(ungs)leistung, Raum(er)wärmungsleistung *f*

room heating efficiency, space heating efficiency, indoor heating efficiency, local heating efficiency Raumheiz(ungs)wirkungsgrad, Raum(er)wärmungswirkungsgrad *m*

room humidification, space humidification, indoor humidification Raum(luft)befeuchtung, Raum(luft)anfeuchtung, Innen(luft)befeuchtung, Innen(luft)anfeuchtung *f*

room humidity detector, room humidity sensor, space humidity detector, space humidity sensor Raumfeuchtigkeitsfühler, Raumfeuchtefühler *m* [*„Sensor", „Detektor" und „Detektionselement" können synonym für „Fühler" verwendet werden*]

R

room illumination, room lighting, space illumination, space lighting, indoor illumination, indoor lighting Raumbeleuchtung, Innenbeleuchtung *f*

room inlet air temperature cascade ((closed) loop) control Raum-Zulufttemperaturkaskadenreg(e)lung *f*

room key, guest('s) key Zimmerschlüssel, Gästeschlüssel *m* [*Hotel*]

room layout, space layout, indoor layout [*e.g. of a heating system to insure proper operation and architectural suitability*] Anordnung im Raum, Aufteilung im Raum, Gliederung im Raum *f* [*Nicht „Raumanordnung" usw., das ist Anordnung der Räume*]

room key, indoor key, space key Raumschlüssel, Innenschlüssel *m*

room load, space load, indoor load Raumlast, Innenlast *f*

room monitoring Raumüberwachung *f* [*Sicherheit*]

room need → room requirement

(room) occupant, space occupant (Raum)Insasse *m*

(room) occupant (Zimmer)Insasse *m*

room ((open) loop) control, space ((open) loop) control, indoor ((open) loop) control Raumsteuerung *f*, Raumsteuern *n* [*Siehe Anmerkung unter „Steuerung"*]

room operation(al) equipment, space operation(al) equipment, indoor operation(al) equipment Raumbedien(ungs)gerät(e) *n,(f)*

room protection, space protection, indoor protection Raumschutz, Innenschutz *m*

room recirculated air system Raumumluftsystem, Innenumluftsystem *n* [*System mit Umluftbehandlung unmittelbar an den Versorgungsstellen*]

room requirement, room need Raumbedarf *m*

rooms Räumlichkeiten, Räume *f*

room security, space security, indoor security Raumsicherung, Innensicherung *f*

room sensor, room detector, space sensor, space detector [*See remark under "sensor"*] Raumsensor, Raumdetektor, Raumfühler *m*, Raumdetektionselement *n*

room service trolley Zimmerwagen *m* [*Hotel*]

room setting, space setting, indoor setting Raumausgestaltung, Innenausgestaltung *f*

room sound, space sound, indoor sound Raumklang, räumliche Klang *m*

room sound damping, room sound absorption, space sound damping, space sound absorption, indoor sound damping, indoor sound absorption Raumschalldämpfung *f*

(room) spray humidifier, space spray humidifier, indoor spray humidifier [*Air humidifier which sprays atomized water directly into the room*] Sprühdüse *f*

room steam humidifier, space steam humidifier, indoor steam humidifier Raumdampfbefeuchter, Raumdampfanfeuchter *m* [*Zum Verteilen des Dampfes in Räumen, z. B. für Holzbearbeitung, Druck, Papierherstellung usw. DIN 30 600 Reg.-Nr. 06099*]

room subdivision, space subdivision, indoor subdivision Raumaufteilung, Raumgliederung *f*

room temperature ((closed) loop) control, space temperature ((closed) loop) control, indoor temperature ((closed) loop) control, ((closed) loop) room temperature control Raumtemperaturreg(e)lung *f*

room temperature ((closed) loop) controller Raumtemperaturregler *m*

room temperature ((open) loop) control, indoor temperature ((open) loop) control, space temperature ((open) loop) control Raumtemperatursteuerung *f*

room temperature sensor, space temperature sensor, indoor temperature sensor [*See remark under "sensor"*] Raumtemperatursensor, Raumtemperaturfühler, Innentemperatursensor, Innentemperaturfühler *m*

room thermostat, space thermostat, indoor thermostat [*It is located in a room so as to respond to room temperature and thereby control heating or cooling devices*] Raum(regel)thermostat, Raumtemperaturregler *m*

room unit, indoor unit, space unit Innengerät *n* [*Raumklimagerät in geteilter Bauweise*]

room user, space user Raum(be)nutzer, Rauminsasse *m*

room ventilation, space ventilation Raumlüftung, Zimmerlüftung, Raumbe- und -entlüftung, Zimmerbe- und -entlüftung *f*

room volume, space volume, indoor volume Raumvolumen *n*, Rauminhalt *m*

room-warm zimmerwarm, raumwarm

root, superuser Systemverwalter *m*

root directory, superuser directory Stammverzeichnis *n* [*Das höchstrangige Verzeichnis auf einer CD-I*]

root locus curve, root locus plot Wurzelort(s)kurve *f* [*Verlauf der Eigenwerte in der komplexen s-Ebene in Abhängigkeit eines freien Systemparameters, der entweder nicht genau bekannt ist oder sich durch Einflüsse ändert, oder für gutes dynamisches Verhalten eingestellt wird*]

root mean square → RMS

root-mean-square value of the sound pressure, r.m.s. value of the sound pressure quadratische energetische Schalldruckmittelwert, energetische quadratische Schalldruckmittelwert *m*

Roots compressor [*Two mating lobed impellers revolve within a cylinder and are prevented from making contact with each other by timing gears mounted outside the cylinder*] Rootskompressor, Rootsverdichter *m*

rope(d) hydraulic elevator (US); rope(d) hydraulic lift (Brit.) Hydraulik-Seilaufzug, Aufzug mit hydraulischem Seil(an)trieb *m*

rope elevator (US) → cable lift

rope lift (Brit.) → cable lift

ROR → release on request

rose(tte) Rosette *f*

rotary act(uat)or, rotating act(uat)or, rotation act(uat)or, rotary final controlling element, rotating final controlling element, rotation final controlling element Drehstellglied *n* [*Siehe Anmerkung unter „Stellglied"*]

rotary armature machine → revolving-armature machine

rotary atomization Drehzerstäubung, Rotationszerstäubung, Drehversprühung, Rotationsversprühung *f*

rotary atomizer Drehzerstäuber, Rotationszerstäuber, Drehversprüher, Rotationsversprüher *m*

rotary-atomizing burner, rotary cup burner Drehzerstäuberbrenner, Rotationszertstäuberbrenner *m* [*Das Öl fließt durch eine schnell rotierende Hohlwelle einem nach der Kesselseite offenen Becher zu. Durch Fliehkraft wird das Öl auf der Becherinnenseite gleichmäßig verteilt und von der Becherkante mit großer Geschwindigkeit abgeschleudert und dabei fein zerstäubt*]

rotary bolt Drehriegel, Rotationsriegel *m* [*Schloss*]

rotary compartmented chamber → drum

rotary control switch → rotary (momentary-contact) control switch

rotary converter Einankerumformer *m*

rotary dehumidifier, rotation dehumidifier, rotating dehumidifier, rotative dehumidifier Rotationsentfeuchter *m*

rotary displacement (gas) meter [*A positive displacement (gas) meter in which the measuring compartment is formed between the walls of a stationary chamber and a rotating element making substantially gas-tight contact with the walls*] Rotorvolumen(gas)zähler *m* [*Nicht verwechseln mit „Drehkolben(gas)zähler"*]

rotary encoder → shaft encoder

R

rotary field tester, rotary field testing device Drehfeldprüfgerät *n*

rotary final controlling element → rotary act(uat)or

rotary (gas) meter [*An inferential meter using a vane type anemometer in the gas stream*] Turbinenrad(gas)zähler, Schraubenrad(gas)zähler *m*

rotary heat exchanger, rotary heat exchange unit, rotary heat interchanger Rotationswärme(aus)tauscher [*In der Industrie so genannt*], Rotationswärmeübertrager *m* [*Siehe Anmerkung unter „Wärme(aus)tauscher"*]

rotary hunting connector (US) Mehrfachleitungswähler, Vielfachleitungswähler, Sammelleitungswähler, Drehwähler *m*

rotary knob Drehknopf *m*

rotary (momentary-contact) control switch Drehschalter *m*

rotary (piston) blower, rotating (piston) blower, rotation (piston) blower Drehkolbengebläse *n*

rotary (piston) compressor, rotary motor compressor [*Positive displacement compressor in which the change in internal volume of the compression chamber(s) is accomplished by the rotary motion of a positive displacement member or positive displacement members*] Dreh(kolben)verdichter, Dreh(kolben)kompressor, Zellenverdichter, Zellenkompressor, Rotations(kolben)verdichter, Rotations(kolben)kompressor *m*

rotary (piston) (door) closer Dreh(kolben)(tür)schließer, Rotations(kolben)(tür)schließer *m*

rotary piston (gas) meter Drehkolben(gas)zähler *m* [*Nicht verwechseln mit „Rotorvolumen(gas)zähler"*]

rotary piston meter Rinkolbenzähler *m*

rotary (piston) (refrigerant) compressor [*A positive displacement compressor in which the change in internal volume of the compression chamber(s) is accomplished by the rotary motion of positive displacement members or member*] Dreh(kolben)(kälte)verdichter, Dreh(kolben)(kälte)kompressor, Rotations(kolben)(kälte)verdichter, Rotations(kolben)(kälte)kompressor *m*

rotary selector switch Drehwahlschalter, Wahldrehschalter, Rotationswahlschalter, Wahlrotationsschalter *m*

rotary stepping relay → stepping relay

rotary timing resistor Einstelldrehwiderstand *m*, Einstellrotationswiderstand, Dreheinstellwiderstand, Rotationseinstellwiderstand

rotary ventilator Drehlüfter *m*

rotary wafer switch Stufendrehschalter, Stufenrotationsschalter, Drehstufenschalter, Rotationsstufenschalter *m*

rotary warded mechanism drehende Riegelgesperre *n* [*Buntbartschloss*]

rotary (water) meter, vane (water) meter, propeller (water) meter, rotating (water) meter, rotation (water) meter [*This type of (water) meter is used to measure the amount of water supplied to bulk consumers. The flow through the meter rotates a vane and registers on a dial*] Wasserzähler, Flügelrad(wasser)zähler *m* [*Nicht verwechseln mit „Wassermesser". DIN 3260*]

rotation Drehen *n* [*Eine Grundfunktion der Bildausschnitt(s)änderung*]

rotation atomizer Rotationszerstäuber *m*

rotations per minute → speed

rough (builder('s)) hardware [*Bolts, nails, spikes, and other metal fittings which are not seen and therefore are not finish hardware*] verdecke (Bau)Beschläge, rohe (Bau)Beschläge *f*

rough file → coarse file

rough fitting → large fitting

roughness → key

rough screen Grobrechen *m*

round (air) duct → circular (air) duct

round bulb lamp Kugelkolbenlampe, Tropfenlampe *f*

round-down key Abstreichtaste(r) *f, (m)*

rounded plunger Kuppenstößel *m* [*Positionsschalter. EN 50047*]

round escalator → circular escalator

round interlocked doors, interlocked round doors Rundschleuse, Türenrundschleuse *f*

round knob Rundolive *f* [*Schloss*]

round moving staircase → circular escalator

round thermostat, round thermorelay Rundthermostat *m*

round trip propagation time [*local area network*] Hin- und Rücklaufzeit, doppelte maximale Laufzeit, maximale doppelte Laufzeit *f*

round trip time Gesamtfahrzeit *f* [*z. B. Aufzug*]

route Leitweg *m* [*DIN 44 302*]

route Trasse *f*

route diagram Streckenbild *n* [*Fernwirktechnik*]

router Rangierer, Rangierbaustein, Rangiermodul *m*

router, level 3 relay Verbindungsglied *n* [*Es verbindet mehrere voneinander entfernte lokale Netzwerke*]

route signal Leitweg(e)signal *n*

R

routine → **(computer) routine**

routine library [*Ordered set of standard and proven computer routines by which problems or parts of them may be solved*] Entstörungsprozedursatz, Ausprüfungsprozedursatz, Fehlerbeseitigungsprozedursatz *m*

routing information protocol, RIP Protokoll über die Wegleitung, Weg(e)leitungsprotokoll *n* [*Die Weg(e)-leitung wird in der Regel mit Tabellen durchgeführt welche an alle beteiligten Rangierer gemeldet und dort ausgewertet werden*]

row Zeile *f*

row address Zeilenadresse *f*

row-address select, RAS Zeilenadress(aus)wahl *f*

row binary (punched) card [*A (punched) card in which the binary data are punched along the rows*] Zeilenbinär(loch)karte *f*

row of luminaires, strip of luminaires, row of light(ing) fittings, strip of light(ing) fittings Leuchtenband, Beleuchtungskörperband *n*

row urinals, urinals row Reihenurinale, Urinalreihe *f*

row vector Zeilenvektor *m*

row-wise zeilenweise, reihenweise

RPE → **remote procedure error**

RPG → **report program generator (US)**

r.p.m. → **speed**

RPOA, recognized private operating agency anerkannte private Betriebsgesellschaft, private anerkannte Betriebsgesellschaft *f*

R-PROM → **reprogrammable read-only memory**

RS-232-D connection; RS-232-C connection [*obsolete*] RS232-Verbindung *f*

RS-232-D interface; RS-232-C interface [*obsolete*] RS232-Schnittstelle *f* [*Die Daten werden seriell, also Bit für Bit übertragen*]

RS-232-D isolation network; RS-232-C isolation network [*obsolete*] RS232-Trennschaltung *f*

RS-232-D signal; RS-232-C signal [*obsolete*] RS232-Signal *n*

RS-232-D transmission; RS-232-C transmission [*obsolete*] RS232-Übertragung *f*

RS flip-flop (circuit), RS bistable element, set-reset flip-flop (circuit), set-reset bistable element, reset-set flip-flop (circuit), reset-set bistable element bistabile Kippschaltung *f* [*Mit Eingängen für Rücksetzen und Setzen*]

RSN → **remote nodal station**

RSSI → **received signal strength indicator**

RST → **radio station terminal**

RSU → **remote subscriber unit**

RT → **reverberation time**

RTC → **real time clock**

RTL, resistor-transistor-logic Rechnerschaltkreis *m* bestehend aus einer Widerstand(s)-Transistor-Kombination, Widerstand(s)-Transistor-Logik *f* [*Logikschaltung*]

RTR → **remote transmisson request**

RTS, ready to send Sendebereit

RTTY, radio teletyper Funkfernschreiber *m*

RTTY method → **radio teletype method**

rubber conduit Gummirohr *n* [*Hartgummirohr (in Stangen) oder Weichgummirohr (in Rollen) zur Verlegung von Leitungen unter Putz*]

rubber (file) → **coarse file**

rubber gasket [*See remark under "gasket"*] Gummi(ab)dicht(ungs)scheibe *f*

rubber grommet Gummitülle *f*

rubber handrail Gummihandlauf *m*

rubber insert Gummieinlage *f*

rubber membran(e), rubber diaphragm Gummimembran(e) *f*

rubber plug Gummistopfen, Gummistöpsel, Gummipfropfen, Gummistecker, Flexostecker *m*

rubber shock absorber Gummistoßdämpfer *m*

rubber vibration dampener Gummischwingungsdämpfer *m*

rubbish → **(building) rubbish**

rub-out signal Irrungssignal *n* [*Fernmeldetechnik*]

rugged robust

ruggedness Robustheit *f*

rule-antecedent part WENN-Teil [*Regel*]

rule base Regelbasis *f*

rule-based regelbasiert

rule-consequent part DANN-Teil, Regel-DANN-Teil *m/n*

rules system → **system of rules**

to run, to operate betreiben, fahren [*Maschine*]

run Lauf, Gang *m*

run, operation, service Betrieb *m*

run Strecke *f*, Weg *m*

run Strang *m*

runaround coil Rundum(rohr)schlange *f*

runaround (type) heat recovery coil → **heat recovery coil of runaround type**

run diagram → **system (flow)chart**

rundown, machine rundown Auslauf(en), Maschinenauslauf(en) *m, (n)*, Austrudeln *n*

rundown battery, depleted battery, empty battery entladene Batterie, erschöpfte Batterie, verbrauchte Batterie, leere Batterie *f*

to run hot heißlaufen

to run in einlaufen [*Maschine*]

to run in steps, to run synchronously synchron laufen, gleichlaufen

run length coding Lauflängenkodierung, Lauflängencodierung *f*

run monitoring Laufüberwachung *f*

runnable ausführbar [*Prozessstatuskode*]

runner → (sliding door) roller

running Betreiben, Fahren *n* [*Maschine*]

running amperage, running current intensity, running current strength Betriebsstromstärke, Laststromstärke *f*

running dry → dry running

running economy, operation(al) economy Betriebswirtschaft *f*

running handrail, mov(e)able handrail bewegliche Handlauf *m* [*Fahrsteig*]

running hour, operating hour Betriebsstunde *f*

running hours counting, operation(al) hours counting Betriebsstundenzählung *f*

running in steps, running synchronously synchron laufend, gleichlaufend

running-in test Probelauf, Prüf(ungs)lauf *m*

running rail Laufschiene *f*

running reserve, power reserve, spring reserve [*clock*] Gangreserve *f* [*Uhr*]

running sample Durchzugsprobe *f*

running through of neutral conductor Durchschleifen *n* des Nulleiters

run(ning) up → start(ing) up

running voltage, operation(al) voltage Betriebsspannung *f*

run-off Abfluss, Ablauf *m*

run-on Nachlauf *m*

run-stall torque Kippmoment *n* [*Schrittmotor*]

to run synchronously → to run in steps

run time Laufzeit, Betriebszeit *f*

run-time based, run-time dependent laufzeitabhängig, betriebszeitabhängig

run-time monitoring Betriebszeitüberwachung, Laufzeitüberwachung *f*

run-time sampling Betriebszeiterfassung, Laufzeiterfassung *f*

run-time system; service routine [*deprecated*] Laufzeitsystem, Betriebszeitsystem *n*

run-time totalization Betriebsstundenzählung, Betriebsstundensummierung, Laufstundenzählung, Laufstundensummierung *f*

to run true rundlaufen, schlagfrei laufen, stoßfrei laufen

rupture member [*A member that will rupture at a predetermined pressure*] Sollbruchglied

rush of air → air rush

rush operation Stoßbetrieb, Hochbetrieb *m*

RWM → read/write memory

S

sabin, absorption unit, open window unit, OW unit, square-foot unit of absorption [*The unit of sound absorption based on 1 square foot (0.09290304 square meter) of perfectly absorbent surface, such as an open door or window, whose absorptivity is taken as 1.00. Named after Prof. W.C. Sabine of Harvard University, a pioneer in the field of architectural acoustics*] Sabin *n*, Sabine-Einheit *f*

sabotage monitoring Zugriff(s)überwachung, Sabotageüberwachung *f*

sabotage-proof sabotagesicher, sabotagegesichert

sabotage security Sabotagesicherung *f*

sabotage signal message Sabotagemeldung *f*

sacrificial anode, sacrificed anode Opferanode *f* [*Kathodenkorrosionsschutz*]

SAD → source address

safe → (fail-)safe

ssafe air cooler Truhenluftkühl(ung(s))gerät *n*

safe (deposit) Geldschrank, Panzerschrank *m;* Tresor *m [umgangssprachlich]*

safe gap Grenzspaltweite *f*

safeguard Sicherheit(s)vorrichtung *f*

safe-keeping, care Verwahrung *f*

safe room → strongroom

safe room door → strongroom door

safe room vault → strongroom vault

safe to touch → contact-protected

safety [*Condition of being safe; freedom from danger or hazard. Not to be confused with "security"*] Sicherheit *f*

safety against intruders, intruder safety Sicherheit gegen Eindringlinge, Eindringlingssicherheit *f*

safety against key duplication, key duplication safety Schlüsselsicherheit *f* [*Schutz gegen unbefugte Beschaffung von Schlüsselkopien zu Missbrauchzwecken*]

safety allowance Sicherheitszuschlag *m*

safety analysis Sicherheitsanalyse *f*

safety badge Sicherheitsausweis *m*

safety belt [*Device, usually worn around the waist, attached to a lanyard and lifeline, or to a structure*] Sicherheitsgurt *m*

safety blow-off valve, blow-off safety valve Sicherheits(ab)blaseventil *n*

safety (builders') fitting Sicherheits(bau)beschlag *m*

safety (builders') fittings, safety (builders') hardware Sicherheits(bau)beschläge *f*

safety central station → central safety station

safety committee Sicherheitsausschuss *m*

safety conscious sicherheitsbewusst

safety consciousness Sicherheitsbewusstsein *n*

safety consultant Sicherheitsberater *m*

safety consultation, safety consultancy Sicherheitsberatung *f*

safety control, safety cut-out [*It causes equipment to cease operation when adverse or excessive conditions occur*] Sicherheitsorgan *n*, Sicherheitsvorrichtung *f*

safety corner valve → corner safety valve

safety cryogen(ic agent) Sicherheitskältemittel, Sicherheitskryomittel *n*

safety cut-out (device) Sicherheitsausschalter, Sicherheitsabschalter *m*

safety(-cylinder) lock Sicherheitsschloss *n*

safety disconnection → safety shutdown

safety door interlock switch Türsicherungsschalter, Sicherungstürschalter *m*

safety earthing (Brit.); safety grounding (US) Sicherheitserdung *f*

safety engineer Sicherheitsingenieur *m*

safety engineering Sicherheitstechnik *f*

safety extra low voltage, SELV Sicherheitskleinstspannung, Kleinstsicherheitsspannung *f*

safety fencing Umwehrung, Umzäunung *f* [*DIN 31 001*]

safety field Sicherheitswesen *n*

safety fitting Sicherheitsarmatur *f* [*Siehe Anmerkung unter „Armatur"*]

safety forward flow, forward safety flow Sicherheitsvorlauf *m*

safety gas hose Sicherheitsgasschlauch *m*

safety gear Fangvorrichtung *f* [*Aufzug*]

safety glazing Sicherheitsverglasung *f*

safety gland Sicherheitsstopfbuchse *f*

safety hardware Sicherheitstechnische Geräte, MSR-Geräte, Mess- und Steuer- und Regelgeräte *f*

safety hook [*Hook with a latch to prevent slings or loads from accidentally slipping off the hook*] Sicherheitshaken *m*

safety interconnection Sicherheitsverbund *m*

safety interruption, safety opening Sicherheitsabschaltung *f* [*Stromkreis*]

safety isolation → safety shutdown

(safety) isolation device, (safety) isolation appliance Freischalt(ungs)gerät *n*, Freischalt(ungs)vorrichtung *f*

safety lack, safety shortage Sicherheitsdefizit *n*, Sicherheitsmangel *m*

safety level Sicherheitsgrad *m*, Sicherheitsstufe, Sicherheitsebene *f*, Sicherheitsniveau *n*

safety lighting, safety illumination Sicherheitsbeleuchtung, Schutz- und Überwachungsbeleuchtung, Überwachungs- und Schutzbeleuchtung *f*

safety limit(ing) switch Sicherheitsgrenzschalter, Grenzsicherheitsschalter *m*

safety line Sicherheitsleitung *f*

safety lock -> safety(-cylinder) lock

safety locking system Sicherheitsschließsystem *n*

safety logic assembly Verknüpfungsschaltung *f* [*Reaktorschutz*]

safety market Sicherheitsmarkt *m* [*Kein Markt der Rohstoffe und Arbeitstechniken gemeinsam hat, sondern seine Aufgabe ist es Risiken und Gefahren für Leib, Leben und Gut zu verhindern und zu vermeiden*]

safety officer Sicherheitsbeauftragte *m*

safety of supply → supply safety

safety opening → safety interruption

S

safety outfit Sicherheitsausrüstung, Sicherheitsausstattung, Sicherheitsbestückung *f*

safety partition Sicherheits-Trennwand *f*

safety pipe [*A vent pipe from a domestic boiler*] Sicherheitsrohr *n*

safety plug → fus(ibl)e plug

safety piston switch Sicherheitsverriegelungsschalter, Sicherheitspositionsschalter *m*

safety power source Sicherheitsstromquelle *f*

safe(ty) pressure limit Sicherheitsdruckgrenze, Drucksicherheitsgrenze *f*

safety pressure limiter, safety pressure limit(ing) device Sicherheitsdruckbegrenzer, Drucksicherheitsbegrenzer *m* [*Er kann nach Abfall der Druckspitze nur von Hand mit einem entsprechenden Werkzeug (z. B. Schraubenzieher) in die Anfangsstellung zurückgestellt werden*]

safety refrigerant Sicherheitskältemittel, Fluor-Chlor-Derivat *n*, $C_mH_nCl_pF_q$

safety (relief) valve → (steam) safety (relief) valve

safety replaceable striking plate Sicherheitsaustauschschließblech, Austauschsicherheitsschließblech *n*

safety requirement, safety need Sicherheits(an)forderung *f* [*Anlage; Gerät*]

safety return (pipe)line, safety return flow line, return safety (pipe)line, return safety flow line Sicherheitsrücklauf(rohr)leitung, SR *f* [*Warmwasserheizung*]

safety rim lock Sicherheitskastenschloss *n*

safety riser (pipe), safety rising pipe Sicherheitsstandrohr, Standsicherheitsrohr, Sicherheitssteig(e)rohr, Steig(e)sicherheitsrohr *n*

safety screen Sicherheitsblende *f*

safety section Sicherheitsprofil *n*

safety sector Sicherheitsbranche *f*, Sicherheitssektor *m*

safety service Sicherheitsdienst *m*

safety shortage → safety lack

safety shutdown, safety disconnection, safety isolation, safety shutoff, safety stop Freischalten *n*, Freischaltung *f* [*VDE 0100, T.200*]

safety shut-off valve → (automatic) safety shut-off valve

safety stopping, safety shutdown Sicherheitsabschaltung *f* [*Maschine*]

safety switch [*IEC 65,348*] Sicherheitsschalter, Sicherheitstrenner *m* [*VDE 0860*]

safety switch → emergency (stop) switch [*elevator*]

safety system Sicherheitssystem *n*

safety temperature limiter, safety temperature limit(ing) device, safety temperature cutout Sicherheitstemperaturbegrenzer *m* [*DIN 3440*]

safety thermostat Sicherheitsthermostat, Sicherheitstemperaturregler *m*

safety transformer Sicherheitstrafo, Sicherheitstransformator *m*

safety valve → (steam) safety (relief) valve

safety vertical pipe → vertical safety pipe

safety zone [*Free space in the front of an escalator*] Stauraum, Freiraum *m*

sales area Verkaufsfläche *f*

sales consultant Verkaufsberater *m*

sales manager Verkaufsleiter *m*

salient corner → arris

salt mist Salznebel *m* [*Klimaprüfung*]

salt solution tank Salzlösebehälter *m* [*Wasserenthärtung*]

to sample entnehmen [*Probe*]

sample Muster *n*, Probe(stück) *f*, *(n)*

sample [*computer engineering*] Abfragewert, Abtastwert *m*

sample conditioning, sample preparation Probenaufbereitung *f*

sample cooler Probenkühler *m*

sampled-data ((closed) loop) control system Abtastreg(e)lung *f*

sampled signal getastete Signal *n* [*In der ersten Stufe einer signalverarbeitenden Maschine werden nach einer Filterung vom Signal in gleichem zeitlichen Abstand Proben gezogen, die nun zeitdiskret sind, aber immer noch jeden beliebigen Wert der Amplitude aufweisen können*]

sample fraction defective, defective sample fraction Fehlereinheitenanteil *m* [*Gütesicherung. Anteil fehlerhafter Einheiten in der Stichprobe*]

sample key Musterschlüssel *m*

sample period, sampling period Abtastperiode *f*

sample randomness Stichprobenzufälligkeit *f*

sample rate Abtastrate *f*

sampling Digitalisieren *n* analoger Audiosignale

sampling Digitalisierung *f* hörbarer Töne und Sprache und Geräusche

sampling, sample taking Probe(n)nahme *f*

sampling Abtastung *f*

sampling cock → inspection cock

sampling element Abtaster *m*

sampling interval, sampling spacing Abtastabstand *m*

sampling period → **sample period**

sampling rate, scanning rate Abtastrate, Abstastfrequenz, Abtasthäufigkeit *f* [*Damit ist gemeint, wie oft ein analoges Eingang(s)signal gemessen wird, um ihm digitale Werte zuzuordnen*]

sampling spacing → **sampling interval**

sampling switch, scanning switch, commutator switch [*A switch that executes a series of switching operations in sequential order*] Taktschalter, Kommutatorschalter *m*

sampling time Abtastzeit *f*

sand catch(er) → **detritor**

sanded up versandet

sand trap, sand separator, sand interceptor Sand(ab)scheider *m*

sandwich (structure) Mehrschichtenaufbau, Vielschichtenaufbau, Mehrlagenaufbau, Viellagenaufbau *m*, Mehrschichtenanordnung, Vielschichtenanordnung, Mehrlagenanordnung, Viellagenanordnung *f*

sanitary accessories Sanitärzubehör *n [selten: m]*

sanitary appliance, sanitary device [*There are two groups of sanitary appliances, viz. soil appliances, which receive human body wastes, that are water closets, urinals, and slop sinks; and ablutionary or waste appliances, that are baths, showers, wash basins, bidets, and, less commonly, scullery sinks, pantry sinks, and drip sinks*] Sanitärvorrichtung *f*, Sanitärgerät *n*

sanitary article → **sanitary object**

sanitary articles → **sanitary ware(s)**

sanitary bracket Sanitärkonsole *f*

sanitary brasswork, sanitary brassware Sanitärmessingware *f*

sanitary building block module, sanitary unitized unit [*"Plumbing" is used for "sanitary" in the USA*] Sanitärinstallationsblock, Sanitär(raum)block, sanitärtechnische Block *m*, Sanitärinstallationszelle, Sanitär(raum)zelle, sanitärtechnische Zelle *f*

sanitary central station → **central sanitary station**

sanitary ceramics Sanitärkeramik *f*

sanitary convenience Bedürfniseinrichtung, Bedürfnisanlage *f*

sanitary core → **sanitary shaft**

sanitary cove (US) → **base (board)**

sanitary cove convector (US) → **base (board) convector**

sanitary cove heater (US) → **base (board) heater**

sanitary drainage, sewage removal, removal of sewage (from buildings) Sanitärentwässerung, Gebäudeentwässerung *f*

sanitary electronics Sanitärelektronik *f*

sanitary engineer Sanitäringenieur, Gesundheitsingenieur *m*

sanitary engineering Gesundheitstechnik, Sanitärtechnik *f*

sanitary engineering industry Sanitärtechnikindustrie, Gesundheitstechnikindustrie *f*

sanitary engineering in the wider sense → **sanitation**

sanitary engineering trade Sanitärtechnikgewerk *n*

sanitary engineer in the wider sense → **sanitation engineer**

sanitary equipment → **sanitary outfit**

sanitary fitter, sanitary plumber Sanitärinstallateur *m*

sanitary fitting [*See remark under "fitting"*] Sanitärarmatur *f*, Sanitärformstück *n*

sanitary house services engineering sanitäre Haustechnik *f*

sanitary (identification) colour (Brit.) Sanitär(kennzeichnungs)farbe *f*

sanitary installation → **plumbing installation**

sanitary object, sanitary article, sanitary product Sanitärgegenstand, Sanitärartikel *m*, Sanitärprodukt, Sanitärerzeugnis *n*

sanitary outfit, sanitary equipment Sanitärausstattung, Sanitärausrüstung, Sanitärbestückung *f*

sanitary pipework, sanitary piping, sanitary pipeline Sanitärrohrleitung *f*

sanitary planning Sanitärplanung *f*

sanitary plumber, sanitary fitter Sanitärinstallateur *m*

sanitary range Sanitärbereich *m [selten: n]*

sanitary room Sanitärraum *m* [*Sanitärräume einer Wohnung sind Badezimmer, Toilettenraum, Küche und Hausarbeitsraum. DIN 18 022*]

sanitary science as applied to buildings, architectural hygiene Bauhygiene *f*

sanitary semiporcelain Sanitärhalbporzellan *n*

sanitary shaft, sanitary core Sanitärschacht, Sanitärkern *m*

sanitary shoe (US) → **congé**

sanitary-side sanitärseitig, sanitärseits

sanitary stoneware Sanitärsteinzeug *n*

sanitary system [*In a building, the plumbing includes the water-supply distributing pipes, the fixtures for using water, the soil, waste and vent pipes, the sanitary building drain, the storm-water building drain, and the building sewer, including all of their devices, appurtenances, and connections, all within or adjacent to the building*] Sanitäreinrichtung *f*, Sanitärsystem *n*

sanitary system manifold, sanitary system central distributor Sanitärverteiler(rohr) *m*, *(n)*

sanitary throughflow controller Sanitärarmatur *f*, Sanitärdurchflussregler *m*

sanitary unitized unit → sanitary building block module

sanitary ventilation system Sanitärlüftungssystem *n*

sanitary (ware) fixture [*See remark under "fixture"*] Sanitäreinbauteil *m,n*

sanitary ware(s), sanitary articles, sanitary products, sanitary items [*"Plumbing ..." is used for "sanitary ..." in the USA*] Sanitärware(n), Sanitärartikel, Sanitärerzeugnisse, Sanitärprodukte *f*, Sanitärzeug *n* [*Badewannen; Ausgussbecken; Klosetts; Waschtische; Wasserleitungsgegenstände usw.*]

sanitary waste incinerator Sanitärabfallverbrenner *m*

sanitary water Sanitärwasser *n*

sanitary works Sanitärarbeiten *f*

sanitation, public health engineering, sanitary engineering in the wider sense Gesundheitstechnik, Sanitärtechnik *f*

sanitation engineer, sanitary engineer in the wider sense, public health engineer Gesundheitsingenieur, Sanitäringenieur *m*

sanitation of a building Gebäude-Sanitäranlagen *f*

SAP → satisfiability problem

SAP → service access point

SAS → single attach station

sash angle Winkel *m*, Scheinecke *f* [*Beschlag*]

sash bolt → fastener

SATAN, storage array tester and analyzer Speicherbereichprüfer *m* und Analysator [*Programmpaket zum Aufspüren von Sicherheitslöchern in Netzwerken*]

satellite computer Satellitenrechner *m* [*Ein mit einer zentralen Datenverarbeitungsanlage in Verbindung stehender kleiner Rechner für Entlastungsarbeiten für die Zentralanlage, z. B. Daten übernehmen, umwandeln und verdichten, Berechnungen ausführen und Datenträger für die Weiterverarbeitung auf den betreffenden Großrechner vorbereiten*]

satellite elevation Satellitenpositionshöhe *f* [*über dem Horizont*]

satellite link Satellitenverbindung *f*

satisfiability Erfüllbarkeit *f*

satisfiability problem, SAP Erfüllbarkeitsproblem *n*

saturable(-core) reactor, transductor [*An iron-core reactor, having an additional control winding that carries direct current whose value is adjusted to change the degree of saturation of the core, thereby changing the reactance that the alternating-current winding offers to the flow of alternating current; with appropriate external circuits, a saturable(-core) reactor can serve as a magnetic amplifier*] sättigbare Drossel, Sättigungsdrossel *f*

saturable current Sättigungsstrom *m* [*In Sperrrichtung fließender Strom einer Sperrschicht, soweit er, im Idealfall, für einen größeren Sperrspannungsbereich von dem Wert der in der Sperrichtung angelegten Spannung unabhängig ist*]

saturated gesättigt, satt

saturated air [*1. Air in which the partial pressure of the water vapo(u)r is equal to the vapo(u)r pressure of water at the existing temperature. 2. Air that holds the maximum water vapo(u)r possible at a specified temperature and pressure*] Sattluft, gesättigte Luft, satte Luft *f*

saturated steam Sattdampf, gesättigte Dampf, satte Dampf *m* [*Flüssigkeitsfreier Dampf auf Siedetemperatur*]

saturated suction temperature Sattsaugtemperatur *f*

saturated vapour (Brit.); saturated vapor (US) [*A vapo(u)r whose temperature equals the temperature of boiling at the pressure existing on it*] gesättigte Wasserdampf, satte Wasserdampf *m*

saturated vapour pressure (Brit.); saturated vapor pressure (US) Wasserdampfsättigungsdruck, Sättigungsdruck des Wasserdampfes *m*

saturated wet steam → wet saturated steam

saturation coefficient → C/B ratio

saturation content Sättigungsgehalt *m* [*Der für eine bestimmte Temperatur höchstmögliche Wasserdampfgehalt der Luft*]

saturation curve, saturation line Sättigungskurve, Sättigungslinie *f*

saturation fall time, saturation decay time Sättigungsabfallzeit, Sättigungsabklingzeit *f*

saturation point → 100% relative humidity

saturation pressure Sättigungsdruck *m*

saturation rise time Sättigungsanstieg(s)zeit *f*

saturation shortage Sättigungsdefizit *n*, Sättigungsmangel *m*

saturation signal [*A radio signal (or radar echo)*] Sättigungssignal *n*

saturation temperature → boiling point

S

saturation vapo(u)r pressure Sättigungswasserdampfdruck *m*

saturation voltage Sättigungsspannung *f*

sauna (bath), sauna installation Saunaanlage *f*, Sauna(bad) *f, (n)*

saver Schoner *m*

saving Einsparung *f*

saving and restoring Sichern *n* und Laden [Programm]

saving key Spartaste *f* [*z. B. um Wasser zu sparen*]

saving lamp → (current-)saving lamp

saving potential (Ein)Sparpotential *n*

saving measure Sparmaßnahme *f*

SAW → surface acoustic wave

saw-cut → kerf

sawmill Sägewerk *n*, Sägemühle *f*

saw-tooth comparison method Sägezahnvergleich(s)verfahren *n*, Sägezahnvergleich(s)methode *f*

saw-tooth voltage Sägezahnspannung *f*

'S' bend gekröpfte Rohr *n*

SBS → building related illness

s.c. → short circuit

scalable skalierbar, skalierfähig

scalding hazard Verbrüh(ungs)gefährdung *f*

scalding protection Verbrüh(ungs)schutz *m*

scalding risk Verbrüh(ungs)risiko *n*

scalding water Verbrüh(ungs)wasser *n*

scale → (boiler) scale

scale Skala, Skalierung *f*

scale Maßstab *m*

scale, hammer scale, forge scale Hammerschlag, Zunder *m*

scale, water scale (Wasser)Stein *m*

scale Waage *f*

scale adjuster, scale setter Skala(ein)steller, Skalen(ein)steller *m*

scale drawing Maßstab(s)zeichnung, maßstab(s)gerechte Zeichnung *f*

scale factor Skalenkonstante *f* [*DIN 1310, Teil 2*]

scale formation, hammer scale formation, forge scale formation, scaling Hammerschlagbildung, Zunderbildung *f*

scale formation, water scale formation, (water) scaling (Wasser)Steinbildung *f*

scale holder Skalahalter, Skalenhalter *m*

scale-limit calibration Skalenendwerteichung *f*

scale mark Teilstrich, Skalenstrich *m*

scale (mark) spacing, graduation Teilabstand *m*, Maßteilung, Gradeinteilung, Skalenteilung, Graduierung *f*

scale-of-two circuit → binary scaler

scale plate Skalenträger *m*, Skalenblech *n*

scaler, counter, scaling circuit [*A circuit that produces an output pulse when a prescribed number of input pulses is received*] Impulszähler *m*

scaler Teiler *m*

scaler chain Teilerkette *f*

scale setter → scale adjuster

scale solvent → disincrustant

scaling Maßstabänderung, Skalierung *f* [*Umwand(e)lung von Dimensionen längs von Achsen durch Multiplizieren mit einem Faktor*]

scaling → scale formation

scaling circuit → scaler

scaling danger Verkrustungsgefahr, Versteinerungsgefahr *f*

SCAM selbstkonfigurierende MC-Schnittstelle *f* [*Siehe Anmerkung unter „MC-Schnittstelle"*]

to scan, to examine sequentially [*Computer engineering. Each item in a list, each record in a file, each point of a display, or each input or output channel of a communication link*] inspizieren in Folge

to scan abtasten; rastern [*Fehlbenennung*] [*Verwandeln von in Datenträgern enthaltenen Informationen in elektrische Impulse*]

to scan, to list [*Computer engineering. Each node in a tree data structure*] abtasten

to scan, to reproduce [*Computer engineering. An image from a document onto a display screen*] abstasten, vervielfältigen

to scan [*Electronics. To sweep over a specified range, field or dimension*] abtasten

to scan [*Telecommunications. To analyse or reproduce an image in a signal-line element by a FAX or television system*] abtasten

to scan [*Telecommunications. To determine the availability of a communication or data channel*] abtasten

to scan, to search [*Telecommunications. To search in a radio for each audible frequency in sequence, remaining tuned to each one for a few seconds before proceeding to the next*] suchen

scanner, scanning element Abtaster *m*, Abtast(ungs)element *n*

scanner Messstellenumschalter, Messstellenwähler *m*

S

scanner driver Abtastertreiber *m*

scanner selector [*In automatic data processing, an electronic device interfacing computer and multiplexers when more than one multiplexer is used*] Abtasterwähler *m*

scanning Abtasten *n*, Abtastung; Rasterung *f* [*Fehlbenennung*] [*Verwandeln von in Datenträgern enthaltenen Informationen in elektrische Impulse*]

scanning, sequential examination [*See remark under "to scan, to examine sequentially"*] Inspizieren *n* in Folge

scanning [*Telecommunications. Determination of the availability of a communication or data channel*] Abtasten *n*, Abtastung *f*

scanning [*Electronics. Sweeping over a specified range, field or dimension*] Abtasten *n*, Abtastung *f*

scanning, reproducing [*Computer engineering. An image from a document onto a display screen*] Abtasten, Vervielfältigen *n*, Abtastung, Vervielfältigung *f*

scanning [*Telecommunications. Analyzing or reproducing an image in a signal-line element by a FAX or television system*] Abtasten *n*, Abtastung *f*

scanning, listing [*Computer engineering. Each node in a tree data structure*] Abtasten *n*, Abtastung *f*

scanning analog(ue)(-to)-digital converter abtastende Analog-Digital-Wandler *m* [*Er benötigt während der Wandlung eine konstante Eingangsspannung*]

scanning ((closed) loop) control Abtast(ungs)reg(e)lung *f*

scanning density Zeilendichte *f* [*Fernkopierer*]

scanning electron microscope [*An instrument that enlarges the image of a specimen by simultaneously and systematically sweeping an electron beam over the specimen and a cathode-ray tube display of the electronic signal created when the beam first strucks the specimen*] Abstastelektronenmikroskop *n*

scanning element → **scanner**

scanning frequency Abtast(ungs)rate, Abtast(ungs)frequenz *f*

scanning interval, scanning spacing Abtast(ungs)abstand *m*

scan(ning) line [*In television a single, continuous narrow strip which is determined by the process of scanning*] Abtast(ungs)zeile *f*

scanning memory Abfragespeicher *m*

scanning moment Abtast(ungs)zeitpunkt *m*

scanning path Abtast(ungs)weg, Abtast(ungs)-pfad *m*

scanning period Abtast(ungs)periode *f*

scanning pulse Abtast(ungs)impuls *m*

scanning rate → **sampling rate**

scanning result Abfrageergebnis *n* [*zyklische Abfrage*]

scanning security Abtast(ungs)sicherung *f*

scanning spacing → **scanning interval**

scanning speed, spot speed [*In facsimile transmission, the speed of the scanning or recording spot within the available line*] Aufzeichnungsgeschwindigkeit *f*

scanning spot Abtast(ungs)punkt *m* [*Fernkopierer*]

scanning switch → **sampling switch**

scanning track Abtast(ungs)spur *f* [*Fernkopierer*]

scanning track direction Abtast(ungs)richtung *f* [*Fernkopierer*]

scanning tunnelling microscope, STM Tunnelstrommikroskop *n* [*Mit einer nahezu atomar spitzen Nadel wird die zu untersuchende Oberfläche in einem Abstand von wenigen Angström (=10^{-10}m) abgetastet. Der Abstand wird über den von der Spitze auf die Oberfläche fließenden Tunnelstrom kontrolliert*]

scaremonging Panikmache *f*

to scatter → **to splash**

scattered light Streulicht *n*

scattered-light particle counter Streulicht-Partikelzähler *m* [*Die Luft wird durch ein scharfes beleuchtetes Messvolumen gesaugt, wobei das Streulicht der Staubteilchen in einen Stromimpuls verwandelt und gemessen wird (Met One, USA, DEHA). Untere noch messbare Größe bei normalem Licht etwa 0,3 µ m (mü m)*]

scattered-light (smoke) detector, scattered-light (smoke) detection device, scattered-light (smoke) detecting device, scattering light (smoke) detection device, scattering light (smoke) detector, scattering light (smoke) detecting device Streulicht(rauch)melder, Streulicht(rauch)meldungsgeber *m*, Streulicht(rauch)meldegerät *n*

scattering Streuung *f*

scattered wave Streuwelle *f*

scattering Streuung *f*

scattering angle Streuwinkel *m* [*Lichttechnik*]

scattering matrix, collision matrix, S matrix Streumatrix *f*

scattering range Streubereich *m* [*selten: n*]

scatter propagation → **(forward) scatter propagation**

scavenging air Spül(ungs)luft *f*

scavenging air blower Spül(ungs)luftgebläse *n*

SCC → **single component computer**

S

SCF → switched capacitor filter

schedule Baureihe *f*, Bauprogramm *n*

schedule (of time) [*Tabulated time for switching control actions*] Schalt(ungs)zeit *f*

schedule ((open) loop) control Zeitplansteuerung *f*

scheduler → sequence ((open) loop) control (system)

schedule setter → setpoint generator

scheduling Terminierung, Ablaufplanung, Terminplanung *f*

schematic diagram → circuit diagram

schematic drawing Prinzipskizze, sinnbildliche Darstellung, schematische Darstellung *f*

scheme → project

Schmitt circuit, Schmitt limiter, Schmitt trigger [*A bistable pulse generator in which an output pulse of constant amplitude exists only as long as the input voltage exceeds a certain value*] Schmittbegrenzer, Grenzwertschalter, Grenzwertmelder, Komparator *m*, Grenzsignalglied *n*

Schottky barrier, hot-carrier Schottkysperre *f*

Schottky (barrier) diode, hot-carrier diode Schottkydiode *f* [*Genannt nach dem Schweizer Physiker Schottky, 1886–1976*]

Schottky contact → metal-semiconductor contact

Schottky defect, Schottky failure, vacancy [*Solid-state physics. A vacancy in a crystal produced by removing an atom from the bulk of the crystal and moving it to the surface*] Schottkydefekt, Schottkymangel *m*

Schottky effect [*Solid-state physics. An effect observed in thermionic emitters in which the work function is reduced when the emitter is subjected to an electric field*] Schottkywirkung *f*

Schottky noise, shot effect, shot noise [*Electronics. Excess voltage generated in a tube by unstable electrons, producing popping sound in a radio broadcast or snow in a television picture*] Schottkygeräusch *n*

Schottky transistor-transistor logic, Schottky TTL Schottky Transistor-Transistor-Logik, Schottky TTL, S-TTL *f* [*Die Schottky-Dioden liegen parallel zu den Basis-Kollektor-Strecken, die Arbeitsgeschwindigkeit wird dadurch gesteigert*]

science of sound → acoustics

scope → (cathode-ray) oscilloscope

scope of delivery, scope of supply, delivery scope, supply scope Lieferumfang *m*

scotch boiler [*Fire-tube boiler consisting of a cylindrical shell with one or more cylindrical internal furnaces in the lower portion, and a bank of tubes attached to both and closures*] schottische Kessel *m* [*Zylindrische Heizröhrenkessel mit Flammrohr und rückkehrenden Heizröhren und innerer Wendekammer*]

scouring Scheuern *n*

SCPC → single-channel-per-carrier

SCR, silicon controlled rectifier Thyristor *m*

scrambler Verwürfler *n* [*Er wandelt Daten nach kryptographischen Verfahren so um, dass sie nur mit einem Entschlüsseler zu nutzen sind*]

scraped-heat exchanger, scraped-surface exchanger [*It has a surface on which a fluid can be solidified and the solid layer scraped off. See remark under "heat exchanger"*] Kratzwärme(aus)tauscher [*In der Industrie so genannt*]; Kratzwärmeübertrager *m*

scraper [*sewage disposal*] Kratzer, Ausräumer *m*

scraper lock → pig lock

scrapping Verschrottung *f*, Verschrotten *n*

scratchpad area, scratch region Schmierbereich *m* [*Speicher*]

scratchpad area, scratch region Notizblockbereich *m* [*selten: n*] [*Speicher*]

scratchpad memory Notizblockspeicher *m*

scratchpad register Hilfsregister *n* [*Mikroprozessor*]

scratch resistance Kratzfestigkeit *f*, Kratzwiderstand *m*

scratch tape Arbeitsband *n* zur Zwischenspeicherung [*Datenverarbeitung*]

scratch test Kratzprobe, Kratzprüfung *f*, Kratzversuch *m*

screed Estrich *m*

screed emulsion Estrichemulsion *f*

to screen [*In automatic data processing, to make a preliminary selection from a set of entities, selection criteria being based on a given set of rules or conditions*] aussortieren

to screen abschirmen, schirmen

screen → baffle

screen → (display) screen

screen [*A covering to give physical protection from light, noise, heat, or flying particles*] Blende *f*

screen [*cable*] Leitschicht *f*

screen → (view(ing)) screen

screen → solar screen

screen brightness [*cathode-ray tube*] Beleuchtungsstärke, Bild(schirm)helligkeit *f*

screen buildup Leucht(bild)schirmanstieg *m*

screen-compatible → (display-)screen-compatible

S

screen continuity Schirmdurchverbindung *f* [*Steckvorrichtung*]

screen designer (Bild)Schirmentwerfer *m*

screen device → **(display) screen device**

screened → **shielded**

screen(ed) mail (Bild)Schirmpost *f*

screen(ed)-on door release Aufschraubtüröffner *m*

screened outlet Siebablauf *m*

(screen) form Maske *f* [*DIN 66 233, Teil 1. Auf dem Bildschirm dargestelltes Schema zur Anzeige und Eingabe von Daten. Diese Maske ist ein spezielles Auswahlmuster*]

screen grid [*A grid placed between a control grid and an anode of an electron tube, and usually maintained at a fixed positive potential, to reduce the electrostatic influence of the anode in the space between the screen grid and the cathode*] Schirmgitter *n*

screen hardcopy (Bild)Schirmausgabe auf Drucker, Drucker-(Bild)Schirmausgabe *f*

screen image, screen picture, display image, display picture (Bild)Schirmdarstellung *f*, Schirmbild, (Bild)Schirmdokument *n*

screen(ing), shield(ing) (Ab)Schirmung *f* [*Geerdeter Schutzmantel, der Wechselwirkungen zwischen Signalleitung und Umgebung verhindert*]

screen(ing) hood (Ab)Schirmhaube *f*

screen instruction (Bild)Schirmbefehl *m*

screen listing (Bild)Schirmprotokoll *n*

screen manager → **(on-)screen display**

screen mask (Bild)Schirmmaske *f*

screen monitoring → **(on-)screen monitoring**

screen printing Siebdrucken *n*

screen saver (Bild)Schirmschoner *m*

screen window (Bild)Schirmfenster *n*

screen workstation → **(display-)screen workstation**

screw compressor → **screw (refrigerant) compressor**

screw Schraube *f*

screw-down tap [*A draw-off tap closed by means of a disc carrying a renewable non-metallic washer which shuts against the water pressure on a seating at right angles to the axis of the screwed spindle which operates it*] Niederschraubhahn *m*

screw-down valve Niederschraubventil *n*

screw(ed) fitting Schraubfitting *m* [*Siehe Anmerkung unter "Fitting"*]

screw(ed) flange Schraubflansch, Gewindeflansch *m*

screw(ed) hose connection, hose screw(ed) connection Schlauchverschraubung *f*

screw(ed) joint, screw(ed) connection, screw(ed) union, screwing Schraub(en)verbindung, (Anschluss)Verschraubung *f*, Schraub(en)anschluss *m*

screw(ed)-on hinge Aufschraubband *n* [*z. B. für Türen*]

screw(ed)-on lock Aufschraubschloss *n*

screw(ed) rose Schraubrosette *f*

screw(ed) (sealing) plug [*A plug, usually of metal, having an external screw thread, intended for sealing the cleaning eye of a trap used with a sanitary appliance*] Verschlussschraube *f*, Schraubverschluss *m*

screw fixing, screw fastening Schraub(en)befestigung *f*

screw head Schraubenkopf *m*

screw hole Schraubenloch *n*

screw hook Schraubenhaken *m*

screw-in fuse Einschraubsicherung *f*

screwless terminal → **self-clamping terminal**

screw pipe hanger Schraubrohrschelle *f*

screw plug, plug screw Verschlussschraube, Stopfenschraube *f*, Schraubverschluss, Schraubstopfen *m*

screw reach → **reach of screw**

screw (refrigerant) compressor, refrigerant screw compressor [*Rotary compressor relying on the intermeshing of two helical rotors to produce compression*] Schrauben(kälte)verdichter *m* mit zwei Rotoren, Schrauben(kälte)kompressor *m* mit zwei Rotoren

screw rose(tte) Schraubenrosette *f*

screw terminal Schraub(en)klemme, Kopfkontaktklemme, Kopfschraub(en)klemme, Schraub(en)kopfklemme, Flachklemme *f*, Schraub(en)anschluss *m*

screw-to-wall bracket angeschraubte Wandkonsole *f*

screw wrench Schraubenschlüssel, Anziehschlüssel *m*

scroll Schwenk(en) des Bildausschnitts, Bildausschnitt(s)schwenk(en) *m*, *(n)*

scrollbar Schiebebalken *m* [*In einer Alarmleiste*]

scrollbar Schiebebalken *m* [*Zum Blättern in einer Meldeleiste*]

scrollbar Rollbalken *m* [*graphische Benutzungsoberfläche*]

scroll (refrigerant) compressor [*Positive-displacement (refrigerant) compressor in which the reduction in internal volume of the compression chamber is accomplished by an orbiting scroll (= involute spiral) within a mating stationary scroll*] Spiral(kälte)verdichter *m*

scrubber Überspannungs-Schutzelement *n*, Druckstoßminderer *m*

scrub board (US) → base (board)

scrub board convector (US) → base (board) convector

scrub board heater (US) → base (board) heater

SCS → sequence ((open) loop) control system

SCS → sensor-controlled sensitivity

SCS → sequence ((open) loop) control system

SCS, sharpness control system automatische Scharfeinstellung *f*

SCSI → small computer system interface

SCSI bus, small computer system interface bus Kleinrechner-Schnittstellenbus *m*

sculptured surface freigeformte (Ober)Fläche, Freiform(ober)fläche *f*

scum [*sewage disposal*] Schwimmschlamm *m*

scutcheon → (e)scutcheon

scutcheon knob → (e)scutcheon knob

SDA → to send data with acknowledge

SDI → subscriber distribution interface

SDL → specification and description language

SDU → service data unit

to seal verschließen

to seal (ab)dichten

seal (Ab)Dichtung *f*

seal → trap seal

seal depth [*trap*] Verschlusshöhe *f*

sealed compression refrigerating system → (hermetically-)sealed compression refrigerating system

sealed compressor → (hermetically-)sealed compressor

sealed condensing unit → (hermetically-)sealed condensing unit

sealed container Tauchrohr *n* [*Dampfdruck-Thermometer*]

sealed design → (hermetically-)sealed design

sealed power consumption Halteleistung *f*

sealed refrigerant pump → (hermetically) sealed refrigerant pump

sealed solution pump → (hermetically-)sealed solution pump

to seal hermetically Dichtschließen

seal-in circuit Selbsthalteschaltung *f*

sealing (Ab)Dichten *n*

sealing Verschließen *n*

sealing bellows Faltenbalg *m*

seal(ing) compound (Ab)Dicht(ungs)masse, Vergussmasse *f*

seal(ing) liquid Sperrflüssigkeit, Verschlussflüssigkeit, (Ab)Dicht(ungs)flüssigkeit, Betriebsflüssigkeit *f* [*Pumpe*]

seal(ing) number Verschlussnummer *f*

seal(ing) oil (Ab)Dicht(ungs)öl *n*

seal(ing) plug Verschlussstopfen, Verschlussstöpsel *m*

seal(ing) rope (Ab)Dicht(ungs)strick *m*

seal(ing) screw Plombierschraube, (Ab)Dicht(ungs)schraube *f*

seal(ing) spring (Ab)Dicht(ungs)feder *f*

seal(ing) surface (Ab)Dicht(ung)fläche *f*

seal(ing) voltage Schließspannung *f* [*magnetisch betätigte Gerät*]

seal(ing) wire Plombendraht, (Ab)Dicht(ungs)draht *m* [*Lampe*]

seal(ing) wrapper (Ab)Dicht(ungs)wickel *m*

seal of approval, approval seal Überwachungssiegel *n* [*(Werk)Stoffprüfung*]

to search → to scan

search Suche *f*

search access [*See remark under "access"*] Suchzugriff *m*

search call Suchruf *m*

search coil Suchspule, Prüf(ungs)spule *f*

search detector Sucher *m*, Suchgerät *n*

search engine Suchmaschine *f* [*WWW*]

searching Suchen *n*

search(ing) machine, e-mail search(ing) machine, electronic mail search(ing) machine [*For obtaining e-mail addresses*] (elektronische) Suchmaschine *f*

search(ing) spot light → adjustable spot light

search key Suchkriterium *n*

searchlight Suchscheinwerfer *m*

search run Suchlauf *m*

search text characteristic Suchtextzeichen *n* [*Ersetzt beliebiges Zeichen: M??er findet Maier, Mayer und Meier. * ersetzt eine beliebige Zahl von Zeichen: M*er findet Maier, Meier und Mayer, aber auch Mair und Mayr. „"findet den Text zwischen den Anführungszeichen. \ steht vor Sonderzeichen, die im Text wie normale Zeichen zu suchen sind. „," bedeutet logisches ODER: So findet die Eingabe „grünes, Gemüse" alle*

S

Dateien die beide Wörter enthalten. & bezeichnet logische UND: Hier findet die Eingabe „grünes & Gemüse" alle Dateien die beide Wörter enthalten. ~ ist das logische NICHT: Dabei schließt der Begriff „~Gemüse" alle Dateien, die „Gemüse" enthalten von der Suche aus]

to season voraltern

seasonal average Jahreszeitliche Durchschnitt *m*

seasonal thermal storage, seasonal heat storage Jahreszeitliche (Wärme)Speicherung *f*

seasonal time-of-day tariff Saisontarif mit Zeitzonen, Zeitzonen-Saisontarif *m*

seasoning Terminieren *n*

seasoning schedule Alterungsschema *n*

seat bench heating Sitzbankheizung *f*

seat body Sitzkörper *m*

seat(ing) [*tap; valve*] Sitz *m*

seat(ing) area Sitzbereich *m [selten: n]*

seat(ing) (bath)tub Sitz(bade)wanne, Stufen(bade)wanne *f*

seating toilet, seating closet Sitztoilette *f*, Sitzklosett *n*, Sitzabtritt, Sitzabort *m*

seat lid → toilet seat lid

secondary air Sekundärluft, Zweitluft *f* [*Durch den Luftstrahl längs des Strahlweges zunehmende, mitgeführte (induzierte) Raumluft*]

secondary air supply Sekundärluftzufuhr, Zweitluftzufuhr *f*

secondary call Sekundärgespräch, Zweitgespräch *n*

secondary call Sekundär(an)ruf, Zweit(an)ruf *m*

secondary circuit -> output circuit

secondary circuit pump Sekundär(kreislauf)pumpe, Zweit(kreislauf)pumpe *f*

secondary circulation [*The circulation of water in the secondary circuit of a hot water system*] Sekundärumwälzung, Sekundärzirkulation, Zweitumwälzung, Zweitzirkulation *f*

secondary ((closed) loop) control → secondary regulation

secondary command Sekundärkommando, Zweitkommando *n*

secondary condenser [*It is cooled by the evaporator of a secondary system*] Sekundärverflüssiger, Sekundärkondensator *m*

secondary controlled variable, corrective variable Hilfsregelgröße *f*

secondary controlled variable feedforwarding, secondary controlled variable feedforward control, corrective variable feedforwarding, corrective variable feedforward control Hilfsregelgrößenaufschaltung *f* [*Eine Aufschaltung zur Verbesserung des statischen und/oder dynamischen Verhaltens der geregelten Größe, wobei eine geeignete Hilfsregelgröße für die Bildung eines Hilfsregelkreises verwendet wird*]

secondary coolant; brine [*deprecated*] Sole *f* [*Bei Temperaturen unter +4 ° C müssen dem Wasser Frostschutzmittel zugesetzt werden. Hierfür wurden früher Salze verwendet, jetzt wegen der geringeren Korrosionsprobleme vorwiegend Glykole. In beiden Fällen spricht man von Sole, obgleich diese Benennung nur für Salzlösungen zutreffend ist*]

secondary data → output data

secondary dial → slave clock

secondary electrode → auxiliary electrode

secondary energy Sekundärenergie, Zweitenergie *f* [*Energieinhalt von Energieträgern, die aus der Umwandlung von Primärenergieträgern oder aus Sekundärenergieträgern gewonnen wurden*]

secondary fan Sekundärventilator, Zweitventilator *m*

secondary forward flow Sekundärvorlauf, Zweitvorlauf *m* [*Warmwasserversorgung*]

secondary forward flow pipe Sekundärvorlaufrohr, Zweitvorlaufrohr *n* [*Warmwasserversorgung*]

secondary light source Sekundärlichtquelle, Zweitlichtquelle *f*

secondary ((open) loop) control Folgesteuerung *f* [*Datenübertragung*]

secondary output voltage → auxiliary output voltage

secondary refrigerant system → double refrigerant system

secondary regulation, secondary ((closed) loop) control [*See remark under "regulation"*] Sekundärreg(e)lung, Sekundärregulierung, Zweitreg(e)lung, Zweitregulierung *f*

secondary return (flow) Sekudärrücklauf, Zweitrücklauf *m* [*Warmwasserversorgung*]

secondary return (flow) pipe Sekundärrücklaufrohr, Zweitrücklaufrohr *n* [*Warmwasserversorgung*]

secondary service, low-voltage service [*up to 600 V*] Niederspannungsbetrieb *m*

secondary sound absorber Sekundärschalldämpfer, Telefonieschalldämpfer, Zweitschalldämpfer *m*

secondary storage (device) → external memory

secondary ventilation Sekundärlüftung, Zweitlüftung *f*

secondary (water) supply Wassernachspeisung *f* [*z. B. in eine Sprinkleranlage*]

second breakdown zweite Durchbruch *m* [*Ein elektrisch-thermischer Regenerationsvorgang bei Leistungstransistoren*]

second level cache asynchrone Zwischenspeicher *m*

second order delay element, delay element of second order Verzögerungsglied zweiter Ordnung, Verzögerungsglied 2. Ordnung, P-T_2-Glied *n* [*Das P-T_2-Glied wir für d < 1 (Schwingungsglied) in seinem Übertragungsverhalten durch die Differenzialgleichung*

$$\frac{1}{w_0^2}\ \dot{v}(t)\ \frac{2d}{w_0}\ \dot{v}(t) + v\,(t) = K\,u\,(t)$$

oder die zugehörige Übertragungsfunktion

$$G\,(s) = \frac{K}{1 + \frac{2d}{w_0}\,s + \frac{1}{w_0^2}\,s^2}$$

Beschrieben E DIN 19 226 T 2]

secret digit, secret figure, secret numeral Geheimziffer *f*

secret lock Geheimschloss *n*

secret number Geheimnummer *f* [*Geheimzahl oder Geheimziffer*]

secret protection Geheimschutz *m*

secret security → security of secrets

section Schnitt(angabe) *m,(f)*

section → profile

section Sektor, Abschnitt *m*, Teil *m,n* [*räumlich*]

section, stretch Teilstrecke *f*

sectional boiler Glieder(heiz(ungs))kessel *m*

sectional cast(-iron) boiler → cast(-iron) sectional boiler

sectional cast(-iron) radiator → cast(-iron) (sectional) radiator

sectional drawing [*Drawing of a surface revealed by an imaginary plane cut through a project, or portion thereof, in such a manner as to show the composition of the surface as it would appear if the part intervening between the cut plane and the eye of the observer were removed*] Schnittzeichnung *f*

(sectional) radiator → radiator

sectional sliding door, sliding sectional door Gliederschiebetür, Schiebegliedertür *f*

sectional tank Gliederbehälter *m*

sectional view Schnittbild *n*

section for interior work → profile for interior work

section frame Profilrahmen *m*

sectioning, shading [*Section lines drawn across the portions of an object cut by a section plane, as on a drawing*] Unterteilung, Zerlegung *f*

section manipulated variable → line manipulated variable

section of conductor → conductor

section output Gliedleistung *f* [*Gliederheiz(ungs)körper*]

section with U(-shape(d)) cross section → channel

sector Branche *f*, Sektor *m*, Wesen *n*

to secure sichern

security [*Not to be confused with "safety"*] Sicherung *f*, Sicherheitsdienst *m* [*Alle Maßnahmen zur Erreichung einer durch den Bedarfsträger bestimmten Sicherheit gegen von ihm unerwünschte, aber vom Täter beabsichtigte Handlungen und/oder technische Unzulänglichkeiten der betroffenen Objekte*]

security access ((closed) loop) control Sicherungszugang(s)reg(e)lung *f*

security access ((open) loop) control Sicherungszugang(s)steuerung *f*

security against intruders → intruder security

security against overcoming obstacles Überwindungssicherung *f*

security against picking, picking security Aufsperr(ungs)sicherung *f*

security alarm Sicherungsalarm *m*

security and monitoring system Sicherungs- und Überwachungssystem *n*

security badge Sicherheitsausweis *m*

security bag Sicherungskoffer *m*

security barrier Sicherungsschranke, Sicherungssperre *f*

security cabinet, security cubicle Sicherungsschrank *m*

security camera Sicherungskamera *f*, Sicherungsphotoapparat *m*

security card Sicherungskarte *f*

security cased lock → cased security lock

security characteristic → security feature

security code Sicherungskode, Sicherungscode *m*

security consultant Sicherungsberater *m*

security consultancy, security consultation Sicherungsberatung *f*

security contractor Sicherungsunternehmer, Sicherungsdienstleister *m*

security (control) centre (Brit.); security (control) center (US); security (control) station Sicherungszentrale, Sicherungswarte, Sicherungsstation *f*

S

security delegate, security officer, authorized security representative Sicherungsbevollmächtigte, Sicherungsbeauftragte *m, f*

security directory Sicherungsadressbuch *n*

security door → armoured door (Brit.)

security engineer Sicherungsingenieur *m*

security engineering Sicherungstechnik *f*

security environment Sicherungsumgebung *f*

security equipment Sicherungsgerät(e) *n, (f)*

security exhibition, security show Sicherungsausstellung *f*

security feature, security characteristic Sicherungsmerkmal *n*

security fencing Sicherungsumzäunung, Sicherungseinzäunung *f*

security fitting Sicherungsarmatur *f*

security gate Sicherungstor *n*

security glazing Sicherungsverglasung *f*

security grill(e) Sicherungsgitter *n*

(security) guard Wächter, Wachmann *m*

(security) (guard) patrol Wachdienst *m*

(security) (guard) patrol company, (security) (guard) patrol firm Wachgesellschaft, Wach- und Schließgesellschaft *f*

(security) (guard) patrol tour (Wächter)Rundgang, Wach(rund)gang, Wachmannrundgang *m*

security hardware [*See remark under "(builders') iron supplies"*] Sicherungs-Kleineisenwaren *f*, Sicherungs-Kleineisen(zeug) *n*

security industry Sicherungsindustrie *f*

security kernel [*Computer programming. A part of an operating system that is protected form unauthorized access and implements basic system security*] Sicherungskern *m*

security level → level of security

security lighting, security illumination Sicherungsbeleuchtung *f*

security lock Sicherungsschleuse *f*

security lock → thief-resistant lock

security management system Sicherungsleit(ungs)-system, Sicherungsführungssystem *n*

security manual, security handbook Sicherungshandbuch *n*

security masonry wall Sicherungsmauer *f*

security measure Sicherungsmaßnahme *f*

security mechanism Sicherungsmechanismus *m*

security memory Sicherungsspeicher *m*

security mirror Sicherungsspiegel *m*

security monitoring Sicherungsüberwachung *f*

security net(work) Sicherungsnetz(werk) *n*

security object Sachwert *m*

security of data → data security

security of objects, object security Sachwertsicherung *f*

security of secrets, secret security Geheimhaltung *f*

security of supply, supply security Versorgungssicherung *f*

security operations Sicherungsbetrieb *m*

security padlock Sicherungs(vor)hangschloss *n* [*DIN 7465*]

security personnel → security staff

security product Sicherungserzeugnis, Sicherungsprodukt *n*

security professional Sicherungsfachmann *m*

security range Sicherungsbereich *m* [*selten: m*] [*Gebäude oder Gebäudeteile in denen sich die zu sichernden Sachen befinden. Ein Sicherungsbereich kann mehrere Meldebereiche umfassen*]

security screen Sicherungsblende *f*

security screw Sicherungsschraube *f*

security seal Sicherungsverschluss *m*

security sector Sicherungsbranche *f*, Sicherungssektor *m*, Sicherungswesen *n*

security service Sicherungsdienst *m*

security show → security exhibition

security software Sicherungsprogramme *f* in der Informationstechnik

security specialist Sicherungsfachmann *m*

security staff, security personnel Sicherungspersonal *n*

security station → security (control) centre (Brit.)

security system Sicherungssystem *n* [*Ein entsprechend realisiertes Sicherheitskonzept*]

security system Werkschutzanlage *f*, Werkschutzsystem *n*

security task Sicherungsaufgabe *f*

security training Sicherungsausbildung *f*

security turnstile Sicherungsdrehkreuz *n*

security wear Sicherungskleidung *f*

security window [*It prevents forcible entry*] Sicherungsfenster *n*

sedimentation plant, sedimentation installation [*water supply*] Absetzanlage *f*

sedimentation tank Absetzbehälter *m*

sediment pan [*A remov(e)able bucket for the collection of detritus or/and oil, if any of them, in a gull(e)y*] Schlammfang *m*

Seebeck effect Seebeck-Effekt *m*, Seebeck-Wirkung *f* [*Er beruht auf der Erscheinung, dass in einem nichtgeschlossenen Kreis zweier verschiedener Leiter, zwischen deren Enden eine Temperaqturdifferenz dT besteht, eine elektromotorische Kraft dE auftritt, die der Metallkombination eigentümlich und der Temperaturdifferenz proportional ist: dE* = e x dT]

seek Positionierung *f* [*Speicherlaufwerk*]

seek area, cylinder [*In automatic data processing, an area of a direct-access storage device, such as a magnetic disk file, assigned to hold records to which rapid access is needed, and located so that the physical characteristic of the device permit such access*] Positionierungszone *f*

seepage Einsickerung, Versickerung, Sickerung *f*

seepage line Sickerleitung, Versickerungsleitung, Einsickerungsleitung *f*

seepage opening, seepage port Sickeröffnung, Versickerungsöffnung, Einsickerungsöffnung *f*

seepage pit, dry well Sickergrube, Versickerungsgrube, Einsickerungsgrube *f*

seepage shaft Versickerungsschacht, Sickerschacht, Einsickerungsschacht *m*

segment [*In automatic data processing, a single section of an overlay program(me) structure, which can be loaded into the main memory when and as needed. ISO 2382*] Segment *n*

segment, dial segment (Zifferblatt)Segment *n*

segmental conductor Segment(strom)leiter *m*

segregating unit, segregation unit [*In automatic data processing, a punched-card machine that selects from a group of punched cards those cards satisfying certain criteria; it has only two output hoppers, one for cards satisfying the criteria and one for those that do not, in contrast with a punched-card sorter which has multiple outputs*] Lochkartenaussonderer, Aussonderer *m*

segregation Trennschottung *f*

segregation, (punched) card segregation Aussonderung, Lochkartenaussonderung *f*

seizing Fressen *n*

seizing load Fresslast *f*

to select (aus)wählen, selektieren

to select → to address

select and execute command (Aus)Wahl- und Ausführungskommando *n* [*Fernwirktechnik*]

selected active area → active selected area

selection → (setpoint) selection

selection (Aus)Wahl *f*

selection end → end of selection

selection manifold, selection central distributor Wahlverteiler(rohr), Umschaltverteiler(rohr) *m*, *(n)*

selection ratio Ansteuer(ungs)verhältnis *n* [*numerische Steuerung*]

selection signal end → end-of-selection signals

selective dump(ing) [*In automatic data programme, an edited or nonedited listing of the contents of selected areas of memory or auxiliary storage*] Archivierung, gezielte Archivierung *f*

selective frequency amplifier frequenzselektive Verstärker *m*

selectively operating circuit breaker Selektivleistungsschalter *m*

selective monitoring Selektivüberwachung *f*

selective radiator [*An object that emits electromagnetic radiation whose spectral energy distribution differs from that of a blackbody with the same temperature*] Selektivstrahler *m*

selective supervisory control Fernwirken *n*

selectivity of resonance, resonance selectivity Resonanzschärfe *f*

selector [*In automatic data processing, a computer device which interrogates a condition and initiates a particular operation dependent upon the report*] Wähler *m*

selector channel [*In automatic data processing, a piece of hardware which allows more than one high-speed input/output device to be accessed by line computer*] Wähl(er)kanal, Wahlkanal *m*

selector disc, selector disk Wähl(er)scheibe, Wahlscheibe *f*

selector handle Wähl(er)griff, Wahlgriff *m*

selector key Wähl(er)taste(r), Wahltaste(r) *f*, *(m)*

selector (switch), multi(ple-)contact switch [*A hand-operated multiposition switch*] Wahlschalter, Wähl(er)schalter *m*

selector valve Wähl(er)ventil, Wahlventil *n*

selenium rectifier Selengleichrichter *m*

self-acting → automatic

self-acting controller, automatic controller [*It generally uses a totally evacuated system with a liquid. The force created by the expansion of the contents in response to a temperature change is used to move a valve or damper spindle. A common example of this type is the thermostatic radiator valve*] selbsttätige Regler, automatische Regler *m*

self-adaptive [*The optimization of certain parameters of a function as a result of the function algorithm itself*] selbstadaptiv

self-adjusting selbstjustierend, selbstverstellend

self alarm Selbstalarm *m* [*Durch Unachtsamkeit ausgelöster unerwünschter Alarm*]

self-atomization Selbstzerstäubung *f*

self-ballasted mercury lamp → mixed-light lamp

self-checking number [*In automatic data processing, a number with a suffix figure related to the figure of the number, used to check the number after it has been transferred from one medium or device to another*] selbstkontrollierende Zahl *f*

self-clamping terminal, screwless terminal schraubenlose Klemme *f*

self-cleaning selbstreinigend

self-closing selbstschließend

self-closing corner valve Selbstschlusseckventil *n*

self-closing device → automatic closing device

self-closing fitting Selbstschlussarmatur *f*

self-closing solenoid valve, self-closing electromagnetic valve, automatic solenoid valve, automatic electromagnetic valve Magnetselbstschlussventil, Selbstschlussmagnetventil *n*

self-closing throughflow controller Selbstschluss(durchfluss)regler, Durchfluss-Selbstschlussregler *m*

self-closing valve → flow valve

self-combined → stand-alone

self-consumption → auto-consumption

self-contained [*Having all essential working parts except energy and control connections so contained in a case or framework that they do not depend on appliances or fastenings outside the machine*] selbstständig, unabhängig, eigenständig

self-contained unit → fan-coil unit

self-convection Eigenkonvektion *f*

self-cooling Eigenkühlung *f* [*Vollklimagerät*]

self-correction → auto-correction

self-draining soap tray, perforated soap tray Lochseifenschale, Seifenlochschale *f*

self drinking water supply, self drinkable water supply, self potable water supply Eigentrinkwasserversorgung; Eigenbrauchwasserversorgung *f* [*DIN 2001*]

self-extinguishing selbstlöschend

self-generating humidifier [*Unit producing steam at or about atmospheric pressure for injection, either directly into the air through a room fan unit, or directly into ducting through an injection manifold*] Eigendampfluftbefeuchter, Eigendampferzeuger zur Luftbefeuchtung *m*

self-ignition [*It results from self-heating*] Selbst(ent)zündung *f*

self-installation, self-mounting Selbsteinbau *m*

self-monitor circuit → watchdog circuit

self-monitoring Selbstüberwachend, eigenüberwachend

self-monitoring Selbstüberwachung, Eigenüberwachung *f*

self-monitoring functional duplication Funktionsbeteiligte Redundanz *f*

self-operated elevator (US); self-operated lift (Brit.) Selbstfahraufzug *m*

self-optimizing selbstoptimierend, lernfähig

self-optimizing algorithm lernfähige Algorithmus, selbstoptimierende Algorithmus *m*

self-powered detector, self-powered detecting device, S.P.D., self-powered detection device Neutronen-Beta-Messer *m* [*Er misst Neutronenflüsse ohne Hilfsenergie*]

self-powered (work(ing)) platform [*Its raising and lowering mechanism is located on its platform*] Selbstantrieb(s)(arbeits)plattform *f*

self-production Eigenerzeugung, Selbsterzeugung *f* [*z. B. von elektrischer Energie*]

self-protected starter Motorstarter *m* mit erhöhtem Kurzschlussschutz

self-regulating controlled system [*IEC 50(351)*], **self-regulating directly controlled member** [*IEC 27-2A*], **self-regulating ((closed) loop) control system, controlled system with self-regulation** Regelstrecke mit Ausgleich, Ausgleich(s)regelstrecke, Strecke mit Ausgleich *f* [*Die Regelgröße strebt einem Höchstwert zu*]

self-regulation value Ausgleich(s)wert *m* [*Reg(e)lung*]

self-resetting loop [*In automatic data processing, a loop whose termination causes the numbers stored in all locations affected by the loop to be returned to the original values which they had upon entry into a loop*] selbsttätige Rückstellschleife, selbsttätige Rücksetzschleife *f*

self-resetting Selbstrückhaltend

self-sealing grommet Würgenippel *m*

self service shoe polish(ing) machine Selbstbedienungs-Schuhputzmaschine *f*

self-sufficient, independent autark, unabhängig

S

self-sufficient heat pump installation → independent heat pump installation

self-superheating Selbstüberhitzen *n*, Selbstüberhitzung *f*

self supply Eigenversorgung, Selbstversorgung *f*

self-test [*The check(ing) each system performs at initialization to verify control logic and the accessibility of the registers*] Selbstkontrolle *f*

self-triggering programm (US); self-triggering programme [*In automatic data processing, a computer program(me) which automatically commences execution as soon as it is fed into the central processing unit*] selbstauslösende Programm *n*

self tuning regulator, STR, self tuning ((closed) loop) controller selbsteinstellende Regler *m*

self-ventilation Selbstlüftung, Fugenlüftung *f* [*Der bei geschlossenen Türen und Fenstern durch deren Undichtheiten (Fugen) infolge des durch Temperaturunterschied und Wind hervorgerufenen Druckunterschiedes zwischen innen und außen entstehende Luftwechsel*]

self-working → automatic

selsyn Drehmelder, Drehfeldgeber, Selsyn *m* [*Wegmessung*]

SELV → safety extra low voltage

semiadjust, half-adjust [*The least significant digit is dropped and, if the least significant digit is one-half or more of the number base, one is added to the next more significant digit and all carries are propagated*] halbaddiert

semianechoic room, semi-free field room, half-anechoic room, half-free field room halbschalltote Raum *m* [*ISO 3475*]

semiassembled representation, half-assembled representation halbzusammenhängende Darstellung *f* [*DIN 40 719, Blatt 1*]

semiautomatic defrosting, half-automatic defrosting [*The defrost(ing) cycle is started manually and ends automatically with automatic resumption of normal refrigerant at the end of defrost(ing) operation*] halbautomatische Entfrostung *f*

semiautomatic hoistway door, semiautomatic hoistway gate, half-automatic hoistway door, half-automatic hoistway gate halbautomatische Fahrschachttür, halbautomatische Aufzugschachttür *f*

semiautomatic mode, half-automatic mode (Betriebsart) Teilautomatik *f* [*Betriebsart, in der nur Teile der Steuerung oder des Programms ohne Bedienereingriff programmiert arbeiten. V DIN 19 237*]

semicarry, half carry, HC Zwischenübertrag *m*

semicoiled winding, half-coiled winding Halbspulenwick(e)lung *f*

semiconducting material, half-conducting material halbleitende (Werk)Stoff, Halbleiter(werk)stoff *m*, halbleitende Material, Halbleitermaterial *n*

semiconductor contactor, half-conductor contactor Halbleiterschütz *n*

semiconductor diode, half-conductor diode Halbleiterdiode *f*, Richtleiter *m*

semiconductor fuse, half-conductor fuse Halbleitersicherung *f*

semiconductor image sensor, half-conductor image sensor Halbleiterbildfühler *m*

semiconductor laser, half-conductor laser Halbleiterlaser *m*

semiconductor noise, half-conductor noise Halbleiterrauschen *n*

semiconductor plate, half-conductor plate Halbleiterplättchen *n*

semiconductor sensor, half-conductor sensor Halbleiterfühler *m*

semiconductor slice → (thin) semiconductor slice

semiconductor transistor, half-conductor transistor Halbleitertransistor *m*

semicontrolled, half-controlled halbgesteuert

semidelayed, half-delayed halbverzögert, mittelträge

semidirect lighting system, semidirect illumination system, half-direct lighting system, half-direct illumination system [*Lighting system or luminaire which directs from 60 to 90 per cent of the light downward*] halbdirekte Beleuchtung, vorwiegend direkte Beleuchtung *f*

semiduplex, half-duplex Halbduplexbetrieb *m* [*Nicht gleichwertige Übertragung von Daten zwischen zwei Stationen in beiden Richtungen (Sendung oder Empfang einer Station)*]

semiduplex interface, half-duplex interface, HDX interface Halbduplexschnittstelle *f*

semiflooded, half-flooded halbüberflutet [*Verdampfer*]

semifree field room → semianechoic room

semiglass door, half-glass door [*A door having glass in the panel above the lock rail*] Halbglastür *f*

semiglossy, half-glossy [*Intermediate between flat and glossy*] halbglänzend

semihermetic(ally-sealed), halfhermetic(ally-sealed), halbhermetisch (abgeschlossen), halbhermetisch gekapselt

S

semihigh-level (flushing) cistern, half-high-level (flushing) cistern Halbhoch(spül)kasten, halbhochhängende (Spül)Kasten *m*

semihermetic (refrigerant) compressor → accessible hermetic (refrigerant) compressor

semi-immersion(-type) rotary (water) meter, half-immersion(-type) rotary (water) meter Trockenläufer *m* [*Bei diesem Wassermesser befindet sich das Zählwerk im Wasser, das darüber liegende Zeigerwerk läuft „trocken"*]

semiindirect lighting system, semiindirect illumination system, half-indirect lighting system, half-indirect illumination system [*Lighting system or luminaire which directs from 60 to 90 per cent of the light upward*] halbindirekte Beleuchtung, vorwiegend indirekte Beleuchtung *f*

semimat, half-mat [*Immediate between glossy and mat, as photographic paper*] halbmatt

semineutral bus duct, half-neutral bus duct Schienenverteiler *m* mit Halbquerschnitt-N-Leiter

semipedestal, half-pedestal Halbsäule *f* [*Sie reicht nicht bis zum (Fuß)Boden*]

semipedestal lavatory, semipedestal (wash)basin, half-pedestel lavatory, half-pedestel (wash)basin Halbstandwaschbecken *n*

semipermanent, half-permanent teilweise bereitgestellt

semiradial pipe fan, half-radial pipe fan Halbradial-Rohrventilator *m*

semireflector, half-reflector [*dual-layer disk*] halbreflektierende Schicht *f*

semiround (roof) gutter, half-round (roof) gutter halbrunde Dach(regen)rinne *f*

semiround wood, half-round wood Halbholz, Breitholz *n* [*Zimmerei*]

semispace key, half-space key Halbschrittaste *f*

semispecular reflection, half-specular reflection halbgerichtete Reflektion, halbspiegelnde Reflektion *f*

semitone, halftone Aufrasterung *f* [*Nicht „Halbton" als photographische Wiedergabe*]

semitone storage cathode ray tube, half-tone storage cathode ray tube Halbtonspeicherröhre, monostabile Speicherröhre *f*

semitrack recording, half-track recording Halbspuraufnahme *f*

semiturning lock → German lock

semiturning (lock) bolt → German (lock) bolt

semi-wave rectifier, half-wave rectifier Halbwellengleichrichter *m*

semiwave voltage doubler, half-wave voltage doubler Greinacherschaltung *f*

semiword i/o buffer [*A buffer, the upper half being used to store the upper half of a word for both input and output characters, the lower half of the buffer being used for purposes such as the storage of constants*] Halbwort-Eingang-Ausgang-Puffer *m*

to send data with acknowledge, SDA senden von Daten mit Quittierung, senden von Daten mit Bestätigung

sending key Sendetaste, Absendetaste *f* [*Rohrpost(anlage)*]

sending readiness Sendebereitschaft *f*

sending station Sendestation, Absendestation *f* [*Rohrpost(anlage)*]

send protocol circuit handler Send(e)-Protokollsteuerung *f* [*Signalisierungssystem Nr. 7*]

send sequence number Send(e)folgenummer *f*

send signal(l)ing input/output handler Eingabe-Ausgabe-Steuerung *f* in Send(e)richtung [*Signalisierungssystem Nr. 7*]

send state variable Send(e)folgezähler *m*

sensation Empfinden *n*

sensation area Hörfläche *f*

sensation temperature Empfindungstemperatur *f*

sense amplifier Leseverstärker *m* [*Integrierte Schaltung, die auf eine Spannung in einem bestimmten Spannungsbereich reagiert und ein digitales Ausgangssignal abgibt*]

sense of vision [*It is based on the eye's ability to absorb and process selectively that portion of the electromagnetic spectrum that is called light*] Sehsinn, Gesichtssinn *m*

sensible air conditioner capacity → (useful) sensible air conditioner capacity

sensible cooler Umluftkühl(ungs)gerät *n*

sensible cooling load fühlbare Kühl(ungs)last, wahrnehmbare Kühl(ungs)last *f* [*Raumlufttechnik. Wärmestrom, der bei konstanter absoluter Luftfeuchte aus dem Raum abgeführt werden muss um eine angestrebte Lufttemperatur aufrechtzuerhalten*]

sensible dry heat → dry sensible heat

sensible heat fühlbare Wärme *f*

sensing Fühlung *f*

sensing area, sensing face aktive Fläche *f* [*Näherungsschalter*]

sensing device → detector

sensing distance → operating distance "S"

sensing element → sensor

S

sensing station → **reading station**

sensitive, susceptible [*Changing easily or quickly in the pretence of some force or agency; very responsive to external conditions, as to light, heat, etc.*] anfällig, empfindlich, sensibel, gefährdet

sensitiveness → **susceptibility**

sensitiveness level → **susceptibility level**

sensitiveness response characteristic → **susceptibility response characteristic**

sensitiveness response curve → **susceptibility response curve**

sensitive switch [*IEC 163*], **sensitive microswitch, quick-make-quick-break switch, fast-make-fast-break switch, rapid-make-rapid-break switch, snap(-action) switch, snap-acting switch, tumbler switch** [*A switch that responds to very small movements of its actuating button or lever and changes rapidly and positively from one contact position to the other; the trademark of one version is Micro Switch*] Schnappschalter, Sprungschalter, Kipp(hebel)schalter *m*

sensor, sensing device, pilot device, sensing element, (primary) detector, detecting element, detection element [*These devices may be responsive to air or water temperature, humidity, steam pressure, or air velocity, and include thermostats, aquastats, static pressure regulators, and similar devices. These are the "eyes and ears" of any control system. They may be electric, pneumatic, electronic, or a combination of any two or all three*] Sensor, Fühler, Detektor *m*, Detektionselement *n* [*Der Begriff „Sensor" ist umfassender als „Mess(wert)fühler", weil er alle Bauelemente zusammenfasst, die auf eine Umweltgröße messbar reagieren und diese in ein elektrisches Signal umsetzen. „Sensor" umfasst die Randbedingungen für Mikroelektronik: 1. Gute Voraussetzungen für eine Anpassung an eine hochentwickelte elektronische Datenaufbereitung bis zur vollständigen Integrierbarkeit. 2. Diese Bedingung wird häufig durch Massenfertigung ermöglicht*]

sensor act(uat)or level Sensor-Aktor-Ebene *f*

sensor cable Sensorkabel *n*

sensor-generated sensorerzeugt

sensor interface Sensorschnittstelle *f*

sensor monitoring Sensorüberwachung *f*

sensor row, sensor line Sensorreihe, Sensorzeile *f*

sensor signal Sensorsignal *n*

sensor signal processing Sensorsignalverarbeitung *f*

sensor switch Sensorschalter *m*

sensor system Sensorsystem *n*

SEP, separating character, separator Trennzeichen *n*

separately excited field fremderregte Feld *n*

separated compressor, split compressor Splitverdichter, Splitkompressor *m* [*Getrennt aufgestellte Verdichter*]

separate network Inselnetzwerk *n*

separate sewer Trennkanal *m*

separate shower Einzelbrause *f*

separate (source) feed, separate (source) supply, external (source) feed, external (source) supply Fremdeinspeisung *f*

separate system (of sewerage), separate system of sewers, separate sewerage system, separate drainage system [*A system of sewers in which sewage and storm water are carried in separate conduits*] Trennentwässerung, Trennkanalisation *f*, Trennsystem *n*

separate urinal Einzelurinal *n*

separating character → **SEP**

separation → **(filter) separation**

separation discharge gap, separation spark(ing) gap Trennfunkenstrecke *f* [*Im Blitzschutz zur Überbrückung von Näherungen eingesetzt*]

separation efficiency → **(filter) separation efficiency**

separation element Trennglied, Trennelement *n*

separation plate Trennblech *n*

separator → **SEP**

separator → **interceptor**

septic tank [*A purifier for sewage where no sewer is available. It is a tank through which sewage flows slowly enough for it to be purified by decomposition. It is divided into two or more chambers separated by scum boards with an opening below*] Faulgrube, Kleinkläranlage *f*, Klärbehälter *m* [*In Deutschland nur für Schmutzwasser zugelassen. DIN 4261*]

to sequence folgen, ablaufen

sequence → **order**

sequence, (work)flow [*In automatic data processing, the sequence in which events take place or operations are carried out*] Folge, Arbeitsfolge *f*, Ablauf, Arbeitsablauf *m*

sequence chain Folgekette, Ablaufkette *f*

sequence chart Ablaufdiagramm, Folgediagramm *n* [*DIN 40 719, Blatt 1*]

sequence check [*In automatic data processing, verification that correct precedence relationships are obeyed, usually by checking for ascending sequence numbers*] Folgekontrolle, Ablaufkontrolle *f*

sequence command Folgekommando, Ablaufkommando *n*

S

sequence computing → (work)flow computing

sequence controlled calculator Sequenzrechner *m* [*von Howard G. Aiken, 1944*]

sequence control signal message Ablaufmeldung, Folgemeldung *f* [*Eine Meldung über den Ablauf der Steuerung, z. B. durch Schrittanzeige*]

sequence error [*In automatic data processing, an error that arises when the arrangement of items in a set, for example, a deck of punch cards, does not follow some specified orders*] Folgefehler, Ablauffehler *m*

sequence instruction Folgebefehl, Ablaufbefehl *m*

sequence number [*In automatic data processing, a number assigned to an item to indicate its relative position in a series of related items*] Satzkennzahl, Satznummer *f*

sequence of bit transmission → order of bit transmission

sequence of operations, sequence of work, operations sequence, work sequence Arbeitsablauf, Betriebsablauf *m*, Arbeitsfolge, Betriebsfolge *f*

sequence of programming instructions → procedure

sequence of trades, trade sequence [*The order in which the various building and service trades carry out their work in a new building. This is the order in which the trade sections occur in a bill of quantities or specification*] Gewerkefolge *f*

sequence ((open) loop) control (system), scs, SCS, sequencing ((open) loop) control system Folgesteuerung, Ablaufsteuerung, ALS *f*

sequence processor Ablaufschaltwerk, Folgeschaltwerk *n* [*DIN 19 237*]

sequencer, sorter [*In automatic data processing, a machine which puts items of information into a particular order, for example, it will determine whether A is greater than, equal to, or less than B, and sort or order accordingly*] Sortierer *m*, Sortiermaschine *f*

sequencer [*A mechanical or electronic device that may be set to initiate a series of events and to make the events flow in given sequences*] Folgesteuer(ungs)gerät, Folgesteuer(ungs)werk, Ablaufsteuer(ungs)gerät, Ablaufsteuer(ungs)werk *n*

(sequence(r)) step Ablaufschritt, Schritt, Folgeschritt *m* [*Die kleinste funktionelle Einheit des Programms von Ablaufsteuerungen. E DIN 19 226, Teil 5*]

sequence selection Ablaufauswahl, Folgeauswahl *f*

sequence table Ablauftabelle, Schalttabelle, (Schalt)Folgetafel, Folgetabelle *f* [*DIN 40 719, Blatt 1*]

sequence timer Zeitablaufglied, Zeitfolgeglied *n*

sequence valve Sequenzventil, Folgeventil *n*

sequential access [*A process that involves reading or writing data sequentially*] sequentielle Zugriff *m*

sequential analysis [*The continuous analysis of data, obtained via sampling, performed as the amount of sampling increases*] sequentielle Analyse *f*

sequential circuit, serial circuit, dynamic digital circuit Sequentielle Schaltung, Ablaufschaltung, serielle Schaltung, Folgeschaltung, dynamische Schaltung *f* [*E DIN 44 300/10.85*]

sequential ((closed) loop) control, sequential closed control Folgereg(e)lung, Folgeregulierung *f*, geschlossene Folgewirkungsablauf *m* [*Siehe Anmerkung unter „Reg(e)lung"*]

sequential color television (US); sequential colour television (Brit.); sequential system [*A colo(u)r television system in which the primary colo(u)r components of a picture are transmitted one after the other; the three basic types are the line-sequential, dot-sequential, and field-sequential colo(u)r television system*] sequentielle Farbfernsehen *n*

sequential data storage sequentielle Datenspeicherung *f*

sequential examination → scanning

sequential function chart, SFC, control system flowchart, CSF, control system function chart Funktionsplan, FUP *m* [*Jede Steuerungsaufgabe (= Funktion) erhält ein entsprechendes Signal*]

sequential key Folgetaste(r) *f*, *(m)*

sequential logic(al) ((open) loop) control logische Ablaufsteuerung, logische sequentielle Steuerung, sequentielle logische Steuerung, logische Folgesteuerung *f*

sequential logic(al) system Schaltwerk *n* [*Die Ausgangssignale sind abhängig vom zeitlich verschiedenen Eintreffen der Eingangssignale*]

sequential number Ablaufnummer, Folgenummer *f*

sequential ((open) loop) control, sequential open control Folgesteuerung, Ablaufsteuerung *f* [*Leistungselektronik*]

sequential operation [*In automatic data processing, the flow of information through a computer in time sequence, using only one digit, word, line or channel at a time*] Folgebetrieb, Ablaufbetrieb *m*

sequential potentiometer → slave potentiometer

sequential processing [*Processing items in a collection of data according to some specified sequence of keys, in contrast to serial processing*] Folgeverarbeitung, Ablaufverarbeitung *f*

sequential switch Folgeschalter *m*

sequential switching Folgeschalten, Ablaufschalten *n*

sequential system → **sequential color television (US)**

serial seriell, Reihen..., Folge...

serial access serielle Zugriff, Reihenfolgezugriff *m*

serial access memory Speicher mit seriellem Zugriff, Reihenfolgezugriffsspeicher *m*

serial active poll(ing) state aktive Serienabfragezustand *m*

serial arsonant, serial fire setter Serienbrandstifter *m*

serial boiler → **sequential boiler**

serial boiler circuit → **sequential boiler circuit**

serial bus serielle Bus *m*

serial data transmission [*Data characters or bytes are transmitted in sequence one at a time over a single path*] serielle Datenübertragung *f*

serial digital computer, digital serial computer [*A digital computer in which the digits are handled serially, although the bits that make up a digit may be handled either serially or in parallel*] serielle Digitalrechner *m*

serial feed [*The method of placing cards in the feed hopper of a punched-card machine in which one of the short edges of the card enters the machine first, so that the conditions of the card are red serially*] serielle Zuführung *f*

serial fire setter → **serial arsonant**

serial interface serielle Schnittstelle, serielle Eingabe/Ausgabe *f* [*Diese Schnittstelle erfordert serielle Übertragung oder einen Nachrichtentransfer, bei dem die Bits, die ein Zeichen bilden, sequentiell gesendet werden. Läuft über einen Einzelkanal. Gegensatz von seriell: parallel*]

serially reusable [*An attribute possessed by a program(me) that can be used for several tasks in sequence without having to be relocated into main memory for each additional use*] seriell wiederverwendbar

serial mode serielle Betrieb *m*

serial mouse serielle Maus *f*

serial operation [*The flow of information through a computer in time sequence, using only one digit, word, line, or channel at a time*] serielle Betrieb *m* [*Eine Betriebsart einer Funktionseinheit, bei der die Aufträge nacheinander abgewickelt werden. DIN 44 300, Teil 9*]

serial parallel [*Descriptive of a device which converts a serial input into a parallel output*] seriell parallel

serial-parallel converter Serien-Parallel-Umsetzer, Serien-Parallel-Umwandler *m*

serial planning serielle Planung *f*

serial printer, character printer, serial writer, character writer serielle Drucker, Zeichendrucker *m*

serial processing [*Processing items in a collection of data in the order that they appear in a storage device, in contrast to sequential processing*] serielle Verarbeitung *f*

serial programming [*In computers, programming in which only one operation is executed at one time*] serielle Programmieren *n*

serial storage [*Computer storage in which time is one of the coordinates used to locate any given bit, character, or word; access time, therefore, includes a variable waiting time, ramping from zero to many word times*] serielle Speicherung *f*

serial storage architecture, SSA serielle Speicherarchitektur *f*

serial transfer [*Transfer of the characters of an element of information in sequence over a single path in a digit computer*] serielle Übergabe, serielle Weitergabe, serielle Übermitt(e)lung *f*

serial transmission [*Transmission of groups of elements of a signal in time intervals that follow each other without overlapping*] serielle Übertragung *f*

serial universal bus → **universal serial bus**

serial writer → **serial printer**

series → **family**

series circuit Reihenschaltung *f*

series compensation, cascade compensation, tandem compensation [*Compensation in which the compensator is placed in series with the forward transfer function*] Serienkompensation *f*

series connection → **connection-in-series**

series element Vorschaltglied *n*

series gap Reihenfunkenstrecke, Löschfunkenstrecke *f*

series multiple [*Type of switchboard jack arrangement in which a single line circuit appears before two or more operations, all appearances being connected in series*] Gruppenschaltung *f*

series of events, event series Ereignisserie *f*

series of jets, jet series [*air washer*] Düsenstock *m*

series-parallel circuit, series-parallel connection [*A circuit in which some of the components or elements are connected in parallel, and one or more of these parallel combinations are in series with other components of the circuit*] Reihenparallelschaltung *f*, Serienparallelkreis *m*

series resonance Reihenresonanz *f*

serious fire, major fire Großbrand *m*, Großfeuer *n* [*Ein Brand, zu dessen Bekämpfung mehr als drei C-Rohre eingesetzt werden (1 B-Rohr = 2 C-Rohre*]

S

series(-wound) generator [*A generator whose armature winding and field winding are connected in series*] Hauptschlussgenerator, Reihenschlussgenerator *m*

serpentine cooler → coil cooler

server Diensterbringer, Erbringer, Server, Dienstanbieter, Anbieter, Dienstleister *m* [*Ein Prozessor, durch den das Netzwerk um einen weiteren Dienst ausgebaut wird. Er besteht aus Hardware und Software. Ein typischer Server ist beispielsweise ein Netzdrucker. DIN 19 245, Teil 2*]

server station, SS (Dienst)Erbringerstation, (Dienst)-Anbieterstation, Serverstation, SS *f*

to service warten

service → operation

service, serv(ic)ing Wartung *f*, Warten *n* [*Vorbeugende oder fehlerbedingte Maßnahme zur Bewahrung des technischen Sollzustandes eines Systems, einer Anlage, usw.*]

service Dienst(leistung) *m,(f)*

serviceability [*The ease within which equipment in buildings can be serviced for maintenance purposes*] Unterhaltungsfreundlichkeit *f*

serviceability → (operational) serviceability

serviceability [*Capability of a structure, assembly, component or product to perform the function(s) for which it was designed and produced over at least a specified time, based on assumed and/or specified levels of use and maintenance*] Brauchbarkeit, Gebrauchsfähigkeit, Funktionstüchtigkeit *f*, Gebrauchsvermögen *n*

serviceable [*In proper service condition*] funktionstüchtig, gebrauchsfähig

service access point, SAP Dienstzugang(s)punkt *m* [*E DIN ISO 7498*]

service agency Kundendienststützpunkt *m*

service attribute Dienstmerkmal *n* [*ISDN*]; Merkmal(s)-attribut *n* [*X.32*]

S

service bit/digit → housekeeping digit

service box → (house) service box

service building Betriebsgebäude, Technikgebäude *n* [*Gebäude mit Betriebsanlagen*]

service cable → connecting cable

service cable duct → connecting cable duct

service cable (entrance) box → connecting cable (entrance) box

service capacity Nennleistung *f* [*Elektromotor*]

service call Dienstgespräch *n*

service chute → (building) service chute

service chute → (inclined) (building) service chute

service chute → (vertical) (building) service chute

service computer Bedien(ungs)rechner *m* [*Vermittlung*]

service connection Betriebsanschluss *m*, Betriebsverbindung *f*

service connector Dienstwähler *m*

service contract Dienstleistungsvertrag *m*

service contractor Dienstleist(ungsunternehm)er *m*

service control center (US); service control centre (Brit.) Sicherungszentrale *f*

service core, mechanical core, building core, utility core Betriebskern, Gebäudekern *m*

service corridor Betriebsflur, Gebäudeflur *m*

service corridor Dienstflur *m* [*Hotel*]

service counter Ausgabetresen *m*

service cylinder, gas bottle, small refrigerant cylinder [*It is used by a service person*] Kältemittel-Kleinzylinder *m*

service data → capacity data

service data unit, SDU Dienstdateneinheit *f*

service distribution cabinet, service distribution cubicle Anschlussverteilerschrank *m*

service drop [*The overhead service wires that serve a building*] Hausanschlussfreileitung *f*

service duct Betriebskanal *m*

service elevator (US); service lift (Brit.) [*It has passenger classification but is primarily used for moving materials and other freight. It is used predominantly in buildings where people other than employees may ride on the elevator*] Betriebsaufzug *m*

service employment Dienstleistungsbeschäftigung *f*

service engineering, serv(ic)ing engineering Wartungstechnik *f*

service enterprise, service undertaking Dienstleistungsunternehmen *n*, Dienstleistungsbetrieb *m*

service entrance point Anschlussstelle *f* [*In einem Haus oder Gebäude zum Anschluss öffentlicher Leitungen*]

service entry Einführung *f*, Hausanschluss *m* [*elektrische Hausanlage*]

service facility Dienst(leistungs)einrichtung *f*

service-friendly wartungsfreundlich

service function Dienstfunktion *f*

service grade → quality of service

service guide Dienstleistungsführer *m*

service incident Dienstvorfall *m*, Dienstvorkommnis *n*

service indicator Dienstekennung *f* in ISDN

service industry Dienstleistungsindustrie *f*

service instruction Wartungsbefehl *m*

service key Betriebstaste *f*

service layer Dienstschicht *f*

service level Dienstniveau *n* [*Beschreibt die Unterschiede verschiedener Gebäude hinsichtlich der Häufigkeit von Reinigungsleistungen, des Sicherungsniveaus des Wachdienstes usw.*]

service level → capacity level

service life → life (cycle)

service life monitoring, life (cycle) monitoring Lebensdauerüberwachung *f*

service lift → dumbwaiter

service management Dienstleistungsverwaltung, infrastrukturelle Gebäudebewirtschaftung *f* [*Leistungen für Werterhaltung, Nutzung und Sicherheit eines Gebäudes*]

service-off duration → loss-of-service duration

service ((open) loop) control Dienststeuerung *f*

service (operation) level Betriebsebene *f*

service operation unit → direct operation unit

service outage duration → loss-of-service duration

service penetration, line penetration Leitungsdurchgang *m*

service personnel → service staff

service pipe (öffentliche) Versorgungsrohr *n*

service pipe → (branch) service pipe

service (pipe)line (öffentliche) Versorgungs(rohr)leitung *f*

service place Dienstleistungsort *m*

(service-)primitive [*An abstract, implementation-independent representation of an interaction between the service-user and the service-provider (ISO TR 8509)*] Dienstelement *n* [*E DIN ISO 7498*]

service provider [*An abstract of the totality of those entities which provide a service to peer service-users (ISO TR 8509)*] Diensterbringer, Dienstleister *m* [*E DIN ISO 7498*]

service quality → quality of service

service range → capacity range

service regulation → capacity regulation

service related dienstbezogen

service request, SRQ Bedienungsaufruf *m*, Bedienungsanforderung *f*

service reservoir → balancing reservoir

service riser [*A gas pipe which rises to supply an upper floor*] Gasstandrohr, Gassteig(e)rohr *n*

service routine [*deprecated*] **→ run time system**

service routine [*A section of a computer code that is used in so many different jobs that it cannot belong to any one job*] Wartungsteilprogramm *n*

service run, service line, service raceway Betriebsleitung *f* [*z. B. für Heizung, Wasser usw.*]

services duct, duct for services Installationskanal, Betriebskanal *m*

service-sealed condensing unit [*A sealed compressor unit in which the housing is sealed by one or more gaskets, and means of access are provided for servicing internal parts in the field*] zugängliche Kondensatoreinheit, zugängliche Verflüssigereinheit *f*

service shaft → (public) utility shaft

service sink → slop sink

service sink grating → slop sink grating

service space Betriebsraum *m* [*Volumen zur Unterbringung einer Betriebsanlage, z. B. eines Entwässerungsrohrleitungssystems in einem Gebäude*]

service staff, service personnel Dienstleistungspersonal *n*, Dienstleistungsbelegschaft *f*

service steam Betriebsdampf *m*

service sticker [*data processing*] Wartungsetikett *n*

service switch → operating switch

service tap (line) Hausanschluss(leitung) *m*, (*f*)

service terminal → operator terminal

service trolley Servierwagen *m*

service undertaking → service enterprise

service user [*An entity in a single open system that makes use of a service through service-access-points (ISO TR 8509)*] Dienstbenutzer *m* [*E DIN ISO 7498*]

service value Gebrauchswert *m*

service valve Betriebsventil *n*

service voltage → utilization voltage

service warm water → warm service water

service water Brauchwasser, Betriebswasser, Gebrauchswasser *n* [*Wasser zum Gebrauch für verschiedene Zwecke; es muss keine Trinkwassergüte haben. DIN 4046*]

service water heating Brauchwasser(er)wärmen, Betriebswasser(er)wärmen, Gebrauchswasser(er)wärmen *n* [*„...erhitzen" und „...erwärmung" können synonym für „...(er)wärmen" verwendet werden*]

service water storage (type) heater → storage (type) service water heater

serv(ic)ing → service

S

serv(ic)ing engineering → service engineering

serving cart (on casters), portable server Servierwagen *m*

servo-act(uat)or Servoaktor *m* [*Siehe Anmerkung unter „Aktor"*]

servo-amplifier [*An amplifier used in a servo-mechanism*] Servoverstärker *m*

servo-assisted valve → servo-operated valve

servo ((closed) loop) control → follow-up ((closed) loop) control

servo ((closed) loop) control system, follow-up ((closed) loop) control system Servoregel(ungs)system, Servosystem *n*

servo(-controlled) drive, actuating drive Servoantrieb, Stell(ungs)antrieb *m*

servo control system → servo ((closed) loop) control system

servo control system → servo ((open) loop) control system

servo-drive Stellantrieb *m* [*Er dient zur direkten Einwirkung auf eine Regelstrecke oder Steuerstrecke. Er verstellt das Stellglied wenn es mechanisch betätigt wird*]

servo-feedback system Servorückführ(ungs)system *n*

servo loop → ((closed) loop) control circuit

servo measuring element, servo measuring mechanism Servomesswerk *n* [*Das Messsignal wird mit elektrischer oder pneumatischer Hilfsenergie verstärkt*]

servo-motor → positioning motor

servo-multiplier [*An electromechanical multiplier in which one variable is used to position one or more gauged potentiometers across which the other variable voltages are applied*] Folgereglermultiplikator *m*

servo ((open) loop) control Servosteuerung *f*

servo ((open) loop) control system, follow-up ((open) loop) control system vorgesteuerte System *n*

servo-operated valve, pilot-controlled valve, servo-assisted valve [*Valve operated by a mechanism that directly responds to some controlled conditions*] vorgesteuerte Schieber *m*

servo-plotting board [*data processing*] Folgereglerregistriertisch *m*

servo-signal → actuating signal

session layer, layer 5 Sitzungssteuer(ungs)schicht, Kommunikationssteuer(ungs)schicht, Kommunikationssteuer(ungs)ebene, fünfte Schicht, fünfte Ebene, Schicht 5, Ebene 5 *f* [*OSI*]

session protocol machine → SPM

session service → SS

session service user → SS user

to set, to adjust, to vary, to control (ein)stellen

to set → to float

set Menge *f* [*Fuzzy-Logik*]

set → assembly

set and plug valve → globe valve

setback Rückstellung *f*

setback [*Reduction of heating or cooling during hours when a building is unoccupied, or during periods when lesser demand is acceptable*] Absenkung *f*

setback heating resistor Absenkung(s)heizwiderstand *m*

set construction method → assembly construction method

set fault switch(ing) off → assembly fault switch(ing) off

set fault switch(ing) over → assembly fault switch(ing) over

set of assembly bolts Anreihverbindungen *f*

set of characters → character set

set of contacts [*IEC 214*] Kontaktsatz *m* [*Trafostufenschalter, VDE 0532, Teil 30*]

set of drawings → drawing(s) set

(set of) fundamental points → basis

set of punched cards → deck

set operator Mengenoperator *m*

setpoint → setpoint (value)

setpoint adjuster, setpoint setter, setpoint device Sollwerteinsteller, Sollwertversteller, Sollwertgeber, Vorgabewerteinsteller, Vorgabewertversteller, Vorgabewertgeber *m*

setpoint compensator Funktionsbildner *m* für Sollwertaufschaltung, Sollwertaufschalt(ungs)glied, Vorgabewertaufschalt(ungs)glied *n*

setpoint control Sollwertführung *f*

setpoint control block, setpoint control module Sollwertführungsbaustein, Vorgabewertführungsbaustein *m*, Sollwertführungsmodul, Vorgabewertführungsmodul *m*

setpoint generator, setpoint device, schedule setter Sollwertgeber, Vorgabewertgeber *m*

setpoint pointer Sollwertzeiger, Vorgabewertzeiger *m*

setpoint potentiometer Sollwertpotentiometer *n*, Sollwertgeber, Sollwerteinsteller *m* [*Gerät, mit dem eine konstante Aufgabengröße (Sollwert) in die Regelvorrichtung oder in den Regler eingegeben wird*]

S

(setpoint) selection (Sollwert)(An)Wahl, Vorgabewert(an)wahl *f*

setpoint setter Sollwertsteller, Vorgabewertsteller *m*

setpoint signal Sollwertsignal, Vorgabewertsignal *n*

setpoint (signal) matching Sollwertanpassung, Vorgabewertanpassung *f*

setpoint station, setpoint device Sollwerteinsteller, Sollwertgeber, Vorgabewerteinsteller, Vorgabewertgeber *m* [*Zur Umwandlung elektrischer Signale eines Prozessrechners in Sollwert für einen automatischen Regler; nach Umschaltung kann der Sollwert von Hand verstellt werden*]

setpoint stepchange Sollwertsprung, Vorgabewertsprung *m*

setpoint (value) Sollwert, eingestellte Führungsgrößenwert, Vorgabewert *m*

set recognition Satzerkennung *f* [*Grammatik*]

set-reset flip-flop (circuit) → RS flip-flop (circuit)

set square Zeichendreieck *n*

setting [*e. g. room setting*] Ausgestaltung *f* [*z. B. Zimmerausgestaltung*]

setting Stellen, Einstellen *n* [*Das Verändern von Masse-, Energie- oder Informationsflüssen durch Stellglieder, Das Stellen kann kontinuierlich oder durch Schaltvorgänge erfolgen. In der Leittechnik werden die Stellglieder zum Prozess gerechnet*]

setting Einstellung, Vorgabe *f*

set(ting) back Rückstellung *f*

setting button (Ein)Stell(ungs)knopf *m*

set(ting) collar (Ein)Stell(ungs)ring *m*

set(ting) command (Ein)Stell(ungs)kommando *n*

set(ting) dial (Ein)Stell(ungs)ziffernblatt *n*

set(ting) instruction (Ein)Stell(ungs)befehl *m*

setting key (Ein)Stell(ungs)schlüssel *m*

set(ting) mark (Ein)Stell(ungs)marke *f*

set(ting) power (Ein)Stell(ungs)kraft *f*

set(ting) range (Ein)Stell(ungs)bereich *m* [*selten: n*]

setting-up mode, initial setting mode Betriebsart *f* Einrichten [*Die Stellgeräte werden einzeln durch Eingriffe des Bedienenden unter Umgehung von Verriegelungen gesteuert. V DIN 19 237*]

set(ting) value (Ein)Stell(ungs)wert *m*

settling basin, settlement basin Absetzbecken *n*

settling time [*ANSI C 85.1*]**, correction time, transient recovery time** [*IEC 478-1*] Beruhigungszeit, Einschwing(ungs)zeit, Ausregel(ungs)zeit *f* [*Zeitspanne, die beginnt, wenn der Wert der Regelgröße nach einem Sprung der Störgröße oder der Führungsgröße einen vorgegebenen Betrag der Sollwertabweichung überschreitet, und die endet, wenn er in den Bereich der zulässigen Sollwertabweichung zum dauernden Verbleib wieder eintritt. E DIN 19 226 T 5*]

set-top-box Zusatzterminal *n*, TV-Aufsatz *m* mit Internetzugang

to set up, to turn up [*The edge of sheet lead*] aufkanten, hochkanten

setup Miniprogramm *n* [*In ihm legt der Rechner seine Konfiguration ab*]

setup Anpassung *f* [*Standardanwendung auf die Hardware*]

setup → facility

setup Anordnung *f*, Aufbau *m*

setup input Vorbereitungseingang *m* [*logische Schaltung*]

setup (interval) [*The ratio between the reference black level and the reference white level in television, both measured from the blanking level; usually expressed as a percentage*] Schwarzabhebung *f*

setups → facilities

setup time [*The time before, after, and between computer machine runs, in which manual tasks are carried out, such as changing tape reels or transporting tapes, cards or supplies to and from the computer equipment to prepare for the new run*] Vorhaltezeit *f*

setup time [*IEC 147*] Vorbereitungszeit *f* [*Zeitunterschied zwischen bestimmten Signalpegeln*]

set value Sollwert *m*

set-value ((closed) loop) control → fixed setpoint ((closed) loop) control

set value controller → fixed setpoint controller

set value ((open) loop) control → fixed setpoint ((open) loop) control

set value regulator → fixed setpoint controller

seven-day dial Sieben-Tage-Zifferblatt *n*

sewage; wastewater (US) Abwasser *n* [*Zum Zweck der Entsorgung in die Kanalisation abgeleitetes Wasser. Man unterscheidet grundsätzlich: Schmutzwasser, Regenwasser (Niederschlagswasser), Schmutz- und Regenwasser (Mischwasser). Zur Erläuterung der Herkunft des Abwassers sind Benennungen wie – häusliche Schmutzwasser, gewerbliche Schmutzwasser, Oberflächenwasser, Kühlwasser usw. anzuwenden*]

sewage analysis; wastewater analysis (US) Abwasseranalyse *f*

sewage charge → sewage rate

S

sewage collection system; wastewater collection system (US) [*The sewer and pumping system used for collection and conveyance of domestic, commercial, and industrial sewage*] Abwassersammelanlage *f*, Abwassersammelsystem *n*

sewage composition; wastewater composition (US) [*The chemical, biological, and physical characteristics of the solid and liquid constituents of sewage, and their relationship*] Abwasseraufbau *m*

sewage condition, sewage state, wastewater condition, wastewater state (US) [*Expressions defining the physical and biological state of sewage as between freshness and staleness, or as between aerobic and anaerobic*] Abwasserbeschaffenheit *f*, Abwasserzustand *m*

sewage decomposition; wastewater decomposition (US) [*Transformations of organic or inorganic materials contained in sewage through chemical or biological processes*] Abwasserzersetzung *f*

sewage disposal; wastewater disposal (US) Abwasserentsorgung *f*

sewage effluent; wastewater effluent (US) Abwasserablauf, Abwasserabfluss, Abwasserauslauf *m*

sewage ejector → ejector

sewage engineering; wastewater engineering (US) Abwassertechnik *f*

sewage facilities; wastewater facilities (US) [*The structures, equipment, and procedures required to collect, convey and treat sewage, and dispose of the effluent and sludge*] Abwasseranlagen *f*

sewage flow; wastewater flow (US) Abwasserabfluss *m*

sewage gas; wastewater gas (US) [*Gas produced during the digestion of sludge*] Faulgas, Biogas *n* [*DIN 4045*]

sewage heat; wastewater heat (US) Abwasserwärme *f*

S

sewage heat recovery; wastewater heat recovery (US) Abwasserwärmerückgewinnung, Abwasserwärmeverwertung, Abwasserwärmenutzung *f*

sewage influent; wastewater influent (US) [*Sewage as it enters a treatment plant or pumping station*] Einlaufabwasser *n*

sewage installation → sewage plant

(sewage) lift → ejector

sewage outfall; wastewater outfall (US) [*The outlet or structure through which sewage is finally discharged*] Abwasserauslauf(bauwerk) *m*, (*n*)

sewage noise; wastewater noise (US) Abwassergeräusch *n*

sewage management; wastewater management (US) [*The collection and treatment of wastewater including such ancillary matters as administration and financing, engineering, planning and design, operation and maintenance, and monitoring and evaluation of overall performance*] Abwasserbewirtschaftung *f*

sewage outlet; wastewater outlet (US) [*The point of final discharge of sewage*] Abwasserablass *m*

sewage oxidation; wastewater oxidation (US) [*The process by which the organic matter is converted to a stable or a mineral form*] Abwasseroxidation *f*

sewage plant, sewage installation; wastewater plant, wastewater installation (US) Abwasseranlage *f*

sewage purification → sewage treatment

sewage purifier; wastewater purifier (US) Kläranlage *f* [*In Deutschland nur für Schmutzwasser zugelassen. DIN 4261*]

sewage rate, sewage charge; wastewater rate, wastewater charge (US) [*A charge, or a schedule of charges, for the collection, or the collection and treatment, of wastewater to users who are connected to the system. It may be based on water consumption, wastewater flow, wastewater strength, number and types of plumbing fixtures, or some condition of these*] Abwassergebühr *f*

sewage reclamation; wastewater reclamation (US) [*Processing of sewage for reuse*] Abwasserwiederaufbereitung *f*

sewage removal → sanitary drainage

sewage renovation; wastewater renovation (US) [*Treatment of sewage for reuse*] Abwasserwiederbehandlung *f*

sewage sewer; wastewater sewer (US) Abwasserkanal *m*, Abwasserleitung *f*

sewage sludge; wastewater sludge (US) Klärschlamm, Abwasserklärschlamm *m*

sewage state → sewage condition

sewage strength; wastewater strength (US) Abwasserkonzentration *f*

sewage survey; wastewater survey (US) [*An investigation of the quality and characteristics of each sewage stream, as in an industrial plant or municipality*] Abwasserstromuntersuchung *f*

sewage system; wastewater system (US) [*Combination of facilities for the collection, treatment, and discharge of sewage*] Abwasseranlage *f*, Abwassersystem *n*

sewage treatment, sewage purification; wastewater treatment, wastewater purification (US) [*Any process to which sewage is subjected in order to remove or alter its objectionable constituents and thus render it* *less offensive or hazardous to human health and the environment*] Abwasserbehandlung *f*, Abwasseraufbereitung, Abwasserreinigung, AWR

sewage treatment plant, sewage treatment system, sewage treatment installation; wastewater treatment plant, wastewater treatment installation, wastewater treatment system (US) [*The central facility of wastewater treatment facilities which contains all treatment processes exclusive of the collection system. "Purification" may synonymously be used for "treatment"*] Abwasseraufbereitungsanlage, Abwasserbehandlungsanlage, Abwasserreinigungsanlage, Kläranlage *f*, Abwasseraufbereitungssystem, Abwasserbehandlungssystem, Abwasserreinigungssystem *n*

sewage type; wastewater type (US) Abwasserart *f*

sewage works; wastewater works (US) [*All-inclusive terms for sewage collection, dumping, treatment and disposal facilities. Terms declining in use*] Abwassersammelanlage *f*, Abwassersammelsystem *n*

sewer Abwasserleitung *f*, Kanal *m*

sewerage Kanalisation *f*

sewer gas [*The foul air and vapo(u)r from sewers*] Kanalgas, Abwasserleitungsgas *n*

SFC → sequential function chart

SFD → simple formattable document

S.G. → specific gravity

shaded pole motor [*Small induction motor having a shaded pole for the purpose of starting a very low starting torque*] Spaltpolmotor, Hilfspolmotor *m*

shader prozedurale Textur *f*

shade temperature Schattentemperatur, Temperatur im Schatten *f*

shading Übergang *m* [*Vom Drahtgittermodell eines dreidimensionalen Objekts zum Objekt mit Außenflächen*]

shading [*Representing shadows on drawings*] Abschattung, Verschattung *f*

shading → sectioning

shading range → range of shading

shadow → multi(ple)-path signal

shadow voltaic wings (multifunktionale) Beschattungs- und Photovoltaiksystem *n*

shaft, building shaft (Gebäude)Schacht *m*

shaft → (escalator) well

shaft → hoistway

shaft [*machine part*] Welle *f*

shaft-angle digitizer (digitalisierende) Drehwinkelmessumformer, Winkelschrittgeber *m*

shaft door → hoistway door

shaft encoder, rotary encoder kodierte Drehgeber, Kodedrehgeber *m* [*Gerät zur Umsetzung analoger Daten, die einen bestimmten Winkel angeben, in kodierte, di-gitale Daten*]

shaft fan Schachtventilator *m*

shaft girder → (escalator) well girder

shaft head Schachtkopf *m*

shaft horsepower → brake horsepower

shaft pit Schachtgrube *f*

shaft seal(ing) Wellen(ab)dichtung *f*

shaft security Schachtsicherung *f*

shaft ventilation, shaft vent(ilat)ing, well ventilation, well vent(ilat)ing [*See remark under "ventilation"*] Schachtlüftung, Schachtventilation *f* [*Der zu lüftende Raum hat einen über Dach geführten Schacht*]

shaft(way) → hoistway

shaft(way) access → elevator hoistway access

shaft(way) air removal → hoistway air removal

shaft(way) biparting door → hoistway biparting door

shaft(way) door → hoistway door

shaft(way) enclosure → hoistway enclosure

shaft(way) gate → hoistway door

shaft(way) machine → hoist(ing) machine

shaft(way) switch (US) → (lift) well switch (Brit.)

shaft(way) pit → hoistway pit

shake → fishing wire

shaking grate Schüttelrost *m*

shallow luminaire, shallow light(ing) fitting [*See remark under "luminaire"*] Flachleuchte *f*, Flachbeleuchtungskörper *m*

shallow pan, shallow bowl Auswaschbecken, Flachspülbecken *n* [*(Spül)Abortbecken*]

shallow pan closet, shallow bowl closet, shallow pan toilet, shallow bowl toilet Auswaschabort, Flachspülabort *m*, Auswaschtoilette, Flachspültoilette *f*, Auswaschklosett, Flachspülklosett *n* [*DIN 1381*]

S & H amplifier → sample-and-hold amplifier

shampoo spray Kopfwaschbrause *f*

shape → profile

shape → configuration

shaped beam antenna (US); shaped beam aerial Strahlformungsantenne *f*

shape for interior work → profile for interior work

shape memory Formgedächtnis *n*

shape-retaining quality, shape-retaining stability Gestalthaltigkeit, Gestalttreue *f*

shaping Gestalten *n*, Gestaltung *f*

shared (computer) memory gemeinsam benutzte (Rechner)Speicher *m*

shared 1:N protection 1:N-Ersatzschaltung *f* [*Ringnetz. Für N genutzte Wege wird ein Ersatzweg bereitgehalten. Auf dem Ersatzweg kann Verkehr mit geringer Priorität geführt werden, der im Fall einer Ersatzschaltung unterbrochen wird, damit der geschützte Verkehr diese Resource nutzen kann*]

shareware (utility) probeweise Hilfsprogramm *n* [*Zuerst ausprobieren und dann bezahlen*]

sharpening Schärfen *n*

sharpness control system → **SCS**

sharpness of resonance, resonance sharpness Resonanzschärfe, Resonanzüberhöhung *f*

sharp/non-sharp switching Scharf/Unscharfschaltung *f*

sharp switchgear Scharfschaltvorrichtung *f*

sharp switching Scharfschalten *n*

sharp switching readiness Scharfschaltbereitschaft *f*

shaver plug Rasiersteckdose *f*

sheath(ed) line, sheathing line Mantelleitung *f* [*Elektrotechnik*]

sheath(ed) thermocouple (assembly) Mantelthermoelement, Mantelthermopaar *n*

sheath(ing) → **jacket**

sheave, drive sheave, driving sheave [*Wheel with grooves which assists in providing traction or driving power for the vertical motion of the car in traction-type elevators*] (Seil)Treibscheibe, (Seil)Antrieb(s)scheibe *f*

sheave, deflection sheave, deflexion sheave [*Wheel with grooves that route the hoist(ing) rope*] (Seil)Ablenkrolle *f*

sheave elevator → **(drive) sheave elevator**

shed, shelter [*It is used for storing materials or for a workshop. It may be open all round*] Schuppen *m*, Bude *f*

shedding Abwurf *m*

sheet Dünnblech *n* [*≤ 5 mm*]

sheet metal cabinet Blechschrank *m*

sheet metal panel Blechtafel *f*

sheet metal waste gas pipe, metal sheet waste gas pipe Blechabgasrohr, Abgasblechrohr *n*

sheet paper Bogenpapier *n*

sheet plate Blechplatte *f*

sheet steel enclosure, steel sheet enclosure, sheet steel housing, steel sheet housing, sheet steel casing, steel sheet casing, sheet steel case, steel sheet case (Stahl)Blechgehäuse *n*

sheffer stroke → **NAND**

shelf Ablage(platte), Ablegeplatte *f*

shell [*shell-and-oil condenser*] Kessel, Schlangenrohrkessel *m*

shell Mantelrohr *n*, Röhrenkessel *m* [*liegende Rohrbündelverflüssiger*]

shell Hülle *f* [*Die äußere Begrenzung von topologischen Elementen im Raum*]

shell Verbindung *f* zwischen Arbeitsstation und Netzwerksystem

shell Befehlsprogramm *n*

shell, carcass Rohbau *m*

shell and coil [*Designation for heat exchangers, condensers, and chillers consisting of a tube coil or pipe coil within a shell or casing*] Mantelrohrschlange *f*

shell and coil (type) condenser Rohrschlangenverflüssiger, Rohrschlangenkondensator *m*

shell and tube (type) condenser Röhrenkesselverflüssiger, Röhrenkesselkondensator *m* [*Er besteht aus einem Mantelrohr mit beiderseits angeschweißten Rohrplatten, in die die Innenrohre eingeschweißt oder eingewalzt sind. Wasser fließt in den Rohren, Kältemittel kondensiert im Mantelraum*]

shell apparatus [*Refrigerant-containing pressure vessel having tubes for the passage of a heating, cooling, or refrigerating fluid*] Kältemittel-Rohrdruckgefäß, Kältemittel-Druckrohrgefäß *n*

shell belt, shell ring; course (US) Mantelschuss, Kessel(mantel)schuss

shell boiler Mantelkessel *m*

shell-form core → **shell(-type) core**

shell-form transformer → **shell(-type) transformer**

shell plate → **(boiler) shell plate**

shell(-type) core, shell-form core (Trafo)Rahmenkern *m*

shell(-type) transformer, shell-form transformer [*Electromagnetism. A transformer constructed such that the magnetic circuit completely surrounds the coils*] Manteltransformator *m*

shelter → **shed**

shelter → **(air-raid) shelter**

shielded, screened, shrouded, guarded (ab)geschirmt

shielded twisted pair → **twisted shielded pair**

shield(ing) → **screen(ing)**

shield standing voltage [*IEC 50(461)*] (Kabel)-Schirmspannung *f*

shift clock serielle Datenwortausleser *m*

shift in, SI Rückschaltung *f* [*Rücksetzen aus der Erweiterung in den normalen Code*]

shift(ing) key Umschalttaste *f*

shift(ing) keying, SK Um(schalt)tasten *n* [*Modulation digitaler Signale*]

to shift left, SL links schieben [*Befehl*]

shift out, SO Dauerumschaltung *f* [*Erweiterung des Datencodes. DIN 66 003*]

shift register, SR, shifting register [*In automatic data processing, a computer hardware element constructed to perform shifting of its contained data*] Schieberegister, Schubregister *n*

to shift right, SR rechts schieben [*Befehl*]

shifts, shift work Schichtarbeit *f*

shift time Schichtzeit *f*

shipping weight Versandgewicht *n*

shn. → **shown**

(shock) absorber (Stoß)Dämpfer *m*

shock absorbing soft storey (Brit.); shock absorbing soft story (US) stoßabsorbierende Etage *f*, stoßabsorbierende Geschoss, stoßabsorbierende Stockwerk *n* [*(Erd)Bebensicherung*]

shock-hazard-protected berührungsgeschützt, kontaktgeschützt

Shockley diode Shockley-Diode, Vierschicht(en)-diode *f*

shock load Stoßlast *f*

shock-proof (computer) memory, shock-proof (computer) storage Pufferspeicher *m*

shoe [*A short length at the foot of a down-pipe bent to direct the flow away from the wall*] Abflussbogen *m* [*DIN 1099*]

shoe [*plumbing*] Auslaufstück *n*

shop drawing → **(work)shop drawing**

shoplifter Ladendieb *m*

shoplifting Ladendiebstahl *m*

shopping arcade, shopping passage [*Strictly a line of arches carried on columns, either free standing or attached to a wall. The term "arcade" was applied during the nine-tenth century to glass-roofed shopping areas modelled on the Crystal Palace in London and is now use for a passage with shops on one or both sides, irrespective of construction*] Einkaufspassage, Einkaufsarkade, Ladenpassage, Geschäftspassage, Verkaufspassage, Ladenarkade, Geschäftsarkade, Verkaufsarkade, Kaufpassage, Kaufarkade *f*

shopping centre (Brit.); shopping center (US); block of grouped shops Einkaufszentrum, Kaufzentrum, Ladenzentrum, Verkaufszentrum *n*

shortage Verknappung, Knappheit *f*

short autowalk, short passenger conveyor Kurzfahrsteig, Kurzrollsteig *m*

short (bath)tub Kurz(bade)wanne *f*

to short-circuit kurzschließen

short-circuit, s.c. [*An electrical defect caused, for example, by the positive and negative wires touching, resulting in a large passage of current which melts the circuit-breakers, as, for instance, fuses*] Kurzschluss *m*

short-circuit alternating current Kurzschlusswechselstrom *m*

short-circuit current Kurzschlussstrom *m*

short-circuit force → **(electromechanical) short-circuit force**

short-circuit impedance [*The input impedance of a circuit when the output terminals are connected together (short-circuited)*] Kurzschlussimpedanz, Synchronimpedanz *f*, Kurzschlusswiderstand, Synchronwiderstand *m*

short-circuiting flow Kurzschlussströmung *f* [*Eine Raumströmungsform*]

short-circuiting section Kurzschlussstrecke *f* [*z. B. Schalter von einer Pumpe auf die andere bei einer Störung*]

short-circuit making capacity Kurzschluss-Einschaltvermögen *n*, Kurzschluss-Einschaltfähigkeit *f*

short-circuit-proof → **(mechanically) short-circuit-proof**

short-circuit-proof end kurzschlussfeste Ausgang *m*, kurzschlussfeste Ende *n* [*Ausgang eines Funktionsgliedes, bei dem der Ausgangsstrom im Kurzschlussfall durch ausreichende innere Impedanz oder Stromreg(e)lung innerhalb zulässiger Grenzen bleibt. DIN 19 237*]

short-circuit protection Kurzschlussschutz *m*

short-circuit to ground (US) → **ground(ing) fault**

short cross garnet, short tee hinge Kurzband *n* [*Es ist kürzer als 1/3 der Türflügelbreite*]

short cycle → **short (work(ing)) cycle**

short-cycling operation Kurztaktbetrieb *m*

short-delayed kurzverzögert

short distance heat Nahwärme *f*

shorted emitter kristallintegrierte Widerstand *m* [*Niederohmige Nebenschluss im Thyristorkristall*]

short flushing Kurzspülen *n*, Kurzspülung *f*

short-haul modem → **baseband modem**

S

short hold modem Deaktivierung *f* einer Verbindung [*Die physikalische Verbindung wird in den Zeiten ohne Datenübertragung abgebrochen*]

'shorting' akustische Kurzschluss *m*

short message service, SMS Datentelegrammdienst *m* [*Versand und Empfang von Datentelegrammen mit bis zu 160 Zeichen mit GSM-Funktelephonen. Die eingehenden Nachrichten werden zum späteren Abruf gespeichert*]

short neck monitor Kurzhalsmonitor *m*

short on and off cycle → short (work(ing)) cycle

short passenger conveyor → short autowalk

short pipe length without socket muffenlose Rohrstück *n*

short-pitch Negativperforation *f* [*Optische Kopierung mit kontinuierlichem Lauf*]

short splice Kurzspleiß *m*

shorts test Kurzschlussprüfung, Kurzschlussprobe *f*, Kurzschlussversuch *m*

short-stroke key kurzhubige Taste(r) *f, (m)*

short tee hinge, short cross garnet Kurzband *n* [*Es ist kürzer als 1/3 der Türflügelbreite*]

short-term kurzzeitig, kurzfristig

short-term enable, short-term EN Kurzzeitfreigabe *f*

short-term monitoring Kurzzeitüberwachung *f*

short-term occupancy kurzfristige Belegung *f*

short-term rating Kurzzeitnennleistung *f*

short-term release key Kurzzeitfreigabetaste(r) *f*, (*m*)

short-time delay Kurzzeitverzögerung *f*, Kurzzeitverzug *m*

short-time operation Kurzzeitbetrieb *m*

short-time storage tank Kurzzeitspeicher *m*

short-time withstand current Haltekurzzeitstrom, Kurzzeithaltestrom *m*

short wedge [*It contains the windings in a slot in a rotor or in the stator core of an electrical machine*] Nutverschlusskeil *m*

short (work(ing)) cycle, short on and off cycle [*Too frequent starting and stopping*] kurze (Arbeits)Gang, kurze (Arbeits)Takt, kurze (Arbeits)Zyklus *m*, kurze (Arbeits)Spiel *n*

shot, photograph Aufnahme, Photographie *f*, Kamerabild *n*

shot effect → Schottky noise

should-be time Sollzeit *f*

showcase → merchandiser

shower, spray shower Brause *f* [*Das Wasser wird in feiner Verteilung gegen den Körper geleitet. Gegenteil: (Strahl)Dusche ((jet) shower), bei der das Wasser in geschlossenem Strahl gegen den Körper geleitet wird*]

shower → (jet) shower

shower bar Brausestange *f*

shower bath [*Shower baths are installed where washing has to be performed quickly, in limited space and with the minimum quantity of water. For this reason they are installed in such places as pit head bath, in gymnasiums, sports pavilions and in factories where work is dirty*] Brausebad *n*

shower (bath)tub Brause(bade)wanne *f*

shower blender → shower mixer (valve)

shower booster Brausedruckverstärker *m*

shower bracket Brausekonsole *f*, Brausewandträger *m*

shower column Brausesäule, Brausestange *f*

shower connection Brauseanschluss *m*, Brauseverbindung *f*

shower controller Brauseregler *m*

shower cubicle → (jet) shower cubicle

shower curtain, spray curtain Brausestandvorhang, Spritzvorhang *m*

shower curtain rail, shower curtain bar, spray curtain rail, spray curtain bar Brausestandvorhangstange, Spritzvorhangstange *f*

shower equipment, shower outfit Brauseausrüstung *f*

shower fitting Brausearmatur *f*

shower (grab) bar, shower (grab) rail Brause(halte)stange, Brause(halte)schiene *f*

shower head Brausekopf *m*

shower holder Brausehalter *m*

shower hose Brauseschlauch *m*

shower jet Brausestrahl *m*

shower mixer (valve), shower mix(ing) valve, shower blender, shower (blending) valve Brausemischbatterie, Mischbrausebatterie *f*

shower ((open) loop) control Brausesteuerung *f*

shower outfit → shower equipment

shower partition Brausetrennwand *f*

shower receptor (US); shower tray, shower pan, shower receiver, shower receptacle Brausetasse, Brausewanne *f*

shower riser (pipe), shower rising pipe, shower ascending pipe Brausestandrohr *n*

shower room (without cubicles) Massenbrause *f*, Brauseraum *m* ohne Zelleneinteilung

S

shower rose Brausekopf *m*

shower screen Brausekabinenwand *f*

shower seat Brausesitz *m*

shower tube Brauseröhre *f*

shower valve → shower mixer (valve)

shown, shn. wie gezeichnet [*Beschriftung im Zeichnungsschriftfeld*]

shredder, shredding machine Aktenvernichtungsanlage *f*, Reißwolf *m*

show window lighting, show window illumination Schaufensterbeleuchtung *f*

shrinkage hose Schrumpfschlauch *m*

shrouded → shielded

shrouded contact (type) socket Kragensteckdose *f*

shrouded knob Schutzkragenknebel, versenkte Knebel *m*

to shunt [*To create a parallel connection*] parallel schalten

shunt, multiple, parallel [*Connected to the same pair of terminals*] im Nebenschluss geschaltet, parallel geschaltet

shunt (connection) → connection-in-parallel

shunt field Nebenschlussfeld *n*

shunt tripping Spannungsauslösung *f*

shunt-wound [*Of a device, having its generator field coils in parallel with the main armature circuit*] nebenschlussgewickelt, mit Nebenschlusswick(e)lung

to shut down → to lay up

shutdown, laying up Stilllegung, Stillsetzung, Außerbetriebnahme, Außerbetriebsetzung *f*, Terminieren *n*

shutdown modus Anzeigeabschaltung *f* im Ruhestand

shutdown on faults, disconnection on faults, fault shutdown, fault disconnection Störabschaltung *f*

to shut off, to stop, to close off [*The flow of fluid in a pipe or a duct*] (ab)sperren

shutoff valve → stop valve

shutter [*Optics. A mechanical device that controls the time during which light is admitted to an optical system*] Verschluss *m*

shutter blind → Venetian blind

shuttering oil → mould oil

shutter ((open) loop) control [*Damper assembly and pressure-operated act(uat)or used on systems with nonunloading-type hermetic compressors in air-cooled condensing unit to maintain minimum required condensing pressure during low ambient operation*] Verschlusssteuerung *f*

shutter rail Rollladenschiene *f*

shut(ting)-off, closing, closure, closeoff (Ab)Sperren, Schließen *n*, (Ab)Sperrung, Schließung *f*

shut(ting)-off ball cock, closing ball cock, closure ball cock, closeoff ball cock (Ab)Sperr(ungs)kugelhahn, Schließkugelhahn *m*

shut(ting)-off (builder's) fittings, shut(ting)-off (builder's) hardware, shut(ting)-off (builder's) furniture, shut(ting)-off (builder's) garnishment [*"Closure", "closing" and "closeoff" may synonymously be used for "shut(ting)-off"*] (Ab)Sperr(ungs)-(bau)beschläge, Schließ(bau)beschläge *f*, (Ab)Sperr(ungs)kleineisen, Schließkleineisen *n*

shut(ting)-off cock, closing cock, closure cock, closeoff cock, stop cock (Ab)Sperr(ungs)hahn, Schließhahn *m*

shut(ting)-off damper, closing damper, closure damper, closeoff damper (Ab)Sperr(ungs)klappe, Schließklappe *f*

shut(ting)-off device, shut(ting)-off appliance, closing device, closing appliance, closure device, closure appliance, closeoff device, closeoff appliance (Ab)Sperr(ungs)gerät, Schließgerät, (Ab-)Sperr(ungs)organ, Schließorgan *n*, (Ab-)Sperr(ungs)vorrichtung, Schließvorrichtung *f*

shut(ting)-off pressure, closing pressure, closure pressure, closeoff pressure (Ab)Sperr(ungs)druck, Schließdruck *m*

shut(ting)-off (slide) valve, closing (slide) valve, closure (slide) valve, closeoff (slide) valve [*A (slide) valve used for turning on or shut(ting) off completely a supply of fluid*] (Ab)Sperr(ungs)schieber, Schließschieber *m* [DIN 3352]

shut(ting)-off valve, closing valve, closure valve, closeoff valve, stop valve [*It operates by screwing down a washer on to a seating*] (Ab)Sperr(ungs)ventil, Schließventil *n*

SI → shift in

SI → subscriber interface

SIC → dielectric constant

sick building krankmachende Gebäude *n*

sick building syndrome → building related illness

side clearance [*The distance separating an object or part from a side*] Seitenfreiheit *f*, Seitenspiel *n*

side cushion, lateral cushion Seitenpolster *n* [*Österreich: m*]

side drive lift (Brit.); side drive elevator (US) [*The machine room is mounted alongside the lift shaft at the top floor*] Seitenantrieb(s)aufzug *m*

side entry, side inlet, lateral entry, lateral inlet Seiteneinlauf *m*

S

to sidegrade → **to update**

sidegraded → **updated**

side hood Seitenhaube *f*

side inlet → **side entry**

side spray shower, lateral spray shower Seitenbrause *f*

side-to-side crosstalk(ing) Übersprechen *n*

side-to-side crosstalk(ing) attenuation Übersprechdämpfung *f*

side view, lateral view Seitenansicht *f*

Siemens voltage multiplier connection Siemensschaltung, Spannungsvervielfachung nach Siemens *f*

sighold Unterbrechung *f* der Signalbehandlung

sight door, sight hatch, view(ing) door, view(ing) hatch, observation door, observation hatch Schauluke, Beobachtungsluke *f*

sight elevator, observation elevator (US); sight lift, observation lift (Brit.) Aussichtsfahrstuhl, Aussichtsaufzug, Panoramafahrstuhl, Panoramaaufzug *m*

sight error, observation error, view(ing) error Beobachtungsfehler, Schaufehler *m*

sight glass, view(ing) glass, observation glass Schauglas, Beobachtungsglas *n*

sight hole, sight port, view(ing) hole, view(ing) port, observation hole, observation port Schauloch, Beobachtungsloch, Sichtloch *n*, Schauöffnung, Beobachtungsöffnung, Sichtöffnung *f*

sight tube, view(ing) tube, observation tube Beobachtungsröhre, Schauröhre *f*

sight unit, view(ing) unit, observation unit Schaueinheit, Beobachtungseinheit *f*

sight window, view(ing) window, observation window Beobachtungsfenster, Sichtfenster *n*

SIGMAL → **structural interactive graphic monitoring application language**

sign Schild *n*

signage Beschilderung *f*

to signal signalisieren

signal Signal *n* [*Die physikalische Darstellung von Nachrichten oder Daten über Größen durch Signalträger, die Parameter enthalten, wobei die Parameterwerte die Zeitfunktionen der Größen abbilden*]

signal and noise and distortion, SINAD Nutzsignal *n* und Rauschsignal und Klirrfaktor *m*, SINAD-Wert *m* [*Das logarithmierte Verhältnis zwischen dem Gesamtsignal und dem aus Rauschen und Oberschwingungen besteheden Störsignal. Es handelt sich um einen Abstand, dessen Wert möglichst groß sein soll.*

$$SINAD = 20 \cdot lg \frac{U_S + U_N + U_D}{U_N + U_o} dB$$]

signal attenuation, signal abatement, signal damping, signal deadening [*See remark under "damping"*] Signaldämpfung, Signalminderung, Signalabschwächung *f*

signal channel [*Communication. A signal path for transmitting electric signals; such paths may be separated by frequency division or time driven*] Signalkanal, Meldekanal, Messkanal *m*

signal condition, signal state Signalbeschaffenheit *f*, Signalzustand *m* [*Jeder der möglichen Werte eines digitalen Signals. DIN 19 226, Teil 5*]

signal conditioning Signalaufbereitung *f*

signal conditioning element, signal conditioner Eingabeglied, Signalaufbereitungselement *n* [*DIN 19 237*]

signal control path → **control path**

signal damping → **signal attenuation**

signal deadening → **signal attenuation**

signal degradation Signalminderung, Signalverschlechterung *f*

signal distance, hamming distance [*Computer programming. In two binary words of the same length, the number of bit positions in which the corresponding bits are different*] Signalabstand, Hammingabstand *m*

signal distortion Signalverzerrung *f*

signal distortion generator [*Electronics. An instrument that generates distorted signals; used to test or adjust communications equipment*] Signalverzerrer *m*

signal disturbance → **noise (distortion)**

signal duration Signallaufzeit *f*

signal earth (Brit.) → **signal ground**

signal edge Signalkante *f*

signal evaluation unit Signalauswert(ungs)einheit *f*

signal flow Signalfluss *m*

signal flow diagram, block diagram Signalflussplan [*DIN 19 221*], Wirkungsplan *m*, Signalflussbild *n*

signal flow path Wirkungslinie *f* [*Signalblock. VDI/VDE 2600*]

signal gain → **gain**

signal generation, signal formation, signal forming, signal shaping Signalbildung, Signalerzeugung, Signalformung *f* [*E DIN 19 226 Teil 5*]

signal generator, test(ing) oscillator [*An electronic test instrument that delivers a sinusoidal output at an accurately calibrated frequency that may be anywhere from the audio to the microwave range; adjustable over a wide area, and the output usually may be amplitude- or frequency-modulated*] Signalgenerator, Signalerzeuger, Messsender, Signalbildner *m*

S

signal glass, alarm glass Verbundglas mit Alarmdrahteinlage, Alarmglas, Signalglas *n* [*Verbundsicherheitsglas mit elektrischen Leitern, welche bei Zerstörung ein Signal auslösen*]

signal (glass) pane → alarm (glass) pane

signal glazing → alarm glazing

signal ground (US); signal earth (Brit.) Betriebserde, Signalerde *f*

signaling rate (US); signalling rate (Brit.) [*Communication. The rate at which signals are transmitted*] Übertragungsgeschwindigkeit *f* [*in Baud*]

signal input Signaleingang *m*, Signaleingabe *f* [*E DIN 19 226 Teil 5*]

signal input card Signaleingang(s)karte, Signaleingabekarte *f*

signal input range Arbeit(s)bereich *m* der Eingang(s)größe [*DIN 44 472*]

signal intensity, signal strength [*The electric-field strength of the electromagnetic wave, transmitting a signal*] Signalstärke, Signalintensität *f*

signal isolation Signalentkopp(e)lung, Potentialfreiheit des Signals *f* [*DIN IEC 381, Tl.2*]

signalization (US); signalisation Signalisierung *f* [*Die Gesamtheit aller Signale, die den Zustand der Steuerung oder der zu steuernden Einrichtungen darstellen. V DIN 19 237*]

signal lamp Signallampe *f*

signal latch Signalspeicher *m*

signal level [*Telecommunications. The difference between the level of a signal at a point in a transmission system and the level of an arbitrarily specified reference signal*] Signalpegel *m*

signal level converter Signalpegelumsetzer *m*

signal line Signalleitung *f* [*Eine Signalleitung überträgt die Messsignale und verbindet Messvorrichtungen auf eine der physikalischen Natur des Messsignals entsprechende Weise. E DIN 19 226, Tl. 5*]

signalling (Brit.); signaling (US) Signalisieren *n*, Signalgabe, Zeichengabe *f*

signal message; signalling message (Brit.); signaling message (US) Meldung *f* [*Sie kennzeichnet einen Zustand oder eine Zustandsänderung der Steuerung oder der zu steuernden Einrichtung und dient vorwiegend zur Information des Menschen. Das zugehörige Signal heißt Meldesignal. D DIN 19 226, Teil 5; V DIN 19 237*]

signal message content Meldungsinhalt, Meldeinhalt *m*

signal message delay Meldungsverzögerung, Meldeverzögerung *f*, Meldungsverzug, Meldeverzug *m*

signal message display Meldungsanzeige, Meldeanzeige *f*

signal message lamp Meldungslampe, Meldelampe *f*

signal message processing Meldeverarbeitung, Meldungsverarbeitung, Meldungsbearbeitung, Meldebearbeitung *f* [*Die Aufbereitung eines Signals oder mehrerer Signale zu einem Messsignal*]

signal message record(ing) Meldungsaufzeichnung, Meldeaufzeichnung *f* [*Die Dokumentation von Meldesignalen (= Meldungen) mittels Datenträger, z. B. Magnetband, Magnetplatte, Papier (Schreiber)*]

signal message registration Meldungsregistrierung, Melderegistrierung *f* [*Die fortlaufende Aufzeichnung von Meldesignalen (= Meldungen)*]

signal message sequence Meldungsfolge, Meldefolge *f*

signal message suppression Meldungsunterdrückung, Meldeunterdrückung *f*

signal message system Meldungssystem, Meldesystem *n*

signal message text Meldungstext, Meldetext *m*

signal message word [*programmable controller*] Meldungswort, Meldewort *n*

signal message zone Meldungszone, Meldezone *f*

signal mixing Aufschalten *n* [*Regelung*]

signal-noise ratio → signal(-to)-noise ratio

signal operation Signalbetrieb *m*

signal output Signal(aus)gabe *f* [*E DIN 19 226 Teil 5; V DIN 19 237*]

signal output Signalausgang *m*

signal path Signal(fluss)weg *m*

signal pit Signalvertiefung *f* [*Laserabtastung. Vom Laser aus gesehen aber Signalerhöhung*]

signal preprocessing Signalvorverarbeitung *f*

signal processing Signalverarbeitung, Signalaufbereitung *f* [*DIN 19 226, Teil 5*]

signal propagation time Signallaufzeit *f* [*Zeitspanne, nach deren Ablauf ein Ausgangssignal auf eine Signaländerung am zugehörigen Eingang funktionsgemäß reagiert. DIN 19 237*]

signal range Signalbereich *m* [*selten: n*]

signal receiver Signalempfänger *m*

signal regeneration, signal reshaping Signalregenerierung *f*

(signal) repeater (Signal)Regenerator *m*

signal replay Rückmeldung *f* [*Sie erfolgt als unmittelbare Auswirkung eines Befehls. DIN 19 235. DIN 19 237*]

S

signal reprocessing Signalwiederverarbeitung *f*

signal selector Signalwähler *m*

signal separation Signal(auf)trennung *f*

signal sequence Signalfolge *f*

signal shaping → **signal generation**

signal source Signalquelle *f*, Signalursprung *m*

signal state → **signal condition**

signal suppression Signalunterdrückung *f*

signal to mask ratio, SMR Verhältnis *n* von Signalamplitude zur minimalen Mithörschwelle

signal(-to)-noise ratio, SNR, S/N ratio (Signal)Störabstand, (Signal)Rauschabstand *m*, (Signal)Rauschverhältnis, Nutz-zu-Rausch-Verhältnis *n* [*Logarithmierte Verhältnis der Nutzsignalleistung zur Rauschsignalleistung bezogen auf die letzte Stelle*]

signal transducer → **measuring transducer**

signal transit time Signallaufzeit *f* [*Zeitspanne, nach deren Ablauf ein Ausgang(s)signal auf eine Signaländerung am zugehörigen Eingang funktionsgemäß reagiert. V DIN 19 237*]

signal transmission, transmission of signals Signalübertragung *f*

signal transmitter Signalgeber *m* [*1.) Die dem Signalglied entsprechende Baueinheit heißt Signalgeber. [E DIN 19 226, Tl. 5] 2.) Gerät, das ein zur Beeinflussung der Steuerung notwendiges Signal gibt. [V DIN 19 237]*]

signal wave → **signal**

sign-bit Vorzeichenbit *n*

sign digit, sign place Vorzeichenziffer *f* [*Plus- oder Minuszeichen, das den positiven oder negativen Charakter einer Koordinate festlegt*]

signed integer ganze Zahl *f* mit Vorzeichen

significant instant Kennzeitpunkt *m*

significant interval Kennintervall *n*

sign-magnitude-code Betragsanzeige *f* mit Vorzeichenbit [*Analog-Digital-Umsetzer*]

sign-magnitude-coding Vorzeichenbetragkodierung *f*

sign-reversing amplifier → **inverting amplifier**

signum generator Signumbildner, Signumerzeuger *m* [*Ein Signumbildner ist ein Glied, dessen Ausgangssignal einen positiven bzw. negativen Nennwert in Abhängigkeit vom Vorzeichen des Eingangssignals annimmt nach der Beziehung* $v(t) = \frac{u(t)}{|u(t)|}$ *E DIN 19 226 T5*]

silencer → **acoustic absorber**

silent alarm stille Alarm *m* [*Telefonbenachrichtigung der Vertrauensperson(en)*]

silica (reine) Siliziumdioxid *n*, SiO_2

silicon Silizium *n*, Si

silicon sensor Siliziumsensor *m*

silicon switching diode Siliziumschaltdiode *f*

silistance Silizium-PTC-Widerstand *m*

silted up verschlufft

silver facing Silberauflage *f*

SIM → **subscriber(s) identity module**

similarity law Ähnlichkeitsgesetz *n*

similarity transformation Änlichkeitstransformation *f*

similitude characteristic value → **characteristic similitude value**

similitude law Ähnlichkeitsgesetz *n*

SIMM → **single-in-line memory module**

simple-cycle (gas) turbine → **gas turbine**

simple formattable document, SFD einfach strukturierbare Dokument *n*

simple goods elevator, simple freight elevator (US); simple goods lift, simple freight lift (Brit.) vereinfachte Güteraufzug *m* [*Er hat höchstens drei Haltestellen. Tragfähigkeit höchstens 2000 kg und Fahrkorbgrundfläche höchstens 2,5 m²*]

simple logic function einfache Verknüpfung *f*

simple type einfache Type, skalare Type, aufzählbare Type *f*

simple vacuum steam heating, simple steam vacuum heating einfache Vakuumdampfheizung *f* [*Konstanter Dampfdruck und unabhängiger Druck in der Kondensatleitung*]

simplex Wechselsprechbetrieb *m*

simplex conference Wechselsprechkonferenz *f*

simplex of reference → **basis**

simplex structure → **nonredundant system**

simplex table Simplextabelle *f*, Simplextableau *n* [*Anordnungsschema aller für die Simplexmethode erforderlichen Daten. Durch jede Simplextransformation wird ein Simplextableau aufgestellt*]

simplex unit Wechselsprechanlage *f*

simulation of homogenous zero-dimensional reaction system nulldimensionale Modell für homogene Gasreaktionssysteme, HOMREA *n*

simulator Simulator *m* [*Ein Rechenautomat, oft ein Analogrechner, oder anderes Gerät, oder ein Programm, das in der Lage ist eine Simulation auszuführen, die so gestaltet wird, dass es einen funktionalen Zusammenhang zwischen dem studierten physikalischen System und dem verwendeten System gibt*]

S

simultaneity Gleichzeitigkeit *f*

simultaneous → concurrent

simultaneous → parallel

simultaneous access → concurrent access

simultaneous factor, simultaneity factor Gleichzeitigkeitsfaktor *m*

simultaneousness Gleichzeitigkeit *f*

simultaneous transfer → parallel transfer

SINAD → signal and noise and distortion

single acknowledg(e)ment Einfachquittieren *n*, Einfachquittierung *f*

single-acting compressor [*Compressor in which compression takes place on but one stroke per revolution in each compressing element*] einfachwirkende Kompressor, einfachwirkende Verdichter *m*

single-action floor spring Drehflügel-(Fuß)Boden-(tür)schließer *m*

single-address system, one-address system [*See remark under "address"*] Ein-Adressensystem *n*

single attach station, SAS Klasse-B-Station *f* [*Rückgratnetzwerk*]

single-ball cock, one-ball cock Einzelkugelhahn *m*

single-bank air heater, one-bank air heater einreihige Lufterhitzer, Lufterhitzer mit einer Rohrreihe *m*

single-bank radiator, one-bank radiator einreihige Gliederheiz(ungs)körper *m*

single basis enclosure, single basis casing, single basis housing, single basis case [*"One-basis" may synonymously be used for "single-basis"*] Einfachleergehäuse *n*

single beep, one-beep Einzelpiepton *m*, kurze Einzelsummerzeichen *n*

single-bit, one-bit binäre Eins *f*, Einser *m*

single-blade damper, one-blade damper Einzelklappenzugregler *m*

single-board computer, one-board computer Ein-Platinen-Rechner *m*

single branch Einfachabzweig *m*

single-busbar substation, one-busbar substation Einfachsammelschienenstation *f*

single-cable sensor, one-cable sensor Einzelkabelsensor *m*

single-cable technique, one-cable technique Einkabeltechnik *f*

single-cell system, one-cell system Einzelzellensystem *n*

single-chamber (air) lock, one-chamber (air) lock Einkammer(luft)schleuse *f*

single-channel time switch, one-channel time switch [*See remark under "time switch"*] Ein-Kanal-Schaltuhr *f*

single chimney, individual chimney Einzelschornstein, Einzelkamin *m*

single-chip computer, one-chip computer Ein-Chip-Rechner *m*

single ((closed) loop) control Einzelreg(e)lung, Einzelregulierung *f*

single compartment sink (bowl), one compartment sink (bowl) [*"Basin" may synonymously be used for "bowl"*] Einbeckenspüle, Einfachspüle *f*

single compartment sink top, one compartment sink top Einbeckenspültisch, Einfachspültisch *m*

single component computer, SCC, one component computer Blockrechner *m*

single connecting line, single connection line Einzelanschlussleitung *f*

single controller Einzelregler *m*

single (convenience) outlet → single socket-outlet

single-day site, one-day site Eintag(es)-Baustelle *f*

single-decade, one-decade eindekadig

single-deck passenger elevator, one-deck passenger elevator (US); single-deck passenger lift, one-deck passenger lift (Brit.) Eindecker-Fahrstuhl, Eindecker-Personenaufzug *m*

single density → frequency modulation

single-digit access, one-digit access Einfingerzugriff *m*

single-digit dialling, one-digit dialling [*telephone*] Einfingerwahl *f*

single-dimensional digital signal processing, one-dimensional digital signal processing eindimensionale Digitalsignalverarbeitung *f*

single-door ((open) loop) control, one-door ((open) loop) control Einzeltürsteuerung *f*

single-ended unsymmetrisch [*Signaltechnik*]

single-duct air conditioning plant, single-duct air conditioning installation, one-duct air conditioning installation, one-duct air conditioning plant [*The air, having been conditioned in a central plant, is distributed to various zones through a single duct*] Ein-Kanal-Klimaanlage, Ein-Kanal-Klimainstallation *f*

single-duct high-pressure air conditioning plant, single-duct high-pressure air conditioning installation, one-duct high-pressure air conditioning plant, one-duct high-pressure air conditioning installation Ein-Kanal-Hochdruck-Klimaanlage, Ein-Kanal-Hochdruck-Klimainstallation, Ein-Kanal-Hochgeschwindigkeits-Klimaanlage, Ein-Kanal-Hochgeschwindigkeits-Klimainstallation *f*

S

single-duct low-pressure air conditioning plant, single-duct low-pressure air conditioning installation, one-duct low-pressure air conditioning plant, one-duct low-pressure air conditioning installation Ein-Kanal-Niederdruck-Klimaanlage, Ein-Kanal-Niederdruck-Klimainstallation, Ein-Kanal-Niedergeschwindigkeits-Klimaanlage, Ein-Kanal-Niedergeschwindigkeits-Klimainstallation *f*

single-duct system, one-duct system Ein-Kanalsystem *n* [*System mit einer Zuluftleitung an den Versorgungsstellen. DIN 1546, Teil 1*]

single-ended amplifier Eintaktverstärker *m*

single-ended output, unbalanced output [*It is grounded on one side*] Eintaktausgang, Einzelausgang *m*

single gas heating → gas single heating

single-gas heating appliance, single-gas heater, one-gas heating appliance, one-gas heater Ein-Gasgerät *n* [*Gasheiz(ungs)gerät für eine Gasart (Gasfamilie)*]

single-generator plant, single-generator installation, one-generator plant, one-generator installation Ein-Generator-Anlage *f*

single glazing Einfachverglasung *f*

single grip (bath)tub/shower installation, one grip (bath)tub/shower installation, single grip bath(ing tub)/shower installation, one grip bath(ing tub)/shower installation Ein-Griff(mischer)-(Bade)Wannenfüll- und Duschbatterie *f* (oder Duschgarnitur)

single grip mixing battery, one grip mixing battery Ein-Griff-Mischbatterie *f*

single grip vanity vertical battery, one grip vanity vertical battery Ein-Griff-Waschtisch-Standbatterie *f*

single-handed calibration, one-handed calibration Einmanneichung *f*

single (hand)grip mixer, single handle mixer, one-(hand)grip mixer, one-handle mixer [*bath*] Eingriffmischer *m*

single handle mixing battery, single handgrip mixing battery, one-handle mixing battery, one-handgrip mixing battery Einhandmischbatterie mit Griff, Griff-Einhandmischbatterie, Ein-Griff(-Misch)-batterie *f* [*Das gemeinsame Stellglied wird durch Griff betätigt*]

single hoistway [*Hoistway for a single elevator or dumbwaiter*] Einfach(aufzug)schacht *m*

single-hole bath/shower mixer (valve), one-hole bath/shower mixer (valve) Ein-Loch-Bad/Brause-Mischbatterie *f*

single-hole single-lever mixer, one-hole one-lever mixer Einloch-Einhebel-Mischer, Einhebel-Einloch-Mischer *m*

single-hole mixer (valve), single-hole blender, single-hole mix(ing) valve, single-hole blending valve [*"One-hole" and "monobloc" may synonymously be used for "single-hole"*] Einloch(misch)batterie *f*

single-hole (wash)basin mixer, one-hole (wash)-basin mixer Einloch-(Wasch)Becken-Mischer *m*

single hot water preparation plant, single hot water preparation installation Einzel-WWB-Anlage, örtliche WWB-Anlage *f*

single hot water supply system Warmwasser-Einzelversorgung(ssystem), WW-Einzelversorgung(ssystem) *f*, (*n*) [*Für die Versorgung einer einzelnen Warmwasserentnahmestelle*]

single-in-line memory module, SIMM Speicherbaustein *m* mit einreihiger Kontaktleiste

single-in-line package, SIP einreihige Steckgehäuse *n*

single-input impedance Eintakteingangsimpedanz *f*

single-input single-output system, SISO, one-input one output system Eingrößensystem *n*

single-latch lock, one-latch lock Ein-Fallenschloss *n*

single-level code → absolute programming

single-lever ball mixer, single-lever globe mixer, one-lever ball mixer, one-lever globe mixer Einhebelkugelmischer *m*

single-lever (bath) mixer (valve), one-lever (bath) mixer (valve) Einhebelbad(e)batterie, Einhebelmischbatterie *f*, Einhebelmischer *m*

single-lever mixer tap, single-lever mixing tap, one-lever mixer tap, one-lever mixing tap Einhebelmischerhahn *m*

single-lever mixer (valve) → lever(-operated) mixer (valve)

single-lever mixing battery Einhandmischbatterie mit Hebel, Hebel-Einhandmischbatterie *f* [*Das gemeinsame Stellglied wird durch Hebel betätigt*]

single-lever ((open) loop) control, one-lever ((open) loop) control Einhebelsteuerung *f*

single-lever shower mixer, one-lever shower mixer Einhebelbrausemischer *m*

single-lever vanity (mixing) battery, one-lever vanity (mixing) battery Einhebelwaschtisch(misch)-batterie, Waschtischeinhebel(misch)batterie *f*

single-line, one-line einzeilig

single-line circuit diagram, one-line circuit diagram Einfachleitung-Stromlaufplan *m*

single-line diagram, one-line diagram einpolige Übersichtsschaltbild *n*

single-line representation, one-line representation einpolige Darstellung *f* [*DIN 40 719, Blatt 1*]

single-loop, one-loop einschleifig

single magnetoresistor Einzelfeldplatte *f*, magnetische Einzelwiderstand *m*

single-management level, one-management level, single-management co-ordinating level, one-management co-ordinating level Einzelleitebene *f* [*Sie fasst alle Teile der Leiteinrichtung zusammen, die unmittelbar über die Stellglieder auf den Prozess einwirken. DIN 19 222*]

single-mode fibre, one-mode fibre; single-mode fiber, one-mode fiber (US) Einmodemfaser *f*

single output impedance Eintaktausgansimpedanz *f*

single ((open) loop) control [*See remark under "((open) loop) control"*] Einzelsteuerung *f*

single-package air-conditioning system → self-contained air-conditioning system

single pass einzige Durchgang, Ein-Durchgang *m*

single-pass method, one-pass method Ein-Durchgang-Verfahren *n*, Ein-Durchgang-Methode *f*

single-phase bridge (connection), one-phase bridge (connection), two-pulse bridge (connection) [*"Circuit" may synonymously be used for "connection"*] Einphasenbrückenschaltung *f*

single-phase non-reversing motor, non-reversing single phase motor, non-reversing one-phase motor, one-phase non-reversing motor Einphasenmotor *m* mit 1 Drehrichtung

single-phase starting switch, one-phase starting switch Einphasenanlassschalter *m*

single-phase-to-earth fault, one-phase-to-earth fault, single-phase-to-ground fault, one-phase-to-ground fault einphasige Erdschluss, einpolige Erdschluss, einsträngige Erdschluss, einphasige Fehler, einpolige Fehler, einsträngige Fehler, Erdschluss einer Phase, Fehler einer Phase *m*

single-phasing sensibility, single-phasing sensitiveness, single-phasing susceptibility, one-phasing sensitivity, one-phasing sensitiveness, one-phasing susceptibility Einphasenempfindlichkeit *f*

single-piece, one-piece einteilig

single-piece boiler, one-piece boiler einteilige Kessel *m*

single-piece construction, one-piece construction einteilige Bauweise *f* [*Einrichtungsgegenstand*]

single-piece section frame, one-piece section frame umlaufende Profilrahmen *m*

single-piece toilet, single-piece w.c., one-piece toilet, one-piece w.c., close-coupled toilet, close-coupled w.c. Tiefspülabort *m*, Tiefspülklosett *n*

single pipe Einzelrohr *n* [*Gegensatz: Rohrbündel*]

single-pipe forced circulation hot water heating system, one-pipe forced circulation hot water heating system Einrohr-Zwangsumlauf-Heißwasserheiz(ungs)anlage *f*

single-pipe forced hot water heating, single-pipe pumped hot water heating, one-pipe forced hot water heating, one-pipe pumped hot water heating [*"Warm water" may synonymously be used for "hot water"*] Einrohr-Pumpenwarmwasserheizung, Einrohr-PWW *f* [*bis 110 Grad C Vorlaufttemperatur*]

single-pipe forced hot water heating system, one-pipe forced hot water heating system [*"Warm" may synonymously be used for "hot"*] Ein-Rohr-Zwangsumlauf-Warmwasser-Heiz(ungs)system *n*

single-pipe heating, single-tube heating, one-pipe heating, one-tube heating Einrohrheizung *f*

single-pipe hot water heating, one-pipe hot water heating [*"Warm" may synonymously be applied for "hot"*] Einrohr-Warmwasserheizung, Einrohr-WWH, Warmwasserheizung im Einrohrsystem, Warmwasser-Einrohrheizung *f* [*Bis 110 Grad C Vorlauftemperatur*]

single-pipe steam heating system, one-pipe steam heating system [*A single main carries steam to the heating units or radiators and also conveys the condensation back to the boiler. Usually there is but one connection from the radiator or heating unit to the main, but sometimes separate supply and return connections are used*] Einrohrdampfheiz(ungs)system *n*

single-pipe steam system, one-pipe steam system [*System in which the condensable vapo(u)r from the supply main passes into a heating unit and returns as condensate through the same supply main*] Einrohr-Dampfsystem *n*

single-piston lift, one-piston lift (Brit.); single-piston elevator, one-piston elevator (US) Einkolbenaufzug *m*

single-point water heater, one-point water heater [*See remark under "water heater"*] Wassererhitzer für Einzelversorgung, Wasser(er)wärmer für Einzelversorgung, Wasserheizer für Einzelversorgung *m*

single pole double-throw switch, SPDT switch einpolige Umschalter, einpolige Zweistellungsschalter *m*

single pole single throw switch, SPST switch Ein/-Aus-Schalter *m*

single-processor computer architecture, one-processor computer architecture Ein-Prozessor-Rechnerarchitektur *f*

single-purpose controller, single-purpose regulator Einfachregler *m* [*Er ist nur für eine bestimmte Regelgröße verwendbar*]

S

single-quadrant converter, one-quadrant converter Einquadrantstromrichter, Einquadrantstromwandler *m*

single-quadrant operation, one-quadrant operation Einquadrantbetrieb *m*

single-range Scherbius system, one-range Scherbius system untersynchrone Scherbiuskaskade *f*

single receptacle-socket-outlet (US) → **single socket-outlet**

single resistance Einzelwiderstand *m* [*Er entsteht durch Richtungs- und Querschnittsänderungen beim Fließen eines Mediums in einer Rohrleitung. Zwei Größen bestimmen die Einzelwiderstände:*
a) die Formziffer ζ *(Zeta), die durch Versuche bestimmt wird und von der Konstruktion abhängig ist (bei WW-Zentralheizungen liegen die Werte zwischen 0,5 und 10).*
b) die Geschwindigkeit des Mediums (in m/s). Man summiert die Formziffern einer Teilstrecke und entnimmt mit der Strömungsgeschwindigkeit aus einer Tabelle den Druckverlust Z.
Dieser Rechengang ist jedoch nur bei einer genauen Nachrechnung notwendig. Bei der vorläufigen Rohrnetzberechnung setzt man für die Einzelwiderstände pauschal 35 bis 50 % des Gesamtwiderstandes ein.]

single-room air conditioning ((open) loop) control, one-room air conditioning ((open) loop) control Einzelraumklimasteuerung *f*

single-room ((closed) loop) control, one-room ((closed) loop) control Einzelraumreg(e)lung, Einzelraumregulierung *f*

single-room controller, one-room controller Einzelraumregler *m*

single-room ((open) loop) control, one-room ((open) loop) control Einzelraumsteuerung *f*

single-rotor (refrigerant) screw compressor, one-rotor (refrigerant) screw compressor Schrauben(kälte)verdichter mit einem Rotor, Schrauben(kälte)kompressor mit einem Rotor *m* [*Angetrieben wird ein Nutenrotor. Verdichtung erfolgt durch eine oder mehrere in die Nuten eingreifende(n) Zahnscheibe(n), die entsprechend der axialen Wandung der Nuten mitrotiert (mitrotieren)*]

single seal siphonic closet (fitment) → **single trap siphonic closet (fitment)**

single-seated (slide) valve Ein-Sitz-Schieber *m*

single-seated valve, one-seated valve Ein-Sitz-Ventil *n*

single-shaft ventilation, one-shaft ventilation Einzelschachtlüftung *f*

single shift → **SS**

single shot → **monoflop**

single-shot pushbutton, one-shot pushbutton Einmal-Druckknopf *m*

single-shot recording, one-shot recording Einzelschrittaufnahme, Einstufenaufnahme *f*

single side band → **SSB**

single side-band modulation → **SS/SSB**

single sink [*See remark under "sink"*] Einfachspüle *f*

single (slide) valve, plain (slide) valve Einzelschieber *m*

single slope converter A/D-Einrampen-Umsetzer, A/D-E.-U., Spannung-Zeit-Umsetzer *m*

single socket-outlet, single receptacle-socket-outlet (US); single receptacle, single (convenience) outlet Einzelsteckdose *f*

single-source mains infeed(ing), one-source mains infeed(ing) Einzel(ein)speisung *f* [*(Ein)Speisung aus nur einer Energiequelle. V DIN 19 237*]

single-speed, one-speed eintourig

single-spray mixing tap, single-spray mixer tap, one-spray mixing tap, one-spray mixer tap Ein-Spritz-Mischerhahn *m*

single-stage compressor, single-step compressor, one-stage compressor, one-step compressor [*Compression from initial to final pressure is completed in a single step or stage*] einstufige Verdichter, einstufige Kompressor *m*

single-stage(d), one-stage(d) einstufig

single-stage selection, one-stage selection Einstufenwahl *f*

single step Einzelschritt *m* [*Eine Mikrorechnerbetriebsart, bei der der Rechner nach jedem ausgeführten Befehl anhält*]

single-step compressor capacity modulation system, single-stage compressor capacity modulation system, one-step compressor capacity modulation system, one-stage compressor capacity modulation system [*Compressor capacity modulation system arranged to reduce capacity in one step. It usually reduces capacity to one-half or full capacity*] Leistungsmodulationssystem für einstufigen Verdichter, Leistungsmodulationssystem für einstufigen Kompressor *n*

single supply Einzelversorgung *f*

single tap lavatory basin, one tap lavatory basin, single tap (wash)basin, one tap (wash)basin [*See remark under "(wash)basin"*] Einzelzapfhahn-(wasch)becken *n*

single teletotalizer Einzelferntotalisator *m*

single terminal Einzelklemme *f*

S

single-throw switching, one-throw switching Einhebelschalten *n*

singleton unscharfe Menge *f* mit einem Wertepaar

single trap siphonic closet (fitment), single seal siphonic closet (fitment) Leersaug(e)abort mit Geruchverschluss, Absaug(e)abort mit Geruchverschluss *m*, Leersaug(e)toilette mit Geruchverschluss, Absaug(e)toilette mit Geruchverschluss *f*

single-tube heating → single-pipe heating

single-tube system, one-tube system Einrohrsystem *n* [*Rohrpost*]

single-turn(ing) lock, one-turn(ing) lock eintourige Schloss, Eintourschloss *n*

single urinal Einzelurinal *n*

single user system Einzelnutzersystem *n* [*Es steht nur einem einzigen Benutzer zur Verfügung*]

single valve Einzelventil *n* [*Im Gegensatz zum Mischventil*]

single vane rotary compressor, one vane rotary compressor Sperrschiebeverdichter *m*

single-walled, one-walled einwandig

single-wall(ed) hot water generator, one-wall(ed) hot water generator einwandige Warmwassererzeuger, einwandige Warmwasserbereiter, einwandige Boiler *m*

single wall metal pipe, one wall metal pipe einwandige Metallrohr *n*

single-way coding, one-way coding Einwegkodierung, Einwegcodierung, Einwegverschlüsselung *f*

single-way servo-valve, one-way servo-valve Einwegstellventil *n*

single-wire, one-wire [*in composite*]; **unifilar** eindrähtig [*Elektrotechnik*]

single-zone detector, one-zone detector, single-zone detecting device, one-zone detecting device, single-zone detection device, one-zone detection device Einzonenmelder *m*, Einzonenmeldegerät *n*

sink [*A sanitary fitting used mainly for culinary and cleaning activities, usually with hot and cold water supply and a trapped waste pipe and gully. It may be made of glazed earthenware, vitreous china, enamelled cast iron or stainless steel. Types vary widely and include domestic sink, commercial kitchen sink, bucket sink and slop sink*] Spüle *f* [*1. Spülbecken 2. Spültisch*]

sink bowl Spülbecken *n*

sink current, sink power Stromsenke *f*

sink faucet (US); sink (water) tap, sink draw(ing) off tap; sink crane [*Scottish term*] Spülenzapfhahn, Spülenentnahmehahn, Spülenkran *m* [*niederdeutsch*], Spülwasserdurchlaufventil *n*

sink-front cabinet, sink-front cubicle Spülenschrank *m*

sink mixer Standbatterie *f*, Standmischer *m*

sink pool Spülloch *n* [*Kleine Senkgrube zum Sammeln des Bodensatzes aus Goßsteinflüssigkeiten*]

sink top Spültisch *m*

sink water heater Spülwasserheizer, Spülwarmwasserbereiter *m*

sinusoidal angular modulation [*The variation in the angle of a sine-wave carrier; particular forms are phase modulation and frequency modulation*] sinusoidale Winkelmodulation *f*

sinusoidal answer Sinusantwort *f* [*Die Sinusantwort ist der zeitliche Verlauf der Ausgangssgröße bei einer Sinusfunktion als Eingangsgröße im eingeschwungenen Zustand. DIN 19 226 Teil 2. Anmerkung: Ein beliebiges Übertragungsglied ist mit der Kenntnis von endlich vielen dieser Antworten nicht vollständig beschreibbar. Die Angabe dieser Antworten kann jedoch nützlich sein, wenn die jeweiligen Eingangsgrößen den charakteristischen Betriebsverhältnissen entsprechen.*]

sinus response Sinusantwort *f*

SIP → single-in-line package

siphon → syphon

siphon(ic) tank, syphon tank Siphonbehälter *m*

siphon(ic) closet, siphon(ic) toilet, siphon(ic) water closet, siphon(ic) WC, washdonn siphon(ic) toilet [*"Syphonic" may synonymously be used for "siphonic"*] Siphonklosett *n*, Siphontoilette *f*, Siphonabort *m* [*„Absaug(e)..." und „Leersaug(e)..." können synonym für „Siphon..." verwendet werden. Zwischen Becken und Rohrleitung befindet sich ein Siphon, der als Wasserverschluss das Eindringen von Kanalgas in die Abortkammer verhindert*]

siren [*A motor-driven siren produces a high-pitched wailing sound, readily distinguished from horn, whistles, or other monotone audible devices and easily recognized through a confusion of other sounds*] Sirene, Motorsirene *f*

SISO → single-input single-output

site control → field control

site key Bauzeitschlüssel *m* [*Er ermöglicht eine zeitlich begrenzte Betätigung von montierten Zylindern während der Bauzeit*]

site landscape lighting, site landscape illumination Grundstücksbeleuchtung *f* [*Nicht verwechseln mit Außenbeleuchtung eines Gebäudes*]

site management → field management

site overview Übersichtsbild *n* [*Bildhierarchie*]

site riveting → riveting on site

S

site wiring → **field wiring**

sitting (type) water closet (fitment) → **western (type) water closet (fitment)**

six-pulse bridge connection → **three-phase bridge connection**

six-wire interface Sechsdrahtschnittstelle *f*

size Größe *f*

size accuracy → **accuracy to size**

sized → **dimensioned**

sized coke; graded coke [*This term is less favoured*] [*Coke in lump of sizes between specified limits obtained by screening*] Siebkoks *m*

size trueness → **accuracy to size**

sizing → **dimensioning**

sizing → **dimensions**

sizing factor, dimensioning factor Bemessungsfaktor, Abmessungsfaktor, Dimensionierungsfaktor *m*

sizing table, dimensioning table Bemessungstabelle, Abmessungstabelle, Dimensionierungstabelle *f*

SK → **shift(ing) keying**

skeleton-key → **pick lock**

skimmer Oberflächen(ab)sauger *m* [*Wasserbecken*]

skew Schräglauf *m*

skew Zeitversatz, Schräglauf *m* [*Band*]

skew Polarisatoreinstellung *f* [*Diese genaue Einstellung ist bei schwenkbaren Antennen unerlässlich und muss getrennt für jeden Kanal durchgeführt werden*]

skew Schrägung *f*, Versatz *m* [*Signal; Bit*]

skew Impulsabstand *m*

skill Können *n*

skinning, lofting Oberflächenüberzug *m* [*Gestaltung eines komplexen dreidimensionalen Objektes. Der Konstrukteur zeichnet mehrere Querschnitte durch ein Objekt – wie die Spanten eines Schiffsbausatzes – und überzieht diese mit einer Oberfläche*]

to skip [*In fixed-instruction-length digital computers, to bypass an ignored one or more instructions in an otherwise sequential process*] übergehen, überspringen, auslassen, überlesen, ausblenden

to skip abbrechen [*Diskette*]

skip Überspringen, Überlesen, Auslassen, Ausblenden, Übergehen *n* [*Eines bestimmten Satzes, bewirkt durch eine kodierte Information*]

skip Sprung *m* [*Überspringen, also keine Abarbeitung, von Befehlen und Programmteilen: a) bedingt (in Abhängigkeit von einer bisherigen Anweisung); b) unbedingt (unabhängig von Kriterien)*]

skip [*In automatic data processing, a device on a card punch that causes columns on a punch in fields where no punching is desired to more rapidly pass the punching station*] Kolonnen(über)sprungvorrichtung *f*

skip code Sprungkode, Sprungcode *m*

skip index Sprungmarkierung *f*

skip instruction → **no-operation instruction**

skip key Kolonnen(über)sprungtaste *f*

skip lot übersprungene Lot *n* [*Qualitätsprüfung*]

skipped code fehlende Kode, fehlende Code, Fehlkode, Fehlcode, Ausfallkode, Ausfallcode *m*

skip scan Anschallung *f* mit Schallstrahlumlenkung

skirt band illumination, skirt band lighting Sockelbandbeleuchtung *f*

skirt cover(ing) section Sockel(ab)deckprofil *n*

skirting (board) (Brit.) → **base (board)**

skirting board convector (Brit.) → **base (board) convector**

skirting (board) heater (Brit.) → **base (board) heater**

skirt(ing) ventilation board → **ventilation skirt(ing) board**

sky vault Himmelsgewölbe *n*

SL → **to shift left**

slab duct Plattenkanal *m*

slab urinal [*A urinal consisting of a flat impervious slab fixed on a wall, the discharge from which is directed into a floor channel fitted immediately beneath the slab. When the slab urinals are fixed in ranges, division slabs may be provided to ensure privacy*] (Wand)Plattenurinal *n*

slack bus Potentialknoten *m* [*Netz*]

slack node → **balancing bus**

slam button Schlagtester *m*

slant(ed) azimuth recording Schrägspuraufzeichnung *f* [*Geneigte Arbeitsspalten der Videoköpfe*]

slant(ed) seat (slide) valve → **inclined-seat (slide) valve**

slant(ed)-seat valve → **inclined-seat valve**

slat → **(blind) slat**

slatted blind, jalousie, louvres [*A screen of spaced parallel slats*] Jalousie *f*

slatted roller blind Rollladen *m*, Rolljalousie *f* [*Besteht aus schmalen Stäbchen, die durch Gurtbänder, Metallketten usw. gelenkig verbunden sind und auf einer Welle hinter dem Fenstersturz mit Hilfe eines Gurtes aufgerollt werden*]

slatted roller blind drive Rollladenantrieb *m*, Rolljalousieantrieb

slave Taktnehmer *m*

slave [*High-level data link control*] antwortende Datenverarbeitungsanlage, ausführende Datenverarbeitungsanlage *f*

slave → (bus) slave

slave Antwortstation *f*

slave zweite Festplatte *f*

slave zweite Laufwerk *n* [*an einem Port*]

slave Binärstufe, BS 2 *f*

slave arm Arbeitsarm *m* [*Manipulator*]

slave clock, secondary dial Nebenuhr, Tochteruhr *f*

slave control unit Unterzentrale *f*

slave distributor Folgeverteiler *m*

slave drive Folge(an)trieb *m*

slave fire (control) panel, slave fire alarm panel Nebenbrandmelderzentrale, Nebenfeuermelderzentrale *f*

slave key Nebenschlüssel *m*

slave operation Folgebetrieb, abhängige Betrieb *m* [*Von Stromerzeugungsgeräten im Verbundbetrieb*]

slave potentiometer, sequential potentiometer Folgepotentiometer *n*

slave series operation Serienbetrieb, Reihenbetrieb *m* [*Von Stromversorgungsgeräten, deren Ausgänge in Reihe geschaltet sind*]

slave station Empfangsstation, Unterstation, datenempfangende Station *f* [*Prozessleitsystem. Diese Station wird zum Datenempfang von Hauptstationen angewählt*]

sleeping stillgelegt [*Prozessstatuskode*]

sleeping zone Schlafbereich *m* [*selten: n*]

sleeptimer Anschaltzeitautomatik *f* [*Zusatzvorrichtung eines Rundfunk- oder Fernsehgerätes für eine programmierbare Anschaltzeit*]

sleeve Muffe *f*

sleeve Hülse *f*

sleeve-bearing motor [*Motor with a sleeve(-type) babbit or bronze bearing requiring oil lubrication*] Gleitlagermotor, Ringlagermotor, Büchsenlagermotor *m*

sleeve-expanding pliers Dreidornzange *f*

sleeve oblique seat valve Muffenschrägsitzventil, Schrägsitzmuffenventil *n*

sleeve terminal Hülsenklemme *f*

slewing momentary contact (control) switch Schwenktast(schalt)er *m*

slew(ing) rate Änderungsgeschwindigkeit *f* [*Rechnerverstärker*]

slice Scheibe *f*

slicer Doppelbegrenzer *m* [*Eingangsbegrenzer für zwei Amplitudengrenzwerte*]

to slide in einschieben

slide-in module Einschubmodul *n*

slider, sliding controller, sliding CTRL [*See remark under "controller"*] Schieberegler, Schubregler *m*, Schieberegelorgan, Schubregelorgan *n*, Schieberegel(ungs)einheit, Schubregel(ungs)einheit *f*

slider-belt(-type) moving walk [*The treadway slides on the supporting surface*] Fahrsteig mit Bandunterstützung durch Gleitunterlagen, Gleitfahrsteig *m* [*Die Oberseite (=Trittfläche) des Bandes ist mit einem Gummibelag mit Längsnuten belegt; die Unterseite gleitet auf Gleitunterlagen und der Rücktrumm auf Querwalzen. Eine Bandunterstützung durch Gleitunterlagen wird nur bei Fahrsteigen mit maximal 20 m Förderbandlänge angewendet, bei größeren Längen wird das Band durch Querwalzen unterstützt*]

slideshow Dia(positiv)vortrag *m*

slide switch, sliding(-dolly) switch Schiebeschalter *m*

slide valve, valve Schieber *m*

slide wire, sliding wire Gleitdraht *m*

sliding bolt → spring bolt

sliding bracket; slide bracket (US) Gleiter *m* [*Brausestange*]

sliding dolly dimmer, slide (type) dimmer Schiebehelligkeitsregler, Helligkeitsregler mit Schieberegler *m*

sliding-dolly switch → slide switch

sliding door, slide door Schiebetür *f*

sliding door lock, slide door lock Schiebetürschloss *n*

sliding door mounting, slide door mounting Schiebetürbeschlag *m*

sliding door partition, slide door partition Schiebe(tür)wand *f* [*Mehrere Schiebetüren zusammengeschoben*]

(sliding door) rail, slide door rail Schiebetürschiene, Laufschiene, Schiene *f*

sliding door release, slide door release Schiebetüröffner *m*

(sliding door) roller, slide door roller, (sliding door) runner, slide door runner Schiebetürrolle, Laufrolle, Rolle *f*

sliding door system, slide door system Schiebetüranlage *f*

sliding fit Gleitsitz *m*

sliding flue damper, slide flue damper Rauchschieber *m*

sliding folding gate, folding sliding gate, slide folding gate, folding slide gate Schiebefalttor, Faltschiebetor *n* [*Ein waag(e)recht bewegtes Hallentor mit durch Bändern verbundenen Faltschiebeflügeln*]

sliding folding gate wing, folding sliding gate wing, slide folding gate wing, folding slide gate wing Schiebefalt(tor)flügel, Faltschiebe(tor)flügel *m* [*Diese Flügel werden zum Öffnen seitlich zusammengefaltet*]

sliding gate, slide gate Schiebetor *n*

sliding hanger, slide hanger Gleitschelle *f*

sliding-in module, slide-in module Einschub *m*

sliding-in reader, sliding-in reading device, slide-in reader, slide-in reading device Einschubleser *m*

sliding limiting value, slide limiting value gleitende Grenzwert *m* [*Eine Funktion, die eine Größe begrenzt, wobei der Grenzwert selber durch ein separates Eingangssignal bereitgestellt wird*]

sliding link terminal, slide link terminal Längstrennklemme, Trennlängsklemme *f*

sliding resistor, slide resistor Schiebewiderstand *m*

sliding screen, slide screen Schiebeblende *f*

sliding seal(ing), slide seal(ing) Gleit(ab)dichtung *f*

sliding sectional door → sectional sliding door

sliding set point (value), curve set point (value), slide set point (value) Sollwertführung, Kennlinie *f*

sliding shut(ting)-off valve, slide shut(ting)-off valve, slide stop valve, sliding stop valve Absperrschieber *m*

sliding sleeve, slide sleeve Schiebehülse *f*

sliding switch, slide switch, sliding-dolly switch Schiebeschalter *m*

sliding switching, slide switching gleitende Schalten *n*

sliding transformer, slide transformer Stelltransformator *m*

sliding(-type) disconnector, slide(-type) disconnector Schubtrenn(schalt)er, Linientrenner *m*

sliding-vane rotary compressor, slide-vane rotary compressor Flügelzellenverdichter, Flügelzellenkompressor *m*

sliding ventilation window → ventilation sliding window

slight cooling Ankühlung *f*

slightly rusted angerostet, leicht gerostet

slightly wet → moist

slim-line fused load-break unit Sicherungstrennerleiste *f*

slim-line lamp, instant-start fluorescent lamp [*It is designed to be started by high voltage without preheating of the electrodes; usually has single-pin base connection*] Sofortstart-Leuchtstofflampe *f*

slip bolt Schubriegel, Schieb(e)riegel *m* [*Falzen oder in einer Kramme auf einem Blech*]

slip gauge; slip gage (US) Endmaß *n*

slip-on connector aufgeschobene Verbinder *m*, Aufsteckverbinder

slip-on marking sleeve aufgeschobene Bezeichnungshülse, Aufsteckbezeichnungshülse *f*

slip-on terminal Flachsteckklemme *f*

slip(ping) speed, slip(ping) r.p.m., slip(ping) rotations per minute Schlupfdrehzahl, Schlupfumdrehungen pro Minute *f*

slip ring → collector ring

slope → fall

slope, inclination Neigung *f*

(sloping) vane [*louver*] geneigte Lamelle *f* [*Luftraster*]

slop sink, service sink [*A rectangular sink set at low level and often provided with a metal, hinged grating on which to stand a bucket. The taps are set high enough on the wall to allow a bucket to be filled standing in the sink or on the grating*] Ausguss(becken) *m*, (*n*)

slop sink water, service sink water Ausgusswasser *n*

slot Steckplatz *m* [*Einbautechnik für elektronische Ausrüstung*]

slot Steckfassung für Karten, Kartensteckfassung *f*

slot [*component of a frame*] Abteil *n*

slot Schlitz *m*

slot burner → (long) slot burner

slot coding Steckplatzkodierung, Steckplatzcodierung *f*, Steckplatzkodieren, Steckplatzcodieren *n*

slot diffuser Schlitz(en)diffusor, Diffusor mit Schlitz(en) *m*

slot for anti-rotation lug, anti-rotation lug slot Nasenausschnitt *m*

slot gutter Schlitzrinne *f*

slot rib Schlitzrippe *f*

slots per module, SPM Einbauplätze *f* je Baugruppe, Einbauraster *m*

slot(ted) (air) diffuser, slot(ted) (air) diffusor, strip (air) diffuser, strip (air) diffusor, linear (air) diffuser, linear (air) diffusor (Luft)Schlitzdiffusor *m*

slot(ted) (air) outlet Schlitz(luft)auslass *m*

S

slot(ted) head screw [*A screw with a groove cut across its head*] Schlitzkopfschraube *f*

slot(ted) (hexagonal) nut Schlitz(sechskant)mutter, Sechskant(schlitz)mutter *f*

slot(ted) mask Schlitzmaske *f*

slot(ted) nut Schlitzmutter *f*

slot(ted) pan head screw Flachkopfschraube *f*

slot(ted) rail air outlet Schlitzschienenluftauslass *m*

slot(ted) Rider (slide) valve → Rider (slide) valve with slots

slot(ted) steel strap Schlitzbandeisen *n*

slot vent Schlitzdurchlass *m*

slow combustion stove Dauerbrandofen *m*

slow filter Langsamfilter *m, n*

slow flashing langsame Blinken *n*

slow ignition valve Langsamzündventil *n*

slow motion Zeitlupe *f*

slow-releasing relay, slow-release relay, SR relay [*A relay in which there is a time delay between a de-energizing of the coil and release of the armature*] abfallverzögerte Relais, rückfallverzögerte Relais, Relais mit Abfallverzögerung, Relais mit Rückfallverzögerung *n*

slow-scan transmission Einzelbildübertragung *f* [*Über das öffentliche Fernsprechnetz werden Einzelbilder in Abständen von 1 bis 10 Sekunden übertragen*]

slow speed (rate) Feingang *m*

SLT → solid logic technics

sludge bucket Schlammeimer *m*

(sludge) digestion Schlamm(aus)faulung *f*

(sludge) digestion plant, (sludge) digestion installation (Aus)Faulanlage, Schlamm(aus)faulanlage *f*

(sludge) digestion tank (Aus)Faulgrube, Schlamm(aus)faulgrube *f*

sludge disposal Schlammentsorgung *f*

sludge solid Schlammfeststoff *m*

sludge treatment Schlammaufbereitung, Schlammbehandlung *f*

small central bore (heating) system, small (bore) pipe central (heating) system [*A pumped, wet, central heating system depending on a sealed pump capable of producing high pressure for its successful operation. As a result of the pressure produced by the pump the diameter of the flow and return pipes can be reduced in size (usually to about 12.5 mm). Such systems are usually installed in small commercial and domestic buildings*] Kleinrohrheiz(ungs)system *n*

small cap Kapitälchen *n* [*Großbuchstabe so hoch wie ein Kleinbuchstabe*]

small central station → midget central station

small current Schwachstrom *m*

small fire Kleinbrand *m*, Kleinfeuer *n*

small fitting → fine fitting

small (private) room → closet (US)

small computer system interface, SCSI parallele Schnittstelle *f*; Protokoll *n* [*Schnittstelle, die parallele oder simultane Übertragung der Bit ermöglicht, die ein Zeichen oder Byte bilden. Dies geschieht entweder über separate Kanäle oder in einem Kanal auf unterschiedlichen Trägerfrequenzen. In der SCSI-2-Norm sind Hardware (parallele (SCSI) Schnittstelle) und Protokoll definiert. An diese Schnittstelle können bis zu 7 Peripheriegeräte angeschlossen werden*]

small goods elevator (US); small goods lift (Brit.) Kleingüteraufzug, Kleinfrachtaufzug *m*

small refrigerant cylinder → service cylinder

small-scale heat pump Kleinwärmepumpe *f*

small scale integration → SSI

small (scale) steam turbine Kleindampfturbine *f*

small sewage treatment plant, small sewage treatment system; small wastewater treatment plant, small wastewater treatment system (US) Kleinkläranlage *f*

small(-sized) furnace Kleinfeuerung, Kleinfeuerstätte *f*

smart card [*Computer engineering. A plastic credit card containing a microprocessor*] Mikroprozessorkarte *f*

smart (peripheral) I/O card → intelligent (peripheral) I/O module

smart power intelligente Leistungshalbleiter *m*

S matrix → scattering matrix

SMD → surface mounted device

SMD technique → surface-mounted technique

smell Gestank *m*

smog Dunstglocke *f* [*über Ballungsräumen*]

smoke Rauch *m*

smoke alarm Rauchalarm *m*

smoke (and fire) vent (cover) → (automatic) smoke (and fire) vent cover

smoke apron Rauchschürze *f*

smoke barrier → smoke stop

smoke box superheater Rauchkammerüberhitzer *m*

smoke cartridge Rauchpatrone *f*

S

smoke control [*Natural or mechanical ventilation for the removal of smoke*] Rauchbekämpfung *f*

smoke control door Rauchbekämpfungstür *f*

smoke crossover Rauchdurchtritt *m*

smoke damper [*Damper arranged to seal off the airflow automatically through a part of an air duct system so as to restrict the passage of smoke*] Rauch(schutz)-klappe *f* [*DIN 30 600, Reg.-Nr. 00 607*]

smoke density meter Rauchdichtemesser *m*

smoke detection Raucherkennung *f*, Rauchnachweis *m*

smoke detection Rauchmeldung *f*

smoke (detection) chamber, smoke detecting chamber Rauchkammer *f*

smoke detector, smoke detecting device, smoke detection device [*A photoelectric system for an alarm when smoke in a chimney or other location exceeds a predetermined density. See remark under "detector"*] Rauchmelder, Rauchmeldungsgeber *m*, Rauchmeldegerät *n*

smoke development Rauchentwick(e)lung *f*

smoke draught regulator, smoke damper, smoke draft regulator Rauchzugregler *m*, Rauch(stell)klappe *f*

smoke duct Rauchkanal *m*

smoke extraction Entrauchen *n*, Rauchentzug *m*

smoke extraction [*Machine or blower-fan device for extracting smoke or gases from burning buildings*] Rauchabzug, Entraucher *m*

smoke-filled → filled with smoke

smoke filter Rauchfilter *m, n*

smoke freedom Rauchfreiheit *f*

smoke gas fan Rauchgasventilator *m*

smoke generator Raucherzeuger *m*

smoke hood Rauchhaube *f*

smokeless fuel rauchlose Brennstoff *m*

smokeless liquid fire rauchlose Flüssigkeitsbrand *m*, rauchlose Flüssigkeitsfeuer *n*

smoke management Rauchführung, Rauchleitung *f*

smoke measuring device Rauchmessgerät *n*

smoke particle Rauchteilchen *n*

smoke penetration Raucheindringung *f*

smoke pipe Rauchrohr *n*

smoke pipe silencer, smoke pipe muffler, smoke pipe sound absorber Rauchrohrschalldämpfer *m*

smoke poisoning Rauchvergiftung *f*

smoke-proof, smoke-resistant rauchdicht, rauchsicher, rauchfest

smoke propagation Rauchausbreitung *f*

smoke proportion Rauchanteil *m*

smoke protection Rauchschutz *m*

smoke protection door, smoke protective door, smoke protecting door Rauch(schutz)tür *f* [DIN 18 095]

smoke removal, smoke exhaust, smoke disposal, smoke discharge, smoke extraction Entrauchung, Rauchbeseitigung, Rauchabführung, Rauchableitung, Rauchentsorgung *f*, Rauchabzug *m*

smoke sensitive (type) (automatic) fire detector [*Alteration of radioactive ionisation determined by photo-electric cells*] automatische Rauchbrandmelder, automatische Rauchfeuermelder, Rauchbrandmeldeautomat, Rauchfeuermeldeautomat *m*

smoke sensitivity, smoke sensitiveness Rauchempfindlichkeit *f*

smoke signal Rauchsignal *n*

smoke signal message Rauchmeldung *f*

smoke stop, smoke barrier [*Partition in corridor or between spaces to retard the passage of smoke, with any opening in such partition protected by a door equipped with a self-closing device*] Rauchsperre *f*, Rauchabschluss *m*

smoke switch Rauchschalter *m*

smoke test Rauchprobe, Rauchprüfung *f*, Rauchversuch *m* [*Dichtheit(s)prüfung*]

smoke tester Rauchprüfer *m*

smoke thinning Rauchverdünnung *f*

smoke tube, fire tube Feuerrohr *n*

smoke tube ferrule, fire tube ferrule Feuerrohrring *m*

smoke tube (steam) boiler → fire tube (steam) boiler

smoke tube superheater, fire tube superheater Rauchrohrüberhitzer *m*

smoke vent Rauchdurchlass *m*

smoke ventilation Rauchentlüftung *f*

smoke washer [*A device for removing particles from smoke by forcing it through a spray of water*] Rauchwäsche(r) *f*, (*m*)

smoking occupancy Raucherbelegung *f*

smoking prohibition Rauchverbot *n*

smoky rauchig

smooth finishing Feinschlichten *n*

smoothness of ride, ride smoothness stoßfreie Fahren, ruckfreie Fahren *n* [*Aufzug*]

smooth operation einwandfreie Betrieb, ruhige Betrieb, fehlerlose Betrieb *m*

S

smooth start(ing) → **soft start(ing)**

smooth surface (sectional) radiator [*See remark under "radiator"*] glatte Radiator, glatte Gliederheiz(ungs)körper *m*

smothering effect Stickwirkung *f* [*Feuerlöschen*]

smouldering [*Very low stage of combustion with little heat or flame and mostly smoke evidenced; phase one of a fire*] Schwelen, Glimmen *n*

smouldering fire Schwelbrand, Glimmbrand *m*, Schwelfeuer, Glimmfeuer *n*

SMPS → **switched mode power supply**

SMR → **signal to mask ratio**

SMS → **storage management services**

SMS → **short message service**

smudge Verschmierung *f*

S/N → **signal(-to)-noise ratio**

to snap einfangen [*von Bildpunkten; CAD*]

to snap [*See remark under "snap action"*] schnappen, kippen, springen [*Näherungsschalter*]

snap action [*Response of a switch to very small movements of its actuating button or lever, changing rapidly and positively from one contact position to the other*] Springen, Kippen, Schnappen *n* [*Näherungsschalter*]

snap(-action) cam Schnappnocken, Sprungnocken, Kippnocken *m*

snap(-action) magnetic amplifier Schnappmagnetverstärker, Kippmagnetverstärker, Sprungmagnetverstärker *m*

snap(-action) ((open) loop) control Sprungsteuerung, Schnappsteuerung, Kippsteuerung *f*

snap(-action) operation → **independent manual operation**

snap(-action) switch → **sensitive switch** [*IEC 163*]

snap gauge; snap gage (US) Rachenlehre *f*

to snap into place einschnappen, zuschnappen

snap-off diode Abreißdiode, Schaltspeicherdiode, Speicherschaltdiode *f*

SNG → **synthetic natural gas**

sniffer Lecksucher, Leckspürer *m*

snow accumulation Schneeansammlung *f*

snow load Schneelast *f*

SNR → **signal(-to)-noise ratio**

snubber Überspannungsschutzelement *n*, Druckstoßminderer *m*

SO → **shift out**

soap bubble test [*Method of detecting leaks by coating with soap solution. Any major leak will cause bubbles to form*] Seifenblasenprüfung, Seifenblasenprobe *f*, Seifenblasenversuch *m*

soap container Seifenbehälter *m*

soap dispenser Seifenspender *m*

soap foam dispenser, foam soap dispenser Seifenschaumspender, Schaumseifenspender *m*

soap grinder Seifenmühle *f*

soap holder Seifenhalter, Seifenträger *m*

soap level indicator [*soap dispenser*] Seifenstandanzeiger *m*

soap tray, soap dish Seifenschale, Seifenablage, Seifenmulde *f*

soap valve Seifenventil *n*

SOC → **state-of-charge**

social service Sozialdienst *m*

socket (Brit.) → **bell**

socket distributor Buchsenverteiler *m*

socket inlet, plug [*An electrical port, fixed generally to the flex of an appliance for connecting it to the supply through a socket outlet, or sometimes through an adaptor*] Stecker *m*

socket outlet; receptacle (US) [*An electrical fixture, usually on a wall, containing two or three holes into which the metal prongs of an adaptor or socket inlet are inserted*] Steckdose, Anschlussdose *f*, Steckkontakt, Anschlusskontakt *m*

socket pipe (Brit.) → **bell pipe**

socket(-ring) → **(door) pan**

socket with earth(ing) contact, two-pole-and-earth socket (Brit.); ground(ing)(-type) receptacle, earth(ing) contact socket Schutzkontaktsteckdose; SCHUKO-Steckdose *f* [*„SCHUKO" ist ein eingetragenes Warenzeichen*]

sodium (discharge) lamp; sodium vapor lamp (US); sodium vapour lamp (Brit.) Natrium(entladungs)-lampe, Natrium(dampf)lampe *f*

soffit, crown [*The highest point of the internal surface of a pipe or culvert at any cross section or of the outlet leg of a trap*] Scheitel(punkt) *m*

soffit (area) Untersicht(fläche) *f*

soft coal Weichkohle *f*

soft copy Sichtgerätausgabe, Bildschirmdarstellung *f*

soft edge unscharfe Kante *f* [*Kantenmodulation*]

to soften weichmachen, enthärten [*Wasser*]

soft(ened) water [*Water free of calcium and magnesium salts*] enthärtete Wasser, Weichwasser *n*

S

softener → **(water) softener**

softening → **(water) softening**

softening filter → **(water) softening filter**

softening resin → **(water) softening resin**

softening system → **(water) softening system**

soft ignition Weichzündung, leise Zündung *f*

soft solder [*A solder in which lead and tin are the principal constituents*] Weichlot *n*

soft soldering Weichlöten *n*, Weichlötung *f*

soft start(ing), reduced-voltage start(ing), cushioned start(ing), smooth start(ing) Sanftanlauf, weiche Anlauf *m*

soft tag [*shoplifting prevention*] Weichetikett *n*

software Rechnerprogramme, Programmierintelligenz, Programme *f*

software engineering Programmtechnik *f* [*Die Kunst, Programme zu schreiben*]

software polling Reg(e)lung *f* der Prioritätsfrage [*Mikroprozessor*]

soft water → **soft(ened) water**

soft welding Widerstands-Rollnaht-Schweißen, Rollnaht-Widerstands-Schweißen *n*

SOH → **start of heading**

soil → **human body waste**

soil drain → **soil sewer**

soil pipe, soil line, closet waste water pipe, closet waste water line, toilet waste water pipe, toilet waste water line [*A pipe collecting from W.C.s and urinals*] Abortrohr, Klosett(abwasser)rohr, Toiletten(abwasser)rohr *n*

soil sewage; soil wastewater (US) Abortabwasser

soil sewer, soil drain; wastewater sewer, wastewater drain (US) Abwasserkanal *m*

soil sewerage; wastewater sewerage (US) Abwasserkanalisation *f*

soil stack, closet soil stack, toilet soil stack Abortstrang, Klosett(abwasser)strang, Toiletten(abwasser)strang *m*

soil wastewater (US) → **soil sewage**

soil/water heat pump Boden-Wasser-Wärmepumpe *f* [*Siehe Anmerkung unter „Wärmepumpe"*]

solar absorber, sun absorber Solarabsorber, Sonnenabsorber *m* [*Die Solarabsorber des Energiedaches sind im Gegensatz zu den Solarkollektoren ganzjährig Tag und Nacht wirksam*]

solar battery, sun battery Sonnenbatterie, Solarbatterie *f* [*Sie wird mit Solarstrom geladen*]

solar brick, sun brick Solarziegel, Sonnenziegel *m*

solar collector, sun collector [*Device designed to absorb incident solar radiation and to transfer the energy to a fluid passing through*] Solarkollektor, Sonnenkollektor *m* [*Nicht verwechseln mit Solarabsorber*]

solar collector system, sun collector system Solarkollektoranlage, Sonnenkollektoranlage *f*

solar central heating station solare Heiz(ungs)zentrale *f*

solar cladding → **solar skin**

solar constant, sun constant [*The rate at which solar radiant energy is received outside the atmosphere on a surface normal to the incident radiation, at the Earth's mean distance from the sun. The value of the mean solar constant is 1.94 gram calories per minute square centimeter*] Solarkonstante, Sonnenkonstante *f*

solar cooling, sun cooling Solarkühlung, Sonnenkühlung *f*

solar curtain wall Solar-Vorhangwand, Sonnen-Vorhangwand *f*

solar detector, solar sensor, sun detector, sun sensor [*It is identical to the outside temperature detector except that it has a special clear plastic front to measure the solar effect. It has no local setting adjustment*] Sonnendetektor, Sonnensensor, Sonnenfühler *m*, Sonnendetektionselement *n*

solar electronics Solarelektronik *f*

solar energy conversion, sun energy conversion Solarenergieumwandlung, Sonnenenergieumwandlung *f*

solar energy storage, sun energy storage Sonnenenergiespeicherung, Solarenergiespeicherung *f*

solar engineering, sun engineering Solartechnik, Sonnenwärmetechnik *f*

solar evaporation, sun evaporation [*The evaporation of water due to the sun's heat*] Solarverdunstung, Sonnenverdunstung *f*

solar fraction solare Deckungsrate *f*

solar glazing, sun glazig Solarverglasung, Sonnenverglasung *f*

solar-heated, sun-heated solarbeheizt, sonnenbeheizt

solar heat gain, sun heat gain [*Solar energy flowing into a building, through windows, doors, openings and structural materials*] Solarwärmegewinn, Sonnenwärmegewinn *m*

solar heat impact, sun heat impact Sonnenwärmeanfall, Solarwärmeanfall *m*

solar heating, sun heating Sonnenheizung, Solarheizung *f*

solar heat(ing) plant, sun heat(ing) plant, solar heat(ing) installation, sun heat(ing) installation Sonnenheiz(ungs)anlage, Solarheiz(ungs)anlage *f* [*DIN 4757*]

solar heat storage, sun heat storage Solarwärmespeicherung, Sonnenwärmespeicherung *f*

solar house, sun house Solarhaus, Sonnenhaus *n*

solar load, sun load Sonnenlast, Solarlast *f*

solar power, sun power Sonnenkraft, Solarkraft *f*

solar-powered, sun-powered solarbetrieben, sonnenbetrieben

solar relay → photo-electric lighting controller

solar roof, sun roof Solardach *n*

solar screen, (sun-)screen (Sonnen)Blende, Solarblende *f*

solar shading, sun shading Solarabschattung, Sonnenabschattung *f*

solar skin, solar cladding Solar-Vorhangfassade, Sonnen-Vorhangfassade *f*

solar tube collector, tube solar collector Röhren(solar)kollektor, Solarröhrenkollektor *m*

solar winter load → winter solar load

solar vacuum tube collector plant, solar vacuum tube collector installation, sun vacuum tube collector plant, sun vacuum tube collector installation Solar-Vakuumröhren-Kollektoranlage, Sonnen-Vakuumröhren-Kollektoranlage *f*

soldered gelötet

soldering Löten *n*, Lötung *f* [*Das Verbinden gleicher oder verschiedener Metalle durch ein Zusatzmetall (= Lot) mit niedriger Schmelztemperatur*]

soldering bit Lötkolben *m*

soldering bush Lötkupp(e)lung *f*

soldering flux Löt(fluss)mittel *n*

soldering lamp → blowlamp

soldering lug Löt(anschluss)fahne *f*

soldering pin Lötstift *m*

soldering stone Lötstein *m*

(soldering) tag Lötfahne, Lötlasche, Lötöse *f*, Lötstützpunkt *m*

(soldering) tag block Lötleiste *f*, Lötstreifenverband *m*

soldering-tight lötdicht

soldering torch Lötbrenner *m*

soldering with bismuth, bismuth soldering Wismutlöten *n*, Wismutlötung *f*

sole → (door) pan

solenoid, electromagnetic coil [*It is wound in the shape of a hollow cylinder or spool, often containing a mov(e)able iron core that is the coil when current flows through the windings, thus allowing it to move other devices such as relays and circuit breakers*] Magnetspule, Zylinderspule, elektromagnetische Spule *f*, Solenoid *n*

solenoid valve, electromagnetic valve [*Magnetically operated valve, generally used to control the flow of liquid to an evaporator. It may also be used where ever off-on control is permissible. Solenoid electrical winding controls the action of the valve*] Magnetventil, Solenoidventil, elektromagnetische Ventil *n*

solid angle Raumwinkel *m*

solid body electronics Festkörperelektronik *f*

solid carbon dioxid(e) [*proprietary term*] **→ dry ice**

solid CO_2 → dry ice

solid concrete floor Vollbetondecke, Decke mit unterem Tragbeton *f*

solid concrete heating floor Vollbetonheiz(ungs)decke, Heiz(ungs)decke mit unterem Tragbeton *f*

solid-expansion thermometer [*It is based on the expansion of a solid substance*] Ausdehnungs-Thermometer *n*

solid (fire) extinguishing agent feste Löschmittel, Trockenlöschmittel *n*

solid floor Massivdecke *f*

solid fuel [*It includes coal, coke, wood, and waste products of industrial and agricultural operations*] Festbrennstoff *m*

solid fuel boiler Festbrennstoffkessel *m*

solid fuel (combustion) chamber Festbrennstoff(verbrennungs)kammer *f*

solid fuel cooker Festbrennstoffkocher *m*

solid fuel domestic boiler, domestic solid fuel boiler Festbrennstoffhauskessel *m*

solid fuel furnace Festbrennstoffeuerung(sanlage), Festbrennstoffeuerungsstätte *f* [*DIN 18160*]

solid geometry Festkörpergeometrie *f*

solid gold Festgold *n*

solid grounding starre Erdung, widerstandslose Erdung *f*

solidification point [*Temperature at which a liquid substance will solidify, but not necessarily crystallize, on removal of heat (usually at standard atmospheric pressure)*] Erstarrungspunkt, Stockpunkt *m*

solid in suspension → suspended solid

solid lubricant Festschmierstoff *m*, Festschmiermittel *n*

solid material → solid stock

solid model, body model Körpermodell *n*

solid pipe → high-pressure pipe

S

solid piston → plunger

solid-state camera Halbleiterkamera *f*, Halbleiterphotoapparat, Halbleiterfotoapparat *m*

solid-state disk, solid-state disc [*Electronic device made with semiconductor components*] Halbleiterplatte *f*

solid-state laser Halbleiterlaser *m*

solid-state ((open) loop) control kontaktlose Steuerung *f*

solid-state physics Festkörperphysik *f*

solid-state switching device Halbleiterschaltgerät, statische Schaltgerät, kontaktlose Schaltgerät *n*

solid-state technique Halbleitertechnik *f*

solid stock, solid material Massivmaterial, Vollmaterial *n*

solid stoker, solid stoking device Festbrennstoffbeschicker *m* [Kessel]

solid surface Feststoffoberfläche *f*

solid thyratron Halbleiterthyratron *n*, Halbleiterschalter *m*

solidus Schrägstrich *m* [*Kode*]

solid waste Müll, feste Abfall(stoff) *m*

solid waste incinerator Müllverbrennungsanlage *f*

solid waste management Müllbewirtschaftung *f*

solid wood Echtholz, Massivholz, Vollholz *n* [*Im Gegensatz zum Sperrholz, Schichtholz usw.*]

solid yoke massive Joch *n*

solitary operation Alleinbetrieb, Inselbetrieb *m*

solubility Löslichkeit *f*

solution Lösung *f*

solution algorithm [*Prescribed set of well-defined rules, or process for the solution of a problem in a finite number of steps, e.g. a full statement of an arithmetical procedure for evaluation sine x to a stated precision*] Lösungsalgorithmus *m*

solution approach Lösungsansatz *m*

to solve a problem lösen einer Aufgabe

solvent Lösungsmittel *n*, Lösungsstoff *m*

solver [*Finite element method*] Analyseprogramm *n*

soot Ruß *m* [*Unverbrennbarer Kohlenstoff aus den Kohlenwasserstoffverbindungen der Brenngase*]

soot blower Rußbläser *m*, Rußgebläse *n*

soot dust Rußstaub *m*

sooting (Ver)Rußen *n*

sooting number [*A measure of the sooting propensity of a gas obtained by determining the volume rate of flow of air required to prevent soot-tailing of the flame of the gas in a standard sooting test burner*] Rußzahl *f*

sooting test Rußprüfung, Rußprobe *f*, Rußversuch *m*

soot meter Rußmesser *m*

soot metering Rußmessung *f*

sootproof, soottight rußdicht, rußsicher, rußfest

soot spot, soot stain Rußfleck *m*

soot tester, soot testing device Rußprüfer *m*, Rußprüfgerät *n*

sophisticated hochentwickelt, technisch ausgereift

sorbent, sorbing agent, sorptive material [*It attracts and holds other gases and liquids*] Sorptionsmittel *n*

sorption [*The binding of one substance to another*] Sorption *f*

sorption heat pump Sorptionswärmepumpe *f*

sorption isotherm [*Materials science. The expression of the hygroscopicity of wood as a curve that relates moisture content with a given relative vapo(u)r pressure*] Sorptionsisotherme *f*

sorption unit (assembly) Sorptionsaggregat *n*, Sorptionseinheit *f*

sorter → sequencer

SOS, silicon on sapphire Silikon-auf-Saphir-Technik *f*

SOU → direct operation unit

to sound [*alarm*] auslösen

sound [*Computer science. Describing a theorem-proving technique or method of reasoning that is guaranted to derive only valid conclusions*] fehlerfrei

sound [*Acoustics. A pressure disturbance that propagates through a medium due to stress or displacement of the medium from its equilibrium state*] Schall *m*

sound [*The auditory perception that is induced by a pressure disturbance that propagates through a medium due to stress or displacement of the medium from its equilibrium state; something heard by the ear*] Klang *m*

sound absorption [*Acoustics. The decay of sound energy as its distance from the sound source increases, due to conversion of the acoustic energy to heat generated in the medium*] Schallabsorption *f*

sound-amplification system → (public-address) (loud)speaker system

sound attenuation Schalldämpfung, Schallabschwächung *f*

sound attenuator [*A device mounted in the ducting of an air conditioning or ventilating system which prevents the transmission of sound from the fan through the ducting to other parts of the building*] Schalldämpfer, Schallabschwächer *m*

to sound a warning akustisch warnen, hörbar warnen

S

sound boom Überschallknall *m*

sound broadcasting Hörfunk *m*

sound confusion → confusion of sounds

sound density Schalldichte *f*

sound depth Klangtiefe *f*

sound detection, sound recognition [*The discrimination of a specific sound of interest separated from background noise*] Schallerkennung *f*, Schallnachweis *m*

sound direction, direction of sound Schallrichtung *f*

sound-dub Nachvertonen *n*

sound energy, acoustic energy [*The energy associated with sound waves, expressed quantitatively by the difference between the total energy of a sound field and the energy that would exist in the same region in the absence of sound*] Schallenergie, Schalleistung *f*

sound energy density Schallenergiedichte *f*

sound energy flux [*Acoustics. The average amount of sound energy transmitted in a specific direction perpendicularly across a unit area at the point considered*] Schalleistung durch ein Oberflächenelement, Schallenergie durch ein Oberflächenelement *f*

sound engineering Schalltechnik *f* [*Die Schalltechnik umfasst die technischen Probleme der Entstehung und Ausbreitung von Schwingungsvorgängen in gasförmigen, flüssigen oder festen Stoffen oder schwingungsfähigen System (Luft-, Wasser- und Körperschall). Schall und Erschütterungen breiten sich in Form von Längs-, Quer- und Biegewellen aus und gehen aus einem Medium in ein anderes nach ähnlichen Gesetzen wie in der Optik über.*]

sound fidelity Klangtreue *f*

sound field Schallfeld *n*

sound film [*Motion-picture film that has an accompanying audio track near the edge of one side of the film*] Tonfilm *m*

sound flanking path, flanking sound path Schallflankenübertragungsweg *m*

sound formation Schallbildung *f*

sound generation Schallerzeugung *f*

sounding test Klangprobe, Klangprüfung *f*, Klangversuch *m*

sound intensity Schallstärke *f*

sound intromission Einschallung *f*

sound level Schallpegel *m*

sound level intensity Schallstärke *f*

sound level meter [*An apparatus that measures sound level intensities; consists of a microphone, an amplifier, frequency-weighting networks, and a meter calibrated in decibels or volume units*] Schallpegelmesser, Schallstärkemesser *m*

sound-mixing table Schallmischpult *n*

soundness test, quality test Güteprüfung, Güteprobe *f*, Güteversuch *m*

sound path Schallweg, Schalllaufweg *m*

sound propagation Schallausbreitung, Schallfortpflanzung *f*

sound protection → protection against sound

sound protection in building construction Schallschutz im Hochbau, Hochbauschallschutz *m* [*DIN 4109*]

sound receiving room → receiving room

sound recognition → sound detection

sound recording Schallaufzeichnung *f*

sound-reduced schallarm, schallgemindert, schallreduziert

sound reduction → (equivalent sound pressure) level difference

sound-reflecting schallhart

sound reinforcement system → (public-address) (loud)speaker system

sound scattering Schallstreuung *f*

sound signal [*Any sound that conveys intelligence, as distinguished from noises*] Schallsignal *n*

sound source room → source room

sound transmission Schallübertragung *f*

sound transmission coefficient [*The ratio of incident sound energy to that of transmission in the event that a sound wave strikes a boundary of different material; the coefficient is dependent upon the material and the incidence angle*] Schallübertragungskoeffizient *m*

sound type, type of sound Schallart *f* [*Man unterscheidet: Körper-, Tritt- und Luftschall*]

sound velocity, sound speed [*About 1100 feet per second at sea level*] Schallgeschwindigkeit *f*

sound wave [*Any of the longitudinal progressive vibrations in an elastic medium, by which sounds are transmitted*] Schallwelle *f*

sour → acidulous

source Träger *m* [*z. B. Wärmeträger*]

source [*The point or object form which something emanates*] Quelle *f*, Ursprung *m*

source address, SAD Quelladresse *f*

source and drain p-leitende Zonen *f* [*Herstellung integrierter Schaltungen*]

source call Quellprogramm *n*

source code Quelltext, Quellkode, Quellcode *m* [*Er wird vom Programmierer formuliert im Gegensatz zum Maschinenkode den der Rechner ausführt bzw. abarbeitet*]

source database Quelldatenbank *f*

source decoding Quellendekodierung, Quellendecodierung, Quellenentkodierung, Quellenentcodierung *f*

source directory Quellenverzeichnis *n*

source encoding Quellenkodierung, Quellencodierung *f*

source-follower amplifier → **common-drain amplifier**

source impedance [*The output impedance of a source in a circuit, such as that of a generator or transducer*] Quellimpedanz *f*

source retaining plate, source retention plate thermische Meldesichtung *f*

source language Quell(en)sprache, Ausgang(s)sprache, Ursprungssprache *f* [*In dieser Sprache liegt eine Befehlsfolge vor, bevor sie einem maschinellen Übersetzungsprozess unterworfen wird*]

source program (US); source programme [*A program(me) coded so that it must be translated into machine language prior to use*] Quellenprogramm *n*

source room, sound source room (Schall)Ausgangsraum *m*

space, gap Lücke, Leerstelle *f*, Zwischenraum *m*

space → **room**

space air condition, space air state Raumluftbeschaffenheit *f*, Raumluftzustand *m*

space air flow Raumluftströmung *f*

space air humidification Raumluftbefeuchtung, Raumluftanfeuchtung *f*

space (character) → **blank (character)**

space cone → **herpolhode cone**

space coordinate Raumkoordinate *f*

space divider Raumteiler *m*

space division Raumaufteilung *f*

space enclosing masonry wall → **enclosing masonry wall**

space enclosing wall → **enclosing wall**

space flow(ing), space stream(ing) Raumströmen *n*, Raumströmung *f*

space gas heater → **gas space heater**

space-heating furnace [*See remark under "furnace"*] Raumheiz(ung(s))feuerung *f*

space line Leerzeile *f*

space management Raumbewirtschaftung *f*

space monitoring Raumüberwachung *f* [*Sicherheitsüberwachung außen und innen*]

spacer (block) [*Any component or device that holds two parts at a distance form each other*] Abstand(s)-halter *m*

space relative raumbezogen

space requirement, space need Platzbedarf (zum Aufstellen), Raumbedarf (zum Aufstellen) *m*

spacer frame Montageflansch *m* [*Ventilator*]

spacer key Leerstellentaste *f*

space-saving, economical of space platzsparend, raumsparend

spacing Ausschluss *m* [*Wortabstand innerhalb einer Zeile; Leerraum zwischen zwei Wörtern; Einrücken am Anfang einer Zeile*]

spacing → **interval**

spacing wave → **back wave**

spackle Spachtel *m* [*Österreich: f*]

spackling [*The process of using a paste (spackle), made from a dry powder and water mix(ture), to fill surface*] (Ver)Spachteln *n*

spandrel (wall) induction unit, breast induction unit, apron induction unit, parapet induction unit; apron wall induction unit (US) [*See remark under "induction unit"*] Brüstungsinduktionsgerät *n* [*In Außenräumen und Fenstern angeordnet*]

Spanish sash bolt → **espagnolette (bolt)**

spanner key Steckschlüssel *m*

spare bit unbelegte Bit *n* [*Zeichengabe*]

sparge pipe [*A horizontal pipe having perforations through which water is sprayed for cleansing purposes, for example, on to a slab (type) urinal or in a tunnel shower*] Sprinklerrohr *n*

spark arrester, spark killer, spark suppressor Funkenlöscher *m*, Funkenlöschvorrichtung *f*

spark chamber [*Nucleonics. An enclosed chamber equipped with a device that generates a series of sparks in an array of spark gaps; used to observe the trajectory of an energetic particle passing through the chamber*] Funkenkammer *f*

spark coil [*A high-voltage induction coil used to produce a spark for ignition systems*] Funkenspule *f*

spark detector, spark detecting device, spark detection device Funkenmelder *m*, Funkenmeldegerät *n*

spark extinguishing, spark quenching, spark arresting, spark killing, spark absorbing, spark blowing out Funkenlöschung *f*

spark gap [*A device consisting of two metal points, tips, or balls separated by a small air gap; a high voltage applied to the electrodes causes a spark or, with an AC voltage, a train of sparks to jump across the gap*] (Schutz)Funkenstrecke *f*

spark ignition Funkenzündung *f*, Zünddurchschlag *m*

spark(ing) voltage [*The voltage required to create an arc across the gap of a spark plug*] Funkenspannung *f*

spark killer → **spark arrester**

spark monitoring system Funkenüberwachungssystem *n*

spark suppressor → **spark arrester**

spark tracking Kriechwegbildung *f* durch Gleitfunken

spark transmitter [*Electronics. A radio transmitter whose source of radio-frequency power is the oscillatory discharge of a capacitor through an inductor and a spark gap*] Funkensender *m*

spatial background Raumhintergrund *m*

spatial brightness Raumhelligkeit *f*

spatial frequency Raumfrequenz, Liniendichte je Längeneinheit *f*

spatial harmonic Raumharmonie *f*

SPD → **serial pole(ing) disenable**

S.P.D. → **self powered detector**

SPDT switch → **single pole double-throw switch**

SPE → **serial poll(ing) enable**

speaker → **(loud)speaker**

speakerphone Freisprechanlage *f*

speaking current, weak electrical oscillation Sprechstrom *m*

speaking transmitter, transmitter insert Sprechkapsel *f*

special character [*Computer programming. 1. A character in a character set other than a letter, digit, or space. 2. In COBOL, a character that is neither numeric nor alphabetic*] Sonderzeichen, Spezialzeichen *n*

special committee Fachkommission *f*, Fachausschuss *m*

special elevator (US); special lift (Brit.) Spezialaufzug, Sonderaufzug *m*

special feature Sonderheit *f*

specialist Fachkraft *f*, Fachmann *m*

special(ist) firm Fachfirma *f*, Fachunternehmen *n*

special knowledge → **know-how**

special(-purpose) language [*Computer programming. A programming language designed for a single objective such as simulation or string processing*] Spezialsprache, Sondersprache *f*

special sewage Sonderabwasser *n*, Spezialabwasser [*Es wird gezielt gereinigt, neutralisiert oder deponiert*]

special system → **(dedicated) special system**

special(ty) lighting, special(ty) illumination Sonderbeleuchtung, Spezialbeleuchtung *f*

special version Sonderausführung, Spezialausführung *f* [*Apparat; Gerät*]

special voltage Sonderspannung, Spezialspannung *f*

species Art(en) *f*

species richness Artenreichtum *m*

specific acoustic impedance, unit-area impedance spezifische Schallimpedanz, Feldimpedanz *f* [*Quotient aus Schalldruck und Schallschnelle*]

specific acoustic resistance [*A real quantity pc, where p is the equilibrium density and c is the sound velocity; the quantity represents an analogy to resistance in electrical circuits*] spezifische Schallwiderstand *m*

specific adhesion allowance bezogene Haftmaß *n*

specific (air) humidity [*It refers to the weight of the vapo(u)r per lb. of the dry air*] spezifische Luftfeuchtigkeit, spezifische Luftfeuchte *f*

specification and description language, SDL Standardbeschreibunssprache *f* [*von der ITU-T entwickelt*]

specification(s), spec(s)., spec(s) technische Beschreibung, technische Daten, technische Merkmale, Kenndaten *f*

specific code → **absolute programming**

specific conductance → **electrical conductivity**

specific enthalpy, specific heat content [*Enthalpy per unit mass of a substance*] spezifische Enthalpie *f*, spezifische Wärmeinhalt *m*

specific gravity, S.G., relative density Wichte, relative Dichte *f*; spezifische Gewicht *n* [*frühere Benennung*]

specific gravity bottle → **density bottle**

specific heat [*Thermodynamics. A partial derivative with respect to temperature; under certain conditions it can be expressed as a ratio of the heat transfer to increase the temperature of a unit mass of a substance by 1 ° to the amount of heat transfer to increase the temperature of a unit mass of water by the same temperature increment, while subject to constant conditions of volume or pressure. Constant-volume specific heat and constant-pressure specific heat are macroscopic properties of a substance*] spezifische Wärme, Eigenwärme *f*

specific heat of air [*The heat (Btu) necessary to raise 1 lb of water 1 F°*] spezifische Luftwärme, Luft-Eigenwärme, Eigen-Luftwärme *f*

specific heat of a subcooled refrigerant spezifische unterkühlte Kältemittelwärme, unterkühlte spezifische Kältemittelwärme *f*

S

specific heat requirement, specific heat need spezifische Wärmebedarf, Eigenwärmebedarf *m*

specific hot water need, specific hot water requirement spezifische Warmwasserbedarf, Warmwassereigenbedarf *m* [*Eine Warmwassermenge in Volumeneinheiten mit bestimmter Temperatur pro Zeit- und Verbrauchereinheiten*]

specific inductive capacity → dielectric constant

specific insulation resistance → volume resistivity

specific memory requirement, specific memory need spezifische Speicherbedarf *m* [*Bei der Programmierung von Funktionen entsteht je nach Gerätetyp ein unterschiedlicher Speicherbedarf. Die für die Programmierung einer Funktion benötigten Speicherplätze werden angegeben. VDI 2880 Blatt 1*]

specific preparation → conditioning

specific true heat → true specific heat

specific volume, SV [*Volume of a unit mass of a material. Usually expressed in cubic feet per pound of cubic meters per kilogram. The reciprocal of density*] spezifische Volumen *n*

specific water separation → water conditioning

to specify, to stipulate vorschreiben

spectrophotometer Spektrophotometer, Spektralphotometer *n*

specular glass reflector → mirror

specular reflector [*A surface capable of reflecting a direct image of a source. A mirror is a specular glass reflector*] Spiegelreflektor *m*

specular reflector luminaire Spiegelleuchte *f*

speech filing system Sprachspeichersystem *n*

speech frequence → voice frequence

speech intelligibility Sprachverständlichkeit *f*

speech quality measurement Sprachgütemessung *f*

speech recognition [*Linguistics. The ability to recognize the communicative meaning of vocally produced sounds*] Spracherkennung *f*

speech recognition [*Artificial intelligence. A process by which a computer can recognize but not necessarily interpret a spoken word, phrase, or sentence*] Spracherkennung *f*

speech station Sprechstelle *f*

speed [*The time rate of change of position without regard to the direction of motion; i.e., the ratio of the distance travel(l)ed to the time elapsed*] Geschwindigkeit *f*

speed, r.p.m., rotations per minute Drehzahl, Umdrehungen pro Minute *f*

speeding-up condenser → acceleration condenser

speed limiter Geschwindigkeitsbegrenzer *m* [*Aufzug*]

speed limit(ing) rope Geschwindigkeitsbegrenzungsseil *n* [*Aufzug*]

speed of light, light speed Lichtgeschwindigkeit *f*

speed of milling out, milling out speed Fräsgeschwindigkeit *f*

speed ((open) loop) control → ((open) loop) speed control

sphere gap Kugelfunkenstrecke *f*

spherical float, globe float Kugelschwimmer *m*, Schwimmerkugel *f*

spherical float valve, globe float valve Kugelschwimmerventil, Schwimmerkugelventil *n*

spherical tank Kugeltank, Kugelbehälter *m*

spill oil → return oil

spill(-over), overflow [*Engineering. An unwanted occurrence in which material becomes released from its container*] Überlauf *m*, Überströmung *f*

spill(-over)level, overflow level Überlaufhöhe, Überströmhöhe *f*

spin [*The rotation of a body about an axis through the body*] Drall *m*

spin ((closed) loop) control Drallreg(e)lung *f*

spindle → stem

spindle extension → stem extension

spindle guide → stem guide

spinning disk atomization, spinning disc atomization Scheibenzerstäubung *f*

spinning disk humidifier, spinning disc humidifier [*Water at the required temperature is atomized and then thrown into the air stream by the spinning disk*] Scheibenluftbefeuchter, Scheibenluftanfeuchter *m*

spiral casing, spiral case, spiral housing Spiralgehäuse, Wendelgehäuse *n*

spiral chute Spiralrutsche, Wendelrutsche *f*

spiral fin, helical fin, strip fin [*Continuous strip of metal helically wound and attached to a tube*] Band(rohr)rippe *f*

spiral fin tube, helical fin tube, strip fin tube Bandrippenröhre *f* [*Die Rippen sind auf das Rohr spiralförmig aufgewickelt*]

spiral four(quad) cable, star quad cable, cable with star quad combination [*A quad cable in which the four conductors are twisted around a common axis; the two sets of opposite conductors are used as pairs*] Sternviererkabel, sternviererverseilte Kabel *n*

S

spiral heat exchanger, spiral heat exchange unit, spiral heat interchanger Spiralwärme(aus)tauscher *m* [*In der Industrie so genannt; Spiralwärmeübertrager. Siehe Anmerkung unter „Wärme(aus)tauscher"*]

spiral nozzle Spiraldüse *f*

spiral pipe [*A pipe having a continuous coiled seam along its length*] Spiralrohr *n*

spiral quad → star quad

spiral rib(bed) pipe, rib(bed) spiral pipe Spiralrippenrohr, Rippenspiralrohr *n*

spiral suction hose, suction spiral hose Spiralsaugschlauch, Saugspiralschlauch *m*

spiral wire brush, wire spiral brush Spiraldrahtbürste, Drahtspiralbürste *f*

spittoon Speibecken, Spuckbecken *m*

SPL → acoustic pressure level

***s*-plane** *s*-Ebene *f*

to splash, to dash, to scatter [*To cause a liquid substance to fly or scatter*] (ver)sprühen

splash(ing) water [*A mass of flying and scattering water*] Sprühwasser *n*

splash lubrication [*Mechanical engineering. A system of engine lubrication in which the connecting-rod bearings dip into oil troughs and splash the oil onto the cylinder and piston rods*] Tauchschmierung *f*

to splash on, to dash on, to scatter on [*To apply a liquid substance on something, so as to wet it*] besprühen

splay Abkantung *f*

splayed (ab)gekantet

splicing (Auf)Spleißen *n*, (Auf)Spleißung *f*

spline Raumkurve *f* [*Sie kann als Bewegungspfad und als Teil eines Extrusionskörpers eingesetzt werden*]

split geteilt

split air-conditioning system [*It consists of equipment provided in more than one assembly or enclosure, usually with supply air distribution equipment housed separately from refrigerant condensing equipment*] Klimaschrankgerät mit getrennt aufgestellter Kältemaschine, Splitgerät *n*

split bolt terminal Mantelklemme *f*

split compressor → separated compressor

split-core (type) current transformer Zangenstromwandler, Anlegestromwandler *m*

split equipment zweiteilige Gerät(e); mehrteilige Gerät(e), vielteilige Gerät(e) *n*, (*f*)

split key Partnerschlüssel *m*

split-level house Halbgeschosshaus, Halbstockwerkhaus, Halbetagenhaus *n*

split(-location air-cooling) unit [*It provides for two locations of the compartment parts: the compressor and condenser are located in the cabinet outside of the house, while the evaporator is located inside of the house, usually in the plenum (chamber) of a forced-air furnace. The refrigerant is piped from the remotely located condenser to the evaporator through well-insulated piping. The furnace blower is utilized for circulating conditioned air through the duct system*] (Kältemittel-)Luftkühl(ungs)anlage *f* in geteilter Bauweise

split-phase switch Hilfsphasenschalter, Spaltphasenschalter *m*

split-range ((closed) loop) control Reg(e)lung *f* mit Bereich(s)aufspaltung

split sliding gate unterteilte Schiebetor *n*

splitter Verteiler, Phasenspalter *m* [*Ein passives Gerät, das am Netzwerk die Verteilerfunktion übernimmt und die Eingangsleistung in zwei oder mehrere Ausgänge geringerer Leistung unterteilt. Kann auch als Kombinator verwendet werden, wenn zwei oder mehrere Leistungseingänge in einem einzelnen Leistungsausgang zusammengeschlossen werden sollen*]

split time Zwischenzeit *f* [*Zeitabschnitt einer Gesamtzeit*]

splitting box Aufteilungsdose *f*, Verteil(er)dose

split treatment Neutralisieren *n* [*Den Kalküberschuss in einem zweiten Becken vom nicht behandelten Rohwasser*]

SPM, session protocol machine Protokollmaschine *f* der Kommunikationssteuer(ungs)schicht

SPM → slots per module

spoilage → decomposition

spoilage, spoiling Verderb *m* [*Nahrungsmittel*]

spoilage, spoiling Beschmutzung *f*, Beschmutzen *n*

spoiler Fleck, Schmutzfleck *m*

sponge filter Schwammfilter *m*, *n*

sponge holder Schwammhalter *m*

sponge tray Schwammschale *f*

spooling Hintergrundprogrammierübergabe *f* [*Übergabe von Druck- oder Zeichendaten an ein Hintergrundprogramm zur Ausgabe an die Endgeräte*]

sports arena Sportstätte *f*

spot, stain Fleck *m*

spot announcement Durchsage *f*

spot check(ing), random check(ing) Stichkontrolle *f*

spot cooling [*Cooling the air of a limited portion of an enclosed space without the use of walls or partitions*] Punktkühlung *f*

spot damped punktgedämpft

S

spot distortion → **coma**

spot light Punktstrahler, Ausstrahler *m*

spot merchandiser Impulstruhe *f* [*Kleine, bewegliche Schautiefkühltruhe*]

spot scanner → **optical scanner**

spot scanning → **optical scanning**

spot size Leuchtfleckgröße *f*

spot speed [*In television, the product of the length (in units of elemental area, that is, in spots) of scanning line by the number of scanning lines in second*] Leuchtfleckgeschwindigkeit, Ablenkgeschwindigkeit *f*

spot speed → **scanning speed**

spray (air) dehumidifier Sprüh(luft)entfeuchter *m* [*DIN 30 600, Reg.-Nr. 06 100*]

spray (air) humidifier Sprüh(luft)anfeuchter *m*, Sprüh(luft)befeuchter [*DIN 30 600, Reg.-Nr. 06 100*]

spray chamber Sprühkammer *f* [*Zentralklimaanlage*]

spray curtain → **shower curtain**

spray curtain rail → **shower curtain rail**

spray deaerator Sprühentlüfter *m*

spray(-filled) water cooling tower [*Water-to-air contact depends entirely on the break up of water by means of pressure-spray nozzles*] Spritzkühlturm *m*

spray freeze drying, spray lyophilization [*Freeze-drying process during which a liquid product is frozen prior to sublimation of the ice*] Spritzgefriertrocknen *n*

spray gun, (compressed-)air gun [*A device which sprays coating material, as, for instance, paint, in minute droplets on to a surface by means of compressed air*] Druckluftspritzpistole, Pressluftspritzpistole, pneumatische Spritzpistole *f*

spray humidifier → **(room) spray humidifier**

spray mixing tap Spritzmischhahn *m*

spray nozzle Spritzdüse *f*

spray nozzle air humidifier Sprühdüsenluftbefeuchter, Sprühdüsenluftanfeuchter *m*

spray nozzle ceiling Spritzdüsendecke *f*

spray plotter → **(ink-)spray plotter**

spray seam Spritznaht *f*

spray shower → **shower**

spray sprinkler Schirmsprinkler, Schirmsprenger *m*

spray washing Spritzwaschen *n*

spray water Spritzwasser *n*

spreading resistance Stromwegwiderstand *m* [*Durch den Halbleiter vom punktförmigen Anschlusskontakt einer Sperrschicht*]

spreadsheet Tabellenkalkulationsprogramm *n*

spring assembly Federgarnitur *f*

spring bolt, catch bolt Riegel mit Feder, Federriegel *m*

spring bias Federspannung *m*

spring bolt, catch bolt Federriegel, Riegel mit (einer) Feder *m*, schießende Falle, Dresd(e)ner Falle *f* [*Sie bewegt sich waag(e)recht*]

spring bolt stock lock → **stock lock with spring bolt**

spring buffer [*Buffer that stores in a spring the kinetic energy of a compressive load*] Federpuffer *m*

spring buffer load rating, spring buffer load output [*Load required to compress the buffer spring an amount equal to its stroke*] Federpuffernennlast *f*

spring buffer stroke [*Distance the contact end of the buffer spring can move under a compressive load until all coils are essentially in contact*] Federpufferhub *m*

spring butt (hinge), spring (hinge and) loop Federangelband, Federeinsetzband, Federeinstemmband *n*

spring-closing federschließend

spring contact Federkontakt *m*

spring contact strip Federleiste *f*

spring core Federkern *m* [*z. B. bei einem Handtuchhalter*]

spring door closer, spring door closing device Federtürschließer *m*, Federtürschließvorrichtung *f*

spring flap lid → **hing(ed) lid**

spring force Federkraft *f*

spring hanger → **spring (pipe) hanger**

spring hinge Federscharnier *n*

spring hood Federhaube *f*

spring return act(uat)or, spring return final controlling element Rückfederung(s)stellglied *n* [*Siehe Anmerkung unter „Stellglied"*]

spring isolator → **spring (vibration) isolator**

spring key of a bolt, bolt spring key Federvorstecker, gespaltene Vorstecker *m*

spring latch Federklinke, Klinke mit Feder *f*

spring-loaded terminal Federdruckklemme, federbelastete Klemme *f*

spring-loaded valve cone federbelastete Ventilkegel, Federdruckventilkegel *m*

spring lock with falling latch, falling latch spring lock, spring stock lock, French lock with latch, latch French lock Schließriegelschloss *n* mit hebender Falle

spring loop → **spring butt (hinge)**

spring manometer, spring pressure gauge; spring pressure gage (US) Federmanometer *n*, Federdruckmesser *m*

S

spring (pipe) hanger Feder(rohr)schelle *f*

spring plate Federteller *m*

spring reserve → running reserve

spring return Rückfederung f

spring return motor [*This motor for two-position control operation is energized electrically, driven to one position, and held there until the circuit is broken when the spring returns the (slide) valve to its normal position*] Federrückzugsmotor, Federrückführungsmotor *m*

spring rod Federstab *m*

spring tension Federspannung *f*

spring (vibration) isolator [*See remark under "isolator"*] Feder(schwingungs)dämpfer *m*

spring vice [*locksmith*] Federhakenzange *f*

spring washer Federunterlegscheibe *f*

spring wire Federdraht *m*

springy federhart

to sprinkle sprinkeln, sprengen

sprinkler Sprinkler, Sprenger *m*, geschlossene Löschdüse *f*

sprinkler alarm Sprinkleralarm, Sprengeralarm *m*

sprinkler distribution station Sprinklerverteilerstation, Sprengerverteilerstation *f*

sprinklered [*Equipped with an approved automatic sprinkler system properly maintained*] sprinklerbestückt, sprengerbestückt

sprinkler guard, sprinkler protection Sprinklerschutz, Sprengerschutz *m*

sprinkler head Sprinklerkopf, Sprengerkopf *m*

sprinkler monitoring Sprinklerüberwachung, Sprengerüberwachung *f*

sprinkler (pipe)line Sprinkler(rohr)leitung, Sprenger(rohr)leitung *f*

sprinkler plant, sprinkler installation, sprinkler system Sprinkleranlage, Sprinklerinstallation, Sprengeranlage, Sprengerinstallation *f* [*Nicht verwechseln mit „Sprühwasserlöschanlage" DIN 14 489*]

sprinkler protection → sprinkler guard

sprinkler pump Sprinklerpumpe, Sprengerpumpe *f*

sprinkler response time Sprinkleransprechzeit, Sprengeransprechzeit *f*

sprinkler spacing [*Distribution of automatic sprinklers to provide the number of square feet of coverage specified for light hazard, ordinary hazard, and extra hazard locations*] Sprinkleranordnung, Sprengeranordnung *f*

sprinkler stopper Sprinklerverschluss, Sprengerverschluss *m*

sprinkler substation Sprinklerunterstation, Sprengerunterstation *f* [*Nicht verwechseln mit „Sprinklernebenstation"*]

sprinkler wet (pipe)line, wet sprinkler (pipe)line Sprinklernassleitung, Sprengernassleitung, Nasssprinklerleitung, Nasssprengerleitung *f*

sprinkler wet system → wet(-pipe) (fire) sprinkler system

sprocket hole Transportloch, Taktloch, Vorschubloch *n* [*Lochstreifen*]

sprung [*A description of timber that has warped*] gesprungen

SPST switch → single pole single throw switch

spud [*A short pipe which serves as connection in a piping system*] Stutzen *m*, Ansatzrohr *n*

spur duct → stub duct

spurious count Störzählimpuls *m*

spurious switch-on pulse suppression, spurious signal suppression, switch-on transient suppression Einschaltfehlimpulsunterdrückung *f*

spurting out, striking out Herausschlagen *n*

sputtering Kat(h)odenstrahlverdampfung *f* im Hochvakuum

square Anschlagwinkel *m*

square (air) duct Quadrat(luft)kanal *m*

square(d) paper Rasterpapier *n*

square equation quadratische Gleichung, Quadratgleichung *f*

square-foot unit of absorption → sabin

square (hinge) flap Quadrat(band)lappen *m*

square joint → butt joint

square-jointed → butt-jointed

square root device Radizierer, Wurzelzieher *m*

square-root element Radizierergiied, Radiziererelement, Wurzelzieherglied, Wurzelzieherelement *n*

square-root extractor block Radizierbaustein, Wurzelzieherbaustein *m*

squaring device Quadrierer *m*

squaring element Quadrierglied *n*

squatting w. c.; Eastern w. c.; Asiatic w. c.; Lascar w. c.; Lascar closet [*deprecated*] [*A w.c. pan with an elongated bowl for installing with its top edge at or near floor level so that the user has to adopt a squatting position*] Hockabort, französische Abort *m*, Hockklosett, französische Klosett *n*

sqawker Mitteltöner *m*

Squid → superconducting quantum interference device

S

SR → shift(ing) register

SR → to shift right

SRAM statische Speicher *m* mit wahlfreiem Zugriff

SRQ → service request

SR relay → slow-releasing relay

SS → (dedicated) special system

SS, single shift einmalige Hin-/Rückschaltung, Einzelumschaltung *f* [*Kode*]

SS, session service Dienst *m* der Kommunikationssteuerung

SS → server station

SSA → serial storage architecture

SSB, single side band Einseitenband *n*

SSI, small scale integration Integration *f* mit bis zu 12 Gatter pro Chip

SS/SSB, single side-band modulation Einseitenbandmodulation *f*

"S" (stench) trap, "S" siphon trap [*A trap with the inlet vertical and the outlet parallel to but not in alignment with the inlet*] (Wasser)(Geruch)Verschluss in S-Form, S-Traps, S-Form(-Geruch)verschluss, Schwanenhals(geruch)verschluss, S-Siphon, S-Geruchverschluss, Knie(geruch)verschluss; Siphon(geruch)verschluss, Syphon(geruch)verschluss *m* [*veraltete Benennungen*]

SS user, session services user Benutzer *m* des Dienstes der Kommunikationssteuerung

stability [*telephone*] Pfeifsicherheit *f*

stability → strength

stabilized power supply → constant-current source

stack Baugruppe *f* [*Leistungselektronik*]

stack → (building) stack

stack, data stack Keller(speicher), Stapelspeicher, Datenkeller(speicher) *m* [*Eine Gruppe von Registern oder Speicherplätzen im Arbeitsspeicher. Sie sind logisch so miteinander verbunden, dass ihre Inhalte in der umgekehrten Reihenfolge der Einspeicherung wieder ausgelesen werden*]

stackable stapelbar, stapelfähig

stack effect → chimney effect

stacking tables → nested tables

stack (pipe) Falleitung *f*, Fallrohr *n* [*Lotrechte Leitung mit oder ohne Verziehung, die durch ein oder mehrere Geschosse führt, über Dach gelüftet wird und das Abwasser einer Grund- oder Sammelleitung zuführt*]

stack pointer Stapelzeiger, Adresszeiger *m* [*Er verweist auf die Adresse des zugänglichen Stapelspeichers, an dem zugefügt oder weggenommen wird*]

staff access, personnel access Personalzugang *m*

staff access control system, personnel access control system Personalzugangssystem *n*

staff checkpoint, personnel checkpoint Personalkontrollstelle *f*

staff data, personnel data Personaldaten, Personaldatei *f*

staff entrance reader, personnel entrance reader Personaleingangsleser, Personalzutrittsleser *m*

staff file, personnel file Personaldatei *f*

staff/goods lift, goods/staff lift, personnel/goods lift, goods/personnel lift (Brit.); staff/goods elevator, goods/staff elevator, personnel/goods elevator, goods/personnel elevator (US) Güteraufzug mit Personenbegleitung, Lastenaufzug, Personal- und Lastenaufzug *m* [*Beförderung von Gütern und Personal (=vom Betreiber des Aufzugs beschäftigte Personen)*]

staff interlocked doors, personnel interlocked doors Personalschleuse *f*

staff lift, personnel lift (Brit.); staff elevator, personnel elevator (US) Personalaufzug, Personalfahrstuhl *m*

staff management, personnel management Personalverwaltung *f*

staff paging system, personnel paging system, staff location system, personnel location system (drahtlose) (Personal)Suchsystem, Personal(aus)rufsystem *n*

staff reader, personnel reader Personalleser *m*

staff restaurant, personnel restaurant Personalrestaurant *n*, Personalkantine *f*

stage lift (Brit.); stage elevator (US) Bühnenaufzug *m*

to stagger versetzen, versetzt anordnen, verziehen

staggered stack (pipe) [*See remark under "stack (pipe)"*] verzogene Fallleitung *f*, verzogene Fallrohr *n* [*Gebäudeentwässerung*]

stagnant air stickige Luft, verbrauchte Luft *f*

stagnant refrigerant unbewegliche Kältemittel *n*

stagnation anemometer Staubklappenanemometer *n*, Staubklappenwindmesser *m*

stain, spot Fleck *m*

stainless steel Edelstahl-Rostfrei, Niro-Stahl, nichtrostende Stahl, Chromstahl *m* [*Je nach korrosiver Beanspruchung wird dieser Stahl als Chromstahl, Chrom-Nickel-Stahl und Chrom-Nickel-Molybdänstahl eingesetzt*]

stainless steel drainer [*See remark under "drainer"*] Edelstahl(ab)tropffläche *f*

S

stainless steel sink Edelstahlspüle *f*

stair bracket Treppenwinkel *m* [*Ein Befestigungsmittel*]

staircase /lift core, staircase /lift well, lift/staircase core, lift/staircase well (Brit.); staircase /elevator core, staircase/ elevator well, elevator/staircase core, elevator/staircase well (US) Aufzug-Treppen-Schacht, Treppen-Aufzug-Schacht *m*

staircase lighting time switch Treppen(haus)licht-Zeitschalter *m*

staircase lock Treppenhausschleuse *f*

staircase pressurization fan Treppenhaus-Druckausgleichventilator *m*

staircase switch Treppenhausschalter *m*

stair lift (Brit.); stair elevator (US) Treppenaufzug *m*

stair nosing Treppen(schutz)leiste *f*

stair rod Treppenläuferstange *f*

stall urinal, urinal stall [*A urinal having a back curved on plan to form a stall for the user. It has an integral floor channel; when stall urinals are fixed in ranges, divisions or closing pieces are provided between each stall*] Einzelurinal, Standurinal *n*, Urinalstand *m* [*DIN 1986*]

stall urinal range, urinal stall range, range of urinal stalls [*See remark under "stall urinal"*] Urinalstandreihen(anlage) *f* [*Einzelurinale in Reihe angeordnet*]

stamping Stempeln *n*, Stemp(e)lung *f*

stand-alone, independent, self-combined unabhängig, autonom, selb(st)ständig, dezentral, Einzelplatz ... , nicht vernetzt, unvernetzt

stand-alone ((closed) loop) control autonome Reg(e)lung, Einzelplatzreg(e)lung *f*

stand-alone controller autonome Regler, Einzelplatzregler *m*

stand-alone ((open) loop) control autonome Steuerung, Einzelplatzsteuerung *f*

stand-alone PV system, remote system Inselsystem *n* [*Autarkes photovoltaisches nicht mit einem Versorgungsnetz verbundenes System*]

standard, norm Norm *f*

standard air [*Air of seventy (70) degrees F. and 29.92 inches of mercury*] Normenluft, Standardluft *f*

standard ambient temperature, ambient standard temperature [*Reference condition for the rating of equipment, usually 68 °F (20 °C)*] Standardumgebungstemperatur, Umgebungsstandardtemperatur *f*

standard broadcast band → broadcast band

standard building heat requirement, standard building heat need Normen-Gebäudewärmebedarf *m*

standard design, standard type, standard version Normalkonstruktion, Normalausführung, Normalbauweise, Normalbauart *f*, Normalaufbau *m*

standard design cast(-iron) (convector) radiator [*See remark under "radiator"*] Gusseisenradiator *m* in Normalausführung [*DIN 4720*]

standard deviation Standardabweichung *f*

standard draft → draft standard

standard driver Standardtreiber *m*

standard external temperature, external standard temperature Normenaußentemperatur *f*

Standard for the exchange of product model data, STEP Schnittstelle *f* für den Austausch geometrischer und produktdefinierender Daten zwischen verschiedenen CAD Systemen [*Diese Schnittstelle ist international genormt*]

Standard for the exchange of product model data – 2-dimensional building subset Geometriedatenschnittstelle *f* zum Austausch geometrischer Informationen [*Diese Schnittstelle wurde auf der Basis von STEP (Standard for the exchange of product model data) speziell auf das Bauwesen abgestimmt*]

standard heat output, standard heat performance [*See remark under "heat output"*] Normenwärmeleistung *f*

standard heat requirement, standard heat demand Normenwärmebedarf *m*

standard interface Standardschnittstelle *f*

standardized signal Einheitssignal *n*

standardized signal regulator, standardized signal ((closed) loop) controller Einheitssignalregler *m* [*Regler, der mit beliebiger Art sowie beliebigem Bereich der Regelgröße arbeiten kann und am Ein- und Ausgang mit genormten analogen Signalen arbeitet. 0 bis 20 mA, 4 bis 20 mA, 0 bis 10 V, 0,2 bis 1,0 bar, usw. DIN 19 225*]

standard make/break contact Gleichschaltglied *n*

standard motor Normenmotor *m* [*Elektromotor*]

standard pilot duty normale Schaltleistung *f*

standard position switch Sicherheitsgrenzschalter *m*

standard product Serienprodukt, Serienerzeugnis *n*

standard profile Standardprofil *n*

standard pump Normenpumpe *f*

standard rating Normennennleistung, Nenn-Normenleistung *f*

standard reflector [*Standard CD-ROM; dual-layer disk*] Standardreflexionsschicht *f*

standards converter Normenwandler *m*

standards institute, standards commission Normenausschuss *m*, Normenkommission *f*

standard specification sheet Normenblatt *n*

standard sprinkler system Regel-Sprinklersystem, Regel-Sprengersystem *n*

standard text AutoText-Eintrag; Textbaustein *m* [*frühere Benennung*]

standard tone Kammerton *m*

standard type → **standard design**

standard ventilation heat requirement, standard ventilating heat requirement [*"Need" may synonymously be used for "requirement"*] Normen-Lüftungswärmebedarf *m*

standard version → **standard design**

standard volume Normenvolumen *n*

stand-by Bereitschaft *f*

stand-by alarm Alarmbereitschaft *f*

stand-by availability level Bereitschaftsgrad *m*

stand-by card Ersatzkarte, Reservekarte *f*

stand-by current, stand-by electricity Notstrom *m*

stand-by current generation, stand-by electricity generation Notstromerzeugung *f*

stand-by current voltage, stand-by electricity voltage Notstromspannung *f*

stand-by electrical service, stand-by electrical supply Notstromversorgung *f*, Notstromdienst *m*

stand-by electricity, stand-by current Notstrom *m*

stand-by fan Reserveventilator, Ersatzventilator *m*

stand-by machine [*It is used as a substitute in times of emergency*] Reservemaschine, Ersatzmaschine *f*

stand-by mode Betriebsart *f* „Bereitschaft"

stand-by operation Bereitschaftsbetrieb *m*

stand-by power load → **emergency (power) load**

stand-by power operation → **backup power operation**

stand-by power supply → **emergency power supply**

stand-by pump Reservepumpe, Ersatzpumpe *f*

stand-by service → **emergency service**

stand-by source → **emergency source**

stand-by state solution stationäre Lösung *f*

stand-by state value, inertia value Beharrungswert, Trägheitswert *m*

stand-by unit Notstromaggregat *n*, Notstromeinheit *f*

standing area → **base**

standing wave ratio, SWR Stehwellenverhältnis, SWV *n* [*Eine Messgröße, aus der die Schall- und Empfangsneigung einer Antenne bezüglich einer Arbeitsfrequenz abgelesen werden kann*]

stand-off mounting Auf-Dach-Montage *f* [*Anordnung von photovoltaischen Modulen mit einem geringem Abstand oberhalb eines geneigten Daches. Die Neigung der Module kann von der Dachneigung abweichen*]

standpipe → **riser**

standpipe for condenser suction → **condenser suction standpipe**

standstill monitoring Stillstandsüberwachung *f*

staple Krampe *f*

staple of a padlock → **padlock staple**

staple of door mounting, door mounting staple (Tür)Schließhaken *m*, (Tür)Öse *f*

staple of window mounting, window mounting staple (Fenster)Schließhaken *m*, (Fenster)Öse *f*

staple plate Hakenblatt, Schließblech mit Riegelhäspchen *n*

star bus Sternbus *m*

starch separator, starch interceptor, starch trap [*See remark under "interceptor"*] Stärke(ab)scheider, Stärkefang, Stärkefänger *m*

star coupler Sternkopp(e)ler *m* [*Lichtwellenleiter*]

star-delta start Stern-Delta-Start, Stern-Delta-Anlauf *m*

star duplex connection → **duplex star connection**

star head earth(ing) switch (Brit.); star head ground(ing) switch (US) Sternerder, Sternerd(ungs)schalter *m*

to start → **to release**

start address Startadresse *f* [*Vorgegebene Speicheradresse, die zuerst abgearbeitet wird*]

start disk, start disc formatierte Systemdiskette, Startdiskette *f*

starter Anlasser, Starter *m* [*Ein Anlassgerät bei dem Stufenschalter und Anlasswiderstandsgerät vereinigt sind*]

starter, triggering electrode [*An auxiliary control electrode used in a gas tube to establish sufficient ionisation to reduce the anode breakdown voltage*] Triggerelektrode, Starterelektrode *f*, Starter, Trigger *m*

starter coil Zündspule, Startspule *f* [*Lampe*]

starter duty Anlasserbetrieb, Starterbetrieb *m*

starter gap Starterentladungsstrecke, Anlasserentladungsstrecke *f*

start(ing) → **start(ing) up**

starting → **release**

start(ing) aid Starthilfe, Zündhilfe *f* [*Lampe*]

start(ing) current Anlaufstrom, Anlassstrom, Startstrom *m* [*Elektromotor*]

start(ing) current Zündstrom, Startstrom *m* [*Lampe*]

start(ing) kick Zündstoß, Startstoß *m* [*Lampe*]

start(ing) master/slave Startlaufwerk *n*

start(ing) motor, pony motor Anwurfmotor *m*

start(ing) sequence Anlauffolge, Anlassfolge, Startfolge *f*

start(ing)–stop system Start-Halt-System *n*

start(ing)-stop transmission Start-Halt-Übertragung *f*

start(ing) store Anfahr(t)speicher *m* [*Rohrpost(anlage)*]

starting torque [*Turning effort of an electric motor on starting*] Anlaufdrehmoment, Anlassdrehmoment *n*

start(ing) up, run(ing) up, start(ing) Starten, Anfahren *n*, Anlauf(en), Hochlauf(en) *m, (n)*

start(ing) winding Anlaufwick(e)lung, Anlasswick(e)lung *f*

start-of-frame sequence Belegt-Token *n*

start of heading, heading start, SOH Zeichenfolgebeginn, Zeichenfolgeanfang, Zeichenfolgestart *m*

start of text, text start Textanfang, Textstart, Textbeginn *m*

star topology [*LAN*] Sternstruktur, Sterntopologie *f* [*Siehe Anmerkung unter „logische Topologie"*]

start sequence → turn-on sequence

to start (up); to run up anfahren, starten, anlaufen, hochlaufen, hochfahren

star washer Sternunterlegscheibe *f*

star wheel [*Automatic data processing*] Rasterrad *n*

state → condition

state, matter phase, phase of matter, matter state, state of matter [*solid; liquid; gas*] Aggregatzustand *m*, Aggregatbeschaffenheit *f*

state change → change of state

state check(ing) function Rückmeldung *f* [*Türöffner*]

state ((closed) loop) control Zustandsreg(e)lung, Zustandsregulierung *f*

state feedback Zustandsrückführung *f*

state graph Zustandsgraph *m* [*Schaltwerk*]

state management Zustandsführung, Zustandsleitung *f* [*Überwachung des Zustandes von Geräten und Maßnahmen, die getroffen werden, um den Betrieb derselben zu ändern, resultierend aus der Analyse überwachter Werte*]

statement [*In automatic data processing, an elementary specification of a computer action or process, complete and not divisible into smaller meaningful units; it is analogous to the simple sentence of a natural language*] Arbeitsanweisung, Arbeitsvorschrift, Anweisung, Vorschrift *f* [*DIN 44 300. Teil 4*]

statement list, STL Anweisungsliste, AWL *f* [*Die Anweisungsliste für eine speicherprogrammierbare Steuerung (SPS) ist die zeilenweise Auflistung aller Steuerungsanweisungen mit entsprechenden, den Steuerungsablauf erläuternden Klartextkommentaren. DIN 19 239*]

state observation, condition observation Zustandsbeobachtung, Beschaffenheitsbeobachtung *f*

state-of-charge, SOC Ladezustand *m* [*Verfügbare Batterieenergie, ausgedrückt als Prozentsatz der Kapazität*]

state of compliance with the order Werksbescheinigung *f* [*Euronorm 21-62*]

state-of(-the)-art (technischer) Entwicklungsstand *m*

state signal message Zustandsmeldung *f*

state strobe, STSTB Zustandsübernahmesignal *n* [*Eine Gehäuseanschlussbezeichnung beim Mikroprozessor 8080. Bei ihm werden die Daten auf dem Bus von Speichern oder vom Mikroprozessor übernommen. Damit werden die Umschaltzeiten unwirksam gemacht, in denen die Signale noch nicht auf Hochpegel angestiegen oder noch nicht auf Niedrigpegel abgesunken sind. Das Zustandsübernahmesignal wird auch zusammen mit dem Rücksetzsignal gegeben und gestattet das Rücksetzen des Mikroprozessors (z. B. beim Anlegen der Betriebsspannung)*]

state table Zustandstabelle, Tabellendarstellung der Logikzustände *f*

state transition matrix Zustandsübergang(s)matrix, Zustandstransitionsmatrix *f*

state transition table Schaltfolgetabelle, Automatentafel *f*

state variable Zustandsgröße, Zwischengröße *f* [*DIN 19 229*]

state vector Zustandsvektor *m*

static allocation, static assignment, static allotment statische Zuordnung, statische Zuteilung, statische Zuweisung *f*

statically balanced statisch (aus)gewuchtet

static balancer, a.c. balancer, AC balancer [*An arrangement of transformers or reactors used to equalise the voltages between the wires of a multi(ple)-wire system*] Mittelpunkttransformator, Spannungsteiler *m*, Ausgleichstransformator *m*

static check(ing) *[Of a computer, one or more test(s) of computing elements, their interconnections, or both, performed under static conditions]* statische Überprüfung, statische Kontrolle *f*

static convector statische Konvektor *m* *[Konvektor mit natürlicher Luftumwälzung]*

static converter → converter

static fan pressure → fan total pressure

static heating statische Heizung *f*

staticizer → serial-(to-)parallel converter

static (pressure) head *[Fluid mechanics. The pressure on a fluid due to the head of the static fluid above it; thus it is also the height to which the fluid will rise in a piezometer]* statische Druckhöhe *f*

static pressure probe statische Drucksonde *f*

static pressure sensor, static pressure sensing element statische Druckfühler *m*

static purge system *[System in which differential pressure is used to remove noncondensables from a centrifugal system]* Differenzdruck(ab)scheider *m*

static regain konstante Druckfall *m* *[Kostenoptimierung von verzweigten Luftkanälen]*

static skew (run) statische Schräglauf *m*

station → (bus) station

stationary automatic firefighting system → automatic stationary firefighting system

stationary fire (grate) bar → fixed fire (grate) bar

stationary laminar premixed flat one-dimensional flame, MIXFLA Modell *n* für eine laminare und eindimensionale und flache Vormischflamme

stationary uniform flow, uniform stationary flow stationäre gleichförmige Fließen, gleichförmige stationäre Fließen *n*

station authorization Stationsberechtigung *f*

station being called, called station angerufene Station, Anrufstation *f*

station panel Stationstafel *f*

statistical tolerance interval statistische Anteilsbereich *m [selten: n]* *[DIN 55 350, T.24]*

stator starter Ständeranlasser, Ständerstarter *m*

status *[building automation]* Gerät(e)meldung *f* *[z. B. gestört]*

status error *[building automation]* Gerät(e)fehler *m*

(statutory) fee schedule for architects and engineers, table of (statutory) fees for architects and engineers Gebührenordnung für Architekten und Ingenieure, Honorarordnung für Architekten und Ingenieure, HOAI *f*

stay → (casement) stay

stay and casement fastener → (combined) stay and casement fastener

stay fastener Feststeller *m*

STE → signalling terminal

steady state ((closed) loop) control accuracy Regel(ungs)genauigkeit *f* im Beharrungszustand

steady-state deviation from desired value, offset, steady-state desired value deviation bleibende Sollwertabweichung *f* *[Die im Beharrungszustand der Regelgröße vorhandene Sollwertabweichung. E DIN 19 226, Teil 5]*

steady-state error signal bleibende Regel(ungs)differenz *f*, bleibende Regel(ungs)unterschied *m*

steady-state level Beharrungszustandshöhe, bleibende Zustandshöhe *f*

steady-state system deviation, steady-state negative deviation Beharrungsregel(ungs)differenz, bleibende Regel(ungs)differenz *f* *[Die im Beharrungszustand vorhandene Regel(ungs)differenz. E DIN 19 226, Teil 5]*

steady-state value Beharrungswert, bleibende Wert *m*

steam Dampf *m*

steam (air) heater *[It consists of a bank of tubes containing the steam and arround which the air passes]* Dampf-Luftheizer, Dampf-Lufterhitzer, Dampf-Luft(er)wärmer *m*

steam (air) heating Dampfheizung *f* *[Erwärmung der Heiz(ungs)luft mit Dampf in Lamellen, Lufterhitzer oder durch Röhrenkessel, mit eigener Dampferzeugung oder mit Fremddampf]*

steam (air) humidification Dampf(-Luft)befeuchtung, Dampf(-Luft)anfeuchtung *f*

steam (air) humidifier *[A humidifier in which steam is injected directly into an airstream]* Dampf(-Luft)-befeuchter, Dampf(-Luft)anfeuchter *m*

steam air mix(ture) heating Dampf-Luft-Gemisch-Heizung *f*

steam atomization, steam atomizing Dampfzerstäubung *f*

steam atomizer Dampfzerstäuber *m*

steam atomizing burner *[Burner in which the firing oil is atomized by steam]* Dampfzerstäubungsbrenner, Dampfzerstäuberbrenner *m*

steam barrel, steam drum Dampftrommel *f*

steam barrel boiler → barrel steam boiler

steam base (board) heating unit Dampf-Sockelheiz(ungs)körper, Dampf-Fußleistenheiz(ungs)körper, Dampf-Scheuerleistenheiz(ungs)körper *m*, Dampf-Heiz(ungs)leiste *f*

S

steam bath Dampfbad *n*

steam blast pipe → blast steam pipe

(steam) blow-off tank Abblasebehälter, Ausblasebehälter, Dampfabblasebehälter, Dampfausblasebehälter *m*

steam calorimeter → throttling calorimeter

steam cavity ceiling heating → steam void ceiling heating

steam central distributor → steam manifold

steam central heating, central steam heating Dampfzentralheizung, Zentraldampfheizung *f*

steam central plant → central steam plant

steam central vacuum heating → vacuum steam central heating

steam chamber Dampfkammer *f*, Dampfraum *m*

to steam-clean reinigen mit Dampfstrahl, dampfstrahlreinigen

steam coil Dampfschlange *f*

steam condensate, steam condensation water, steam condens(at)ing water, water of steam condensation Dampfkondensat, Dampfkondenswasser *n*

steam condenser [*A device that condenses exhaust steam from an engine or turbine into water*] Dampfverflüssiger, Dampfkondensator *m*

steam connection Dampfanschluss *m*, Dampfverbindung *f*

steam consumer Dampfverbraucher *m*

steam conveyance, steam transport(ation) Dampfförderung, Dampfüberführung *f*, Dampftransport *m*

steam cooler → injection steam cooler

(steam-)cured, steamed (dampf)gehärtet [*Beton; Mörtel*]

steam cushion Dampfpolster *n* [*Österreich: m*]

steam distribution Dampfverteilung *f*

steam distribution (pipe)line Dampfverteil(ungs)-(rohr)leitung *f*

steam district heating → district steam heating

steam district heating network → district steam heating network

steam district heating (pipe)line → district steam heating (pipe)line

steam district heating plant, steam district heating installation Dampfstadtheiz(ungs)anlage, Dampffernheiz(ungs)anlage *f*

steam drum → steam barrel

steam dryer, steam drier Dampftrockner *m*

steam dryness Dampftrockenheit *f*

steam duct [*See remark under "duct"*] Dampfkanal *m*

steam engine Dampf(kraft)maschine *f*, Dampfmotor *m*

steam engineer Dampfingenieur *m*

steam engineering Dampftechnik *f*

steam filtering, steam filtration Dampffilterung, Dampffiltration *f*

steam flow(ing), steam streaming Dampfströmung *f*, Dampfströmen *n*

steam for heating → heating steam

steam-generating furnace → boiler furnace

steam generation, steam generating, steam raising, raising of steam, generating of steam, generation of steam, steam production, production of steam Dampferzeugung *f*

steam generation plant, steam raising plant, steam raising installation, steam generation installation Dampferzeugungsanlage, Dampferzeugungsinstallation *f*

steam-heated dampfbeheizt

steam heater Dampfheizer *m*

steam heating Dampfheizung *f*

steam heating coil Dampfheiz(ungs)schlange *f*

steam heating plant, steam heating installation Dampfheiz(ungs)anlage, Dampfheiz(ungs)installation *f*

steam heat trace system → heat trace steam system

steam high-pressure (pipe)line → high-pressure steam (pipe)line

steam (injection) cooler → injection steam cooler

steam (injection) humidification Dampfbefeuchtung, Dampfanfeuchtung *f*, Dampfbefeuchten, Dampfanfeuchten *n*

steam (injection) humidifier [*Moisture in the form of steam is injected directly into the air stream*] Dampf(luft)befeuchter, Damp(luft)anfeuchter *m*

steam injector Dampf(kessel)injektor *m* [*Pumpe, die Wasser in einen Dampfkessel einspritzt*]

steam inlet branch Dampfeinlassstutzen *m*

steam input Dampfzufuhr, Dampfeingabe *f*

steam jacket Dampfmantel *m*

steam jet Dampfstrahl *m*

steam-jet refrigerating system, steam-jet chiller [*System in which high-pressure steam, supplied through a nozzle and acting to eject water vapo(u)r from the evaporator, maintains the requisite low pressure on one side and produces a high pressure on the other by virtue of compression in a following diffusion passage*] Dampfstrahl(kompressor)kältesystem, Dampfstrahlverdichterkältesystem *n*

S

steam lance Dampflanze *f*

steam leakage Dampfleckage *f*

steam main (line), main steam line Dampfhauptleitung, Hauptdampfleitung *f*

steam manifold, steam central distributor, central steam distributor Dampfverteiler(rohr), Dampfverteil(ungs)rohr *m, (n)*, Dampfverteil(er)stück, Dampfverteil(ungs)stück *n*, Dampfverteil(er)batterie, Dampfverteilungsbatterie *f*

steam meter Dampfmesser *m*

steam metering Dampfmessen *n*

steam network Dampfnetz *n*

steam nozzle Dampfdüse *f*

steam–operated dampfbetätigt

steam output, steam performance Dampfleistung *f*

steam parallel slide shut(ting) off valve, steam parallel slide stop valve Dampfparallel(ab)sperrschieber *m*

steam pipe Dampfrohr, schwere Gewinderohr *n* [*DIN 2441*]

steam power plant, steam power installation Dampfkraftwerk *n*

steam pressure Dampfdruck *m* [*Fehlbenennung: Dampfspannung f*]

steam pressure condensate return feeder plant, steam pressure condensate return feeder installation Dampfdruck-Kondensat-Rückspeiseanlage *f* [*Sie dient zur Kesselspeisung und zur Kondensatförderung in einem höherliegenden Behälter*]

steam pressure curve (Sättigungs)Dampfdruckkurve *f* [*Graphische Darstellung des Sättigungsdampfdruckes einer Flüssigkeit*]

steam (pressure) gauge; steam (pressure) gage (US) Dampfmanometer *n*, Dampfdruckmesser *m*

S

steam pressure limiter, steam pressure limiting device Dampfdruckbegrenzer *m*

steam pressure monitoring Dampfdrucküberwachung *f*

steam pressure ((open) loop) control Dampfdrucksteuerung *f*

steam pressure relief pipe Dampfdruckentlastungsrohr *n*

steam production → steam generation

steam purifier → steam separator

steam raising → steam generation

steam raising plant → steam generation plant

steam return (flow) Dampfrücklauf *m*

steam riser, steam rising line Dampf-Steig(e)leitung, Steig(e)-Dampfleitung *f*

(steam) safety (relief) valve [*A spring-loaded, pressure-actuated valve that allows steam to escape from a boiler at a pressure slightly above the safe working level of the boiler; fitted by law to all boilers*] Sicherheitsventil, Entlastungsventil, Dampfsicherheitsventil, Dampfentlastungsventil *n*

steam (sectional) radiator [*See remark under "radiator"*] Dampfradiator, Dampfgliederheiz(ungs)körper *m*

steam separator, steam purifier [*Device used to remove moisture from steam after it leaves the boiler*] Dampffeuchteabscheider, Dampffeuchtigkeitsabscheider, Wasserabscheider *m*

steam side Dampfseite *f*

steam soot blower Dampfrußbläser *m*

steam space Dampfraum *m*

steam space boiler Dampfraumkessel *m*

steam storage device Dampfspeicher *m* [*Dampf wird in Wasser eingeblasen, kondensiert und wärmt das Speicherwasser bei steigendem Druck auf*]

steam streaming → steam flow(ing)

steam superheater [*A steam generator component that adds additional heat to the steam after it has been vapo(u)rized*] Dampfüberhitzer *m*

steam supply valve Dampfentnahmeventil *n*

steam temperature ((closed) loop) control Dampftemperaturreg(e)lung *f*

steam throttle lever Dampfdrosselhebel *m*

steam throughput Dampfdurchsatz *m*

steam transport(ation) → steam conveyance

steam trap [*Device for allowing the passage of condensate and preventing the passage of steam, or for allowing the passage of air as well as condensate*] Kondens(at)topf, Entwässserungstopf, Kondenswasserabscheider, Kondenswassersammler *m*

steam turbine Dampfturbine *f*

steam turbine condenser cooling water Kaltwasser *n* [*Kühlwasser des Kondensators des Dampfturbinenprozesses mit Vorlauftemperaturen von 25 °C bis 35 °C*]

steam turbine heating power plant, steam turbine heating power installation Dampfturbinen-Heizkraftwerk *n*

steam valve sizing chart → valve sizing chart for steam

steam void ceiling heating, steam cavity ceiling heating Dampf-Hohlraumdeckenheizung, Hohlraumdecken-Dampfheizung *f*

steam/water mixing valve, steam/water mixer (valve) Dampf-Wasser-Mischventil *n*

steam whistle Dampfpfeife, Sicherheitspfeife *f* [*Dampfkessel*]

steel baseboard heating element Stahlsockelheiz(ungs)körper, Stahlfußleistenheiz(ungs)körper *m*, Stahlheiz(ungs)leiste *f*

steel cabinet, steel cubicle Stahlschrank *m*

steel case(work) → **steel housing**

steel central distributor → **steel manifold**

steel channels U-Stahl *m*, Stahl-U-Profile *f*

steel conduit Stahl(panzer)rohr, Staparohr *n* [*Elektroinstallation*]

steel (convector) radiator, steel sectional radiator [*See remark under "radiator"*] Stahlradiator, Stahlgliederheiz(ungs)körper; Stahlheiz(ungs)körper *m* [*Fehlbenennung. DIN 4722*]

steel element, seamless steel pipe nahtlose Stahlrohr *n*

steel frame building Stahlrahmengebäude *n*

steel heating boiler Stahlheiz(ungs)kessel *m*

steel housing, steel case(work), steel casing Stahlgehäuse *n*

steel jacket pipe Stahlmantelrohr *n*

steel manifold, steel central distributor Stahlverteiler(rohr) *m*, (*n*), Stahlverteilungsrohr, Stahlverteilerstück, Stahlverteilungsstück *n*, Stahlverteilerbatterie *f*

steel plate (bath)tub Stahlblech(bade)wanne *f*

steel plate cabinet, steel plate cubicle Stahlblechschrank *m*

steel plate enclosure, steel plate casing, steel plate case, steel plate housing Stahlblechgehäuse *n*

steel plate portal Stahlblechportal *n*

steel (sectional) radiator → **steel (convector) radiator**

steel sheet casing → **sheet steel enclosure**

steel sheet enclosure → **sheet steel enclosure**

steel sheet housing → **sheet steel enclosure**

steel sheet plate, sheet steel plate Stahlblechplatte *f*

steel support(ing) frame(work) Stahltragrahmen *m*

steel track Stahlschiene *f* [*Türanlage*]

(steel) truss [*escalator*] Stahlrohrfachwerk(konstruktion) *n*, (*f*)

steel tube (convector) radiator [*See remark under "radiator"*] Stahlröhrenradiator *m*

steel (wall) bracket Stahlkonsole *f*, Stahlwandträger *m*

steel wool filter Stahlwollefilter *m*, *n*

steep slope, steep gradient Steilgefälle *n*

Stefan-Boltzmann law, fourth power law [*Total energy radiated from a blackbody is proportional to the fourth power of the temperature of the body*] Gesetz von Stefan und Boltzmann, Stefan-Boltzmann-Gesetz *n*

stem, spindle Spindel *f* [*Schieber*]

stem extension, spindle extension Spindelverlängerung *f* [*Schieber*]

stem number Sachnummer *f*

stem valve → **spindle valve**

stench trap → **trap**

step Stufe *f*

step Schritt *m*

step → **bit(t)**

step → **(sequence(r)) step**

STEP → **Standard for the exchange of product model data**

step band, band of steps Stufenband *n* [*Rolltreppe*]

step band locking [*escalator*] Stufenbandverrieg(e)lung *f*

step brass (Zapfenband)Pfanne, Bandpfanne *f*

step by step, stepwise, in steps stufenweise, Stufe für Stufe

step-by-step action, stepwise action, in steps action Weiterschaltung, Stufenschaltung *f*

step chain [*escalator*] Stufenkette *f*

step comb [*escalator*] Stufenkamm *m*

step controller, stepping controller [*Multi(ple)-switch assembly in which a moving element trips multi(ple) output steps successively*] Stufenregler *m*

step display Stufenanzeige *f*

step-down method → **decay method**

step enabling condition, stepping condition, progression condition Weiterschalt(ungs)bedingung *f* [*DIN 19 226 Teil 5. „Fortschaltbedingung" und „Prozesskriterium" sind hier zu vermeiden (DIN 19 237)*]

step-forced response [*ANSI C81.5*] → **step response** [*IEC 50(351)*]

step generator [*Electronics. A device used to test the linearity of an amplifier by applying a step wave to the input and observing wave form at the output with an oscilloscope*] Stufenerzeuger, Stufengenerator *m*

step guide [*escalator*] Stufenführung *f*

step lowering monitoring [*escalator*] Stufen(ab)senküberwachung *f*

stepped down, reduced [*voltage*] abgespannt, herunter transformiert, reduziert, verringert

S

step(ped) grate external furnace [*See remark under "furnace"*] Treppenrostvorfeuerung *f*

step(ped) grate furnace → furnace with step(ped) grate

step(ped) key → bit(ted) key

step(ped) piston compound compressor, differential piston compressor [*Single-acting compressor for compound operation, the lower stage of which is compressed in a chamber in which the larger diameter of the piston moves and the higher stage is compressed in a chamber in which the smaller diameter of the piston moves*] Differenzialkolbenverdichter, Differenzialkolbenkompressor *m*

stepper [*hard disk*] magnetische Motor *m*

stepping condition → step enabling condition

stepping controller → step controller

stepping motor, stepper (motor), pulse motor Schrittmotor *m*

stepping relay, rotary stepping relay [*A relay whose contact arm may rotate through 360° but not in one operation*] Schrittrelais, Stufenrelais *n*

stepping stone method Primal-Dual-Algorithmus von Dantzig, Dantzig-Algorithmus *m* [*Ein weit verbreiteter Lösungsalgorithmus der Transportoptimierung*]

step response [*IEC 50(351)*], **step-forced response** [*ANSI C 81.5*] Sprungantwort *f* [*Der zeitliche Ablauf der Ausgangsgröße bei einer Sprungfunktion als Eingangsgröße. Hat eine Sprungantwort für alle $t \geq t_0$ dasselbe Vorzeichen und strebt sie für t° ° gegen einen von Null verschiedenen endlichen Grenzwert an, so lassen sich für diese die folgenden Kennwerte festlegen und bestimmen E DIN 19 226 T2*]

step roller [*escalator*] Stufenrolle *f*

step run guard [*escalator*] Stufenlaufführung *f*

step sequencer Schritt(fort)schaltwerk *n*

step sequencing, progression to (the) next step Schritt(fort)schaltung *f* [*programmierbare Steuer(ungs)gerät*]

step setting mode (Betriebsart *f*) Schrittsetzen *n* [*Betriebsart bei Ablaufsteuerungen, in der durch Eingriff des Bedieners die Ablaufkette auf den beliebigen Schritt gesetzt wird. V DIN 19 237*]

step speed → travelling speed

stepwise → step by step

steradiance → radiance

stick switch → (wobble) stick switch

stiffness Steifigkeit, Steife *f*

still air, motionless air stehende Luft, ruhende Luft *f*

still picture → (telephony) still picture

stimulation threshold Reizschwelle *f*

stink trap → trap

to stipulate → to specify

stipulated value → specified value

stirrer → agitator

stirring → agitation

stirrup handle Bügelgriff *m*

stirrup(-type) guard rail Umgehungsbügel *m* [*Er stellt sicher, dass beim Betreten der Fahrtreppe die Lichtschranke betätigt und dass die Kontaktmatte nicht umgangen wird*]

STL → statement list

STM → scanning tunnelling microscope

stock → on-hand quantity

stock lock with spring bolt, spring bolt stock lock, French lock with spring bolt Riegelschloss *n* mit schießender Falle

stock size Lagergröße, Bestand(s)größe *f*

stokehole Heiz(ungs)loch, Schürloch *n*

stoker, fireman, boiler operator Kesselwärter, Heizer, Kesselbediener *m*

stoneware sink → Belfast sink

stop Anschlag *m*

stop Halt *m* [*Aufzug*]

stop angle valve → angle stop valve

stop bit, end bit, E Endbit *n* [*Letztes Bit einer Bitfolge; es zeigt das Ende einer Information an*]

stop bracket, stop angle Anschlagwinkel *m*

stop controller Ausschaltregler *m*

stop disc, stop disk Anschlagscheibe *f*

stop instruction → program stop (US)

stopped, traced angehalten, im Einzelschrittmodus ausgeführt [*Prozessstatuskode*]

stop pin Anschlagstift *m*

stopping, shut-down Abschalten, Stillsetzen, Anhalten *n* [*Maschine*]

stopping point Haltestelle *f* [*Aufzug*]

stopping switch Halteschalter *m* [*Aufzug*]

stop time → deceleration time

stop valve → shutting-off valve

storage [*This term refers to magnetic drums, disks, cores, tapes, punched cards, or solid-state devices outside of the computer*] Speicherzubehör *n* [*selten: m*]

storage → incorporating

storage → **(computer) memory**

storage array tester and analyzer → **SATAN**

(storage) battery; accumulator (Brit.); reversible cell, ACC [*A register in which the results of arithmetic, logical and I/O operations can be stored. The contents of the accumulator can be cleared, checked, complemented, incremented, dercremented, or shifted either to the right or left*] Akku(mulator) *m*, Sammler-(batterie) *m*, (*f*), Akkubatterie, Akkumulator(en)batterie *f*

storage block, storage element Speicherblock *m* [*Nachtstromspeicherheizung*]

storage boiling water heater → **boiling storage water heater**

storage cabinet, storage cubicle Aufbewahrungsschrank *m* [*Küche*]

storage capacity → **memory capacity**

(storage) cistern → **cistern**

storage device → **data carrier**

storage direct heating → **direct storage heating**

storage display screen, display storage screen Speicher(bild)schirm *m*

storage drinking water heater → **drinking water storage heater**

storage element → **storage block**

storage floor heating → **floor storage heating**

storage gas Lagergas, Speichergas *n*

storage gas (water) heater → **gas storage (water) heater**

storage heater Speicher(er)wärmer, Speicherheizer *m*

storage heating (system) Speicherheizung *f*

storage location Speicherstelle *f*, Speicherplatz *m*

storage management services, SMS Datenmigration, Datenauslagerung, Datenwanderung *f* [*Novell Netware 4*]

storage mass Speichermasse *f*

storage matrix Speichermatrix *f* [*von Magnetbändern*]

storage medium [*e.g. of a magnetic drum*] Speicherträger *m*, Speichermedium *n*

storage oscilloscope Speicheroszilloskop *n*

storage pool nicht zugewiesene Speicherbereiche *f* [*Diese Bereiche sind kaum prozesszugewiesen und werden vom Zuweisungssystem dazu benutzt Anforderungen zu erfüllen. Ein von einem Prozess freigegebener Speicher wird für eine Neuzuweisung zurückgegeben*]

storage potable water heater → **drinking water storage heater**

storage protection → **memory protection**

storage quantity Lagermenge *f*

storage space Bevorratungsraum, Speicherraum, Vorratsraum, Lagerraum *m*

storage structure Speicherstruktur *f*

storage system Speichersytem *n*

storage tank Lagertank, Vorratstank, Speicher(tank), Bevorratungstank *m*

storage temperature Speichertemperatur *f*

storage time Speicherzeit *f*

storage tube Speicher(bild)röhre *f*

storage (type) electrical water heater, electrical storage (type) water heater Elektro-Speicherwasserheizer, Elektro-Vorratswasserheizer *m*

storage (type) service water heater, service water storage (type) heater Brauchwasserspeicherheizer, Brauchwasservorratsheizer, Betriebswasserspeicherheizer, Betriebswasservorratsheizer *m*

storage unit Speichereinheit *f*

storage vessel Speicher(behälter), Vorratsbehälter *m*

storage water Speicherwasser, Vorratswasser *n*

storage (water) heater [*Water heater that heats at a thermostatically controlled temperature, for delivery on demand*] Vorratswasserheizer, Warmwasserspeicher, Vorratswarmwasserbereiter, Speicherwasserheizer, Speicherwasser(er)wärmer, Speicher(wassererhitzer) *m*, Speichergerät *n*

to store → **to incorporate**

store Rollvorhang *m*

store-and-forward messaging Mitteilungsdienst *m*

store cooling Lagerkühlung, Speicherkühlung *f*

stored and delayed instruction gespeicherte und verzögerte Befehl *m*

stored data Speicherdaten, Vorratsdaten *f*

stored gas Gasvorrat *m*, Vorratsgas *n*

stored instruction Speicherbefehl, gespeicherte Befehl *m*

stored program logic (US); stored programme logic (Brit.) [*Program(me) that is stored in a memory unit containing logical commands in order to perform the same processes and all problems*] gespeicherte Programmlogik, Speicherprogrammlogik *f*

stored routine; stored program (US); stored programme (Brit.) [*In computers. a series of instructions in storage to direct the step-by-step operation of the machine*] gespeicherte Programm, Speicherprogramm *n*

stored warmth Speicherwärme, gespeicherte Wärme *f*

S

stored word [*The actual linear combination of letters (or their machine equivalents) to be placed in the machine memory; this may be physically quite different from a dictionary word*] gespeicherte Wort, Speicherwort *n*

store entrance Ladeneingang, Geschäftseingang *m*

store exit Ladenausgang, Geschäftsausgang *m*

store merchandise Ladenware, Geschäftsware *f*

store program(me) control switching system programmgesteuerte Vermittlung *f*

store-programmed speicherprogrammiert

store security Lagersicherung, Speichersicherung *f*

storey → floor

storing Archivieren *n*, Archivierung *f*

storing → incorporating

storm, thunderstorm Gewitter *n*

storm drain → rainfall drain

storm overvoltage, thunderstorm overvoltage Gewitterüberspannung *f*

storm(water) [*The water which drains off a catchment area during and after a fall of rain, snow, or hail*] abfließende Niederschlagswasser *n*

storm(water) drain [*A pipe or channel used for conveying rainwater, groundwater, subsurface water, condensate, cooling water, or other similar discharge to a storm sewer or combined sewer*] Niederschlagswasserdrän *m*

storm(water) overflow Niederschlagswasserüberlauf *m*

story (US) → floor

storyboarder Konzeptionist *m* [*Seine Aufgabe ist die inhaltliche Gestaltung und Konzeption interaktiver Medien*]

STP → twisted shielded pair

STR → self-tuning regulator

straight-blade(d) centrifugal fan → radial centrifugal fan

straight expansion piece, straight compensating piece Kompensator, Ausgleicher *m* [*Rohrleitung*]

straight fitting [*See remark under "fitting"*] g(e)rade Armatur, egale Armatur *f*

straight joint → butt joint

straight lock → rim lock

straight pipe Geradrohr, Strangrohr *n*

straight pipes g(e)rade Rohre, Strangware, Strangrohre *f*

straight(-type) return (flow) temperature limiter Durchgangs-Rücklauftemperaturbegrenzer *m*

straight(-way) adjusting valve, adjusting straight(-way) valve (Ein)Stelldurchgangsventil, Verstelldurchgangsventil, Justierdurchgangsventil, Durchgangs(ein)stellventil, Durchgangsverstellventil, Durchgangsjustierventil *n*

straight(-way) ((closed) loop) control valve, ((closed) loop) control straight(-way) valve Regeldurchgangsventil, Durchgangsregelventil *n*

straight(-way) float(ing) valve, floating straight (-way) valve Durchgangsschwimmerventil, Schwimmerdurchgangsventil *n*

straight(-way) matrix Durchgangsmatrix *f*

straight(-way) nonreturn valve, nonreturn straight(-way) valve Durchgangsrückschlagventil, Rückschlagdurchgangsventil *n*

straight(-way) pressure reducing valve, pressure reducing straight(-way) valve Druckminderdurchgangsventil, Durchgangsdruckminderventil *n*

straight(-way) seat(ing) G(e)radsitz(form), Durchgang(sform) *m, (f)* [*Schieber; Ventil*]

straight(-way) throttling valve, throttling straight(-way) valve Drosseldurchgangsventil, Durchgangsdrosselventil *n*

straight(-way) valve Durchgangsventil *n*

strain Dehnung *f*

strain bearing element Zugentlastungselement *n* [*Gummischlauchleitung*]

strainer → dirt trap

strainer (type) (air) filter → dry (medium) (air) filter

stranding [*cable*] Verseilung *f*

s-transform pair LAPLACE-Transformationspaar *n*

strap hanger Laschenschelle *f*

strategic facility management strategische Gebäudebewirtschaftung *f*

stratified flow(ing), stratified streaming Schichtenströmung, geschichtete Strömung *f*, Schichtenströmen, geschichtetes Strömen *n*

stray current, leakage current, current from external source Streustrom, Erdstrom, Irrstrom, Ableit(ungs)strom, Fremdstrom *m*

stray heat Streuwärme *f* [*Siehe Anmerkung unter „Hitze"*]

stray magnetic field, magnetic stray field Magnetstreufeld, Streumagnetfeld, magnetisches Streufeld *n*

stray radiation Streustrahlung *f*

to stream → to flow

stream Strom *m*, Vollduplex-Datenverbindung *f* [*Zwischen einem Anwenderprogramm und dem Betriebssystem*]

stream → **flow**

streamer Magnetkassettenspeicher, Kassettenmagnetspeicher *m*

streamer Laufwerk, Bandlaufwerk *n*

streamer tape Kassettenmagnetband, Magnetkassettenband *n*

stream head Stromkopf *m* [*Er bildet die Schnittstelle zwischen einer Vollduplex-Datenverbindung (= Strom) und dem Anwenderprogramm*]

streamline (flow) → **laminar flow**

street [*Not to be confused with "road"*] Straße, Stadtstraße *f*

street door Haustür, Straßentür *f*

street drainage (Stadt)Straßenentwässerung *f*

street duct (Stadt)Straßenkanal *m*

street folding door zweiflüg(e)lige Haustür, zweiflüg(e)lige Straßentür *f*

street gutter Straßenrinne *f*

street level Straßenhöhe , Straßenkote *f*, Straßenniveau *n*

street sewer Straßenabwasserleitung *f*

strength Stärke *f* [*Akustik*]

strength, stability Festigkeit, Beständigkeit, Kraft *f* [*z. B. Haftfestigkeit oder Festigkeit gegen Säuren*]

stretch → **section**

stretcher lift (Brit.); stretcher elevator (US) Krankenbahrenfahrstuhl *m*

stretching Dehnung *f*

strike plate → **(lock) striking plate**

striker, trip cam Auslösenocken *m*

strike(r) → **(lock) striking plate**

striker armature Schlaganker *m*

striking out → **spurting out**

striking plate → **(lock) striking plate**

string, character string Zeichenkette, Zeichenfolge, Zeichenreihe *f*

string character Kettenzeichen, Folgezeichen, Reihenzeichen *n*

string of numbers, number string Nummernfolge, Nummernkette, Nummernreihe *f*

strip Streifen *m*

strip Leiste *f*

strip (air) diffuser → **slot(ted) (air) diffuser**

strip conductor electrode → **ground(ing) strip**

strip fin → **spiral fin**

strip heating ceiling, heating strip ceiling Streifenheiz(ungs)decke *f*

strip mask Streifenmaske *f*

strip of luminaires → **row of luminaires**

stripping Datenaufteilung auf Laufwerke, Datenverteilung auf Laufwerke, Feld-Funktionalität *f*

stripping column → **dephlegmator**

strip (type) radiant heater, radiant strip (type) heater Streifenstrahlungsheizer, Streifenstrahlungserhitzer, Streifenstrahlungs(er)wärmer *m*

strobe pulse Strobe-Impuls, (Aus)Tastimpuls, Stroboskopimpuls, Übernahmeimpuls, Markier(ungs)-impuls, Auswerteimpuls, Abfrageimpuls, Abtastimpuls *m* [*DIN IEC 469, Teil 1. Er bewirkt, dass ein vorbereiteter Vorgang ausgelöst wird bzw. dass eine mit ihm verknüpfte Information während seiner Dauer zum Abtasten zu Verfügung steht*]

stroke [*The linear distance travel(l)ed in either direction by a piston or rod in an engine. Generally, the term is used for all consecutive up and down movements of any object*] Hub *m* [*DIN 1940*]

strongroom, safe room Tresorraum, Wertschutzraum, Panzerraum *m*

strongroom door, safe room door Tresorraumtür, Panzerraumtür, Wertschutzraumtür *f*

strongroom vault, safe room vault Tresorraumgewölbe, Panzerraumgewölbe, Wertschutzraumgewölbe *n*

strong solution [*absorption refrigeration*] reine Lösung *f*

structural conversion Umbau *m*

structural fabric Bausubstanz *f*

structural fire protection [*Those features in the layout and/or construction of a building which are intended to reduce the effect of a fire*] bauliche Brandschutz, bauliche Feuerschutz *m*

structural floor → **floor**

structural interactive graphic monitoring application language, SIGMAL blockstrukturierte graphische Programmiersprache, graphische blockstrukturierte Programmiersprache *f* [*Sie ermöglicht das Erstellen verwickelter Funktionsabläufe am Bildschirm*]

structural renewal konstruktive Erneuerung *f*

structural steel frame(work) Baustahlrahmen *m*

structure, building structure Bauwerk *n*, bauliche Anlage *f*

STSTB → **status-strobe**

stub duct, spur duct, branch duct Stichkanal, (Ab)Zweigkanal *m*

S-tube → **bend offset**

S

stub line → branch line

stuck-at failure, stuck-at outage Haftfehler, Haftausfall *m*, Haftversagen *n* [*"Stuck-at" bedeutet, dass ein Signalpegel entgegen seiner Sollfunktion ständig auf 0 (stuck-at 0 oder Sa0] oder 1 (stuck-at 1 oder Sa1) liegt. Auch offene Eingänge und Kurzschlüsse äußern sich als Haftfehler*]

stuck open nichtleitend [*CMOS-Schaltung*]

stuck short permanent leitend [*CMOS-Schaltung*]

stud plattierte Nagel, Beschlagnagel *m*

stud-bolt Schraubenbolzen *m* [*Mit Gewinde an beiden Enden*]

stuffingless → pack(ing)less

to style → to design

style sheet Druckformat *n*

styling → design

stylus (Ab)Taster *m* [*mechanischer (Ab)Taster für Prüfstücke*]

stylus printer → matrix printer

stylus print head → matrix print head

stylus printing → matrix printing

subassembly, assembly group Baugruppe *f*

subband coding Tiefbandkodierung, Tiefbandcodierung *f*

subband signal Tiefbandsignal *n*

subcarrier Hilfsträger *m*

sub-circuit distribution board Stromkreislauf *m*

subcode Hilfskode, Hilfscode *m*

subcontractor Unterauftragnehmer, Subunternehmer *m*

to subcool, to undercool, to supercool unterkühlen, tiefkühlen

subcooler → supercooler

subcooling → supercooling

S

subdirectory Unterverzeichnis *n*

subdistribution box Nebenverteilerkasten *m*

sub-exchange [*telephone*] Unterzentrale *f*

subfloor heating → underfloor heating

subframe Unterrahmen *m*

subharmonic [*Acoustics. Specifically, a sound wave whose frequency is the inverse of a whole number of the frequency of a specified fundamental; thus, the subharmonics of a frequency of 4000 are 2000, 1000, 500, 250, 125 and so on*] Unterschwingung, Unterharmonie, harmonische Unterschallwelle *f*

subindex Kennung *f* [*Elektrizitätszähler*]

subject characteristic Sachmerkmal *n*

subject characteristics data Sachmerkmaldaten *f* [*DIN 4000 Teil 1*]

subject index Sach(wort)index *m*

subject key Sach(wort)schlüssel *m*

subject to permission erlaubnispflichtig, genehmigungspflichtig

subayer Teilschicht, Teillage *f* [*ISO-Referenzmodell*]

sublim(at)ed sublimiert

sublim(at)ed coating thermal transfer line printing Thermotransfer-Farb-Sublimations-Drucken *n*

Sub MAIN Hauptroutine *f* [*Adressverwaltung*]

submaster controller [*1. Controller whose set point is automatically readjusted from a remote location over a predetermined range by variations in an applied signal from a master. 2. Controller whose set point may be automatically reset at the command of either a conventional or master controller*] Nebenregler *m*

submaster dimmer Nebenabblendregler *m*

submaster station Knotenstation *f* [*Fernwirktechnik*]

submersible, immersible (ein)tauchfähig, (ein)tauchbar

submeter Zwischenzähler, Zwischenmesser *m*

submetering individuelle Verbrauchsmessung, individuelle Verbrauchszählung, Verbrauchsmessung einer Wohneinheit, Verbrauchszählung einer Wohneinheit, Zwischenmessung, Zwischenzählung *f* [*Sie wird vom Abrechnungsspezialisten durchgeführt*]

submode Unterbetriebsart *f*

submodule Untermodul *m*, Platine *f*

sub-mounting platform vorgezogene Geräteplatte *f*

Sub Name DER Routine Unterroutine *f* [*Adressenverwaltung*]

subnet(work) Unternetz *n*

subpanel Untertafel *f*

subquery Teilanfrage *f*

subrack → card rack

subrack tier → card rack tier

subroutine [*A subprogram(me) that is called by name from another program(me) or subroutine and may have arguments passed to it in the call. The term "subprogram(me)" as a synonym for "subroutine" is not correct, since "routine" means a sequence of instructions directing a computer to carry out a set of operations that form part of a program(me) and are not the program(me) itself*] Unterprogramm *n*

subscriber Teilnehmer *m* [*Fernsprechwesen*]

subscriber distribution interface, SDI Teilnehmerverteilerschnittstelle *f*

subscriber interface, SI Teilnehmerschnittstelle *f*

subscriber's identity module, SIM Telefonkarte, Fernsprechkarte *f*

subsriber's line → post-office line

subsriber's station [*automatic telephony*] Sprechstelle *f*

subsidiary ventilation, subsidiary vent(ilat)ing [*It occurs through the cracks around doors and the frames of windows*] Fugenlüftung, Selbstlüftung *f* [*DIN 1946, Teil 1*]

subsignal Befehlszusatz *m*

subsonic [*deprecated*] **→ infrasonic**

substance, matter Substanz, Materie *f*

substation, utility substation [*power engineering*] Unterwerk, Hilfswerk, Außenwerk, Übergabewerk, Umspannwerk *n*, Unterstation, Hilfstation, Außenstation, Unterstelle, Hilfsstelle, Übergabestation, Übergabestelle, Außenstelle *f*

to substitute → to replace

substitute Ersatz(werk)stoff *m*, Ersatzmaterial *n*

substitute element lag Ersatztotzeit, Totersatzzeit *f* [*Glieder höherer Ordnung werden vielfach zur vereinfachten Behandlung durch die Reihenschaltung eines* T_t*- und eines P-*T_1*–Gliedes ersetzt. Die Totzeit des* T_t*-Gliedes heißt dann Ersatztotzeit* T_{tE}*.*
Anmerkung: Wird statt eines P-T_t*-Gliedes ein Glied höherer Ordnung als Ersatz verwendet, so setzt sich die Ersatztotzeit* T_{tE} *aus der Totzeit des* T_t*-Gliedes und der Verzugszeit* T_u *des Ersatzgliedes zusammen. DIN 19 226, Teil 2*]

substitute fuel Ersatzbrennstoff *m*

substitute refrigerant Ersatzkältemittel *n*

subsystem Untersystem *n*

subtractor [*A circuit that performs arithmethic subtraction operations on binary numbers*] Subtrahierwerk *n*, Subtrahierer *m*, Subtrahierschaltung *f*

subventilation Nebenlüftung *f*

subway [*A (pipe) duct below ground level large enough to walk through*] begehbare (Rohr)Kanal *m*

subwoofer Basslautsprecher *m* [*Lautsprecher für Töne unter ungefähr 200 Hz*]

sub-zero temperature; low temperature (US) Minustemperatur *f*

succeeding unit [*cascade control*] nachgeschaltete Einheit *f*, Folgeregler *m*

successive approximation allmähliche (An)Näherung *f*

successive converter allmähliche Umsetzer *m*

sucking (in), suction Einsaugen, Saugen *n*

suction, sucking (off) Absaugen, Saugen *n*

suction airtube conveyor, suction airtube system Saugrohrpost *f*

suction connection Sauganschluss *m*, Saugverbindung *f*

suction draught; suction draft (US) Saugzug *m*

suction duct Saugkanal *m*

suction filter Saugfilter *m,n*

suction fire detector, suction fire detecting device Saugfeuermelder, Saugbrandmelder *m* [*Er besteht aus einem Sauggerät mit Rauchdetektor und Luftstromüberwachung sowie einem Rohrnetz mit bis zu 12 Saugöffnungen*]

suction gas [*Gas entering the suction side of the compressor*] Sauggas *n*

suction gas-cooled sauggasgekühlt

suction gas line Sauggasleitung *f* [*Verbindungsleitung im Kältemittelkreislauf zwischen Ausgang des Verdampfers und Verdichter*]

suction gas manifold, suction gas central distributor [*Device to distribute suction gas equally from a common suction line to multiple compressors connected in parallel*] Sauggasverteiler(rohr) *m, (n)*, Sauggasverteilungsrohr, Sauggasverteilerstück, Sauggasverteilungsstück *n*, Sauggasverteilerbatterie, Sauggasverteilungsbatterie *f*

suction gas riser [*Vertical tube or pipe that carries suction gas from an evaporator on a lower level to compressor on a higher level*] Sauggassteig(e)leitung *f*

suction gas temperature Sauggastemperatur *f* [*Eintrittstemperatur des dampfförmigen Kältemittels in den Verdichter*]

suction gauge; suction gage (US) Saugdruckmesser *m*

suction lance Sauglanze *f*

suction-line filter [*Filter installed in the suction line of a refrigerating system for the purpose of removing solid containments of a shell with fittings and internal filtering element*] Saugleitungsfilter *m, n*

suction-line valve, suction stop valve [*Shutoff valve fitted to the compressor, or in the suction line adjacent to it, to isolate the compressor from the suction line*] Saugabsperrventil *n*

suction noise Sauggeräusch *n*

suction nozzle Saugdüse *f*

suction opening Saugöffnung *f*

suction pipe Saugrohr *n*

suction (pipe)line Saug(rohr)leitung *f*

suction (pipe)line valve Saug(rohr)leitungsventil *n*

S

suction piston pump, piston suction pump Saugkolbenpumpe, Kolbensaugpumpe *f*

suction pressure Saugdruck *m*

suction pressure ((closed) loop) control Saugdruckreg(e)lung *f*

suction pressure ((closed) loop) controller Saugdruckregler *m*

suction pressure ((open) loop) control Saugdrucksteuerung *f*

suction/pressure pump Saug- und Druckpumpe *f*

suction pump [*A pump which raises water, or draws water in plumbing, by producing a partial vacuum within a pipe*] Saugpumpe *f*

suction shaft Saugschacht *m*

suction side [*1. Side of a compressor connected to low pressure. 2. The outlet side of an evaporator*] Saugseite *f*

suction-side, on the suction side saugseitig, auf der Saugseite

suction spiral hose → spiral suction hose

suction stop valve → suction-line valve

suction stream ((open) loop) control Saugstromsteuerung *f*

suction stroke [*Movement of a piston in a compressor or engine cylinder during which fluid is driven into the cylinder*] Saughub *m*

suction tank [*Gravity water supply system*] Saugbehälter, Saugtank *m*

suction urinal Saugurinal *n*

suction valve [*In a compressor, the valve which allows refrigerant to enter the cylinder from the suction line and prevents return flow*] Saugventil *n*

suction vapor (US) → (refrigerant) suction vapor

sufficient erschöpfend [*Schätzung*]

Suffolk latch (Brit.) → thumb latch

S

suitability Eignung, Betätigung *f*

suitable for networking vernetzgeeigent, vernetzfähig, vernetzbar

suite Garnitur *f* [*z. B. Klosettgarnitur*]

sulphuric acid dew point (Brit.); sulfuric acid dew point (US) Schwefelsäuretaupunkt *m* [*Er darf 150 °C bei Abkühlung der Abgase nicht überschreiten*]

sum compilation → total compilation

summation amplifier Summierverstärker, Summenverstärker *m*

summer air conditioning [*It is carried out primarily when outside temperature and humidity are above those to be maintained in the conditioned space*] Sommerklimatisierung *f*

summer/winter thermostat → heating/cooling thermostat

sum-min inference SUM-MIN-Inferenz, unscharfe Schlussfolgerungsmethode *f*, unscharfe Schlussfolgerungsverfahren *n*

sump [*A tank or pit which receives the discharge from drains and which must be emptied by mechanical means*] Sumpf *m*

sump pump Sumpfpumpe *f*

sun angle declination → declination of sun angle

sun blind ((open) loop) control Rolladensteuerung, Sonnenschutzsteuerung, Solarschutzsteuerung *f*

sunk hydrant, below-grade hydrant Unterflurhydrant *m*

sun relay → photoelectric lighting controller

sun shadow Sonnenschatten *m*

sun slate schindelförmige PV-Modul *n*

superatmospheric pressure Überdruck *m*

superbeam Satellitenstrahl *m* [*Er ist sehr stark gebündelt und energiereich*]

superbrightness → superintensity

superchilling [*Chilling a product to a temperature very near or sometimes below its freezing point*] Unterkühlen *n*, Unterkühlung *f*

superconducting, superconductive supraleitend, superleitend

superconducting gyro(scope) → cryogenic gyro(scope)

superconducting quantum interference device, SQUID superleitende Quanteninterferometer, supraleitende Quanteninterferometer *n*

superconductivity Superleitfähigkeit, Supraleitfähigkeit *f*, Superleitvermögen, Supraleitvermögen *n*

to supercool → to subcool

supercooler, undercooler, subcooler Unterkühler, Tiefkühler *m*

supercooling, undercooling, subcooling [*A condition in which a pure substance is cooled below its freezing point, condensation point, or sublimation point, without experiencing the corresponding change of state that its cooling would normally bring about*] Unterkühlen, Tiefkühlen *n* ,Unterkühlung, Tiefkühlung *f*

superfast passenger lift (Brit.); superfast passenger elevator (US) Schnellstfahrstuhl, Schnellstpersonenaufzug, Personenschnellstaufzug *m*

superficial content → area

superfine filter cartridge Feinstfilterpatrone *f*

to superheat [*Thermodynamics. To bring a substance to a temperature above that at which a change of state would normally occur, without this change taking place; e.g. to heat liquid above its normal boiling point without producing vapo(u)rization*] überhitzen

superheated [*Thermodynamics. At a temperature above which a change of state would normally occur (e.g. the change of water from a liquid to a vapo(u)r, without this change having taken place)*] überhitzt, heiß

superheated air Heißluft, überhitzte Luft *f*

superheated steam [*Thermodynamics. Water vapo(u)r that has been heated, at constant pressure and out of constant pressure with the water from which it was formed, to a temperature significantly above its saturation point, so that it does not recondense to water as readily as ordinary steam; used to provide increased efficiency in steam engines*] Heißdampf, überhitzte Dampf *m*

superheated steam slide valve Heißdampfschieber *m*

superheated steam valve Heißdampfventil *n*

superheated vapor(US); superheated vapour (Brit.) [*Thermodynamics. A vapo(u)r that is heated to a temperature that exceeds the boiling point of the liquid phase*] Heißwasserdampf, überhitzte Wasserdampf *m*

superheated water Heißwasser, überhitzte Wasser *n* [*Nicht verwechseln mit „Warmwasser"*]

superheated water heater Heißwasserbereiter, Heißwasserheizer, HW-Bereiter, HW-Heizer *m*

superheater [*A coil or other device through which steam from boiler passes and is heated above its saturation temperature*] Überhitzer *m*

superheater chamber in smoke box Überhitzerfeuerkammer *f*

superheater coil → zig-zag superheating tube

superheating [*Heating of a substance above the temperature at which a change of state would ordinarily take place without such a change of state occurring, for example, the heating of a liquid above its boiling point without boiling taking place; this results in a metastable state*] Überhitzen *n*, Überhitzung *f*

superheating degree Überhitzungsgrad *m*

superheavy → extraheavy

superimposed hopper [*boiler*] Aufbaufüll(ungs)-schacht *m*

superintensity, superbrightness, superluminosity Superlichtstärke, Höchstlichtstärke, Superhelligkeit, Höchsthelligkeit *f*

superior übergeordnet

supermarket Verbrauchermarkt, Supermarkt *m*

to supersaturate übersättigen

superseded by ersetzt durch

supersensitive, supersusceptible höchstempfindlich

supersonic (turbo) compressor [*The fluid flow, in certain parts of the machine, attains supersonic velocity*] Überschallkompressor, Überschallverdichter *m*

supersonic velocity Überschallgeschwindigkeit *f*

supersonic wind tunnel Überschallwindkanal *m*

superuser → root

superuser directory → root directory

supervision Beaufsichtigung, Aufsicht *f*

supervisor Leiter *m*

supervisor Organisationsprogramm *n*

supervisory control level Bedienstation *f* [*früher: Gesamtleitebene. Oberste Ebene in der Systemhierarchie*]

supervisory program (US); supervisory programme (Brit.) Arbeitsablaufprogramm *n* [*Zur Organisierung und Regulierung des Arbeitsablaufes in einem automatischen Datenverarbeitungssystem*]

supplementary energy → additional energy

supplementary interface, additional interface Zusatzschnittstelle *f*

supplied [*Installed, furnished or provided by the owner or operator*] bauseits, bauseitig

supplier Lieferant *m*

supplier → provider

supplies Lieferungen *f*

supply → (entering) supply

supply Lieferung *f*, (gelieferte) Mittel *n*

supply Versorgung *f*

supply air → inlet air

supply bottleneck Lieferengpass *m*

supply cable → connecting cable

supply cable duct → connecting cable duct

supply cable (entrance) box → connecting cable (entrance) box

supply condition Lieferbedingung, Versandbedingung *f*

supply conductor, supply lead, lead wire Zuleitung *f* [*Leiter*]

supply current Versorgungsstrom *m*

supply duct Versorgungskanal *m*

supply fan → (air) supply fan

supply group Versorgungsgruppe *f*

S

supply inside wall outlet [*See remark under "supply outlet"*] Innenwandzuluftöffnung *f*

supply installation → **supply plant**

supply lead → **supply conductor**

supply line, supply main Versorgungsleitung *f*

supply line flow pressure, supply main flow pressure Versorgung(fließ)druck *m* [*Fließdruck in der Versorgungsleitung*]

supply mains, supply net(work), supply lines Versorgungsnetz *n*

supply module, voltage module Versorgungsmodul, Spannungsmodul, Versorgungsbaustein, Spannungsbaustein *m*

supply network Versorgungsleitungsnetz *n*

(supply) order Bestellauftrag *m*, Bestellung *f*

supply outlet [*An opening through which air is delivered into a conditioned space. Outlets are grouped under the broad classification of registers and diffusers. Although the difference between registers and diffusers is not sharply defined, a register is commonly considered as an outlet which discharges the air in a confined jet. On the other hand, a diffuser discharges the air in a spreading jet*] Zuluftöffnung *f*, Zuluftdurchlass *m*

supply pipe Versorgungsrohr *n*

supply pipe from source → **(water) supply pipe from source**

supply pipeline, supply piping Versorgungsrohrleitung *f*

supply plant, supply installation Versorgungsanlage, Versorgungsinstallation *f*

supply point Versorgungsstelle *f* [*Bereich, der von der kleinsten Einheit einer RLT-Anlage versorgt wird (z. B. von einem Induktionsgerät, Raumgerät, Luftdurchlass)*]

supply pressure Versorgungsdruck *m*

S

supply range Versorgungsbereich *m* [*selten: n*]

supply riser, supply standpipe Versorgungsstandrohr, Standversorgungsrohr *n*

supply safety, safety of supply Versorgungssicherheit *f*

supply scope → **scope of supply**

supply security → **security of supply**

supply value Entnahmeventil *n*

supply voltage rejection ratio, SVRR Betriebsspannungsunterdrückung *f* [*Operationsverstärker*]

supply zone Versorgungszone *f*

support(ing) steelwork → **(escalator) truss**

support of membership function, membership function support Träger *m*, Stützmenge einer unscharfen Menge *f* [*Fuzzy-Logik*]

to suppress entstören

to suppress → **to disable**

suppression, disablement Unterdrückung, Entaktivierung *f*

suppression Entstörung *f*

suppression of radio interference, radio interference suppression Funkentstörung *f*

suppression reactor Entstör(ungs)drossel *f*

suppressor, suppressing element Entstörer *m*, Entstör(ungs)glied *n*

suppressor Trägerstauwirkung-Schutz(be)schaltung *f*

suppressor circuit (Überspannungs-)Schutzschaltung, Entstör(ungs)(be)schaltung *f*

suppressor diode Löschdiode *f*

surface acoustic wave, SAW, acoustic surface wave akustische Oberflächenwelle, AOW *f*

surface adherence → **adsorption**

surface area, area of surface (Ober)Flächeninhalt *m*

surface-barrier-transistor Oberflächensperrschichttransistor, Randschichttransistor *m*

surface burner, cooking burner [*kitchen range*] Kochplatte *f*

surface clearance Bodenfreiheit *f*, Bodenabstand *m* [*Abstand zwischen Stelloberfläche und Objektunterseite*]

(surface) condensation, sweating Schwitzwasserbildung, Tauwasserbildung, Kondenswasserbildung, Kondensatbildung, Kondensation *f* [*Niederschlag der Luftfeuchtigkeit, z. B. auf Wänden und Decken, an Glasscheiben oder Rohrleitungen, wenn durch die Abkühlung die Feuchtigkeitsaufnahmefähigkeit der Luft sinkt und infolge zu geringer Porigkeit des Untergrundes die Feuchtigkeit nicht aufgesaugt wird*]

surface condenser [*A heat-transfer device in which exhaust is condensed by contact with metal surfaces cooled by a flow of cold water on the sides opposite the condensing surface*] Oberflächenverflüssiger, Oberflächenkondensator *m*

surface cooler Oberflächenkühler *m* [*Berippte Röhrenbündel werden innen vom Kühlmittel durchflossen, außen von Luft umspült*]

surface cooling [*Cooling air or other gas by passing it over a cold surface*] Oberflächenkühlung *f*

surface cylinder lock → **cylinder surface lock**

surface evaporation Verdunstung *f* [*Diffundieren von Wasserdampf bei Temperaturen < 100 °C aus einer Wasseroberfläche oder einer befeuchteten Fläche in die Luft, z. B. bei Kontakt- oder Rieselbefeuchtern*]

surface hydrant → above-grade hydrant

surface-mounted aufputzmontiert

surface(-mounted) dead bolt lock aufliegendes Riegelschloss, Aufsatzriegelschloss, Aufliegeriegelschloss *n*

surface(-mounted) device, SMD platinenverlötete Bauteil *m, n* [*Dieses Bauteil wird nicht in Lochraster gesteckt und verlötet, sondern auf der Oberfläche der Platine verlötet*]

surface(-mounted) lock Aufsatzschloss, Aufliegerschloss, aufliegendes Schloss *n*

surface(-mounted) technique, SMD technique Oberflächenmontagetechnik *f*

surface mounting Aufputzmontage *f*

surface mounting base Aufputz(montage)sockel *m*

surface mounting housing, surface mounting casing, surface mounting case Aufputz(montage)-gehäuse *n*

surface protecting wire Flächenschutzdraht *m*

surface raceway → wireway

surface radiant burner, radiant surface burner Oberflächenstrahlungsbrenner, Strahlungsoberflächenbrenner *m*

surface retention → adsorption

surface storage in buildings oberirdische Lagerung *f* in Gebäuden

surface (type) thermostat [*It has its sensing element clamped to the surface of a tank or pipe and is an expedient substitute for an insertion or immersion thermostat*] Oberflächenthermostat, Oberflächentemperaturregler *m*

surge [*Electricity. A momentary, sudden increase or change in the current or voltage of a circuit*] Stoß, nicht periodische Stoß *m* [*DIN 41 785*]

surge arrester, transient arrester Überspannungsleiter *m*

surge tank Pufferspeicher *m* [*Um unterschiedliche Wassermengen zu fördern müssen die Pumpenkreisläufe getrennt werden. Der hierzu verwendete Pufferspeicher wird zwischen Wärmepumpenkreislauf und Heizkreislauf eingebaut*]

surge ventilation, transient ventilation Stoßlüftung *f*

surge voltage → transient voltage

surrounding, environment Umgebung *f*

surrounding → ambient

surrounding area Umschließungsfläche *f* [*z. B. eine Wand*]

surrounding zone Außenzone *f* [*An die Fassade(n) angrenzender Raum oder angrenzende Räume*]

surveillance Sicherheitsüberwachung *f*

surveillance plant, surveillance installation Sicherheitsüberwachungsanlage *f*

survival(s) Bestand *m*, Bestände *f* [*Gütesicherung*]

survival(s) care Bestandspflege *f* [*Gütesicherung*]

susceptibility, sensitiveness, sensitivity Anfälligkeit, Empfindlichkeit *f*

susceptibility level, sensitiveness level, sensitivity level Anfälligkeitsstand, Empfindlichkeitsstand *m*

susceptibility response characteristic, sensitiveness response characteristic, sensitivity response characteristic Anfälligkeitscharakteristik, Empfindlichkeitscharakteristik *f*

susceptibility response curve, sensitiveness response curve, sensitivity response curve Anfälligkeitskurve, Empfindlichkeitskurve *f*

susceptible → sensitive

suspended, hung, pendant, hanging abgehängt, hängend, (auf)gehängt

suspended (bathroom) cabinet, hung (bathroom) cabinet, pendant (bathroom) cabinet (Badezimmer)Hängeschrank *m*

suspended ceiling, hanging ceiling, hung ceiling, dropped ceiling, suspension ceiling, pendant ceiling [*It is suspended below the overhead structure slab or from the structural elements of a building and not bearing on the walls*] Decke, Hängedecke, abgehängte Decke *f*

suspended dry sprinkler, dry suspended sprinkler, pendant dry sprinkler, hung dry sprinkler, dry hung sprinkler, dry pendant sprinkler Trockenhängesprinkler, Hängetrockensprinkler, hängende Trockensprinkler, abgehängte Trockensprinkler *m* [*„Sprenger" kann synonym für „Sprinkler" verwendet werden*]

suspended electrode, suspension electrode, hanging electrode, hung electrode Hängeelektrode *f*

suspended glass bulb sprinkler, pendant glass bulb sprinkler, hung glass bulb sprinkler Hängeschirmsprinkler, Hängeschirmsprenger *m*

suspended concrete heating ceiling, hung concrete heating ceiling, pendant concrete heating ceiling Betonrohrheiz(ungs)decke *f* [*z. B. unterhalb einer Hohlsteintragdecke*]

suspended heating ceiling, hanging heating ceiling, hung heating ceiling Heiz(ungs)hängedecke, Hängeheiz(ungs)decke *f*

suspended matter → suspended solid

S

suspended mounting, suspension mounting, hanging mounting, pendant mounting, hung mounting Hängemontage *f*

suspended mounting channel, hung mounting channel, pendant mounting channel, hanging mounting channel, suspension mounting channel Pendel-Montageschiene, Montage-Pendelschiene *f* [*Deckenleuchte*]

suspended panel(s) ceiling heating, suspension panel(s) ceiling heating, pendant panel(s) ceiling heating, hanging panel(s) ceiling heating, hung panel(s) ceiling heating Strahl(ungs)plattendeckenheizung, Deckenstrahl(ungs)plattenheizung *f*

suspended radiant heating ceiling, hung radiant heating ceiling, pendant radiant heating ceiling, suspension radiant heating ceiling Hänge-Strahl(ungs)heiz(ungs)decke *f*

suspended socket, hanging socket, hung socket, suspension socket, pendant socket Hängesteckdose, Kupp(e)lungssteckdose *f*

suspended solid, suspended matter, solid in suspension, matter in suspension Schweb(e)stoff, ungelöste feste Bestandteil *m*

suspended switch, (pull-)cord switch, hanging switch, cord(-operated) switch, pull(ing) switch, hung switch Schnurschalter, Zugschalter, Hängeschalter *m* [*VDE 0632*]

suspended tubular concrete heating ceiling, hanging tubular concrete heating ceiling, pendant tubular concrete heating ceiling (Rohr-)Betonheiz(ungs)decke *f* [*Sie wird an der Tragdecke aufgehängt*]

suspension ... → suspended ...

suspension cable, suspension rope Hängekabel, Hängeseil *n*

suspension channel, suspension duct [*vent(ilat)ed ceiling*] Hängekanal *m*

suspension insulator → chain insulator

SV → specific volume

SVRR → supply voltage rejection ratio

swaging Sickung, rinnenförmige Biegung, Kehlung *f*

swap data Auslagerungsdaten *f*

swapping Auslagern *n* [*Von Daten auf die Festplatte*]

swapping Auslagern *n* [*Zwischenzeitliches Auslagern eines Prozesses auf einen Hintergrundspeicher*]

swap (region) Auslagerungsbereich *m [selten: n]* [*Definierter Festplattenbereich zur zwischenzeitlichen Auslagerung von Prozessen*]

to sweat anschwitzen [*Löten*]

sweating, (surface) condensation Schwitzwasserbildung, Tauwasserbildung, Kondenswasserbildung, Kondensatbildung *f* [*Niederschlag der Luftfeuchtigkeit, z.B. auf den Wänden und Decken, an Glasscheiben oder Rohrleitungen, wenn durch die Abkühlung die Feuchtigkeitsaufnahmefähigkeit der Luft sinkt und infolge zu geringer Porigkeit des Untergrundes die Feuchtigkeit nicht aufgesaugt wird*]

to sweat (together) feuerlöten, heißlöten

sweep(ing) Erzeugen *n* dreidimensionaler Geometrie [*Durch Führen einer Linie entlang einer räumlichen Leitkurve*]

swift setting mortar, place-and-finish mortar Schnellmörtel *m*

(swimming) pool Schwimmbecken *n*

(swimming) pool care system (Schwimmbecken)-Wasserpflegesystem *n*

(swimming) pool hall Hallen(schwimm)bad *n*

(swimming) pool surrounding area, (swimming) pool surrounding zone (Schwimm)Beckenbereich *m, [selten: n]*, (Schwimm)Beckenzone *f*

swing-through butterfly valve durchschlagendes Drosselventil *n*

swipe reader, push-through reader Durchzugsleser *m*

to switch → to change

switch Schalter *m*, Weiche *f* [*Nicht „Elektroschalter"*]

switch, electric switch Schalter, Elektroschalter *m*

switchable schaltbar, schaltfähig

switchable plane circular interpolation Kreisinterpolation *f* in verschiedenen Ebenen

switch accessories Schalterzubehör *n [selten: m]*

switch actuation → switching

switch act(uat)or Schaltgriff *m*

to switch back (zu)rückschalten

switch block Schalterblock *m*

switchboard Schalttafel *f*

switchboard diagram, switchboard chart Schalttafelplan *m*, Schalttafelschema, Schalttafelbild, Schalttafeldiagramm *n*

switchboard panel Schalttafelfeld *n*

switch clock → timing element

switch (closed) loop control Umschaltreg(e)lung *f*

switch cover plate Schalterdeckel *m*

(switch) disconnector [*IEC 265; BS 4727*], **(switch)isolator, load interrupter, interrupter switch, quick-break switch, isolating switch, disconnect(ing) switch; load-break switch** [*deprecated*] Lasttrenn(schalt)er, Trenn(schalt)er, TS, Trennschutzschalter *m* [*VDE 0670*]

S

switched capacitor filter, SCF Schalter-Kondensator-Filter *m, n*

switched communication link, dial(l)ed communication link geschaltete Kommunikationsstrecke, gewählte Kommunikationsstrecke, vorübergehend aufgebaute Kommunikationsstrecke *f*

switched connection Wählverbindung, Wahlverbindung *f*

switched mains unit Schalt(ungs)signal *n*

switched mode power supply, SMPS Schaltnetzteil *m, n*

switched network vermittelte Netz *n*

switched neutral schaltbare Nullleiter *m*

switched off abgeschaltet, ausgeschaltet

switched on eingeschaltet

switched(-type) power supply unit, SWPS unit, chopper(-type) power supply unit Schalternetzteil *m, n*

switched video-broadcast service vermittelte Videoverteil(ungs)dienst *m*

switchfuse [*It consists of a separate switch and fuse enclosed in the same case*] Sicherungsschalter *m*

switchgear Schalt(ungs)vorrichtung *f*

switchgear cubicle, switchgear cabinet Schaltschrank *m*

switchgear cubicle for measuring and open control and closed control, measuring open control/closed control switchgear cubicle [*"Cabinet" may synonymously be used for "cubicle"*] MSR-Schaltschrank, Schaltschrank für Messen/Steuern/Regeln *m*

switching [*telephony*] Vermitteln *n*, Vermittlung *f*

switching, switch actuation Schalten *n*, Schalt(erbetätig)ung *f*

switching act(uat)or, three-point act(uat)or, triple-point act(uat)or [*See remark under "act(uat)or"*] Schalt(ungs)stellglied *n*

switching capacity, switching performance, switching duty, switching service, making capacity, making performance, making duty, making service Schalt(ungs)leistung, Schalt(ungs)kapazität *f*

switching center (US); switching centre (Brit.) [*telephony*] Vermittlungsstelle *f*

switching command Schalt(ungs)kommando *n* [*Fernwirktechnik*]

switching controller, on/off controller Schalt(ungs)-regler *m*

switching diagram, switching chart Schalt(ungs)-schema, Schalt(ungs)bild, Schalt(ungs)diagramm *n*, Schalt(ungs)plan *m*, Schaltung *f* [*Die Darstellung der Wirkungsweise von Geräten usw., des Wirkungsweges oder der Verbindungsleitungen durch Symbole. Nicht verwechseln mit „Stromlaufplan"*]

switching diode Schalt(ungs)diode *f*

switching duty → **switching capacity**

switching element, switching constant Schaltungsglied, Schaltungselement *n*, Stromkreiskonstante *f*

switching frequency → **switching rate**

switching function Schalt(ungs)funktion *f* [*Eine Funktion, bei der jede Argumentvariable und die Funktion selbst endlich viele Werte aufnehmen können. Wird eine Schaltfunktion mit Hilfe eines Operationssymbols dargestellt, so ist das eine Verknüpfung. E DIN 44 300, Teil 5*]

switching-in → **switching-on**

switching installation Schalt(ungs)anlage *f*

switching interval → **switching (time) interval**

switching-in water level, switching-on water level Einschalt(ungs)wasserstand *m*

switching load Schalt(ungs)last *f*

switching matrix Koppelvielfache *n*

switching net(work) section Koppelnetzbereich *m* [*selten: n*] [*Ein Funktionsbereich eines Vermittlungsmoduls*]

switching off → **cutout**

switching-off circuit → **deenergizing circuit**

switching-off timer → **cutout timer**

switching-off torque → **cutout torque**

switching-off water level Ausschalt(ungs)wasserstand *m*

switching-on, switching-in Einschalten *n*, Einschaltung *f*

switching-on time, switching-in time Einschalt-(ungs)zeit *f*

switching over Umschalten *n*, Umschaltung *f*

switching over valve Umschalt(ungs)ventil *n*

switching overvoltage Schalt(ungs)überspannung *f*

switching panel Schalt(ungs)tafel *f*

switching performance → **switching capacity**

switching point [*telephony*] Verteil(ungs)fernamt *n*

switching point Schalt(ungs)punkt *m*

switching position Schalt(ungs)stellung *f*

switching rate, switching frequency, operating frequency, operating rate, frequency of operation, frequency of operating cycles Schalt(ungs)frequenz, Schalt(ungs)häufigkeit *f* [*EN 50 032, DIN IEC 147-1 D*]

switching room Schalt(ungs)raum *m*

switching sensitivity, switching sensitiveness, switching susceptibility Schalt(ungs)empfindlichkeit *f*

switching section Schalt(ungs)feld *n*

switching sequence Schalt(ungs)folge *f*

switching service → **switching capacity**

switching station Schalt(ungs)warte, Schalt(ungs)-zentrale, Schalt(ungs)station *f*

switching test(ing) run Schalt(ungs)probelauf *m*

switching (time) interval Schalt(ungs)(zeit)abstand *m*

switching (time) optimization Schalt(ungs)(zeit)-optimierung *f*

switching transistor Schalt(ungs)transistor *m*

switching tree Schalt(ungs)baum *m*

switching unit Schalt(ungs)einheit *f*

switching valve Schalt(ungs)ventil *n*

switching variable Schalt(ungs)größe *f*

switching voltage Schalt(ungs)spannung *f*

switching wire Schalt(ungs)draht *m*

switch-in point → **cut-in point**

switch-in pressure → **cut-in pressure**

switch interlock(ing) (system) Schalterkombination *f*

switch key Schaltertaste *f*

switch logic Schalterlogik *f* [*Darstellung der Logikfunktionen mit Schaltersymbolen*]

switch module Schalterbaustein *m*, Schaltermodul *n*

switch mounting Schaltermontage *f*

to switch off, to switch out, to turn off, to turn out, to cut off, to cut out, to deenergize, to deactivate, to disconnect abschalten, ausschalten, stromlos machen

to switch over umschalten, wechseln

switchover Umschalten, Wechseln *n*, Umschaltung *f*, Wechsel *m*

switch position detector, switch position detecting device, switch position detection device Schalterstellungsmelder, Schalterstellungsgeber *m*, Schalterstellungsmeldegerät *n*

switch position signal message; switch position signalling message (Brit.); switch position signaling message (US) Schalterstellungsmeldung *f*

switch socket-outlet Schaltersteckdose, Steckdose mit Schalter *f*

switch starter Anlassschalter *m*

swivel damper, revolving damper Dreh(rauch)schieber, Rauchschieber *m*, Dreh(rauch)klappe, Rauchklappe *f*

swivel hook Wirbel *m*

swivel(ling) elbow; swivel(ing) elbow (US) Winkeleinstell(ungs)verschraubung *f*

swivel(ling) head; swivel(ing) head (US) Schwenkrosette *f* [*Dusche; Brause*]

swivel(ling) outlet (Brit.); swivel(ing) outlet (US) Schwenkauslauf, Schwenkablauf, Schwenkauslass, Schwenkablass *m*

swivel(ling) tubing spout (Brit.); swivel(ing) tubing spout (US) schwenkbare Rohrauslauf, schwenkbare Rohrablauf *m*

swivel pivot and socket centre hinge (Brit.) [*US spelling "center"*] Körnerpfanne *f*

SWPS unit → **switched(-type) power supply unit**

SWR → **standing wave ratio**

symbol Sinnbild, Symbol, Kurzzeichen *n*

symbolic address Symboladresse, Sinnbildadresse, Kurzzeichenadresse *f* [*Sie hat nicht die für das Steuerwerk unmittelbar verständliche Form, sie kann z. B. als Variablenbezeichnung vorliegen*]

symbolic language Symbolsprache, Sinnbildsprache, Kurzzeichensprache *f*

symbol interference Symbolinterferenz, Sinnbildinterferenz, Kurzzeicheninterferenz *f* [*Die Überlagerung einer Informationseinheit bit (= Symbol) durch mehrere Nachbarbits (= Nachbarsymbole)*]

symbol recognition Symbolerkennung, Sinnbilderkennung, Kurzzeichenerkennung *f*

synchronism, synchronous operation Gleichlauf, Synchronismus *m*

synchronization bit Synchronisationsbit *n*

synchronized handrail monitoring [*escalator*] Handlaufüberwachung *f*

synchronous binary communication(s) → **binary synchronous communication(s)**

synchronous data link control, SDLC Kommunkationsprotokoll *n* von IBM in Verbindung mit SNA

synchronous (electric) heating → **(electric) synchronous heating**

synchronous generator → **alternator**

synchronous speed, synchronous revolutions per minute, synchronous rpm. Synchrondrehzahl, Synchronumdrehungen pro Minute *f*

sync separator → **impulse separator**

syntax check(ing) Syntaxkontrolle *f*

syntax error Syntaxfehler *m* [*Er entsteht durch Missachtung der Regeln einer Programmiersprache*]

synthesizer [*Any or various instruments that integrate simple elements or functions to produce more complex entities*] Zusammensetzer *m*

synthetic natural gas, SNG hochkalorige Syntheseerdgas *n*

syphon, siphon Geruchsverschluss, Siphon *m*

system → controlled system [*IEC 50(351)*]

system [*A set or arrangement of things so related or connected as to form a unit or organic whole: as, e.g., an irrigation system, supply system and the like*] Anlage *f*, (technische) System *n*

system access ((open) loop) control Systemzugriffssteuerung *f* [*Überwachung des Zugriffs von Bedienern, um nur autorisierten die Überwachung eines BMS und dessen Parameteränderungen zu gestatten.*]

system activity logbook Systemprotokoll *n* [*Aufzeichnung (üblicherweise eine elektronische) mit Zeit und Datum, von Ereignissen in Systemen, d. h. welche Bediener auf das Gebäudebetriebssystem zugegriffen und welche Maßnahmen sie getroffen haben (z. B. Alarme, Einträge usw.)*]

system architecture Systemarchitektur *f*

(systematic) procedure → process

system blocking Systemblockierung, System(ab)sperrung *f*

system data Systemdaten *f* [*Daten zur Darstellung der für die Prozessleitung relevanten Größen der technischen Anlage. DIN 66 201, Teil 1*]

system data file Systemdatei *f* [*Siehe Anmerkung unter „Datei"*]

system design, system type, system version Systemaufbau *m*, Systemkonstruktion, Systemausführung, Systembauweise, Systembauart *f*

system designation Systembezeichnung *f*

system designer Systemkonstrukteur *m*

system deviation [*IEC 50(351)*], **deviation** [*ANSI C85.1*] Regelabweichung *f* [*Unterschied zwischen Regelgröße und Führungsgröße oder Sollwert (negative Regelunterschied)*]

system diagnostics Systemdiagnose *f* [*Funktionen, die bei der Diagnose von Systemfehlern zu Hilfe genommen werden*]

system dull → dull of system

system electronics Systemelektronik *f*

system encoding Systemkodierung, Systemcodierung *f*

system engineering Ingenieurarbeit, technische Bearbeitung und Inbetriebnahme *f* [*In der Gebäudetechnik gehören dazu im wesentlichen: Koordinierung verschiedener Möglichkeiten der Energiebedarfsdeckung, Aufstellung und Optimierung konkreter Konzeptvarianten, Ausschöpfung der Möglichkeiten zur Mehrfachnutzung der Einsatzenergien, Untersuchung von wirtschaftlich realisierbaren Energierückgewinnungsmaßnahmen, Ermittlung der günstigsten Energieträger, ggf. in kombinierter Anwendung, Anwendungsmöglichkeiten alternativer und additiver Energiequellen, usw.*]

(system) extension (System)Erweiterung *f*

system (flow) chart, run diagram [*It emphasizes system component operations, including those performed either by machines or by people*] Systemflussdiagramm, Systemflussschema, Systemflussbild *n*, Systemflussplan *m*

system hum Netzbrummen *n*

system integration Systemintegration, Systemvernetzung *f*

system integrator Systemintegrierer, Systemintegrator *m*, integrierte Systemglied, System-I-Element *n*

system interface Systemschnittstelle *f*

system isolation System(ab)trennung *f*

system level Systemebene *f*

system level controller Systemebene(n)regler *m*

system maintenance Systemunterhaltung *f*

system management Systemverwaltung, Systemführung *f* [*Überwachung des Leistungsstatus des Systems als Ganzes, gefolgt von einer Analyse und – falls erforderlich – von Korrekturmaßnahmen*]

system management screen Systemverwaltungs(bild)schirm, Systemführungs(bild)schirm *m*

system maximum continuous operating voltage höchste Netzbetriebsspannung *f*

system of rules, rules system Regelwerk, Regelsystem *n*

system of units, units system Einheitensystem, Einheitenwerk *n*

S

system ((open) loop) control Systemsteuerung *f*

system operation Systembetrieb *m*

system operator Systembetreiber *m*

system optimization → optimization

system planner Systemplaner *m*

system pressure Systemdruck *m*

system printer, system writer Systemdrucker *m*

system regulation → system ((closed) loop) control

system report Logbuch *n*

system report entry Logbucheintrag(ung) *m*, *(f)*

system response time Systemreaktionszeit *f* [*Zeitspanne zwischen dem Zeitpunkt, zu dem ein Auftrag an ein System als erteilt gilt und dem Zeitpunkt, zu dem die Entscheidung über seine Annahme als mitgeteilt gilt.*]

system selection System(aus)wahl *f*

system shape → system configuration

system shaping, system forming Systemgestalten *n*, Systemgestaltung, Systemformen, Systemformung *f*

system time constant Systemzeitkonstante *f*

system type → system design

system unit Systemeinheit *f*

system variable Systemgröße *f*

system version → system design

system voltage, transformer voltage [*Voltage from the power company*] Transformatorspannung, Trafospannung, Netzspannung *f*

system writer → system printer

T

T → tee (branch)

tabbed tabelliert, tabuliert

table Tabelle *f*

table of (statutory) fees for architects and engineers → (statutory) fee schedule for architects and engineers

table-service restaurant [*Patrons are served only at tables*] Tischrestaurant *n*

tablespace logischer Teil *m*, *n* einer Datenbank [*Speicherung von Tabellen oder Indizes*]

tachometer generator, tachogenerator, pilot generator, tachometric generator Drehzahlgenerator *m*

tachometric relay, tachorelay Drehzahlrelais *n*

tachoswitch, tachometric switch Drehzahlwächter *m*

tack → clasp nail

TACS → talker active state

tactile touch Druckpunkt *m*

tag Anhängezettel *m*

tag Feldkennung *f*

tag, identification tag Etikett *n* [*Zur maschinellen Interpretation von Daten*]

tag Marke *f*, Bezeichner *m* [*von Datenobjekten*]

tag Ringzunge *f* [*Elektrozähler*]

tag → (soldering) tag

TAG → talker address group

to tag etikettieren

to tag markieren [*Mit einem Vermerk versehen*]

tag applicator, tag device Etikettiergerät *n*

tag block → (soldering) tag block

tagging Etikettieren *n*, Etikettierung *f* [*Zur Interpretation von Daten*]

tagging → (time) tagging

tag (pin) Senkelstift *m*

tail door spring, rod door spring (Tür)Zuschlagfeder *f*

tailored → customized

tailor-made zugeschnitten [*Problemlösung*]

to take a reading, to read off ablesen

to take off entnehmen

take-off Entnahme *f*

to take over übernehmen

take-over Übernahme *f*

talk Gespräch *n*

talker Sprecher *m* [*Funktionselement zum Informationsaustausch. DIN IEC 625*]

talker active state, TACS aktive Sprecherzustand *m* [*programmierbare Messgerät. DIN IEC 625*]

talker address group, TAG Sprecheradressengruppe *f*

talker idle state, TIDS Sprecherruhezustand *m*, Sprecherruhebeschaffenheit *f*

talker/listener Sender/Empfänger *m* [*Er kann wahlweise Daten senden oder empfangen, z. B. Magnetspeicher*]

talking clock sprechende Uhr, Sprechuhr *f*

talk recording Gesprächsaufzeichnung *f*

tam, television-audience measurement Feststellung *f* der Anzahl von Fernsehempfängern und empfangenen Programmen

tamper → misuse

tandem compensation → series compensation

tandem compressor → double compressor

tandem elevator, double deck elevator (US); tandem lift, double deck lift (Brit.) [*The bottom cab serves odd-numbered floors and the top cab stops at even ones. Inside the entrance, the visitors are directed to a lower level for elevators to odd-numbered floors and to an upper level for those to even ones*] Doppelaufzug, Tandemaufzug *m*

tank → (liquid) tank

tank crown [*fuel oil tank*] Behälterdecke, Tankdecke *f*

tank filling manifold, tank filling central distributor Tankfüll(ungs)verteiler(rohr), Behälterfüll(ungs)-verteiler(rohr) *m, (n)*

tankless heater [*A coil in a hot water heating boiler for heating domestic hot water*] behälterlose Heizer *m*

tank(-type) soap dispenser Behälterseifenspender *m*

tantalum electrolytic capacitor, electrolytic tantalum capacitor Tantal-Elektrolytkondensator *m*

to tap anbohren

to tap schneiden [*Gewinde*]

to tap, to draw off (ab)zapfen, anzapfen

tap Basisband *n*

tap Breitband *n* [*Ein passives Gerät, das einen Teil der Signalleistung vom Kabelzweig abnimmt und diesen an die Ausgangsleitung weitergibt, um den Eingang/Ausgang zum Netz zu bilden.*]

tap → pickoff

tap → (water) tap

tap and die Gewindebohrer und -schneider *m*

tap comparison Stellungsvergleich *m* [*Trafostufenschalter*]

tape → (data) (magnetic) tape

tape → (paper) tape

tape barbed wire → barbed wire tape

tape drive → (magnetic) tape drive

tape operating system, TOS Bandbetriebssystem *n*

tape punch → (paper) tape punch

tape reader → (paper) tape reader

tape recorder (Magnet)Tonbandgerät *n*

taper(ed) → conical

taper(ed) pendulum → conical pendulum

taper(ed) pipe, diminshing pipe, reducer, reducing pipe [*A pipe of diminshing diameter*] Übergangsstück, Reduzierstück, Reduktionsstück, Übergangsrohr, Reduzierrohr, Reduktionsrohr *n*

taper(ed) (pipe) fitting thread → conical (pipe) fitting thread

taper(ed) plug → plug

tape routine end → end-of-tape routine

tape streamer Magnetband *n*

tape transmitter → (perforated) tape transmitter

TAPI → telephony application programming interface

tap (line) Anschluss(leitung) *m, (f)*

tapped resistor (An)Zapfwiderstand, Abzapfwiderstand *m*

tapping Anbohren *n*

tapping, draw(ing) off (An)Zapfung, Abzapfung *f*, (An)Zapfen, Abzapfen *n*

tap(ping) Entnahme(stelle) *f* [*Messblende; Drosselgerät*]

tap(ping) circuit → regulating circuit

tap(ping) clip Anbohrstelle *f*

tap(ping) hole, draw(ing) (off) hole (An)Zapfloch, Abzapfloch *n*

tap(ping) line, draw(ing) (off) line (An)Zapfleitung, Abzapfleitung *f*

tap(ping) output, draw(ing) off output, tap(ping) performance, draw(ing) off performance (An)Zapfleistung, Abzapfleistung *f*

tap(ping) peak, draw(ing) off peak (An)Zapfspitze, Abzapfspitze *f*

tap(ping) pipe, draw(ing) off pipe (An)Zapfrohr, Abzapfrohr *n*

tap(ping) point, draw(ing) off point (An)Zapfstelle, Abzapfstelle *f*

tap(ping) spindle, thread spindle Gewindespindel *f*

tap(ping) temperature, draw(ing) off temperature (An)Zapftemperatur, Abzapftemperatur *f*

tap(ping) valve, draw(ing) off valve (An)Zapfventil, Abzapfventil *n* [*Eine Entnahmevorrichtung für Flüssigkeiten aus Zulaufleitungen*]

tap(ping) valve, draw(ing)-off valve, blow-off valve Ablassventil *n* [*Für Schlamm bei der Kesselreinigung*]

tap selector Stufenwähler *m* [*Trafo*]

tap water → mains water

target → destination

target address → destination address

target check(ing) Zielkontrolle *f*

target directory Zielverzeichnis *n*

target language Zielsprache *f*

target network → destination network

target room temperature setpoint Ziel-Raumtemperatur-Sollwert, Raumtemperatur-Ziel-Sollwert *m*

target setpoint, final setpoint, end setpoint Zielsollwert, Endsollwert, Zielvorgabewert, Endvorgabewert *m*

tariff counter Gebührenzähler, Tarifzähler *m*

tariff dependent switching tarifabhängige Schalten, gebührenabhängige Schalten *n*

tariff-rate changeover impulse receiver [*electric storage heating*] Rundsteuer(ungs)empfänger *m*

tar rope Teerstrick *m*, Teerseil *n*

tar yield Teerausbeute, Teerergiebigkeit *f*

task Aufgabe *f*

task Programmaufgabe *f* [*Eine in sich geschlossene Aufgabe, repräsentiert durch einen Programmteil, ein vollständiges Programm oder eine Programmaufgabe. DIN 19 243 Teil 4*]

task light, task lamp Arbeitsplatzlampe *f*

task loading Netzknotenladen *n* von Anwendungsprogrammen

task-related, task-orien(ta)ted aufgabenorientiert, aufgabenbezogen

TAV → CA

TAYLOR series TAYLOR-Reihe *f*

TBF → time between failures

t-conorm, triangular conorm triangulare Konorm, Dreieck-Konorm, t-Conorm

TCP/IP, Transmission Control Protocol/Internet Protocol Internetprotokoll *n* [*Das Protokoll, über welches sich die Rechner im Internet verständigen. Dieses Protokoll ist plattformunabhängig. Auch örtliche Netzwerke können mit ihm arbeiten*]

Td → derivative time

TDM, time division multiplexing Zeitbündelung *f* [*Methode zur effizienten Nutzung einer gegebenen Kanalkapazität, wobei jeder Knoten einem kurzen Zeittakt unterliegt, während dem eine Nachricht oder ein Datenpaket übertragen werden kann. Die Konten erhalten einen zeitlich begrenzten Anschluss, haben dabei allerdings die Berechtigung zum Senden von Nachrichten. Nachrichtenwünsche mehrerer Knoten werden in einer Warteschlange zusammengefasst und in der entsprechenden Reihenfolge übertragen.*]

TDR → (time-)delay relay

TE, totally enclosed ganzgekapselt, vollgekapselt

teach in automatische Übernahme *f*, Einlernen *n* [*Die gewünschten Positionen werden mit der Maschine angefahren und mit Tastendruck automatisch in den Programmspeicher übernommen*]

teamwork Gruppenarbeit *f*

tear gas Tränengas *n*

tear-off cleat mechanism Reiß-Halte-Mechanismus *m* [*Papierhandtuchspender*]

tearproof (zer)reißfest

tea service trolley Tee(servier)wagen *m*

technical aid technische Hilfsmittel *n*

technical building system technische Gebäudesystem, TGS *n*, technische Ausbau *m*

technical facility management technische Gebäudebewirtschaftung *f*

technical services installations, technical services plants betriebstechnische Anlagen, BTA, BTI, betriebstechnische Installationen *f*

technical supply condition technische Lieferbedingung *f*

technical support technische Betreuung *f* [*von Kunden*]

technical support manager technische Betreuer, technische Leiter *m* [*von Kunden*]

technique Technik, technische Ausführung *f*

technological language, TECLA natürliche Sprache, Umgangssprache *f*

technology, engineering science [*Not to be confused with "technique" and "engineering"*] Technologie, Ingenieurwissenschaft *f* [*Lehre und Erforschung technischer Verfahren zur Kenntniserweiterung*]

TECLA → technological language

tee (branch), T(ee) (branch) [*A branch fitting used to connect to a main pipe a branch pipe which is at an angle of 90° to the main pipe*] T(-Stück) *n*

t(ee) coupling piece, T(ee) coupling piece T-Verschraubung *f*

t(ee)-square, T(ee)-square Reißschiene *f*

TEFC → totally enclosed fan-cooled

tele ... [*Prefix meaning "from a distance"*] Fern ...

tele ... → remote ...

telecold, remote cold Fernkälte *f*

telecommunication Fernübertragung, Fernmeldung *f*

telecommunication circuit Fernmeldeleitung *f*

telecommunication code Fernmeldekode *m*

telecommunication plant, telecommunication installation Fernmeldeanlage, Schwachstromanlage *f* [*Fernsprecher-, Telegraphen-, Signal-, Feuermelde-, Klingel- und Rundfunkanlage*]

telecommunications Fernmeldewesen *n*

telecommunication safety Fernmeldesicherheit *f*

T

telecommunication security Fernmeldesicherung *f*

telecommunication service Fernmeldedienst *m*

teleconference [*A conference between persons remote from one another but linked by a telecomunications system*] Fernkonferenz *f*

telecontrol Fernwirktechnik *f*

telegraph alphabet, telegraph code Telegraphenalphabet, Telegrafenalphabet *n*, Telegraphenkode, Telegraphencode, Telegrafenkode, Telegrafencode *m*

telegraph and communications central station Fernmeldezentrale, Fernmeldezentralstation *f*

telegraph cable Telegraphenkabel, Telegrafenkabel *n*

telegraph circuit [*In a telegraph system, the complete radio or wire circuit over which signal currents move between transmitting and receiving devices*] Telegraphenleitung, Telegrafenleitung *f*

telegraph repeater [*A repeater inserted at intervals in long telegraph lines to amplify weak code signals, with or without reshaping of pulses, and to retransmit them automatically over the next section of the line*] Telegraphenübertrager, Telegrafenüberträger *m*

telegraph signal distortion Telegraphiesignalverzerrung, Telegrafiesignalverzerrung *f*

telegraph transmitter Telegraphiesender, Telegrafiesender *m*

telegraphy [*Communication at a distance by means of code signals consisting of current pulses sent over wires or by radio*] Telegraphie, Telegrafie *f*

teleline Fernleitung *f*

telemetering [*The operation of instruments that provide data readout to a location remote from their sensing device(s)*] Fernmessen *n* [*Gegensatz: Telemetrie f*]

telemetry Telemetrie, Messdatenübertragung *f* [*Gegensatz: Fernmessen n*]

telemonitoring Fernüberwachung *f*

teleoperation Fernbedienung *f*

telephone arrestor Glimmsicherung *f*

telephone call recording system Telefonmitschnittsystem, Fernsprechermitschnittsystem *n*

telephone cell, telephone booth Fernsprecherzelle, Telefonzelle *f*

telephone closet Telefonschrank, Fernsprecherschrank *m*

telephone cord outlet → **telephone outlet (box)**

telephone directory Fernsprechbuch, Telefonbuch *n*

telephone hood Fernsprechhaube, Telefonhaube *f*

telephone intercom(munication) system Wechselsprechanlage *f*

telephone line Fernsprechleitung, Telefonleitung *f*

telphone manhole Fernsprecheinsteigschacht, Telefoneinsteigschacht *m*

telephone meter (Fernsprech)Gebührenzähler, Telefongebührenzähler *m*

telephone modem [*See remark under "modem"*] Fernsprechmodem, Telefonmodem *n*

telephone network Fernsprechnetz, Telephonnetz *n*

telephone outlet (box), telephone cord outlet Fernsprechanschlussdose, Telefonanschlussdose *f*

telephone pilot Fernsprechader, Telefonader *f*

telephone plug [*A standard plug having a $^3/_4$-inch-diameter (19-mm-diameter) shank, used with headphones, microphones, and other audio equipment; usually designed for use with either two or three conductors*] Fernsprechstecker, Telefonstecker, Fernsprechstöpsel, Telefonstöpsel *m*

telephone point [*e. g. in a wall or floor*] Fernsprechanschluss, Telefonanschluss *m*

telephone receiver Fernsprechhörer, (Telefon)Hörer *m*

telephone relay Fernsprechrelais, Telefonrelais *n*

telephone selector Telefonwählgerät, Fernsprechwählgerät *n*

telephone service Fernsprechdienst, Telefondienst *m*

telephone service Fernsprechanschluss, Telefonanschluss *m*

telephone set Telephon(apparat) *n,(m)*, Fernsprecher *m*

telephone subscriber Fernsprechteilnehmer, Telefonteilnehmer *m*

telephone switchboard Vermittlungsschrank *m*

telephone user part → **TUP**

telephony engineering Telefontechnik, Fernsprechtechnik *f*

(telephony) still picture (Fernsprech)Einzelbild, (Fernsprech)Festbild, Telefoneinzelbild, Telefonfestbild *n*

telephoto lens Teleobjektiv *n*

teleprinter, remote printer Ferndrucker *m*

teleprocessing, remote processing Fernverarbeitung *f*

telereading, remote reading Fernablesung *f*

telescopic bolt lock, bolt telescopic lock Teleskopriegelschloss, Riegelteleskopschloss *n*

telescopic ladder → **extending ladder**

telescopic sliding gate, telescoping sliding gate Teleskopschiebetor *n*

telescopic suction lance, telescoping suction lance Teleskopsauglanze *f*

telescoping sliding door, telescopic sliding door Teleskopschiebetür *f*

telescoping (utility) rod, telescopic utility rod [*A bathroom accessory*] Teleskopstange *f*

telesensor → **remote sensor**

to teleswitch → **to remote-switch**

teleswitch, remote switch Fernschalter *m*

telethermometer [*It transmits its data readout to a remote location*] Fernthermometer *n*

teletransmission of data → **data remote transmission**

to teletype (US); to teleprint (Brit.) fernschreiben

teletypewriter (US); TWX machine; teleprinter (Brit.) [*An electric typewriter that emits coded electric signals corresponding to manually typed characters and automatically types messages upon reception of coded signals from another machine*] Fernschreiber *m*

teletyping (US); teleprinting (Brit.) Fernschreiben *n*

television-audience measurement → **tam**

television broadcast Fernsehsendung *f*

television disturbance, television interference Fernsehstörung *f*

television monitoring Fernsehüberwachung *f*

television receiver Fernsehempfänger *m*

television screen Fernseh(bild)schirm *m*

teleworker Heimarbeiter am PC, Telearbeiter *m*

teleworking Heimarbeit am PC, Telearbeit *f*

teller assist unit → **CA**

telltale → **watchman time detector**

telltale (board) Anzeige(r)tafel *f*

telltale lamp Meldelampe, (Ausfall)Warnlampe *f*

temperature adap(ta)tion Temperaturanpassung *f*

temperature adjustment Temperaturverstellung, Temperatur(ein)stellung, Temperaturjustierung *f*

temperature booster Temperaturerhöher, Temperaturverstärker *m*

temperature calculation Temperaturberechnung *f*

temperature ((closed) loop) control → **((closed) loop) temperature control**

temperature-controlled switch → **heat switch**

temperature controller → **thermostat**

temperature detector, temperature detecting device (Maximal)Temperaturmelder, (Maximal)Temperaturdetektor *m*, (Maximal)Temperaturmeldegerät *n*

temperature difference, temperative gradiant [*Thermal "pressure", indoors to out*] Temperaturgefälle *n*, Temperaturunterschied *m*

temperature drop, temperature lowering [*As applied to water systems, the difference in temperature of water leaving and returning to the boiler*] Temperaturgefälle *n*, Temperaturrückgang, Temperaturunterschied, Temperaturabfall *m*, Temperaturabnahme, Temperatur(ab)senkung *f*

temperature first element → **temperature primary element**

temperature fluctuation, temperature variation [*The temperature varies around a mean value*] Temperaturschwankung *f*

temperature fuse → **thermofuse**

temperature galvanometer → **hoop drop relay**

temperature humidity index → **comfort index**

temperature immersion controller → **immersion thermostat**

temperature increment(al step) Temperaturinkrement *n*, Temperaturzunahmestufe *f* [*Betrag, um den eine Temperatur zunimmt*]

temperature layering → **temperature stratification**

temperature level Temperaturniveau *n*, Temperaturpegel, Temperaturstand *m*

temperature limit switch Temperaturgrenzwertschalter *m*

temperature lowering → **temperature drop**

temperature measuring device, temperature measuring appliance Temperaturmessgerät *n*, Temperaturmessvorrichtung *f*, Temperaturmesser *m*

temperature on the Kelvin scale → **K**

temperature ((open) loop) control [*See remark under "((open) loop) control"*] Temperatursteuerung *f*

temperature preselection Temperaturvorwahl *f*

temperature primary detector → **temperature sensor**

temperature primary element, temperature first element Temperatur-Messwertgeber *m*

temperature profile [*Graph representing the distribution of temperature in a plane section of a body or a space, or over a period of time*] Temperaturprofil *n*

temperature range Temperaturbereich *m* [*selten: n*]

temperature recorder → **thermograph**

temperature regulation → **temperature ((closed) loop) control**

temperature regulator Temperaturregler *m*

temperature-related temperaturbezogen

temperature rise [*The difference in temperature of the return air and the cooled air delivered*] Temperaturanstieg *m*, Temperaturzunahme *f*

temperature selector Temperaturwähler *m*

temperature selector handle Temperaturwähl(er)-griff, Temperaturwahlgriff *m*

temperature-sensitive paper, heat-sensitive paper Thermopapier *n*

temperature sensor, temperature sensing element, temperature primary detector [*See remark under "sensor"*] Temperatursensor, Temperaturfühler *m*

temperature setpoint setter Temperatursollwertsteller *m*

temperature setting Temperatureinstellung *f*

temperature shock Temperaturstoß *m*

temperature spread [*radiator*] Temperaturspreizung *f*

temperature stratification, temperature layering Temperaturschichtung *f*

temperature stress Temperaturbeanspruchung *f*

temperature submersible controller → immersion thermostat

temperature switch Temperaturschalter *m*

temperature transmitter Temperaturübertrager *m*

temperature variation → temperature fluctuation

tempered temperiert, temperaturgeregelt

template Vorlage *f* [*Dokument*]

template, picture, mask Schablone, Maske *f* [*Eine vom Programmierer für das jeweilige Programm definierte oder vom Hersteller festverdrahtete Systemfolge, die je nach Aufgabenstellung unterschiedliches Aussehen haben kann*]

temporary error signal zeitweilige Regelunterschied, vorübergehende Regelunterschied *m*, zeitweilige Regeldifferenz, vorübergehende Regeldifferenz *f* [*DIN 19 226, Teil 5*]

temporary overvoltage zeitweilige Überspannung, vorübergehende Überspannung *f*

temporary personnel and material hoist [*It is a special kind of elevator whose drive machine is located on the car. Multiple hydraulic motors drive pinion gears engaging a full-height rack over the distance travel(l)ed. The hoist is installed to assist the construction of buildings*] Zahnstangenaufzug *m*

temporary threshold shift, TTS zeitweilige Hörschwellenverschiebung, vorübergehende Hörschwellenverschiebung *f* [*durch starken Lärm*]

tenant Mieter *m*

tenant('s) management Mieterverwaltung *f*

tenant('s) meter Mieterzähler *m*

tenon hole... → mortice... (Brit.)

tension band Spannband *n* [*Drehspulinstrument*]

TENV, totally enclosed nonvent(ilat)ed ganzgekapselt und unbelüftet, vollgekapselt und unbelüftet

term → label

terminal [*An electrically conductive element, attached to the end of a conductor or piece of equipment for connection to an external conductor*] Klemme, Anschlussklemme *f*

terminal [*Conductor point at which an electrical circuit or component may be connected to another electrical component or circuit*] Anschluss *m*

terminal [*Point in a system or communication net(work) at which data can either enter or leave*] Endplatz *m*, Terminal *n*, Endstelle, Endstation *f*

terminal, data terminal Netzendgerät, (Daten)Endgerät *n* [*Für den Dialog mit der Zentraleinheit*]

terminal adapter [*rarely: adaptor*] Anpassungseinheit *f*

terminal allocation, terminal assignment Klemmenzuweisung, Klemmenzuordnung, Klemmenzuteilung *f*

terminal assembly Rahmenverteiler, Klemmenstreifen *m*

terminal block → (modular) terminal block

terminal cabinet → connection cabinet

terminal clamp → clamp(ing) terminal

terminal cover Klemmendeckel *m*

terminal diagram → terminal (switching) diagram

terminal down Terminal *n* außer Betrieb, Terminalausfall *m*, Terminalpanne *f*

terminal element [*Means by which the energy from a system is finally delivered, e.g. registers, diffusors, radiators, lighting fixtures, faucets and the like*] Endglied, Schlussglied *n*

terminal (elevator) landing, end (elevator) landing (US); terminal (lift) landing, end (lift) landing (Brit.); terminal stop (Aufzug)Endhalt(estelle) *m*, (*f*)

terminal (equipment), terminals Endgerät(e) *n*, (*f*)

terminal equipment controller → application-specific controller

terminal exists Terminal vorhanden

terminal expander Terminalerweiterer *m*

terminal gate valve, terminal pipeline valve Endschieber *m*

terminal insert Klemmeneinsatz *m*

terminal interface unit, following unit Endschnittstelle, Endanpassungsschaltung *f*

terminal landing → terminal (elevator) landing

T

terminal leg, terminal stub [*Piece of cable that comes with a cable terminal for splicing into the main cable*] Klemmenkabel *n*

terminal loop configuration Terminalschleifenaufbau *m*

terminal marking Klemmenbezeichnung, Anschlussklemmenbezeichnung *f* [*DIN EN 50 011*]

terminal pair [*An associated pair of accessible terminals, such as the input or output terminals of a device or net(work)*] (Anschluss)Klemmenpaar *n*

terminal pipeline valve → **terminal gate valve**

terminal pressure Enddruck *m*

terminal repeater Endverstärker *m*

terminal room [*In telephone practice, a room associated with a central office, private branch exchange, or private exchange, which contains distributing frames, relays, and similar apparatus, except that mounted in the switchboard section*] Wählerraum *m*

terminal sequence Klemmenfolge *f*

terminal server Netzwerkanschlussgerät *n* [*Es schließt mehrere (Daten)Endgeräte an ein Netzwerk an*]

terminal shroud Klemmenabdeckung, Klemmenbedeckung *f*

terminal station [*Receiving equipment and associated multiplex equipment used at the ends of a radio-relay system*] Endstelle *f*

terminal stop → **terminal (elevator) landing**

terminal strip Klemmenleiste *f*

terminal stub → **terminal leg**

terminal (switching) diagram Anschluss(schalt)plan *m* [*DIN 40 719 Blatt 1*]

terminal-to-earth fault Klemmenerdschluss *m*

terminal trunk exchange Fernsprechendamt, End(fern)amt, EF *n*

terminal unit Luftnachbehandlungsgerät *n* [*Aggregat, das in einem lokalen Bereich eingesetzt wird, um eine HLK-Aufgabe (HLK = Heizung, Lüftung, Klima) zu erfüllen (z. B. Gebläsekonvektor).*]

terminal up Terminal *n* in Betrieb

terminal velocity [*Velocity of an airstream at the end of the motion*] Endgeschwindigkeit *f*

terminal window Terminalfenster *n*

terminating resistor → **matching resistor**

termination date Ablaufzeitpunkt *m*, Ablaufdatum *n*

terms of trade, trade terms reale Austauschrelation *f* [*Verhältnis zwischen einem für die Ausfuhr- und einem für die Einfuhrgüter einer Volkswirtschaft repräsentativen Preisindex*]

terrain, ground(s) Gelände, Terrain *n*

tertiary air Drittluft, Tertiärluft *f*

tertiary flow Drittströmung, Tertiärströmung *f*

terylene ladder web Teryleneleiterkordel *f* [*Jalousie*]

terylene tape Terylene(aufzug)band *n* [*Raffjalousie*]

test area manager übergeordnete Prüf(ungs)system *n* [*Zur Steuerung und Überwachung aller z. B. in einem Prüffeld zusammengefassten Prüfautomaten*]

testbed Prüf(ungs)umgebung *f* [*Mit Werkzeugen (tools) speziell für kommunikative Einrichtungen*]

test circuit Prüf(ungs)schaltung *f*

test current Prüf(ungs)strom *m*

test duct Prüf(ungs)kanal *m*

tester, test(ing) device, test(ing) appliance Prüf(ungs)gerät, Prüf(ungs)organ *n*, Prüf(ungs)vorrichtung *f*

test(ing) agency Prüf(ungs)stelle *f*

test(ing) authority Prüf(ungs)behörde *f*

test(ing) campaign, test(ing) series Prüf(ungs)serie, Prüf(ungs)reihe *f*

test(ing) certificate Prüf(ungs)zeugnis *n*

test(ing) characteristic Prüf(ungs)zeichen *n*

test(ing) coil Prüf(ungs)spule *f*

test(ing) device → **tester**

test(ing) flow Prüf(ungs)ablauf *m*, Prüf(ungs)folge *f*

test(ing) gas Prüf(ungs)gas *n*

test(ing) language Prüf(ungs)sprache *f*

test(ing) level Prüf(ungs)niveau *n*, Prüf(ungs)schärfe *f*

test(ing) oscillator → **signal generator**

test(ing) pattern Prüf(ungs)muster *n*

test(ing) plug Prüf(ungs)stecker *m*

test(ing) pressure Prüf(ungs)druck *m*

test(ing) principle Prüf(ungs)grundsatz *m*

test(ing) probe Klemmprüf(ungs)spitze *f*

test(ing) release Prüf(ungs)auslösung *f*

test(ing) screw Prüf(ungs)schraube *f*

test(ing) series → **test(ing) campaign**

test(ing) set Prüf(ungs)garnitur *f*

test(ing) signal Prüf(ungs)signal *n* [*Ein Signal, bei dessen Benutzung das Eingangssignal eines Gliedes oder Systems geprüft wird bzw. das Glied oder System geprüft werden*]

test(ing) transmitter Prüf(ungs)übertrager *m*

test(ing) unit Prüf(ungs)einheit *f*

test(ing) water meter Prüf(ungs)wassermesser *m*

tetrad Tetrade, vier Bitstellen *f*

tetrode, four-electrode electron tube [*It contains an anode, a cathode, a control electrode, and an electrode that is ordinarily a grid*] Tetrode, Vierpolröhre *f*

tetrode thyristor, thyristor tetrode Thyristortetrode, PUT *f* [*Ein bistabiler, steuerbarer Halbleiterschalter, der über zwei Steuerelektroden wahlweise mit Strömen beider Polaritäten ein- und ausgeschaltet wird*]

TEV → thermostatic expansion valve

TEWC, totally enclosed and watercooled vollgekapselt und wassergekühlt, ganzgekapselt und wassergekühlt

text → language

text body Textkörper *m*

text character Textzeichen *n*

text data file Textdatei *f*

text end → end of text

text end character → end-of-text character

textile braid Textilgeflecht *n* [*Gummischlauchleitung*]

textile floor Textil(fuß)boden *m*

text image format, TIF Textabbildungsformat, Textbildformat *n*

text interchange format, TIF Textübermittelungsformat *n*

text line Textzeile *f*

text module, text block Textmodul, Textbaustein *m*

text network Textnetz(werk) *n*

text processing Textverarbeitung *f*

text programming Textprogrammierung *f*

text recognition Texterkennung *f*

text start → start of text

text station Textstation *f*

texture mapping Oberflächenbelegung, Textur(en)-belegung *f* [*Belegung einer Oberfläche mit einer graphischen Bildinformation*]

T-flipflop (circuit), T bistable multivibrator [*See remark under "bistable multivibrator"*] bistabile T-Multivibrator *m*, Auslöse-Triggerschaltung, Trigger-Auslöseschaltung, bistabile T-Kippstufe *f*

TFT, thin-film-transistor, insulated-gate field-effect transistor Dünnschichttransistor, Isolierschicht-Transistor *m*

TG width → timing generator width

thawing (off) Auftauen, Abtauen *n*

thawing (off) (temperature) regulator Abtau(temperatur)regler, Auftau(temperatur)regler *m*

theft Diebstahl *m*

theft alarm Diebstahlalarm *m*

theft attempt Diebstahlversuch *m*

theft prevention Diebstahlverhütung *f*

theftproof, thiefproof diebstahlsicher, diebstahlgesichert

theorem Satz, Lehrsatz *m*

theoretical compressor work [*Enthalpy difference along an isentrope*] theoretische Verdichterarbeit, theoretische Kompressorarbeit *f*

theory of fuzzy sets Theorie der unscharfen Mengen, unscharfe Mengentheorie *f*

therapeutic (bathing) pool Therapie(bade)becken *n*

thermal comfort thermische Behaglichkeit *f*, thermische Wohlbefinden *n*

thermal conductivity, heat conductivity Wärmeleitfähigkeit *f*, Wärmeleitvermögen *n*

thermal energy → heat

thermal foil wärmeisolierende Folie *f*

thermal internal resistance → internal thermal resistance

thermal nominal output → nominal heat output

thermal storage wall, heat storage wall Wärmespeicherwand *f*

thermically controlled [*closed loop control*] thermisch geregelt

thermically controlled [*open loop control*] thermisch gesteuert

thermic compressor thermische Kompressor, thermische Verdichter *m*

thermic drive thermische (An)Trieb, Thermo(an)-trieb *m*

thermistor Thermistor *m* [*Ein Widerstand, dessen Widerstandswert sich in Abhängigkeit der Umgebungstemperatur nicht linear ändert.*]

thermoacoustics Geräuschemission *f* von Wärmeerzeugern

thermochemical thermochemisch

thermochemical energy storage thermochemische Energiespeicherung *f*

thermocircuit Thermostromkreis *m*

thermo-controlled [*close control*] thermisch geregelt

thermo-controlled [*close control*] thermisch gesteuert

thermocouple Thermopaar, Thermoelement *n* [*Zwei an einem Ende miteinander verschweißte Drähte*]

thermodynamic cycle [*e.g. the Carnot cycle*] thermodynamische Prozess *m*

thermodynamic cryogenic production, thermodynamic production of low temperature thermodynamische Kälteerzeugung *f*

thermodynamic open system → open thermodynamic system

thermodynamic production of low temperature → thermodynamic cryogenic production

thermodynamic property thermodynamische Eigenschaft *f*

thermodynamics [*Science of the relation of heat to other forms of energy*] Thermodynamik *f*

thermodynamic steam trap [*See remark under "steam trap"*] thermodynamische Kondens(at)topf *m*

thermodynamic system thermodynamische System *n* [*Ein wasserdichtes thermodynamisches System wird als ein geschlossenes, ein durch einen Kontrollraum begrenztes als offenes definiert*]

thermoelectric battery, thermoelectric refrigerating module thermoelektrische Batterie, Peltierbatterie *f*

thermoelectric cooling thermoelektrische Kühlung *f*, thermoelektrische Kühlen *n*

thermoelectric heat pump [*It is based on the Peltier effect*] thermoelektrische Wärmepumpe *f*

thermoelectric refrigeration thermoelektrische Kälteerzeugung *f*

thermoelectric thermometer [*It measures temperatures by inserting one thermocouple into a constant-temperature bath and employing another as a measuring junction*] thermoelektrische Thermometer *n*

thermofuse, temperature fuse thermische Sicherung, Thermosicherung *f*

thermograph, recording thermometer, temperature recorder [*It senses, measures and records the temperature of the atmosphere*] Thermograph, schreibende Temperaturmesser, Temperaturschreiber, Thermograf *m*

thermograph [*In optics. A far-infrared image-forming device that provides a thermal photography by scanning a far-infrared image of an object or scene*] Thermograph, Thermograf *m*

thermography [*A method of measuring surface temperature by using luminescent materials; the main types are contact thermography and projection thermography*] Thermographie, Thermografie *f*

thermomagnetic heat pump thermomagnetische Wärmepumpe *f*

thermomechanical thermomechanisch, wärmemechanisch

thermophysical wärmephysikalisch, thermophysikalisch

thermoplotter Thermostrichzeichner *m*

thermostat, temperature controller [*Automatic control device responsive to temperature used to maintain constant (static) temperature*] Thermostat, Temperaturregler *m*

thermostat heater valve Thermostatheiz(ungs)körperventil *n*

thermostatically controlled thermostatgesteuert

thermostat(ically controlled) valve [*Valve in which the position of the plunger in relation to the seat is determind by the deviation of temperature from the set value*] Thermostatventil *n*

thermostatic ((closed) loop) control Thermostatreg(e)lung *f*

thermostatic (convector) radiator valve thermostatische Radiatorventil, thermostatische (Glieder-) Heiz(ungs)körperventil; thermostatische Heiz(ungs)-körperventil *n* [*Fehlbenennung*]

thermostatic expansion valve, TEV thermostatische Expansionsventil *n* [*Es besteht aus einem thermisch gesteuerten und einem druckgesteuerten Teil*]

thermostatic fan speed open loop control thermostatische Ventilatordrehzahlsteuerung *f*

thermostatic mixer, thermostatic blender thermostatische Mischbatterie, Thermostat-Mischbatterie, Temperaturfühler-Mischbatterie *f*

thermostatic radiator regulator thermostatische Heiz(ungs)körperregler *m*

thermostatic (refrigerant) expansion valve [*Controlling device for regulating flow of refrigerant into a cooling unit, actuated by the changes in evaporator pressure and superheat of the refrigerant leaving the cooling unit. The basic response is to the superheat*] thermostatische (Kältemittel)Expansionsventil *n*

thermostat lever, thermostatic lever Thermostathebel *m*

thermostat set(ting), thermostatic set(ting) Thermostateinstellung *f*

thermostress Thermospannung, Thermobeanspruchung *f*

thermoswitch → heat switch

thermosyphon [*Water cooling system. Water circulates through the jackets due to difference in density between the cold and warm outlet water*] Thermosiphon *m*

thermo transfer plotter Farb(en)zeichenmaschine *f* [*Die Farbe wird durch Erwärmen einer Folie auf Papier übertragen*]

THI → comfort index

thick-film (integrated) circuit Dickschichtschaltung, Dickfilmschaltung *f* [*Eine Weiterentwicklung und Verkleinerung der Leiterplatte*]

T

thick-film resistor Dickschichtwiderstand, Dickfilmwiderstand *m*

thick-walled dickwandig; starkwandig [*Fehlbenennung*] [*Rohr*]

thiefproof diebstahlsicher, diebessicher

thief-resistant lock, security lock [*British Standard 3621*] diebstahlsichere Schloss, Sicherungsschloss, diebessichere Schloss *n*

thin-film (integrated) circuit Dünnschichtschaltung, Dünnfilmschaltung *f*

thin-film memory Dünnschichtspeicher, Dünnfilmspeicher *m*

thin-film transistor → TFT

(thin) semiconductor slice, (thin) half-conductor slice Halbleiterscheibchen *n*, Halbleiterscheibe *f*

third octave band filter Terzfilter *m, n*

third party zone Fremdzone, Drittenzone *f*

Thomson effect → (Joule-)Thomson effect

thorn lock → pin lock

thread(ed) connection Gewindeanschluss *m*, Gewindeverbindung *f*

thread(ed) cover Gewindedeckel *m*

thread(ed) glass globe Gewindeglas *n* [*Leuchte*]

thread(ed) list Verkettungsstrukturliste *f* [*Ihr sind Verkettungsstrukturen (threads) hinzugefügt, um Durchgänge in besonderen Reihenfolgen zu ermöglichen*]

thread(ed) nipple Gewindenippel *m*

thread(ed) pipe Gewinderohr *n*

thread(ed) rod Gewindestange *f*

thread(ed) tube Gewinderöhre *f*

threading starke Riefenbildung *f* [*Kommutator*]

threadless gewindelos, ohne Gewinde

thread seal(ing) Gewinde(ab)dichtung *f*

thread spindle → tap(ping) spindle

threat analysis Bedrohungsanalyse *f* [*Bei dieser Analyse werden Bedrohungen festgestellt und untersucht: Schutzziele, mutmaßliche Täterbild, örtliche Gegebenheiten, Verfügbarkeit der hilfeleistenden Kräfte und vertretbares Restrisiko*]

three-address system, triple-address system [*See remark under "address"*] Drei-Adressen-System *n*

three-bank air heater, triple-bank air heater dreireihige Lufterhitzer, Lufterhitzer mit drei Rohrreihen *m*

three bolts lock → lock with three bolts

three-burner range, triple-burner range Dreiplattenherd *m*

three chamber septic tank, triple chamber septic tank Dreikammerfaulgrube, Dreikammerkleinkläranlage *f*, Dreikammerklärbehälter *m* [*In Deutschland nur für Schmutzwasser zugelassen. DIN 4261*]

three compartment sink (bowl), triple compartment sink (bowl) [*"Basin" may synonymously be used for "bowl"*] Dreibeckenspüle *f*

three-component ((closed) loop) control, triple-component ((closed) loop) control Dreikomponentenreg(e)lung *f* [*Aufschaltung eines Signals proportional des Unterschiedes von Störgrößen und Stellgrößen auf den Regeleingang. Diese Reg(e)lung kann auch als Verhältnisreg(e)lung (z/y) mit aufgeschalteter Regelgröße x angesehen werden*]

three-dimensionality, triple-dimensionality Dreidimensionalität *f*

three-hole mixer (valve), three-hole blender, three-hole mix(ing) valve, three-hole blending valve, triple-hole mixer (valve), triple-hole mix(ing) valve, triple-hole blender, triple-hole blending valve Drei-Loch-Mischbatterie, Misch-Drei-loch-Batterie *f*

three-hole vanity fitting, triple-hole vanity fitting Drei-Loch-Waschtischarmatur *f*

three-hole (wash)basin mixer, triple-hole (wash-) basin mixer Dreiloch-(Wasch)Becken-Mischer *m*

three-latch lock, triple-latch lock Dreifallenschloss *n*

three-layer diode → trigger diode

three-phase bridge, triple-phase bridge, six-pulse bridge Drehstrombrückenschalter, DB-Schalter *m*

three-phase bridge connection, triple-phase bridge connection, six-pulse bridge connection Drehstrombrückenschaltung, DB-Schaltung *f*

three-phase current, triple-phase current Drehstrom *m* [*Drei Wechselströme, deren Verlauf jeweils um ein Drittel der Periode zeitlich verschoben ist. Amplitude der Wechselströme nahezu konstant und gleich hoch (symmetrisch)*]

three-phase current outside-rotor motor, triple-phase current outside-rotor motor Drehstromaußenläufermotor *m*

three-phase rectifier connection, triple-phase rectifier connection Drehstromgleichrichterschaltung *f*

three-phase star connection, three-phase Y-connection, triple-phase star connection, triple-phase Y-connection Drehstromsternschaltung, DS *f*

three-phase transformer bank, triple-phase transformer bank Dreiphasentransformatorgruppe, Dreiphasentransformatorbank *f*, Dreiphasentransformatorsatz *m*

three-piece, triple-piece dreiteilig

three-piece bidet mixer, triple-piece bidet mixer dreiteilige Bidetmischer, dreiteilige Sitzwaschbeckenmischer *m*

three-piece set, triple-piece set Dreiersatz *m*

three-piece (wash)basin mixer, three-piece lavatory mixer, triple-piece (wash)basin mixer, triple-piece lavatory mixer dreiteilige Waschbeckenmischer *m*

three-pin pliers, triple-pin pliers Dreidornzange *f*

three-pin plug, triple-pin plug [*B.S. 1778*] Dreifachstecker, dreipolige Stecker, Dreipolstecker *m*

three-pipe air conditioning system, triple-pipe air conditioning system [*See remark under "air conditioning installation". "Plant" and "installation" may synonymously be used for "system"*] Dreirohrklimaanlage *f* [*"System" und "Installation" können synonym für „Anlage" verwendet werden*]

three-piston key, triple-piston key Dreistellungstaste *f*

three-point act(uat)or → switching act(uat)or

three-point regulator, triple-point regulator Dreipunktregler *m*

three-point suspension, triple-point suspension Dreipunktauflagerung *f*

three-pole fuse, triple-pole fuse Dreipolsicherung *f*

three-position ((closed) loop) control, triple-position ((closed) loop) control, three-position regulation, triple-position regulation [*"step..." may synonymously be used for "position..."*] Dreipunkt(schritt)reg(e)lung, Dreipunkt(schritt)regulierung *f*

three-position ((closed) loop) controller → three-step ((closed) loop) controller

three-position element, triple-position element Dreipunktglied *n*

three-pot sink, triple-pot sink Dreifachspüle *f*

three-row coil, triple-row coil dreireihige Schlange, Dreireihenschlange *f*

T

three-stage switch, triple-stage switch Dreistufenschalter *m*

three-state logic, three-condition logic, triple-state logic, triple-condition logic Dreizustandslogik, Dreibeschaffenheitslogik *f*

three-step (action) element, triple-step (action) element Dreipunktglied *n* [*E DIN 19 226 T 2*]

three-step ((closed) loop) control, three-position ((closed) loop) control, triple-step ((closed) loop) control, triple-position ((closed) loop) control Dreipunktreg(e)lung *f*

three-step ((closed) loop) controller, three-step regulator, three-position ((closed) loop) controller, three-position regulator, triple-step ((closed) loop) controller, triple-step regulator, triple-position ((closed) loop) controller, triple-position regulator Dreipunktregler *m* [*Unsteter Regler ohne Rückführung, der bei Überschreiten einer bestimmten Regelabweichung das Ausgangssignal sprungweise in drei Stufen ändert. DIN 19 225*]

three-tank dish washer, three-tank dishwashing machine, triple-tank dish washer, triple-tank dishwashing machine Dreitank-Geschirrwaschmaschine, Dreibehälter-Geschirrwaschmaschine *f*

three-tier terminal, triple-tier terminal Dreistockklemme *f*

three-tube air conditioning plant, triple-tube air conditioning plant Dreirohrklimaanlage *f*

three-unit, triple-unit dreilängig [*Schienenleuchte*]

three-way cock, triple-way cock Dreiwegehahn *m* [*DIN 30 600, Reg.-Nr. 00601*]

three-way mixer, triple-way mixer Dreiwegemischer *m*

three-way mixing valve, triple-way mixing valve Dreiwegemischventil *n*

three-way pressure reducing valve, triple-way pressure reducing valve Dreiwegedruckminderventil *n*

three-way switch, triple-way switch [*A three-terminal switch that connects c-a or c-b where c is common*] Dreiwegeschalter *m*

three-way tap, triple-way tap Dreiwegehahn *m*

three-way temperature controller, triple-way temperature controller Dreiwegetemperaturregler *m*

three-way throttling valve, triple-way throttling valve Dreiweg(e)-Regelventil, Dreiweg(e)-Drosselventil *n* [*DIN 30 600, Reg.-Nr. 06162*]

three-way valve, triple-way valve [*Valve having either a single inlet and two outlets (diverting) or two inlets and a single outlet (mixing), in which either one or the other is open. Service valve for dual-mounted safety relief valves*] Dreiwegeventil *n*

three-wire circuit, triple-wire circuit Dreileiterschaltung *f*

threshold comb [*Toothed portion of a threshold plate designed to mesh with a grooved treadway surface of a moving walk or moving ramp*] Kammplatte *f*, Schwellenkamm *m*

threshold (level), threshold value Schwelle(nwert) *f*, (*m*) [*Kleinste Signalstärke ab der ein Empfänger ein brauchbares Nutzsignal liefert*]

threshold plate [*Portion at the entrance or exit to the treadway of a moving walk or moving ramp consisting of one or more stationary or slightly mov(e)able plates*] Plattform *f* [*Fahrsteig; Fahrrampe*]

threshold voltage Schwellenspannung *f* [*Diode; Thyristor*]

threshold zone, threshold member Einsichtstrecke *f*

throat Horneingang *m* [*Beschallung*]

to throttle, to restrict [*The flow of fluid*] drosseln, einschränken, abdrosseln, beschränken

throttle bore, throttle boring Drosselbohrung *f* [*z. B. bei einem Nippel*]

throttle ((closed) loop) control Drosselreg(e)lung *f*

throttled flow → variable flow

throttling → modulating

throttling calorimeter, steam calorimeter, throttle calorimeter Dampfkalorimeter *n*, Dampfwärmemesser *m*

throttling cone, throttle cone Drossel(ungs)kegel *m*

throttling curve, throttle curve Kennlinie, V-H-Linie, Drossel(ungs)kurve *f*

throttling element, throttle ement Drossel(ungs)-organ *n*

throttling loss, throttle loss Drossel(ungs)verlust *m*

throttling nozzle, throttle nozzle Drossel(ungs)-düse *f*

throttling range [*The amount of change in the controlled variable required to cause the controller to move the controlled device from one extreme to the other. It can be adjusted to meet job requirements*] Drossel-(ungs)bereich *m* [*selten: n*]

throttling straight(-way) valve → straight(-way) throttling valve

throttling valve, throttle valve [*Valve used to (throttle) the flow of fluid*] Drossel(ungs)ventil *n*

through-burning stove Durchbrandofen *m*

through fault capacity Eigenkurzschlussfestigkeit *f*

throughflow Durchfluss *q m* [*Differentialquotient aus durchgeflossener Fluidmenge und Zeit*]

throughflow anti-siphoning pipe, throughflow puff pipe, throughflow A.S.P., throughflow back siphonage pipe [*See remark under "anti-siphonage pipe"*] Durchflussrohrbelüfter *m* [*Diese Belüfter werden in Geräten, die unmittelbar mit der Trinkwasserleitung und der Entwässerungsleitung verbunden sein müssen, z. B. Geschirrspül- und Waschmaschinen, sowie an Auslaufventilen mit Schlauchverschraubungen eingebaut*]

throughflow ((closed) loop) controller (Durchfluss)Regler *m* [*Ausrüstungsteil von Rohrleitungen zur Reg(e)lung des Durchflusses*]

throughflow cross-section Durchflussquerschnitt *m*

throughflow meter Durchflussmesser *m*, Durchflussmessgerät *n*

throughflow pattern Durchflussbild *n*

throughflow pickup, throughflow measured-value transmitter Durchflussmesswertgeber *m*

throughflow quantity counter Durchflussmengenzähler *m*

throughflow rate Durchflussrate *f* [*Durchflussmenge oder Durchflussgeschwindigkeit*]

throughflow resistance Durchflusswiderstand *m*

throughflow tank Durchflussbehälter *m*

throughput Durchsatz *m*

throughput class Durchsatzklasse *f*

throughput class assignment, throughput class allocation Zuordnung der Durchsatzklasse, Zuweisung der Durchsatzklasse, Zuteilung der Durchsatzklasse *f*

throughput class negotiation Absprache *f* der Durchsatzklasse [*Leistungsmerkmal*]

through-the-wall (air) conditioner [*Two-compartment air conditioner (split-system); its elements are mounted on each side of a wall and in line with each other*] (Außen)Wand(klima)gerät, Kasten(klima)-gerät für Wandeinbau, Wandeinbau(klima)gerät *n*

through-ventilation Durchlüftung *f*

through(way) shut-off cock Durchgangsabsperrhahn *m*

through(way) shut-off valve Durchgangsabsperrventil *n*

through(way) wiring Durch(gangs)verdrahtung, durchgehende Verdrahtung *f*

throwaway filter → disposable filter

throwaway seal → disposable seal

throw of jet → jet throw

throwover Kippen, Umschlagen *n*

throwover switch, changeover switch, transfer switch, double throw switch Umschalter *m*

thrust block Festpunkt *m* [*Rohrleitung*]

thruster → (centrifugal) thruster

thrust load Schublast *f*

thumb grip fastener (Daumen)Griffverschluss *m*

thumb grip handle (Daumen)Knebelgriff *m*

thumb latch, Norfolk latch; Canadian latch, Suffolk latch, Garden City latch (Brit.); lift latch (US) (Daumen)Drückerfalle *f*

thumb-lock (Daumen)Federschloss *n* [*Durch Druck zu öffnen*]

thumb-plate (Daumen)Drücker *m*

T

thyristor Thyristor, steuerbare Gleichrichter *m*, Vierschichtendiode *f* [*Er hat eine separate Zündelektrode, die bei der Shockley-Diode fehlt*]

TIC Temperatur anzeigen und regeln [*Kennzeichnet die MSR-Funktion*]

TIDS → talker idle state

tie → (hollow) wall tie

to tie down abgleichen [*Messtechnik*]

tie-down point Abgleichpunkt *m* [*Messtechnik*]

tight building Gebäude *n* mit eingeschränkter Spaltlüftungsrate

tight building syndrome → building related illness

tight-closing dichtschließend

tight seat Dichtsitz *m*

tile Mosaikbaustein *m* [*Für Anlagenfließbilder; heute von Bildschirmen verdrängt*]

tile(d) wall Fliesenwand, geflieste Wand *f*

tile stove Kachelofen, gekachelte Ofen *m*

tile stove hot air heating Kachelofen-Warmluftheizung, Warmluft-Kachelofenheizung *f*

tilt(ing) head Kippkopf *m*

tilt(ing) mirror Kippspiegel *m*

timber window casement turn, timber window casement fastener Holzfenstervorreiber *m*

to time → to clock

time → ((closed) loop) time

time acquisition Zeiterfassung *f*

time amount → amount of time

time authorization Zeitberechtigung *f*

time base Zeitbasis *f*

time base sweep Zeitbasisablenkung *f*

time behaviour (Brit.); time behavior (US) Zeitverhalten *n*

time between failures, TBF Betriebszeit, Betriebsdauer *f*

T

time block → time module

time card Abwesenheit(s)zeitkarte, Zeitkarte *f*

time card clock Zeitkartenuhr *f*

time ((closed) loop) control Zeitreg(e)lung *f*

time code Einzelbildkennung *f* [*Auf einem Videoband*]

(time) comparison gate (Zeit)Vergleich(s)gatter *n*

(time) compressing (Zeit)Raffung *f*

time constant Zeitkonstante *f*

time continuous [*e.g. a control*] zeitkontinuierlich, zeitstetig

time control → time ((closed) loop) control

time control → time ((open) loop) control

time-control(led switching) Zeitschalten *n*, Zeitschaltung *f*

time-controlled operating device zeitgesteuerte Schaltvorrichtung *f* [*Sie lässt erst nach Ablauf der eingestellten Sperrzeit weitere Schaltvorgänge zu. DIN VDE 0833*]

time-control(led switching) routine; time-control(led switching) program (US); time-control(led switching) programme (Brit.) Zeitschalt(ungs)programm *n*

time controller, time regulator Zeitplanregler *m*

time counter Zeitzähler *m*

time current characteristic Zeit/Strom-Kennlinie *f*

timed zeitlich (ein)gestellt, zeit(ein)gestellt

timed → clocked

time delay, time lag Zeitverzögerung *f*, Zeitverzug *m*

(time-)delay fuse [*It opens some time after current exceeds the fuse rating*] träge Sicherung *f*, Zeitsicherung, Verzögerungssicherung, Verzugsssicherung

(time-)delay interval relay → interval (time-)delay relay

(time-)delay relay, timing relay, TDR, (time-)lag relay [*Relay whose activation is delayed for a predetermined period of time. It may be either time-delay opening (TDO) or time-delay closing (TDC)*] träge Relais, Zeitrelais *n*, Verzögerungsrelais, Verzugsrelais

(time-)delay switch, (time-)lag switch, timing switch Verzögerungsschalter *m*, Verzugsschalter, Zeitschalter, träge Schalter

time-dependent sequential ((open) loop) control zeitabhängige Ablaufsteuerung *f*

time distribution system [*IEC 50(35)*] Zeitdienstsystem *n*

time division Zeit(auf)teilung *f*

time division multiple access, TDMA zeitliche (Auf-)Teilung *f* mit mehrfachem Zugriff

time division multiplexing → TDM

time domain Zeitbereich *m* [*selten: n*]

time domain reflectometry Reflexionszeitmessung *f*

time(d) sequence programme; time(d) sequence program (US) Zeitschaltprogramm *n*, Schaltzeitprogramm

time element Zeitglied *n*, binäre Zeitglied, Zeitschaltwerk *n* [*DIN 19 237*]

time expenditure Zeitaufwand *m*

time frame Zeitrahmen *m*

time interval Zeitabstand *m*

time-invariant zeitinvariant, zeitunveränderlich

time lag → time delay

(time-)lag interval relay → interval (time-)delay relay

time lapse mode Langzeitaufzeichnungsmodus, Langzeitbetrieb *m [Gegensatz: Echtzeitaufzeichnungsmodus]*

time lapse recorder Langzeitrekorder, Langzeitrecorder *m*

time lapse video cassette recorder Langzeitvideokassettenrekorder, Langzeitvideokassettenrecorder *m*

time limited instruction zeitlich begrenzte Befehl *m [Er ist nicht gespeichert]*

time lock Zeitschloss *n*

time management Zeitwirtschaft, Zeitverwaltung *f*

time module, time block Zeitbaustein, Zeitblock *m*

time monitoring Zeitüberwachung *f*

time of current → ON time

time of day, day time Tageszeit *f*

time ((open) loop) control Zeitsteuerung *f*

time order zeitliche Ablauf, zeitliche Verlauf *m*

time-orien(ta)ted sequential control zeitabhängige Ablaufsteuerung, zeitgeführte Ablaufsteuerung *f [Die Weiterschaltbedingungen sind nur von der Zeit abhängig. E DIN 19 226 Teil 5; V DIN 19 237]*

time out delay Auszeitverzögerung *f*, Auszeitverzug *m*

time out signal Bereitsignal *n [Mikroprozessor]*

time out status Zeitfehlerstatus, Zeitfehlerzustand *m*

time overtravel, time overshoot Zeitnachlauf *m*

time period Zeitspanne *f*, Zeitraum *m*

time program (US); time programme (Brit.) Zeitprogramm *n [Eine Reihe von Zeitpunkten, die festlegen, wann Anlagekomponenten ein- und ausgeschaltet werden. Ein solches Zeitprogramm erstreckt sich üblicherweise über 24 Stunden und enthält mehrere Ein-/Ausschaltpunkte.]*

time pulse generator → emitter

time pulse metering Zeitimpulsmessung *f*

timer → timing element

time range Zeitbereich *m [selten: n]*

time range preselector Zeitbereichsvorwahlschalter *m*

time range selection Zeitbereichswahl *f*

timer controlled, timing element controlled schaltuhrgesteuert, zeit(glied)gesteuert, zeitgebergesteuert

time recording Zeitaufzeichnen *n*, Zeitaufzeichnung *f*

time regulator → time controller

time/response programme (Brit.); time/response program (US); time/response routine Zeit-Reaktions-Programm *n*

timer motor Zeitgebermotor, Taktgebermotor *m*

timer ((open) loop) control, ((open) loop) timer control Schaltuhrsteuerung *f*

timer relay Zeitgeberrelais, Taktgeberrelais *n*

timer program (US); timer programme (Brit.); timer routine Zeitschaltprogramm *n*

timer with binary outlet, timing element with binary outlet Zeitglied *n* mit binärem Ausgang *[Dieses Schaltglied gibt binäre Eingangssignale zeitlich verändert weiter. E DIN 19 226 Teil 5; V DIN 19 237]*

time scale Zeitmaßstab *m*

time scattering Zeitstreuung *f*

time schedule *[Record of the desired changes of a variable or operational status with respect to time]* Zeitplan *m*

time schedule ((closed) loop) control Zeitplanreg(e)lung *f*

time schedule ((closed) loop) controller Zeitplanregler *m*

time schedule ((open) loop) control Zeitplansteuerung *f*

time scheduling Zeitplanung *f*

time selectivity Zeitselektivität *f*

time sequence chart, time sequence diagram Zeitablaufdiagramm *n [DIN 40 719, Blatt 1]*

time sequence table Zeitablauftabelle *f [DIN 40 719, Blatt 1]*

time series Zeitreihe *f [Eine Reihe von Beobachtungsdaten, Messwerten usw., aufgenommen über eine lange Zeit, meistens in gleichen Abständen]*

time set(ting) Zeiteinstellung *f*

time-sharing operating system, TSOS Teilzeitbetriebssystem, TBS *n*

time-sharing system → TSS

time-sheets calculation Zeitabrechnung *f*

time signal Zeitsignal *n*

time signal generator Zeitgeber *m [Ein Zeitsignal erzeugende Baueinheit. E DIN 19 226]*

times Italic Kursivschrift, Schrägschrift *f*

time slice distributor Zeittaktverteiler *m*, ZTV

time-slicing Aufteilung *f* der Rechenzeit in (Zeit)-Takte

time slot Zeitschlitz *m*

time slot code, TSC Zeitschlitzkode, Zeitschlitzcode *m*

times Roman Normalschrift *f*

time stamp Zeitstempel *m* [*Die mit dem Auftreten eines Ereignisses in Zusammenhang stehende Zeit und das Datum. Der Zeitstempel wird üblicherweise als Teil der Ereignismeldung übermittelt.*]

time switch → timing element

time switches Rundsteuer(ungs)anlage *f* [*Speicherofenheizung*]

time switching, clock(-controlled) switching Zeitschalten *n*, Zeitschaltung *f*

(time) tagging [*IEC 50(371)*] Absolutzeiterfassung *f* [*Fernwirktechnik*]

time target Zeitvorgabe *f*

time trigger Zeitauslöser, Zeitanstoß *m*

time waste → waste of time

timework Zeitarbeit *f*

time zone Zeitzone *f*

timing Takten *n*, Taktung, Zeitabstimmung *f*

timing block → timing module

timing circuit Zeitgeberschaltung, Taktleitung *f* [*Datenübertragung*]

timing element, timer, time switch, switch clock, clock(-controlled) switch [*For opening or closing a circuit at one or more predetermined time(s)*] Zeitglied, Zeitschaltwerk, binäre Zeitglied *n*, Zeitschalter, Zeitgeber, Uhrschalter *m*, Schaltuhr *f* [*DIN 19 237*]

timing element with binary outlet → timer with binary outlet

timing extraction, timing gain Zeitgewinn(ung) *m*, *(f)*

timing gear, timing mechanism Taktlaufwerk *n*, Taktmechanismus *m*

timing generator width, TG width Zeitmarkendauer *f* [*Dauer einer Zeitmarke des Prüfautomaten innerhalb eines Prüfzyklus. Siehe Anmerkung unter „Zeitmarkenerzeuger"*]

timing machine Taktwaage *f*

timing mark Taktmarke *f*

timing mechanism → timing gear

timing module, timing block Taktmodul, Taktbaustein *m*

timing-pulse ((open) loop) control → ((open) loop) timing-pulse control

timing range Takteinstellbereich *m [selten: n]*

timing relay → (time-)delay relay

timing signal Taktsignal *n*

timing tract Taktspur *f*

tinsmith Klempner; Spengler *m* [*Süddeutschland*]

title block → (drawing) title block

title domain Gültigkeitsbereich einer Bezeichnung, Gültigkeitsbereich eines Namens *m [selten: n]* [*E DIN ISO 7498*]

title domain name Gültigkeitsbereichsname *m*, Gültigkeitsbereichsbezeichnung *f*

TIV system → protection multiple earthing

TL → acoustic transmission loss

TLI → transport level interface

TN → twisted nematic

***t*-norm, triangular norm** triangulare Norm, Dreiecknorm, *t*-Norm *f*

tobacco smoke Tabakrauch *m*

t.o.c.d. → twist-on connecting device

toe bolt → bent bolt

toilet, closet, water toilet, water closet, flush(ing) (tank) closet, flush(ing) (tank) toilet, W.C., w.c. [*The term "lavatory" is also used as a euphemism for "toilet" or "closet"*] Abort, Abritt, Spülabort, Wasserabort, Spülabtritt *m*, Toilette *f*, Klosett, W.C., Spülklosett, Wasserklosett *n*

toilet-and-shower room, closet-and-shower room Abort-Dusch-Raum, Toiletten-Dusch-Raum, Abtritt-Dusch-Raum, Klosett-Dusch-Raum *m*

toilet area, closet area Toilettenbereich, Abortbereich, Klosettbereich, Abtrittbereich *m [selten: n]*

toilet basin → toilet bowl

toilet block, closet block Toilettentrakt, Klosett-Trakt, Abtritt-Trakt, Abort-Trakt *m*

toilet bowl, toilet pan, toilet basin, water toilet bowl, water toilet pan, water toilet basin, (water) closet bowl, (water) closet pan, (water) closet basin, (W.C.) bowl, (W.C.) pan, (W.C.) basin, flush(ing) bowl, flush(ing) pan, flush(ing) basin Toiletten(spül)becken, W.C.-(Spül)Becken, Klosett(spül)becken, Abortbecken, Spülabortbecken, Spülklosettbecken, Spülabtrittbecken, Wasserabortbecken, Wassertoilettenbecken, Wasserklosettbecken, Wasserabtrittbecken, Spültoilettenbecken, (Abort)-Spülbecken, Abtritt(spül)becken *n*, Klosettkörper, Abortkörper *m*, Klosettschüssel *f* [*Bei Aborten ohne Wasserspülung bildet das Abortbecken nur eine Erweiterung der Abfallrohrleitung. Bei Abortbecken mit Wasserspülung ist zwischen Becken und Rohrleitung ein Siphon vorhanden*]

toilet brush, closet brush Toilettenbürste, Klosettbürste, Abortbürste, Abtrittbürste *f*

T

toilet carrier, water toilet carrier, (water) closet carrier, W.C. carrier, w.c. carrier [*An iron or steel frame to support a water closet that hangs from a wall*] Toilettenhalter(ung), (Spül)Aborthalter(ung), Wasserklosetthalter(ung), Wasseraborthalter(ung), Abtritthalter(ung) *m*, *(f)*

toilet cistern, (closet) cistern, (W.C.) flush(ing) tank, (W.C.) flush(ing) cistern [*"Box" may synonymously be used for "cistern". A small vessel above a WC that holds water to flush the WC pan*] Toilettenspülkasten, (Abort)Spülkasten, Klosettspülkasten, Spül(abort)kasten, Sitzabortkasten, Spül(abort)-apparat, Abtrittspülkasten, Abortspülapparat *m*

toilet cubicle, toilet compartment, toilet enclosure, closet cubicle, closet compartment, closet enclosure Toilettenkabine, Toilettenzelle, Abortkabine, Abortzelle, Klosettkabine, Klosettzelle, Abtrittkabine, Abtrittzelle *f*

toilet exhaust duct, closet exhaust duct Toilettenabzugkanal, Abortabzugkanal, Klosettabzugkanal, Antrittabzugkanal *m*

toilet facility, closet facility Toilettenanlage, Abortanlage, Klosettanlage, Abtrittanlage *f*

toilet fan, closet fan Abortventilator, Klosettventilator, Toilettenventilator, Abtrittventilator *m*

toilet fixing, closet fixing Toilettenbefestigung, Klosettbefestigung, Abtrittbefestigung, Abortbefestigung *f*

toilet flushometer, W.C. flushometer (US); toilet flush(ing) valve, W.C. flush(ing) valve (Brit.) [*See remark under "flush(ing) valve"*] Klosett(druck)-spüler, Abort(druck)spüler, Toiletten(druck)spüler, Abtritt(druck)spüler *m*

toilet pan → toilet bowl

toilet fitment, water toilet fitment, (water) closet fitment, W.C. fitment, flush(ing) (tank) closet fitment, flush(ing) tank toilet fitment, w.c. fitment, W.C. fitment [*It comprises W.C. pan, flushing apparatus and connecting pipework*] Wasseraborteinrichtung, Wasserklosetteinrichtung, Spülaborteinrichtung, Spülklosetteinrichtung, Wassertoiletteneinrichtung *f*

toilet flush(ing), toilet wash, closet flush(ing), closet wash, flush(ing), wash (Abort)Spülung, Klosettspülung, Toilettenspülung, Druckspülung, Abtrittspülung *f*

toilet flush(ing) apparatus, closet flush(ing) apparatus, toilet wash apparatus, closet wash apparatus (Abort)Spüler, Klosettspüler, Toilettenspüler, Abtrittspüler *m*

toilet flush(ing) pipe, toilet wash(ing) pipe, (closet) flush(ing) pipe, (closet) wash(ing) pipe [*B.S. 1125. A pipe for the conveyance of flush(ing) water from a flush(ing) cistern or flush(ing) valve to a W.C. pan, slop hopper or urinal*] Spül(ungs)(abort)-rohr, Klosettspül(ungs)rohr, Abortspül(ungs)rohr, Toilettenspül(ungs)rohr, Druckspül(ungs)rohr *n*

toilet flush(ing) valve, (water) closet flush(ing) valve (Brit.); toilet flushometer (valve), (water) closet flushometer (valve) (US) [*See remark under "flush(ing) valve"*] Abort(druck)spüler, Toiletten-(druck)spüler, Klosett(druck)spüler, Abtritt(druck)-spüler *m*

toilet flush(ing) water, toilet wash water, closet flush(ing) water, closet wash water, flush(ing) water, wash water (Abort)Spülwasser, Klosettspülwasser, Toilettenspülwasser, Abtrittspülwasser *n*

toilet paper dispenser, closet paper dispenser Klosettpapierspender, Toilettenpapierspender, Abortpapierspender, Abtrittpapierspender *m*

toilet paper holder, closet paper holder [*"Roll" may synonymously be used for "holder"*] Toilettenpapierhalter, Abortpapierhalter, Abtrittpapierhalter, Klosettpapierhalter *m*

toilet paper roll holder, closet paper roll holder Toilettenpapier-Rollenhalter, Klosettpapier-Rollenhalter, Abortpapier-Rollenhalter, Abtrittpapier-Rollenhalter *m*

toilet pipeline, closet pipeline Abortrohrleitung, Klosettrohrleitung, Toilettenrohrleitung, Abtrittrohrleitung *f*

toilet privy, water toilet privy, water closet privy, W.C. privy, w.c. privy englische Abtritt, englische Abort *m*, englische Klosett *n*, englische Toilette *f*

toiletries Toilettenartikel *f*

toilet room, (water) closet (room), (water) toilet room [*It should be properly screened, have adequate daylight and ventilation and open off a ventilated lobby, not directly off habitable rooms, workshops or the like*] Spülabortkammer, Spülklosettkammer, Spülabtrittkammer, Abortkammer, Klosettkammer, Toilettenkammer, Abtrittkammer *f*, Spülabortraum, Spülklosettraum, Spülabtrittraum, Abortraum, Klosettraum, Toilettenraum, Abtrittraum *m* [*DIN 18 022*]

toilet seat, closet seat, WC seat, water closet seat [*B.S. 1254*] Toilettensitz, Klosettsitz, Abortsitz, Abtrittsitz, W.C.-Sitz *m*, Klosettbrille, Abortbrille, Abtrittbrille, WC-Brille, Toilettenbrille *f*

toilet seat lid, (closet) seat lid Toilettendeckel, Abortdeckel, Klosettdeckel, Abtrittdeckel *m*

toilet soil stack → soil stack

toilet suite, closet suite Klosettgarnitur, Abortgarnitur, Toilettengarnitur, Abtrittgarnitur *f* [*Sie besteht aus Becken, Spülkasten und Zubehör*]

T

toilet tissue dispenser, closet tissue dispenser Toilettenpapierspender, Klosettpapierspender *m*

toilet waste water pipe → soil pipe

toilet trailer, mobile toilet unit, closet trailer Toilettenwagen, Klosettwagen, Abtrittwagen, Abortwagen *m*

toilet unit, water toilet unit, (water) closet unit [*e.g. a water closet, a urinal or a lavatory*] Toiletteneinheit, Aborteinheit, Abtritteinheit, Klosetteinheit *f*

toilet valve, water toilet valve, water closet valve [*It controls the flushing cycle of a tank(-type) water closet*] Toilettenspülventil, Abortspülventil, Klosettspülventil, Abtrittspülventil *n*

toilet wash apparatus → toilet flush(ing) apparatus

toilet water, closet water Toilettenspülwasser, Klosettspülwasser, Abortspülwasser, Abtrittspülwasser *n*

toilet with attached flushing tank, toilet with attached cistern Klosettspüler mit angeformtem Spülkasten *m*, Einstückanlage, Einstückgarnitur *f*

toilet with top-mounted flush(ing) tank, toilet with top-mounted cistern Klosettspüler mit aufgesetztem Spülkasten *m*, Zweistückanlage, Zweistückgarnitur *f*

token (Sende)Berechtigungsmarke *f*, Token *n* [*Ein ständig in einem Netzwerk umlaufendes und die zu übertragenden Daten transportierendes Bitmuster*]

token Schlüsselsymbol, Schlüsselsinnbild, Schlüsselzeichen *n*

token passing [*A method of sharing access to a type of LAN. Only the communicating device in current possession of a single control (permission) token is allowed to transmit a message on the medium. All the devices wishing to transmit messages are connected in the form of a logical ring and, after a device receives the token and transmits any waiting messages, it passes the token on to the next device in the ring*] (Sende)-Berechtigungsweitergabe *f*

token ring [*LAN*] Ringnetz mit Sendeberechtigungsmarkierung *n*, Z-Ring, Token-Ring *m*

tonal components Tonhaltigkeit *f*

tonal nuisance, tonal annoyance Tonbelästigung, Tonlastigkeit *f*

ton-day of refrigeration [*Heat removed by a ton of refrigeration operating for a day; 288,000 Btu. It is approximately equal to the latent heat of fusion or melting of 1 ton (2,000 lb) of ice, from and at 32°F*] Kälteerzeugung *f* pro Tagestonne Kältemittel

tongs große Zange *f*

tongue-and-groove flanged joint Feder-Nut-Flansch(en)(ab)dichtung *f*

ton of refrigeration, refrigeration ton [*1. Useful refrigerating effect equal to 12,000 Btu/hour; 200 Btu/minute. 2. Removal of heat at a rate of 12,000 Btu/hour*] Kälteerzeugungstonne, Kälteerzeugungs-Tonne *f*

tool box Werkzeugkasten *m*

tool kit Werkzeugsatz *m*

toothed threshold plate [*moving ramp; moving walk*] Kammplatte *f*

top-down-method Verfahren *n* zur strukturierten Programmentwicklung, Methode *f* zur strukturierten Programmentwicklung

top energy efficiency → upper energy efficiency

top event unerwünschte Ereignis *n* [*Zeichnerische Wiedergabe eines Fehlerbaumes*]

top-hat rail, DIN rail (DIN-)Hut(profil)schiene *f* [*Nach DIN EN 50 032, 35x7,5 ausgeführte Schiene, auf der entsprechend gestaltete Reihen-Einbaugeräte (REG) aufgeschnappt werden*]

top hinge frog → upper hinge frog

top hood, upper hood Oberhaube *f*

topographical (re)presentation lagerichtige Darstellung *f* [*DIN 40 719, Blatt 1*]

topology Topologie, Netzwerkarchitektur *f* [*Die physikalische Auslegung der Verkabelung eines Netzes und der Geräte*]

top overtravel, top overshoot, top overrun [*Distance the car floor can travel above the level of the upper terminal landing until the car is stopped by the terminal stopping device*] obere Durchfahrt, obere Überfahrt *f*, obere Durchfahren, obere Überfahren *n*

top (terminal) landing obere (End)Haltestelle *f* [*Aufzug*]

top track obere Laufschiene *f* [*Tor*]

top-up charge Nachfüllung *f*

top valve part → upper valve part

top view Draufsicht *f*

torquemeter Drehmomentmesser *m*

torr → pascal

torsional wave Torsionswelle *f* [*Schalltechnik*]

torsion-free verwindungsfrei

TOS → tape operating system

total absorbing power of the walls and materials in a room (Schall)Schluckvermögen *n* eines Raumes [*Es wird als äquivalente Schallschluckfläche (Schallabsorptionsfläche) A angegeben. Es ist dies die Fläche mit vollkommener Schluckung, die den gleichen Anteil an Schallenergie schlucken würde wie die gesamte Oberfläche des Raumes und die in ihm befindlichen Gegenstände und Personen tatsächlich schlucken*]

T

total actual volumetric efficiency, volumetric efficiency, charge efficiency, (overall) actual volumetric efficiency, (overall) real volumetric efficiency Liefergrad *m*

total air conditioner capacity → (useful) total air conditioner capacity

total bit rate, overall bit rate, complete bit rate Summenbitrate, Summenbinärzeichenrate *f*

total compilation, sum compilation, overall compilation, accrued compilation Summenbildung *f*

total consumption, overall consumption, complete consumption Gesamtverbrauch, Totalverbrauch, Vollverbrauch, Ganzverbrauch *m*

total cooling effect, overall cooling effect [*1. Difference between the total enthalpy of the dry air and the water vapo(u)r mix(ture) entering and leaving the cooler. 2. Amount of sensible and latent heat removed from the conditioned space*] Gesamtkühlleistung, Totalkühlleistung, Vollkühlleistung, Ganzkühlleistung *f*

total co-ordination, overall co-ordination übergeordnete Koordination *f*

total depth, overall depth, leading depth Tiefe über alles, Gesamttiefe, Totaltiefe, Volltiefe, Ganztiefe *f*

total dimension, overall dimension, leading dimension Maß über alles, Gesamtmaß, Totalmaß, Vollmaß, Ganzmaß *n*

total disconnection, total switching off, overall disconnection, overall switching off Gesamtabschaltung, Totalabschaltung, Vollabschaltung, Ganzabschaltung *f*

total efficiency, overall efficiency Gesamtwirkungsgrad, Totalwirkungsgrad, Vollwirkungsgrad, Ganzwirkungsgrad *m*

total energy system, total energy plant, overall energy system, overall energy plant Total-Energie-Anlage, Gesamt-Energie-Anlage, Voll-Energie-Anlage, Ganz-Energie-Anlage *f* [*Eine Klein-Kraftanlage, die bei weitgehender Ausnutzung der Abwärme annähernd den gesamten Wärme-, Kälte- und Elektroenergiebedarf eines Gebäudes oder eines Industriekomplexes deckt, so dass nur eine Art Primärenergie benötigt wird (einschienige Energieversorgung)*]

total fan pressure → fan total pressure

total flow meter, quantity meter, overall flow meter [*It shows the total quantity in volume units that has passed through a conduit (e.g. a pipe) between two successive readings*] Volumenmesser, Durchflussmesser *m*

(total) heat content → enthalpy

(total) heat content ((closed) loop) control → enthalpy ((closed) loop) control

(total) heat (content) controller → enthalpy controller

total heat control → enthalpy control

total heat gain, overall heat gain, complete heat gain Gesamtwärmeabgabe, Totalwärmeabgabe, Vollwärmeabgabe, Ganzwärmeabgabe *f*

total heating value → higher heating value

total illumination, total lighting, overall illumination, overall lighting Vollbeleuchtung, Totalbeleuchtung, Gesamtbeleuchtung, Ganzbeleuchtung *f*

total(ized) hours worked, overall hours worked Gesamtarbeitsstunden, Totalarbeitsstunden, Vollarbeitsstunden, Ganzarbeitsstunden *f*

totalizer Summierer *m*, Summier(mess)gerät *n*

totalizer pulse, totalizing pulse Mengenimpuls, Summier(ungs)impuls *m*

totalizer pulse relay, totalizing pulse relay Impulssummier(ungs)relais, Impulsmengenrelais *n*

total level, overall level Gesamtpegel, Totalpegel, Vollpegel, Ganzpegel *m*

total length, length overall, overall length [*On drawings abbreviated: LOA*] Länge über alles, Gesamtlänge *f*

totally enclosed, TE, completely enclosed ganzgekapselt, vollgekapselt, totalgekapselt, gesamtgekapselt

totally enclosed fan-cooled, TEFC ganzgekapselt und ventilatorgekühlt, vollgekapselt und ventilatorgekühlt, gesamtgekapselt und ventilatorgekühlt, totalgekapselt und ventilatorgekühlt

totally enclosed nonvent(ilat)ed, TENV ganzgekapselt und unbelüftet, vollgekapselt und unbelüftet, totalgekapselt und unbelüftet, gesamtgekapselt und unbelüftet

totally enclosed water-cooled, TEWC ganzgekapselt und wassergekühlt, vollgekapselt und wassergekühlt, totalgekapselt und wassergekühlt, gesamtgekapselt und wassergekühlt

totally integrated, fully integrated, completely integrated gesamtintegriert, vollintegriert, totalintegriert, ganzintegriert

total monitoring, complete monitoring, overall monitoring Vollüberwachung, Ganzüberwachung, Gesamtüberwachung, Totalüberwachung *f*

total monitoring range, overall monitoring range, complete monitoring range Gesamtüberwachungsbereich, Totalüberwachungsbereich, Vollüberwachungsbereich, Ganzüberwachungsbereich *m* [*selten: n*]

total noise Gesamtgeräusch, Totalgeräusch, Vollgeräusch, Ganzgeräusch *n*

T

total outage, total failure (action), total breakdown, accrued outage, accrued failure (action), accrued breakdown, overall failure (action), overall breakdown Gesamtausfall, Totalausfall, Vollausfall, Ganzausfall *m*, Gesamtversagen, Totalversagen, Vollversagen, Ganzversagen *n*, Gesamtpanne, Totalpanne, Vollpanne, Ganzpanne *f*

total opening time, overall opening time Gesamtausschalt(eigen)zeit, Totalausschalt(eigen)zeit, Vollausschalt(eigen)zeit, Ganzausschalt(eigen)zeit *f*

total pitch error → accumulated pitch error

total pressure, overall pressure Gesamtdruck, Totaldruck, Volldruck, Ganzdruck *m*

total project, total scheme, overall project, overall scheme Gesamtprojekt, Totalprojekt, Vollprojekt, Ganzprojekt *n*

total refrigeration capacity, gross refrigeration capacity, overall refrigeration capacity [*Total rate of heat removal from all sources by the evaporator of a refrigerating system at stated conditions. It is numerically equal to the system refrigerating effect*] Bruttokälteleistung, Gesamtkälteleistung, Totalkälteleistung, Vollkälteleistung, Ganzkälteleistung *f*

total residence time, overall residence time gesamte Verweilzeit, totale Verweilzeit, ganze Verweilzeit, volle Verweilzeit *f*

total response time, overall response time Gesamtreaktionszeit, Totalreaktionszeit, Vollreaktionszeit, Ganzreaktionszeit *f* [*Verstärker*]

total room heat, overall room heat totale Raumwärme, gesamte Raumwärme, ganze Raumwärme, volle Raumwärme *f*

total solar radiation, overall solar radiation Totalsonnenstrahlung, Vollsonnenstrahlung *f*

total safety, overall safety Gesamtsicherheit, Totalsicherheit, Ganzsicherheit, Vollsicherheit *f*

total switching off → total disconnection

total shielding, overall shielding Gesamtabschirmung, Totalabschirmung, Vollabschirmung, Ganzabschirmung *f*

total thermal resistance, overall thermal resistance, overall heat resistance, total heat resistance, overall resistance to heat transfer, total resistance to heat transfer Gesamtwärmewiderstand, Totalwärmewiderstand, Vollwärmewiderstand, Ganzwärmewiderstand *m*

total thermal transmittance value for building fabrics, 'O' value, overall thermal transmittance value for building fabrics [*In Btu/sq ft./°F./hr*] Wärmedurchlasszahl (λ) *f*, kcal/m²h°

total waterflow meter, water quantity meter, waterflow volume meter, overall waterflow meter Wasserzähler *m* [*Gerät zur Totalisation der durchfließenden Wassermenge, d.h. des über einen bestimmten Zeitraum integrierten Volumenstromes*]

totem-pole Gegentakt(ausgangs)stufe *f*

to the atmosphere ins Freie

touch dimmer, sensor dimmer Tasthelligkeitsschalter, Helligkeitstastschalter *m*

touch input Tasteingabe, Berührungseingabe *f* [*Bildschirm*]

touch keyboard → diaphragm keyboard

touch probe → tracer

touch-safe → contact-protected

touch(-sensitive) screen, touch(-sensitive) CRT Tast(bild)schirm, Berührungs(bild)schirm, berührungsempfindliche (Bild)Schirm *m*

touch-tone dialling (Brit.); touch-tone dialing (US) [*automatic telephony*] Tastatur-Tonwahl *f*

touch-up paint Ausbesserungsfarbe, Nachbesserungsfarbe *f*

touch voltage Berührungsspannung *f*

tour → (guard) tour

towel airer, towel dryer; towel drier (US) Handtuchtrockner *m*

towel bar, towel rail Handtuchstange, Handtuchschiene *f*

towel basket Handtuchkorb *m*

towel cabinet, towel cubicle Handtuchschrank *m*

towel dispenser Handtuchspender *m*

towel dryer → towel airer

towel holder Handtuchhalter *m*

towel hook Handtuchhaken *m*

towel ladder Handtuchleiter *f*

towel pull-out rod Handtuchausziehstange *f*

towel rack Handtuchständer, Handtuchstand *m*

towel rail → towel bar

towel ring Handtuchring *m*

towel warmer Handtuch(er)wärmer, Handtuchwärmekörper *m*

tower Turm *m*

town drainage → city drainage

town gas, city gas Stadtgas; Leuchtgas *n* [*frühere Benennung*] [*Ein Gemisch aus Wasserstoff (H_2), Methan (CH_4), Kohlenmonoxid (CO), schweren Kohlenwasserstoffen (C_nH_m), Kohlendioxid (CO_2), Stickstoff (N_2) und Sauerstoff (O_2)*]

T

town gas cock, city gas cock Stadtgashahn *m*

town gas heating → heating by town gas

town water → city water

TPDDI → twisted pair distributed data interface

tpi → tracks per inch

trace Ablaufverfolgung *f*

traced → stopped

trace electric heat system → heat trace electric system

trace heat system → heat trace system

tracer, touch probe Taster *m* [*numerische Steuerung; Kopierfühler*]

tracer Spurenstoff, Markierungsstoff *m*

tracer gas Spurengas, Markierungsgas *n*

tracer gas constant concentration method Spurengasmessverfahren mit konstanter Konzentration, Markierungsgasmessverfahren mit konstanter Konzentration *n*

tracer heating → (pipe) tracer heating

trace steam heat system → heat trace steam system

tracing [*A graphic record produced by lines visible through a transparent paper*] Pause, Durchzeichnungskopie *f*

track → (trolley) track

track Spur *f*

track → (feed) track

track → printed(-circuit) board

track(er) ball [*A peripheral ball used to position a cursor on a video terminal. Similar to a mouse*] Rollkugel, Steuer(ungs)kugel *f*

track(er) wire → equipment wire

track following, track finding Spurnachführung *f*, Spurfolgen, Spurnachführen *n*

track index Spurindex *m*

tracking Nachführung *f*, Folgen, Nachführen *n* [*numerische Steuerung*]

tracking error Nachführ(ungs)fehler, Folgefehler *m*

track(-mounted) luminaire → channel(-mounting) luminaire

tracks per inch, tpi Spuren *f* pro Zoll [*Diskette*]

track-suspended luminaire → channel(-mounting) luminaire

track suspension device [*IEC 570*], **busway hanger** Stromschienenaufhängung, Sammelschienenaufhängung, Verteilerschienenaufhängung *f* [*„Hänger" kann synonym für „Aufhängung" verwendet werden*]

track wire → equipment wire

tract Spur *f*

traction drive [*The lift car is moved by traction developed between the ropes and the specially designed sheave*] Seil(an)trieb, Kabel(an)trieb *m*

traction (driven) lift (Brit.); traction (driven) elevator (US) Seil((an)trieb)aufzug, Kabel((an)trieb)-aufzug *m*

traction (lift) drive, traction (lift) machine (Brit.); traction (elevator) drive, traction (elevator) machine (US) [*The car is moved by the traction developed between the ropes and the groove(d) sheave*] Seil(an)triebwerk, Kabel(an)triebwerk *n* [*Aufzug*]

trade Gewerbe *n*

trade fair Fachmesse *f*

trade-integrating → integrating of trades

trade laundry gewerbliche Wäscherei, Waschanstalt *f*

trade-mark → (registered) trade-mark

trade premise Gewerbestätte *f*

trade sequence → sequence of trades

trade sewage; trade wastewater (US) gewerbliche Abwasser *n*

trade waste Gewerbeabfall(stoff), gewerbliche Abfall(stoff) *m*

traffic installation autowalk, traffic installation passenger conveyor [*As opposed to a commercial centres autowalk*] Verkehrsanlagenrollsteig, Verkehrsanlagenfahrsteig *m*

traffic installation escalator [*As opposed to a commercial centres escalator*] Verkehrsanlagenrolltreppe, Verkehrsanlagenfahrtreppe *f*

traffic volume Verkehrsaufkommen *n*, Verkehrsmenge *f*

trafo, transformer Transformator, Trafo *m*

trafoscape Trafowächter *m*

trailing pointer → drag pointer

training Synchronisation(svorgang) *f*, (*m*) [*Vorgang der Synchronisation zweier Modems beim Verbindungsaufbau*]

training Ausbildung, Schulung *f*

training course Lehrgang, Ausbildungskursus, Schulungskursus *m*

train printer → chain printer

transceiver Übertrager *m* [*Ein Gerät, das in einem Basisband-Netzwerk erforderlich ist, um die digitalen Signale vom Rechner (oder Terminal) aufzunehmen und auf ein Basisbandmedium zu übertragen.*]

transceiver, transmitter-receiver [*A radio transmitter and receiver combined in one unit and having switching arrangements such as to permit use of one or more tubes for both transmitting and receiving*] Sender-Empfänger *m*, Sende-Empfangs-Gerät *n*

transceiver → (bus) transceiver

transceiver cable Übertragungskabel *n* [*Kabel zwischen Übertrager und Netzwerksteuereinheit; es ermöglicht die Entnahme von Knoten von einem Basisbandmedium*]

transceiver data link, transmitter-receiver data link [*Integrated data processing by means of punched cards, using transceivers as terminal equipment; the transmission path can be wire or radio*] Sender-Empfänger-Datenverbindung, Sende-Empfangs-Gerät-Datenverbindung *f*

to transcribe umschreiben

transcriber [*The equipment used to convert information from one form to another, as for converting computer input data to the medium and language used by the computer*] Umschreiber *m*

trans-critical vapour compression cycle (Brit.); trans-critical vapor compression cycle (US) transkritische CO_2-Prozess *m* [*Warmwasserbereitung*]

transducer → magnetic amplifier

transducer → measuring transducer

transducer, signal transducer [*Any device or element which converts an input signal into an output signal of a different form; examples include the microphone, phonograph pickup, (loud)speaker, barometer, photoelectric cell, automobile horn, doorbell, and underwater sound transducer*] Signalumsetzer, Signalwandler *m*

transducer [*Device that converts power of one type into power of another, for instance, a loudspeaker*] Umsetzer, Wandler *m*

transductor → saturable(-core) reactor

transfer [*relay*] Umschlagen *n*, Umschlag *m*

transfer [*computer*] Verzweigung *f*

transfer Übergabe, Weitergabe, Übermitt(e)lung, Übertragung *f*, Transfer *m*

transfer check(ing) [*Check(ing) (usually automatic) on the accuracy of the transfer of a word in a computer operation*] Übergabekontrolle, Weitergabekontrolle, Übermitt(e)lungskontrolle, Transferkontrolle, Übertragungskontrolle *f*

transfer complete time Sendebeendigungszeit, Sendeschlusszeit *f*

transfer forward signal → forward transfer signal

transfer function → ((closed) loop) transfer function

transfer instruction Übergabeanweisung, Weitergabeanweisung, Übermitt(e)lungsanweisung *f*, Übergabebefehl, Weitergabebefehl, Übermitt(e)lungsbefehl *m*

transfer matrix Übergabematrix, Weitergabematrix, Übermitt(e)lungsmatrix *f*

transfer pump Umfüll(ungs)pumpe *f*

transfer rate [*The speed at which data are moved from a direct-access device to a central processing unit*] Übergabegeschwindigkeit, Weitergabegeschwindigkeit, Übermitt(e)lungsgeschwindigkeit *f*

transfer sequence switch Übergabefolgeschalter, Weitergabefolgeschalter, Übermitt(e)lungsfolgeschalter *m*

transfer signal [*telephony*] Umlegungszeichen *n*

transfer station Übergabestation, Weitergabestation, Übermitt(e)lungsstation *f*

transfer switch → throwover switch

transfer syntax [*That concrete syntax used in the transfer of data between open systems (ISO 7498)*] Übergabesyntax, Weitergabesyntax, Übermitt(e)lungssyntax *f*

transfer (to another number) → (call) transfer (to another number)

to transform umwandeln, umformen, transformieren

to transform [*electrical engineering*] umspannen

transformation → conversion

transformer, trafo Transformator, Trafo *m*

transformer read-only store [*In computers, read-only store in which the presence or absence of mutual inductance between two circuits determines where a binary 1 or 0 is stored*] Umsetzer-Nur-Lesespeicher, Wandler-Nur-Lesespeicher *m*

transformer switch Transformatorschalter, Trafoschalter *m*

transformer vault Transformatorkeller, Trafokeller *m*

transformer voltage → system voltage

transient [*Pulse or other temporary phenomenon occurring in a system prior to reaching a steady-state condition*] vorübergehende Änderung, zeitweilige Änderung, zeitweise Änderung *f*

transient-absorption-zener diode spitzenabsorbierende Z-Diode *f*

transient but stay resident, TSR speicherresident [*Speicherresidente Programme stehen sofort zur Verfügung*]

transient overvoltage, surge overvoltage Stoßüberspannung *f*

transient error signal vorübergehende Regel(ungs)unterschied *m*

T

transient recovery time → **settling time**

transient response Anwort im Zeitbereich, Zeitbereich(s)antwort *f*

transient value tolerance Einschwing(ungs)toleranz *f*

transient ventilation → **surge ventilation**

transient voltage, surge voltage transiente Spannung, Ausgleichsspannung, Stoßspannung, Einschwing(ungs)spannung, freie Spannung *f*, Spannungswischer *m* [*Sie wird bei plötzlichem Anlegen der Elektronen mit dem vollen Spannungswert noch ausgehalten*]

transistor [*An active component of an electronic circuit consisting of a small block of semiconducting material to which at least three electrical contacts are made, usually two closely spaced rectifying contacts and one ohmic (nonrectifying) contact; it may be used as an amplifier, detector, or switch*] Transistor *m*

transistor amplifier [*An amplifier in which one or more transistors provide amplification comparable to that of electron tubes*] Transistorverstärker *m*

transistor array Transistorfeld *n*

transistor chopper Transistorzerhackerschaltung *f*

transistor circuit [*An electric(al) circuit in which a transistor is connected*] Transistorschaltung *f*

transistor cutoff region Transistorsperrbereich *m* [*selten: n*]

transistor magnetic amplifier [*A magnetic amplifier together with a transistor preamplifier, the latter used to make the signal strong enough to change the flux in the core of the magnet completely during a half-cycle of the power supply voltage*] Transistormagnetverstärker *m*

transistor preamplifier Transistorvorverstärker *m*

transistor-transistor logic, TTL Transistor-Transistor-Logik, TTL *f* [*Bei dieser Bausteinfamilie werden als aktive Elemente nur am Sättigungsbereich betriebene Transistoren verwendet*]

transition card [*In reading a deck of punched cards by a computer, a card that causes the computer to stop reading cards*] Lesesperrkarte, Sperrlesekarte *f*

transition connection branch Übergangsstutzen *m*

transition counting Pulswechselverfahren *n*, Pulswechselmethode *f*

transition pipe Übergangsrohr *n*

transition zone Übergangsstrecke *f*

transit through-connect signal → **TTC**

to translate [*language*] übersetzen

translation [*language*] Übersetzung *f*

translation [*carrier frequency engineering*] Umsetzung *f*

translator Übersetzer(programm) *m*, (*n*) [*Programm zur Überführung von Programmen aus einer Sprache in eine andere, meistens maschinenorientierte, Sprache*]

translator [*carrier frequency engineering*] Umsetzer *m*

to transliterate umkodieren, umcodieren

transmission Durchlicht *n*

transmission → **transmittance**

transmission [*The process of transferring a signal, message, picture or other form of intelligence from one location to another location by means of wire lines, radio, light beams, infrared beams, or other communication systems*] Übertragung *f*

transmission bit order → **bit order of transmission**

transmission block, function(al) block Übertragungsblock *m* [*Rechteck. Kennzeichnung der wirkungsmäßigen (kausalen) Abhängigkeit der Ausgangsgröße von der Eingangsgröße*]

transmission block end → **end of transmission block**

transmission channel Übertragungskanal *m*

transmission clarity Übertragungsklarheit, Übertragungsreinheit *f*

transmission code Übertragungskode, Übertragungscode *m*

transmission communication protocol/internet protocol Protokoll *n* TCP/IP

transmission control character → **transmission ((open) loop) control character**

Transmission control Protocol/Internet Protocol → **TCP/IP**

transmission device Übertragungsgerät *n*

transmission electron microscopy Durchstrahlungselektronenmikroskopie *f*

transmission element Übertragungsglied *n* [*Bauglied oder Gerät der Messung, Steuerung oder Reg(e)lung, das eine physikalische Größe in eine andere überträgt*]

transmission end → **end of transmission**

transmission engineering Übertragungstechnik *f* [*als Wissenschaft*]

transmission error Übertragungsfehler *m*

transmission heat Übertragungswärme *f*

transmission heat gain [*The transmission heat gain through walls, windows and ceiling surfaces results from the temperature difference between indoor air and outdoor air, and is largely dependent upon the construction of the barrier between indoor air and outdoor air*] Übertragungswärmegewinn *m*

T

transmission identification, transmission ID Übertragungsidentifizierung, Übertragungskenn(zeichn)ung *f*

transmission interface Übertragungsschnittstelle *f*

transmission interface converter [*A device that converts data to or from a form suitable for transfer over a channel connecting two computer systems or connecting a computer with its associated data terminals*] Übertragungsschnittstellenwandler *m*

transmission line, transmission path Übertragungsstrecke *f*, Übertragungsweg *m*

transmission-line attenuation [*The decrease in power of a transmission-line signal from one point to another, expressed as a ratio or in decibels*] Übertragungsleitungsabschwächung, Übertragungsleitungsdämpfung *f*

transmission-line cable [*The coaxial cable which forms a transmission line; a number of standard types have been specified by size and materials*] Übertragungsleitungskabel *n*

transmission-line constant, transmission-line parameter, line(ar electrical) parameter [*The quantity which is necessary to specify the impedance per unit length of a transmission line, and the admittance per unit length between various conductors of the line*] Übertragungsleitungskonstante *f*

transmission-line transducer loss [*The ratio of the power delivered by a transmission line to a load to that produced at the generator, expressed in decibels; equal to the sum of the attenuation of the line and the mismatch loss*] Übertragungsleitungsverlust *m*

transmission-line voltage [*The work which would be required to transport a unit electrical charge between the specified conductors of a transmission line at a given instant*] Übertragungsleitungsspannung *f*

transmission loss [*The actual power that is lost in transmitting a signal from one point to another through a medium or along a line*] Übertragungsverlust *m*

transmission material, transmit(ting) material [*They are classified as clear transparent, primatic, as in clear ribbed glass or plastic; or diffusing, as in white glass or plastic. They can also be classified as high transmission or low transmission. This refers to the percent of incident energy transmitted through the material*] Durchlicht(werk)stoff *m*, Durchlichtmaterial *n*

transmission modulation [*Amplitude modulation of the reading-beam current in a charge storage tube as the beam passes through apertures in the storage surface; the degree of modulation is controlled by the stored charge pattern*] Übertragungsmodulation *f*

transmission monitor [*video*] Hauptmonitor *m*

transmission net(work), transmission system Übertragungsnetz, Übertragungssystem *n*

transmission of signals → signal transmission

transmission performance Übertragungsleistung *f*

transmission processing Übertragungsverarbeitung *f*

transmission protocol Übertragungsprotokoll *n*

transmission rate → rate of transmission

transmission request Übertragungsaufforderung *f*

transmission route, transmission path Übertragungsweg *m* [*Fernwirktechnik*]

transmission safety Übertragungssicherheit *f*

transmission security [*Component of a communications security which results from all measures designed to protect transmissions from unauthorized interception, traffic analysis, and imitative deception*] Übertragungssicherung *f*

transmission sequence Übertragungsfolge *f*

transmission system → transmission net(work)

transmission technique Übertragungstechnik *f* [*als Anwendung*]

transmission trouble [*audible communication*] schlechte Übertragung *f*

transmittance, transmission [*In electromagnetism, the radiant power transmitted by a body divided by the total radiant power incident upon the body*] Strahlendurchlässigkeit *f*

transmitted data Sendedaten *f*

transmitted light detector, transmitted light detecting device, transmitted light detection device Durchlichtmelder *m*, Durchlichtmeldegerät *n*

transmitted-light smoke detector, transmitted-light smoke detecting device Durchlichtrauchmelder *m*, Durchlichtrauchmeldegerät *n* [*Der Lichtstrahl einer Quelle wird in zwei Strahlen aufgeteilt. Ein Strahl wird durch einen gegen außen hermetisch abgeschlossenen Referenzkanal geleitet, der andere Strahl durch einen Raummesskanal. Der Raummesskanal kontrolliert die Raumluft kontinuierlich auf Rauchteilchen. Der Referenzkanal dient zur Temperatur- und Alterungskompensation. Beide Strahlen werden von Lichtempfängern gemessen, verstärkt und der digitalen Signalverarbeitung mit Signalprozessor und hinterlegten Rechenalgorithmen zugeführt*]

transmitter Übertrager *m*

transmitter, radio transmitter [*Telecommunications. Equipment that produces and modulates radio-frequency energy for the purpose of radio communication*] Sender, Funksender *m*

transmitter insert → speaking transmitter

transmitter valve, transmitting valve Senderöhre *f*

transmit(ting) frequency Sendefrequenz *f*

transmit(ting) material → transmission material

transonic barrier → acoustic barrier

transparent cover(ing) Klarsicht(ab)deckung *f*

transparent door Klarsichttür *f*

transparent paper Pauspapier, Transparentpapier *n*

transpiration cooling [*Cooling produced by evaporation of water lost by a body*] Schwitzkühlung *f*

transport(ation) → conveyance

transport(ation) lag → distance/velocity lag

transport(ation) of valuables Werttransport *m*

transport(ation) plant, transport(ation) installation Transportanlage, Förderanlage *f* [*In einem Gebäude*]

transport delay Transportverzögerung *f*, Transportverzug *m* [*Signal*]

transport entity Transportgröße *f* [*Zwischen den Bezugspunkten von meist getrennten Netzelementen*]

transport layer, layer 4 vierte Schicht, vierte Ebene, Transportschicht, Transportebene, Schicht 4, Ebene 4 *f* [*OSI*]

transport level interface, TLI Werkzeug *n* zur Netzwerkentwick(e)lung

transverse-vent(ilat)ed → cross-vent(ilat)ed

transverse window ventilation → cross window ventilation

to trap verschließen mit Geruchverschluss

trap, drain trap, stench trap, U-bend, fixture trap, siphon trap, water trap [*A bend in a sanitary pipe, so formed that it is always filled with water, to trap the air in the drains, and so prevent it from escaping. Traps are named after the letter they represent: D-trap, P-trap, S-trap*] (Wasser)(Geruch)Verschluss *m* [*Die Benennungen „Syphon", „Siphon" und „Traps" sind veraltet. DIN 1986*]

trap → interceptor

trap [*A scaffold board projecting over a putlog so as to be a source of danger*] Kraggerüstbrett *n*

trap → condensate separator

trapezoidal (bath(room)) cabinet Trapez(badezimmer)schrank *m*

trapezoidal membership function trapezförmige Zugehörigkeitsfunktion *f*

trap monitoring Fallenüberwachung *f* [*Von Bereichen, die der Täter mit großer Wahrscheinlichkeit betritt (Überwachung durch Tretmatten, Schranken, Kontaktfallen an Türen usw.)*]

trap outlet, overflow connection Auslauf(stutzen), Ablauf(stutzen) *m* [*Geruchverschluss*]

trapped air → (en)trapped air

trapped gull(e)y, entrapped gull(e)y [*A gull(e)y containing a trap*] Ablauf mit Verschluss, Verschlussablauf, Auslauf mit Verschluss, Verschlussauslauf *m*

trapping sieve → dirt trap

trap protection [*This consists of fitting detection devices at strategic points within the building – literally setting traps for the unwary*] Fallenschutz *m*

trap seal, (water) seal [*The water in a trap which acts as a barrier to the passage of air through the trap*] Wasserverschluss *m*

travel, path Weg, Pfad *m*

travel, rise [*Of an elevator, escalator, etc., the vertical distance between the bottom terminal landing and the top terminal landing*] Förderhöhe, Fahrt, Fahrhöhe *f*

travel control → PC

travel(ing) path, travel(ing) route (US); travel(ling) path, travel(ling) route (Brit.) Fahrweg *m*, Rollweg [*Fahrtreppe; Fahrsteig*]

travel(ling) speed (Brit.); traveling speed (US); step speed Fahrgeschwindigkeit, Betriebsgeschwindigkeit *f* [*Rolltreppe*]

travel(ling) speed (Brit.); travel(ing) speed (US) Fahrgeschwindigkeit, Betriebsgeschwindigkeit *f*

tray Pritsche *f*

tray soapholder Schalenseifenhalter, Seifenhalter mit Schale *m*

tread plate section [*escalator*] Trittrostprofil, Tretrostprofil *n*

treadway [*Exposed passenger-carrying member of a moving walk or moving ramp*] Trittfläche, Tretfläche *f*

treated air [*Not to be confused with "conditioned air"*] behandelte Luft, aufbereitete Luft *f*

treated potable water → warm drinking water

treatment plant, treatment installation Aufbereitungsanlage *f*

treechart Baumdiagramm *n*

tree topology Baumstruktur, Baumtopologie *f* [*Sonderform des Feldbusses, welche durch Verzweigungen einer Buslinie entsteht. Sie ermöglicht eine besonders gute Anpassung an räumlich verteilte Signalerfassungen*]

trembler Kontakthammer *m*

trend Entwick(e)lungsrichtung *f*

trespass, unauthorized entry Eindringen, unbefugte Betreten *n*

trial and error Erprobung *f*, Versuch *m* und Irrtum *m*

triangle equation → angle equation

triangular circuit Dreieckschaltung *f*

triangular conorm → ***t*-conorm**

triangular duct Dreieckkanal *m*

triangular membership function dreieckige Zugehörigkeitsfunktion *f*

triangular norm → ***t*-norm**

triangular scale rule Dreikantmaßstab *m*

tribological tribologisch

tribophysics, friction physics Reibungsphysik *f*

trichloro(mono)fluoromethane → refrigerant-11

trichlorotrifluoromethane → refrigerant-113

trickle charge Ladungserhaltung *f* [*Geringfügige Ladung, welche die Verluste einer Selbstentladung ausgleicht und damit eine Zelle oder Batterie in vollem Ladezustand erhält*]

trickling deaerator Rieselentlüfter *m*

trickling filter → biofilter

to trigger [*In automatic data processing, to execute a jump to the first instruction of a program(me) after it has been loaded into the computer*] triggern, ansteuern

to trigger, to initiate [*1. In electricity, to initiate an action, which then continues for a period of time, as by applying a pulse to a trigger circuit. 2. To cause a procedure to be executed*] auslösen, triggern, hervorrufen

trigger Grenzwertglied *n* [*Es vergleicht den Wert eines Eingangssignals mit einem Grenzwert. Wird dieser Grenzwert über- oder unterschritten so ändert sich der Zustand des binären Ausgangssignals. DIN 19 226 Teil 5*]

trigger → trigger (pulse)

trigger Auslöser, Trigger *m*

trigger action → triggering

trigger circuit [*Electronics. A circuit that depends on the application of an input initiating signal or stimulus to perform a specific predetermined function; examples include bistable multivibrators, monostable multivibrators, time delays, and sweep waveform generators*] Triggerschaltung, Auslöserschaltung *f*

trigger control → trigger ((open) loop) control

trigger diode, three-layer diode, triple-layer diode, diac [*Electronics. A two-terminal, three-layer avalanche semiconductor diode characterized by its ability to conduct symmetrically with voltages of either polarity; used mainly in Ac power-control applications*] Triggerdiode, Auslöserdiode *f*

triggered pulsed attenuator multiplier Multiplikator *m* mit triggerimpulsgesteuertem Dämpfungsglied

triggered spark gap [*A fixed spark gap in which the discharge passes between two electrodes but is initiated by an auxiliary trigger electrode to which low-power pulses are applied at regular intervals by a pulse amplifier*] angesteuerte Funkenstrecke, Auslöserfunkenstrecke, Triggerfunkenstrecke *f*

trigger equipment, trigger set [*IEC 146*] Ventilsteuer(ungs)vorrichtung *f* [*VDE 0558, Teil 1*], Ventilsteuer(ungs)satz *m*

trigger flip-flop → bistable multivibrator

trigger extension [*A metal(lic) item attached to a trigger in order to extend its length*] Triggerverlängerung, Auslöserverlängerung *f*

trigger gap Triggerentladungsstrecke, Auslöserentladungsstrecke *f*

trigger hold-off Triggersperre, Auslösersperre *f* [*Oszilloskop*]

triggering, initiating action, initiation, release, trigger action Auslösung, Triggerung, Ansteuerung, Taktgabe, Taktung, Impulsansteuerung, Freigabe *f*, Triggern *n* [*DIN IEC 469, Teil 1*], auslösende Eingriff *m*

triggering electrode → starter

triggering (frequency) range Trigger(frequenz)-bereich, Auslöser(frequenz)bereich *m* [*selten: n*] [*Oszilloskop*]

triggering of paging system, paging system triggering Ansteuerung *f* einer Personenrufanlage [*Signal, das an eine Personenrufanlage gesendet wird, um ein Personenrufsignal an einen entsprechenden Empfänger auszulösen.*]

triggering spark gap Triggerfunkenstrecke, Auslöserfunkenstrecke *f*

triggering threshold Triggeransprechschwelle, Auslöseransprechschwelle *f* [*Oszilloskop*]

triggering uncertainty [*counter*] Einschalt(ungs)-ungenauigkeit *f*

trigger ((open) loop) control [*Electronics. The on/off control of thyristors and other gas-filled electron tubes in such a way that current flow may be started or stopped, but not regulated as to rate*] Triggersteuerung, Auslösersteuerung *f*

trigger oscillator Auslöserimpulsgenerator, Triggerimpulsgenerator *m*

trigger (pulse) [*Electronics. An electrical pulse used as a trigger*] Triggerimpuls, Auslöserimpuls *m*

trigger qualifier Triggerkennzeichner, Auslöserkennzeichner *m*

trigger rate source Synchronimpulsgenerator *m*

trigger set → trigger equipment

trigger switch *[A switch that is actuated by pulling a trigger, and is usually mounted in a pistol-grip handle]* Triggerschalter, Auslöserschalter *m*

trigger timing pulse Triggerzeitkontrollimpuls, Auslöserzeitkontrollimpuls *m*

trigger tube *[A cold-cathode gas-filled electron tube in which one or more auxiliary electrodes initiate the anode current but do not control it]* (Glimm)Relaisröhre *f*

trigger unit Triggereinheit, Auslösereinheit *f [Meistens elektrische Schaltung zur Ansteuerung von Motorschützen, elektronischen Leistungsgruppen, mit oder ohne Rückmeldung, und Signalisierung]*

trigger window Triggerfenster, Auslöserfenster *n*

trigonometric function Winkelfunktion, trigonometrische Funktion *f*

to trim abgleichen, trimmen

trim → profile

trim for interior work → profile for interior work

trimming capacitor Trimmkondensator, Abgleichkondensator *m [Einmalige Einstellung mittels Werkzeug]*

trimming circuit, coordinate trimming system Trimmschaltung, Abgleichschaltung *f [numerische Steuerung]*

trimming resistor, trimmer resistor *[A small rheostat, usually connected to adjust the value of a fixed resistor]* Abgleichwiderstand, Trimmwiderstand *m*

trimming the plates, plate trimming Beschneiden der Bleche, Blech(e)beschneiden *n*

triode field effect transistor, field effect triode transistor Feldeffekttransistortriode, Transistorfeldeffekttriode *f*

to trip, to free, to release *[To put into mechanical action]* auslösen, entsperren, freigeben, freisetzen

triple *[Having three parts, being of three types, being three times as great, and so on]* dreifach, dreimalig

triple bolts lock → lock with three bolts

triple coding Dreifachkodierung, Dreifachcodierung *f*

triple-layer diode → trigger diode

triple point *[Particular temperature and pressure at which three different phases of one substance can coexist in equilibrium. Water is an example of a substance that has a well-know triple point]* Tripelpunkt, Dreifachpunkt *m*

triple-pole, three-pole dreipolig

triple precision *[Computer programming. Of or relating to processing data that has been stored into three machine words in order to increase precision]* Dreifachpräzision *f*

triple (thread) screw *[It has three parallel threads]* Dreigewindeschraube *f*

tripper → trip(ping) mechanism

tripping → releasing

trip(ping) band, releasing band Auslöseband, Entsperr(ungs)band *n*

trip(ping) characteristic, releasing characteristic Auslösekennlinie, Auslösecharakteristik, Entsperr(ungs)kennlinie, Entsperr(ungs)charakteristik *f*

trip(ping) command, releasing command Auslösekommando, Entsperr(ungs)kommando *n [Siehe Anmerkung unter „Kommando"]*

trip(ping) contact, releasing contact Auslösekontakt, Entsperr(ungs)kontakt *m*

trip(ping) factor Auslösefaktor, Entsperr(ungs)faktor *[Messgerät]*; Ausschalt(ungs)faktor *m*

tripping independent of line voltage, releasing independent of line voltage netzspannungsunabhängige Auslösung, netzspannungsunabhängige Entsperrung *f*

trip(ping) key, release key, releasing key Freigabetaste, Auslösetaste, Entsperr(ungs)taste *f*

trip(ping) latch, releasing latch Auslöseklinke, Entsperr(ungs)klinke *f*

trip(ping) limit, releasing limit Auslösgrenze, Entsperr(ungs)grenze *f*

trip(ping) linkage, releasing linkage Auslösegestänge, Entsperr(ungs)gestänge *n*

trip(ping) magnet, releasing magnet Auslösemagnet, Entsperr(ungs)magnet *m*

trip(ping) mechanism, releasing device, releasing mechanism, trip, trip(ping) device, tripper Auslösevorrichtung, Entsperr(ungs)vorrichtung *f*, Auslöser, Entsperrer *m*

trip(ping) relay, releasing relay Auslöserelais, Entsperr(ungs)relais *n*

trip(ping) segment, releasing segment *[dial]* Auslösesegment, Entsperr(ungs)segment *n*

trip(ping) signal, releasing signal Auslösesignal, Entsperr(ungs)signal *n*

trip(ping) speed, releasing speed, trip(ping) r.p.m., releasing r.p.m., trip(ping) rotations per minute, releasing rotations per minute Abschaltdrehzahl, Ausschaltdrehzahl, AUS-Drehzahl *f*

trip(ping) temperature, releasing temperature, opening temperature Auslösetemperatur, Öffnungstemperatur, Entsperr(ungs)temperatur *f [Sprinkler]*

trip(ping) time, releasing time, opening time Auslösezeit, Entsperr(ungs)zeit, Öffnungszeit *f [Sprinkler]*

trip-wire contact Fadenzugkontakt *m*, Fadenzugberührung *f*

trip-wire switch Fadenzugschalter *m*

trislot rotor Dreinutenläufer *m*

tristate driver hochohmige Treiber(baustein), hochohmige Puffer(baustein) *m*

tristate logic, TSL hochohmige Logik *f*

tristate output Gegentaktausgänge *f* [*Sie können 3 Zustände annehmen: H-Pegel, L-Pegel und hochohmig (neutral). Sie passen Schaltungen an ein Bausystem an*]

troffer (luminaire) → (ceiling) troffer (luminaire)

trolley Laufrolle *f*

trolley busway Schienenverteiler mit Stromabnehmerwagen, Schienenverteiler mit fahrbarem Stromabnehmer *m* [*VDE 0660, Teil 502*]

trolley collector Stangenstromabnehmer, Stromstangenabnehmer *m*

(trolley) track, overhead track Laufschiene, Deckenschiene *f* [*Tür; Tor*]

trouble → fault

trouble shooting, failure shooting, fault shooting Fehlerbehebung, Fehlerkorrektur, Fehlerbeseitigung, Fehlerberichtigung *f*

trouble signal message → failure signal message

trough Wanne *f*, Trog *m*

trough lavatory, lavatory trough Waschwanne *f*, Waschtrog *m*

trough luminaire Wannenleuchte, Leuchtwanne *f*, Trogleuchte, Leuchttrog *m*

trough (roof) gutter → box(ed) (roof) gutter

trough urinal → urinal trough

trouser piece → breeches pipe

throw-away filter → disposable filter

true alarm → valid alarm

True Color (US); True Colour (Brit.) Farbtiefe *f* 24 Bit

true colour (Brit.); true color (US) naturgetreue Farbe *f*

true current → active current

true indication Istanzeige *f*

trueness Richtigkeit *f* [*Qualitätssicherung. DIN 55 350. Teil 13*]

trueness → accuracy

trueness of the mean → accuracy of the mean

trueness to size → accuracy to size

true specific heat, specific true heat wahre spezifische Wärme(kapazität), spezifische wahre Wärme(kapazität) *f*

trumpet shape, trumpet form Trompetenform, Trompetengestalt *f*

trunk-and-branch duct arrangement Hauptkanal-Abzweigkanal-Führung *f*

trunk cable Hauptleitungskabel, Fernkabel, Sammelkabel *n* [*Koaxialkabel zur Verteilung von RF-Signalen über große Entfernungen*]

trunk circuit Amtsleitung *f* [*Ein verbindender Übertragungskanal zwischen einer Vermittlungsstelle an einem Ort und einer Vermittlungsstelle an einem benachbarten Knoten*]

trunk distribution frame, trunk distributing frame (US); main distribution frame, main distributing frame, MDF, principal distribution frame, principal distributing frame Hauptverteiler *m* [*Fernsprechwesen*]

trunking, trunk duct [*Where a very large number of cables have to follow the same route it is often better to use trunking in place of many conduits; trunking consists of a metal trough with detachable lid and conduit or m.i.c.c. cables may be connected to it*] Installationskanal *m*

trunking → (air) ductwork

trunking luminaire → channel(-mounting) luminaire

trunking potentiometer Abgleichpotentiometer, Justierpotentiometer, Abgleichpotenziometer, Justierpotenziometer *n*

trunking unit Schienenkasten *m* [*Installationskanal*]

trunk line Fernleitung *f*

trunk piston compressor Tauchkolbenverdichter, Tauchkolbenkompressor *m*

truss → (steel) truss

truss → (escalator) truss

truth function Wahrheitsfunktion, Logikfunktion *f*

truth table Wahrheitstabelle, Logiktabelle *f*

TSC → time slot code

TSL → tristate logic

TSOS → time-sharing operating system

T-square → tee-square

TSR → transient but stay resident

TSS, time-sharing-system Teilzeitsystem *n*

TTC, transit through-connect signal Transit-Durchschalte-Kennzeichen [*dezentrale Zeichengabe*], Antwortkennzeichen für Transitgleichschaltung *n* [*zentrale Zeichengabe*]

TTL → transistor-transistor logic

TTS → temporary threshold shift

tub *[May be a bath tub or wash tub]* Wanne *f*

tube Röhre *f*

tube bundle, tube nest Röhrenbündel, Röhrennest *n*

tube bundle condenser, tube nest condenser Röhrenbündelkondensator, Bündelröhrenkondensator *m*

tube bundle heat exchanger, tube bundle heat exchange unit, tube bundle heat interchanger *["Nest" may synonymously be used for "bundle"]* Röhrenbündelwärme(aus)tauscher, Bündelröhrenwärme(aus)tauscher, Röhrenbündelwärmeübertrager, Bündelröhrenwärmeübertrager *m [Siehe Anmerkung unter „Wärme(aus)tauscher (im weiteren Sinne)"]*

tube bundle pass → pass

tube clip Röhrenschelle *f*

tube coil Röhrenschlange *f*

tube condenser, tube condens(at)ing unit, tube condensation unit, tube refrigerant condenser Röhrenkondensator, Röhrenverflüssiger *m [Siehe Anmerkung unter „Kondensator"]*

tube conveyor system Rohrpostanlage *f*, Rohrpostsystem *n*

tube efficiency Röhrenwirkungsgrad *m*

tube floor heating, floor tube heating Deckenröhrenheizung, Röhrendeckenheizung *f*

tube heater, tube heating unit Röhrenheiz(ungs)-körper *m*

tube heat exchanger, tube heat exchange unit, tube heat interchanger Röhrenwärme(aus)tauscher *[In der Industrie so genannt]*; Röhrenwärmeübertrager *m [Siehe Anmerkung unter „Wärme(aus)tauscher"]*

tube-in-tube condenser → double-pipe condenser

tube lamp *[B.S. 1853]* Röhrenlampe *f*

tube motor Röhrenmotor *m [für Rolläden und Markisen]*

tube nest, tube bundle Röhrenbündel, Röhrennest *n*

tube noise Röhrenrauschen *n*

tube pull(ing) space Röhrenziehraum *m*

tube (sectional) radiator *[See remark under "radiator"]* Rohrradiator, Röhrenradiator, Rohrgliederheiz(ungs)körper, Röhrengliederheiz(ungs)körper; Rohrheiz(ungs)körper, Röhrenheiz(ungs)körper *m [Fehlbenennungen] [Besonders in Hochdruckanlagen für Hochhäuser und Fernheizungen verwendet. Diese Radiatoren bestehen aus nahtlosen senkrecht angeordneten runden oder profilierten Rohren, die oben und unten in je ein gemeinsames Sammelrohr eingeschweißt sind. Die senkrechten Röhren sind entweder glatt oder es sind zur Vergrößerung der Heiz(ungs)fläche verschiedene Stahlbleche elektrisch angeschweißt]*

tube solar collector → solar tube collector

tube system Röhrensystem *n*

tube waste heat boiler Röhrenabhitzekessel, Abhitzeröhrenkessel *m*

tub filler → (bath)tub filler

tub mixer → (bath)tub filler

tubular ceiling plaster heating, ceiling plaster tubular heating Rohrdeckenheizung, Deckenputzheizung *f [Heizrohre im Deckenputz]*

tubular concrete floor heating, concrete floor tubular heating Rohrdeckenheizung, Tragdeckenheizung *f [Heizrohre im Beton der Tragdecke]*

tubular floor heating Rohr(fuß)bodenheizung *f*

tubular frame door Rohrrahmentür *f*

tubular frame latch lock Rohrrahmenfallenschloss *n*

tubular frame rose(tte) Rohrrahmenrosette *f*

tubular frame three-latch lock, tubular frame triple-latch lock Rohrrahmendreifallenschloss, Dreifallenrohrrahmenschloss *n*

tumbler lock → French lock

tumbler pin Zuhaltungsstift *m*

tumbler switch → sensitive switch

tumbling bay manhole → backdrop manhole

tunable abstimmbar

tundish Auffangtrichter *m [Druckvor(er)wärmer]*

tuner Kanalwähler *m [Fernseher]*

tuner UKW-Empfangsteil *m, n*

tuner Abstimm(ungs)vorrichtung *f*

tungsten filament lamp *[B.S. 232]* Wolframdrahtlampe *f*

tuning *[The process of adjusting the inductance or the capacitance or both in a tuned circuit, for example, in a radio, television, or radar receiver or transmitter, so as to obtain optimum performance at a selected frequency]* Abstimmen *n*, Abstimmung *f*

tuning Konfigurieren *n [z. B. eine Festplatte]*

tuning capacitor *[A variable capacitor used for tuning purposes]* Abstimmkondensator *m*

tuning circuit, tuned circuit *[A circuit whose components can be adjusted to make the circuit responsive to a particular frequency in a tuning range]* Abstimmschaltung *f*

tuning core *[A ferrite core that is designed to be moved in and out of a coil or transformer to vary the inductance]* Abstimmkern *m*

tuning diode Abstimmdiode *f*

tuning fork Stimmgabel *f*

T

tuning memory Abstimmspeicher *m*

tuning parameter Abstimmparameter *m [Siehe Anmerkung unter „Parameter"]*

tuning range Abstimmbereich *m [selten: n]*

tuning range power ratio Leistungsverhältnis *n* im Abstimmbereich *[Oszillationsröhre, DIN IEC 235, Teil 1]*

tunnel Tunnel *m*, Verknüpfung *f [lokale Netz]*

tunnel burner *[A burner in which combustion takes place in a refractory duct called a "burner tunnel"]* Tunnelbrenner *m*

tunnel detector, tunnel detecting device, tunnel detection device Tunnelmelder *m*, Tunnelmeldegerät *n*

tunnel diode, Esaki diode Tunneldiode, Esakidiode *f*

tunnel freezer, freezing tunnel, blast-freeze tunnel *[Elongated enclosure provided with rapid cold air circulation for the freezing of foodstuffs]* Gefrierkanal, Gefriertunnel *m*

tunnel reader Tunnelleser *m [Er wird aktiviert sobald das Gepäck durch einen Tunnel mit elektromagnetischem Feld geschleust wird. Der Kofferanhänger strahlt die gespeicherten Daten im Radiofrequenzbereich ab]*

tunnel shower bath *[A form of shower bath consisting of a passageway with impervious walls and a drained floor and having a succession of shower heads, or a sparge pipe, along its length controlled by one or more mixing valves]* Tunnelbrausebad *n*

tunnel terminal Buchsenklemme *f*

tunnel test → (wind) tunnel test

TUP, telephone user part Fernsprechbenutzerteil *m, n [Signalisierungssystem Nr. 7]*

turbine, turbomachine *[Fluid-energized acceleration machine for generating rotary mechanical power from the energy in a fluid stream]* Turbine *f*

turbine blade Turbinenschaufel *f*, Turbinenblatt *n*

turbo cold water set Turbokaltwassersatz *m*

turbo compressor Turboverdichter, Turbokompressor *m*

turbomachine → turbine

turbulence device → (air) turbulence device

turbulence boundary layer turbulente Grenzschicht *f*

turbulent buoyancy jet Freistrahl *m*

turbulent flow, eddy flow, turbulence, eddy current, turbulence current *[Fluid flow in which the velocity varies in magnitude and direction in an irregular manner throughout the mass]* turbulente Strömung, ungeordnete Strömung, Wirbelströmung, wirbelnde Strömung *f*, Wirbelstrom, turbulente Strom, wirbelnde Strom, ungeordnete Strom *m*

turbulent noise Turbulenzgeräusch *n*

turn Windung *f*

turnaround time Durchlaufzeit, Umschalt(e)zeit *f [Zeitspanne zwischen dem Zeitpunkt, bei welchem ein Auftrag an eine Instanz als erteilt gilt, und dem Zeitpunkt, zu dem das Auftragsergebnis am bestimmten Ort vorliegt. E DIN 44 306, Teil 7]*

turnbuckle, turn button *[Rotary bolt or fastening made in various forms. The common form is a simple bar secured by a screw in the center on which it rotates. In another form this bar is mounted on a circular plate. The terms are also applied to a catch having a sliding bolt operated by a rotating knob or T-handle]* Vorreiber, Wirbel *m*

turned clockwise nach rechts gedreht, im Uhrzeigersinn gedreht

turned off → switched off

turned-reed frequency meter → (vibrating-)reed frequency meter

turning joint, hinge joint Scharnier *n*

turning vane, turning baffle, turning deflector *[Adjustable device, usually in a duct, to change direction of a stream of liquid, gas, or air]* Drehschikane, (ein)stellbare Schikane *f*, Drehumlenker, (ein)stellbare Umlenker *m*

turnkey schlüsselfertig

turn knob *[Small knob with a spindle attached for operating the dead bolt of a mortise lock]* Drehknopf *m*

to turn off → to switch off

to turn on einschalten

turn-on sequence, start sequence, beginning sequence Beginn(ungs)folge, Anfang(s)folge *f*

turnstile; turnpike (US) Drehkreuz *n*, Drehsperre *f*

to turn up → to set up

tutorial Lehrbuch *n*

TÜV agency Technischer Überwachungsverein *m [Bundesrepublik Deutschland]*

tweeter ((loud) speaker) Hochtonlautsprecher, Hochtöner *m*

twice-turning lock, double turn lock, lock with double turn, dual turn lock, twin turn lock zweitourige Schloss, Zweitourschloss, doppeltourige Schloss, Doppeltourschloss *n*

twilight control, dawn control Dämmerungssteuerung *f [Rolladen]*

twilight sensor, dawn sensor Dämmerungssensor *m*

twilight switch, dawn switch Dämmerungsschalter *m*

twin ... → double ...

to twist verdrillen, verdrallen, verwinden

twist Drall *m*, Verwindung, Drillung, Verdrillung *f*

twisted nematic, TN verdrillt nematische Zelle, TN-Zelle *f*

twisted pair distributed data interface, TPDDI „Kupfervariante" *f* des LWL

twisted pair (wires), twisted wire pair Paar verdrillter Drähte, verdrillte Drahtpaar, verdrallte Drahtpaar, verdrillte Leiterpaar, verdrallte Leiterpaar, Paar verdrallter Drähte *n*, verdrallte Drähte, verdrillte Leitung, verdrillte Drähte, verdrallte Leitung *f*

twisted shielded pair, shielded twisted pair Paar verdrillter abgeschirmter Drähte, abgeschirmtes verdrilltes Drahtpaar, verdrilltes abgeschirmtes Drahtpaar, STP-Kabel, Paar verdrallter abgeschirmter Drähte *n*

twistfree drallsicher

twist-on connecting device, t.o.c.d. Drehklemme [*EVDE 0613, Teil 3*], Verbindungskappe *f*

twist outlet Drallauslass *m*

two-link international call Durchgangsverbindung über zwei internationale Leitungen, Einfachtransitverbindung, Transiteinfachverbindung *f*

two-pole-and-earth socket (Brit.) → socket with earth(ing) contact

two-position controller, two-position regulator Zweipunktregler *m* [*Regler zur Reg(e)lung einer Prozessgröße durch Ein-/Ausschalten einer Anlage oder eines Anlagenteils unter Berücksichtigung von zu parametrierenden Sollwert und Hysterese.*]

two-position PID ((closed) loop) controller, two-position proportional-plus-integral derivative ((closed) loop) controller [*"Two-state" may synonymously be used for "two-position"*] Zweipunkt-PID-Regler *m* [*Zweipunktregler mit PID-Verhalten durch eine verzögerte und nachgebende Rückführung hinter dem Vergleicher einwirkend. DIN 19 225*]

two-position proportional-plus-derivative ((closed) loop) controller [*"Two-state" may synonymously be used for "two-position"*] Zweipunkt-PD-Regler *m* [*Zweipunktregler mit PD-Verhalten, durch eine verzögerte Rückführung hinter dem Vergleicher einwirkend. DIN 19 225*]

two-pulse bridge connection → single-phase bridge connection

twos complement binary Zweierkomplement, B-Komplement *n* [*Das Zweierkomplement ermöglicht auch die Darstellung von negativen Werten und ist eine sehr häufig verwendete Form. Dabei werden positive Zahlen und die Null durch gewöhnliche Binärcodes dargestellt, negative Zahlen werden durch bitweise Inversion des entsprechenden positiven Wertes und anschließender Addition von 1 dargestellt. Die Zweierkomplementdarstellung steht in engem Zusammenhang mit Binärdarstellung mit Offset. Beispiel: Die Zahl 182 wird als 000010110110, die Zahl −182 als 111101001010 dargestellt.*]

twos complement coding Zweierkomplementkodierung *f*

two-step action controller Zweipunktregler *m*, Zweipunktregelvorrichtung *f*

TWX machine → teletypewriter

tymbre crest kronenartige Schornsteinaufsatz *m*, kronenartige Schornsteinhaube *f* [*„Kamin" kann synonym für „Schornstein" verwendet werden*]

Tyndall effect [*Visible scattering of light along the path of a light beam as it passes through a system containing discontinuities, such as the surface of colloidal particles in a colloidal solution*] Tyndalleffekt *m*, Tyndallwirkung *f*

type → design

type approval, type permission Typenzulassung, Typengenehmigung *f*

type designation Typenbezeichnung *f*

type face → font

type font Schriftart *f*

type H trip(ping) characteristic, type H operating characteristic Typ H Auslösekennlinie, Auslösekennlinie des Typs H *f*

type mnemonic Typkürzel *n*

type L trip(ping) characteristic, type L operating characteristic Typ L Auslösekennlinie, Auslösekennlinie des Typs L *f*

type of lock → lock type

type output Ausgabemeldung *f*

typetested typengeprüft, (bau)mustergeprüft

typewriter paper, typing paper Schreibmaschinenpapier *n*

typing pool, typewriter pool Schreibmaschinensaal *m*

U

UART → universal asynchronous receiver transmitter

U-bend → trap

UC, universal current Universalstrom *m* [*Gleich- und Wechselstrom*]

UCP → unit control panel

UDI → unrestricted digital information

UDM → universal driver model

U duct → rectangular duct

UJT → unijunction transistor

ultrahigh frequency ultrahohe Frequenz, Höchstfrequenz *f* [*470-790 MHz*]

ultralow temperature Tiefsttemperatur, Ultratieftemperatur *f*

ultralow voltge Tiefstspannung, Ultratiefspannung *f*

ultrarapid, ultra-quick, ultra-fast superschnell, ultraschnell

ultrasonic alarm, ultrasound alarm Ultraschallalarm *m*

ultrasonic appliance → ultrasonic device

ultrasonic atomizer, ultrasound atomizer Ultraschallzerstäuber *m*

ultrasonic barrier, ultrasound barrier Ultraschallschranke *f*

ultrasonic device, ultrasonic appliance, ultrasound device, ultrasound appliance Ultraschallgerät *n*

ultrasonic motion detector, ultrasound motion detector, ultrasonic motion detecting device, ultrasound motion detecting device, ultrasonic motion detection device, ultrasound motion detection device Ultraschallbewegungsmelder *m*, Ultraschallbewegungsmeldegerät *n*

ultrasonic pilot transmitter, ultrasound pilot transmitter Ultraschallpilotsender, Pilotultraschallsender *m*

ultrasonic receiver, ultrasound receiver Ultraschallempfänger *m*

ultrasonic scanning, ultrasound scanning Ultraschallabtastung *f*

ultrasonic space monitoring, ultrasound space monitoring Ultraschallraumüberwachung *f*

ultrasonic system, ultrasound system Ultraschallsystem *n*

ultrasonic transmitter, ultrasound transmitter Ultraschallsender *m*

ultrasound Ultraschall *m*

U

ultraviolet flame detection system UV-Flammenmelde-System, Ultraviolett-Flammenmelde-System *n*

ultraviolet flame detector, ultraviolet flame detection device, ultraviolet flame detecting device [*See remark under "flame detector"*] Ultraviolettflammenmelder, UV-Flammenmelder *m*, Ultraviolettflammenmeldegerät, UV-Flammenmeldegerät *n*

ultraviolet image converter, ultraviolet imaging converter Ultraviolett-Bildwandlerröhre *f* [*Siehe Anmerkung unter „Bildwandlerröhre"*]

ultraviolet light erasing Löschen *n* mit ultraviolettem Licht [*Vorgang, der die Daten eines EPROMs mittels UV-Licht mit genau definierter Wellenlänge und Intensität löscht. Danach können neue Programme in das EPROM einprogrammiert werden*]

unacknowledged data transfer → data-transfer-unacknowledged

unassigned unbelegt, unbesetzt, frei verfügbar [*z. B. Kanal; Klemme*]

unattended unbedient, bedienerlos

unauthorized unbefugt, unberechtigt

unauthorized entry → trespass

unauthorized intervention unbefugte Eingriff, unberechtigte Eingriff *m*

unauthorized person Unbefugter, Unberechtigter *m* [*unauthorized male person*], Unbefugte, Unberechtigte *f* [*unauthorized female person*]

unbalanced output → single-ended output

unbored key → French key

UNC → outstation

unchokable → non-chokable

unconditioned unklimatisiert, nichtklimatisiert

unconditioned air, raw air unbehandelte Luft, Rohluft *f*

unconditioned jump unbedingte Sprung *m* [*Er ist an keine Bedingung gebunden*]

unconfirmed data-transfer → data-transfer unacknowledged

unconserved (air) conditioning plant, unconserved (air) conditioning installation Klima(tisierungs)-anlage *f* ohne Wasserrücknahme

unsolicited message telegram, USM telegram Spontanmeldung *f* einer Inselzentrale an die Leitzentrale

uncontrolled adiabatic air washer ungeregelte adiabatische Luftwäsche(r), adiabatische ungeregelte Luftwäsche(r) *f, (m)*, Düsen-Stockbefeuchter, Düsen-Stockanfeuchter *m*

to uncover → to bare

to undercool → to subcool

undercooler → supercooler

undercooling → supercooling

undercounter, (base) cabinet Unterschrank, Unterbau *m*

undercounter appliance, (base) cabinet appliance, undercounter device, (base) cabinet device [*kitchen*] Unterbaugerät, Unterschrankgerät *n* [*Es ist unter einer durchgehenden Arbeitsfläche integriert*]

undercounter dishwasher, (base) cabinet dishwasher Unterbau-Geschirrspüler, Unterschrank-Geschirrspüler *m*, Unterbau-Geschirrspülmaschine, Unterschrank-Geschirrspülmaschine *f*

undercounter (kitchen) range, (base) cabinet (kitchen) range Unterbau(küchen)herd, Unterschrankherd *m*

undercounter refrigerator, (base) cabinet refrigerator Unterbaukühlschrank, Tischkühlschrank, Kühlunterschrank *m* [*Er wird unter der Arbeitsplatte neben anderen Unterschränken aufgestellt*]

undercounter storage heater, (base) cabinet storage heater Unterbauspeicherheizer, Unterschrankspeicherheizer *m*

underdamped system PT2-System *n* mit d<1 [*Schwingfeld*]

underflow Unterlauf *m*, Unterschreitung *f* [*Bereichsüberschreitung bei einer arithmetischen Operation nach unten*]

underground cable Untergrundkabel *n*

underground cable entry Kabeleinführung *f* im Keller [*elektrische Hausanlage*]

underground cable service, cable main service Kabelhausanschluss *m*

underground heating oil storage unterirdische Heizöllagerung *f*

underground hydrant Unterflurhydrant *m*

underground space center (US); underground space centre (Brit.) Untertage-(Welt)Raumfahrtzentrum *n*

underground storage unterirdische Lagerung *f*

underlayed unterlagert

to underline unterstreichen

under lock and key verschließbar

under min operator Unter-MIN-Operator *m* [*Alle t-Normen außer MIN*]

undershoot Unterschwingen *n* [*Regelverstärker*]

undersized unterdimensioniert, unterbemessen

undersizing Unterbemessung, Unterdimensionierung *f*

under tension → **alive**

under-the-window fan coil (unit), under-the-window fan convector (unit) [*See remark under "fan coil (unit)"*] Ventilatorkonvektor unter Fenster, Gebläsekonvektor unter Fenster *m*

undertravel Unterfahrt *f* [*Aufzug*]

undervoltage [*Voltage is lower than rated*] Unterspannung *f*

undervoltage protection Unterspannungsschutz *m*

undervoltage release [*IEC 157-1*], **undervoltage opening release** Unterspannungsauslöser *m*

underwater television examination Unterwasserfernsehuntersuchung *f*

under-window air conditioner, under-window air conditioning unit [*See remark under "air conditioner"*] Unterfensterklima(tisierungs)gerät *n*

under-window cooler Unterfensterkühl(ungs)gerät *n*, Unterfensterkühler *m*

undesired alarm → **unwanted alarm**

undesired signal → **unwanted signal**

undesired sound, unwanted sound, parasitic sound, spurious sound Stör(ungs)schall *m*

undirectional breakdown diode, undirectional zener diode Einwegzenerdiode *f*

to undo rückgängig machen, widerrufen

undo Rückgängigmachen *n*, Widerruf *m*

undo rückgängig, widerrufen

undrilled key → **French key**

unequal → **diversified**

unequal load → **diversified load**

uneven-numbered ung(e)radzahlig, ung(e)rade

uneven parity ung(e)rade Parität *f*

unfinished unfertig, nicht fertig

ungrounded (US); unearthed (Brit.) ungeerdet, nicht geerdet

unheated ungeheizt, unbeheizt, nicht geheizt, nicht beheizt [*Raum*]

unhibited → **disabled**

unhygienic → **unsanitary**

unifilar → **single-wire**

uniflow (type) compressor Gleichstromverdichter, Gleichströmungsverdichter *m*

uniformity of illumination, uniformity of lighting Gleichmäßigkeit g1 und g2, Einheitlichkeit g1 und g2 *f*

Uniform Resource Locator → **URL**

uniform stationary flow → **stationary uniform flow**

unijunction transistor, UJT Zweizonentransistor, Transistor mit einem pn-Übergang, Doppelbasis-Transistor *m* [*Ein Siliziumtransistor mit Schalteigenschaften ähnlich der Vierschichtdiode*]

unilet Verteilerkasten *m*, Leergehäuse *n* [*Elektrotechnik*]

unimpeded → **unobstructed**

uninterrupted, continuous unterbrechungsfrei, stetig, ununterbrochen

uninterrupted operation → continuous operation

uninterrupted power supply, UPS unterbrechungsfreie Stromversorgung, ununterbrochene Stromversorgung, stetige Stromversorgung *f*

union operation Vereinigungsoperation bei Mengen, Mengen-Vereinigungsoperation, ODER-Operation *f*

unipolar (field effect) transistor Unipolartransistor, einpolige Transistor *m* [*Ein Transistor, an dessen Verstärkungsmechanismus nur eine Art von Ladungsträgern beteiligt ist*]

unipolar machine → acyclic machine

unipolar phototransistor Unipolarphototransistor, Photofeldeffekttransistor, Sperrschichtphototransistor, einpolige Phototransistor *m*

unit → (basic) unit

unit [*An assembly or device capable of independent operation, such as a radio receiver, cathode-ray oscilloscope, or computer subassembly that performs the same inclusive operation or function*] Einheit *f*, Einzelgerät, Aggregat *n*

unit → profile

unit → making/breaking unit

unit area Einheitsfläche *f*

unit-area impedance, specific acoustic impedance spezifische Schallimpedanz, Feldimpedanz *f* [*Quotient aus Schalldruck und Schallschnelle*]

unitary controller (US) → application-specific controller

unit circle Einheitskreis *m*

unit construction method → assembly construction method

unit consumption Einheitsverbrauch *m*

unit controller Einheitsregler *m* [*Diese Regler arbeiten an ihren Ein- und Ausgängen mit genormten Signalen (0...20mA, 4...20mA, bzw. 0,2...1,0 bar)*]

unit control panel, UCP [*More accessory control panels are grouped to serve a single remote system. Contains both accessory devices and local loop controls*] Steuer(ungs)tafelgruppe, Gruppensteuer(ungs)tafel *f*

unit data service Übertragung *f* ohne Verbindungsauf- und -abbau

unit current signal Einheitsstromsignal *n*

unit for interior work → profile for interior work

unit-guided dialogue exchange geführte Dialogverkehr *m*

unit heater [*It is used for industrial space-heating either with H.P.H.W. (high-pressure hot water), or steam. It is a forced convector consisting of a heater battery through which air is forced by means of an electrically driven fan with sufficient velocity to give a directional air stream. These heaters are made in sizes up to 500,000 Btu/hr*] Industriegebläsekonvektor *m*

unit-impulse function Einheitsimpulsfunktion *f*

unit-impulse response Einheitsimpulsantwort *f*

unitized components assembly, modular components assembly Zusammenbau *m* gleichförmiger (Bau)Elemente

unitized system → modular system

unit matrix Einheitsmatrix *f*

unit operation → process

unit priority system → (basic) unit priority system

unit pulse rate switch block → Baudot rate switch block

unit-ramp function Einheitsanstiegsfunktion *f*

unit-ramp response Einheitsanstiegsantwort *f*

unit record → input-output medium

units connection → connection of units

unit sign, unit symbol Einheitszeichen, Einheitssymbol *n*

units system → system of units

unit-step function Einheitssprungfunktion *f*

unit surface power → power per unit surface

unit time Einheitszeit *f*

unit wiring diagram Gerät(e)verdrahtungsplan *m* [*DIN 40 719, Blatt 1*]

unit without auxiliary power → (basic) unit without auxiliary power

unity Eins *f*

unity-feedback (closed) control system Regel(ungs)-system *n* mit direkter Gegenkopp(e)lung

unity gain frequency Transitfrequenz *f* [*Operationsverstärker*]

universal access number Service 130 *m*

universal asynchronous receiver transmitter, UART Universal-Sende- und -Empfängerbaustein *m* [*Ein integrierter Schaltkreis, der parallel anliegende Daten in einen seriellen Datenstrom wandelt*]

universal boiler, all-fuel boiler, all-purpose boiler Umstellbrandkessel, Universalbrandkessel *m* [*Er ist für eine Verfeuerung von festen, flüssigen und gasförmigen Brennstoffen geeignet. Der eine Kesselzubehör muss abgebaut und der andere aufgebaut werden. Nicht verwechseln mit „Wechselbrandkessel"*]

universal ((closed) loop) control, all-purpose ((closed) loop) control Universalreg(e)lung *f*

universal computer, all-purpose computer Universalrechner *m*, elektronische Großrechneranlage *f*

universal contactor, all-purpose contactor Universal(schalt)schütz *n*

universal controller, all-purpose controller Universalregler *m*

universal current → **UC**

universal device → **a.c.-d.c. device**

universal device → **allrounder**

universal digital computer, digital universal computer, all-purpose digital computer, digital all-purpose computer Universaldigitalrechner, universelle Digitalrechner *m* [*Er gliedert sich, grob gesehen, in Zentralprozessor, Zentralspeicher und Eingabe-Ausgabewerk. An das letztere sind Ausgabegeräte, Eingabegeräte und periphere Speicher angeschlossen*]

universal driver model, UDM, all-purpose driver model Universaltreibermodell *n*

universal functionality, all-purpose functionality umfassende Funktionalität, Einheitsfunktionalität, Universalfunktionalität *f*

universal governor → **all-speed governor**

universal limitation, all-purpose limitation Universalbegrenzung *f*

universal limit switch, all-purpose limit switch Universalbegrenzungsschalter *m*

universal network, all-purpose network Universalnetz(werk) *n*

universal network controller → **outstation**

universal receiver → **ac/dc receiver**

universal serial bus, serial universal bus, USB, all-purpose serial bus, serial all-purpose bus serielle Universalbus, serielle universale Bus, universale serielle Bus *m*

universal shunt, Arton shunt, all-purpose shunt [*A shunt used to increase the range of a galvanometer without changing the damping*] Mehrfachnebenwiderstand, Vielfachnebenwiderstand *m*

universal switch → **AC-DC switch**

universal sychronous/asychronous receiver/transmitter, USART [*"All-purpose" may synonymously be used for "universal"*] universelle Synchron/Asynchron-Empfänger/Sender-Baustein *m* [*Mikroprozessor*]

universal time → **Greenwich mean time**

universal time coordinated → **UTC**

unknown quantity Dunkelziffer; Dunkelzahl *f*

to unlatch aufklinken [*Tür*]

unless otherwise specified wenn nicht anders angegeben [*Beschriftung im Zeichnungsschriftfeld*]

to unlime → **to decalcify**

to unlist entadressieren [*Die Funktionen eines als Hörer eingestellten Gerätes aufheben*]

unloaded entlastet, unbelastet

unloader valve Leerlaufventil *n* [*Von Hand oder automatisch steuerbares Ventil, das in geöffnetem Zustand den ganzen Förderstrom einer Verdrängungspumpe nahezu drucklos von der Druckseite zur Saugseite oder zum Saugbehälter fließen lässt, sodass sich die Pumpe im Leerlauf befindet*]

to unlock, to deactivate entriegeln, aufschließen, aufsperren [*Schloss*]

unlocked, deactivated entriegelt, aufgesperrt, aufgeschlossen [*Schloss*]

unlocking → **(interlock) deactivation**

unlocking → **(door) unlocking**

unlocking security Aufsperr(ungs)sicherung, Entriegel(ungs)sicherung, Aufschließsicherung *f*

unmodified instruction → **basic instruction**

unobstructed, unimpeded, free [*e.g. passage*] unbehindert, frei [*z. B. Durchgang*]

unoccupancy Nichtbelegung, Nichtbesetzung *f*

unoccupied, nonoccupied unbelegt, leerstehend, unbesetzt, nicht besetzt, nicht belegt [*Gebäude*]

unoccupied mode → **night (setback) mode**

unplastered unverputzt, ungeputzt, roh

unplasticized weichmacherfrei, ohne Weichmacher

unpressurized, pressureless, depressurized drucklos, ohne Druck

unrestricted digital information, UDI uneingeschränkte digitale Information *f*

unsanitary, unhygienic unhygienisch

unscaled unit Grundeinheit, Basiseinheit *f* [*Messgerät*]

unscreened un(ab)geschirmt, nicht(ab)geschirmt

unshielded twisted pair, UTP un(ab)geschirmte verdrillte Kabelelement *n* [*Die Benennung unshielded twisted pair ist irreführend,weil neben einer paarweisen Verseilung auch eine Sternvierer-Konfiguration realisiert werden kann, und weil UTP-Kabel genauso wie STP-Kabel als ganzes mittels Geflechtschirm oder Metallfolie geschirmt sein können.*]

unsigned vorzeichenlos

unsigned integer vorzeichenlose ganze Zahl, ganze vorzeichenlose Zahl, ganze Zahl ohne Vorzeichen *f*

to untalk aufheben [*Sprecherfunktion*]

unthreaded gewindelos

untimed ohne Zeitzählung, zeitlos

untreated sewage → crude sewage

unvent(ilat)ed lüftungslos

unvent(ilat)ed water heater → pressure (type) water heater

unwanted alarm, undesired alarm, parasitic alarm, spurious alarm unerwünschte Alarm *m*

unwanted signal, parasitic signal, spurious signal, undesired signal unerwünschte Signal *n*

unwanted sound → undesired sound

unweighted signal-to-noise ratio Fremdspannungsabstand *m* [*Unterschied zwischen dem Pegel der Summe aus Nutzsignal und Störsignal (=Gesamtsignal) und dem Pegel des Störsignals*]

unzoned nichtzoniert

up-and-over gate Schwingtor *n* [*Ein senkrecht öffnendes Tor, meistens ein Garagentor*]

upangle Aufwärtswinkel *m*

upblow air curtain Aufwärtsluftstromschleier, Aufwärtsluftstromvorhang *m*

upcounter Aufwärtszähler, Vorwärtszähler, Hochzähler *m*

upcounting Hochzählen, Vorwärtszählen, Aufwärtszählen *n*

to update, to upgrade, to sidegrade aktualisieren, überarbeiten, modernisieren, nachführen

update, upgrade, sidegrade modernisierte Programmversion, aktualisierte Programmversion, überarbeitete Programmversion, nachgeführte Programmversion *f* [*Ersetzt eine Vorgängerversion*]

updated, upgraded, sidegraded geschult [*Personal*]

updating, upgrading, sidegrading Aktualisieren, Modernisieren, Überarbeiten, Nachführen *n*

updating in place Überschreiben *n* [*mit optischem Doppelsystem*]

up-down counter → bidirectional counter

upfeed(ing) (pipe) riser, upfeed(ing) (pipe) rising line Anstiegspeise(rohr)leitung *f*

upflow furnace [*Furnace in which the circulating air flows upward, discharging at or near the top of the casing*] Steigluftfeuerung, Steigluftfeuerstätte *f*

to upgrade → to update

upgraded → updated

upgrading → updating

upkeep → maintenance

to upload laden zum System, heraufladen [*Datei*]

upload(ing) [*The process of transferring an executable program(me) image or a database from a remote device in such a manner as to allow subsequent download*] Laden zum System, Heraufladen *n*

upper energy efficiency, top energy efficiency obere Wirkungsgrad *m* [*Diffusionskennzahl*]

upper hinge frog, top hinge frog Oberpfanne *f* [*Türangel*]

upper latch → latch

upper limit(ing) value → high limit(ing) value

upper valve part, top valve part Ventiloberteil *m,n*

upright (discharge) valve, pillar (discharge) valve Stand(ablauf)ventil, Steh(ablauf)ventil *n*

upright glass bulb sprinkler Stehschirmsprinkler, Standschirmsprinkler, Stehschirmsprenger, Standschirmsprenger *m*

upright valve, vertical valve Standventil, Stehventil *n*

UPS, uninterrupted power supply unterbrechungsfreie Stromversorgung *f*

upstream vorgeschaltet

upstream einspeiseseitig

upstream [*Electrically speaking in the direction toward(s) the power source*] stromaufwärts, quell(en)-seitig

upstream pressure [*fluid in motion*] Vordruck *m*

uptake → principal flue

uptime Betriebszeit *f* [*Zeit in der ein Rechner arbeitet oder betriebsbereit ist*]

upward air distribution [*Air is introduced at low level and extracted at the top*] Aufwärtsluftverteilung *f*

upward compatibility Aufwärtsverträglichkeit *f*

upward compatible aufwärtsverträglich [*Ein größerer Rechner verarbeitet Programme eines kleineren Rechners desselben Herstellers*]

upward travel → rise

urinal (bowl), bowl urinal Urinal(becken), Pissoirbecken, PP-Becken, Urinierbecken *n* [*DIN 1986*]

urinal flush(ing) cistern → flush(ing) urinal cistern

urinal flush(ing) device, urinal flush(ing) appliance Urinalspüler *m*, Urinalspülvorrichtung *f*, Urinalspülgerät *n*

urinal gull(e)y, urinal inlet Urinalgully, Urinalablauf, Urinaleinlauf, Urinalsinkkasten *m*

urinal gull(e)y valve, urinal inlet valve Urinalablaufventil, Urinalgullyventil, Urinaleinlaufventil, Urinalsinkkastenventil *n*

U

urinal installation Pissoiranlage, Urinalanlage *f* [*DIN 1986*]

urinal installation flush(ing) valve (Brit.); urinal installation flushometer (valve) (US) [*See remark under "flush(ing) valve"*] Urinalanlagen(druck)spüler, Pissoiranlagen(druck)spüler *m*

urinal outlet Urinalablauf, Pissoirablauf *m*

urinal partition Urinaltrennwand, Pissoirtrennwand *f*

urinal room Urinalraum, Pissoirraum *m*

urinal sewage Urinalabwasser, Pissoirabwasser *n*

urinal sparge pipe [*A perforated water pipe for flushing a urinal*] Urinalspülrohr, Pissoirspülrohr *n*

urinals row → **row urinals**

urinal stall → **stall urinal**

urinal stall range → **stall urinal range**

urinal (stench) trap, urinal water(-seal) trap Urinal(geruch)verschluss, Pissoir(geruch)verschluss *m*

urinal trough, trough urinal, urinal channel, channel urinal Urinalrinne, Pissoirrinne *f*, Rinnenurinal, Rinnenpissoir *n*

urinal valve Urinalventil, Pissoirventil *n*

urinal wall Urinalwand, Pissoirwand *f*

urinary calculus Urinstein *m*

URL, Uniform Resource Locator deskriptive Ressourcenadressen, Adressen im Internet, IP-Adressen *f* [*Sie beinhalten den jeweils angesprochenen Dienst (Web, E-Mail, News)*]

usable → **applicable**

USART → **universal synchronous/asynchronous receiver/transmitter**

USB → **universal serial bus**

use → **application**

use change → **change of use**

use frequency → **frequency of use**

useful energy Nutz(ungs)energie *f* [*Technische Form der Energie, welche der Verbraucher für den jeweiligen Zweck verwendet, z. B. Wärme, Kälte, Licht, mechanische Energie usw.*]

useful function Nutz(ungs)funktion *f*

useful latent air conditioner capacity, net latent air conditioner capacity [*Available refrigerating capacity of an air conditioner for removing latent heat from the space being conditioned*] Klimagerätnutzleistung *f* für latenten Wärmeentzug

usefulness Nützlichkeit *f*

(useful) sensible air conditioner capacity, net sensible air conditioner capacity [*Available refrigerating capacity of an air conditioner for removing sensible heat from the space being conditioned*] Klimagerätnutzleistung *f* zur Abführung fühlbarer Wärme

useful signal Nutzsignal *n*

useful signal level Nutz(signal)pegel *m*

(useful) total air conditioner capacity, net total air conditioner capacity [*Available refrigerating capacity of an air conditioner for removing sensible and latent heat from the space being conditioned*] gesamte Klimagerätnutzleistung *f*

useful sound Nutzschall *m*

useful water Nutzwasser *n* [*Es umfasst Trinkwasser und Brauchwasser*]

useful water heating by steam Nutzwasser(er)wärmung *f* mit Dampf

use identification → **(network) user identification**

useless imitation → **dummy**

user Benutzer, Anwender, Teilnehmer, Teilhaber, Verwender *m*

user Nutzer, Nutznießer *m*

user → **(network) user**

user address Benutzeradresse, Verwenderadresse, Anwenderadresse, Teilnehmeradresse, Teilhaberadresse *f* [*Grundnorm: DIN V 6779 T1, Fachnorm: DIN V 32 734 §6.1, Memotechnischer Text als alphanumerische Zeichenfolge zur Bezeichnung von Informationen in Menüs, Dialogen und für Kommandos. Die Benutzeradresse soll im Gesamtsystem eindeutig sein und in allen Ebenen (Feld bis Management) Verwendung finden. Die Adress-Struktur kann nach DIN V 32 734 auf Basis des Kennzeichnungssystems DIN V 6779 T1 (9/92) gewählt werden. Die Festlegung des Benutzeradress-Systems für ein Bauobjekt kann bereits mit der Bearbeitung der Informationsliste nach VDI 3814 Bl.2 (1993) bei der Planung erfolgen.*]

user applicability Anwenderfreundlichkeit, Benutzerfreundlichkeit *f*

user charge Benutzergebühr, Anwendergebühr, Verwendergebühr *f*

user-defined benutzerdefiniert, anwenderdefiniert, verwenderdefiniert

user-dependent benutzerabhängig, anwenderabhängig, verwenderabhängig

user environment Benutzerumgebung, Anwenderumgebung, Verwenderumgebung *f*

user error Benutzerfehler, Anwenderfehler, Verwenderfehler *m*

user facility Benutzerleistungsmerkmal, Anwenderleistungsmerkmal *n*

user friendliness Benutzerfreundlichkeit, Anwenderfreundlichkeit, Verwenderfreundlichkeit *f*

user-friendly, user-orien(ta)ted anwenderfreundlich, benutzergerecht, benutzerfreundlich, anwendergerecht, verwenderfreundlich, verwendergerecht

user group Benutzergruppe, Anwendergruppe, Verwendergruppe *f*

user guidance Benutzerführung, Anwenderführung, Verwenderführung *f*

user identification Benutzeridentifizierung, Benutzerkenn(zeichn)ung, Anwenderidentifizierung, Anwenderkenn(zeichn)ung, Verwenderidentifizierung, Verwenderkenn(zeichn)ung *f*

user interface Benutzerschnittstelle, Anwenderschnittstelle, Verwenderschnittstelle *f*

user language Benutzersprache, Verwendersprache, Anwendersprache *f*

user magnetic card Benutzermagnetkarte, Verwendermagnetkarte, Anwendermagnetkarte *f*

user memory → principal memory

user-orient(at)ed → user-friendly

user program (US); user programme (Brit.) Anwenderprogramm, Benutzerprogramm , Verwenderprogramm *n* [*Spezialprogramm zur Lösung einer Anwenderaufgabe. Gegensatz: Universal- oder Betriebsprogramm*]

user-related benutzerbezogen, anwenderbezogen, verwenderbezogen

user's choice, free choice freie Verfügung *f*

user terminal Benutzerstation, Anwenderstation *f*

user time Benutzungszeit, Nutzungszeit, Verwendungszeit, Anwendungszeit *f*

user(-written) program (US); user(-written) programme (Brit.) Benutzerprogramm, Anwenderprogramm, Verwenderprogramm *n* [*Gegensatz: Systemprogramm*]

use setpoint → utilization setpoint

U(-shape(d)) kitchen U-förmige Küche *f* [*Eine stirnseitig verbundene zweizeilige Küche mit umlaufender Arbeitsfläche*]

U

U(-shape(d)) section → channel

U(-shape(d)) tube manometer, double-legged manometer, dual-legged manometer, two-legged manometer [*"Differential pressure ga(u)ge" may synonymously be used for "manometer". The term "glass" may precede all terms given*] U-Rohr-Manometer *n*, U-Rohr-Differenzdruckmesser *m* [*Der in einem Schenkel (leg) des U-Rohres wirkende positive Überdruck drückt die Flüssigkeitssäule nach unten; um den gleichen Betrag steigt sie im anderen Schenkel nach oben*]

USM telegram → unsolicited message telegram

UT → Greenwich mean time

UTC, universal time coordinated, coordinated universal time europäische Sommerzeit *f* plus zwei Stunden [*Eine Weltzeit besonders für den Kurzwellenempfang*]

utility Dienst(leistungs)programm *n* [*Zur Erledigung von Aufgaben auf Betriebssystemebene. Es erleichtert das Verwalten von Dateien (löschen, erzeugen, editieren, sortieren usw.) und wird vom Betriebssystem bereitgestellt oder arbeitet als separates Programm.*]

utility Hilfsprogramm *n*

utility (company) → (public) utility (company)

utility core → service core

utility gas Netzgas *n*

utility line → utility service

utility metering Messung *f* öffentlicher Versorgungsleitungen

utility outlet Steckdose *f* für öffentliche Versorgung

utility power Netzstrom *m*

utility program (US); utility programme (Brit.) Dienstprogramm *n* [*Allgemeines Hilfsprogramm, das den Umgang mit einem Datenverarbeitungssystem erleichtert und verarbeitet*]

utility service, utility line, utility (run) (öffentliche) Versorgungsleitung *f*

utility shaft → (public) utility shaft

utility steam öffentliche Dampf *m*

utility substation → substation

utilizability Nutzbarkeit *f*

utilization break, utilization interruption Nutzungspause, Nutzungsunterbrechung *f*

utilization change → change of use

utilization coefficient → coefficient of utilization

utilization setpoint, use setpoint Nutz(ungs)sollwert *m*

utilization voltage, service voltage Netzbetriebsspannung, Betriebsspannung des Netzes, Gebrauchsspannung *f*

UTP → unshielded twisted pair

V

VA → **volt-ampere**

vacancy → **Schottky defect**

vacant connection freie Anschluss *m*, freie Verbindung *f*

vacuum Vakuum *n* [*Druck Null im leeren Raum*]

vacuum axial pump → **axial vacuum pump**

vacuum bottle → **Dewar flask**

vacuum breaker, air-admittance valve [*A valve that prevents the formation of a vacuum in a pipe system, such as a soil stack, where the vacuum might break the water seal*] Vakuumbrecher *m*

vacuum chamber, evacuated chamber Vakuumkammer *f*

vacuum chlorinator Vakuum-Chlorgasdosier(ungs)-anlage *f* [*DIN 19 606*]

vacuum ((closed) loop) control valve Vakuumregelventil *n*

vacuum condensate return feeder plant, vacuum condensate return feeder installation Vakuum-Kondensat-Rückspeiseanlage *f* [*Das erste Arbeitsspiel beginnt mit dem Einströmen von Dampf in den Behälter*]

vacuum disc, vacuum disk Vakuumscheibe *f*

vacuum electronics Vakuumelektronik *f*

vacuum fluorescent display, fluorescent vacuum display Vakuumfluoreszenzanzeigeeinheit, Fluoreszenz-Vakuum-Anzeigeeinheit *f*

vacuum freeze dryer, vacuum freeze drier [*A type of indirect batch dryer used to dry materials that would be destroyed by the loss of volatile ingredients or by drying temperatures above the freezing point*] Vakuumgefriertrockner *m*

vacuum freezing [*Freezing of a substance by lowering pressure to induce vapo(u)rization of a part of the solvent (usually water)*] Vakuumgefrieren *n*

vacuum gauge, vacuum tester, vacuum meter; vacuum gage (US) Vakuummessgerät, Vakuummeter *n*, Vakuummesser *m* [*„Unterdruck" kann synonym für „Vakuum" verwendet werden*]

vacuum heating [*A two-pipe steam heating system in which a vacuum pump is used to maintain a suction in the return piping, thus creating a positive return flow of air and condensate*] Vakuumdampfheizung, Dampfvakuumheizung *f*

vacuum insulation → **evacuated (thermal) insulation**

vacuum low-pressure steam heating, low-pressure vacuum steam heating Vakuum-Niederdruckdampfheizung, Niederdruck-Vakuumdampfheizung, Vakuum-ND-Dampfheizung, ND-Vakuumdampfheizung *f*

vacuum meter → **vacuum gauge**

vacuum porous insulation → **evacuated (thermal) porous insulation**

vacuum powder insulation → **evacuated (thermal) powder insulation**

vacuum pump [*A compressor for exhausting air and noncondensable gases from a space that is to be maintained at subatmospheric pressure*] Vakuumpumpe *f*

vacuum steam central heating, steam central vacuum heating Vakuumdampfzentralheizung, Dampfvakuumzentralheizung *f*

vacuum steam heating Vakuumdampfheizung, Dampfvakuumheizung *f*

vacuum steam heating plant, vacuum steam heating installation Vakuumdampfheiz(ungs)anlage, Vakuumdampfheiz(ungs)installation *f*

vacuum switch Vakuumschalter *m*

vacuum tester → **vacuum gauge**

vacuum tube collector Vakuumröhrenkollektor *m* [*Solarsystem*]

vaguely patterned leichtgemustert, schwachgemustert

val. → **(e)valuation**

valence Valenz *f* [*Klassifizierung der Zugriff(e)rechte in die vier Kategorien: LZ-Exklusivität, LZ-Vorrang, IZ-Exklusivität, IZ-Vorrang*]

valet towel holder [*Unit that holds wash cloths, face towels, bath towels, and a bath mat. Hooks on the bottom hold shoe shine cloths or other items. It is made in two-, three-, or four-guest sizes*] Handtuchhalter *m*

valid alarm, true alarm, danger alarm echte Alarm, Gefahrenalarm *m* [*Gegensatz: Falschalarm*]

valley gutter; centre gutter (Brit.); center gutter (US) [*A (roof) gutter having a flat sole and two sides angled to suit the slopes of adjacent pitched roofs*] (Dach)Kehl(rinn)e *f*

valuating → **(e)valuating**

valuation → **(e)valuation**

value Wert *m*, Größe *f*

value allocation, value assignment Wertzuordnung, Wertzuweisung, Wertzuteilung *f*

value call → **call by value**

value change → **change of value**

value(d) item, value(d) article, value(d) product Wertartikel *m*, Werterzeugnis, Wertprodukt *n*

value indicator Messwertanzeiger *m* [*Gerät, das die Messgröße mit der vorgegebenen Einheit als Messwert analog oder digital anzeigt*]

value of a variable, variable value Wert einer Größe, Größenwert *m*

value of property management → **property value management**

value stage Wertstufe *f*

value theorem Wertsatz *m*

valve Ventil *n*

valve → **slide valve**

valve arrangement Ventilanordnung *f*

valve authority → **authority (of a valve)**

valve body Ventilkörper *m*

valve bonnet Ventilhaube *f*, Ventilaufsatz *m*

valve chamber, slide valve chamber Schieberkammer *f*

valve chamber drain(ing) pipe, slide valve chamber drain(ing) pipe Schieberkammer-Entwässerungsrohr *n*

valve characteristic (curve) Ventilkennlinie, Ventilkennkurve, *kv*-Linie, *kv*-Kurve *f* [*Durchgang(s)ventil. Sie enthält den Durchfluss der Flüssigkeit in Abhängigkeit vom Hub bei einem konstanten Druckabfall von 1 bar. Je nach Kegelausbildung gibt es verschiedene Kennlinien; lineare (Tellerventile), quadratische, logarithmische (gleichprozentige) und andere*]

valve cone Ventilkegel *m*

valve control → **((closed) loop) valve control**

valve control → **((open) loop) valve control**

valve-controlled → **((closed) loop) valve-controlled**

valve-controlled → **((open) loop) valve-controlled**

valve controller Ventilregler *m*

valve (convector) radiator [*See remark under "radiator"*] Ventilradiator, Ventilgliederheiz(ungs)körper *m*

valved ventilbestückt

valve dimensioning → **valve sizing**

valve drive Schieberantrieb *m*

valve ((open) loop) control Schiebersteuerung *f*

valve operation platform Schieberbedien(ungs)bühne *f*

valve rangeability [*The ratio of the maximum controllable flow (fully open) to the minimum controllable flow. This minimum will be when the valve (Schieber) is not quite shut. There are always tolerances to consider but most manufacturers use a ratio of 40 or 50 to 1*] Fließverhältnis *n*

valve rose(tte) Ventilrosette *f*

valve sizing, valve dimensioning, slide valve sizing, slide valve dimensioning Schieberbemessung, Schieberabmessung, Schieberdimensionierung *f*

valve sizing chart for steam, steam (valve) sizing chart, slide valve sizing chart for steam, steam slide valve sizing chart Schieberbemessungstabelle für Dampf, Schieberdimensionierungstabelle für Dampf *f*

valve sizing chart for water, water valve sizing chart, slide valve sizing chart for water, water slide valve sizing chart [*To choose the correct valve size for hot or cold water the flow rate is required in litres per second or cubic metres per hour*] Schieberbemessungstabelle für Wasser, Schieberdimensionierungstabelle für Wasser *f*

valve stroke ((open) loop) control Ventilhubsteuerung *f*

valve taking non-steady flow into account Ventil *n* mit instationärer Gasströmung

valve(-type) arrester, valve(-type) surge diverter Kat(h)odenfallableiter *m*

vandal-proof, vandal-resistant zerstörungssicher, beschädigungssicher

van der Waals equation [*An empirical equation of state which takes into account the finite size of the molecules and the attractive force between them: p = [RT/ (v – b)] – (a/v2), where p is the pressure, v is the volume per mole, T is the absolute temperature, R is the gas constant, and a and b are constants*] van der Waals-Gleichung *f*

vane → **(sloping) vane**

vane, baffle, deflector Schikane *f*, Umlenker *m*

vane anemometer Flügelradanemometer *n*, Flügelradwindmesser *m*

vane ((closed) loop) control [*fan*] Drallreg(e)lung *f*

vane (water) meter → **rotary (water) meter**

vane wheel → **blade wheel**

vane wheel (inferential) meter Flügelradzähler, Läufer; Flügelradmesser *m* [*veraltete Benennung*]

vane wheel scanning Flügelradabtastung *f*

vanity Waschtisch *m*

vanity (base) cabinet → **vanity undercounter**

vanity battery Waschtischbatterie *f*

V

vanity bowl Waschtischbecken *n*

vanity corner-slab → corner vanity slab

vanity fitting Waschtischarmatur *f*

vanity pedestal Waschtisch(stand)säule *f*

vanity single-hole mixer (valve), vanity single-hole mix(ing) valve, vanity single-hole blender, vanity single-hole blending valve [*"One-hole" and "monobloc" may synonymously be used for "single-hole" and "battery" for "valve"*] Waschtisch-Einloch(misch)batterie *f*

vanity single-lever mixer, vanity one-lever mixer Waschtisch-Einhebelmischer, Einhebel-Waschtischmischer *m*

vanity slab Waschtischplatte *f*

vanity soap dispenser Waschtischseifenspender, Tischseifenspender *m*

vanity undercounter, vanity (base) cabinet, vanity (base) cubicle Waschtischunterbau, Waschtischunterschrank *m*

vapor (US); vapour (Brit.) [*An air dispersion of molecules of a substance that is liquid or solid in its normal state, i. e., at standard temperature and pressure. Examples are water vapo(u)r and benzene vapo(u)r. Vapo(u)rs of organic liquids are also loosely called "fumes"*] Dampf *m*

vapor concentration (US) → absolute humidity

vaporization (US) → evaporation

vapour (Brit.) → (water) vapour

vapour barrier (Brit.); vapor barrier (US) [*A moisture-impervious layer or coating, such as special paint, or a membrane on roofing felt or on building paper or on the inner (warm) surface of a concrete wall or floor, which prevents the absorption and condensation of moisture*] Dampfsperre *f*

vapour compression (Brit.); vapor compression (US) Wasserdampfverdichtung *f*

vapour compression cycle (Brit.), vapor compression cycle (US) Wasserdampfverdichtungszyklus *m*

vapour compression system (Brit.); vapor compression system (US) Wasserdampfverdichtungssystem *n*

vapourizing burner (Brit.); vaporizing burner (US) [*It is designed to utilize the temperature of the heated combustion chamber to vapo(u)rize the liquid fuel into such chamber*] Verdampfungsbrenner, Schichtungsbrenner *m* [*Nicht verwechseln mit „Zerstäubungsbrenner"*]

vapourizing temperature (Brit.); vaporizing temperature (US) Verdampfungstemperatur *f*

vapour precipitation (Brit.); vapor precipitation (US) Wasserdampfniederschlag *m*, Niederschlagen von Wasserdampf *n* [*aus der Luft*]

vapour pressure Reid (Brit.) → Reid vapour pressure

vapour retarder (Brit.); vapor retarder (US) [*These retarders are classified as rigid, flexible, or coating*] Wasserdampfverzögerer *m*

VAr controller → reactive-power controller

variable veränderlich

variable → adjustable

variable [*Anything which in concept or in practice is measurable or quantifiable, as a physical variable such as pressure, length, voltage, power, enthalpy, temperature, speed. "Quantity" often means variable but usually refers to an amount or numerical size. "Parameter" can be variable during design but is constant after construction and during use of equipment.*] Größe, Veränderliche *f*

variable access Lesen *n* und Schreiben von Veränderlichen [*azyklisch und zyklisch*]

variable air flow air conditioning plant, variable air flow air conditioning installation Klimaanlage mit veränderlichem Luftstrom, Klimainstallation mit veränderlichem Luftstrom *f*

variable air volume stream veränderliche Luftvolumenstrom *m* [*System mit lastabhängig geregeltem Luftvolumenstrom an den Versorgungsstellen*]

variable boiler temperature gleitend gefahrene Kesseltemperatur *f*

variable capacitance diode Kapazitätsdiode *f*, spannungsabhängige Kondensator *m*

variable clearance → adjustable clearance

variable flow, throttled flow Drosselfließen *n*, Drosselströmen, Drosselströmung *f*, Drosselfluss *m*

variable key → extension key

variable output [*For each element of a control system, a physical quantity whose changes are governed ultimately by the functions of the element*] veränderliche Ausgabe *f*, veränderliche Ausgang *m*

variable parameter veränderliche Parameter *m*

variable-reactance amplifier → parametric amplifier

variable refrigerant volume, VRV veränderliche Kältemittelvolumen *n*

variable resistor → (voltage) variable resistor

variable slide resistor, variable sliding resistor Stellschiebewiderstand *m*

variable-speed drehzahlgeregelt, drehzahlveränderlich

variable-speed drive electric act(uat)or, VSD act(uat)or drehzahlveränderliche Stellglied *n* [*Siehe Anmerkung unter „drehzahlgeregeltes Stellglied"*]

V

variable-speed starter Anlasser *m* mit n-Einstellung

variable tariff meter Mehrtarifzähler, Vieltarifzähler *m*

variable value → **value of a variable**

variable volume single-duct air conditioning system, variable volume single-duct air conditioning plant, variable volume single-duct air conditioning installation, variable volume one-duct air conditioning system, variable volume one-duct air conditioning plant, variable volume one-duct air conditioning installation Einkanalklimaanlage mit veränderlichem (Luft-)Volumenstrom, VVS-Anlage *f*

variance → **discrimination**

variance comparison, discrimination comparison Soll-Ist-Vergleich, Ist-Soll-Vergleich *m*

variance ratio → **discrimination ratio**

varipath throat besonders ausgeformte Horneingang *m* [*Beschallung*]

varistor → **(voltage) variable resistor**

varmeter, reactive volt-ampere meter Blindleistungsmesser, Blindleistungszähler *m*

varnish pin Lackstift *m* [*Zum Ausbessern von Lackschäden*]

to vary → **to adjust**

VBA → **Visual Basic for Applications**

VBD → **voice band data**

VCA, voltage-controlled amplifier spannungssteuerbare Verstärker *m*

VCO → **voltage-controlled oscillator**

VDE → **voice data entry**

VDR, voltage-dependent-resistor spannungsabhängige Widerstand, Varistor *m*

VDU → **CRT unit**

vector ((closed) loop) control feldorientierte Reg(e)lung *f*

vector graphics Vektorgraphik, Vollgraphik *f*

vector group Schaltgruppe *f* [*Trafo*]

vector interrupt independence gerichtete Unterbrechung *f* [*Unterbrechung, bei der jeder anfordernde Baustein ein eigenes Programm enthält*]

vector of estimation errors Schätz(ungs)fehlervektor *m*

vector of observer errors Beobachtungsfehlervektor *m*

vector plotter Vektorzeichenmaschine *f*

(vector) refresh(ing) display (Vektor)Wiederholanzeige *f*

(vector) refresh(ing) tube (Vektor)Wiederhol(bild)-röhre *f*

vector with magnitude, magnitude vector Vektor mit Betrag, Betragsvektor *m*

V(ee)-belt Keilriemen *m*

V(ee) reciprocating compressor, V(ee) piston compressor [*Two cylinders are unequally disposed on a common crank*] V-Kolbenverdichter, V-Kolbenkompressor *m*

vegetable (preparation) sink [*See remark under "sink"*] Gemüsespüle *f* [*1. Gemüsespülbecken. 2. Gemüsespültisch*]

vehicle (access) barrier, vehicle control barrier Fahrzeugsperre, Fahrzeugschranke *f*

vehicle check(ing) Fahrzeugkontrolle *f*

vehicle control barrier → **vehicle (access) barrier**

veiling reflection [*window glazing*] Schleierreflexion *f*

velocity [*The time rate at which a body changes its position vector; velocity is a vector quantity whose magnitude is expressed in units of distance over time, such as miles per hour. (From the Latin word for "speed")*] Geschwindigkeit *f* [*Die Geschwindigkeit ist ein Vektor*]

velocity energy Geschwindigkeitsenergie *f*

velocity head [*The square of the speed of flow of a fluid divided by twice the acceleration of gravity; it is equal to the static pressure head corresponding to a pressure equal to the kinetic energy of the fluid per unit volume*] Geschwindigkeitshöhe *f*

velocity potential [*Of a sound field in an isotropic fluid medium at rest or in a state of irrotational flow. A scalar potential function, the gradient of which is the negative of the particle velocity at a point in the field.*] Potential der Schallschnelle, Schnellepotential *n* [*Funktion, deren Gradient die Schallschnelle ist*]

velocity pressure Geschwindigkeitsdruck *m*

velocity pressure → **wind pressure**

velocity pressure side → **wind pressure side**

velocity profile Geschwindigkeitsprofil *n*

velocity triangle Geschwindigkeitsdreieck *n* [*Ventilator*]

vending machine Verkaufsautomat *m*

Venetian blind, jalousie, shutter blind [*An outside blind formed with slats like louvres that open and close*] Jalousie *f*, Rolladen *m*

vent [*Air-filled piping that prevents siphonage of trap seals or the bubbling of air through trap seals*] Verschluss *m*

vent [*Any opening designed to allow air, water or pressure to enter or escape from a confined space, as in a building or mechanical system*] Durchlass *m*

vent [*Pipe designed to convey the products of combustion from an appliance to a flue or chimney*] Verbrennungsabzug *m*

vented condensate return, open condensate return offene Kondensatrücklauf *m*

ven(tila)ted [*Provided with a means to permit circulation of air sufficient to remove an excess of heat, fumes, or vapo(u)rs*] entlüftet

ven(tila)ted [*Provided with a means to supply outside fresh air to a space*] belüftet

ven(tila)ted air heating plant, ven(tila)ted air heating installation Luftheiz(ungs)anlage mit Lüftung, Lüftungsheiz(ungs)anlage *f*

ventilating system, ventilating equipment [*It is used for supplying air to or removing air from room or space*] Lüftungsanlage *f*

ventilation [*The process of supplying outside fresh air to a space*] Belüftung, Lüftung *f*

ventilation [*The process of removing air in a space to the outside*] Entlüftung, Lüftung *f*

ventilation ability → ventilation power

ventilation air Lüftungsluft *f*

ventilation air cooling system Lüftungs- und Luftkühl(ungs)anlage *f* [*Die Anlage arbeitet mit angesaugter Außenluft*]

ventilation air heating system Lüftungs- und Luftheiz(ungs)anlage *f* [*Die Anlage arbeitet mit angesaugter Außenluft*]

ventilation aperture, ventilation opening, ventilation orifice Lüftungsöffnung *f*

ventilation appliance → ventilation device

ventilation assembler → ventilation fitter

ventilation assembly, vent(ilat)ing assembly [*"Set" and "kit" may synonymously be used for "assembly"*] Lüftungsgarnitur, Lüftungsgruppe, Lüftungskombination *f*, Lüftungs(bau)satz *m*, Lüftungsaggregat *n*

ventilation board Lüftungsleiste *f*

ventilation capability → ventilation power

ventilation cavity, vent(ilat)ing cavity Lüftungshohlraum *m*

ventilation ceiling [*Suspended ceiling containing many small apertures through which air, at low pressure, is forced downward from an overhead plenum dimensioned by the concealed space between the suspended ceiling and the floor or roof above*] Lüftungs(hänge)decke, Hängelüftungsdecke *f*

ventilation chimney, ventilating chimney Lüftungsschornstein, Lüftungskamin *m*, Lüftungsesse *f*

ventilation ((closed) loop) control Lüftungsreg(e)lung *f*

ventilation control centre (Brit.) → central ventilation station

ventilation controller Lüftungsregler *m*

ventilation degree day Lüftungsgradtag *m*

ventilation device, ventilation appliance Lüftungsgerät *n*, Lüftungsvorrichtung *f*

ventilation direct heating device Lüftungs-Direktheiz(ungs)gerät *n*

ventilation duct, ventilating duct Lüftungskanal *m*

ventilation duct detector, ventilation duct detecting device, ventilating duct sensing device Lüftungskanalmelder, Lüftungskanaldetektor *m*, Lüftungskanalmeldegerät *n*

ventilation duct humidifier Lüftungskanalbefeuchter, Lüftungskanalanfeuchter *m*

ventilation effectiveness Lüftungswirksamkeit *f* [*Nicht verwechseln mit „Lüftungswirkungsgrad"*]

ventilation engineer Lüftungsingenieur *m*

ventilation engineering Lüftungstechnik *f*

ventilation equipment [*See remark under "equipment"*] Lüftungsgerät(e) *n*, *(f)*

ventilation fitter, ventilation assembler Lüftungsmonteur, Lüftungsschlosser *m*

ventilation flat roof, flat roof with air circulation [*A roof where the layer of thermal insulation is installed near the ceiling rather than near the outside. The air in the roof space is therefore unheated by the building and will be cold in winter. Cold roofs are prone to condensation and need protection by vapo(u)r barriers and ventilation*] zweischalige (Flach)Dach, Kaltdach, doppelschalige (Flach)Dach, durchlüftete (Flach)Dach *n*

ventilation for human occupancy → comfort ventilation

ventilation grill(e), ventilation grate, ventilation grating Lüftungsgitter *n*

ventilation heat Lüftungswärme *f*

ventilation heat demand, ventilating heat requirement Lüftungswärmebedarf *m*

ventilation heat gain Lüftungswärmegewinn *m*

ventilation/heating/heat recovery system Lüftungs-/Heiz(ungs)-/Wärmerückgewinnungssystem *n*

ventilation heating system Lüftungs-/Heiz(ungs)-system *n*

ventilation heat loss Lüftungswärmeverlust *m*

ventilation hole, ventilation port Lüftungsloch *n*, Lüftungsöffnung *f*

ventilation hood Lüftungshaube *f*

ventilation installation → ventilation plant

ventilation jet Lüftungsstrahl *m*

ventilation lay-in ceiling Lüftungseinlegedecke, Einlegelüftungsdecke *f*

ventilation line Luftleitung, Lüftungsleitung *f* [*DIN 4741 Tab. 1, 2 und 5; DIN 4740 Tab. 1,2 und 5*]

ventilation luminaire Lüftungsleuchte *f*, Lüftungsbeleuchtungskörper *m*

ventilation nozzle, air nozzle [*See remark under "nozzle"*] Luftdüse, Be- und Entlüftungsdüse *f*

ventilation of building → building ventilation

ventilation opening → ventilation aperture

ventilation ((open) loop) control Lüftungssteuerung *f*

ventilation orifice → ventilation aperture

ventilation pipe [*A pipe (not being a drain) open to the external air at its highest point, which ventilates a drainage system either by connection to a drain or to a soil pipe or waste pipe and does not convey any soil water, wastewater, or rainwater. Code of Practice C.P.304: 1968 defines a vent(ilating) pipe as "a pipe provided to facilitate the circulation of air within the system and protect trap seals from excessive pressure fluctuations"*] Lüftungsrohr *n*

ventilation plant, ventilation installation Lüftungsanlage *f* [*Luft wird durch einen Zuluftventilator in einen Raum gefördert und durch einen Abluftventilator aus einem Raum abgesaugt*]

ventilation plant with air treatment, ventilation installation with air treatment Lüftungsanlage *f* mit Luftbehandlung

ventilation plant with dehumidification, ventilation installation with dehumidification Lüftungsanlage *f* mit Entfeuchtung

ventilation plant with heating and cooling, ventilation installation with heating and cooling Lüftungsanlage *f* mit Heizung und Kühlung

ventilation plant with humidification, ventilation installation with humidification Lüftungsanlage mit Befeuchtung, Lüftungsanlage mit Anfeuchtung *f*

ventilation plug, ventilation stopper Lüftungsstopfen, Lüftungspfropfen, Lüftungsstöpsel *m*

ventilation port → ventilation hole

ventilation power, ventilation ability, ventilation capability, ventilation capacity Lüftungsvermögen *n*, Lüftungsfähigkeit *f*

ventilation power, ventilation output, ventilation performance Lüftungsleistung *f*

ventilation rate Luftwechselrate *f*

ventilation rate procedure [*ASHRAE-Standard 62-1989*] Luftwechselratenverfahren *n*, Luftwechselratenmethode *f*

ventilation screw Lüftungschraube *f*

ventilation selector Lüftungswähler *m*

ventilation sensitivity, ventilation sensitiveness, ventilation susceptibility Lüftungsempfindlichkeit *f*

ventilation shaft, ventilation well Lüftungsschacht *m*

ventilation skirt(ing) board, skirt(ing) ventilation board Lüftungsscheuerleiste, Scheuerlüftungsleiste *f*

ventilation sliding window, sliding ventilation window Lüftungsschiebefenster, Schiebelüftungsfenster *n*

ventilation slot Lüftungsschlitz *m*

ventilation stack [*A vertical pipe installed primarily for the purpose of providing circulation of air to or from any part of the building-drainage system, and to protect the water seals of the traps from syphonage and back pressure*] Lüftungs(steig(e))leitung *f*

ventilation station Lüftungsstation *f*

ventilation stopper → ventilation plug

ventilation system [*All equipment intended or installed for the purpose of supplying air to or removing air from a room or space by mechanical means, other than equipment that is a portion of a comfort heating, cooling, absorption, or evaporative cooling system*] Lüftungsanlage *f*, Lüftungssystem *n*

ventilation time optimizer Lüftungszeitoptimierer *m*

ventilation unit [*A device, usually connected to ductwork, to move air, which also may clean and condition the air*] Lüftungseinheit *f*

ventilation unit for ceiling void mounting Lüftungseinheit *f* in Flachbauweise

ventilation valve, air valve [*A device to control flow and volume in air distribution*] Luftventil, (Be- und Ent-)Lüftungsventil, LVT *n*

ventilation valve [*battery*] Zellenventil *n*

ventilation vane Lüftungsfahne *f*

ventilation window Lüftungsfenster *n*

ventilator Lüfter *m* [*Er entlüftet Räume durch thermischen Luftauftrieb, also ohne Motorkraft. In der Praxis werden „Ventilatoren" (fans) oft „Lüfter" genannt*]

vent system [*Gas vent or chimney and vent connector assembled to form a continuous unobstructed passageway from the gas appliance to the outside atmosphere for the purpose of removing vent gases*] Gasabführ(ungs)anlage *f*

venture capital Risikokapital, Wagniskapital *n*

Venturi burner Venturibrenner *m*

Venturi flue Venturizug *m*

Venturi gas washer Venturigaswäsche(r) *f*,(*m*) [*Arbeitet nach dem Prinzip des Venturirohres*]

Venturi meter [*An instrument for efficiently measuring fluid flow rate in a piping system; a nozzle section increases velocity and is followed by an expanding section for recovery of kinetic energy*] Venturimesser *m*

Venturi nozzle Venturidüse *f*

Venturi tee, Venturi T-piece [*It directs water through the convector branches*] Venturi-T(-Stück) *n*

Venturi tube [*A constriction that is placed in a pipe and causes a drop in pressure as fluid flows through it, consisting essentially of a short straight pipe section or throat between two tapered sections; it is used to measure fluid flow rate (a venturi meter), or to draw fuel into the main flow stream, as in a carburet(t)or*] Venturiröhre *f*

Venturi tube [*misnomer*] → **burner (mixing) tube**

venue Veranstaltungsort *m*

verifiable nachweisbar, nachprüfbar

verification Bewahrheitung, Verifikation, Feststellung der Wahrheit *f*, Nachweis der Richtigkeit, Richtigkeit(s)nachweis *m*

verification tone Bewahrheitungston, Verifikationston, Nachweiston der Richtigkeit, Richtigkeit(s)nachweiston *m*

vernier scale Nonius, verschiebbare Messstabzusatz *m*

versatility Vielseitigkeit, Mehrseitigkeit *f*

version → **design**

vertex, node [*Of a graph. By the duality principle, edges may also be viewed as nodes; in network application, vertices are called nodes and edges are called branches*] Knoten *m*

vertical bar chart, vertical bar diagram senkrechte Balkendiagramm, vertikale Balkendiagramm *n*

vertical air heating equipment → **standing air heating equipment**

(vertical) (building) service chute [*A vertical tube or shaft that conveys and controls the fall of objects such as mail, laundry, and garbage to a lower level*] Schlucker *m*

vertical cable, downcomer [*lightning protection system*] Senkrechtkabel, Vertikalkabel *n*

vertical convector Standkonvektor, Stehkonvektor, Senkrechtkonvektor, Vertikalkonvektor *m*

vertical conveyance Senkrechtförderung, Vertikalförderung *f*, Senkrechttransport, Vertikaltransport *m*

vertical damper Senkrechtschieber, Vertikalschieber *m*, Senkrechtklappe, Vertikalklappe *f*

vertical frequency, field frequency Vertikalfrequenz, Bildwiederholfrequenz, Senkrechtfrequenz *f* [*Sie wird in Hertz angegeben und bezeichnet die pro Sekunde dargestellten Bilder. Je höher der Wert, desto geringer der Flimmereindruck*]

vertical grate Stehrost, Standrost *m*

vertical laundry chute Wäscheschlucker, Wäscheabwurfschacht *m*

vertically-unified cells, VUC zwei Zellen übereinander [*Elektrospeicher*]

vertical mail chute Postschlucker, Postabwurfschacht *m*

vertical mounting Standmontage, Stehmontage *f*

vertical outflow valve, outflow vertical valve Standauslaufventil, Standablaufventil, Standausflussventil, Standabflussventil, Standausfließventil *n* [*siehe Anmerkung unter „Auslaufventil"*]

vertical (passenger) transport(ation) [*For moving large numbers of people from one level to another*] Senkrechtpersonen(be)förderung, Vertikalpersonen(be)förderung *f*

vertical (pipe)line Fall(rohr)leitung, senkrechte (Rohr)Leitung, vertikale (Rohr)Leitung *f*

vertical reciprocating compressor, vertical piston compressor [*The cylinder or cylinders are arranged vertically*] Senkrechtkolbenverdichter, Senkrechtkolbenkompressor, Vertikalkolbenverdichter, Vertikalkolbenkompressor *m*

vertical redundancy check(ing), VRC Querparitätskontrolle *f* [*Magnetbandspeicher. „Kreuzsicherung" genannt, wenn spalten- und zeilenweise bei einem Datenblock angewendet*]

vertical safety pipe, safety vertical pipe Sicherheitsstandrohr, Standsicherheitsrohr *n*

vertical shell-and-tube condenser [*The water flows down the inner surface of the tubes in a single pass, without filling them. Its main advantages are economy of floor space and ease of cleaning, which can be carried out under load*] Senkrechtröhrenverflüssiger, Senkrechtröhrenkondensator, Vertikalröhrenverflüssiger, Vertikalröhrenkondensator *m*

vertical soap dispenser Standseifenspender *m*

vertical steam boiler Standdampfkessel, Senkrechtdampfkessel *m*

vertical transport(ation) Senkrechtförderung, Vertikalförderung *f*, Senkrechttransport, Vertikaltransport *m* [*in einem Gebäude*]

vertical transport(ation) engineering Senkrechtfördertechnik, Senkrechttransporttechnik, Vertikalfördertechnik, Vertikaltransporttechnik *f*

vertical transport(ation) plant, vertical transport(ation) installation Senkrechtförderanlage, Senkrechttransportanlage, Vertikalförderanlage, Vertikaltransportanlage *f* [*in einem Gebäude*]

vertical(-type) evaporator [*It consists of a series of parallel tubes arranged vertically*] Steilrohrverdampfer *m*

vertical valve → upright valve

very high speed integrated circuit hardware description language, VHDL Sprachformat *n* zur Beschreibung der Funktionseigenschaften von elektrischen Schaltungen

very large scale integration → VLSI

vessel Gefäß *n*

VF → voice frequency

VFC → voltage(-to)-frequency converter

VFD → virtual field device support

VHDL → very high speed integrated circuit hardware description language

vibration Vibration, Erschütterung *f*

vibration(al) severity Vibrationsschärfe, Erschütterungsschärfe *f*

vibration contact Erschütterungskontakt, Vibrationskontakt *m*

vibration detector, vibration detecting device, vibration detection device Erschütterungsmelder, Erschütterungsmeldungsgeber, Erschütterungsdetektor, Vibrationsmelder, Vibrationsmeldungsgeber, Vibrationsdetektor *m*, Erschütterungsmeldegerät, Vibrationsmeldegerät *n*

vibration eliminator mounting vibrationsfreie Aufstellung, vibrationssichere Aufstellung, erschütterungsfreie Aufstellung, erschütterungssichere Aufstellung *f*

vibration isolator Vibrationsdämpfer, Erschütterungsdämpfer *m*

vibratory detector, automatic detector, vibratory detecting device, automatic detecting device, vibratory detection device, automatic detection device automatische Melder, Vibrationsmelder, Erschütterungsmelder *m*, automatische Meldegerät, Vibrationsmeldegerät, Erschütterungsmeldegerät *n* [*Eine federnd gelagerte Masse wird durch Vibration zum Schwingen gebracht. Hierbei öffnet und schließt sich ein Kontakt mehrmals und unterbricht die Meldelinie*]

video Video *n* [*Unter dieser Benennung werden alle Verfahren zusammengefasst, die es ermöglichen, Bildinformationen aufzuzeichnen und wiederzugeben. Es wird unterschieden zwischen Videorecordern (einschließlich zugehörigen Komponenten, z. B. Videokamera) und Bildwiedergabegeräten, wie der Bildplatte*]

video amplifier Videoverstärker *m*

video camera Videokamera *f*

video channel Videokanal *m*

video clarity Sehdeutlichkeit *f*

video cross-bar Videokreuzschiene *f*

video data digital processing [*Digital processing of video signals for pictures transmitted over a television link; the computer compares each scanned line with adjacent lines and eliminates extreme changes caused by electromagnetic interference*] digitale Videodatenaufbereitung *f*

video digitizer Bilddigitalisierer *m* [*Gerät, das Bilder analoger Videosignale digitalisiert, sodass diese im Rechner bearbeitet und/oder auf Papier ausgedruckt werden können*]

videodisk, videodisc Bildscheibe *f*

video display unit → monitor

video frequency [*One of the frequencies existing in the output of a television camera when an image is scanned; it may be any value from almost zero to well over 4 megahertz*] Videofrequenz *f*

video (frequency) response Video(frequenz)gang *m*

video image Videobild *n*

video input Videoeingabe *f*

video intercom(munication) system Bildsprechanlage *f*, Bildsprechsystem *n*

video look-up table, VLUT Farbkodierer, Farbcodierer *m* [*graphische Datenverarbeitung*]

video luminaire, video light(ing) fitting, video light(ing) fixture Videoleuchte *f*

video monitoring Videoüberwachung *f*

video motion detection [*The electronic ability to visually monitor a designated area, make intelligent decisions as to whether motion has occurred and generate corrent responses based on the interpreted data*] Videobewegungsmeldung, Videobewegungsdetektion *f*

video motion detector, video motion detecting device, video motion detection device [*"Movement" may synonymously be used for "motion"*] Videobewegungsmelder, Videobewegungsdetektor, Videobewegungsmeldungsgeber *m*, Videobewegungsmeldegerät *n*

video on demand → pay television

video output Bildausgabe *f* [*Bildschirmgerät*]

video recorder [*A magnetic tape recorder capable of storing video signals for a television program(me) and feeding them back later to a television transmitter or directly to a receiver*] Videorecorder, Videorekorder *m*

video rectifier Videogleichrichter *m*

video remote transmission, remote video transmission Videofernübertragung *f*

video signal Videosignal *n*

video signal changeover switch, video signal transfer switch, video signal double throw switch, video signal throw-over switch Videosignalumschalter *m*

video tape [*A heavy-duty magnetic tape designed primarily for recording the video signals of a television program(me)*] Videoband *n*

video telemonitoring Videofernüberwachung *f*

video telephone [*A communication instrument which transmits visual images along with the attendant speech*] Bildtelefon *n*, Bildfernsprecher *m*

video telephony Bildfernsprechen, Bildtelefonieren *n*

video terminal → CRT unit

videotext → (broadcast) videotext

video transmission system, video transmission net(work) Videoübertragungssystem, Videoübertragungsnetz *n*

view Bearbeitungsauszug *m* [*Der Anwender definiert einen räumlichen und/oder thematischen Ausschnitt des Datenbestandes als eine logische Einheit, die er mit einem Namen versehen und unter diesem Namen abrufen kann*]

view Datenzusammenfassung *f* [*Zusammenfassung der Daten die in mehreren Tabellen gespeichert sind*]

view Ansicht *f*

viewer Projektor *m*

viewer Betrachter *m*

viewfinder Sucher *m*

(view(ing)) screen (Bild)Schirm *m* [*Datensichtgerät mit einfarbiger oder mehrfarbiger Darstellung*]

VIP Board [*VIP = VLB-ISA-PCI*] Hauptplatine *f* mit Steckplätzen für VLB (=Visa-Local-Bus) und PCI

virtual virtuell, scheinbar [*Als virtuell bezeichnete Daten oder Vorrichtungen bestehen physikalisch nicht, sie werden in einer Datenverarbeitungsanlage simuliert*]

virtual address [*A 32-bit integer identifying a byte location in virtual address space. The memory management hardware translates a virtual address to a physical address. The term "virtual address" may also refer to the address used to identify a virtual block on a mass storage device*] virtuelle Adresse, scheinbare Adresse *f*

virtual basic function virtuelle Grundfunktion, scheinbare Grundfunktion *f*

virtual circuit virtuelle Verbindung *f*, scheinbare Verbindung [*Ein verbindungsähnlicher Dienst im Rahmen der Software-Protokolle eines Netzwerkes, durch das zwei Endpunkte miteinander kommunizieren können, so als wäre eine physische Verbindung gegeben (logische, über Software realisierte Verbindung). Die wesentlichen Vorteile virtueller Verbindungen im Vergleich zu physischer Verbindung liegen in der Auflösung von Geschwindigkeitsabweichungen, in der Wiederholung von Übertragungen im Falle von Übertragungsfehlern und in der Umwandlung von Nachrichten, die über die Netzwerk-Verbindung laufen.*]

virtual (computer) memory virtuelle (Rechner)Speicher *m*, scheinbare (Rechner)Speicher

virtual field device support, VFD Statusabfrage *f* und Identifikation *f* [*Der für die Kommunikation sichtbare Teil eines Feldgerätes oder einer komplexen Funktion*]

virtual information address → address of virtual information

virtual memory, VM, virtual storage [*The set of storage locations in physical memory and on disk that are referred to by virtual addresses. From the programmer's viewpoint, the secondary storage locations appear to be locations in physical memory. The size of virtual memory in any system depends on the amount of physical memory available and the amount of disk storage used for non-resident virtual memory.*] virtuelle Speicher, scheinbare Speicher *m*

virus scanner Virensucher *m*

virus scanner program (US); virus scanner programme (Brit.) Virensuchprogramm *n*

virus tester Virenprüfer *m*

viscosity index [*Empirical number evaluated by comparing the viscosity of a substance with that of a standard substance*] Viskositätsindex, Zähflüssigkeitsindex *m*

viscous filter [*It consists of a series of oil coated plates or baffles so that the direction of the air is continually changed as it passes through the filter, and the dirt or dust is left sticking to the surface*] Kleb(e)filter *m*, *n*

visibility Sicht(barkeit) *f*

visibility distance, visible distance Sichtweite *f*

visibility factor → display loss

visible sichtbar

visible mounting Sichtmontage *f*

visible radiation Sichtstrahlung, sichtbare Strahlung *f* [*Frequenz* v_1]

vision field → field of vision

visitor location register, VLR Besucherregister, alphabetische Besucherverzeichnis *n*

visitor management Besucherverwaltung *f*

visual acuity → acuity

visual alarm (signal) [*See remark under "alarm (signal)"*] Sichtalarm(signal) *m*, (*n*), Sichtstör(ungs)signal *n*

visual alarm (signal) unit Sichtalarm(signal)anlage, Sichtstör(ungs)signalanlage *f*

visual announcement [*See remark under "announcement"*] Sichtmitteilung, Sichtankündigung *f*

visual area Blickfeld, Sichtfeld *n*

Visual Basic for Applications, VBA Programmiersprache *f* [*Sie ist in Access 7.0, Exced 7.0 und Visual Basic 4.0 integriert*]

visual call system Sichtrufsystem *n*

visual check(ing) Sichtkontrolle, Blickkontrolle *f*

visual communication Sichtkommunikation, optische Kommunikation *f*

visual environment Sichtumgebung, Blickumgebung *f*

visual field Sehfeld, Blickfeld *n*

visual indication Sichtanzeige, optische Anzeige *f*

visual indicator Sichtanzeiger, optische Anzeiger *m*

visualization Visualisierung, Sichtbarmachung *f*

visually prominent lighting system sichtbar hervortretende Beleuchtungssystem *n*

visually subordinate lighting system sichtbar unterlagerte Beleuchtungssystem *n*

visual performance Sehvermögen, Blickvermögen *n*, Sehfähigkeit, Blickfähigkeit, Sehleistung, Blickleistung *f*

visual plane Blickebene, Sichtebene *f*

visual signal Sichtsignal, Blicksignal *n*

visual smoke detector, visual smoke detecting device [*"Optical" may synonymously be used for "visual"*] Sichtrauchmelder, Sichtrauchdetektor, Rauchsichtmelder, Rauchsichtdetektor, optische Rauchmelder, optische Feuermelder, optische Brandmelder, Sichtfeuermelder, Sichtbrandmelder, Sichtbranddetektor, Sichtfeuerdetektor *m*

visual state indicator Sichtzustandsanzeiger *m*

visual system Sichtsystem *n*

visual transmission line optische Übertragungsstrecke *f*

visual water level gauge; visual water level gage (US) Sichtwasserstandsmesser *m*

vitiated air, foul air Schlechtluft, verbrauchte Luft *f*

vitiated air flue → foul air flue

VLR → visitor location register

VLSI, very large scale integration Integration mit mehr als 500 Gatter pro Chip, höchstintegrierte Sicherheit *f*

VLUT → video look-up table

VM → virtual memory

VMOS → vertical metal oxid(e) semiconductor

VOC → volatile organic compound

voice answer Sprachantwort, Sprechantwort *f* [*Datenübertragung*]

voice band data, VBD Sprachbanddaten, Daten im Sprachband, Sprechbanddaten, Daten im Sprechband *f*

voice-controlled, voice-actuated sprachgesteuert, sprechgesteuert

voice data entry, VDE Spracheingabe, Sprecheingabe *f*

voice frequency, VF, speech frequency, low frequency Sprechfrequenz, Tonfrequenz, Sprachfrequenz, Niederfrequenz, NF *f*

voice level, voice intensity, voice volume, voice loudness [*In sound recording, the relative intensity of a person's voice*] Stimmen(laut)stärke, Stimmenlautheit, Stimmenempfindungsstärke *f*

voice mail hörbare Post, Stimmenpost *f* [*Wenn der Hörer nicht abgenommen wird, werden ankommende Gespräche im zentralen Telefonrechner gespeichert und können dort, z. B. nach Eingabe einer persönlichen Schlüsselzahl, abgerufen werden*]

voice(-operated) switch, VOX sprechbetätigte Schalter, sprachbetätigte Schalter *m*

voice recognition Spracherkennung, Sprecherkennung *f*

voice trace Spracheingabegerät, Sprecheingabegerät *n*

void ungültig

void farbfreie Stelle *f* [*Im Zeichen bei der Datenverarbeitung*]

void → air space

void ceiling heating, cavity ceiling heating Hohlraumdeckenheizung, Deckenhohlraumheizung *f* [*Die Heiz(ungs)rohre liegen im Zwischenraum von Tragdecke (floor) und Zwischendecke (false ceiling)*]

volatile (computer) memory, volatile (computer) storage (device) nicht permanente (Rechner)Speicher, flüchtige (Rechner)Speicher, spannungsabhängige (Rechner)Speicher, Nichtdauer(rechner)speicher *m* [*Sein Inhalt geht bei Ausfall oder Abschalten des Stromes verloren*]

volatile organic compound, VOC, organic volatile compound flüchtige organische Verbindung, organische flüchtige Verbindung *f*

volatilization heat → vaporization heat (US)

voltage amplification → voltage gain

voltage backup Spannungs(unter)stützung, Spannungsreserve *f*

voltage circuit, potential circuit Spannungsschaltung *f*

voltage clamp Spannungsbegrenzer, Spannungsklemmer *m* [*Siehe Anmerkung unter „Begrenzer"*]

voltage ((closed) loop) control, voltage regulation [*The ability of a transformer to maintain secondary voltage under different load conditions*] Spannungsreg(e)lung *f*

voltage comparator, voltage comparing element Spannungsvergleicher *n*, Spannungskomparator *m*, Spannungskoinzidenzeinheit *f*

voltage-controlled [*close control*] spannungsgeregelt

voltage-controlled [*open control*] spannungsgesteuert

voltage-controlled amplifier → VCA

voltage-controlled oscillator, VCO spannungsgesteuerte Oszillator *m*

voltage-dependent spannungsabhängig

voltage-dependent-resistor → VDR

voltage divider Spannungsteiler *m*

voltage divider probe (Spannungs-)Tastteiler *m*

voltage driver Spannungs-Treiber *m*

voltage drop [*The decrease in voltage in the conductors of an electric circuit resulting from their resistance*] Spannungsabfall *m*, Spannungsabnahme *f*

voltage-fed spannungsgespeist, spannungsführend

voltage-free → floating

voltage-frequency converter → voltage(-to)-frequency converter

voltage gain, voltage amplification Spannungsverstärkung *f*, Spannungsgewinn *m*

voltage gradient → electric field intensity

voltage gradient level → electric field intensity level

voltage input Spannungseingabe *f*

voltage level Spannungsebene *f*, Spannungspegel *m*

voltage measuring range Spannungsmessbereich *m* [*selten*: *n*]

voltage module → supply module

voltage multiplier connection Spannungsvervielfacherschaltung *f*

voltage ((open) loop) control Spannungssteuerung *f*

voltage-operated earth-leakage circuit breaker Fehlerspannungs-Schutzschalter *m*

voltage output Spannungsausgangsleistung, Spannungsabgabeleistung, Spannungsleistungsabgabe, abgegebene Spannungsleistung *f*

voltage overshoot Spannungsüberschwingen *n*, Spannungsüberhöhung, Spannungsüberschwingweite *f*

voltage peak Spannungsspitze *f*

voltage range Spannungsbereich *m* [*selten*: *n*]

voltage regulation → voltage ((closed) loop) control

voltage regulator Spannungsregler *m*

voltage relay [*Device that is voltage sensitive with the ability to sense between two or more voltage levels to operate a set of contacts*] spannungsempfindliche Relais *n*

voltage return Spannungswiederkehr *f*

voltage signal Spannungssignal *n*

voltage source Spannungsquelle *f*

voltage stabilizer → (automatic) voltage regulator

voltage supply Spannungsversorgung *f*

voltage(-to)-frequency converter, VFC Spannungs-Frequenz-Umsetzer, Spannungs-Frequenz-Wandler *m*

voltage tolerance Spannungssicherheit *f*

voltage transfer Spannungsübertritt *m*

(voltage) variable resistor, varistor [*Electronics. A two-electrode resistor made of semiconductive material and having voltage-dependent nonlinear resistance that drops with an increase in applied voltage*] Stellwiderstand, veränderliche Widerstand *m*

voltage waveform, voltage waveshape Spannungsform *f*

volt-ampere, VA Voltampere *n*

volt-ampere acceptance Leistungsaufnahme *f* in VA [*(Schmelz)Sicherung*]

voltmeter changeover switch Spannungsmesser-Umschalter, Spannungsmesser-Wechselschalter *m*

volume → intensity [*acoustics*]

volume act(uat)or → (air) volume act(uat)or

volume ((closed) loop) control system [*An electronic system that regulates the signal amplification or limits the output of a circuit, such as a volume compressor or volume expander*] Volumenstärkeregel(ungs)system *n*

volume controller [*A potentiometer used to vary the loudness of a reproduced sound by varying the audio frequency signal voltage at the input of the audio amplifier*] Lautstärkeregler *m*

volume damper [*Device that, when installed, will restrict, retard, or direct the flow of air in any duct or the products of combustion in any heat-producing equipment, vent connector, vent, or chimney therefrom*] Volumenklappe *f*

volume expander → **(automatic) volume expander**

volume expansion Volumen(aus)dehnung, Raum(aus)dehnung *f*

volume flow → **(air) volume flow**

volume flow, volumetric flow Volumenstrom *m*, Volumenströmung *f* [*Querschnitt mal Strömungsgeschwindigkeit*]

volume flow ((closed) loop) controller → **(air) volume flow ((closed) loop) controller**

volume flow increase Volumenstromerhöhung, Volumenstromzunahme, Volumenströmungserhöhung, Volumenströmungszunahme *f*

volume flow regulator Volumenstromregler, VR *m*

volume flow sensor Volumenströmungsfühler, Volumenstromfühler, VF *m*

volume indicator, volume unit meter [*A standardized instrument for indicating the volume of a complex electric wave such as that corresponding to speech or music; the reading in volume units is equal to the number of decibels above a reference level which is realized when the instrument is convected across a 600-ohm resistor that is dissipating a power of 1 milliwatt at 100 hertz*] Volumeneinheitsmesser *m*

volume label Dateikennsatz *m*

volume lifetime [*Solid state physics. Average time between the generation and recombination of minority carriers in a homogeneous semiconductor*] Trägerlebensdauer *f*

volume meter → **bulk meter**

volume model Volumenmodell *n*

volume range [*Engineering acoustics. The difference expressed in decibels, between the maximum and minimum volumes of a complex audio-frequency signal occurring over a specified period of time*] Dynamikumfang, Dynamikbereich *m* [*selten*: *n*], Lautstärke(n)umfang, Lautstärke(n)bereich *m*

volume resistivity, specific insulation resistance [*Electrical resistance between opposite faces of a 1-centimeter cube of insulating material, commonly expressed in ohm-centimeters*] spezifische Raumwiderstand, spezifische Durchgangswiderstand, spezifische Innen-Isolationswiderstand *m*

volume system → **(air) volume system**

volume throughflow Volumendurchfluss *qv m* [*Die betrachtete Menge ist das Volumen*]

volume trailer label Datenträgerendekennzeichen *n*

volumetric compressor → **positive displacement compressor**

volumetric efficiency [*1. Ratio obtained by dividing the actual gas volume delivered by the displacement. 2. Ratio of the capacity of the compressor to the displacement of the compressor. The term does not apply to centrifugal compressors*] Volumenwirkungsgrad *m*

volumetric flow → **volume flow**

volumetric (gas) meter → **positive displacement (gas) meter**

volumetric protection [*This consists of filling the interior of the building with sonic, ultrasonic, or radio waves so that a standing pattern is related which, when disturbed by anyone or anything moving within the monitored area, will sense the movement and give the alarm*] Bewegungsschutz *m*

volume unit, VU [*A unit for expressing the audio-frequency power level of a complex electric wave, such as that corresponding to speech or music; the power level in volume units is equal to the number of decibels above a reference level of 1 milliwatt as measured with a standard volume indicator*] Volumeneinheit *f*

volume unit meter → **volume indicator**

volume velocity [*The rate of flow of a medium through a specified area due to a sound wave*] Schalldurchfluss *m*

vortex breakdown Axialströmung *f* mit ausgeprägter Rückströmzone und heftigem Stoffaustausch

vortex dynamics Wirbeldynamik *f*

vortex tube Wirbelröhre *f*

voting logic Bewertungslogik *f*

VOX → **voice operated switch**

VRC → **vertical redundancy check(ing)**

VRV → **variable refrigerant volume**

VSD act(uat)or → **variable-speed drive electric act(uat)or**

VU → **volume unit**

VUC → **vertically-unified cells**

vulnerable to fire → **fire vulnerable**

VxD virtuelle Gerätetreiber, scheinbare Gerätetreiber *m*

VXO spannungsgesteuerte Quarzoszillator *m*

V

W

wafer [*A (thin) semiconductor slice on which matrices of microcircuits can be fabricated, or which can be cut into individual slices for fabricating single transistors and diodes*] Siliziumscheibchen *n*

wafer switch Stufenschalter *m*

wafer tag [*See remark under "tag"*] Siliziumscheibchenetikett *n*

waffle slab Kassettenplatte *f*

wage snatch Entreißdiebstahl *m*

waist level turnstile; waste level turnpike (US) Taillenhöhedrehkreuz *n*

wait Warten *n*, „Warte" [*Eine Gehäuseanschlussbezeichnung beim Mikroprozessor 8080. Dieses Signal zeigt an, dass sich der Mikroprozessor in der Wartestellung befindet*]

waiting interval Wartezeit *f* [*Fahrstuhl*]

waiting position Halteposition, Warteposition *f* [*Anruf*]

waiting time music Warte(schlange)musik *f* [*Fernsprecher*]

wait request, wait demand Warteanforderung *f*

walk-in duct begehbare Kanal *m*

walk-in pit begehbare Grube *f* [*Aufzugschacht*]

walkway Gehweg *m*

wall → **hoistway wall (US)**

wall accent luminaire Wandakzentleuchte, Akzentwandleuchte *f*

wall air cooler, (air) wall cooler Wandluftkühler *m*

wall air grill(e) → **(air) wall grill(e)**

wall air inlet, (air) wall inlet Wandlufteinlass, Luftwandeinlass *m*

wall air outlet, (air) wall outlet Wandluftauslass, Luftwandauslass *m*

wall (area) heating Wand(flächen)heizung *f*

wall ash tray Wandaschenbecher *m*

wall (basin) faucet (US) → **wall (water) tap**

wall box Wandkasten *m*

(wall) bracket (Wand)Konsole *f*

wall breakthrough, wall opening, wall penetration Wanddurchbruch *m*, Wanddurchführung *f*

wall built-in casing, wall inbuilt casing, wall built-in case, wall inbuilt case Wandeinbaugehäuse *n*

wall cabinet → **wall (stock kitchen) cabinet**

wall connection Wandanschluss *m*, Wandverbindung *f*

wall convector → **wall(-hung) convector**

wall cooler → **wall air cooler**

wall crane [*Scottish term*] → **wall (water) tap**

wall diffuser, wall diffusor Wanddiffusor *m*

wall distance Wandabstand *m*

wall draw(ing) off tap → **wall (water) tap**

wall duct Wandkanal m

wall fan Wandventilator *m*

wall filter Wandfilter *m, n*

wall finish Wandoberflächengestaltung *f*

wall (fire) sprinkler [*Sprinkler designed to be installed in piping along the sides of a room instead of with normal spacing. See remark under "sprinkler"*] Wandsprinkler, Wandsprenger *m*

(wall) gas heater → **(recessed) (wall) gas heater**

wall grill(e) → **(air) wall grill(e)**

wall heating Wandheizung *f*

wall heating surface Wandheiz(ungs)fläche *f*

wall (heat) unit, wall heater [*Either of two major types. 1. Electrical wiring embedded in glass or ceramic tile to radiate heat. 2. Fixture recessed in a wall deep enough to enclose resistance coils and sometimes fans. Note: This term may apply also to a gas heater*] Wandheizer *m*, Wandheiz(ungs)gerät *n*

wall hook Wandhaken *m*

wall(-hung) air cooler, wall-mounted air cooler Wandluftkühler *m*

wall(-hung) air heater, wall-mounted air heater Wandlufterhitzer *m*, Wandluftheizer, Wandluft(er)wärmer, Wandluftheiz(ungs)gerät *n*

wall(-hung) bidet, wall-mounted bidet Wandbidet, wandhängende Sitzwaschbecken, Wandsitzwaschbecken, wandhängende Bidet *n*

wall(-hung) boiler, wall-mounted boiler Therme *f*, Wandheiz(ungs)kessel *m*

wall(-hung) casing, wall-mounted casing Wandgehäuse *n*

wall(-hung) closet, wall(-hung) toilet, wall(-hung) W. C.; wall(-hung) water closet, wall-mounted closet, wall-mounted toilet, wall-mounted W. C., wall-mounted water closet Wandabort, Wandabtritt *m*, Wandtoilette *f*, Wandklosett, Wand-W. C. *n* [*freischwebend*]

wall(-hung) convector, wall-mounted convector Wandkonvektor *m*

wall(-hung) design, wall-mounted design wandhängende Ausführung *f*

wall(-hung) discharge valve, wall-mounted discharge valve, wall mounted drain(ing) valve, wall(-hung) drain(ing) valve Wandauslaufventil, Wandablaufventil *n*

wall(-hung) (door) buffer, wall-mounted (door) buffer Wand(tür)puffer *m*

wall(-hung) gas air heater, wall-mounted gas air heater Wand-Gas-Luftheizer, Wand-Gas-Luft(er)wärmer, Wand-Gas-Lufterhitzer *m*

wall(-hung) gas geyser, wall-mounted gas geyser Wandgasbadeofen *m*

wall(-hung) grab bar, wall-mounted grab bar Wandstange, Wandgreifstange *f*

wall(-hung) heating panel, wall(-hung) heater panel, wall-mounted heating panel, wall-mounted heater panel Wandplattenheiz(ungs)körper, Wandplattenheizer *m*

wall(-hung) holder, wall-mounted holder Wandhalter(ung) *m*, (*f*)

wall(-hung) hydrant, wall-mounted hydrant Wandhydrant *m*

wall(-hung) hydrant cabinet, wall-mounted hydrant cabinet, wall(-hung) hydrant cubicle, wall-mounted hydrant cubicle Wandhydrantschrank *m*

wall(-hung) hygrometer, wall-mounted hygrometer Wandhygrometer *n*, Wand-Luftfeuchtigkeitsmesser *m*

wall(-hung) mixer fitting, wall-mounted mixer fitting Wand(misch)batterie *f* [*Diese wandmontierte Batterie mit zwei Zulaufanschlüssen für Warm- und Kaltwasser wird zur ungemischten und gemischten Wasserentnahme verwendet*]

wall(-hung) mixing cock for cold and hot water, wall-mounted mixing cock for cold and hot water Wandbatterie *f*

wall(-hung) model, wall-mounted model Wandmodell *n*

wall(-hung) oil-fired boiler, wall-mounted oil-fired boiler Öltherme *f*, Ölwandheiz(ungs)kessel *m*

wall(-hung) outflow valve, wall-mounted outflow valve Wandauslaufventil, Wandablaufventil, Wandabflussventil, Wandausflussventil, Wandausfließventil *n* [*Siehe Anmerkung unter „Auslaufventil"*]

wall(-hung) room heater, wall-mounted room heater Wand-Raumheizer, Wand-Raum(er)wärmer, Wand-Raumerhitzer *m*

wall(-hung) safe, wall-mounted safe Wandtresor *m*

wall(-hung) shower, wall-mounted shower Wanddusche *f*

wall(-hung) soap dispenser, wall-mounted soap dispenser Wandseifenspender *m*

wall(-hung) switch, wall-mounted switch Wandschalter *m*

wall(-hung) thermostat, wall-mounted thermostat Wandthermostat *n*

wall(-hung) toilet, wall-mounted toilet, wall-mounted closet, wall-mounted WC, wall (-hung) closet, wall (-hung) WC Wandtoilette *f*, Wandklosett *n*, Wandabort *m*

wall(-hung) toilet cistern, wall-mounted toilet cistern, wall(-hung) (closet) cistern, wall-mounted (closet) cistern, wall(-hung) flush(ing) cistern, wall-mounted flush(ing) cistern Wand(toiletten)-spülkasten, Wandklosettspülkasten, Wandabortspülkasten *m*

wall(-hung) urinal (bowl), wall-mounted urinal (bowl) Wandurinal(becken) *n*

wall(-hung) wash basin, wall (-hung) lavatory basin, wall-mounted wash basin, wall-mounted lavatory basin Wandwaschbecken *n*

wall(-hung) (water) (valve), wall(-hung) (water) mix(ing) valve, wall(-hung) blender, wall(-hung) blending valve [*"Wall-mounted" may synonymously be used for "wall(-hung)"*] Wandmischbatterie *f*, Wandmischventil *n*, Wandmischer(ventil) *m*, (*n*)

wall(-hung) WC pan, wall-mounted WC pan, wall(-hung) water closet pan, wall-mounted water closet pan Wand(toiletten)spülbecken, Wandabortspülbecken, Wandklosettspülbecken *n*

wall hydrant Wandhydrant *m*

wall illumination → wall lighting

wall inbuilt casing → wall built-in casing

(wall) inset box (Wand)Einbaukasten *m*

wall installation → wall mounting

wall-installed urinal flush(ing) device, wall-installed urinal flush(ing) unit Wandeinbauurinalspüler *m*, Wandeinbauurinalspülvorrichtung *f*

wall-integrated cabinet, wall-integrated cubicle Wandschrank *m*

wall kitchen cabinet → wall (stock kitchen) cabinet

wall lift → roof cradle

wall lift machine → roof cradle machine

wall lighting, wall illumination Wandbeleuchtung *f*

wall luminaire, wall light(ing) fitting, wall light(ing) fixture [*See remark under „luminaire"*] Wandleuchte *f*, Wandbeleuchtungskörper *m*

W

wall-mounted → **wall-hung**

wall mounting, wall installation Wandinstallation, Wandmontage, Wandanbringung *f*

wall opening → **wall breakthrough**

wall outlet Wandauslauf, Wandablauf *m*

wall oven Wandbackofen, Wandbratofen *m* [*Küche*]

wall panel radiant heating, panel wall radiant heating Wandstrahl(ungs)plattenheizung, Strahl(ungs)wandplattenheizung *f*

wall pedestal Wandsäule *f* [*Waschtisch*]

wall penetration → **wall breakthrough**

wall plaster Wandputz *m*

wall radiant heating, radiant wall heating Wandstrahl(ungs)heizung *f*

wall radiator [*Sections are joined parallel to the length of the radiator, allowing a thickness of not more than 2 inch wide*] Wandradiator, Wandgliederheiz(ungs)körper *m*

wall rose(tte) Wandrosette *f*

wall shower holder Wandbrausehalter *m*

wall sleeve [*An opening in a wall having a shroud insert to accept the chassis of an air-conditioning unit*] Wandöffnungskragen *m*

wall slot Wandschlitz *m*

wall space Wandraum *m*

wall sprinkler → **wall (fire) sprinkler**

wall (stock kitchen) cabinet, wall stock cabinet, wall kitchen cabinet (Küchen)Wandschranktisch *m*

wall support Wandarm *m*

wall surface installation Vorwandinstallation *f* [*Gegensatz: Inwandinstallation*]

wall tie → **(hollow) wall tie**

wall washer Wandanleuchtung *f* mit versenktem Schrägstrahler

wall (water) tap, wall draw(ing) off tap; wall crane [*Scottish term*]; **wall (basin) faucet (US)** Wand(wasser)(zapf)hahn, Wandentnahmehahn *m*, Wanddurchlaufventil; Wandkorn *n* [*niederdeutsch*]

wall wiring Wandverdrahtung *f*

wall (wiring) conduit Elektro-Wandleerrohr, (Elt-)-Wandleerrohr *n*

WAN → **wide area network**

warble tone Wobbelton *m*

ward, guards Einstrich *m*, Besatzung *f*, Gewirre, Eingerichte *n* [*Schloss*]

ward bit Schlüsselbarteinschnitt *m*

warded rising mechanism → **rising warded mechanism**

ward lock Schloss *n* mit Besatzung und Eingerichte

wardrobe hook Garderobe(n)haken *m*

warehouse lift (Brit.); warehouse elevator (US) Magazinaufzug *m*

to warm, to become warm warm werden

to warm, to make warm, to warm up [*To raise the temperature moderately*] (er)wärmen, aufwärmen

warm, hot warm

warm air, hot air Warmluft *f*

warm air fan-assisted furnace → **mechanical warm air furnace**

warm air hand dryer → **(blown) warm air hand dryer**

warm air (mechanical) furnace → **mechanical warm air furnace**

warm drinking water → **hot drinking water**

warm forced air → **forced warm air**

warming coil → **(air) heating coil**

warming plant, warming installation (Er)Wärmungsanlage, (Er)Wärmungsinstallation *f*

warming room [*Room designed to warm foodstuffs coming out of a cold room and thus avoid surface condensation on produce*] Wärmeraum *m*

warming (up), warm-up Erwärmen, (Auf)Wärmen *n*, Erwärmung, (Auf)Wärmung *f*

warm(ing-)up allowance [*Addition to the capacity of a heating system (as calculated for heat loss) to provide quick warm(ing-) up in the morning*] Aufwärm(ungs)zuschlag *m*

warm(ing)-up thermostat, morning warm(ing)up thermostat Aufwärm(ungs)thermostat, Aufwärm(ungs)temperaturregler *m*

warm roof → **non-ventilated (flat) roof**

warm service water, service warm water Brauchwarmwasser, Warmbrauchwasser, Betriebswarmwasser, Warmbetriebswasser, Gebrauchswarmwasser, Warmgebrauchswasser *n*

warmth [*As an acoustic quality, the liveness of the bass, or ful(l)ness of bass tone relative to that of the mid-frequency tone*] Klangwärme *f*

warmth [*As felt by the human body*] Wärme *f*

warmth to touch, mild heat, background warmth [*As felt by the human body*] milde Wärme, behagliche Wärme, Fühlwärme, wohlige Wärme *f*

warm to the tread, tread-warm fußwarm, trittwarm

to warm (up) → **to warm**

warm-up → **warming (up)**

warm-up switch Aufwärm(ungs)schalter *m*

warning Warnung *f*

Warning! Achtung!

warning device, warning appliance Warngerät *n*

warning lamp Warnlampe *f*

warning message Warnnachricht, Warnmitteilung *f* [*Eine Stör(ungs)meldung über mögliche sich anbahnende Gefahr(en) für Mensch, Umwelt, Erzeugnis oder/und Anlage. DIN 19 235*]

warning sign Warn(ungs)schild *n*

warning signal Warn(ungs)signal *n*

warning signal message Warn(ungs)meldung *f* [*Siehe Anmerkung unter „Meldung"*]

to wash, to flush spülen

wash appliance, wash device, ablutionary appliance, ablutionary device [*Bath; shower, wash basin; bidet, and less commonly, scullery sink, pantry sink, and drip sink*] Waschvorrichtung *f*, Waschgerät *n*

(wash)basin, lavatory (basin) [*A sanitary appliance primarily intended for washing the upper parts of the body. It has a waste connection and a water supply and is generally of ceramic ware but can also be of metal or plastics*] (Wasch)Becken *n* [*Diese Benennung umfasst Handwaschbecken und Waschtisch*]

(wash)basin mixer, lavatory (basin) mixer (Wasch)Beckenmischer *m*, (Wasch)Beckenbatterie *f*

(wash) basins in range, basin range, lavatory basins in range, lavatories in range [*An installation of multiple washbasins, as used in public washrooms*] Wasch(becken)reihe, Wasch(becken)batterie, Reihenwaschanlage, Reihenwaschbecken, Beckenreihe, Beckenbatterie, Beckenanlage *f*

(wash)basin soap dish, lavatory basin soap dish (Wasch)Beckenseifenschale *f*

(wash)basin unit, lavatory basin unit (Wasch)Beckenanlage *f*

(wash)basin-vanity unit, lavatory basin-vanity unit (Wasch)Becken-Waschtisch-Einheit *f*

washboard (US) → **base (board)**

washboard convector (US); base (board) convector

washboard heater (US) → **base (board) heater**

wash boiler Waschkessel *m*

wash device → **wash appliance**

to wash down herunterspülen

washdown siphon(ic) toilet → **siphon(ic) closet**

washdown (water) closet, washdown WC Tiefspülklosett *n* [*Die Fäkalien fallen in einen wassergefüllten Trichter und versinken zum Teil bis zum Spülen*]

washer Unterlegscheibe *f*

washing benches Reihenwaschanlage, Waschreihe *f*

wash(ing) boat → **(window) cradle machine**

wash(ing) cradle → **(window) cradle machine**

wash(ing) fixture, wash(ing) facility Waschanlage *f* [*In der Wohnung, meist im Badezimmer, Waschbecken und Badewanne, im Duschraum Dusche, in Schlafräumen, insbesondere in Hotels, Waschnische, in Fabriken, Sporthallen und ähnlichen Bauten kombinierte Umkleide- und Waschräume mit Waschrinnen oder Waschbrunnen*]

wash(ing) fountain Waschbrunnen *m*

wash(ing) pipe → **toilet flush(ing) pipe**

washout method → **decay method**

washout valve [*A valve for draining a water main*] Entwässerungsventil *n*

washout (water) closet, washout water toilet, washout WC Auswaschklosett, Flachspülklosett, Ausspülklosett *n*, Auswaschabort, Flachspülabort, Ausspülabort *m*

washroom, lavatory [*A room equipped with basin(s) for washing hands and face*] Waschraum *m*

washroom soap dispenser → **lavatory soap dispenser**

wash tube [*A large deep sink with a sloping front having a roll edge, used for washing clothes*] Waschwanne *f*

washup basket → **dishwashing basket**

washup machine → **dishwasher**

washup waste water → **dishwashing waste water**

wash water → **toilet flush(ing) water**

wastage Verschwendung *f*

waste Ablauf *m* [*als Gegenstand*]

waste Abfall(stoff) *m*

waste air Abluft *f*

waste air duct Abluftkanal *m*

waste air line Abluftleitung *f*

waste air lock Abluftschleuse *f*

waste air valve Abluftventil *n*

waste basket Abfall(stoff)korb *m*

waste brine Ablauge *f*

waste cadmium plating abfallarme galvanische Kadmierung, galvanische abfallarme Kadmierung *f*

waste chute Abfall(stoff)rutsche *f*

waste coal → **refuse coal**

waste combustion [*In boilers and industrial furnaces*] (nutzbringende) Abfallverbrennung *f*

W

waste data → **data waste**

waste disposal Abfallbeseitigung, Abfallentsorgung *f* [*Im allgemeinen Sprachgebrauch auch „Abfallwirtschaft" genannt*]

waste disposal unit → **disposal unit**

waste food → **food waste**

waste food grinder → **disposal unit**

waste food incinerator → **food waste(s) incinerator**

waste gas Abgas *n*

waste gas branch Abgasstutzen *m* [*Feuerstätte*]

waste gas chimney Abgasschornstein, Abgaskamin *m* [*DIN 18 160 Blatt 1*]

waste gas connection Abgasanschluss *m*, Abgasverbindung *f*

waste gas damper Abgasklappe, Abgas(ab)sperrvorrichtung *f* [*DIN 33 88*]

waste gas desulphurization; waste gas desulfurization (US) Abgasentschwefelung *f*

waste gas discharge Abgasabführung, Abgasableitung f

waste gas duct Abgaskanal *m*

waste gas engineering Abgastechnik *f*

waste gas fan Abgasventilator *m*

waste gas flow, waste gas stream Abgasstrom *m*

waste gas generator abgasbeheizte Austreiber, abgasbeheizte Kocher, abgasbeheizte Generator *m* [*Siehe Anmerkung unter „Austreiber"*]

waste gas handling Abgasumgang *m*

waste gas heat Abgaswärme *f*

waste gas heat exchanger, waste gas heat interchanger, waste gas heat exchange unit Abgaswärme(aus)tauscher *m*

waste gas heat recovery plant, waste gas heat recovery installation Abgaswärmerückgewinnungsanlage *f*

waste gas hood Abgashaube *f*

waste gas line Abgasleitung *f* [*Rohr oder Schacht in Wand, Decke oder (Fuß)Boden*]

waste gas loss Abgasverlust *m*

waste gas monitoring Abgasüberwachung *f*

waste gas muffler → **waste gas silencer**

waste gas pipe Abgasrohr *n* [*Zwischen Feuerstätte und Abgasschornstein. DIN 18 160 Blatt 2*]

waste gas plant, waste gas installation Abgasanlage, Abgasinstallation *f*

waste gas removal Abgasabführung *f*

waste gas return Abgasrückführung *f*

waste gas sensor [*See remark under "sensor"*] Abgasfühler *m*

waste gas shaft Abgasschacht *m*

waste gas shut(ting) off device, waste gas shut(ting) off appliance, waste gas stop device, waste gas stop appliance, waste gas closing device, waste gas closing appliance Abgas(ab)sperrorgan, Abgasschließorgan, Abgasschließgerät, Abgas(ab)sperrgerät *n*, Abgas(ab)sperrvorrichtung, Abgasschließvorrichtung *f*

waste gas silencer, waste gas muffler, waste gas sound absorber Abgasschalldämpfer *m*

waste gas temperature regulator Abgastemperaturregler *m*

waste gas temperature sensor Abgastemperaturfühler *m*

waste gas test Abgasprüfung, Abgasprobe *f*, Abgasversuch *m*

waste gas thermometer Abgasthermometer *n*

waste gas utilization Abgasverwendung, Abgasnutzung *f*

waste heat [*Unused heat rejected from a system, usually a heat engine or combustion furnace, to its surroundings*] Abwärme, Sekundärwärme *f*

waste heat boiler [*A steam-raising boiler utilizing the sensible heat contained in the waste gases from a carbonizing plant. Producer gas and additional air may be added to the waste gases at the inlet to the boiler to increase the steam output of the boiler. This is known as augmentation*] Abwärmekessel, Sekundärwärmekessel *m*

waste heat drying, drying by waste heat Abwärmetrocknung, Trocknung mit Abwärme, Sekundärwärmetrocknung *f*

waste heat engine Abwärmemotor, Sekundärwärmemotor *m*

waste heat exchanger Abwärme(aus)tauscher, Sekundärwärme(aus)tauscher *m*

waste heat heating → **heating by waste heat**

waste heat plant, waste heat installation Abwärmeanlage, Abwärmeinstallation, Sekundärwärmeanlage *f*

waste heat reclamation district heating [*Use of collected heat from an available process as a source for a district heating system*] Abwärmefernheizung, Abwärmestadtheizung, Sekundärwärmefernheizung, Sekundärwärmestadtheizung *f*

waste heat steam generator Abwärmedampferzeuger, Sekundärwärmedampferzeuger *m*

waste heat management Abfall(stoff)wirtschaft *f*

waste of time, time waste Zeitverschwendung, Zeitvergeudung *f*

waste oil Abfallöl, Altöl *n*

waste opening, waste outlet Ablauföffnung *f*

waste paper → **paper waste**

waste pipe → **downspout**

waste pipe [*A pipe collecting from baths, basins and sinks*] Bade- und Küchenschmutzwasserrohr *n*

waste plug and chain Ablaufpfropfen *m* und Kette *f*

waste receptacle Abfall(stoff)behälter *m*

wastes, house wastes [*Wastes from all plumbing in the house, including kitchen sinks, food waste disposal units, laundry trays, washing machines, discharge into the sewage system piping and then to the septic tank*] (Haus)Abfälle, (Haus)Abfallstoffe *f*

waste shredder Abfallzerkleinerer, Abfallschnitzler *m*

waste steam, exhaust steam Abdampf *m*

waste steam, absorption chiller, waste steam absorption refrigerating machine, exhaust steam absorption chiller, exhaust steam absorption refrigerating machine Abdampf-Absorptionskältemaschine, Absorptions-Abdampfkältemaschine *f*

waste steam heating, exhaust steam heating Abdampfheizung *f* [*Abdampf einer Dampfkraftanlage wird entweder direkt oder nach Druckminderung als Wärmeträger in Niederdruckdampfheizungen verwendet*]

waste steam heat, exhaust steam heat Abdampfwärme *f*

waste steam injector, exhaust steam injector Abdampfinjektor *m*

waste steam residue, exhaust steam residue Abdampfrückstand *m*

waste steam turbine, exhaust steam turbine Abdampfturbine *f*

waste superheat, high-temperature waste heat Abhitze, Abwärme hoher Temperatur *f*

waste treatment Abfallbehandlung *f*

waste water [*In a legal sense, not contaminated water that is not needed*] nichtbenötigte nichtverunreinigte Wasser *n*

waste water [*In a legal sense, not contamined water which has been used and is permitted to escape, or which unavoidably escapes from conduits or reservoirs of the lawful owners of such structures*] Abflusswasser, Ablaufwasser, abfließende Wasser, ablaufende Wasser *n* [*Aus Leitungen und Becken*]

waste(water) → **foul water**

wastewater (US) → **sewage**

wastewater... (US) → **sewage...**

wastewater drain (US) → **soil sewer**

wastewater sewer (US) → **soil sewer**

wastewater sewerage (US) → **soil sewerage**

waste wood Abfallholz *n*

watchdog, self-monitor Wächter, Netzausfalldetektor, Systemselbstüberwacher *m* [*Er verhindert bei Stromausfall dass keine unkontrollierten Daten in die Speicher geschrieben werden und dass danach der Prozesszyklus wieder richtig einsetzt*]

watchdog circuit, self-monitor circuit Wächterschaltung *f*

watchman Wächter, Wachmann *m*

watchman clock system Wächteruhrenanlage, Wachmannuhrenanlage *f*

watchman monitoring system Wächterüberwachungssystem, Wachmannüberwachungssystem *n*

watchman service Wachdienst, Wächterdienst, Wachmanndienst *m*

watchman('s) (supervisory) system [*It requires the watchmen on the various tours to produce a record in the superintendent's or chief guard's quarters at the start and at the finish of each tour*] Wachmannkontrollsystem, Wächterkontrollsystem *n*

watchman time detector, telltale Kontrolluhr, Wächteruhr, Wachmannuhr *f*

water absorption Wasseraufnahme, Wasserabsorption *f*

water aerator Wasserbelüfter *m*

water/air conditioning plant Wasser-Luft-Klima(tisierungs)anlage *f*

water/air heater [*It consists of a bank of tubes containing the heating water and around which the air passes*] Wasser-Luft-Erhitzer, Wasser-Luft-(Er-)Wärmer, Wasser-Luft-Heizer *m*

water/air heating Wasser-Luft/-Heizung *f* [*Luftheizung mit Warm- oder Heißwasser als Zwischenmedium*]

water/air heat pump Wasser/Luft-Wärmepumpe *f*

water/air system Wasser-Luft-Anlage *f*, Wasser-Luft-System *n*

water alarm Wassermangelsicherung, Wassermangelpfeife *f* [*Kessel. Sie muss über der Höchstdrucklinie liegen*]

water alarm valve Wassermangelventil *n*

water ascending pipe → **water riser pipe**

water atomizer Wasserzerstäuber, Wasserversprüher, Wasservernebler *m*

water atomizing nozzle Wasserzerstäubungsdüse, Wasserversprühungdüse, Wasservernebelungsdüse *f*

W

water atomization Wasserzerstäubung, Wasserversprühung, Wasservernebelung *f*

water authority Wasserbehörde *f*, Wasseramt *n*

water barrier → **water stop**

water booster installation, water booster plant Wasserdruckerhöhungsanlage *f*

water-borne sound Wasserschall *m* [*Sich im Wasser fortpflanzender Schall*]

water care agent Wasserpflegemittel *n*

water central distributor → **water manifold**

water charge Wassergeld *n*, Wassergebühr *f*

water chemistry → **aquatic chemistry**

water chiller Kaltwassersatz *m*

water chute → **water slide**

water circulating pump, water circulation pump Wasserumwälzpumpe, Wasserumlaufpumpe, Wasserzirkulationspumpe *f*

water circulation Wasserumlauf *m*, Wasserumwälzung, Wasserzirkulation *f*

water ((closed) loop) control Wasserreg(e)lung *f*

water closet → **toilet**

water closet seat → **toilet seat**

water closet suite → **WC suite**

water coil Wasserschlange *f*

water column → **water head**

water company → **water (supply) company**

water condition, water state Wasserbeschaffenheit *f*, Wasserzustand *m*

water conditioning, specific water preparation Wasserkonditionierung *f*

water connection Wasseranschluss *m*, Wasserverbindung *f*

water consumer, water user Wasserbezieher, Wasserabnehmer, Wasserverbraucher, Wasserkonsument *m*

water consumption, water use Wasserverbrauch *m*

water control (slide) valve Wassersteuer(ungs)schieber *m*

water conveyance, water transport(ation) Wasserförderung *f*, Wassertransport *m*

water-cooled condenser [*Condenser in which heat of the refrigerant is removed by water flowing over condensing surfaces*] wassergekühlte Kondensator, wassergekühlte Verflüssiger *m*

water cooler [*An electric, refrigerated drinking fountain*] Elektrotrinkbrunnen *m*

water cooler Wasserkühler *m*

water cooling [*The cooling of hot bodies by cold water*] Wasserkühlung *f*

water cooling ceiling Wasserkühl(ungs)decke *f*

water cooling coil Wasserkühl(ungs)schlange *f*

water cooling machine Wasserkühl(ungs)maschine *f*

water cup Wassertasse *f* [*Wassergefülltes offenes Gefäß zur (Ab)Dichtung eines unter Vakuum stehenden Raumes*]

water curtain Wasserschleier, Regenvorhang, Wasservorhang *m*, Regenwand, Wasserwand *f* [*Brandbekämpfung*]

water deficiency → **water shortage**

water defrosting [*Defrosting in which water is sprayed or poured over the frosted surface*] Wasserentfrostung, Wasserenteisung, Wasserabtauung *f*

water demineralization → **demineralization**

water discharge Wasserabfluss, Wasserablauf *m* [*Der einen Abflussquerschnitt durchfließende Wasserstrom*]

water dispense point Wasser(ab)zapfstelle *f*

water distribution Wasserverteilung *f*

water distribution installation Wasserverteilungsanlage *f* [*DIN 19 630*]

water distribution pipe, distribution water pipe Wasser-Gebäuderohr, Gebäude-Wasserrohr, Wasserverteilungsrohr, Verteilungswasserrohr *n*

water draught (Brit.); water draft (US) Wasserentnahme *f*

water drop Wassertropfen *m*

water duct Wasserkanal *m*

water ejector [*See remark under "ejector"*] Wasserstrahlpumpe *f*, Wasserejektor *m*

water engineering Wassertechnik *f*

water evaporation; water vaporization (US); water vapourization (Brit.) [*See remark under "evaporation"*] Wasserverdampfung *f*

water evapotranspiration Wasserverdunstung *f*

water expansion Wasserausdehnung, Wasserexpansion *f*

water extraction Wassergewinnung *f*

water failure → **water shortage**

water feature, water effect device Beckenattraktion, Wasserbeckenattraktion *f*

water feed(ing) Wasserspeisung *f*

water-filled radiator → **(electric) water-filled radiator**

water film Wasserfilm *m*

water (fire) extinguisher, gas pressure (fire) extinguisher, compressed gas (fire) extinguisher Nass(feuer)löscher, Nassbrandlöscher, Wasser(feuer)löscher, Wasserbrandlöscher *m*

water fitting [*See remark under "fitting"*] Wasserarmatur *f*, Wasserpassstück *n*, Wasserfitting *m*, Wasserformteil *n, m*

waterflow Wasservorlauf *m*

waterflow, waterstream Wasserfluss, Wasserstrom *m*, Wasserfließen *n*

waterflow volume meter → total waterflow meter

water flushing Wasserspülung *f*

water fog Wassernebel *m*

water for general use → mains water

water fountain → ablution fountain

water fountain, architectural fountain Zierbrunnen, Schmuckbrunnen, Dekor(ations)brunnen *m*

water gauge; water gage (US); wg Wassermesser *m*

water hammer → hydraulic shock

water hazard Wassergefährdung *f*

water head, water column, hydraulic head, hydraulic column Druckhöhe; Wassersäule *f* [*veraltete Benennung. Ein Pumpenbetriebswert*]

water heat Wasserwärme *f*

water heater, boiler, hot water generator [*Closed vessel in which water is heated by the combustion of fuels or electricity, and is withdrawn for use external to the system at pressures not exceeding 160 psig [1100 kPa (gage)], including the apparatus by which heat is generated, and all controls and devices necessary to prevent water temperatures from exceeding 210 °F (99 °C)*] Wasserheizer, Warmwasserbereiter, Boiler, Warmwassererzeuger, Speicher, Warmwasserspeicher *m*

water heating Wasserheizung *f*

water heating plant, water heating installation Wasserheiz(ungs)anlage *f*

water holding Wasserfassung *f*

water holding capacity Wasserfassungsvermögen *n*, Wasserfassungsfähigkeit *f*

water humidification (of air) Wasserbefeuchtung (der Luft), Wasseranfeuchtung (der Luft) *f*

water injection Wassereinspritzung *f*

water injection valve Wassereinspritz(ungs)ventil *n*

water installation Wasserinstallation *f*

(water) intake, (water) inlet (Wasser)Einlauf, (Wasser)Zufluss *m*

water jet pump Wasserstrahlpumpe *f*

water leakage Wasserleckage, Wasserundichtigkeit *f*

water level Wasserstand *m*, Wasserhöhe *f*, Wasserniveau *n*

water level controller, water level regulator Wasserstandsregler *m*

water level gauge; water level gage (US) Wasserstandsmesser *m*

water level gauge glass; water level gage glass (US) Wasserstandsglas *n*

water level indicator Wasserstandsanzeiger *m*

water level limiter, water level limiting device Wasserstandsbegrenzer *m*

water level mark Wasserstandsmarke *f*

water level monitoring Wasserstandsüberwachung *f*

water level regulator → water level controller

water line Wasserleitung *f*

water line → water (pipe)line

water loss Wasserverlust *m*

water main [*Underground water supply pipe for public or community use*] Hauptwasserrohr, Wasserhauptrohr *n*

water management, water regimen, water resources policy Wasserwirtschaft *f*, Wasserhaushalt *m* [*DIN 4046*]

water manifold, water central distributor, central water distributor Wasserverteiler(rohr) *m*, (*n*), Wasserverteil(ungs)rohr, Wasserverteil(er)stück, Wasserverteilungsstück *n*, Wasserverteil(er)batterie, Wasserverteilungsbatterie *f*

water meter Wassermesser *m* [*Zur Durchflussmessung, als Messung des zeitbezogenen Volumenstromes. Nicht verwechseln mit „Wasserzähler"*]

water meter Wasserzähler *m*

water metering Wassermessung *f* [*Nicht verwechseln mit „Wasserzählung"*]

water mist Wasser(sprüh)nebel *m*

water need → water requirement

water of condensation → condensate

water of steam condensation → steam condensate

water-only four-pipe system Vierrohr-Nur-Wasser-system, Nur-Wasser-Vierrohrsystem *n* [*System mit zwei Vorlaufrohren unterschiedlicher Temperatur und zwei Rücklaufrohren an den Versorgungsstellen*]

water-only single-pipe system Einrohr-Nur-Wasser-system, Nur-Wasser-Einrohrsystem *n* [*System mit einem gemeinsamen Vor- und Rücklaufrohr an den Versorgungsstellen*]

W

water-only system Nur-Wasser-System *n* [*Energiezufuhr an den Versorgungsstellen nur durch Wasser. DIN 1946, Teil 1*]

water ((open) loop) control Wassersteuerung *f*

water origin → water source

water outlet Wasserablauf, Wasserauslauf *m*

water physics → aquatic physics

water pipe Wasserrohr *n*

water pipe burst Wasserrohrbruch *m*

water (pipe)line Wasser(rohr)leitung *f* [*DIN 19 630*]

water (pipe (line)) installation, water (pipe (line)) plant Wasser(rohr)leitungsanlage, Wasser(rohr)leitungsinstallation *f* [*in einem Gebäude*]

water (pipe)line noise Wasser(rohr)leitungsgeräusch *n*

water pipework, water piping, water pipe system Wasser(rohr)leitungssystem *n*

water pollutant [*Substance in potable water creating an objectionable condition, but not creating a health hazard*] Wasserverunreinigungsstoff *m*

water-post → (fire) hydrant

water pressure Wasserdruck *m*

water pressure boosting Wasserdruckerhöhung *f*

water pressure reducer Wasserdruckminderer *m*

waterproof case, waterproof casing, waterproof enclosure, waterproof housing Wetter(schutz)gehäuse *n*

waterproofing [*The term "waterproofing" is sometimes restricted to mean "waterproofing against ground water" whereas the term "dampproofing" means protection against any moisture. In spite of this, the term "waterproofing" is generally applied, especially in combined words, as, for instance, "waterproofing agent" etc.*] (Ab)Dichtung, Sperrung, Isolierung *f* [*Zur Verhinderung der Durchfeuchtung der Baustoffe und Bauteile werden diese durch geeignete Stoffe entweder selbst wasserundurchlässig gemacht oder mit einem wasser- und unter Umständen auch chemikalienbeständigen Überzug versehen. Nicht verwechseln mit „Dämmung", dem Schutz gegen Wärme- und Schalldurchgang*]

water propeller metering → water rotary metering

water purification [*obsolete*] **→ water treatment**

water purification facility [*obsolete*] **→ water treatment facility**

water purification plant [*obsolete*] **→ water treatment plant**

water purification works [*obsolete*] **→ water treatment works**

water purifier Wasserreiniger *m*

water quantity meter → total waterflow meter

water radiator → water (sectional) radiator

water rate Wassertarif *m*

water recooler Wasserrückkühler *m*

water recooling Wasserrückkühlung *f*

water recooling plant, water recooling installation Wasserrückkühl(ungs)anlage *f*

water recooling pump Wasserrückkühl(ungs)pumpe *f*

water refeeding Wassernachspeisen *n*, Wassernachspeisung *f*

water regimen → water management

water removal Wasserentsorgung, Wasserbeseitigung *f*

water renewal Wassererneuerung *f*

water requirement, water need Wasserbedarf *m*

water return (flow) Wasserrücklauf *m*

water riser pipe, water rising pipe, water ascending pipe Wassersteigrohr, Steigwasserrohr *n*

water rotary metering, water vane metering, water propeller metering Wasserzählung, Flügelrad(wasser)messung *f* [*Nicht verwechseln mit „Wassermessung"*]

water sale Wasserverkauf *m*

water sampling, water sample taking Wasserprobe(n)nahme *f*

water saving Wassereinsparung *f*

water scale → scale

water scale formation → scale formation

(water) seal → trap seal

water seal(ing) head → head of water seal(ing)

water (sectional) radiator [*See remark under "radiator"*] Wasserradiator, Wassergliederheiz(ungs)körper *m*

water separator Wasser(ab)scheider *m*

water service pipe [*That part of a building main installed by, or under the jurisdiction of, a water department or company*] Wasserwerkrohr *n*

water services [*Hot and cold water piping and equipment*] Warm- und Kaltwasseranlagen *f*

watershield Wasserschild *m* [*Sprinkler*]

water shortage, water deficiency, deficiency of water, shortage of water, water failure, failure of water Wasserknappheit, Wasserklemme *f*, Wassermangel, Wasserfehlbedarf *m*

water shutoff, water shutting off Wasserabsperren *n*, Wasserabsperrung *f*

waterside Wasserseite *f*

waterside Wasserseite, Rohrinnenseite *f*

water sieve Wassersieb *n*

water slide, water chute Wasserrutsche *f*

(water) softener, (water) softening agent (Wasser)Enthärter, (Wasser)Weichmacher *m*

(water) softening (Wasser)Enthärtung *f*, (Wasser)-Enthärten, (Wasser)Weichmachen *n*

(water) softening filter (Wasser)Enhärterfilter, (Wasser)Enthärtungsfilter *m, n*

(water) softening resin, resin (water) softener (Wasser)Enthärterharz, (Wasser)Enthärtungsharz *n*

water source, water origin Wasserherkunft *f*, Wasserursprung *m*

water space Wasserraum *m*

water splash Wasserspritzer *m*

water state → water condition

water stop, water barrier Wasservorlage *f* [*Geruchverschluss*]

water storage Wasserspeicherung *f* [*DIN 4046*]

water storage vessel Wasserspeicher *m* [*Unbeheizte Speicherbehälter. Betriebsmäßig mit einer Wärmepumpe und gegebenenfalls mit einer Elektrozusatzheizung und/oder mit Wärmeaustauschern für andere Heiz(ungs)systeme ausgerüstet oder verbunden. DIN 8947*]

waterstream → waterflow

water supplier Wasserversorger *m*

water supply Wasserzufuhr *f* [*z. B. in eine Sprinkleranlage*]

water supply, building water supply (Gebäude)Wasserversorgung, (Gebäude)Bewässerung *f* [*DIN 4046*]

water supply and distribution system, water supply and distribution network Wasserversorgungs- und -verteilungsnetz *n*

water supply area Wasserversorgungsgebiet *n*

water (supply) company, water (supply) undertaking, water (supply) utility Wasserversorgungsunternehmen, Wasserwerk, WVU *n*

water supply engineer Wasserversorgungsingenieur *m*

water supply engineering Wasserversorgungstechnik *f*

water supply flow Wasserzufluss *m*

water supply installation, water supply plant Wasserversorgungsanlage, Wasserversorgungsinstallation, Bewässerungsanlage, Bewässerungsinstallation *f* [*In einem Gebäude. DIN 1988 und DIN 4179*]

water supply line, building water supply line (Gebäude)Wasserversorgungsleitung, (Gebäude)Bewässerungsleitung *f*

water supply mains, water supply network Wasserversorgungsnetz *n*

(water) supply pipe from source Rohwasser(zufluss)-rohr *n*

water supply pipe [*to systems*] Wasserzuführungsrohr *n*

water supply piping system, water supply piping network, water supply pipeline system, water supply pipeline network Wasserleitungssystem *n*, Wasserleitungsanlage *f*

water (supply) undertaking → water (supply) company

water surge Wasserschwall *m*

water switch Wasserschalter *m*

water tank Wasserbehälter, Wassertank *m*

(water) tap, draw(ing) off tap; crane [*Scottish term*]; **(basin) faucet (US)** [*As a sanitary term, a valve with a free outlet used as a draw-off or delivery point*] Wasserdurchlaufventil *n*; (Wasser)(Zapf)Hahn, Entnahmehahn, Kran *m* [*niederdeutsch*]

water temperature controller Wassertemperaturregler *m*

water temperature switch Wassertemperaturschalter *m*

water test Abdrückversuch *m*, Abdrückprobe, Abdrückprüfung *f* [*Abdrücken einer Rohrleitung mit Wasser*]

water thermal stop system, water thermal saving system Wasser-Wärme-Energie-Spar-System *n*

water throughput Wasserdurchsatz *m*

watertight tanking weiße Wanne *f* [*Ein wasserundurchlässiges Bauwerk bei dem die Betonkonstruktion eine tragende und raumabschließende Funktion und Abdichtung gegen drückendes Wasser übernimmt*]

(water)toilet → toilet

water transport(ation) → water conveyance

water trap → trap

water treatment; water purification [*obsolete*] [*A series of chemical and physical processes to remove dissolved and suspended solids from raw water to produce potable water for distribution and use*] Trinkwasseraufbereitung, Brauchwasseraufbereitung *f* [*Für Wasserwerke bedeutet „Trinkwasseraufbereitung" das gesamte Verfahren von der Gewinnung des Grundwassers bis zur Trinkwasserversorgung. Der Installateur versteht unter „Trinkwasseraufbereitung" einen Filter oder eine Enthärtungsanlage beim Bedarfsträger und diese Aufbereitung heißt jetzt „Wassernachbehandlung". DIN 2000*]

water treatment facility; water purification facility [*obsolete*] Trinkwasseraufbereitungsanlage, Brauchwasseraufbereitungsanlage *f*

water treatment plant, water treatment installation; water purification plant, water purification installation [*obsolete*] [*The central facility for the production of potable water; it contains all treatment proc-esses and appurtenances exclusive of the distribution system*] Trinkwasseraufbereitungsanlage, Brauchwasseraufbereitungsanlage *f*

water treatment works; water purification works [*obsolete*] [*A group or assemblage of processes, devices and structures used for the treatment or conditioning of water*] Wasserwerk *n*

water tube Wasserröhre, Siederöhre, Verdampfungsröhre *f* [*„Rohr" für „Röhre" ist ungenau und sollte deshalb als Synonym vermieden werden*]

water-tube boiler Wasserröhrenkessel, Siederöhrenkessel *m*

water-tube steam generator Wasserröhrendampferzeuger, Siederöhrendampferzeuger, Verdampfungsröhrendampferzeuger *m*

water undertaking → water (supply) company

water use → water consumption

water user → water consumer

water utility → water (supply) company

water valve Wasserventil *n*

water (valve) sizing chart → valve sizing chart for water

water vane metering → water rotary metering

water vaporization (US) → water evaporation

(water) vapour (Brit.); (water) vapor (US) Wasserdampf *m*

water vapour permeability, water vapour perviousness (Brit.); water vapor permeability, water vapor perviousness (US) Wasserdampfdurchlässigkeit *f*

water vent Wasserdurchlass *m*

water verification paste Wassernachweispaste *f*

water volume ((closed) loop) controller, water volume regulator Wasservolumenregler, Volumenwasserregler *m*

water waste perventer Auslauf mit Schwimmerventil, Schwimmerventilauslauf *m*

water/water heat pump Wasser/Wasser-Wärmepumpe, W-W-WP *f* [*Siehe Anmerkung unter „Wärmepumpe"*]

water-wetted wasserbenetzt, wasserberührt

waterworks Wasserwerk(e) *n*, (*f*)

water yield Wasserertrag *m*

WATS, wide area telephone service Fernsprechfernverkehr *m* zu Festgebühr

wattage Leistungsaufnahme, Wattzahl *f*

watt(ful) current → active current

watt-hour meter Wattstundenzähler, Wattstundenmesser *m*

wattless current → reactive current

wattmeter Wattmesser, Wattzähler *m*

watt(s) loading Wattbelastung *f*

wave filter [*Transducer for separating waves on the basis of their frequency*] Wellenfilter *m*, *n* Wellensieb *n*

waveform Wellenform *f* [*Darstellung von Tonfrequenzen*]

wave interference → interference

wavelength constant → phase constant

wave machine pool Wellenbecken *n* [*Schwimmbad*]

Wave Metrics Messdatenverarbeitung *f* IGOR

W.C. → toilet

WC suite, water closet suite komplette WC-Anlage *f*

WE → write enable

weak → dilute

weak current → light current

weak current (distribution) network, light current (distribution) network
Schwachstrom(verteiler)netz, Schwachstromverteilungsnetz *n*

weak electrical oscillation → speaking current

weakening of current → current weakening

weak point Schwachstelle *f*

weak point analysis Schwachstellenanalyse *f* [*Untersuchung im Rahmen einer Sicherheitsberatung über Schwachstellen in einem Schutzobjekt zur Projektierung von Sicherungsvorrichtungen und Erstellung eines Sicherheitskonzepts*]

wear(ing) failure Verschleißausfall, Fehlerausfall, Versagensausfall m

weather-actuated ((open) loop) control wettergeführte Steuerung, witterungsgeführte Steuerung *f*

weather barrier, weather divider [*Between the exterior and interior working parts of an incremental conditioner*] Wetterteiler *m*

weather change Witterungsumschwung, Witterungswechsel *m*

weather data base Wetterdatenbank *f*

weather head → lead-in bell

weather measure value transmitter Witterungsmesswertgeber *m*

weather penetration Witterungseindringung *f*

weatherproofing Klimaschutz, Wetterschutz *m*

weather protection grill(e) Wetterschutzgitter, Klimaschutzgitter *n*

weather report Wetterbericht, Klimabericht *m*

weatherstrip → air lock strip (US)

weblock Klappzungenschloss *n*

web plate Stegblech *n*

wedge bonding Keilbonden *n* [*Siehe Anmerkung unter „Bonden"*]

wedge for punches → punching wedge

week day block generation [*EIB range*] Wochentagsblockbildung *f*

week day controlled wochentagsgesteuert

week day dial → day-of-week dial

week-end peak Wochenendspitze *f*

weekend switch(ing) off Wochenend(e)abschaltung *f*

week(ly) dial Wochenzifferblatt *n*

week(ly) schedule Wochenplan *m*

week timer Wochenschaltuhr *f*

weight Masse *f* [*frühere Benennung: Gewicht n*]

weighted signal-to-noise ratio Geräuschspannungsabstand *m* [*Unterschied zwischen dem Pegel der Summe aus Nutzsignal und Störsignal (= Gesamtsignal) und dem Störsignalpegel, beide gemessen hinter einem Bewertungsfilter*]

weight(-load)ed gewichtsbelastet, gewichtet

welded bracket Schweißkonsole *f*

welded mesh fence, mesh welded fence Schweißmaschenzaun, Maschenschweißzaun *m*

welded-on pin Anschweißstift *m*

welded (pipe) elbow Schweiß(rohr)bogen *m*

welding burner insert Schweißbrennereinsatz *m*

welding rectifier Schweißgleichrichter *m*

welding symbol Schweißsinnbild, Schweißsymbol *n*

welding torch Schweißbrenner *m*

well → hoistway

well → (escalator) well

well girder → (escalator) well girder

well → (water) well

well access → elevator hoistway access

well air removal → hoistway air removal

well in advance rechtzeitig

well potable water supply → non-public potable water supply

well-proven praxisbewährt

well riser Brunnensteigleitung, Steigbrunnenleitung *f*

well vent(ilat)ing → shaft ventilation

well water Brunnenwasser *n*

well water heat pump Brunnenwasserwärmepumpe *f*

well water meter Brunnenwasserzähler *m*

wet air cooler → wet(-type) air cooler

wet alarm valve Nassalarmventil, Alarmnassventil *n* [*Sprinkleranlage*]

wet alarm valve station Alarmnassventilstation, Nassalarmventilstation *f*, NAV-Station [*Sprinkleranlage*]

wet and dry bulb thermometer → psychrometer

wet area Nassbereich *m* [*selten: n*]

wet bar [*A bar with water and sink provisions for minor washing of glass*] Gläserspülbar *f*

wet bulb approach Feuchtkugelannäherung *f*

wet bulb temperature Feuchtkugeltemperatur *f* [*DIN 33 403*]

to wet-clean nassreinigen

wet condensate (pipe)line nasse Kondensat(rohr)leitung, tiefliegende Kondensat(rohr)leitung *f* [*„Druckzone" ist der Bereich in der Kondensatleitung über dem Kesselwasserstand. In diesem Bereich sind alle mit dem Kesselwasser direkt in Verbindung stehenden Leitungen geflutet. Sie werden als „nass" bezeichnet*]

wet (condensate) return (flow) tiefliegende (Kondensat)Rücklauf, Nass(kondensat)rücklauf *m* [*Dampfheizung. Das Kondensat wird unterhalb der Druckzone zum Kessel geführt*]

wet cooler Nasskühler *m* [*In Bauart identisch mit Luftbefeuchter. Kühlung der Luft ist zwangsläufige Nebenerscheinung der Befeuchtung*]

wet cooling Nasskühlung *f*

wet (cooling) tower Nasskühlturm *m*

wet (cooling) tower with induced draught fan; wet (cooling) tower with induced draft fan (US) Nasskühlturm *m* mit (An)Saugventilator

wet-dry (cooling) tower Nass-Trocken-(Kühl)Turm *m*

wet-dry (cooling) tower with natural draft, wet-dry (cooling) tower with natural draught Nass-Trocken-(Kühl)Turm mit natürlichem Zug, Naturzug-Nass-Trocken-(Kühl)Turm *m*

wet-dry firefighting installation Nass-Trocken-Feuerlöschanlage, Feuerlöschanlage nass/trocken *f*

W

wet/dry riser, wet/dry rising pipe(line) Nass-Trocken-Steig(e)leitung, Steig(e)leitung nass/trocken *f* [*Löschwasserleitung die im Bedarfsfall durch Fernbetätigung von Armaturen mit Wasser aus dem Netz gespeist wird. DIN 14 462, Teil 1*]

wet fixture [*See remark under "fixture"*] Nasseinbauteil *m, n*

wet (fire) extinguisher Nass(feuer)löscher *m*

wet (gas) meter, drum (type) (gas) meter [*A meter in which the volume of gas is measured by admitting the gas to a drum having compartments of known volume sealed by water or other liquid. The drum rotates under the influence of the gas pressure differential and the gas is displaced from the compartments by the sealing liquid*] Nass-Gaszähler, Trommel(gas)zähler, nasse Gaszähler, Gas-Trommelzähler *m*

wet heating → hot water heating

wet(-pipe) (fire) sprinkler system, (fire) sprinkler wet system, wet(-pipe) automatic system [*Piping contains water under pressure for immediately release when valve of sprinkler head is activated*] Nass(rohr)-Sprinkleranlage, Nass(rohr)-Sprengeranlage *f*

wet (pipe)line, wet piping Nass(rohr)leitung *f* [*Brandbekämpfung*]

wet (pipe) sprinkler system [*Sprinkler system in which all piping is maintained filled with water and connected to an approved source or sources of water supply*] Nass(-Sprinkler)anlage, Nass-Sprengeranlage *f*

wet reed relay [*Reed-type relay containing mercury at the relay contacts to reduce arcing and contact bounce*] Nassschutzgas(kontakt)relais, Nassreedrelais *n*

wet return (flow(ing)) Nassrücklauf, Nassrückfluss *m*, Nassrückfließen, Nassrücklaufen *n*

wet riser, wet rising pipe(line) nasse Steig(e)leitung *f* [*Ständig unter Druck stehende Löschwasserleitung*]

wet-room general-diffuse luminaire, wet-room non-cutoff luminaire Feuchtraumfreistrahler *m*, freistahlende Feuchtraumleuchte *f*

wet saturated steam, saturated wet steam [*Steam at the saturation temperature corresponding to the pressure and containing water particles in suspension*] Nasssattdampf, Sattnassdampf *m*

wet separator, wet interceptor, wet trap [*See remark under "interceptor"*] Nass(ab)scheider *m*

wet sprinkler (pipe)line → sprinkler wet (pipe)line

wet standpipe Nassstandrohr, Standnassrohr *n*

wet steam Nassdampf *m* [*Gemisch aus siedender Flüssigkeit und gesättigtem Dampf*]

wet sump, pump sump Pumpensumpf *m*

wet system alarm valve station Nassalarmventilstation *f*, Strömungsschalter *f, mpl* [*Mehrzahl*] [*Sprinkleranlage*]

wet(ted) (dust) filter [*The filtering material is coated with mineral oil or other viscous material so that the dust sticks to it*] Nass(staub)filter *m, n*

wetting [*Electricity. The coating of a contact surface with an adherent film of mercury*] Benetzen *n*, Benetzung *f*

wetting Benetzung *f*, Benetzen, Nassmachen *n* [*Ausbreitung einer Flüssigkeit auf einer Festkörperfläche unter Vergrößerung der Grenzfläche Flüssigkeit/Festkörper*]

wet trap → wet separator

wet(-type) electronic (air) filter Elektroniknass(luft)filter *m, n*

wet vane wheel (inferential) meter Nassläufer, Nassflügelradzähler *m*

wg → water gauge

WG → work(ing) group

wet(-type) air cooler, wet(-type) air cooling unit, wet(-type) air unit cooler [*Unit bringing air in contact with liquid by bubbling through the liquid or spraying liquid into it*] Nassluftkühler *m*

wheel(ed) basket [*shopping store*] Einkauf(s)wagen *m*

wheel lock → letter(-keyed) lock

wheel printer [*In automatic data processing, a line printer that prints its characters from the rim of a wheel around which is the type for the alphabet, numerals, and other characters*] Raddrucker *m*

whirlpool Wasserstrudel *m*

whirlpool (bath)tub Strudel(wasser)(bade)wanne *f*

whirl(-type) air outlet Drall-Luftauslass *m*

whistle alarm Pfeifenalarm *m*

white balance Weißabgleich *m* [*Farbkamera*]

white noise [*A noise whose spectral density is substantially constant over a significant frequency range. White noise need not necessarily be random*] weiße Rauschen *n* [*Rauschen, dessen spektrale Intensitätsdichte über dem interessierenden Frequenzbereich konstant ist*]

white room [*obsolete*] **→ clean room**

Whitworth thread Whitworth(rohr)gewinde *n* [*DIN 2999. Ein zylindrisches Innen- oder konisches Außengewinde*]

whole body vibration Ganzkörperschwingung, GKS *f*

whole number → **integer**

whole-room heat Ganzzimmerwärme *f*

whole tone Ganzton *m* [*Akustik*]

wick feed oil cup Dochtöler *m*

wide-angle lens Breitwinkelobjektiv, Weitwinkelobjektiv *n*

wide area telephone service → **WATS**

wide band → **broad band**

wide band integrated services digital network → **B-ISDN**

wide-pitch fluting, coarse-pitch fluting [*radiator panel*] Grobprofilierung *f*

widebeam, broadbeam Breitabstrahlbündel, Abstrahlbreitbündel *n*

wide tolerance [*A tolerance greater than standard tolerance*] Übertoleranz *f*

wideviewing angle Breitsichtwinkel *m*

width; breadth [*deprecated*] [*One of two horizontal dimensions, normally the lesser*] Breite *f*

Wiedemann-Franz law, Lorenz relation [*The law that the ratio of the thermal conductivity of a metal to its electrical conductivity is a constant, independent of the metal, times the absolute temperature*] Wiedemann-Franz-Lorenzsche Gesetz *n*

Wiegand badge (card), Wiegand identity card Wiegandausweis(karte) *m*, (*f*) [*Im Ausweis ist ein Kunststoffstreifen eingeschweißt der zwei Eisendrähte enthält. Diese Drähte verändern ihren magnetischen Zustand beim Einbringen in den Leser*]

(Wien's) displacement law, Wien's radiation law [*A law for blackbody radiation which states that the wavelength at which the maximum amount of radiation occurs is a constant equal to approximately 2898 times the product of 1 micrometer and 1 kelvin*] Wiensche Verschiebungsgesetz *n*

wildcard [*A symbol, such as an asterisk, that is used in place of a file name, file type, directory name, or version number in a file specification to indicate "all" for the given field*] Joker *m*, Ersatzzeichen *n*

wilful vorsätzlich, absichtlich

Winchester disk, Winchester disc Winchesterplatte *f* [*Festplatten-Speichersystem mit hoher Aufzeichnungsdichte und geringer Zugriff(s)zeit*]

wind chill factor → **chill factor**

wind detector, wind sensor [*It works on a different principle to the outside temperature detection. It is not a resistance changing device but measures the cooling effect of the wind on a heated element*] Winddetektor, Windsensor, Windfühler *m*, Winddetektionselement *n*

wind direction pickup, wind direction measured value transmitter Windrichtungsmesswertgeber *m*

wind direction sensor Windrichtungsfühler, Windrichtungssensor *m*

wind effect Windwirkung *f*

wind guard Windschutz, Windschützer *m* [*als Vorrichtung*]

wind occurrence Windanfall *m*

wind guard Windschutz *m*

window Fenster *n* [*Teil eines Bildschirm(e)s zur Darstellung bestimmter Informationen*]

window accessories Fensterzubehör *n* [*seltener: m*]

window air conditioner [*Factory assembled unit designed to be installed in window openings, providing temperature control, cooling cycles, heating cycles, ventilation and filtered air*] Fenster(raum)klimagerät *n*

window air cooling device, window air cooling appliance Fenster-Luftkühlgerät *n*

window air curtain Fensterluftschleier, Fensterluftvorhang *m*

window air outlet Fensterluftauslass *m*

window cable duct Fensterkabelkanal *m*

window casement fastener → **casement fastener**

window cleaning equipment, window washing equipment Fensterreinigungsgerät(e), Fensterputzgerät(e) *n*, (*f*)

window discriminator [*See remark under "discriminator"*] Fensterdiskriminator *m* [*Er analysiert eine anliegende Eingangsspannung auf Überschreiten eines oberen und Unterschreiten eines unteren Grenzwertes*]

(window) cradle machine, (window) cleaning boat, (window) cleaning cradle, (window) wash(ing) boat, (window) wash(ing) cradle [*Motorized carriage running vertically on rails let into the window frames. "Gondola" may also be used*] (Fenster)Putzwagen, (Fenster)Reinigungswagen *m*

window fan Fensterventilator *m*

window fitting, window mounting Fensterbeschlag *m* [*Zum Einbau eines beweglichen Fensterflügels in Blendrahmen*]

window handle, window handgrip Fenster(hand)griff *m*

window hinge Fensterband *n*

window lead-in Fensterdurchführung *f*

window lock Fensterschloss *n*

window mounting kit Fenstereinbausatz *m*

window mounting staple → **staple of window mounting**

window sash knob Fenster(aufzieh)knopf *m*

window (sectional) radiator [*See remark under "radiator"*] Fensterradiator, Fenstergliederheiz(ungs)körper *m*

window security Fenstersicherung *f*

window sill duct system Fensterbankkanalanlage *f*

windows open services architecture, WOSA Richtlinien *f* von Microsoft für den Aufbau von Fenster-APIs

window state monitoring Fensterzustandsüberwachung *f*

window temperature sensor Fenstertemperaturfühler, Fenstertemperatursensor *m*

window unit Fensterklimagerät *n*

window ventilation, window vent(ilat)ing Fensterlüftung, Stoßlüftung, Fensterbe- und -entlüftung *f* [*Luftwechsel durch Öffnen von Fenstern*]

window ventilator Fensterlüfter *m*

(window) washing boat → (window) cradle machine

window washing equipment → window cleaning equipment

wind path Windweg *m*

wind pressure, velocity pressure [*Total force excerted on a structure by wind*] Windangriff, Staudruck *m*, Anströmung *f*

wind pressure side, velocity pressure side Anströmseite, Staudruckseite, Windangriffseite *f*

wind sensor → wind detector

wind speed meter → anemometer

wind tie Windkammer *f* [*Satteldach*]

(wind) tunnel test (Wind)Kanalprüfung, (Wind)-Kanalprobe *f*, (Wind)Kanalversuch *m*

windup [*Relative movement due to deflection under load*] Torsionsspiel *n* [*Werkzeugmaschine*]

wind vane Windfahne *f*

wind vane relay Windfahnenrelais *n*

wind ventilation Windlüftung *f* [*Der durch Wind entstehende Luftwechsel. Wärmegewinnung ist nicht möglich*]

winter (air) conditioning [*Heating, humidification, air distribution, and air cleaning where outside temperatures are below the inside or room temperature*] Winterklimatisierung, Winterklimareg(e)lung *f*

winter (air) conditioning plant, winter (air) conditioning installation Winterklima(tisierungs)anlage, Winterklimareg(e)lungsanlage *f*

winter cooling load Winterkühl(ungs)last *f*

winter cycle Winterzyklus *m*

winter heating load Winterheiz(ungs)last *f*

winter service Winterbetrieb *m* [*Heizung*]

winter solar load, winter sun load, solar winter load, sun winter load Solarwinterlast, Sonnenwinterlast, Wintersonnenlast, Wintersolarlast *f*

wiping cloth → cleaning coth

to wire verdrahten

wire Ader *f*, Draht *m*

wire bell Drahtklingel *f*

wire bonding (method) Drahtbonden, Drahtbondverfahren *n*, Drahtbondmethode *f* [*Siehe Anmerkung unter „Bonden"*]

wirebound drahtgebunden

wire breaking → wire rupture

wire conductor → conductor

wire cryoton Drahtkryoton, Drahttieftemperaturschaltelement *n*

wired verdrahtet

wired AND verdrahtete UND *n* [*Die Parallelschaltung mehrerer Bausteine mit offenem Kollektor entspricht einem logischen UND. Dieses verdrahtete UND ist von großem Vorteil wenn viele Ausgänge durch ein logisches UND verknüpft werden*]

wired clock system Drahtuhrensystem *n*

wire detector, wire detecting device Drahtmelder, Drahtdetektor, Drahtmeldungsgeber *m*, Drahtmeldegerät *n*

wired for connection, wired for being connected anschlussfertig verdrahtet, betriebsbereit verdrahtet

wired-in festverdrahtet

wired logic control, WLC verdrahtungsprogrammierte Schützensteuerung *f*

wire entry guide, wire entry aid Drahteinführungshilfe *f*

wire glass Drahtglas *n*

wire guard Drahtschutz(gitter) *m*, (*n*)

wire gutter top [*down-pipe connection*] Korbseiher *m* [*Regenrohranschluss*]

wire heating Drahtheizung *f*

wire jumper, wire link Drahtbrücke *f*

wire memory → (plated) wire memory

wire mesh fence → mesh wire fence

wire printer → matrix printer

wire print head → matrix print head

wire printing → matrix printing

wire routing Drahtführung *f*

wire rupture, wire breaking, wire breakage Drahtbruch *m*

wire spiral brush → **spiral wire brush**

wire thermocouple (assembly) Drahtthermoelement, Drahtthermopaar *n*

wireway, surface raceway Verdrahtungskanal *m*

wire(-wound) resistance Drahtwiderstand *m*

wire-wrap connection Drahtwickelanschluss *m*

wire-wrap technique Drahtwickeltechnik *f*

wiring Schaltung *f*

wiring Verdrahtung, Verlegung von Drähten, Drahtverlegung *f*, Verdrahten, Drahtverlegen *n*

wiring accessories Verdrahtungszubehör *n, m* [*Elektrotechnik*]

wiring aid Verdrahtungshilfe *f*, Verdrahtungshilfsmittel *n*

wiring board → **plug board**

wiring by the Owner bauseitige Verdrahtung *f*

wiring cable Installationskabel *n*

wiring chamber, wiring closet Verdrahtungskammer *f*

wiring chase, wiring groove Verdrahtungsrille *f*

wiring colour code (Brit.); wiring color code (US) Verdrahtungsfarb(en)kennzeichnung *f*

wiring configuration Verdrahtungskonfiguration *f*

wiring device, wiring appliance [*Receptacle, switch, pilot light, small dimmer, or any device that is wired in a branch circuit*] verdrahtete Gerät *n*

wiring diagram → **circuit diagram**

wiring harness Kabelbaum, Fertigkabelsatz, vorgefertigte Kabelsatz *m*

wiring manual, wiring handbook Schaltungs(hand)-buch *n*

wiring monitoring Verdrahtungsüberwachung *f*

wiring retainer, wiring holder Verdrahtungshalter *m*

wiring run Verdrahtungszug *m*

wiring schedule Verdrahtungsliste *f*

wiring schematic, wiring scheme Beschaltungsplan, Verdrahtungsplan *m*, Beschaltungsschema, Verdrahtungsschema *n*

wiring sequence Verdrahtungs(reihen)folge *f*

wiring space Verdrahtungsraum *m*

wiring system Verdrahtungssystem *n*

wiring topology Verdrahtungstopologie *f*

with communication capacity → **capable of communication**

to withdraw, to evacuate [*To remove something, especially gases and vapo(u)rs, from an enclosure, such as from the envelope of an electron tube, or from a well or the like*] abziehen

to withdraw herausziehen

withdrawable circuit breaker Einschubleistungsschalter *m*

(withdrawable) zone module Linieneinschub *m* [*Feuermeldeanlage*]

with nonreversible poles polunverwechselbar

without gear → **gearless**

without pack(ing) → **pack(ing)less**

witness line Maßhilfslinie *f*

WLC → **wired logic control**

WLL → **radio local loop**

WNT → **radio net(work) termination**

wobble plate Taumelscheibe *f*

(wobble) stick switch Knüppelschalter, Taumelstabschalter *m*

Woltman (water) meter Woltmann(wasser)zähler *m* [*DIN 19 625*]

women's dormitory Frauenschlafsaal *m*

wood(en) draining board, wood(en) drainer (Ab)-Tropfbrett *n* [*Teil eines Spültisches*]

wood(en)-frame(d) mirror Holz(rahmen)spiegel *m*

wood(en) louver (US); wood(en) louvre (Brit.) Holzlüftungsklappe *f*

woo(en) (stock kitchen) cabinet, wood(en) stock cabinet, wood(en) kitchen cabinet (Küchen)Holzschranktisch *m*

wood heating Holzheizung *f*

woodworking home shop, home woodworking shop Holz-Hauswerkstatt, Holz-Heimwerkstatt *f*

woofer Tieftöner *m*

word Wort *n* [*Gibt an wieviel zweiwertige Ziffern (Bits) ein Rechner gleichzeitig verarbeiten kann*]

word address Wortadresse *f* [*Programmiermethode für Lochstreifen bei numerischer Steuerung*]

word gap → **word space**

word length Wortlänge *f* [*Entsprechend dem Anwendungsgebiet werden mehrere Bits zu Worten unterschiedlicher Länge (4, 8, 12, 16 und 32 Bits) zusammengefasst. Ein Wort mit 8 Bit wird als Byte bezeichnet*]

word processing Wortverarbeitung *f*

word processor wortverarbeitende Gerät *n*

words block → **(computer) words block**

word space, word gap Wortzwischenraum *m*, Wortlücke, Leerstelle *f*

workbook Arbeitsbuch *n*

work/break relation(ship) Betrieb-Pause-Verhältnis *n* [*Gebäudenutzung durch Menschen*]

work cell → production cell

work engineering → job engineering

(work)flow → sequence

(work)flow computing, sequence computing Arbeitsflussautomatisierung, (Arbeits)Ablaufautomatisierung *f* [*Zwischen miteinander vernetzten Abteilungen*]

work function (elektronische) Austrittt(s)arbeit *f* [*Energie, um ein Leitungselektron aus einem Metall heraus im Vakuum zu bringen*]

work(ing) area Arbeitsfläche *f*

work(ing) chamber, work(ing) hearth [*The part furnace in or on which the stock is placed during heat treatment*] Arbeitskammer *f*

work(ing) characteristic → work(ing) feature

work(ing) clear space → clear (work(ing)) space

work(ing) current release Arbeitsstromauslösung *f*

work(ing) data Arbeitsdaten *f*

work(ing) directory Arbeitsverzeichnis *n*

work(ing) drawing Arbeitszeichnung, Ausführungszeichnung *f*

work(ing) efficiency Arbeitswirkungsgrad *m*

work(ing) equipment Arbeitsgerät(e) *n*, (*f*)

work(ing) feature, work(ing)) characteristic Arbeitsmerkmal *n*

work(ing) file Arbeitsdatei *f*

work(ing) fluid [*e. g. steam*] Arbeitsströmungsmedium *n*

work(ing) gas Arbeitsgas *n*

work(ing) group, WG Arbeitsgruppe *f*

work(ing) hearth → work(ing) chamber

work(ing) layout Arbeitsvorbereitung *f*

work(ing) layouter Arbeitsvorbereiter *m*

work(ing) manual Arbeitshandbuch *n*

work(ing) memory, work(ing) storage Arbeitsspeicher, Hauptspeicher *m* [*In engerer Bedeutung wird nur ein Teil des Hauptspeichers als Arbeitsspeicher bezeichnet, um die Bereiche für Eingabe, Ausgabe, Programm und jeweils notwendige Zwischenspeicherungen (Arbeitsspeicher) voneinander abzugrenzen*]

work(ing) mode signal message Arbeitsartenmeldung *f* [*Eine Meldung über die vorgewählte Arbeitsart*]

work(ing) oil level Arbeitsölstand *m*

work(ing) plane Arbeitsebene *f*

work(ing) posture Arbeitshaltung *f*

work(ing) pressure [*Pressure indicated by a ga(u)ge when the system is in normal operation*] Arbeitsdruck *m*

work(ing) principle Arbeitsgrundsatz *m*

work(ing) range, work(ing) region Arbeitsbereich *m* [*selten: n*]

work(ing) register Arbeitsregister *n* [*Zwischenergebnisse speichernder und einen schnellen Zugriff gestattender Speicher*]

work(ing) safety Arbeitssicherheit *f*

work(ing) space Arbeitsraum *m*

work(ing) speed Rollgeschwindigkeit, Fahrgeschwindigkeit *f* [*Rolltreppe*]

work(ing) substance [*e.g. the refrigerant in a refrigeration system*] Arbeitsmittel *n*

work(ing) temperature Arbeitstemperatur *f*

work(ing) time → operation time

work(ing) time acquisition system Arbeitszeiterfassungssystem *n*

work(ing) time acquisition terminal Arbeitszeiterfassungs-Terminal, AZET *n*

work(ing) time and attendance, working T/A Arbeitszeiterfassung *f*

workmanship Arbeitsausführung *f*

workplace shell, WPS OS/2-Arbeitsoberfläche *f*

works certificate Werkprüf(ungs)zeugnis *n*

work sequence → sequence of operations

worksheet Arbeitsblatt *n*

workshop (Praxis)Kurs(us), Lehrgang *m*

workshop Werkstatt *f*

(workshop) fitter (Betriebs)Schlosser *m*

(work)shop drawing Werk(statt)zeichnung *f*

workstation (computer), desk(top) computer, personal computer, PC Arbeitsplatzrechner, APR, Einzelplatzrechner *m*

work system → production system

world clock Weltuhr *f*

World Wide Web, WWW, www weltweite Netz, Internet *n*

WORM → write once read many memory

WORM disc → write once read many (magneto-optical) disc

worm eccentric pump → eccentric worm pump

worm geared reduction unit, geared machine [*lift*] Schneckenuntersetzungsgetriebe *n*

worm-to-worm gear reduction Schneckengetriebeuntersetzung *f*

worst case ungünstigste Fall, schlechteste Fall *m*

worst case analysis ungünstigste Toleranzanalyse *f* [*Toleranzanalyse bei der unterstellt wird, dass in einer Schaltung die Bauelemente jeweils die ungünstigsten, gerade noch innerhalb der Toleranz liegenden, Parameterwerte haben. Untersucht wird wieweit die Zielfunktion der Schaltung bei den minimalen bzw. maximalen Bauelementen von ihrem Sollwert abweicht*]

wort Rechnerwort *n*

WOSA → windows open services architecture

wow and flutter Gleichlaufschwankung *f*

WPS → workplace shell

wrapping Bandage *f*, Wickel *m*

wrapping Bandagierung, Umwick(e)lung *f*, Bandagieren, Umwickeln *n*

writeable → printable

write control memory Schreib-Lese-Speicher *m*, Pufferspeicher

write enable, WE Einschreibfreigabe *f*

write instruction, writing instruction Schreibbefehl *m*, Schreibanweisung *f*

write once read many (magneto-optical) discs, write once read many (magneto-optical) disks, WORM discs, WORM disks einmal beschreibbare magneto-optische Platte *f*

write once read many memory, WORM Speicher immer wieder lesbar, dauerlesbare Speicher *m*

to write out → to print out

write protection Schreibschutz *m* [*Diskette*]

write protection pin Schreibschutzstift *m*

writer → printer

write-read head Schreib-Lese-Kopf *m*

write-read-store Schreib-Lese-Speicher *m* [*DIN 19 226, Teil 5*]

to write to beschreiben [*Schreib-Lese-Speicher*]

wrong operation → maloperation

WWW → World Wide Web

X

XBS → exchange baseband station

xenon beacon, xenon flash instrument Xenonblinker *m*, Xenonblinklicht *n*

xerography [*A printing method developed by the Xerox Corporation; a negative image is formed by a resinous powder on an electrically charged plate, and this image is transferred and thermally fixed onto a paper as a positive*] Trockenschrift *f*

X-Tal-Oscillator Quarzoszillator *m*

xylene-toluene mix(ture), xylol-toluene mix(ture), dimethylbenzene tuluol mix(ture) Weingeist *m*, Xylol-Toluol-Mischung *f*, Xylol-Toluol-Gemisch *n*

x-y-plotter, X-Y-plotter x-y-Zeichengerät, X-Y-Zeichengerät *n*, x-y-Planzeichner, X-Y-Planzeichner, Zweikoordinatenschreiber *m* [*Der Stift bewegt sich in beiden Richtungen des Koordinatensystems (x-y)*]

Y

Yale lock Yaleschloss *n*

Y-amplifier Y-Verstärker *m*

yard drainage Hofentwässerung *f*

yard gull(e)y → gull(e)y

yard trap → gull(e)y trap

Y-attenuator Y-Abschwächer *m*

yearly jährlich

year-round, yearly ganzjährig

year-round air-conditioning system [*It provides ventilation and heating, humidification in winter and dehumidification in summer, and also provides the desired degree of air motion and cleanliness*] Jahresklimaanlage *f*

yellow cable Koaxialkabel *n* 50 Ohm [*Ethernet IEEE 802.3*]

yield Ertrag *m*

yield stress → flow stress

Y piece → breeches pipe

Z

ZBR → zoned bit recording

z-clipping räumliche Abschneiden *n*

Zener diode [*Electronics. A P-N diode that shows avalanche breakdown at a specified value of reverse bias; used for voltage reference purposes and for voltage stabilization*] Zenerdiode, Z-Diode *f* [*Nach dem Erfinder Zener genannt*]

Zener point Zenerknie *n* [*Zenerdiode*]

Zener voltage → breakdown voltage

zenithal light → diffuse daylight

zero adjustment Nullabgleich *m*, Nullanpassung *f*

zero curve Wirbelfreiheit *f*

zero-dimensional nulldimensional

zero-emission schadstofffrei

zero-energy band Nullenergieband *n*

zero energy house Nullenergiehaus *n*

zero growth Nullwachstum *n*, Nullwuchs *m*

zero input response Lösung *f* der homogenen Zustandsgleichung

zero lux condition Null-Lux-Zustand *m*

zero mean crossing distribution Nulldurchgangsverteilung *f*

zero offset error Nullpunktfehler *m*

zero order hold, ZOH Halt(e)glied nullter Ordnung, nullte Halt(e)glied *n*

zero output signal Nullausgangssignal *n*

zero point adjuster, zero point adjustor Nullpunktversteller *m*

zero position → position of rest

zero position interlock → off position interlock

zero shift Nullpunktverschiebung *f*

zero signal current Ruhestrom *m*

zero speed Drehzahlnull, Nulldrehzahl *f*, Stillstand *m*

zero suppression Null(en)unterdrückung *f*

zeroth order hold(ing) element, hold(ing) zeroth order element Halteglied *n* nullter Ordnung

zeroth sound nullte Schall *m*

zero voltage, (electrical) ground [*Conduction connection between an electrical circuit or equipment and the earth or some conduction body that serves in place of the earth*] Erd(an)schluss *m*, Nullspannung, Erde *f*

zinc (bath)tub Zink(bade)wanne *f*

zinc sheet metal Zinkblech *n*

ZOH → zero order hold

zone Zone *f* [*Teil eines Bereiches, dessen Versorgung gemeinsam mit anderen Zonen geregelt oder gesteuert wird*]

zero adjuster, zero adjustor Null(ein)steller *m*, Null(ein)stellgerät *n*

zig-zag heating tube, heater coil Heiz(ungs)schlange *f*

zig-zag superheating tube, superheater coil schlangenförmige Überhitzerröhre, Schlangenüberhitzerröhre, Überhitzerschlange *f*

zinc-coated verzinkt

zinc coating Verzinkung *f*

ZOH → zero order hold

zone [*Section of a heating, cooling, ventilating o air conditioning system separately controllable*] Zone *f*

zone [*An area composed of a building , a portion of a building, or a group of buildings, affected by a device or piece or equipment and subject to an open and/or closed control strategy*] Zone *f*, Bereich *m* *[selten: n]*

zone afterheater Zonennacherhitzer *m* [*Einkanalklima(tisierungs)anlage*]

zone air conditioner, zone air conditioning unit [*It conditions the air of adjacent rooms having similar requirements. See remark under "air conditioning unit"*] Zonenklimagerät *n*

zone air conditioning Zonenklimatisierung *f*

zone ((closed) loop) control Zonenreg(e)lung *f*

zone configuration → **zoning**

zone controller Zonenregler *m*

zoned zoniert

zone damper Zonenklappe *f*

zone(d) bit recording, ZBR, zone(d) density recording, ZDR Zonendichteaufzeichnung *f* [*Platte*]

zone(d) recirculated air system Zonenumluftsystem *n* [*System mit Umluftbehandlung getrennt für die Zonen des Versorgungsbereiches*]

zone identification sheet support plate Dechiffriereinsatz, Entchiffriereinsatz, Entschlüsselungseinsatz *m*

zone isolation (slide) valve Zonentrennschieber *m*

zone module → **(withdrawable) zone module**

zone ((open) loop) control Zonensteuerung *f*

zone recirculated air Zonenumluft *f* [*Abluft, die in derselben Zone als Zuluft wiederverwendet wird. DIN 1946, Teil 1*]

zone thermostat Zonenthermostat *m*

zone valve [*Valve that modulates water flow to a thermal zone in response to a downstream temperature signal*] Zonenventil *n*

zone-wise zonenweise

zoning, zone configuration Zonierung, Zoneneinteilung *f*

zoom(ing) Bildausschnittänderung *f* [*Vergrößern oder Verkleinern*]

Z